VOLUME	EDITOR-IN-CHIEF	PAGES	
20	CHARLES F. H. ALLEN	113	*Out of print*
21	THE LATE NATHAN L. DRAKE	120	*Out of print*
22	LEE IRVIN SMITH	114	*Out of print*
23	LEE IRVIN SMITH	124	*Out of print*
24	THE LATE NATHAN L. DRAKE	119	*Out of print*
25	THE LATE WERNER E. BACHMANN	120	*Out of print*
26	THE LATE HOMER ADKINS	124	*Out of print*
27	R. L. SHRINER	121	*Out of print*
28	H. R. SNYDER	121	*Out of print*
29	CLIFF S. HAMILTON	119	
Collective Vol. 3	A revised edition of Annual Volumes 20–29 E. C. HORNING, *Editor-in-Chief*	890	
30	ARTHUR C. COPE	115	
31	R. S. SCHREIBER	122	
32	RICHARD T. ARNOLD	119	
33	CHARLES C. PRICE	115	
34	WILLIAM S. JOHNSON	121	
35	T. L. CAIRNS	122	
36	N. J. LEONARD	120	
37	JAMES CASON	109	
38	JOHN C. SHEEHAN	120	
39	MAX TISHLER	114	
Collective Vol. 4	A revised edition of Annual Volumes 30–39 NORMAN RABJOHN, *Editor-in-Chief*	1036	
40	MELVIN S. NEWMAN	114	
41	JOHN D. ROBERTS	118	
42	VIRGIL BOEKELHEIDE	118	

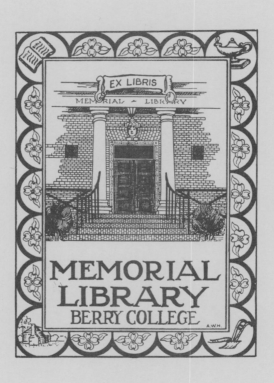

EX LIBRIS

MEMORIAL · LIBRARY

MEMORIAL
LIBRARY
BERRY COLLEGE

ORGANIC SYNTHESES

ROGER ADAMS

ORGANIC SYNTHESES

Collective Volume 4

A REVISED EDITION OF
ANNUAL VOLUMES 30–39

JOHN WILEY & SONS, INC.

New York • Chichester • Brisbane • Toronto • Singapore

ISBN 0 471 70470 9

Library of Congress Catalog Card Number: 32–11772

PRINTED IN THE UNITED STATES OF AMERICA

THIS BOOK IS DEDICATED TO
PROFESSOR ROGER ADAMS,
ONE OF THE FOUNDERS OF ORGANIC SYNTHESES, INC.,
AND ITS LEADER DURING THE LAST FORTY-ODD YEARS.

PREFACE

Collective Volume 4 of *Organic Syntheses* follows the plan established for the preceding three collected works. The material in annual volumes 30–39 has been combined and revised where necessary. Errors found in the original printings have been corrected, calculations and references have been checked, and modifications and improvements which have been brought to the attention of the Editorial Board have been included. New procedures for the preparation of 2,3-diphenyl-succinonitrile, ethyl azodicarboxylate, and mucobromic acid have been added.

The section of Methods of Preparation for each procedure has been revised to include material of preparative value recorded in the literature covered by *Chemical Abstracts* through Volume 54 (January–June, 1960). Later references have been supplied in a few cases. The *Chemical Abstracts* indexing name for each preparation is given as a subtitle when that name is different from the title. A fair number of changes have been made in the subtitles which were used in the annual volumes.

The practice has been continued of noting which of the compounds, whose preparations have been described, have become commercially available. This information is indicated by an asterisk after the title of the preparation.

A considerable number of warning notices have been added to the procedures in this volume on the basis of information supplied by users of the annual volumes. This suggests that many of the procedures and chemicals described may be hazardous and should be treated with due caution, as dictated by one's experience and knowledge.

Two innovations have been made in this collective volume. They are the use of the first reference number in each procedure to indicate the laboratory in which the directions were developed, and the addition of an author index. This index is included in Collective Volume 4 along with the previously developed Type of Reaction Index, Type of Compound Index, Formula Index, Preparation or Purification of Solvents and Reagents Index, Apparatus Index, and General Index.

The practice of solicitation of new synthetic procedures of current interest, instigated several years ago, accounts for many of the directions in the present volume.

Illustrations of several special pieces of equipment are shown in this collective volume. The preparations of benzoylacetanilide and oleoyl chloride illustrate the use of the general aspects of laboratory-sized continuous reactors. A low-temperature distillation column is described in the procedure for monovinylacetylene, and directions for the preparation and use of an ion-exchange column are given in the ε-aminocaproic acid procedure. A number of helpful suggestions in regard to modifications of standard pieces of equipment, as well as apparatus for specialized synthetic procedures, are illustrated and may be located by consulting the Apparatus Index.

The editors of *Organic Syntheses* are indebted to the contributors and users of these volumes and appreciate the assistance that they have supplied during the many years that the series has been published. Additional suggestions, comments, and corrections will be welcomed and should be sent to the Secretary to the Editorial Board. Submissions of directions for the annual volumes also should be forwarded to the Secretary. Further details about submissions may be obtained by consulting the preface to the most recent annual volume.

I wish to take this opportunity to express my gratitude to my graduate students who helped with the checking of calculations, references, and proof.

NORMAN RABJOHN

December, 1962

CONTENTS

ABIETIC ACID.................... 1
3-ACETAMIDO-2-BUTANONE.......... 5
N-(p-ACETYLAMINOPHENYL)-
  RHODANINE.................... 6
9-ACETYLANTHRACENE.............. 8
α-ACETYL-δ-CHLORO-γ-
  VALEROLACTONE................ 10
1-ACETYLCYCLOHEXANOL........... 13
2-p-ACETYLPHENYLHYDRO-
  QUINONE..................... 15
δ-ACETYL-n-VALERIC ACID......... 19
ACROLEIN ACETAL................. 21
ALLOXAN MONOHYDRATE............ 23
ALLOXANTIN DIHYDRATE........... 25
2-AMINO-4-ANILINO-6-(CHLORO-
  METHYL)-s-TRIAZINE............ 29
p-AMINOBENZALDEHYDE............ 31
2-AMINOBENZOPHENONE............ 34
ε-AMINOCAPROIC ACID............. 39
2-AMINO-3-NITROTOLUENE.......... 42
3-AMINOPYRIDINE................ 45
p-AMINOTETRAPHENYLMETHANE...... 47
ARYLUREAS...................... 49
  p-BROMOPHENYLUREA............ 49
  p-ETHOXYPHENYLUREA........... 52
DL-ASPARTIC ACID................ 55
ATROLACTIC ACID................ 58
AZELANITRILE................... 62
1,1'-AZO-bis-1-CYCLOHEXANENITRILE.. 66
BENZENEBORONIC ANHYDRIDE....... 68
BENZHYDRYL β-CHLOROETHYL
  ETHER...................... 72
BENZOFURAZAN OXIDE............. 74
BENZOGUANAMINE................ 78
BENZOYLACETANILIDE............. 80
BENZOYLCHOLINE IODIDE AND
  CHLORIDE.................... 84
3-BENZOYLPYRIDINE.............. 88
2-BENZYLAMINOPYRIDINE........... 91
3-BENZYL-3-METHYLPENTANOIC
  ACID....................... 93
BENZYLTRIMETHYLAMMONIUM
  ETHOXIDE.................... 98
BISCHLOROMETHYL ETHER.......... 101
N-BROMOACETAMIDE.............. 104
β-BROMOETHYLPHTHALIMIDE........ 106
4-BROMO-2-HEPTENE............. 108
p-BROMOMANDELIC ACID.......... 110
2-BROMO-3-METHYLBENZOIC ACID..... 114
n-BUTYLACETYLENE.............. 117
sec-BUTYL α-n-CAPROYLPROPIONATE.. 120
n-BUTYL GLYOXYLATE............. 124
tert-BUTYL HYPOCHLORITE......... 125
2-BUTYN-1-OL.................. 128
BUTYRCHLORAL.................. 130

β-CARBETHOXY-γ,γ-DIPHENYL-
  VINYLACETIC ACID.............. 132
β-(o-CARBOXYLPHENYL)PROPIONIC
  ACID....................... 136
CETYLMALONIC ESTER............. 141
CHLOROACETONITRILE............. 144
3-(o-CHLOROANILINO)PROPIONITRILE.. 146
CHLORO-p-BENZOQUINONE.......... 148
N-CHLOROBETAINYL CHLORIDE...... 154
trans-2-CHLOROCYCLOPENTANOL...... 157
1-CHLORO-2,6-DINITROBENZENE...... 160
2-CHLORO-2-METHYLCYCLOHEXANONE
  AND 2-METHYL-2-CYCLOHEXENONE . 162
2-CHLORONICOTINONITRILE......... 166
α-CHLOROPHENYLACETIC ACID....... 169
o-CHLOROPHENYLCYANAMIDE........ 172
α-(4-CHLOROPHENYL)-γ-PHENYL-
  ACETOACETONITRILE............ 174
1-(p-CHLOROPHENYL)-3-PHENYL-2-
  PROPANONE................... 176
p-CHLOROPHENYL SALICYLATE....... 178
o-CHLOROPHENYLTHIOUREA......... 180
2-CHLOROPYRIMIDINE............. 182
2-CHLORO-1,1,2-TRIFLUOROETHYL
  ETHYL ETHER................. 184
β-CHLOROVINYL ISOAMYL KETONE.... 186
Δ⁴-CHOLESTEN-3,6-DIONE........... 189
Δ⁴-CHOLESTEN-3-ONE.............. 192
CHOLESTEROL, Δ⁵-CHOLESTEN-3-
  ONE AND Δ⁴-CHOLESTEN-3-ONE.... 195
COUMALIC ACID.................. 201
CREOSOL...................... 203
N-2-CYANOETHYLANILINE.......... 205
CYANOGEN IODIDE................ 207
3-CYANO-6-METHYL-2(1)-PYRIDONE . . . 210
1-CYANO-3-PHENYLUREA............ 213
1,2-CYCLODECANEDIOL............. 216
CYCLODECANONE................. 218
CYCLOHEPTANONE................ 221
1,2-CYCLOHEXANEDIONE DIOXIME.... 229
CYCLOHEXENE SULFIDE........... 232
CYCLOHEXYLIDENECYANOACETIC
  ACID AND 1-CYCLOHEXENYLACE-
  TONITRILE.................... 234
CYCLOPENTADIENE AND 3-
  CHLOROCYCLOPENTENE.......... 238
DIACETYL-d-TARTARIC ANHYDRIDE.... 242
2,5-DIAMINO-3,4-DICYANOTHIOPHENE.. 243
2,4-DIAMINO-6-HYDROXYPYRIMIDINE . . 245
DIAMINOURACIL HYDROCHLORIDE.... 247
DIAZOMETHANE.................. 250
DIBROMOACETONITRILE............ 254
4,4'-DIBROMOBIPHENYL........... 256
DI-n-BUTYLDIVINYLTIN........... 258
DI-tert-BUTYL MALONATE.......... 261

xi

4,4'-Dichlorodibutyl Ether...... 266
1,1-Dichloro-2,2-difluoro-
ethylene..................... 268
2,2-Dichloroethanol............. 271
1,1'-Dicyano-1,1'-bicyclohexyl..... 273
1,2-Di-1-(1-cyano)cyclohexyl-
hydrazine.................... 274
Dicyanoketene Ethylene Acetal.. 276
Dicyclopropyl Ketone........... 278
1-Diethylamino-3-butanone...... 281
N,N'-Diethylbenzidine 283
Diethyl Benzoylmalonate........ 285
Diethyl 1,1-Cyclobutanedi-
carboxylate.................. 288
Diethyl Δ²-Cyclopentenyl-
malonate.................... 291
Diethyl Ethylidenemalonate..... 293
Diethyl Mercaptoacetal........ 295
Diethyl Methylenemalonate..... 298
Diethyl γ-Oxopimelate........... 302
Diethyl cis-Δ⁴-Tetrahydro-
phthalate and Diethyl cis-
Hexahydrophthalate........... 304
Diethylthiocarbamyl Chloride... 307
3,4-Dihydro-2-methoxy-4-methyl-
2H-pyran.................... 311
9,10-Dihydrophenanthrene....... 313
9,10-Dihydroxystearic Acid...... 317
1,4-Diiodobutane.............. 321
1,6-Diiodohexane................ 323
Diisopropyl Methylphosphonate.. 325
2,3-Dimethoxycinnamic Acid...... 327
Dimethyl Acetylenedicarboxylate......................... 329
p-Dimethylaminobenzaldehyde.... 331
β-Dimethylaminoethyl Chloride
Hydrochloride................ 333
2-(Dimethylamino)pyrimidine...... 336
4,6-Dimethylcoumalin.............. 337
N,N-Dimethylcyclohexyl-
methylamine.................. 339
Dimethylfurazan................. 342
β,β-Dimethylglutaric Acid........ 345
Dimethylketene................. 348
5,5-Dimethyl-2-n-pentyltetrahy-
drofuran..................... 350
3,5-Dimethylpyrazole............. 351
2,2-Dimethylpyrrolidine......... 354
5,5-Dimethyl-2-pyrrolidone...... 357
N,N-Dimethylselenourea......... 359
asym-Dimethylurea............... 361
2,6-Dinitroaniline............... 364
p,p'-Dinitrobibenzyl............. 367
1,4-Dinitrobutane................ 368
3,4-Dinitro-3-hexene............. 372
Diphenylacetaldehyde............ 375
Diphenylacetylene............... 377
1,4-Diphenyl-5-amino-1,2,3-triazole
and 4-Phenyl-5-anilino-1,2,3-
triazole.................... 380
N,N'-Diphenylbenzamidine....... 383
α,β-Diphenylcinnamonitrile....... 387
Diphenyl Succinate............. 390

2,3-Diphenylsuccinonitrile........ 392
p-Dithiane..................... 396
trans-2-Dodecenoic Acid.......... 398
Ethanedithiol.................... 401
Ethoxyacetylene................. 404
Ethyl α-Acetyl-β-(2,3-dimethoxy-
phenyl)propionate.............. 408
Ethyl Azodicarboxylate.......... 411
Ethyl Benzoylacetate............ 415
Ethyl tert-Butyl Malonate....... 417
N-Ethyl-p-chloroaniline......... 420
Ethyl Chlorofluoroacetate...... 423
Ethyl Diazoacetate.............. 424
Ethyl Diethoxyacetate........... 427
Ethyl Enanthylsuccinate........ 430
Ethylenimine.................... 433
2-Ethylhexanonitrile............. 436
Ethyl Isocyanide................ 438
β-Ethyl-β-methylglutaric Acid.... 441
4-Ethyl-2-methyl-2-octenoic Acid.. 444
5-Ethyl-2-methylpyridine........ 451
Ethyl α-Nitrobutyrate.......... 454
Ethyl Orthocarbonate........... 457
Ethyl β,β-Pentamethyl-
eneglycidate.................. 459
Ethyl Phenylcyanoacetate....... 461
Ethyl (1-Phenylethylidene)-
cyanoacetate.................. 463
Ethyl N-Phenylformimidate...... 464
Ethyl α-(1-Pyrrolidyl)propionate. 466
Ethyl Pyruvate.................. 467
1,1'-Ethynylene-bis-
cyclohexanol.................. 471
Ferrocene....................... 473
Flavone......................... 478
9-Fluorenecarboxylic Acid....... 482
5-Formyl-4-phenanthroic Acid..... 484
Fumaronitrile................... 486
Furfural Diacetate.............. 489
2-Furfuryl Mercaptan........... 491
2-Furoic Acid................... 493
Glutaric Acid and Glutarimide... 496
Guanylthiourea.................. 502
d-Gulonic-γ-Lactone.............. 506
Hemimellitene................... 508
Hendecanedioic Acid............. 510
n-Heptamide.................... 513
3-n-Heptyl-5-cyanocytosine...... 515
Hexahydro-1,3,5-tripropionyl-s-
triazine..................... 518
Hexamethylbenzene.............. 520
Hexamethylene Diisocyanate..... 521
n-Hexyl Fluoride................ 525
4-Hydroxy-1-butanesulfonic
Acid Sultone.................. 529
6-Hydroxynicotinic Acid......... 532
3-Hydroxytetrahydrofuran....... 534
Indazole........................ 536
Indole-3-aldehyde............... 539
Iodocyclohexane................. 543
2-Iodothiophene................. 545
4-Iodoveratrole................. 547

ISODEHYDROACETIC ACID AND ETHYL
ISODEHYDROACETATE.............. 549
ISOPHORONE OXIDE................. 552
ITACONYL CHLORIDE................ 554
6-KETOHENDECANEDIOIC ACID....... 555
β-KETOISOÖCTALDEHYDE DIMETHYL
ACETAL....................... 558
LAURONE......................... 560
LAURYLMETHYLAMINE............... 564
2-MERCAPTO-4-AMINO-5-CARBETHOXY-
PYRIMIDINE AND 2-MERCAPTO-4-
HYDROXY-5-CYANOPYRIMIDINE...... 566
2-MERCAPTOBENZIMIDAZOLE......... 569
METHANESULFONYL CHLORIDE....... 571
o-METHOXYPHENYLACETONE......... 573
p-METHOXYPHENYLACETONITRILE.... 576
METHYL p-ACETYLBENZOATE........ 579
o-METHYLBENZYL ALCOHOL.......... 582
2-METHYLBENZYLDIMETHYLAMINE.... 585
O-METHYLCAPROLACTIM............. 588
3-METHYLCOUMARONE.............. 590
METHYL CYCLOPENTANE-
CARBOXYLATE.................... 594
METHYL CYCLOPROPYL KETONE..... 597
2-METHYL-2,5-DECANEDIOL......... 601
N-METHYL-2,3-DIMETHOXYBENZYL-
AMINE........................ 603
N-METHYL-1,2-DIPHENYLETHYLAMINE
AND HYDROCHLORIDE............. 605
trans-2-METHYL-2-DODECENOIC ACID.. 608
METHYLENECYCLOHEXANE AND N,N-
DIMETHYLHYDROXYLAMINE
HYDROCHLORIDE................. 612
2-METHYLENEDODECANOIC ACID..... 616
1-METHYL-3-ETHYLOXINDOLE....... 620
9-METHYLFLUORENE............... 623
5-METHYLFURFURYLDIMETHYLAMINE.. 626
3-METHYL-2-FUROIC ACID AND 3-
METHYLFURAN.................. 628
β-METHYLGLUTARIC ANHYDRIDE..... 630
METHYLGLYOXAL-ω-PHENYLHYDRA-
ZONE......................... 633
METHYL HYDROGEN
HENDECANEDIOATE............... 635
4-METHYL-6-HYDROXYPYRIMIDINE.... 638
1-METHYLISOQUINOLINE............. 641
METHYLISOUREA HYDROCHLORIDE.... 645
METHYL 3-METHYL-2-FUROATE...... 649
METHYL γ-METHYL-γ-NITRO-
VALERATE..................... 652
3-METHYL-4-NITROPYRIDINE-1-OXIDE . 654
3-METHYLOXINDOLE............... 657
3-METHYL-1,5-PENTANEDIOL........ 660
β-METHYL-β-PHENYL-α,α'-DICYANO-
GLUTARIMIDE.................. 662
β-METHYL-β-PHENYLGLUTARIC
ACID......................... 664
1-METHYL-3-PHENYLINDANE........ 665
METHYL 2-THIENYL SULFIDE........ 667
METHYL β-THIODIPROPIONATE....... 669
3-METHYLTHIOPHENE.............. 671
METHYL p-TOLYL SULFONE......... 674
β-METHYL-δ-VALEROLACTONE....... 677

MONOBENZALPENTAERYTHRITOL...... 679
MONOBROMOPENTAERYTHRITOL...... 681
MONOVINYLACETYLENE............. 683
MUCOBROMIC ACID................. 688
1-NAPHTHALDEHYDE............... 690
NAPHTHALENE-1,5-DISULFONYL
CHLORIDE...................... 693
1,5-NAPHTHALENEDITHIOL........... 695
1,4-NAPHTHOQUINONE.............. 698
α-NAPHTHYL ISOTHIOCYANATE....... 700
NEOPHYL CHLORIDE................ 702
NICOTINAMIDE-1-OXIDE............. 704
NICOTINONITRILE................. 706
o-NITROACETOPHENONE............. 708
9-NITROANTHRACENE............... 711
o- AND p-NITROBENZALDIACETATE.... 713
m-NITROBENZAZIDE................ 715
m-NITROBIPHENYL................. 718
o-NITROCINNAMALDEHYDE.......... 722
1-NITROÖCTANE................... 724
1-(p-NITROPHENYL)-1,3-BUTADIENE... 727
trans-o-NITRO-α-PHENYLCINNAMIC
ACID......................... 730
m-NITROSTYRENE................. 731
6-NITROVERATRALDEHYDE.......... 735
NORBORNYLENE.................... 738
OLEOYL CHLORIDE................. 739
PARABANIC ACID.................. 744
1,4-PENTADIENE.................. 746
PENTAERYTHRITYL TETRABROMIDE... 753
4-PENTYN-1-OL................... 755
PHENANTHRENEQUINONE............ 757
PHENYLACETAMIDE................. 760
PHENYLACETYLENE................ 763
γ-PHENYLALLYLSUCCINIC ACID....... 766
N-PHENYLBENZAMIDINE............ 769
trans-1-PHENYL-1,3-BUTADIENE 771
α-PHENYL-α-CARBETHOXYGLU-
TARONITRILE................... 776
α-PHENYLCINNAMIC ACID........... 777
2-PHENYLCYCLOHEPTANONE.......... 780
PHENYLDICHLOROPHOSPHINE........ 784
4-PHENYL-m-DIOXANE.............. 786
o-PHENYLENE CARBONATE.......... 788
α-PHENYLGLUTARIC ANHYDRIDE...... 790
1-PHENYL-1-PENTEN-4-YN-3-OL....... 792
1-PHENYLPIPERIDINE............... 795
3-PHENYL-1-PROPANOL............. 798
PHENYLPROPARGYLALDEHYDE
DIETHYL ACETAL................ 801
PHENYLSUCCINIC ACID............. 804
o-PHTHALALDEHYDE................ 807
α-PHTHALIMIDO-o-TOLUIC ACID....... 810
PROPIOLALDEHYDE................. 813
PSEUDOPELLETIERINE.............. 816
PUTRESCINE DIHYDROCHLORIDE...... 819
2,3-PYRAZINEDICARBOXYLIC ACID..... 824
PYRIDINE-N-OXIDE................. 828
2-PYRROLEALDEHYDE.............. 831
2-(1-PYRROLIDYL)PROPANOL......... 834
QUINACETOPHENONE MONOMETHYL
ETHER....................... 836
SEBACIL....................... 838

SEBACOIN........................ 840
SODIUM NITROMALONALDEHYDE
  MONOHYDRATE................... 844
SODIUM β-STYRENESULFONATE AND
  β-STYRENESULFONYL CHLORIDE.... 846
STEAROLIC ACID.................. 851
STEARONE....................... 854
cis-STILBENE.................... 857
trans-STILBENE OXIDE............ 860
α-SULFOPALMITIC ACID............ 862
SYRINGIC ALDEHYDE............... 866
TETRAACETYLETHANE.............. 869
dl-4,4',6,6'-TETRACHLORODIPHENIC
  ACID......................... 872
TETRACYANOETHYLENE............. 877
TETRAETHYLTIN.................. 881
1,2,3,4-TETRAHYDROCARBAZOLE..... 884
ar-TETRAHYDRO-α-NAPHTHOL....... 887
cis-Δ⁴-TETRAHYDROPHTHALIC
  ANHYDRIDE.................... 890
TETRAHYDROTHIOPHENE............ 892
TETRALIN HYDROPEROXIDE......... 895
α-TETRALONE.................... 898
β-TETRALONE.................... 903
2,2,6,6-TETRAMETHYLOLCYCLO-
  HEXANOL...................... 907
TETRAPHENYLARSONIUM CHLORIDE
  HYDROCHLORIDE................ 910
TETRAPHENYLETHYLENE............ 914
2-THENALDEHYDE................. 915
3-THENALDEHYDE................. 918
3-THENOIC ACID................. 919
3-THENYL BROMIDE............... 921
THIOBENZOIC ACID............... 924

THIOBENZOPHENONE............... 927
THIOLACETIC ACID............... 928
o-TOLUALDEHYDE................. 932
p-TOLUENESULFENYL CHLORIDE.... 934
p-TOLUENESULFINYL CHLORIDE..... 937
p-TOLUENESULFONIC ANHYDRIDE.... 940
p-TOLYLSULFONYLMETHYLNITRO-
  SAMIDE....................... 943
2,4,6-TRIBROMOBENZOIC ACID...... 947
TRICHLOROMETHYLPHOSPHONYL
  DICHLORIDE................... 950
p-TRICYANOVINYL-N,N-DIMETHYL-
  ANILINE...................... 953
TRIETHYL PHOSPHITE............. 955
2,4,4-TRIMETHYLCYCLOPENTANONE... 957
α,β,β-TRIPHENYLPROPIONIC ACID.... 960
α,α,β-TRIPHENYLPROPIONITRILE..... 962
TRIPTYCENE..................... 964
TRITHIOCARBODIGLYCOLIC ACID..... 967
10-UNDECYNOIC ACID............. 969
VANILLIC ACID.................. 972
VINYL LAURATE AND OTHER VINYL
  ESTERS....................... 977
2-VINYLTHIOPHENE............... 980
o-XYLYLENE DIBROMIDE........... 984
TYPE OF REACTION INDEX........ 987
TYPE OF COMPOUND INDEX........ 995
FORMULA INDEX.................. 1007
INDEX TO PREPARATION OR PURI-
  FICATION OF SOLVENTS AND
  REAGENTS..................... 1011
APPARATUS INDEX................ 1013
AUTHOR INDEX.................. 1015
GENERAL INDEX................. 1021

ABIETIC ACID *

Wood rosin + HCl → Isomerized wood rosin

Isomerized wood rosin + $(C_5H_{11})_2NH$ →

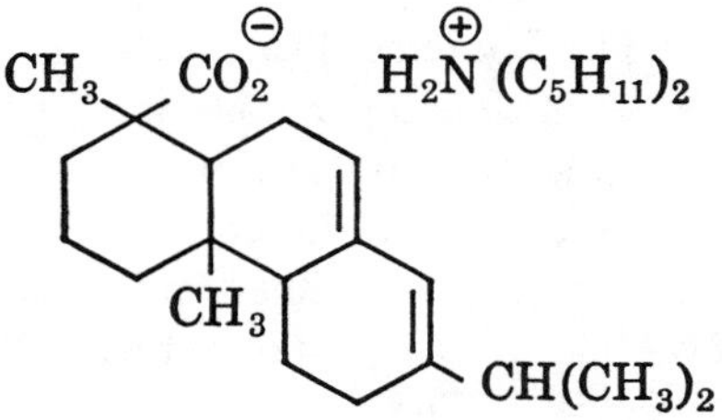

Amine salt + CH_3CO_2H →

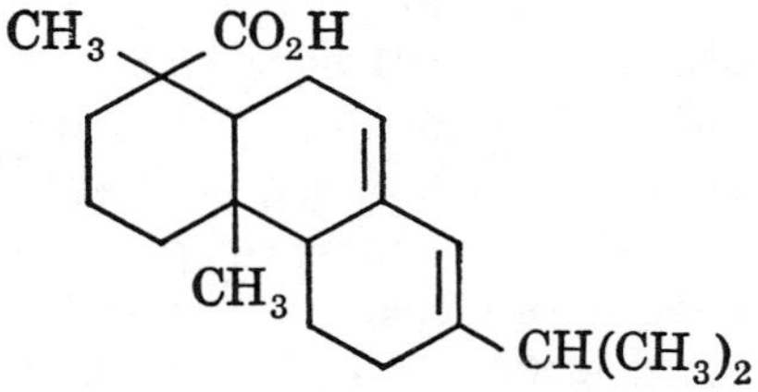

Submitted by G. C. HARRIS and T. F. SANDERSON.[1]
Checked by R. T. ARNOLD and K. L. LINDSAY.

1. Procedure

In a 2-l. round-bottomed flask fitted with a 35-cm. reflux condenser are placed 250 g. (0.74 mole) of N-grade wood rosin (Note 1), 740 ml. of 95% ethanol, and 42 ml. of hydrochloric acid (sp. gr. 1.19). A stream of carbon dioxide is passed over the surface of the solution by means of a glass tube which extends downward through the condenser during this reaction (Note 2). The mixture is boiled under reflux for 2 hours (Note 3). At the

* See Preface for explanation of asterisks.

1

end of this time, the ethanol and acid are removed by steam distillation and the water is decanted. The residue is cooled to room temperature and dissolved in 1 l. of ether. The ether solution is extracted with water and dried over 200 g. of anhydrous sodium sulfate. The bulk of the ether is evaporated on the steam bath, and the last traces are removed by fusing the rosin over a free flame and under a vacuum furnished by a water aspirator. The molten rosin, blanketed continuously with carbon dioxide, is most conveniently handled by being poured into a paper boat; yield 245 g.; $[\alpha]_D^{24}$ $-35°$ (Note 4).

The isomerized rosin, 245 g. (0.72 mole) (Note 1), is placed in a 1-l. Erlenmeyer flask and dissolved in 375 ml. of acetone by heating the mixture on a steam bath. To this solution, at incipient boiling, is added slowly and with vigorous agitation (Note 5) 127 g. (0.81 mole) of diamylamine [2] (Note 6). Upon cooling to room temperature, crystals appear in the form of rosettes. The mass is agitated, cooled well in an ice bath, and filtered by suction. The crystalline salt is washed on a Büchner funnel with 150 ml. of acetone and dried in a vacuum oven at 50° for 1 hour. The optical rotation of this material is $[\alpha]_D^{24}$ $-18°$ (Note 4). The solid is recrystallized four times from acetone. Each time a sufficient quantity (20 ml. per g.) of acetone is used to obtain complete solution, and the solvent is evaporated until incipient precipitation of the salt occurs. The yield of product is 118 g.; $[\alpha]_D^{24}$ $-60°$ (Note 4). An additional 29 g. of product, having the same rotation, can be recovered from filtrates of the previous crystallizations.

The amine salt (147 g.) is placed in a 4-l. Erlenmeyer flask and dissolved in 1 l. of 95% ethanol by heating the mixture on a steam bath. To the solution, which has been cooled to room temperature (Note 7), is added 39 g. (35.8 ml.) of glacial acetic acid, and the solution is stirred. Water (900 ml.) is added cautiously at first and with vigorous agitation until crystals of abietic acid begin to appear; the remainder of the water is then added more rapidly. The abietic acid is collected on a Büchner funnel (Note 8) and washed with water until the acetic acid has been removed completely as indicated by tests with indicator paper. Recrystallization can be effected by dissolving the crude product in 700 ml. of 95% ethanol, adding 600 ml. of water as described above, and cooling the solution. The yield of abietic acid is 98 g. (40% based on the weight of isomerized rosin; $[\alpha]_D^{24}$ $-106°$ (Notes 4 and 9). The ultraviolet absorption spectrum shows a maximum at 241 mμ; $\alpha = 77.0$ (Note 10).

2. Notes

1. The calculation of molar quantities is based on an acid number of 166 for N-grade wood rosin as obtained from Hercules Powder Company, Wilmington, Delaware. Acid number is the number of milligrams of potassium hydroxide required to neutralize 1 g. of sample.

2. Blanketing the rosin in solution or in the molten state with carbon dioxide serves to keep it out of contact with air to avoid oxidation.

3. The maximum negative optical rotation, $[\alpha]_D^{24}$ $-35°$, is obtained with a minimum reflux time of 2 hours.

4. Rotations are reported as those of 1% solutions in absolute ethanol.

5. The addition of the amine to the hot solution is necessary for the formation of the salt. However, it must be done slowly and with rapid stirring because of the resulting vigorous exothermic reaction.

6. Commercial diamylamine, a mixture of isomers, purchased from Sharples Chemicals Company, Philadelphia, Pennsylvania, was employed.

7. The acid is added to a cooled solution of the salt in ethanol to minimize the chance for isomerization of the liberated abietic acid.

8. An early filtration is desirable for the purpose of removing the abietic acid from the acidic solution where isomerization can take place. Washing with a large volume of water and recrystallizing assures the complete removal of acetic acid.

9. The pure acid is dried in a vacuum desiccator over sodium hydroxide or calcium sulfate and stored in an oxygen-free atmosphere. Undue exposure to higher temperatures will result in isomerization, and contact with oxygen will result in oxidation.

10. The absorption spectrum data were obtained from measurements made with a Beckman Ultraviolet Spectrophotometer.

The formulas employed in making the calculations use the term α, specific absorption coefficient.

$$\alpha = \frac{\log_{10} I_0/I}{cl}$$

where I_0 = intensity of radiation transmitted by the solvent; I = intensity of radiation transmitted by the solution; c = concentration of solute in grams per liter; l = length in centimeters of solution through which the radiation passes.

3. Methods of Preparation

Abietic acid has usually been prepared [2] from rosin through the acid sodium salt $(3C_{20}H_{30}O_2 \cdot C_{19}H_{29}CO_2Na)$ with the subsequent formation and recrystallization of the diamylamine salt. The acid is regenerated from the pure salt by decomposition of the latter with a weak acid such as acetic acid. In addition, it has been purified [3] through the potassium, piperidine, and brucine salts, as well as through abietic anhydride and trityl abietate. The acid is regenerated from the pure salts by decomposition of the latter with a weak acid such as acetic acid, and from the pure acid derivatives by treatment with potassium hydroxide.

Two improvements have been introduced in the first method [2] by the procedure described above: (1) the abietic content of the rosin is increased by isomerization, and (2) a much better recovery of acid is obtained by applying the amine salt technique directly to the isomerized rosin, thus eliminating the step involving the acid sodium salt.

[1] Hercules Powder Company, Wilmington, Delaware.
[2] Palkin and Harris, *J. Am. Chem. Soc.*, **56**, 1935 (1934).
[3] Lombard and Frey, *Bull. soc. chim. France*, **1948**, 1194.

3-ACETAMIDO-2-BUTANONE *

(2-Butanone, 3-acetamido-)

$$CH_3CH(NH_2)CO_2H + 2(CH_3CO)_2O \xrightarrow{\text{Pyridine}}$$
$$CH_3CH(NHCOCH_3)COCH_3 + CO_2 + 2CH_3CO_2H$$

Submitted by RICHARD H. WILEY and O. H. BORUM.[1]
Checked by R. S. SCHREIBER and B. D. ASPERGREN.

1. Procedure

A mixture of 156.6 g. (159 ml., 1.98 moles) of pyridine (Note 1), 239.9 g. (224 ml., 2.35 moles) of acetic anhydride (Note 2), and 35.1 g. (0.39 mole) of vacuum-dried alanine (Notes 3 and 4) is heated with stirring (Note 5) on the steam bath for 6 hours after solution is complete (Note 6). The excess pyridine and acetic anhydride, and the acetic acid, are removed at reduced pressure. The residue is distilled through a 15-cm. column, packed with glass helices, to give 41.5–47.5 g. of crude product, boiling at 110–125°/3 mm. Refractionation gives 41–45 g. (81–88%) of 3-acetamido-2-butanone; b.p. 102–106°/2 mm.; n_D^{25} 1.4558–1.4561 (Note 7).

2. Notes

1. A commercial C.P. grade can be used. The checkers used Merck A.R. grade.
2. A commercial grade, 95% minimum assay, can be used. The checkers used Merck A.R. grade.
3. Any good commercial grade material appears to be satisfactory.
4. Reducing the molar ratio of pyridine or anhydride to the amino acid reduces the yield.
5. Without stirring the yield is 46%.
6. With other amino acids, notably glycine and sarcosine, it is necessary to reflux the reactants 1–6 hours.
7. The checkers found it necessary to heat the column to obtain the maximum available product.

3. Methods of Preparation

This method, an adaptation of a previously described procedure,[2-4] has been used with a variety of amino acids and anhydrides to give the following products: 1-phenyl-1-propionamido-2-butanone (75%);[5] acetamidoacetylacetone (60%);[5] N-methyl-acetamidoacetone;[6] 1-phenyl-2-acetamido-3-butanone (79%);[7] 1-phenyl-2-propionamido-3-pentanone (41%);[7] 1-phenyl-2-butyramido-3-hexanone (27%);[7] α-benzamidopropiophenone (42%);[7] α-benzamido-β-phenylpropiophenone (44%);[7] 1-phenyl-1-acetamidoacetone (72–90%);[8,9] 1-phenyl-1-benzamidoacetone (65%);[8] 1-phenyl-2-benzamido-3-butanone (78%);[8] 3-benzamido-2-butanone (65–88%);[8] and 3-acetamido-5-methyl-2-hexanone (73%).[10]

[1] University of Louisville, Louisville, Kentucky.

[2] Dakin and West, *J. Biol. Chem.*, **78**, 91, 757 (1928).

[3] Levene and Steiger, *J. Biol. Chem.*, **74**, 689 (1927); **79**, 95 (1928).

[4] Wiley, *J. Org. Chem.*, **12**, 43 (1947).

[5] Wiley and Borum, *J. Am. Chem. Soc.*, **70**, 2005 (1948).

[6] Wiley and Borum, *J. Am. Chem. Soc.*, **72**, 1626 (1950).

[7] Cleland and Niemann, *J. Am. Chem. Soc.*, **71**, 841 (1949).

[8] Searles and Cvejanovich, *J. Am. Chem. Soc.*, **72**, 3200 (1950).

[9] Rondestvedt, Manning, and Tabibian, *J. Am. Chem. Soc.*, **72**, 3183 (1950).

[10] Borum, Ph.D. Thesis, University of North Carolina, 1949.

N-(p-ACETYLAMINOPHENYL)RHODANINE

[Rhodanine, 3-(p-acetamidophenyl)-]

$$CH_3CONH\!\!-\!\!\bigcirc\!\!-\!\!NH_2 + (HO_2C\!-\!CH_2S)_2CS \xrightarrow{H_2O}$$

$$CH_3CONH\!\!-\!\!\bigcirc\!\!-\!\!N\!\!-\!\!C\!\!=\!\!O + HSCH_2CO_2H$$

Submitted by R. E. Strube.[1]
Checked by John D. Roberts and Stanley L. Manatt.

1. Procedure

In a 2-l. round-bottomed flask fitted with a mechanical stirrer

and a reflux condenser are placed 30.0 g. (0.20 mole) of *p*-amino-
acetanilide (Note 1) and 400 ml. of water. The mixture is heated
on a steam bath with stirring, and to the clear solution is added
at once a hot solution of 45.2 g. (0.20 mole) of trithiocarbodiglycolic acid (p. 967) in 500 ml. of water. Heating and stirring are
continued for 5 hours (Note 2). The steam bath is then replaced
by an ice bath, and the reaction mixture is cooled to 20–25°. The
precipitate is removed by suction filtration. The solid is transferred to a 500-ml. Erlenmeyer flask containing 200 ml. of water.
The mixture is heated on the steam bath to 70–75° while the
lumps are crushed by a glass rod to obtain a homogeneous mixture. The mixture is filtered with suction while hot, and the flask
is cleaned by rinsing it with small amounts of hot water. The
solid on the filter is sucked as dry as possible and then transferred
to a 2-l. round-bottomed flask fitted with a reflux condenser.
Glacial acetic acid (1.5 l.) is added and the mixture is heated in an
oil bath to vigorous reflux for 5 minutes (Note 3). A small
amount of solid does not dissolve, and this is removed by filtration while hot (Note 4). The filtrate is stirred mechanically and
cooled to 15–20° by an ice bath and kept at this temperature for 1
hour. The slightly yellow crystals are collected by suction filtration, washed successively with 25 ml. of glacial acetic acid, 100 ml.
of ethanol, and 100 ml. of ether. The yield of air-dried material
is 26–28 g. (49–53% yield). The compound decomposes on heating above 240° (Note 5).

2. Notes

1. *p*-Aminoacetanilide (white label) supplied by Eastman
Kodak Company was used.
2. Within 10 minutes a precipitate is formed; the greater part
of the reaction product is present after 2 hours' heating.
3. The purification should be carried out in a hood, since gas
escapes during the heating and hot acetic acid is irritating to the
eyes. The checkers used a 2-l. heating mantle instead of an oil
bath.
4. The filtration of the hot acetic acid solution should be done
with care. The flask was surrounded by a towel and rubber
gloves were worn. The filtration can best be done in two steps.
Approximately half of the hot acetic acid solution is filtered

through a large, fluted filter paper; the other half is heated again to reflux and then filtered through another fluted filter paper. Filtration through a steam-heated Büchner funnel may sometimes be troublesome, since the suction accelerates crystallization causing plugging of the funnel stem.

5. Analytical values: Calcd. for $C_{11}H_{10}N_2O_2S_2$: C, 49.62; H, 3.78; N, 10.52; S, 24.08. Found: C, 49.76; H, 3.76; N, 10.36; S, 24.07.

3. Methods of Preparation

This procedure is based on the method of Holmberg[2] for preparing N-substituted rhodanines. The synthesis of N-(p-acetylaminophenyl)rhodanine has not yet been reported in the literature.

[1] The Upjohn Company, Kalamazoo, Michigan.
[2] Holmberg, *J. prakt. Chem.*, 81, 451 (1910).

9-ACETYLANTHRACENE *

(Ketone, 9-anthryl methyl)

Submitted by CHARLES MERRITT, JR., and CHARLES E. BRAUN.[1]
Checked by WILLIAM S. JOHNSON and RALPH F. HIRSCHMANN.

1. Procedure

Fifty grams (0.28 mole) of purified anthracene (Note 1) is suspended in 320 ml. of anhydrous benzene and 120 ml. (1.68 moles) of reagent grade acetyl chloride contained in a 1-l. three-necked flask. The flask is fitted with a thermometer which is immersed in the suspension, a calcium chloride drying tube, an efficient motor-driven sealed stirrer, and a rubber addition tube to which a 125-ml. Erlenmeyer flask containing 75 g. (0.56 mole) of anhydrous aluminum chloride is attached.[2]

The flask is surrounded by an ice-calcium chloride cooling mixture, and the aluminum chloride is added in small portions from the Erlenmeyer flask at such a rate that the temperature is maintained between $-5°$ and $0°$. After the addition is complete, the mixture is stirred for an additional 30 minutes, and the temperature is then allowed to rise slowly to $10°$. The red complex which forms is collected with suction on a sintered-glass funnel and washed thoroughly with dry benzene (Note 2). The complex is added in small portions by means of a spatula with stirring to a 600-ml. beaker nearly filled with a mixture of ice and concentrated hydrochloric acid. The mixture is then allowed to come to room temperature, and the crude ketone is collected on a suction filter.

The product is digested under reflux for about 20 minutes with 100–150 ml. of boiling 95% ethanol. The suspension (Note 3) is cooled quickly almost to room temperature and filtered rapidly with suction to remove any anthracene. The 9-acetylanthracene, which separates in the filtrate, is redissolved by heating and allowed to crystallize by slowly cooling the solution (finally to 0–$5°$ in an icebox) (Note 4). A second recrystallization from 95% ethanol yields 35–37 g. (57–60%) of light-tan granules of 9-acetylanthracene melting at 75–76° (Note 5).

2. Notes

1. The Eastman Kodak Company grade melting at 214–215° is satisfactory. Technical grade anthracene can be purified by codistillation with ethylene glycol. (See Fieser, *Experiments in Organic Chemistry*, 2nd ed., p. 345, footnote 13, D. C. Heath and Company, 1941.)

2. A regular Büchner funnel fitted with a mat of glass wool can be employed successfully. The filtration should be carried out as rapidly as possible, and the hydrolysis should be performed immediately thereafter if the humidity is high to minimize reaction on the funnel.

3. Most of the unreacted anthracene remains undissolved as a brown fluffy residue.

4. If the product has a tendency to separate as an oil, the addition of more solvent followed by heating to redissolve the material

and subsequent cooling will usually yield a crystalline product.

5. Lüttringhaus and Kacer [3] reported the melting point as ca. 80°, but May and Mosettig [4] have found it to be 74–76°.

3. Methods of Preparation

The procedure described is essentially that of Lüttringhaus and Kacer [3] except for the method of isolation of the product, which is due to May. [5]

[1] University of Vermont, Burlington, Vermont.
[2] Fieser, *Experiments in Organic Chemistry*, 3rd ed., p. 265, Fig. 46.4, D. C. Heath and Company, 1955.
[3] Lüttringhaus and Kacer, Ger. pat. 493,688 [*C. A.*, **24**, 2757 (1930)].
[4] May and Mosettig, *J. Am. Chem. Soc.*, **70**, 686 (1948).
[5] May, Private communication.

α-ACETYL-δ-CHLORO-γ-VALEROLACTONE

(Valeric acid, 2-acetyl-5-chloro-4-hydroxy-, γ-lactone)

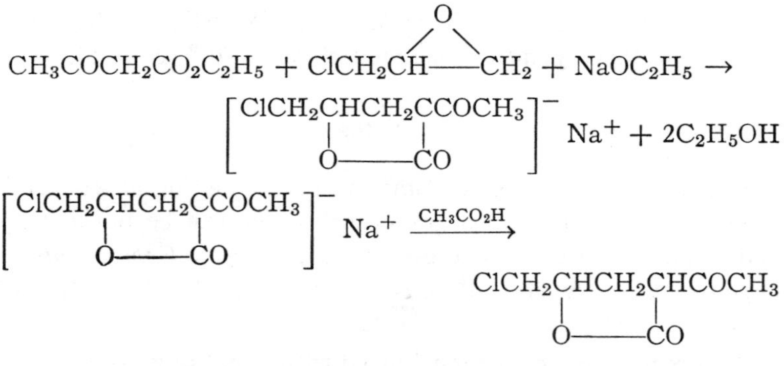

$$CH_3COCH_2CO_2C_2H_5 + ClCH_2CH\overset{O}{\overbrace{\qquad}}CH_2 + NaOC_2H_5 \rightarrow$$

$$\left[\begin{array}{c} ClCH_2CHCH_2CCOCH_3 \\ | \qquad\qquad | \\ O\text{———}CO \end{array}\right]^{-} Na^+ + 2C_2H_5OH$$

$$\left[\begin{array}{c} ClCH_2CHCH_2CCOCH_3 \\ | \qquad\qquad | \\ O\text{———}CO \end{array}\right]^{-} Na^+ \xrightarrow{CH_3CO_2H}$$

$$ClCH_2CHCH_2CHCOCH_3 \\ | \qquad\qquad | \\ O\text{———}CO$$

Submitted by G. D. Zuidema, E. van Tamelen, and G. Van Zyl.[1]
Checked by William S. Johnson and Herbert I. Hadler.

1. Procedure

A 1-l. three-necked round-bottomed flask is equipped with a sealed stirrer, a thermometer, a dropping funnel, and an efficient condenser, the upper end of which is protected with a calcium chloride drying tube. In this flask 23 g. (1 g. atom) of lustrous sodium (Note 1) is dissolved in 400 ml. of absolute ethanol (Note 2). The sodium is cut into about 25 pieces, and the entire amount is added at one time. It may be necessary to cool the flask in a cold-water bath if the reaction becomes violent. When all the sodium has dissolved, the solution is cooled to 50° and 130 g. (127 ml., 1 mole) of ethyl acetoacetate (Note 3) is added dropwise while the temperature is maintained between 45° and 50°. The resulting solution is cooled to about 35°, and 92.5 g. (78.4 ml., 1 mole) of epichlorohydrin (Note 4) is added dropwise with stirring over a period of 20 minutes. The temperature is then raised to 45° and is kept at 45–50° for 18 hours. The clear red-orange solution is cooled to 15°, and chilled glacial acetic acid (60–65 ml.) is added with stirring until the solution is just acid to litmus; a mush of sodium acetate crystals precipitates. The dropping funnel is replaced by a capillary tube, and the condenser is set for distillation. About three-fourths of the ethanol is removed under reduced pressure while air is bubbled into the mixture through the capillary tube (Note 5). Care is taken that the internal temperature does not exceed 100°.

The mushy residue is shaken with 250–300 ml. of water until the sodium acetate dissolves. The oily layer of lactone is separated, and the aqueous phase is extracted with two 100-ml. portions of ether. The combined oil and ether extracts are washed with 150 ml. of water and dried overnight over anhydrous sodium sulfate. The ether is removed under reduced pressure, and the product is distilled from a modified Claisen flask. The fraction boiling at 160–170°/11 mm. is collected; refractionation yields 107–114 g. (61–64%) of product boiling at 164–168°/11 mm. or 151–156°/8 mm.; n_D^{25} 1.4815–1.4830 (Notes 6 and 7).

2. Notes

1. The sodium must be present in an equivalent amount for best results. When 0.2 g. atom of sodium was used, the yield was only 10%.

2. It is necessary to maintain strictly anhydrous conditions in this reaction. The apparatus should be carefully predried and the absolute ethanol freshly prepared either by the diethyl phthalate method [2] or by the magnesium ethoxide method.[3]

3. Eastman Kodak Company white label quality ethyl aceto-acetate (b.p. 78–79°/11 mm.) was used.

4. Epichlorohydrin may be prepared from glycerol-α,γ-dichlo-rohydrin.[4] It is also commercially available.

5. Unless this precaution is taken, there is considerable bump-ing due to the presence of the solid sodium acetate in the mixture.

6. The product may become slightly colored upon standing.

7. This reaction is typical of those between the following epox-ides and ethyl acetoacetate:

Epoxide	Boiling Point of Product	% Yield of Product
Butadiene monoxide	148–151°/32 mm.	54
Propylene oxide	138–141°/26 mm.	49
Styrene oxide	164–167°/3 mm.	60
Ethyl glycidyl ether	160–163°/14–15 mm.	46
Phenyl glycidyl ether	195–197°/1 mm.	77

3. Methods of Preparation

α-Acetyl-δ-chloro-γ-valerolactone has been prepared only by the condensation of epichlorohydrin with ethyl acetoacetate. The preparation described is based on the method of Traube and Lehman.[5]

[1] Hope College, Holland, Michigan.

[2] Manske, *J. Am. Chem. Soc.*, **53**, 1106 (1931), footnote 9.

[3] Lund and Bjerrum, *Ber.*, **64**, 210 (1931). See Fieser, *Experiments in Organic Chemistry*, 3rd ed., p. 286, D. C. Heath and Company, Boston, Massachusetts, 1955.

[4] *Org. Syntheses Coll. Vol.* **1**, 233 (1941).

[5] Traube and Lehman, *Ber.*, **34**, 1980 (1901).

1-ACETYLCYCLOHEXANOL *

(Ketone, 1-hydroxycyclohexyl methyl)

Submitted by GARDNER W. STACY and RICHARD A. MIKULEC.[1]
Checked by JOHN C. SHEEHAN, GEORGE A. MORTIMER, and
NORMAN A. NELSON.

1. Procedure

In a 1-l. three-necked round-bottomed flask, equipped with a sealed stirrer, a reflux condenser, a thermometer, and a dropping funnel, is dissolved 5 g. of mercuric oxide (Note 1) in a solution of 8 ml. of concentrated sulfuric acid and 190 ml. of water. The solution is warmed to 60°, and 49.7 g. (0.40 mole) of 1-ethynyl-cyclohexanol (Note 2) is added dropwise over a period of 1.5 hours. After the addition has been completed, the reaction mixture is stirred at 60° for an additional 10 minutes and allowed to cool. The green organic layer which settles is taken up in 150 ml. of ether, and the aqueous layer is extracted with four 50-ml. portions of ether (Note 3). The combined ethereal extracts are washed with 100 ml. of saturated sodium chloride solution (Note 4) and dried over anhydrous sodium sulfate. The drying agent is removed, the ether is evaporated, and the residue is distilled under reduced pressure through a 15-cm. column packed with glass helices. The 1-acetylcyclohexanol is collected at 92–94°/15 mm. as a colorless liquid, n_D^{25} 1.4670, d_4^{25} 1.0248 (Note 5). The yield is 37–38 g. (65–67%).

2. Notes

1. Mallinckrodt mercuric oxide red (analytical reagent or N.F. 1x grade) was used.
2. 1-Ethynylcyclohexanol is available commercially. It may be prepared as reported by Saunders.[2]

3. To facilitate subsequent extractions, the solid material remaining after separation of as much of the aqueous phase as possible should be removed by gentle suction filtration and washed with 25 ml. of ether.

4. The sodium chloride solution removes the green color from the ether extract, leaving a yellow solution.

5. The checkers found b.p. 100°/21 mm., n_D^{25} 1.4662–1.4665, d_4^{25} 1.0235–1.0238. Others have reported b.p. 92–94°/12 mm., d_0^{20} 1.0256;[3] b.p. 91°/11 mm., n_D^{11} 1.4726, d_4^{11} 1.1033;[4] b.p. 88.0–88.6°/12 mm., n_D^{28} 1.4712.[5]

Establishing a criterion for the purity of the product is of particular importance because of the known tendency of ethynyl-carbinols to undergo rearrangement.[5, 6] The authors have reported that consecutive small fractions of the distillate possess a constant boiling point and refractive index. Furthermore, representative fractions, treated with periodic acid and subsequently with 2,4-dinitrophenylhydrazine, give cyclohexanone 2,4-dinitrophenylhydrazone in 83% over-all yield in a high state of purity.

3. Methods of Preparation

1-Acetylcyclohexanol has been prepared by the hydrolysis of 1-bromo-1-acetylcyclohexane[3] and of 1-acetoxy-1-acetylcyclohexane oxime,[7] by the hydration of 1-ethynylcyclohexanol,[4, 5, 8–12] by the treatment of 1-hydroxycyclohexanecarboxylic acid with methyllithium[13, 14] and by the hydrolysis of 1-(isopropoxyethoxy)-1-(1-iminoethyl)cyclohexane[15] and 1-(2-tetrahydropyranoxy)-1-(1-iminoethyl)cyclohexane.[16] The present procedure is based upon that of Stacy and Mikulec for the preparation of 1-acetyl-cyclopentanol.[6]

[1] State College of Washington, Pullman, Washington.

[2] *Org. Syntheses Coll. Vol.* **3**, 416 (1955).

[3] Favorskii, *J. Russ. Phys. Chem. Soc.*, **44**, 1339 (1912) [*C. A.*, **7**, 984 (1913)].

[4] Locquin and Wouseng, *Compt. rend.*, **176**, 516 (1923).

[5] Newman, *J. Am. Chem. Soc.*, **75**, 4740 (1953).

[6] Stacy and Mikulec, *J. Am. Chem. Soc.*, **76**, 524 (1954).

[7] Wallach, *Ann.*, **389**, 191 (1912).

[8] Bergmann, Brit. pat. 640,477 [*C. A.*, **45**, 1622 (1951)]; U. S. pat. 2,560,921 [*C. A.*, **46**, 3072 (1952)].

[9] Stacy and Hainley, *J. Am. Chem. Soc.*, **73**, 5911 (1951).

[10] Newman (to Ohio State University Research Foundation), U. S. pat. 2,853,520 [*C. A.*, **54**, 345 (1960)].

[11] Papa, Ginsberg, and Villani, *J. Am. Chem. Soc.*, **76**, 4441 (1954).

[12] Hennion and Watson, *J. Org. Chem.*, **23**, 656 (1958).

[13] Billimoria and MacLagan, *Nature*, **167**, 81 (1951); *J. Chem. Soc.*, **1951**, 3067.

[14] MacLagan and Billimoria, Brit. pat. 742,571 [*C. A.*, **50**, 16845 (1956)].

[15] Tchoubar, *Compt. rend.*, **237**, 1006 (1953).

[16] Elphimoff-Felkin, *Bull. soc. chim. France*, **1955**, 784.

2-*p*-ACETYLPHENYLHYDROQUINONE

(Acetophenone, 4′-(2,5-dihydroxyphenyl)-)

Submitted by GEORGE A. REYNOLDS and J. A. VANALLAN.[1]
Checked by R. S. SCHREIBER and A. H. NATHAN.

1. Procedure

In a 500-ml. beaker equipped with a mechanical stirrer are placed 27 g. (0.2 mole) of p-aminoacetophenone (Note 1), 100 g. of chopped ice, and 53 ml. of concentrated hydrochloric acid (sp. gr. 1.19). To the stirred mixture is added, over a period of 5 minutes, a solution of 13.8 g. (0.2 mole) of sodium nitrite dissolved in 75 ml. of water. The stirring is continued for 15 minutes, during which all the insoluble amine hydrochloride reacts to form the soluble diazonium compound (Note 2).

In a 4-l. beaker, equipped with a high-speed stirrer (Note 3), are placed 20 g. (0.185 mole) of quinone (Note 4), 34 g. (0.4 mole) of sodium bicarbonate, 50 g. of chopped ice, and 500 ml. of water. About 10 ml. of the above diazonium salt solution is added (Note 5). After the frothing has subsided (Note 6), the diazonium salt solution is added in 10- to 20-ml. portions over a period of about an hour (Note 7). The temperature of the reaction mixture is kept below 15° during this period by the addition of ice. After the diazonium salt solution has been added, the mixture is allowed to warm to room temperature, and the stirring is continued for an additional hour. The precipitate of 2-p-acetylphenylquinone is collected on a Büchner funnel and washed thoroughly with approximately 1 l. of water. The yield of crude yellow-brown solid is 40–41 g. (96–98%). The melting point ranges from 125–135° to 134–136° (Note 8).

The crude quinone is dissolved in 250 ml. of chloroform (Note 9) and added to a solution of 40 g. of sodium hydrosulfite in 300 ml. of water. The mixture is shaken for 10 minutes, and the light-tan 2-p-acetylphenylhydroquinone which precipitates from solution is collected on a Büchner funnel and dried. The yield of crude hydroquinone is 32–37 g. (78–92%). The melting point ranges from 175–180° to 184–194° (Note 10).

A suspension of 35 g. (0.153 mole) of 2-p-acetylphenylhydroquinone in 77 ml. of acetic anhydride is treated with 0.5 ml. of concentrated sulfuric acid (sp. gr. 1.84). The hydroquinone goes into solution immediately with the evolution of much heat. The dark-colored solution is allowed to stand at room temperature overnight; then it is poured into 400 ml. of water. The acetylated material is collected by suction filtration and dried. The crude

2-*p*-acetylphenylhydroquinone diacetate is distilled at reduced pressure (b.p. 236–241°/1 mm. or 182–190°/0.1 mm.), and the hot distillate is poured into 20 ml. of *n*-butyl alcohol (Note 11). The product immediately separates as a colorless, crystalline mass, which is collected by suction filtration and dried. The yield is 32–35 g. (67–73%), m.p. 104–105°.

To a 300-ml. three-necked round-bottomed flask, equipped with a sealed stirrer, a condenser, and a gas inlet tube, is added a solution of 34 g. (0.11 mole) of 2-*p*-acetylphenylhydroquinone diacetate in 140 ml. of hot methanol. The solution is cooled to room temperature, causing some of the hydroquinone diacetate to crystallize. A slow stream of nitrogen is passed through the suspension, and 70 ml. of methanol containing 6.1 g. of anhydrous hydrogen chloride is added. The reaction mixture is stirred at room temperature for 2 hours under nitrogen, during which period the hydroquinone diacetate gradually dissolves. The pale yellow solution is poured onto 500 g. of chopped ice, and the colorless or faintly yellowish solid is collected by suction filtration and dried. The yield of *p*-acetylphenylhydroquinone melting at 193–194° is 24.8 g. (quantitative). The over-all yield of product based on quinone is 50–66%.

2. Notes

1. The purest grade of *p*-aminoacetophenone supplied by the Eastman Kodak Company was used without further purification.

2. This reaction has been successfully carried out on a 3-mole scale.

3. A "Lightnin" mixer (manufactured by the Mixing Equipment Company, Rochester, N. Y.) equipped with a propeller stirrer was used. If rapid stirring is not maintained, the reaction does not go to completion.

4. The checkers used a practical grade of quinone obtainable from the Eastman Kodak Company.

5. If nitrogen is not evolved immediately, the reaction may be initiated by the addition of a small amount of hydroquinone.

6. If the foaming becomes too violent, a few drops of octyl alcohol are added.

7. The checkers found it convenient to add the diazonium salt

solution slowly from a dropping funnel. The time of addition was 25–45 minutes.

8. Recrystallization from butanol gives material melting at 139–140°. The pure substance is reported to melt at 152–153°.[2]

9. Ethanol can also be used as a solvent but has the disadvantage of more readily dissolving the hydroquinone, thus making it necessary to evaporate the solution nearly to dryness.

10. It is difficult to purify the crude hydroquinone by recrystallization; therefore the remainder of the procedure is recommended in order to obtain a highly purified product.

11. The checkers found it convenient to crystallize the viscous distillate from 125 ml. of methanol by chilling a hot solution in the refrigerator overnight. A small second crop amounting to about 2 g. may also be obtained by concentration of the mother liquors.

3. Methods of Preparation

This procedure is a modification of the method described for the preparation of 2-chlorophenylhydroquinone.[3] 2-p-Acetylphenylquinone has been prepared by carrying out the coupling in alcohol solution in the presence of sodium acetate instead of sodium bicarbonate.[2] Reduction by zinc, acetic acid, and a small amount of concentrated hydrochloric acid yielded 2-p-acetylphenylhydroquinone.[2]

[1] Eastman Kodak Company, Rochester, New York.

[2] Kvalnes, *J. Am. Chem. Soc.*, 56, 2478 (1934).

[3] B.I.O.S., Report 1146 (1946). [Reports obtainable from British Intelligence Objectives Subcommittee, 32 Bryanston Sq., London, W. 1.]

δ-ACETYL-*n*-VALERIC ACID

(Heptanoic acid, 6-oxo-)

$$\text{H}_2\text{C} \overset{\displaystyle\text{CH}_2}{\underset{\displaystyle\text{CH}_2}{\begin{array}{c}\text{CHCH}_3\\[4pt]\text{CHOH}\end{array}}} + 2\text{CrO}_3 + 3\text{H}_2\text{SO}_4 \rightarrow$$

$$\text{CH}_3\overset{\text{O}}{\underset{\|}{\text{C}}}(\text{CH}_2)_4\text{CO}_2\text{H} + 4\text{H}_2\text{O} + \text{Cr}_2(\text{SO}_4)_3$$

Submitted by J. R. SCHAEFFER and A. O. SNODDY.[1]
Checked by RICHARD T. ARNOLD and H. W. TURNER.

1. Procedure

A solution of 368 g. of 96% sulfuric acid in 664 ml. of water is cooled to room temperature and placed in a 3-l. three-necked flask provided with a mechanical stirrer, a thermometer, and a dropping funnel. To this acid solution is added 114 g. (1 mole) of 2-methylcyclohexanol (Note 1). A mixture of 220 g. (2.2 moles) of chromic oxide (Note 2) in 368 g. of 96% sulfuric acid and 664 ml. of water is added from the dropping funnel to the 2-methylcyclohexanol suspension at such a rate that the temperature of the mixture remains at 30 ± 2° (Note 3). Good agitation and an ice bath are necessary to control the temperature in this range. The mixture is stirred at 30 ± 2° for 1 hour and then at room temperature until all the chromic oxide is consumed (Note 4). The sulfuric acid solution is extracted with ether until the returns from the ether extractions fall to an insignificant amount. Approximately 10 extractions with 200-ml. portions of ether are required (Note 5). The ether extracts are combined, and the ether is removed by distillation on the steam bath. The resulting crude δ-acetyl-*n*-valeric acid is a yellow liquid with a sharp odor and amounts to about 130 g. The crude acid is purified by distillation through a 30-in. Vigreux column, using a variable take-off, a reflux ratio of 3:1, and a pressure of 1 mm. A fore-run of approximately 30 g. of material distilling up to 122°/1 mm. is obtained. The main fraction which distils at 122–123°/1 mm. is pure δ-acetyl-*n*-valeric acid and amounts to 66–79 g. (46–55%). The pure acid is a colorless crystalline hygroscopic solid which melts (sealed capillary) at 34–35° and is miscible with water in all proportions. The literature records the melting point of δ-acetyl-*n*-valeric acid as ranging from 31° to 42°.[2–4]

2. Notes

1. Eastman Kodak Company practical grade 2-methylcyclohexanol was used in this preparation.

2. Technical grade chromic oxide (99.5% CrO_3) in flake form was used.

3. An alcohol-Dry Ice bath is very convenient for this purpose; with this bath only about 45 minutes is needed for the addition of the chromic oxide solution. A water-ice bath can be used, but a longer time will be required for the addition of the chromic oxide solution.

4. The chromic oxide content of the mixture at any time may be determined by titrating a test portion against standard ferrous ammonium sulfate solution. If the mixture is allowed to stand overnight at room temperature without stirring, it will be free from chromic oxide. A convenient procedure is to perform the oxidation in the afternoon and the extraction the next day.

5. A liquid-liquid continuous extractor is convenient for extracting the crude acid from the aqueous solution. With such apparatus, it is possible to extract all the crude δ-acetyl-*n*-valeric acid from the aqueous acid in 6–8 hours.

6. The fore-run and the still residue contain some δ-acetyl-*n*-valeric acid. These fractions may be combined and redistilled to yield an additional 5–10% of δ-acetyl-*n*-valeric acid, but the low cost of the starting materials and the ease of preparing the crude δ-acetyl-*n*-valeric acid scarcely justify the labor unless a considerable number of batches are being prepared.

3. Methods of Preparation

δ-Acetyl-*n*-valeric acid has been prepared by the oxidation of 1-methylcyclohexene with potassium permanganate;[5] by the oxidation of 2-methylcyclohexanone with chromic oxide and sulfuric acid,[6, 7] with neutral potassium permanganate,[8] and by air in the presence of adipic acid and manganese nitrate;[9] by the reaction of methylzinc iodide on the ethyl ester of adipic acid chloride, followed by saponification of the ester so obtained;[10] by the saponification of the ethyl ester of diacetylvaleric acid;[2] by the hydrolysis of ethyl α-acetyl-δ-cyanovalerate with boiling 20% hydrochloric acid;[3] by the permanganate oxidation of 1-methyl-1,2-cyclopentanediol;[11] and by the alkaline cleavage of 2-acetylcyclopentanone.[12]

[1] Procter and Gamble Company, Ivorydale, Ohio.
[2] Perkin, *J. Chem. Soc.*, **57**, 229 (1890).
[3] Derick and Hess, *J. Am. Chem. Soc.*, **40**, 551 (1918).
[4] Blaise and Kohler, *Compt. rend.*, **148**, 490 (1909).
[5] Wallach, *Ann.*, **329**, 371, 376 (1903).
[6] Wallach, *Ann.*, **359**, 300 (1908).
[7] Ruzicka, Seidel, Schinz, and Pfeiffer, *Helv. Chim. Acta*, **31**, 422 (1948).
[8] Acheson, *J. Chem. Soc.*, **1956**, 4232.
[9] Badische Anilin- und Soda-Fabrik, Ger. pat. 812,073 [*C. A.*, **47**, 2769 (1953)].
[10] Blaise and Koehler, *Bull. soc. chim.*, [4] **7**, 222 (1910).
[11] Adkins and Roebuck, *J. Am. Chem. Soc.*, **70**, 4041 (1948).
[12] Hauser, Swamer, and Ringler, *J. Am. Chem. Soc.*, **70**, 4023 (1948).

ACROLEIN ACETAL *

(Acrolein diethyl acetal)

$$CH_2=CHCHO + HC(OC_2H_5)_3 \xrightarrow{NH_4NO_3} CH_2=CHCH(OC_2H_5)_2 + HCO_2C_2H_5$$

Submitted by J. A. VanAllan.[1]
Checked by T. L. Cairns and R. E. Benson.

1. Procedure

A warm solution of 3 g. of ammonium nitrate in 50 ml. of anhydrous ethanol is added to a mixture of 44 g. (52.4 ml., 0.79 mole) of acrolein and 144 g. (160 ml., 0.97 mole) of ethyl orthoformate, and the mixture is allowed to react at room temperature for 6–8 hours (Note 1). The light-red solution is filtered, 4 g. of sodium carbonate is added, and the reaction mixture is distilled from the sodium carbonate through a good column (Note 2).

The fraction boiling at 120–125° is acrolein acetal and weighs 73–81 g. (72–80%); n_D^{25} 1.398–1.407 (Note 3).

2. Notes

1. Refluxing causes the formation of some resinous material. The solution remains warm for about 1.5 hours after mixing.

2. The column described by Pingert [2] is suggested.

3. This reaction appears to be of general application; crotonaldehyde diethyl acetal is formed in 68% yield; b.p. 145–147°; n_D^{25} 1.409. (In this preparation the amount of ethyl orthoformate used is reduced to exactly one equivalent since it distils at 142–143°. For this particular acetal, it is preferable to use ethyl orthosilicate according to Helferich.[3]) Tiglylaldehyde diethyl acetal is formed in 79% yield; b.p. 158–160°; n_D^{25} 1.419.

3. Methods of Preparation

These have been reviewed previously.[2,4] The procedure described above is an adaptation of a method reported in a German patent.[5] It has been claimed that the reaction of acrolein with ethanol in the presence of hydrochloric acid [2] produces a mixture of substances from which no acrolein acetal can be isolated.[6] More recently it has been reported [7] that acrolein acetal is formed in 62% yield from acrolein and ethanol in the presence of p-toluenesulfonic acid.

[1] Eastman Kodak Company, Rochester, New York.

[2] Org. Syntheses, **25**, 1 (1945).

[3] Helferich, Ger. pat. 404,256 (1924) [Frdl., **14**, 228 (1921–1925)].

[4] Org. Syntheses Coll. Vol. **2**, 17 (1943).

[5] Schmidt, Ger. pat. 553,177 (1932) [Frdl., **19**, 229 (1932)].

[6] Hall and Stern, Chem. & Ind. (London), **1950**, 775.

[7] Weisblat, Magerlein, Myers, Hanze, Fairburn, and Rolfson, J. Am. Chem. Soc., **75**, 5893 (1953).

ALLOXAN MONOHYDRATE *

(Barbituric acid, 5,5-dihydroxy-)

$$\begin{array}{ccc}
\text{HN—CO} & & \text{HN—CO} \\
| \quad | & \xrightarrow{\text{CrO}_3} & | \quad |\diagup\text{OH} \\
\text{OC} \quad \text{CH}_2 & & \text{OC} \quad \text{C} \\
| \quad | & & | \quad |\diagdown\text{OH} \\
\text{HN—CO} & & \text{HN—CO}
\end{array}$$

Submitted by A. V. HOLMGREN and WILHELM WENNER.[1]
Checked by T. L. CAIRNS and R. W. UPSON.

1. Procedure

In a 2-l. three-necked, round-bottomed flask with glass joints are placed 850 g. of commercial glacial acetic acid and 100 ml. of water. The flask is fitted with a stirrer. One of the side necks carries a reflux condenser and a thermometer reaching to the bottom of the flask; the other is provided with a stopper which can be replaced by a powder funnel. The flask is surrounded by a water bath. At room temperature 156 g. (1.53 moles) of 98–99% chromium trioxide (Note 1) is added, and the mixture is stirred for about 15 minutes to effect solution of the oxidizing agent.

One hundred and twenty-eight grams (1 mole) of barbituric acid is added in the course of about 25 minutes in portions approximating 15–20 g. The temperature of the mixture rises from about 25–30° at the beginning of the reaction to 50° and is held at that value until all the barbituric acid has been added (Note 2). During the addition, alloxan monohydrate begins to crystallize. The temperature of the solution is held at 50° for 25–30 minutes after completion of the addition of barbituric acid. Then the reaction slurry, which contains the major amount of alloxan monohydrate in crystalline form, is cooled to 5–10° and filtered through a 5-in. Büchner funnel fitted with a piece of filter cloth. The product is washed while still on the funnel with cold glacial acetic acid until the washings are practically colorless. In order to effect rapid drying, the acetic acid is finally washed out of the filter cake by means of 100–200 ml. of ether. The yellow alloxan monohydrate weighs 120–125 g. (75–78%) after drying; m.p. 254° (dec.). It is pure enough for most purposes (Note 3).

2. Notes

1. This amount was found to give best yields.

2. It is very important that the temperature does not rise above 50°. If the addition of barbituric acid is carried out too rapidly, the temperature rise cannot be checked satisfactorily and the yield may drop considerably.

3. If entirely pure alloxan monohydrate is desired, this material is recrystallized according to the directions in an earlier volume of this series.[2]

3. Methods of Preparation

The methods for the preparation of alloxan have been reviewed earlier.[2] The present method is essentially that of Wenner.[3]

[1] Hoffmann-La Roche, Inc., Nutley, New Jersey.

[2] *Org. Syntheses Coll. Vol.* **3**, 37, 39 (1955).

[3] Wenner, U. S. pat. 2,445,898 [*C. A.*, **43**, 2227 (1949)].

ALLOXANTIN DIHYDRATE [*]

$$\begin{array}{c}\text{HN—C=O} \\ | \quad | \\ \text{O=C} \quad \text{C} \overset{\text{OH}}{\underset{\text{OH}}{\diagdown}} \\ | \quad | \\ \text{HN—C=O}\end{array} + H_2S \rightarrow \begin{array}{c}\text{HN—C=O} \\ | \quad | \\ \text{O=C} \quad \text{CHOH} \\ | \quad | \\ \text{HN—C=O}\end{array} + S + H_2O$$

$$\begin{array}{c}\text{HN—C=O} \\ | \quad | \\ \text{O=C} \quad \text{C} \overset{\text{OH}}{\underset{\text{OH}}{\diagdown}} \\ | \quad | \\ \text{HN—C=O}\end{array} + \text{HOHC} \begin{array}{c}\text{O=C—NH} \\ | \quad | \\ \text{C=O} \\ | \quad | \\ \text{O=C—NH}\end{array} \xrightarrow{H_2O}$$

$$\begin{array}{cc}\text{HN—C=O} & \text{O=C—NH} \\ | \quad | \diagup^{\text{H}} & \text{HO} \diagdown | \\ \text{O=C} \quad \text{C———O———C} \quad \text{C=O} (2H_2O) \\ | \quad | & | \quad | \\ \text{HN—C=O} & \text{O=C—NH}\end{array}$$

Submitted by R. STUART TIPSON.[1]
Checked by T. L. CAIRNS and F. S. FAWCETT.

1. Procedure

An apparatus is assembled in the hood, as shown in Fig. 1. In the 2-l. globe-shaped separatory funnel H (stopcock J closed) is placed 1.3 l. of deaerated water (Note 1). Three-holed rubber stopper G (bearing the stem of a 125-ml. separatory funnel C with stopcock D closed, a long inlet tube E, and a short outlet tube F) is inserted in the neck of H. In the neck of C, a rubber stopper B, provided with an inlet tube A, is inserted. H is flushed with nitrogen (Note 2) admitted at E. To the water is added 16.0 g. (0.1 mole) of alloxan monohydrate (Note 3), and the mixture is stirred by the flow of nitrogen through E until the alloxan monohydrate has dissolved. The nitrogen flow is discontinued, and hydrogen sulfide (Note 4) is passed in at E until the mixture is saturated with this gas and the aqueous solution is free from opalescence (about 2 hours). Carbon disulfide (100 ml.) is now added, and the mixture is agitated for 5 minutes by means of the hydrogen sulfide gas stream. The carbon disulfide layer is then cautiously withdrawn through J and discarded, and the aqueous solution is washed once more with 100 ml. of carbon disulfide which is separated and discarded. The hydrogen sulfide flow is discontinued, and nitrogen is passed in at E for about 2 hours or until the gas emerging at F gives no more than a faint test for hydrogen sulfide with lead acetate paper.

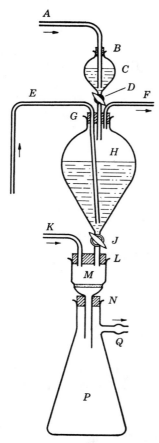

FIG. 1. Apparatus for preparation of alloxantin dihydrate.

Tubes E and F are now closed, stopper B is loosened, and C is flushed out with nitrogen admitted through tube A. Deaerated water (100 ml.) is placed in C, which is then flushed with nitrogen. To this water is added alloxan monohydrate (16.0 g.; 0.1 mole), and the mixture is stirred with A, while the nitrogen stream is continued, until the solid has dissolved. B is now pushed down to give a tight fit, a slight pressure of nitrogen is applied (at A), F and D are opened, and the solution is allowed to pass from C into H. To wash out traces of alloxan, a further 10 ml. of water is placed in C and run into H. D is closed, and nitrogen is passed in at E and out of F until the solutions are thoroughly mixed. E and F are now closed, and the mixture in H is allowed to stand until crystallization is complete (overnight). In the meantime, the stem of funnel H is inserted in one hole of the two-holed rubber stopper L (also bearing inlet tube K); L is inserted in the mouth of M (a 150-ml., Pyrex Büchner funnel with coarse, fritted-glass septum); and the stem of M is inserted in the rubber stopper N placed in the neck of a 2-liter Büchner flask P (side arm, Q). To flush out M and P, nitrogen is passed in at K and out of Q. K and Q are now closed, F is opened, and a slight pressure of nitrogen is applied at F. Q and J are opened, and, if necessary, slight suction is applied at Q. When all the suspension has passed out of H, the nitrogen stream is continued for a few minutes, to remove as much as possible of the liquid clinging to the precipitate in M. Then F, J, and Q are closed. The Büchner funnel M and its contents are quickly removed, placed in a vacuum desiccator (preflushed with nitrogen), and dried to constant weight, at room temperature, over phosphorus pentoxide and soda-lime. The yield is 27–27.5 g. (84–85%) (Notes 5, 6, and 7).

2. Notes

1. Deaerated water is prepared as follows. A boiling stone is added to distilled water which is then boiled under reflux for at least 5 minutes; it is cooled in ice to room temperature under an atmosphere of oxygen-free nitrogen.

2. For preparation of moist, oxygen-free nitrogen, the commercial gas is passed through (a) a 500-ml. Drechsel bottle containing a fresh solution of 25 g. of sodium hydroxide plus 5 g. of pyrogallol in 250 ml. of deaerated water, (b) a reversed, empty 500-ml. bottle, and (c) a 500-ml. bottle containing 250 ml. of deaerated water.

3. Alloxan monohydrate from Eastman Kodak Company is satisfactory. It is dried to constant weight over soda-lime and phosphorus pentoxide in the vacuum desiccator at room temperature. It should be colorless, and readily and completely soluble in 5 volumes of cold water. The sample employed assayed 99–100% alloxan monohydrate (p. 23) by Tipson and Cretcher's method.[2]

4. Commercial hydrogen sulfide is passed through a (reversed) empty 500-ml. Drechsel gas-washing bottle and then through 250 ml. of deaerated water in a similar bottle (not reversed).

5. The solubility of alloxantin dihydrate in water at room temperature is about 0.29 g. per 100 ml. of solution.[3] An additional 4 g. of product may be obtained by evaporation of the mother liquor to dryness at 25° under reduced pressure (nitrogen atmosphere).

6. The yield is slightly less if traces of crystals are left adhering to the inner walls of funnel H. The alloxantin obtained by

dehydration of the product at 120–150° under reduced pressure for 2 hours melts with decomposition in 0.5 to 1 minute when placed in a block heated to 245°.[4]

7. The submitter reports that with minor modification the above hydrogen sulfide reduction procedure can be applied to the preparation of the *dialuric acid monohydrate* intermediate. The apparatus is assembled as in Fig. 1 with the exception that funnel *C* and its accessories are deleted. The above reduction procedure is followed initially, employing 500 ml. of deaerated water and 50 g. of alloxan monohydrate instead of the quantities shown above. After the saturation with hydrogen sulfide (determined by weighing) and the first agitation with carbon disulfide have been conducted as above, the funnel is assembled for filtration in an atmosphere of hydrogen sulfide (rather than nitrogen), and the suspension in *H* is filtered through *M* by manipulations analogous to those described. The colorless crystals of dialuric acid monohydrate are washed on the filter with an additional 100 ml. of carbon disulfide added portionwise via *H*, and, while wet with carbon disulfide and hydrogen sulfide, the crystals and funnel *M* are transferred to a shielded vacuum desiccator and dried over soda-lime and phosphorus pentoxide under high vacuum (Dry Ice-cooled trap). The yield is 44–44.5 g. (87–88%). Even at 300°, the compound exhibits no true melting or gas evolution. Heated at 2° per minute in an aluminum block (initial temperature, 150°) it appears unchanged at 200°, turns faintly pink at 203–206°, and gradually becomes reddish brown (229–232°) and then purplish black at about 270°.

3. Methods of Preparation

Methods of preparation are reviewed in *Org. Syntheses Coll. Vol.* **3**, 42 (1955).

[1] Mellon Institute of Industrial Research, Pittsburgh, Pennsylvania.

[2] Tipson and Cretcher, *Anal. Chem.*, **22**, 822 (1950).

[3] Thunberg, *Skand. Arch. Physiol.*, **33**, 217 (1916) [*C. A.*, **11**, 456 (1917)]; Biilmann and Bentzon, *Ber.*, **51**, 522 (1918).

[4] *Org. Syntheses Coll. Vol.* **3**, 42 (1955).

2-AMINO-4-ANILINO-6-(CHLOROMETHYL)-*s*-TRIAZINE

[*s*-Triazine, 2-amino-4-anilino-6-(chloromethyl)-]

$$\underset{\text{C}_6\text{H}_5\text{NH}\overset{||}{\text{C}}\text{NH}\overset{||}{\text{C}}\text{NH}_2}{\overset{\text{NH} \quad \text{NH}}{}} + \text{ClCH}_2\text{CO}_2\text{C}_2\text{H}_5 \rightarrow$$

$$+ \text{C}_2\text{H}_5\text{OH} + \text{H}_2\text{O}$$

Submitted by C. G. OVERBERGER and FRANCIS W. MICHELOTTI.[1]
Checked by B. C. McKUSICK and F. W. STACEY.

1. Procedure

Methanol (225 ml.) is placed in a 500-ml. two-necked flask equipped with a mechanical stirrer and a reflux condenser. Sodium (6.8 g., 0.30 g. atom) in small pieces is dropped down the condenser into the stirred methanol. The resultant solution is cooled to room temperature, and 64 g. (0.30 mole) of 1-phenyl-biguanide hydrochloride (Note 1) is added. The mixture is stirred at room temperature for 20 minutes. The sodium chloride that precipitates is separated on a Büchner funnel and washed with 25 ml. of methanol.

The combined filtrates, which contain 1-phenylbiguanide as the free base, are placed in a 500-ml. three-necked flask equipped with a mechanical stirrer, a drying tube containing calcium chloride, and a dropping funnel, and 36.8 g. (0.30 mole) of ethyl chloroacetate (Note 2) is added at room temperature with stirring. The mixture is stirred at room temperature for 14 hours, during which time 2-amino-4-anilino-6-(chloromethyl)-*s*-triazine precipitates as a white solid. After being separated by filtration and air-dried, the triazine weighs 37–40 g. and melts at 138–140°. The methanol filtrate is added to 500 ml. of cold water. The mixture is cooled in an ice bath with stirring for 2 hours and filtered to remove an additional 10–12 g. of gray triazine, m.p. 140–142°. The total yield of crude product is 47–52 g.

The triazine is purified by recrystallizing it from 250 ml. of dioxane, using 2 g. of decolorizing carbon and filtering hot. The recrystallized triazine is dried for 5 hours at 60° (1–5 mm. pressure) in a vacuum oven (Note 3). It then weighs 31–33 g. (44–47%) (Note 4), m.p. 142–143° (Note 5).

2. Notes

1. 1-Phenylbiguanide hydrochloride can be obtained from American Cyanamid Company. If 1-phenylbiguanide itself is available, the triazine can be prepared in the same way by dissolving 53 g. (0.30 mole) of 1-phenylbiguanide in 250 ml. of methanol, adding 36.8 g. of ethyl chloroacetate, and proceeding as before. The same yield is obtained whether one starts with the free base or its hydrochloride.

2. Ethyl chloroacetate from Fisher Scientific Company was used.

3. The checkers found that less rigorous drying failed to remove all the dioxane.

4. An additional 3–5 g. of the triazine, m.p. 141–143°, can be obtained by concentrating the dioxane filtrate to about 60 ml. and cooling the concentrate.

5. 2-Chloromethyl-4,6-diamino-s-triazine can be prepared in 82% yield by stirring a mixture of biguanide and ethyl chloroacetate in methanol in the same way.

3. Methods of Preparation

2-Amino-4-anilino-6-(chloromethyl)-s-triazine has been prepared from 1-phenylbiguanide and ethyl chloroacetate in the presence of methanol [2] or sodium methoxide at −40°.[3]

[1] Polytechnic Institute of Brooklyn, Brooklyn, New York.

[2] Schuller, U. S. pats. 2,822,364 and 2,848,452 [C. A., **52**, 6807, 19169 (1958)].

[3] Shapiro and Overberger, J. Am. Chem. Soc., **76**, 97 (1954).

p-AMINOBENZALDEHYDE *

(Benzaldehyde, *p*-amino-)

$$\underset{\substack{\\ NO_2}}{\overset{\substack{CH_3 \\ \\}}{\bigcirc}} \xrightarrow[\text{H}_2\text{O}]{\text{Na}_2\text{S}_x + \text{NaOH}} \underset{\substack{\\ NH_2}}{\overset{\substack{CHO \\ \\}}{\bigcirc}}$$

Submitted by E. Campaigne, W. M. Budde, and G. F. Schaefer.[1]
Checked by Cliff S. Hamilton and R. C. Rupert.

1. Procedure

To 600 ml. of distilled water in a 1-l. beaker are added 30 g. (0.125 mole) of crystalline sodium sulfide nonahydrate (Note 1), 15 g. (0.47 g. atom) of flowers of sulfur, and 27 g. (0.67 mole) of sodium hydroxide pellets. The mixture is heated on a steam bath for 15–20 minutes with occasional stirring and then poured into a 2-l. round-bottomed flask containing a hot solution of 50 g. (0.36 mole) of *p*-nitrotoluene (Note 2) in 300 ml. of 95% ethanol. A reflux condenser is attached, and the mixture is heated under reflux for 3 hours. The resulting clear but deep red solution is rapidly steam-distilled until about 1.5–2 l. of condensate has been collected (Note 3). The distillate should be clear when the distillation is stopped. The residue in the 2-l. flask should have a volume of 500–600 ml.; if less, it should be diluted to this volume with boiling water. The solution is rapidly chilled in an ice bath with occasional vigorous shaking and stirring to induce crystallization. After 2 hours in the ice bath the golden yellow crystals of *p*-aminobenzaldehyde are collected on a Büchner funnel and washed with 500 ml. of ice water to remove sodium hydroxide (Note 4). The product is immediately placed in a vacuum desiccator over solid potassium hydroxide pellets for 24 hours. The yield of *p*-aminobenzaldehyde, m.p. 68–70°, amounts to 18–22 g. (40–50%). The product contains some impurities but is pure enough for most purposes (Note 5). It should be stored in a sealed bottle (Note 6).

2. Notes

1. Merck's reagent grade of sodium sulfide nonahydrate was used. Since sodium sulfide decomposes on contact with air, a freshly opened bottle should be employed. "Sodium Sulfhydrate" (Hooker Electrochemical Company hydrated sodium hydrosulfide) is also satisfactory; the amount should be based upon the formula $NaHS \cdot 2H_2O$, and an equivalent amount of sodium hydroxide in excess of the 27 g. is required.

2. The p-nitrotoluene used was Eastman Kodak Company practical grade.

3. The steam distillation should be carried out as rapidly as possible. The distillate contains ethanol, p-toluidine, and some unchanged p-nitrotoluene.

It has been reported [2] that a large amount of a dark oily tar may be present at this stage. Presumably it consists of Schiff's base polymers which have formed during the time necessary for reflux and steam distillation. The clear solution may be decanted from the oil, and the expected orange-yellow crystals of p-aminobenzaldehyde are obtained on cooling the solution.

The oily tar may be dissolved in boiling acetic anhydride, and upon dilution of the reaction mixture with water and partial concentration, crude p-acetamidobenzaldehyde separates. The latter may be purified by dissolving it in hot sodium bisulfite solution and fractionally precipitating the aldehyde by the addition of sodium hydroxide solution. From 12.3 g. of intractable tars there were obtained a first fraction which consisted of a dark sludge which was discarded, a second fraction which weighed 5.2 g., m.p. 150°, and a third which weighed 1.9 g., m.p. 147°. The melting point of p-acetamidobenzaldehyde is recorded [3] as 153°.

4. It is sometimes necessary to suspend the precipitate in about 200 ml. of ice water, stir it vigorously, and filter again to remove all traces of alkali.

5. The chief impurities are the polymeric condensation products of p-aminobenzaldehyde with itself. No satisfactory method for recrystallization has been found. If the melting point is high and a pure product is desired, it is best to extract with boiling water until the filtrate is clear, and extract the

monomer from the water with ether. This procedure gives recoveries of 25–30%.

Readily purified aldehyde derivatives may be prepared in good yields from the crude polymer mixture. The oxime melts at 124°, the azine at 245°, and the phenylhydrazone at 175°.[4] If these derivatives are hydrolyzed, the same crude *p*-aminobenzaldehyde of broad melting range results.

6. Care must be taken to exclude all traces of acid fumes from *p*-aminobenzaldehyde, since they catalyze its self-condensation.

3. Methods of Preparation

p-Aminobenzaldehyde has been prepared by the action of sodium polysulfide upon *p*-nitrotoluene [5–8] on which the method described is based. It can be prepared also from *p*-nitrobenzyl alcohol and sodium sulfide,[6] by heating *p*-nitrobenzaldehyde with sodium bisulfite and decomposing the addition product with hydrochloric acid,[9] by the reduction of *p*-nitrobenzaldoxime with ammonium sulfide [3, 10] and subsequent hydrolysis of the amino oxime, and by the decomposition of the benzenesulfonylhydrazide of *p*-aminobenzoic acid in the presence of powdered glass and sodium carbonate.[11]

[1] Indiana University, Bloomington, Indiana.

[2] Overberger, Private communication.

[3] Gabriel and Herzberg, *Ber.*, **16**, 2003 (1883).

[4] Walther and Kausch, *J. prakt. Chem.*, [2] **56**, 97 (1897).

[5] Geigy, Ger. pat. 86,874 (1895) [*Frdl.*, **4**, 136 (1894–1897)]; Friedländer and Lenk, *Ber.*, **45**, 2087 (1912).

[6] Beard and Hodgson, *J. Chem. Soc.*, **1944**, 4.

[7] Mukherjee, Indian pat. 43,527 [*C. A.*, **46**, 7121 (1952)].

[8] DeGarmo and McMullen (to Monsanto Chemical Co.) U. S. pat. 2,795,614 [*C. A.*, **51**, 16542 (1957)].

[9] Meister Lucius and Brüning, Ger. pat. 106,590 (1898) [*Chem. Zentr.*, I **71**, 1084 (1900)].

[10] Cohn and Springer, *Monatsh.*, **24**, 87 (1903).

[11] Newman and Caflish, *J. Am. Chem. Soc.*, **80**, 862 (1958).

2-AMINOBENZOPHENONE *

(Benzophenone, 2-amino-)

$$\text{(benzene-CO}_2\text{H, NH}_2) + RSO_2Cl \,^\dagger + Na_2CO_3 \rightarrow$$

$$\text{(benzene-CO}_2\text{Na, NHSO}_2R) + NaCl + CO_2 + H_2O$$

$$\text{(benzene-CO}_2\text{Na, NHSO}_2R) + HCl \longrightarrow \text{(benzene-CO}_2\text{H, NHSO}_2R) + NaCl$$

$$\text{(benzene-CO}_2\text{H, NHSO}_2R) + PCl_5 \longrightarrow \text{(benzene-COCl, NHSO}_2R) + POCl_3 + HCl$$

$$\text{(benzene-COCl, NHSO}_2R) + C_6H_6 \xrightarrow{\text{AlCl}_3} \text{(benzene-COC}_6\text{H}_5\text{, NHSO}_2R) + HCl$$

$$\text{(benzene-COC}_6\text{H}_5\text{, NHSO}_2R) \xrightarrow{\text{H}_2\text{SO}_4} \text{(benzene-COC}_6\text{H}_5\text{, } \overset{\oplus}{N}H_3 \; \overset{\ominus}{H}SO_4) + RSO_3H$$

$$\text{(benzene-COC}_6\text{H}_5\text{, } \overset{\oplus}{N}H_3 \; \overset{\ominus}{H}SO_4) \xrightarrow{\text{NH}_4\text{OH}} \text{(benzene-COC}_6\text{H}_5\text{, NH}_2) + (NH_4)_2SO_4$$

Submitted by H. J. SCHEIFELE, JR., and D. F. DETAR.[1]
Checked by R. T. ARNOLD and JOHN D. JONES.

1. Procedure

A. *p-Toluenesulfonylanthranilic acid.* In a 5-l. three-necked flask equipped with a stirrer and a thermometer extending to the bottom of the flask are placed 1.5 l. of water and 260 g. (2.4 moles) of technical grade dry sodium carbonate (Note 1). While the mixture is warmed, 137 g. (1 mole) of anthranilic acid is

† R = $p\text{-CH}_3\text{C}_6\text{H}_4$—.

added in three portions, and the temperature is raised to 70° to effect complete solution. The solution is allowed to cool to about 60°, and 230 g. (1.2 moles) of technical *p*-toluenesulfonyl chloride is added in 5 portions over a period of about 20 minutes (Note 2). When all the *p*-toluenesulfonyl chloride has been added, the reaction mixture is maintained at 60–70° for an additional 20 minutes. The temperature is raised to about 85°, 10 g. of Norit is added cautiously, and the solution is filtered by suction through a previously heated Büchner funnel.

In a 4-l. beaker equipped with a stirrer which can be operated above the liquid level to break the foam are placed 250 ml. of 12*N* hydrochloric acid and 250 ml. of water. The filtrate obtained above is cooled to about 50° and is added to the hydrochloric acid in small portions and at such a rate that the mixture does not foam over. If efficient stirring is used in the foam layer, this addition can be carried out in 5 minutes. The product is isolated by filtration through a Büchner funnel and is washed on the filter, first with a 250-ml. portion of dilute hydrochloric acid (prepared by diluting 50 ml. of 12*N* hydrochloric acid to about 250 ml.) to remove anthranilic acid, and then with 500 ml. of water. The product is sucked as dry as possible and is then spread in a thin layer and allowed to air dry for about 15 hours. When easily pulverizable, the material is transferred to an oven and dried for 3 hours at 100–120°.

There is obtained 257–265 g. (88–91%) of *p*-toluenesulfonylanthranilic acid as a pale lavender powder with a neutral equivalent of 294–300, which indicates a purity of 97–99% (Note 3). This product is suitable for conversion to 2-aminobenzophenone, but it may be recrystallized by dissolving in hot 95% ethanol (10 ml. per g.) and then adding water (4 ml. per g.). The recovery in the first crop is about 75% of material melting at 229–230° and having a neutral equivalent of 295.

B. *2-Aminobenzophenone.* In a dry 3-l. three-necked flask equipped with a stirrer, a reflux condenser connected to a hydrogen chloride trap, and a thermometer extending to the bottom of the flask are placed 146 g. (0.5 mole) of dry *p*-toluenesulfonylanthranilic acid, 1.5 l. of thiophene-free benzene, and 119 g. (0.57 mole) of phosphorus pentachloride. The mixture is stirred and heated at about 50° for 30 minutes. The murky solution (Note 4) is then cooled to 20–25°, and 290 g. (2.2 moles) of anhydrous aluminum chloride is added in 4 portions. When addition is complete, the dark mixture is heated with stirring at about 80–90° for 4 hours. The mixture is cooled to room temperature and poured onto a mixture of 500 g. of ice and 40 ml. of $12N$ hydrochloric acid in a 5-l. round-bottomed flask. The benzene is best removed by vacuum distillation using a water aspirator (Note 5). The grainy, brown, crude product is separated by filtration on a Büchner funnel and washed thoroughly with dilute hydrochloric acid, with water, then with two 500-ml. portions of 5% sodium carbonate (to remove anthranilic acid and starting material), and finally with three 500-ml. portions of water (Note 6). The filter cake is sucked reasonably dry.

The crude, moist sulfonamide is dissolved in 1.6 l. of concentrated sulfuric acid by warming on the steam bath for 15 minutes. The sulfuric acid solution is divided into two equal parts, each of which is placed in a 4-l. beaker. The beakers are cooled in ice baths while 1.6 kg. of ice is added slowly and with stirring to the contents of each beaker. During the addition of the ice, phenyl p-tolyl sulfone separates. A total of 50 g. of Norit is added, and the solution is filtered (Notes 7 and 8).

The filtrates are best neutralized separately. Two 5-gal. crocks are half filled with crushed ice, and one-half of the total filtrate is poured into each. For neutralization, commercial $12N$ ammonium hydroxide is added slowly with stirring; a total of 4.8 l. is required. The solid is collected on a Büchner funnel, washed with water, and air-dried.

The product is obtained in the form of bright yellow crystals, m.p. 103–105°. The yield is 68–71 g. (69–72% based on p-toluenesulfonylanthranilic acid). This material is dissolved in 1 l. of hot 95% ethanol, treated with 15 g. of Norit, and filtered. The hot solution is diluted with 700 ml. of hot water and cooled. After a second recrystallization, the yield of hexagonal yellow plates is 47 g.; m.p. 105–106°. Another 6 g. of pure aminoketone can be recovered from the filtrate. The total yield of recrystallized 2-aminobenzophenone is 53 g. (54%) (Note 9).

2. Notes

1. If sodium hydroxide is used, the main product is the *p*-toluenesulfonic acid salt of anthranilic acid. This salt has properties quite similar to those of the desired *p*-toluenesulfonylanthranilic acid but is useless for the preparation of 2-aminobenzophenone.

2. It is advisable to have sodium hydroxide solution available in case carbon dioxide is evolved indicating that the amount of sodium carbonate used was insufficient.

There is a tendency for salts to precipitate from the mixture and for some foaming to occur if much less water is used.

3. The melting point of the *p*-toluenesulfonylanthranilic acid is not a good criterion of purity because the *p*-toluenesulfonic acid salt of anthranilic acid has about the same value. The neutral equivalents are widely different: 154 for the salt and 291 for *p*-toluenesulfonylanthranilic acid. The compound obtained in this preparation gives a negative test for anthranilic acid on diazotization and treatment with alkaline β-naphthol solution. The probable impurity is the sodium salt of *p*-toluenesulfonylanthranilic acid.

4. When recrystallized *p*-toluenesulfonylanthranilic acid is used, the solution is clear at this point. The crude acid gives rise to a dark solution containing a small amount of suspended solid. The yield of 2-aminobenzophenone is the same in either case.

5. Steam distillation may also be used but should not be prolonged. If the contents of the flask are kept below 80°, the crude product is obtained as a fine powder. If the temperature becomes too high, the material melts and anthranilic acid is not easily removed from the solid mass obtained on cooling.

6. It is convenient to keep the wash solutions at 75–80°, but the temperature should not exceed 85° or part of the organic material will melt and clog the filter. It is advisable to transfer the solid to a 2-l. beaker to permit thorough washing. Most of the wash solution can be separated by decantation.

7. Phenyl *p*-tolyl sulfone may be isolated at this point by filtering the acid solution before using Norit. It can be purified by recrystallization from 95% ethanol; m.p. 125°.

8. The temperature of the solution should be 30–35°. If too

cold some of the product will be retained on the filter, and if too hot the filter paper will be attacked by the acidic solution. 2-Aminobenzophenone is a weak base and separates as the free base from sulfuric acid solutions below about $4N$.

9. 4'-Methyl-2-aminobenzophenone can be prepared similarly by substituting toluene for benzene. The yield of crude material, m.p. 85–88°, is 70%. On recrystallization from 95% ethanol, using 5 ml. per g., there is obtained, in two crops, a 70% recovery of 4'-methyl-2-aminobenzophenone, m.p. 92–93°. Because of the higher temperature required in the steam distillation (cf. Note 5), the sulfonamide is obtained in a form difficult to purify. As a result the crude aminoketone usually contains 1–2 g. of aluminum oxide.

3. Methods of Preparation

The above procedure is essentially that of Ullmann and Bleier.[2] 2-Aminobenzophenone has also been prepared by reduction of 2-nitrobenzophenone,[3] by the Hofmann reaction of the amide of o-benzoylbenzoic acid with sodium hypobromite,[4] by the action of an excess of benzoyl chloride on aniline at 220°,[5] and by hydrolysis of the acetyl derivative which is obtained by the action of phenylmagnesium bromide on 2-methyl-3,1,4-benzoxaz-4-one (from anthranilic acid and acetic anhydride).[6] Various methods for the preparation of 2-aminobenzophenones have been summarized critically by Simpson, Atkinson, Schofield, and Stephenson.[7]

[1] Cornell University, Ithaca, New York.

[2] Ullmann and Bleier, *Ber.*, **35**, 4273 (1902); cf. also Stoermer and Finche, *Ber.*, **42**, 3118 (1909).

[3] Geigy and Koenigs, *Ber.*, **18**, 2400 (1885); Tatschaloff, *J. prakt. Chem.*, [2] **65**, 308 (1902); Gabriel and Stelzner, *Ber.*, **29**, 1300 (1896).

[4] Graebe and Ullmann, *Ann.*, **291**, 8 (1896); Hewett, Lermit, Openshaw, Todd, Williams, and Woodward, *J. Chem. Soc.*, **1948**, 292. (The submitters have found that the Curtius procedure gives better results than the method of Hofmann; cf. P. A. S. Smith, in Adams, *Organic Reactions*, Vol. 3, p. 337, John Wiley & Sons, New York, 1946.)

[5] Chattaway, *J. Chem. Soc.*, **85**, 386 (1904).

[6] Lothrop and Goodwin, *J. Am. Chem. Soc.*, **65**, 363 (1943).

[7] Simpson, Atkinson, Schofield, and Stephenson, *J. Chem. Soc.*, **1945**, 646.

ε-AMINOCAPROIC ACID *

(Hexanoic acid, 6-amino-)

$$(CH_2)_5 \begin{array}{c} NH \\ | \\ C \end{array} \diagdown O + H_2O \xrightarrow{HCl} \overset{\ominus}{Cl} \ \overset{\oplus}{H_3N}(CH_2)_5CO_2H$$

$$\overset{\ominus}{Cl} \ \overset{\oplus}{H_3N}(CH_2)_5CO_2H \xrightarrow{Resin} H_2N(CH_2)_5CO_2H$$

Submitted by CAL Y. MEYERS and LEONARD E. MILLER.[1]
Checked by RICHARD T. ARNOLD and WILLIAM R. HASEK.

1. Procedure

Into a 500-ml. round-bottomed flask containing 50 g. (0.44 mole) of 2-ketohexamethylenimine (ε-caprolactam) is poured a solution containing 45 ml. of concentrated hydrochloric acid (sp. gr. 1.19) dissolved in 150 ml. of water. The mixture is boiled for 1 hour, and the resulting yellow solution is decolorized with Norit and evaporated to dryness under reduced pressure on a steam bath (Note 1).

The resulting ε-aminocaproic acid hydrochloride is converted into the amino acid by means of a column containing Amberlite IR-4B resin (Note 2):

1. Construct the column as shown in Fig. 2.

2. *Exhaustion.* Pass a 1% aqueous hydrochloric acid solution through the column (downflow) until the pH of the solution leaving the column decreases from 5.5–6.5 to about 2.

3. *Regeneration.* Now pass a 1% aqueous sodium hydroxide solution through the column (downflow) until the solution leaving the column is strongly alkaline.

4. *Classification.* Wash the resin (upflow) with 10 l. of distilled water.

5. Wash the resin with distilled water (downflow) (Note 3) until the salts are all washed out and the pH of the washings is 5.6–6.5. The column is now ready for use (Note 4).

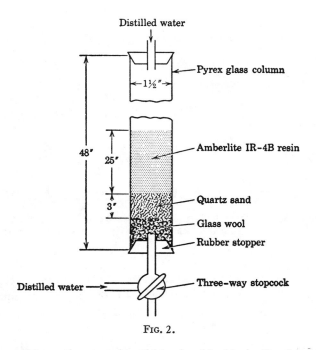

FIG. 2.

The solid ε-aminocaproic acid hydrochloride is dissolved in 1 l. of distilled water. This solution is passed through the column (downflow) and followed by at least 2 l. of distilled water (Note 5).

The collected solution (pink) is concentrated by distillation, under reduced pressure, to a volume of about 100 ml. (Note 6), and the resulting orange-colored solution is decolorized with Norit. After the addition of 300 ml. of absolute ethanol and 500 ml. of ether, followed by vigorous shaking, a white solid forms within a few minutes. The ε-aminocaproic acid is collected on a Büchner funnel and dried in a vacuum desiccator until no more ether-alcohol odor is detected. A yield of 52 g. (90%) is obtained; m.p. 202–203°.

2. Notes

1. This hydrolysis is similar to that utilized previously.[2]
2. Obtained from the Resinous Products Division, Rohm and Haas Company, Philadelphia, Pennsylvania. The checkers found that this resin sometimes liberates carbon dioxide when

treated with 1% hydrochloric acid. In such cases the total resin sample may be pretreated in a beaker with 1% hydrochloric acid until no more gas is evolved, and then a 30-in. (rather than 25-in.) column containing 1% hydrochloric acid is packed with resin. Further treatment of the resin is continued as described by the submitters.

3. Intermittent silver nitrate tests will indicate whether the solution is free of salts.

4. The liquid level of the column should always be above the resin when the column is not in use.

5. To assure the user that the resin is functioning, the eluant should be tested frequently for pH and the presence of salts. If the test for chloride ion becomes positive, or if the pH falls below 5.6, regeneration procedures are necessary; generally the column is good for two runs.

6. Excessive heating and excessive evaporation may result in peptide formation.

3. Methods of Preparation

ε-Aminocaproic acid has been prepared by the hydrolysis of ε-benzoylaminocapronitrile,[3] by the hydrolysis of diethyl ω-phthalimidobutylmalonate,[4] from cyclohexanone oxime by rearrangement and hydrolysis,[5] by hydrochloric acid hydrolysis of ε-caprolactam and removal of the acid by the use of litharge, silver oxide, etc.,[2] by alkaline hydrolysis of ε-caprolactam[6] or by hydrolysis of ε-caprolactam in the presence of a cation exchange resin,[7] by the reduction of δ-cyanovaleric acid[8] or the corresponding ethyl ester,[9] and by the hydrogenation of ε-oximinocaproic acid.[10]

[1] University of Illinois, Urbana, Illinois.

[2] *Org. Syntheses Coll. Vol.* **2**, 28 (1943).

[3] von Braun and Steindorff, *Ber.*, **38**, 117 (1905); von Braun, *Ber.*, **40**, 1839 (1907); Ruzicka and Hugoson, *Helv. Chim. Acta*, **4**, 479 (1921); Marvel, MacCorquodale, Kendall, and Lazier, *J. Am. Chem. Soc.*, **46**, 2838 (1924); Sumitomo and Hachihama, *Chem. High Polymers (Japan)*, **8**, 332 (1951) [*C. A.*, **48**, 593 (1954)].

[4] Gabriel and Maass, *Ber.*, **32**, 1266 (1899).

[5] Wallach, *Ann.*, **312**, 188 (1900); Eck and Marvel, *J. Biol. Chem.*, **106**, 387 (1934); Algemeene Kunstzijde Unie N.V., Ger. pat. 812,076 [*C. A.*, **48**, 6464 (1954)]; Nagasawa et al., Jap. pat. 7577 (1954) [*C. A.*, **50**, 10764 (1956)].

[6] Shpital'nyĭ, Shpital'nyĭ, and Yablochnik, *Zhur. Priklad. Khim.*, **32**, 617 (1959) [*C. A.*, **53**, 16005 (1959)].

[7] Itin and Kahr (to Inventa A.-G. für Forschung und Patentverwertung), Brit. pat. 774,468 [*C. A.*, **52**, 2056 (1958)].

[8] Shono and Hachihama, *Chem. High Polymers (Japan)*, **8**, 504 (1951) [*C. A.*, **48**, 10581 (1954)].

[9] Chrétien, *Ann. chim. (Paris)*, [13] **2**, 682 (1957).

[10] Chiusoli and Minisci, *Gazz. chim. ital.*, **88**, 261 (1958).

2-AMINO-3-NITROTOLUENE *

(*o*-Toluidine, 6-nitro-)

$$
\underset{}{\text{CH}_3}\text{C}_6\text{H}_3\text{NH}_2 + (\text{CH}_3\text{CO})_2\text{O} \longrightarrow \underset{}{\text{CH}_3}\text{C}_6\text{H}_3\text{NHCOCH}_3 + \text{CH}_3\text{CO}_2\text{H}
$$

$$
\underset{}{\text{CH}_3}\text{C}_6\text{H}_3\text{NHCOCH}_3 + \text{HNO}_3 \longrightarrow \underset{}{\text{CH}_3}\text{C}_6\text{H}_3(\text{NHCOCH}_3)(\text{NO}_2) + \text{H}_2\text{O}
$$

$$
\underset{}{\text{CH}_3}\text{C}_6\text{H}_3(\text{NHCOCH}_3)(\text{NO}_2) + \text{H}_2\text{O} \xrightarrow{\text{HCl}} \underset{}{\text{CH}_3}\text{C}_6\text{H}_3(\text{NH}_2)(\text{NO}_2) + \text{CH}_3\text{CO}_2\text{H}
$$

Submitted by JOHN C. HOWARD.[1]
Checked by CHARLES C. PRICE and JOSEPH D. BERMAN.

1. Procedure

A 1-l. three-necked flask is fitted with a sealed Hershberg stirrer, a reflux condenser, and a dropping funnel. The flask is charged with 650 ml. of acetic anhydride, and 107 g. (107 ml., 1 mole) of *o*-toluidine (Note 1) is introduced from the dropping funnel. The mixture becomes very warm. After the amine has been completely added, the solution is cooled to 12–13° in an ice-salt bath (Note 2). During the cooling, the dropping funnel and condenser are replaced by another dropping funnel containing 126 ml. (2 moles) of 70% nitric acid and a thermometer which can be read to within 0.5° in the range from 10° to 20° (Note 3).

The nitric acid is added drop by drop to the cold slurry at a rate which maintains the temperature carefully within the limits of 10–12° (Note 4). If the temperature persists in dropping, the addition is stopped after about 5 minutes. The ice bath is removed until the temperature rises 0.5°, the ice-salt bath is replaced, and addition is continued. As the reaction progresses, the acetotoluide which may have precipitated redissolves, and the solution becomes deeply colored. The addition is complete in 1–2 hours, and the nitro compounds may start to separate.

The solution is poured, with stirring, into 3 l. of ice water. The mixture of 4- and 6-nitroacetotoluides precipitates as a cream-colored solid which is collected on a large Büchner funnel. After thorough washing with four 500-ml. portions of ice water, the precipitate is partly dried by suction (Note 5). The moist product is then placed in a steam-distillation apparatus (Note 6), covered with 300 ml. of concentrated hydrochloric acid, and heated until the mixture boils. The acetotoluides are rapidly hydrolyzed, and the solution becomes dark red. Steam is then introduced, and the distillation is thus continued until 36 l. of distillate has been collected (Notes 7 and 8). The 2-amino-3-nitrotoluene, which separates as bright orange needles when the distillate is cooled, is collected on a large Büchner funnel. The dried product amounts to 75–84 g. (49–55%), m.p. 92–94°. The product may be further purified by a second steam distillation. Ten grams of the amine is distilled from 150 ml. of water, and 3 l. of distillate is collected, yielding 8.7 g. of 2-amino-3-nitrotoluene, m.p. 95–96° (cor.).

2. Notes

1. Commercially available *o*-toluidine, b.p. 75–77°/10 mm., is suitable. Redistillation of this material gave no significantly better results. The checkers obtained a 42% yield of 2-amino-3-nitrotoluene using practical grade *o*-toluidine directly, and a 57% yield after redistillation.

2. The flask should be immersed up to the neck in the slurry of ice and salt. During the cooling, the acetotoluide may suddenly precipitate, immobilizing the stirrer; a few turns manually break up the mass of crystals and allow the stirring to be continued.

3. A low-temperature thermometer with a range from $-15°$ to $+50°$ is suitable.

4. If the temperature is allowed to rise above 18°, violent if not explosive decomposition may ensue.

5. The precipitate can be air-dried to a constant weight of 150–160 g.

6. An efficient steam-distillation apparatus such as that described by Fieser [2] is recommended. A 12-l. round-bottomed flask cooled in a tub of ice serves as the receiver, which is equipped with an auxiliary vertical condenser attached to a gas absorption trap [3] to accommodate the hydrogen chloride which distils first.

7. The third 12-l. portion yields about 20 g. of 2-amino-3-nitrotoluene. The residue in the steam-distillation flask, about 20 g. of crude 2-amino-5-nitrotoluene, solidifies when cooled and may be separated by filtration. It can be recrystallized from 2 l. of hot water, yielding 14–15 g. of yellow plates, m.p. 130–131° (cor.).

8. Instead of separating the mixture of isomers by the slow steam distillation, one may employ the procedure of Wepster and Verkade.[4] In the latter, the product from the nitration of o-methylacetanilide is treated with the Witt-Utermann solution,[5] which consists of a water-alcohol solution of potassium hydroxide. 2-Acetylamino-5-nitrotoluene is insoluble in this solution, while the 3-nitro isomer is soluble and may be recovered in high yield and a good state of purity by acidification of the red filtrate. Hydrolysis of the acetyl derivatives affords 2-amino-5-nitro- and 2-amino-3-nitrotoluene.

3. Methods of Preparation

2-Amino-3-nitrotoluene has been prepared by the nitration of oxalotoluide [6] and by the nitration of o-acetotoluide in acetic acid with fuming nitric acid,[7] with a mixture of nitric and sulfuric acid,[8] or with metal nitrates.[9]

[1] Cornell University, Ithaca, New York.

[2] Fieser, *Experiments in Organic Chemistry*, 3rd ed., p 257 (Figs. 45.4 and 45.5), D. C. Heath and Company, Boston, Massachusetts, 1955.

[3] *Org. Syntheses Coll. Vol.* **2**, 4 (1943).

[4] Wepster and Verkade, *Rec. trav. chim.*, **68**, 77 (1949).

[5] Witt and Utermann, *Ber.*, **39**, 3901 (1906).

[6] Hadfield and Kenner, *Proc. Chem. Soc.*, **30**, 253 (1914).

[7] Cohen and Dakin, *J. Chem. Soc.*, **79**, 1127 (1901).
[8] McGookin and Swift, *J. Soc. Chem. Ind.*, **58**, 152 (1939).
[9] Kyryacos and Schultz, *J. Am. Chem. Soc.*, **75**, 3597 (1953).

3-AMINOPYRIDINE *

(Pyridine, 3-amino-)

$$+ Br_2 + 4NaOH \rightarrow$$

$$+ Na_2CO_3 + 2NaBr + 2H_2O$$

Submitted by C. F. H. ALLEN and CALVIN N. WOLF.[1]
Checked by CLIFF S. HAMILTON and MARJORIE DEBRUNNER.

1. Procedure

In a 2-l. beaker equipped with a mechanical stirrer and immersed in an ice-salt bath is placed a solution of 75 g. (1.87 moles) of sodium hydroxide in 800 ml. of water. To the solution is added, with stirring, 95.8 g. (30.7 ml., 0.6 mole) of bromine. When the temperature of the solution reaches 0°, 60 g. (0.49 mole) of nicotinamide (Note 1) is added all at once with vigorous stirring. After being stirred for 15 minutes, the solution is clear. The ice-salt bath is replaced by a bath containing water at 75°, and the solution is stirred and heated at 70–75° for 45 minutes.

The solution is cooled to room temperature, saturated with sodium chloride (about 170 g. is required), and extracted with ether in a continuous extractor (Note 2). The extraction time is 15–20 hours. The ether extract is adjusted to a volume of 1 l., dried over 4–5 g. of sodium hydroxide pellets, and filtered, and the ether is removed by distillation from a steam bath. The residue crystallizes on cooling. The yield of dark red crystals melting at 61–63° is 39–41 g. (85–89%).

The crude product is dissolved in a mixture of 320 ml. of benzene and 80 ml. of ligroin (b.p. 60–90°) and heated on a steam bath with 5 g. of Norit and 2 g. of sodium hydrosulfite for 20 minutes. The hot solution is filtered by gravity, allowed to cool slowly to room temperature, and then chilled overnight in a refrigerator. The product is isolated by gravity filtration (Note 3), washed on the filter with 25 ml. of ligroin, and dried in a vacuum desiccator. The yield of white crystals melting at 63–64° amounts to 28–30 g. (61–65%). By concentrating the combined filtrate and washings to a volume of 150 ml., an additional 2–3 g. of pale yellow crystals melting at 62–64° can be obtained. The total yield of 3-aminopyridine is 30–33 g. (65–71%).

2. Notes

1. The nicotinamide should be finely powdered to facilitate rapid solution.

2. The continuous extractor described by Pearl [2] was used. If the material is extracted in a separatory funnel, four 800-ml. portions and ten 500-ml. portions of ether are required to give the above yield.

3. Since 3-aminopyridine is somewhat hygroscopic, it tends to liquefy if collected on a suction filter.

3. Methods of Preparation

3-Aminopyridine has been prepared by heating nicotinamide in an alkaline potassium hypobromite solution at 70°; [3,4] by hydrolysis of β-pyridylurethane with oleum; [5] by heating 3-aminopyridine-2-carboxylic acid at 250°; [6] by reduction of 3-nitropyridine with zinc and hydrochloric acid; [7] by heating 3-bromopyridine with ammonia and copper sulfate in a sealed tube, [8,9] and by the hydrolysis of benzyl 3-pyridylcarbamate, prepared from nicotinic acid hydrazide through the corresponding azide. [10]

[1] Eastman Kodak Company, Rochester, New York.
[2] Pearl, *Ind. Eng. Chem., Anal. Ed.*, **16**, 62 (1944).
[3] Camps, *Arch. Pharm.*, **240**, 354 (1902).
[4] Philips, *Ann.*, **288**, 263 (1895).
[5] Curtius and Mohr, *Ber.*, **31**, 2494 (1898).
[6] Gabriel and Colman, *Ber.*, **35**, 2833 (1902).
[7] Binz and Räth, *Ann.*, **486**, 95 (1931).

[8] Maier-Bode, *Ber.*, **69**, 1534 (1936).

[9] Gitsels and Wibaut, *Rec. trav. chim.*, **60**, 176 (1941).

[10] Sugasawa, Akahoshi, Toda, and Tomisawa, *J. Pharm. Soc. Japan*, **72**, 192 (1952) [*C. A.*, **47**, 6418 (1953)].

p-AMINOTETRAPHENYLMETHANE *

(*p*-Toluidine, α,α,α-triphenyl-)

$$C_6H_5NH_2 \cdot HCl + (C_6H_5)_3COH \rightarrow$$
$$p\text{-}NH_2C_6H_4C(C_6H_5)_3 \cdot HCl + H_2O$$

$$p\text{-}NH_2C_6H_4C(C_6H_5)_3 \cdot HCl + NaOH \rightarrow$$
$$p\text{-}NH_2C_6H_4C(C_6H_5)_3 + NaCl + H_2O$$

Submitted by Benjamin Witten and E. Emmet Reid.[1]
Checked by Richard T. Arnold and Jerome J. Rosenbaum.

1. Procedure

Into a 1-l. round-bottomed flask equipped with a reflux condenser are introduced 100 g. (0.385 mole) of technical grade triphenylcarbinol (Note 1), 105 g. (0.81 mole) of dry aniline hydrochloride (Note 2), and 250 ml. of glacial acetic acid. The mixture is heated at the reflux temperature for 3 hours. During the period of reflux a clear brown homogeneous solution is formed. The solution while still hot is poured with stirring into a 4-l. beaker containing 2 l. of water. *p*-Aminotetraphenylmethane hydrochloride, which is not very soluble in water, separates as a light-brown solid. It is collected on a Büchner funnel and washed with 1 l. of water. The solid is then put back into the beaker, and a solution of 40 g. of sodium hydroxide in 2 l. of water is added. The mixture is heated to boiling for 1 hour to convert the hydrochloride to the free base (Note 3), which likewise is not very soluble in water. The mixture is allowed to cool to room temperature and is filtered with suction through a Büchner funnel. The solid material is washed with 500 ml. of water and is dried in an oven at 110–120°. The crude substance melts at 243–247°. It is purified by crystallization from 1.7 l. of toluene. The purified product (90–95 g., 70–74%) melts at 249–250° (Note 4).

2. Notes

1. Technical grade triphenylcarbinol is satisfactory, provided it is dry. The checkers obtained a final product having a higher melting point by starting with Eastman Kodak Company purest grade triphenylcarbinol.

2. The aniline hydrochloride must be dry if a good yield of product is to be obtained. Aniline hydrochloride can be prepared conveniently by mixing 75 g. of aniline and 80 ml. of concentrated hydrochloric acid in an evaporating dish and evaporating to dryness. The aniline hydrochloride should be dried in an oven at 110–120° before use. Aniline hydrochloride (Merck) which has been washed with ether and dried at 110–120° can be employed satisfactorily.

3. The mixture tends to bump during the period of heating. This bumping can be overcome by stirring the solution mechanically.

4. A product melting at 256–257° (uncor.) [2] was obtained by the checkers (Note 1).

3. Methods of Preparation

The procedure given is similar to one described by Ullmann and Münzhuber,[2] except that one-half as much aniline hydrochloride and two-thirds as much glacial acetic acid are used, and the time of reflux is reduced from 6 to 3 hours. p-Aminotetraphenylmethane can be prepared from triphenylchloromethane and aniline hydrochloride, following the same procedure outlined for triphenylcarbinol and aniline hydrochloride, except that a reaction time of 1 hour is sufficient. It has been prepared also from triphenylchloromethane and aniline.[3]

[1] Chemical Warfare Center, Edgewood Arsenal, Maryland.
[2] Ullmann and Münzhuber, *Ber.*, **36**, 407 (1903).
[3] MacKenzie and Chuchani, *J. Org. Chem.*, **20**, 336 (1955).

ARYLUREAS

I. CYANATE METHOD

p-BROMOPHENYLUREA

[Urea, (*p*-bromophenyl)-]

$$\text{Br}\langle\bigcirc\rangle\text{NH}_2 + \text{HNCO} \rightarrow \text{Br}\langle\bigcirc\rangle\text{NHCONH}_2$$

Submitted by FREDERICK KURZER.[1]
Checked by RICHARD T. ARNOLD and LESTER C. KROGH.

1. Procedure

In a 2-l. beaker, 86 g. (0.5 mole) of *p*-bromoaniline is dissolved in 240 ml. of glacial acetic acid and 480 ml. of water at 35°. This solution is treated with a solution of 65 g. (1 mole) of sodium cyanate (Note 1) in 450 ml. of water at 35° (Note 2). About 50 ml. of the sodium cyanate solution is added slowly with stirring until a white crystalline precipitate of the product appears. The rest is then added quickly with vigorous agitation (Note 3). The very rapid separation of the product is accompanied by a rise in the temperature to 50–55°. The thick, paste-like suspension is stirred for another 10 minutes, allowed to stand at room temperature for 2–3 hours, and diluted with 200 ml. of water. After cooling to 0°, the material is filtered with suction, washed with water, drained thoroughly, and dried. The yield of crude *p*-bromophenylurea, a white crystalline powder, is 95–100 g. (88–93%). The product is sufficiently pure for further synthetic work, but it can be recrystallized from aqueous ethanol (12 ml. of ethanol and 3 ml. of water per gram of crude material) to give a 65% recovery of lustrous white prisms of *p*-bromophenylurea, m.p. 225–227° (Note 4). This method is suitable for the preparation in excellent yields of a large number of arylureas (Note 5).

2. Notes

1. Comparable results are obtained with an equivalent quantity (81 g.) of potassium cyanate.

2. Instead of the solution, a well-stirred suspension of the cyanate in 150 ml. of water may be used with equal success.

3. Considerable frothing occurs with loss of some isocyanic acid (faint smell resembling that of sulfur dioxide). The foam collapses readily on stirring.

4. Melting points varying between 220° and 278° have been reported for *p*-bromophenylurea (see Methods of Preparation). It has been shown [2] that the thermal conversion of arylureas to the corresponding diarylureas takes place extremely rapidly, even below the melting point. This is particularly true of arylureas containing certain substituents in the *para* position to the ureido grouping. Melting points of such compounds are therefore liable to be indefinite and correspond to mixtures of the *mono* and *sym* disubstituted urea, especially if the temperature is raised slowly.

Reproducible values for the melting points of such compounds, including *p*-bromophenylurea, have now been determined by the following simple procedure: After the approximate softening temperature or melting range of the urea derivative has been found, the bath temperature of the apparatus is raised a further 10–20° above that point and samples are inserted into the slowly cooling bath until a temperature is reached at which a specimen just fails to liquefy. Insertion of a further specimen at a temperature 1° higher causes instantaneous fusion and can be taken as the "melting point" of the urea derivative under examination. Under these conditions practically no conversion to a diarylurea occurs, and samples withdrawn immediately after fusion show nitrogen contents corresponding to the original arylurea. This method gives reproducible physical constants for urea derivatives whose properties preclude slow heating.

5. A list of substituted arylureas prepared by the method above is given in the accompanying table, which records slight variations in the quantities of starting materials employed and in the yields obtained.

The quantities of water recorded in column III are used for preparing the arylamine acetate solutions. The sodium cyanate is dissolved or suspended in the appropriate volume of water as detailed in the synthesis of *p*-bromophenylurea. The yields refer to the quantity of crude product. The substituted ureas thus obtained are usually sufficiently pure for further syntheses but can be recrystallized from ethanol.

SUBSTITUTED ARYLUREAS

I Arylurea	II Aryl- amine, moles	III Water, ml.	IV Acetic Acid, ml.	V Sodium Cyanate, moles	VI Initial Tempera- ture, °C	VII Yield, %
p-Tolylurea	0.5	500	50	1.0	25	96
m-Tolylurea	0.5	400	75	1.0	30	86
o-Tolylurea	0.5	400	75	1.0	30	94
1-(4-Biphenylyl)urea	0.3	500	350	0.6	40	80
1-(2-Biphenylyl)urea	0.25	400	400	0.5	30	75
p-Methoxyphenylurea	0.3	150	30	0.6	40	94
p-Ethoxyphenylurea	0.5	500	75	1.0	25	95
m-Ethoxyphenylurea	0.3	500	120	0.6	20	54
o-Ethoxyphenylurea	0.5	500	170	1.0	20	88
m-Bromophenylurea	0.5	300	150	1.0	20	90
o-Bromophenylurea	0.5	100	250	1.0	25	92
o-Chlorophenylurea	0.5	100	200	1.0	25	92

3. Methods of Preparation

p-Bromophenylurea has been prepared by the bromination of phenylurea, using glacial acetic acid,[3, 4] ethanol,[4] or chloroform [5] as solvents; by the action of potassium cyanate on p-bromoaniline hydrochloride; [4, 6] and by the interaction of p-bromoaniline and ethyl allophanate.[7]

[1] University of London, London, England.

[2] Kurzer, J. Chem. Soc., **1949**, 2292.

[3] Pinnow, Ber., **24**, 4172 (1891).

[4] Wheeler, J. Am. Chem. Soc., **51**, 3653 (1929).

[5] Desai, Hunter, and Khalidi, J. Chem. Soc., **1934**, 1186.

[6] Scott and Cohen, J. Chem. Soc., **121**, 2034 (1922).

[7] Dains and Wertheim, J. Am. Chem. Soc., **42**, 2303 (1920).

II. UREA METHOD

p-ETHOXYPHENYLUREA

(Dulcin)

(Urea, *p*-ethoxyphenyl-)

$$C_2H_5O\langle\underset{}{\bigcirc}\rangle NH_2 \cdot HCl + NH_2CONH_2 \rightarrow$$

$$C_2H_5O\langle\underset{}{\bigcirc}\rangle NHCONH_2 + NH_4Cl$$

1. Procedure

A mixture of 870 g. (5 moles) of *p*-phenetidine hydrochloride and 1.2 kg. (20 moles) of urea is placed in a 12-l. round-bottomed flask (Note 1). To this mixture are added 2 l. of water, 40 ml. of concentrated hydrochloric acid, and 40 ml. of glacial acetic acid, and the well-shaken suspension is heated to boiling. The dark purple solution thus obtained is boiled vigorously for 45–90 minutes until the reaction is complete. The liquid remains clear during the first half of the heating period. Separation of the product begins during the last half and proceeds with increasing rapidity until the entire contents of the vessel suddenly set to a solid mass. The source of heat is immediately withdrawn at this point (Note 2).

After cooling to room temperature, the product is broken up with the addition of 1–1.5 l. of water, filtered with suction, washed with cold water, drained, and dried. The crude *p*-ethoxyphenylurea is obtained in a yield of 740–810 g. (82–90%) as a nearly white to pale yellow solid (Note 3). This material may be purified by recrystallization from boiling water (Note 4), when minute white plates, m.p. 173–174°, are obtained. The method is applicable to the preparation of other substituted arylureas (Note 5).

2. Notes

1. An enameled-steel vessel of approximately 2-gal. capacity is suitable for carrying out this reaction.

2. After removal of the heat source, a vigorous reaction may

continue for a few minutes, and the reaction mixture tends to froth somewhat. It eventually sets to a sponge-like formation of a crystalline mass. In smaller-scale experiments the final stage of the reaction is more easily controlled. The checkers employed one-tenth the scale and conventional equipment.

3. The crude product contains varying small quantities of the symmetrical disubstituted compound, di-(p-ethoxyphenyl)urea $[(p\text{-}C_2H_5OC_6H_4NH)_2CO]$. This substance is removed in the crystallization from water (Note 4).

4. To 1 l. of boiling water, 35 g. of crude p-ethoxyphenylurea is added. The bulk of the urea dissolves readily, and the solution is decolorized by the addition of 3 g. of activated charcoal, boiled for 5 minutes, and quickly filtered with suction through a preheated Büchner funnel. The colorless filtrate is slowly cooled to 0°, when lustrous minute plates of p-ethoxyphenylurea separate. An 80% recovery of material having m.p. 173–174° is obtained. Prolonged boiling of the solution should be avoided, since slow conversion to sym-di-(p-ethoxyphenyl)urea occurs under these conditions.

5. p-Anisidine hydrochloride (0.5 mole), when boiled with the proportionate quantities of urea and other reagents for 1 hour, gives 80–85% yields of p-methoxyphenylurea. Owing to the greater solubility of this product in water, less of the material separates during the heating period, but satisfactory crystallization occurs when the reaction liquid is slowly cooled to 0°. 1-Amino-2-naphthol hydrochloride gives 72–87% yields of 1-(2-hydroxy-1-naphthyl)urea.

3. Methods of Preparation

p-Ethoxyphenylurea has been prepared by the action of potassium cyanate on p-phenetidine hydrochloride [1] or p-phenetidine acetate,[2] and by the interaction of p-phenetidine with the following agents: phosgene in benzene or toluene and treatment of the product with ammonia;[3] urethan;[4] urea salts;[5] acetylurea;[6] and a mixture of urea and ammonium chloride.[7] It has been obtained from the reaction of phenetidine salts (usually the hydrochloride) with urea,[5,8] or with a mixture of sodium cyanide and sodium hypochlorite or peroxide.[9] p-Ethoxyphenylurea has also been prepared by heating *p*-ethoxyphenylurethan and ammonia to 100–180°;[4] by heating di-(*p*-ethoxyphenyl)urea with urea, ammonium carbamate, commercial ammonium carbonate,[10] or ethanol and ammonia;[11] by treating ammonium *p*-ethoxyphenyldithiocarbamate with lead carbonate in alcoholic solution;[12] by ethylating *p*-hydroxyphenylurea;[13] and by the action of ammonia on *p*-ethoxyphenyl isocyanate.[14]

[1] Berlinerblau, *J. prakt. Chem.*, [2] **30**, 103 (1884).

[2] Sonn, Ger. pat. 399,889 (1924) [*Chem. Zentr.*, II **95**, 1513 (1924)].

[3] Berlinerblau, Ger. pat. 63,485 [*Frdl.*, **3**, 906 (1890–1894)].

[4] Riedel, Ger. pat. 77,420 [*Frdl.*, **4**, 1269 (1894–1897)].

[5] Riedel, Ger. pat. 76,596 [*Frdl.*, **4**, 1268 (1894–1897)].

[6] Riedel, Ger. pat. 79,718 [*Frdl.*, **4**, 1270 (1894–1897)]; Roy and Ray, *Quart. J. Indian Chem. Soc.*, **4**, 339 (1927).

[7] Loginov and Polyanskii, U.S.S.R. pat. 65,779 [*C. A.*, **40**, 7234 (1946)].

[8] Roura, *Industria y quimica Buenos Aires*, **3**, 160 (1941); Volynkin, *Zhur. Obshcheĭ Khim.*, **27**, 483 (1957) [*C. A.*, **51**, 15437 (1957)].

[9] Riedel, Ger. pat. 313,965 [*Frdl.*, **13**, 1049 (1916–1921)].

[10] Riedel, Ger. pat. 73,083 [*Frdl.*, **3**, 907 (1890–1894)].

[11] Riedel, Ger. pat. 77,310 [*Frdl.*, **4**, 1271 (1894–1897)].

[12] Heller and Bauer, *J. prakt. Chem.*, [2] **65**, 379 (1902).

[13] Riedel, Ger. pat. 335,877 (1921) [*Chem. Zentr.*, IV **92**, 1324 (1921)].

[14] Sah and Chang, *Ber.*, **69**, 2762 (1936).

DL-ASPARTIC ACID *

$$\left[\begin{array}{c} \text{(phthalimido)} \end{array} NC(CO_2C_2H_5)_2 \right]^{-} Na^{+} + ClCH_2CO_2C_2H_5 \longrightarrow$$

$$\text{(phthalimido)} NC(CO_2C_2H_5)_2CH_2CO_2C_2H_5 + NaCl$$

$$\text{(phthalimido)} NC(CO_2C_2H_5)_2CH_2CO_2C_2H_5 + 5H_2O \xrightarrow[CH_2CO_2H]{HCl}$$

$$HO_2CCH_2CH(NH_2)CO_2H + \text{(benzene)}\begin{array}{c}-CO_2H\\-CO_2H\end{array} + 3C_2H_5OH + CO_2$$

Submitted by M. S. Dunn and B. W. Smart.[1]
Checked by H. T. Clarke and W. Pearlman.

1. Procedure

A. *Triethyl α-phthalimidoethane-α,α,β-tricarboxylate.* Three hundred and twenty-seven grams (1.0 mole) of diethyl sodium phthalimidomalonate [2] and 735 g. (6.0 moles) of ethyl chloroacetate (b.p. 144–145°) are placed in a 2-l. Claisen flask fitted with a reflux condenser and rubber stoppers. The mixture is heated under reflux in an oil bath at 150–160° for 2.25 hours. The excess ethyl chloroacetate is removed by distillation at 30 mm. until the heating bath temperature reaches 150° and no more distillate is obtained (Note 1). The brown residual mass is cooled and then extracted with 1250 ml. of ether. The oil dissolves, leaving a solid residue which is separated by filtration and washed with 750 ml. of ether. The combined ether extracts are distilled to remove ether, and the residual oil is heated on a steam bath under reduced pressure (35 mm.) to remove traces of ethyl chloroacetate. The yield of triethyl α-phthalimidoethane-α,α,β-tricarboxylate, dried at 45° for 48 hours, is 373–389 g. (95–99%) (Note 2).

B. DL-*Aspartic acid.* A mixture of 383 g. of the above crude product, 1 l. of concentrated hydrochloric acid, 1 l. of glacial acetic acid, and 1 l. of water is boiled under reflux in a 5-l. round-bottomed flask for 2–3 hours. The reflux condenser is then replaced by a fractionating column, and the mixture is slowly distilled until the temperature at the head of the column has risen to 108°. This requires about 13 hours. The distillate amounts to 1.5 l. (Note 3).

The residual mixture is allowed to cool, and the phthalic acid which crystallizes is removed by filtration and washed with 350 ml. of 1% hydrochloric acid (Note 4). The combined filtrate and washings are distilled nearly to dryness on a steam bath under reduced pressure; the bulk of the hydrochloric and acetic acids remaining is removed by slowly adding 300 ml. of water through a dropping funnel while the distillation under reduced pressure is continued. The dark brown residue is warmed on a steam bath with 700 ml. of water, is allowed to cool, and is filtered to remove a small amount of black insoluble matter. The filtrate is decolorized with 2 g. of Norit, 200 ml. of hot water being used to wash the Norit. The volume of the combined filtrate and washings, amounting to about 1.2 l., is measured accurately, and a small portion is analyzed for chloride (Note 5). An amount of pyridine corresponding exactly to the chloride content is added, diluted with 500 ml. of 95% ethanol. The DL-aspartic acid, which crystallizes at once, is separated by filtration after the mixture has stood for 24 hours at room temperature and is washed with 50–100 ml. of cold water (Note 6).

The crude DL-aspartic acid, amounting to 58–60 g., is recrystallized from 600 ml. of hot water and yields 54 g. of pure DL-aspartic acid. The mother liquors on evaporation to about 90 ml. yield an additional 2–3 g. (Note 7). The total yield of pure colorless DL-aspartic acid is 56–57 g. (42–43%) (Note 8).

2. Notes

1. From 490 to 536 g. of ethyl chloroacetate (b.p. 144–145°) is recovered by the distillation.

2. Although this product cannot be purified by distillation, it contains almost the theoretical amount of nitrogen as shown by Kjeldahl analysis.

3. During the first few hours the distillate contains ethyl acetate; the distillate obtained during the first hour, amounting to 137 ml., distils below 99° and on saturation with sodium chloride yields 115 ml. of crude ethyl acetate.

4. The phthalic acid so obtained is brown and weighs 140–150 g.

5. The total amount of chloride found should be less than 1 mole.

6. The mother liquor contains too little DL-aspartic acid to justify its recovery. When the filtrate and washings are evaporated to a syrup and treated with 500 ml. of 95% ethanol, the pyridine hydrochloride dissolves completely, leaving 8–9 g. of crude glycine which yields little or no sparingly soluble DL-aspartic acid on treatment with a minimum quantity of cold water.

7. The final mother liquor from the recrystallization of DL-aspartic acid yields a small quantity (about 0.5 g.) of glycine.

8. The purity of the recrystallized DL-aspartic acid was established by nitrogen analysis by the Kjeldahl and Van Slyke methods. The decomposition point of this product is 325–348°.

3. Methods of Preparation

The above method for the preparation of DL-aspartic acid is a modification of one described by Dunn and Smart.[3] Other methods are: the decomposition of acid ammonium malate by heat;[4] the racemization of active aspartic acid[5] and active asparagine;[6] the reaction of maleic and fumaric acids with ammonia in a closed tube;[7] the reduction of oxalacetic ester oxime;[8] the reduction of silver fumarate by hydroxylamine hydrochloride;[9] the reduction of nitrosuccinic ester;[10] the catalytic reduction and amination of oxalacetic acid; [11] the hydrolysis of tri-ethyl α-aminoethane-α,α,β-tricarboxylate; [12] the hydrolysis of β,β-dicarbethoxy-β-acetylaminopropionic acid; [13] the hydrolysis of dimethyl (carbethoxy)formamidomalonate; [14] the oxidation of β-(2-furyl)-α-alanine by potassium permanganate; [15] and the hydrolysis of diethyl α-acetyl-α-aminosuccinate. [16]

[1] University of California, Los Angeles, California.
[2] *Org. Syntheses Coll. Vol.* **2**, 384 (1943).
[3] Dunn and Smart, *J. Biol. Chem.*, **89**, 41 (1930).
[4] Dessaignes, *Compt. rend.*, **30**, 324 (1850); **31**, 432 (1850); Wolff, *Ann.*, **75**, 293 (1850).
[5] Michael and Wing, *Ber.*, **17**, 2984 (1884); *Am. Chem. J.*, **7**, 278 (1885).
[6] Piutti, *Ber.*, **19**, 1691 (1886).
[7] Engel, *Compt. rend.*, **104**, 1805 (1887); **106**, 1734 (1888); Stadnikoff, *Ber.*, **44**, 44 (1911).
[8] Piutti, *Gazz. chim. ital.*, **17**, 519 (1887).
[9] Tanatar, *Ber.*, **29**, 1477 (1896).
[10] Schmidt and Widmann, *Ber.*, **42**, 497 (1909).
[11] Knoop and Oesterlin, *Z. physiol. Chem.*, **148**, 294 (1925).
[12] Keimatsu and Kato, *J. Pharm. Soc. Japan*, **49**, 111 (1929) [*C. A.*, **24**, 70 (1930)].
[13] Warner and Moe (to General Mills, Inc.), U. S. pat. 2,523,744 [*C. A.*, **45**, 5718 (1951)].
[14] Meek, Minkowitz, and Miller, *J. Org. Chem.*, **24**, 1397 (1959).
[15] Takahashi, *Nippon Kagaku Zasshi*, **76**, 735 (1955) [*C. A.*, **51**, 17758 (1957)].
[16] Takizawa, *Science Repts. Research Insts. Tôhoku Univ.*, Ser. A, **4**, 316 (1952) [*C. A.*, **48**, 1959 (1954)].

ATROLACTIC ACID *

$$C_6H_5COCH_3 + NaCN + HCl \longrightarrow C_6H_5\underset{\underset{CN}{|}}{C}(OH)CH_3 + NaCl$$

$$C_6H_5\underset{\underset{CN}{|}}{C}(OH)CH_3 + H_2O \xrightarrow{HCl} C_6H_5\underset{\underset{CONH_2}{|}}{C}(OH)CH_3$$

$$C_6H_5\underset{\underset{CONH_2}{|}}{C}(OH)CH_3 + NaOH \longrightarrow C_6H_5\underset{\underset{CO_2Na}{|}}{C}(OH)CH_3 + NH_3$$

$$C_6H_5\underset{\underset{CO_2Na}{|}}{C}(OH)CH_3 + HCl \longrightarrow C_6H_5\underset{\underset{CO_2H}{|}}{C}(OH)CH_3 + NaCl$$

Submitted by Ernest L. Eliel and Jeremiah P. Freeman.[1]
Checked by Arthur C. Cope, William F. Gorham, and Roscoe A. Pike.

1. Procedure

Caution! This preparation must be conducted in a hood to avoid exposure to the poisonous hydrogen cyanide that is evolved.

In a 1-l. three-necked round-bottomed flask equipped with a Hershberg stirrer, a thermometer, and a 250-ml. dropping funnel are placed 80 g. (0.67 mole) of acetophenone, 60 ml. of ether, and 100 ml. of water. The apparatus is assembled in a well-ventilated hood, the flask is surrounded by an ice-salt bath, and 82 g. (1.67 moles) of granulated sodium cyanide is added all at once with vigorous stirring. When most of the sodium cyanide has dissolved and the temperature of the mixture has fallen to 5°, 140 ml. (1.7 moles) of concentrated hydrochloric acid is added from the dropping funnel at such a rate that the temperature remains between 5° and 10°. The addition requires about 1.7 hours. After all the acid has been added, the cooling bath is removed and vigorous stirring is continued for 2 hours. The mixture is allowed to settle, and the liquid portion is decanted into a 1-l. separatory funnel. The water layer is returned to the reaction flask, and 100 ml. of water is added to dissolve the salts. The aqueous solution is extracted with four 50-ml. portions of ether, and the original ether layer and extracts are combined in a 500-ml. round-bottomed flask. The ether is distilled at 20–30 mm. pressure (water aspirator) (Note 1), and the residual oil is poured slowly with stirring into 160 ml. of concentrated hydrochloric acid in a 1-l. round-bottomed flask. The mixture then is saturated with hydrogen chloride gas in the hood (Notes 2 and 3) and allowed to stand overnight.

Part of the excess hydrogen chloride is removed by blowing (or drawing) air through the solution for 1 hour. The solution then is made alkaline by the slow addition of 50% aqueous sodium hydroxide (Note 4) with mechanical stirring and cooling in an ice bath. Solid sodium hydroxide (24 g.) is added, and the mixture is steam-distilled until no more ammonia and acetophenone pass into the distillate (Note 5). About 3–4 l. of distillate is collected (Note 6).

Water is added to the residue if necessary to make the volume 700 ml., and the solution is treated with 2 g. of Norit and filtered with suction (Note 7). The filtrate is extracted with 100 ml. of ether (which is discarded), and acidified by the addition of 80 ml. of concentrated hydrochloric acid. After thorough chilling, preferably overnight in a refrigerator, the precipitated atrolactic acid is collected on a suction filter (Note 8) and air-dried at a temperature not exceeding 65°.

The crude product weighs somewhat more than 70 g. and contains water and sodium chloride. It is dissolved in 500 ml. of boiling benzene, the solution is filtered by gravity, and the solid in the flask and funnel is rinsed with 50 ml. of boiling benzene. The filtrate is concentrated by distillation until no more water collects in the distillate, and the residual solution (300–400 ml.) is cooled. The atrolactic acid which crystallizes is collected on a suction filter and washed with 25 ml. of cold benzene and 25 ml. of commercial pentane (b.p. 30–40°). After air-drying, the cream-colored product amounts to 42.9–43.5 g., m.p. 88–90° with shrinking at 82° (Notes 9 and 10). It is pure enough for most purposes but is intermediate in composition between the anhydrous acid and hemihydrate and contains about 3% sodium chloride. This product is dissolved in 200 ml. of boiling water and treated with 10 g. of Norit. The solution is filtered and cooled overnight at 0–5°. The pure, colorless crystals of atrolactic acid hemihydrate that separate are collected on a suction filter and air-dried. The yield is 33.5–34.7 g. (29–30%), m.p. 88–90° with softening beginning at 75°. The anhydrous acid can be obtained by drying the hemihydrate at 55°/1–2 mm.; m.p. 94.5–95°.

2. Notes

1. By maintaining ebullition with a capillary air inlet and cooling the receiver in a bath containing Dry Ice and trichloro-ethylene, the distillation can be completed in 1 hour or less.

2. The solution becomes homogeneous when it is nearly saturated. An exothermic reaction begins within a few minutes, and some hydrogen chloride gas escapes.

3. The preparation should be carried to this point in 1 day; the time required is about 8 hours. All the operations must be conducted in a well-ventilated hood.

4. About 150 g. of the 50% sodium hydroxide solution is required.

5. The distillation flask should be heated so as to maintain a liquid volume of 600–700 ml.

6. Acetophenone (15–16.5 g., b.p. 92–94°/20 mm.) can be recovered by extracting the distillate with 100 ml. of pentane (b.p. 30–40°). The extract is dried over sodium sulfate, concentrated, and distilled.

7. Only a small amount of solid, in addition to the Norit, should be retained by the filter. If a copious precipitate of sodium atrolactate forms, it should be dissolved by the addition of more water.

8. Only small amounts of crude atrolactic acid are recovered by extracting the filtrate with ether.

9. If the yield is low, the mother liquors should be concentrated to a volume of 50 ml. A second crop of crystals can sometimes be obtained in that way.

10. The submitters have obtained the same yields in preparations on five times this scale.

3. Methods of Preparation

Atrolactic acid has been prepared by the oxidation of hydratropic acid with alkaline permanganate,[2] by hydrolysis of α-chloro- or α-bromohydratropic acid,[3,4] by sodium amalgam reduction of β,β-dibromoatrolactic acid,[5] from α-aminohydratropic acid and nitrous acid,[6,7] by permanganate oxidation of 2,5-dihydroxy-2,5-diphenyl-3-hexyne,[8] by reaction of ethyl phenylglyoxylate with methylmagnesium iodide followed by hydrolysis,[9] from pyruvic acid and phenylmagnesium bromide,[10] and from α,α-dibromopropiophenone and sodium hydroxide.[11] The last method could not be adapted to the preparation of the acid in large quantities by the submitters.

The synthesis of atrolactic acid through acetophenone cyanohydrin was first described by Spiegel [12] and has since been used by several other investigators.[6,13–17] The above preparation is adapted from the methods of McKenzie and Clough [15] and Freudenberg, Todd, and Seidler.[17]

[1] University of Notre Dame, South Bend, Indiana.
[2] Ladenburg and Rugheimer, *Ber.*, **13**, 375 (1880); Ladenburg, *Ann.*, **217**, 107 (1883).
[3] McKenzie and Clough, *J. Chem. Soc.*, **97**, 1022 (1910).
[4] Merling, *Ann.*, **209**, 21 (1881); Fittig and Wurster, *Ann.*, **195**, 153 (1879); Fittig and Kast, *Ann.*, **206**, 24 (1880).
[5] Bottinger, *Ber.*, **14**, 1238 (1881).
[6] Tiemann and Kohler, *Ber.*, **14**, 1976 (1881).
[7] McKenzie and Clough, *J. Chem. Soc.*, **101**, 397 (1912).
[8] Dupont, *Compt. rend.*, **150**, 1524 (1910); *Ann. chim. Paris*, [8] **30**, 532 (1913).
[9] Grignard, *Compt. rend.*, **135**, 628 (1902); *Ann. chim. Paris*, [7] **27**, 556 (1902).
[10] Peters, Griffith, Briggs, and French, *J. Am. Chem. Soc.*, **47**, 453 (1925).
[11] Levine and Stephens, *J. Am. Chem. Soc.*, **72**, 1642 (1950).
[12] Spiegel, *Ber.*, **14**, 1352 (1881).
[13] Staudinger and Ruzicka, *Ann.*, **380**, 289 (1911).
[14] Smith, *J. prakt. Chem.*, **84**, 731 (1911).
[15] McKenzie and Clough, *J. Chem. Soc.*, **101**, 393 (1912).
[16] McKenzie and Wood, *J. Chem. Soc.*, **115**, 833 (1919).
[17] Freudenberg, Todd, and Seidler, *Ann.*, **501**, 213 (1932).

AZELANITRILE *

$$HO_2C(CH_2)_7CO_2H + 2NH_3 \xrightarrow[\text{gel}]{\text{Silica}} NC(CH_2)_7CN + 4H_2O$$

Submitted by ARTHUR C. COPE, ROBERT J. COTTER, and LELAND L. ESTES.[1]
Checked by N. J. LEONARD and R. W. FULMER.

1. Procedure

Caution! This preparation should be conducted in a hood to avoid exposure to ammonia.

The reaction is carried out in the apparatus shown in Fig. 3. *A* is a Pyrex combustion tube, 54 cm. long and 3.5 cm. in diameter, having a glass star seal or indentations 14 cm. from the lower end to support the catalyst, and fitted with ground-glass joints. *B* is a Pyrex tube, 21 cm. long and 3.5 cm. in diameter, which has a 6-cm. length of 6-mm. capillary tubing attached at the lower end. The capillary tube is attached to a ground-glass joint and an inlet tube, *C*, is inserted above the end of the capillary tube. The top section of the apparatus (above the first ground-glass joint) is covered with asbestos and wound with 14 ft. of No. 22 Chromel A resistance wire. *D* is a 500-ml. round-bottomed two-necked flask with ground-glass joints and a 4-mm.

stopcock sealed to the bottom. An efficient condenser is attached
to the flask and connected to a gas absorption trap.[2] The hot
junction of a pyrometer is placed in contact with the glass combustion tube A at its center, and the tube is wrapped with a thin

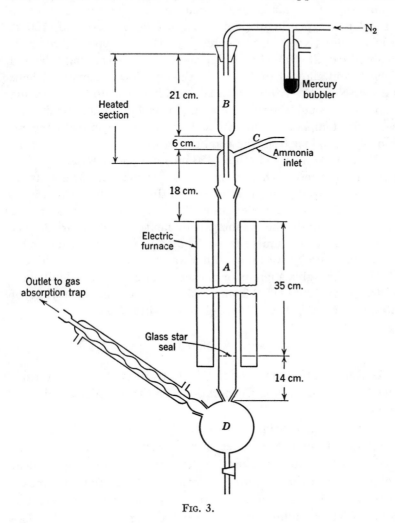

FIG. 3.

layer of asbestos paper. An electric furnace with a 33-cm.
heated section which is rated at 6.8 amp. (110 v.) is used to heat
the combustion tube.

A piece of glass wool is placed on the star seal, and the combustion tube is filled with 74 g. (110 ml.) of 14- to 20-mesh silica gel (Note 1). The tube is heated to 500°, and nitrogen is passed through the column for 30 minutes to activate the catalyst. Anhydrous ammonia is then passed through the column at a rate of 3.9 moles per hour (Note 2). Molten azelaic acid, 100 g. (0.53 mole) (Note 3), is poured into the reservoir B, which is maintained at 108–112° by means of the electrically heated jacket, and allowed to drop onto the hot silica gel over a 4-hour period (Note 4). After the azelaic acid has been added, ammonia is passed through the tube for an additional period of 30 minutes, with the temperature maintained at 500°, to complete the removal of product from the combustion tube.

Water (100 ml.) and ether (350 ml.) are added to the condensate in the receiver D, and the mixture is shaken. The aqueous layer is removed, and the ethereal solution is washed with 75 ml. of $6N$ sodium hydroxide solution. The ethereal solution is washed with water until the washings are neutral and is dried over anhydrous magnesium sulfate. The ether is removed under reduced pressure by warming with a water bath, and the residue is distilled through a Vigreux or packed column. The azelanitrile is collected at 120–121°/0.2 mm., 175–176°/6 mm., in a yield of 50–54 g. (63–68%), n_D^{25} 1.4443–1.4448 (Notes 5 and 6).

2. Notes

1. Refrigeration grade silica gel, 14- to 20-mesh, obtained from the Davison Chemical Corporation, Baltimore, Maryland, was used.

2. A manometer-type flow meter [3] was calibrated by passing ammonia through the meter into a standard solution of $4N$ hydrochloric acid containing a few drops of phenolphthalein solution. The time required for exact neutralization of a measured volume of acid was recorded, together with the pressure differential, in millimeters, between the manometer arms. The logarithm of the rate of flow in moles per hour plotted against the logarithm of the pressure, for various rates of flow, gave a straight-line plot which was used to determine flow rates for other pressure differentials.

3. Azelaic acid obtained from Emery Industries, Inc., Cincinnati, Ohio, was recrystallized from water to a melting point of 101–102°.

4. Comparable results have been obtained when the acid was added over a 2-hour period. It is desirable to maintain a nitrogen atmosphere over the acid at a pressure slightly above that of the ammonia flow to prevent reduction of the acid flow by salt formation in the capillary.

5. Cooling the receiver D may help prevent loss of product by mechanical carry-over in the gas flow.

6. This preparation illustrates a general method for producing nitriles from monocarboxylic [4] and dicarboxylic acids.[5]

3. Methods of Preparation

Azelanitrile has been prepared in 80% yield by treating 1,7-dibromoheptane with potassium cyanide;[6] by treating 1,7-diiodoheptane with potassium cyanide;[7] from azelaic acid through the intermediate acid chloride and diamide;[8] in 60–70% yield by the dehydration of the diamide of azelaic acid;[9] and by the action of sodium amide and acetonitrile on the 1,5-dihalopentanes.[10]

[1] Massachusetts Institute of Technology, Cambridge, Massachusetts.

[2] *Org. Syntheses Coll. Vol.* **2**, 4 (1943).

[3] Altieri, *Gas Analysis and Testing of Gaseous Materials*, American Gas Association, Inc., New York, N. Y., 1945, p. 57.

[4] Mitchell and Reid, *J. Am. Chem. Soc.*, **53**, 321 (1931).

[5] Lazier, U. S. pat. 2,144,340 [*C. A.*, **33**, 3398 (1939)]; Miwa, Veno, and Fujimura, Jap. pat. 5514 (1954) [*C. A.*, **49**, 16271 (1955)].

[6] von Braun and Danziger, *Ber.*, **45**, 1971 (1912).

[7] Dionneau, *Ann. chim.*, [9], **3**, 248 (1915).

[8] Trunel, *Ann. chim.*, [11], **12**, 112 (1939).

[9] Solonia, *J. Russ. Phys. Chem. Soc.*, **29**, 410 (1897) [*Chem. Zentr.*, [2] **68**, 848 (1897)].

[10] Paul and Tchelitcheff, *Bull. soc. chim. France*, **1949**, 470.

1,1'-AZO-*bis*-1-CYCLOHEXANENITRILE

(Cyclohexanecarbonitrile, 1,1'-azodi-)

$$\begin{array}{c} CH_2\!-\!CH_2 \quad CN \qquad NC \quad CH_2\!-\!CH_2 \\ CH_2 \qquad\quad C\!-\!NHNH\!-\!C \qquad\quad CH_2 + Br_2 \rightarrow \\ CH_2\!-\!CH_2 \qquad\qquad\quad CH_2\!-\!CH_2 \end{array}$$

$$\begin{array}{c} CH_2\!-\!CH_2 \quad CN \qquad NC \quad CH_2\!-\!CH_2 \\ CH_2 \qquad\quad C\!-\!N\!=\!N\!-\!C \qquad\quad CH_2 + 2HBr \\ CH_2\!-\!CH_2 \qquad\qquad\quad CH_2\!-\!CH_2 \end{array}$$

Submitted by C. G. OVERBERGER, PAO-TUNG HUANG, and M. B. BERENBAUM.[1]
Checked by N. J. LEONARD and E. H. MOTTUS.

1. Procedure

In a 600-ml. beaker equipped with a stirrer, thermometer, and dropping funnel are placed 24.6 g. (0.1 mole) of finely powdered 1,2-di-1-(1-cyano)cyclohexylhydrazine (p. 274) and 130 ml. of 90% ethanol. To this mixture is added slowly, with cooling, 45 ml. of concentrated hydrochloric acid. The beaker is placed in an ice bath, and, after the suspension has been cooled to 10°, bromine is added at such a rate that the temperature does not rise above 15°. About 16–17 g. (about 0.1 mole) of bromine is required to reach the end point characterized by a permanent orange-yellow color. The reaction mixture is poured into 80 ml. of ice water. After 15 minutes the suspension is filtered with the aid of a Büchner funnel, washed with 250 ml. of water, and pressed dry. The solid is transferred to a 500-ml. Erlenmeyer flask, 120 ml. of boiling 95% ethanol is added, and the crude product is dissolved as rapidly as possible while being heated on a steam bath (Note 1). The solution is filtered through a fluted filter in a heated funnel, and the filtrate is placed in a refrigerator overnight. The solid is collected on a Büchner funnel and dried in a vacuum desiccator over calcium chloride. The yield of product is 20.5–22.0 g. (84–90%); m.p. 113.5–115.5° (Notes 2 and 3).

2. Notes

1. Prolonged heating of the solution will cause excessive decomposition of the azo compound.

2. This compound is stable indefinitely if stored at room temperature. Prolonged heating at temperatures of 80° or higher, however, will result in rapid decomposition involving possible hazards.[2]

3. 2,2'-Azo-*bis*-isobutyronitrile can be prepared in a similar manner. The product after recrystallization from 95% ethanol is obtained in a yield of 85–90%; m.p. 102–103°. This compound must be regarded as an explosive.[2]

3. Methods of Preparation

1,1'-Azo-*bis*-1-cyclohexanenitrile has been prepared in a similar manner by Hartman.[3] The method has been substantiated by Overberger, O'Shaughnessy, and Shalit [4] and is a modification of that used originally by Thiele and Heuser [5] for the synthesis of 2,2'-azo-*bis*-isobutyronitrile.

[1] Polytechnic Institute of Brooklyn, Brooklyn, New York.
[2] Carlisle, *Chem. Eng. News*, **27**, 150 (1949); **28**, 803 (1950).
[3] Hartman, *Rec. trav. chim.*, **46**, 150 (1927).
[4] Overberger, O'Shaughnessy, and Shalit, *J. Am. Chem. Soc.*, **71**, 2661 (1949).
[5] Thiele and Heuser, *Ann.*, **290**, 1 (1896).

BENZENEBORONIC ANHYDRIDE *

(Boroxin, triphenyl-)

$$C_6H_5MgBr + B(OCH_3)_3 \rightarrow [C_6H_5B(OCH_3)_3]^-MgBr^+$$

$$[C_6H_5B(OCH_3)_3]^-MgBr^+ + 3H_2O \rightarrow$$

$$C_6H_5B(OH)_2 + 3CH_3OH + Mg(OH)Br$$

$$2Mg(OH)Br + H_2SO_4 \rightarrow MgBr_2 + MgSO_4 + 2H_2O$$

$$3C_6H_5B(OH)_2 \rightarrow C_6H_5{-}B \begin{array}{c} O{-}B{-}C_6H_5 \\ \diagup \quad \diagdown \\ \quad\quad O \\ \diagdown \quad \diagup \\ O{-}B{-}C_6H_5 \end{array} + 3H_2O$$

Submitted by ROBERT M. WASHBURN, ERNEST LEVENS, CHARLES F. ALBRIGHT, and FRANKLIN A. BILLIG.[1]
Checked by B. C. McKUSICK and H. C. MILLER.

1. Procedure

Caution! Benzeneboronic acid and its anhydride are toxic substances and may irritate mucous tissues such as those of the eyes. In case of contact, carefully wash exposed parts of the body with soap and water (Note 1).

The apparatus consists of a four-necked 5-l. round-bottomed Morton flask [2] fitted with a 500-ml. graduated dropping funnel with a pressure-equalizing side arm, a 1-l. graduated dropping funnel of the same type, a thermometer, an efficient mechanical stirrer (Note 2), and an inlet for dry nitrogen. The apparatus is thoroughly swept with dry nitrogen, and the reaction flask is charged with 1.5 l. of anhydrous ether, dry nitrogen (Note 3) being used for pressure transfer.

Three hundred thirty-six milliliters (312 g., 3 moles) of methyl borate is distilled directly into the 500-ml. dropping funnel shortly before starting the reaction (Note 4). One liter (544 g., 3 moles) of a $3M$ ethereal solution of phenylmagnesium bromide is pressure-transferred with dry nitrogen to the 1-l. dropping funnel (Note 5). During subsequent operations until the hydrolysis step, a positive pressure of 10–20 mm. of nitrogen is maintained in the closed system by means of a mercury bubbler to prevent access of atmospheric moisture. The ether is cooled to below $-60°$ by a bath of Dry Ice and acetone and is kept below $-60°$ all during the reaction (Note 6). The reactants are added to the well-stirred reaction mixture alternately in small portions, first 10 ml. of methyl borate and then 30 ml. of phenylmagnesium bromide, the rate of addition being as rapid as is possible without the temperature of the mixture rising above $-60°$ (Note 7). Stirring is continued for an additional 20 minutes below $-60°$ after the addition of the reagents is completed.

The stirred mixture, maintained at or below $0°$, is hydrolyzed by the addition of 200 ml. of distilled water during 5 minutes. It is then neutralized by addition of a solution of 84 ml. of concentrated sulfuric acid in 1.7 l. of distilled water during 15 minutes. The mixture is transferred to a 5-l. separatory funnel, the ether layer is separated, and the aqueous layer is extracted with three 250-ml. portions of ether.

The combined ether layer and extracts are transferred to a 5-l. round-bottomed flask equipped with a Hershberg stirrer,[3] a dropping funnel, a Claisen head with a water-cooled condenser, an electric heating mantle, and an ice-cooled receiver (Note 8). After approximately one-half of the ether has been removed by distillation from the stirred mixture, 1.5 l. of distilled water is added slowly while the distillation is continued until a head temperature of $100°$ is reached (Note 9).

While stirring is continued, the aqueous distilland is cooled in an ice bath (Note 10). The benzeneboronic acid, which separates as small white crystals, is collected on a Büchner funnel and washed with petroleum ether. The petroleum ether removes traces of dibenzeneborinic acid, which are seen in the hot mother liquor as globules of brown oil and which may color the product. The acid is dehydrated to benzeneboronic anhydride by heating it in an oven at $110°$ and atmospheric pressure for 6 hours (Note 11). Benzeneboronic anhydride is obtained as a colorless solid, weight 240–247 g. (77–79%) (Note 12), m.p. 214–216°.

2. Notes

1. A summary of the physiological activity of benzeneboronic acid may be found in reference 4a.

2. The submitters found that for a preparation of this size a 1-inch Duplex Dispersator (Premier Mill Corp., Geneva, New York) operating at 7500 r.p.m. provided excellent agitation of the heterogeneous reaction mixture. For smaller preparations (1-l. flask) they found that a Stir-O-Vac (Labline, Inc., 217 N. Desplainer St., Chicago 6, Illinois) operating at 5000 r.p.m. was satisfactory. The type of agitation is very important for, whereas the submitters obtained yields of around 91%, the checkers obtained yields of only 77–80% with either a Morton stirrer [2] (excessive splashing deposited some of the reaction mixture on the warm upper walls of the flask) or a Polytron dispersion mill type of stirrer (there was too much hold-up in the stirrer housing).

3. Tank nitrogen was dried with phosphorus pentoxide.

4. Methyl borate (b.p. 68°) forms a 1:1 azeotrope (b.p. 54.6°) with methanol (b.p. 64°).[5] Since the presence of even a small amount of methanol reduces the yield considerably more than would be expected from the stoichiometry,[4, 6] methyl borate stocks should be freshly distilled through a good column to remove as fore-run any methyl borate-methanol azeotrope which may have been formed by hydrolysis during storage.

5. Mallinckrodt analytical reagent grade ether, dried over sodium, was used. The methyl borate was the commercial product of American Potash and Chemical Corporation containing 99% ester as received. The phenylmagnesium bromide was purchased as a 3.0M solution in ether from Arapahoe Special Products, Inc., Boulder, Colorado.

6. The yield of benzeneboronic anhydride is highly dependent upon the reaction temperature, as the following data of the submitters show. At a reaction temperature of 15° the yield was 49%; at 0°, 76%; −15°, 86%; −30°, 92%; −45°, 92%; −60°, 99%. The yields are based on the combined first and second crops of benzeneboronic acid.

7. At a given temperature, the maximum yield of benzeneboronic acid and the minimum amount of by-product dibenzeneborinic acid are obtained when neither reagent is present in excess. The addition of small increments of reactants is a convenient approximation imposed by the difficulty of adjusting stopcocks to small rates of flow. Alternatively, the Hershberg dropping funnel [7] or other metering device may be used to maintain the

stoichiometry. Addition times, which depend upon the efficiency of stirring and heat transfer, vary from about 1 hour at −60° to 15 minutes at 0°.

8. Stirring is helpful during the ether distillation to prevent superheating.

9. Small amounts of benzene, phenol, and biphenyl, which may be formed in the reaction, are removed by the steam distillation. Enough water has been added to ensure solution of all of the product.

10. The product crystallizes at 43° with a temperature rise to 45°. The solubility of benzeneboronic acid in water (g./100 g. of water) is approximately 1.1 at 0° and 2.5 at 25°; the solubility-temperature relationship is linear to at least 45°.

11. If benzeneboronic acid rather than its anhydride is desired, it can be obtained by air-drying the moist acid in a slow stream of air nearly saturated with water. The yield of acid is 282–332 g. One can readily convert the anhydride to the acid by recrystallizing it from water. Benzeneboronic acid gradually dehydrates to the anhydride if left open to the atmosphere at room temperature and 30–40% relative humidity. The melting point observed is that of the anhydride because the acid dehydrates before it melts.

12. The submitters report a yield of 91% and state that an additional 27 g. (9%) of acid can be obtained from the aqueous mother liquor.

3. Methods of Preparation

The procedure described [4] is a modification of the method of Khotinsky and Melamed,[8] who first reported the preparation of boronic acids from Grignard reagents and borate esters. Benzeneboronic acid and the corresponding anhydride also have been prepared by reaction of phenylmagnesium bromide with boron trifluoride;[9] by the reaction of phenyllithium with butyl borate;[10] by the reaction of diphenylmercury with boron trichloride;[11] by the reaction of benzene with boron trichloride in the presence of aluminum chloride;[12] and by the reaction of triphenylborane with boric oxide.[13]

The present procedure is also applicable to the synthesis of substituted benzeneboronic acids.[4a] Benzeneboronic acid and its anhydride are of use as starting materials for the synthesis of phenylboron dichloride [14] and of various substituted boronic and borinic acids and esters.[6, 15]

[1] American Potash and Chemical Corporation, Whittier, California.
[2] Morton, *Ind. Eng. Chem., Anal. Ed.*, **11**, 170 (1939); Morton and Redman, *Ind. Eng. Chem.*, **40**, 1190 (1948).
[3] *Org. Syntheses Coll. Vol.* **2**, 117 (1943).
[4] (*a*) Washburn, Levens, Albright, Billig, and Cernak, *Advances in Chem. Ser.*, **23**, 102 (1959); (*b*) Washburn, Billig, Bloom, Albright, and Levens, *Advances in Chem. Ser.*, **32**, 208 (1961).
[5] Schlesinger, Brown, Mayfield, and Gilbreath, *J. Am. Chem. Soc.*, **75**, 213 (1953).
[6] Seaman and Johnson, *J. Am. Chem. Soc.*, **53**, 711 (1931).
[7] *Org. Syntheses Coll. Vol.* **2**, 129 (1943).
[8] Khotinsky and Melamed, *Ber.*, **42**, 3090 (1909).
[9] Krause and Nitsche, *Ber.*, **55B**, 1261 (1922); Krause, German pat. 371,467 (1923) [*C. A.*, **18**, 992 (1924)].
[10] Brindley, Gerrard, and Lappert, *J. Chem. Soc.*, **1955**, 2956.
[11] Michaelis and Becker, *Ber.*, **15**, 180 (1882).
[12] Muetterties, *J. Am. Chem. Soc.*, **82**, 4163 (1960).
[13] McCusker, Hennion, Ashby, and Rutowski, *J. Am. Chem. Soc.*, **79**, 5194 (1957).
[14] Dandegaonker, Gerrard, and Lappert, *J. Chem. Soc.*, **1957**, 2893.
[15] Lappert, *Chem. Revs.*, **56**, 987, 1013 (1956).

BENZHYDRYL β-CHLOROETHYL ETHER

(Ether, benzohydryl 2-chloroethyl)

$$\underset{C_6H_5}{\overset{C_6H_5}{\diagdown}}CHOH + HOCH_2CH_2Cl \xrightarrow{H_2SO_4}$$

$$\underset{C_6H_5}{\overset{C_6H_5}{\diagdown}}CHOCH_2CH_2Cl + H_2O$$

Submitted by SHIGEHIKO SUGASAWA and KUNIO FUJIWARA.[1]
Checked by N. J. LEONARD, P. D. THOMAS, and L. A. MILLER.

1. Procedure

In a 500-ml. three-necked round-bottomed flask equipped with a sealed stirrer, a reflux condenser, and a dropping funnel are placed 36 g. (0.45 mole) of ethylene chlorohydrin (Note 1), 5 ml. of concentrated sulfuric acid, and 35 ml. of benzene. The mixture is warmed on a water bath, and to it is added, with efficient stirring, a solution of 55 g. (0.30 mole) of benzhydrol (Note 2) in 65 ml. of benzene (Note 3) during 30–50 minutes. The reaction mixture is heated at the reflux temperature for an additional 4 hours with stirring. To the cooled mixture is added about 35 ml. of benzene, and the combined benzene layer is washed with water and dried over calcium chloride. The drying agent is removed, the benzene is evaporated, and the residue is distilled under reduced pressure. The benzhydryl β-chloro-ethyl ether is collected at 144–148°/1.0 mm. (174–177°/4 mm.) as a colorless, viscous oil, n_D^{30} 1.5651, which should be removed from the receiver to a beaker or Erlenmeyer flask immediately after the distillation. The oil solidifies to a hard white mass, m.p. 27.4–27.8°, when kept in an ice chest (Note 4). The yield is 60.0–65.3 g. (81–88%).

2. Notes

1. Commercial ethylene chlorohydrin is dried over anhydrous sodium sulfate and distilled before use; b.p. 126–127°/743 mm. Excess is used to avoid the formation of dibenzhydryl ether as a by-product.

2. Eastman Kodak Company benzhydrol, m.p. 67–67.5°, can be used directly.

3. It is necessary to warm the mixture in order to complete the solution of benzhydrol in the benzene.

4. The checkers found this product to be analytically pure without recourse to further purification.

3. Methods of Preparation

This method is based on the process of the submitters.[2]

[1] Pharmaceutical Institute, Medical Faculty, University of Tokyo, Tokyo, Japan.
[2] Sugasawa and Fujiwara, J. Pharm. Soc. Japan, 71, 365 (1951) [C. A., 46, 951h (1952)]; Jap. pat. 184,243 (Aug. 12, 1949).

BENZOFURAZAN OXIDE *

(Benzofurazan 1-oxide)

I. HYPOCHLORITE OXIDATION OF *o*-NITROANILINE

$$
\text{(structure: benzene ring with NH}_2\text{ and NO}_2\text{)} + \text{NaOCl} \xrightarrow{\text{KOH}} \text{(benzofurazan oxide structure)} \text{O} + \text{NaCl} + \text{H}_2\text{O}
$$

Submitted by F. B. MALLORY.[1]
Checked by T. L. CAIRNS and H. E. SIMMONS.

1. Procedure

A. *Sodium hypochlorite solution.* A solution of sodium hypochlorite [2] is prepared immediately before it is to be used. A mixture of 50 g. (1.25 moles) of sodium hydroxide and 200 ml. of water is swirled until the solid dissolves. The solution is cooled to 0°, and 100 g. of crushed ice is added. The flask is then placed in an ice bath, and chlorine gas from a tank is bubbled through the solution until 41 g. (0.58 mole) is absorbed. An excess of chlorine should be avoided. The solution of sodium hypochlorite is kept in the dark at 0° until needed.

B. *Benzofurazan oxide.* A mixture of 21 g. (0.32 mole) of potassium hydroxide and 250 ml. of 95% ethanol in a 1-l. Erlenmeyer flask is heated on a steam bath until the solid dissolves (Note 1). *o*-Nitroaniline (40 g., 0.29 mole) (Note 2) is dissolved in the warm alkali solution. The resulting deep red solution is then cooled to 0°, and the sodium hypochlorite solution from part A is added slowly with good stirring over the course of 10 minutes (Note 3). The flocculent yellow precipitate is collected on a large Büchner funnel, washed with 200 ml. of water, and air-dried. The crude product weighs 36.0–36.5 g. and melts at 66–71° (Note 4). The product is purified by recrystallization from a solution made up from 45 ml. of 95% ethanol and 15 ml. of water. Material insoluble in the hot solvent is removed by filtration, and the hot filtrate is allowed to cool to room temperature. The yield of yellow benzofurazan oxide is 31.6–32.5 g. (80–82%), m.p. 72–73°.

2. Notes

1. A small residue of insoluble carbonate may be ignored.

2. Eastman Kodak white label grade material, melting at 71.5–73.5°, was used.

3. The temperature of the mixture should be kept close to 0° to avoid decomposition of the sodium hypochlorite and prevent formation of tarry materials that occurs at 10–12°. A Dry Ice-acetone bath was found convenient by the checkers.

4. There may be some material that does not melt under 100°, which is not present after recrystallization.

II. DECOMPOSITION OF o-NITROPHENYLAZIDE

$$\text{(o-}NH_2\cdot HCl\text{, }NO_2\text{ ring)} + HONO \longrightarrow \text{(o-}N_2Cl\text{, }NO_2\text{ ring)} + 2H_2O$$

$$\text{(o-}N_2Cl\text{, }NO_2\text{ ring)} + NaN_3 \xrightarrow{HCl} \text{(o-}N_3\text{, }NO_2\text{ ring)} + NaCl + N_2$$

$$\text{(o-}N_3\text{, }NO_2\text{ ring)} \xrightarrow{\Delta} \text{(benzofurazan oxide)} + N_2$$

Submitted by P. A. S. SMITH and J. H. BOYER.[3]
Checked by ARTHUR C. COPE, DAVID J. MARSHALL, and DOUGLAS S. SMITH.

1. Procedure

A. *o-Nitrophenylazide.* A mixture of 28 g. (0.2 mole) of o-nitroaniline (Note 1), 80 ml. of water, and 45 ml. of concentrated hydrochloric acid is placed in a 500-ml. three-necked flask equipped with a stirrer, a thermometer, and a dropping funnel. The stirrer is started, and the flask is cooled in an ice-salt bath until the temperature of the mixture is 0–5°. After this temperature has been reached, the amine hydrochloride is diazotized by adding dropwise a solution of 14.5 g. of reagent grade sodium nitrite in 50 ml. of water. Stirring is then continued for 1 hour at 0–5°. The yellow-green solution is filtered from traces of insoluble impurities and poured into a 2-l. beaker surrounded by an ice bath. With stirring, a solution of 13 g. (0.2 mole) of sodium azide in 50 ml. of water is added (Note 2). Almost immediately the *o*-nitrophenylazide begins to precipitate as a light-cream to colorless solid, which is collected on a Büchner funnel after the nitrogen evolution has ceased (15–20 minutes). The yield of *o*-nitrophenylazide, m.p. 52–55°, is 31–32 g. (94–97%). This crude product can be used for the preparation of benzofurazan oxide in Part B (Note 3).

The impure azide is dissolved in 110–120 ml. of 95% ethanol at 50–55° (Note 4), and 2 g. of activated carbon is added to aid in the removal of impurities. After being filtered through a steam-heated funnel, the warm solution is allowed to cool to room temperature, whereupon 14–15 g. of the product precipitates as light-yellow prisms, m.p. 53–55°. Concentration (Note 5) of the mother liquor to 30–40 ml. by evaporation at room temperature under an air stream causes the separation of an additional 7–8 g. of material, m.p. 52–54° (Note 6). The total yield of purified *o*-nitrophenylazide is 63–69%.

B. *Benzofurazan oxide.* A mixture of 16.4 g. (0.1 mole) of *o*-nitrophenylazide and 30 ml. of reagent grade toluene is placed in a 100-ml. round-bottomed flask equipped with a reflux condenser and is heated on a steam cone. Moderate nitrogen evolution commences immediately and continues for about 3 hours. When there are no more visible signs of gas evolution, the solution is cooled in an ice bath. After a few minutes, precipitation of light straw-colored clusters of prisms commences. About 6 g. of pure product, m.p. 70–72°, is obtained in this manner. Evaporation of the mother liquor yields another 4.5–5.5 g. of the oxide, slightly darker in color, m.p. 69–71°, which may be purified by recrystallization from 15 ml. of 70% ethanol to give material having a melting point of 70–71°. The total yield is 10.5–11.5 g. (77–85%).

2. Notes

1. If the *o*-nitroaniline is contaminated with *p*-nitroaniline, as it is likely to be, the yield and quality of the *o*-nitrophenylazide are lowered. The submitters obtained yields of 72–80% from *o*-nitroaniline melting at 70–71° obtained from the Eastman Kodak Company. The yields reported were obtained with *o*-nitroaniline melting at 72–73.5°.

2. A large container is necessary for this reaction because of excessive frothing which accompanies the nitrogen evolution. This step should be conducted in a hood to avoid possible exposure to hydrazoic acid.

3. A somewhat lower yield of benzofurazan oxide, m.p. 67–70°, is obtained if non-purified azide is used.

4. At this temperature there is no danger of decomposition. Loss of nitrogen commences at 80°.

5. Recrystallization of *o*-nitrophenylazide from a smaller volume of ethanol gives a considerably higher recovery, but it may be rather difficult to avoid separation of the product as an oil.

6. The submitters report that similar yields of phenylazide can be obtained from aniline in the same manner. Phenylazide must be distilled rather than recrystallized.[4]

3. Methods of Preparation

Benzofurazan oxide is prepared most conveniently by the hypochlorite oxidation of *o*-nitroaniline according to the method described, which is adapted from the procedure of Green and Rowe.[5]

Benzofurazan oxide also has been prepared by thermal decomposition of *o*-nitrophenylazide;[6,7] by oxidation of the dioxime of *o*-benzoquinone with dilute nitric acid or potassium ferricyanide in alkaline solution;[6] and by the oxidation of *o*-nitroaniline with phenyl iodosoacetate.[8]

The present synthesis of benzofurazan oxide from *o*-nitrophenylazide is a modification of the methods of Noelting and Kohn[7] and of Zincke and Schwarz.[6] The hypochlorite oxidation method has been used in the synthesis of various substituted benzofurazan oxides.[9]

o-Nitrophenylazide has been prepared by the action of sodium azide or hydrazine on o-nitrobenzenediazonium sulfate; [10] ammonia on o-nitrobenzenediazonium perbromide; [6,11] O-benzylhydroxylamine hydrochloride on o-nitrobenzenediazonium acetate; [12] sodium nitrite and hydrochloric acid on o-nitrophenylhydrazine; [6] and aqueous alkali on o-nitrobenzenediazo-4-semicarbazinocamphor. [13]

[1] California Institute of Technology, Pasadena, California.
[2] A similar procedure is given in *Org. Syntheses Coll. Vol.* **1**, 309 (1941).
[3] University of Michigan, Ann Arbor, Michigan.
[4] *Org. Syntheses Coll. Vol.* **3**, 710 (1955).
[5] Green and Rowe, *J. Chem. Soc.*, **101**, 2452 (1912).
[6] Zincke and Schwarz, *Ann.*, **307**, 28 (1899).
[7] Noelting and Kohn, *Chem. Ztg.*, **18**, 1095 (1894).
[8] Pausacker, *J. Chem. Soc.*, **1953**, 1989.
[9] Gaughran, Picard, and Kaufman, *J. Am. Chem. Soc.*, **76**, 2233 (1954).
[10] Noelting and Michel, *Ber.*, **26**, 86, 88 (1893).
[11] Noelting, Grandmougin, and Michel, *Ber.*, **25**, 3328 (1892).
[12] Bamberger and Renauld, *Ber.*, **30**, 2288 (1897).
[13] Forster, *J. Chem. Soc.*, **89**, 233 (1906).

BENZOGUANAMINE *

(s-Triazine, 2,4-diamino-6-phenyl-)

Submitted by J. K. SIMONS and M. R. SAXTON.[1]
Checked by T. L. CAIRNS and A. K. SCHNEIDER.

1. Procedure

Five grams of potassium hydroxide (85% KOH) is dissolved in 100 ml. of methyl Cellosolve (Note 1) in a 500-ml. flask (Note 2) fitted with a mechanical stirrer, reflux condenser, and a heating mantle. Dicyandiamide (50.4 g.; 0.6 mole) (Note 3) and benzonitrile (50 g.; 0.48 mole) are added, and the mixture is stirred and heated. A solution is formed, and, when the temperature reaches 90–110°, an exothermic reaction begins and the product separates as a finely divided white solid. The vigor of the reaction is kept under control by the refluxing of the solvent (Note 4).

When the exothermic reaction is ended, the slurry of product is stirred and refluxed for 5 hours to ensure complete reaction (Note 5). The mixture is then cooled and filtered. The product is washed by suspension in hot water (Note 6), filtered, and dried. The yield is 68–79 g. (75–87%). The product melts at 227–228° (Note 7).

1. Notes

1. The commercial solvent is used without purification. Other primary alcohols of similar or higher boiling point are suitable solvents.

2. The large flask is chosen to allow for better stirring and heat transfer.

3. The reaction may be carried out with exactly molar equivalents of reactants. Slightly better yields are obtained by using the 0.2 to 0.25 molar excess of dicyandiamide.

4. Boiling of the methyl Cellosolve dissipates the heat of reaction, which begins at about 90°. The submitters state that, in one preparation using 100 times the indicated quantities (21-l. flask), two condensers were used and intermittent cooling of the flask with a small stream of water kept the boiling from becoming too violent.

5. Yields reported were obtained after 5 hours under reflux. Preparation of other guanamines indicates an optimum yield after 2.5 hours under reflux.

6. Thorough washing with hot water serves to remove any dicyandiamide or melamine which may be present in the crude product. Benzoguanamine is slightly soluble in hot water.

7. The Cellosolve filtrate may be evaporated to obtain further quantities of product, m.p. 227–228°, raising the total yield to 90–95%.

3. Methods of Preparation

Benzoguanamine has been prepared by the reaction of dicyandiamide with benzonitrile in a sealed tube at 220–230°,[2] with an excess of benzonitrile in the presence of piperidine and potassium carbonate,[3] with benzonitrile in a solvent in the presence of a basic catalyst,[4] with benzonitrile in liquid ammonia,[5] and with benzamidine hydrochloride at elevated temperatures.[6] It also has been prepared from the reaction of biguanide acetate with benzamidine hydrochloride,[7] or of biguanide sulfate with benzoyl chloride in an alkaline medium,[8] by the distillation of guanidine benzoate,[6a] and from benzonitrile, urea, and ammonia.[9]

[1] Plaskon Division, Libby-Owens-Ford Glass Company, Toledo 6, Ohio.

[2] Ostrogovich, *Atti accad. Lincei*, [5] **20**, I, 251 [*C. A.*, **5**, 2099 (1911)].

[3] DeBell, Goggin, and Gloor, *German Plastics Practice*, p. 267, DeBell and Richardson, Springfield, Massachusetts, 1946.

[4] Zerweck and Brunner (vested in the Alien Property Custodian) U. S. pat. 2,302,162 [*C. A.*, **37**, 2016 (1943)]; Thrower and Pinchin (to British Oxygen Co.) Brit. pat. 758,601 [*C. A.*, **51**, 10593 (1957)]; Jones (to British Oxygen Co.) U. S. pat. 2,735,850 [*C. A.*, **50**, 15598 (1956)].

[5] Bann, Grimshaw, Jones, and Pinchin, *Compt. rend. 27° congr. intern. chim. ind. Brussels*, **1954**, 3; *Industrie chim. belge*, **20**, Spec. No. 342 (1955) [*C. A.*, **50**, 11053 (1956)].

[6] (a) Ostrogovich, *Atti accad. Lincei*, [5] **20**, I, 185 [*C. A.*, **5**, 2099 (1911)]; (b) Birtwell, *J. Chem. Soc.*, **1952**, 1279.

[7] Reference 6a, p. 252.

[8] Rackmann, *Ann.*, **376**, 181 (1910).

[9] Mackay (to American Cyanamid Co.), U. S. pat. 2,527,314 [*C. A.*, **45**, 2513 (1951)].

BENZOYLACETANILIDE *

(Acetanilide, 2-benzoyl-)

$$C_6H_5COCH_2CO_2C_2H_5 + C_6H_5NH_2 \rightarrow$$
$$C_6H_5COCH_2CONHC_6H_5 + C_2H_5OH$$

Submitted by C. F. H. Allen and W. J. Humphlett.[1]
Checked by Max Tishler and R. Connell.

1. Procedure

A mixture of 105.7 g. (0.55 mole) of ethyl benzoylacetate and 46.6 g.

(0.5 mole) of aniline (Note 1) is placed in the dropping funnel *D* (Fig. 4) at the top of the continuous reactor (Notes 2 and 3) after the column has been heated to 135° (transformer set at 80 volts) (Notes 4 and 5). The reactants are then admitted to the column during about 15 minutes (this corresponds to a rate of amide formation of 396–400 g. per hour). Alcohol distils (Note 5) noticeably from the column during the addition and collects in flask *G*. At the completion of the reaction, 100 ml. of xylene is passed through the hot column to rinse out the residual amide (Note 6). An additional 200 ml. of xylene is added to the receiver *C*, and solution is effected by warming. After the solution has been cooled enough to induce crystallization, 100 ml. of petroleum ether (b.p. 35–60°) is added with manual stirring. The mixture is chilled to 15°, then the crystalline product is separated by suction filtration and washed with 300–400 ml. of petroleum ether. The yield is 99–100 g. (83–84%), m.p. 106–106.5° (Note 7). The melting point of this product is not altered by recrystallization from benzene (Notes 8–10).

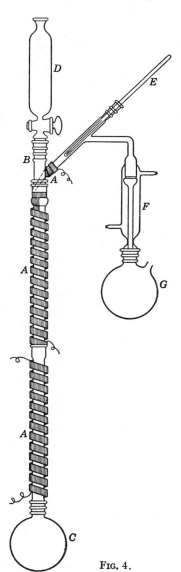

FIG. 4.

2. Notes

1. Commercial grades of ethyl benzoylacetate and aniline were freshly redistilled before use.

2. The continuous reactor shown in Fig. 4 is of general utility for reactions that proceed at a rapid rate. Optimum conditions must be determined by experiment for each new product, but a high yield may often be secured. To determine a yield, it is necessary to run a given

weight of the components through, rinse the column, and work up the combined products.

3. The reactor is built from stock pieces of glassware having 29/42 and 24/40 standard taper joints. It consists of a column 100 cm. long, of 2 cm. inside diameter, in two 50-cm. sections. The column is packed with $\frac{1}{8}$-in. glass helices such as are used for packing distillation columns. The column is heated by two 275-watt flexible heating tapes A, 6 ft. long and $\frac{1}{2}$ in. wide, with lead wires connected to variable transformers, which are attached to a source of 110-volt power. The heating elements may be covered with asbestos tape if desired. At the bottom of the column is a 500-ml. flask C for receiving the product. At the top of the column is a 250-ml. addition funnel D (a cylindrical shape is convenient for calibration of volume or for winding with heating tape in instances where melting a solid or heating a solution is required). The addition funnel is connected to the column through a section B having a side arm provided with a thermometer E and a downward condenser F leading to a 500-ml. flask G, which has an outlet to the atmosphere for effluent gas.

4. It is always desirable to use a "wet" column; once a column has been used, this condition prevails. It is advantageous to admit a little xylene while regulating the heater, before addition of the reactants.

5. With two 275-watt heating tapes, the required temperature inside the column is secured at a voltage setting of about 80. If alcohol does not distil noticeably during the reaction, the internal temperature is not high enough and the transformer should be adjusted, or more time should be allowed for preheating the column.

6. Part of the product solidifies in the receiver.

7. This compares with 74–76% yield by the batch process.[2]

8. Recrystallization from benzene produces a whiter product in 92% recovery.

9. This procedure may be used to prepare other substituted anilides. If one of the components is a solid, it can be dissolved in an excess of the other. For instance, 46 g. of 2-amino-5-nitroanisole in 285 ml. of hot ethyl benzoylacetate is passed through in 30 minutes, with a transformer setting of 70 volts; then the column is rinsed with 50 ml. of the ester. The product crystallizes in the receiver and is separated by filtration; the filtrate is used to make up more of the reacting component mixture. The work-up yields 75–76 g. (89%) of amide with the correct melting point (178.5–180°) and color. When

the same amide is made by a batch procedure, the yield is only 81% of a product melting at 130–150°. This illustrates the advantage of short time of exposure to heat in the continuous reactor.

10. This reactor has also been used with other types of reactions.

	Volt-age	Time, hr.	Mole Ester	Yield, %
Ethyl benzalmalonate (batch) [3]		18	0.63	90.8
(cont.)	65	0.5	0.69	82
3-Carboxy-4-hydroxyquinoline [4]	90	0.66		35
4-Benzal-2-phenyl-5-oxazolone (batch) [5]				64
(an azlactone) (cont.)	60	0.25		73

3. Methods of Preparation

This preparation and oleoyl chloride (p. 739) illustrate the use of the general form of a laboratory-sized continuous reactor.[6] This device has many advantages over the commonly used flasks (batch procedure). In particular, the short time of exposure to heat results in a better quality of product, as shown by less color, fewer side reactions, and better melting point, often unchanged by recrystallization. Furthermore, the unlimited capacity, very short reaction time, and use of concentrated solutions permit a larger output with no increase in size of apparatus and less delay required for removal of solvents.

The continuous reactor is most useful with reactions that take place at a relatively rapid rate. Its wide versatility enables it to be used in many types of reactions. Benzoylacetanilide was selected as an example because it has previously appeared in *Organic Syntheses*.[2]

[1] Eastman Kodak Company, Rochester, New York.
[2] *Org. Syntheses Coll. Vol.* **3**, 108 (1955).
[3] *Org. Syntheses Coll. Vol.* **3**, 377 (1955).
[4] Gould and Jacobs, *J. Am. Chem. Soc.*, **61**, 2893 (1939).
[5] *Org. Syntheses Coll. Vol.* **2**, 490 (1943).
[6] Allen, Byers, Humphlett, and Reynolds, *J. Chem. Educ.*, **32**, 394 (1955).

BENZOYLCHOLINE IODIDE AND CHLORIDE *

(Choline, chloride benzoate, and Choline, iodide benzoate)

$$C_6H_5COCl + HOCH_2CH_2Cl \rightarrow C_6H_5CO_2CH_2CH_2Cl + HCl$$

$$C_6H_5CO_2CH_2CH_2Cl + NaI \rightarrow C_6H_5CO_2CH_2CH_2I + NaCl$$

$$C_6H_5CO_2CH_2CH_2I + (CH_3)_3N \rightarrow C_6H_5CO_2CH_2CH_2\overset{+}{N}(CH_3)_3I^-$$

$$C_6H_5CO_2CH_2CH_2\overset{+}{N}(CH_3)_3I^- + AgCl \rightarrow$$
$$C_6H_5CO_2CH_2CH_2\overset{+}{N}(CH_3)_3Cl^- + AgI$$

Submitted by A. H. FORD-MOORE.[1]
Checked by R. L. SHRINER and CALVIN N. WOLF.

1. Procedure

A. *2-Chloroethyl benzoate.* In a 500-ml. round-bottomed flask attached to a 100-cm. air condenser by a ground-glass joint are placed 80.5 g. (66 ml., 1 mole) of redistilled ethylene chlorohydrin (b.p. 128–129°) and 140.5 g. (115.5 ml., 1 mole) of benzoyl chloride. The apparatus is set up in a good hood, and the mixture is warmed gently with a low flame until the reaction starts (Note 1). The source of heat is withdrawn until the reaction moderates and is then again applied for an additional 30 minutes, during which time the temperature rises to about 200–215°. The flask is fitted with a short column (about 20 cm.) and arranged for distillation. After volatile material has been removed by evacuation with a water pump at a bath temperature of 100–110° the residual liquid is fractionated under reduced pressure. The yield of 2-chloroethyl benzoate boiling at 101–104°/2 mm. is 165–168 g. (89–91%), n_D^{19} 1.5298.

B. *2-Iodoethyl benzoate.* A mixture of 170 g. of anhydrous sodium iodide and 1.2 l. of methyl ethyl ketone (Note 2) is heated on a steam bath for 1 hour with occasional shaking in a 3-l. round-bottomed flask fitted with a water-cooled reflux condenser. 2-Chloroethyl benzoate (162 g., 0.88 mole) is added to the mixture, and heating is maintained for an additional 22–24 hours with occasional shaking. The mixture is cooled to room temperature and filtered through a 15-cm. Büchner funnel with suction. The inorganic salts on the filter are washed with 200 ml. of methyl ethyl ketone, and the filtrate is concentrated by distillation of about 1 l. of the solvent. The residue is poured into 1 l. of water contained in a separatory funnel, which is shaken, and the lower layer is withdrawn. The latter is washed successively with 200 ml. of 10% sodium bisulfite solution, 200 ml. of 5% sodium bicarbonate solution, and 100 ml. of water. It is dried with anhydrous magnesium sulfate (5–7 g.) and fractionated under reduced pressure. The yield of material boiling at 133–136°/2.5 mm., n_D^{15} 1.5820, is 190–196 g. (78–81%).

C. *Benzoylcholine iodide.* A solution of 194 g. (0.70 mole) of 2-iodoethyl benzoate in 200 ml. of dry acetone is treated with 270 ml. of a 19.5% solution of trimethylamine in acetone (Note 3) in a 1-l. Pyrex reagent bottle which is closed with a tightly fitting rubber stopper wired in place. The solution is allowed to stand at room temperature for 24 hours (Note 4), and at the end of this time the quaternary salt is separated by filtration with suction, washed with 200 ml. of dry acetone, and air-dried (Note 5). The weight of the quaternary iodide melting with decomposition at 247° is 200–210 g. (85–90%) (Note 6).

D. *Benzoylcholine chloride.* Silver chloride is prepared by dissolving 160 g. (0.94 mole) of silver nitrate in 500 ml. of boiling distilled water and adding 120 ml. of analytical reagent hydrochloric acid (sp. gr. 1.18) from a dropping funnel in a period of 15 minutes, with continuous stirring. The silver chloride is washed by decantation with three 300-ml. portions of boiling distilled water. The moist silver chloride is suspended in 750 ml. of water warmed to 50–60° in a 2-l. beaker, and 210 g. (0.63 mole) of benzoylcholine iodide is added in a period of 1 hour, with good mechanical stirring. After the addition is completed, stirring is continued for an additional 30 minutes without the application of heat. The mixture is cooled and filtered with suction. The silver salts on the filter are washed with 200 ml. of hot water (Note 7), and the combined filtrates are evaporated to dryness under reduced pressure (water pump). The residue is dried by twice distilling to dryness with 250 ml. of absolute ethanol and then once with 250 ml. of dry acetone, the last of the acetone being removed under reduced pressure. The product is recrystallized by dissolving it in 240 ml. of isopropyl alcohol (Note 8) and allowing the solution to cool in a refrigerator. It is filtered and dried, first at 100° and then in a vacuum desiccator over silica gel. The yield of pure product, m.p. 207° (dec.), is 125–132 g. (82–87%) (Note 9).

2. Notes

1. Usually the reaction starts at a temperature of 55° to 60° as evidenced by liberation of hydrogen chloride. If the reaction becomes too vigorous it may be moderated by applying a wet towel to the flask.

2. A purified grade of methyl ethyl ketone should be used. The technical material may be purified by allowing it to stand over solid anhydrous calcium chloride for 24 hours, decanting from the syrupy layer and solid through a filter, and then distilling; b.p. 79–80°.

3. The trimethylamine may be generated by the action of alkali on trimethylamine hydrochloride and dissolved in acetone.[2] The submitter prepared trimethylamine by the method of Sommelet and Ferrand[3] and obtained a 65% yield by the interaction of ammonia, formaldehyde, and formic acid. The checkers found that a commercial 25% solution of trimethylamine in methanol (210 ml.) gave the same yields as the acetone solution.

4. Increasing the reaction period to 48 hours gives an additional 5 to 7 g. of product.

5. If this preparation is carried out during periods of high humidity it is best to place the product in a vacuum desiccator, which is evacuated several times in order to remove the solvent. Anhydrous calcium chloride may be placed in the bottom of the desiccator.

6. This product is quite pure as shown by titration of iodide ion.

7. The silver residues are saved for the recovery of silver[4] and iodine.

8. The submitter states that the compound may be recrystallized by boiling with acetone under an upright condenser and adding ethanol cautiously down the condenser until the solid just dissolves. The substance is appreciably soluble in the cold sol-

vent mixture. It is necessary to distil the mother liquor and recrystallize the residue from acetone-ethanol, for otherwise a considerable loss of product will occur. The checkers used isopropyl alcohol for crystallization.

9. Benzoylcholine chloride prepared by this method is pure as shown by titration of ionic chlorine. It is somewhat hygroscopic, though much less so than choline chloride. It is not appreciably hydrolyzed by boiling with water for 1 hour, although more prolonged heating leads to the formation of benzoic acid. Benzoylcholine chloride can be characterized as the picrate, m.p. 177°, formed by treating a strong aqueous solution with an appropriate amount of $0.5N$ calcium picrate and crystallizing the product from methyl ethyl ketone.

3. Methods of Preparation

2-Chloroethyl benzoate has been prepared from benzoyl chloride and ethylene chlorohydrin; [5-7] from benzoic acid, ethylene glycol, and hydrogen chloride [8] at 100°; from ethylene oxide and benzoyl chloride; [9] from benzoyl chloride and dioxane in the presence of titanium tetrachloride; [10] from benzoic acid, ethylene, and chlorine in the presence of various catalysts.[11] It has also been obtained by hydrolysis of 2-chloroethyl benzimidate; [12] by the action of bromomagnesium benzoate on 2-chloroethyl p-toluene-sulfonate; [13] and by the action of sodium benzoate on bis-(2-chloroethyl) sulfate.[14]

2-Iodoethyl benzoate has been obtained by the action of alcoholic sodium iodide on 2-chloroethyl benzoate.[5]

Benzoylcholine chloride has been prepared by heating choline chloride with benzoyl chloride [15] and by the action of trimethylamine on 2-chloroethyl benzoate.[16]

[1] Chemical Defence Experimental Station, Wilts, England.
[2] Org. Syntheses Coll. Vol. 1, 528 (1941).
[3] Sommelet and Ferrand, Bull. soc. chim. France, [4] 35, 446 (1924).
[4] Inorg. Syntheses, 1, 2 (1939).
[5] Zaki, J. Chem. Soc., 1930, 2271.
[6] Kirner, J. Am. Chem. Soc., 48, 2751 (1926).
[7] Jones and Major, J. Am. Chem. Soc., 49, 1535 (1927).
[8] Simpson, Ann., 113, 120 (1860).
[9] Altwegg and Landrivon, U. S. pat. 1,393,191 [C. A., 16, 422 (1922)].

[10] Goldfarb and Smorgonskii, *J. Gen. Chem. U.S.S.R.*, **8**, 1516 (1938) [*C. A.*, **33**, 4593 (1939)].

[11] Brit. pat. 460,720 [*C. A.*, **31**, 4675 (1937)].

[12] Gabriel and Neumann, *Ber.*, **25**, 2384 (1892).

[13] Gilman and Beaber, *J. Am. Chem. Soc.*, **45**, 842 (1923).

[14] Suter and Evans, *J. Am. Chem. Soc.*, **60**, 537 (1938).

[15] Nothnagel, *Arch. Pharm.*, **232**, 267 (1894).

[16] Fourneau and Page, *Bull. soc. chim. France*, [4] **15**, 552 (1914).

3-BENZOYLPYRIDINE *

(Ketone, phenyl 3-pyridyl)

Submitted by FRANK J. VILLANI and MARY S. KING.[1]
Checked by MAX TISHLER and MATTHEW A. KOZLOWSKI.

1. Procedure

In a 2-l. three-necked flask, fitted with a sealed mechanical stirrer (Note 1), a reflux condenser protected with a calcium chloride tube, and a dropping funnel, is placed 123 g. (1 mole) of nicotinic acid (Note 2). The stirrer is started, and 500 ml. (818 g., 6.9 moles) of distilled thionyl chloride is added in a slow stream over a period of 15–20 minutes (Note 3). After the addition is complete, the mixture is heated on the steam bath with continuous stirring for 1 hour; then the reflux condenser is replaced by one set for downward distillation, and the excess thionyl chloride is removed by distillation at reduced pressure as heating on the steam bath is continued (Notes 1 and 3). After most of the thionyl chloride has been distilled, 200 ml. of anhydrous benzene is added, and the benzene is distilled at reduced pressure. An additional 500 ml. of anhydrous benzene is added; then the flask is fitted with a thermometer and a reflux condenser and is placed in an ice-salt bath. The stirrer is started, and 330 g. (2.5 moles) of anhydrous aluminum chloride is added in portions over a period of 1 hour as the internal temperature is held between 5° and 10°. The ice bath is removed, and the flask is permitted to warm to room temperature and is finally heated under reflux for 6 hours.

The dark red-brown reaction mixture is cautiously poured onto a mixture of 2 kg. of ice and 200 ml. of concentrated hydrochloric acid. The organic layer is separated and discarded. The acid solution is extracted with three 500-ml. portions of ether (Note 4), which are discarded; then it is treated with 50% aqueous sodium hydroxide until the aluminum hydroxide which first forms redissolves (Note 5). After cooling, the organic material is extracted with five 300-ml. portions of chloroform. The combined chloroform extracts are washed with water, the solvent is removed by distillation on the steam bath, and the product is distilled. The yield of 3-benzoylpyridine (Note 6), b.p. 107–110°/0.3 mm. or 141–145°/4 mm., is 165–175 g. (90–96%), n_D^{26} 1.6088.

2. Notes

1. It is convenient to use a sealed stirrer, such as the Trubore stirrer, which may be left in the flask during the distillation of thionyl chloride and benzene at reduced pressure; however, the stirrer cannot be left running during this distillation, for it is stopped by the cake of acid chloride hydrochloride. The distillations are accomplished most conveniently if the dropping funnel is removed and replaced by a capillary and the stirrer is left in place but not running during the distillations.

2. Satisfactory results were obtained with nicotinic acid from either Eastman Organic Chemicals or Matheson, Coleman and Bell.

3. The submitters used thionyl chloride from Hooker Electrochemical Company. It was distilled and collected over a 1° range (78–79°). The first few drops of thionyl chloride are added cautiously as the initial reaction may be quite vigorous. Recovered thionyl chloride may be used for subsequent runs.

4. The ether extractions remove any diphenyl sulfoxide that is formed.

5. About 800–1000 ml. of 50% sodium hydroxide is required.

6. 4-Benzoylpyridine can be obtained by this procedure from iso-

nicotinic acid in yields of 87–90%. This product is recrystallized from hexane, m.p. 72–73°.

3. Methods of Preparation

The described method of preparing 3-benzoylpyridine is a modification of that of Wolffenstein and Hartwich.[2] Other methods of preparing this compound are by the addition of phenylmagnesium bromide to 3-cyanopyridine,[3] the addition of 3-pyridyllithium to benzonitrile,[4] the chromic acid oxidation of phenyl-3-pyridylcarbinol,[5] and the decarboxylation of β-benzoylpicolinic acid obtained from quinolinic acid anhydride and benzene.[6]

[1] Schering Corporation, Bloomfield, New Jersey.
[2] Wolffenstein and Hartwich, *Ber.*, **48**, 2043 (1915).
[3] LaForge, *J. Am. Chem. Soc.*, **50**, 2486 (1928).
[4] French and Sears, *J. Am. Chem. Soc.*, **73**, 469 (1951); Wibaut, deJonge, van der Voort, and Otto, *Rec. trav. chim.*, **70**, 1054 (1951).
[5] Kleipool and Wibaut, *Rec. trav. chim.*, **69**, 1041 (1950).
[6] Bernthsen and Mettegang, *Ber.*, **20**, 1209 (1887).

2-BENZYLAMINOPYRIDINE *

(Pyridine, 2-benzylamino-)

$$\underset{N}{\underset{NH_2}{\bigcirc}} + C_6H_5CH_2OH \xrightarrow{KOH} \underset{N}{\bigcirc}-NHCH_2C_6H_5 + H_2O$$

Submitted by YAÏR SPRINZAK.[1]
Checked by MAX TISHLER and MATTHEW A. KOZLOWSKI.

1. Procedure

A 500-ml. Claisen flask with a 35-mm. indented side arm (Note 1) is attached downward to a Liebig condenser for distillation. A thermometer, held by a cork stopper, is inserted through the neck of the flask and adjusted so that its bulb is close to the bottom. The flask is heated by means of an electric mantle or air bath. To the flask are charged 94 g. (1.0 mole) of 2-aminopyridine (Note 2), 150 g. (1.4 moles) of benzyl alcohol, and 9 g. of 85% potassium hydroxide. The mixture is heated to boiling and boiled vigorously enough to cause slow distillation of water accompanied by as little benzyl alcohol as possible (Note 3). The temperature of the boiling mixture rises gradually from 182° to 250° during a period of 30 minutes. The mixture is maintained at 250° for 3 minutes, then allowed to cool. The distillate amounts to 19–20 ml. of a water-rich layer and 2–4 ml. of a benzyl alcohol-rich layer.

The residual product is cooled to about 100° (Note 4) and poured into 250 ml. of water. The crystallized solid is crushed and collected on a 12-cm. Büchner funnel. Slight suction is applied at first, but, after most of the mother liquor has been removed, the crystals are pressed down with strong suction. The product is then washed thoroughly with water. After drying, the yield of colorless 2-benzylaminopyridine (Notes 5 and 6), m.p. 95–96°, amounts to 180–183 g. (98–99% of the theoretical amount). The product may be recrystallized from isopropyl alcohol with 90% recovery. For each gram of amine 3 ml. of solvent is employed. The melting point of recrystallized material is 96.0–96.7° (cor.), lit.[2] m.p. 97–98°.

2. Notes

1. A short Vigreux column or any other short column for distillation may be used.

2. 2-Aminopyridine was obtained from Matheson, Coleman and Bell, Inc., East Rutherford, New Jersey.

3. If the reaction mixture is heated too strongly or the vapors are inadequately fractionated, correspondingly greater amounts of benzyl alcohol will distill, with concomitant loss of yield. The distillate should be clear, not milky.

4. If the product has partially solidified, it should be melted for easy handling.

5. By essentially the same procedure N,N'-dibenzyl-p-phenylenediamine has been obtained in 92% yield. The heating period requires 1 hour, and the final temperature is 260°.

6. N-Benzylaniline has been obtained in 90–94% yield by an appropriate modification[3] of this method.

3. Methods of Preparation

For 2-benzylaminopyridine the methods of preparation of significance are condensation of 2-pyridinesulfonic acid and benzylamine,[4] condensation of the alkali metal salts of 2-aminopyridine with benzyl chloride[5] or benzyl alcohol,[6] reductive alkylation of 2-aminopyridine in the presence of benzaldehyde and formic acid,[2] oxidation of N-benzyl-N-pyridylaminoacetonitrile or N-benzyl-N-pyridylaminoacetaldoxime,[7] and the method described here modified by use of an inert solvent.[8]

[1] The Weizmann Institute of Science, Rehovoth, Israel.

[2] Kaye and Kogon, Rec. trav. chim., 71, 309 (1952).

[3] Sprinzak, J. Am. Chem Soc., 78, 3207 (1956).

[4] Mangini and Colonna, Gazz. chim. ital., 73, 313 (1943) [C. A., 41, 1224 (1947)].

[5] Huttrer, Djerassi, Beears, Mayer, and Scholz, J. Am. Chem. Soc., 68, 1999 (1946).

[6] Géczy, Magyar Kém. Folyóirat, 62, 162 (1956) [C. A., 52, 10075 (1958)].

[7] Bristow, Charlton, Peak, and Short, J. Chem. Soc., 1954, (616).

[8] Hirao and Hagashi, J. Pharm. Soc. Japan, 74, 853 (1954) [C. A., 49, 10308 (1955)].

3-BENZYL-3-METHYLPENTANOIC ACID

(Valeric acid, 3-benzyl-3-methyl-)

$$C_2H_5-CO-CH_3 + NC-CH_2CO_2C_2H_5 \xrightarrow{\beta\text{-Alanine}}$$

$$\begin{array}{c} CH_3 \\ | \\ C_2H_5-C{=}C-CO_2C_2H_5 + H_2O \\ | \\ CN \end{array}$$

$$\begin{array}{c} CH_3 \\ | \\ C_2H_5-C{=}CCO_2C_2H_5 + C_6H_5CH_2MgCl \xrightarrow{(1)} \xrightarrow{(2)\ H_2O} \\ | \\ CN \end{array}$$

$$\begin{array}{cc} CH_3 & CN \\ | & | \\ C_6H_5CH_2-C\text{------}CH-CO_2C_2H_5 + Mg(OH)Cl \\ | \\ C_2H_5 \end{array}$$

$$\begin{array}{cc} CH_3 & CN \\ | & | \\ C_6H_5CH_2-C\text{------}CH-CO_2C_2H_5 + 2KOH \rightarrow \\ | \\ C_2H_5 \end{array}$$

$$\begin{array}{c} CH_3 \\ | \\ C_6H_5CH_2-C-CH_2-CN + C_2H_5OH + K_2CO_3 \\ | \\ C_2H_5 \end{array}$$

$$\begin{array}{c} CH_3 \\ | \\ C_6H_5CH_2-C-CH_2CN + H_2O \xrightarrow[(2)\ HCl]{(1)\ KOH} \\ | \\ C_2H_5 \end{array}$$

$$\begin{array}{c} CH_3 \\ | \\ C_6H_5CH_2-C-CH_2CO_2H + NH_4Cl + KCl \\ | \\ C_2H_5 \end{array}$$

Submitted by F. S. PROUT, R. J. HARTMAN, E. P.-Y. HUANG,
C. J. KORPICS, and G. R. TICHELAAR.[1]
Checked by JAMES CASON, K. C. DEWHIRST, E. J. GAUGLITZ, JR., and
WILLIAM G. DAUBEN.

1. Procedure

A. *Ethyl sec-butylidenecyanoacetate.* In a 1-l. round-bottomed flask fitted with a 24/40 joint are placed 0.45 g. of β-alanine, 113 g. (106 ml., 1.0 mole) of ethyl cyanoacetate (Note 1), 87 g. (108 ml., 1.2 moles) of butanone, 20 ml. of glacial acetic acid, and 100 ml. of benzene. A Barrett-type water separator (Note 2) and a condenser are attached to the flask, and the mixture is heated briskly under reflux until water ceases to be collected in the trap (7–12 hours).

The reaction mixture is decanted into a 500-ml. round-bottomed flask which is attached to a fractionating column (Note 3). The solvent is removed at atmospheric pressure while the oil bath is heated finally at 160°. The residue is distilled at reduced pressure to furnish four fractions: (a) acetic acid and other materials boiling below 95°/16 mm.; (b) ethyl cyanoacetate, b.p. 95–110°/16 mm.; (c) intermediate, b.p. 110–124°/16 mm.; and (d) ethyl *sec*-butylidenecyanoacetate, b.p. 124–126°/16 mm., n_D^{25} 1.4640–1.4648. Fraction d amounts to 117–122 g., and refractionation of fraction c yields an additional 18–24 g.; total yield, 135–146 g. (81–87.5%) (Note 4).

B. *Ethyl 3-benzyl-2-cyano-3-methylpentanoate.* A 2-l. three-necked round-bottomed flask, fitted with a tantalum wire Hershberg stirrer, a condenser, and a separatory funnel, is arranged for use of a nitrogen atmosphere.[2] Magnesium (19.2 g., 0.79 g. atom) and 100 ml. of dry ether [3] are placed in the flask, and a solution of 100 g. (91 ml., 0.79 mole) of benzyl chloride in 500 ml. of dry ether is added in a period of 1.5–2.0 hours, with stirring, while the mixture boils spontaneously. The mixture is boiled for 15 minutes after completion of the addition, then a solution of 110 g. (0.66 mole) of ethyl *sec*-butylidenecyanoacetate in 130 ml. of benzene is added over a 30-minute period with spontaneous reflux. The reaction mixture is stirred and heated under reflux for an additional hour. A precipitate separates after about 30 minutes.

The reaction mixture is poured onto about 400 g. of cracked ice and is made acidic with 20% sulfuric acid. After two clear phases have formed the mixture is poured into a separatory funnel, and the lower layer is removed. This aqueous layer is extracted with two 100-ml. portions of benzene and discarded. The three organic extracts are washed separately and successively with 125 ml. of water and 125 ml. of saturated sodium chloride solution, then filtered successively through a layer of anhydrous sodium sulfate.

The combined extract (about 1 l.) is flash-distilled at atmospheric pressure from a 250-ml. Claisen flask. After the solvent and a small amount of fore-run (ca. 15 g., b.p. 45°/3 mm.) have been removed, the product is distilled to yield 157–162 g. (92–95%), b.p. 150–162°/3 mm. (bath temperature, 180–190°), n_D^{25} 1.5053–1.5063 (Notes 5, 6, and 7).

C. *3-Benzyl-3-methylpentanenitrile.* Sixty-seven grams (1 mole) of potassium hydroxide (85%) is dissolved by heating in 360 ml. of ethylene glycol and is added to a 1-l. round-bottomed flask containing 155 g. (0.6 mole) of ethyl 3-benzyl-2-cyano-3-methylpentanoate (above). A condenser is attached with a rubber stopper, and the mixture is heated under gentle reflux for 3 hours (Note 8). The resulting two-phase mixture is cooled, diluted with 350 ml. of water, and extracted with three portions of ether (250 ml., 100 ml., 100 ml.). The three extracts are washed successively with 100 ml. of water and 100 ml. of saturated sodium chloride solution, then filtered through a layer of anhydrous sodium sulfate (Note 9). The combined extracts are flash-distilled at atmospheric pressure from a 250-ml. Claisen flask to remove the ether. The residue is distilled at reduced pressure to furnish 102–105 g. (91–93%) of nitrile, b.p. 150–160°/11 mm. (bath temperature, 190–200°), n_D^{25} 1.5111–1.5128 (Notes 10 and 11).

D. *3-Benzyl-3-methylpentanoic acid.* A solution of 112 g. (1.6 moles) of potassium hydroxide (85%) in 400 ml. of ethylene glycol is added to 93.6 g. (0.5 mole) of 3-benzyl-3-methylpentanenitrile in a 1-l. round-bottomed copper or stainless-steel flask. A condenser with a rubber stopper is attached, and the solution is heated under brisk reflux for 6 hours (Note 12). The reaction mixture is cooled, diluted with 400 ml. of water, and extracted with three portions of ether (250 ml., 100 ml., 100 ml.). The ether extracts are washed successively with two 75-ml. portions of water and then discarded (Note 13).

The combined aqueous phases are acidified to Congo red with 200 ml. of concentrated hydrochloric acid and extracted with three portions of benzene (200 ml., 75 ml., 75 ml.). The benzene extracts are washed successively with 100 ml. of water and 100 ml. of saturated sodium chloride solution, then filtered through anhydrous sodium sulfate. The combined extract is flash-distilled from a 250-ml. Claisen flask at atmospheric pressure (bath temperature, up to 160°). The residue is distilled at reduced pressure to give 94–96 g. (91–93%) of acid; b.p. 173–177°/7 mm. (bath temperature, 207–220°), n_D^{25} 1.5160–1.5163 (Notes 14 and 15).

2. Notes

1. Ethyl cyanoacetate was obtained from Kay-Fries Chemicals, 180 Madison Avenue, New York, New York.

2. The submitters used a Barrett Distilling Receiver, Corning No. 3622, Corning Glass Works, Corning, New York.

3. The submitters used a 60-cm. heated Vigreux column to effect this fractionation. The checkers used a similar column with partial take-off head.

4. Fractions b–d consist entirely of ethyl cyanoacetate and the product. Pure ethyl cyanoacetate and ethyl sec-butylidenecyanoacetate have n_D^{25} 1.4151 and 1.4650, respectively. The purity of fractions b–d can be estimated by their indexes of refraction, which are proportional to the weight per cent.

5. The pure product obtained by fractional distillation has n_D^{25} 1.5052. The product obtained by distillation from a Claisen flask is contaminated mainly with bibenzyl, b.p. 122–125°/3 mm., f.p. 44°. The purity of the product can be estimated by determination of the saponification equivalent in ethanol.

6. The use of dibenzylcadmium gave no improvement in yield.

7. Phenylmagnesium bromide gives a 79% yield of product, b.p. 178–180°/11 mm., n_D^{25} 1.5063; and n-propylmagnesium bromide gives 33–42% yields, b.p. 150–153°/22 mm., n_D^{25} 1.4429, of alkylation product when essentially the same procedure is used. The yield obtained with n-propylmagnesium bromide depends upon the efficiency of separation from the reduction product, ethyl sec-butylcyanoacetate, b.p. 126°/22 mm., n_D^{25} 1.4277.[4]

8. After 30 minutes of reflux the second phase begins to

separate. The formation of this nitrile layer is probably complete after 2.5–3.0 hours. A small amount of solid, presumably ammonium carbonate, collects in the condenser during the heating.

9. When the combined aqueous washes from three runs were acidified, extracted, and distilled, there was obtained 1.8 g. of 3-benzyl-3-methylpentanoic acid, n_D^{25} 1.5158.

10. This nitrile is contaminated with some lower-boiling bibenzyl and some higher-boiling amide. The pure nitrile, obtained by fractional distillation, has n_D^{25} 1.5110.

11. Hydrolyses of ethyl 2-cyano-3-methyl-3-phenylpentanoate and ethyl 2-cyano-3-ethyl-3-methylhexanoate (cf. Note 7) by essentially this procedure gave 71% (b.p. 149–151°/16 mm., n_D^{25} 1.5149) and 68% (b.p. 103–104°/31 mm., n_D^{25} 1.4291) yields of nitriles, respectively. In the second case, about 12% additional yield of nitrile could be obtained from the acidic fraction, which contains some undecarboxylated product, 2-cyano-3-ethyl-3-methylhexanoic acid.

12. The two-phase solution becomes homogeneous after 1.5–2.0 hours of boiling. This alkaline solution is very corrosive, and a glass flask can be used only a few times in this reaction. A stainless-steel or copper flask is preferable.

13. If emulsions are encountered, the addition of a few milliliters of saturated aqueous sodium chloride clears them readily. The combined ether extracts contain 4–5 g. of solid, neutral material. This product is mainly bibenzyl, b.p. 138–143°/7 mm., f.p. 40°.

14. The best sample of this acid obtained by fractional distillation had n_D^{25} 1.5160; neut. equiv., 207.3 (calcd., 206.3).

15. Hydrolyses of 3-methyl-3-phenylpentanenitrile and 3-ethyl-3-methylhexanenitrile (cf. Note 11) by the described procedure gave 88% and 95% yields of 3-methyl-3-phenylpentanoic acid (b.p. 190–194°/26 mm., n_D^{25} 1.5182) and 3-ethyl-3-methylhexanoic acid (b.p. 136–137°/15 mm., n_D^{25} 1.4377), respectively.

3. Method of Preparation

Ethyl sec-butylidenecyanoacetate has been prepared by this condensation using various amino acids,[5] ammonium acetate,[6] sodium sulfate-piperidine,[7] and zinc chloride-aniline.[8]

The ethyl 3-benzyl-2-cyano-3-methylpentanoate, 3-benzyl-3-methylpentanenitrile, and 3-benzyl-3-methylpentanoic acid preparations follow the procedure given by Prout, Huang, Hartman, and Korpics.[9]

[1] De Paul University, Chicago 14, Illinois.

[2] Cason and Rapoport, *Laboratory Text in Organic Chemistry*, 2nd ed., p. 452, Prentice-Hall, Englewood Cliffs, New Jersey, 1962; cf. *Org. Syntheses Coll. Vol.* **3**, 601 (1955).

[3] Cason and Rapoport, *Laboratory Text in Organic Chemistry*, 2nd ed., p. 460, Prentice-Hall, Englewood Cliffs, New Jersey, 1962.

[4] Prout, *J. Am. Chem. Soc.*, **74**, 5915 (1952).

[5] Prout, *J. Org. Chem.*, **18**, 928 (1953).

[6] Cope, Hofmann, Wyckoff, and Hardenbergh, *J. Am. Chem. Soc.*, **63**, 3452 (1941).

[7] Cowan and Vogel, *J. Chem. Soc.*, **1940**, 1528.

[8] Scheiber and Meisel, *Ber.*, **48**, 259 (1915).

[9] Prout, Huang, Hartman, and Korpics, *J. Am. Chem. Soc.*, **76**, 1911 (1954).

BENZYLTRIMETHYLAMMONIUM ETHOXIDE

(Ammonium, benzyltrimethyl-, ethoxide)

$$C_6H_5CH_2Cl + N(CH_3)_3 \xrightarrow{C_2H_5OH} C_6H_5CH_2\overset{+}{N}(CH_3)_3\overset{-}{Cl}$$

$$2C_2H_5OH + 2Na \xrightarrow{C_2H_5OH} 2NaOC_2H_5 + H_2$$

$$C_6H_5CH_2\overset{+}{N}(CH_3)_3\overset{-}{Cl} + NaOC_2H_5 \xrightarrow{C_2H_5OH}$$

$$C_6H_5CH_2\overset{+}{N}(CH_3)_3\overset{-}{O}C_2H_5 + NaCl$$

Submitted by W. J. CROXALL, MARIAN F. FEGLEY, and H. J. SCHNEIDER.[1]
Checked by JOHN C. SHEEHAN and M. GERTRUDE HOWELL.

1. Procedure

A 3-l. three-necked flask (flask *A*) is fitted with a nitrogen inlet, a reflux condenser protected by a soda-lime tube, and a reflux condenser fitted with a dropping funnel protected by a soda-lime tube.

A second 3-l. three-necked flask (flask *B*) is fitted with a gas-tight modified Hershberg stirrer (Note 1), a gas inlet tube, and an appropriately designed gas outlet tube bearing a thermometer and connections leading to a soda-lime tube and an open-end mercury manometer. All rubber stoppers and connections are wired in place with 16-gauge copper wire.

Sodium (69 g., 3 g. atoms) is introduced into flask A, which had been flushed previously with nitrogen (Note 2). Ethanol (1.2 kg.) (Note 3) is added at such a rate that a continuous reflux is maintained. After the sodium is completely dissolved, the solution is allowed to cool to room temperature. During this period the benzyltrimethylammonium chloride is prepared.

A solution of 379.5 g. (3 moles) of benzyl chloride (Note 4) dissolved in 750 g. of anhydrous ethanol (Note 3) is placed in flask B. The system is flushed with trimethylamine previously dried by passage through a soda-lime tower. The gas outlet is closed and connected to the manometer. Over a period of 80 minutes, 195 g. (3.3 moles) of trimethylamine (Note 5) is introduced with stirring. The reaction is exothermic and must be cooled to keep the temperature below 50°. After the amine addition is complete, the solution is kept at 50° under an amine pressure of 5 cm. of mercury above atmospheric pressure for 1 hour. The mixture is then cooled to room temperature under an amine atmosphere.

The inlet tube of flask B is replaced by a rubber stopper bearing a short glass outlet tube and a glass inlet tube which is connected to an appropriately designed glass siphon outlet tube extending to the bottom of flask A. Gentle suction applied to the outlet tube of flask B draws the ethoxide solution slowly into flask B. The benzyltrimethylammonium chloride solution is stirred throughout the addition of the sodium ethoxide. After the transfer of the sodium ethoxide is completed, the finely divided precipitate of sodium chloride is allowed to settle overnight. When the sodium chloride has completely settled, the stopper bearing the inlet and outlet tubes is replaced by a stopper bearing a 15-in. length of 19-mm. glass tubing which is attached in turn by a gum rubber connection to a clean, dry, 1-gal. bottle fitted with an inlet tube and an outlet tube protected by soda-lime. The bottle is flushed with nitrogen before use. The 15-in. tube is adjusted so that the bottom is approximately ½ in. above the level of the precipitated sodium chloride. With the open-end manometer sealed off by a screw clamp, nitrogen pressure is applied through the outlet tube of the flask until 1.7–1.9 kg. of supernatant liquor is forced from the flask into the bottle (Note 6). This solution contains 24–30% benzyltrimethylammonium ethoxide (2.1–2.7 moles), as determined by titration with $0.1N$ hydrochloric acid, using methyl red as indicator. This represents a yield of 70–90%. An additional 270–400 g. of solution is obtained by filtration of the residual mixture under nitrogen. To ensure rapid filtration, a filter aid, such as Filtercel (Note 7), must be employed. The filtrate contains 24–30% benzyltrimethylammonium ethoxide (0.3–0.7 mole). The total yield is 89–100%. The solutions are stored under nitrogen and refrigeration in bottles sealed with rubber stoppers which are wired in place (Note 8).

Three hundred and thirty-five grams of the 25% ethanolic solution of benzyltrimethylammonium ethoxide (0.43 mole) is placed under nitrogen in a 3-l. three-necked flask equipped with a gas inlet tube, a gas-tight modified Hershberg stirrer (Note 1), and a gas outlet tube fitted with a thermometer. All stoppers and rubber connections are wired in place with 16-gauge copper wire. There is obtained by evaporation at 40° (Note 9) under reduced pressure (Note 10) 97 g. (0.40 mole) of the ethoxide containing an equivalent of ethanol. The vacuum is broken with dry nitrogen.

2. Notes

1. The stirrer is of the type designed by Hershberg,[2] but it has a single paddle of 16-gauge Nichrome wire.

2. A bubble counter is employed for indicating nitrogen flow.

3. The water content of the ethanol, determined by Karl Fischer analysis, is 0.02–0.10%.

4. Benzyl chloride is redistilled before use. The fraction collected at 40–41°/6 mm. is used.

5. The flask is disconnected periodically from the gas train and stirring mechanism for weighing.

6. In small-scale preparations (approximately 0.5 mole) centrifugation of the total mixture is a convenient method for separation of the product from sodium chloride.

7. Filtercel is supplied by Johns-Manville Corporation.

8. This method is more satisfactory than storage of the benzyltrimethylammonium ethoxide monoethanolate. The excess ethanol is removed as necessary.

bis(CHLOROMETHYL) ETHER

HAZARD NOTE

Very high carcinogenic activity has been reported for bis(chloromethyl) ether when administered to rats by inhalation and by subcutaneous injection. This compound should be handled with great care.

Reported by B. L. Van Duuren, A. Sivak, B. M. Goldschmidt, C. Katz, and S. Melchionne, *J. Nat, Cancer Inst.*, **43,** 481 (1969).

Please insert this sheet opposite page 101 of *Org. Syn., Coll. Vol.* 4 (1963).

9. The temperature is maintained by surrounding the flask with a warm water bath. Higher temperatures are not recommended because of the tendency of the quaternary ammonium ethoxide to decompose.

10. The bulk of the ethanol was removed by using a water pump (40–50 mm.) and collected in 1-l. and 200-ml. Dry Ice traps connected in series. The remaining ethanol was removed by using an oil pump protected with two 300-ml. Dry Ice traps.

3. Methods of Preparation

The procedure described is adapted from the preparation outlined by Meisenheimer.[3] The method has been applied by the authors to the preparation of benzyltrimethylammonium methoxide, tetramethylammonium ethoxide, dibenzyldimethylammonium methoxide, bisisopropylbenzyltrimethylammonium methoxide and benzyltriethylammonium ethoxide.

[1] Rohm and Haas Company, Philadelphia 37, Pennsylvania.
[2] Hershberg, *Ind. Eng. Chem., Anal. Ed.*, 8, 313 (1936).
[3] Meisenheimer and Bratring, *Ann.*, **397**, 295 (1913).

BISCHLOROMETHYL ETHER

[Ether, bis(chloromethyl)-]

$$2CH_2O + 2HCl \rightarrow ClCH_2OCH_2Cl + H_2O$$
$$2CH_2O + 2ClSO_3H + H_2O \rightarrow ClCH_2OCH_2Cl + 2H_2SO_4$$

Submitted by SAUL R. BUC.[1]
Checked by CHARLES C. PRICE, FREDERICK V. BRUTCHER, and JEROME COHEN.

1. Procedure

Caution! It has been reported that exposure to bischloromethyl ether in low concentrations has caused severe irritation of the eyes and respiratory tract, and perhaps pulmonary edema.

Proper safeguards should be taken to prevent exposure to this material both during and after its preparation. It is recommended that all operations involving this compound be carried out in a good hood.

In a 1-l. three-necked flask immersed in an ice bath and provided with a stirrer, thermometer, and dropping funnel are placed 168 ml. (200 g.) of concentrated (37–38%) hydrochloric acid (2 moles) and 240 g. of paraformaldehyde (effectively 8 moles of formaldehyde). While the temperature is maintained below 10°, 452 ml. (6.9 moles) of chlorosulfonic acid is added dropwise at such a rate that gaseous hydrogen chloride is not lost from the mixture. This requires about 5.5 hours. The mixture is stirred for 4 hours in the melting ice bath and comes to room temperature. It may be allowed to stand overnight. The layers are separated, and the product (upper layer) is washed twice with ice water. Ice is added to the product, and 250 ml. of 40% sodium hydroxide is then added to the mixture slowly with vigorous agitation until the aqueous phase is strongly alkaline (Note 1). The product is separated and dried rapidly over potassium carbonate and then over potassium hydroxide, keeping the product cold during drying (Note 2). After separation of the drying agent by filtration, 350–370 g. (76–81%) of product sufficiently pure for many purposes is obtained. On distillation there is obtained 330–350 g. (72–76%) of bischloromethyl ether, boiling at 100–104°. Approximately 95% boils at 101–101.5°, n_D^{25} 1.4420 (Note 3).

2. Notes

1. Local overheating must be carefully avoided during the alkaline washing, since it may result in vigorous decomposition.

2. The washing and drying are carried out as rapidly as possible to avoid hydrolysis of the product.

3. The submitter reports that operation of this procedure on a scale of 72 moles of paraformaldehyde gave a crude yield of 3601 g. (87%) and a distilled yield of 3519 g. (85%).

3. Methods of Preparation

Bischloromethyl ether has been prepared by saturation of formalin with dry hydrogen chloride,[2–4] by the reaction of paraformaldehyde with phosphorus trichloride[5,6] or phosphorus oxychloride,[7,8] by solution of paraformaldehyde in concentrated sulfuric acid and treatment with ammonium chloride or dry hydrogen chloride,[9] and by suspension of paraformaldehyde in seventy [10] or eighty [11] percent sulfuric acid and treatment with chlorosulfonic acid. It is formed together with the asymmetrical isomer when methyl ether is chlorinated [12] and when paraformaldehyde is treated with chlorosulfonic acid.[13] It also is obtained when chloromethyl methyl ether is chlorinated by means of chlorine [14] or sulfuryl chloride.[15] The present method has been published.[16]

[1] Central Research Laboratory, General Aniline and Film Corporation, Easton, Pennsylvania.

[2] Tishchenko, *Zhur. Russ. Fiz. Khim. Obshchestva*, **19**, 464 (1887); *Ber.*, **20**, 701 (1887).

[3] Litterscheid and Thimme, *Ann.*, **334**, 1 (1904).

[4] Stephen, Short, and Gladding, *J. Chem. Soc.*, **117**, 510 (1920).

[5] Descudé, *Bull. soc. chim. Paris*, [3] **35**, 953 (1906).

[6] Beeby and Mann, *J. Chem. Soc.*, **1949**, 1799.

[7] Löbering and Fleischmann, *Ber.*, **70B**, 1680 (1937).

[8] Backès, *Bull. soc. chim. France*, [5] **9**, 60 (1942).

[9] Schneider, *Z. angew. Chem.*, **51**, 274 (1938); Zahn, Dietrich, and Gerstner, *Chem. Ber.*, **88**, 1737 (1955).

[10] Norris, *Ind. Eng. Chem.*, **11**, 827 (1919).

[11] Vorozhtov and Yuruigina, *Zhur. Obshcheĭ Khim.*, **1**, 49 (1931) [*C. A.*, **25**, 4521 (1931)].

[12] Salzberg and Werntz (to E. I. du Pont de Nemours and Co.), U. S. pat. 2,065,400 [*C. A.*, **31**, 1046 (1937)]; Evans (to Dow Chemical Co.), U. S. pat. 2,811,485 [*C. A.*, **52**, 2886 (1958)]; Badische Anilin- & Soda Fabrik Akt.-Ges., Ger. pat. 857,949 [*C. A.*, **52**, 5448 (1958)].

[13] Fuchs and Katscher, *Ber.*, **60B**, 2288 (1927).

[14] Evans and Gray, *J. Org. Chem.*, **23**, 745 (1958).

[15] Böhme and Dörries, *Chem. Ber.*, **89**, 723 (1956).

[16] Buc (to General Aniline and Film Corp.), U. S. pat. 2,704,299 [*C. A.*, **50**, 1891 (1956)].

N-BROMOACETAMIDE *

(Acetamide, N-bromo-)

$$CH_3CONH_2 + Br_2 + KOH \rightarrow CH_3CONHBr + KBr + H_2O$$

Submitted by EUGENE P. OLIVETO and CORINNE GEROLD.[1]
Checked by RICHARD T. ARNOLD and CARL G. KRESPAN.

1. Procedure

Twenty grams of acetamide (0.34 mole) is dissolved in 54 g. of bromine (0.34 mole) contained in a 500-ml. Erlenmeyer flask, and the solution is cooled to 0–5° in an ice bath. An ice-cold aqueous 50% potassium hydroxide solution is added in small portions with swirling and cooling until the color becomes a light yellow. Approximately 33–34 ml. of the caustic solution is required. The nearly solid reaction mixture is allowed to stand at 0–5° for 2–3 hours.

The mixture is treated with 40 g. of salt and 200 ml. of chloroform and warmed on the steam bath with vigorous swirling. After 2–3 minutes the clear red chloroform layer is decanted from the semisolid lower layer, and the extraction is repeated twice more with 200- and 100-ml. portions of chloroform respectively (Note 1). The combined extracts are dried over sodium sulfate, the solution is filtered by gravity through a fluted filter into a 2-l. Erlenmeyer flask, and 500 ml. of hexane is added with swirling. White needles of N-bromoacetamide begin to form at once (Note 2). After chilling for 1–2 hours, the crystals are collected with suction, washed with hexane, and air-dried. The yield is 19–24 g. (41–51%), m.p. 102–105° (Note 3), purity 98–100% (Notes 4 and 5).

2. Notes

1. Six additional extractions, using 50-ml. portions of chloroform, may produce an increase in yield of 4–5 g.

2. Occasionally it may be necessary to add seed crystals to promote crystallization.

3. Material melting as much as 10° lower is sometimes obtained. However, it still has a purity of better than 96% as determined by thiosulfate titration (Note 4). This may indicate the presence of small amounts of N,N-dibromoacetamide.

4. The purity is determined by titration with standard sodium thiosulfate solution. An accurately weighed sample of about 200 mg. is dissolved in water, and a solution of approximately 1 g. of potassium iodide in 10 ml. of water is added. The solution is acidified with 10 ml. of 10% sulfuric acid and titrated with 0.1N thiosulfate to the starch end point.

$$\% \text{ N-bromoacetamide} = \frac{\text{ml. } S_2O_3^= \times \text{normality} \times 69 \times 100}{\text{weight of sample (mg.)}}$$

5. The product is unstable and should be stored in a cool place protected from light.

3. Methods of Preparation

N-Bromoacetamide has been prepared from acetamide and bromine in the presence of potassium hydroxide,[2,3] zinc oxide,[4] or calcium carbonate.[4]

[1] Schering Corporation, Bloomfield, New Jersey.
[2] Behrend and Schreiber, *Ann.*, **318**, 371 (1901).
[3] Hofmann, *Ber.*, **15**, 407 (1882).
[4] Likhosherstov and Alekseev, *J. Gen. Chem. U.S.S.R.*, **3**, 927 (1933).

β-BROMOETHYLPHTHALIMIDE *

(Phthalimide, N-2-bromoethyl-)

$$\text{(phthalic anhydride)} + H_2NCH_2CH_2OH \rightarrow$$

$$\text{(N-hydroxyethylphthalimide)} \; NCH_2CH_2OH + H_2O$$

$$3 \; \text{(N-hydroxyethylphthalimide)} \; NCH_2CH_2OH + PBr_3 \rightarrow$$

$$3 \; \text{(N-bromoethylphthalimide)} \; NCH_2CH_2Br + H_3PO_3$$

Submitted by T. O. SOINE and M. R. BUCHDAHL.[1]
Checked by CLIFF S. HAMILTON and JOHN D. SCULLEY.

1. Procedure

Caution! This preparation should be carried out in a hood.

In a 1-l. round-bottomed flask are placed 74 g. (0.5 mole) of phthalic anhydride and 30 ml. (0.5 mole) of freshly distilled monoethanolamine. The mixture is heated on a steam bath for 30 minutes; the initial reaction is vigorous (Note 1). The reaction mixture is cooled to room temperature, and a reflux condenser is attached to the flask. To the cooled reaction mixture is added slowly, with shaking, 32 ml. (91.3 g., 0.34 mole) of freshly distilled phosphorus tribromide. The reaction flask is then placed on a steam bath and heated under reflux with occasional shaking for 1.25 hours (Note 2). The hot liquid reaction mixture is poured with stirring onto 750 g. of crushed ice. When the ice has melted completely, the crude β-bromoethylphthalimide is collected on a Büchner funnel, washed with cold water, and allowed to dry for a few minutes. The crude product (Note 3) is dissolved in 1.2 l. of aqueous ethanol (50% by volume) with the aid of heat. If necessary a small amount of 95% ethanol is added to effect complete solution. The hot solution is filtered and cooled in a refrigerator. A white crystalline product weighing 94–99 g. is obtained. Concentration of the mother liquor to 400 ml. yields an additional 1–3 g. of product. The total yield of product is 95–102 g. (75–80%); m.p. 80–82°.

2. Notes

1. It is not necessary to isolate the intermediate β-hydroxyethylphthalimide before going on to the next step. However, recrystallization of the product from 250 ml. of boiling water yields an initial crop of white crystals; m.p. 128°. The mother liquors will deposit further material until a yield of 95% may be obtained.

2. The final reaction mixture should contain no undissolved material.

3. This product weighs approximately 110 g. when dry.

3. Methods of Preparation

β-Bromoethylphthalimide has been prepared by the method of Gabriel [2] as recorded by Salzberg and Supniewski.[3] Illg and Smoliński have carried out this reaction in a specially designed apparatus.[4] The procedure outlined above is a modification of the method given by Soine.[5]

[1] University of Minnesota, Minneapolis, Minnesota.
[2] Gabriel, *Ber.*, **20**, 2224 (1887); **21**, 566 (1888); **22**, 1137 (1889).
[3] *Org. Syntheses Coll. Vol.* **1**, 119 (1941).
[4] Illg and Smoliński, *Roczniki Chem.*, **23**, 426 (1949) [*C. A.*, **45**, 6173 (1951)].
[5] Soine, *J. Am. Pharm. Assoc.*, **33**, 141 (1944).

4-BROMO-2-HEPTENE

(2-Heptene, 4-bromo-)

$$CH_3CH_2CH_2CH_2CH{=}CHCH_3 + \underset{\overset{|}{CH_2-C}}{\overset{CH_2-C}{\diagdown}} \overset{O}{\underset{O}{\underset{\diagup}{NBr}}} \rightarrow$$

$$CH_3CH_2CH_2\underset{\overset{|}{Br}}{CH}CH{=}CHCH_3 + \underset{\overset{|}{CH_2-C}}{\overset{CH_2-C}{\diagdown}} \overset{O}{\underset{O}{\underset{\diagup}{NH}}}$$

Submitted by F. L. Greenwood, M. D. Kellert, and J. Sedlak.[1]
Checked by John D. Roberts and A. T. Bottini.

1. Procedure

In a 500-ml. round-bottomed flask fitted with a stirrer, nitrogen inlet tube, and reflux condenser are placed 40 g. (0.41 mole) of 2-heptene, 48.1 g. (0.27 mole) of N-bromosuccinimide, 0.2 g. of benzoyl peroxide, and 250 ml. of carbon tetrachloride (Note 1). The reaction mixture is stirred and heated under reflux in a nitrogen atmosphere for 2 hours (Note 2). The succinimide is removed by suction filtration, washed twice with 15-ml. portions of carbon tetrachloride and the carbon tetrachloride washings are combined with the filtrate (Note 3). The carbon tetrachloride solution is transferred to a 500-ml. Claisen flask modified so that the distilling arm carries a 25 x 300 mm. section packed with glass helices. The capillary is attached to a source of nitrogen and the carbon tetrachloride removed at 36–38°/190 mm. (Note 4).

The residue is transferred to a 125-ml. Claisen flask modified so that the distilling arm carries an 18 x 180 mm. section packed with glass helices. Nitrogen is led into the capillary, and, after a forerun of 1–3 g., there is collected 28–31 g. (58–64%) of 4-bromo-2-heptene, b.p. 70–71°/32 mm., n_D^{25} 1.4710–1.4715 (Note 5). A residue of 7–10 g. remains in the distilling flask (Note 6).

2. Notes

1. The 2-heptene was the pure grade material purchased from Phillips Petroleum Company, Bartlesville, Oklahoma. This olefin is comparable to material prepared by a Boord synthesis. The N-bromosuccinimide was obtained from Arapahoe Chemicals, Inc., Boulder, Colorado. The benzoyl peroxide was used as received from Distillation Products, Rochester, New York. The carbon tetrachloride was reagent grade, from J. T. Baker Chemical Company, Phillipsburg, New Jersey.

2. The reaction is not rapid, and benzoyl peroxide is necessary to effect reaction. Longer reflux times lead to darkening of the reaction mixture.

3. The succinimide recovered corresponded to 97–98% of the theoretical amount and analyzed for 0.4% active bromine.

4. Removal of the carbon tetrachloride at a lower pressure results in loss of product. Distillation at a pressure much above 200 mm. causes considerable darkening of the liquid. Carbon tetrachloride removed at the higher pressure gives no precipitate with aqueous silver nitrate; this indicates the absence of product.

5. When first distilled, the product is nearly colorless. On standing under nitrogen in the refrigerator for several days, the material acquires a pale yellow color. Evidence for the identity of the product as 4-bromo-2-heptene is outlined in Reference 3.

6. This includes the liquid wetting the helices as well as the small amount of dark residue in the flask.

3. Methods of Preparation

Ziegler and coworkers [2] indicated that allylic methylene groups undergo bromine substitution more readily than allylic methyl groups. This has been shown [3] to be true, and the treatment of 2-heptene with N-bromosuccinimide gives rise to 4-bromo-2-heptene.

[1] Tufts University, Medford, Massachusetts.
[2] Ziegler et al., *Ann.*, **551**, 80 (1942).
[3] Greenwood and Kellert, *J. Am. Chem. Soc.*, **75**, 4842 (1953).

p-BROMOMANDELIC ACID *

(Mandelic acid, *p*-bromo)

$$p\text{-}BrC_6H_4COCH_3 + 2Br_2 \rightarrow p\text{-}BrC_6H_4COCHBr_2 + 2HBr$$
$$p\text{-}BrC_6H_4COCHBr_2 + 3NaOH \rightarrow$$
$$p\text{-}BrC_6H_4CHOHCO_2Na + 2NaBr + H_2O$$
$$p\text{-}BrC_6H_4CHOHCO_2Na + HCl \rightarrow$$
$$p\text{-}BrC_6H_4CHOHCO_2H + NaCl$$

Submitted by J. J. KLINGENBERG.[1]
Checked by R. T. ARNOLD and C. D. WRIGHT.

1. Procedure

A. *p,α,α-Tribromoacetophenone.* In a 1-l. three-necked flask (Note 1) equipped with an efficient mechanical stirrer, a dropping funnel, and a gas outlet leading to a hood or trap are placed 100 g. (0.5 mole) of *p*-bromoacetophenone (Note 2) and 300 ml. of glacial acetic acid. The resulting solution is stirred and cooled to 20°, and a solution of 26 ml. (0.5 mole) of bromine in 100 ml.

of glacial acetic acid is added dropwise (Note 3). Crystals of the mono-α-brominated derivative separate during the addition, which requires about 30 minutes. When the addition is completed, a second solution of 26 ml. (0.5 mole) of bromine in 100 ml. of glacial acetic acid is added dropwise. Slight heating may be necessary to keep the reaction proceeding, as indicated by decolorization of the bromine, but the temperature should be kept as near 20° as possible. During the addition, which requires about 30 minutes, the solid dissolves and crystals of the di-α,α-brominated derivative appear toward the end of the addition. The flask is heated to dissolve the contents, which are transferred, preferably in a hood, to a 1-l. beaker and cooled rapidly by means of an ice-water bath (Note 4). The mixture is filtered with suction (Note 5), and the solid is washed with 50% ethanol until colorless. The air-dried product has a slight pink cast and melts at 89–91°. The yield is 130–135 g. (73–75%) (Note 6). A pure, white solid melting at 92–94° is obtained by recrystallization from ethanol, but the initial product is sufficiently pure for the next step.

B. *p-Bromomandelic acid*. In a Waring-type blender are placed 89 g. (0.25 mole) of *p*,α,α-tribromoacetophenone and 100–150 ml. of cold water. The mixture is stirred for 10–15 minutes, and the contents are transferred to a 1-l. wide-mouthed bottle. The mixing vessel is rinsed with 150–200 ml. of ice-cold water. The material from the rinse is combined with the mixture in the bottle, and sufficient crushed ice is added to bring the temperature below 10°. One hundred milliliters of a chilled aqueous solution containing 50 g. of sodium hydroxide is added slowly while the bottle is rotated (Note 7). The contents are stored for approximately 4–5 days in a refrigerator (5°) and are shaken occasionally. During this time most of the solid dissolves, but a slight amount remains as a yellow sludge and the liquid assumes a yellow to amber color. The mixture is filtered, and the insoluble material is discarded. An excess of concentrated hydrochloric acid is added to the filtrate. The entire resulting mixture containing a white solid is extracted with three 200-ml. portions of ether. The ether extracts are combined, dried over anhydrous sodium sulfate, and filtered into a 1-l. flask. The ether is carefully removed by distillation using a hot-water bath to give a yellow oil

which solidifies when cooled. The product is recrystallized from
500 ml. of benzene. The crystals are collected by filtration and
washed with benzene until the filtrate is colorless. The air-dried
product (Note 8) weighs 40–48 g. (69–83% based on p,α,α-tri-
bromoacetophenone), m.p. 117–119° (Note 9). A second recrystallization from 500 ml. of benzene is sometimes necessary.

2. Notes

1. The use of ground-glass equipment is desirable but not
necessary.

2. The preparation of p-bromoacetophenone is described in
Organic Syntheses.[2] The compound is also available from Eastman Kodak Company.

3. Sometimes the initation of the reaction is slow. The reaction may be started by heating the solution until the bromine is
decolorized (approximately 45°), after which the reaction will
proceed normally at 20°.

4. The checkers found that maximum yields of product were
obtained when cooling was carried to the point where crystallization of the solvent commenced. The trace of crystalline solvent
quickly melts during the filtering procedure.

5. These compounds are lachrymatory and should be kept
away from the eyes.

6. An additional quantity of less pure material can be isolated
from the glacial acetic acid mother liquid and alcohol filtrates by
evaporation of the solvent. This, after recrystallization from
ethanol, amounts to 15–30 g.

7. It is important that the reaction mixture be kept cold at
this point. The amount of sludge and colored material increases
as the temperature increases.

8. The last traces of benzene leave very slowly and if present
lower the melting point. The product should be thoroughly
dried in air or on a clay plate before the melting point is taken.

9. Under similar conditions, p-chloromandelic acid melting
at 119–120° was obtained from p-chloroacetophenone in 54%
yield and p-iodomandelic acid melting at 135–136° from p-iodo-
acetophenone in 21% yield.

3. Methods of Preparation

p-Bromomandelic acid has been prepared by the bromination of *p*-bromoacetophenone followed by alkaline hydrolysis;[3] by the alkaline hydrolysis of the product formed by the addition of chloral to *p*-bromophenylmagnesium bromide;[4] and by the condensation of bromobenzene and ethyl oxomalonate in the presence of stannic chloride followed by hydrolysis and decarboxylation.[5] *p*-Bromomandelic acid is a valuable reagent in the analyses of zirconium and hafnium.[6]

[1] Xavier University, Cincinnati, Ohio.

[2] *Org. Syntheses Coll. Vol.* **1**, 109 (1941).

[3] Collet, *Bull. soc. chim. France*, [3] **21**, 67 (1899).

[4] Hebert, *Bull. soc. chim. France*, [4] **27**, 45 (1920).

[5] Riebsomer, Baldwin, Buchanan, and Burkett, *J. Am. Chem. Soc.*, **60**, 2974 (1938).

[6] Hahn, *Anal. Chem.*, **23**, 1259 (1951); Klingenberg and Papucci, *Anal. Chem.*, **24**, 1861 (1952).

2-BROMO-3-METHYLBENZOIC ACID

(*m*-Toluic acid, 2-bromo-)

$$+ \text{KCN} + 2\text{H}_2\text{O} \rightarrow$$

$$+ \text{KNO}_2 + \text{NH}_3$$

Submitted by J. F. BUNNETT and M. M. RAUHUT.[1]
Checked by JOHN D. ROBERTS and MARC S. SILVER.

1. Procedure

A. *2-Bromo-4-nitrotoluene.* In a 200-ml. three-necked, round-bottomed flask provided with an efficient reflux condenser bearing a suitable trap for absorbing hydrogen bromide, a 100-ml. separatory funnel, and a ball-joint or mercury-sealed mechanical stirrer are placed 68.5 g. (0.5 mole) of *p*-nitrotoluene (Note 1) and 1.0 g. of iron powder. The mixture is heated to 75–80° by means of a water bath, vigorous stirring is begun, and 30.5 ml. (94.8 g., 0.59 mole) of bromine is added over the course of 30 minutes. After the addition of bromine is complete, the reaction mixture is maintained at 75–80° with continuation of stirring for an additional 1.5 hours.

The reaction mixture is poured with vigorous stirring into 750 ml. of ice-cold 10% sodium hydroxide solution, the solid is allowed to settle, and the supernatant liquid is decanted. To the residue is added 250 ml. of glacial acetic acid, and the mixture is heated until the solid is completely melted. The two liquid phases are thoroughly mixed by stirring, the mixture is cooled to 5° in an ice bath, and the supernatant liquid is decanted. The product is then heated with 500 ml. of 10% acetic acid until molten, stirred thoroughly, and cooled to room temperature. The aqueous liquor is decanted, and the cycle is repeated with 500 ml. of 1% sodium hydroxide solution (Note 2). The solid 2-bromo-4-nitrotoluene is collected on a Büchner funnel and thoroughly washed with water. The moist product may be used directly in the next stage of the synthesis. It can be dried to yield 93–97 g. (86–90%) of light-brown material melting at 75–76°.

B. *2-Bromo-3-methylbenzoic acid. Caution! This procedure must be carried out in a hood with a good draft, because poisonous hydrogen cyanide is evolved.* In a 5-l. round-bottomed flask are placed 90 g. of potassium cyanide, 900 ml. of 2-ethoxyethanol (Note 3), 850 ml. of water, and the moist 2-bromo-4-nitrotoluene obtained above. A reflux condenser is attached, and the mixture is boiled for 16 hours (Note 4). To the hot, dark-red solution is then added 1.5 l. of water, and the mixture is acidified with concentrated hydrochloric acid. (*Caution! Hydrogen cyanide is evolved.*) The acidified mixture is boiled for 15 minutes to expel hydrogen cyanide and then allowed to cool to 35–40°. Five grams of diatomaceous earth is stirred in, and the mixture is filtered through a Büchner funnel precoated with a little diatomaceous earth. The solid is discarded, and the filtrate is extracted three times with 200-ml. portions of chloroform. The chloroform extracts are combined and extracted with three 100-ml. portions of 5% ammonium carbonate solution. The basic extracts are combined, acidified with concentrated hydrochloric acid, and cooled in an ice bath. The oil which first forms soon crystallizes.

The tarry solid is collected on a Büchner funnel, washed with 50 ml. of water, and dried. The dried solid is pulverized and boiled under reflux for 3 hours with 500 ml. of petroleum ether (b.p. 90–100°). The hot mixture is then filtered (fluted filter paper), and the solid is discarded. The filtrate is allowed to cool to room temperature, and the 2-bromo-3-methylbenzoic acid is collected by filtration. The white acid when dry weighs 7.5–8.5 g. (7–8%, based on the *p*-nitrotoluene) and melts at 134–136° (Note 5).

2. Notes

1. Eastman Kodak Company practical grade *p*-nitrotoluene was used.

2. The hot mixture of 2-bromo-4-nitrotoluene and 1% sodium hydroxide solution should be stirred vigorously during cooling in order to avoid obtaining the product as a solid cake. If stirring is omitted, subsequent treatment is less convenient.

3. Commercial Cellosolve was used.

4. An electric heating mantle is a convenient heat source.

5. The submitters report 11–14 g. (10–13%) of material having m.p. 132–135°, which yields, after crystallization from benzene, 9–13 g. of product having m.p. 135–137°.

3. Methods of Preparation

Although 2-bromo-4-nitrotoluene has been obtained by several routes, it is most easily prepared by bromination of *p*-nitrotoluene.[2] The procedure given is adapted from that described by Cavill.[3] 2-Bromo-3-methylbenzoic acid has not been prepared by any other means; evidence for its structure is presented elsewhere.[4]

[1] University of North Carolina, Chapel Hill, North Carolina.

[2] Scheufelen, *Ann.*, **231**, 152 (1885); Lucas and Scudder, *J. Am. Chem. Soc.*, **50**, 244 (1928); Frejka and Vitha, *Publs. fac. sci. univ. Masaryk*, **20** (1925) [*Chem. Zentr.*, **96**, II, 1153 (1925)]; Higginbottom, Hill, and Short, *J. Chem. Soc.*, **1937**, 263; Truce and Amos, *J. Am. Chem. Soc.*, **73**, 3013 (1951).

[3] Cavill, *J. Soc. Chem. Ind.* (*London*), **65**, 124 (1926).

[4] Bunnett and Rauhut, *J. Org. Chem.*, **21**, 934 (1956).

n-BUTYLACETYLENE *

(1-Hexyne)

$$2HC{\equiv}CH + 2Na \xrightarrow{\text{liq. NH}_3} 2HC{\equiv}CNa + H_2$$

$$HC{\equiv}CNa + n\text{-}C_4H_9Br \xrightarrow{\text{liq. NH}_3} n\text{-}C_4H_9C{\equiv}CH + NaBr$$

Submitted by Kenneth N. Campbell and Barbara K. Campbell.[1]
Checked by R. S. Schreiber, M. F. Murray, and A. C. Ott.

1. Procedure

Caution! This preparation should be conducted in a hood to avoid exposure to ammonia.

A 5-l. three-necked flask is fitted with an efficient stirrer, mounted through a short glass bushing, and a long gas inlet tube which dips below the surface of the liquid ammonia. The third neck carries a device for holding sodium; this consists of a short piece of 10–12 mm. glass tubing, bent at a 45° angle, through which is passed a 12-in. piece of stout, flexible iron wire (picture wire is satisfactory). The lower end of the wire is attached to a stout iron fish-hook. (Sneck Hook No. 5/0 is satisfactory.)

The flask is charged with about 3 l. of liquid ammonia (Note 1), the stirrer is started, and a rapid stream of acetylene gas (about 5 bubbles per second) is passed in for about 5 minutes to saturate the ammonia. The acetylene from a tank is sufficiently purified by passage through a sulfuric acid wash bottle; a safety trap also should be inserted in the line. Sodium (92 g., 4 g. atoms) is cut in strips (about ½ by ½ by 3 in.) so that they can be inserted through the side neck of the flask. One of these pieces of sodium is attached to the fish-hook and is gradually lowered into the liquid ammonia while a rapid stream of acetylene is passed in. The sodium should be added at such a rate that the entire solution does not turn blue. If it does, the sodium should be raised above the level of the ammonia until the color is discharged (Note 2). The rest of the sodium is added in a similar manner; the addition requires about 45 minutes, depending on the rate of passage of the acetylene.

The acetylene is shut off, and the gas inlet tube and iron wire are removed, but not the bent glass tubing. A potassium hydroxide tower (Note 3) is attached to the end of this tubing, and a dropping funnel (Note 4) is mounted in the other neck of the flask. *n*-Butyl bromide (548 g., 428 ml., 4.0 moles) is added dropwise with stirring over a period of 45–60 minutes (Note 5). The mixture is stirred for an additional 2 hours. Ammonium hydroxide (500 ml.) is then added dropwise, followed by about 1–1.5 l. of distilled water. When the frost on the outside of the flask is loose and can be pulled off easily, the contents of the flask are transferred to a large separatory funnel, and the lower aqueous layer is removed. The organic layer is washed with 100 ml. of distilled water, then with about 100 ml. of $6N$ hydrochloric acid (the aqueous layer should be tested to make sure that it remains acid; if not, another washing with acid should be carried out), and finally with about 100 ml. of 10% sodium carbonate solution. The material is dried over potassium carbonate, a small amount of solid sodium carbonate is added, and the liquid is fractionated through a helix-packed column of about 10–14 plates. The yield of pure *n*-butylacetylene is 230–252 g. (70–77%); b.p. 71–72°, n_D^{20} 1.3984–1.3990 (Notes 6 and 7).

2. Notes

1. More liquid ammonia should be added from time to time to maintain approximately the same level. Liquid ammonia can be handled in ordinary apparatus, and it is not necessary to use a Dewar flask or to cool the reaction flask in a Dry Ice bath, as a thick frost forms on the outside and partially insulates the contents. In making very volatile alkylacetylenes, such as ethylacetylene and propylacetylene, it is advisable to cool the apparatus in a Dry Ice bath, to minimize loss by entrainment, and to use a Dry Ice condenser. With the higher alkylacetylenes the use of a Dry Ice condenser, as recommended by Henne and Greenlee,[2] does not improve the yields enough to justify the extra trouble.

2. In order to see inside the flask a little alcohol may be poured over the outside.

3. This is not absolutely necessary except in damp weather,

for the ammonia escaping through the outlet tube prevents the entrance of appreciable amounts of moisture.

4. A calibrated funnel is convenient, as the rate of addition may be judged better.

5. If the addition is carried out more slowly, the yield of product is lowered, unless a Dry Ice condenser is used.

6. The reaction may be carried out on a smaller scale without much loss in yield.

7. Other alkylacetylenes can be made in the same way from primary alkyl bromides. With *n*-propylacetylene the time of addition of *n*-propyl bromide should be about 45–60 minutes, and the yield is lower (40–50%), owing to losses by entrainment unless a Dry Ice condenser is used (b.p. 39–40°; n_D^{20} 1.3850). *n*-Amylacetylene (b.p. 98°; n_D^{20} 1.4088) and isoamylacetylene (b.p. 91–92°; n_D^{20} 1.4060) can be prepared in 70–80% yields by this method; the time of addition of the halide is 1.5–2.0 hours. *n*-Hexylacetylene (b.p. 76–77°/150 mm.; n_D^{20} 1.4157) can be obtained in 65% yields if a 1-mole excess of sodium acetylide is used. The halide is added during the course of 1 hour, and the mixture is stirred for an additional 3 hours before hydrolysis.

The method is not satisfactory for methyl- and ethylacetylenes or with secondary and tertiary alkyl halides or with primary alkyl halides above hexyl.

3. Methods of Preparation

n-Butylacetylene has been prepared from *n*-butyl bromide and sodium acetylide in liquid ammonia [2, 3] or a mixture of tetrahydrofuran and dimethylformamide,[4] from *n*-butyl bromide and a slurry of calcium carbide and potassium hydroxide in dibutyl carbitol,[5] and from *n*-butyl bromide and ethynylmagnesium bromide at 80–90°.[6] It also has been obtained from the reaction of alcoholic potassium hydroxide on dibromohexanes,[7, 8] and sodamide in xylene on 2-bromo-1-hexene.[9] *n*-Butylacetylene is formed, along with 2-ethoxy-1-hexene when 1-bromo-2-ethoxyhexane is caused to react with sodium in liquid ammonia.[10] It has been found [11] that a mixture of xylene and dimethylformamide is an excellent medium in which to carry out the alkylation of sodium acetylide with alkyl bromides.

The method described here is a modification of the one published by Vaughn, Vogt, Hennion, and Nieuwland.[3]

[1] University of Notre Dame, Notre Dame, Indiana.
[2] Henne and Greenlee, *J. Am. Chem. Soc.*, **67**, 484 (1945).
[3] Vaughn, Vogt, Hennion, and Nieuwland, *J. Org. Chem.*, **2**, 1 (1937); Marszak and Guermont, *Mém. services chim. état (Paris)*, **34**, 423 (1948) [*C. A.*, **44**, 5794 (1950)].
[4] Badische Anilin- & Soda-Fabrik Akt.-Ges., Ger. pat. 944,311 [*C. A.*, **52**, 16194 (1958)].
[5] Lyon and Rutledge (to Air Reduction Co., Inc.), U. S. pat. 2,724,008 [*C. A.*, **50**, 4193 (1956)].
[6] Grignard, Lapayre, and Faki, *Compt. rend.*, **187**, 519 (1928); **188**, 520 (1929).
[7] Welt, *Ber.*, **30**, 1494 (1897).
[8] van Risseghem, *Bull. soc. chim. Belg.*, **35**, 356 (1926).
[9] Bourguel, *Ann. chim.*, [10] **3**, 222, 380 (1925).
[10] Eglinton, Jones, and Whiting, *J. Chem. Soc.*, **1952**, 2873.
[11] Rutledge, *J. Org. Chem.*, **24**, 840 (1959).

sec-BUTYL α-*n*-CAPROYLPROPIONATE

(Octanoic acid, 2-methyl-3-oxo-, *sec*-butyl ester)

$$CH_3(CH_2)_4CN + \underset{\underset{C_2H_5}{|}}{\underset{\underset{CH_3}{|}}{Br}CHCO_2CHCH_3} \xrightarrow{Zn}$$

$$\underset{\underset{CH_3}{|}}{CH_3(CH_2)_4}\overset{NZnBr}{\overset{\|}{C}}CHCO_2\underset{\underset{C_2H_5}{|}}{CHCH_3}$$

$$\underset{\underset{CH_3}{|}}{CH_3(CH_2)_4}\overset{NZnBr}{\overset{\|}{C}}CHCO_2\underset{\underset{C_2H_5}{|}}{CHCH_3} \xrightarrow[H_2O]{H^+} \underset{\underset{CH_3}{|}}{CH_3(CH_2)_4}\overset{O}{\overset{\|}{C}}CHCO_2\underset{\underset{C_2H_5}{|}}{CHCH_3}$$

Submitted by KENNETH L. RINEHART.[1]
Checked by JOHN C. SHEEHAN and J. IANNICELLI.

1. Procedure

A 2-l. three-necked round-bottomed flask is fitted with a condenser arranged for distillation, a 200-ml. dropping funnel, and a mercury-sealed mechanical stirrer. Connections are most conveniently made with ground-glass joints, and the flask is heated on a steam cone. Cupric bromide (0.4 g.), 39.2 g. (0.60 g. atom) of freshly sand-papered zinc foil cut into narrow strips, and 1.2 l. of benzene previously dried over sodium are added to the flask. To dry the apparatus and contents, 300 ml. of benzene is slowly distilled from the flask with stirring. Heating is interrupted, and the condenser is quickly arranged for reflux and fitted with a calcium chloride drying tube. To the flask are added rapidly with stirring 38.9 g. (0.40 mole) of *n*-capronitrile (Note 1) and 125.4 g. (0.60 mole) of *sec*-butyl α-bromopropionate (Note 2). Refluxing is resumed, and after a 3–5 minute induction period blackening of the zinc surface and clouding of the solution are noted as the first signs of reaction. Heating is maintained a total of 45 minutes, after which the solution is cooled for 15 minutes in an ice bath (Note 3).

To the cooled solution is added with vigorous stirring 400 ml. of ice-cold $12N$ sulfuric acid. The ice bath is removed, and stirring is continued at room temperature for a total of 2 hours (Note 4). The reaction mixture is poured into a 2-l. separatory funnel; after separation, the aqueous lower layer is drained into a 2-l. separatory funnel containing 800 ml. of water which has been used to wash the reaction flask. The aqueous layer is extracted twice with 200-ml. portions of benzene which have also been employed to rinse the reaction flask. The original organic layer and the combined benzene extracts are kept separate and are washed successively with 400-ml. portions of water, saturated sodium bicarbonate solution, and again with water. The two organic portions are combined and allowed to stand over anhydrous sodium sulfate until clear. The solvent is removed at atmospheric pressure by flash distillation through a fractionating column (Note 5). For this operation, a side-arm distilling flask equipped with a dropping funnel is heated by an oil bath whose temperature is maintained at 140–150°. After all the solvent has been added, the dropping funnel is replaced by a capillary, and the last of the solvent is removed at reduced pressure furnished by a water pump. Finally, an oil pump is attached and the product is fractionated at about 5 mm. pressure (Note 6). After a small fore-run, *sec*-butyl α-*n*-caproylpropionate is obtained as a clear liquid; b.p. 112–114°/4.5 mm. (134–136°/12 mm.), n_D^{25}

1.4293 (Notes 7 and 8). The yield is 46.0–52.9 g. (50–58% based on *n*-capronitrile) (Note 9).

2. Notes

1. Commercial *n*-capronitrile (Eastman Kodak white label grade) is distilled through a fractionating column before use; b.p. 161–163°, n_D^{25} 1.4052.

2. α-Bromoesters are severe lachrymators, and it is necessary to conduct reactions involving them in a well-ventilated hood. *sec*-Butyl α-bromopropionate is prepared from commercially available ethyl α-bromopropionate (Sapon Laboratories, Inc., 543 Union Street, Brooklyn 15, New York) by transesterification with *sec*-butyl alcohol using sulfuric acid catalyst. The product boils at 97–98°/39 mm. and has n_D^{25} 1.4420. Darkening of the α-bromoester is prevented by storage over a few drops of mercury.

3. Unreacted zinc usually remains at this point, and the solution is yellow to brown. Most of the color is discharged on subsequent addition of the acid.

4. With 3-methylnonanoicnitrile it is necessary to continue the stirring in sulfuric acid for an additional period; a total of 19 hours is sufficient.

5. A 50 cm. by 8 mm. simple Podbielniak column [2] with partial-reflux head is adequate for this and subsequent distillations.

6. Fractionation at higher pressures and temperatures leads to some thermal decomposition of the product.

7. An analytical sample of a redistilled center cut had n_D^{25} 1.4302.

8. β-Ketoesters with alkyl substituents in the α-position give a blue-violet color with ferric chloride.[3] Copper chelates of the α-alkyl-β-ketoesters could not be isolated. The corresponding 5-pyrazolones, although often difficultly crystallizable, appear to be the most suitable derivatives. The 5-pyrazolone from *sec*-butyl α-*n*-caproylpropionate is prepared according to von Auwers and Dersch [4] and has a melting point of 80.4–82.9°.

9. This method is suitable for the preparation of mono- and di-α-substituted β-ketoesters. Bromoacetates fail in this reaction. Yields with ethyl α-bromopropionate are considerably lower (30–

36% with capronitrile); however, ethyl esters are useful for higher-molecular-weight compounds whose *sec*-alkyl esters are cracked by distillation. With 3-pentyl α-bromopropionate, the yields are slightly higher (53–60% with capronitrile). Both aromatic and aliphatic nitriles are suitable; benzonitrile gives yields comparable to those obtained with capronitrile. Alkyl substitution in the α- and β-positions (cf. Note 4) of aliphatic nitriles lowers the yield to 29% and 38%, respectively; γ-substitution has no effect. Recently it has been shown that a significant improvement in the yields of ketoesters is obtained by the use of a mixture of benzene and ether as a solvent.[5]

3. Methods of Preparation

The procedure is essentially that of Horeau and Jacques [6] as modified by Cason, Rinehart, and Thornton.[7] The reaction described is that discovered by Blaise [8] and is more extensively discussed in the second-named paper. Ethyl α-*n*-caproylpropionate has been prepared by the alkylation of ethyl *n*-caproylacetate with methyl iodide in the presence of sodium ethoxide.[9] This method would not be expected to yield a pure product owing to contamination with starting ketoester and disubstituted ketoester.

[1] University of California, Berkeley, California.

[2] Podbielniak, *Ind. Eng. Chem., Anal. Ed.*, **3**, 177 (1931); **5**, 119 (1933).

[3] Henecka, *Ber.*, **81**, 188 (1948).

[4] von Auwers and Dersch, *Ann.*, **462**, 115 (1928).

[5] Cason and Fessenden, *J. Org. Chem.*, **22**, 1326 (1957).

[6] Horeau and Jacques, *Bull. soc. chim. France*, **1947**, 58.

[7] Cason, Rinehart, and Thornton, *J. Org. Chem.*, **18**, 1594 (1953).

[8] Blaise, *Compt. rend.*, **132**, 478, 978 (1901); Blaise and Courtot, *Bull. soc. chim. France*, [3] **35**, 599 (1906).

[9] Locquin, *Bull. soc. chim. France*, [3] **31**, 596 (1904).

n-BUTYL GLYOXYLATE

(Glyoxylic acid, n-butyl ester)

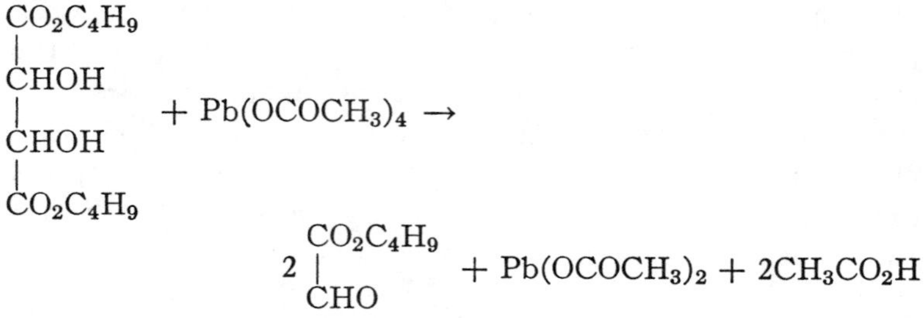

$$
\begin{array}{l}
CO_2C_4H_9 \\
| \\
CHOH \\
| \qquad + Pb(OCOCH_3)_4 \rightarrow \\
CHOH \\
| \\
CO_2C_4H_9
\end{array}
$$

$$
\begin{array}{l}
\qquad CO_2C_4H_9 \\
2 \; | \qquad\qquad + Pb(OCOCH_3)_2 + 2CH_3CO_2H \\
\qquad CHO
\end{array}
$$

Submitted by FRANK J. WOLF and JOHN WEIJLARD.[1]
Checked by N. J. LEONARD and L. A. MILLER.

1. Procedure

In a 3-l. three-necked round-bottomed flask provided with a Hershberg stirrer and a thermometer are placed 1.25 l. of reagent-grade benzene and 325 g. (1.24 moles) of di-n-butyl d-tartrate (Note 1). The mixture is stirred rapidly, and 578 g. (1.30 moles) of lead tetraacetate (Note 2) is added over a period of about 25 minutes. The temperature is maintained below 30° by occasional cooling with cold water. After the addition is complete, the mixture is stirred for 1 hour, during which time the gummy salts become crystalline. The salts are removed by filtration with suction and washed with 500 ml. of benzene. The benzene and acetic acid are removed from the filtrate by distillation at 65°/50 mm. The residue is distilled at 20 mm. under nitrogen introduced through an ebullator, using a Vigreux column (2 by 30 cm.), and the fraction boiling between 65° and 79°/20 mm. (main portion at 68–74°) is collected as product. The crude n-butyl glyoxylate, which weighs 247–280 g. (77–87%), n_D^{20} 1.442–1.443, d_4^{25} 1.085, is satisfactory for most purposes (Note 3).

2. Notes

1. The di-n-butyl d-tartrate used was the purest grade from Eastman Kodak Company, m.p. 20–22°.

2. Lead tetraacetate was obtained from the G. Frederick Smith Chemical Company, Columbus, Ohio. It may also be prepared by the procedure described in *Inorganic Syntheses*.[2]

3. The product undergoes autoxidation and should be stored under nitrogen. Further purification may be effected by a second fractional distillation under nitrogen at reduced pressure. The product decomposes on boiling (159–161°) at atmospheric pressure.

3. Methods of Preparation

n-Butyl glyoxylate has not been described previously. *Anal.* Calcd. for $C_6H_{10}O_3$: C, 55.37; H, 7.75. Found: C, 54.95; H, 7.83. Ethyl glyoxylate has been prepared in good yield by oxidation of ethyl tartrate with red lead oxide [3] or sodium bismuthate.[4] These papers describe isolation of ethyl glyoxylate as carbonyl derivatives.

[1] Merck and Company, Rahway, New Jersey.
[2] *Inorg. Syntheses*, **1**, 47 (1939).
[3] Hamamura, Suzumoto, and Hayashima, *J. Agr. Chem. Soc. Japan*, **22**, 25 (1948) [*C. A.*, **46**, 10108 (1952)].
[4] Rigby, *Nature*, **164**, 185 (1949).

tert-BUTYL HYPOCHLORITE

$$(CH_3)_3COH + Cl_2 + NaOH \rightarrow (CH_3)_3COCl + NaCl + H_2O$$

Submitted by H. M. Teeter and E. W. Bell.[1]
Checked by Charles C. Price and T. C. Schwann.

1. Procedure

Caution! This preparation should be carried out in a good hood. The product should be protected from strong light, overheating, or exposure to rubber to avoid vigorous decomposition.

A solution of 80 g. (2 moles) (Note 1) of sodium hydroxide in about 500 ml. of water is prepared in a 2-l. three-necked round-bottomed flask equipped (Note 2) with a gas inlet tube reaching nearly to the bottom of the flask, a gas outlet tube, and a mechanical stirrer. The flask is placed in a water bath at 15–20°

(Note 3), and, after the contents have cooled to this temperature, 74 g. (96 ml., 1 mole) of *tert*-butyl alcohol (Note 4) is added together with enough water (about 500 ml.) to form a homogeneous solution. With constant stirring, chlorine is passed into the mixture for 30 minutes at a rate of approximately 1 l. per minute (Note 5) and then for an additional 30 minutes at a rate of 0.5–0.6 l. per minute.

The upper oily layer is then separated with the aid of a separatory funnel (Note 6). It is washed with 50-ml. portions of 10% sodium carbonate solution until the washings are no longer acidic to Congo red. It is finally washed 4 times with an equal volume of water and dried over calcium chloride. The yield is 78–107 g. (72–99%) (Note 7); d_{20}^{20} 0.910; n_D^{20} 1.403. The product is best stored under an inert atmosphere (Note 8) in sealed bottles kept in the dark in a refrigerator (Note 9).

2. Notes

1. The submitters have successfully carried out this preparation in quantities up to 7 moles (519 g.) of *tert*-butyl alcohol using 2.25 hours for the addition of chlorine at each rate.

2. *tert*-Butyl hypochlorite reacts violently with rubber. The apparatus should therefore be assembled by means of ground-glass joints. Synthetic plastic tubing (Tygon) may be used instead of rubber tubing.

3. For 7-mole runs, a satisfactory bath consists of a 10-gal. earthenware jar fitted with a water inlet and an overflow device. The desired temperature is obtained by adjusting the rate of flow of tap water through the jar. The reaction can also be carried out successfully using an ice-water bath.

4. The *tert*-butyl alcohol was a commercial product obtained from the Shell Chemical Corporation, New York, New York.

5. Rates of flow are conveniently measured with the usual U-tube and capillary using one of the liquid Arochlors as the indicating fluid. The checkers did not use a flow meter but passed in chlorine rapidly at first and then slowly for 30 minutes after the initial rapid absorption.

6. In 7-mole runs carried out by the submitters, the bulk of the aqueous layer was removed conveniently by siphoning through the gas inlet tube.

7. This product, which is sufficiently pure for most purposes, contains about 2% free chlorine. It may be purified by distillation in an all-glass apparatus heated by a steam bath. The yield of pure product is 75–104 g. (69–96%); b.p. 77–78°/760 mm. Active chlorine assay indicates 97–98% purity for the crude product, 98–100% purity for the distilled material.

8. The inert atmosphere helps to minimize any tendency of vapors to ignite during sealing of the bottle. Filled bottles should be cooled in solid carbon dioxide before sealing.

9. When exposed to light, *tert*-butyl hypochlorite decomposes with formation of acetone and methyl chloride. When induced by radiation from an ultraviolet lamp, this decomposition proceeds rapidly enough to raise the temperature of the hypochlorite to the boiling point. The decomposition does not continue after irradiation is stopped. Customary room illumination does not induce the decomposition to a noticeable extent during ordinary handling of this material. However, a sealed glass bottle of hypochlorite should not be allowed to stand in light for a prolonged time as pressure sufficient to burst the bottle may be built up gradually.

3. Methods of Preparation

tert-Butyl hypochlorite has been prepared by the action of chlorine upon alkaline solutions of *tert*-butyl alcohol.[2-5] Solutions of *tert*-butyl hypochlorite have been prepared by shaking a solution of the alcohol in carbon tetrachloride,[6] fluorotrichloromethane (Freon 11), and other solvents [7] with aqueous hypochlorous acid. The procedure described is that of Teeter et al.[4]

[1] Northern Regional Research Laboratory, U. S. Department of Agriculture, Peoria, Illinois.

[2] Chattaway and Backeberg, *J. Chem. Soc.*, **1923**, 2999.

[3] Irwin and Hennion, *J. Am. Chem. Soc.*, **63**, 858 (1941).

[4] Teeter, Bachmann, Bell, and Cowan, *Ind. Eng. Chem.*, **41**, 848 (1949).

[5] Deanesly, U. S. pat. 1,938,175 [*C. A.*, **28**, 1053 (1934)].

[6] Taylor, MacMullin, and Gammal, *J. Am. Chem. Soc.*, **47**, 395 (1925).

[7] Fort and Denivelle, *Bull. soc. chim. France*, **1954**, 1109.

2-BUTYN-1-OL

$$CH_3C\!\!=\!\!CHCH_2Cl \xrightarrow[\text{H}_2\text{O}]{\text{Na}_2\text{CO}_3} CH_3C\!\!=\!\!CHCH_2OH$$
$$\qquad | \qquad\qquad\qquad\qquad\qquad | $$
$$\qquad Cl \qquad\qquad\qquad\qquad\qquad Cl$$

$$CH_3C\!\!=\!\!CHCH_2OH \xrightarrow[\text{NaNH}_2]{\text{Liq. NH}_3} CH_3C\!\!\equiv\!\!CCH_2ONa$$
$$\qquad | $$
$$\qquad Cl$$

$$CH_3C\!\!\equiv\!\!CCH_2ONa \xrightarrow{\text{NH}_4\text{Cl}} CH_3C\!\!\equiv\!\!CCH_2OH$$

Submitted by P. J. Ashworth, G. H. Mansfield, and M. C. Whiting.[1]
Checked by John C. Sheehan, George Buchi, and Walfred S. Saari.

1. Procedure

Caution! The experimental procedure involving liquid ammonia should be conducted in a hood.

In a 3-l. three-necked round-bottomed flask fitted with a reflux condenser and a mercury-sealed stirrer, 250 g. (2 moles) of 1,3-dichloro-2-butene (Note 1) and 1.25 l. of 10% sodium carbonate are heated at reflux temperature for 3 hours. The 3-chloro-2-buten-1-ol is extracted with three 300-ml. portions of ether, which are then dried over anhydrous magnesium sulfate. The ether is removed by distillation through a 20-cm. Fenske column, and the residue is distilled from a 250-ml. Claisen flask, yielding 134 g. (63%) of 3-chloro-2-buten-1-ol, b.p. 58–60°/8 mm., n_D^{20} 1.4670.

A solution of sodium amide in liquid ammonia is prepared according to the procedure described on p. 763 using a 4-l. Dewar flask equipped with a plastic cover (Note 2) and a mechanical stirrer. Anhydrous liquid ammonia (3 l.) is introduced through a small hole in the plastic cover, and 1.5 g. of hydrated ferric nitrate is added followed by 65 g. (2.8 g. atoms) of clean, freshly cut sodium. The mixture is stirred until all the sodium is converted into sodium amide, after which 134 g. (1.26 moles) of 3-chloro-2-buten-1-ol is added over a period of 30 minutes. The mixture is stirred overnight, then 148 g. (2.8 moles) of solid ammonium chloride is added in portions at a rate that permits control of the exothermic reaction. The mixture is transferred to a metal bucket (5-l., preferably of stainless steel) and allowed to stand overnight in the hood while the ammonia evaporates. The residue is extracted thoroughly with five 250-ml. portions of ether, which is removed by distillation through a 20-cm. Fenske column. Distillation of the residue yields 66–75 g. (75–85%) of 2-butyn-1-ol, b.p. 55°/8 mm., n_D^{20} 1.4550 (Note 3).

2. Notes

1. 1,3-Dichloro-2-butene was obtained from Eastman Kodak Company. Redistillation of the crude material did not appreciably alter the yield in the first stage of the reaction.

2. The checkers used an ordinary flask, well insulated, for this reaction.

3. Others have reported b.p. 91–93°/50 mm., n_D^{25} 1.4630; [2] b.p. 160.5°, n_D^{25} 1.4635; [3] b.p. 160–161°, n_D^{25} 1.4635, for 3-chloro-2-buten-1-ol, and b.p. 87–88°/100 mm., n_D^{25} 1.4520; [2] b.p. 140–141°, n_D^{25} 1.4517 [4] for 2-butyn-1-ol.

3. Methods of Preparation

2-Butyn-1-ol has been prepared as above for the first stage [2–4] but using aqueous sodium hydroxide [2] and alcoholic potassium hydroxide [4] for the second stage. The reaction between propynylmagnesium bromide and formaldehyde also has been employed. [5–7]

[1] Victoria University of Manchester, Manchester, England.
[2] Hatch and Nesbitt, J. Am. Chem. Soc., 72, 729 (1950).
[3] Hatch and Hudson, J. Am. Chem. Soc., 72, 2505 (1950).
[4] Hatch and Chiola, J. Am. Chem. Soc., 73, 360 (1951).
[5] Yvon, Compt. rend., 180, 748 (1925).
[6] Hurd and Cohen, J. Am. Chem. Soc., 53, 1074 (1931).
[7] Schulte and Reiss, Chem. Ber., 87, 964 (1954).

BUTYRCHLORAL

(Butyraldehyde, 2,2,3-trichloro-)

$$CH_3CH{=}CHCHO + Cl_2 + H_2O \rightarrow$$
$$CH_3CHOHCHClCHO + HCl$$
$$CH_3CHOHCHClCHO \rightarrow CH_3CH{=}CClCHO + H_2O$$
$$CH_3CH{=}CClCHO + Cl_2 \rightarrow CH_3CHClCCl_2CHO$$

Submitted by Gus A. Ropp, W. E. Craig, and Vernon Raaen.[1]
Checked by T. L. Cairns and H. N. Cripps.

1. Procedure

Crotonaldehyde (Note 1) (70 g., 1 mole) and 300 ml. of water are stirred with a glass-enclosed bar magnet while chlorine gas is passed in from a cylinder. The reaction vessel, equipped with a water-cooled reflux condenser, is cooled in an ice bath to maintain the temperature at about 10°, and the introduction of chlorine is continued for about 2 hours until the temperature does not rise rapidly when the bath is removed. At this point, the total weight increase is 70–80 g. and the upper oil layer has been converted entirely to a lower layer of viscous white oil. The reaction mixture is heated for 30 minutes at reflux temperature with slow stirring to dehydrate the chlorohydrin. The light-brown oil layer is extracted with chloroform, and the extract (Note 2) is washed twice with water and dried thoroughly over anhydrous magnesium sulfate in the refrigerator. The dry solution is filtered into a thoroughly dried flask (Note 3) equipped with a thermometer, a reflux condenser, and a bubbler tube for the introduction of dry chlorine. Over a period of 2.5 hours, chlorine, passed first through a drying tube filled with Drierite, is introduced. The temperature in the reaction mixture is kept at 0° to 10° by an ice bath, and chlorination is continued until the weight increase indicates that 1 mole of chlorine has been added. The reaction mixture is stirred for 1 hour longer in the ice bath. Then dry carbon dioxide (or nitrogen) is bubbled through the solution at room temperature to remove excess chlorine. The reaction vessel is fitted to a dry 20-mm. by 12-in.

Vigreux column wrapped with glass-wool insulation and equipped with a vacuum-jacketed, total take-off-type still head and a water-cooled condenser. The receiver is a dry tared flask with side arm leading to a manometer and, through a Drierite-filled U-tube, to a water pump. Dry carbon dioxide is passed through the still pot during the distillation. The chloroform is distilled (Note 4) at atmospheric pressure. The pressure is then decreased, and the pale yellow butyrchloral is distilled. The yield is 91–93 g. (52–53%); b.p. 57–60°/23 mm.; $n_D^{25} = 1.4712$–1.4740. The oil has a persistent and characteristic but not unpleasant odor (Note 5).

2. Notes

1. Best yields were obtained from Eastman's best grade croton-aldehyde which had been distilled immediately before use.

2. When this extract is dried and distilled, a high yield of α-chlorocrotonaldehyde, b.p. 147–148°, is obtained. This alde-hyde is a powerful lachrymator.

3. Care is necessary to ensure absolute dryness of all glass-ware in which butyrchloral is contained since the aldehyde quickly forms a solid, insoluble hydrate.

4. If only a moderately pure sample of butyrchloral is needed as a reaction intermediate, the concentrate remaining after evaporation of the chloroform may be used.

5. For best yields, the entire series of reactions should be com-pleted within 1 or 2 days. Distillation at higher pressures tends to cause some decomposition. Analysis of the product showed % Cl, 60.40, 60.64 (calculated % Cl, 60.63).

3. Methods of Preparation

Butyrchloral has been prepared by chlorination of acetalde-hyde [2] and paraldehyde.[3] Butyrchloral hydrate has been pre-pared by treatment of α,β-dichlorobutyraldehyde with chlorine and water.[4] Butyrchloral has also been prepared [5] by treatment of crotonaldehyde with hydrogen chloride followed by chlorina-tion. Brown and Plump have used a procedure similar to the one described here.[4]

[1] Oak Ridge National Laboratory, Oak Ridge, Tennessee.

[2] Kraemer and Pinner, *Ber.*, **3**, 383 (1870).

[3] Reicheneder and Zoebelein, Ger. pat. 814,594 [*C. A.*, **52**, 1204 (1958)].

[4] Brown and Plump (to Pennsylvania Salt Manufacturing Company), U. S. pat. 2,351,000 (1944) [Brit. pat. 576,435 (1946)].

[5] High (to Udylite Corporation), U. S. pat. 2,280,290 (1942).

β-CARBETHOXY-γ,γ-DIPHENYLVINYLACETIC ACID

(Succinic acid, α-benzhydrylidene-, α-ethyl ester)

$$(C_6H_5)_2C{=}O + \overset{\displaystyle CH_2CO_2C_2H_5}{\underset{\displaystyle CH_2CO_2C_2H_5}{|}} + KOC(CH_3)_3 \rightarrow$$

$$(C_6H_5)_2C{=}\overset{\displaystyle CH_2CO_2K}{\underset{}{\overset{|}{C}CO_2C_2H_5}} + HOC(CH_3)_3 + C_2H_5OH$$

$$(C_6H_5)_2C{=}\overset{\displaystyle CH_2CO_2K}{\overset{|}{C}CO_2C_2H_5} + HCl \rightarrow (C_6H_5)_2C{=}\overset{\displaystyle CH_2CO_2H}{\overset{|}{C}CO_2C_2H_5} + KCl$$

Submitted by WILLIAM S. JOHNSON and WILLIAM P. SCHNEIDER.[1]
Checked by ARTHUR C. COPE and MALCOLM CHAMBERLAIN.

1. Procedure

Caution!   See Note 3 concerning the safe handling of potassium.

The reaction is conducted in a 500-ml. round-bottomed flask attached (by a ground-glass joint) to a Pyrex reflux condenser, the top of which is connected to a three-way stopcock leading to (a) a source of nitrogen and a mercury trap and (b) a water aspirator (Fig. 5). The flask and condenser are dried by warming with a free flame while the system is under reduced pressure (stopcock turned to (b) to engage aspirator). Dry nitrogen (Note 1) is then admitted to the apparatus by turning the stopcock slowly to the position indicated in Fig. 5 while nitrogen is bubbling through the mercury trap. The cooled flask is quickly charged with 45 ml. of dry *tert*-butyl alcohol (Note 2) and 2.15 g. (0.055 g. atom) of potassium (Note 3) and is then reconnected to the apparatus. The flow of nitrogen is stopped, the screw clamp is closed, and the mixture is boiled under reflux until the potassium is dissolved (Note 4), hydrogen being liberated through the mercury trap.   The solution is then cooled to room temperature while nitrogen is admitted to equalize the pressure. The flask is quickly disconnected just long enough for the addition of 9.11 g. (0.05 mole) of benzophenone (Note 5) and 13.05 g. (0.075 mole) of diethyl succinate (Note 5).   The system is then evacuated (until the alcohol begins to boil) and filled with nitrogen.   With the stopcock as shown in Fig. 5 and the screw clamp

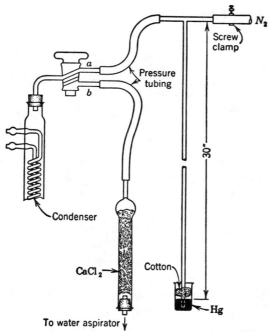

FIG. 5.   Apparatus for alternately evacuating and introducing nitrogen into the reaction vessel.

closed, the mixture is refluxed gently for 30 minutes (Note 6). It is then chilled, acidified with about 10 ml. of cold 1:1 hydrochloric acid, and distilled under reduced pressure (water aspirator) until most of the alcohol is removed.   Water is added to the residue, which is extracted thoroughly with ether, and the combined extracts are washed with successive portions of 1N ammonium hydroxide until a test portion gives no precipitate on acidification.   The combined alkaline solutions are washed once with a fresh portion of ether and then added slowly with stirring to an excess of cold dilute hydrochloric acid. When the addition is complete the mixture should still be acidic to Congo red. The pale tan crystalline half-ester is separated on a suction funnel, washed well with water, and dried. The yield is 14.0–14.5 g. (92–94%), m.p. 120–124°. If a purer material is desired the product may be recrystallized by dissolving it in about 50 ml. of warm benzene, filtering, and adding an equal volume of petroleum ether (b.p. 40–60°). Upon cooling, 13.0–13.4 g. of almost colorless half-ester crystallizes, m.p. 123–124.5°.

2. Notes

1. Ordinary tank nitrogen is dried satisfactorily by passage through a train consisting of (a) a trap, (b) a wash bottle containing concentrated sulfuric acid, and (c) a drying tube containing fresh soda lime.

2. Commercial *tert* -butyl alcohol is dried by refluxing with sodium (about 3 g. per 100 ml.) until the metal is about two-thirds dissolved, and then distilling. It may be necessary to add fresh sodium in order to have free metal present throughout the distillation.

3. The following procedure is recommended for the safe handling of potassium. The metal may be cut conveniently under xylene (which has been dried over sodium wire) contained in a mortar. A beaker or crystallizing dish should *not* be used because it is too fragile. Each scrap obtained in cutting off the outer oxide-coated surface of the metal should be immediately transferred with tweezers to a second deep mortar containing dry xylene, where the accumulated residues are decomposed as described below as soon as the cutting operation is completed. In order to weigh the freshly cut metal it may be removed with tweezers, blotted rapidly with a piece of filter paper, and introduced into a tared beaker containing dry xylene. The weighed potassium is then introduced into the reaction mixture, the proper precautions, such as exclusion of air and moisture and rate of addition being taken, depending on the nature of the reaction involved. *Caution!* It is the small scraps of metal that adhere to the knife or float on top of the xylene that are most likely to start a fire.

Danger! Potassium residues have been known to explode even under a protective liquid. It is therefore important that all such residues be decomposed *immediately;* under no circumstances should they be stored. The mortar containing the scraps is moved to the rear of the hood and *tert* -butyl (not methyl or ethyl) alcohol is added in small portions from a medicine dropper or beaker at such a rate that the reaction does not become too vigorous. A square sheet of asbestos large enough to cover the mortar should be at hand. If the liquid should catch fire it may be extinguished easily by covering the mortar with the asbestos sheet. There should be no other inflammable material or flames in the hood during this treatment. Sufficient *tert* -butyl alcohol must be employed to ensure *complete* decomposition of all the potassium. Small specks of potassium usually remain in the first mortar used for the cutting operation; they should be decomposed in the hood by cautious addition of small amounts of *tert* - butyl alcohol as described above.

4. If the alcohol and apparatus have been properly dried, the dissolution of the potassium will be slow, requiring more than 4 hours of refluxing.

5. Eastman Kodak Company grade material is satisfactory if dried by redistillation.

6. The potassium salt of the half-ester may precipitate during the period of heating.

3. Methods of Preparation

β-Carbethoxy-γ,γ-diphenylvinylacetic acid has been prepared by the condensation of benzophenone with diethyl succinate in the presence of sodium ethoxide [2] or sodium hydride.[3] The procedure described here is a modification involving the use of potassium *tert*-butoxide as the condensing agent.[4]

[1] University of Wisconsin, Madison, Wisconsin.
[2] Stobbe, *Ann.*, **308**, 89 (1899).
[3] Daub and Johnson, *J. Am. Chem. Soc.*, **70**, 418 (1948).
[4] Johnson, Petersen, and Schneider, *J. Am. Chem. Soc.*, **69**, 74 (1947).

β-(o-CARBOXYPHENYL)PROPIONIC ACID

(Hydrocinnamic acid, o-carboxy-)

$$\underset{\text{OH}}{\text{naphthol}} \xrightarrow{\text{CH}_3\text{CO}_3\text{H}} \underset{\substack{\text{CO}_2\text{H}\\\text{CH=CHCO}_2\text{H}}}{\text{arene}} \xrightarrow[\text{NaOH}]{\text{Ni-Al}} \underset{\substack{\text{CO}_2\text{H}\\\text{CH}_2\text{CH}_2\text{CO}_2\text{H}}}{\text{arene}}$$

Submitted by G. A. PAGE and D. S. TARBELL.[1]
Checked by WILLIAM S. JOHNSON, SHIRLEY ROSENBERG, and
ROBERT D. EBERHARDT.

1. Procedure

A. *o-Carboxycinnamic acid.* Eighty-eight grams (78 ml., 0.46 mole of peracid) of 40% peracetic acid (Note 1) is placed in a 250-ml. Erlenmeyer flask which is immersed in a water bath maintained at 25–30°. A 150-ml. dropping funnel is mounted so that the stem enters the flask to within about 4 cm. of the liquid surface. With mechanical stirring (Note 2), a cold solution of 20 g. (0.14 mole) of β-naphthol (Note 3) in 100 ml. of glacial acetic acid is added dropwise over a period of 4 hours to the peracid. With the appropriate rate of addition, the temperature of the reaction mixture slowly rises to 30–35° and should not exceed 40°. Solid material begins to separate from the orange solution when one-third or more of the naphthol solution is introduced. When the addition is complete, the mixture is stirred for 1 hour and the flask is allowed to stand in the water bath until the exothermic reaction ceases (usually 6–8 hours), then at room temperature for 4 days (Note 4). The solid material is collected by suction filtration and washed on the filter with sufficient (10–20 ml.) acetic acid to remove colored impurities. Drying in the air gives 19.6–20.1 g. of crude *o*-carboxycinnamic acid as a pale yellow crystalline solid (Notes 4 and 5).

The crude acid is purified by dissolving in 360–400 ml. of cold 5% sodium bicarbonate solution, filtering, and acidifying the filtrate with sufficient excess of mineral acid to turn Congo red paper blue. The product is separated by suction filtration, washed with water to remove mineral acid, and air-dried. Material thus obtained weighs 17.9–18.7 g. (67–70% yield). It melts generally (Note 6) between 202° and 205°, and is sufficiently pure (Note 7) for most practical purposes.

B. *β-(o-Carboxyphenyl)propionic acid.* In an open 1-l. wide-mouthed round-bottomed flask are placed 18 g. (0.094 mole) of *o*-carboxycinnamic acid and 550 ml. of 10% sodium hydroxide solution. The mixture is warmed to 90° (Note 8) on a steam bath and stirred mechanically. The steam bath is then removed while 54 g. (Note 9) of nickel-aluminum alloy (Raney catalyst) powder is added through the open neck of the flask in small portions (from the end of a spatula) at frequent intervals (Note 10). When addition of the alloy is complete (about 50 minutes), the mixture is stirred and maintained at 90–95° for 1 hour by warming on a steam bath. Distilled water is added as needed to maintain the total volume at approximately 550 ml. The hot mixture is filtered with suction, and the metallic residue is washed with 50 ml. of hot 10% sodium hydroxide solution and two 50-ml. portions of hot water in such a manner that the solid is always covered with liquid (Note 11). The cooled filtrate and washings are added dropwise with mechanical stirring to 300 ml. of concentrated hydrochloric acid (sp. gr. 1.19) in an open 2-l. beaker at such a rate that the temperature does not exceed 80–85° (Note 12). Separation of crystalline material begins almost immediately and is complete when the beaker contents have cooled to room temperature. The β-(o-carboxyphenyl)propionic acid is separated by suction filtration, washed with water, and air-dried (Note 13). The yield is 16.8–17.3 g. (92–95%), m.p. 165.5–167° (Note 14).

2. Notes

1. Commercial 40% (w/w) peracetic acid is available from the Becco Sales Corporation, Buffalo 7, New York. The use of a 3.3 molecular proportion of the peracid results in slightly higher

and more consistent yields of product than when the theoretical 3.0 proportion is employed. The procedure gives the same yield (percentage) of product when using proportionately smaller quantities of reactants.

2. The operator should be protected by means of a safety shield. A glass (propeller-blade) stirrer passing through the open neck of the Erlenmeyer flask is convenient; rapid stirring is not essential.

3. β-Naphthol of c.p., u.s.p., or n.f. grade has been used with equal success.

4. After 15 hours' standing, 15.0–16.8 g. of crude o-carboxy-cinnamic acid may be recovered by filtration, washing, and drying.

5. The filtrate, either on concentration under reduced pressure or upon dilution with water, fails to yield more o-carboxycinnamic acid, but a crystalline by-product, presumably 4-(o-carboxy-phenyl)-5,6-benzocoumarin,[2] may be encountered in small yield.

6. On melting, o-carboxycinnamic acid cyclizes to give the lactone of β-hydroxy-β-(o-carboxyphenyl)propionic acid (phtha-lideacetic acid), m.p. 153°. If the melting point is taken too slowly, or if the diacid is not washed completely free of mineral acid, it may therefore melt considerably below 200°. The reduction step (part B), however, proceeds normally with such material.

7. The acid may be recrystallized from aqueous ethanol to give small, white, felted prisms, m.p. 205°.

8. It is advisable to insert a thermometer only at intervals since the alkaline mixture attacks glassware to an appreciable extent on prolonged contact.

9. Using less than 50 g. of the alloy results in the same yield of final product, which, however, contains small amounts of unchanged o-carboxycinnamic acid. Equally good results are obtained when proportionately smaller quantities of reactants are used.

10. If excessive foaming is encountered it may be controlled as required by the addition of a few drops of octyl alcohol.

11. The metallic residue may ignite if allowed to dry on the filter. Disposal can be carried out by dissolving the residue in dilute nitric acid. (*Caution! Vigorous reaction.*)

12. With this order of addition, aluminum salts remain in solution, thus simplifying the procedure. External cooling may be applied in order to save time.

13. Extraction of the filtrate with ether gives an additional 0.4–0.5 g. of the crude acid after removal of solvent by distillation. It may be purified by conventional means to give an additional 0.25–0.35 g. of the pure product.

14. β-(o-Carboxyphenyl)propionic acid may be recrystallized from hot water (about 20 ml./g.), giving material m.p. 166.5–167.5°.

3. Methods of Preparation

o-Carboxycinnamic acid has been prepared by the hydrolysis of o-carboxycinnamonitrile,[3] by the opening of the lactone ring in phthalideacetic acid,[4,5] and by the dehydration of metallic salts of β-hydroxy-β-(o-carboxyphenyl)propionic acid.[4] It has also been obtained from β-naphthol by reaction with the following oxidizing agents: potassium permanganate in neutral or alkaline solutions;[6] 30% hydrogen peroxide in acetic acid;[7] and peracetic acid in acetic acid.[8] β-Naphthoquinone may be oxidized to give o-carboxycinnamic acid, with 30% hydrogen peroxide[9] or perbenzoic acid.[10] Naphthalene also yields this acid on oxidation with peracetic acid.[11] The procedure described here is essentially that of Böeseken and Königsfeldt and of Greenspan.[8]

β-(o-Carboxyphenyl)propionic acid has been prepared by the action of reagents (mostly potassium permanganate as an oxidant) upon di- and tetrahydronaphthalenes and their derivatives.[12] o-Carboxycinnamic acid has been reduced by means of sodium amalgam,[13] and cinnam-o-hydroxamic acid has been reduced catalytically,[14] to give the propionic acid. The dialdehyde of β-(o-carboxyphenyl)propionic acid has been oxidized to the diacid, using potassium permanganate in sodium carbonate solution.[15] The diacid has also been prepared by the action of heat upon o-carboxybenzylmalonic acid,[16] and from o-cyanohydrocinnamonitrile,[17] β-(o-cyanophenyl)propionic acid,[18] and ethyl o-cyanobenzylmalonate[19] by procedures involving hydrolysis. Hydrolytic cleavage of bis(1-keto-2-hydrindylidenemethyl)hydroxylamine and of 2-cyanohydrindone-1[20] gives β-(o-carboxyphenyl)propionic acid. It has been obtained also by the nitric acid oxidation of α-tetralone.[21]   The present procedure has been published [22] and is adapted from a general method for reducing cinnamic acids.[23]

[1] University of Rochester, Rochester, New York. The submitters wish to thank the National Institutes of Health for a grant under which this work was done.

[2] Bader, *J. Am. Chem. Soc.*, **73**, 3731 (1951); Dischendorfer and Danziger, *Monatsh.*, **48**, 325 (1927).

[3] Beckmann and Liesche, *Ber.*, **56**, 7 (1923); Edwards, *J. Chem. Soc.*, **1926**, 816.

[4] Gabriel and Michael, *Ber.*, **10**, 2203 (1877).

[5] Leupold, *Ber.*, **34**, 2834 (1901); Rowe, Haigh, and Peters, *J. Chem. Soc.*, **1936**, 1104; Titley, *J. Chem. Soc.*, **1928**, 2576.

[6] Ehrlich and Benedikt, *Monatsh.*, **9**, 528 (1888); Leman and Deremaux, *Bull. soc. chim. France*, **9**, 165 (1942).

[7] Raacke-Fels, Wang, Robins, and Christensen, *J. Org. Chem.*, **15**, 627 (1950).

[8] Böeseken and Königsfeldt, *Rec. trav. chim.*, **54**, 316 (1935); Greenspan, *Ind. Eng. Chem.*, **39**, 847 (1947).

[9] Böeseken, Lichtenbelt, Milo, and van Marlen, *Rec. trav. chim.*, **30**, 146 (1911); Böeseken and Slooff, *Rec. trav. chim.*, **49**, 91 (1930).

[10] Karrer and Schneider, *Helv. Chim. Acta*, **30**, 859 (1947).

[11] Böeseken and Slooff, *Rec. trav. chim.*, **49**, 100 (1930).

[12] Bamberger and Bammann, *Ber.*, **22**, 967 (1889); Bamberger and Helwig, *Ber.*, **22**, 1915 (1889); Bamberger and Kitschelt, *Ber.*, **23**, 1562 (1890); Straus and Ekhard, *Ann.*, **444**, 158 (1925); Straus and Lemmel, *Ber.*, **46**, 239 (1913); Straus and Rohrbacher, *Ber.*, **54**, 66 (1921).

[13] Beckmann and Liesche, *Ber.*, **56**, 7 (1923); Gabriel and Michael, *Ber.*, **10**, 2204 (1877).

[14] Neunhoeffer and Kölbel, *Ber.*, **68**, 262 (1935).

[15] von Braun and Zobel, *Ber.*, **56**, 2140 (1923).

[16] Wislicenus, *Ann.*, **242**, 39 (1887).

[17] Snyder and Poos, *J. Am. Chem. Soc.*, **71**, 1395 (1949).

[18] Linstead, Rowe, and Tuey, *J. Chem. Soc.*, **1940**, 1076; Mayer, Philipps, Ruppert, and Schmitt, *Ber.*, **61**, 1971 (1928).

[19] Mitchell and Thorpe, *J. Chem. Soc.*, **1910**, 2271.

[20] Johnson and Shelberg, *J. Am. Chem. Soc.*, **67**, 1757 (1945).

[21] Balaceanu and de Radzitzky (to Institut Français du petrole, des carburants, et lubrifiants), Brit. pat. 778,311 [*C. A.*, **52**, 1245 (1958)]; de Radzitzky and Balaceanu, *Compt. rend. 27ᵉ congr. intern. chim. ind., Brussels*, **1954**, 2; *Ind. chim. belge*, **20**, *Spec. No. 182* (1955) [*C. A.*, **50**, 15462 (1956)].

[22] Page and Tarbell, *J. Am. Chem. Soc.*, **75**, 2053 (1953).

[23] Schwenk, Papa, Whitman, and Ginsberg, *J. Org. Chem.*, **9**, 175 (1944).

CETYLMALONIC ESTER

(Malonic acid, cetyl-, diethyl ester)

$$C_{17}H_{35}CO_2C_2H_5 + (CO_2C_2H_5)_2 \xrightarrow{NaOC_2H_5}$$

$$C_{16}H_{33}\underset{\underset{COCO_2C_2H_5}{|}}{CH}CO_2C_2H_5 + C_2H_5OH$$

$$C_{16}H_{33}\underset{\underset{COCO_2C_2H_5}{|}}{CH}CO_2C_2H_5 \rightarrow C_{16}H_{33}CH(CO_2C_2H_5)_2 + CO$$

Submitted by Don E. Floyd and Sidney E. Miller.[1]
Checked by James Cason and George A. Gillies.

1. Procedure

A solution of sodium ethoxide in ethanol is prepared by adding 23 g. (1 g. atom) of freshly cut sodium metal, as ¼-in. cubes, in portions of 3–4 g., to 300 ml. of absolute ethanol, contained in a 1-l. three-necked flask, fitted with a thermometer and an upright, water-cooled condenser, the open end of which is capped with a drying tube filled with a mixture of Drierite and coarse lime (Note 1). The sodium is added through the third neck of the flask, which is otherwise kept stoppered.

When all the sodium has dissolved, the condenser is removed and 584 g. (4 moles) of diethyl oxalate (Note 2) and 312 g. (1 mole) of ethyl stearate (Note 3) are quickly added. The condenser is replaced by a still head connected to a condenser and receiver arranged for distillation at reduced pressure. If a water aspirator is used, a drying tube filled with Drierite should be inserted in the line between the receiver and aspirator. The receiver is cooled in a bath of ice water. The reaction mixture is heated at 50°, and the pressure in the system is reduced to about 100 mm. The temperature is gradually raised to 60°, with the system at a pressure of 100 ± 10 mm., to remove the ethanol used as solvent and formed as by-product. This distillation requires 2–3 hours (Note 4).

The receiver is changed and the excess diethyl oxalate is distilled under reduced pressure in the range of 76–82°/15 mm.

(Note 5). The residue in the reaction flask, a viscous, red-brown mass, consists of the sodio derivative of ethyl α-ethoxalylstearate (Note 6). To it is gradually added, without cooling, 66 g. (1.1 mole) of glacial acetic acid. As the mixture is stirred by hand, 1 l. of water is added. Finally, the mixture is stirred mechanically for a few minutes, then allowed to separate into layers. The organic (upper) layer is taken up in 300 ml. of ether and washed well with sodium bicarbonate solution and with water; then the ether is removed by distillation from a steam bath.

The residual α-ethoxalyl ester is decarbonylated to cetyl-malonic ester by heating at 160–170° under reduced pressure. A water aspirator is sufficient to facilitate removal of the carbon monoxide formed during the heating. The decarbonylation requires about 1–1.5 hours (Note 7). The cetylmalonic ester which remains as a dark liquid is fractionated under reduced pressure (Note 8).

A fore-run, consisting of ethyl stearate and an intermediate fraction, amounts to about 60 g., b.p. 185–204°/2 mm. (Note 8). The product is a colorless liquid amounting to 265–275 g. (68.5–71% yield), b.p. 204–208°/2 mm., n_D^{25} 1.4433 (Notes 9 and 10).

2. Notes

1. The evolved hydrogen should be led into a hood or out of a window to minimize the explosion hazard.

2. Diethyl oxalate can be dried satisfactorily by distilling a small portion from the bulk of the material. The water is removed with the distillate.

3. The ethyl stearate used by the checkers was prepared by esterifying commercial stearic acid (Armour Neo-Fat 1–65 or General Mills Aliphat 7) with 8 equivalents of absolute ethanol containing 10% by weight of concentrated sulfuric acid. The washed and dried product was distilled directly through a 60-cm. Vigreux column with a heated jacket and partial reflux head to give an 80% yield of material, b.p. 186.5–189°/3.5 mm., m.p. 32.2–34.2° (cor.). The purest grade of ethyl stearate as supplied by the Eastman Kodak Company is also satisfactory.

4. A column is not needed during distillation of the ethanol. When low-molecular-weight esters are substituted for ethyl

stearate a suitable indented column should be used at this point to prevent loss of the ester. The reaction temperature should not be allowed to exceed 60°, because decomposition of the condensation product may result.

5. The contents of the flask should not be heated to a temperature higher than 90° during removal of the excess oxalate or some decomposition of the product may result. The recovered diethyl oxalate contains small amounts of ethanol and ethyl carbonate. It can readily be purified by fractional distillation; however, it is quite suitable for reuse in repeat preparations without purification.

6. This sodio derivative can be dissolved in a solvent such as toluene and employed as an intermediate for other reactions (alkylation, acylation, etc.).

7. The carbon monoxide evolved should be led into a hood to avoid the dangers of poisoning and explosion. No difficulties have been encountered at this stage, but precautions are advisable.

8. A 60-cm. column of the Vigreux type, with heated jacket and partial reflux head, was found satisfactory by the checkers. Ethyl stearate may solidify in the condenser if a total reflux head is used. Refractionation of the fore-run gives an additional 30–40 g. (10–13%) of pure product. Material containing a few per cent of ethyl stearate can be obtained in 85–90% yield by simply distilling the total crude product from a Claisen flask.

9. Cetylmalonic acid, m.p. 115.5–120.5°, is obtained by saponification of the ester and one crystallization from acetone.

10. This general procedure is applicable to lower fatty acid esters.

3. Methods of Preparation

This method is based on the general procedure previously described by Floyd and Miller.[2] Cetylmalonic ester has also been prepared by condensation of ethyl stearate with ethyl carbonate[3] and by alkylation of sodiomalonic ester with cetyl iodide[4] or cetyl bromide.[5]

[1] General Mills Laboratory, Minneapolis, Minnesota.
[2] Floyd and Miller, *J. Am. Chem. Soc.*, **69**, 2354 (1947).
[3] Wallingford, Homeyer, and Jones, *J. Am. Chem. Soc.*, **63**, 2056 (1941).

[4] Kraft, *Ber.*, **17**, 1630 (1884); Guthzeit, *Ann.*, **206**, 357 (1881).
[5] Phillips and Mumford, *J. Chem. Soc.*, **1931**, 1736.

CHLOROACETONITRILE *

(Acetonitrile, chloro-)

$$3\text{ClCH}_2\text{CONH}_2 + \text{P}_2\text{O}_5 \rightarrow 3\text{ClCH}_2\text{CN} + 2\text{H}_3\text{PO}_4$$

Submitted by D. B. Reisner and E. C. Horning.[1]
Checked by R. L. Shriner and Calvin N. Wolf.

1. Procedure

In a 3-l. round-bottomed three-necked flask fitted with an efficient mechanical stirrer, a reflux condenser, and a thermometer are placed 170 g. (1.2 moles) of phosphorus pentoxide, 187 g. (2 moles) of chloroacetamide [2] (Note 1), and 800 ml. of dry technical trimethylbenzene (Note 2). The mixture is refluxed gently with vigorous stirring for 1 hour. The reaction mixture is then allowed to cool to about 100° with continuous stirring, and the reflux condenser is replaced with a distilling adapter fitted with a thermometer and a water-cooled condenser.

The crude product and part of the solvent are distilled at atmospheric pressure (Note 3). The yield of crude product boiling at 124–128° is 121–131 g. (80–87 %) (n_D^{25} 1.441–1.444). In order to obtain a pure product, the crude chloroacetonitrile is mixed with 10 g. of phosphorus pentoxide and redistilled through an efficient packed fractionating column (Note 4). The yield of pure chloroacetonitrile distilling at 123–124° is 93–106 g. (62–70%) (Note 5).

2. Notes

1. The practical grade of chloroacetamide obtainable from the Eastman Kodak Company can be used.

2. Technical trimethylbenzene with a boiling range of 166–174° is satisfactory.

3. About 200 ml. of solvent can be recovered. The remainder is left in the flask to facilitate removal of the residue.

4. A Fenske column packed with glass helices previously described in *Organic Syntheses* [3] is satisfactory. The product has n_D^{20} 1.426, d_4^{20} 1.1896, in good agreement with reported values.[4] When a Vigreux column was used the distillate had d_4^{20} 1.072 and n_D^{25} 1.430–1.436, indicating incomplete separation from the trimethylbenzenes (1,3,5-trimethylbenzene has d_4^{20} 0.86; n_D^{25} 1.494).

5. The product can also be distilled under reduced pressure; b.p. 60–61°/100 mm.; 30–32°/15 mm.

3. Methods of Preparation

Practical syntheses of chloroacetonitrile depend upon dehydration of chloroacetamide with phosphorus pentoxide. The present method uses a liquid reaction medium; in previous procedures the dry reagents were heated in the absence of solvent or liquid medium. [5,6]

[1] University of Pennsylvania, Philadelphia, Pennsylvania.
[2] *Org. Syntheses Coll. Vol.* **1**, 153 (1941).
[3] *Org. Syntheses*, **25**, 2 (1945).
[4] Rogers, *J. Am. Chem. Soc.*, **69**, 457 (1947).
[5] School, *Ber.*, **29**, 2417 (1896).
[6] Steinkopf, *Ber.*, **41**, 2540 (1908).

3-(o-CHLOROANILINO)PROPIONITRILE *

(Propionitrile, 3-o-chloroanilino-)

Submitted by S. A. HEININGER.[1]
Checked by JOHN C. SHEEHAN and ALMA M. BOSTON.

1. Procedure

A 500-ml. three-necked flask equipped with a stirrer, reflux condenser, and thermometer is charged with 255 g. (2.0 moles) of o-chloroaniline, 106 g. (2.0 moles) of acrylonitrile, and 10.2 g. (4.0% by weight of the amine) of cupric acetate monohydrate (Note 1). The mixture is then stirred and heated to reflux, beginning at about 95°. Refluxing is continued for 3 hours (Note 2), with the pot temperature rising to about 130° (Note 3).

The dark mixture is then transferred to a 500-ml. distilling flask fitted with a 15.2-cm. modified Vigreux column and the unchanged acrylonitrile (17–20 g.) collected at 100 mm. (water pump). The distillation is continued (vacuum pump) and the unchanged o-chloroaniline (110–120 g.), b.p. 57–60°/0.5 mm., is recovered. The 3-(o-chloroanilino)propionitrile (182–192 g.) is obtained as a colorless, somewhat viscous liquid, b.p. 139–141°/0.3 mm., n_D^{25} 1.5728–1.5735 (Note 4).

A pot residue of 30–35 g. remains (Note 5). The conversion of o-chloroaniline to 3-(o-chloroanilino)propionitrile is 50.5–53%, with a yield of 90–95% based on o-chloroaniline, and 53–65% based on acrylonitrile (Note 6).

2. Notes

1. The commercially available monohydrate form of cupric acetate was used. Anhydrous cupric acetate gives the same results. From 2% to 5% of catalyst by weight of the amine employed gives good yields of cyanoethylated products from a variety of anilines.

2. Slightly improved yields may be obtained by use of longer reaction times.

3. Maintaining the temperature at 100–110° for the same period of time gives equivalent results.

4. Physical constants for pure 3-(o-chloroanilino)propionitrile are: b.p. 139–141°/0.3 mm., n_D^{25} 1.5734, d_{25}^{25} 1.2103.

5. The residue consists mainly of polyacrylonitrile and copper or copper salts. It is slowly soluble in acetone, more readily soluble in polyacrylonitrile solvents such as dimethylformamide or dimethyl sulfoxide, especially when warmed.

6. An attempt to prepare 3-(o-chloroanilino)propionitrile by the Cymerman-Craig procedure (o-chloroaniline hydrochloride, diethylamine, and acrylonitrile) (p. 205) gave no isolable product, with recovery of 75% of the o-chloroaniline as its acetyl derivative, m.p. 86–87°; reported m.p. 87–88°.[2] Thus representative *ortho*-substituted anilines can be cyanoethylated in much better yields by use of the cupric acetate catalyst than by the Cymerman-Craig route, which is known to be subject to steric interferences.[3] Comparative yields for cyanoethylation of o-toluidine substantiate this conclusion; cupric acetate gave 62%, whereas Cymerman-Craig reported 25%.[3] Bulky N-substituents appear to affect yields similarly: methyl-, ethyl-, *n*-propyl-, and isopropylanilines gave yields of 65%, 41%, 17.5%, and 0.5% by the exchange reaction,[3] whereas, with cupric acetate, *n*-butylaniline was cyanoethylated in 68% yield.

3. Methods of Preparation

Cupric acetate is an efficient catalyst for the cyanoethylation of all but nitro-substituted aromatic amines. It is particularly effective with anilines which give poor yields by known methods, i.e., those with substituents on the nitrogen atom or in the *ortho* position.[4]

Other known catalysts for cyanoethylation of aromatic amines include acetic acid,[5,6] acetic acid-cuprous chloride mixtures,[7,8] aniline salts,[9] and choline.[10] 3-Anilinopropionitriles may also be prepared by an exchange reaction between the aniline hydrochloride and diethylaminopropionitrile (p. 205).[3]

[1] Research Department, Research and Engineering Division, Monsanto Chemical Company, Dayton, Ohio.

[2] Beilstein and Kurbatow, *Ann.*, **182**, 100 (1876).

[3] Bates, Cymerman-Craig, Moyle, and Young, *J. Chem. Soc.*, **1956**, 388; and earlier papers.

[4] Heininger, *J. Org. Chem.*, **22**, 1213 (1957).

[5] Braunholtz and Mann, *J. Chem. Soc.*, **1952**, 3046.

[6] Cookson and Mann, *J. Chem. Soc.*, **1949**, 67.

[7] Smith and Yu, *J. Am. Chem. Soc.*, **74**, 1096 (1952).

[8] Braunholtz and Mann, *J. Chem. Soc.*, **1953**, 1817.

[9] Bekhli and Serebrennikov, *J. Gen. Chem. U.S.S.R.*, **19**, 1553 (1949); [*C. A.*, **44**, 3448 (1950)].

[10] Pietra, *Gazz. chim. ital.*, **86**, 70 (1956).

CHLORO-*p*-BENZOQUINONE

(*p*-Benzoquinone, chloro-)

$$o\text{-ClC}_6\text{H}_4\text{NO}_2 \xrightarrow[\text{Electrolytic}]{\text{[H]}} o\text{-ClC}_6\text{H}_4\text{NHOH}$$

Submitted by R. E. Harman.[1]
Checked by N. J. Leonard and R. W. Fulmer.

1. Procedure

A. *Apparatus.* The reduction is carried out in a 400-ml. beaker. The anode, a cylinder of sheet lead 3 in. in height and 2.25 in. in diameter, rests on the bottom of the beaker. Clamped inside the anode and extending almost to the bottom of the beaker is the catholyte chamber, a porous cup (Note 1) 5 in. deep and 1.75 in. in diameter. Clamped securely inside the cup and about 0.25 in. above its inside bottom surface is the cathode, a cylinder of 25-gauge sheet copper 6 in. long and 1.25 in. in diameter. It is provided with a row of seven vertical slots 0.25 in. wide and 1.25 in. long, evenly spaced around its circumference; the lower ends of the slots are about 2.5 in. from the bottom of the cathode (Note 2). Both electrodes are provided with binding posts for connection to the circuit. A stirrer is constructed with two double propeller blades attached to the shaft, one about 1.5 in. above the other, at a pitch that will direct the flow of liquid downward. The mechanically driven stirrer is set to run deep inside the cathode (Note 3).

The current may be drawn from a commercially available battery charger (Note 4) or from a storage battery, each capable of operating at about 6 volts. Experiment has shown that the considerable ripple in the output of the charger has no adverse effect on the reduction. A transformer in the input to the charger, or a variable resistance in the battery circuit, and a 0–3 ampere range d.c. ammeter complete the apparatus.

B. *Reduction of o-chloronitrobenzene.* The stirrer is set in position carefully inside the cathode, which is clamped in place at the top. The porous cup, previously impregnated with the electrolyte, is charged with 11.5 g. (0.073 mole) of *o*-chloronitrobenzene (Note 5) and about 80 ml. of a mixture of acetic acid (70 ml.), concentrated sulfuric acid (22 ml.), and water (8 ml.) (Note 6). The cup is then clamped securely in position, and the beaker containing the anode is supported just clear of the bottom of the porous cup and filled with the same solution of aqueous sulfuric and acetic acids to the same level as the liquid inside the cup. The stirrer is started, the current turned on, and the system observed for a few minutes until the current has become stable. It is then adjusted at some convenient value no greater than 2 amperes.

The temperature will rise slowly and should be maintained at 30–45° throughout the reduction, with cooling as necessary. The system requires no attention while in operation except for occasional inspection and possibly the addition of a few milliliters of 90% acetic acid to maintain the surface of the catholyte at the desired level (Note 7).

C. *Oxidation of 4-amino-3-chlorophenol.* When the reduction is complete (Note 8), the system is disassembled, the catholyte poured into a flask, and the apparatus rinsed with hot water into the same flask. The combined catholyte and washings are extracted with ether in a continuous-type liquid-liquid extractor to remove the acetic acid almost completely (Note 9). The residual aqueous acid solution of the aminophenol (Note 10) is made up to $4N$ in sulfuric acid by adding 20 ml. of concentrated sulfuric acid and diluting with water to 400 ml.

At room temperature there is added in one portion a solution of 11 g. (0.037 mole, 50% excess over the theoretical amount) of sodium dichromate dihydrate in 20 ml. of water. A rise in temperature of some 6–7° will be observed; it is desirable to moderate the temperature by external cooling if it should rise above 35°.

After the mixture has been allowed to stand at room temperature for at least an hour (Note 11), a few grams of clean sand is added and the dark red mixture is filtered with suction. The filtrate is extracted with ether (100 ml., then 4–5 portions of about 30 ml. each), and each extract is used in turn to extract the filter cake.

The ether extracts are combined; the ether is removed by distillation (Note 12), and the dark red tarry residue subjected to steam distillation. Some 40–50 ml. of distillate is collected after solid quinone no longer appears in the condenser (total volume about 150 ml.).

There is obtained by filtration 5.0–5.5 g. (48–53%) of yellow chloro-*p*-quinone of m.p. 53–55°; ether extraction of the filtrate yields a further 1.0 g., which brings the total yield to 58–63% (Notes 12 and 13). Crystallization from a small volume of aqueous ethanol (85–90%) raises the melting point to 55–56°.

2. Notes

1. A Norton dense grade Alundum extraction thimble has been found suitable. Deposition of silicic acid in the interstices of the cup to decrease transfer of material is desirable. This is accomplished by soaking the cup for an hour in 20% water glass, draining off the excess, and then keeping the cup totally immersed for 3 hours in 20% aqueous sulfuric acid.[2] Finally, the cup is stored completely immersed in the electrolyte.

2. This cathode design, with the stirring described, is regarded as preferable to the combination used by Lukens [3] and much simpler in construction. It provides better dispersion of the organic material, a feature which is particularly necessary if these reductions are carried out in aqueous solution (Note 6).

3. Very rapid and efficient stirring is required to strip the intermediate *o*-chlorophenylhydroxylamine from the cathode so that the acid-catalyzed rearrangement to 4-amino-3-chlorophenol may occur rather than reduction of the substituted hydroxylamine to *o*-chloroaniline.

4. A Battery Booster, type 6-AC-4, manufactured by P. R. Mallory and Company, Inc., Indianapolis, Indiana, gave good service.

5. Material equivalent to Eastman Kodak Company white label grade was used. It has been found that the same results are obtained using *m*-chloronitrobenzene. Likewise, either *o*- or *m*-nitrotoluene may be used in the preparation of toluquinone (Note 13).

6. The submitter states that lower yields of quinones (50–65%) are obtained by using as the electrolyte 75% by weight aqueous sulfuric acid. The catholyte at the end of the reduction is diluted to about 500 ml. ($4N$ in sulfuric acid) and oxidized directly. The low solubility of the nitro compounds in the aqueous acid makes exceedingly efficient stirring a necessity, and the cell must be maintained at 50–60° during the reduction to keep the nitro compound molten and so promote the formation of a fine emulsion. Solid aminophenol sulfate sometimes crystallizes near the end of the reduction, and evolution of hydrogen near the end of the run may force the pasty material out the top of the porous cup. In general, these difficulties are more troublesome than the

continuous extraction necessitated by the use of acetic acid in the catholyte.

7. In operation, a flood of electrolyte should be observable pouring over the lower ends of the slots in the cathode.

8. Complete reduction of the nitro compound is assured by the use of about 1.2 equivalents of current (9.4 ampere-hours), but a larger excess of current is no disadvantage.

9. It has been found satisfactory to continue the extraction for 30 minutes after the volume of the aqueous phase no longer decreases visibly. Saturated aqueous sodium bicarbonate washes may be used to effect removal of the acetic acid from ether solutions of alkyl quinones. The cascade distribution apparatus devised by Kies and Davis [4] is useful for this purpose. As halogenated quinones have been found to be unstable to bicarbonate, the acetic acid must be removed before oxidation of the corresponding aminophenols.

10. It has been found that the sulfates of the aminophenols may be obtained in yields about the same as those recorded for the quinones by concentrating the aqueous catholytes under reduced pressure after removal of the acetic acid.

11. It is recommended that the oxidation mixture be stored in a refrigerator if more than 6–8 hours must elapse before the work-up is completed.

12. Prolonged heating after removal of the ether may lead to sublimation of the product, with resulting lower yield.

13. Essentially the same procedure, with alteration only in the temperature during the oxidation, is stated by the submitter to be satisfactory for the preparation of several other p-benzoquinones. For p-benzoquinone itself and for toluquinone (both in 80% yield) the oxidation is best carried out at 5–10°; for 3-chlorotoluquinone (70% yield), at 15–20°; for 2,5-dichloroquinone (50% yield), at 55–60°. In the last case, some of the weakly basic aminophenol is ether extracted along with the acetic acid; it may be recovered by distilling most of the ether and acetic acid under reduced pressure, extracting the dark residue with two 25-ml. portions of $1N$ sodium hydroxide, and making the alkaline extract neutral or slightly acid with sulfuric acid. Normal oxidation is carried out after combining this suspension of aminophenol with the original aqueous solution.

The low vapor pressure of 2,5-dichloroquinone precludes the use of steam distillation; however, the quinone precipitates directly from the oxidation mixture as a tan solid and may be purified readily by sublimation, followed by crystallization from ethanol. An excellent sublimation apparatus that will easily handle 10-g. quantities of this quinone has been described.[5]

This method has given only 25% yield of methoxyquinone which is highly sensitive to acid and requires a special isolation procedure.[6]

3. Methods of Preparation

The present method is the result of a study by Cason, Harman, Goodwin, and Allen.[6, 7] The sequence of electrolytic reduction followed by oxidation has been used for the preparation of 5-bromotoluquinone,[8] 5-chlorotoluquinone,[9] and 3-chlorotoluquinone.[10] The preparation of an intermediate *p*-aminophenol from the corresponding aromatic nitro compound by electrolytic reduction is a useful general method.[11-18] Chloro-*p*-quinone has been prepared by acid dichromate oxidation of chlorohydroquinone [19-22] or 2-chloro-4-aminophenol.[23, 24] It has been shown that pure chloro-*p*-quinone is obtained only with some difficulty when chlorohydroquinone is used.[7] Potassium bromate also has been used to oxidize chlorohydroquinine to chloro-*p*-quinone.[25]

[1] University of California, Berkeley, California.

[2] Swann, "Electrolytic Reductions," in Weissberger, *Technique of Organic Chemistry*, Vol. 2, 2nd ed., p. 385, Interscience Publishers, New York, New York, 1956.

[3] Lukens, *Ind. Eng. Chem.*, **13**, 562 (1921).

[4] Kies and Davis, *J. Biol. Chem.*, **189**, 637 (1951).

[5] Morton, Mahoney, and Richardson, *Ind. Eng. Chem., Anal. Ed.*, **11**, 460 (1939).

[6] Harman and Cason, *J. Org. Chem.*, **17**, 1047, 1058 (1952).

[7] Cason, Harman, Goodwin, and Allen, *J. Org. Chem.*, **15**, 860 (1950).

[8] Gattermann, *Ber.*, **27**, 1931 (1894).

[9] Raiford, *Am. Chem. J.*, **46**, 445 (1911).

[10] Cason, Allen, and Goodwin, *J. Org. Chem.*, **13**, 403 (1948).

[11] Gattermann, *Ber.*, **26**, 1844 (1893).

[12] Gattermann and Koppert, *Chem. Ztg.*, **17**, 210 (1893).

[13] Piccard and Larsen, *J. Am. Chem. Soc.*, **40**, 1090 (1918).

[14] Brigham and Lukens, *Trans. Electrochem. Soc.*, **61**, 281 (1932).

[15] Fieser and Martin, *J. Am. Chem. Soc.*, **57**, 1840 (1935).

[16] Mann, Montonna, and Larian, *Trans. Electrochem. Soc.*, **69**, 367 (1936).

[17] Dey, Govindachari, and Rajagopalan, *J. Sci. Ind. Research* (*India*), **4**, 574 (1946) [*C. A.*, **40**, 4965 (1946)].

[18] Dey, Maller, and Pai, *J. Sci. Ind. Research (India)*, **7B**, 113 (1948) [*C. A.*, **43**, 3730 (1949)].

[19] Levy and Schultz, *Ann.*, **210**, 145 (1881).

[20] den Hollander, *Rec. trav. chim.*, **39**, 481 (1920).

[21] Eckert and Endler, *J. prakt. Chem.*, (2) **104**, 83 (1922).

[22] Conant and Fieser, *J. Am. Chem. Soc.*, **45**, 2201 (1923).

[23] Kollrepp, *Ann.*, **234**, 14 (1886).

[24] van Erp, *Ber.*, **58**, 663 (1925).

[25] Grinev and Terent'ev, *Vestnik Moskov. Univ. Ser. Mat., Mekh., Astron., Fiz., Khim.*, **12**, No. 6, 147 (1957) [*C. A.*, **53**, 3187 (1959)].

N-CHLOROBETAINYL CHLORIDE

(Ammonium chloride, [(chloroformyl)methyl]trimethyl-)

$$(CH_3)_3N(Cl)CH_2CO_2H + SOCl_2 \rightarrow$$
$$(CH_3)_3N(Cl)CH_2COCl + SO_2 + HCl$$

Submitted by B. VASSEL and W. G. SKELLY.[1]
Checked by JOHN C. SHEEHAN and J. IANNICELLI.

1. Procedure

Caution! This preparation should be conducted in a good hood.

A 1-l. round-bottomed flask is equipped with an internal thermometer, a sealed stirrer, and a reflux condenser, the upper end of which is protected with a calcium chloride drying tube. In this flask are placed 307 g. (2 moles) of dry, pulverized betaine hydrochloride (Note 1) and 285 g. (174 ml., 2.4 moles) of thionyl chloride. The mixture is stirred and heated slowly. When the internal temperature reaches 68° copious evolution of sulfur dioxide and hydrogen chloride occurs, and the mass becomes pasty. The temperature is maintained with stirring at 68–70° for 1.5 hours (Note 2).

Warm (80°) dry toluene (150 ml.) is added to the melt, and stirring is continued for 5 minutes. The entire mass is quickly poured into a dry beaker (Note 3) and slowly stirred, manually, until the entire mass has crystallized (Note 4). The cool toluene is decanted rapidly, and 150 ml. of warm toluene is added. The mixture is heated sufficiently to melt the crystals (about 68°), then allowed to cool again with stirring. The toluene is decanted once more, and 150 ml. of hot (60°) dry benzene is added. The mass is melted once more and cooled with stirring. The crystalline mass, with the benzene layer still covering it, is transferred rapidly to a Büchner funnel sufficiently large to hold all the contents of the beaker. The funnel is immediately covered with a rubber diaphragm, and suction is applied (Note 5). The crystals are quickly covered with 150 ml. of dry methylene chloride, dried with suction, and quickly transferred into a glass vacuum oven at 50° (Note 6). The yield of N-chlorobetainyl chloride is 337–344 g. (98–100%) of 98–100% purity if moisture was rigorously excluded (Note 7).

2. Notes

1. The betaine hydrochloride was obtained from International Minerals and Chemical Corporation. It was pulverized, dried at 105° for 3 hours, ground again, and dried once more at 105° for 3 hours.

2. The mixture should be fluid after about the first 20 minutes of heating. If at this stage the betaine hydrochloride has not completely lost its granular appearance it is probable that it had not been adequately dried or that the reaction temperature had been below 68°. If it is considered that the betaine hydrochloride had not been dry enough the reaction may be completed by adding an additional 41 g. (25 ml., 0.34 mole) of thionyl chloride and continuing the stirring and heating for 1.25 hours.

3. Success in the isolation of the pure acid chloride depends upon the rigorous exclusion of moisture. The acid chloride hydrolyzes to the hydrochloride with great rapidity when exposed to even traces of moist air. During the stirring of the warm melt, care must be exercised that a layer of toluene covers the melt at all times.

4. If the melt is permitted to crystallize without stirring, a hard glasslike layer forms which cannot be broken up without exposure to air.

5. The rubber diaphragm is sold by Fisher Scientific Company as dental dam. It is held in place by two strong rubber bands. The suction flask is protected from moisture by attachment of a calcium chloride drying tube to the vacuum line if a water aspirator is used.

6. The dried acid chloride is conveniently stored in about 25 bottles with tightly fitting plastic screw caps which are kept in a desiccator over phosphorus pentoxide. In this manner the acid chloride is exposed to a minimum of atmospheric moisture when reactions are run which require only part of the preparation.

7. The same yield and purity were obtained when 1229 g. (8 moles) of betaine hydrochloride was used. In this case the volumes of toluene, benzene, and methylene chloride do not have to be increased proportionally if narrow, tall beakers are used. About 300 ml. of each is sufficient.

3. Methods of Preparation

N-Chlorobetainyl chloride has been prepared by treating betaine hydrochloride with either thionyl chloride,[2] phosphorus pentachloride in acetyl chloride,[3,4] or phosphorus pentachloride in phosphorus oxychloride.[5] None of the patents mentions the instability of the acid chloride towards moisture or describes a method for the isolation of the pure product.

[1] Central Research Laboratories, International Minerals and Chemical Corporation, Skokie, Illinois.

[2] Linch, U. S. pats. 2,359,862; 2,359,863 [C. A., **39**, 2076, 2077 (1945)]; Byrne, U. S. pat. 2,888,383 [C. A., **53**, 15492 (1959)].

[3] Society of Chemical Industry of Basel, Brit. pats. 589,232; 590,727 [C. A., **42**, 230, 210 (1948)].

[4] Ruzicka and Plattner, U. S. pat. 2,429,171 [C. A., **42**, 930 (1948)].

[5] Plattner and Geiger, *Helv. Chim. Acta*, **28**, 1362 (1945).

trans-2-CHLOROCYCLOPENTANOL

(Cyclopentanol, 2-chloro-, *trans*-)

$$NH_2CONH_2 + Cl_2 \longrightarrow NH_2CONHCl + HCl$$

$$CaCO_3 + 2HCl \longrightarrow CaCl_2 + CO_2 + H_2O$$

$$NH_2CONHCl + H_2O \xrightarrow{CH_3CO_2H} NH_2CONH_2 + HOCl$$

$$\begin{array}{c} CH_2\text{---}CH \\ | \quad\quad || \\ CH_2 \quad CH \\ \diagdown \diagup \\ CH_2 \end{array} + HOCl \longrightarrow \begin{array}{c} CH_2\text{---}CHOH \\ | \quad\quad\quad | \\ CH_2 \quad CHCl \\ \diagdown \diagup \\ CH_2 \end{array}$$

Submitted by HUGH B. DONAHOE and CALVIN A. VANDERWERF.[1]
Checked by ARTHUR C. COPE and ELBERT C. HERRICK.

1. Procedure

A mixture of 150 g. (2.5 moles) of urea, 125 g. (1.25 moles) of reprecipitated calcium carbonate, and 150 ml. of water in a 2-l. three-necked flask is tared and cooled in an ice-salt bath. The flask is equipped with a thermometer which extends into the re-action mixture, a gas inlet tube, an outlet tube leading to a gas-absorption trap,[2] and a slip- or mercury-sealed mechanical stirrer which will disperse chlorine gas below the surface of the liquid (Note 1).

A rapid stream of chlorine gas is bubbled into the mixture at 0–15° with vigorous stirring until an increase in weight of about 95 g. has occurred (30–40 minutes) (Note 2). A 250-ml. portion of water at room temperature is added to the suspension, which is then filtered by suction through rather porous filter paper on a 16-cm. Büchner funnel. The filtrate is removed and cooled in an ice bath, and the filter cake is washed on the funnel with a 250-ml. portion of water at room temperature. The filtrate is poured back in the funnel and sucked through the filter cake repeatedly until no more solid appears to dissolve, and then is combined with the original filtrate (Note 3).

The cold filtrates (solutions of monochlorourea) are trans-ferred to a 3-l. two-necked flask immersed in an ice-salt bath.

The flask is equipped with a slip- or mercury-sealed mechanical stirrer and an efficient reflux condenser. To the flask are added 500 g. of ice, 100 ml. of glacial acetic acid, and 136 g. (2.0 moles) of cyclopentene (or 1.43 times the weight increase in grams during introduction of the chlorine) (Note 4). Mechanical stirring is begun and is continued while the flask is kept packed in ice until the cyclopentene (the top layer) disappears and a heavy oil settles to the bottom (Note 5).

The solution is saturated with sodium chloride and distilled with steam until all the 2-chlorocyclopentanol is collected, which requires distillation of a volume of about 2 l. The distillate is saturated with sodium chloride, the oily layer separated, and the aqueous layer extracted four times with 300-ml. portions of ether. The ether extracts are added to the oil, and the solution is washed with a saturated sodium chloride solution and dried over anhydrous sodium sulfate. The ether is removed by distillation, and the product is distilled under reduced pressure through a total condensation, variable take-off, 15 by 1.5 cm. column packed with glass helices. A trap cooled with Dry Ice is placed in the vacuum line between the column and the pump. Low-boiling fractions, b.p. 43–81°/15 mm., amount to 19–40 g. (Note 6). The *trans*-2-chlorocyclopentanol is collected at 81–82°/15 mm. in a yield of 126–135 g. (52–56%); n_D^{25} 1.4770 (Note 7).

2. Notes

1. Rapid absorption of chlorine depends on efficient stirring and dispersal of the gas through the liquid phase. A stirrer which disperses the gas through the solution by vigorous agitation may be used (p. 891) or the gas may be introduced through the stirrer.[3]

2. The actual weight of chlorine absorbed is equal to the sum of the weight increase noted plus the weight of carbon dioxide formed minus the relatively small weight of carbon dioxide that remains dissolved in the reaction mixture.

3. The combined filtrates may be titrated with $1N$ sodium thiosulfate solution to determine the yield of monochlorourea.[4] This preparation of monochlorourea is a modification of procedures previously described.[4,5]

4. Best yields result when an excess of cyclopentene is used, as specified. The weight increase should be roughly 95 g., and the amount of cyclopentene should be varied proportionately by using 1.43 times the weight increase in grams.

5. The stirring time is about 12–15 hours. It is advantageous to allow the ice to melt and the reaction mixture to come to room temperature during the last 2–3 hours.

6. The fractions boiling at 43–48°/15 mm. and 48–81°/15 mm. contain a mixture of 1,2-dichlorocyclopentane and *cis*- and *trans*-2-chlorocyclopentanol.[6, 7]

7. The yield is based on the weight of cyclopentene, although this reagent is used in slight excess.

3. Methods of Preparation

2-Chlorocyclopentanol has been prepared by the reaction of dry hydrogen chloride[8] or thionyl chloride[9] with 1,2-cyclopentanediol, and by the addition of hypochlorous acid to cyclopentene.[6–8, 10]

[1] University of Kansas, Lawrence, Kansas.

[2] *Org. Syntheses Coll. Vol.* **2**, 4 (1943).

[3] Russell and Vanderwerf, *Ind. Eng. Chem., Anal. Ed.*, **17**, 269 (1945).

[4] McRae, Charlesworth, and Alexander, *Can. J. Research*, **21 B**, 1 (1943).

[5] Detoeuf, *Bull. soc. chim. France*, [4] **31**, 102 (1922).

[6] Rothstein and Rothstein, *Compt. rend.*, **209**, 761 (1939).

[7] Godchot, Mousseron, and Granger, *Compt. rend.*, **200**, 748 (1935).

[8] Meiser, *Ber.*, **32**, 2052 (1899).

[9] Mousseron, Winternitz, and Mousseron-Canet, *Bull. soc. chim. France*, **1953**, 737.

[10] Mousseron, Granger, Winternitz, and Combes, *Bull. soc. chim. France*, **1946**, 610.

1-CHLORO-2,6-DINITROBENZENE *

(Benzene, 2-chloro-1,3-dinitro-)

$$O_2N-C_6H_3(NH_2)(NO_2) \xrightarrow[H_2SO_4]{NaNO_2} O_2N-C_6H_3(\overset{\oplus}{N_2}\ HSO_4^{\ominus})(NO_2)$$

$$O_2N-C_6H_3(\overset{\oplus}{N_2}\ HSO_4^{\ominus})(NO_2) + HCl \xrightarrow{Cu_2Cl_2} O_2N-C_6H_3(Cl)(NO_2) + N_2 + H_2SO_4$$

Submitted by F. D. GUNSTONE and S. HORWOOD TUCKER.[1]
Checked by ARTHUR C. COPE, DAVID J. MARSHALL, and RONALD M. PIKE.

1. Procedure

Concentrated sulfuric acid (160 ml.) is placed in a 1-l. three-necked flask fitted with a glass stirrer and a thermometer, and 15.2 g. (0.22 mole) of solid sodium nitrite is added over a period of 10–15 minutes with stirring. After the addition is completed, the temperature is raised to 70° and the mixture is stirred until all the sodium nitrite dissolves. The solution is cooled to 25–30° with an ice bath, and a solution of 36.6 g. (0.2 mole) of 2,6-dinitroaniline (p. 364) in 400 ml. of hot glacial acetic acid is added slowly, with stirring, at such a rate that the temperature remains below 40° (Note 1). After the addition is completed, the solution is stirred at 40° for 0.5 hour. A solution of 44 g. (0.44 mole) of cuprous chloride in 400 ml. of concentrated hydrochloric acid is prepared in a 2-l. beaker and cooled in an ice bath, and the solution of the diazonium salt is added in portions over a period of about 5 minutes, with manual stirring, at a rate which keeps the effervescence from becoming too vigorous. The mixture becomes hot during the addition, and it is stirred intermittently while being cooled in an ice bath until the effervescence slackens. It is then heated on a steam bath with occasional stirring until the temperature reaches 80°. After about 20 minutes at that temperature the effervescence ceases, and then an equal volume of water is added and the mixture is cooled in an ice bath. After several hours the yellow, crystalline 1-chloro-2,6-dinitrobenzene is collected on a suction filter, washed with water, and dried (Note 2). The product, which is sufficiently pure for most purposes without recrystallization, is obtained in a yield of 28.7–30 g. (71–74%); m.p. 86–88°. The product can be recrystallized from 90% (by volume) acetic acid (2 ml. per g.) or by dissolving it in hot benzene (1.5 ml. per g.) and adding petroleum ether (3 ml. per g.); m.p. 86–87°.

2. Notes

1. The temperature of diazotization is critical. Lower yields are obtained if the temperature rises above 40°.

2. 1-Chloro-2,6-dinitrobenzene is a skin irritant, and contact with it should be avoided.

3. Methods of Preparation

1-Chloro-2,6-dinitrobenzene has been prepared from 2,6-dinitroaniline by the Sandmeyer reaction,[2-4] from 2,6-dinitrophenol and phosphorus oxychloride in the presence of N,N-diethylaniline,[5] and from the mixture of isomers (in which 1-chloro-2,4-dinitrobenzene is present in largest amount) obtained by nitrating o-nitrochlorobenzene.[3, 6] The mixture of 2,4- and 2,6-dinitrochlorobenzene has been separated by the use of solution of sodium hydroxide in ethanol and water.[7]

[1] The University, Glasgow, Scotland.
[2] Welsh, J. Am. Chem. Soc., 63, 3276 (1941).
[3] Hodgson and Dodgson, J. Chem. Soc., 1948, 1006.
[4] Gunstone and Tucker, J. Appl. Chem. (London), 2, 204 (1952).
[5] Boothroyd and Clark, J. Chem. Soc., 1953, 1504.
[6] Borsche and Rantscheff, Ann., 379, 152 (1911).
[7] Molard and Vaganay, Mêm. poudres, 39, 111 (1957) [C. A., 52, 19989 (1958)].

2-CHLORO-2-METHYLCYCLOHEXANONE

(Cyclohexanone, 2-chloro-2-methyl-)

and

2-METHYL-2-CYCLOHEXENONE

(2-Cyclohexen-1-one, 2-methyl-)

Submitted by E. W. WARNHOFF, D. G. MARTIN, and WILLIAM S. JOHNSON.[1]
Checked by M. S. NEWMAN, G. R. KAHLE, and D. E. REID.

1. Procedure

A. *2-Chloro-2-methylcyclohexanone.* A 3-l. three-necked flask, fitted with a sealed mechanical stirrer with glass blade, a dropping funnel, and an outlet tube connected to a gas-absorption trap,[2] is charged with a solution of 224 g. (2.0 moles) of 2-methylcyclohexanone (Note 1) in 1 l. of dry carbon tetrachloride. A solution of 179 ml. (297 g., 2.2 moles) of sulfuryl chloride (Note 2) in 300 ml. of dry carbon tetrachloride is added from the dropping funnel over a 1-hour period with stirring. The slightly exothermic reaction is moderated by cooling the flask with a bath of water at room temperature. After the addition is complete, stirring is continued for 2 hours. The yellow solution is then washed successively with three 300-ml. portions of water, two 200-ml. portions of saturated sodium bicarbonate solution, and one 200-ml. portion of saturated salt solution, and finally dried over anhydrous magnesium sulfate.

The solvent is removed by distillation through a 15-cm. Vigreux column, first at atmospheric pressure and finally at reduced pressure (water aspirator). The residue is satisfactory for the preparation of 2-methyl-2-cyclohexenone described below (part B). Distillation through the column gives, after a small fore-run, 243–248 g. (83–85%) of colorless 2-chloro-2-methylcyclohexanone, b.p. 94–96°/27 mm., n_D^{25} 1.4672, d_4^{25} 1.088 (Note 3).

B. *2-Methyl-2-cyclohexenone.* (*a*) *Collidine method.* The crude (undistilled) 2-chloro-2-methylcyclohexanone prepared as described above (part A) is transferred to a 1-l. three-necked flask fitted with a stout sealed Hershberg wire stirrer and two efficient reflux condensers, one attached to each side neck; and 290 ml. (266 g., 2.2 moles) of 2,4,6-collidine (Note 2) is added rapidly through one of the condensers with stirring. The flask is heated to 145–150° (bath temperature) with an oil bath until there ensues a sudden exothermic reaction which results in vigorous boiling of residual carbon tetrachloride. The reaction is essentially complete within 1 minute, and the mixture becomes very viscous with suspended collidine hydrochloride. The heating is discontinued, and, as the reaction mixture cools and becomes viscous enough to impede stirring, a total of 500 ml. of benzene is added cautiously (with vigorous boiling) through a condenser in order to maintain fluidity. The collidine hydrochloride is collected by suction filtration on a 10-cm. sintered glass suction filter, then transferred to a beaker, triturated with 300 ml. of benzene, and refiltered. After repetition of this treatment, the weight of residual salt is about 303 g. (96%).

The combined dark-brown filtrates are washed (Note 4) with two 300-ml. portions of 10% hydrochloric acid saturated with sodium chloride, with one 300-ml. portion of saturated sodium bicarbonate solution, and with one 300-ml. portion of saturated salt solution, and finally dried over anhydrous magnesium sulfate. The benzene is removed by distillation through a 15-cm. Vigreux column, and then all material boiling at 70–97°/56 mm. is collected (Note 5). Fractionation of this material through a 20-cm. heated column packed with steel saddles gives, after a fore-run of benzene, 100–109 g. (45–49% yield from 2-methylcyclohexanone) of colorless 2-methyl-2-cyclohexenone, b.p. 98–101°/77 mm., n_D^{25} 1.4830–1.4835, d_4^{25} 0.972, λ_{max}^{alc} 234 mμ (ϵ 9660) (Note 6).

(*b*) *Lithium chloride method.* A 1-l. three-necked flask, fitted with a sealed Hershberg wire stirrer, a thermometer, and a tube leading to a source of nitrogen, is charged with 26 g. of lithium chloride, 250 ml. of dimethylformamide (Note 7) and crude 2-chloro-2-methylcyclohexanone prepared as described above (part A, half-scale) from 112 g. of 2-methylcyclohexanone. The stirrer is started, the air swept out with nitrogen, and the flask immersed in an oil bath maintained at 100°. The temperature of the reaction mixture rises to 100° in about 10 minutes, then to 112° and back to 105° in another 25 minutes. At the end of this time the mixture is cooled, 1-l. each of ether and of 2.5% sulfuric acid is added, and the mixture is stirred for about 4 hours to hydrolyze the dimethylformamide. The aqueous layer is separated, saturated with sodium chloride, and extracted with four 150-ml. portions of ether. These extracts are combined with the original ether layer, washed with saturated sodium chloride solution and saturated sodium bicarbonate solution, and finally dried over anhydrous sodium sulfate. The ether is removed by distillation through a 15-cm. Vigreux column, and then a fraction boiling at 79–107°/28 mm. is collected. Redistillation through a 20-cm. heated column packed with steel saddles gives 47–50 g. (43–45% yield from 2-methylcyclohexanone) of colorless 2-methyl-2-cyclohexenone, b.p. 83–85.5°/35 mm., n_D^{25} 1.4833–1.4840, λ_{max}^{alc} 234 mμ (ϵ 9680) (Note 6).

2. Notes

1. 2-Methylcyclohexanone is available commercially. In the present work it was prepared as follows. A 3-l. three-necked flask, fitted with a Hershberg wire stirrer, dropping funnel, and a thermometer, is charged with a solution of 228 g. (2.0 moles) of 2-methylcyclohexanol (Eastman Kodak Company practical grade) in 1 l. of benzene. A solution of 238 g. of sodium dichromate dihydrate in 1 l. of water containing 324 ml. of concentrated sulfuric acid and 100 ml. of acetic acid is added from the funnel over a period of 2.5 hours with stirring. The temperature of the reaction mixture is maintained (ice bath) at 10° or slightly below during the addition and also during a 3-hour stirring period after the addition. The aqueous layer is separated, diluted with 250 ml. of water, and extracted with two 300-ml. portions of benzene. These extracts are combined with the original benzene layer, and the whole is washed in sequence with 500 ml. of water, 400 ml. of saturated sodium bicarbonate solution, and 400 ml. of saturated salt solution. After the solution has been dried over anhydrous magnesium sulfate, the benzene is removed by distillation through a 20-cm. Vigreux column. Distillation of the residue through a 20-cm. column packed with steel saddles gives 193–200 g. (85–88% yield) of 2-methylcyclohexanone, b.p. 162.5–163.5°/742 mm., n_D^{25} 1.4459.

2. Eastman Kodak Company practical grade.

3. The homogeneity of this product has been demonstrated.[3] It is stable for long periods if stored in a brown bottle over a little magnesium oxide.

4. Excessive washing with aqueous solution is avoided as the product has appreciable solubility in water.

5. The dark oily distillation residue (about 72 g.) distils at 140–200°/1 mm. and consists mainly of the dimer of 2-methylenecyclohexanone which can be isolated in 23% yield.[3]

6. On standing, the ketone gradually turns yellow, and the refractive index increases. Redistillation of such material gives pure ketone.

7. Supplied by Matheson, Coleman and Bell.

3. Methods of Preparation

The only published method for producing pure 2-chloro-2-methylcyclohexanone is by the action of sulfuryl chloride on 2-methylcyclohexanone.[3] Direct chlorination gives mixtures of the 2- and 6-chloro compounds.[3]

2-Methyl-2-cyclohexenone has been prepared by the action of nitrosyl chloride on 1-methylcyclohexene, followed by dehydrohalogenation with sodium methoxide[4] or sodium acetate,[5] and hydrolysis of the resulting oxime; in an impure condition by several methods;[6–12] by dehydration of the ketol produced by the reaction of methylmagnesium iodide with 1,2-cyclohexanedione;[7b, 13] by reduction of the enol methyl ether produced by the reaction of diazomethane with 1,2-cyclohexanedione;[14] by bromination of 2-methylcyclohexanone with N-bromosuccinimide, followed by dehydrobromination with pyridine or with 2,4-dinitrophenylhydrazine;[15] by treatment of 5-heptenoic acid chloride with stannic chloride;[16] and by the present method.[3] The use of lithium chloride in dimethylformamide for the dehydrohalogenation is an adaptation of the method of Holysz.[17]

[1] University of Wisconsin, Madison, Wisconsin.

[2] Org. Syntheses Coll. Vol. 2, 4 (1943).

[3] Warnhoff and Johnson, J. Am. Chem. Soc., 75, 494 (1953).

[4] Wallach, Ber., 35, 2822 (1902); Ann., 359, 303 (1908).

[5] Haworth, J. Chem. Soc., 103, 1242 (1913).

[6] Godchot and Bedos, Compt. rend., 181, 919 (1925).

[7] (a) Urion, Compt. rend., 199, 363 (1934); (b) Butz, Davis, and Gaddis, J. Org. Chem., 12, 122 (1947).

[8] Mousseron, Jacquier, and Winternitz, Compt. rend., 224, 1230 (1947).

[9] Mousseron, Winternitz, and Jacquier, Compt. rend., 224, 1062 (1947).

[10] Dupont, *Bull. soc. chim. Belg.*, **45**, 57 (1936).

[11] Whitmore and Pedlow, *J. Am. Chem. Soc.*, **63**, 758 (1941).

[12] Birch, *J. Chem. Soc.*, **1946**, 593.

[13] Godchot and Gauquil, *Bull. soc. chim.*, [5]**2**, 1100 (1935).

[14] Bornstein, Pappo, and Szmuszkovicz, *Bull. Research Council Israel*, **2**, 273 (1952) [*C. A.*, **48**, 9933 (1954)].

[15] Rinne, Deutsch, Bowman, and Joffe, *J. Am. Chem. Soc.*, **72**, 5759 (1950).

[16] Riobé, *Compt. rend.*, **248**, 2774 (1959).

[17] Holysz, *J. Am. Chem. Soc.*, **75**, 4432 (1953).

2-CHLORONICOTINONITRILE

(Nicotinonitrile, 2-chloro-)

$$+ 2PCl_5 \xrightarrow{POCl_3} + 2POCl_3 + 3HCl$$

Submitted by E. C. TAYLOR, JR.,[1] and ALDO J. CROVETTI.[2]

Checked by CHARLES C. PRICE and WALTER A. SCHROEDER.

1. Procedure

Caution! This preparation should be conducted in a good hood.

In a 1-l. round-bottomed flask are placed 85.0 g. (0.62 mole) of nicotinamide-1-oxide (p. 704) and 180.0 g. (0.86 mole) of phosphorus pentachloride (Note 1), and the solids are thoroughly mixed. Two hundred and forty-three milliliters of phosphorus oxychloride is added slowly with shaking. A spiral condenser provided with a drying tube is attached to the flask which is then placed in an oil bath preheated to 60–70°. The temperature is slowly (20–25 minutes) raised to 100°, during which time the reaction mixture is occasionally shaken. In the range 100–105° the evolution of hydrogen chloride gas increases, and a spontaneous, vigorous refluxing of the phosphorus oxychloride begins. The reaction flask is removed from the oil bath, and the rate of refluxing is controlled by the application of an ice-water bath (Note 2). After the vigorous reaction has subsided (about 5 minutes) the oil bath is replaced, and heating under reflux is continued at 115–120° for 1.5 hours.

After the reaction mixture has been cooled, the excess phosphorus oxychloride is distilled under reduced pressure (80–100 mm.) (0.5–1.0 hour). Near the end of the distillation the product begins to sublime into the still head. The residual dark-brown oil is poured with stirring into an 800-ml. beaker containing 280–300 g. of crushed ice (Notes 3 and 4). The volume of the ice-water mixture is brought to 600 ml. and allowed to stand at 5° overnight. The crude light-brown product is filtered by suction and washed with water.

The solid is suspended in 300 ml. of 5% sodium hydroxide at 15° (Note 5). The mixture is stirred for 30 minutes, and the solid is filtered by suction and washed with water until the filtrates are no longer alkaline. The procedure is repeated, but stirring is continued for 0.75–1.0 hour. After the solid has been filtered by suction, washed, and pressed as dry as possible, it is dried under reduced pressure over phosphorus pentoxide for 12–16 hours.

The solid is transferred to a Soxhlet thimble (45 × 125 cm.) containing a 5-cm. layer of anhydrous sodium carbonate on the bottom (Note 6), and the solid is extracted for 2 to 3 hours with anhydrous ether (700–800 ml.). The total volume of ether is brought to 800–900 ml. The ethereal solution is treated with charcoal and boiled for 10–15 minutes under reflux; then the solution is filtered by suction (Note 7). After evaporation of solvent, 30–33 g. (35–39%) of white 2-chloronicotinonitrile is obtained, m.p. 105–106° (Note 8).

2. Notes

1. *To avoid exposure to irritating fumes, the phosphorus pentachloride and oxychloride are handled in the hood. The heating under reflux is also carried out in the hood.*

2. The reaction becomes very exothermic and if uncontrolled there is serious flooding of the condenser. The reaction mixture also becomes dark red to black when the reaction temperature is not controlled.

3. It is difficult to transfer all the oil before solidification starts in the flask. The residual solid is removed by melting on the steam bath and pouring on ice. By repeating this procedure several times, almost all the product can be removed from the flask. Any remaining solid is removed by adding cold water to the flask, breaking the solid with a spatula, and swirling the mixture out of the flask.

4. It is important to stir as rapidly as possible to break the product

into small pieces. The beaker should be secured by a large clamp. If the stirring is too slow or the oil is poured too fast, large clumps result which do not solidify completely.

5. Most of the acidic impurities are removed in this step.

6. The sodium carbonate retains any residual moisture and acidic impurities. After the extraction, a brown, gummy, hygroscopic mass remains in the thimble.

7. The checkers found a sintered-glass filter funnel suitable.

8. This product may be recrystallized from ligroin-acetone with 80–85% recovery. Analysis of the product after such recrystallization gave the following analytical results: Calcd. for $C_6H_3ClN_2$: C, 51.98; H, 2.16; Cl, 25.63; N, 20.21. Found: C, 52.41; H, 2.39; Cl, 25.43; N, 19.60.

3. Methods of Preparation

The preparation described here is a modification of that reported by Taylor and Crovetti.[3] 2-Chloronicotinonitrile has also been prepared by the dehydration of 2-chloronicotinamide [4] and by the Sandmeyer reaction on 3-amino-2-chloropyridine.[5]

[1] Princeton University, Princeton, New Jersey.
[2] University of Illinois, Urbana, Illinois.
[3] Taylor and Crovetti, *J. Org. Chem.*, **19**, 1633 (1954).
[4] Späth and Koller, *Ber.*, **56**, 880 (1923).
[5] von Schickh, Binz, and Schulz, *Ber.*, **69**, 2593 (1936).

α-CHLOROPHENYLACETIC ACID

(Acetic acid, chlorophenyl-)

$$C_6H_5CHOHCO_2H + C_2H_5OH \xrightarrow{HCl} C_6H_5CHOHCO_2C_2H_5$$

$$C_6H_5CHOHCO_2C_2H_5 + SOCl_2 \rightarrow$$
$$C_6H_5CHClCO_2C_2H_5 + HCl + SO_2$$

$$C_6H_5CHClCO_2C_2H_5 + CH_3CO_2H \xrightarrow{HCl}$$
$$C_6H_5CHClCO_2H + CH_3CO_2C_2H_5$$

Submitted by Ernest L. Eliel, Milton T. Fisk, and Thomas Prosser.[1]
Checked by James Cason, Lois J. Durham, and Gerhard J. Fonken.

1. Procedure

A. *Ethyl mandelate.* To 152 g. (1.0 mole) of mandelic acid and 200 ml. of absolute ethanol in a 1-l. round-bottomed flask equipped with a reflux condenser, there is added 100 ml. of absolute ethanol containing about 10 g. of anhydrous hydrogen chloride (Note 1). The solution is heated under reflux on a steam bath for 5 hours, then poured into 1 l. of ice water in a 3-l. beaker (Note 2). A saturated aqueous solution of sodium bicarbonate is added until the mixture is faintly alkaline (Note 3). It is then extracted with two 300-ml. portions of ether in a 2-l. separatory funnel. The ether extracts are washed with a 200-ml. portion of water and dried over 50 g. of anhydrous sodium sulfate. The dried ether solution is concentrated by distillation from a 250-ml. Claisen flask, and the residue is distilled at reduced pressure. There is obtained 147–154 g. (82–86%) of ethyl mandelate, b.p. 144–145°/16 mm. The ester may crystallize upon standing for a prolonged period. It melts at 30.5–31.5°.

B. *Ethyl α-chlorophenylacetate.* Ethyl mandelate (135 g., 0.75 mole) is dissolved in 98 g. (59 ml., 0.82 mole) of thionyl chloride (Note 4) contained in a 500-ml. round-bottomed flask equipped with a reflux condenser capped with a drying tube. The apparatus is allowed to stand in a hood overnight (about 16 hours), at the end of which time the solution is heated under reflux for 30 minutes on a steam bath. The solution is then poured into 750 ml. of ice water contained in a 2-l. separatory funnel (Note 5), and the mixture is extracted with two 300-ml. portions of ether. The combined ether extracts are washed with two 250-ml. portions of saturated aqueous sodium bicarbonate solution and one 250-ml. portion of water. The washed extracts are dried over 45 g. of anhydrous sodium sulfate and concentrated by distillation. The residue is distilled from a 250-ml. Claisen flask at reduced pressure. The yield of ethyl α-chlorophenylacetate is 121–127 g. (81–85%), b.p. 134–136°/15 mm., n_D^{20} 1.5149.

C. *α-Chlorophenylacetic acid.* A solution of 119 g. (0.6 mole) of ethyl α-chlorophenylacetate in 238 ml. of glacial acetic acid and 119 ml. of concentrated hydrochloric acid, contained in a 1-l. round-bottomed flask, is heated under reflux in a hood for 1.5 hours (Note 6). At the end of the heating period the solution is concentrated by heating in an oil bath at 100° at reduced pressure (15–20 mm.) until no further material is distilled (Note 7). The residue is allowed to cool to room temperature and poured slowly, with stirring, into 1-l. of ice-cold saturated sodium bicarbonate solution contained in a 2-l. beaker. Solid sodium bicarbonate is added in small portions until the solution becomes neutral to universal indicator paper (Note 8). The solution is then extracted with two 200-ml. portions of ether in a 2-l. separatory funnel (Notes 9, 10). The aqueous phase is placed in a 3-l. beaker and acidified cautiously with ice-cold 12N sulfuric acid until the mixture is acid to Congo red paper (Note 11). The oily suspension is extracted with two 200-ml. portions of ether in a 2-l. separatory funnel. The ether extracts are washed with two 100-ml. portions of water and dried over 45 g. of anhydrous sodium sulfate. The dried ether extract is transferred to a 1-l. Erlenmeyer flask and concentrated on a steam bath until ether is no longer distilled. To the residue there is added 500 ml. of warm (50–60°) concentrated hydrochloric acid (in a hood), and the suspension is allowed to cool with occasional swirling (Note 12). Crystallization is completed by chilling in ice, and the product is collected on a sintered-glass funnel. After the product has been dried as much as possible on the funnel it is dried to constant weight in a vacuum desiccator over solid potassium hydroxide. The yield of dry acid is 82–84 g. (80–82%), m.p. 77.5–79.5°. It is satisfactory for most purposes. If very pure material is desired the acid may be recrystallized from three volumes of hexane to give material, m.p. 78.5–79.5°, in 90–95% recovery.

2. Notes

1. The hydrogen chloride may be generated by the method described in *Organic Syntheses*, Coll. Vol. **1**, 293 (1941). It should be dried by passage through concentrated sulfuric acid. The hydrogen chloride may be taken directly from a cylinder, if one is available.

2. The flask should be rinsed with a small portion of water.

3. The amount of bicarbonate solution varies, depending on the weight of hydrogen chloride used. There is usually required 250–350 ml.

4. The submitters used Matheson thionyl chloride without further purification.

5. The flask should be rinsed with a small portion of ether.

6. The solution becomes homogeneous after a few minutes of heating.

7. The checkers found that about 250–270 ml. of distillate is obtained under these conditions.

8. About 90–130 g. of sodium bicarbonate is usually required.

9. When the combined ether layers are dried over sodium sulfate and concentrated, and the residue is distilled at reduced pressure, about 11 g. (10%) of unchanged ethyl α-chlorophenylacetate, b.p. 134–137°/15 mm., is recovered.

10. The solution should be kept ice-cold during the extraction, which must be carried out rapidly lest hydrolysis of the chlorophenylacetic acid to mandelic acid take place.

11. About 70 ml. of acid is required. A 25-ml. excess of $12N$ sulfuric acid is recommended in order to ensure the acidity of the mixture.

12. The objective of the hydrochloric acid treatment is to remove small amounts of mandelic acid from the product. If seed crystals are available, the slurry should be seeded as it is cooled.

3. Methods of Preparation

α-Chlorophenylacetic acid has been prepared from mandelonitrile and hydrochloric acid in a sealed tube,[2,3] from mandelic acid and hydrochloric acid in a sealed tube,[4] from α-nitrostyrene and hydrochloric acid in a sealed tube,[5] from phenylglycine, hydrochloric acid, and sodium nitrite,[6] from mandelic acid and phosphorus pentachloride (to give the acid chloride which is then hydrolyzed),[7] and, in poor yield, from mandelic acid and thionyl chloride.[8] In the method described, ethyl mandelate is prepared according to Fischer and Speier.[9] The conversion to the chloroester and the acid hydrolysis step are modifications of a preparation described by McKenzie and Barrow.[8]

[1] University of Notre Dame, Notre Dame, Indiana.
[2] Spiegel, *Ber.*, **14**, 239 (1881).
[3] Meyer, *Ann.*, **220**, 41 (1883).
[4] Radziszewski, *Ber.*, **2**, 208 (1869).
[5] Priebs, *Ann.*, **225**, 336 (1884).
[6] Jochem, *Z. physiol. Chem.*, **31**, 123 (1900).
[7] Bischoff and Walden, *Ann.*, **279**, 122 (1894).
[8] McKenzie and Barrow, *J. Chem. Soc.*, **99**, 1916 (1911).
[9] Fischer and Speier, *Ber.*, **28**, 3252 (1895).

o-CHLOROPHENYLCYANAMIDE

(Carbanilonitrile, o-chloro-)

$$+ (CH_3CO_2)_2Pb \rightarrow$$

$$+ PbS + 2CH_3CO_2H$$

Submitted by FREDERICK KURZER.[1]
Checked by CLIFF S. HAMILTON and DAVID B. CAPPS.

1. Procedure

To a suspension of 37.4 g. (0.2 mole) of o-chlorophenylthiourea (p. 180) in 300 ml. of water at 100°, contained in a 3-l. beaker, is added a boiling solution of 132 g. (2 moles) of 85% potassium hydroxide in 300 ml. of water. The resulting solution is immediately treated with a hot saturated solution of 83.5 g. (0.22 mole) of lead acetate trihydrate, added as rapidly as possible and with good stirring (Note 1). The reaction mixture, from which large quantities of lead sulfide separate instantly, is boiled for 6 minutes and cooled to 0°, and the lead sulfide is filtered with suction by means of a large Büchner funnel (Note 2). The colorless filtrate is acidified at 0–5° (Note 3) by the slow addition with stirring of 120–140 ml. of glacial acetic acid. The white crystalline precipitate of nearly pure *o*-chlorophenylcyanamide which separates is collected by filtration on a suction filter and is washed with six 150-ml. portions of ice water. The crystalline mass of white plates is filtered, drained, and dried (Note 4). The yield of product, melting at 100–104°, is 26–28 g. (85–92%).

Recrystallization from benzene-light petroleum ether (8 ml. and 4 ml. respectively per gram of the dried precipitated material) gives lustrous needles of *o*-chlorophenylcyanamide, m.p. 105–106° (60–70% recovery) (Note 5). The above method is generally applicable and affords an excellent route to arylcyanamides (Note 6).

2. Notes

1. The exothermic reaction may cause the contents of the beaker to froth vigorously if the lead acetate solution is added too quickly. The reaction is readily controlled by adding the liquid in a rapid, thin stream with good stirring.

2. The use of a hardened filter paper or two thicknesses of ordinary filter paper is recommended. The black filter cake of lead sulfide may be extracted once again with 100 ml. of boiling 4% potassium hydroxide solution, and the extracts combined with the main filtrate.

3. The temperature of the solution is kept below 5° by the addition of suitable quantities of clean ice.

4. The product is sufficiently pure for use in further syntheses.

5. In benzene solution or in the solid state, *o*-chlorophenylcyanamide does not polymerize on storage for several months at 20–30°.

6. The desulfurization of arylthioureas can be used generally for preparing arylcyanamides in excellent yields. The submitter

reports that α-naphthylcyanamide, melting at 124–128°, can be prepared in yields of 77–90% by this method.

3. Methods of Preparation

o-Chlorophenylcyanamide has been prepared by the action of lead acetate on o-chlorophenylthiourea.[2] The method is based on the analogous preparation of phenylcyanamide.[3]

[1] University of London, London, England.
[2] Kurzer, *J. Chem. Soc.*, **1949**, 3033.
[3] Rathke, *Ber.*, **12**, 772 (1879); Krall et al., *J. Indian Chem. Soc.*, **19**, 343 (1942); **23**, 373 (1946).

α-(4-CHLOROPHENYL)-γ-PHENYLACETOACETONITRILE

[Acetoacetonitrile, 2-(p-chlorophenyl)-4-phenyl-]

$$p\text{-}ClC_6H_4CH_2CN + C_6H_5CH_2CO_2C_2H_5 \xrightarrow{NaOC_2H_5}$$
$$[p\text{-}ClC_6H_4C(CN)COCH_2C_6H_5]^{\ominus}Na^{\oplus}$$

$$[p\text{-}ClC_6H_4C(CN)COCH_2C_6H_5]^{\ominus}Na^{\oplus} + HCl \rightarrow$$
$$p\text{-}ClC_6H_4CH(CN)COCH_2C_6H_5 + NaCl$$

Submitted by STEPHEN B. COAN and ERNEST I. BECKER.[1]
Checked by CHARLES C. PRICE and G. VENKAT RAO.

1. Procedure

A solution of 11.5 g. (0.5 g. atom) of sodium is prepared in 150 ml. of absolute ethanol (Note 1) in a 500-ml. three-necked flask equipped with a stirrer, a condenser, and a dropping funnel. While this solution is refluxing with stirring, a mixture of 37.8 g. (0.25 mole) of 4-chlorophenylacetonitrile and 50.8 g. (0.31 mole) of ethyl phenylacetate is added through the dropping funnel over a period of 1 hour. The solution is refluxed for 3 hours, cooled, and poured into 600 ml. of cold water. The aqueous alkaline mixture is extracted three times with 200-ml. portions of ether and the ether extracts discarded. The aqueous solution is acidified with cold 10% hydrochloric acid and extracted three times with 200-ml. portions of ether. The ether solution is then extracted once with 100 ml. of water, twice with 100 ml. each of

10% sodium bicarbonate solution, and once with 100 ml. of water, the aqueous extracts being discarded in turn. The organic phase is dried with anhydrous sodium sulfate, filtered through a fluted filter, and the ether removed by distillation. The yield of α-(4-chlorophenyl)-γ-phenylacetoacetonitrile is 58–62 g. (86–92%), m.p. 128.5–130.0°. For many purposes, this crude product may be used without further purification. If a purer product is desired, however, it may be recrystallized from methanol or aqueous methanol (Note 2) to yield 50–55 g. (74–82%), m.p. 131.0–131.2° (Note 3).

2. Notes

1. Comparable results were obtained when fresh commercial grade (2-B) anhydrous ethanol was used.

2. For recrystallization, a solution of the nitrile in 3–4 ml. of hot methanol per gram was treated with Darco G-60 and filtered on a fluted paper. Distilled water was added to the hot solution until incipient crystallization was observed. In general the final solvent was approximately 90% methanol.

3. The above method has been used in the preparation of other ring-substituted diphenylacetoacetonitriles. The method is equally successful when applied to the condensation of phenylacetonitrile with the ethyl ester of a 4-substituted phenylacetic acid. The table summarizes the results reported by the submitters.

DIPHENYLACETOACETONITRILES

Acetoacetonitrile	Yield, %	M.P.
From Phenylacetonitrile and an Ethyl 4-Substituted Phenylacetate		
α-Phenyl-γ-phenyl-	82	79.4–80.0°
α-Phenyl-γ-(4-methylphenyl)-	84	88.0–89.0°
α-Phenyl-γ-(4-methoxyphenyl)-	81	69.5–70.4°
α-Phenyl-γ-(4-bromophenyl)-	80	94.0–95.0°
α-Phenyl-γ-(4-methylthiophenyl)-	85	85.0–85.2°
From 4-Substituted Phenylacetonitrile and Ethyl Phenylacetate		
α-(4-Fluorophenyl)-γ-phenyl-	75	111.8–112.0°

3. Methods of Preparation

The above procedure is a modification of that described by Walther and Hirschberg.[2] α,γ-Diphenylacetoacetonitrile has been prepared by the condensation of phenylacetonitrile with ethyl phenylacetate in the presence of sodium ethoxide without solvent.[3]

[1] Polytechnic Institute of Brooklyn, Brooklyn 2, New York.
[2] Walther and Hirschberg, *J. prakt. Chem.*, [2] **67**, 377 (1903).
[3] Walther and Schickler, *J. prakt. Chem.*, [2] **55**, 305 (1897).

1-(p-CHLOROPHENYL)-3-PHENYL-2-PROPANONE

[2-Propanone, 1-(p-chlorophenyl)-3-phenyl-]

$$p\text{-ClC}_6\text{H}_4\text{CH(CN)COCH}_2\text{C}_6\text{H}_5 \xrightarrow[\text{H}_2\text{SO}_4]{\text{H}_2\text{O}}$$

$$p\text{-ClC}_6\text{H}_4\text{CH}_2\text{COCH}_2\text{C}_6\text{H}_5$$

Submitted by STEPHEN B. COAN and ERNEST I. BECKER.[1]
Checked by CHARLES C. PRICE and G. VENKAT RAO.

1. Procedure

In a 250-ml. three-necked flask equipped with a stirrer and a condenser are placed 75 ml. of 60% sulfuric acid and 25 g. (0.093 mole) of α-(4-chlorophenyl)-γ-phenylacetoacetonitrile (p. 174). While being stirred, the mixture is heated at reflux until the evolution of carbon dioxide ceases (Notes 1 and 2). The mixture is cooled, poured into 200 ml. of ice water, and extracted three times with 150-ml. portions of ether. The ether solution is washed once with 50 ml. of water, twice with 100-ml. portions of 10% sodium hydroxide solution, and then with 50 ml. of water. After drying over sodium sulfate and filtering, the ether is distilled on a steam bath, affording 15–17 g. (66–75%) of 1-(p-chlorophenyl)-3-phenyl-2-propanone, m.p. 34.5–35.5°. Recrystallization from 160 ml. of petroleum ether (b.p. 40–60°) gives 12–13 g. (53–57%) of product, m.p. 35.9–36.5° (Note 3).

2. Notes

1. The hydrolysis and decarboxylation of the nitrile require from 18 to 24 hours.

2. The evolution of carbon dioxide is conveniently observed by passing the effluent gases through a saturated solution of barium hydroxide.

3. The submitters report that a similar procedure has been used in the preparation of other monosubstituted dibenzyl ketones from α-phenyl-γ-(4-substituted phenyl)acetoacetonitriles.

Product	Yield, %	M.P.
1,3-Diphenyl-2-propanone	65–71	34.2–34.5°
1-(*p*-Tolyl)-3-phenyl-2-propanone	66	30.8–31.2°
1-(*p*-Bromophenyl)-3-phenyl-2-propanone	50	53.8–54.2°
1-(*p*-Fluorophenyl)-3-phenyl-2-propanone	50	36.0–36.5°
1-(*p*-Methylthiophenyl)-3-phenyl-2-propanone	40	43.9–44.2°
1-(*p*-Methoxyphenyl)-3-phenyl-2-propanone [a]	19	46.6–47.4°

[a] It is necessary to employ 5 ml. of glacial acetic acid and 5 ml. of 20% aqueous hydrochloric acid, instead of the sulfuric acid, per gram of keto-nitrile as reaction solvent in order to obtain the desired product.

3. Methods of Preparation

This method of preparation of 1-(*p*-chlorophenyl)-3-phenyl-2-propanone has been reported by Coan and Becker.[2] The method utilized is a modification of that described for the formation of 1,3-diphenyl-2-propanone from α,γ-diphenylaceto-acetonitrile.[3, 4]

[1] Polytechnic Institute of Brooklyn, Brooklyn 2, New York.
[2] Coan and Becker, *J. Am. Chem. Soc.*, **76**, 501 (1954).
[3] Meyer, *J. prakt. Chem.*, [2] **52**, 81 (1895).
[4] Walther and Schickler, *J. prakt. Chem.*, [2] **55**, 305 (1897).

p-CHLOROPHENYL SALICYLATE

(Salicylic acid, p-chlorophenyl ester)

$$3 \; o\text{-HOC}_6\text{H}_4\text{CO}_2\text{H} + 3 \; p\text{-HOC}_6\text{H}_4\text{Cl} + \text{POCl}_3 \rightarrow$$

$$3 \underset{\text{OH}}{\bigcirc}\!\!-\text{CO}_2\!\!-\!\!\bigcirc\!\!\text{Cl} + \text{H}_3\text{PO}_4 + 3\text{HCl}$$

Submitted by Norman G. Gaylord and P. M. Kamath.[1]
Checked by R. T. Arnold and J. John Brezinski.

1. Procedure

To a mixture of 138.1 g. (1 mole) of salicyclic acid and 128.6 g. (1 mole) of p-chlorophenol in a 2-l. round-bottomed flask fitted with a thermometer reaching to the bottom of the flask and a reflux condenser with a drying tube (Note 1) is added 58.3 g. (0.38 mole) of phosphorus oxychloride. The mixture is heated with occasional swirling, and the temperature is maintained at 75–80°. At the end of 4 hours the reactants have been reduced to a molten mass, and this is poured slowly, with vigorous stirring, into a solution of 120 g. of sodium carbonate in 800 ml. of water. The precipitated ester is collected on a filter and washed with four 200-ml. portions of water. The yield of crude, air-dried, p-chloro-phenyl salicylate is 174–189 g. (70–76%); m.p. 65–66°. Recrystallization from absolute ethanol yields 136–154 g. (55–62%) of pure product; m.p. 69.5–70.5°. A second crop may be obtained by concentration of the filtrate from the first crop or by the addition of water (Note 2).

2. Notes

1. It is advisable to attach the drying tube to a water trap [2] in order to prevent the escape of hydrogen chloride into the atmosphere.
2. The above method has been used in the preparation of other substituted phenyl salicylates to give the following yields of recrystallized products with the indicated melting points.

	Yield, (%)	M.P.
o-Chlorophenyl ester	67	52.0– 52.4°
p-Nitrophenyl ester	53	151.5–151.9°
p-tert-Butylphenyl ester	42	63.0– 63.4°
p-Phenylphenyl ester	66	109.2–109.6°
o-Phenylphenyl ester	70	89.2– 89.5°

These esters, with the exception of the *p*-nitrophenyl derivative, can be recrystallized from absolute ethanol. The nitro compound is recrystallized from dioxane.

3. Methods of Preparation

Substituted phenyl salicylates can be prepared by heating salicylic acid and the appropriate phenol in the presence of phosphorus oxychloride,[3–5] phosphorus trichloride,[4, 5] phosphorus pentachloride,[4, 6] phosgene,[4] or thionyl chloride,[4] or by heating the phenol and salol.[7]

[1] Polytechnic Institute of Brooklyn, Brooklyn, New York.

[2] *Org. Syntheses Coll. Vol.* **1**, 97 (1941).

[3] Seifert, *J. prakt. Chem.*, [2] **31**, 472 (1885); Nencki and Heyden, Ger. pats. 38,973 [*Ber.*, **20R**, 351 (1887)] and 43,713 [*Ber.*, **21R**, 554 (1888)]; Walther and Zipper, *J. prakt. Chem.*, [2] **91**, 399 (1915); Krauz and Remenec, *Collection Czechoslov. Chem. Communs.*, **1**, 610 (1929) [*C. A.*, **24**, 1365 (1930)]; Kolloff and Page, *J. Am. Chem. Soc.*, **60**, 948 (1938).

[4] Nencki and Heyden, Ger. pat. 70,519 [*Ber.*, **26R**, 967 (1893)].

[5] Harris and Christiansen, *J. Am. Pharm. Assoc.*, **24**, 553 (1935); U. S. pat. 2,141,172 [*C. A.*, **33**, 2655 (1939)].

[6] Tozer and Smiles, *J. Chem. Soc.*, **1938**, 1897.

[7] Cohn, *J. prakt. Chem.*, [2] **61**, 550 (1900).

o-CHLOROPHENYLTHIOUREA

[Urea, 1-(o-chlorophenyl)-2-thio-]

$$Cl-C_6H_4-NH_2 + HCl + NH_4SCN \rightarrow Cl-C_6H_4-NHCSNH_2 + NH_4Cl$$

Submitted by Frederick Kurzer.[1]
Checked by Cliff S. Hamilton and David B. Capps.

1. Procedure

To a suspension of 38.3 g. (31.5 ml., 0.30 mole) of o-chloro-aniline in 300 ml. of warm water is added, with stirring, 27.5 ml. (0.33 mole) of concentrated hydrochloric acid (12N). The resulting solution is placed in a 500-ml. porcelain evaporating dish, 25 g. (0.33 mole) of ammonium thiocyanate is added (Note 1), and the mixture is heated on the steam bath for 1 hour (Note 2). The liquid, from which a mass of large needles of o-chloro-aniline thiocyanate separates, is allowed to cool, set aside at room temperature for 1 hour (Note 3), and then evaporated slowly to dryness over a period of 2–3 hours. The crystalline residue is crushed finely, 300 ml. of water is added, and again the mixture is evaporated slowly. The dry grayish white residual powder is heated finally on the steam bath for 4–5 hours.

The resulting mixture of crude o-chlorophenylthiourea and ammonium chloride (58–62 g.) is powdered finely and suspended in 300 ml. of water. The mixture is warmed slowly to 70° with mechanical stirring, then allowed to cool to 35°, and the solid is filtered with suction. The yield of crude o-chlorophenylthio-urea, melting at 140–144°, is 30–35 g. (54–63%).

The crude material is dissolved in 60 ml. of absolute ethanol, the solution boiled with decolorizing carbon for a few minutes, and the clear, nearly colorless filtrate (Note 4) diluted with 100 ml. of hot benzene and 20 ml. of light petroleum ether (b.p. 60–80°). The white crystalline mass of o-chlorophenylthiourea, which separates gradually on cooling and standing, is separated by filtration under reduced pressure, washed with light petroleum ether, and dried, m.p. 144–146°. The yield of purified material is 20–24 g. (36–43%). Evaporation of the mother liquors and crystallization of the residue from a proportionally smaller volume of solvents yields a second crop (6–8 g.) (Note 5).

2. Notes

1. A good commercial grade of o-chloroaniline and pure ammonium thiocyanate are satisfactory for this preparation.

2. Comparable results are obtained when three times the quantities specified are used.

3. Uninterrupted evaporation of the initial reaction mixture sometimes tends to give a partly oily product from which only smaller yields can be obtained.

4. The filtration is best effected by the use of reduced pressure employing a preheated Büchner funnel and filter flask.

5. According to the submitter the method is generally applicable to the synthesis of aromatic thioureas. For example, phenylthiourea may be prepared in yields of 37–42%.

3. Methods of Preparation

o-Chlorophenylthiourea has been prepared from o-chlorophenylisothiocyanate and ammonia,[2] by the interaction of o-chloroaniline hydrochloride[3] or sulfate[4] with sodium thiocyanate in chlorobenzene, or with ammonium thiocyanate in aqueous solution.[4, 5]

[1] University of London, London, England.
[2] Dyson and George, J. Chem. Soc., 125, 1705 (1924).
[3] Dalgliesh and Mann, J. Chem. Soc., 1945, 900.
[4] Elderfield and Short, J. Org. Chem., 18, 1092 (1953).
[5] Kurzer, J. Chem. Soc., 1949, 3033.

2-CHLOROPYRIMIDINE

(Pyrimidine, 2-chloro-)

$$\underset{\text{NH}_2}{\underset{\big|}{\left(\text{N} \diagdown \diagup \text{N}\right)}} \xrightarrow[\text{(2) NaOH}]{\text{(1) HNO}_2,\text{ HCl}} \underset{\text{Cl}}{\underset{\big|}{\left(\text{N} \diagdown \diagup \text{N}\right)}}$$

Submitted by IRVING C. KOGON, RONALD MININ, and C. G. OVERBERGER.[1]
Checked by CHARLES C. PRICE and T. L. V. ULBRICHT.

1. Procedure

Caution! This procedure should be carried out in a good hood.

In a 3-l. three-necked round-bottomed flask fitted with a stirrer and a low-temperature thermometer is placed 500 ml. of concentrated hydrochloric acid (6.0 moles), and the solution is cooled to 0°. To the cooled solution, 142 g. (1.5 moles) of 2-aminopyrimidine (Note 1) is added portionwise with stirring until a homogeneous solution is obtained. The solution is cooled to −15° (Note 2), and a 500-ml. dropping funnel is fitted to the flask. A cold solution of 207 g. (3.0 moles) of sodium nitrite in 375 ml. of water is then added dropwise with stirring over a period of 55 minutes, the reaction temperature being maintained at −15° to −10° (Note 3). The solution is stirred an additional hour, and the temperature is allowed to rise to −5°. The mixture is then carefully neutralized to about pH 7 with a 30% solution of sodium hydroxide (about 3.0 moles), care being taken not to allow the temperature to rise above 0° (Note 4). The solid which forms, consisting of 2-chloropyrimidine and sodium chloride, is collected by filtration and washed thoroughly with ether to dissolve all the 2-chloropyrimidine. The cold solution is extracted with four 75-ml. portions of ether (Note 5). The combined extracts are dried over anhydrous sodium sulfate, the solvent is removed, and the residue is recrystallized from isopentane to give white crystals of 2-chloropyrimidine. The yield is 44–46 g. (26–27%), m.p. 64.5–65.5°.

2. Notes

1. Purchased from the Matheson, Coleman and Bell Company, Norwood, Ohio.
2. Cooling below $-15°$ causes the mixture to solidify.
3. Care should be exercised since at this point nitrogen oxides are being evolved. Addition should be started cautiously, as there tends to be a rapid initial rise in temperature.
4. Yields are appreciably reduced if the temperature is allowed to rise above $0°$.
5. Filtration and extraction should be performed immediately or extensive decomposition occurs.

3. Methods of Preparation

2-Chloropyrimidine has been prepared by Howard[2] and by Sperber, Papa, Schwenk, Sherlock, and Fricano[3] by a similar procedure. The compound also has been obtained from 2-hydroxypyrimidine hydrochloride by treatment with a mixture of phosphorus pentachloride and phosphorus oxychloride[4] or by treatment with phosphorus oxychloride alone.[5] The present procedure has been published.[6]

[1] Polytechnic Institute of Brooklyn, Brooklyn 2, New York.

[2] Howard, U. S. pat. 2,477,409 [C. A., **43**, 8105 (1949)].

[3] Sperber, Papa, Schwenk, Sherlock, and Fricano, J. Am. Chem. Soc., **73**, 5752 (1951).

[4] Matsukawa and Ohta, J. Pharm. Soc. Japan, **69**, 491 (1949) [C. A., **44**, 3456 (1950)].

[5] Copenhaver and Kleinschmidt, Brit. pat. 663,302 [C. A., **46**, 10212 (1952)].

[6] Overberger and Kogon, J. Am. Chem. Soc., **76**, 1065 (1954).

2-CHLORO-1,1,2-TRIFLUOROETHYL ETHYL ETHER *

(Ether, 2-chloro-1,1,2-trifluoroethyl ethyl)

$$CFCl{=}CF_2 + C_2H_5OH \xrightarrow{C_2H_5ONa} HCFClCF_2OC_2H_5$$

Submitted by BRUCE ENGLUND.[1]
Checked by R. S. SCHREIBER and BURRIS D. TIFFANY.

1. Procedure

The apparatus used is shown in Fig. 6 (Note 1). A solution of sodium ethoxide prepared by dissolving 2.5 g. (0.11 g. atom) of clean sodium in 230 g. (292 ml., 5 moles) of absolute ethanol under anhydrous conditions is added to the reaction tube C.

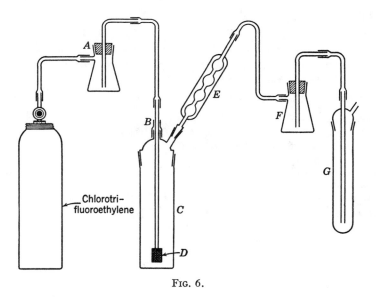

Chlorotri-
fluoroethylene

FIG. 6.

Tube C thus charged is weighed and placed in position with the gas inlet tube, fitted with fritted-glass dispersion cylinder D, extending nearly to the bottom. Several inches of ethanol is placed in tube G, which serves as a flow indicator. The traps A and F are provided to protect against suck-back if the gas flow is interrupted. The seal at B is a sleeve of rubber tubing. Any efficient reflux condenser E is satisfactory.

Chlorotrifluoroethylene (Note 2) is introduced at such a rate, controlled by the needle valve, that it is essentially all absorbed in the reaction tube C, as indicated by the escape of little or no gas through tube G. At this rate, 233 g. (2 moles) (Note 3) is absorbed in 2–2.5 hours, during which the temperature rises to the point of reflux (Note 4).

When the required amount of chlorotrifluoroethylene has been absorbed, the reaction mixture is poured into 500 ml. of water. The product separates as a light-yellow oil, which is separated, washed with 250 ml. of water, and dried over 20 g. of anhydrous calcium chloride. From 233 g. of chlorotrifluoroethylene (2 moles), the yield of crude product is 300–315 g. (92–97%). This material is suitable for use in preparation of ethyl chlorofluoroacetate (p. 423). It may be fractionated through an efficient column to give 285–300 g. (88–92%) of pure chlorotrifluoroethyl ethyl ether, b.p. 87–88°, n_D^{25} 1.3427.

2. Notes

1. The reaction tube C may be of any convenient size. The tube used by the submitter was 300 by 55 mm., fitted with a 55/35 standard taper ground-glass joint. The checkers used a 500-ml. calibrated dropping funnel which was convenient for following the increase in volume during the absorption of the chlorotrifluoroethylene. For safety, a heavy grease such as Cello-Grease was used in the stopcock.

2. The chlorotrifluoroethylene used was inhibited polymerization grade, supplied by the Kinetic Chemicals Division, Organic Chemicals Department, E. I. du Pont de Nemours and Company, Wilmington, Delaware.

3. The amount of chlorotrifluoroethylene absorbed is determined by weighing the reaction tube C and contents. A convenient alternative is to note the increase in volume of the reaction mixture, which in this case amounts to 165 ml.

4. The rate of absorption is slow at first but increases as the temperature rises.

3. Methods of Preparation

2-Chloro-1,1,2-trifluoroethyl ethyl ether has been prepared by the base-catalyzed addition of ethanol to 1-chloro-1,2,2-trifluoroethylene.[2-4]

[1] Chemical Department, Experimental Station, E. I. du Pont de Nemours and Company, Wilmington, Delaware.

[2] Hanford and Rigby, U. S. pat. 2,409,274 [*C. A.*, **41**, 982 (1947)].

[3] Park, Vail, Lea, and Lacher, *J. Am. Chem. Soc.*, **70**, 1550 (1948).

[4] Barr, Rapp, Pruett, Bahner, Gibson, and Lafferty, *J. Am. Chem. Soc.*, **72**, 4480 (1950).

β-CHLOROVINYL ISOAMYL KETONE

(1-Hepten-3-one, 1-chloro-6-methyl-)

$$(CH_3)_2CHCH_2CH_2COCl + HC{\equiv}CH \xrightarrow{AlCl_3}$$
$$(CH_3)_2CHCH_2CH_2COCH{=}CHCl$$

Submitted by CHARLES C. PRICE and JOSEPH A. PAPPALARDO.[1]
Checked by R. S. SCHREIBER, WM. BRADLEY REID, JR., and R. W. JACKSON.

1. Procedure

Caution! This preparation should be carried out in a good hood.

A 1-l. three-necked flask fitted with a mercury-sealed stirrer, a gas inlet tube, and a gas outlet tube protected by a calcium chloride drying tower (Note 1) is surrounded by an ice-water bath. The system is flushed slowly with acetylene for 3 minutes (Note 2). Carbon tetrachloride (260 g.) (Note 3) is poured into the flask, and acetylene is bubbled through at a saturation rate for 3 minutes. Aluminum chloride (98 g., 0.74 mole) (Note 4) is added, and acetylene is bubbled continuously through the mixture with stirring for 5 minutes.

The gas inlet tube is replaced by a dropping funnel protected by a calcium chloride drying tube. Isocaproyl chloride (84.7 g., 0.63 mole) (Note 5) is added to the reaction mixture with stirring over a period of 20 minutes. The dropping funnel is replaced by the gas inlet tube (inlet tube wiped with a dry towel), and, with stirring, acetylene is bubbled just below the surface of the mixture at a rate noticeably above the saturation rate. After 15 minutes to 1 hour the rate of absorption of acetylene suddenly becomes very rapid, and the acetylene is passed through as rapidly as it is absorbed (Note 6). The introduction of acetylene is continued for 30 minutes after the rapid absorption (which lasts 1–2 hours) has subsided.

The reaction mixture is poured with stirring onto a mixture of 700 g. of crushed ice and 300 ml. of a saturated solution of sodium chloride. The organic layer is separated, and the aqueous layer is extracted with three 100-ml. portions of ether. To the combined organic layers is added 2 g. of hydroquinone, and the mixture is dried over calcium chloride (Note 7).

The liquid is decanted from the solid, and the solid is washed with 50 ml. of carbon tetrachloride. The carbon tetrachloride layers are combined and 2 g. of hydroquinone is added. The solution is then distilled from a Claisen flask at such a pressure as to maintain a distillation temperature of about 30°. When most of the solvent has been removed it is discarded, and the residue is distilled as rapidly as possible until the temperature reaches about 90° at 5 mm. Redistillation from a Claisen flask with a short Vigreux side arm gives a colorless liquid; yield 55–65 g. (54–64%) (Note 8); b.p. 96–98°/20 mm.; n_D^{25} 1.4619 (Note 9).

2. Notes

1. A rubber tube with an eye-dropper is attached to the tower. The tip of the eye-dropper is immersed in mineral oil to indicate the absorption of acetylene.

2. The reaction is carried out in a well-ventilated hood to remove excess acetylene. All subsequent operations are also carried out in the hood because the product of the reaction has an objectionable odor. The acetylene used is passed through a train consisting of an empty 300-ml. bottle, a 300-ml. bottle containing 150 ml. of concentrated sulfuric acid (through which acetylene passes by means of a gas dispersion tube), a pressure-release valve, and another empty 300-ml. bottle.

3. Carbon tetrachloride (c.p. grade) decanted from calcium chloride was used.

4. Baker and Adamson Company powdered aluminum chloride was used.

5. Eastman Kodak Company white label grade isocaproyl chloride was redistilled, and the middle 80% portion was used.

6. To keep the temperature in the immediate vicinity of the flask from rising, the ice bath should be stirred occasionally.

7. The combined organic layers should be allowed to stand in a separatory funnel about 30 minutes before drying because some additional water separates on standing.

8. The checkers consistently obtained yields of 75–76%.

9. This preparation works equally well for the isobutyl and isohexyl homologs. For β-chlorovinyl methyl ketone, which is both a lachrymator and a vesicant, the initial rapid rate of absorption of acetylene begins only after 3–9 hours.

3. Methods of Preparation

Carpmael,[2] Yakubovich and Merkulova,[3] and Price and Pappalardo [4] have used this general reaction to make β-chlorovinyl ketones from a variety of acid chlorides, employing various solvents and acidic catalysts. Bayer and Nelles [5] made β-chlorovinyl ketones by the reaction of an acid chloride and vinyl chloride.

[1] University of Notre Dame, Notre Dame, Indiana.

[2] Carpmael, Brit. pat. 461,080 [C. A., **31**, 4676 (1937)].

[3] Yakubovich and Merkulova, J. Gen. Chem. U.S.S.R., **16**, 55 (1946) [C. A., **41**, 91 (1947)].

[4] Price and Pappalardo, J. Am. Chem. Soc., **72**, 2613 (1950).

[5] Bayer and Nelles, U. S. pat. 2,137,664 [C. A., **33**, 1758 (1939)].

Δ⁴-CHOLESTEN-3,6-DIONE

(Cholest-4-ene-3,6-dione)

Submitted by LOUIS F. FIESER.[1]
Checked by WILLIAM S. JOHNSON and DAVID G. MARTIN.

1. Procedure

A 500-ml. Erlenmeyer flask containing 64 g. (10 oxygen equivalents) of sodium dichromate dihydrate and 225 ml. of acetic acid is heated and swirled until all the solid has dissolved, and then is cooled in an ice bath to bring the temperature of the solution to 15°. In a 1-l. Erlenmeyer flask 25 g. (0.065 mole) of commercial cholesterol is dissolved in 225 ml. of benzene by brief warming, and the solution is cooled to 20°; 225 ml. of acetic acid is added, the solution is cooled to 15°, and the dichromate solution (at 15°) is poured in, whereupon a thick orange paste of cholesteryl chromate, $(C_{27}H_{45}O)_2CrO_2$, separates. The flask is immersed in an ice-water bath that is allowed to stand in a refrigerator for 40–48 hours. No stirring or other attention is required in this period; the temperature soon drops to 0° (Note 1), and the chromate dissolves in a few hours.

The resulting brown solution is poured into a separatory funnel and shaken with 225 ml. of 30–60° petroleum ether. After brief standing, the mixture when viewed against a strong light can be seen to have separated into a reddish upper hydrocarbon layer and a smaller, very dark lower layer containing chromium compounds and acetic acid. The lower layer is drawn off and discarded. Then 50 ml. of water is added to the hydrocarbon layer, and the mixture is shaken and allowed to settle. Another lighter-colored, lower layer is drawn off and discarded, and the process is repeated with 50 ml. more water. The hydrocarbon layer, now light in color, is then shaken with 110 ml. of Claisen's alkali (Note 2); this first portion neutralizes the residual acetic acid and acidic oxidation products and extracts some of the enedione as the yellow enolate. The funnel is allowed to stand, with occasional twirling, until the lower layer has settled to a clear yellow solution; the upper hydrocarbon layer acquires a dirty red-brown color. The extract is drawn off into a wide-mouthed 2-l. separatory funnel charged with 200 ml. of water, 600 g. of ice, 200 ml. of 36% hydrochloric acid, and 300 ml. of ether (Note 3). The hydrocarbon layer is extracted in the same way with five more 100-ml. portions of Claisen's alkali and then discarded (Note 4).

After the last alkaline extract has been run into the receiving funnel, this is shaken and the aqueous layer is drawn off and discarded. The light-yellow ethereal layer is run into a smaller funnel and shaken with 100 ml. of 5% sodium carbonate solution and 30 ml. of saturated sodium chloride solution (Note 5). This extraction is repeated a second time (Note 6). The carbonate extract is either discarded or worked up for recovery of acidic oxidation products (Note 7). The ethereal solution is finally shaken with 100 ml. of saturated sodium chloride solution, filtered into a round-bottomed flask (Note 8) by gravity through a paper containing 25 g. of anhydrous magnesium sulfate, and evaporated to dryness. The residue is initially an oil, but when the last traces of solvent are removed by evacuating the flask (Note 8), reheating, and re-evacuating, it is obtained as a yellow solid (about 13 g.). This is taken up in 125 ml. of boiling methanol, and crystallization is allowed to proceed, first at room temperature and then at 0°. Δ^4-Cholesten-3,6-dione separates in glistening, thin, yellow plates, m.p. 124–125°; yield in the first crop, 9.0–9.3 g. Concentration of the filtrate affords a second crop of 0.7–1.2 g., melting in the range 118–121.5°; total yield, 10.0–10.2 g. (39–40%).

2. Notes

1. The temperature of the oxidation is highly critical. When

the mixture, initially at 15°, is allowed to stand in a refrigerator at an air temperature of 4–8°, the temperature of the reaction mixture during the first 7 hours is in the range 18–5° (exothermic); yields of enedione after varying reaction periods are as follows: 7 hours, 27%; 15–22 hours, 32.5% (less-pure product). At 25° the enedione is oxidized completely to acids in about 18 hours.

2. Sufficient Claisen's alkali for one run is made by dissolving 175 g. of potassium hydroxide pellets in 125 ml. of distilled water, cooling to room temperature, adding 500 ml. of methanol, and again cooling; this gives 655–665 ml. of solution.

3. $Δ^4$-Cholesten-3,6-dione is sensitive to air oxidation under certain conditions, particularly in alkaline solution. If the protective procedure specified is not followed the reaction product may be deep red and tarry.

4. The residual solution contains a little cholesterol, 0.2–0.3 g. of $Δ^4$-cholesten-3,6-dione, and small amounts of products derived from the oxidation of cholestanol and other companions.

5. Sodium chloride speeds up the separation of layers.

6. The sodium carbonate extract contains only a few milligrams of the enedione.

7. The oxidation procedure described affords a mixture of acids from which no component is easily isolable. When the oxidation is allowed to proceed in the temperature range 18-5° for 15–22 hours, as described in Note 1, isolation of the Diels acid (3,4-seco-$Δ^5$-cholesten-3,4-dioic acid) is easily accomplished as follows. The carbonate extract is acidified and shaken with ether, and the clear aqueous layer is discarded. The ethereal solution, which may contain some suspended Diels acid, is not dried but is run into a flask and diluted with an equal volume of acetone. The mixture is evaporated to a volume of 15–20 ml. and cooled, and the Diels acid is collected as a white powder, m.p. 280–285°; yield 0.5 g.

8. A round-bottomed flask is recommended in order to withstand the evacuation process.

3. Methods of Preparation

Mauthner and Suida [2] isolated, as one of three neutral products resulting from the oxidation of cholesterol with aqueous chromic

acid in acetic acid solution, the substance later identified as
Δ^4-cholesten-3,6-dione. The present procedure is based upon
results of a reinvestigation of the oxidation by a low-temperature,
non-aqueous procedure.[3]

[1] Harvard University, Cambridge 38, Massachusetts.

[2] Mauthner and Suida, *Monatsh.*, **17**, 579 (1896); see also Windaus, *Ber.*, **39**, 2249
(1906); Ross, *J. Chem. Soc.*, **1946**, 737.

[3] Fieser, *J. Am. Chem. Soc.*, **75**, 4386 (1953).

Δ^4-CHOLESTEN-3-ONE

(Cholest-4-en-3-one)

Submitted by JEROME F. EASTHAM and ROY TERANISHI.[1]
Checked by JAMES CASON, JAMES JIU, and ELMER J. REIST.

1. Procedure

To a 5-l. three-necked flask, equipped with a sealed mechanical
stirrer, a dropping funnel, and a take-off reflux condenser, are
added 2 l. of sulfur-free toluene and two boiling chips. The
openings of the dropping funnel and the condenser are protected
by drying tubes containing Drierite. A portion (200 ml.) of the
toluene is distilled from the flask (take-off cock open) in order to
dry the system by azeotropic distillation; then 100 g. (0.26 mole)
of cholesterol and 500 ml. of cyclohexanone (Note 1) are added

to the flask. After an additional 50 ml. of toluene has been distilled, a solution of 28 g. (0.14 mole) of aluminum isopropoxide (Note 2) in 400 ml. of dry toluene (Note 3) is added dropwise over a period of approximately 30 minutes. During this time toluene is distilled at a rate slightly greater than the rate of addition of catalyst solution, so that when the addition is complete about 600 ml. of toluene has distilled. An additional 300 ml. of toluene is distilled, and the murky orange-colored reaction mixture is then allowed to cool to room temperature.

Four hundred milliliters of a saturated aqueous solution of potassium-sodium tartrate (Note 4) is added to the mixture, and the organic layer becomes clear and orange. The stirrer assembly is removed, and the mixture is steam-distilled until about 6 l. of distillate has been collected. The residual mixture is cooled and extracted successively with one 300-ml. portion and two 100-ml. portions of chloroform. The combined extracts are washed with two 100-ml. portions of water and dried over anhydrous magnesium sulfate. The chloroform is removed by distillation on the steam bath at reduced pressure (water aspirator). The residual viscous amber oil (Note 5) is dissolved by heating in 150 ml. of methanol. When the solution has cooled to about 40°, seeds of cholestenone are added, and the flask is wrapped with a small towel to ensure slow cooling (Note 6). After the bulk of the material has crystallized, which requires several hours, the mixture is stored at 0° overnight (Note 7). The product is collected by suction filtration, washed with 40–50 ml. of methanol previously cooled in an ice-salt bath, then dried at reduced pressure, first at room temperature and finally at 60°. The yield of light-cream-colored Δ⁴-cholesten-3-one is 81–93 g. (81–93%), m.p. 76–79°. Recrystallization from methanol gives material melting at 79.5–80.5° in 90% recovery (Note 8).

2. Notes

1. Commercial cholesterol, m.p. 146–149°, is satisfactory if dried at reduced pressure at 100° for 48 hours. The cyclohexanone is simply distilled before use, b.p. 153–155°.

2. Directions for the preparation of aluminum isopropoxide are given by Young, Hartung, and Crossley.[2] This compound is also

commercially available. Turbidity of the catalyst solution is to be expected and is not detrimental.

3. Toluene is dried satisfactorily by distilling about 20% of it and using the residue, which is cooled with protection from moisture.

4. Either Rochelle or Seignette salt is satisfactory. The tartrate ion serves to keep the aluminum ion in solution.

5. Prolonged heating is not necessary, since the chloroform remaining with the residual oil does not hamper the crystallization.

6. Seed crystals may be obtained by transferring a few drops of the solution to a small test tube, then cooling and scratching.

Ordinarily the cholestenone will separate as an oil from a solution saturated much above 50°. In this event, rather than working with larger volumes of methanol, small portions of chloroform may be added to the solution to lower the saturation temperature to a point at which seed crystals do not turn to oil. The best yield of good crystalline product is realized by inducing crystallization at the highest possible temperature in order to obtain a large initial crop of crystals. A second crop is difficult if not impossible to obtain by direct crystallization.

The towel should be placed around the flask only after crystallization, rather than oil formation, is assured. The towel is carefully removed occasionally during the first hour to ascertain whether oil is forming on top of the crystals. If so, the solution is warmed slightly in order to redissolve the oil but not the crystals. The product is quite impure if it first separates as an oil which later crystallizes.

7. If initial crystallization is not induced at 40° or above, a further period of cooling in an ice-salt bath is necessary to obtain the best yield.

8. In order to avoid large volumes, it is recommended that 700 ml. of methanol be employed to recrystallize 90 g. of cholestenone and that either chloroform or acetone be used to increase the solubility of the cholestenone to the required amount.

3. Methods of Preparation

The methods of preparation of Δ^4-cholesten-3-one have been summarized in an earlier volume.[3] In addition, it has been ob-

tained by the deacetylation of 3-acetoxy-3,5-cholestadiene.[4]
The present modification is less laborious than the earlier [3]
method.

[1] University of California, Berkeley, California.
[2] Young, Hartung, and Crossley, *J. Am. Chem. Soc.*, **58**, 100 (1936).
[3] *Org. Syntheses Coll. Vol.* **3**, 207 (1955).
[4] Dory, Geri, Szabó, and Oposzky, *Med. Prom. S.S.S.R.*, **13**, No. 10, 14 (1959)
[*C. A.*, **54**, 12192 (1960)].

CHOLESTEROL,* Δ⁵-CHOLESTEN-3-ONE,* AND Δ⁴-CHOLESTEN-3-ONE *

(Cholesterol, Cholest-5-en-3-one, and Cholest-4-en-3-one)

Submitted by LOUIS F. FIESER.[1]
Checked by WILLIAM S. JOHNSON and DAVID G. MARTIN.

1. Procedure

A. *Cholesterol dibromide.* In a 4-l. beaker 150 g. (0.39 mole)
of commercial cholesterol (Note 1) is dissolved in 1 l. of absolute

ether by warming on the steam bath and stirring with a stout glass rod; the solution is then cooled to 25°. A second solution is prepared by adding 5 g. of powdered anhydrous sodium acetate (0.06 mole) (Note 2) to 600 ml. of acetic acid, stirring the mixture and breaking up the lumps with a flat stirring rod; 68 g. (0.4 mole) of bromine is then added, and the solution is poured with stirring into the cholesterol solution. The solution turns yellow and promptly sets to a stiff paste of dibromide. The mixture is cooled in an ice bath to 20°, and then the product is collected on a 16-cm. Büchner funnel (Note 3). The cake is pressed down and washed with acetic acid until the filtrate is completely colorless; 500 ml. is usually sufficient. A second crop of satisfactory dibromide is obtained by adding 800 ml. of water to the combined filtrate and washings, collecting the precipitate, and washing it with acetic acid until colorless. Dibromide moist with acetic acid is satisfactory for most transformations; dry dibromide, even when highly purified by repeated crystallization, begins to decompose (darkens) within a few weeks. When material prepared as described is dried to constant weight at room temperature, it is obtained as the 1:1 dibromide/acetic acid complex. Yields obtained in the first and second crops, respectively, are: 171–186 g. and 13–25 g., total yield 197–199 g. (84–85%) (Note 1).

B. *Cholesterol from the dibromide.* The acetic acid-moist filter cake of dibromide from 150 g. of cholesterol is transferred (Note 3) to a 3-l. round-bottomed flask and covered with 1.2 l. of U.S.P. ether, and the suspension is stirred mechanically above a bucket of ice and water that can be raised as required (Notes 4, 5). Forty grams (0.61 g. atom) of fresh zinc dust is added in the course of 5 minutes. The first 5- to 10-g. portion is added without cooling; when the reaction has started, as evidenced by solution of part of the dibromide and by ebullition, the cooling bath is raised during the remainder of the addition. At the end, the ice bath is lowered, and the mixture, which soon sets to a paste of white solid (Note 6), is stirred for 15 minutes longer. Then 50 ml. of water is added to dissolve the white solid, and the ethereal solution is decanted into a separatory funnel and washed with 400 ml. of water containing 25 ml. of 36% hydrochloric acid. After three more washings with 400-ml. portions of water, the solution is shaken thoroughly with 300 ml. of water and 150 ml.

of 25% sodium hydroxide solution, and the ethereal solution is tested with moist blue litmus paper to make sure that all the acetic acid is removed (Note 7). The solution is then dried over magnesium sulfate and evaporated to a volume of 600 ml., methanol (600 ml.) is added, and the solution is evaporated to the point where crystallization just begins (about 1 l.). After standing at room temperature and then at 0–4°, the main crop of cholesterol is collected and dried at room temperature; yield 108–110 g., m.p. 149.5–150° (Note 8). A second crop of 8.4–10.4 g., m.p. 148–149°, is obtained after evaporation of the mother liquor to a volume of 250 ml. (Note 9); total yield 117–120 g. (78–80% from commercial cholesterol) (Note 10).

C. *5α,6β-Dibromocholestan-3-one.* The moist dibromide from 150 g. of cholesterol (part A) is suspended in 2 l. of acetic acid in a 5-l. round-bottomed flask equipped with a stirrer and mounted over a bucket of ice and water that later can be raised to immerse the flask (Note 5). The suspension is stirred at room temperature (25–30°), and a solution, preheated to 90°, of 80 g. (2 oxygen equivalents) of sodium dichromate dihydrate in 2 l. of acetic acid is poured in through a funnel (Note 3). The mixture reaches a temperature of 55–58° during the oxidation, and all the solid dissolves in 3–5 minutes. After another 2 minutes the ice bucket is raised until the flask is immersed; stirring is then stopped, and the mixture is allowed to stand in the ice bath without disturbance for 10 minutes to allow the dibromoketone to separate in easily filterable crystals. With stirring resumed, the temperature is brought to 25° and then, after addition of 400 ml. of water, to 15°. The product is collected on a 21-cm. Büchner funnel, and the filter cake is drained until the flow of filtrate amounts to no more than 25 drops per minute. The suction is released, the walls of the funnel are washed down with methanol, and 200 ml. of methanol is added. After a few minutes' standing, suction is applied and the crystals are drained thoroughly of solvent before they are washed in the same way with 200 ml. of fresh methanol. The last drops of filtrate should be completely colorless. Dried to constant weight at room temperature in a dark cupboard, the dibromoketone consists of shiny white crystals, m.p. 73–75° (dec.), $[\alpha]_D^{25}$ −47° chloroform (c = 2.11) (Note 11); yield about 171 g. (96% in the oxidation or 81% from cholesterol).

D. Δ^5-*Cholesten-3-one.* (Note 12.) The moist 5α,6β-dibromo-cholestan-3-one from 150 g. of cholesterol is transferred to a 3-l. round-bottomed flask and covered with 2 l. of U.S.P. ether and 25 ml. of acetic acid. The suspension is stirred mechanically, an ice bath is raised into position (Note 5), and the temperature is brought to 15°. The ice bath is then lowered, and 5 g. of fresh zinc dust is added. As soon as the exothermic reaction of de-bromination sets in, the temperature is controlled to 15–20° by cooling during the addition (in about 5 minutes) of 35 g. more zinc dust. The ice bath is then lowered, and the ethereal solution containing suspended zinc dust is stirred for 10 minutes longer. With continued stirring, 40 ml. of pyridine is added; this pre-cipitates a white zinc salt (Note 13). The mixture is filtered through a Büchner funnel, and the filter cake is washed well with ether. The colorless filtrate is washed with three 600-ml. portions of water and then shaken thoroughly with 600 ml. of 5% aqueous sodium bicarbonate solution until free from acetic acid as in-dicated by testing the ethereal solution with moist blue litmus paper. The solution is dried over magnesium sulfate and evap-orated to a volume of about 1 l.; 500 ml. of methanol is added, and the evaporation is continued until the volume is approxi-mately 1.2 l. Crystallization is allowed to proceed at room tem-perature, then at 0–4°, and the large colorless prisms are collected by suction filtration; yield in the first crop 87–94 g., melting point in the range 124–129° (camphorlike odor), $[\alpha]_D^{25}$ −2.5° chloro-form ($c = 2.03$), no selective absorption at 242 mμ. Concentra-tion of the mother liquor gives a second crop of 12–19 g. melting in the range 117–125° and suitable for conversion to Δ^4-cholesten-3-one; total yield 106–108 g. (71–72% from cholesterol).

E. Δ^4-*Cholesten-3-one.* A mixture of 100 g. of Δ^5-cholesten-3-one (0.26 mole), 10 g. (0.11 mole) of anhydrous oxalic acid (Note 14), and 800 ml. of 95% ethanol is heated on the steam bath until all the solid is dissolved (15 minutes) and for 10 min-utes longer, and then is allowed to stand at room temperature. If crystallization has not started after a period of several hours, the solution is seeded or scratched. After crystallization has pro-ceeded at room temperature and then at 0–4°, the large, colorless, prismatic needles that separate are collected by suction filtration; yield in the first crop 88–92 g., m.p. 81–82°, $[\alpha]_D^{25}$ 92° chloroform

$(c = 2.01)$; $\lambda_{\text{max.}}^{\text{ethanol}}$ 242 mμ (ϵ = 17,000). A second crop (5.0–7.5 g., melting in the range 78–82°) is obtained after concentration of the mother liquor to a volume of about 100 ml., and a third crop (3–4 g., low melting) by dilution with water. Recrystallization of these crops from 95% ethanol gives a total of 6.8–8.1 g. of satisfactory material, m.p. 81–82°; total yield 96–98 g. (68–69% over-all yield from cholesterol).

2. Notes

1. Cholesterol of high quality and of recent production was employed. Cholesterol undergoes slow autoxidation in the solid state, and samples that have been in storage for a few years give lower yields of dibromide. The checkers used U.S.P. material, m.p. 149–150°, as supplied by Wilson Company, Chicago, Illinois.

2. The yield of dibromide dropped from 84% to 73% when no buffering sodium acetate was used. No improvement resulted from doubling the quantity of sodium acetate specified. Buffered bromination in a stirred suspension of acetic acid (no ether) at 20° raised the yield to 89%, but this material on debromination afforded sterol of low melting point (145–147°) containing halogen (Beilstein test).

3. The operation should be done in a hood, and the hands should be protected with Neoprene gloves.

4. If dry dibromide is used, 25 ml. of acetic acid is added to the suspension.

5. The ice bucket is conveniently mounted on an automobile jack.

6. The solid appears to be a complex of cholesterol and a zinc salt.

7. If the acetic acid is not removed some cholesteryl acetate may be formed during the evaporation.

8. The melting-point determination should be done in an evacuated capillary tube. In an open tube autoxidation occurs readily enough to lower the melting point when the bath is heated very slowly.

9. The residual mother liquor contains about 4 g. of material containing bromine not removed by repetition of the treatment with zinc dust.

10. Purification of cholesterol through the dibromide completely eliminates cholestanol, 7-dehydrocholesterol, and lathosterol (Δ^7-cholestenol). The first crop of material from methanol-ether is also free from cerebrosterol (24-hydroxycholesterol) and 25-hydroxycholesterol, a product of autoxidation present in cholesterol that has been stored in the crystalline state for a few years with access to air. When material of highest purity is desired, only first-crop dibromide should be employed, since debromination of second-crop material gives sterol melting at 146–147° and giving a positive Beilstein test.

11. Dibromocholestanone sometimes begins to decompose (turns purplish) after standing in the dark for a few hours; it rapidly darkens when dried at 70° or when exposed to bright sunlight. Hence it is advisable to use the material moist with methanol directly after preparation.

12. For success in the preparation of this labile non-conjugated ketone in high yield and purity, the intermediates, cholesterol dibromide and dibromocholestanone, should be processed further in the solvent-moist state as soon as prepared. The three reactions can be completed easily in one day.

13. If the bulk of the ionic zinc is not precipitated at this point it will cause troublesome emulsions when the solution is washed with water.

14. When isomerization of 100-g. batches of non-conjugated ketone was effected in ethanol under catalysis by either hydrochloric acid or sodium hydroxide (followed by neutralization of the yellow enolate solution with acetic acid), a permanent yellow coloration developed, the first-crop material was yellowish and melted at 78–80°, and the second-crop material was very impure.

3. Methods of Preparation

Cholesterol dibromide has been prepared by unbuffered [2] and buffered [3] bromination of cholesterol and oxidized to 5α,6β-dibromocholestan-3-one with acid permanganate,[2] chromic acid,[4-6] and sodium dichromate.[3] Regeneration of sterols, or more usually of sterol acetates, from their dibromides has been accomplished by use of zinc dust in boiling acetic acid,[2] sodium iodide,[7] ferrous chloride,[8] and chromous chloride,[9] and by the method [3] here described. Δ^5-Cholesten-3-one has been prepared by debromination of dibromocholestanone with zinc dust in boiling ethanol or methanol [5] and by zinc dust in ether containing a little acetic acid.[3]

The Oppenauer oxidation of cholesterol to Δ^4-cholesten-3-one of m.p. 77–79° in 70–93% yield has been reported in these volumes (p. 192).[10] Isomerization of Δ^5-cholesten-3-one by a mineral acid or a base has been conducted satisfactorily only on a micro scale; [5] the method of isomerization with oxalic acid has been reported.[3]

[1] Harvard University, Cambridge 38, Massachusetts.
[2] Windaus, *Ber.*, **39**, 518 (1906).
[3] Fieser, *J. Am. Chem. Soc.*, **75**, 5421 (1953).
[4] Ruzicka, Brüngger, Eichenberger, and Meyer, *Helv. Chim. Acta*, **17**, 1413 (1934).
[5] Butenandt and Schmidt-Thomé, *Ber.*, **69**, 882 (1936).
[6] Inhoffen, *Ber.*, **69**, 1134 (1936).
[7] Schoenheimer, *Z. physiol. Chem.*, **192**, 86 (1930); *J. Biol. Chem.*, **110**, 461 (1935).
[8] Bretschneider and Ajtai, *Monatsh.*, **74**, 57 (1943).
[9] Julian, Cole, Magnani, and Meyer, *J. Am. Chem. Soc.*, **67**, 1728 (1945).
[10] *Org. Syntheses Coll. Vol.* **3**, 207 (1955).

COUMALIC ACID *

$$2HO_2CCHOHCH_2CO_2H \xrightarrow[H_2SO_4]{SO_3}$$

Submitted by RICHARD H. WILEY and NEWTON R. SMITH.[1, 2]
Checked by C. F. H. ALLEN and GEORGE A. REYNOLDS.

1. Procedure

In a 2-l. round-bottomed flask are placed 200 g. (1.49 moles) of powdered malic acid (Note 1) and 170 ml. of concentrated sulfuric acid. To this suspension are added three 50-ml. portions of 20–30% fuming sulfuric acid at 45-minute intervals. After the evolution of gas has slackened, the solution is heated on a water bath for 2 hours with occasional shaking. The reaction mixture is then cooled and poured slowly onto 800 g. of crushed ice with stirring. After standing 24 hours, the acid is filtered on a Büchner funnel, washed with three 50-ml. portions of ice-cold water, and dried on a water bath. The yield of crude acid, melting at 195–200°, is 75–80 g. (Notes 2 and 3).

One-half of the crude product is dissolved in five times its weight of hot methanol, and the solution is boiled with 3 g. of Norit or decolorizing carbon. The solution is filtered while hot and cooled in an ice bath. The precipitate is collected on a filter and washed with 25 ml. of cold methanol. The mother liquors are used to recrystallize the remaining crude material. The yield of bright yellow coumalic acid, melting at 206–209°, is 68–73 g. (65–70%) (Note 4).

2. Notes

1. A technical free-flowing powder, melting at 126–128°, was used.

2. This washing is essential to remove the mineral acid and to avoid partial esterification that otherwise takes place during the methanol recrystallization step.

3. The submitters state that an additional 10–12 g. of crude acid can be obtained from the filtrate by extraction with ether in a continuous extractor.

4. Depending on the color of the crude acid, several additional recrystallizations may be required to obtain a colorless product.

3. Methods of Preparation

This procedure is essentially that of von Pechmann.[3] Esters of coumalic acid may be obtained by heating the sulfuric acid solution with the appropriate alcohol.[4]

[1] University of Louisville, Louisville, Kentucky.

[2] The submitters wish to thank the Research Corporation for a grant under which this work was done.

[3] von Pechmann, *Ann.*, **264**, 272 (1891).

[4] Campbell and Hunt, *J. Chem. Soc.*, **1947**, 1176; Gilman and Burtner, *J. Am. Chem. Soc.*, **55**, 2903 (1933); Ruzicka, *Helv. Chim. Acta*, **4**, 504 (1921).

CREOSOL

OH   OH
$\qquad$OCH$_3$ Zn(Hg) $\qquad$OCH$_3$
$\xrightarrow{\text{Zn(Hg)}}$
$\xrightarrow{\text{HCl}}$
CHO   CH$_3$

Submitted by R. SCHWARZ and H. HERING.[1]
Checked by R. L. SHRINER and C. L. FURROW, JR.

1. Procedure

In a 5-l. three-necked flask fitted with a stopper, a reflux condenser, and a 250-ml. dropping funnel are placed 1.5 kg. (22.9 g. atoms) of amalgamated zinc (Note 1) and 800 ml. of concentrated hydrochloric acid. The flask is heated to cause gentle refluxing to occur. A solution of 152 g. (1.0 mole) of vanillin in 450 ml. of 95% ethanol and 1.5 l. of concentrated hydrochloric acid (Note 2) is added dropwise through the dropping funnel over an 8-hour period (Note 3). After the addition of the vanillin is complete the mixture is refluxed for 30 minutes more. The liquid, consisting of two layers, is decanted from the zinc (Note 4) into a 4-l. separatory funnel. The aqueous layer is removed and washed three times with 200-ml. portions of benzene, and the benzene extracts are combined with the crude creosol separated initially. The residual amalgamated zinc is washed twice with 100-ml. portions of benzene, and this benzene solution is added to the combined extracts. The extracts are washed twice with 200-ml. portions of 5% sodium bicarbonate solution, and the resultant precipitate of metallic salts is removed by filtration. The solution is washed once with 200 ml. of water.

The benzene is removed under reduced pressure by distillation on a steam bath. The residue is distilled under reduced pressure, the fraction boiling at 78–79°/4 mm., at 104–105°/15 mm., or at 219–222.5°/760 mm. being collected. A yield of 83–92.5 g. (60–67%) of colorless or pale yellow creosol is obtained, d_4^{25}, 1.092; n_D^{20} 1.5354 (Notes 5 and 6).

2. Notes

1. The amalgamated zinc may be prepared by adding 1500 g. of clean granulated zinc to 600 ml. of 5% mercuric chloride solution. After standing for 2 hours with occasional shaking, the liquid is decanted and the zinc is used immediately.

Alternatively,[2] 1500 g. of granulated zinc is added to a solution of 62.5 g. of mercuric chloride and 62.5 ml. of concentrated hydrochloric acid in 1875 ml. of water. The mixture is shaken for about 5 minutes, the liquid decanted, and the zinc used immediately.

2. The vanillin is first dissolved in ethanol by warming gently, and then 1.5 l. of concentrated hydrochloric acid ($d = 1.19$) is added. A clear yellow solution is obtained, which soon becomes blue-green.

3. It is advisable to avoid overheating the upper parts of the flask to prevent some brown-red tarry masses from forming.

4. The recovered zinc (900–950 g.) can be used for further preparations by adding more zinc and renewing amalgamation.

5. The submitters report yields up to 103 g. (75%), but this yield could not be consistently obtained by the checkers.

6. The reduction of vanillin by the Clemmensen method using toluene as an auxiliary solvent has been reported.[3] By following these directions, a yield of 49% of creosol was obtained with large amounts of tar.

3. Methods of Preparation

Creosol (also called 2-methoxy-p-cresol, 4-methylguaiacol, and 3-methoxy-4-hydroxytoluene) has been obtained by the fractionation of beach creosote tar,[4] by the reduction of vanillin by electrolytic methods,[5, 6] by hydrogen and palladium on charcoal or barium sulfate,[7, 8] with hydrazine,[9] and by amalgamated zinc and hydrochloric acid.[3, 10, 11] It has also been prepared by methylation of 4-methylcatechol with methyl iodide [12, 13] or with methyl sulfate [14] and is reported to be formed by the distillation of the calcium salt of 3-methoxy-4-hydroxyphenylacetic acid.[15]

[1] Pharmaceutical-Chemical Institute, Vienna, Austria.

[2] Martin, *Org. Reactions*, **1**, 155 (1942).

[3] Fletcher and Tarbell, *J. Am. Chem. Soc.*, **65**, 1431 (1943).

[4] Mendelsohn, Dissertation, Berlin, 1877, p. 13.

[5] Schepss, *Ber.*, **46**, 2571 (1913).

[6] Shima, *Mem. Coll. Sci. Kyoto*, **A12**, 79 (1929).

[7] Rosenmund and Jordan, *Ber.*, **58**, 162 (1925).

[8] St. Pfau, *Helv. Chim. Acta*, **22**, 550 (1939).

[9] Wolff, *Ann.*, **394**, 100 (1912); Lock, *Monatsh.*, **85**, 802 (1954).

[10] Kawai and Sugiyama, *Ber.*, **72B**, 367 (1939).

[11] Buu-Hoi, Hiong-Ki-Wei, and Royer, *Bull. soc. chim. France*, **12**, 866 (1945).

[12] Ono and Imoto, *J. Chem. Soc. Japan*, **57**, 112 (1936); *Bull. Chem. Soc. Japan*, **11**, 127 (1936).

[13] Steinkopf and Klopfer, *Ber.*, **64B**, 990 (1931).

[14] Fahlberg, Ger. pat. 258,105 [*Frdl.*, **11**, 891 (1912)].

[15] Tiemann and Nagai, *Ber.*, **10**, 206 (1877).

N-2-CYANOETHYLANILINE

(Propionitrile, 3-anilino-)

$$C_6H_5NH_2 \cdot HCl + (C_2H_5)_2NH + CH_2{=}CHCN \rightarrow$$
$$C_6H_5NHCH_2CH_2CN + (C_2H_5)_2NH \cdot HCl$$

Submitted by J. CYMERMAN-CRAIG and M. MOYLE.[1]
Checked by WILLIAM S. JOHNSON and DUFF S. ALLEN, JR.

1. Procedure

A mixture of 12.95 g. (0.10 mole) of aniline hydrochloride (Note 1), 6.6 g. (0.12 mole) of acrylonitrile, and 9.1 g. (0.12 mole) of diethylamine (Note 2) is placed in a 100-ml. round-bottomed flask fitted with an efficient reflux condenser and heated for 2.5 hours in a bath maintained at 180°.

The melt is cooled to 0°, 50 ml. of 10% aqueous sodium hydroxide solution is added, and the mixture is extracted with four 50-ml. portions of chloroform. The combined chloroform extracts are washed with two 25-ml. portions of water, and these in turn are extracted with 10 ml. of chloroform. The organic layers are combined and dried partially over anhydrous sodium sulfate. The solvent is removed by distillation on the steam bath, and the residue is distilled at reduced pressure from a 50-ml. distilling flask. After a fore-run of about 4 g. (Note 3), b.p. 60–70°/1.5 mm. (bath temperature taken up to 125°), the cyanoethylaniline is collected at 115–120°/0.01 mm. The product solidifies in the form of colorless plates, m.p. 48–51° (Note 4). The yield is 10.5–11.4 g. (72–78%) (Note 5).

2. Notes

1. An equimolar quantity of aniline benzenesulfonate may be used in place of the hydrochloride.

2. No reaction occurs if the diethylamine is omitted.

3. In a typical run this fraction weighed 4.4 g. and contained 1.6 g. of aniline (estimated as acetanilide) and 2.8 g. of 3-diethylaminopropionitrile.

4. Reported properties are b.p. 178–186°/16 mm. and m.p. 51.5°.[2]

5. The submitters have found that other arylamines may be employed in a similar manner in place of aniline. Thus N-2-cyanoethyl-*p*-anisidine was obtained in 76% yield as plates, m.p. 62–64°, b.p. 130–140°/0.01 mm.; N-2-cyanoethyl-*m*-chloroaniline, in 42% yield as needles, m.p. 44–46°, b.p. 125–130°/0.01 mm.; N,N′-bis-2-cyanoethyl-*o*-phenylenediamine, in 70% yield as needles, m.p. 116–118°, b.p. 190–200°/0.01 mm.; and N,N′-bis-2-cyanoethyl-*p*-phenylenediamine in 22% yield as plates, m.p. 138–139°.

3. Methods of Preparation

N-2-Cyanoethylaniline has been prepared (accompanied by much of the N,N′-bis-2-cyanoethyl compound) by heating aniline, acrylonitrile, and acetic acid in an autoclave,[2,3] or at refluxing temperature for 10 hours in the presence of various inorganic catalysts.[4] The substance also has been obtained, free of the N,N′-bis-2-cyanoethyl compound from aniline salts and β-diethylaminopropionitrile.[5-7] A number of other cyanoethylated compounds have been heated with aniline and water to form N-2-cyanoethylaniline,[8] and a study has been made of the conditions for the addition of aromatic amines to acrylonitrile.[9]

Heininger [10] has shown that cupric acetate is a superior catalyst for the cyanoethylation of aniline; N-2-cyanoethylaniline has been obtained in 73% yield by this method. The use of cupric acetate as a catalyst in cyanoethylation is demonstrated in the procedure for the preparation of 3-(*o*-chloroanilino)propionitrile (p.146).

[1] The University of Sydney, Sydney, Australia.

[2] Cookson and Mann, *J. Chem. Soc.*, **1949**, 67.

[3] Braunholtz and Mann, *J. Chem. Soc.*, **1952**, 3046; Strepikheev and Bogdanov, *Sbornik Stateĭ Obshcheĭ Khim.*, **2**, 1462 (1953) [*C. A.*, **49**, 4551 (1955)].

[4] Braunholtz and Mann, *J. Chem. Soc.*, **1953**, 1817; Gurvich and Terent'ev, *Sbornik Stateĭ Obshcheĭ Khim.*, Akad. Nauk S.S.S.R., **1**, 409 (1953) [*C. A.*, **49**, 1048 (1955)].

[5] Bauer and Cymerman, *Chem. & Ind. (London)*, **1951**, 615.

[6] Bauer, Cymerman, and Sheldon, *J. Chem. Soc.*, **1951**, 3311.

[7] Bekhli, *Zhur. Obshcheĭ Khim.*, **21**, 86 (1951) [*C. A.*, **45**, 7540 (1951)].

[8] Butskus and Denis, *Nauch. Doklady Vyssheĭ Shkoly, Khim. i Khim. Tekhnol.*, **1958**, No. 1, 130, 743 [*C. A.*, **53**, 3082, 8028 (1959)].

[9] Bekhli, *Doklady Akad. Nauk S.S.S.R.*, **113**, 558 (1957) [*C. A.*, **51**, 14577 (1957)].

[10] Heininger, *J. Org. Chem.*, **22**, 1213 (1957).

CYANOGEN IODIDE

(Iodine cyanide)

$$NaCN + I_2 \rightarrow ICN + NaI$$

Submitted by B. BAK and A. HILLEBERT.[1]
Checked by R. T. ARNOLD and H. E. FRITZ.

1. Procedure

Caution! Cyanogen iodide is relatively volatile and highly toxic. Therefore, these operations should be conducted in a good hood.

A three-necked 500-ml. flask is surrounded by an ice-water bath and provided with a stirrer and thermometer. Twenty-seven grams (0.55 mole) of sodium cyanide is dissolved in 100 ml. of water, added to the reaction flask, and cooled to 0°. To this, in 3- to 4-g. portions, is added with good stirring a total of 127 g. (0.50 mole) of iodine over a period of 30–40 minutes. A given portion of iodine is not added until the preceding one has reacted almost completely. Ten minutes after the addition of iodine is completed, 120 ml. of peroxide-free ether is added and the mixture is stirred for a few minutes until the precipitated cyanogen iodide has dissolved in the ethereal layer. The entire contents are then transferred to a previously cooled separatory funnel, and the aqueous layer is separated. This aqueous solution is again extracted successively with 100-ml. and 80-ml. portions of cold, peroxide-free ether. The combined ethereal extracts are poured into a 500-ml. round-bottomed flask, and the ether is evaporated under reduced pressure at room temperature. To the slightly brown crude product, which weighs about 90 g., is added 120 ml. of water. A slightly diminished pressure ($\frac{1}{2}$ atm.) is maintained while the contents in the closed flask are heated at 50° for 15 minutes with occasional vigorous shaking (Note 1). The mixture is then cooled to 0°, and the crystalline cyanogen iodide is separated from the light yellow mother liquor by suction on a sintered-glass funnel or filter plate (Note 2). The crude product is washed with six 25-ml. portions of ice water, removed from the sintered-glass funnel, and air-dried (in a good hood) for 1 hour at room temperature. Colorless cyanogen iodide weighing about 59 g. (77%) is obtained; m.p. 141–144° (Note 3).

Cyanogen iodide of highest purity may be produced in the following way. The above crude product is dissolved in 150 ml. of boiling chloroform, and the solution is filtered through a plug of glass-wool on a hot-water funnel into a 250-ml. Erlenmeyer flask. This solution, after being cooled at room temperature for 15 minutes, is placed in an ice-salt bath and cooled to −10° (Note 4). By means of suction filtration, the crystalline product is collected on a sintered-glass funnel, washed with three 15-ml. portions of cold chloroform (0°), and freed from the last traces of solvent by being placed on a watch glass and exposed to the atmosphere (in a good hood) at room temperature for 1 hour. Practically colorless needle-shaped crystals weighing 45 g. (59%) are obtained; m.p. 146–147° (Notes 3 and 5).

Removal of 100 ml. of chloroform from the above filtrate by means of evaporation under reduced pressure at room temperature and subsequent cooling permits isolation of an additional 2 g. of cyanogen iodide; m.p. 146–147°. The total yield thus becomes 47 g. (62% based on iodine).

2. Notes

1. In this way sodium iodide, soluble in the solution of cyanogen iodide in ether [complex formation of $NaI_2(CN)$], is removed. This complex is avoided in the procedure by Ketelaar and Kruyer [2] in which chlorine is used. The method adopted here

is faster and simpler and gives almost the same yield of purified cyanogen iodide.

2. Contact with filter paper must be avoided.

3. Determinations of the melting point of cyanogen iodide must be made using a sealed capillary which is kept totally immersed in the heating bath.

4. When the chloroform solution is cooled, a small aqueous layer is observed which finally separates as ice. The ice is filtered with cyanogen iodide but melts on the filter plate and is removed with the chloroform used as washing liquid.

5. Owing to the volatility of cyanogen iodide, the yield is slightly dependent on the speed of operation. By the above method sublimation as a means of purification is avoided. If, however, sublimation is desirable, it can be accomplished with appreciable speed only under reduced pressure and at temperatures at which cyanogen iodide is slowly decomposed into iodine and cyanogen. The vacuum must be constantly renewed during the operation.

3. Methods of Preparation

Cyanogen iodide has been prepared from mercuric cyanide and iodine,[3] potassium cyanide and iodine,[4] and sodium cyanide and iodine.[5]

[1] Universitetets Kemiske Laboratorium, Copenhagen K, Denmark.
[2] Ketelaar and Kruyer, *Rec. trav. chim.*, **62**, 550 (1943).
[3] Seubert and Pollard, *Ber.*, **23**, 1062 (1890); Woolf, *J. Chem. Soc.*, **1953**, 4121.
[4] Grignard and Crouzier, *Bull. soc. chim. France*, [4] **29**, 214 (1921).
[5] Moller, *Kgl. Danske Videnskab. Selskab, Math.-fys. Medd.*, **14**, 3 (1936).

3-CYANO-6-METHYL-2(1)-PYRIDONE

(Nicotinonitrile, 1,2-dihydro-6-methyl-2-oxo-)

$$CH_3COCH_3 + HCO_2C_2H_5 + CH_3ONa \rightarrow$$

$$[CH_3COCH{=}CHO]^{\ominus}Na^{\oplus} + C_2H_5OH + CH_3OH$$

$$[CH_3COCH{=}CHO]^{\ominus}Na^{\oplus} + CH_2(CN)CONH_2 \xrightarrow[\text{then } CH_3CO_2H]{\text{Piperidine acetate,}}$$

$$+ CH_3CO_2Na + H_2O$$

Submitted by RAYMOND P. MARIELLA.[1]
Checked by ARTHUR C. COPE, JAMES E. KRUEGER, and DAVID J. MARSHALL.

1. Procedure

In a 2-l. three-necked flask fitted with a Hershberg stirrer sealed by a lubricated rubber sleeve, a dropping funnel, and a reflux condenser attached to a calcium chloride drying tube are placed 46.5 g. (0.86 mole) of sodium methoxide (Note 1) and 1 l. of ether (dried over sodium wire). The flask is cooled in an ice bath, and a mixture of 46.4 g. (0.8 mole) of acetone (Note 2) and 59.2 g. (0.8 mole) of ethyl formate (Note 3) is added through the dropping funnel at a rate of about 2 drops per second, with stirring, during a period of about 1 hour. Stirring is continued 15 minutes longer with the ice bath in place and then 1 hour after it is removed. The reflux condenser is replaced by a condenser set for distillation, and the ether is distilled by heating the mixture in a water bath at a temperature which is not allowed to rise above 70°. Stirring is continued as long as possible during the distillation. The last of the ether is removed by distillation under reduced pressure with the aid of a water aspirator (Note 4).

To the solid residue of sodium formylacetone remaining in the flask are added a solution of 67 g. (0.8 mole) of cyanoacetamide (Note 5) in 400 ml. of water and piperidine acetate (prepared by adding piperidine to 8 ml. of glacial acetic acid in 20 ml. of water until the solution is just basic to litmus). The flask is equipped with a reflux condenser, and the mixture is heated under reflux for 2 hours. At the end of this time 200 ml. of water is added, and the solution is acidified (to litmus) with acetic acid, causing separation of the product as a voluminous yellow precipitate. The mixture is cooled in an ice bath for 2 hours, and the product is collected on a suction filter, washed on the filter with three 100-ml. portions of ice water, and dried (Note 6). The yield of 3-cyano-6-methyl-2(1)-pyridone is 59–67 g. (55–62%); m.p. 292–294° (dec., cor.) (Notes 7, 8, and 9).

2. Notes

1. Commercial sodium methoxide obtained from the Mathieson Alkali Works was used. If the commercial product is not available, sodium methoxide can be prepared by dissolving clean sodium in dry methanol and removing the excess methanol by distillation and finally by heating the residue to 200° under good vacuum furnished by an oil pump (protected by Dry Ice traps).[2]

2. Reagent grade acetone is dried over potassium carbonate and distilled.

3. Commercial ethyl formate was purified by the procedure described earlier.[3]

4. If not all the ether is removed, it will be impossible to obtain the reflux temperature needed for the subsequent condensation, in which case the remainder of the ether must be removed by a preliminary distillation after adding the aqueous solution of cyanoacetamide and piperidine acetate.

5. Either Eastman Kodak Company white label grade cyanoacetamide or a product prepared according to the procedure described previously [4] can be used.

6. The product retains water tenaciously and is best dried in a vacuum oven at 70–100° at a pressure of 30 mm. or lower.

7. Melting points determined in the ordinary way are unsatisfactory because gradual decomposition occurs over a broad temperature range. Reproducible melting points were obtained by placing the sample in a melting-point tube and displacing the air in the tube with nitrogen introduced through a fine capillary (prepared by drawing out a piece of glass tubing, which then was at-

tached to a nitrogen cylinder through a T-tube dipping into 1–2 cm. of mercury). The nitrogen-filled melting-point tube was sealed quickly, and the melting point was determined by placing the tube in a bath 10° below the melting point and raising the temperature 2° per minute.

8. Analysis of the crude product is approximately 0.9% low in carbon. Analytically pure 3-cyano-6-methyl-2(1)-pyridone, m.p. 296.5–298.5° (dec., cor., under nitrogen), can be obtained by one recrystallization from 50% (by volume) ethanol, using 66 ml. per g. of product and treating the hot solution with Darco. Recovery of the product is 60%, and concentration of the mother liquor yields impure material.

9. A similar procedure can be used for preparing other 2(1)-pyridones. For example, 3-cyano-6-isobutyl-2(1)-pyridone can be obtained from the sodium salt of formylmethyl isobutyl ketone, and 3-cyano-5,6-dimethyl-2(1)-pyridone can be prepared from the sodium salt of α-formylethyl methyl ketone.[5]

3. Methods of Preparation

The procedure used for preparing the sodium salt of formylacetone is a modification of a previously described procedure.[2] 3-Cyano-6-methyl-2(1)-pyridone has been prepared by the condensation of β-ethoxycrotonaldehyde diethyl acetal[6] or acetoacetaldehyde bismethylacetal[7] with cyanoacetamide, and by the condensation of the sodium salt of formylacetone with cyanoacetamide.[8]

[1] Northwestern University, Evanston, Illinois.
[2] Johnson, Woroch, and Mathews, J. Am. Chem. Soc., **69**, 570 (1947).
[3] Org. Syntheses Coll. Vol. **2**, 180 (1943).
[4] Org. Syntheses Coll. Vol. **1**, 179 (1941).
[5] Mariella, J. Am. Chem. Soc., **69**, 2670 (1947).
[6] Dornow, Ber., **73B**, 153 (1940).
[7] Franke and Kraft, Chem. Ber., **86**, 797 (1953).
[8] Perez-Medina, Mariella, and McElvain, J. Am. Chem. Soc., **69**, 2574 (1947).

1-CYANO-3-PHENYLUREA

(Urea, 1-cyano-3-phenyl-)

$$C_6H_5NCO + NH_2CN + NaOH \rightarrow C_6H_5NHCONNaCN + H_2O$$

$$C_6H_5NHCONNaCN + HCl \rightarrow C_6H_5NHCONHCN + NaCl$$

Submitted by FREDERICK KURZER and J. ROY POWELL.[1]
Checked by JOHN C. SHEEHAN and ERNEST R. GILMONT.

1. Procedure

To a solution of 16.8 g. (0.40 mole) of cyanamide in 50 ml. of water (Note 1) contained in a 400-ml. flask or beaker is added 50 ml. of aqueous $3N$ sodium hydroxide (0.15 mole). To the resulting solution, cooled to 15–18°, is added, in 2-ml. portions with shaking over a 15–18 minute period, 23.8 g. (22.0 ml., 0.2 mole) of phenyl isocyanate. The isocyanate dissolves rapidly, while the temperature rises slightly. The mixture is maintained at 20–25° by occasional external cooling in ice water. When half the phenyl isocyanate has been introduced, a second portion of 50 ml. (0.15 mole) of $3N$ sodium hydroxide is added to keep the reaction mixture strongly alkaline throughout the experiment. When addition is complete (Note 2), the slightly turbid liquid is diluted with 40 ml. of water and is filtered immediately under reduced pressure (Note 3) to remove undissolved impurities and traces of separated diphenylurea (Note 4). The cyanourea is precipitated from the clear colorless filtrate by the slow addition of concentrated hydrochloric acid with stirring until a permanent turbidity just appears (Note 3). Cracked ice (30–40 g.) is added to lower the temperature to 18–20°. Precipitation is then completed in this temperature range by the alternate addition of concentrated hydrochloric acid (total volume required, approximately 30 ml.) and cracked ice until the suspension is acid to Congo red. The crude cyanourea forms a microcrystalline, white precipitate, which, after storage at 0° for 3 hours, is collected by filtration under reduced pressure (Note 5) and is washed with two portions of cold water (20–25 ml.) (Note 6). The crude product is drained thoroughly, air-dried at room temperature, and finally dried to constant weight in a desiccator over phosphorus pentoxide. The yield of crude 1-cyano-3-phenylurea (Note 7), m.p. 122–126° with decomposition (Note 8), varies between 29 and 30.5 g. (90–95%). The dried material is purified as follows. A solution of the crude product in 100 ml. of boiling acetone is diluted slowly with gentle swirling with 30–40 ml. of petroleum ether (boiling range 40–60°). As crystallization proceeds, an additional 20–30 ml. of petroleum ether is added carefully at such a rate that the supernatant liquid does not become turbid. After 15 minutes at room temperature, the mixture is set aside at 0°. The product is collected by filtration under reduced pressure and washed successively with 50 ml. of an acetone-petroleum ether mixture (1:3) and 50 ml. of petroleum ether. 1-Cyano-3-phenylurea thus obtained forms colorless lustrous needles, m.p. 127–128° with decomposition (Note 8), yield 20–21.5 g. (62–67%). Slow dilution of the filtrates with petroleum ether to a total volume of 350–400 ml. and storage at 0° affords a second crop, m.p. 123–126° (dec.), of satisfactory purity; yield 4–5 g. (12–16%). The synthesis is generally applicable to the preparation of 1-cyano-3-arylureas (Note 9).

2. Notes

1. The checkers used 16.8 g. of Eastman Kodak Company cyanamide (P1995) without further purification. A convenient method of preparing cyanamide from commercial calcium cyanamide has been described (p. 645). According to the submitters, an aqueous solution of crude cyanamide is satisfactory in the present synthesis and is obtained by adapting this published procedure (p. 645) as follows.

The residual crude cyanamide remaining after evaporation of the ethereal extracts (which need not be dried previously) is dissolved readily in the appropriate volume of cold water. A small quantity of water-insoluble oily or semisolid by-products is removed by shaking the solution with carbon and filtering the liquid through a small ordinary filter, followed by rinsing with a few milliliters of water. The clear filtrate is suitable for the subsequent operation.

In order to allow for small losses in the filtration and for the presence of impurities (such as solvent and oily by-products) an excess of approximately 10–15% by weight of crude cyanamide is allowed.

2. The reaction is completed when the liquid no longer has the odor of phenyl isocyanate; shaking is continued until this stage is reached.

3. The reaction mixture is not cooled during the addition of the last two or three portions of phenyl isocyanate, so that the final temperature is near 25°; this procedure prevents separation of the sodium salt of cyanophenylurea, which crystallizes readily at low temperatures. For the same reason, the filtered solution of the salt is not precooled, but rather is cooled during the precipitation of the free cyanourea.

4. The residue on the filter is rinsed with 5–10 ml. of water, in order to redissolve any small quantities of the sodium salt of cyanophenylurea that may have been collected.

5. The aqueous filtrate does not deposit any further material on storage at 0° and is discarded.

6. For the purpose of washing, the filter cake on the funnel is covered with ice water. After being allowed to remain in contact for a few seconds, the liquid is quickly drained under reduced pressure. The second portion of washing water should no longer be acid to Congo red.

7. According to the submitters the crude material may turn very pale pink on drying, but the product is again colorless after recrystallization. The checkers found that the crude product could be dried to less than 0.2% moisture by drawing air through the filter cake for 3 hours.

8. The decomposition temperature is somewhat influenced by the rate of heating. The material does not form a clear melt during the decomposition.

9. According to the submitters 1-cyano-3-α-naphthylurea [2] is obtained similarly from α-naphthyl isocyanate in 85–90% yields. Crystallization from acetone-petroleum ether (12 and 6 ml., respectively, per gram of crude product; recovery approximately

60% per crystallization) yields lustrous prisms, m.p. 148–149° with decomposition.

3. Methods of Preparation

1-Cyano-3-phenylurea, first obtained by the alkaline hydrolysis of 5-anilino-3-*p*-toluyl-1,2,4-oxadiazole,[3] has been prepared by the condensation of phenyl isocyanate and the sodium salt of cyanamide.[4] However, in these publications an incorrect structural assignment for the product was made. 1-Cyano-3-phenylurea is obtained also, together with other products, by warming gently 1-cyano-3-phenylthiourea with caustic soda in the presence of ethylene chlorohydrin,[5] or by gradually adding caustic potash to a boiling solution of 1-phenyldithiobiuret and ethylene chlorohydrin in ethanol.[5]

[1] Royal Free Hospital School of Medicine, University of London, London, England.

[2] Kurzer and Powell, *J. Chem. Soc.*, **1955**, 1500.

[3] Böeseken, *Rec. trav. chim.*, **16**, 350 (1897).

[4] Böeseken, *Rec. trav. chim.*, **29**, 279 (1910).

[5] Fromm and Wenzl, *Ber.*, **55B**, 809 (1922).

1,2-CYCLODECANEDIOL

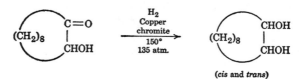

(*cis* and *trans*)

Submitted by A. T. BLOMQUIST and ALBERT GOLDSTEIN.[1]
Checked by N. J. LEONARD and F. H. OWENS.

1. Procedure

A mixture of 140 g. (0.82 mole) of sebacoin (p. 840) (Note 1), 50 g. of copper chromite catalyst [2] (Note 2), and 230 ml. of 95% ethanol is placed in an 800-ml. high-pressure hydrogenation bomb (Note 3). Hydrogen is admitted to the bomb at 135 atm., and the bomb is heated to 150°. When this temperature is reached, the shaker is started. The temperature is stabilized at 150° after its initial rapid fluctuation, and the hydrogenation is allowed to proceed until the mixture ceases to absorb hydrogen (2–4 hours). The heating and shaking are discontinued, and the bomb is allowed to cool to room temperature. The excess hydrogen is vented, and the bomb is dismantled.

The cis-1,2-cyclodecanediol will have crystallized out of solution, while the trans-diol remains in the ethanol. The entire mixture is washed out of the bomb with 95% ethanol (about 1 l.). The cis-glycol is redissolved by heating the ethanolic mixture at reflux temperature. Filter aid ("Celite") is added to the mixture, and the hot mixture is filtered through a bed of filter aid on a Büchner funnel to remove the catalyst. The ethanol is removed from the filtrate by distillation on a steam bath under water-pump pressure. The residue is dissolved in a minimum of hot 1:1 benzene-ethanol solution (about 250 ml.), and the cis-diol crystallizes upon cooling to room temperature. The crystals are collected by filtration, and the mother liquor is concentrated to dryness on a steam bath under water-pump pressure. The residue is again dissolved in a minimum of hot 1:1 benzene-ethanol (about 100 ml.), and additional cis-diol crystallizes upon cooling in a refrigerator. The process is repeated using 50 ml. of 1:1 benzene-ethanol solvent. The total yield of cis-1,2-cyclodecanediol is 68–73 g. (48–52%), m.p. 137–138°.

The trans-diol remains in the mother liquor and may be recovered by complete evaporation of the solvent followed by recrystallization of the residue from pentane. The yield of trans-1,2-cyclodecanediol is 38–45 g. (27–32%), m.p. 53–54°.

2. Notes

1. A sebacoin-sebacil mixture may be used.

2. A commercial catalyst was employed by the submitters: Harshaw Chemical Company, CU-0202P; 556–002.

3. A suitable apparatus is the "Aminco" high-pressure hydrogenation apparatus, manufactured by the American Instrument Company, Silver Spring, Maryland.

3. Methods of Preparation

1,2-Cyclodecanediol has been prepared by the hydrogenation of sebacoin in the presence of Raney nickel [3] or platinum,[4] by the reduction of sebacoin with aluminum isopropoxide [4] or lithium aluminum hydride,[4] and by the oxidation of cyclodecene with osmium tetroxide and pyridine.[3]

[1] Cornell University, Ithaca, New York.

[2] *Org. Syntheses Coll. Vol.* **2**, 142 (1943), Note 11.

[3] Prelog, Schenker, and Günthard, *Helv. Chim. Acta*, **35**, 1598 (1952); Prelog, Urech, Bothner-By, and Würsch, *Helv. Chim. Acta*, **38**, 1095 (1955).

[4] Blomquist, Burge, and Sucsy, *J. Am. Chem. Soc.*, **74**, 3636 (1952).

CYCLODECANONE *

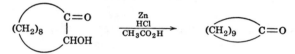

Submitted by ARTHUR C. COPE, JOHN W. BARTHEL, and RONALD DEAN SMITH.[1]

Checked by N. J. LEONARD and J. C. LITTLE.

1. Procedure

A 1-l. round-bottomed three-necked flask is fitted with a sealed stirrer (Note 1), a dropping funnel, and a reflux condenser, through which a thermometer extends nearly to the bottom of the flask. In the flask are thoroughly mixed 40.5 g. (0.62 g. atom) of zinc dust (Note 2) and 100 ml. of glacial acetic acid, and to this mixture is added 42.5 g. (0.25 mole) of sebacoin (p. 840) (Note 3). The mixture is stirred rapidly, and 90 ml. of concentrated c.p. hydrochloric acid is added dropwise during a period of 5 to 10 minutes, or as fast as control of foaming and temperature permits. The temperature must be kept between 75 and 80° (Note 4), and cooling by a water bath may be necessary during the addition of the hydrochloric acid. Stirring is continued for 1.5 hours at 75–80°. Thirty minutes after the initial addition of hydrochloric acid, and again 30 minutes later, 90-ml. portions of concentrated hydrochloric acid are added to the mixture while the temperature is maintained at 75–80°. After the reaction is complete, the remaining zinc is separated from the cooled mixture by decantation (Note 5). The liquid phase is diluted with 700 ml. of saturated aqueous sodium chloride solution and extracted with four 250-ml. portions of ether, each of which is first used to wash the residual zinc (Note 6). The ether extracts are combined and washed with 250 ml. of saturated sodium chloride solution, three 250-ml. portions of 10% sodium carbonate solution (foaming!), and finally 250 ml. of saturated sodium chloride solution. The ethereal solution is dried over anhydrous magnesium sulfate (about 25 g. is needed). After the drying agent has been removed by filtration and the solvent by distillation, the residue is distilled at reduced pressure through an efficient column (Note 7). After a small fore-run consisting mostly of cyclodecane, cyclodecanone is collected at 99–101°/8 mm. The yield is 29–30 g. (75–78%), n_D^{25} 1.4808–1.4810 (Note 8).

2. Notes

1. A metal stirrer must not be used. A simple glass stirrer with a ball-joint seal is satisfactory.

2. Mallinckrodt technical grade may be used.[2] If Mallinckrodt analytical reagent zinc dust is used, the reaction temperature must be maintained at 50–55° instead of 75–80°.

3. Pure sebacoin gives a colorless product. A sebacoin-sebacil mixture must first be purified by recrystallization from pentane as described (p. 841). The sebacil apparently is not reduced completely according to the accompanying directions and thus may contaminate the product (see Note 7).

4. The reaction temperature is important. At temperatures below 75° some sebacoin remains unreduced, while at temperatures above 80° considerable cyclodecane is formed. The submitters report that the reaction run at the reflux temperature gives cyclodecanone in 27% yield and cyclodecane in 32% yield.

5. The product should be isolated and distilled as quickly as possible inasmuch as the unreacted sebacoin is readily oxidized to sebacil, which cannot be separated from the cyclodecanone by simple distillation.

6. The residual zinc may be pyrophoric.

7. For efficient separation of cyclodecanone from cyclodecane, a 60-cm. column of the simple Podbielniak type [3] may be used. Removal of sebacil cannot be accomplished readily by fractional distillation, since cyclodecanone and sebacil have virtually identical boiling points.

8. Cyclodecanone regenerated from its semicarbazone, m.p. 203.5–205.5°, has n_D^{25} 1.4806.

3. Methods of Preparation

The procedure described is a modification of the directions of Prelog, Frenkiel, Kobelt, and Barman.[4] Cyclodecanone has been prepared by the dehydration of sebacoin followed by catalytic hydrogenation,[5] by the pyrolysis of the thorium or yttrium salt of nonane-1,9-dicarboxylic acid,[6] and by the ring enlargement of cyclononanone,[7] as well as by the reduction of sebacoin.[8]

[1] Massachusetts Institute of Technology, Cambridge 39, Massachusetts.

[2] Brown and Borkowski, *J. Am. Chem. Soc.*, **74**, 1901 (1952).

[3] Cason and Rapoport, *Laboratory Text in Organic Chemistry*, 2nd ed., p. 294, Prentice-Hall, Englewood Cliffs, New Jersey, 1962.

[4] Prelog, Frenkiel, Kobelt, and Barman, *Helv. Chim. Acta*, **30**, 1741 (1947).

[5] Stoll, *Helv. Chim. Acta*, **30**, 1837 (1947).

[6] Ruzicka, Stoll, and Schinz, *Helv. Chim. Acta*, **9**, 249 (1926); **11**, 670 (1928).

[7] Kohler, Tishler, Potter, and Thompson, *J. Am. Chem. Soc.*, **61**, 1057 (1939).

[8] Blomquist, Burge, and Sucsy, *J. Am. Chem. Soc.*, **74**, 3636 (1952).

CYCLOHEPTANONE *

I. NITROUS ACID METHOD

Submitted by Hyp J. Dauben, Jr., Howard J. Ringold, Robert H. Wade, David L. Pearson, and Arthur G. Anderson, Jr.[1]

Checked by Arthur C. Cope, Warren N. Baxter, and Robert J. Cotter.

1. Procedure

Caution! The sodium salt of 1-(nitromethyl)cyclohexanol may be an explosive (Note 1).

A solution of sodium ethoxide is prepared by adding 57.5 g. (2.5 g. atoms) of clean sodium to 1.2 l. of absolute ethanol (Note 2) in a 3-l. three-necked flask equipped with an Allihn reflux condenser fitted with a drying tube, a large sturdy sealed Hershberg stirrer, and a dropping funnel. After the sodium has dissolved, the solution is cooled to 40° and the condenser is replaced by a thermometer extending into the liquid. A mixture of 245.5 g. (258.5 ml., 2.5 moles) of redistilled cyclohexanone and 198 g. (175 ml., 3.25 moles) of redistilled nitromethane (Note 3) is added dropwise with vigorous stirring over the course of about 3 hours at a rate that maintains an internal temperature of 45 ± 3° (Note 4). After addition is complete, the white, pasty mass is stirred for an additional 3 hours without cooling or heating and

then is allowed to stand overnight. The resulting suspension
is cooled with an ice bath, and the white sodium salt of 1-(nitromethyl)cyclohexanol is collected on a 25-cm. Büchner funnel and
dried by suction for about 1 hour (Note 1). The sodium salt
cake is broken up and transferred to a 4-l. beaker equipped with
a Hershberg stirrer and immersed in an ice bath. A cold solution of 184 g. (175 ml., 3 moles) of glacial acetic acid in 1250 ml.
of water is added in a single portion, and the mixture is stirred
for 10–30 minutes to complete dissolution. The oily layer of 1-
(nitromethyl)cyclohexanol is separated, and the aqueous layer is
extracted with three 100-ml. portions of ether. The ether extracts and the 1-(nitromethyl)cyclohexanol are combined, dried
briefly over magnesium sulfate, and concentrated by distillation
from a steam bath with a water aspirator at 20–35 mm. to remove
ether and excess nitromethane. The crude, undistilled 1-(nitromethyl)cyclohexanol (Note 5) is dissolved in 450 ml. of glacial
acetic acid in a 2-l. externally cooled stainless-steel hydrogenation
bottle (Note 6). Three heaping teaspoonfuls of W-4 Raney nickel
catalyst [2] are added, and the mixture is shaken with hydrogen at
40–45 p.s.i., with cooling to maintain the temperature below 35°,
until about 90% of the theoretical amount (7.5 moles) is taken up
and absorption ceases (Note 7). The catalyst is separated by
filtration with suction through Filter-Cel, and the filtrate (sometimes green or tan) containing the 1-(aminomethyl)cyclohexanol
is used directly in the next step (Note 8).

The filtrate is transferred to a 5-l. round-bottomed flask, immersed in an ice-salt mixture and equipped with a Hershberg
stirrer, a thermometer, and a dropping funnel. The solution is
diluted with 2.3 l. of ice water; then an ice-cold solution of 290 g.
(4.2 moles) of sodium nitrite in 750 ml. of water is added dropwise
during a period of about 1 hour with stirring and cooling (ice-
salt bath) to maintain the temperature at −5°. The mixture is
stirred for an additional period of 1 hour and then allowed to
come to room temperature overnight as the ice in the cooling
bath melts. The acetic acid in the reaction mixture is neutralized
by the addition of small portions of solid sodium bicarbonate,
and the neutral (to litmus paper) solution is then steam-distilled
until about 2 l. of distillate is collected. The oily cycloheptanone
layer is separated, the aqueous layer is extracted with three

100-ml. portions of ether, and the combined organic layers are dried briefly over magnesium sulfate. Most of the ether is removed by distillation through a 17 by 2.5 cm. glass helix-packed column at atmospheric pressure (Note 9). The residue is then distilled through the same column, and the fraction boiling at 80–85°/30 mm. is collected. The yield of cycloheptanone is 112–118 g. (40–42%), n_D^{25} 1.4600 (Notes 10 and 11).

2. Notes

1. It has been reported [3] that when the crude product from the condensation of 2 moles of nitromethane (as the sodium salt in absolute ethanol) with 1 mole of 1,1,3,3-tetramethylcyclobutane-2,4-dione (dimethylketene dimer) was filtered, and the filter cake was dried on a porous clay plate, a violent explosion occurred upon addition of the dried powder to crushed ice.

This experience suggests that it may be inadvisable to dry completely the sodium salt of 1-(nitromethyl)cyclohexanol, and that the material should be handled with considerable caution. Also, in other procedures where salts of nitro compounds are involved, and the extent of the reaction and the nature of the products are not known with certainty, the salts probably should not be filtered, but should be neutralized in the reaction mixtures and the free nitro alcohols separated by suitable means.

2. Commercial absolute ethanol is used without additional drying. The submitters state that the use of absolute ethanol may be avoided by employment of the alternative condensation procedure of Wood and Cadorin [4] using 5 mole per cent of sodium hydroxide in aqueous methanol at 15–20° for 6 hours, but the yield of 1-(nitromethyl)cyclohexanol is appreciably lower (51%).

3. Cyclohexanone is dried over magnesium sulfate or calcium sulfate (Drierite) for 1 day and distilled; the fraction boiling at 152.5–154° (uncor.) is used. Nitromethane is dried in the same manner and distilled; small quantities of acidic impurities are removed in the fore-run, and the fraction boiling at 101.5–102.5° (uncor.) is used. Drying is frequently unnecessary when good grades of cyclohexanone and nitromethane are used.

4. Efficient stirring of the pasty reaction mixture is necessary to obtain maximum yield.

5. Removal of residual amounts of acetic acid in the crude product is unnecessary; their presence is actually preferable to the presence of traces of bases, such as any sodium bicarbonate remaining from neutralization, which reverse the condensation on attempted distillation of 1-(nitromethyl)cyclohexanol.[4] The submitters state that, in preparations conducted on twice the scale specified, fractional distillation of the residue yielded 620–670 g. (78–84%) of 1-(nitromethyl)cyclohexanol, b.p. 129–132°/19 mm., n_D^{25} 1.4835.

6. A Pyrex bottle can be used; it must be enclosed to prevent possible injury by fragments of glass in case of explosion.

7. Catalytic hydrogenation of 1-(nitromethyl)cyclohexanol in acetic acid solvent is a markedly exothermic reaction, and, unless the temperature is moderated to about 35°, low yields of 1-(aminomethyl)cyclohexanol result owing to hydrogenolysis and deactivation of the catalyst. Cooling can be accomplished by running cold water through a copper coil surrounding the hydrogenation bottle or by periodically adding ice to the container in which the bottle is placed. Considerable cooling is necessary during initial stages of the hydrogenation, but cooling below 25° greatly retards the reduction. Absorption of hydrogen is usually complete in 15–18 hours under these conditions, but longer times may be required, depending on the temperature and pressure and on the amount and activity of the catalyst.

8. 1-(Aminomethyl)cyclohexanol can be isolated as its acetic acid salt by the addition of 2 volumes of ether to the acetic acid solution or to the residue after removal of the acetic acid under reduced pressure, followed by trituration and refrigeration overnight. After filtration of the first crop of crystals, m.p. 118–121°, a second crop of the salt, m.p. 113–116°, is obtained on removal of the dissolved nickel from an aqueous solution of the concentrated filtrate by saturation with hydrogen sulfide, concentration, and treatment with ether. Total yields of the air-dried crude acetic acid salt of 1-(aminomethyl)cyclohexanol obtained from 100–350 g. lots of 1-(nitromethyl)cyclohexanol in this manner ranged from 69% to 94%.

9. Distillation of the product from a modified Claisen flask and collection of the fraction boiling at 67–77°/20 mm. (largely

69–72°) gave 140–149 g. of impure cycloheptanone, n_D^{25} 1.4570–1.4590, containing about equal amounts (10–15 g.) of lower-boiling and higher-boiling (mainly 1-(nitromethyl)cyclohexanol) contaminants, which were separated by distillation through the glass helix-packed column.

10. The submitters state that the yields of cycloheptanone obtained from 60- to 700-g. quantities of the crude acetic acid salt of 1-(aminomethyl)cyclohexanol (isolated by the procedure of Note 8) were 57–65%.

11. Small quantities of cycloheptanone are prepared more conveniently by the diazomethane method [5] or by dry distillation of suberic acid salts [6-8] or suberic acid admixed with iron filings and barium hydroxide.[9]

II. DIAZOMETHANE METHOD

$$p\text{-}CH_3C_6H_4SO_2N(NO)CH_3 + C_2H_5OH \xrightarrow{\text{aq. KOH}}$$
$$p\text{-}CH_3C_6H_4SO_2OC_2H_5 + CH_2N_2 + H_2O$$

Submitted by TH. J. DE BOER and H. J. BACKER.[10]
Checked by JAMES CASON, JOHN B. ROGAN, and WM. G. DAUBEN.

1. Procedure

Caution! Diazomethane is very toxic; therefore the operations must be carried out in a well-ventilated hood. (See p. 250.)

In a 500-ml. round-bottomed flask, provided with a mechanical stirrer (Note 1), a thermometer, and a dropping funnel, are placed 49 g. (0.5 mole) of cyclohexanone (Note 2), 125 g. (0.58 mole) of p-tolylsulfonylmethylnitrosamide (p. 943), 150 ml. of 95% ethanol, and 10 ml. of water (Note 3). The nitroso compound is largely undissolved. The stirrer is adjusted so that only the upper portion of solution is stirred and the precipitate moves slightly. The thermometer bulb is placed in the liquid.

The reaction mixture is cooled to about 0° with an ice-salt bath; then with gentle stirring a solution of 15 g. of potassium hydroxide in 50 ml. of 50% aqueous ethanol is added dropwise very slowly from the dropping funnel. After the addition of 0.5–1 ml. of the alkaline solution a brisk evolution of nitrogen commences and the temperature rises. The rate of addition is adjusted so that the temperature is maintained at 10–20° (Note 4). The addition of the alkali requires about 2 hours, during which the nitroso compound gradually disappears. The orange-yellow solution is stirred for an additional 30 minutes, then 2N hydrochloric acid (about 50 ml.) is added until the solution is acidic to litmus paper.

A solution of 100 g. of sodium bisulfite (Note 5) in 200 ml. of water is added as stirring is continued. After a few minutes a thick precipitate separates. The mixture is stirred or preferably shaken mechanically at room temperature with exclusion of air for 10 hours. The bisulfite addition product is separated by suction filtration, washed with ether until colorless, and decomposed in a flask with a lukewarm solution of 125 g. of sodium carbonate in 150 ml. of water. The ketone layer is separated, the aqueous layer is extracted with four 25-ml. portions of ether, and the combined organic layers are dried over anhydrous sodium sulfate. Most of the ether is removed by distillation at atmospheric pressure, and the residual oil is distilled at reduced pressure (Note 6).

After a few drops of fore-run, practically all the liquid distils at 64–65°/12 mm. The yield of cycloheptanone is 18.5–20.2 g. (33–36%) (Note 6).

2. Notes

1. If it is desired to determine the amount of nitrogen evolved, then it is necessary to use a sealed stirrer and a discharge tube for the nitrogen; however, observing the amount of nitrogen evolved affords no particular advantage. The temperature exerts sufficient control on the reaction rate, and dissolution of the nitroso compound indicates the progress of the reaction.

2. Cyclohexanone was distilled before use, b.p. 154–155°.

3. The water keeps in solution the potassium p-toluenesulfonate, which is formed by partial saponification of the ester.

4. In some experiments, the temperature was allowed to rise temporarily to 35° without mishap, but the lower temperature is preferred for reasons of safety.

5. Cyclooctanone, which is a by-product, is removed at this stage since it does not form an adduct with bisulfite.

6. The checkers found it desirable to distil the product through a fractionating column in order to effectively remove the small fore-run and a smaller after-run. The pattern obtained in a 50-cm. column of the simple Podbielniak type follows: fraction 1, 1.13 g., b.p. up to 63°/12 mm., n_D^{25} 1.4549; fraction 2, 19.44 g., b.p. 63–64°/12 mm., n_D^{25} 1.4592; fraction 3, 0.55 g., b.p. > 64°/12 mm., n_D^{25} 1.4604. A center cut of fraction 2 had n_D^{25} 1.4590.

3. Methods of Preparation

Cycloheptanone has been prepared by variations of two general routes, ring closure and ring enlargement. Ring-closure methods that have been employed are: (a) dry distillation of calcium [6] (35–50% yield), thorium [7] (45% yield), cerium [7] (45% yield), and zinc or magnesium [8] (55–60% yield) salts of suberic acid; (b) pyrolysis of a mixture of suberic acid, iron filings, and barium hydroxide [9] (40% yield) or pyrolysis of suberic acid in the presence of catalytic amounts of thorium oxide; [11] (c) Dieckmann condensation of diethyl suberate with sodium ethoxide in ether [12] or potassium tert-butoxide in xylene under high dilution conditions; [13] (d) Thorpe-Ziegler condensation of suberonitrile using preferably sodium methylanilide in ether by a high-dilution technique followed by hydrolysis and decarboxylation (80–85% yield); [14] and (e) dehydrohalogenation of suberyl chloride using triethylamine by a high-dilution technique (33% yield). [15] Ring-enlargement methods that have been used are: (a) diazomethane on cyclohexanone (33–36% yield) [5] and (b) catalytic reduction of cyclohexanone cyanohydrin [16] or electrolytic reduction of 1-(nitromethyl)cyclohexanol [17] and treatment of the resulting 1-(aminomethyl)cyclohexanol with nitrous acid, and hydrogenation of tetrahydrobenzonitrile (from acrylonitrile and butadiene) to cyclohexylmethylamine which then is caused to react with nitrous acid. [18]

Procedure I represents an improved modification of the proce-

dure of Dauben, Ringold, Wade, and Anderson,[19] and Procedure
II is a modification of earlier diazomethane methods and employs
the relatively stable *p*-tolylsulfonylmethylnitrosamide (p. 943).

[1] University of Washington, Seattle, Washington.
[2] Pavlic and Adkins, *J. Am. Chem. Soc.*, **68**, 1471 (1946).
[3] Rosowsky, Private communication.
[4] Wood and Cadorin, *J. Am. Chem. Soc.*, **73**, 5504 (1951).
[5] Kohler, Tishler, Potter, and Thompson, *J. Am. Chem. Soc.*, **61**, 1059 (1939); Mosettig and Burger, *J. Am. Chem. Soc.*, **52**, 3460 (1930); Meerwein, Ger. pat. 579,309 [*C. A.*, **27**, 4546 (1933)]; de Boer and Backer, *Proc. Koninkl. Ned. Akad. Wetenschap.*, **55B**, 444 (1952) [*C. A.*, **48**, 9903 (1954)].
[6] Day, Kon, and Stevenson, *J. Chem. Soc.*, **1920**, 642.
[7] Ruzicka, Brugger, Pfeiffer, Schinz, and Stoll, *Helv. Chim. Acta*, **9**, 515 (1926).
[8] Böeseken and Derx, *Rec. trav. chim.*, **40**, 530 (1921); Derx, *Rec. trav. chim.*, **41**, 338 (1922).
[9] Vogel, *J. Chem. Soc.*, **1928**, 2032.
[10] De Rijks-Universiteit, Groningen, The Netherlands.
[11] Badische Anilin- & Soda-Fabrik Akt.-Ges. (by Schlichting, Dörries and Gehm), Ger. pat. 1,025,872 [*C. A.*, **54**, 9799 (1960)].
[12] Dieckmann, *Ann.*, **317**, 49 (1901).
[13] Leonard and Schimelpfenig, *J. Org. Chem.*, **23**, 1708 (1958).
[14] Ziegler, Eberle, and Ohlinger, *Ann.*, **504**, 120 (1933); Ziegler and Aurnhammer, *Ann.*, **513**, 57 (1934).
[15] Blomquist and Spencer, *J. Am. Chem. Soc.*, **70**, 30 (1948).
[16] Tchoubar, *Bull. soc. chim. France*, **1949**, 160, 164, 169; Blicke, Doorenbos, and Cox, *J. Am. Chem. Soc.*, **74**, 2924 (1952).
[17] Blicke (to University of Michigan), U. S. pat. 2,846,474 [*C. A.*, **53**, 2120 (1959)].
[18] Schmid, *Kunststoffe—Plastics*, **3**, 165 (1956) [*C. A.*, **52**, 11006 (1958)].
[19] Dauben, Ringold, Wade, and Anderson, *J. Am. Chem. Soc.*, **73**, 2359 (1951).

1,2-CYCLOHEXANEDIONE DIOXIME *

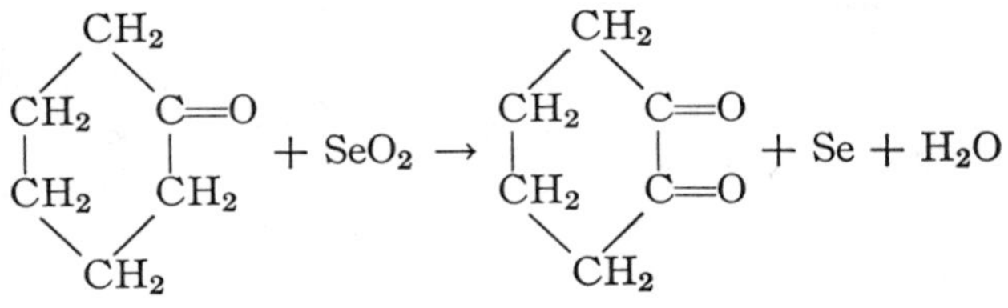

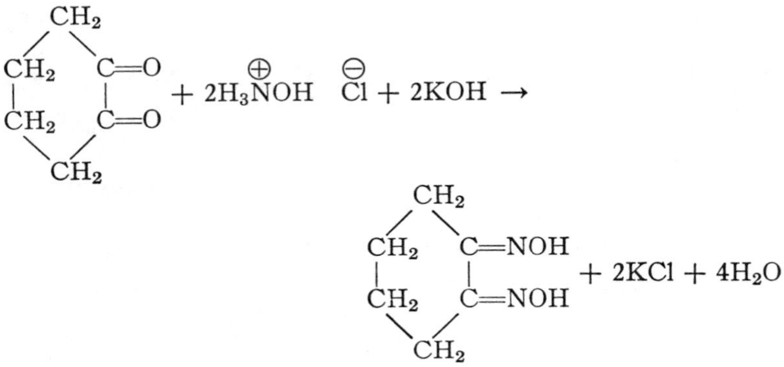

$$
\begin{array}{c}
CH_2 \\
CH_2 \quad C=O \\
CH_2 \quad CH_2 \\
CH_2
\end{array}
+ SeO_2 \rightarrow
\begin{array}{c}
CH_2 \\
CH_2 \quad C=O \\
CH_2 \quad C=O \\
CH_2
\end{array}
+ Se + H_2O
$$

$$
\begin{array}{c}
CH_2 \\
CH_2 \quad C=O \\
CH_2 \quad C=O \\
CH_2
\end{array}
+ 2H_3\overset{\oplus}{N}OH \quad \overset{\ominus}{Cl} + 2KOH \rightarrow
$$

$$
\begin{array}{c}
CH_2 \\
CH_2 \quad C=NOH \\
CH_2 \quad C=NOH \\
CH_2
\end{array}
+ 2KCl + 4H_2O
$$

Submitted by CLIFFORD C. HACH, CHARLES V. BANKS, and HARVEY DIEHL.[1]
Checked by R. T. ARNOLD and PHILIP N. GORDON.

1. Procedure

A. 1,2-*Cyclohexanedione.* A 3-l. round-bottomed flask, fitted with stirrer and dropping funnel, is placed in a water bath containing a copper coil through which cooling water may be circulated. In the 3-l. flask is placed 1708 g. (17.4 moles, 1.8 l.) of cyclohexanone (Note 1). Tap water is circulated through the cooling coil (Note 2), and a solution containing 387 g. (3 moles) of selenious acid (H_2SeO_3) (Note 3), 500 ml. of dioxane, and 100 ml. of water is added dropwise and with stirring to the cyclohexanone over a period of 3 hours. The reaction mixture immediately turns yellow, and red amorphous selenium gradually appears. Stirring is continued for 5 additional hours at water-bath temperatures and then for 6 more hours at room temperature. Removal of the bulky, amorphous selenium is accomplished with the aid of a 6-in. Büchner funnel. The selenium is returned to the reaction flask and extracted with 300 ml. of boiling 95% ethanol for 1 hour (Note 4). The solution, obtained by decantation from the compact gray selenium, is combined with the above filtrate in a 4-l. distilling flask. Distillation under reduced pressure gives two fractions. The lower-boiling fraction (25–60°/16 mm.) consists mainly of ethanol, water, dioxane, and cyclohexanone; the higher-boiling one (60–90°/16 mm.) contains cyclohexanone and 1,2-cyclohexanedione with traces of water and dioxane. The yield of crude product is approximately 322 g.

The higher-boiling fraction is redistilled (Note 5), and again two fractions, boiling at 25–75°/16 mm. and 75–79°/16 mm., are collected. The latter fraction is essentially pure 1,2-cyclohexanedione and crystallizes at 34° to ice-like crystals which become light yellow-green when exposed to the air; yield 202.5 g. (60% based on selenous acid). A considerable amount of light-brown, clear, resinous residue remains in the distilling flask.

B. *1,2-Cyclohexanedione dioxime.* In a 1-l. Erlenmeyer flask is placed 200 ml. of water and 100 g. of cracked ice. To this ice-water mixture is added 86.9 g. (1.25 moles) of hydroxylammonium chloride. An ice-cold basic solution is prepared by dissolving 82.4 g. (1.25 moles) of 85% potassium hydroxide in 50 ml. of water and then adding 150 g. of cracked ice. The ice-cold potassium hydroxide solution is added to the hydroxylammonium chloride solution, and the mixture is thoroughly shaken. To the mixture is added about 0.5 g. of nioxime (1,2-cyclohexanedione dioxime) with stirring (Note 6). The solution turns red owing to the reaction of the nioxime with quantities of iron and other impurities in the reagents. Four grams of Norit is added; the mixture is thoroughly shaken and then filtered with the aid of a 5-in. Büchner funnel. After this operation the filtrate should be water white.

The cold solution is transferred to a 1-l. Erlenmeyer flask and is placed in an ice bath. To the stirred solution is added slowly 56 g. (0.5 mole) of melted 1,2-cyclohexanedione. Precipitation of the 1,2-cyclohexanedione dioxime should take place almost immediately. If not, the solution may be seeded to initiate rapid precipitation. The mixture is stirred for 30 minutes after the addition of the 1,2-cyclohexanedione, and the precipitate is then collected on a 5-in. Büchner funnel. The precipitate is thoroughly washed with water to remove inorganic salts. The 1,2-cyclohexanedione dioxime is partially dried by suction and finally dried in a vacuum desiccator to give snow-white crystals; yield 52.5 g. (74%); m.p. 185–188° (darkening at 170°) (Note 7). The crude 1,2-cyclohexanedione dioxime is recrystallized from 550 ml. of water using 2.5 g. of iron-free Norit; yield 39.3 g. (55%); m.p. 186–188°.

2. Notes

1. The cyclohexanone need not be freshly distilled. Commercial cyclohexanone, obtained from the Barrett Division, Allied Chemical & Dye Corporation, New York, New York, gave practically the same yield as carefully fractionated cyclohexanone.

2. If too much selenous acid is added at once, or the cooling discontinued, the solution will heat up and the reaction will become extremely vigorous with subsequent decrease in yield.

3. Selenous acid or selenium dioxide can apparently be used interchangeably.

4. The selenium filtered from the reaction mixture is refluxed with 3 l. of 95% ethanol for 1 hour; this converts the red amorphous form to the gray hexagonal form and frees it of organic matter. This metallic selenium is removed by filtration, washed with water, and converted to the dioxide by the method of Baker and Maxson.[2]

5. A distilling head such as J-1104 obtained from Scientific Glass Apparatus Company, Bloomfield, New Jersey, was used.

6. This treatment must obviously be omitted on the first preparation; the product will be light pink rather than snow white, as obtained when the reagents are purified in this manner.

7. This material is 96.5% pure as determined by precipitation of the nickel compound and is satisfactory as an analytical reagent.

3. Methods of Preparation

1,2-Cyclohexanedione has been prepared by brominating cyclohexanone and treating the resulting 2,6-dibromocyclohexanone with aqueous potassium hydroxide to obtain the dihydroxy compound which loses water to yield the dione;[3] by heating divinyl glycol with copper;[4] and by oxidizing cyclohexanone with selenium dioxide in an ethanolic solution.[5, 6]

1,2-Cyclohexanedione dioxime has been prepared by oximating 1,2-cyclohexanedione with hydroxylammonium chloride in aqueous potassium hydroxide solution;[3, 6] by oximating 2-isonitrosocyclohexanone with hydroxylammonium chloride;[7, 8] by oximating sodium 2-isonitrosocyclohexanone with hydroxylammonium chloride in methanolic solution;[6] and by the reaction of hydroxylamine with 2-chloro-[9] or 2-bromocyclohexanone,[10] and with perhydro-4a,9a-epoxydibenzo-p-dioxin-5a,10a-diol.[11]

[1] Iowa State University, Ames, Iowa.

[2] Baker and Maxson, *Inorg. Syntheses*, 1, 119–120 (1939).

[3] Wallach and Weissenborn, *Ann.*, **437**, 172 (1924).

[4] Urion, *Compt. rend.*, **192**, 1662 (1931).

[5] Riley, Morley, and Friend, *J. Chem. Soc.*, **1932**, 1875.

[6] Rauh, Smith, Banks, and Diehl, *J. Org. Chem.*, **10**, 199 (1945).

[7] Treibs and Dinelli, *Ann.*, **517**, 160 (1935).

[8] Jaeger and van Dijk, *Proc. Acad. Sci. Amsterdam*, [1] **39**, 384, 392 (1936).

[9] Tokura and Oda, *Bull. Inst. Phys. Chem. Research* (*Tokyo*), **22**, 844 (1943) [*C. A.*, **43**, 2176 (1949)].

[10] Belcher, Hoyle, and West, *J. Chem. Soc.*, **1958**, 2743.

[11] Svoboda and Krátký, Czech. pat. 87,466 [*C. A.*, **54**, 8674 (1960)].

CYCLOHEXENE SULFIDE *

(7-Thiabicyclo[4.1.0]heptane)

Submitted by EUGENE E. VAN TAMELEN.[1]
Checked by CHARLES C. PRICE and PAUL F. KIRK.

1. Procedure

Ninety-eight grams (1 mole) of cyclohexene oxide (Note 1) is divided into two approximately equal portions; one portion is added to a solution of 121 g. (1.25 moles) of potassium thio-cyanate in 100 ml. of water and 75 ml. of 95% ethanol. After standing for 3–4 hours (Note 2) the clear solution is transferred to a 1-l. flask equipped with a mechanical stirrer. The second portion of oxide is added, and the resulting solution is stirred vigorously for 36 hours at room temperature. The supernatant layer and the aqueous phase are then decanted from the precipitated potassium cyanate into a 1-l. separatory funnel. The potassium cyanate is rinsed with 50 ml. of ether, which is subsequently added to the separatory funnel and used to extract the cyclohexene sulfide. The ether extract is washed twice with 50-ml. portions of saturated sodium chloride solution and then dried over anhydrous sodium sulfate. The excess ether is removed on the steam bath, and the residual liquid is distilled under reduced pressure through an 18-in. Vigreux column while the distillate is being cooled in ice. The main fraction boils at 71.5–73.5°/21 mm. (69–71°/19 mm.) (Note 3); n_D^{25} 1.5306–1.5311. A fore-run, boiling up to 71.5°/21 mm., yields more of the product on re-distillation. The total yield of cyclohexene sulfide is 81.5–83.5 g. (71–73%).

2. Notes

1. The cyclohexene oxide (b.p. 129–134°) was prepared from 2-chlorocyclohexanol.[2]
2. During this time a temperature rise of about 5° occurs.
3. Cyclohexene sulfide can be stored at about 5° in a closed container for at least a month without apparent decomposition.

3. Methods of Preparation

Cyclohexene sulfide has been prepared by the action of thiourea, potassium thiocyanate, or ammonium thiocyanate on cyclohexene oxide,[3] by the hydrolysis of S-(*trans*-2-hydroxycy-clohexyl)thiuronium sulfate in sodium carbonate solution,[4] and by the alkaline hydrolysis of 2-mercaptocyclohexyl acetate or 2-acetylmercaptocyclohexyl acetate.[5]

The method described above is a modification of that of Snyder, Stewart, and Ziegler.[6]

[1] University of Wisconsin, Madison, Wisconsin.
[2] *Org. Syntheses Coll. Vol.* **1**, 185 (1941).
[3] Culvenor, Davies, and Pausacker, *J. Chem. Soc.*, **1946**, 1050.
[4] Bordwell and Andersen, *J. Am. Chem. Soc.*, **75**, 4959 (1953).
[5] Harding, Miles, and Owen, *Chem. & Ind. (London)*, **1951**, 887.
[6] Snyder, Stewart, and Ziegler, *J. Am. Chem. Soc.*, **69**, 2672 (1947).

CYCLOHEXYLIDENECYANOACETIC ACID AND 1-CYCLOHEXENYLACETONITRILE

($\Delta^{1,\alpha}$-Cyclohexaneacetic acid, α-cyano-, and 1-Cyclohexene-1-acetonitrile)

Submitted by ARTHUR C. COPE, ALFRED A. D'ADDIECO, DONALD EDWARD WHYTE, and SAMUEL A. GLICKMAN.[1]
Checked by T. L. CAIRNS and R. E. HECKERT.

1. Procedure

In a 500-ml. round-bottomed flask equipped with a side arm through which a capillary tube is inserted (Note 1) are placed 108 g. (1.1 moles) of cyclohexanone (Note 2), 85 g. (1.0 mole) of cyanoacetic acid (Note 3), 3.0 g. (0.04 mole) of ammonium acetate (Note 4), and 75 ml. of benzene. The flask is attached to a modified Dean and Stark constant water separator,[2, 3] which in turn is attached to an efficient reflux condenser. The mixture is heated in an oil bath at 160–165° so that a vigorous reflux is maintained, and the water that collects in the separator is removed at intervals. The theoretical amount of water (18 ml.) is collected in the course of 2 hours, and the mixture is heated under reflux for an additional 1 hour. At this point, Part A is followed for the isolation of cyclohexylidenecyanoacetic acid, or Part B for the preparation of 1-cyclohexenylacetonitrile.

A. *Cyclohexylidenecyanoacetic acid.* The benzene solution is diluted with an additional 100 ml. of hot benzene and transferred to a 1-l. separatory funnel. The solution is allowed to cool until it is slightly above room temperature, and then 200 ml. of ether is added, small portions being used to rinse the reaction flask. After the solution has cooled to room temperature, it is washed with two 50-ml. portions of cold water (Note 5). The emulsion which normally forms at this point is broken by slow filtration through a Büchner funnel. The ether is removed, and the benzene solution is concentrated to approximately 300 ml. by distillation under reduced pressure. The solution is allowed to cool slowly to room temperature and then is cooled to about 10° in a refrigerator (Note 6). Cyclohexylidenecyanoacetic acid crystallizes as colorless prisms (Note 7). It is collected on a Büchner funnel, washed with two 100-ml. portions of cold benzene (10°), and dried in a vacuum desiccator to constant weight (88–92 g.). The filtrate and washings are concentrated by distillation under reduced pressure to about 150 ml. and cooled as in the first crystallization. The second crop of crystals is separated by filtration, washed with two 50-ml. portions of cold benzene (10°), and dried in a vacuum desiccator to constant weight (21–25 g.). Further concentration of the mother liquor and washings to a volume of about 75 ml. followed by cooling, filtering, washing with two 10-ml. portions of cold benzene, and drying yields an additional 2–5 g. The total yield of cyclohexylidenecyanoacetic acid, m.p. 110–110.5°, is 108–126 g. (65–76%).

B. *1-Cyclohexenylacetonitrile.* The benzene solution is allowed to cool to about 50°, and the flask is attached to a Vigreux column. The benzene is removed under reduced pressure, whereupon the residual cyclohexylidenecyanoacetic acid solidifies. The flask is then heated slowly in an oil bath to 165–175° while the system is evacuated with a water pump to a pressure of 35–45 mm. (not lower). The acid melts, decarboxylation occurs very rapidly, and the crude 1-cyclohexenylacetonitrile distils at 100–120°/35–45 mm.

The crude product is diluted with 50 ml. of ether, washed with 10 ml. of 5% sodium carbonate solution then with 10 ml. of water, and dried over anhydrous sodium sulfate. The ether is removed by distillation, and the residue is distilled under reduced pressure. 1-Cyclohexenylacetonitrile is collected as a colorless liquid, b.p. 74–75°/4 mm. (110–112°/25 mm.), n_D^{25} 1.4769, in a yield of 92–110 g. (76–91%).

2. Notes

1. The capillary aids ebullition in the distillation under reduced pressure in Part B and may be omitted if cyclohexylidene-cyanoacetic acid is to be prepared according to Part A.

2. Commercial cyclohexanone obtained from the Barrett Division of the Allied Chemical and Dye Corporation was used.

3. Good quality commercial cyanoacetic acid was used. It may be purchased from the Benzol Products Company, Newark, New Jersey.

4. The amount of ammonium acetate specified permits completion of the condensation in a relatively short reaction period.

5. The product is washed with water to remove small amounts of ammonium acetate and acetamide which are formed from the ammonium acetate during the condensation.

6. If the rate of cooling is too fast at this point, very small crystals difficult to wash are formed. The rate and time of cooling also control the proportions of product found in the three fractions.

7. The solvated crystals effloresce upon drying, leaving a white solid which is easily powdered.

3. Methods of Preparation

Cyclohexylidenecyanoacetic acid has been prepared by the condensation of cyclohexanone and cyanoacetic acid in the presence of piperidine [4, 5] or basic ion-exchange resins,[6] and by the hydrolysis of ethyl cyclohexylidenecyanoacetate.[4]

1-Cyclohexenylacetonitrile has been prepared by the decarboxylation of cyclohexylidenecyanoacetic acid; [4, 5] by the dehydration of 1-cyclohexenylacetamide; [5] by the condensation of cyclohexanone and cyanoacetic acid in the presence of piperidine; [7] by the condensation of cyclohexanone and ethyl cyanoacetate in the presence of sodium ethoxide; [4, 8] and by the condensation of cyclohexanone and cyanoacetic acid in the presence of ammonium acetate followed by decarboxylation.[9] Ammonium acetate also has been used as a catalyst for the condensation of ketones with ethyl cyanoacetate.[3, 10]

In a number of instances the decarboxylation of α,β-unsaturated (conjugated) cyanoacetic acids has been found to yield β,γ-unsaturated (unconjugated) nitriles.[5, 11]

[1] Massachusetts Institute of Technology, Cambridge, Massachusetts.
[2] Dean and Stark, *Ind. Eng. Chem.*, **12**, 486 (1920).
[3] Cope, Hofmann, Wyckoff, and Hardenbergh, *J. Am. Chem. Soc.*, **63**, 3452 (1941).
[4] Harding, Haworth, and Perkin, *J. Chem. Soc.*, **93**, 1959 (1908).
[5] Kandiah and Linstead, *J. Chem. Soc.*, **1929**, 2142, 2145.
[6] Astle and Gergel, *J. Org. Chem.*, **21**, 493 (1956).
[7] Shemyakin and Trakhtenberg, *Compt. rend. acad. sci. U.R.S.S.*, **24**, 763 (1939) [*C. A.*, **34**, 3676 (1940)].
[8] Birch and Kon, *J. Chem. Soc.*, **123**, 2444 (1923).
[9] Whyte and Cope, *J. Am. Chem. Soc.*, **65**, 2002 (1943).
[10] *Org. Syntheses Coll. Vol.* **3**, 399 (1955).
[11] von Auwers, *Ber.*, **56**, 1172 (1923).

CYCLOPENTADIENE AND 3-CHLOROCYCLOPENTENE

(Cyclopentene, 3-chloro-)

$$C_{10}H_{12} \xrightarrow{\Delta} 2 \begin{array}{c} HC\text{——}CH \\ \| \quad\quad \| \\ HC \quad\quad CH \\ \diagdown \quad \diagup \\ CH_2 \end{array}$$

$$\begin{array}{c} HC\text{——}CH \\ \| \quad\quad \| \\ HC \quad\quad CH \\ \diagdown \quad \diagup \\ CH_2 \end{array} + HCl \rightarrow \begin{array}{c} HC\text{====}CH \\ | \quad\quad |_! \\ H_2C \quad\quad CHCl \\ \diagdown \quad \diagup \\ CH_2 \end{array}$$

Submitted by ROBERT BRUCE MOFFETT.[1]
Checked by R. T. ARNOLD and GEORGE P. SCOTT.

1. Procedure

A. *Cyclopentadiene.* Two hundred milliliters (195 g.) of technical dicyclopentadiene (Note 1) is placed in a 500-ml. two-necked round-bottomed flask equipped with thermometer and an upright Friedrichs-type condenser (through which water at 50° (Note 2) is circulated). The ground-glass (Note 3) outlet of the Friedrichs condenser is connected to the side arm of a simple distilling head fitted with a thermometer and attached to an efficient water-cooled condenser held in a vertical position. At the lower end of this condenser is a receiver which consists of a carefully weighed 500-ml. two-necked round-bottomed flask immersed in a Dry Ice bath (Note 4) and protected from the air by a calcium chloride drying tube.

The flask containing dicyclopentadiene is now heated by means of an electric heating mantle or oil bath to approximately 160°, or until cyclopentadiene distils smoothly at 38–46° and a little dicyclopentadiene refluxes from the cold-finger (Friedrichs) condenser. After two-thirds of the dicyclopentadiene has been pyrolyzed (during the course of 4–5 hours), the residue in the flask may tend to become viscous and a higher temperature for the pyrolysis will be required in order to obtain rapid distillation of cyclopentadiene. In such an event it is desirable to discard the residue while it is still hot and mobile.

Cyclopentadiene dimerizes rapidly at room temperature and should be used immediately (Note 5) or stored at Dry Ice temperatures. As obtained above, the product has a refractive index of about 1.433 at 25° and is quite satisfactory as a starting material for the following preparation (Note 6). The yield, which is determined by weighing the receiving flask plus product, depends upon the quality of dicyclopentadiene employed (Note 7).

B. *3-Chlorocyclopentene.* The flask containing cyclopentadiene is weighed and the quantity of cyclopentadiene determined (Note 8). A thermometer and gas inlet tube are passed through a two-holed rubber stopper which is fitted to the center neck of the flask; the side neck is fitted with a calcium chloride drying tube. While the flask containing cyclopentadiene is being cooled in a Dry Ice bath (Note 9), dry hydrogen chloride (Note 10) is passed in rapidly. During this operation the temperature of the reaction mixture must be kept below 0° and the flask swirled to ensure good mixing. From time to time the flask is detached, wiped dry, and weighed quickly in order to determine the amount of hydrogen chloride that has been added. An excess is to be avoided, and it is advisable to stop the addition about 10% short of the theoretical quantity.

For many purposes, this crude 3-chlorocyclopentene, either as such or in solution (Note 5), may be used without purification (Notes 11 and 12). If a purer product is desired, however, it may be distilled according to the following procedure.

The flask containing the crude product is equipped with a capillary tube and distilling head and surrounded by a water bath, which may be heated by a hot plate or steam cone. A water-cooled condenser connects the distilling head with the center neck of a two-necked receiver which is surrounded by a Dry Ice bath. The outer neck of the receiver is fitted with a Dry Ice condenser arranged in such a way that vapors which first escape the receiver are condensed and returned to it. This apparatus is connected through a second Dry Ice trap and soda-lime tower (Note 13) to a vacuum pump. A fore-run is removed at a bath temperature of 20° and a pressure of about 15 mm. The receiver is changed, and the product is distilled at 18–25°/5 mm. The temperature of the water bath should not exceed 30°.

3-Chlorocyclopentene is obtained as a colorless liquid in a yield of 70–90% based on cyclopentadiene; n_D^{26} 1.4708.

2. Notes

1. Coarse iron filings or turnings may be added to speed up the rate of depolymerization,[2] but they are of questionable value.

2. Water at 50° may be obtained by carefully mixing streams of hot and cold water taps. A more elegant method is to circulate water from a thermostatically controlled constant-temperature bath.

It has been suggested [3] that a modification of the equipment described for the preparation of dehydroacetic acid,[4] but without the fractionating column, is very convenient for the depolymerization of dicyclopentadiene. The partial condenser is filled with methylene chloride or other liquid with a boiling point in this range, and attached directly to the flask which contains the dicyclopentadiene. This arrangement eliminates the inconvenience of maintaining a supply of water at 50°.

3. It is desirable to use ground-glass jointed equipment throughout.

4. An ice-salt bath may be used but is not so effective. A very convenient fluid for use in Dry Ice baths is ethylene glycol monomethyl ether (methyl Cellosolve, Carbide and Carbon Chemicals Corporation, New York, New York).

5. If the cyclopentadiene is to be used in solution, it is convenient to collect it directly in a flask containing a weighed amount of the desired solvent (e.g., toluene or ether).

6. Further purification, if desired, may be accomplished by distillation at about 20 mm. while collecting the product in a Dry Ice cooled receiver.

7. The checkers used Eastman Kodak Company blue lable grade dicyclopentadiene and obtained yields approximating 87%.

8. This reaction should be carried out with undiluted cyclopentadiene if reasonably pure distilled 3-chlorocyclopentene is to be isolated. However, if a distilled product is not required, a solution of cyclopentadiene (Note 5) may be used. 3-Chlorocyclopentene is somewhat more stable in solution.

9. If Dry Ice is not available, an ice-salt bath may be used

but then the hydrogen chloride must be added at a much slower rate.

10. Hydrogen chloride gas from a cylinder is most convenient; however, it may be generated if desired.[2]

11. If the temperature has been kept sufficiently low and an excess of hydrogen chloride avoided, the product will be a clear (or only slightly turbid) colorless, mobile liquid. If an excess of hydrogen chloride seems to be present, or if the product is dark, distillation is advisable.

12. 3 Chlorocyclopentene is unstable at room temperature and soon polymerizes to a black tar which is very difficult to remove from the apparatus. It should be used at once or stored at temperatures well below 0°. All apparatus must be cleaned as soon as possible after completion of the experiment. Ground-glass joints, if not separated soon after the reaction is completed, may become cemented together.

13. In spite of the Dry Ice cooled trap and soda-lime tower, some low-boiling material (as evidenced by the odor) often reaches the pump. It is therefore recommended that an old pump be used and the oil changed immediately after the experiment.

3. Methods of Preparation

The only practical laboratory preparation of cyclopentadiene is by the depolymerization of dicyclopentadiene.[5-8] 3-Chlorocyclopentene has been prepared by the addition of hydrogen chloride to cyclopentadiene.[9-13]

[1] The Upjohn Company, Kalamazoo, Michigan.
[2] Org. Syntheses Coll. Vol. 1, 293, 534 (1941).
[3] Kittila, Private communication.
[4] Org. Syntheses Coll. Vol. 3, 231 (1955).
[5] Perkins and Cruz, J. Am. Chem. Soc., 49, 518 (1927).
[6] Stobbe and Reuss, Ann., 391, 151 (1912).
[7] Khambata and Wasserman, Nature, 138, 368 (1936).
[8] Ward, U. S. pat. 2,372,237 [C. A., 39, 3312 (1945)].
[9] Kraemer and Spilker, Ber., 29, 552 (1896).
[10] Noeldechen, Ber., 33, 3348 (1900).
[11] Noller and Adams, J. Am. Chem. Soc., 48, 2444 (1926).
[12] Buu-Hoï and Cagniant, Bull. soc. chim. France, [5] 9, 99 (1942).
[13] Guest and Stansbury, Brit. pat. 790,842 [C. A., 53, 5157 (1959)]; U. S. pat. 2,891,888 [C. A., 53, 18372 (1959)].

DIACETYL-*d*-TARTARIC ANHYDRIDE

(Tartaric anhydride, diacetate of *d*-)

HOCHCO$_2$H   CH$_3$CO$_2$CH—CO

$\xrightarrow[\text{H}_2\text{SO}_4]{\text{(CH}_3\text{CO)}_2\text{O}}$

HOCHCO$_2$H   CH$_3$CO$_2$CH—CO

O

Submitted by R. L. SHRINER and C. L. FURROW, JR.[1]
Checked by N. J. LEONARD and R. R. SAUERS.

1. Procedure

In a 500-ml. three-necked round-bottomed flask fitted with a liquid-sealed stirrer and two reflux condensers (Note 1) is placed 40 g. (0.27 mole) of anhydrous, powdered *d*-tartaric acid (Note 2). A solution of 1.2 ml. of concentrated sulfuric acid in 136 g. (126 ml., 1.33 moles) of acetic anhydride is added, and the stirrer is started. The mixture warms up, and the tartaric acid goes into solution. The solution is heated gently (Note 1) under reflux with stirring for 10 minutes. The solution is poured into a beaker and cooled for 1 hour in an ice bath. The crude crystalline product is collected on a 15-cm. Büchner funnel (Note 3), washed twice with 20-ml. portions of dry benzene, stirred mechanically with 175 ml. of cold absolute ether, filtered, and placed in a vacuum desiccator over phosphorus pentoxide and paraffin shavings for 24 hours. The yield of diacetyl-*d*-tartaric anhydride is 41–44.5 g. (71–77%), m.p. 133–134° (Note 4), $[\alpha]_D^{20}$ 97.2° in dry chloroform ($c = 0.47$).

2. Notes

1. The reaction may be quite vigorous at its start, and the use of a large flask with two condensers is advised.
2. The anhydrous *d*-tartaric acid was obtained from Matheson, Coleman and Bell, East Rutherford, New Jersey.
3. Additional but lower-grade product may be acquired by pouring the mother liquor into petroleum ether and filtering the mixture. The recovered product is washed twice with absolute

ether, filtered, and dried. About 7 g. of product, m.p. 129–131°, is thus obtained.

4. The product is not stable and should be prepared only as needed. It may be kept in a vacuum desiccator over phosphorus pentoxide and paraffin, but the melting point drops about 1 degree during the first 4 days and then remains constant at approximately 132–134°. If placed in an ordinary stoppered bottle, the product becomes gummy and the melting point falls to about 100° within 3 days. Attempts to recrystallize the anhydride invariably led to decomposition and lowered melting point.

3. Methods of Preparation

The acetylation of d-tartaric acid with acetic anhydride has been effected by means of sulfuric acid,[2,3] hydrogen chloride,[4] or 85% phosphoric acid.[5]

[1] State University of Iowa, Iowa City, Iowa.
[2] Chattaway and Parkes, *J. Chem. Soc.*, **123**, 663 (1923).
[3] Roberts, *J. Chem. Soc.*, **1952**, 3315.
[4] Lucas and Baumgarten, *J. Am. Chem. Soc.*, **63**, 1653 (1941).
[5] Fuchs (to Emulsol Corp.), Brit. pat. 674,710 [*C. A.*, **47**, 4904 (1953)].

2,5-DIAMINO-3,4-DICYANOTHIOPHENE

(3,4-Thiophenedicarbonitrile, 2,5-diamino-)

$$(NC)_2C\!=\!C(CN)_2 + 2H_2S \xrightarrow{\text{Pyridine}} \underset{\underset{S}{H_2N}}{\overset{NC}{\underset{}{}}}\!\!\!\!\overset{CN}{\underset{NH_2}{}} + S$$

Submitted by W. J. Middleton.[1]
Checked by James Cason and Ralph J. Fessenden.

1. Procedure

Caution! Since carbon disulfide is highly flammable and hydrogen sulfide highly toxic, this reaction should be carried out in a hood, with due precaution against fire. It is also recommended that tetracyanoethylene not be allowed to come into contact with the skin.

A 1-l. three-necked flask is fitted with a sealed mechanical stirrer, a condenser protected by a drying tube, a thermometer, and an inlet tube extending to the bottom of the flask. A solution of 25.6 g. (0.2 mole) of recrystallized tetracyanoethylene (p. 877) in 300 ml. of acetone is placed in the flask, and 300 ml. of carbon disulfide is added. The flask and its contents are cooled to 0° by means of a salt-ice bath. With good stirring, hydrogen sulfide is passed into the reaction mixture at a moderate rate while the temperature is maintained at 0–5°. The solution becomes milky after a few minutes owing to the formation of colloidal sulfur. The hydrogen sulfide addition is continued for about 30 minutes, or until the solution is thoroughly saturated.

The hydrogen sulfide addition is temporarily suspended, and 100 ml. of pyridine is added rapidly in one portion through the condenser as the solution is stirred vigorously (Note 1). The solution becomes clear, and then 2,5-diamino-3,4-dicyanothiophene begins to precipitate immediately. The hydrogen sulfide addition is resumed and is continued for about 30 minutes while the temperature of the reaction mixture is maintained at 0–5°. Finally, the reaction mixture is stirred for an additional 30 minutes at 0–5°, then the yellow precipitate of the thiophene is collected on a Büchner funnel, thoroughly washed with about 500 ml. of acetone, and dried in the air or in a vacuum desiccator. The yield of crude product of yellow or buff color amounts to 30–31 g. (92–95%).

This material is sufficiently pure for most purposes. If a purer product is desired, the crude material is dissolved in 300 ml. of dimethylformamide, 10 g. of activated alumina (48–100 mesh) is added, and the mixture is filtered. The filtrate is heated to 80–90° on a steam bath, then 1 l. of boiling water is added immediately (Note 2). The resultant mixture is cooled in an ice bath, and the light buff crystals of 2,5-diamino-3,4-dicyanothiophene that separate are collected on a Büchner funnel and thoroughly washed with 500 ml. of acetone; weight 26–28 g. (79–85%). The product has no definite melting point but sublimes with some decomposition when heated above 250°.

2. Notes

1. Unless the pyridine is added quite rapidly, some of the product will begin to precipitate before all of the sulfur has dissolved, and the final product will be contaminated with sulfur. 2. This operation should be carried out as rapidly as possible, since prolonged heating in dimethylformamide results in loss of product.

3. Methods of Preparation

2,5-Diamino-3,4-dicyanothiophene has been prepared only by the action of hydrogen sulfide or sodium sulfide on tetracyanoethylene or tetracyanoethane.[2] Unlike most aminothiophenes, 2,5-diamino-3,4-dicyanothiophene is very stable and can be stored indefinitely. Its amino groups show the normal reactivity of aromatic amines. For example, they readily condense with aromatic aldehydes to form highly colored bis-anils.[2] Hot 10% sodium hydroxide rearranges 2,5-diamino-3,4-dicyanothiophene to 2-amino-3,4-dicyano-5-mercaptopyrrole.[2]

[1] Contribution No. 483 from Central Research Department, Experimental Station, E. I. du Pont de Nemours & Co., Wilmington, Delaware.
[2] Middleton, Englehardt, and Fisher, *J. Am. Chem. Soc.*, **80**, 2822 (1958).

2,4-DIAMINO-6-HYDROXYPYRIMIDINE *

(4-Pyrimidinol, 2,6-diamino-)

$$
\underset{\substack{\text{HN}=\text{C}\\|\\\text{NH}_2}}{\overset{\text{NH}_2}{}} + \underset{\substack{\text{CH}_2\\|\\\text{CN}}}{\overset{\text{C}_2\text{H}_5\text{OC}=\text{O}}{}} \xrightarrow{\text{NaOC}_2\text{H}_5} \underset{\substack{\text{H}_2\text{N}-\text{C}\\\|\\\text{N}}}{} \underset{\substack{\text{C}-\text{H}\\\|\\\text{C}-\text{NH}_2}}{\overset{\text{N}=\text{C}-\text{OH}}{}} + \text{C}_2\text{H}_5\text{OH}
$$

Submitted by J. A. VANALLAN.[1]
Checked by CLIFF S. HAMILTON and JOHN D. SCULLEY.

1. Procedure

A solution of sodium ethoxide is prepared from 23 g. (1 g. atom) of sodium and 250 ml. of anhydrous ethanol in a 1-l. round-bottomed flask fitted with a reflux condenser carrying a calcium chloride drying tube. After the sodium has dissolved, the solution is cooled and 113 g. (1 mole) of ethyl cyanoacetate is added (Note 1). This mixture is allowed to stand while a second solution of sodium ethoxide of the same volume and concentration is prepared. To this solution is added 97 g. (1.02 moles) of guanidine hydrochloride. The sodium chloride is separated by filtration, and the clear filtrate containing guanidine is added to the solution of ethyl sodiocyanoacetate. This mixture is heated for 2 hours under reflux and is then evaporated to dryness at atmospheric pressure. The solid product is dissolved in 325 ml. of boiling water and acidified with 67 ml. of glacial acetic acid. Upon cooling of the solution, 101–103 g. (80–82%) of yellow needles separates; m.p. 260–270° (dec.) (Note 2).

2. Notes

1. Eastman Kodak Company white label grade of ethyl cyanoacetate was used by the checkers.

2. Analyses carried out by the checkers showed that the product is quite pure.

3. Methods of Preparation

This procedure is a modification of the method of Traube,[2] which has been studied by Berezovskiĭ and Strel'chunas.[3] The reaction has been carried out in the presence of alkali or alkali earth hydroxides.[4]

[1] Eastman Kodak Company, Rochester, New York.

[2] Traube, *Ber.*, **33**, 1371 (1900); Ger. pat. 135,371 [*Frdl.*, **6**, 1192 (1900–1902)].

[3] Berezovskiĭ and Strel'chunas, *Trudy Vsesoyuz. Nauch. Issledovatel. Vitamin Inst.*, **5**, 28 (1954) [*C. A.*, **51**, 7379 (1957)].

[4] Heinrich and Buth (to Wickhen Products Inc.), U. S. pat. 2,673,204 [*C. A.*, **49**, 1825 (1955)]; Sugino, Jap. pat. 2733 (1953) [*C. A.*, **49**, 2527 (1955)].

DIAMINOURACIL HYDROCHLORIDE

(Uracil, 5,6-diamino-, hydrochloride)

Submitted by Wm. R. Sherman and E. C. Taylor, Jr.[1]
Checked by T. L. Cairns and D. S. Acker.

1. Procedure

In a 3-l., three-necked flask (Note 1) equipped with a reflux condenser and an efficient stirrer is placed 1 l. of absolute (99.8%) ethanol. To this is added 39.4 g. (1.72 g. atom) of sodium, and, after solution is complete (Note 2), 91.5 ml. (97.2 g., 0.86 mole) of ethyl cyanoacetate (Note 3) and 51.5 g. (0.86 mole) of urea are added. The mixture is heated under reflux on a steam bath with vigorous stirring for 4 hours. After about 2 hours, the reaction mixture becomes practically solid, and the stirrer may have to be stopped. At the end of the reaction time, 1 l. of hot (80°) water is added to the reaction mixture, and stirring is resumed. After complete solution has taken place, the stirred mixture is heated at 80° for 15 minutes and is then neutralized to litmus with glacial acetic acid (Note 4). Additional glacial acetic acid (75 ml.) is added, followed by cautious addition of a solution of 64.8 g. (0.94 mole) of sodium nitrite dissolved in 70 ml. of water. The rose-red nitroso compound separates almost immediately as an expanded precipitate which almost stops the stirrer. After a few minutes the nitroso compound is removed by filtration and washed twice with a small amount of ice water. The moist material is transferred back to the 3-l. flask, and 430 ml. of warm water (50°) is added.

Caution! This procedure should be conducted in a good hood. The slurry is stirred while being heated on a steam bath, and solid sodium hydrosulfite is added until the red color of the nitroso compound is completely bleached (Note 5). Then an additional 30 g. of sodium hydrosulfite is added; the light tan suspension is stirred with heating for 15 minutes more and is allowed to cool. The dense diaminouracil bisulfite is filtered from the cooled solution, washed well with water, and partially dried.

The crude product is readily purified by conversion to its hydrochloride salt. The bisulfite salt is transferred to a wide-mouthed 1-l. flask, and concentrated hydrochloric acid is added until the consistency of the resulting mixture is such as to permit mechanical stirring (100 to 200 ml. of acid). The slurry is heated on a steam bath with stirring for 1 hour (*Hood!*). The tan diaminouracil hydrochloride is filtered on a sintered glass funnel, washed well with acetone, and vacuum-dried over phosphorus pentoxide. The yield of diaminouracil hydrochloride is 104–124 g. (68–81%) (Notes 6 and 7).

2. Notes

1. Since the reaction mixture becomes almost solid after nitrosation, one of the necks of the flask should be of large diameter to facilitate removal of the product.

2. The usual precautions must be observed with respect to the hydrogen evolved. The reflux condenser is capped with a drying tube after hydrogen evolution ceases. The sodium ethoxide must be used immediately after its preparation, for it discolors rapidly and in this state leads to an impure product.

3. Eastman Kodak white label grade, Dow Chemical, and Kay-Fries ethyl cyanoacetate were all used with equal success.

4. Caution must be exercised in the addition of the glacial acetic acid in order to avoid frothing of the hot solution. The frothing becomes most vigorous as the 6-aminouracil begins to precipitate from the solution. The heating and subsequent neutralization assure cyclization of the initially formed cyanoacetylurea to 6-aminouracil. The nitrosation is carried out on the two-phase (solid-liquid) system.

5. The amount of sodium hydrosulfite used depends on its age and quality. The submitters never had to use more than 250 g. per run.

6. The preparation may be interrupted after the nitroso compound has been separated or after the crude bisulfite salt has been isolated. This preparation has been satisfactorily carried out on a scale seven times that given.

7. If placed in a preheated melting-point block, the product melts with decomposition in the range 300–305°. Diaminouracil hydrochloride in $0.1N$ hydrochloric acid has a well-defined absorption peak at 260 mμ, log $\epsilon = 4.24$. Satisfactory analyses for nitrogen and chlorine are difficult to obtain with this type of compound although good results are obtained for carbon and hydrogen.

3. Methods of Preparation

The procedure for the formation of diaminouracil bisulfite is slightly modified from that of Cain, Mallette, and Taylor,[2] which in turn is derived from preparations of Bogert and Davidson,[3] and Traube.[4] The sulfate salt may be formed in lower yield than the hydrochloride described here by dissolving the bisulfite salt in aqueous base and precipitating with sulfuric acid.[3,4] The hydrochloride is appreciably soluble in water, while the sulfate salt is only slightly soluble.

Other methods of reducing the nitroso compound include the use of ammonium sulfide [4] and hydrogenation utilizing Adams catalyst.[5]

Bredereck, Hennig, and Pfleiderer [6] describe a method for the formation of diaminouracil from uric acid which involves acetylation and subsequent hydrolysis of the acetyl derivative. This preparation was attempted on a large scale by the submitters without success (even when the acetylation step was carried out twice on the same material).

[1] University of Illinois, Urbana, Illinois.
[2] Cain, Mallette, and Taylor, *J. Am. Chem. Soc.*, **68**, 1996 (1946).
[3] Bogert and Davidson, *J. Am. Chem. Soc.*, **55**, 1668 (1933).
[4] Traube, *Ber.*, **33**, 1371 (1900).
[5] E. C. Taylor, Jr., unpublished results.
[6] Bredereck, Hennig, and Pfleiderer, *Ber.*, **86**, 321 (1953).

DIAZOMETHANE

(Methane, diazo-)

$$p\text{-}CH_3C_6H_4SO_2N(NO)CH_3 + ROH \xrightarrow{KOH}$$
$$CH_2N_2 + p\text{-}CH_3C_6H_4SO_2OR + H_2O$$

Submitted by TH. J. DE BOER and H. J. BACKER.[1]
Checked by JAMES CASON, MAX J. KALM, and R. F. PORTER.

1. Procedure

Caution! Diazomethane is toxic and prone to cause development of specific sensitivity; in addition, it is potentially explosive. Hence one should wear heavy gloves and goggles while performing this experiment and should work behind a safety screen or a hood door with safety glass. Also, it is recommended that ground joints and sharp surfaces be avoided. Thus all glass tubes should be carefully fire-polished, connections should be made with rubber stoppers, and separatory funnels should be avoided, as should etched or scratched flasks. Furthermore, at least one explosion of diazomethane has been observed at the moment crystals (sharp edges!) suddenly separated from a supersaturated solution. Stirring by means of a Teflon-coated magnetic stirrer is greatly to be preferred to swirling the reaction mixture by hand, for there has been at least one case of a chemist whose hand was injured by an explosion during the preparation of diazomethane in a hand-swirled reaction vessel.

It is imperative that diazomethane solutions not be exposed to direct sunlight or placed near a strong artificial light, because light is thought to have been responsible for some of the explosions that have been encountered with diazomethane. Particular caution should be exercised when an organic solvent boiling higher than ether is used. Because such a solvent has a lower vapor pressure than ether, the concentration of diazomethane in the vapor above the reaction mixture is greater and an explosion is more apt to occur.

Most diazomethane explosions take place during its distillation. Hence diazomethane should not be distilled unless the need justifies it. An ether solution of diazomethane satisfactory for many uses can be prepared as described by Arndt,[2] where nitrosomethylurea is added to a mixture of ether and 50% aqueous potassium hydroxide

*and the ether solution of diazomethane is subsequently decanted
from the aqueous layer and dried over potassium hydroxide pellets
(not sharp-edged sticks!). When distilled diazomethane is required,
the present procedure is particularly good because at no time is much
diazomethane present in the distilling flask.*

*The hazards associated with diazomethane have been discussed by
Gutsche,[3] and LeWinn [4] has reported on a fatal case of diazomethane
poisoning.*

A 125-ml. distilling flask is fitted with a condenser set for distillation and with a long-stem dropping funnel. The condenser
is connected by means of an adapter to a 250-ml. Erlenmeyer
flask. Through a second hole in the stopper of the Erlenmeyer
flask is placed an outlet tube bent so as to pass into and nearly to
the bottom of a second Erlenmeyer flask which is not stoppered.
Both receivers are cooled in an ice-salt mixture; in the first is
placed 10 ml. of ether (Note 1), and in the second 35 ml. of ether.
The inlet tube passes below the surface of the ether in the second
flask.

In the distilling flask are placed a solution of 6 g. of potassium
hydroxide dissolved in 10 ml. of water, 35 ml. of Carbitol (Note 2),
10 ml. of ether (Note 1), and the "Teflon"-coated bar of a magnetic stirrer (Note 3). The dropping funnel is attached and adjusted so that the stem is just above the surface of the solution in
the distilling flask. There is placed in the dropping funnel a solution of 21.5 g. (0.1 mole) of *p*-tolylsulfonylmethylnitrosamide
(p.943) in 125 ml. of ether (Note 4). The distilling flask is heated
in a water bath (Note 3) at 70–75°, the stirrer is started, and the
nitrosamide solution is added at a regular rate during 15–20 minutes. As soon as all the nitrosamide solution has been added,
additional ether (Note 1) is placed in the dropping funnel and
added at the previous rate until the distillate is colorless. Usually 50–100 ml. additional of ether is required. The distillate
contains 2.7–2.9 g. (64–69%) of diazomethane, as determined by
titration [2] (Notes 5 and 6).

2. Notes

1. If an alcohol-free solution of diazomethane is required,
absolute ether should be used throughout this preparation.

2. The Carbitol (monoethyl ether of diethylene glycol) was the Carbide and Carbon Chemicals Company product, which was distilled before use, b.p. 192–196°. It is a suitable solvent to render the reactants mutually soluble. Aqueous alkali with an ether solution of the nitrosamide does not yield diazomethane.

3. The same results may be obtained by an occasional careful agitation of the flask by hand; however, an explosion during this agitation by hand would be unfortunate. If the flask is placed in contact with the bottom of a beaker containing the heated water, and the magnetic stirring unit is placed in contact with the beaker, the bar may be spun satisfactorily. The checkers used the magnetic stirring apparatus, no. 9235-R, supplied by the A. H. Thomas Company, Philadelphia, Pa., and heated the water bath with the thermostated electric immersion heater, "Chill Chaser," model S-1005, supplied by the Still-Man Company, New York 56, N. Y. Satisfactory results should also be obtained by use of a combined heater and magnetic stirrer, such as no. 25210T supplied by the Will Corporation, New York 12, N. Y.

4. Nitrosamide prepared as described (p. 943), not recrystallized, is suitable for the present preparation. The solubility of the nitrosamide in ether drops sharply with temperature; below 20°, more than the specified amount of ether may be required, especially if recrystallized nitrosamide is used.

5. If an entirely dry solution of diazomethane is required, round pellets of potassium hydroxide should be used.[2]

6. The submitters reported that, in cases where the presence of alcohol in diazomethane solutions is not objectionable, they utilized ethanol as the solvent for the reaction mixture in which the diazomethane was prepared. They believe that ethanol has an advantage over Carbitol in that mixing of the reactants is achieved readily during the distillation, since the mixture is not viscous (see *Caution*).

They used the same apparatus as described in the present procedure with the following exceptions: (1) the reaction flask was of 200-ml. capacity, (2) the first receiver was an empty 500-ml. Erlenmeyer flask, and the second flask of 100-ml. capacity contained 40 ml. of ether, and (3) a magnetic stirrer was not employed. In the reaction flask was placed a solution of 10 g. of potassium hydroxide in 15 ml. of water and 50 ml. of 95% ethanol. The water bath was heated to 60–65° and a solution of 43 g. (0.2 mole) of p-tolylsulfonylmethylnitrosamide (p. 943) in 200 ml. of ether was added from the dropping funnel in 45 minutes. After all had been added, ether (about 30 ml.) was introduced through the dropping funnel until the condensing ether became colorless. The combined solutions in the receivers contained 5.9–6.1 g. (70–73%) of diazomethane..

3. Methods of Preparation

The more important methods of preparation of diazomethane include those from nitrosomethylurea,[2] nitrosomethylurethan,[5] N-nitroso-β-methylaminoisobutyl methyl ketone,[6] and 1-methyl-1-nitroso-3-nitroguanidine.[7, 8] It also has been prepared from N-nitrosoacetamide,[9] from N,N'-dimethyl-N,N'-dinitrosoöxamide and methylamine [10] or other alkaline reagents,[11] from nitrosyl chloride and methylamine,[12] from the reaction of bis-(N-methyl-N-nitroso)terephthalamide with alkali,[13] and by treatment of the sodium salt of formaldehyde oxime with chloramine.[14]

The advantage of the present method [15] resides in the stability of the starting material and the manipulative advantage resulting from its solubility in organic solvents.

[1] Organisch Chemisch Laboratorium der Rijks-Universiteit, Groningen, The Netherlands.

[2] Org. Syntheses Coll. Vol. 2, 165 (1943).

[3] Gutsche, Org. Reactions, 8, 391 (1954).

[4] LeWinn, Am. J. Med. Sci., 218, 556 (1949).

[5] Org. Syntheses Coll. Vol. 3, 119 (1955).

[6] Org. Syntheses Coll. Vol. 3, 244 (1955).

[7] McKay, J. Am. Chem. Soc., 70, 1974 (1948); 71, 1968 (1949).

[8] McKay, Ott, Taylor, Buchanan, and Crooker, Can. J. Research, 28B, 683 (1950).

[9] Heyns and Woyrsch, Chem. Ber., 86, 76 (1953).

[10] Fawcett (to E. I. du Pont de Nemours and Co.), U. S. pat. 2,675,378 [C. A., 49, 1777 (1955)].

[11] Reimlinger, Chem. Ber., 94, 2547 (1961).

[12] Phrix-Werke Akt.-Ges. (by Müller and Rundel), Ger. pat. 1,033,671 [C. A., 54, 10861 (1960)]; (by Müller, Haiss, and Rundel), Ger. pat. 1,104,518 [C. A., 56, 11445 (1962)].

[13] Org. Syntheses, 41, 16 (1961).

[14] Rundel, Angew. Chem., 74, 469 (1962).

[15] de Boer and Backer, Rec. trav. chim., 73, 229, 582 (1954).

DIBROMOACETONITRILE *

(Acetonitrile, dibromo-)

$$NCCH_2COOH + 2 \begin{array}{c} CH_2-CO \\ \diagdown \\ \diagup \\ CH_2-CO \end{array} NBr \xrightarrow{H_2O}$$

$$NCCHBr_2 + 2 \begin{array}{c} CH_2-CO \\ \diagdown \\ \diagup \\ CH_2-CO \end{array} HN + CO_2$$

Submitted by JAMES W. WILT and JAMES L. DIEBOLD.[1]
Checked by B. C. McKUSICK and H. E. KNIPMEYER.

1. Procedure

A solution of 63.8 g. (0.75 mole) of cyanoacetic acid (Note 1) in 750 ml. of cold water is placed in a 2-l. beaker. N-Bromosuccinimide (267 g., 1.5 moles) (Note 2) is added in portions with good mechanical stirring over a period of about 6 minutes (Note 3). The slightly exothermic reaction which attends the separation of the dibromoacetonitrile as a heavy oil is completed in about 20 minutes, after which time the beaker is placed in an ice bath and allowed to cool for 2 hours (Note 4).

The precipitated succinimide is collected on a large Büchner funnel atop a 2-l. filter flask and is washed with six 50-ml. portions of methylene chloride. The lower organic layer in the filtrate is separated from the aqueous phase, which is extracted with two 25-ml. portions of methylene chloride. The organic layer and the extracts are combined, washed vigorously with 50 ml. of a 5% sodium hydroxide solution (Note 5) and three 80-ml. portions of water, and dried over 10 g. of anhydrous sodium sulfate for several hours in a flask wrapped with aluminum foil (Note 6).

The colorless dried oil is distilled through a 45-cm. Widmer column (Note 7). Most of the methylene chloride is removed by heating the contents of the distillation pot to 75° at atmospheric pressure. The pressure is then lowered to about 20 mm., and 112–129 g. (75–87%) of dibromoacetonitrile is collected as a colorless oil, b.p. 70–72°/20 mm., n_D^{25} 1.540–1.542, d_4^{20} 2.369 (Notes 8 and 9).

2. Notes

1. Commercial cyanoacetic acid (Eastman Kodak Company), m.p. 67–71.5°, about 98% pure, was found satisfactory for this preparation.

2. N-Bromosuccinimide obtained from Matheson, Coleman and Bell, Inc., as well as from Arapahoe Chemicals, Inc., was used as received. The N-chlorosuccinimide (see Note 9) was commercial material from Arapahoe Chemicals, Inc.

3. The N-bromosuccinimide should be added as rapidly as possible consistent with the foaming produced by evolution of carbon dioxide.

4. About one-third of the succinimide (50–53 g.) precipitates from the solution during the cooling period, thus rendering the subsequent purification of the nitrile easier. Sometimes the solution must be seeded to start the precipitation. The checkers found it necessary to store the mixture in a refrigerator overnight in order to obtain 50–53 g. of succinimide.

5. The basic aqueous phase becomes pink through action of the base on dibromoacetonitrile.

6. Pure dibromoacetonitrile is fairly stable to air and light, but traces of basic impurities cause it to become discolored (see Note 5). Protective shielding retards this coloration.

7. Unless an efficient column is used, some nitrile distils over with the methylene chloride. The checkers used a 120-cm. spinning-band column.

8. The product shows good shelf stability, but, as a precautionary measure, it is best stored under nitrogen in a sealed brown vessel. Dichloroacetonitrile (Note 9) is less sensitive and may be stored in an ordinary glass-stoppered brown bottle.

9. In exactly the same fashion, from 63.8 g. (0.75 mole) of cyanoacetic acid and 200.3 g. (1.5 moles) of N-chlorosuccinimide there may be obtained 45–54 g. (55–65%) of colorless dichloroacetonitrile, b.p. 110–112°/760 mm., n_D^{25} 1.439, d_4^{20} 1.369. Since this reaction is slightly slower, 30 minutes should be allowed for the reaction before chilling.

3. Methods of Preparation

Dibromoacetonitrile has been prepared by the dehydration of dibromoacetamide with phosphorous pentoxide [2,3] and by the present method.[4] An early report that dibromoacetonitrile can be obtained from cyanoacetic acid by treatment with bromine,[5] a method similar to that described here, was later shown to be wrong.[2] Dibromoacetonitrile has been prepared also by the bromination of ethyl cyanoacetate in the presence of magnesium oxide.[6]

[1] Loyola University, Chicago, Illinois.
[2] Steinkopf, *Ber.*, **38**, 2694 (1905).
[3] Ghigi, *Gazz. chim. ital.*, **71**, 641 (1941).
[4] Wilt, *J. Org. Chem.*, **21**, 920 (1956).
[5] van't Hoff, *Ber.*, **7**, 1382, 1571 (1874).
[6] Felton, *J. Chem. Soc.*, **1955**, 515.

4,4'-DIBROMOBIPHENYL *

(Biphenyl, 4,4'-dibromo-)

$$C_6H_5-C_6H_5 + 2Br_2 \rightarrow Br\langle\bigcirc\rangle-\langle\bigcirc\rangle Br + 2HBr$$

Submitted by ROBERT E. BUCKLES and NORRIS G. WHEELER.[1]
Checked by R. S. SCHREIBER, WM. BRADLEY REID, JR., and ROBERT W. JACKSON.

1. Procedure

In a 15-cm. evaporating dish is placed 15.4 g. (0.10 mole) of finely powdered biphenyl (Note 1). The dish is set on a porcelain rack in a 30-cm. desiccator with a 10-cm. evaporating dish under the rack containing 39 g. (12 ml., 0.24 mole) of bromine. The desiccator is closed, but a very small opening is provided for the escape of hydrogen bromide (Note 2). The biphenyl is left in contact with the bromine vapor for 8 hours (or overnight). The orange solid is then removed from the desiccator and allowed to stand in the air under a hood for at least 4 hours (Note 3). At this point, the product weighs about 30 g. and has a melting point in the neighborhood of 152°. The crude 4,4'-dibromobiphenyl is dissolved in 75 ml. of benzene, filtered, and cooled to 15°. The resulting crystals are filtered, giving a yield of 23.4–24.0 g. (75–77%) of 4,4'-dibromobiphenyl, m.p. 162–163° (Note 4).

2. Notes

1. The checkers used Eastman Kodak Company white label grade of biphenyl.

2. If a vacuum desiccator is used, the stopcock can be opened slightly to allow for the escape of hydrogen bromide.

3. The standing period allows hydrogen bromide and bromine to escape from the crystals.

4. 4,4'-Dibromobibenzyl can be prepared in the same manner. Eighteen grams (0.10 mole) of finely divided bibenzyl is left in contact with the vapor from 39 g. (12 ml., 0.24 mole) of bromine for 24 hours. The desiccator is put under an opaque cover to keep out light, which causes the formation of α,α'-dibromobibenzyl. The somewhat sticky reaction product is allowed to stand overnight. The crude product is dissolved in 300 ml. of isopropyl alcohol, filtered, and cooled. A yield of 15.0–17.0 g. (44–50%) of 4,4'-dibromobibenzyl, m.p. 113–114°, is obtained.

3. Methods of Preparation

This method is a modification of that of Buckles, Hausman, and Wheeler.[2] 4,4'-Dibromobiphenyl has also been prepared by the bromination of biphenyl in water,[3] carbon disulfide,[4] and glacial acetic acid;[5] by the bromination of a mixture of biphenylsulfonic acids in dilute sulfuric acid;[6] by the action of sodium carbonate[7] or ethanol[8] on the perbromide obtained from the reaction of diazotized benzidine with bromine water; and by passing p-dibromobenzene vapor through a red-hot tube.[9]

4,4'-Dibromobibenzyl has been prepared by the bromination of bibenzyl in water;[10] by the reaction of p-bromobenzyl bromide with zinc dust in water;[11] and by the reaction of bromobenzene with ethylene oxide in the presence of anhydrous aluminum chloride.[12]

[1] State University of Iowa, Iowa City, Iowa.

[2] Buckles, Hausman, and Wheeler, *J. Am. Chem. Soc.*, **72**, 2494 (1950).

[3] Fittig, *Ann.*, **132**, 204 (1864).
[4] Carnelley and Thomson, *J. Chem. Soc.*, **47**, 586 (1885).
[5] Scholl and Neovius, *Ber.*, **44**, 1087 (1911).
[6] Datta and Bhoumik, *J. Am. Chem. Soc.*, **43**, 306 (1921).
[7] Griess, *J. Chem. Soc.*, **20**, 91 (1867).
[8] Loh and Turner, *J. Chem. Soc.*, **1955**, 1274.
[9] Meyer and Hofmann, *Monatsh.*, **38**, 141 (1917).
[10] Stelling and Fittig, *Ann.*, **137**, 267 (1866).
[11] Errera, *Gazz. chim. ital.*, **18**, 236 (1888).
[12] Smith and Natelson, *J. Am. Chem. Soc.*, **53**, 3476 (1931).

DI-*n*-BUTYLDIVINYLTIN

(Tin, dibutyldivinyl-)

$$CH_2{=}CHBr + Mg \xrightarrow{\text{Tetrahydrofuran}} CH_2{=}CHMgBr$$

$$2CH_2{=}CHMgBr + (n\text{-}C_4H_9)_2SnCl_2 \rightarrow$$

$$(n\text{-}C_4H_9)_2Sn(CH{=}CH_2)_2 + MgCl_2 + MgBr_2$$

Submitted by DIETMAR SEYFERTH.[1]
Checked by MELVIN S. NEWMAN and S. RAMACHANDRAN.

1. Procedure

In a 2-l. three-necked flask, equipped with a Dry Ice-acetone reflux condenser, a mechanical stirrer, and a 250-ml. dropping funnel, is placed 29.2 g. (1.2 g. atoms) of magnesium turnings. Enough tetrahydrofuran (THF) (Note 1) to cover the magnesium is added, stirring is begun, and about 5 ml. of vinyl bromide (Note 2) is added. After the reaction has started (Note 3), an additional 350 ml. of the THF is added. The rest of the vinyl bromide (140 g., 1.3 moles, total), dissolved in 120 ml. of THF, is added at such a rate that a moderate reflux is maintained. After the addition has been completed, the solution is refluxed for 30 minutes (Note 4). The Grignard solution is then cooled to room temperature, and the Dry Ice-acetone condenser is replaced with a water condenser which is fitted with a Drierite-filled drying tube. A solution of 135 g. (0.44 mole) of di-*n*-butyltin dichloride (Note 5) in 250 ml. of THF (Note 6) is then added, with stirring, at such a rate that a moderate reflux is maintained. After the addition has been completed, the reaction mixture is refluxed for

20 hours. The mixture is cooled to room temperature and is hydrolyzed by the slow addition of 150 ml. of a saturated ammonium chloride solution (Note 7). The organic layer is then decanted and the residual salts are washed thoroughly with 3 portions of ether, the washings being added to the organic layer. The ether and the THF are stripped off at atmospheric pressure; a Claisen distillation head is used. The residue is distilled at reduced pressure using a vacuum-jacketed Vigreux column equipped with a total-reflux partial take-off head to give 95–116 g. (74–91%) of di-*n*-butyldivinyltin, b.p. 60°/0.4 mm., n_D^{25} 1.4797 (Notes 8 and 9).

2. Notes

1. Tetrahydrofuran, obtained from the Electrochemicals Department of E. I. du Pont de Nemours & Co., was distilled from lithium aluminum hydride prior to use. It is not advisable to leave THF purified in this manner standing around for longer periods, since, in the absence of the inhibitor present in the commercial material, peroxides form fairly rapidly. (See also p. 793 ; Note 2).

2. Vinyl bromide, obtained from the Matheson Company, was redistilled prior to use. The distillate was collected in a receiver cooled with a Dry Ice-acetone mixture and protected from daylight.

3. In most cases the formation of the Grignard reagent began in the absence of any initiator. In cases where the reaction did not begin within a few minutes, 0.5 ml. of methyl iodide served to initiate attack on the magnesium.

4. In small-scale preparations of vinylmagnesium bromide it is advisable to carry out the reaction in an atmosphere of dry nitrogen in order to prevent hydrolysis and oxidation of the Grignard reagent. In larger-scale preparations such as the one described here, where a considerable excess of Grignard reagent is used, such precautions are not necessary.

5. Di-*n*-butyltin dichloride is a commercial product of Metal and Thermit Corporation, Rahway, New Jersey.

6. The checkers found that 250 ml. of dry ether was equally effective.

7. Enough saturated ammonium chloride solution is added to cause coagulation of the inorganic salts to a particle size of about 2–5 mm. in diameter; the volume of solution required varies but averages about 100–120 ml. per mole of Grignard reagent. If the hydrolysis is stopped at this point, a clear, essentially dry organic layer results, and in most instances no further drying is required before distillation.

8. This general procedure has been used to prepare [2,3] a large number of vinyltin compounds, including:

$CH_2\!\!=\!\!CHSnR_3$ $(R = CH_3, C_2H_5, n\text{-}C_3H_7, n\text{-}C_4H_9, C_6H_5)$

$(CH_2\!\!=\!\!CH)_2SnR_2$ $(R = CH_3, n\text{-}C_4H_9, (CH_3)_3SiCH_2, C_6H_5)$

$(CH_2\!\!=\!\!CH)_3SnR$ $(R = n\text{-}C_4H_9, C_6H_5)$

$(CH_2\!\!=\!\!CH)_4Sn$

9. Grignard reagents other than vinylmagnesium bromide may be used in this general procedure. The initial use of a Dry Ice-acetone condenser, is then not required. Use of the THF solvent provides a distinct advantage over the method recently described in detail (p. 881), in which ether is used as a solvent, since fewer steps are required.

3. Methods of Preparation

The above procedure is essentially that described previously by the author.[2] Di-n-butylvinyltin has been prepared by the reaction between vinylmagnesium chloride with either di-n-butyltin dichloride or di-n-butyltin oxide.[4] The preparation of vinylmagnesium bromide was first described by Normant.[5]

[1] Massachusetts Institute of Technology, Cambridge, Massachusetts.

[2] Seyferth and Stone, *J. Am. Chem. Soc.*, **79**, 515 (1957).

[3] Seyferth, *J. Am. Chem. Soc.*, **79**, 2133 (1957).

[4] Rosenberg, Gibbons, and Ramsden, *J. Am. Chem. Soc.*, **79**, 2137 (1957).

[5] Normant, *Compt. rend.*, **239**, 1510 (1954).

DI-*tert*-BUTYL MALONATE

(Malonic acid, di-*t*-butyl ester)

I. ISOBUTYLENE METHOD

$$CH_2(CO_2H)_2 + 2(CH_3)_2C{=}CH_2 \underset{}{\overset{H_2SO_4}{\rightleftharpoons}} CH_2[CO_2C(CH_3)_3]_2$$

Submitted by ALLEN L. MCCLOSKEY, GUNTHER S. FONKEN,
RUDOLPH W KLUIBER, and WILLIAM S. JOHNSON.[1]
Checked by JAMES CASON, GERHARD J. FONKEN, and WILLIAM G. DAUBEN.

1. Procedure

A 500-ml. Pyrex heavy-walled narrow-mouthed pressure bottle is charged with 100 ml. of ether (Note 1), 5 ml. of concentrated sulfuric acid, 50.0 g. (0.48 mole) of malonic acid, and approximately 120 ml. (about 1.5 moles) of isobutylene (Note 2), which is liquefied by passage into a large test tube immersed in a Dry Ice-acetone bath. The bottle is closed with a rubber stopper which is clamped or wired securely in place (Note 3) and is shaken mechanically at room temperature until the suspended malonic acid dissolves (Note 4). The bottle is chilled in an ice-salt bath and opened; then the contents are poured into a separatory funnel containing 250 ml. of water, 70 g. of sodium hydroxide, and 250 g. of ice. The mixture is shaken (carefully at first), the layers are separated, and the aqueous portion is extracted with two 75-ml. portions of ether. The organic layers are combined, dried over anhydrous potassium carbonate, and filtered into a dropping funnel attached to the neck of a 125-ml. modified Claisen flask (Note 5). The flask is immersed in an oil bath at about 100°, and the excess isobutylene and ether are removed by flash distillation effected by allowing the solution to run in slowly from the dropping funnel. The dropping funnel is then removed, and the residue is distilled at reduced pressure. The fraction boiling at 112–115°/31 mm. is collected. The yield of colorless di-*tert*-butyl malonate is 60.0–62.0 g. (58–60%), n_D^{25} 1.4158–1.4161, freezing point −5.9 to −6.1° (Notes 6 and 7).

2. Notes

1. Increase in the concentrations of reactants and product by elimination of the solvent shifts the equilibrium to the right and thus increases the yield of ester. In several runs by the checkers in which the described procedure was followed except that ether was omitted, isobutylene was increased to 240 ml. (3 moles), and shaking was continued for 12–15 hours to effect solution, yields of 88–91% were obtained. When ether was used as solvent, the larger amount of isobutylene raised the yield to only 73%. The submitters, however, have found that in the procedure without solvent the yield is more variable (in the range 69–92%), complete solution of the acid sometimes fails to occur, and in runs requiring long shaking for complete solution there is formed a lower-boiling substance the separation of which requires fractional distillation. Without solvent, there is usually an exothermic reaction as the mixture warms up. In the size run described this is no disadvantage, but in larger runs the heat evolved might be sufficient to cause the reaction to get out of control.

2. Technical grade isobutylene supplied by the Matheson Company was used.

3. The pressure during reaction on this scale does not exceed 40 p.s.i.

4. Solution is usually complete within 6 hours, but sometimes as long as 12 hours may be required.

5. The flask should be carefully washed with alkali before rinsing and drying, to ensure the removal of traces of acid which will catalyze the decomposition of the ester on warming to give isobutylene, carbon dioxide, and acetic acid. Once this decomposition begins, as evidenced by severe foaming, it is autocatalyzed (by the acetic acid formed) and cannot be prevented from continuing at an accelerated rate except by rewashing the product and apparatus with alkali. The addition of some solid potassium carbonate or magnesium oxide before distillation has been used with some *tert*-butyl esters to aid in inhibiting incipient decomposition. This treatment, however, does not appear to be necessary in the present preparation.

6. This preparation has been carried out by H. C. Dehm on a larger scale. From 150 g. of malonic acid, 200 ml. of ether, 10 ml.

of concentrated sulfuric acid, and 375 ml. of isobutylene, there was obtained after shaking for 22 hours in a 1-l. bottle 201.3 g. (64% yield) of ester, $n_D^{24.2}$ 1.4161.

7. Other esters that have been prepared by this general procedure are: *tert*-butyl acetate, 50% yield, b.p. 94–97°/738 mm., n_D^{25} 1.3820; *tert*-butyl chloroacetate, 63% yield, b.p. 56–57°/16–17 mm., n_D^{25} 1.4204–1.4210 (carried out by R. C. Hunt); *tert*-butyl bromoacetate, 65% yield, b.p. 74–76°/25 mm., n_D^{25} 1.4162; *tert*-butyl α-chloropropionate, yield 63%, b.p. 52–53°/12 mm., n_D^{25} 1.4163 (carried out by J. S. Belew); *tert*-butyl *o*-benzoylbenzoate, 70% yield, m.p. 65–69°; di-*tert*-butyl succinate (dioxane was used instead of ether as solvent), 52% yield, b.p. 105-107°/7 mm., m.p. 31.5–35°; di-*tert*-butyl glutarate, 60% yield, b.p. 113–119°/9 mm., n_D^{25} 1.4215; di-*tert*-butyl β,β-dimethylglutarate, 67% yield, b.p. 72–75°/1 mm., n_D^{25} 1.4246.

II. ACID CHLORIDE METHOD

$$CH_2(CO_2H)_2 + 2SOCl_2 \rightarrow CH_2(COCl)_2 + 2SO_2 + 2HCl$$
$$CH_2(COCl)_2 + 2(CH_3)_3COH + 2C_6H_5N(CH_3)_2 \rightarrow$$
$$CH_2[CO_2C(CH_3)_3]_2 + 2C_6H_5N(CH_3)_2 \cdot HCl$$

Submitted by CHITTARANJAN RAHA.[2]
Checked by WILLIAM S. JOHNSON and RUDOLPH W. KLUIBER.

1. Procedure

A. *Malonyl dichloride.* In a 250-ml. Erlenmeyer flask (Note 1) fitted by a ground-glass joint to a reflux condenser capped with a calcium chloride drying tube are placed 52 g. (0.5 mole) of finely powered, dry malonic acid (Note 2) and 120 ml. (about 1.65 mole) of thionyl chloride (Note 3). The flask is warmed for 3 days in a heating bath kept at 45–50° (Note 4). During this period the mixture, which is agitated occasionally by gentle swirling, gradually darkens to a deep brownish red or sometimes a blue color. Finally the mixture is heated at 60° for 5–6 hours. After cooling, it is transferred to a 125-ml. modified Claisen flask and distilled at reduced pressure (water aspirator). A calcium chloride guard tube is inserted between the vacuum line and the apparatus, and the flask is heated with a bath rather than a free flame. After a small fore-run of thionyl chloride, the malonyl chloride distils at 58–60°/28 mm. The pale yellow product amounts to 50.5–60 g. (72–85% yield), n_D^{29} 1.4572.

B. *Di-tert-butyl malonate*. A 1-l. three-necked flask is fitted with a thermometer, a mercury- or rubber sleeve-sealed mechanical stirrer, a reflux condenser protected by a calcium chloride guard tube, and a dropping funnel (either pressure-equalized or protected by a calcium chloride guard tube). A mixture of 100 ml. (about 1 mole) of *tert*-butyl alcohol, dried by distillation from sodium (p. 134, Note 2), and 80 ml. (0.63 mole) of dry dimethylaniline (Note 5) is placed in the flask, the stirrer is started, and a solution of 28.0 g. (0.2 mole) of malonyl dichloride in about 60 ml. of dry, alcohol-free chloroform (Note 6) is added slowly from the dropping funnel while the reaction flask is cooled in an ice bath. The reaction is strongly exothermic, and the rate of dropping is regulated so that the temperature of the mixture does not exceed 30°. After the addition is complete (about 30 minutes) the light-greenish mixture is heated under reflux for 4 hours. The mixture is then cooled, 150 ml. of ice-cold 6N sulfuric acid is added with stirring, and the product is extracted with three 250-ml. portions of ether (Note 7). The combined ether extracts are washed once with 6N sulfuric acid, twice with water, twice with 10% potassium carbonate, and once with saturated sodium chloride, and are finally dried over anhydrous sodium sulfate to which a small amount of potassium carbonate is added. The ether is removed by distillation at reduced pressure (water aspirator), and the residue (to which a pinch of magnesium oxide is added) is distilled at reduced pressure from a modified Claisen flask (Note 8). The yield of colorless di-*tert*-butyl malonate, distilling at 65–67°/1 mm., 110–111°/22 mm., is 35.8–36.2 g. (83–84%), n_D^{25} 1.4159, m.p. about −6°.

2. Notes

1. Better results are obtained by using a flat-bottomed flask, which permits the insoluble malonic acid to be distributed over a greater surface.

2. The reaction mixture is heterogeneous at first, and if the acid is not finely powdered, some of it remains unreacted. Attempts to carry out the reaction on a larger scale resulted in some charring and lower yields.

3. Eastman Kodak Company white label quality thionyl chloride is satisfactory.

4. The temperature range is critical, and yields are lower if it is not controlled carefully. The use of pyridine as a catalyst is not recommended as it produces charring even after relatively short reaction periods.

5. J. T. Baker dimethylaniline (purified grade) is satisfactory without distillation.

6. The chloroform was dried over and distilled from anhydrous calcium chloride just before use.

7. The dimethylaniline may be recovered from the aqueous layer where it is dissolved as the salt.

8. Di-*tert*-butyl malonate, like most *tert*-butyl esters, decomposes readily on heating in the presence of traces of acids. It is therefore desirable to give all glassware to be used for distillation of the material an alkali rinse before use. The addition of a small amount of magnesium oxide also helps to inhibit the decomposition during distillation.[3] When decomposition starts, foaming is generally observed. In this event the addition of glass wool to the distillation flask helps to keep the product from foaming over.

3. Methods of Preparation

Procedure I is a modification [3] of the method of Altschul [4] for preparing *tert*-butyl esters. Di-*tert*-butyl malonate also has been prepared by the reaction of malonyl dichloride and *tert*-butyl alcohol in the presence of a base,[5, 6] and by the reaction of carbon suboxide with *tert*-butyl alcohol.[7] Procedure II is based on the former [5] method and developed from studies initiated by P. C. Mukharji of the University College of Science and Technology, Calcutta.

Malonyl dichloride has been prepared from malonic acid and thionyl chloride,[5, 8-12] and from carbon suboxide and anhydrous hydrogen chloride.[13] The present procedure is adapted from that of Staudinger and Bereza [10] and of Backer and Homan.[5]

[1] University of Wisconsin, Madison, Wisconsin.
[2] Rose Research Institute, Calcutta, India.
[3] Fonken and Johnson, J. Am. Chem. Soc., 74, 831 (1952).
[4] Altschul, J. Am. Chem. Soc., 68, 2605 (1946).
[5] Backer and Homan, Rec. trav. chim., 58, 1048 (1939).
[6] Backer and Lolkema, Rec. trav. chim., 57, 1234 (1938).
[7] Hagelloch and Feess, Chem. Ber., 84, 730 (1951).
[8] Auger, Ann. chim. et phys., [6] 22, 347 (1891).
[9] Asher, Ber., 30, 1023 (1897).
[10] Staudinger and Bereza, Ber., 41, 4463 (1908).
[11] von Auwers and Schmidt, Ber., 46, 477 (1913).
[12] McMaster and Ahmann, J. Am. Chem. Soc., 50, 145 (1928).
[13] Diels and Wolf, Ber., 39, 696 (1906).

4,4′-DICHLORODIBUTYL ETHER *

[Ether, bis(4-chlorobutyl)]

$$\underset{\displaystyle \overset{\displaystyle CH_2}{|}\;\;\;\overset{\displaystyle CH_2}{|}}{\underset{O}{CH_2 \diagdown \diagup CH_2}} \xrightarrow[H_2SO_4]{POCl_3} Cl(CH_2)_4O(CH_2)_4Cl$$

Submitted by KLIEM ALEXANDER and H. V. TOWLES.[1]
Checked by ARTHUR C. COPE, MALCOLM CHAMBERLAIN, and MARK R. KINTER.

1. Procedure

Caution! This preparation should be conducted in a good hood because some hydrogen chloride is evolved.

In a 2-l. three-necked flask fitted with a mercury-sealed stirrer (Note 1), a reflux condenser connected to a calcium chloride tube, and a thermometer is placed 360 g. (406 ml., 5 moles) of dry tetrahydrofuran (Note 2). The flask is surrounded by an ice bath, stirring is started, and 256 g. (153 ml., 1.67 moles) of phosphorus oxychloride is added rapidly. The mixture is cooled to 10–15°, and 50 ml. of concentrated sulfuric acid (sp. gr. 1.84) is added during the course of 3–10 minutes at a rate that does not cause the temperature to rise above 40°. The ice bath is then removed and the mixture is heated cautiously over a low luminous flame until an exothermic reaction becomes evident at about 88–90° (Note 3). By moderate cooling or warming as may be required the temperature is maintained at 90–100° until the exothermic reaction ceases, as indicated by the increased rate of heating required to maintain the reaction temperature, and thereafter for an additional 10 minutes (Note 4). Six hundred milliliters of water is added, the mixture is heated under reflux for 30 minutes and then distilled through a downward condenser until the vapor temperature reaches 99–100° (Note 5).

The dark reaction mixture is cooled to room temperature, transferred to a separatory funnel, and extracted with 225 ml. of ether. The ether extract is washed with four 100-ml. portions of water and dried over anhydrous sodium sulfate or magnesium sulfate. The mixture is filtered, the ether is removed by distillation, and the residual liquid is fractionated under reduced pressure from a modified Claisen flask. The yield of colorless 4,4'-dichlorodibutyl ether, b.p. 84–86°/0.5 mm. (116–118°/10 mm.), n_D^{25} 1.4562, d_4^{25} 1.0690, is 257–268 g. (52–54% based on tetrahydrofuran) (Note 6).

2. Notes

1. Efficient stirring is necessary to permit good control of the reaction temperature.

2. The same yield of 4,4'-dichlorodibutyl ether was obtained from redistilled tetrahydrofuran dried over sodium or a good-quality commercial grade obtained from the Electrochemicals Department of the E. I. du Pont de Nemours and Company and dried over Drierite.

3. About 30 minutes is required for the temperature to reach 88–90°. The temperature rises rapidly to 76°, at which point refluxing of tetrahydrofuran accompanied by evolution of some hydrogen chloride occurs. Thereafter refluxing gradually diminishes and the temperature increases approximately as follows: 0 minutes, 76°; 5 minutes, 77°; 10 minutes, 78°; 15 minutes, 80°; 20 minutes, 85°; 25 minutes, 94°; 27 minutes, 100°.

4. The temperature is easily kept within the range 90–100° by occasional cooling with an ice-water bath. Above 100° the reaction tends to become violent, resulting in excessive evolution of hydrogen chloride and a lower yield of the product. The exothermic phase of the reaction is usually complete in 15–20 minutes.

5. Refluxing with water serves to decompose phosphorus-containing complexes and facilitates isolation of the product. The aqueous distillate contains small amounts of 1,4-dichlorobutane and unchanged tetrahydrofuran.

6. The product as obtained with these physical constants is analytically pure.

3. Methods of Preparation

The procedure described is based on one reported by Alexander and Schniepp.[2] 4,4'-Dichlorodibutyl ether also has been prepared by the action of thionyl chloride on tetrahydrofuran [3] or 1,4-butanediol,[4] and by heating 4-chlorobutanol and hydrogen chloride under pressure.[5]

[1] Northern Regional Research Laboratory, U. S. Department of Agriculture, Peoria, Illinois.

[2] Alexander and Schniepp, *J. Am. Chem. Soc.*, **70**, 1839 (1948).

[3] Krzikalla and Maier, PB 631, Office of Technical Services, U. S. Department of Commerce; Lutkova, Kutsenko, and Itkina, *Zhur. Obshcheĭ Khim.*, **25**, 2102 (1955) [*C. A.*, **50**, 8584 (1956)].

[4] Haga et al. (to Mitsubishi Chemical Industries Co.), Jap. pats. 1268, 1271 (1953) [*C. A.*, **48**, 2086 (1954)]; Jap. pat. 5124 (1952) [*C. A.*, **48**, 8812 (1954)].

[5] Trieschmann, U. S. pat. 2,245,509 [*C. A.*, **35**, 5914 (1941)].

1,1-DICHLORO-2,2-DIFLUOROETHYLENE *

(Ethylene, 1,1-dichloro-2,2-difluoro-)

$$CCl_3—CClF_2 + Zn \xrightarrow{CH_3OH} CCl_2{=}CF_2 + ZnCl_2$$

Submitted by J. C. Sauer.[1]
Checked by Charles C. Price and Maseh Osgan.

1. Procedure

In a 500-ml. three-necked round-bottomed flask equipped with a 100-ml. separatory funnel, a thermometer, and a short fractionating column (Note 1) leading through a condenser to a 100-ml. tared receiver are charged 150 ml. of methanol, 42.2 g. (0.65 g. atom) of powdered zinc, and 0.2 g. of zinc chloride. Acetone cooled in a Dry Ice-acetone bath is circulated through the condenser, and the distillation receiver is immersed in ice water (Note 2). The mixture in the flask is heated to 60–63°, and a 10–15-ml. portion of a solution consisting of 122.4 g. (0.6 mole) of 1,1,1,2-tetrachloro-2,2-difluoroethane (Note 3) in 50 ml. of methanol is added dropwise over a period of a few minutes. The reaction generally becomes moderately vigorous at this point, and refluxing of 1,1-dichloro-2,2-difluoroethylene part-way up the column is observed. The heating bath is removed at this time. Addition of the ethane derivative is continued at such a rate that the temperature of the refluxing liquid at the head of the column is maintained at 18–22°. The flask may require occasional shaking to prevent the zinc dust from agglomerating. The take-off rate of the ethylene derivative is adjusted to about one-half the rate of input of the ethane derivative. During this addition, the temperature in the distillation flask drops to 45–50°. The addition of the ethane derivative requires 45 minutes to 1 hour. The heating bath is replaced, and another hour is required to complete the dehalogenation and to distil the last of the product. The temperature in the distillation flask during this period gradually rises to 69–70°. There is collected in the receiver 71–76 g. (89–95% yield) (Note 4) of 1,1-dichloro-2,2-difluoroethylene, distilling at 18–21° (Note 5), n_D^0, 1.3730–1.3746.

2. Notes

1. A 16-in. column packed with platinum gauze and fitted with a variable take-off was used by the submitter. The checkers used a 20 × ¾-in. jacketed column packed with ⅟₁₆-in. glass helices.

2. The checkers used a Dry Ice-acetone-cooled cold-finger condenser fitted with a partial take-off adapter.

3. Tetrachlorodifluoroethane (CCl_3—$CClF_2$) is sold under the trade name "Genetron-131" by the General Chemical Division, Allied Chemical and Dye Corporation, 40 Rector Street, New York 6, N. Y. The sample, as received, melted at 38–40°. According to infrared data, the compound was 95–99% pure. Distillation through a high-precision, spinning band column showed about 10% distilling at 86–90° (mostly 90°), another 10% at 90–90.5°, and the remainder at 91°. All fractions were essentially

the same by infrared analysis. The first fraction melted at about 35°, and the main fraction melted at 40.5°.[2]

4. This product contains 1–1.7% methanol and a trace of the starting material.

5. This figure is slightly above the true boiling point of 1,1-dichloro-2,2-difluoroethylene. Upon redistillation in a low-temperature still fitted with a thermocouple, about 95% of this material distilled at 17°,[3] n_D^3 1.3710.

3. Methods of Preparation

1,1-Dichloro-2,2-difluoroethylene has been prepared by the zinc dehalogenation method in ethanol.[2,3] This compound also has been prepared by dehydrochlorination of 1,2,2-trichloro-1,1-difluoroethane by sodium hydroxide,[4] by a solution of potassium hydroxide in aqueous methanol,[5] by passing it over active carbon containing barium chloride or strontium chloride,[6] or by thermal cleavage at 590°.[7] In addition, it has been obtained by the thermal cleavage of 1,1,2,2-tetrachloro-3,3,4,4-tetrafluorocyclobutane at 650° over aluminum fluoride.[8]

[1] Contribution No. 285 from the Chemical Department, Experimental Station, E. I. du Pont de Nemours and Company, Wilmington, Delaware.

[2] Locke, Brode, and Henne, *J. Am. Chem. Soc.*, **56**, 1726 (1934).

[3] Henne and Wiest, *J. Am. Chem. Soc.*, **62**, 2051 (1940).

[4] Frederick (to B. F. Goodrich Co.), U. S. pat. 2,709,181 [*C. A.*, **50**, 5017 (1956)].

[5] McBee, Hill, and Bachman, *Ind. Eng. Chem.*, **41**, 70 (1949).

[6] Alexander Wacker Gesellschaft für electrochemische Industrie G.m.b.H., Brit. pat. 723,715 [*C. A.*, **50**, 5720 (1956)].

[7] Padbury and Tarrant (to American Cyanamid Company), U. S. pat. 2,566,807 [*C. A.*, **46**, 2561 (1952)].

[8] Miller and Calfee (to Allied Chemical and Dye Corp.), U. S. pat. 2,674,631 [*C. A.*, **49**, 4007 (1955)].

2,2-DICHLOROETHANOL *

(Ethanol, 2,2-dichloro-)

$$2CHCl_2COCl + LiAlH_4 \rightarrow LiAlCl_2(OCH_2CHCl_2)_2$$
$$2LiAlCl_2(OCH_2CHCl_2)_2 + 3H_2SO_4 \rightarrow$$
$$4CHCl_2CH_2OH + 2LiCl + 2HCl + Al_2(SO_4)_3$$

Submitted by C. E. Sroog and H. M. Woodburn.[1]
Checked by R. S. Schreiber and C. J. Lintner, Jr.

1. Procedure

Caution! This procedure should be carried out in a good hood, and a spark-proof motor should be used.

A 1-l. three-necked flask having ground-glass joints is fitted with an efficient reflux condenser, a dropping funnel, and a mercury-sealed stirrer, and all exits are protected by drying tubes (Note 1). In the flask are placed 300 ml. of rigorously dried ether (Note 2) and 13.6 g. (0.36 mole) of pulverized lithium aluminum hydride (Note 3). After the mixture has been stirred for 15 minutes, a milky suspension is produced. A solution of 88.6 g. (0.60 mole) of dichloroacetyl chloride in 75 ml. of dry ether is added from the dropping funnel (Note 4) at a rate such as to produce gentle reflux. The process is completed in about 2.5 hours (Note 5).

Agitation is continued for 30 minutes after all the chloride has been introduced, and the excess hydride is then destroyed by the careful, dropwise addition of water to the stirred and cooled reaction mixture. The hydrolysis is accompanied by the formation of a white curdy mass of aluminum hydroxide, and the mixture has a semisolid consistency when the hydride has been completely decomposed (Note 6). With constant stirring, 500 ml. of 10% sulfuric acid is then added slowly to the mass. Stirring is continued for 30 minutes, and the solution becomes clear during this period.

The ether layer is separated, and the solvent is removed by distillation under atmospheric pressure, and the dark-colored residue is fractionated under reduced pressure through a 25-cm. column packed with glass helices. The fraction boiling at 37–38.5°/6 mm. is collected; it weighs 44–45 g. (64–65%); n_D^{25} 1.4626; d_4^{25} 1.404 (Note 7).

2. Notes

1. To prevent partial hydrolysis of the lithium aluminum hydride, the checkers found it desirable to dry the glassware by flaming while dry nitrogen was being swept through the apparatus.

2. Anhydrous ether is recommended since a solvent containing moisture will react to coat the hydride with oxides which will retard its rate of solution. The checkers used ether dried over sodium wire.

3. The hydride can be pulverized rapidly and safely by breaking the large pieces with a spatula, followed by careful crushing with a mortar and pestle.

4. The acid chloride was obtained from the Eastman Kodak Company and was used without prior treatment. The submitters report[2] that dichloroacetic acid or ethyl dichloroacetate may be substituted for dichloroacetyl chloride but that the reaction appears to be smoother with the acid chloride.

5. The acid chloride should be added slowly since the reaction appears to have a short induction period and it is undesirable to accumulate a large quantity of unreacted material.

6. The decomposition of the hydride is accompanied by the vigorous evolution of hydrogen. Cessation of hydrogen formation is evidence that the hydride is completely decomposed.

7. The following constants have been reported in the literature for 2,2-dichloroethanol: d_4^{15} 1.415;[2] d_4^{19} 1.416; $n_D^{17.3}$ 1.4752.[3]

3. Methods of Preparation

2,2-Dichloroethanol has been prepared by the reaction of dichloroacetaldehyde with zinc dialkyls[4] and aluminum alkoxides.[3]

[1] University of Buffalo, Buffalo, New York.

[2] Sroog, Chih, Short, and Woodburn, *J. Am. Chem. Soc.*, **71**, 1710 (1949).

[3] Boeseken, Tellegen, and Plusje, *Rec. trav. chim.*, **57**, 73 (1938).

[4] de Lacre, *Compt. rend.*, **104**, 1184 (1887).

1,1'-DICYANO-1,1'-BICYCLOHEXYL *

([Bicyclohexyl]-1,1'-dicarbonitrile)

Submitted by C. G. OVERBERGER and M. B. BERENBAUM.[1]
Checked by N. J. LEONARD and E. H. MOTTUS.

1. Procedure

In a 200-ml. flask equipped with a reflux condenser are placed 20 g. (0.082 mole) of 1,1'-azo-*bis*-1-cyclohexanenitrile (p. 66) and 50 ml. of toluene. The solution is heated under gentle reflux for 8 hours, and the reaction mixture is placed in a refrigerator overnight. The product is collected on a Büchner funnel and air-dried; yield 11.5–12.2 g. (65–69%); m.p. 224.5–225.5° (Note 1).

2. Notes

1. Tetramethylsuccinonitrile can be prepared from 2,2'-azo-*bis*-isobutyronitrile in a similar manner.[2] In this case petroleum ether (b.p. 60–70°) is added to the reaction mixture before cooling it overnight. The yield is 75–81% of the theoretical amount; m.p. 168–170°. Sublimation at 20–50 mm. raises the melting point to 170.5–171.5° with very little loss of material.

3. Methods of Preparation

1,1'-Dicyano-1,1'-bicyclohexyl has been prepared previously by Hartman [3] in a similar manner. The procedure has been substantiated by Overberger, O'Shaughnessy, and Shalit.[4]

[1] Polytechnic Institute of Brooklyn, Brooklyn, New York.
[2] Thiele and Heuser, *Ann.*, **290**, 1 (1896).
[3] Hartman, *Rec. trav. chim.*, **46**, 150 (1927).
[4] Overberger, O'Shaughnessy, and Shalit, *J. Am. Chem. Soc.*, **71**, 2661 (1949).

1,2-DI-1-(1-CYANO)CYCLOHEXYLHYDRAZINE

(Cyclohexanecarbonitrile, 1,1′-hydrazodi-)

$$
2 \begin{array}{c} O \\ \parallel \\ C \\ \diagup \diagdown \\ CH_2 \quad CH_2 \\ | \quad\quad | \\ CH_2 \quad CH_2 \\ \diagdown \diagup \\ CH_2 \end{array} + 2NaCN + H_2NN\overset{\oplus}{H}_3 \quad H\overset{\ominus}{SO}_4 \rightarrow
$$

$$
\begin{array}{c} CH_2\text{—}CH_2 \quad CN \\ CH_2 \diagup \quad\quad | \\ \diagdown CH_2\text{—}CH_2 \diagup C\text{—}NHNH\text{—} \end{array}
\begin{array}{c} NC \quad CH_2\text{—}CH_2 \\ | \diagup \quad\quad \diagdown CH_2 \\ C \diagdown CH_2\text{—}CH_2 \diagup \end{array}
$$

$$+ \, Na_2SO_4 + 2H_2O$$

Submitted by C. G. OVERBERGER, PAO-TUNG HUANG, and M. B. BERENBAUM.[1]
Checked by N. J. LEONARD and E. H. MOTTUS.

1. Procedure

Caution! These operations should be conducted in an efficient hood.

In a 600-ml. screw-capped bottle are placed 15.4 g. (0.32 mole) of sodium cyanide, 20.5 g. (0.16 mole) of hydrazine sulfate (Note 1), and 400 ml. of ice water. The bottle is capped to prevent loss of hydrogen cyanide (Note 2) and held in an ice bath for 15 minutes. To the cooled mixture is added 29.4 g. (0.3 mole) of cyclohexanone. The bottle is recapped and cooled for an additional 15 minutes. The bottle is then shaken intermittently over a period of 6 hours and allowed to stand an additional 14 hours. The bottle is cooled again before opening. The suspension is filtered by means of suction, and the cake is washed thoroughly with 250 ml. of ice water. After the crude product has been pressed dry on the filter (Note 3), it is transferred to a 500-ml. Erlenmeyer flask and 150 ml. of boiling 95% ethanol is added. The suspension is brought into solution as rapidly as possible by warming on a hot plate (Note 4) and is filtered quickly through a prewarmed 7.5-cm. Büchner funnel. An additional 10 ml. of hot 95% ethanol is used to dissolve any organic residue on the filter, and the combined filtrates are warmed to redissolve any precipitate. The solution is allowed to stand undisturbed for 6 hours in an icebox. The product is collected on a Büchner funnel, washed with 15 ml. of cold 95% ethanol, and then dried over solid calcium chloride in a vacuum desiccator. The yield of product is 24.5–26 g. (66–70%); m.p. 147–149° (Note 5).

2. Notes

1. Technical hydrazine sulfate of known purity is suitable.

2. Both hydrazine and hydrogen cyanide are toxic, and appropriate precautions should be taken.

3. The checkers have used the dry, crude product directly in the conversion to 1,1′-azo-*bis*-1-cyclohexanenitrile (see p. 66) with an over-all yield of 72% based upon cyclohexanone.

4. Excessive or prolonged heating will result in a decreased yield owing to decomposition of the product.

5. 1,2-Di-2-(2-cyano)propylhydrazine (2,2′-hydrazo-*bis*-isobutyronitrile) may be prepared in a similar manner from acetone. The crude product (m.p. 89–91°; yield 88–93%) can be oxidized directly to 2,2′-azo-*bis*-isobutyronitrile. The dried impure material can be recrystallized from ether; m.p. 91.5–92.5°; (72–77%).

3. Methods of Preparation

1,2-Di-1-(1-cyano)cyclohexylhydrazine has been prepared by a procedure similar to that used by Hartman.[2] The procedure has been substantiated by Overberger, O'Shaughnessy, and Shalit.[3]

[1] Polytechnic Institute of Brooklyn, Brooklyn, New York.

[2] Hartman, *Rec. trav. chim.*, **46**, 150 (1927).

[3] Overberger, O'Shaughnessy, and Shalit, *J. Am. Chem. Soc.*, **71**, 2661 (1949).

DICYANOKETENE ETHYLENE ACETAL

(1,2-Dioxolane-$\Delta^{2,\alpha}$-malononitrile)

$$\begin{matrix} NC \\ \\ NC \end{matrix} C=C \begin{matrix} CN \\ \\ CN \end{matrix} \quad + \quad \begin{matrix} HO-CH_2 \\ | \\ HO-CH_2 \end{matrix} \xrightarrow{\text{Urea}}$$

$$\begin{matrix} NC \\ \\ NC \end{matrix} C=C \begin{matrix} O-CH_2 \\ | \\ O-CH_2 \end{matrix} \quad + \quad 2HCN$$

Submitted by C. L. DICKINSON and L. R. MELBY.[1]
Checked by JAMES CASON, EDWIN R. HARRIS, and WILLIAM T. MILLER.

1. Procedure

Caution! This preparation must be carried out in a good hood because hydrogen cyanide is evolved. It is inadvisable to allow contact of tetracyanoethylene with the skin.

Urea (4.0 g., 0.067 mole) is dissolved in 50 ml. of distilled ethylene glycol (Note 1) contained in a 125-ml. Erlenmeyer flask. Finely divided recrystallized tetracyanoethylene (p. 877) (25.6 g., 0.20 mole) is added, and the flask is heated on a steam bath at 70–75° with frequent stirring by hand with a thermometer until solution is complete (about 15 minutes). The resultant brownish yellow solution is then cooled in ice water, and the precipitated dicyanoketene ethylene acetal is collected on a Büchner funnel. The acetal is first washed with two 25-ml. portions of cold ethylene glycol and then washed thoroughly with cold water to remove the ethylene glycol. The dicyanoketene ethylene acetal, which may be dried in air or in a vacuum desiccator, is obtained in the form of large slightly pink needles, m.p. 115–116.5° (Note 2); yield 21–23 g. (77–85%).

2. Notes

1. Moisture in the ethylene glycol leads to lowered yields.

Satisfactory results are obtained with glycol collected at 199.5–201° from a simple distillation.

2. The color may be removed by recrystallization from ethanol after treatment with decolorizing carbon; however, the melting point is not improved and occasionally is found to be lowered.

3. Methods of Preparation

The synthesis of dicyanoketene ethylene acetal described here is a slight modification of one published recently.[2] The procedure has been applied successfully to the synthesis of dicyanoketene dimethyl acetal and dicyanoketene diethyl acetal.[2]

Dicyanoketene ethylene acetal reacts with tertiary amines to give quaternary ammonium inner salts.[2] Similarly, it reacts with sulfides to give sulfonium inner salts.[2] These products are generally solids that can be used to characterize tertiary amines and sulfides. Dicyanoketene acetals can be converted to pyrimidines, pyrazoles, or isoxazoles in one step.[3]

[1] Contribution No. 481 from Central Research Department, Experimental Station, E. I. du Pont de Nemours & Co., Wilmington, Delaware.
[2] Middleton and Engelhardt, *J. Am. Chem. Soc.*, 80, 2788 (1958).
[3] Middleton and Engelhardt, *J. Am. Chem. Soc.*, 80, 2829 (1958).

DICYCLOPROPYL KETONE *

(Ketone, dicyclopropyl)

Submitted by OMER E. CURTIS, JR., JOSEPH M. SANDRI,
RICHARD E. CROCKER, and HAROLD HART.[1]
Checked by V. BOEKELHEIDE, R. TABER, and D. S. TARBELL.

1. Procedure

A solution of sodium methoxide is prepared from 50 g. (2.17 g. atoms) of freshly cut sodium and 600 ml. of absolute methanol (Note 1) in a 3-l. three-necked flask placed on a steam bath and equipped with a sealed stirrer (Note 2), dropping funnel, and a condenser set downward for distillation (Note 3). To the stirred solution is added in one portion 344 g. (4.0 moles) of γ-butyrolactone (Note 4), and the flask is heated until methanol distils at a rapid rate. After 475 ml. of methanol is collected, a filter flask or other suitable device equipped with a side arm is connected to the condenser. This receiver is surrounded by an ice bath, and reduced pressure from an aspirator is applied cautiously (frothing) with continuous stirring. An additional 50–70 ml. of methanol is collected in this way. The residue presumably is dibutyrolactone (Note 5).

The condenser is set for reflux, and the steam bath is replaced with a more potent source of heat (electric heating mantle, oil bath, or direct flame). Concentrated hydrochloric acid is added with stirring, cautiously at first because a considerable amount of carbon dioxide is evolved. A total of 800 ml. of acid is added in about 10 minutes (Note 6). The mixture is heated under reflux with stirring for 20 minutes, then cooled in an ice bath (Note 7). A solution of 480 g. of sodium hydroxide in 600 ml. of water is added to the stirred mixture as rapidly as possible, without allowing the temperature to go above 50° (Note 8). The mixture is then heated under reflux for an additional 30 minutes.

The condenser is arranged for downward distillation, and a total of 650 ml. of ketone-water mixture is collected as distillate. Sufficient potassium carbonate is added to saturate the aqueous layer, and about 130 ml. of ketone is separated. The aqueous layer is extracted with three 100-ml. portions of ether, and the combined ether and ketone layers are dried over 25 g. of anhydrous magnesium sulfate. The product remaining after removal of the ether is distilled through an efficient column. The yield of dicyclopropyl ketone boiling at 72–74°/33 mm., n_D^{25} 1.4654, is 114–121 g. (52–55%) (Note 9).

2. Notes

1. Commercial sodium methoxide (117 g.) in 520 ml. of methanol may be used instead of metallic sodium.

2. The stirrer should be sturdy and capable of giving vigorous agitation.

3. It is desirable to have the condenser arranged for reflux during preparation of the methanolic sodium methoxide, if it is made from sodium metal.

4. Commercial lactone (available from General Aniline and Film Corporation, 435 Hudson Street, New York 14, New York) should be redistilled, b.p. 88–90°/12 mm., before use.

5. Dibutyrolactone can be isolated as a crystalline solid, m.p. 86–87°, from this residue.[2] The preparation can be interrupted at this point without jeopardizing the yield.

6. The color of the mixture changes from yellow through dark orange to dark reddish brown.

7. At this point, the following procedure may be used to prepare 1,7-dichloro-4-heptanone. To the cooled, stirred solution is added 200 ml. of ether, which brings the dense dichloroketone to the upper layer. The latter is separated, and the acid layer is extracted with two 100-ml. portions of ether. The combined

ether layers are dried over 25 g. of anhydrous calcium chloride. After removal of the solvent, the residue is distilled through an efficient column. The yield of 1,7-dichloro-4-heptanone, b.p. 106–110°/4 mm., n_D^{25} 1.4713, is 263–278 g. (72–76%). The material takes on a purple cast rapidly and should be stored in a refrigerator.

8. Considerable salt separates at this point but does not interfere with the subsequent steps.

9. According to the submitters, a similar procedure can be applied to substituted lactones; di-(2-methylcyclopropyl) ketone, b.p. 65–67°/7 mm., n_D^{25} 1.4600, has been made from γ-valerolactone in 50% yield.

3. Methods of Preparation

This procedure is a modification of one recently described in the literature.[3] The first step is based on early work of Fittig [4] and Volhard [5] as modified by Spencer and Wright.[2] The third step, ring closure of a γ-haloketone,[5] is well known (p. 597). Dicyclopropyl ketone was reported to form in small amounts from the decarboxylation of cyclopropanecarboxylic acid over thoria,[6] but there is some doubt [3] about the product.

[1] Michigan State University, East Lansing, Michigan.
[2] Spencer and Wright, J. Am. Chem. Soc., 63, 1281 (1941).
[3] Hart and Curtis, Jr., J. Am. Chem. Soc., 78, 112 (1956).
[4] Fittig, Ann., 256, 50 (1889); Fittig and Stöm, Ann., 267, 191 (1892).
[5] Volhard, Ann., 267, 78 (1892).
[6] Michiels, Bull. soc. chim. Belges, 24, 396 (1910) [Chem. Zentr., 82, I, 66 (1911)].

1-DIETHYLAMINO-3-BUTANONE *

(2-Butanone, 4-diethylamino-)

$$CH_3COCH_3 + (C_2H_5)_2\overset{+}{N}H_2\overset{-}{Cl} + HCHO \rightarrow$$

$$CH_3COCH_2CH_2\overset{+}{N}H(C_2H_5)_2\overset{-}{Cl} + H_2O$$

$$CH_3COCH_2CH_2\overset{+}{N}H(C_2H_5)_2\overset{-}{Cl} + KOH \rightarrow$$

$$CH_3COCH_2CH_2N(C_2H_5)_2 + H_2O + KCl$$

Submitted by Alfred L. Wilds, Robert M. Nowak, and Kirtland E. McCaleb.[1]
Checked by William S. Johnson and Duane Zinkel.

1. Procedure

In a 3-l. round-bottomed flask equipped with a reflux condenser (Note 1) are placed 176 g. (1.60 moles) of diethylamine hydrochloride (Note 2), 68 g. (2.26 moles) of paraformaldehyde, 600 ml. (8.2 moles) of acetone, 80 ml. of methanol, and 0.2 ml. of concentrated hydrochloric acid. The mixture is heated for 12 hours at a moderate to vigorous rate of reflux (Note 3). The light-yellow solution, in which a small amount of gelatinous solid remains, is cooled, and a cold solution of 65 g. of sodium hydroxide in 300 ml. of water is added. The mixture is extracted with three 200-ml. portions of ether, the combined extracts are washed with two 150-ml. portions of saturated sodium chloride solution, and the washes are re-extracted with two 150-ml. portions of ether.

The combined ether solutions are dried overnight with about 80 g. of anhydrous sodium sulfate, filtered, and then distilled under reduced pressure (5 to 12 mm.) (Note 4) through a 20-cm. asbestos-wrapped Vigreux distilling column, with an efficient water-cooled condenser (Note 5). After the solvent and a small fore-run have been distilled, 150–171 g. (66–75%) of 1-diethylamino-3-butanone is collected as a light-yellow to nearly colorless liquid, b.p. 63–67°/7mm. (75–77°/15mm.), n_D^{25} 1.4300–1.4310. The product may contain a small amount of 1,1-bis(diethylaminomethyl)acetone (the bis-Mannich base), which can interfere with some uses of this product. Refractionation gives relatively pure material, 142–161 g. (62–70%), b.p. 72–75°/10mm., n_D^{25} 1.4301–1.4307, d_4^{25} 0.8626, M_D (found) 43.2–43.3, M_D (calcd.) 43.1 (Note 6).

2. Notes

1. Ground-glass joints are desirable.

2. A good grade of commercial diethylamine hydrochloride (Eastman Kodak white label grade or Matheson, Coleman and Bell) is satisfactory without purification.

3. At first some bumping may occur if the heating is too vigorous; mechanical stirring may reduce this, but does not improve the yield. The submitters found an electrically heated oil bath or a steam bath to be satisfactory, but *not a heating mantle*.

4. If the temperature of distillation is too high, or if a heating mantle is used, decomposition to methyl vinyl ketone may occur. The submitters used an electrically heated oil bath and prefer pressures below 12 mm. to minimize decomposition. A more elaborate fractionating column necessitating a higher bath temperature or prolonged heating is also undesirable. If the material stands more than one or two days before distillation it may decompose.

5. Unless the condenser is efficient, some product will be lost; a Dry Ice-cooled trap located between the receiver and the pump is recommended.

6. This product gave satisfactory analytical values: Calcd. for $C_8H_{17}NO$: C, 67.1; H, 12.0. Found: C, 67.2; H, 11.9. The neutral equivalent of various samples, titrated potentiometrically with standard hydrochloric acid solutions, ranged between 144 and 145 (calcd. 143.2).

3. Methods of Preparation

1-Diethylamino-3-butanone has been prepared by the Mannich reaction,[2-7] and by the condensation of methyl vinyl ketone with diethylamine hydrochloride in the presence of acetic anhydride.[8] The present procedure is a modification of that described by Wilds and Shunk.[5]

[1] University of Wisconsin, Madison, Wisconsin.

[2] Mannich, *Arch. Pharm.*, **255**, 261 (1917).

[3] du Feu, McQuillin, and Robinson, *J. Chem. Soc.*, **1937**, 53.

[4] Tuda, Hukusima, and Oguri, *J. Pharm. Soc. Japan*, **61**, 69 (1941) [*C. A.*, **36**, 3154 (1942)].

[5] Wilds and Shunk, *J. Am. Chem. Soc.*, **65**, 469 (1943).

[6] Spaeth, Geissman, and Jacobs, *J. Org. Chem.*, **11**, 399 (1946).

[7] Halsall and Thomas, *J. Chem. Soc.*, **1956**, 2431.

[8] Swaminathan and Newman, *Tetrahedron*, **2**, 88 (1958).

N,N'-DIETHYLBENZIDINE

(Benzidine, N,N'-diethyl-)

$$H_2NC_6H_4C_6H_4NH_2 + 2C_2H_5OH \xrightarrow[Ni]{Raney}$$
$$C_2H_5NHC_6H_4C_6H_4NHC_2H_5 + 2H_2O$$

Submitted by RIP G. RICE and EARL J. KOHN.[1]
Checked by MAX TISHLER, W. H. JONES, and W. F. BENNING.

1. Procedure

In a 5-l. three-necked flask, fitted with an efficient stirrer (Note 1), a stopper, and a reflux condenser, are placed, in order, 184.2 g. (1 mole) of benzidine (Note 2), 500 ml. of commercial absolute ethanol, about 125 g. of Raney nickel,[2] and 500 ml. of ethanol. The mixture is heated under reflux with stirring for a total of 15 hours (Note 3). The volume is brought to 3 l. with 95% ethanol, and about 150 g. of filter aid ("Super-Cel") is added with stirring. The mixture is heated to boiling, filtered rapidly through a ¼-in. layer of filter aid on a Büchner funnel into a 4-l. filter flask (Notes 4 and 5), and the nickel is washed well with hot 95% ethanol (Note 6). The filtrate is concentrated to a volume of 1.5 l., cooled slowly to room temperature, then chilled in a refrigerator and filtered. The yield is 185–202 g. (77–84%) of light gray or purplish gray flakes, m.p. 105–115°.

The crude product is dissolved in 1.5 l. of hot 95% ethanol, the solution is treated with 5 g. of activated carbon, and the mixture is filtered rapidly with suction. The filtrate is made up to 1.5 l. with 95% ethanol and heated to dissolve the solid. The solution is allowed to cool slowly to room temperature, then chilled in a refrigerator and filtered. The yield is 143–161 g. (60–67%) of colorless flakes melting at 115–116° (Note 7).

2. Notes

1. Efficient stirring is necessary to keep the nickel uniformly distributed throughout the reaction mixture; otherwise, the yield is decreased considerably.

2. The submitters used benzidine obtained by neutralization of an aqueous solution of C.P. benzidine dihydrochloride (Matheson, Coleman and Bell) with 20% sodium hydroxide solution, followed by crystallization from 60–70% ethanol. The benzidine prepared in this manner was in the form of light tan flakes. Freshly prepared benzidine should be used for the best results.

3. The refluxing time does not have to be continuous. The submitters heated the reaction mixture for 7 hours, allowed the mixture to stand overnight, and continued the refluxing the next day.

4. This operation should be carried out in a well-ventilated hood.

5. To prevent clogging of the Büchner funnel during filtration it is necessary to preheat the funnel and keep it hot during filtration.

6. The nickel is pyrophoric and must be kept moist to prevent spontaneous ignition.

7. The method has been applied successfully by the submitters [3] to the preparation of N,N'-dibutylbenzidine (61% crude yield), N-ethyl-, N-propyl-, N-butyl-, N-amyl-, and N-benzylanilines (80–84% yields), N-hexylaniline (72%), and N-isobutyl- and N-isoamylanilines (41, 49%). N-Propylbutylamine was obtained from propyl alcohol and butylamine in 57% yield. The submitters state that no alkylation of benzidine took place with methyl alcohol or of aniline with methyl, isopropyl, or sec-butyl alcohol. The method has been applied to the synthesis of other N-alkyl and N-cycloalkyl aromatic amines.[4,5]

3. Methods of Preparation

N,N'-Diethylbenzidine has been prepared by heating ethyl iodide, benzidine, and ethanol in a pressure tube at water-bath temperature,[6,7] and by the reaction of diethylzinc on benzenediazonium chloride.[7] The method described here is a modification of that of Shah, Tilak, and Venkataraman.[8]

[1] Naval Research Laboratory, Washington, D. C.
[2] Org. Syntheses Coll. Vol. 3, 181 (1955).
[3] Rice and Kohn, J. Am. Chem. Soc., 77, 4052 (1955).
[4] Ainsworth, J. Am. Chem. Soc., 78, 1636 (1956).

[5] Kao, Tilak, and Venkataraman, *J. Sci. Ind. Research India*, **14B**, 624 (1955).
[6] Hofmann, *Ann.*, **115**, 365 (1860).
[7] Bamberger and Tichvinsky, *Ber.*, **35**, 4179 (1902).
[8] Shah, Tilak, and Venkataraman, *Proc. Indian Acad. Sci.*, **28A**, 145 (1948).

DIETHYL BENZOYLMALONATE

(Malonic acid, benzoyl-, diethyl ester)

$$\underset{CO_2C_2H_5}{\overset{CO_2C_2H_5}{CH_2}} + Mg + C_2H_5OH \rightarrow \underset{CO_2C_2H_5}{\overset{CO_2C_2H_5}{C_2H_5OMgCH}}$$

$$C_6H_5CO_2H + N(C_2H_5)_3 + ClCO_2C_2H_5 \rightarrow$$
$$C_6H_5CO_2CO_2C_2H_5 + N(C_2H_5)_3 \cdot HCl$$

$$C_6H_5CO_2CO_2C_2H_5 + C_2H_5OMgCH(CO_2C_2H_5)_2 \rightarrow$$
$$C_6H_5COCH(CO_2C_2H_5)_2 + CO_2 + Mg^{+2} + 2OC_2H_5{}^-$$

Submitted by JOHN A. PRICE and D. S. TARBELL.[1]
Checked by T. L. CAIRNS and C. L. DICKINSON.

1. Procedure

A. *Ethoxymagnesiummalonic ester* (*Note 1*). In a 250-ml. three-necked flask equipped with a dropping funnel and an efficient reflux condenser fitted with a calcium chloride drying tube are placed 5.0 g. (0.2 g. atom) of magnesium turnings (Grignard), 5 ml. of absolute alcohol (Note 2), 0.2 ml. of carbon tetrachloride, and 6 ml. of a mixture of 32.0 g. (30.2 ml., 0.2 mole) of diethyl malonate (Note 3) and 16 ml. of absolute alcohol. The reaction will proceed in a few minutes and may require occasional cooling before the addition of the remainder of the diethyl malonate solution. The addition should be controlled so that the reaction goes at a fairly vigorous rate. When the reaction mixture has cooled to room temperature, 60 ml. of ether dried over sodium wire is cautiously (Note 4) added. When the reaction again appears to subside, gentle heating by means of a steam bath is begun and continued until nearly all the magnesium has disappeared (Note 5). The alcohol and ether are removed by distillation, first at atmospheric pressure and then at reduced pressure secured with a water pump. To the partially crystalline product is added 60 ml. of dry benzene, and the solvent is again removed by distillation at atmospheric and then reduced pressure (Note 6). The residue is dissolved in 60 ml. of dry ether to await the completion of the mixed carbonic-carboxylic anhydride preparation.

B. *Mixed benzoic-carbonic anhydride* (*Note 7*). In a 500-ml. three-necked flask, equipped with a low-temperature thermometer, an efficient sealed stirrer, and an adaptive joint carrying a drying tube and a dropping funnel, is placed a solution of 24.4 g. (0.2 mole) of benzoic acid (Note 8) and 20.2 g. (0.2 mole) of triethylamine (Note 9) in 200 ml. of dry toluene. The solution is cooled below 0° by means of an ice-salt mixture, and 21.7 g. (0.2 mole) of ethyl chlorocarbonate (Note 10) is added at such a rate that the temperature does not rise above 0° (approximate time for addition is 25–30 minutes). Triethylamine hydrochloride precipitates both during the addition and while the mixture is stirred for 15–25 minutes thereafter.

C. *Diethyl benzoylmalonate.* The dropping funnel used for the chlorocarbonate addition is replaced by another into which the ethereal solution of the ethoxymagnesium compound has been transferred. Approximately 30 ml. of dry ether is used to rinse the flask, and this is also added to the dropping funnel. The ether solution is added to the mixed anhydride with stirring, as the temperature is held at −5° to 0°. After the mixture has been allowed to stand overnight and to come to room temperature during this time, it is treated cautiously with 400 ml. of 5% sulfuric acid; then the aqueous solution is separated and extracted once with ether. The two organic layers are combined, washed once with dilute sulfuric acid and then with a concentrated sodium bicarbonate solution until no further benzoic acid is obtained from acidification of the bicarbonate extracts (Note 11). The organic layer is washed with water and dried over anhydrous sodium sulfate. After removal of the sodium sulfate by filtration, the solvent is removed at water-pump pressure from a water bath held at about 50°. The resulting product is purified by distillation through a 30-cm. Vigreux column, and the fraction boiling at 144–149°/0.8 mm. is collected (Note 12). The yield is 35.8–39.4 g. (68–75%), n_D^{25} 1.5063–1.5066.

2. Notes

1. The described procedure is essentially the same as that reported

by Lund and Voigt.[2] A similar preparation has been described by Reynolds and Hauser (p. 708).

2. A commercial grade is satisfactory for this preparation.

3. Commercial malonic ester was redistilled at reduced pressure to give material with n_D^{20} 1.4047.

4. The ether dissolves the crystalline cake which has formed on cooling. This releases unreacted material and vigorous reaction again sets in.

5. From 6 to 8 hours is required for this operation.

6. The benzene removes any residual alcohol which may interfere with the subsequent acylation.[3]

7. The mixed carbonic anhydride procedure [4-6] has been useful in the preparation of amide linkages and thiol esters. Mixed carbonic anhydrides have successfully acylated, under very mild conditions, the carbanions derived from diethyl ethylmalonate and diethylcadmium.[7] The latter gives as a product the corresponding ketone. Mixed anhydrides derived from acetic and acetylsalicylic acids give results similar to those described here.[7]

8. A good reagent grade of benzoic acid is satisfactory.

9. Redistillation of a commercial grade gave material boiling at 87.0–87.2°.

10. Redistillation of a commercial grade gave material boiling at 90.8–92.0°.

11. If the benzoic acid is not all removed at this stage, it is troublesome during the final distillation.

12. The fore-run, if giving a positive test with ferric chloride reagent, may be redistilled to give increased yields.

3. Methods of Preparation

Diethyl benzoylmalonate has been prepared by treatment of the copper derivative of ethyl benzoylacetate with ethyl chlorocarbonate.[8] It has also been obtained by the action of benzoyl chloride on the magnesium derivative of diethyl malonate,[9] and on a mixture of malonic ester and sodium methoxide [10, 11] or sodium.[12, 13] This compound has been found to be obtainable in higher yield by the reaction of benzoyl chloride and the ethoxymagnesium derivative.[14] The present method has been described in a previous communication and is of interest as an illustration of the use of mixed carbonic anhydrides as acylating agents.[7]

[1] University of Rochester, Rochester, New York.
[2] *Org. Syntheses Coll. Vol.* **2**, 594 (1943).
[3] Riegel and Lilienfeld, *J. Am. Chem. Soc.*, **67**, 1273 (1945).
[4] Vaughan, *J. Am. Chem. Soc.*, **73**, 3547 (1951).
[5] Boissonnas, *Helv. Chim. Acta*, **34**, 874 (1951).
[6] Wieland and Bernhard, *Ann.*, **572**, 190 (1951).
[7] Tarbell and Price, *J. Org. Chem.*, **21**, 144 (1956); **22**, 245 (1957).
[8] Bernhard, *Ann.*, **282**, 165 (1894).
[9] Gagnon, Boivin, and Laflamme, *Can. J. Chem.*, **34**, 530 (1956).
[10] Claisen and Falk, *Ann.*, **291**, 72 (1896).
[11] Bülow and Hailer, *Ber.*, **35**, 934 (1902).
[12] King, King, and Thompson, *J. Chem. Soc.*, **1948**, 552.
[13] Borsche and Wannagat, *Ber.*, **85**, 193 (1952).
[14] Lund, *Ber.*, **67**, 935 (1934).

DIETHYL 1,1-CYCLOBUTANEDICARBOXYLATE *

(1,1-Cyclobutanedicarboxylic acid, diethyl ester)

$$CH_2(CO_2C_2H_5)_2 + BrCH_2CH_2CH_2Cl + 2NaOC_2H_5 \rightarrow$$

$$
\begin{array}{c}
CH_2 \\
\diagup \quad \diagdown \\
CH_2 \qquad C(CO_2C_2H_5)_2 + NaBr + NaCl + 2C_2H_5OH \\
\diagdown \quad \diagup \\
CH_2
\end{array}
$$

Submitted by RAYMOND P. MARIELLA and RICHARD RAUBE.[1]
Checked by WILLIAM S. JOHNSON, WILLIAM DeACETIS, and HERBERT TITLE.

1. Procedure

A solution of sodium ethoxide is prepared by adding 138 g. (6.0 g. atoms) of fresh-cut sodium in small pieces to 2.5 l. of absolute ethanol in a 5-l. round-bottomed flask fitted with an efficient reflux condenser capped with a calcium chloride drying tube (Note 1). In a three-necked 5-l. round-bottomed flask, equipped with a reflux condenser capped with a calcium chloride tube, a rubber-sealed mechanical stirrer, and an inlet tube for addition of the sodium ethoxide solution, are mixed 480 g. (3.0 moles) of diethyl malonate (Note 2) and 472 g. (3.0 moles) of trimethylene chlorobromide (Note 3). The mixture is heated to 80° and vigorously stirred while the sodium ethoxide solution is slowly forced into the flask by means of dry air pressure. The rate of addition is regulated so that the reaction mixture refluxes smoothly. After the addition is complete (this requires about 1.5 hours), the mixture is refluxed, with continued stirring (Note 4), for an additional 45 minutes. Upon completion of the reflux period, the alcohol is removed by distillation (Note 4), 90–95% of the alcohol being recovered. The reaction mixture is cooled, and 900 ml. of cold water is added. After the sodium halides are completely dissolved, the organic layer is separated and the aqueous layer is extracted with three 500-ml. portions of ether. The organic layer and the ether extracts are combined, shaken with 50 ml. of saturated salt solution, and dried over 100 g. of anhydrous sodium sulfate. The solution is filtered, the ether is removed by distillation on a steam bath, and the residue, which weighs 600–625 g., is distilled through a short Vigreux column. The yield of product boiling at 91–96°/4 mm. is 320–330 g. (53–55%) (Note 5).

2. Notes

1. It is important to maintain strictly anhydrous conditions throughout this reaction. The equipment should be carefully predried and the absolute ethanol freshly prepared (preferably by the magnesium ethoxide method [2]) and distilled directly into the reaction flask. If the volume of ethanol is less than 2.5 l. the sodium ethoxide may not remain in solution. It is convenient to employ a three-necked flask carrying two condensers for this operation and to add the sodium through the third neck, which is otherwise kept stoppered.

2. Material boiling at 95.2–95.8°/14 mm., n_D^{25} 1.4120, was used.

3. Trimethylene chlorobromide can be obtained commercially. Material boiling at 141–142°/755 mm., n_D^{25} 1.4843, was used.

4. The stirring must be continued during this operation; otherwise the mixture will bump badly.

5. This material, n_D^{25} 1.433–1.434, d_{20}^{25} 1.042–1.044, is of fair purity and can be hydrolyzed and decarboxylated [3,4] to give cyclobutanecarboxylic acid in more than 80% yield. In order to obtain a highly pure product it may be necessary to fractionally distil the material. For example, on slow redistillation of a

product prepared as described above through a 3-ft. Vigreux column at a high reflux ratio, the checkers obtained 8.5% of forerun n_D^{25} 1.4287–1.4328, 43% of diethyl 1,1-cyclobutanedicarboxylate, n_D^{25} 1.4332–1.4335, and 2.5% of higher-boiling material n_D^{25} 1.4362–1.4427. The pure product is reported to boil at 104.6°/12 mm., 85.2°/3.5 mm., or 60.0°/0.5 mm., n_D^{25} 1.4336, d_{20}^{25} 1.0470.[5]

3. Methods of Preparation

Diethyl 1,1-cyclobutanedicarboxylate has been prepared by the alkylation of diethyl sodiomalonate with trimethylene dibromide,[3,4,6,7] trimethylene diiodide,[7] or trimethylene chlorobromide.[8,9] It is claimed that the yield of product from the latter reaction may be increased by the use of a mixture of benzene and ethanol as a solvent.[10] Diethyl 1,1-cyclobutanedicarboxylate also may be obtained by the peroxide-catalyzed addition of hydrogen bromide to diethyl allylmalonate followed by intramolecular alkylation.[11] The procedure described here is that of Mariella and Raube.[8]

[1] Northwestern University, Evanston, Illinois.

[2] Lund and Bjerrum, *Ber.*, **64**, 210 (1931). See Fieser, *Experiments in Organic Chemistry*, 3rd ed., p. 286, D. C. Heath and Company, Boston, Massachusetts, 1955.

[3] *Org. Syntheses Coll. Vol.* **3**, 213 (1955).

[4] Cason and Allen, *J. Org. Chem.*, **14**, 1036 (1949).

[5] Perkin, *J. Chem. Soc.*, **51**, 1 (1887).

[6] Rupe, *Ann.*, **327**, 183 (1903).

[7] Gol'mov, *Zhur. Obshcheĭ Khim.*, **22**, 1944 (1952) [*C. A.*, **47**, 9268 (1953)].

[8] Mariella and Raube, *Bol. col. quím. Puerto Rico*, **8**, 24 (1951) [*C. A.*, **46**, 4491h (1952)].

[9] Kishner, *J. Russ. Phys. Chem. Soc.*, **37**, 507 (1905) [*Chem. Zentr.*, [2] **76**, 761 (1905)]; Favorskaya and Yakovlev, *Zhur. Obshcheĭ Khim.* (*J. Gen. Chem.*), **22**, 122 (1952) [*C. A.*, **46**, 11118 (1952)]; Gol'mov and Kazanskiĭ, *Akad. Nauk. S.S.S.R. Inst. Org. Khim. Sintezy Org. Soedineniĭ Sbornik*, **I**, 93 (1950) [*C. A.*, **47**, 8003 (1953)].

[10] Raik and Kazanskiĭ, *Vestnik Moskov. Univ.*, **8**, No. 3, Ser. Fiz.-Mat. i Estestven. Nauk, No. 2, 125 (1953) [*C. A.*, **49**, 3833 (1955)].

[11] Walborsky, *J. Am. Chem. Soc.*, **71**, 2941 (1949).

DIETHYL Δ²-CYCLOPENTENYLMALONATE *

(2-Cyclopentene-1-malonic acid, diethyl ester)

$$\underset{\underset{CH_2}{\diagdown\diagup}}{\overset{\displaystyle CH\!=\!\!=\!\!CH}{\underset{\displaystyle CH_2 \quad CH}{\vert \qquad \vert}}}\!-\!Cl + Na^{\oplus}\ [CH(CO_2C_2H_5)_2]^{\ominus} \rightarrow$$

$$\underset{\underset{CH_2}{\diagdown\diagup}}{\overset{\displaystyle CH\!=\!\!=\!\!CH}{\underset{\displaystyle CH_2 \quad CH}{\vert \qquad \vert}}}\!-\!CH(CO_2C_2H_5)_2 + NaCl$$

Submitted by Robert Bruce Moffett.[1]
Checked by R. T. Arnold and George P. Scott.

1. Procedure

Nine hundred and twenty-five milliliters of absolute ethanol (Note 1) is placed in a 2-l. three-necked round-bottomed flask, fitted with a mercury or glycerin-sealed stirrer (Note 2), dropping funnel, and reflux condenser. To this is added 46 g. (2 g. atoms) [2] of freshly cut sodium, a few pieces at a time and at such a rate that the reaction proceeds rapidly but the solvent does not reflux too vigorously. When most of the sodium has dissolved, a calcium chloride drying tube is fitted to the top of the condenser and 320 g. (2 moles) of redistilled diethyl malonate is added from the dropping funnel. Then 205 g. (2 moles) of 3-chlorocyclopentene (p. 42) (Note 3) is added at such a rate that a gentle reflux is maintained. Toward the end of the addition, it is desirable to test the reaction mixture with pH test paper, and the addition should be stopped if the solution becomes acidic.

When the addition is complete, the condenser is set downward for distillation, the stirring is continued, and most of the ethanol is removed by distillation on a steam bath (Note 4). After cooling, the reaction mixture is diluted with sufficient water to dissolve the salt, and the layers are separated. The aqueous layer is extracted with 50 ml. of ether. The ethereal solution is added to the ester, and the resulting solution is washed with saturated salt solution and dried over anhydrous magnesium sulfate. The ethereal solution is transferred to a Claisen flask, the solvent is evaporated, and the product is distilled under reduced pressure. The fraction boiling at about 85–140°/12 mm. is collected and redistilled through an efficient fractionating column (Note 5). The product distils at 130°/12 mm. (Note 6). The yield is about 276.5 g. (61%); n_D^{20} 1.4536.

2. Notes

1. The preparation of absolute ethanol has been described earlier.[3]

2. The glycerin-sealed stirrer has been described earlier.[2]

3. It is not necessary to use redistilled 3-chlorocyclopentene; however, if large amounts of impurities are present, difficulty may be encountered in obtaining a pure product.

4. If a glycerin-sealed stirrer is used, the rate of solvent evaporation may be accelerated by means of gentle water-pump suction.

5. Careful fractionation is necessary in order to secure a pure product. The submitter used a 12-in. column packed with ⅛-in. glass helices. The fore-run should be redistilled to separate any ester it may contain.

6. Other boiling points are 140°/18 mm.; 97°/1 mm.

3. Methods of Preparation

Diethyl Δ^2-cyclopentenylmalonate has been prepared by the alkylation of malonic ester with 3-chlorocyclopentene.[4–8]

[1] The Upjohn Company, Kalamazoo, Michigan.
[2] *Org. Syntheses Coll. Vol.* **3**, 368 (1955).
[3] *Org. Syntheses Coll. Vol.* **1**, 259 (1941).
[4] Noller and Adams, *J. Am. Chem. Soc.*, **48**, 2444 (1926).
[5] Perkins and Cruz, *J. Am. Chem. Soc.*, **49**, 518 (1927).
[6] Wagner-Jauregg and Arnold, *Ann.*, **529**, 274 (1937).
[7] Horclois, *Chim. & ind. (Paris)*, **31**, 357 (Special No., April, 1934).
[8] Buu-Hoï and Cagniant, *Bull. soc. chim. France*, [5] **9**, 99 (1942).

DIETHYL ETHYLIDENEMALONATE *

(Malonic acid, ethylidene-, diethyl ester)

$$(CH_3CHO)_3 + 3CH_2(CO_2C_2H_5)_2 + 3(CH_3CO)_2O \rightarrow$$
$$3CH_3CH{=}C(CO_2C_2H_5)_2 + 6CH_3CO_2H$$

Submitted by WILLIAM S. FONES.[1]
Checked by ARTHUR C. COPE, HARRIS E. PETREE, and E. R. TRUMBULL.

1. Procedure

In a 1-l. three-necked flask, equipped with a thermometer and a reflux condenser, are placed 60 g. of paraldehyde (0.45 mole, equivalent to 1.35 moles of acetaldehyde) and 100 ml. (1.06 moles) of acetic anhydride. Ice water is circulated through the condenser, and the reaction mixture is protected from atmospheric moisture by a drying tube containing Drierite. The temperature of the mixture is raised slowly by heating with an electric mantle to 125°, at which point gentle refluxing begins. Then 100 g. (0.62 mole) of diethyl malonate is added in 15-ml. portions at a rate of 1 portion every 30 minutes. During the addition of diethyl malonate, the temperature gradually drops to about 100°, and the mixture is heated so as to maintain a reflux rate of 30–60 drops per minute. After the addition is complete, the reaction mixture is heated under reflux for 4 hours at the specified rate.

The reflux condenser is replaced by a Claisen distillation head, and the reaction mixture is distilled until the temperature of the vapor reaches 140°. The residue is transferred to a smaller flask and fractionated through a 30-cm. column packed with glass helices. A low-boiling fraction containing ethylidene diacetate and diethyl malonate is collected first, followed by 79–89.5 g. (68–77%) of diethyl ethylidenemalonate; b.p. 102–106°/10 mm.; n_D^{25} 1.4394 (Notes 1 and 2).

2. Notes

1. The submitter obtained a yield of 70% when three times

the above quantities were used, in which case the diethyl malonate was added at a rate of 90 ml. per hour.

2. Horton [2] has reported that a higher yield of diethyl ethylidenemalonate has been obtained by the following procedure. A mixture of 50 g. of redistilled malonic ester, 50 g. of acetic anhydride and 28.5 g. of acetaldehyde, enclosed in a hydrogenation bomb, was mixed and heated at 100° for 24 hours without shaking. The bomb contents were transferred with the aid of benzene, and three pooled runs were distilled through a 15-in. Vigreux column at 18–20 mm. A fore-run of diethyl malonate was removed, and then 149.8 g. (86.2%) of diethyl ethylidenemalonate was collected at 106–109°/13 mm. The product was stored at 5°.

3. Methods of Preparation

Diethyl ethylidenemalonate has been prepared by heating acetaldehyde, diethyl malonate, and acetic anhydride; [3,4] by heating the same reagents plus zinc chloride; [5] by treating acetaldehyde and diethyl malonate with sodium ethoxide or piperidine; [6] and by heating diethyl malonate, ethylidene bromide, and ethanolic sodium ethoxide. [7]

[1] National Institutes of Health, Bethesda, Maryland.
[2] Horton, Private communication.
[3] Komnenos, *Ann.*, **218**, 157 (1883).
[4] Goss, Ingold, and Thorp, *J. Chem. Soc.*, **123**, 3353 (1923).
[5] von Auwers and Eisenlohr, *J. prakt. Chem.*, [2] **84**, 101 (1911).
[6] Higginbotham and Lapworth, *J. Chem. Soc.*, **123**, 1622 (1923).
[7] Loevenich, Losen, and Dierichs, *Ber.*, **60**, 957 (1927).

DIETHYL MERCAPTOACETAL

(Acetaldehyde, mercapto-, diethyl acetal)

$$\text{BrCH}_2\text{CH} \begin{smallmatrix} \text{OC}_2\text{H}_5 \\ \\ \text{OC}_2\text{H}_5 \end{smallmatrix} \quad + \text{Na}_2\text{S}_3 \xrightarrow[\text{H}_2\text{O}]{\text{C}_2\text{H}_5\text{OH}}$$

$$\begin{smallmatrix} \text{C}_2\text{H}_5\text{O} \\ \\ \text{C}_2\text{H}_5\text{O} \end{smallmatrix} \text{CHCH}_2\text{—S}_n\text{—CH}_2\text{CH} \begin{smallmatrix} \text{OC}_2\text{H}_5 \\ \\ \text{OC}_2\text{H}_5 \end{smallmatrix} \xrightarrow[\text{NH}_3]{\text{Na}}$$

$$\text{NaSCH}_2\text{CH} \begin{smallmatrix} \text{OC}_2\text{H}_5 \\ \\ \text{OC}_2\text{H}_5 \end{smallmatrix} \xrightarrow[\text{CO}_2]{\text{NH}_4\text{Cl, HCl}} \text{HSCH}_2\text{CH} \begin{smallmatrix} \text{OC}_2\text{H}_5 \\ \\ \text{OC}_2\text{H}_5 \end{smallmatrix}$$

Submitted by WILLIAM E. PARHAM and HANS WYNBERG.[1]
Checked by MAX TISHLER and GEORGE PURDUE.

1. Procedure

A. *1,1,1′,1′-Tetraethoxyethyl polysulfide* (Note 1). In a 5-l.
round-bottomed flask, equipped with a stirrer, a reflux condenser,
and an addition funnel, 540 g. (2.25 moles) of sodium sulfide
enneahydrate (Note 2) is dissolved in 2.7 l. of boiling 95%
ethanol. Sulfur (144 g., 4.5 g. atoms) is added, and the solution is heated under reflux until it has assumed a deep-red
color (10 minutes). The source of heat is removed, and 588 g.
(3 moles) of diethyl bromoacetal (Note 3) is added over a period
of about 30 minutes at such a rate that the solution boils gently.
The mixture is heated under reflux for an additional 4 hours.
During the last hour of heating, ethanol is allowed to distil from
the reaction flask, and after the first 700 ml. of distillate is collected sodium bicarbonate (135 g.) (Note 4) is added to the reaction mixture. A total of 2.0–2.5 l. of ethanol is collected.
Water (1.5 l.) is added, and the upper layer is separated. The
aqueous layer is extracted with two 250-ml. portions of ether, and

the combined organic layers are washed successively with two 100-ml. portions of 10% sodium hydroxide solution, three 100-ml. portions of water, and two 250-ml. portions of saturated sodium chloride solution, and are dried over potassium carbonate. The ether is removed by distillation, and the residue is brought to constant weight by heating for about 1 hour on the steam bath at 10–15 mm. pressure. The orange liquid obtained in this way (Note 5) weighs 450–500 g. and has n_D^{25} 1.515–1.517.

B. *Diethyl mercaptoacetal.* A 12-l. three-necked flask, equipped with an efficient Hershberg stirrer, a Dry Ice condenser, and a stopper, is placed in an enameled container (Note 6). The flask is charged with 500 g. (1.38 moles, calculated as tetrasulfide) of 1,1,1′,1′-tetraethoxyethyl polysulfide and 500 ml. of anhydrous ether. The condenser is filled with acetone and Dry Ice, and the contents of the flask are cooled by means of acetone and Dry Ice kept at $-35°$ to $-45°$. Liquid ammonia (Note 7) is introduced into the stirred solution until 4.5–5.0 l. has been added (Note 8). The solution is stirred for several minutes to ensure homogeneity, and 200 g. (8.7 g. atoms) of sodium is added in 2- to 5-g. lumps to the mixture over a period of 30–45 minutes. The last 2–10 g. of sodium causes the solution to turn deep blue. The mixture is stirred for 30 additional minutes, and ammonium chloride (about 200 g.) is cautiously added with stirring until the blue color is discharged. The cooling bath is removed and replaced by a bath of hot water into which steam may be introduced. The mixture is stirred while the ammonia is removed by distillation or evaporation (the distillation should be continued until the residue can no longer be stirred or is difficult to stir). The soft, semisolid mass containing small amounts of ammonia is cooled in ice, and 1 l. of ice-cold water is added with stirring. When most of the solid has dissolved, ice-cold hydrochloric acid is added cautiously, with frequent checks of the pH, until a pH of 8.0–8.5 is reached. The pH is then adjusted to 7.8–8.0 (preferably by a pH meter) by passing carbon dioxide into the cold solution (Note 9). The oil is separated from the aqueous layer, which is extracted with two 300-ml. portions of ether. The combined organic layers are washed successively with 100 ml. of water and two 100-ml. portions of saturated sodium chloride, and are dried over magnesium sulfate. The ether is removed by distillation, and the residue is distilled from a 500-ml. flask using a small Vigreux column or a Claisen head. The product (Note 10) weighs 320–372 g. (77–90% yield, based on polysulfide; 71–83% yield, based on bromoacetal), boils at 62–64°/12 mm., and has n_D^{25} 1.4391–1.4400. The residue, 23–27 g., n_D^{25} 1.4702–1.4730, consists mainly of 1,1,1',1'-tetraethoxyethyl sulfide.

2. Notes

1. These directions have been used equally successfully with twice, one-third, and one-fifth the amounts specified. The reaction of chloro- or bromoacetal with sodium disulfide results in the formation of a considerable quantity of the corresponding monosulfide which is not subsequently reduced to mercaptan. Polysulfides are, however, easily reduced to mercaptans. 1,1,1',1'-Tetramethoxyethyl polysulfide has been prepared from commercially available dimethyl chloroacetal in a similar fashion. A 10-hour heating period and the addition of 5 g. of potassium iodide per 100 g. of acetal are recommended in the latter preparation.

2. The sodium sulfide ($Na_2S \cdot 9H_2O$) should be crystalline and finely divided.

3. The procedure used for the preparation of diethyl bromoacetal (b.p. 78–79°/24 mm., n_D^{20} 1.4418) is that of Bedoukian.[2]

4. Di- and polysulfides are cleaved by strong alkali. The bicarbonate is added as a buffer.

5. Analysis of this material showed it to have the average composition calculated for a tetrasulfide; the yield of product, calculated as tetrasulfide, is 83–92%.

6. In order to use as little coolant as possible the flask should fit the container snugly.

7. All normal precautions should be observed during the handling of these large quantities of ammonia and sodium. A well-ventilated hood, a gas mask, and a bucket of sand were available to the submitters. The only hazard in this reaction may arise if too much ammonia escapes and the sodium does not react properly as a consequence. The addition of 0.5–1.0 l. of ammonia to the reaction mixture after the sodium has been added is a useful precaution.

8. The submitters measured the liquid ammonia in a 1-l. graduated cylinder.

9. Complete neutralization of the salt of mercaptoacetal with mineral acid may result in hydrolysis of the acetal.

10. Mercaptoacetal is sensitive to acid and oxygen. The product should be stored in a dry, alkali-washed bottle under nitrogen in a refrigerator.

3. Methods of Preparation

Diethyl mercaptoacetal has been prepared by treating diethyl bromoacetal with potassium hydrosulfide;[3] by the reduction of 1,1,1',1'-tetraethoxyethyl disulfide[3] with lithium aluminum hydride; by reduction of 1,1-diethoxyethyl benzyl sulfide,[3, 4] 1,1,1',1'-tetraethoxyethyl disulfide,[3] and 1,1,1',1'-tetraethoxyethyl polysulfide[3] with sodium and liquid ammonia. The method described is adapted from the last-named preparation. Dimethyl mercaptoacetal has been prepared by the same methods.[3]

[1] University of Minnesota, Minneapolis, Minnesota.
[2] Bedoukian, *J. Am. Chem. Soc.*, **66**, 651 (1944).
[3] Parham, Wynberg, and Ramp, *J. Am. Chem. Soc.*, **75**, 2065 (1953).
[4] Hesse and Jörder, *Ber.*, **85**, 924 (1952).

DIETHYL METHYLENEMALONATE

(Malonic acid, methylene-, diethyl ester)

$$[C_2H_5OCH{=}C(CO_2C_2H_5)_2 + H_2] \xrightarrow{\text{Raney Ni}}$$

$$[C_2H_5OCH_2CH(CO_2C_2H_5)_2]$$

$$[C_2H_5OCH_2CH(CO_2C_2H_5)_2] \xrightarrow{\Delta}$$

$$CH_2{=}C(CO_2C_2H_5)_2 + C_2H_5OH$$

Submitted by WAYNE FEELY and V. BOEKELHEIDE.[1]
Checked by MAX TISHLER, BARBARA P. BIRT, and ARTHUR A. PATCHETT.

1. Procedure

A solution of 108 g. (0.5 mole) of diethyl ethoxymethylenemalonate (Note 1) in 100 ml. of commercial absolute alcohol is placed in an apparatus for high-pressure hydrogenation together with 10 g. of Raney nickel catalyst (Note 2). The pressure in the bomb is raised to 1000–1500 lb. with hydrogen, and the temperature is adjusted to 45° (Note 3). The bomb is shaken, and the reaction is allowed to proceed for 12–20 hours, during which time 0.5 mole of hydrogen is absorbed.

The apparatus is allowed to cool to room temperature, the pressure is released, and the catalyst is removed by filtration. Concentration of the filtrate under reduced pressure at room temperature yields a colorless oil (Note 4).

The residual oil is transferred to a distillation flask (Note 5), and the flask is carefully warmed with a small flame (Note 6). After a fore-run of ethanol is collected, the variable transformer controlling the heating tape is set at a voltage to give an internal temperature in the distilling head of 80–100° before distillation. The flask is then heated to effect slow distillation at atmospheric pressure. About 10–13 g. of fore-run is collected before the temperature of the distillate vapor reaches 200°. The main fraction (68–71 g., 79–82%) is collected between 200° and 216° as a colorless oil. For most purposes, the fore-run is sufficiently pure that it can be combined with the main fraction to give an over-all yield of 78–81 g. (91–94%) (Note 7). Diethyl methylenemalonate polymerizes on standing to a white solid, from which the monomer can be recovered by slow distillation. The diethyl methylenemalonate should be distilled just prior to use (Note 8).

2. Notes

1. Diethyl ethoxymethylenemalonate can be prepared by the method of Parham and Reed.[2] The submitters used diethyl ethoxymethylenemalonate obtained commercially from Kay-Fries Chemicals Inc., 180 Madison Avenue, New York 16, New York.

2. Raney nickel catalyst can be prepared by the method of Mozingo.[3] The submitters used Raney nickel catalyst obtained commercially from the Gilman Paint and Varnish Company, Chattanooga, Tennessee.

3. At higher temperatures (about 70°) the diethyl ethoxy-

methylmalonate formed tends to eliminate ethanol, forming diethyl methylenemalonate which is hydrogenated to diethyl methylmalonate.

4. This oil, n_D^{20} 1.4254, is presumably diethyl ethoxymethylmalonate, as evidenced by its infrared and ultraviolet absorption spectra. It is relatively stable and can be stored at room temperature without change. If the concentration of the filtrate under vacuum is carried out by heating on a steam bath instead of by keeping it at room temperature, the diethyl ethoxymethylmalonate undergoes elimination of ethanol to some extent, giving diethyl methylenemalonate directly. The yield of diethyl methylenemalonate obtained eventually is not altered by this procedure.

5. The distillation apparatus is shown in Fig. 7. The upper

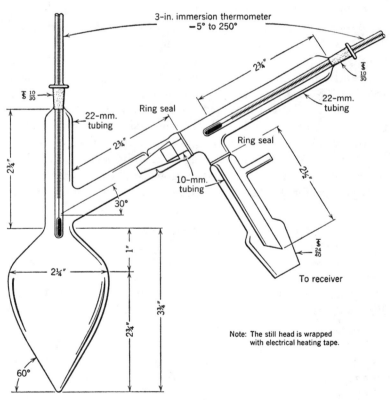

Fig. 7.

part of the flask and the short column are wrapped in electrical heater tape, which is operated during distillation of the product.

6. The elimination of ethanol is slightly exothermic. It is advisable to heat the flask cautiously with a small flame until the reaction starts and then to remove the flame until the reaction subsides.

7. The fore-run has a refractive index (usually about n_D^{22} 1.4154) lower than that of the main fraction (about n_D^{22} 1.4250). The combined fractions show refractive indices in the range n_D^{22} 1.4210–1.4259.

8. When the solid polymer from the above preparation was slowly distilled to recover diethyl methylenemalonate, there was obtained 57 g. of colorless oil, b.p. 210–216°/730 mm., n_D^{20} 1.4220.

3. Methods of Preparation

Diethyl methylenemalonate was prepared first by Perkin [4] from formaldehyde and malonic ester, and more recently a study has been made of the effect of various catalysts on this system.[5] An alternative procedure involving the reaction of methylene chloride or iodide with sodiomalonic ester has been developed by Tanatar.[6] Diethyl methylenemalonate also is formed when hexaethyl pentane-1,1,3,3,5,5-hexacarboxylate is treated with methylene iodide and sodium ethoxide,[7] or when methyl chloromethyl ether is caused to react with sodiomalonic ester.[8] Diethyl methylenemalonate has been obtained from paraformaldehyde [9] and vapor and liquid reactions of formaldehyde with diethyl malonate in the presence of various catalysts.[10]

[1] University of Rochester, Rochester, New York.

[2] Org. Syntheses Coll. Vol. 3, 395 (1955).

[3] Org. Syntheses Coll. Vol. 3, 181 (1955).

[4] Perkin, Ber., 19, 1053 (1886).

[5] Takagi and Asahara, J. Chem. Soc. Japan, Ind. Chem. Sect., 56, 901 (1953) [C. A., 49, 6836 (1955)].

[6] Tanatar, Ann., 273, 48 (1893).

[7] Bottomley and Perkin, J. Chem. Soc., 77, 294 (1900).

[8] Vasiliu and Barbulescu, Analele univ. C. I. Parhon Bucureşti, Ser. stiint. nat., No. 16, 99 (1957) [C. A., 53, 1238 (1959)].

[9] Sakurai, Midorikawa, and Aoyama, J. Sci. Research Inst. (Tokyo), 52, 112 (1958).

[10] Bachman and Tanner, J. Org. Chem., 4, 493 (1939).

DIETHYL γ-OXOPIMELATE *

(Heptanedioic acid, 4-oxo-, diethyl ester)

$$\text{CH}{=}\text{CHCO}_2\text{H} + 2\text{C}_2\text{H}_5\text{OH} \xrightarrow{\text{HCl}}$$

$$\text{C}_2\text{H}_5\text{O}_2\text{C}(\text{CH}_2)_2\text{CO}(\text{CH}_2)_2\text{CO}_2\text{C}_2\text{H}_5$$

Submitted by W. S. EMERSON and R. I. LONGLEY, JR.[1]
Checked by WILLIAM S. JOHNSON and I. A. DAVID.

1. Procedure

A 3-l. three-necked flask equipped with a stirrer, reflux condenser, and gas inlet tube is charged with 476 g. (3.45 moles) of furylacrylic acid [2] and 1580 g. (about 33 moles) of 95% ethanol (Note 1). The mixture is heated to boiling, and anhydrous hydrogen chloride is introduced at such a rate that the mixture becomes saturated after 90 minutes. The gas inlet tube is replaced by a stopper, and a 2-ft. Vigreux column is substituted for the reflux condenser. About 250 ml. of solvent is removed by distillation at atmospheric pressure; then another 300 ml. is removed while the pressure is slowly reduced (water aspirator). The residue is cooled and stirred with a solution of about 260 g. of sodium carbonate in water (Note 2). The mixture is extracted with two 250-ml. portions of benzene, and the combined extracts are washed with 100 ml. of water (Note 3). Distillation of the organic portion through a 2-ft. Vigreux column yields, after a small fore-run, 579–657 g. (73–83%) of diethyl γ-oxopimelate, b.p. 116–121°/0.3 mm., n_D^{25} 1.4395–1.4400.

2. Notes

1. If absolute ethanol is used the yield is much lower.
2. A slight excess of sodium carbonate is used. The amount required depends on the amount of hydrogen chloride remaining after the distillation.
3. Sodium chloride may be added if the layers do not separate.

3. Methods of Preparation

Diethyl γ-oxopimelate has been prepared by saturating an ethanol solution of furylacrylic acid [3-7] or γ-oxopimelic acid dilactone with hydrogen chloride [5] or sulfuric acid.[8] It was found as a by-product in the esterification of furylacrylic acid with ethanol in the presence of *p*-toluenesulfonic acid.[9] A mechanism has been proposed for the formation of esters of γ-oxopimelic acid from furylacrylic acid.[10] The present procedure is a modification of the original Marckwald process.[3, 4]

[1] Monsanto Chemical Company, Dayton, Ohio.

[2] *Org. Syntheses Coll. Vol.* **3**, 425 (1955).

[3] Marckwald, *Ber.*, **20**, 2811 (1887).

[4] Marckwald, *Ber.*, **21**, 1398 (1888).

[5] Volhard, *Ann.*, **253**, 206 (1889).

[6] Chichibabin, *Chim. & ind. (Paris)*, **27**, 563 (1932).

[7] Komppa, *Ann. Acad. Sci. Fennicae*, **A51**, No. 3 (1938) [*C. A.*, **34**, 2335 (1940)].

[8] Reppe et al., *Ann.*, **596**, 80 (1955).

[9] Murahashi, *Bull. Inst. Phys. Chem. Research (Tokyo)*, **22**, 476 (1943) [*C. A.*, **42**, 1205 (1948)].

[10] Gavăt and Glineschi, *Rev. chim. (Bucharest)*, **7**, 575 (1956) [*C. A.*, **52**, 1991 (1958)].

DIETHYL cis-Δ⁴-TETRAHYDROPHTHALATE AND DIETHYL cis-HEXAHYDROPHTHALATE

(4-Cyclohexene-1,2-dicarboxylic acid, diethyl ester, cis-, and 1,2-Cyclohexanedicarboxylic acid, diethyl ester, cis-)

Submitted by ARTHUR C. COPE and ELBERT C. HERRICK.[1]
Checked by CHARLES C. PRICE and GEORGE A. CYPHER.

1. Procedure

A. *Diethyl cis-Δ⁴-tetrahydrophthalate.* In a 2-l. round-bottomed flask are placed 228 g. (1.5 moles) of *cis*-Δ⁴-tetrahydrophthalic anhydride (p. 890), 525 ml. (9 moles) of commercial absolute ethanol, and 2.5 g. of *p*-toluenesulfonic acid monohydrate. The flask is connected to a reflux condenser and heated in an oil bath maintained at 95–105° for 12–16 hours.

At this time 270 ml. of toluene is added to the mixture and the condenser is changed for distillation. An azeotropic mixture of ethanol, toluene, and water is distilled at 75–78° with the bath at

105–110°. When the temperature begins to drop (Note 1), 525 ml. of commercial absolute ethanol is added and the mixture is again heated under reflux for 12-16 hours (Note 2). Again 270 ml. of toluene is added, and the azeotropic mixture is distilled until the vapor temperature falls to 68°. After the residue is cooled, the system is evacuated to 25–35 mm. and the remaining ethanol and toluene are distilled.

The residual liquid is diluted with 200 ml. of ether. The ether solution is washed twice with 100-ml. portions of 3% sodium carbonate solution (Note 3). The combined carbonate solutions are extracted three times with 100-ml. portions of ether, and the ether solutions are combined and washed with 100 ml. of distilled water. The ether solution is dried over magnesium sulfate, filtered, and concentrated, and the residue is distilled under reduced pressure. The yield of product boiling at 129–131°/5 mm., n_D^{25} 1.4605–1.4610, is 280–292 g. (83–86%) (Note 4).

B. *Diethyl cis-hexahydrophthalate.* The reaction is carried out in a low-pressure catalytic hydrogenation apparatus.[2] In a 500-ml. Pyrex centrifuge bottle are placed 0.5 g. of Adams platinum oxide catalyst (Note 5) and 20 ml. of commercial absolute ethanol (Note 6). The bottle is connected to a calibrated low-pressure hydrogen tank and alternately evacuated and filled with hydrogen twice. Hydrogen is then admitted to the system until the pressure is 1–2 atmospheres (15–30 lb.), and the bottle is shaken for 20–30 minutes to reduce the platinum oxide. The shaker is stopped, the bottle is evacuated, and air is admitted. Two hundred and twenty-six grams (1 mole) of diethyl cis-Δ⁴-tetrahydrophthalate is placed in the bottle. The container in which the ester was weighed is rinsed with 10 ml. of absolute ethanol, which is added to the ester. The bottle is alternately evacuated and filled with hydrogen twice. Hydrogen is admitted to the system until the pressure is 25–30 lb. (approximately 2 atmospheres), and the bottle is shaken until the pressure drop indicates that the theoretical amount (1 mole) of hydrogen has been taken up and the absorption ceases (3–5 hours). The bottle is evacuated and then air is admitted. The catalyst is removed by filtration through a small Hirsch funnel. The bottle is washed with 15 ml. of alcohol, which is poured through the funnel. Most of the solvent is distilled at 25–35 mm. Distillation of the residue under reduced pressure yields 215–219 g. (94–96%) of diethyl cis-hexahydrophthalate, b.p. 130–132°/9 mm., n_D^{25} 1.4508–1.4510 (Note 7).

2. Notes

1. A period of 5–8 hours is required for this distillation and for the similar subsequent distillation.

2. If the second period of reflux and azeotropic distillation is omitted the yield is decreased to 66%.

3. The product should be washed with sodium carbonate solution until the aqueous solution remains basic.

4. Dimethyl cis-Δ^4-tetrahydrophthalate can be prepared by a similar procedure. cis-Δ^4-Tetrahydrophthalic anhydride (228 g., 1.5 moles) is heated under reflux with 364 ml. (9 moles) of commercial anhydrous methanol and 2.5 g. of p-toluenesulfonic acid monohydrate for 12–16 hours. At this time 270 ml. of toluene is added and the mixture is distilled. When the distillation temperature drops from 68–70° to 45°, after about 4–6 hours, 364 ml. of absolute methanol is added and the mixture again is heated under reflux for 12–16 hours. An additional 270 ml. of toluene is added, and distillation is continued for 4–6 hours. The residual liquid is purified by a procedure similar to the one described for the ethyl ester. The yield of dimethyl cis-Δ^4-tetrahydrophthalate, boiling at 120–122°/5 mm., n_D^{25} 1.4700, is 239 g. (80%).

5. The platinum oxide catalyst was obtained from Baker and Company, Newark, New Jersey. Platinum oxide may also be obtained from the American Platinum Works, Newark, New Jersey.

6. The reduction also may be carried out using 1 g. of palladium on carbon,[3] in which case no solvent is required and prereduction of the catalyst is unnecessary.

7. Dimethyl cis-hexahydrophthalate also may be prepared by a similar reduction of dimethyl cis-Δ^4-tetrahydrophthalate. With 0.5 g. of prereduced Adams platinum oxide catalyst, 198 g. (1 mole) of dimethyl cis-Δ^4-tetrahydrophthalate was reduced to give 196 g. (98%) of dimethyl cis-hexahydrophthalate, b.p. 110–112°/5 mm., n_D^{25} 1.4570.

3. Methods of Preparation

Diethyl *cis*-hexahydrophthalate has been prepared from *cis*-hexahydrophthalic acid, absolute ethanol, and sulfuric acid,[4,5] and from *cis*-hexahydrophthalic anhydride, absolute ethanol, and sulfuric acid.[6] Diethyl *cis*-Δ^4-tetrahydrophthalate has been prepared from *cis*-Δ^4-tetrahydrophthalic acid or its anhydride, ethanol, and sulfuric acid.[7] Dimethyl *cis*-Δ^4-tetrahydrophthalate and dimethyl *cis*-hexahydrophthalate have been synthesized by the procedures of this preparation.[8]

[1] Massachusetts Institute of Technology, Cambridge, Massachusetts.
[2] Suitable apparatus is described in *Org. Syntheses Coll. Vol.* **1**, 61 (1941).
[3] *Org. Syntheses Coll. Vol.* **3**, 385 (1955).
[4] von Auwers and Ottens, *Ber.*, **57**, 437 (1924).
[5] Hückel and Goth, *Ber.*, **58**, 447 (1925).
[6] Price and Schwarcz, *J. Am. Chem. Soc.*, **62**, 2891 (1940).
[7] Brooks and Cardarelli, U. S. pat. 1,824,069 [*C. A.*, **26**, 152 (1932)].
[8] Cope and Herrick, *J. Am. Chem. Soc.*, **72**, 983 (1950).

DIETHYLTHIOCARBAMYL CHLORIDE *

(Carbamoyl chloride, diethylthio-)

Submitted by R. H. Goshorn, W. W. Levis, Jr., E. Jaul, and E. J. Ritter.[1]
Checked by T. L. Cairns and H. E. Cupery.

1. Procedure

A 200-ml. three-necked flask, equipped with a mechanical stirrer arranged to permit escape of gas, a thermometer, and a gas-inlet tube 10 mm. in diameter (Note 1), is placed in a vessel to which cooling water may be added. The entire apparatus is placed in a well-ventilated hood (Note 2). The flask is charged with 74 g. (0.25 mole) of dry (Note 3), molten (70°) tetraethyl-thiuram disulfide (Note 4). The molten mass is stirred vigorously, and chlorine is passed through a safety trap and is introduced below the surface of the liquid through the inlet tube. The reaction is exothermic, and the temperature is held at 70–75° by adjusting the rate of chlorine addition and by surrounding the reaction flask with cold water. After approximately 6 g. of chlorine has been absorbed, the temperature is lowered to 50–55° and held in this range for the remainder of the chlorination (Note 5). When about 90% of the theoretical amount of chlorine has been added, sulfur begins to precipitate and the reaction mass changes from a clear yellow-to-red solution to a cloudy yellow mixture. The reaction is considered complete when 18 g. (0.25 mole) of chlorine has been absorbed (measured by gain in weight of the reaction mixture). The time required for the chlorination is about 40 minutes.

The crude reaction product at 50° consists of a reddish yellow upper layer of diethylthiocarbamyl chloride saturated with sulfur (Note 6) and a viscous lower layer of amorphous sulfur saturated with diethylthiocarbamyl chloride. The mechanical stirrer is replaced by a 6-in. glass-helix-packed column arranged for distillation. A magnetic stirrer is used throughout the distillation. The diethylthiocarbamyl chloride is distilled under reduced pressure, b.p. 80–85°/1 mm. (Notes 7 and 8), m.p. 48–51°. The yield is 71–72 g. (94–95%) (Notes 9 and 10).

2. Notes

1. A tube of smaller diameter may be used, but a tube 10 mm. in diameter is recommended to minimize plugging.

2. The hood serves to carry away unabsorbed chlorine and diethylthiocarbamyl chloride vapors. The fumes are irritating to the eyes, nose, and throat and may have other injurious physiological effects.

3. The quality of the product depends to a large extent upon the purity of the tetraethylthiuram disulfide used. Since the yield is decreased in proportion to the quantity of water present,

reagents used should be dry. Dry Sharples Tetraethylthiuram Disulfide No. 163 was found to be satisfactory without further purification.

4. An alternative method is to add chlorine to a solution or suspension of tetraethylthiuram disulfide in an inert liquid medium such as carbon tetrachloride. If the quantity and nature of the solvent are such as to dissolve the diethylthiocarbamyl chloride, most of the liberated sulfur can be separated mechanically and the chloride isolated by distillation after evaporation of the solvent.

5. As the chlorination progresses, the melting point of the reaction mass becomes lower so that after about one-third of the theoretical quantity of chlorine has been added the temperature may be dropped to 50°, which is approximately the melting point of the final product. The low temperature is desirable because there is less danger of overheating at the point of entry of the chlorine.

6. If the crude product is held at 50–55° for 6–8 hours, approximately 70% of the theoretical amount of free sulfur precipitates and may be separated by decantation and filtration. The remaining 30% of the theoretical quantity of sulfur remains in solution in the diethylthiocarbamyl chloride.

7. Good fractionation is not necessary, and a better product can usually be obtained by a rapid distillation than by a slow distillation at a high reflux ratio. A column with a 6- or 8-in. packed section gives a sufficiently pure product for most purposes.

The distillation may be carried out at higher pressures with only slightly more decomposition. At a pressure of 13–14 mm. diethylthiocarbamyl chloride distils at 117–120°. If the pot temperature is above 140° decomposition becomes appreciable, and at 160–190° a vigorous decomposition occurs, especially when a relatively large amount of sulfur is present.

8. The crude diethylthiocarbamyl chloride need not be distilled if sulfur does not interfere in the reaction for which the chloride is to be used and if sulfur can be readily separated from the product. At approximately 105° all the sulfur liberated by the chlorination of tetraethylthiuram disulfide will dissolve in the diethylthiocarbamyl chloride, forming a homogeneous product

which may be used on the basis of 100% conversion of the disulfide to the chloride.

9. The submitters used a scale 10 times larger than that given here and obtained yields of 89–94%.

10. The submitters have prepared the following thiocarbamyl chlorides by the chlorination of the corresponding disulfides: dimethylthiocarbamyl chloride, b.p. 90–95°/0.5 mm., m.p. 42.5–43.5°, carbon tetrachloride reaction medium; diisopropylthiocarbamyl chloride, m.p. 69–71°, benzene reaction medium; diisobutylthiocarbamyl chloride, m.p. 46–48°, no solvent.

3. Methods of Preparation

Billeter prepared diethylthiocarbamyl chloride [2] and other carbamyl chlorides [3,4] by the reaction of the appropriate amine hydrochloride with thiophosgene. Free amines also have been used in this reaction.[5] Diethylthiocarbamyl chloride has been obtained from the reaction of trichloromethanesulfenyl chloride with sodium diethyl dithiocarbamate.[6] The present procedure represents an adaptation of the Ritter process.[7]

[1] Pennsylvania Salt Manufacturing Company, Wyndmoor, Pennsylvania.
[2] Billeter, *Ber.*, **26**, 1681 (1893).
[3] Billeter, *Ber.*, **20**, 1629 (1887).
[4] Mazzara, *Gazz. chim. ital.*, **23**, I, 37 (1893) [*Chem. Zentr.*, **64**, I, 647 (1893)].
[5] American Cyanamid Co., Brit. pat. 683,141 [*C. A.*, **48**, 1446 (1954)].
[6] Zbirovský and Ettel, *Chem. listy*, **52**, 95 (1958) [*C. A.*, **52**, 16335 (1958)].
[7] Ritter, U. S. pat. 2,466,276 [*C. A.*, **43**, 5038 (1949)].

3,4-DIHYDRO-2-METHOXY-4-METHYL-2H-PYRAN

(2H-Pyran, 3,4-dihydro-2-methoxy-4-methyl-)

$$
\begin{array}{ccc}
\text{CH}_3 & & \text{CH}_3 \\
| & & | \\
\text{CH} & & \text{CH} \\
\end{array}
$$

$$
\begin{array}{ccccc}
\text{HC} & & \text{CH}_2 & & \\
| & + & \| & \rightarrow & \\
\text{HC} & & \text{CHOCH}_3 & & \\
& \text{O} & & &
\end{array}
$$

Submitted by RAYMOND I. LONGLEY, JR., WILLIAM S. EMERSON, and ALBERT J. BLARDINELLI.[1]
Checked by T. L. CAIRNS and T. E. YOUNG.

1. Procedure

In a high-pressure autoclave, arranged for agitation by shaking or rocking, are placed 286 g. (336 ml., 4.08 moles) of croton-aldehyde, 294 g. (5.06 moles) of methyl vinyl ether, and 1.1 g. of hydroquinone (Notes 1 and 2). The autoclave is heated to 200° (Note 3) and held there for 12 hours. The autoclave is cooled and vented, and the black product is distilled through a 1 by 60 cm. helix-packed column. The yield of 3,4-dihydro-2-methoxy-4-methyl-2H-pyran is 270–297 g. (52–57%), b.p. 42–50°/19 mm., n_D^{25} 1.4349–1.4374 (Notes 4, 5, and 6).

2. Notes

1. The purer grade of crotonaldehyde supplied by the Eastman Kodak Company was used. Methyl vinyl ether was obtained from the Matheson Chemical Company.
2. The submitters condensed the methyl vinyl ether in the aldehyde cooled to 0° and then charged this mixture to a pre-cooled autoclave. The checkers cooled the autoclave containing the crotonaldehyde to −70°, evacuated, and condensed the required amount of methyl vinyl ether directly into the autoclave.
3. The autoclave should be capable of withstanding a pressure

of 3000 p.s.i. This provides a margin of safety, since at 220° the pressure is about 2600 p.s.i.

4. Pure 3,4-dihydro-2-methoxy-4-methyl-2H-pyran boils at 135–138°/760 mm. and at 79–80°/100 mm. and has n_D^{25} 1.4370.

5. The submitters used approximately three times the quantities reported here and obtained yields of 82–83%.

6. Under comparable conditions the submitters found that the corresponding dihydropyran derivatives were similarly obtained by the condensation of acrolein with methyl vinyl ether in 80–81% yield, with ethyl vinyl ether (77–85% yield), with n-butyl vinyl ether (82% yield), with ethyl isopropenyl ether (50% yield), and with n-butyl cyclohexenyl ether (40% yield). Other α,β-unsaturated carbonyl compounds that have thus been condensed with ethyl vinyl ether are crotonaldehyde (87% yield), methacrolein (40% yield), α-ethyl-β-n-propylacrolein (54% yield), cinnamaldehyde (60% yield), β-furylacrolein (85% yield), methyl vinyl ketone (50% yield), benzalacetone (75% yield), and benzalacetophenone (74% yield).

3. Methods of Preparation

3,4-Dihydro-2-methoxy-4-methyl-2H-pyran has been prepared only by the addition of methyl vinyl ether to crotonaldehyde.[2-5]

[1] Monsanto Chemical Company, Dayton 7, Ohio.

[2] Longley and Emerson, J. Am. Chem. Soc., 72, 3079 (1950).

[3] Smith, Norton, and Ballard, J. Am. Chem. Soc., 73, 5267 (1951).

[4] Smith, Norton, and Ballard, U. S. pat. 2,514,168.

[5] N. V. de Bataafsche Petroleum Maatschappij, Brit. pat. 653,764 [C. A., 47, 5452 (1953)].

9,10-DIHYDROPHENANTHRENE *

(Phenanthrene, 9,10-dihydro-)

Submitted by DONALD D. PHILLIPS.[1]
Checked by WILLIAM S. JOHNSON and DAVID C. REMY.

1. Procedure

A. *Purification of phenanthrene.* 1. By azeotropic distillation.[2] A mixture of 300 g. of commercial phenanthrene (Note 1), 90 g. of maleic anhydride, and 600 ml. of xylene, contained in a 2-l. round-bottomed flask, is heated under reflux for 20 hours (Note 2). The initially yellow solution rapidly turns to a dark brown on heating. This solution is cooled to room temperature and filtered by suction to remove any insoluble adduct. The filtrate is then extracted with two 100-ml. portions of dilute sodium hydroxide, and the basic extracts are discarded. The organic phase is next washed with water and saturated sodium chloride solution, and finally is filtered through a layer of anhydrous magnesium sulfate. The excess xylene is removed by distillation, first at atmospheric pressure; then the final portions are removed at reduced pressure. The residue, while still hot, is poured into a large mortar and, after solidification, is powdered to a convenient size. The yield of crude phenanthrene is 230–240 g.

A solution of 52 g. of the crude phenanthrene in 400 ml. of diethylene glycol (Note 3) is azeotropically distilled through a small column (Note 4). A fore-run of approximately 50 g. is collected at 155–165°/100 mm., followed by the main fraction of 390–400 g., b.p. 140–141°/21 mm. (Note 5). The fore-run contains considerable fluorene and should be discarded. The main fraction is added to five times its volume of water, and the precipitated hydrocarbon is collected by suction filtration and washed well with water to remove the last traces of diethylene glycol. The colorless product (41–43 g.) is heated under reflux for 3 hours with about 450 ml. of 95% ethanol containing approximately 9 g. of Raney nickel catalyst. The hot solution is filtered with slight suction through a sintered-glass funnel. On being concentrated to about 250 ml. and cooled, the filtrate deposits 33–35 g. (63-67%) of colorless phenanthrene, m.p. 97.5–98°.

2. *By sodium treatment.*[3] Commercial phenanthrene (Note 1) is treated with maleic anhydride as described above (part 1), and 170 g. of the residue is added to a 1-l. three-necked flask equipped with a Hershberg mercury-sealed Nichrome stirrer,[4] an air condenser, and a thermometer. Ten grams of sodium is added, and the mixture is vigorously stirred at 190–200° for 6 hours. The dark residue is cooled to about 80°, and 300 ml. of benzene is added. The mixture is brought to reflux with stirring and, while still hot, is *cautiously* filtered through a coarse sintered-glass funnel with *gentle* suction (Note 6). The benzene is removed by distillation at atmospheric pressure, and the residual phenanthrene is distilled through a small column (Note 4) adapted to the distillation of solids to give 125–130 g. (74–76%) of colorless phenanthrene, b.p. 183–183.5°/15 mm. This product is heated under reflux for 3 hours with about 1.2 l. of 95% ethanol containing approximately 9 g. of Raney nickel catalyst, and the hot solution is filtered as described under paragraph 1. The filtrate on cooling deposits 115–120 g. (68–71%) of colorless phenanthrene, m.p. 97–98°.

B. *Catalytic reduction.* A hydrogenation bomb of approximately 300 ml. total capacity is charged with 29.5 g. (0.17 mole) of purified phenanthrene (Note 7); then 70 ml. of cyclohexane (Note 8) and 1.5 g. of copper chromium oxide catalyst (Note 9) are added. The bomb is filled with hydrogen to an initial pressure of 2000 p.s.i. at 20° and heated with shaking to 150° (maximum pressure about 2900 p.s.i.). The hydrogenation proceeds rapidly under these conditions, and about 85% of the theoretical uptake is complete within 1.75–2 hours. The reaction is interrupted at this point (Note 10), and the catalyst is removed by centrifugation or filtration. The cyclohexane is evaporated, and the residue is distilled through a small column (Note 4). After a small fore-run (0.2–0.3 g.) distilling below 182°, there is collected 21–23 g.(70–77% yield) of 9,10-dihydrophenanthrene, b.p. 183–184°/25 mm., n_D^{25} 1.6401–1.6416. The residue consists of 4.5–5.0 g. (15–17%) of phenanthrene, m.p. 96.5–98°, which may be recycled.

2. Notes

1. Technical grade phenanthrene (80–90%) is satisfactory for this preparation.

2. An electric heating mantle is convenient for this operation.

3. Technical diethylene glycol may be used with satisfactory results.

4. The submitter used a 65-cm. Podbielniak type column equipped with partial reflux head.[5] For distillation of the sodium-treated phenanthrene the checkers employed a 6-in. Vigreux column. For the fractionation of the dihydrophenanthrene, the checkers employed a 15-cm. spinning-band column obtainable from Nester and Faust, Exton, Pennsylvania.

5. At these concentrations, the azeotrope is solid and adequate heating of the condenser and receivers must be provided by an infrared lamp or similar device. The use of twice this amount of diethylene glycol is reported [2] to give a liquid azeotrope but requires the distillation of proportionately larger amounts, for the azeotrope has nearly the same boiling point as diethylene glycol.

6. The finely divided sodium presents a serious fire hazard, and as much of it as possible should be retained in the flask. This may be accomplished by careful decantation. The material that is collected on the funnel should always be covered with a layer of solvent and should not be allowed to become dry. The residues may be safely destroyed by placing the funnel and flask in a large pail and adding about 1 l. of isopropyl alcohol. This operation is best conducted out-of-doors.

7. Phenanthrene purified by the sodium treatment was found superior to that from the azeotropic distillation, but both products gave satisfactory results. A good grade of commercially available phenanthrene ("white label" grade supplied by the Eastman Kodak Company), although recrystallized and treated with Raney nickel, resisted hydrogenation under the described conditions.

8. Cyclohexane as supplied by Matheson Company was used without further purification. The use of ethanol as solvent [6] gave inconsistent results, and the yield of 9,10-dihydrophenanthrene never exceeded 50%. Erratic results were also obtained when the solvent was omitted.

9. Copper-chromium oxide (HJS2) was prepared as reported by Adkins and coworkers.[7]

10. If the hydrogenation is allowed to proceed to completion, the product is contaminated with considerable polyhydrogenated material, as indicated by its low refractive index. The optimum time for obtaining about 85% hydrogenation may vary with the purity of the phenanthrene and activity of the catalyst. The purest 9,10-dihydrophenanthrene is obtained when the lower limits of hydrogen uptake are realized, although the yield is correspondingly lower.

3. Methods of Preparation

9,10-Dihydrophenanthrene has been prepared from 2,2'-bis(bromomethyl)biphenyl and sodium;[8] from the reduction of 2,2'-diiodobibenzyl in the presence of 1% palladium on barium carbonate catalyst;[9] by the hydrogenation of phenanthrene in the presence of nickel [8] or copper-chromium oxide catalyst;[3,6,10] by the coupling of 2,2'-bis(bromomethyl)biphenyl with lithium phenyl;[11] and by the thermal decomposition of 1,1-(o,o'-biphenylenebismethylene)-2-p-toluenesulfonylhydrazine.[12]

[1] Cornell University, Ithaca, New York.

[2] Feldman, Pantages, and Orchin, J. Am. Chem. Soc., 73, 4341 (1951).

[3] Fieser and Johnson, J. Am. Chem. Soc., 61, 168 (1939).

[4] Org. Syntheses Coll. Vol. 2, 117 (1943).

[5] Cason and Rapoport, Laboratory Text in Organic Chemistry, 2nd ed., p. 293, Prentice-Hall, Englewood Cliffs, New Jersey, 1962.

[6] Durland and Adkins, J. Am. Chem. Soc., 59, 135 (1937).

[7] Adkins, Burgoyne, and Schneider, J. Am. Chem. Soc., 72, 2626 (1950).

[8] Schroeter, Müller, and Huang, Ber., 62, 645 (1929).

[9] Busch and Weber, J. prakt. Chem., 146, 1 (1936).

[10] Burger and Mosettig, J. Am. Chem. Soc., 58, 1857 (1936).

[11] Hall, Lesslie, and Turner, J. Chem. Soc., 1950, 711.

[12] Carpino, Chem. & Ind. (London), 1957, 172.

9,10-DIHYDROXYSTEARIC ACID

(low-melting isomer)

(Octadecanoic acid, 9,10-dihydroxy-)

$$CH_3(CH_2)_7CH\!\!=\!\!CH(CH_2)_7CO_2H + H_2O_2 + HCO_2H \rightarrow$$
Oleic acid

$$CH_3(CH_2)_7\underset{\underset{OH}{|}}{CH}\!\!-\!\!\underset{\underset{OCHO}{|}}{CH}(CH_2)_7CO_2H$$

$$CH_3(CH_2)_7\underset{\underset{OH}{|}}{CH}\!\!-\!\!\underset{\underset{OCHO}{|}}{CH}(CH_2)_7CO_2H + 2NaOH \rightarrow$$

$$CH_3(CH_2)_7\underset{\underset{OH}{|}}{CH}\!\!-\!\!\underset{\underset{OH}{|}}{CH}(CH_2)_7CO_2Na + HCO_2Na$$

$$CH_3(CH_2)_7\underset{\underset{OH}{|}}{CH}\!\!-\!\!\underset{\underset{OH}{|}}{CH}(CH_2)_7CO_2Na + HCl \rightarrow$$

$$CH_3(CH_2)_7\underset{\underset{OH}{|}}{CH}\!\!-\!\!\underset{\underset{OH}{|}}{CH}(CH_2)_7CO_2H + NaCl$$

Low-melting isomer

Submitted by DANIEL SWERN, JOHN T. SCANLAN,
and GERALDINE B. DICKEL.[1]
Checked by JOHN D. ROBERTS and EDGAR F. KIEFER.

1. Procedure

To a well-stirred mixture of 141 g. (0.5 mole) of oleic acid (Note 1) and 425 ml. of formic acid (Note 2) in a 1-l. three-necked flask at 25° is added the appropriate amount (Note 3) of 30% (100 volume) hydrogen peroxide (approximately 60 g.) over a 15-minute period (Note 4). The reaction becomes mildly exothermic after a lag of about 5–10 minutes, and homogeneous after about 20–30 minutes. The temperature is maintained at 40° with a cold water bath at the beginning, and with a warm water bath or heating mantle toward the end, of the reaction. After about 3 hours or after analysis has indicated that the peroxide has been consumed (Note 5), the formic acid is removed by distillation under reduced pressure (b.p. 50°/125 mm.) in a stream of gas (carbon dioxide or nitrogen) to prevent bumping (Note 6). The residue in the flask, which consists of hydroxy-formoxystearic acids, is heated for 1 hour at 100° with an excess of 3N aqueous sodium hydroxide, and the hot, amber-colored soap solution is cautiously poured into an excess of 3N hydrochloric acid with stirring. The oil which separates is allowed to solidify, and the aqueous layer is discarded. The tan-colored solid is remelted on the steam bath by addition of hot water and stirred well to remove residual salts and water-soluble acids (Note 7). When the oil has solidified, the aqueous layer is discarded, and the solid is broken into small pieces and dissolved in 400 ml. of 95% ethanol by heating on the steam bath. After crystallization at 0° for several hours, the product is collected on a filter and dried under vacuum. The yield of crude 9,10-dihydroxystearic acid is 75–80 g., m.p. 85–90°. After a second recrystallization from 250 ml. of 95% ethanol, the product weighs about 60–65 g. and melts at about 90–92°. A third recrystallization may be necessary to produce a pure product melting at 94–95°. The over-all yield is 55–60 g. (50–55%, based on the available oleic acid) (Note 8).

2. Notes

1. The checkers used commercial u.s.p. oleic acid, which has an iodine number of about 60–70 and contains 65–75% oleic acid. The submitters report that, if highly purified oleic acid is used, the yield of fairly pure 9,10-dihydroxystearic acid is almost quantitative, but the purification procedure for oleic acid [*Biochem. Preparations*, **2**, 100 (1952)] is more lengthy and inconvenient than the purification of the hydroxylation product. The over-all yield is approximately the same in either case.

2. The formic acid employed is the 98–100% grade. The submitters report that the 90% grade of acid is satisfactory, but the reaction mixture remains heterogeneous throughout. They also state that, instead of formic acid, an equal volume of glacial acetic acid containing 2.5% by weight of concentrated sulfuric acid may be employed. With acetic acid-sulfuric acid, a 6-hour reaction time is required. However, the yield of 9,10-dihydroxystearic acid is slightly lower than the yield obtained when formic

acid is employed and the iodine number of the crude reaction product is about 6–9.

3. If commercial oleic acid is used, the iodine number should be determined beforehand and the quantity of hydrogen peroxide adjusted accordingly. The hydrogen peroxide should be assayed immediately before use; "100 volume peroxide" usually contains about 30% hydrogen peroxide by weight. This determination is conveniently carried out by weighing 0.2–0.3 g. of the hydrogen peroxide solution into an Erlenmeyer flask with a ground-glass stopper and adding 20 ml. of a glacial acetic acid-chloroform solution (3:2 by volume). Two milliliters of saturated aqueous potassium iodide solution is added, and the mixture is allowed to stand for 5 minutes. Distilled water (75 ml.) is added and the liberated iodine titrated with 0.1N sodium thiosulfate solution to a starch end point. This procedure is also satisfactory for determining the peroxide content of the oxidation mixture, except that 1–2 g. samples are taken [cf. Wheeler, *Oil & Soap*, **9**, 89 (1932)].

4. The submitters state that in one-tenth scale preparations the hydrogen peroxide solution can be added in 1 portion. In larger runs the addition may require 30 minutes to 1 hour.

5. The reaction time ranges from 1.5 to about 4 hours. Progress of the reaction should be followed by determining the peroxide content of the oxidation mixture at half-hour intervals after all the hydrogen peroxide has been added. Approximately all the peroxide should be consumed before distillation is attempted.

6. Instead of removing the formic acid by distillation, the reaction mixture may be poured into a large quantity of water and the oily layer dissolved in ether. The ether solution is washed free of formic acid and then subjected to distillation to remove the ether; hydroxyformoxystearic acids are left as a residue. The submitters found that, in larger-scale operations (five or more times the size of the run described), no ether was required and the oily layer was washed with water until free of formic acid. When acetic acid containing sulfuric acid was employed as the solvent, the reaction mixture was poured into hot water, and the oil which formed was separated mechanically or by extraction with ether.

7. The pH of the wash water should be below 6 in order to be certain that all soap in the product has been converted to free acid. If the pH is above 6, a small quantity of $3N$ hydrochloric acid should be added and the stirring continued for several minutes.

8. The submitters report that the high-melting isomer of 9,10-dihydroxystearic acid can be prepared from elaidic acid [*Biochem. Preparations*, **3**, 118 (1953)] by essentially the procedure described for oleic acid. With elaidic acid, instead of removing the formic acid by distillation, the reaction mixture may be poured into hot water and the oil which forms separated mechanically. The product is not readily soluble in ether. When acetic acid containing sulfuric acid is employed as the solvent, the reaction mixture is poured into *hot* water with thorough mixing, allowed to cool to room temperature, and filtered. The subsequent procedure (saponification and acidification) is the same as that described for the hydroxylation of oleic acid except that the crude dihydroxystearic acid, obtained after acidification of the soap, cannot be melted with hot water during the washing but is merely stirred well at 95–100° on the steam bath with a large quantity of hot water (Note 7). About 5 ml. of ethanol per gram of solute should be used in the recrystallization. The pure product melts at 130–131°. The yield depends on the purity of the starting material; if highly purified elaidic acid is used, the yield is about 80% after one recrystallization.

3. Methods of Preparation

The procedures described have been published.[2] Other procedures, which are not so satisfactory as the ones described, have also been published.[3, 4]

[1] U. S. Dept. of Agriculture, Eastern Utilization Research and Development Division, Philadelphia 18, Pennsylvania.

[2] Swern, Billen, Findley, and Scanlan, *J. Am. Chem. Soc.*, **67**, 1786 (1945).

[3] Hilditch, *J. Chem. Soc.*, **1926**, 1828; Hilditch and Lea, *ibid.*, **1928**, 1576.

[4] Scanlan and Swern, *J. Am. Chem. Soc.*, **62**, 2305 (1940).

1,4-DIIODOBUTANE *

(Butane, 1,4-diiodo-)

$$\begin{matrix} CH_2\text{---}CH_2 \\ | \qquad | \\ CH_2 \quad CH_2 \\ \diagdown\diagup \\ O \end{matrix} + 2KI + 2H_3PO_4 \rightarrow$$

$$ICH_2CH_2CH_2CH_2I + H_2O + 2KH_2PO_4$$

Submitted by HERMAN STONE and HAROLD SHECHTER.[1]
Checked by CLIFF S. HAMILTON and R. C. RUPERT.

1. Procedure

Tetrahydrofuran (36 g., 0.5 mole) (Note 1) is added to a mixture of potassium iodide (332 g., 2 moles), 85% orthophosphoric acid (231 g., 135 ml., 2 moles), and phosphoric anhydride (65 g.) (Notes 2, 3, and 4) in a 1-l. three-necked flask equipped with a sealed mechanical stirrer, a reflux condenser, and a thermometer. The mixture is stirred and heated at its reflux temperature for 3 hours, during which time a dense oil separates from the acid layer. The stirred mixture is cooled to room temperature, and 150 ml. of water and 250 ml. of ether are added (Note 5). The ether layer is separated, decolorized with dilute aqueous sodium thiosulfate solution, washed with cold saturated sodium chloride solution, and dried over anhydrous sodium sulfate. The ether is removed by distillation on a steam bath, and the residue is distilled under reduced pressure from a modified Claisen flask. The portion boiling at 108–110°/10 mm. is collected. The yield of colorless 1,4-diiodobutane (n_D^{20} 1.615; d_4^{20} 2.300) (Note 6) is 143–149 g. (92–96%).

2. Notes

1. Tetrahydrofuran was obtained by the submitters from E. I. du Pont de Nemours and Company.

2. The specified mixture of commercial 85% orthophosphoric acid and phosphoric anhydride corresponds to a 95% orthophosphoric acid solution. The phosphoric anhydride is placed in the dry flask, and the 85% orthophosphoric acid is added with stir-

ring. After the mixture has cooled to room temperature, solid potassium iodide is added. The solution should be cooled, before addition of the potassium iodide, to prevent evolution of hydrogen iodide and formation of iodine. After the tetrahydrofuran is added, the mixture can be heated as desired since the hydrogen iodide reacts as rapidly as it is formed.

3. Orthophosphoric acid of 95% concentration is most efficient for effecting cleavage of tetrahydrofuran. Commercial orthophosphoric acid (85%) may be used; however, the yield drops to 82% and approximately 10% of the tetrahydrofuran is recovered. Anhydrous orthophosphoric acid and tetraphosphoric acid cannot be employed conveniently because of the limited solubility of hydrogen iodide in these reagents.

4. This procedure has been used successfully to convert simple aliphatic ethers into their corresponding iodides. Yields of iodides obtained in the reaction of di-n-butyl ether and diisopropyl ether with potassium iodide and 95% orthophosphoric acid were 81 and 90% respectively. Small quantities of the corresponding alcohols were also isolated as products from these reactions.

5. Usually one extraction with ether is sufficient to decolorize the acid layer; if this fails, an additional extraction with 100 ml. of ether is recommended.

6. The checkers obtained values of: n_D^{25} 1.619; d_4^{26} 2.349. The product darkens slowly on standing.

3. Methods of Preparation

1,4-Diiodobutane has been prepared in 51% yield by the reaction of phosphorus, iodine, and tetrahydrofuran.[2] It has also been prepared by the reaction of hydriodic acid with phenoxybutyl iodide [3, 4] and with the diisoamyl ether of 1,4-butanediol.[5] Sulfuric acid has been used in place of phosphoric acid in the reaction described in the present procedure.[6]

[1] Ohio State University, Columbus, Ohio.
[2] Heisig, *J. Am. Chem. Soc.*, **61**, 525 (1939).
[3] von Braun and Beschke, *Ber.*, **39**, 4357 (1906).
[4] Marvel and Tännenbaum, *J. Am. Chem. Soc.*, **44**, 2650 (1922).
[5] Hamonet, *Compt. rend.*, **132**, 345 (1901).
[6] Bräuniger and Mengering, *Pharmazie*, **14**, 191 (1959) [*C. A.*, **54**, 548 (1960)].

1,6-DIIODOHEXANE

(Hexane, 1,6-diiodo-)

$$HOCH_2(CH_2)_4CH_2OH + 2KI + 2H_3PO_4 \rightarrow$$
$$ICH_2(CH_2)_4CH_2I + 2KH_2PO_4 + 2H_2O$$

Submitted by HERMAN STONE and HAROLD SHECHTER.[1]
Checked by T. L. CAIRNS, B. C. MCKUSICK, and G. V. MOCK.

1. Procedure

In a 1-l. three-necked flask, equipped with a short reflux condenser, a sealed mechanical Hershberg stirrer, and a thermometer, is placed 65 g. (0.46 mole) of phosphoric anhydride, and 231 g. of 85% orthophosphoric acid (135 ml., 2 moles) is added (Note 1). After the stirred mixture has cooled to room temperature, 332 g. (2 moles) of potassium iodide and 59 g. (0.5 mole) of recrystallized 1,6-hexanediol (Notes 2, 3, and 4) are added. The mixture is stirred and heated at 100–120° for 3–5 hours, during which time the homogeneous solution separates into two phases, and finally a dense oil settles through the acid layer. The stirred mixture is cooled to room temperature, and 150 ml. of water and 250 ml. of ether are added (Note 5). The ether layer is separated, decolorized by shaking with 50 ml. of 10% sodium thiosulfate solution, washed with 200 ml. of cold saturated sodium chloride solution, and dried with 50 g. of anhydrous sodium sulfate. The ether is removed by distillation on a steam bath, and the product is distilled from a modified Claisen flask under reduced pressure. The fraction boiling at 123–128°/4 mm. is collected. The yield of 1,6-diiodohexane is 140–144 g. (83–85%), n_D^{15} 1.585, m.p. 10° (Notes 6 and 7).

2. Notes

1. The specified mixture of commercial 85% orthophosphoric acid and phosphoric anhydride corresponds to 95% orthophosphoric acid. Ninety-five per cent orthophosphoric acid is recommended for this reaction. If 85% orthophosphoric acid is used, the reaction proceeds more slowly and the yield is reduced.

2. 1,6-Hexanediol,[2] m.p. 40–41°, was prepared by catalytic reduction of diethyl adipate with hydrogen over copper chromite catalyst. It can also be purchased from Columbia Organic Chemicals Company, Inc.

3. The solution must be cool before the potassium iodide is added to avoid the evolution of hydrogen iodide and formation of iodine. After the 1,6-hexanediol has been added, the mixture can be heated as desired since the hydrogen iodide reacts as rapidly as it is formed.

4. This procedure has been used successfully for conversion of various aliphatic and alicyclic alcohols to the corresponding iodides. Yields of iodides from 1-propanol, 2-methyl-1-propanol, 2-methyl-2-propanol, and cyclohexanol were 95, 88, 90, and 79.5%, respectively.

5. Usually one extraction of the reaction product with ether is sufficient to remove the color from the acid layer.

6. Slightly yellow 1,6-diiodohexane crystallizes as white needles when cooled in an ice-water mixture. The addition of a few drops of mercury to the yellow product produces a nearly colorless liquid.

7. The submitters reported yields of 93–95% and a melting point of 8.5–9.0°.

3. Methods of Preparation

1,6-Diiodohexane has been prepared in 73% yield by the reaction of 1,6-hexanediol, red phosphorus, and iodine.[3] It has also been prepared by reactions of hydrogen iodide and 1,6-diphenoxyhexane [4] and 1,6-diethoxyhexane,[5] respectively. Physical constants have been reported by Dionneau.[6] The method described here has been published.[7]

[1] Ohio State University, Columbus, Ohio.

[2] *Org. Syntheses Coll. Vol.* **2**, 325 (1943).

[3] Müller and Rölz, *Ber.*, **61**, 571 (1928).

[4] Salonina, *Ber.*, **26**, 2988 (1893); Gol'mov, *Zhur. Obshcheĭ Khim.* (*J. Gen. Chem.*), **22**, 809 (1952) [*C. A.*, **47**, 3251 (1953)].

[5] Farmer, Laroia, Switz, and Thorpe, *J. Chem. Soc.*, **1927**, 2951.

[6] Dionneau, *Ann. chim.*, [9] **3**, 257 (1915).

[7] Stone and Shechter, *J. Org. Chem.*, **15**, 491 (1950).

DIISOPROPYL METHYLPHOSPHONATE

(Phosphonic acid, methyl-, diisopropyl ester)

$$[(CH_3)_2CHO]_3P + CH_3I \rightarrow$$
$$[(CH_3)_2CHO]_2P(CH_3)O + (CH_3)_2CHI$$

Submitted by A. H. Ford-Moore and B. J. Perry.[1]
Checked by William S. Johnson and James Ackerman.

1. Procedure

A 2-l. round-bottomed flask containing 284 g. (113 ml., 2 moles) of methyl iodide [2] is fitted with an efficient water-cooled condenser and a dropping funnel which is charged with 416 g. (453 ml., 2 moles) of triisopropyl phosphite (Note 1). A few pieces of porous plate are added to the methyl iodide, and about 50 ml. of the phosphite is introduced. The mixture is heated over a gauze with a free flame until an exothermic reaction begins. The flame is then withdrawn and the remainder of the phosphite is added at such a rate that the mixture keeps boiling briskly. Towards the end of the addition it may be necessary to reapply heat. After the addition is complete, the mixture is boiled under reflux for 1 hour. The condenser is replaced by a 50–75 cm. Vigreux column attached to a condenser set for distillation, and the bulk of the isopropyl iodide is distilled at 85–95° (atmospheric pressure). The residue is transferred to a pear-shaped flask for distillation through a 75-cm. Vigreux column under reduced pressure. The remainder of the isopropyl iodide is distilled at water-pump pressure, a Dry Ice trap being interposed between the receiver and the pump in order to effect complete condensation. A total of 310 g. (91%) of isopropyl iodide is thus recovered. The residue is then fractionated at vacuum-pump pressure. Except for a small fore-run and residue, the product distils almost entirely at 51°/1.0 mm. (46°/0.8 mm.). The yield of colorless product is 308–325 g. (85–90%); n_D^{20} 1.4101, n_D^{25} 1.4081; d_4^{24} 0.985, d_4^{10} 0.997 (Note 2).

2. Notes

1. The triisopropyl phosphite is prepared according to the procedure for triethyl phosphite (p. 955) and should be free from any diisopropyl hydrogen phosphite. The latter substance does not enter into the reaction but is difficult to remove from the final product. The starting material was supplied to the submitters by Messrs. Albright, Wilson and Company, Oldbury, Birmingham, England.

2. Diisopropyl ethylphosphonate can be obtained by a similar procedure, using the appropriate amount of ethyl iodide in place of methyl iodide. Ethyl iodide is less reactive, and it is necessary to apply heat during the addition of the phosphite and to allow the mixture to reflux for 7 hours after the addition. On a $2M$ scale the yield is 354 g. (91%), b.p. 61°/0.7 mm., n_D^{25} 1.4108, d_4^{25} 0.968. The recovery of isopropyl iodide is 317 g. (93%).

Diethyl ethylphosphonate may be obtained by refluxing 332 g. (348 ml., 2 moles) of triethyl phosphite and 250 g. (1.6 moles) of ethyl iodide for 3 hours. After distillation of 231 g. (92%) of ethyl iodide, the residue is fractionated under reduced pressure, giving 329 g. (98.5%) of product, b.p. 56°/1 mm. (58.5°/1.8 mm.); n_D^{25} 1.4141, n_D^{20} 1.4161; d_4^{25} 1.022.

Diethyl methylphosphonate may be prepared similarly by refluxing one molar equivalent of triethyl phosphite with one mole of methyl iodide, but it is very difficult to separate the product from the small amount of diethyl ethylphosphonate that is formed simultaneously by the interaction of the phosphite with the ethyl iodide liberated in the reaction. The pure substance boils at 51°/1 mm., n_D^{25} 1.4117, d_4^{25} 1.050.

3. Methods of Preparation

Diisopropyl methylphosphonate has been prepared from diisopropylethyl phosphite and methyl iodide,[3] and by treating sodium diisopropylphosphonate with methyl chloride.[4] The method described here for the preparation of diisopropyl methylphosphonate is a modification of the Arbusov rearrangement.[5]

[1] Chemical Defence Experimental Station, Porton, Nr. Salisbury, Wilts, England.

[2] *Org. Syntheses Coll. Vol.* **2**, 404 (1943).

[3] Landauer and Rydon, *J. Chem. Soc.*, **1953**, 2224.

[4] Smith, U. S. pats. 2,853,507 [*C. A.*, **53**, 7989 (1959)], 2,880,224 [*C. A.*, **53**, 15978 (1959)].

[5] Arbusov, *Chem. Zentr.*, [II] **77**, 1640 (1906); Ford-Moore and Williams, *J. Chem. Soc.*, **1947**, 1465.

2,3-DIMETHOXYCINNAMIC ACID *

(Cinnamic acid, 2,3-dimethoxy-)

Submitted by J. Koo, M. S. Fish, G. N. Walker, and J. Blake.[1]
Checked by T. L. Cairns and A. E. Barkdoll.

1. Procedure

In a 3-l. round-bottomed flask (Note 1), fitted with a reflux condenser and a thermometer, are placed 208 g. (2 moles) of malonic acid (Note 2), 166 g. (1 mole) of 2,3-dimethoxybenzaldehyde (Note 3), and 400 ml. of pyridine. The malonic acid is dissolved by shaking and warming on a steam bath (Note 4). Piperidine (15 ml.) is then added, the reflux condenser and thermometer are fitted into place (Note 5), and the mixture is heated to 80°. About 30 minutes should be allowed for this rise in temperature. An internal temperature of 80–85° is maintained for 1 hour, and the material is finally heated under reflux (109–115°) for an additional 3 hours (Note 6).

After being cooled the reaction mixture is poured into a large beaker containing 4 l. of cold water. The mixture is acidified by slowly adding with stirring 500 ml. of concentrated hydrochloric acid; it should be strongly acidic at this point. The light-brown crystals are separated by suction filtration and washed 4 times with 150-ml. portions of cold water. The crude acid is dissolved in a solution of 80 g. of sodium hydroxide in 3 l. of water. The resulting solution is filtered, diluted with an additional 1.2 l. of water, and acidified by adding with stirring 600 ml. of 1:1 hydrochloric acid. The mixture is filtered, and the crystalline material is washed with three 150-ml. portions of cold water. The product is dried at 60–70° (Note 7). The yield is 180–205 g. (87–98%), m.p. 174–178° (Note 8). Further purification is usually not necessary, but it may be accomplished by recrystallization from methyl ethyl ketone, using 12 ml. of solvent per gram of acid. The hot solution is filtered rapidly through a steam-heated Büchner funnel and chilled for several hours. A recovery of 70% of product, m.p. 179–180°, may be obtained.

2. Notes

1. A large flask is preferred to ensure against possible loss by foaming.

2. An excess of malonic acid is necessary for high yields. An equimolecular amount of malonic acid results in yields as low as 50%.

3. A practical grade of 2,3-dimethoxybenzaldehyde gives satisfactory results. It is convenient to weigh and transfer this material as a liquid.

4. If the malonic acid is not in solution before addition of the piperidine, the reaction cannot be controlled properly. It is advisable to heat the mixture to 50° to effect solution.

5. The thermometer may be suspended in the mixture through the condenser by means of a long wire.

6. Evolution of carbon dioxide begins at about 55–60°. The prescribed temperatures are necessary to prevent undue foaming.

7. The product should be dried to constant weight in an oven. Drying for several days is usually required.

8. This method is a general one. It can be used with a variety of substituted aromatic aldehydes.

3. Methods of Preparation

2,3-Dimethoxycinnamic acid has been prepared by heating 2,3-dimethoxybenzaldehyde with acetic anhydride and sodium acetate at 200° [2] and by the condensation of 2,3-dimethoxybenzaldehyde and ethyl acetate with sodium, followed by hydrolysis.[3] The present preparation represents an adaption of the Doebner reaction.

[1] University of Pennsylvania, Philadelphia, Pennsylvania.
[2] von Krannichfeldt, *Ber.*, **46**, 4021 (1913).
[3] Perkin and Robinson, *J. Chem. Soc.*, **105**, 2387 (1914).

DIMETHYL ACETYLENEDICARBOXYLATE *

(Acetylenedicarboxylic acid, dimethyl ester)

$$KO_2CC\equiv CCO_2H + H_2SO_4 \rightarrow HO_2CC\equiv CCO_2H + KHSO_4$$

$$HO_2CC\equiv CCO_2H + 2CH_3OH \text{ (excess)} \xrightarrow{H_2SO_4}$$
$$CH_3O_2CC\equiv CCO_2CH_3 + 2H_2O$$

Submitted by E. H. HUNTRESS, T. E. LESSLIE, and J. BORNSTEIN.[1]
Checked by T. L. CAIRNS and M. J. HOGSED.

1. Procedure

To 400 g. (510 ml., 12.5 moles) of methanol (commercial grade) in a 2-l. round-bottomed flask is added in small portions with cooling 200 g. (111 ml., 2.04 moles) of concentrated sulfuric acid. To this cooled solution is added 100 g. (0.66 mole) of the potassium acid salt of acetylenedicarboxylic acid (Note 1). The flask is fitted with a stopper holding a calcium chloride drying tube and allowed to stand with occasional swirling for 4 days at room temperature.

The liquid in the flask is then decanted from the inorganic salt, which is washed with 500 ml. of cold water. The solutions are combined and extracted with five 500-ml. portions of ether. The ether extracts are combined and washed successively with 200 ml. of cold water, 150 ml. of saturated sodium bicarbonate solution (Note 2), and 200 ml. of cold water and then dried over anhydrous calcium chloride. After removal of the ether by distillation from a steam bath, the ester is distilled under reduced pressure from a modified Claisen flask. The yield of ester boiling at 95–98°/19 mm. is 67–82 g. (72–88%) (Notes 3 and 4); n_D^{25} 1.4444–1.4452.

2. Notes

1. The potassium acid salt of acetylenedicarboxylic acid is commercially obtainable from the National Aniline Division, Allied Chemical and Dye Corporation, New York, New York. Directions for the preparation of the free acid are given in earlier volumes.[2, 3]

2. If the ether extract is not washed with sodium bicarbonate solution, considerable loss occurs during the distillation of the ester because of decomposition in the flask.

3. Dimethyl acetylenedicarboxylate is a powerful lachrymator and vesicant; it should be handled with extreme care. Even traces of ester on the skin should be washed off at once with 95% ethanol followed by washing with soap and water.

4. The same general method has been used by the submitters to prepare diethyl acetylenedicarboxylate. In this case absolute ethanol was used, and the ether extract was dried over anhydrous magnesium sulfate. The yield of diethyl ester from 100 g. of the acid potassium salt of acetylenedicarboxylic acid was 57–59 g. (51–53%); b.p. 96–98°/8 mm.; n_D^{25} 1.4397.

3. Methods of Preparation

Dimethyl acetylenedicarboxylate has been prepared by refluxing the acid potassium salt of acetylenedicarboxylic acid with methanol and sulfuric acid.[4, 5] The method described here is a substantial improvement over the method of Moureu and Bongrand,[6] who prepared it from acetylenedicarboxylic acid, absolute methanol, and sulfuric acid.

[1] Massachusetts Institute of Technology, Cambridge, Massachusetts.
[2] *Org. Syntheses Coll. Vol.* **2**, 10 (1943).
[3] *Org. Syntheses*, **18**, 3 (1938).
[4] Baudrowski, *Ber.*, **15**, 2694 (1882).
[5] Curtius and Heynemann, *J. prakt. Chem.*, [2] **91**, 66 (1915).
[6] Moureu and Bongrand, *Ann. chim.*, [9] **14**, 11 (1920).

p-DIMETHYLAMINOBENZALDEHYDE *

(Benzaldehyde, *p*-dimethylamino-)

$$(CH_3)_2NCHO + POCl_3 \rightarrow [(CH_3)_2NCHO \cdot POCl_3]$$

$$[(CH_3)_2NCHO \cdot POCl_3] + (CH_3)_2NC_6H_5 \rightarrow$$

$$\left[\begin{array}{c} OPOCl_2 \\ \diagup \\ p\text{-}(CH_3)_2NC_6H_4CH \qquad \cdot HCl \\ \diagdown \\ N(CH_3)_2 \end{array} \right]$$

$$\left[\begin{array}{c} OPOCl_2 \\ \diagup \\ p\text{-}(CH_3)_2NC_6H_4CH \qquad \cdot HCl \\ \diagdown \\ N(CH_3)_2 \end{array} \right]$$

$$+ 3H_2O + 4CH_3CO_2Na \rightarrow$$

$$p\text{-}(CH_3)_2NC_6H_4CHO + 3NaCl + NaH_2PO_4$$

$$+ CH_3CO_2NH_2(CH_3)_2 + 3CH_3CO_2H$$

Submitted by E. Campaigne and W. L. Archer.[1]
Checked by N. J. Leonard and R. W. Fulmer.

1. Procedure

In a 2-l. three-necked round-bottomed flask, equipped with a sealed stirrer, dropping funnel, and a reflux condenser topped by a calcium chloride tube, is placed 440 g. (6 moles) of dimethylformamide (Note 1). While the flask is carefully cooled in an ice bath, 253 g. (1.65 moles) of phosphorus oxychloride is added dropwise with stirring. An exothermic reaction occurs with the formation of the phosphorus oxychloride-dimethylformamide complex. When all the phosphorus oxychloride has been added, and the heat of the reaction has subsided, 200 g. (1.65 moles) of dimethylaniline (Note 2) is added dropwise with stirring. When the addition of the dimethylaniline is complete, a yellow-green precipitate begins to form. The reaction mixture is heated on a steam bath, and stirring is continued for 2 hours. The yellow-green precipitate redissolves when heating is begun. The mixture is then cooled and poured over 1.5 kg. of crushed ice in a 5-l.

beaker. Any precipitate that remains in the flask may be washed into the ice mixture with cold water. The solution is neutralized to pH 6–8 (Universal Test Paper) by the dropwise addition of approximately 1.5 l. of saturated aqueous sodium acetate with vigorous stirring (Note 3). p-Dimethylaminobenzaldehyde begins to precipitate soon after the addition of the sodium acetate is begun. The neutral mixture (total volume about 4.5 l.) is stored in the refrigerator overnight (Note 4). The greenish-tinted crystalline precipitate is filtered by suction, with the aid of a rubber dam, and washed several times with water on the filter. The green color is readily removed during the washing. The very light-yellow to nearly colorless product, after air-drying, weighs 198–208 g. (80–84%) and melts at 73–74°. It is essentially pure and useful for most purposes as obtained (Note 5).

2. Notes

1. The dimethylformamide is available as technical grade DMF from the Grasselli Chemicals Department of E. I. du Pont de Nemours and Company. Dimethylformamide can be prepared by the method of Mitchell and Reid [2] from dimethylamine and formic acid.

2. Dimethylaniline free from monomethylaniline (Eastman Kodak Company) is used.

3. It is possible to neutralize the acid solution partially with sodium hydroxide before the sodium acetate is added, but it is more difficult to avoid localized heating by this method. It is important to keep the reaction mixture below 20° during the neutralization, by the addition of ice if necessary, since any excessive increase in temperature of the aqueous solution leads to the formation of greenish blue dyestuffs, which are very difficult to remove from the product.

4. The mixture may turn orange-colored when allowed to stand overnight.

5. If a purer product is desired, the aldehyde may be purified by the method described by Adams and Coleman.[3]

3. Methods of Preparation

p-Dimethylaminobenzaldehyde has been prepared from dimethylaniline, formaldehyde, and p-nitrosodimethylaniline in 56–59% yield,[3] by the formylation of dimethylaniline with N-methylformanilide in approximately 50% yield,[4] by the formylation of dimethylaniline with dimethylformamide,[5] and by the condensation of methyl formate with p-dimethylaminophenylmagnesium chloride in tetrahydrofuran.[6]

[1] Indiana University, Bloomington, Indiana.
[2] Mitchell and Reid, *J. Am. Chem. Soc.*, **53**, 1879 (1931).
[3] *Org. Syntheses Coll. Vol.* 1, 214 (1941).
[4] Vilsmeier and Haack, *Ber.*, **60**, 119 (1927).
[5] Brit. pat. 607,920 [*C. A.*, **43**, 2232 (1949)].
[6] Ramsden (to Metal & Thermit Corp.), Brit. pat. 806,710 [*C. A.*, **54**, 2264 (1960)].

β-DIMETHYLAMINOETHYL CHLORIDE HYDROCHLORIDE *

(Ethylamine, 2-chloro-N,N-dimethyl-, hydrochloride)

$$(CH_3)_2NCH_2CH_2OH + SOCl_2 \rightarrow (CH_3)_2NCH_2CH_2Cl \cdot HCl + SO_2$$

Submitted by LUTHER A. R. HALL, VERLIN C. STEPHENS, and
J. H. BURCKHALTER.[1]
Checked by RICHARD T. ARNOLD and WILLIAM LEE.

1. Procedure

Caution! This preparation should be conducted in a good hood.

In a dry 1-l. flask fitted with a sealed mechanical stirrer, an efficient reflux condenser, and a 500-ml. dropping funnel is placed 290 g. (2.44 moles) of thionyl chloride (Note 1). The reaction flask must be cooled in an ice bath throughout the entire period of operation, as the reaction is very exothermic. β-Dimethylaminoethanol (210 g., 2.35 moles) (Note 2) is added dropwise through the funnel to the cooled thionyl chloride (Note 3) over a period of an hour, during which time there is a copious evolution of sulfur dioxide (Note 4). After all the β-dimethylaminoethanol has been added, the ice bath is removed and the reaction mixture is stirred for another hour (Note 5). The temperature of the mixture is 35–50°. At this point the reaction mixture consists of a brown semisolid slush of the desired product together with a slight excess of thionyl chloride.

The entire contents of the reaction flask are transferred to a 2-l. beaker (or wide-mouthed Erlenmeyer flask) containing approximately 1 l. of absolute ethanol (Note 6). The resulting brown solution is heated to boiling on a hot plate, during which time there is a copious evolution of gases (Note 6). The solution is filtered hot, leaving a small amount of insoluble material. Upon cooling of the filtrate in a salt-ice bath, the desired product is obtained as beautiful white crystals which are collected on a Büchner funnel and dried in a vacuum desiccator over phosphorus pentoxide (Note 7). The yield of pure product melting at 201.5–203° is 227–272 g. (67–80%).

Upon evaporation of the last filtrate to one-third of its volume and cooling in a salt-ice bath, an additional 33–69 g. (10–20%) of good-quality product is obtained. The total yield is 296–305 g. (87–90%).

2. Notes

1. Eastman Kodak Company practical grade thionyl chloride is satisfactory.

2. A good commercial grade (Eastman Kodak Company or Union Carbide and Carbon Corporation) of β-dimethylaminoethanol is satisfactory.

3. Continued and efficient cooling of the reaction vessel is needed to prevent too vigorous an evolution of sulfur dioxide and a subsequent loss of thionyl chloride through trapping of this reagent by effluent gases. Cooling also prevents too high a reaction temperature. The reaction proceeds more smoothly if the temperature is kept below 50°.

4. Care should be taken that the dropping funnel inlet does not become clogged with solid product. If the tip of the dropping funnel is in such a position that the drops of β-dimethylaminoethanol fall directly into the thionyl chloride and do not drain down the walls of the flask, mechanical difficulties are reduced markedly. The reaction must be carried out in an efficient hood or with a suitable trap in order to remove the noxious sulfur dioxide formed.

5. The reaction mixture may be stirred for a longer time and allowed to stand overnight without affecting the yield.

6. The ethanol not only converts the excess thionyl chloride to gaseous by-products (sulfur dioxide, hydrogen chloride, and ethyl chloride) but also serves as the recrystallizing solvent for the desired product. The checkers found that about 80% of this thick product can be poured directly into 800 ml. of ethanol. Two hundred milliliters of warm ethanol should be used to decompose the product remaining in the reaction flask. This is combined with the main portion.

7. The product is somewhat hygroscopic, especially in humid weather. It should be dried in a vacuum desiccator to prevent the formation of hydrated forms.

3. Methods of Preparation

β-Dialkylaminoethyl bromide hydrobromides have been known for many years. However, the standard method of preparation requires large volumes of hydrobromic acid.[2] The less expensive analogous chlorides are preferred since their preparation is simpler and their reactivity is sufficient for the synthesis of well-known drugs.[3] Ordinarily β-dialkylaminoalkyl chloride hydrochlorides are prepared in good yield by treatment of β-dialkylaminoalkanols with an excess of thionyl chloride in chloroform or benzene.[4] An article on the German commercial preparation of Atabrine refers to the action of thionyl chloride on β-diethylaminoethanol hydrochloride without solvent.[5] The present method has been published.[6]

[1] University of Kansas, Lawrence, Kansas.

[2] Org. Syntheses Coll. Vol. 2, 92 (1943).

[3] Huttrer et al., J. Am. Chem. Soc., 68, 1999 (1946).

[4] Burger, J. Am. Pharm. Assoc., Sci. Ed., 36, 372 (1947); Tchoubar and Letellier-Dupré, Bull. soc. chim. France, 1947, 792; Elderfield et al., J. Am. Chem. Soc., 68, 1579 (1946); Marechal and Bagot, Ann. pharm. franç., 4, 172 (1946); Giral and Cascajares, Ciencia (Mex.), 5, 105 (1944) [C. A., 41, 4892 (1947)]; Ward, U. S. pat. 2,072,348 [C. A., 31, 2614 (1937)]; Mannich and Baumgarten, Ber., 70, 210 (1937); Brit. pat. 456,338 [C. A., 31, 2230 (1937)]; French pat. 802,416 [C. A., 31, 1824 (1937)]; Slotta and Behnisch, Ber., 68, 754 (1935); Gough and King, J. Chem. Soc., 1928, 2436; Meister, Lucius, and Brüning, Brit. pat. 167,781 [Brit. Abstracts, 122, 529 (1922)].

[5] Greene, Am. J. Pharm., 120, 39 (1948).

[6] Burckhalter, Stephens, and Hall, J. Am. Pharm. Assoc., 39, 271 (1950).

2-(DIMETHYLAMINO)PYRIMIDINE

(Pyrimidine,2-dimethylamino-)

$$\text{Cl-pyrimidine} + 2(CH_3)_2NH \rightarrow \text{N(CH}_3)_2\text{-pyrimidine} + (CH_3)_2NH \cdot HCl$$

Submitted by C. G. OVERBERGER, IRVING C. KOGON, and RONALD MININ.[1]
Checked by CHARLES C. PRICE and T. L. V. ULBRICHT.

1. Procedure

In a 250-ml. three-necked flask equipped with a reflux condenser and a gas-inlet tube are placed 45.6 g. (0.4 mole) of 2-chloropyrimidine (p. 182) and 150 ml. of absolute ethanol. The mixture is refluxed for 6 hours while anhydrous dimethylamine is bubbled into the solution (Note 1). The solution is cooled, and 100 ml. of ethanol is removed by distillation using a water aspirator. The residue is chilled in an ice bath for 1 hour, and 75 ml. of ether is added to cause precipitation of dimethylamine hydrochloride. After the removal of dimethylamine hydrochloride and solvent, the residue is distilled at reduced pressure from a Claisen flask (Note 2). The fraction boiling at 85–86°/28 mm. is collected; yield 40–42.5 g. (81–86%), n_D^{25} 1.5420 (Note 3).

2. Notes

1. Anhydrous dimethylamine may be conveniently prepared by allowing 25% aqueous dimethylamine to drop onto solid potassium hydroxide, the gas evolved being dried by passage over solid potassium hydroxide.

2. The compound is hygroscopic, and care should be taken to prevent exposure to air.

3. N-Methylaminopyrimidine is similarly prepared; b.p. 96–98°/28 mm., m.p. 57.5–58.5° (65% yield).

3. Methods of Preparation

Similar procedures for this preparation have been reported by Brown and Short [2] and by Copenhaver and Kleinschmidt.[3]

[1] Polytechnic Institute of Brooklyn, Brooklyn 2, New York.
[2] Brown and Short, *J. Chem. Soc.*, **1953**, 331.
[3] Copenhaver and Kleinschmidt, Brit. pat. 663,303 [*C. A.*, **46**, 10212 (1952)].

4,6-DIMETHYLCOUMALIN

(Sorbic acid, 5-hydroxy-3-methyl, δ-lactone)

Submitted by NEWTON R. SMITH and RICHARD H. WILEY.[1,2]
Checked by R. S. SCHREIBER and H. H. FALL.

1. Procedure

In a 125-ml. Claisen flask, equipped with a capillary for vacuum distillation and a thermometer, are placed 50 g. (0.3 mole) of isodehydroacetic acid (p. 549) and 2 g. of copper powder (Note 1). A 125-ml. simple distilling flask, cooled by a water jet, is used as a vacuum receiver and is attached to a water aspirator. The Claisen flask is heated at atmospheric pressure in an oil bath at 230–235° for 45 minutes or until the decarboxylation has ceased. The pressure on the system is then slowly reduced, and the dimethylcoumalin is distilled directly from the reaction flask (Note 2). The crude dimethylcoumalin (34–35 g., 92–95%) is redistilled from a Claisen flask. The yield of 4,6-dimethylcoumalin is 30–32 g. (81–87%); b.p. 140–142°/35 mm. (Note 3); m.p. 50–51°.

2. Notes

1. Copper chromite catalyst may be substituted.
2. The submitters recommend 3 hours for decarboxylation. At 45-minute intervals the pressure on the system is reduced

and the dimethylcoumalin distilled directly from the reaction flask. After the distillation slackens, the pressure is returned to atmospheric and the decarboxylation is continued. However, the checkers found that the decarboxylation is virtually completed during the first 45-minute period.

3. The checkers observed a boiling point of 134–136°/35 mm. for 4,6-dimethylcoumalin. The melting point, however, was identical with that reported by the submitters.

3. Methods of Preparation

4,6-Dimethylcoumalin has been prepared by the decarboxylation of isodehydroacetic acid in sulfuric acid or by heating,[3] and by the distillation of 4-methyl-2-pyrone-6-acetic acid.[4]

[1] University of Louisville, Louisville, Kentucky.

[2] The submitters wish to thank the Research Corporation for a grant under which this work was done.

[3] Hantzsch, *Ann.*, **222**, 17 (1883).

[4] Rice and Vogel, *Chem. & Ind.* (*London*), **1959**, 992.

N,N-DIMETHYLCYCLOHEXYLMETHYLAMINE

(Cyclohexanemethylamine, N,N-dimethyl-)

$$\text{(cyclohexane)}CO_2H + SOCl_2 \rightarrow \text{(cyclohexane)}COCl + SO_2 + HCl$$

$$\text{(cyclohexane)}COCl + 2HN(CH_3)_2 \rightarrow$$

$$\text{(cyclohexane)}CON(CH_3)_2 + (CH_3)_2NH \cdot HCl$$

$$\text{(cyclohexane)}CON(CH_3)_2 \xrightarrow{\text{LiAlH}_4} \text{(cyclohexane)}CH_2N(CH_3)_2$$

Submitted by ARTHUR C. COPE and ENGELBERT CIGANEK.[1]
Checked by WILLIAM E. PARHAM and ROBERT KONCOS.

1. Procedure

A. *N,N-Dimethylcyclohexanecarboxamide.* In a 1-l. three-necked flask equipped with a reflux condenser and a dropping funnel, both carrying drying tubes, is placed 128 g. (1.0 mole) of cyclohexanecarboxylic acid (Note 1). Thionyl chloride (179 g., 1.5 moles) (Note 1) is added during 5 minutes to the acid, with stirring by a magnetic stirrer. The flask is placed in an oil bath and heated at a bath temperature of 150° for 1 hour. The reflux condenser is then replaced by a distillation head (Note 2), 200 ml. of anhydrous benzene is added, and the mixture is distilled until the temperature of the vapors reaches 95°. The mixture is cooled, another 200 ml. of anhydrous benzene is added, and the distillation is continued until the temperature of the vapors again reaches 95°. The cooled residual acid chloride is transferred with a little benzene to a dropping funnel which is attached to a 2-l. three-necked flask. The flask is fitted with an efficient mechanical stirrer and a drying tube and is immersed in an ice bath. A solution of 135 g. (3.0 moles) of anhydrous dimethylamine (Note 1) in 150 ml. of anhydrous benzene is introduced into the flask. The acid chloride is added very slowly from the dropping funnel to the vigorously stirred solution, the addition taking about 2 hours. The mixture is then stirred at room temperature overnight. Two hundred milliliters of water is added, the layers are separated, and the aqueous phase is extracted with two 100-ml. portions of ether. The extracts are combined with the benzene layer, washed with saturated sodium chloride solution, and dried over 100 g. of anhydrous magnesium sulfate. Most of the solvent is removed by distillation through a 20-cm. Vigreux column at atmospheric pressure, and the residual liquid is distilled through the column under reduced pressure. The fraction boiling at 85–86°/1.5 mm. is collected (Note 3). The yield of N,N-dimethylcyclohexanecarboxamide is 133–138 g. (86–89%), n_D^{25} 1.4800–1.4807.

B. *N,N-Dimethylcyclohexylmethylamine.* In a 3-l. three-necked flask equipped with a reflux condenser and a dropping funnel, both protected by drying tubes, is placed a suspension of 32 g. (0.85 mole) of lithium aluminum hydride (Note 4) in 400 ml. of anhydrous ether (Note 5). The mixture is stirred with a magnetic stirrer using a 40-mm. Teflon-covered stirring bar. A solution of 133 g. (0.86 mole) of N,N-dimethylcyclohexanecarboxamide in 300 ml. of anhydrous ether (Note 5) is added at such a rate as to maintain gentle reflux. The addition requires about 1 hour. The flask is then placed in an electric heating mantle, and the mixture is stirred and heated under reflux for 15 hours. The heating mantle is replaced by an ice bath, and the flask is fitted with an efficient mechanical, sealed stirrer. Water (70 ml.) is added slowly with vigorous stirring. Stirring is continued for 30 minutes after the addition of water is complete. A cold solution of 200 g. of sodium hydroxide in 500 ml. of water is added at once, and the flask is fitted for steam distillation. The mixture is steam-distilled until the distillate is neutral; about 1.5 l. is collected. The distillate is acidified by careful addition, with water cooling, of 95 ml. of concentrated hydrochloric acid. The two layers are separated and the ether layer washed with 50 ml. of 10% hydrochloric acid. The combined acidic solutions are concentrated until no more distillate comes over at steam bath temperature and 20 mm. pressure. The residue is dissolved in 200 ml. of water, the solution cooled, and 110 g. of sodium hydroxide pellets is added slowly, with stirring and external cooling with ice. The two layers are separated, and the aqueous phase is extracted with three 100-ml. portions of ether (Note 6). The combined amine layer and ether extracts are dried over 40 g. of potassium hydroxide pellets for 3 hours. The drying agent is separated by decantation, and the solvent is removed by distillation through a 20-cm. Vigreux column. The residue, on distillation under reduced pressure, yields 106–107 g. (88%) of N,N-dimethylcyclohexylmethylamine, b.p. 76°/29 mm., n_D^{25} 1.4462–1.4463.

2. Notes

1. The material as supplied by the Eastman Kodak Company (white label grade) may be used without further purification.

2. No fractionating column was used.

3. In some runs, small amounts of sulfur-containing compounds distilled together with the amide. These impurities did not affect the yield and purity of the N,N-dimethylcyclohexylmethylamine obtained in the subsequent reduction with lithium aluminum hydride.

4. Lithium aluminum hydride as supplied by Metal Hydrides Inc., Beverly, Massachusetts, may be used without prior pulverization.

5. Mallinckrodt absolute ethyl ether (reagent grade) may be used without further drying.

6. The checkers added enough water to dissolve most of the solid before the second and third ether extractions.

3. Methods of Preparation

N,N-Dimethylcyclohexylmethylamine has been prepared by reduction of N,N-dimethylcyclohexanecarboxamide with lithium aluminum hydride;[2,3] by the action of dimethylformamide on cyclohexanecarboxaldehyde;[4] by methylation of cyclohexylmethylamine[3,5] and of N-methylcyclohexylmethylamine by the Clarke-Eschweiler method (treatment with formaldehyde and formic acid) and by the action of dimethylamine on cyclohexylmethyl bromide.[6]

N,N-Dimethylcyclohexanecarboxamide has been prepared by the action of dimethylamine on cyclohexanecarbonyl chloride.[2, 3, 7] The experimental procedure described is a modification of the method reported by Mousseron, Jacquier, Mousseron-Canet, and Zagdoun [2] and by Baumgarten, Bower, and Okamoto.[3]

[1] Massachusetts Institute of Technology, Cambridge 39, Massachusetts. Supported by the Office of Ordnance Research, U. S. Army, under Contract No. DA-19-020-ORD-4542.

[2] Mousseron, Jacquier, Mousseron-Canet, and Zagdoun, *Bull. soc. chim. France*, **1952**, 1042.

[3] Baumgarten, Bower, and Okamoto, *J. Am. Chem. Soc.*, **79**, 3145 (1957).

[4] Mousseron, Jacquier, and Zagdoun, *Bull. soc. chim. France*, **1952**, 197.

[5] Cope, Bumgardner, and Schweizer, *J. Am. Chem. Soc.*, **79**, 4729 (1957).

[6] Dunn and Stevens, *J. Chem. Soc.*, **1934**, 279.

[7] Bernhard, *Z. physiol. Chem.*, **248**, 256 (1937).

DIMETHYLFURAZAN *

(Furazan, 3,4-dimethyl-)

$$
\begin{array}{c}
CH_3C\!-\!-\!-\!CCH_3 \\
\| \quad\quad \| \\
NOH \quad NOH
\end{array}
+
\begin{array}{c}
CH_2\!-\!CO \\
| \qquad\quad \diagdown \\
| \qquad\qquad O \rightarrow \\
| \qquad\quad \diagup \\
CH_2\!-\!CO
\end{array}
$$

$$
\begin{array}{c}
CH_3\!-\!-\!-\!CH_3 \\
\| \quad\quad \| \\
N \quad\quad N \\
\diagdown \;\; \diagup \\
O
\end{array}
+
\begin{array}{c}
CH_2CO_2H \\
| \\
CH_2CO_2H
\end{array}
$$

Submitted by LYELL C. BEHR and JOHN T. BRENT.[1]
Checked by T. L. CAIRNS and J. E. CARNAHAN.

1. Procedure

One hundred grams (1 mole) of succinic anhydride and 116 g. (1 mole) of dimethylglyoxime (Note 1), both finely ground, are intimately mixed and introduced into a 1-l. three-necked flask which is equipped with a sealed mechanical stirrer, a thermometer reaching nearly to the bottom of the flask, and an outlet tube connected to a water-cooled condenser arranged for distillation. The mixture is heated slowly with an electric mantle or oil bath, and stirring is commenced as soon as practicable (Note 2). The mixture liquefies at about 100°, and a rapid reaction begins at 150–170° accompanied by a sudden rise in temperature. This initial rapid reaction can be controlled readily by removing the heater when the temperature reaches 170° and applying a cooling bath until the inside temperature is 150° (Note 3). Heat is then applied again, and distillation of the product begins at a flask temperature of 160° and continues until the temperature reaches 200°. The stirrer is stopped, the receiver is changed, and, after the flask and contents have cooled to about 120°, 50 ml. of water is added. The thermometer is replaced by an inlet tube, and steam is passed in until no more insoluble material passes over. Usually collection of about 200–300 ml. of distillate is sufficient. The distillate is extracted with two 100-ml. portions of ether, and the extracts are combined with the dimethylfurazan obtained by direct distillation (Note 4). The ether solution is dried for a short time over anhydrous magnesium sulfate. The drying agent is removed by filtration, and the ether is evaporated on a steam bath. The residue is distilled at atmospheric pressure through a short column, and after a fore-run, consisting chiefly of biacetyl, the dimethylfurazan distils at 154–159° as a colorless liquid, n_D^{25} 1.4234–1.4243, m.p. -7.2 to $-6.6°$. The yield is 59–63 g. (60–64%).

2. Notes

1. A good grade of both reagents should be used. The better-quality products supplied by the Eastman Kodak Company are satisfactory.

2. The mixture may be difficult to stir mechanically below its melting point unless a powerful motor is used.

3. The temperature usually climbs to about 190° during this interval, and some dimethylfurazan distils rapidly.

4. The product obtained by direct distillation contains some water, but the dimethylfurazan can be readily separated.

3. Methods of Preparation

Dimethylfurazan has been obtained from dimethylglyoxime by heating with water, aqueous ammonia, or aqueous sodium hydroxide.[2,3] The usual acid dehydrating agents fail.

[1] Mississippi State University, State College, Mississippi. Work supported in part by a gift from the Research Corporation.

[2] Wolff, *Ber.*, **28**, 69 (1895).

[3] Rimini, *Gazz. chim. ital.*, **25**, II, 266 (1895).

β,β-DIMETHYLGLUTARIC ACID *

(Glutaric acid, 3,3-dimethyl-)

$$\underset{\displaystyle (CH_3)_2C}{\overset{\displaystyle CH_2-C}{\diagup}}\overset{\displaystyle O}{\diagdown} \quad CH_2 + 3NaOCl \rightarrow$$

$$(CH_3)_2C \diagup \overset{CH_2-C-ONa}{\underset{CH_2-C-ONa}{}} + CHCl_3 + NaOH$$

$$(CH_3)_2C \diagup \overset{CH_2-C-ONa}{\underset{CH_2-C-ONa}{}} + 2HCl \rightarrow$$

$$(CH_3)_2C \diagup \overset{CH_2-C-OH}{\underset{CH_2-C-OH}{}} + 2NaCl$$

Submitted by WALTER T. SMITH and GERALD L. McLEOD.[1]
Checked by WILLIAM S. JOHNSON and DONALD D. CAMERON.

1. Procedure

A solution of 218 g. (5.45 moles) of sodium hydroxide in 300 ml. of water in a 3-l. three-necked flask is cooled to room temperature. To this solution 1250 g. of ice is added, and a stream of chlorine is passed in rapidly through a delivery tube having a small opening and extending almost to the bottom of the liquid. The passage of the chlorine is continued until 161 g. (2.27 moles) has been absorbed. The flask is then fitted with a mechanical stirrer, a thermometer, and a 500-ml. separatory funnel.

Seventy grams (0.5 mole) of methone [2] is dissolved in a solution of 65 g. (1.16 moles) of potassium hydroxide in 525 ml. of water. The solution is cooled to room temperature, poured into the separatory funnel, and run slowly with stirring into the sodium hypochlorite solution. The temperature rises gradually to 35–40° during the addition. After the addition has been completed, the solution is stirred for 6–8 hours until the temperature drops to room temperature.

Without interrupting the stirring, 50 g. of sodium sulfite is added to decompose the excess sodium hypochlorite, and the solution is acidified to Congo red by adding concentrated hydrochloric acid slowly with stirring to avoid foaming. The acid solution is then concentrated by distillation until salts just begin to precipitate (Note 1).

The mixture is then cooled to room temperature, 300 ml. of ether and enough water are added to dissolve all of the precipitate, and the whole is transferred to a 3-l. separatory funnel. The layers are separated, and the aqueous portion is extracted with three 200-ml. portions of ether. The ether extracts are combined and dried for several hours over 15–20 g. of anhydrous magnesium sulfate. The ether is then removed by distillation. This may be conveniently carried out by fitting a 250-ml. Claisen flask with a separatory funnel in order to add the solution as the ether distils. When only 150–200 ml. of solution remains in the flask, the distillation is stopped and the residue is poured into a small beaker. The remaining ether is removed by heating on a steam bath, and the residue solidifies on cooling. The colorless, crystalline β,β-dimethylglutaric acid is dried in air. The yield is 73–77 g. (91–96%), m.p. 97–99°. Crystallization from 100–125 ml. of benzene gives 65–73 g. (81–91%) of acid, m.p. 100–102° (Note 2).

2. Notes

1. The chloroform formed in the reaction comes over during the early stages of this distillation. The precipitation of salts usually begins after the solution has been concentrated to about one-half the original volume.

2. The melting points given in the literature are 101°,[3] 101–102°,[4] 103–104°,[5] and 98–100°.[6]

3. Methods of Preparation

β,β-Dimethylglutaric acid has been prepared by heating dimethylpropanetricarboxylic acid above its melting point;[3] by hydrolysis of the condensation product of ethyl cyanoacetate and ethyl β,β-dimethylacrylate;[7] by the action of sulfuric acid on diethyl β,β-dimethyl-α,α'-dicyanoglutarate;[8] by hydrolysis of the nitrile obtained by the action of calcium cyanide on β,β-dimethylbutyrolactone;[4] by the action of sulfuric acid on β,β-dimethyl-α,α'-dicyanoglutarimide;[9] and by the action of sodium hypobromite on methone.[5] The present procedure is essentially that of Walker and Wood.[6]

[1] State University of Iowa, Iowa City, Iowa.

[2] *Org. Syntheses Coll. Vol.* 2, 200 (1943).

[3] Perkin and Goodwin, *J. Chem. Soc.*, 69, 1472 (1896).

[4] Blaise, *Compt. rend.*, 126, 1153 (1898).

[5] Guareschi, *Atti reale accad. sci. Torino*, [1] 36, 261 (1900–1901) [*Chem. Zentr.*, [1] 72, 821 (1901)]; Fredga and Sikström, *Arkiv Kemi*, 8, 433 (1955).

[6] Walker and Wood, *J. Chem. Soc.*, 89, 598 (1906).

[7] Perkin and Thorpe, *J. Chem. Soc.*, 75, 48 (1899).

[8] Komppa, *Ber.*, 33, 3531 (1900).

[9] Komppa, *Ber.*, 32, 1423 (1899); Benica and Wilson, *J. Am. Pharm. Assoc.*, 39, 451 (1950); Benkeser and Bennett, *J. Am. Chem. Soc.*, 80, 5414 (1958).

DIMETHYLKETENE

(Ketene, dimethyl-)

$$(CH_3)_2CHCO_2H + Br_2 \xrightarrow{P} (CH_3)_2CBrCOBr \xrightarrow{Zn}$$
$$(CH_3)_2C{=}C{=}O$$

Submitted by C. W. Smith and D. G. Norton.[1]
Checked by T. L. Cairns and J. C. Sauer.

1. Procedure

A. *α-Bromoisobutyryl bromide.* To a mixture of 250 g. (2.85 moles) of isobutyric acid and 35 g. (0.28 mole) of red phosphorus in a 1-l. three-necked flask, fitted by ground-glass joints to a dropping funnel, mechanical stirrer, and reflux condenser, is added, dropwise with stirring, 880 g. (5.5 moles) of bromine. After the addition is complete, the solution is warmed to 100° over a period of 6 hours. The unreacted bromine and hydrogen bromide are removed under reduced pressure (30 mm.). The α-bromoisobutyryl bromide is decanted from the phosphorous acid and fractionated through a short helices-packed column. After a considerable fore-cut, the main fraction, 493–540 g. (75–83%), is collected at 91–98° (100 mm.).

B. *Dimethylketene* (Note 1). The apparatus for this preparation consists of a 500-ml. flask equipped with an inlet tube for nitrogen and a dropping funnel and fitted to a 6-in. modified Claisen still head leading to a tared spiral inlet trap having stopcocks on the inlet and exit sides and cooled in Dry Ice-acetone (Note 2). This trap is connected to a vacuum line, and the reaction is carried out at 300 mm. pressure. After 40 g. (0.61 g. atom) of zinc turnings and 300 ml. of ethyl acetate have been placed in the flask and the system has been flushed with nitrogen (free of oxygen and moisture) and heated to incipient boiling, 111 g. (0.48 mole) of α-bromoisobutyryl bromide is added dropwise at such a rate that the ethyl acetate boils gently. A slow stream of nitrogen is continued throughout the reaction. Dimethylketene distils along with ethyl acetate and is obtained in 46–54% yield as a 9–10% solution in ethyl acetate (15–18 g. of dimethylketene in 190–200 ml. of ethyl acetate) (Note 3).

2. Notes

1. This ketene reacts rapidly with oxygen to form an explosive peroxide. Drops of solution allowed to evaporate in air may detonate. Washing with water is an efficient means of decontamination.

2. If simple traps are used, it is necessary to use two in series to condense all the dimethylketene and ethyl acetate.

3. The concentration of dimethylketene is determined by titration of an aliquot at ice temperatures with $0.1N$ sodium hydroxide using phenolphthalein indicator. Under these conditions, blank determinations indicate that ethyl acetate is not hydrolyzed.

The identity of the dimethylketene may be determined (and an approximate check made on the concentration) by adding 35 g. of dimethylketene solution to 15 g. of aniline in 75 ml. of ether. After 2–3 minutes, the ether solution is washed with dilute hydrochloric acid, dilute potassium carbonate, then water, and the ether is evaporated. Isobutyroanilide, m.p. 102–103° (103–104° after one recrystallization, no melting-point depression in mixture with an authentic sample), is obtained in about 90% yield based on the concentration of dimethylketene indicated by titration.

3. Methods of Preparation

The preparation of ketenes has been discussed by Hanford and Sauer in *Organic Reactions*.[2] Dimethylketene has been prepared by the treatment of α-bromoisobutyryl bromide with zinc,[3] and by the pyrolysis of isobutyrylphthalimide,[4] dimethylmalonic anhydride,[5] or α-carbomethoxy-α,β-dimethyl-β-butyrolactone.[6] Dimethylketene dimer has been prepared by heating isobutyryl chloride with a tertiary amine. Pyrolysis of the dimer yields dimethylketene.[7]

α-Bromoisobutyryl bromide has been prepared in a two-step process involving the bromination of isobutyric acid to α-bromoisobutyric acid followed by treatment with phosphorus tribromide.[8] A one-step process utilizing the Hell-Volhard-Zelinsky reaction [9] is more satisfactory.

[1] Shell Development Company, Emeryville, California.
[2] Hanford and Sauer, in Adams, *Organic Reactions*, Vol. 3, p. 108, John Wiley & Sons, 1946.
[3] Staudinger and Klever, *Ber.*, **39**, 968 (1906).
[4] Hurd and Dull, *J. Am. Chem. Soc.*, **54**, 2432 (1932).
[5] Staudinger, *Helv. Chim. Acta*, **8**, 306 (1925).
[6] Ott, *Ann.*, **401**, 159 (1913).
[7] See reference 2, p. 136.
[8] Taufen and Murray, *J. Am. Chem. Soc.*, **67**, 754 (1945).
[9] Volhard, *Ann.*, **242**, 161 (1887).

5,5-DIMETHYL-2-n-PENTYLTETRAHYDROFURAN

(Furan, tetrahydro-2,2-dimethyl-5-pentyl-)

$$(CH_3)_2\overset{\displaystyle OH}{\underset{\displaystyle |}{C}}(CH_2)_2CHOH(CH_2)_4CH_3 \xrightarrow{H_3PO_4}$$

$$(CH_3)_2C\underset{\displaystyle O}{\overset{\displaystyle CH_2\text{——}CH_2}{\diagdown\diagup}}CH(CH_2)_4CH_3 + H_2O$$

Submitted by J. COLONGE and R. MAREY.[1]
Checked by V. BOEKELHEIDE and H. KAEMPFEN.

1. Procedure

In a 100-ml., three-necked, round-bottomed flask, fitted with a sealed mechanical stirrer, a reflux condenser, and a thermometer reaching to the bottom of the flask, are placed 37.6 g. (0.2 mole) of 2-methyl-2,5-decanediol (p. 601) and 17 g. of 85% phosphoric acid. The limpid liquid obtained is heated and maintained at 125° for 40 minutes. Then the acidic lower layer is discarded, and the organic layer is washed with three or four 50-ml. portions of lukewarm distilled water.

Distillation of the resulting crude oil using a simple fractionating column gives 32–33 g. (94–97%) of pure 5,5-dimethyl-2-n-pentyltetrahydrofuran as a colorless liquid boiling at 31–33°/1.5 mm.; n_D^{25} 1.4257 (Note 1).

2. Note

1. The submitters have also prepared 5,5-dimethyl-2-heptyltetrahydrofuran, n_D^{25} 1.4360, by a similar dehydration of 2-methyl-2,5-undecanediol obtained from the reaction of methylmagnesium bromide and γ-undecanoic acid lactone.

3. Methods of Preparation

There is no report on the preparation of 5,5-dimethyl-2-n-pentyltetrahydrofuran.

[1] École de Chimie Industrielle de Lyon and Établissement Descollonges Frères (Lyon).

3,5-DIMETHYLPYRAZOLE *

(Pyrazole, 3,5-dimethyl-)

$$CH_3COCH_2COCH_3 + H_2NNH_2 \rightarrow$$

Submitted by RICHARD H. WILEY and PETER E. HEXNER.[1]
Checked by WILLIAM S. JOHNSON and ROBERT J. HIGHET.

1. Procedure

Sixty-five grams (0.50 mole) of hydrazine sulfate (Note 1) is dissolved in 400 ml. of 10% sodium hydroxide in a 1-l. round-bottomed flask, fitted with a separatory funnel, a thermometer, and a stirrer. The flask is immersed in an ice bath and cooled. When the temperature of the mixture reaches 15° (Note 2), 50 g. (0.50 mole) of acetylacetone (Note 3) is added dropwise with stirring while the temperature is maintained at about 15°. The addition requires about 30 minutes to complete, and the mixture is stirred for 1 hour at 15° (Note 4). The contents of the flask are diluted with 200 ml. of water to dissolve precipitated inorganic salts, transferred to a 1-l. separatory funnel, and shaken with 125 ml. of ether. The layers are separated, and the aqueous layer is extracted with four 40-ml. portions of ether. The ether extracts are combined, washed once with saturated sodium chloride solution, and dried over anhydrous potassium carbonate. The ether is removed by distillation, and the slightly yellow residue of crystalline 3,5-dimethylpyrazole obtained by drying at reduced pressure (approximately 20 mm.) weighs 37–39 g. (77–81%), m.p. 107–108°. This product, which is of good quality, can be recrystallized from about 250 ml. of 90–100° petroleum ether without significant change in appearance or melting point. The yield after drying in a vacuum desiccator containing paraffin chips is 35–37 g. (73–77%) (Note 5).

2. Notes

1. Hydrazine sulfate supplied by the Eastman Kodak Company is satisfactory, or it may be prepared by a previously described procedure.[2]

2. A precipitate of sodium sulfate may form at this point.

3. Union Carbide and Carbon Corporation technical 2,4-pentanedione was used without purification.

4. The 3,5-dimethylpyrazole precipitates during this period.

5. Recrystallization from methanol or ethanol gives practically colorless material of the same melting point, but it is more difficult to obtain good recovery owing to the high solubility of the pyrazole in these solvents.

3. Methods of Preparation

3,5-Dimethylpyrazole has been prepared from acetylacetone and hydrazine hydrate in ethanol[3] or hydrazine sulfate in aqueous alkali.[4–6] The latter method is preferred, because the reaction with hydrazine hydrate is sometimes violent.[3,4] 3,5-Dimethylpyrazole also has been prepared by hydrolysis and decarboxylation of the 1-carbamido- or 1-carboxamidine derivatives, obtained by reaction of semicarbazide[7] or aminoguanidine[8] with acetylacetone, and from 1,2-pentadien-4-one and hydrazine hydrate.[9]

[1] University of Louisville, Louisville, Kentucky.

Org. Syntheses Coll. Vol. 1, 309 (1941).
Rothenberg, J. prakt. Chem., [2] 52, 50 (1895); Ber., 27, 1097 (1894).
Rosengarten, Ann., 279, 237 (1894).
Morgan and Ackerman, J. Chem. Soc., 123, 1308 (1923).
Zimmerman and Lochte, J. Am. Chem. Soc., 60, 2456 (1938).
Posner, Ber., 34, 3980 (1901).
Thiele and Dralle, Ann., 302, 294 (1898).
Bertrand, Compt. rend., 245, 2306 (1957).

2,2-DIMETHYLPYRROLIDINE

(Pyrrolidine, 2,2-dimethyl-)

$$4 \quad \underset{CH_3}{\overset{CH_3}{\diagdown}} \underset{N}{\overset{CH_2-CH_2}{\diagdown}} \underset{H}{\overset{CH_2-CH_2}{\diagdown}} C=O + 3LiAlH_4 \rightarrow$$

$$\left[\underset{CH_3}{\overset{CH_3}{\diagdown}} \underset{N}{\overset{CH_2-CH_2}{\diagdown}} \underset{CH_2}{\overset{CH_2-CH_2}{\diagdown}} \right]_4 LiAl + 2LiAlO_2 + 4H_2$$

$$\left[\underset{CH_3}{\overset{CH_3}{\diagdown}} \underset{N}{\overset{CH_2-CH_2}{\diagdown}} \underset{CH_2}{\overset{CH_2-CH_2}{\diagdown}} \right]_4 LiAl + 8HCl \rightarrow$$

$$4 \quad \underset{CH_3}{\overset{CH_3}{\diagdown}} \underset{N}{\overset{CH_2-CH_2}{\diagdown}} \underset{H}{\overset{CH_2-CH_2}{\diagdown}} CH_2 + AlCl_3 + LiCl$$

$$\cdot HCl$$

$$\underset{CH_3}{\overset{CH_3}{\diagdown}} \underset{N}{\overset{CH_2-CH_2}{\diagdown}} \underset{H}{\overset{CH_2-CH_2}{\diagdown}} CH_2 + NaOH \rightarrow$$

$$\cdot HCl$$

$$\underset{CH_3}{\overset{CH_3}{\diagdown}} \underset{N}{\overset{CH_2-CH_2}{\diagdown}} \underset{H}{\overset{CH_2-CH_2}{\diagdown}} CH_2 + NaCl + H_2O$$

Submitted by ROBERT BRUCE MOFFETT.[1]
Checked by N. J. LEONARD and J. W. CURRY.

1. Procedure

A 3-l. three-necked round-bottomed flask is placed on a steam bath and fitted with a mercury-sealed Hershberg stirrer, a dropping funnel, and an efficient reflux condenser topped with a tube containing soda lime and calcium chloride. In this flask are placed 38 g. (1 mole) of pulverized lithium aluminum hydride (Note 1) and 400 ml. of dry tetrahydrofuran (Note 2). The mixture is heated under reflux with stirring for 15 minutes or until most of the lithium aluminum hydride has dissolved. A solution of 90.5 g. (0.8 mole) of 5,5-dimethyl-2-pyrrolidone (p.357) in 200 ml. of dry tetrahydrofuran (Note 2) is added slowly at such a rate that the solvent refluxes gently without external heating. When the addition is complete and the initial reaction subsides, the mixture is stirred and heated at gentle reflux for 8 hours.

The condenser is then set for downward distillation, and, while the mixture is stirred, about 450 ml. of solvent is distilled (Notes 3 and 4). The condenser is reset in the reflux position, and 300 ml. of ether (commercial anhydrous) is added slowly from the dropping funnel with vigorous stirring. This is followed by 50 ml. of ethyl acetate added very slowly with vigorous stirring and finally by 500 ml. of 6N hydrochloric acid added in the same manner.

The condenser is again set downward, and the dropping funnel is replaced by a tube reaching nearly to the bottom of the flask. Steam is passed in, and the distillation is continued for several minutes after the boiling point reaches 100° (Note 5). The distillate is discarded. The mixture in the flask is cooled, and to it is added carefully 350 ml. of 12N sodium hydroxide with stirring (Note 6). The alkaline mixture is then steam-distilled until the distillate is no longer basic (Note 7).

The 2,2-dimethylpyrrolidine may be recovered from the aqueous distillate in two ways: (a) the distillate can be extracted continuously with ether;[2] or (b) the distillate can be acidified with hydrochloric acid and concentrated to dryness under reduced pressure to give crude 2,2-dimethylpyrrolidine hydrochloride. The base is then liberated by adding an excess of saturated aqueous sodium hydroxide solution. The oily layer is separated. The aqueous layer and salt are extracted several times with ether, which is combined with the amine. In either case the ether solution is dried thoroughly over anhydrous potassium carbonate (Note 8).

The drying agent is removed by filtration, and the ether is stripped through a short helices-packed column. The residue is fractionally distilled at 103–105°/745 mm.; n_D^{20} 1.4330; n_D^{25} 1.4304; d_4^{25} 0.8211. The yield is 53–62 g. (67–79%).

2. Notes

1. The hydride can be pulverized rapidly and safely by breaking the large pieces with a spatula, followed by careful crushing with a mortar and pestle. Caution must be observed because the solid may inflame on prolonged grinding or abrasion. The hydride dust is caustic and irritating.

2. Tetrahydrofuran from E. I. du Pont de Nemours and Company can be dried conveniently by adding to it lithium aluminum hydride in small portions until no further reaction (evolution of hydrogen) ensues. After the mixture has been stirred for a few minutes, most of the tetrahydrofuran is distilled from it with stirring (to prevent bumping) (Note 3), and collected in a receiver protected from moisture by a calcium chloride tube.

3. Care must be taken in distilling solutions of lithium aluminum hydride. Explosions have been reported [3] toward the end of distillations of such solutions, especially if they contained carbon dioxide. It is therefore recommended that these distillations be carried out behind a shield and that not more than *three-fourths* of the solvent be removed.

4. If this tetrahydrofuran is collected in a receiver protected from moisture it may be used in subsequent runs.

5. This steam distillation removes ether, tetrahydrofuran, and other volatile neutral products. If too much water accumulates in the flask, it may be heated in an electric heating mantle after most of the ether has been removed.

6. At this point the mixture should be a very strongly basic, mobile, milky slurry.

7. From time to time an aliquot of the distillate being collected can be titrated with standard acid to determine whether significant amounts of amine are distilling.

8. Sufficient drying agent should be used so that no aqueous liquid phase appears.

3. Methods of Preparation

2,2-Dimethylpyrrolidine has been prepared by the hydrogenation of 5-amino-2,2-dimethylpyrroline-N-oxide or 5-imino-2,2-dimethylpyrrolidine in the presence of Raney nickel [4] or by reduction with sodium and alcohol.[4] The present procedure has been published.[5]

[1] The Upjohn Company, Kalamazoo, Michigan.
[2] *Org. Syntheses Coll. Vol.* **1**, 277 (1941).
[3] Barbaras, Barbaras, Finholt, and Schlesinger, *J. Am. Chem. Soc.*, **70**, 877 (1948).
[4] Buckley and Elliott, *J. Chem. Soc.*, **1947**, 1508.
[5] Moffett and White, *J. Org. Chem.*, **17**, 407 (1952).

5,5-DIMETHYL-2-PYRROLIDONE

(2-Pyrrolidinone, 5,5-dimethyl-)

$$(CH_3)_2CCH_2CH_2CO_2CH_3 + 3H_2 \xrightarrow{Ni}$$

$$\overset{|}{NO_2}$$

$$(CH_3)_2CCH_2CH_2CO_2CH_3 + 2H_2O$$

$$\overset{|}{NH_2}$$

$$(CH_3)_2CCH_2CH_2CO_2CH_3 \rightarrow$$

$$\overset{|}{NH_2}$$

Submitted by R. B. MOFFETT.[1]
Checked by N. J. LEONARD, W. E. SMITH, and B. L. RYDER.

1. Procedure

To a solution of 148 g. (0.845 mole) of methyl γ-methyl-γ-nitrovalerate (p. 652) in 500 ml. of commercial absolute ethanol (total volume about 632 ml.) in a 2.5-l. rocking high-pressure bomb is added 12.5–25.0 g. (Note 1) of W-5 [2] Raney nickel catalyst (Note 2), previously rinsed with absolute ethanol. The bomb head and fittings are placed in position, including a thermocouple attached to a semi-automatic heating control (Micromax). Hydrogen is introduced into the bomb until the pressure reaches 1000 lb. per sq. in. (Note 3).

The bomb is rocked and the temperature of the solution is raised carefully to 55° during the course of 0.5–1.0 hour (Note 4). The hydrogen uptake begins at 40–51°, and during the reaction period the temperature is held at 55–60°. The rate of pressure drop is 50–100 lb. per sq. in. each 15 minutes. The rocking of the bomb is continued, and the temperature is maintained at 60° until the pressure reading is constant for 1 hour, in order to ensure completion of the reaction.

After the bomb has been cooled the contents are removed and allowed to stand until the catalyst has settled (Note 5). The mixture is filtered, and the filtrate is transferred to a Claisen flask placed in an oil bath. The solvent is removed by distillation at atmospheric pressure, and the oil-bath temperature is raised to 200°. After temporary cooling, the residue in the flask is distilled under reduced pressure. The 5,5-dimethyl-2-pyrrolidone boils at 126.5–128.5°/12 mm. and solidifies in the receiver. The yield is 84–92 g. (88–96%); m.p. 42–43° (Note 6).

2. Notes

1. It is unnecessary and may in fact be dangerous to use a larger amount of catalyst.

2. The checkers have found that commercial grade Raney nickel (Gilman Paint and Varnish Company) is a satisfactory substitute for W-5 catalyst. The yields obtained with the two catalysts are identical, but the hydrogenation requires 2–3 hours with commercial catalyst compared with 1–1.5 hours for W-5 catalyst.

3. The theoretically required drop in hydrogen pressure for a free space of 1868 ml. and equivalent to 2.535 moles of hydrogen is 507 lb. per sq. in. The theoretical pressure drop will vary with the free space when bombs of different capacity are used. The observed pressure drop usually exceeds the theoretical by about 10%.

4. Since batches of Raney nickel may vary in activity, caution must be exercised during the heating period. The temperature should not exceed 60° at any time.

5. Filtration through Hiflo Super-Cel (Johns-Manville Company) speeds the operation but lowers the final yield.

6. The product can be recrystallized from petroleum ether (b.p. 30–38°).

3. Methods of Preparation

5,5-Dimethyl-2-pyrrolidone has been prepared by the hydrolysis of 5-imino-2,2-dimethylpyrrolidine in the presence of Raney nickel [3] or by the hydrogenation of 5-amino-2,2-dimethylpyrroline-N-oxide in the presence of Raney nickel.[3] The preparation by this method has been published.[4]

[1] The Upjohn Company, Kalamazoo, Michigan.
[2] Adkins and Billica, *J. Am. Chem. Soc.*, **70**, 695 (1948).
[3] Buckley and Elliott, *J. Chem. Soc.*, **1947**, 1508.
[4] Moffett and White, *J. Org. Chem.*, **17**, 407 (1952).

N,N-DIMETHYLSELENOUREA

(Urea, 1,1-dimethyl-2-seleno-)

$$(CH_3)_2NCN + H_2Se \rightarrow (CH_3)_2NCNH_2$$
$$\overset{\|}{Se}$$

Submitted by Frank Bennett and Ralph Zingaro.[1]
Checked by C. F. H. Allen and K. C. Kennard.

1. Procedure

Caution! Hydrogen selenide is very toxic. This preparation should be carried out in a well-ventilated hood. The operator should wear rubber gloves. The apparatus should be screened from bright light.

A mixture of 37 g. (0.5 mole) of dimethylcyanamide (Note 1), 75 ml. of concentrated ammonium hydroxide, and 25 ml. of water is placed in a 250-ml. round-bottomed three-necked flask fitted with a stirrer, thermometer, and glass delivery tube (Note 2). The solution is stirred slowly, while hydrogen selenide (Note 3) is slowly bubbled in, maintaining the temperature at 20–30° by occasional external cooling. At the end of the reaction (4–5 hours) when there is no more gas evolution from the generator (4–5 hours), the flask and contents are cooled to 5–10°

and stirred vigorously for one-half hour. The solid is then collected on a 9-cm. Büchner funnel and washed with 250 ml. of ice-cold 95% ethanol. The crude, gray product weighs 65–73 g. (81–91%).

For purification, it is dissolved in 9–9.5 l. of boiling benzene (Note 4). The solution is filtered rapidly, using a 32-cm. folded filter in an 8-in. short-stem glass funnel. The recovery is 59–70 g. The fine, white crystals, which melt at 169–170° (Note 5), are stored in a brown glass bottle under nitrogen (Notes 6, 7, 8).

2. Notes

1. The practical grade, b.p. 162–164°, obtained from the Eastman Kodak Company, was used.

2. A tube 8 mm. in diameter is used. A sintered-glass gas addition tube is quickly clogged by selenium.

3. Hydrogen selenide is generated from iron or aluminum selenide and a mineral acid. In a 2-l. three-necked flask, fitted with sealed stirrer, reflux condenser, and dropping funnel are placed 135 g. (1 mole) of powdered iron selenide (Note 9) and 350 ml. of water. The generator is heated on a steam cone, while 350 ml. of concentrated hydrochloric acid is admitted slowly with stirring, so that the hydrogen selenide is evolved at a steady rate. (The generation of gas is regulated both by the rate of heating and the rate of addition of acid, so that the gas is absorbed almost entirely; it is led from the top of the condenser to the delivery tube.) The amount of selenide used is sufficient. The residual liquid in the generator is left open to the air overnight in a hood. Then sufficient 50% sodium hydroxide solution is added to make it basic, and the whole is flushed down the sink, using a large amount of water.

4. Recrystallization from benzene affords a product that has superior storage qualities. A liter dissolves 5–6 g. at the boiling point.

5. The melting point is unchanged by further recrystallization.

6. The product turns pink, and eventually gray, if exposed to air for an extended period. It should be stored under nitrogen.[2]

7. The submitters indicate that N,N-diethylselenourea may be prepared similarly, using diethylcyanamide. In this case, a

solution of 50 ml. each of concentrated aqueous ammonium hydroxide and ethanol is used as a solvent. The reaction is carried out at 60°, the solvent being replenished as needed by a solution containing 80 ml. of ethanol and 20 ml. of concentrated ammonium hydroxide. The yield of crude material is 65–80%. Recrystallization from benzene gives a white product, m.p. 117–118°.

8. The procedure gives a commensurate yield when carried out on twice the scale.

9. Iron selenide can be obtained from the Canadian Copper Refiners Ltd., Montreal, Quebec.

3. Methods of Preparation

The procedure used is that of Zingaro, Bennett, and Hammar.[2]

[1] Camera Works Division, Eastman Kodak Company, Rochester 17, New York.
[2] Zingaro, Bennett, and Hammar, *J. Org. Chem.*, **18**, 292 (1953).

asym-DIMETHYLUREA

(Urea, 1,1-dimethyl-)

$$(CH_3)_2NH + H_2NCONHNO_2 \rightarrow$$
$$(CH_3)_2NCONH_2 + N_2O + H_2O$$

Submitted by FREDERICK KURZER.[1]
Checked by WILLIAM S. JOHNSON and WILLIAM T. TSATSOS.

1. Procedure

In a 1.5-l. beaker (Note 1), 191 ml. (180 g.) of aqueous 25% dimethylamine solution (1.0 mole) is diluted with 64 ml. of water and treated with 116 g. (1.1 moles) of nitrourea.[2] The temperature of the resulting brownish liquid rises spontaneously to 35–42°. The solution is warmed to 56–60°, and a reaction sets in vigorously with evolution of nitrous oxide. External cooling with water is applied when required; the reaction temperature is maintained below 70° during the first 5–7 minutes and below 85° during the second period of 5–7 minutes. After a total of 10–15 minutes, the effervescence slackens and the reaction mixture is kept at 90–100° until the evolution of gas has completely ceased. This usually requires an additional 15–20 minutes.

The resulting liquid is heated with about 1 g. of activated carbon and is filtered with suction while hot; the clear, faintly colored filtrate is transferred to an evaporating dish and heated on a steam bath to remove most of the water. The residual, somewhat viscous, liquid (about 120 ml.) is then transferred to a beaker; the evaporating dish is rinsed with 10 ml. of water, and this solution is added to the contents of the beaker. To this is added 95% ethanol (50 ml.), and the mixture is warmed to effect solution. On cooling, large crystals separate which, at 0°, almost fill the bulk of the solution. The crystalline mass is broken up with a glass rod, collected on a suction filter, and washed quickly, while on the filter, with two successive portions of ice water (Note 2), drained, and air-dried (Note 3). The yield of large, colorless prismatic crystals of *asym*-dimethylurea is 35–40 g. (40–45%); m.p. 182–184° (Note 4).

The combined filtrates, on further evaporation and dilution with 95% ethanol as above, yield an additional 15–20 g. of material of approximately the same quality, bringing the total yield up to 57–68%. In a series of preparations, yields can be further increased a few per cent by carrying over mother liquors to subsequent batches.

2. Notes

1. The reaction mixture froths considerably in the initial stages, and the use of a sufficiently large reaction vessel is important.

2. Since *asym*-dimethylurea is appreciably soluble in water, enough ice water is added to just cover the crystals on the filter, and suction is applied immediately.

3. A slight odor of dimethylamine, which persists, disappears when the product is dried at 80° for 1–2 hours.

4. The use of nitrourea is generally applicable to the preparation of urea derivatives and usually affords the required carbamide in excellent yield.[3, 4]

3. Methods of Preparation

asym-Dimethylurea has been prepared by the interaction of dimethylamine sulfate and potassium cyanate [5] or dimethylamine and nitrourea.[3] It is also obtained by the hydrolysis of dimethylcyanamide in acid [6] and alkaline [7] media. Other reactions yielding the product include the action of dimethylamine on methyl γ-methylallophanate [8] or diethoxymethyleneimine,[9] and the hydrolysis of 1,1,2-trimethylisourea [10] or methyl dimethylthiocarbamate.[11]

[1] School of Medicine, University of London, London, England.
[2] Org. Syntheses Coll. Vol. 1, 417 (1941).
[3] Davis and Blanchard, J. Am. Chem. Soc., 51, 1798 (1929).
[4] Buck and Ferry, J. Am. Chem. Soc., 58, 854 (1936).
[5] Franchimont, Rec. trav. chim., 2, 122 (1883); 3, 222 (1884).
[6] Diels and Gollmann, Ber., 44, 3165 (1911).
[7] von Braun and Röver, Ber., 36, 1197 (1903).
[8] Biltz and Jeltsch, Ber., 56, 1920 (1923).
[9] Schenck, Arch. Pharm., 249, 467 (1911); Z. physiol. Chem., 77, 368 (1912).
[10] McKee, Am. Chem. J., 42, 25 (1909).
[11] Delepine and Schving, Bull. soc. chim. France, [4] 7, 900 (1910).

2,6-DINITROANILINE *

(Aniline, 2,6-dinitro-)

$C_6H_5Cl + 2H_2SO_4 + 2KNO_3 \rightarrow$

$$O_2N-\underset{SO_3K}{\overset{Cl}{\bigcirc}}-NO_2 + KHSO_4 + 3H_2O$$

$$O_2N-\underset{SO_3K}{\overset{Cl}{\bigcirc}}-NO_2 + 2NH_4OH \rightarrow$$

$$O_2N-\underset{SO_3K}{\overset{NH_2}{\bigcirc}}-NO_2 + NH_4Cl + 2H_2O$$

$$O_2N-\underset{SO_3K}{\overset{NH_2}{\bigcirc}}-NO_2 + H_2O \xrightarrow{H_2SO_4} O_2N-\overset{NH_2}{\bigcirc}-NO_2 + KHSO_4$$

Submitted by HARRY P. SCHULTZ.[1]
Checked by ARTHUR C. COPE and DOUGLAS S. SMITH.

1. Procedure

In a 1-l. round-bottomed flask fitted with a mechanical stirrer are placed 50 ml. (55.4 g., 0.49 mole) of chlorobenzene (Note 1), 300 ml. of concentrated sulfuric acid (sp. gr. 1.84), and 50 ml. (92 g.) of fuming sulfuric acid (containing approximately 25% free sulfur trioxide). The mixture is stirred and heated on a steam bath for 2 hours and then cooled to room temperature. The stirrer is removed from the reaction flask and replaced with a thermometer. To the clear solution is added 170 g. (1.68 moles) of potassium nitrate in 4 portions. The temperature of the mixture during this time is held at 40–60° by cooling the flask and its contents in ice water. After the mixture has been swirled briefly in the reaction flask to dissolve most of the potassium nitrate, it is heated to 110–115° (Note 2) and held at that temperature for 20 hours. The hot contents of the flask are poured onto 2 kg. of cracked ice. After the ice has melted, the yellow precipitate is filtered with suction and pressed as dry as possible.

Without further drying, the potassium 4-chloro-3,5-dinitrobenzenesulfonate is recrystallized from 600 ml. of boiling water (Note 3). Insoluble material is removed by decantation and filtration of the hot solution. The solution is cooled to 5–10° for 12 hours, and the crystalline potassium salt is collected on a suction filter, pressed as dry as possible, and placed at once in a solution of 400 ml. of concentrated ammonium hydroxide (sp. gr. 0.90) in 400 ml. of water. The solution is boiled for 1 hour under a reflux condenser which has been connected to a gas absorption trap,[2] and then is cooled at 5–10° for 12 hours. The orange, crystalline potassium 4-amino-3,5-dinitrobenzenesulfonate is filtered with suction and pressed as dry as possible on a 10-cm. Büchner funnel.

The damp salt is placed in a solution of 200 ml. of concentrated sulfuric acid (sp. gr. 1.84) and 200 ml. of water in a 1-l. round-bottomed flask, and the mixture is boiled vigorously under reflux for 6 hours (Note 4). The hot acid solution is poured onto 1 kg. of cracked ice, filtered on a 7.5-cm. Büchner funnel, slurried twice with 100-ml. portions of water, and pressed as dry as possible on the funnel. The damp, impure 2,6-dinitroaniline is dissolved in 500 ml. of hot 95% ethanol, and the solution is boiled under reflux for 10 minutes with 3 g. of Norit and 3 g. of filter aid. The hot ethanol solution is filtered through a heated funnel (Note 5) and cooled slowly to room temperature. Light-orange needles of 2,6-dinitroaniline separate and are collected on a suction filter and air-dried. The yield is 27.4–32.3 g. (30–36%) (Note 6), m.p. 139–140°.

2. Notes

1. The best grade of Eastman Kodak Company chlorobenzene was used.

2. Since the reaction is moderately exothermic during the first 4 hours, the temperature of the reaction mixture must be controlled carefully. A gas trap [2] may be used to absorb the small amount of nitrogen dioxide evolved, or the reaction may be carried out in a hood. Excessive fuming is avoided if the temperature is kept in the range 110–115°.

3. If the potassium 4-chloro-3,5-dinitrobenzenesulfonate is not recrystallized before ammonolysis very impure 2,6-dinitroaniline is obtained.

4. The condenser should be cleared occasionally with a small glass rod to remove the 2,6-dinitroaniline that may collect there.

5. The funnel must be heated to avoid crystallization during filtration.

6. The solubility of pure 2,6-dinitroaniline in 95% ethanol at room temperature is about 0.4 g. per 100 ml.

3. Methods of Preparation

2,6-Dinitroaniline has been prepared by the ammonolysis of 2,6-dinitroanisole,[3] 2,6-dinitroiodobenzene,[4] 2,6-dinitrophenyl 4-nitrobenzyl ether,[5] and 2,6-dinitrochlorobenzene;[6] by the rearrangement of o-nitrophenylnitramine;[7] and by the desulfonation of potassium 4-amino-3,5-dinitrobenzenesulfonate.[8-10] The method described above is based on the procedures of Ullmann [8] and Welsh.[9]

[1] University of Miami, Coral Gables, Florida.
[2] Org. Syntheses Coll. Vol. 2, 4 (1943).
[3] Salkowski, Ann., 174, 273 (1874).
[4] Koerner, Gazz. chim. ital., 4, 324 (1874).
[5] Kumpf, Ann., 224, 118 (1884).
[6] Borsche and Rantscheff, Ann., 379, 162 (1911).
[7] Hoff, Ann., 311, 108 (1900).
[8] Ullmann, Engi, et al., Ann., 366, 102 (1909).
[9] Welsh, J. Am. Chem. Soc., 63, 3276 (1941).
[10] Fisher and Joullie, J. Org. Chem., 23, 1944 (1958).

p,p'-DINITROBIBENZYL *

(Bibenzyl, 4,4'-dinitro-)

$$2CH_3\langle\bigcirc\rangle NO_2 + \tfrac{1}{2}O_2 \rightarrow$$

$$O_2N\langle\bigcirc\rangle CH_2CH_2\langle\bigcirc\rangle NO_2 + H_2O$$

Submitted by HERBERT O. HOUSE.[1]
Checked by JOHN C. SHEEHAN and J. IANNICELLI.

1. Procedure

In a 3-l. three-necked flask equipped with a mechanical stirrer and an inlet tube extending to the bottom of the flask is placed 2 l. of 30% methanolic potassium hydroxide (Note 1). The flask is immersed in an ice bath, and, when the solution has cooled to 10°, 100 g. (0.73 mole) of p-nitrotoluene (Note 2) is added to the flask. Vigorous stirring is begun, and a rapid stream of air from a compressed-air source is passed through the inlet tube. After 3 hours the ice bath is removed and the passage of air through the mixture is continued with uninterrupted, vigorous stirring for an additional 5 hours. The reaction mixture is immediately filtered with suction (Note 3), and the solid, while still on the filter, is washed with 2 l. of boiling water followed by 300 ml. of 95% ethanol at room temperature. The product is allowed to dry thoroughly in air and then is dissolved in a minimum quantity of boiling benzene (Note 4). The hot solution is filtered to remove a small amount of insoluble red-orange material and is allowed to cool. The p,p'-dinitrobibenzyl crystallizes as orange needles, m.p. 178–180°. The yield is 73–75 g. (74–76%). A second recrystallization from benzene gives yellow needles, m.p. 179–180°.

2. Notes

1. Thirty per cent methanolic potassium hydroxide may be prepared by dissolving 680 g. of c.p. (minimum 85%) potassium hydroxide pellets in 2 l. of methanol.

2. A good grade of p-nitrotoluene, m.p. 51–52°, such as supplied by the Eastman Kodak Company, was used.

3. A double layer of ordinary filter paper is satisfactory for this filtration.

4. Two to three liters of benzene is required. The checkers found the use of a heated funnel to be advantageous.

3. Methods of Preparation

p,p'-Dinitrobibenzyl has been prepared by the nitration of bibenzyl;[2] by the action of alkaline zinc chloride on p-nitrobenzyl chloride;[3] by the action of alkali on p-nitrotoluene;[4] by the oxidation of α,α-bis(p-nitrobenzyl)hydrazine with mercuric oxide;[5] and by the present method.[6] The course of the oxygen absorption in the latter reaction has been followed kinetically.[7]

[1] University of Illinois, Urbana, Illinois.

[2] Rinkenbach and Aaronson, *J. Am. Chem. Soc.*, **52**, 5040 (1930); Tsekhanskiĭ, *Izvest. Vysshykh Ucheb. Zavedeniĭ, Khim. i Khim. Tekhnol.*, **1958**, No. 4, 61 [*C. A.*, **53**, 6150 (1959)].

[3] Roser, *Ann.*, **238**, 363 (1887).

[4] Green, Davies, and Horsfall, *J. Chem. Soc.*, **1907**, 2076.

[5] Busch and Weiss, *Ber.*, **33**, 2701 (1900).

[6] Fuson and House, *J. Am. Chem.-Soc.*, **75**, 1325 (1953).

[7] Tsuruta, Nagatomi, and Furukawa, *Bull. Inst. Chem. Research, Kyoto Univ.*, **30**, 46 (1952) [*C. A.*, **48**, 11369 (1954)].

1,4-DINITROBUTANE

(Butane, 1,4-dinitro-)

$$ICH_2CH_2CH_2CH_2I + 2AgNO_2 \rightarrow$$
$$O_2NCH_2CH_2CH_2CH_2NO_2 + 2AgI$$

Submitted by HENRY FEUER and GERD LESTON.[1]
Checked by JOHN C. SHEEHAN and J. IANNICELLI.

1. Procedure

The reaction is carried out in a 1-l. three-necked round-bottomed flask fitted with a ball-sealed mechanical stirrer, a reflux condenser, and a dropping funnel. The openings of the condenser and dropping funnel are protected from moisture by drying tubes. In the flask, which is protected from light (Note 1), are placed 170 g. (1.1 moles) of silver nitrite and 300 ml. of absolute ether. The silver nitrite is suspended by vigorous stirring, and the mixture is cooled to 0° by an ice bath. Then 155 g. (0.5 mole) of 1,4-diiodobutane (p. 321) is added dropwise over a period of 3 hours. The temperature is maintained at 0° for an additional 2 hours, and then the reaction mixture is allowed to come slowly to room temperature (25°) by permitting the ice in the cooling bath to melt (Note 2). Twenty-four hours after the addition of the diiodobutane has been completed, the solution is tested for unreacted iodide (Note 3). If the test is negative, the mixture is filtered and the silver iodide washed with a total of 200 ml. of benzene (Note 4). The ethereal solution and the benzene washings are combined, and the solvents are distilled on a steam bath, the pressure being reduced (water aspirator) toward the end of the distillation.

In a 500-ml. three-necked flask equipped with a mechanical stirrer, a dropping funnel, and a thermometer is placed 200 ml. of concentrated sulfuric acid. The flask is immersed in an ice-salt bath, and the acid is cooled to 0–5°. The crude dinitrobutane is added dropwise with vigorous stirring at such a rate that the temperature does not exceed 8°. Stirring is continued for an additional 10 minutes after completion of the addition. The solution is poured cautiously onto 1 kg. of crushed ice with manual stirring. The ice is allowed to melt, and the product is separated by suction filtration, washed with water, and air-dried. Recrystallization from methanol at −70°, using a Dry Ice-methylene chloride cooling bath, yields 30–34 g. (41–46%) of 1,4-dinitrobutane, m.p. 33–34° (Notes 5, 6, 7, and 8).

2. Notes

1. All light should be excluded. It is most convenient to run the reaction in a dark room.

2. The cooling bath should not be removed, or the temperature of the mixture will rise above room temperature.

3. The Beilstein test is carried out in the following manner:

A copper wire is cleaned in the flame of a Bunsen burner and allowed to cool. The stirring is stopped, and the wire is inserted carefully into the clear ether solution so as not to touch the silver iodide at the opening and at the bottom of the flask. The wire is withdrawn and held in the reducing part of the flame. A green color constitutes a positive test, and stirring is resumed until the test is negative.

4. An alternative method of purification is to wash the silver iodide with 250 ml. of methanol instead of benzene. The ether is evaporated, and the residue is combined with the methanol solution. The product is crystallized at $-70°$. Recrystallization of this crude product from methanol at $-70°$ gives the same yields as the other method of purification.

5. The methanol solution may be treated with charcoal if a colorless product is not obtained.

6. Starting with 169 g. of 1,6-diiodohexane (p. 323), 1,6-dinitrohexane may similarly be obtained in 46–48% yield, m.p. 36.5–37.5°. The alternative procedure of isolation described in Note 4 may also be used.

7. 1,3-Dinitropropane may be prepared in a similar manner starting with 148 g. of 1,3-diiodopropane (supplied by the Eastman Kodak Company and by the Eastern Chemical Corporation). However, the dinitro compound is a liquid and has to be purified in the following manner: The crude 1,3-dinitropropane is extracted from the aqueous acid layer with four 150-ml. portions of benzene. The benzene is removed by distillation at atmospheric pressure, and the residue is distilled from a 50-ml. Claisen flask, b.p. 108–110°/1 mm., n_D^{20} 1.465. The yield is 24–25 g. (36–37%). It is colorless during the distillation but rapidly turns yellow on storage. If the aqueous layer is extracted continuously with benzene or ether for 24 hours an additional 3 g. of product may be obtained.

8. 1,5-Dinitropentane may be prepared in a similar manner starting with 162 g. of 1,5-diiodopentane prepared from tetrahydropyran according to the directions for 1,4-diiodobutane (p. 321). The dinitro compound is a liquid and is obtained by extracting the aqueous acid layer with three 125-ml. portions of benzene. The benzene is removed by distillation at atmospheric pressure, and the residue is distilled from a 50-ml. Claisen flask.

The fraction, b.p. 134°/1.2 mm., amounts to 36.6 g. (45% yield), n_D^{20} 1.461. The distillate is colorless but rapidly turns yellow.

3. Methods of Preparation

1,3-Dinitropropane,[2,3] 1,4-dinitrobutane,[4,5] 1,5-dinitropentane,[4,5] and 1,6-dinitrohexane [5] have been prepared by the method described here, which is that of Victor Meyer. 1,4-Dinitrobutane also has been obtained by the hydrolysis of dipotassium α,α'-dinitroadiponitrile.[6]

[1] Purdue University, Lafayette, Indiana.
[2] Keppler and Meyer, *Ber.*, **25**, 1710 (1892).
[3] Kispersky, Hass, and Holcomb, *J. Am. Chem. Soc.*, **71**, 516 (1949).
[4] von Braun and Sobecki, *Ber.*, **44**, 2528 (1911).
[5] McElroy, Ph.D. Thesis, Purdue University, 1943.
[6] Feuer and Savides, *J. Am. Chem. Soc.*, **81**, 5826 (1959).

3,4-DINITRO-3-HEXENE

(3-Hexene, 3,4-dinitro-)

$$2CH_3CH_2CHClNO_2 + 2KOH \rightarrow CH_3CH_2\overset{\displaystyle NO_2}{\underset{\displaystyle NO_2}{C}}\!\!=\!\!CCH_2CH_3$$

Submitted by D. E. Bisgrove, J. F. Brown, Jr., and L. B. Clapp.[1]
Checked by John C. Sheehan, Richard L. Wasson, and Herbert O. House.

1. Procedure

This procedure involves the possibility of an explosion and therefore must be conducted with caution.

In a 1-l. three-necked flask equipped with a mechanical stirrer, a thermometer, and a dropping funnel, and cooled externally with an ice-salt bath, is placed a solution of 118 g. (1.8 moles) of u.s.p. 85% potassium hydroxide (Note 1) in 300 ml. of water. The temperature of the solution is maintained between 0° and 10° (with the addition of ice to the flask if necessary) while 247 g. (205 ml., 2.0 moles) of 1-chloro-1-nitropropane (Note 2) is added from a dropping funnel over a 20-minute period. The cooling bath is removed, and concentrated hydrochloric acid is added dropwise (Note 1) until the momentary green coloration produced by the addition of each drop of acid spreads rapidly throughout the solution (near a pH of 9). The temperature of the solution rises to about 70° with the separation of a deep-green oily layer. Stirring is continued until the reaction mixture reaches room temperature (about 3 hours).

After the green oil has been separated, it is washed with 75 ml. of a warm (*Caution! Note 3*) 20% solution of potassium hydroxide in water to remove, as its potassium salt, the 1,1-dinitropropane formed as a by-product. The remaining 100–110 g. of oil is diluted with 90 ml. of 95% ethanol, and the green solution is cooled in an ice-salt bath to −5° to −10°. The crystalline product is collected on a cold (Note 4) 5.5-cm. Büchner funnel and washed with two 5-ml. portions of ice-cold alcohol. The yield is 50 g. of impure crystals.

Distillation of the alcoholic filtrate (*Caution! Note 5*) under reduced pressure and in an atmosphere of nitrogen permits the isolation of an additional 8–10 g. of 3,4-dinitro-3-hexene. The blue oil boiling below 75°/20 mm. is discarded. To the undistilled residue is added 25 ml. of 95% ethanol, the resulting solution is cooled in an ice-salt bath, and the crystals are collected on a cold Büchner funnel and washed with 5 ml. of ice-cold alcohol. The combined crops of crystals are recrystallized from 80 ml. of 95% ethanol. The pure 3,4-dinitro-3-hexene separates as light-yellow needles, m.p. 31–32°, weight 50–55 g. (29–32%) (Note 6).

2. Notes

1. Although slightly less than an equivalent amount of potassium hydroxide is used, the last of the 1-chloro-1-nitropropane dissolves slowly, and the pH drops sufficiently to allow rapid reaction only after a variable (usually about 3 hours) period of standing. The reaction may also be started by heating one spot on the container with a jet of steam, but it is more convenient and reliable to initiate reaction by cautious addition of acid in the manner described.

2. The 1-chloro-1-nitropropane used was the commercial grade obtained from Commercial Solvents Corporation. The yield was unchanged when a distilled sample, b.p. 143°, was used.

3. *The temperature of the potassium hydroxide solution should not be above 35° when used.* The submitters report the isolation of 15–20 g. (9–12%) of potassium 1-nitropropylnitronate from the cooled extract. The checkers, having been advised that *this product is a hazardous explosive*, discarded the warm alkaline extract. If 1,1-dinitropropane is not desired (its preparation has been described [2]), it is recommended that it be extracted as its more soluble sodium salt by washing the green oil with sodium hydroxide solution rather than potassium hydroxide solution.

4. It is necessary to keep the Büchner funnel cold since the crystals melt near room temperature. An external cooling jacket for the funnel can be fabricated from a metal can by cutting a hole in the bottom of the can of such size that it fits high enough on a rubber stopper to allow a tight fit between the rubber stopper and the suction flask. The cooling jacket is filled with crushed Dry Ice. The checkers employed a 60-mm. sintered glass funnel surrounded by a 400-ml. beaker, in the bottom of which was a hole large enough to accommodate a No. 3 rubber stopper surrounding the stem of the funnel.

5. *Distillation behind safety glass in a nitrogen atmosphere appears advisable in view of the nature of polynitro compounds* although the submitters have not had an explosion in the preparation of 3,4-dinitro-3-hexene.

6. 2,3-Dinitro-2-butene may be prepared from 1-chloro-1-nitroethane by the same procedure, in 30% yield. The compound melts at 28–28.5° and has a boiling point of 135°/11 mm. Commercially available 1-chloro-1-nitroethane contains about 10% 1,1-dichloro-1-nitroethane and 2-chloro-2-nitropropane which cannot be separated by distillation, but these impurities do not interfere with the preparation. *Distillation of 2,3-dinitro-2-butene behind safety glass in a nitrogen atmosphere is advisable.* The submitters, in preparing this compound, have had one explosion over a period of ten years.

3. Methods of Preparation

The action of a sodium bicarbonate solution or a 10% sodium hydroxide solution on 1-chloro-1-nitropropane will produce 3,4-dinitro-3-hexene. The procedure described here is a modification of that described by Nygaard and Noland.[3]

[1] Brown University, Providence, Rhode Island. This work was supported in part by Office of Ordnance Research Contract DA-19-020-ORD-592 at Brown University.

[2] ter Meer, *Ann.*, **181**, 1 (1876); Belew, Grabiel, and Clapp, *J. Am. Chem. Soc.*, **77**, 1110 (1955).

[3] Nygaard and Noland (to Socony-Vacuum Oil Company), U. S. pat. 2,396,282 [*C. A.*, **40**, 3126 (1946)].

DIPHENYLACETALDEHYDE *

(Acetaldehyde, diphenyl-)

$$\text{C}_6\text{H}_5 \quad \text{O} \quad \text{H}$$

$$\text{C}\text{---}\text{C} \xrightarrow{\text{BF}_3 \cdot \text{O}(\text{C}_2\text{H}_5)_2} (\text{C}_6\text{H}_5)_2\text{CHCHO}$$

$$\text{H} \qquad \text{C}_6\text{H}_5$$

Submitted by DONALD J. REIF and HERBERT O. HOUSE.[1]
Checked by M. S. NEWMAN and W. H. POWELL.

1. Procedure

In a 1-l. separatory funnel is placed a solution of 39.2 g. (0.2 mole) of *trans*-stilbene oxide (Note 1) in 450 ml. of reagent benzene. To the solution is added 13.2 ml. (0.1 mole) of boron trifluoride etherate (Note 2). The solution is swirled, allowed to stand for 1 minute (Note 3), and then washed with two 300-ml. portions of water. The organic layer is separated, and the benzene is removed by distillation (Note 4). The residual crude aldehyde is purified by distillation under reduced pressure. The product, collected at 115–117°/0.6 mm., amounts to 29–32 g. (74–82%), n_D^{25} 1.5875–1.5877 (Note 5).

2. Notes

1. The *trans*-stilbene oxide (p. 860) should be free of *trans*-stilbene, since the stilbene, if present, is not altered by the reaction conditions and will contaminate the final product.

2. A practical grade of boron trifluoride etherate, purchased from Eastman Kodak Company, was redistilled before use. The pure etherate boils at 126°.

3. Longer reaction times result in a marked decrease in the yield of diphenylacetaldehyde.

4. The submitters found that distillation of the benzene solution is necessary to obtain an anhydrous product. If the benzene solution is dried over magnesium sulfate and the benzene removed under reduced pressure, the diphenylacetaldehyde is contaminated with water.

5. The product obtained by this procedure, when treated with 2,4-dinitrophenylhydrazine,[2] produced the 2,4-dinitrophenylhydrazone of diphenylacetaldehyde, m.p. 146.8–147.8°, in 94% yield.

3. Methods of Preparation

Diphenylacetaldehyde has been prepared by the isomerization of 1,2-dihydroxy-1,2-diphenylethane either thermally [3] or in the presence of sulfuric acid,[4-6] oxalic acid,[7] or acetic anhydride.[8] The aldehyde has also been produced by the reaction of 2,2-diphenyl-2-hydroxyethyl ether with sulfuric acid [6,7] or oxalic acid; [7,9] by the reaction of hydrochloric acid with 2-amino-1,1-diphenylethanol; [10] by the reaction of hydrobromic [11] or hydrochloric [12] acid with 2-diethylamino-1,1-diphenylethanol; by the hydrolysis of β,β-diphenylvinyl ethyl ether; [13] by the thermal rearrangement of deoxybenzoin; [14] by the hydrolysis and decarboxylation of the glycidic ester obtained from ethyl chloroacetate and benzophenone; [15] by passing a mixture of diphenylacetic and formic acid over thorium oxide on pumice at 450°; [16] and by the hydrolysis of diphenylacetaldehyde enol methyl ether (obtained from benzophenone and methoxymethylenetriphenylphosphorane).[17]

Diphenylacetaldehyde also has been prepared by the isomerization of trans-stilbene oxide in the presence of sodium bisulfite [18] or lithium diethylamide,[19] or by the isomerization of either cis- or trans-stilbene oxide in the presence of boron trifluoride etherate.[2] The procedure chosen illustrates the ready isomerization of substituted ethylene oxides to carbonyl compounds. The procedure is applicable to substituted epoxides in which one of the carbon atoms of the oxirane ring is bonded either to two other carbon atoms or to an aromatic nucleus or a carbon-carbon double bond.

[1] Massachusetts Institute of Technology, Cambridge, Massachusetts.
[2] House, J. Am. Chem. Soc., 77, 3070 (1955).
[3] Ramart-Lucas and Salmon-Legagneur, Compt. rend., 186, 1848 (1928).
[4] Henze and Leslie, J. Org. Chem., 15, 901 (1950).
[5] Tiffeneau, Compt. rend., 142, 1537 (1906); Ann. chim. (Paris), [8] 10, 322 (1906).
[6] Stoermer, Ber., 39, 2288 (1906).
[7] Danilov and Venus-Danilova, Ber., 59B, 1032 (1926).
[8] Tiffeneau, Compt. rend., 150, 1181 (1910).

[9] Behal and Sommelet, *Bull. soc. chim. France*, [3] **31**, 300 (1904).

[10] Thomas and Bettzieche, *Z. physiol. Chem.*, **140**, 261 (1924).

[11] Sou, *Bull. fac. sci. univ. franco-chinoise Peiping*, **1935**, No. 5, 1 [*C. A.*, **30**, 4463 (1936)].

[12] Bersch, Meyer, and Hubner, *Pharm. Zentralhalle*, **96**, 381 (1957) [*C. A.*, **52**, 13684 (1958)].

[13] Buttenberg, *Ann.*, **279**, 324 (1894).

[14] Brueur and Zincke, *Ann.*, **198**, 141 (1879).

[15] Ecary, *Ann. chim. (Paris)*, [12] **3**, 445 (1948).

[16] Haarmann & Reimer, Chemische Fabrik zu Holzminden G.m.b.H., Ger. pat. 825,085 [*C. A.*, **49**, 11713 (1955)].

[17] Wittig and Knauss, *Angew. Chem.*, **71**, 127 (1959).

[18] Klages and Kessler, *Ber.*, **39**, 1753 (1906).

[19] Cope, Trumbull, and Trumbull, *J. Am. Chem. Soc.*, **80**, 2844 (1958).

DIPHENYLACETYLENE *

(Acetylene, diphenyl-)

$$C_6H_5COCOC_6H_5 + 2NH_2NH_2 \cdot H_2O \rightarrow \underset{\underset{H_2NN}{\|} \quad \underset{NNH_2}{\|}}{C_6H_5C\text{---}CC_6H_5} + 4H_2O$$

$$\underset{\underset{H_2NN}{\|} \quad \underset{NNH_2}{\|}}{C_6H_5C\text{---}CC_6H_5} + 2HgO \rightarrow$$

$$C_6H_5C\equiv CC_6H_5 + 2Hg + 2N_2 + 2H_2O$$

Submitted by ARTHUR C. COPE, DOUGLAS S. SMITH, and ROBERT J. COTTER.[1]
Checked by CHARLES C. PRICE and THOMAS F. McKEON, JR.

1. Procedure

A solution of 105.1 g. (0.5 mole) of benzil (Note 1) in 325 ml. of *n*-propyl alcohol is prepared in a 1-l. round-bottomed flask which is fitted with an efficient reflux condenser. To this solution 76 g. (1.30 moles) of 85% hydrazine hydrate (Note 2) is added, and the mixture (Note 2) is heated under reflux for 60 hours. The solution is cooled with an ice bath, and the benzil dihydrazone is separated by suction filtration. The crystals are washed with 200 ml. of cold, absolute ethanol and dried (Note 3) on the suction filter for 1 hour. The yield of benzil dihydrazone is 99–106 g. (83–89%), m.p. 150–151.5°.

The benzil dihydrazone is added to 480 ml. of reagent grade benzene in a 1-l. three-necked flask fitted with a reflux condenser and a sealed stirrer. A small amount of yellow mercuric oxide (2–4 g.) is added to the mixture with stirring to keep the benzil dihydrazone suspended, and the mixture is warmed slightly on a steam bath. Nitrogen is evolved, and the mixture turns gray. Additional yellow mercuric oxide is then introduced in small portions so as to keep the reaction mixture gently refluxing until a total of 240 g. (1.11 moles) has been added. The mixture is stirred for 1 hour and allowed to stand overnight. It is then filtered, and the residue (mercury and mercuric oxide) is washed with 100 ml. of benzene, which is combined with the original red benzene filtrate. After drying over anhydrous sodium sulfate, the benzene is removed by distillation under reduced pressure by heating with a water bath. The residue is distilled from a flask connected to a short distillation head at 95–105°/0.2–0.3 mm. and yields 60–65 g. (67–73% from benzil) of diphenylacetylene, m.p. 59–60°. The product can be recrystallized from 100 ml. of 95% ethanol, m.p. 60–61° (Note 4).

2. Notes

1. Eastman Kodak Company white label grade benzil or material prepared by the procedure described in *Organic Syntheses* [2] is satisfactory.

2. Hydrazine hydrate (85%) as supplied by the Edwal Laboratories was used. On addition of this reagent to the benzil solution, the monohydrazone of benzil precipitates, but it redissolves readily on heating.

3. Benzil dihydrazone should not be dried in a vacuum desiccator, for it sublimes easily.

4. Diphenylacetylene prepared by this method has the advantage of being uncontaminated with stilbene, with which it forms a solid solution not readily separable.[3] Di-*p*-tolylacetylene[4] and α-naphthylphenylacetylene[5] have been prepared by the same method in high yields.

3. Methods of Preparation

In addition to the methods mentioned in a previous preparation,[6] diphenylacetylene has been prepared by the action of potassium hydroxide on 5,5-diphenyl-3-nitroso-2-oxazolidone (100%),[7] on *meso*-stilbene dibromide in triethylene glycol,[8] and on stilbene tetrabromide in butyl alcohol; [9] by the condensation of bromobenzene with sodium phenylacetylide in liquid ammonia,[10] by the oxidation of benzil dihydrazone with silver trifluoroacetate in triethylamine,[11] and by the pyrolysis of α-benzoylbenzylidenetriphenylphosphorane.[12] The present procedure is a modification of that of Schlenk and Bergmann.[13]

[1] Massachusetts Institute of Technology, Cambridge, Massachusetts.
[2] *Org. Syntheses Coll. Vol.* 1, 87 (1941).
[3] Pascal and Normand, *Bull. soc. chim. France*, [4] 13, 151 (1913).
[4] Kastner and Curtius, *J. prakt. Chem.*, [2] 83, 225 (1911).
[5] Ruggli and Reinert, *Helv. Chim. Acta*, 9, 67 (1926).
[6] *Org. Syntheses Coll. Vol.* 3, 350 (1955).
[7] Newman and Kutner, *J. Am. Chem. Soc.*, 73, 4199 (1951).
[8] Fieser, *J. Chem. Educ.*, 31, 291 (1954).
[9] Drefahl and Plötner, *Chem. Ber.*, 91, 1280 (1958).
[10] Scardiglia and Roberts, *Tetrahedron*, 3, 197 (1958).
[11] Newman and Reid, *J. Org. Chem.*, 23, 665 (1958).
[12] Trippett and Walker, *J. Chem. Soc.*, 1959, 3874.
[13] Schlenk and Bergmann, *Ann.*, 463, 76 (1928).

1,4-DIPHENYL-5-AMINO-1,2,3-TRIAZOLE

(1H-1,2,3-Triazole, 5-amino-1,4-diphenyl-)

AND

4-PHENYL-5-ANILINO-1,2,3-TRIAZOLE

[1H-1,2,3-Triazole, 4-phenyl-5-(phenylamino)-]

$$
C_6H_5N_3 + C_6H_5CH_2CN \xrightarrow{\text{NaOCH}_3}
$$

$$
\xrightarrow[\text{Heat}]{C_5H_5N}
$$

Submitted by EUGENE LIEBER, TAI SIANG CHAO, and C. N. RAMACHANDRA RAO.[1]
Checked by T. L. CAIRNS and E. L. MARTIN.

1. Procedure

A. *1,4-Diphenyl-5-amino-1,2,3-triazole.* A 500-ml. three-necked flask is equipped with a sealed stirrer, a thermometer well, and a dropping funnel which is protected by a drying tube and has a pressure-equalizing side arm. A mixture of 35.7 g. (0.3 mole) of phenyl azide (Note 1) and 38.6 g. (0.33 mole) of phenylacetonitrile (Note 2) is placed in the flask. The flask is immersed in an ice-water mixture contained in a 1-gal. Thermos flask. After the reaction mixture has cooled to about 2°, a solution of 24.3 g. (0.45 mole) of sodium methoxide (Note 3) in 150 ml. of absolute ethanol is added dropwise during the course of 2 hours. The reaction mixture is then stirred at 2–5° in the ice-water bath for a period of 48 hours (Note 4). After the cooling bath has been removed and the flask allowed to warm spontaneously to room temperature, the mixture is filtered by suction on a sintered glass funnel, and the collected product is washed with three 50-ml. portions of absolute ethanol.

The dried product weighs 62–65 g. (88–92%) and consists of white, fine platelike crystals, m.p. 169–171°. Recrystallization from benzene does not alter the melting point (Note 5).

B. *4-Phenyl-5-anilino-1,2,3-triazole.* Six grams (0.025 mole) of 1,4-diphenyl-5-amino-1,2,3-triazole is dissolved in 20 g. of dry pyridine (distilled from solid sodium hydroxide) and heated under reflux for 24 hours (Note 6). The reaction mixture (Note 7) is poured into 1 l. of ice water. The product separates as a slightly yellowish milky oil which is converted to white needle-like crystals by stirring the mixture and scratching the beaker with a glass rod. The product is collected by suction filtration, washed with water, suction-dried, and recrystallized from aqueous ethanol (Note 8). The yield is 5.5–5.6 g. (92–93%) of fine white needle-like crystals, m.p. 167–169° (Note 9), soluble in hot water and ether, but difficultly soluble in benzene.

2. Notes

1. Prepared by the method of Lindsay and Allen, *Org. Syntheses,* Coll. Vol. **3**, 710 (1955). The phenyl azide used had b.p. 41–43°/5 mm., $n_D^{25.5}$ 1.5567. The boiling point deviates slightly from that given by Lindsay and Allen, namely 49–50°/5 mm.; however, it agrees fairly well with the other value given by these authors, namely 66–68°/21 mm., and with the values of Darapsky,[2] and the vapor pressure determinations by Carothers,[3] as shown by a plot of log p vs. $1/T$.

2. The Eastman product was used without purification.

3. Anhydrous sodium methoxide from Matheson Chemical Corporation was used.

4. The yield and purity of the product, i.e., with respect to decreased content of acidic isomer (4-phenyl-5-anilino-1,2,3-triazole), depends upon maintaining a low temperature throughout the entire reaction.

5. The product is essentially pure. 1,4-Disubstituted-5-amino-1,2,3-triazoles are readily isomerized;[4] accordingly, care must be exercised in the recrystallization of such products from solvents. It has been found by experiment that the best practice, in order to avoid isomerization, is to heat the benzene to boiling before addition of the product for recrystallization. Repeated tests have shown that a single careless recrystallization of the product from benzene can increase the content of acidic isomer by as much as 4%. Polar solvents must be avoided. The purity of the product can be determined [5] by titration in glacial

acetic acid, using perchloric acid (in glacial acetic acid) as titrant and methyl violet (0.2 g. of methyl violet in 100 ml. of chlorobenzene) as visual indicator (the first appearance of blue color is taken as the end point).

6. This is more than enough time to allow for the complete irreversible isomerization. The extent of isomerization is checked by removing a small quantity of the reaction mixture, isolating the product by dilution with water, and testing its solubility in dilute potassium hydroxide solution. It should be completely soluble.

7. If any solid material is present, it should be removed by filtration.

8. The acidic isomer can be recrystallized from ethanol without the formation of any of the basic isomer. The checkers used for each gram of product 5 ml. of ethanol and 2.5 ml. of water. After solution of the product, treatment with Darco and filtration, 2.5 ml. of water was added to the hot solution.

9. The purity of the acidic isomer is best determined [5] by titration against sodium methoxide in dimethylformamide, using potentiometric indicator. This test showed the present product to be free of basic isomer.

3. Methods of Preparation

The present method is a modification of that first reported by Dimroth.[6] The method of irreversible isomerization in boiling pyridine was first reported by Dimroth [7] for the conversion of 1-phenyl-5-amino-1,2,3-triazole to 5-anilino-1,2,3-triazole.

[1] De Paul University, Chicago, Illinois.
[2] Darapsky, *Ber.*, **40**, 3038 (1907).
[3] Carothers, *J. Am. Chem. Soc.*, **45**, 1734 (1923).
[4] Lieber, Chao, and Rao, *J. Org. Chem.*, **22**, 654 (1957).
[5] Lieber, Rao, and Chao, *Anal. Chem.*, **29**, 932 (1957).
[6] Dimroth and Werner, *Ber.*, **35**, 4058 (1902).
[7] Dimroth, *Ann.*, **364**, 183 (1909).

N,N'-DIPHENYLBENZAMIDINE

(Benzamidine, N,N'-diphenyl-)

$$C_6H_5CONHC_6H_5 \xrightarrow{PCl_5} C_6H_5\overset{\overset{\displaystyle Cl}{|}}{C}\!\!=\!\!NC_6H_5 \xrightarrow{C_6H_5NH_2}$$

$$C_6H_5\overset{\overset{\displaystyle NHC_6H_5}{|}}{C}\!\!=\!\!NC_6H_5 + HCl$$

Submitted by ARTHUR C. HONTZ and E. C. WAGNER.[1]
Checked by R. S. SCHREIBER and WM. BRADLEY REID, JR.

1. Procedure

In a 1-l. three-necked round-bottomed flask (Note 1) are placed 90.0 g. (0.456 mole) of benzanilide [2] (Note 2) previously dried in an oven at 120° (Note 3) and 95 g. (0.456 mole) (Note 4) of phosphorus pentachloride. The solids are mixed by shaking, and lumps are reduced by manipulation with a rod. A short reflux condenser and a small dropping funnel are attached.

The mixture is heated in an electric mantle or oil bath at 110° for 30 minutes and heated under reflux at 160° for 90 minutes or until active evolution of hydrogen chloride ceases (Note 5). To the mixture are added slowly through a dropping funnel first 36.4 g. (0.46 mole) of pyridine (Note 6) previously dried over pellet-form potassium hydroxide, and then 42.4 g. (0.456 mole) of freshly distilled aniline. The contents of the flask are mixed by swirling. The mixture is heated at 160° for about 20 minutes or until the red color is discharged, at which point the flask is removed from the source of heat. The mixture is cooled to about 90°, and 250 ml. of water is added slowly through the dropping funnel with agitation to ensure separation of the solid product in granular form. After the mixture has cooled to room temperature, the solid is collected on a Büchner funnel and air-dried.

The crude diphenylbenzamidine hydrochloride is transferred to a 1-l. beaker and treated with 500 ml. of 28% ammonia water (hood). The mixture is stirred mechanically and warmed very gently for an hour. The diphenylbenzamidine is collected on a

Büchner funnel and air-dried. The melting point of the product at this stage is 130–136°.

To purify the product it is recrystallized from 80% ethanol (Note 7), using 8–10 ml. per gram of diphenylbenzamidine. The small insoluble residue of unconverted hydrochloride which may remain (Note 8) is removed by filtering the hot solution. The solution is chilled in an ice-salt bath, and the crystalline product is collected on a Büchner funnel, pressed dry, and finally dried in the air or in an oven at 100°. The yield is 91–100 g. (73–80%) (Note 9) of product having a melting point of 142–144° (Notes 10 and 11). Recrystallization of 100 g. of this material from 800 ml. of 80% ethanol gives 87 g. (87% recovery) of pure N,N'-diphenylbenzamidine, m.p. 144–145°.

2. Notes

1. An all-glass apparatus is desirable as cork connections are attacked during formation of the imido chloride. The checkers used mechanical stirring.

2. The benzanilide should be of good quality, or tarry material will form and interfere with purification of the product. Benzanilide is readily made by the Schotten-Baumann procedure from aniline, 10% aqueous sodium hydroxide, and benzoyl chloride in the proportions 6:30:5, and after crystallization from 95% ethanol is sufficiently pure.

3. The use of benzanilide dried by fusion [3] did not improve the yield.

4. The use of one-fourth an equivalent of phosphorus pentachloride, in an attempt to utilize fully the dehydroxylating capacity of both phosphorus pentachloride and the derived phosphorus oxychloride, led to a low yield of diphenylbenzamidine (50%) and to formation of tarry material that interfered with purification. By use of phosphorus oxychloride [4] instead of the pentachloride, a temperature of 170° was required to keep the mixture liquid, tarry material was considerable, and the operation was generally unsatisfactory.

5. Isolation of the imido chloride by distillation under reduced pressure offers no advantage.

6. Pyridine serves to make all the aniline available. It does not prevent combination of hydrogen chloride with the N,N'-diphenylbenzamidine even when two equivalents of pyridine are used. In the presence of pyridine the reaction mixture is a suspension and is easily handled. In the absence of pyridine it solidifies to a cake, which must be pulverized to permit liberation of the amidine from the hydrochloride.

7. A series of tests showed ethanol of 78–82% by weight (sp. gr. 0.8344–0.8442 at 25.5°; 84–87% by volume) to be the best solvent. To dissolve 1.0 g. of N,N'-diphenylbenzamidine at the boiling point required 15.5 ml. of 94% (by weight) ethanol, 7.24 ml. of 80% ethanol, and 11 ml. of 71% ethanol, and the recoveries on chilling were respectively 45%, 73%, and 45%; ten intermediate concentrations of ethanol gave results consistent with these. The solubility of N,N'-diphenylbenzamidine in methanol is greater than in ethanol, but recovery is relatively low.

8. The undissolved residue is unchanged hydrochloride and is usually small. If considerable it may be re-treated with ammonia and the free base recovered.

9. By chilling to 20° the yield is about 68% of pure material that melts at 144–145°. A second crop of less pure material can be obtained from the mother liquor by concentrating and chilling. By chilling in a Dry Ice-ethanol bath the yield is about 80% but the product is less pure. Admixed tarry material may be removed in large part by extraction with cold ether, in which the tar dissolves readily.

10. The checkers used a Fisher-Johns block.

11. This method is capable of extension to the preparation of other N,N'-disubstituted amidines.[5] In some preparations it may be advantageous to remove phosphorus oxychloride by distillation under reduced pressure before addition of the amine. The method is not wholly satisfactory for preparation of N,N'-diarylformamidines, which are better made by the orthoformic ester method.[6] During preparation of diphenylacetamidine considerable gluey material formed by decomposition of the intermediate N-phenylacetimidochloride [7] is an impediment to the isolation of the product.[8] This amidine is better prepared by the method of Sen and Ray.[9]

3. Methods of Preparation

N,N'-Diphenylbenzamidine and closely related amidines have been made by several procedures which involve interaction of amines with N-substituted imido chlorides either preformed [10] or formed *in situ* from an acylamine by action of phosphorus trichloride,[11] phosphorus oxychloride,[4] or phosphorus pentachloride.[5] Diphenylbenzamidine is formed from aniline and benzanilidochloroiodide,[12] from aniline and phenyl benzimido ether,[13] and from aniline hydrochloride and N-phenylbenzamidine.[14] It is obtained also from carbanilide and benzoyl chloride,[15] from carbodiphenylimide and phenylmagnesium bromide,[16] from aniline hydrochloride and benzonitrile at 220–240°,[17] and from benzanilide and phenyl isocyanate.[18] Amidines, including diphenylbenzamidine, are obtainable from nitriles by heating with ammonium or amine salts of sulfonic acids,[19] by heating benzotrichloride with amines,[20] and from Schiff bases by action of *tert*-amyl hypochlorite.[21] Good yields are claimed for a patented process using an acylamine, benzenesulfonyl chloride, and amine in the presence of pyridine.[22] The method described is based on the procedures of Wallach [3] and Hill and Cox.[5]

[1] University of Pennsylvania, Philadelphia, Pennsylvania.
[2] *Org. Syntheses Coll. Vol.* **1**, 82 (1941).
[3] Wallach, *Ann.*, **184**, 79 (1877); Holljes and Wagner, *J. Org. Chem.*, **9**, 43 (1944).
[4] Sidiki and Shah, *J. Univ. Bombay*, **6**, II, 132 (1937).
[5] Hill and Cox, *J. Am. Chem. Soc.*, **48**, 3214 (1926).
[6] von Walther, *J. prakt. Chem.*, **53**, 473 (1896).
[7] von Braun, Jostes, and Heymons, *Ber.*, **60**, 93 (1927).
[8] W. F. Tomlinson, University of Pennsylvania, 1947, unpublished results.
[9] Sen and Rây, *J. Chem. Soc.*, **1926**, 646.
[10] Gerhardt, *Ann.*, **108**, 219 (1858).
[11] Hofmann, *Z. Chem.*, **1866**, 165.
[12] Lander and Laws, *J. Chem. Soc.*, **85**, 1696 (1904).
[13] Lossen and Kobbert, *Ann.*, **265**, 155 (1891).
[14] Bernthsen, *Ann.*, **184**, 355 (1877).
[15] Dains, *J. Am. Chem. Soc.*, **22**, 190 (1900).
[16] Busch and Hobein, *Ber.*, **40**, 4297 (1907).
[17] Bernthsen, *Ann.*, **184**, 349 (1877).
[18] Wiley, *J. Am. Chem. Soc.*, **71**, 3746 (1949).
[19] Oxley and Short, *J. Chem. Soc.*, **1946**, 147; U. S. pat. 2,433,489 [*C. A.*, **42**, 3780 (1948)].
[20] Joshi, Khanolkar, and Wheeler, *J. Chem. Soc.*, **1936**, 793.
[21] Fusco and Musante, *Gazz. chim. ital.*, **66**, 258 (1936).
[22] Brit. pat. 577,478 [*C. A.*, **42**, 7321 (1948)].

α,β-DIPHENYLCINNAMONITRILE

(Acrylonitrile, 2,3,3-triphenyl-)

$$C_6H_5CH_2CN + NaNH_2 \rightarrow [C_6H_5CHCN]^-Na^+ + NH_3$$

$$[C_6H_5CHCN]^-Na^+ + (C_6H_5)_2C{=}O \rightarrow$$

$$(C_6H_5)_2C{=}C(C_6H_5)CN + NaOH$$

Submitted by STANLEY WAWZONEK and EDWIN M. SMOLIN.[1]
Checked by ARTHUR C. COPE and MALCOLM CHAMBERLAIN.

1. Procedure

Caution! This preparation should be conducted in a hood to avoid exposure to ammonia.

A solution of sodium amide in liquid ammonia is prepared according to a procedure previously described (Note 1) in a 2-l. three-necked round-bottomed flask fitted with a reflux condenser attached to a soda-lime tower which is connected to a gas absorption trap,[2] a mercury-sealed mechanical stirrer, and an inlet tube. Anhydrous liquid ammonia (350 ml.) is introduced through the inlet tube, and about 0.3 g. of hydrated ferric nitrate and 13.7 g. (0.59 g. atom) of freshly cut sodium are added (Notes 1 and 2).

After the conversion of sodium to sodium amide is complete, the inlet tube is replaced with a 500-ml. dropping funnel and the flask is cooled in a bath of Dry Ice and trichloroethylene. Benzyl cyanide (69 g., 0.59 mole) (Note 3) is added dropwise with stirring during about 20 minutes. The cooling bath is removed, and the solution is stirred for an additional period of 20 minutes, after which 700 ml. of dry ether is added slowly through the dropping funnel. The solution is allowed to stand or is warmed gently with a water bath until it comes to room temperature. The rate of addition of ether and subsequent warming are controlled so that the ammonia which is vaporized passes through the gas absorption trap rather than escaping in part through the mercury seal of the stirrer. When most of the ammonia has been removed, an additional 300-ml. portion of dry ether is added

(Note 4). The flask is then heated on a hot water bath. By turning off the cooling water of the reflux condenser for a short time, a small amount of ether is allowed to distil out of the reaction mixture in order to remove as much of the ammonia as possible (Note 5). The condenser water is then turned on again, and dry nitrogen lines under a positive pressure of 2–3 cm. of mercury (maintained by a T-tube dipping into mercury) are attached to the top of the dropping funnel and the reflux condenser. A solution of 200 g. (1.1 moles) of benzophenone (Notes 6 and 7) in 300 ml. of dry ether (Note 8) is added through the dropping funnel during a period of 15 minutes, and the mixture is heated under reflux with stirring for 24 hours. At the end of this period the solution contains a reddish brown precipitate. The mixture is allowed to cool, and 250 ml. of water is added slowly through the dropping funnel. The aqueous and ether layers are separated in a 2-l. separatory funnel, and the water layer is filtered directly through a 15-cm. Büchner funnel. The aqueous filtrate is discarded. The ether layer also is run into the funnel, and the filtrate is concentrated to a volume of about 80 ml. The solid that separates is added to the solid previously collected on the filter. The crude reddish brown product is recrystallized from 600 ml. of glacial acetic acid. α,β-Diphenylcinnamonitrile separates as white crystals which are collected on a suction filter, washed with 100 ml. of water, and dried at 110° for 24 hours. The yield of a pure product melting at 166–167° is 83–110 g. (50–66%).

2. Notes

1. The procedures for preparing sodium amide and the sodium derivative of benzyl cyanide are described in *Organic Syntheses*.[3] Another procedure for preparing sodium amide (p. 763) is convenient and should be equally satisfactory.

2. If vaporization of ammonia reduces the liquid volume to less than 250 ml. before the conversion of sodium to sodium amide is complete, more ammonia should be added through the inlet tube.

3. The benzyl cyanide used should be washed with warm 50% sulfuric acid to remove benzyl isocyanide [4] and redistilled.

4. Some of the ether is lost through the reflux condenser during vaporization of the ammonia.

5. Traces of ammonia remain in the reaction mixture after this procedure.

6. A smaller excess of benzophenone results in a lowered yield.

7. The benzophenone should be dried by redistillation or storage over phosphorus pentoxide in a vacuum desiccator.

8. The mixture of benzophenone and ether should be warmed gently until the solid dissolves.

3. Methods of Preparation

α,β-Diphenylcinnamonitrile has been prepared by the condensation of benzophenone with the sodium [5] or lithium [6] derivative of benzyl cyanide, or from benzophenone, benzyl cyanide, and sodium ethoxide.[7] It has been obtained also by heating benzyl cyanide with benzophenone dichloride at 200°.[8]

[1] State University of Iowa, Iowa City, Iowa.

[2] Org. Syntheses Coll. Vol. 2, 4 (1943).

[3] Org. Syntheses Coll. Vol. 3, 219 (1955).

[4] Org. Syntheses Coll. Vol. 1, 108 (1941).

[5] Bodroux, Bull. soc. chim. France, [4] 9, 758 (1911).

[6] Ivanov and Vasilev, Compt. rend. acad. bulgare sci., 10, No. 1, 53 (1957) [C. A., 52, 5353 (1958)].

[7] Stobbe and Zeitschel, Ber., 34, 1967 (1901).

[8] Heyl and Meyer, Ber., 28, 1798, 2785 (1895).

DIPHENYL SUCCINATE *

(Succinic acid, diphenyl ester)

$$\begin{matrix} CH_2CO_2H \\ | \\ CH_2CO_2H \end{matrix} + 2C_6H_5OH + POCl_3 \rightarrow$$

$$\begin{matrix} CH_2CO_2C_6H_5 \\ | \\ CH_2CO_2C_6H_5 \end{matrix} + HPO_3 + 3HCl$$

Submitted by Guido H. Daub and William S. Johnson.[1]
Checked by James Cason, Robert A. Wessman, and William G. Dauben.

1. Procedure

A mixture of 118 g. (1 mole) of succinic acid, 188 g. (2 moles) of phenol, and 138 g. (83 ml., 0.9 mole) of phosphorus oxychloride (Note 1) is placed in a 2-l. round-bottomed flask fitted with an efficient reflux condenser capped with a calcium chloride tube (Notes 2 and 3). The mixture is heated on a steam bath in a hood (Note 3) for 1.25 hours, 500 ml. of benzene is added, and the refluxing is continued for an additional hour. The hot benzene solution is decanted from the red syrupy residue of phosphoric acid and filtered by gravity into a 1-l. Erlenmeyer flask. The syrupy residue is extracted with two 100-ml. portions of hot benzene, which are also filtered into the Erlenmeyer flask. The combined benzene solutions are concentrated to a volume of about 600 ml. (Note 4), and the pale yellow solution is allowed to cool, whereupon the diphenyl succinate separates as colorless crystals. It is filtered with suction on a Büchner funnel, washed with three 50-ml. portions of ether, and dried on a porous plate at 40°. The yield of diphenyl succinate, m.p. 120–121°, is 167–181 g. (62–67%) (Note 5).

2. Notes

1. The use of a larger proportion of phosphorus oxychloride failed to improve the yield and in general gave an inferior product.

2. Ground-glass joints are preferred; however, burnt-cork stoppers may be used.

3. Provision should be made to dispose of the hydrogen chloride which is evolved during the reaction.

4. If the mother liquor is concentrated further before removal of phenol and phosphorus oxychloride a dark, impure product is obtained.

5. If desired, a second crop may be obtained by the following procedure: The mother liquor is extracted in a 1-l. separatory funnel with eight 50-ml. portions of 5% potassium hydroxide solution. Each extract is extracted in turn with a 50-ml. portion of benzene. The two benzene solutions are washed in turn with a 50-ml. portion of water and two 50-ml. portions of saturated sodium chloride solution. All the benzene solutions are combined, dried over anhydrous sodium sulfate, filtered, and concentrated to a volume of 150 ml. Cooling yields a second crop of diphenyl succinate, which is collected, washed, and dried as described for the first crop. The yield is 13–27 g. (5–10%), m.p. 118–120°.

3. Methods of Preparation

The above procedure is a modification of that described by Rasinski.[2] Diphenyl succinate has also been prepared by the reaction of phenol and succinic acid in the presence of phosphorus pentoxide in toluene [3] and by the treatment of phenol with succinyl chloride.[4]

[1] University of Wisconsin, Madison, Wisconsin.

[2] Rasinski, *J. prakt. Chem.*, **26**, 63 (1882). See also Bischoff and Hedenstrom, *Ber.*, **35**, 4073 (1902).

[3] Bakunin, *Gazz. chim. ital.*, **30**, 358 (1900).

[4] Weselsky, *Ber.*, **2**, 519 (1869).

2,3-DIPHENYLSUCCINONITRILE

(Succinonitrile, 2,3-diphenyl-)

I. METHOD A

$$C_6H_5CHO + C_6H_5CH_2CN + NaCN \rightarrow$$

$$C_6H_5 - \underset{\underset{CN}{|}}{CH} - \underset{\underset{CN}{|}}{CH} - C_6H_5 + NaOH$$

Submitted by R. B. Davis and J. A. Ward, Jr.[1]
Checked by B. C. McKusick and R. L. Morgan.

1. Procedure

Caution! These operations should be conducted in a good hood.

A 1-l. three-necked flask is fitted with a stirrer, a condenser, a dropping funnel, and a heating mantle. Sixty-one grams (1.25 moles) of sodium cyanide and 100 ml. of distilled water are added, and the mixture is warmed with stirring for about 15 minutes while the sodium cyanide dissolves. Then 400 ml. of methanol (Note 1) is introduced, and the mixture is rapidly heated to gentle reflux. Fifty grams (0.42 mole) of benzyl cyanide is added all at once, followed by dropwise addition of a solution of 53 g. (0.50 mole) of benzaldehyde and 30 g. (0.26 mole) of benzyl cyanide (Note 2). The addition requires about 40 minutes, during which time the reaction mixture turns from yellow to green and crystalline 2,3-diphenylsuccinonitrile begins to separate (Note 3). The reaction mixture is stirred at gentle reflux for 30 minutes after the addition is complete. The reaction mixture, now blue in color, becomes brown as it is cooled to room temperature.

The nitrile is separated from the cooled reaction mixture by suction filtration, washed successively with 150-ml. portions of 75% aqueous methanol, water, 75% aqueous methanol, and ether, and dried in air overnight. It is obtained as a colorless solid that weighs 84–89 g. (72–77%) and melts over a range of 1–2° between 202° and 206° (Note 4). It is the *meso*-isomer of

2,3-diphenylsuccinonitrile mixed with a small amount of the *dl*-isomer.   Recrystallization from glacial acetic acid (about 27 ml. per g.), followed by washing with 150-ml. portions of water, 75% aqueous methanol, and ether, yields 79–84 g. (68–72%) of the pure *meso*-isomer, m.p. 238–239° (Note 5).

2. Notes

1. Reagent grade methanol, very low in acetone content, is used; that available from Rascher and Betzold or Baker is satisfactory.

2. The benzyl cyanide, chlorine-free benzaldehyde, and reagent grade granular sodium cyanide obtainable from Matheson, Coleman and Bell, are satisfactory.

3. Initial precipitation, occurring during the course of the addition, is exothermic.   Hence gentle reflux is advised, and the operator should be prepared to remove the heating mantle for a few minutes should the refluxing become vigorous.   Early seeding, if seed is available, is helpful in this regard.

4. Prepared mixtures of the isomeric nitriles have been reported to melt over a range of 1–2° between 206° and 221°.[2]

5. The checkers used a calibrated thermometer and observed a melting point of 229–230° for an analytically pure sample which was heated in an open capillary in an oil bath.

II. METHOD B

$$C_6H_5-CH{=}\overset{\displaystyle |}{\underset{\displaystyle CN}{C}}-C_6H_5 + HCN \rightarrow C_6H_5-\overset{\displaystyle |}{\underset{\displaystyle CN}{CH}}-\overset{\displaystyle |}{\underset{\displaystyle CN}{CH}}-C_6H_5$$

Submitted by J. A. McRae and R. A. B. Bannard.[3]
Checked by R. T. Arnold and Stuart W. Fenton.

1. Procedure

Caution!   This preparation should be conducted in a good hood.

Two hundred and five grams (1 mole) of α-phenylcinnamonitrile,[4] 2250 ml. of methanol, and 750 ml. of ether are placed in a 5-l. round-bottomed flask fitted with a two-necked adapter sup-

porting a 1-l. separatory funnel and reflux condenser. The
α-phenylcinnamonitrile is dissolved by gentle heating, and the
solution is heated under reflux. A solution of 274 g. (4 moles)
of 95% potassium cyanide in 600 ml. of water (which has been
preheated to 45°) is added rapidly from the separatory funnel
(Note 1). A small amount of potassium cyanide precipitates.
The solution is heated under reflux for 1 hour, after which a
solution of 154 g. (2 moles) of ammonium acetate in 250 ml. of
water is added rapidly from the separatory funnel (Note 2).
Heating is discontinued, and the mixture is allowed to stand for
24 hours. The slightly orange-yellow crystalline product is
collected on a Büchner funnel and washed, first with two 300-ml.
portions of water at 60° to remove inorganic salts, then with
800 ml. of 70% methanol for removal of water and some unreacted α-phenylcinnamonitrile, and finally with two 300-ml.
portions of ether for further drying. The colorless needles thus
obtained consist of a mixture of the stereoisomeric α,α'-diphenyl-
succinonitriles and, after drying at room temperature for 6 hours,
weigh 213–218 g. (92–94%). This mixture of stereoisomers
melts over a 1–2° range in the region of 202–206° (Note 3).

The mixed dinitrile can be recrystallized from glacial acetic
acid (using 27 ml. per g.) to give pure *meso*-α,α'-diphenylsuc-
cinonitrile. Residual acetic acid is removed readily by washing
the crystalline product with small portions of water, aqueous
methanol (70%), and ether; yield 95–98% (based on mixed
dinitrile); m.p. 240–241.5° (cor.).

2. Notes

1. Care must be exercised to avoid superheating of the solution before addition of the cyanide; otherwise ether and α-phenylcinnamonitrile will be lost through the condenser. Superheating
can be avoided by adding a boiling chip periodically as the
solution approaches its boiling point. No difficulty with superheating occurs after the cyanide has been added.

2. Addition of the ammonium acetate solution may be omitted;
the yield is then 203–208 g. (87.5–89.6%).

3. The mixture of dinitriles has been reported to melt over a
1–2° range in the region of 206–221°.[2] Analysis of a sample of

the mixture of dinitriles produced by the present method showed that no impurities are present.

3. Methods of Preparation

A mixture of the stereoisomeric 2,3-diphenylsuccinonitriles has been obtained by the interaction of benzyl cyanide, benzal chloride, and potassium cyanide in aqueous ethanolic solutions;[2] by warming α-phenylcinnamonitrile with excess potassium cyanide in aqueous ethanolic solution with,[5] or without,[6, 7] subsequent addition of dilute acetic acid; by heating under reflux an aqueous ethanolic solution of α-phenylcinnamonitrile and excess potassium cyanide with subsequent addition of saturated ammonium or magnesium chloride solution;[8] by the action of sunlight on a mixture of benzyl cyanide and benzophenone;[9] and by treatment of benzyl cyanide with sodium methylate and ethereal iodine solution.[10]

In addition, 2,3-diphenylsuccinonitriles have been prepared by treating α-bromo- or α-chlorobenzyl cyanide with sodium iodide in acetone solution,[11] and by the condensation of benzyl cyanide, benzaldehyde, and sodium cyanide.[12] The mixture of stereoisomeric 2,3-diphenylsuccinonitriles has been separated into the *meso* compound and the racemic form.[13]

Method I is more practical than Method II, which had been developed earlier, since it affords 2,3-diphenylsuccinonitrile in a single procedure and in good yields from readily available materials.

[1] University of Notre Dame, Notre Dame, Indiana.
[2] Chalanay and Knoevenagel, *Ber.*, **25**, 289 (1892).
[3] Queens University, Kingston, Ontario, Canada.
[4] *Org. Syntheses Coll. Vol.* **3**, 715 (1955).
[5] Lapworth and McRae, *J. Chem. Soc.*, **121**, 1699 (1922).
[6] Lapworth, *J. Chem. Soc.*, **83**, 995 (1903).
[7] Knoevenagel and Schleussner, *Ber.*, **37**, 4067 (1904).
[8] I. G. Farbenind., A.-G., Ger. pat. 427,416 [*Chem. Zentr.*, [2] **97**, 1100 (1926)].
[9] Paterno, *Gazz. chim. ital.*, **44**, 237 (1914).
[10] Heller, *J. prakt. Chem.*, **120**, 193 (1929).
[11] Tronov and Aksenenko, *Zhur. Obshchei Khim.*, **26**, 1393 (1956) [*C. A.*, **50**, 14636 (1956)].
[12] Davis, *J. Am. Chem. Soc.*, **80**, 1752 (1958); U. S. pat. 2,851,477 [*C. A.*, **53**, 3153 (1959)]; University of Notre Dame, Brit. pat. 824,605 [*C. A.*, **54**, 7657 (1960)].
[13] Coe, Gale, Linstead, and Timmons, *J. Chem. Soc.*, **1957**, 123.

p-DITHIANE *

$$\begin{array}{c} CH_2SH \\ | \\ CH_2SH \end{array} + 2C_2H_5ONa \rightarrow \begin{array}{c} CH_2SNa \\ | \\ CH_2SNa \end{array} + 2C_2H_5OH$$

$$\begin{array}{c} CH_2SNa \\ | \\ CH_2SNa \end{array} + \begin{array}{c} CH_2Br \\ | \\ CH_2Br \end{array} \rightarrow \begin{array}{c} S \\ / \quad \backslash \\ CH_2 \quad CH_2 \\ | \qquad | \\ CH_2 \quad CH_2 \\ \backslash \quad / \\ S \end{array} + 2NaBr$$

Submitted by RICHARD G. GILLIS and A. B. LACEY.[1]
Checked by B. C. McKUSICK and R. J. HARDER.

1. Procedure

In a 3-l. round-bottomed flask fitted with a mechanical stirrer and a reflux condenser is placed 2.0 l. of anhydrous ethanol. To this is added 11.5 g. (0.5 g. atom) of sodium cut into small pieces. When the sodium is completely dissolved, 23.6 g. (21.0 ml., 0.25 mole) of 1,2-ethanedithiol (p. 401) is added, followed by 47.0 g. (21.7 ml., 0.25 mole) of ethylene dibromide. The mixture is stirred and heated under reflux for 4 hours, cooled, and filtered to remove some sodium bromide mixed with polyethylene sulfide. The solid is washed with 100 ml. of ethanol, and the combined filtrates are distilled with stirring. When bumping becomes troublesome, as it generally does when 1.3–1.5 l. of distillate has been collected, the hot reaction mixture is filtered to remove sodium bromide, and the sodium bromide is washed with 100 ml. of hot ethanol.

The combined filtrates are returned to the reaction vessel, and distillation with stirring is continued until virtually all the ethanol has been removed. The distillation is stopped when crystals of p-dithiane appear in the condenser or when dilution of the distillate with water causes a milky appearance or the formation of a small quantity of crystals. One liter of water is added to the residue, and the stirred mixture is distilled, using the apparatus of Fig. 8 (Note 1), until no more p-dithiane solidifies in the condenser.

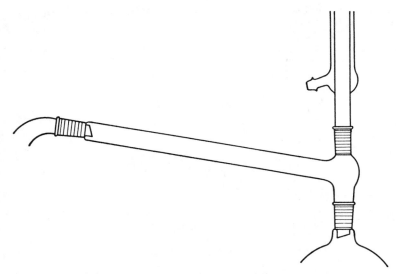

FIG. 8. Apparatus for steam distillation of a solid.

The dithiane is filtered and dried over phosphorus pentoxide or sodium hydroxide in a desiccator at atmospheric pressure. It melts at 112–113° and weighs 16.5–18.1 g. (55–60%).

2. Note

1. The apparatus illustrated is convenient for the steam distillation of compounds which solidify in the condenser. By having the water condenser vertical, it can easily be cleared with a glass rod. No solidification occurs in the side arm, which behaves as a short air condenser. The adapter shown need not be specially constructed but may be assembled from commercially available components; the dimensions and joint sizes are not critical.

3. Methods of Preparation

The procedure described is essentially that of Victor Meyer.[2] *p*-Dithiane has also been obtained from the pyrolysis of the polymer formed by the reaction of ethylene dibromide and potassium sulfide, either alone [3-5] or in phenol,[6] and by the treatment of 2-mercaptoethanol with a cation exchange resin.[7]

[1] Australian Defence Scientific Service (Defence Standards Laboratories, Department of Supply, Melbourne, Australia).
[2] Meyer, *Ber.*, **19**, 3259 (1886).
[3] Crafts, *Ann.*, **124**, 110 (1862).
[4] Husemann, *Ann.*, **126**, 281 (1863).
[5] Masson, *J. Chem. Soc.*, **49**, 234 (1886).
[6] Mansfeld, *Ber.*, **19**, 697 (1886); Fuson, Lipscomb, McKusick, and Reed, *J. Org. Chem.*, **11**, 513 (1946).
[7] Swistak, *Compt. rend.*, **240**, 1544 (1955).

trans-2-DODECENOIC ACID *

$$n\text{-}C_8H_{17}CH_2CH_2CH_2CO_2H \xrightarrow[PCl_3]{Br_2}$$

$$n\text{-}C_8H_{17}CH_2CH_2CHBrCO_2H \xrightarrow{KOC(CH_3)_3}$$

$$n\text{-}C_8H_{17}CH_2CH{=}CHCO_2K + n\text{-}C_8H_{17}CH{=}CHCH_2CO_2K \xrightarrow[C_2H_5OH]{H_2SO_4}$$

$$n\text{-}C_8H_{17}CH_2CH{=}CHCO_2H + n\text{-}C_8H_{17}CH{=}CHCH_2CO_2C_2H_5$$

Submitted by C. Freeman Allen and Max J. Kalm.[1]
Checked by William S. Johnson and Alan D. Lourie.

1. Procedure

Caution! The bromination step should be carried out in a hood, and appropriate precautions should be employed in handling potassium (Note 1).

A dry 125-ml. three-necked flask fitted (glass joints) with a sealed mechanical stirrer, an addition funnel, and a reflux condenser capped with a calcium chloride drying tube, is charged with 30.0 g. (0.15 mole) of dodecanoic acid (Note 2) and 0.6 ml. (0.007 mole) of phosphorus trichloride. The mixture is heated at 90–95° (bath temperature), and 8.5 ml. (0.165 mole) of dry bromine (Note 3) is added in one portion with stirring. After stirring for 3 hours at 90–95°, an additional 7.7 ml. (0.15 mole) of dry bromine is added, and the heating and stirring are continued for an additional 7 hours. The dark reaction mixture is then cooled, dissolved in about 100 ml. of carbon tetrachloride, and shaken vigorously with two 100-ml. portions of water. The organic solution is filtered through anhydrous sodium sulfate, and the solvent and excess bromine are removed by distillation at steam-bath temperature and reduced pressure (water aspirator). The residue of bromo acid, which is pale orange in color, is slowly and cautiously added at room temperature to a solution of potassium *tert*-butoxide which has been prepared from 14.7 g. (0.375 g. atom) of potassium (Note 1) and 350 ml. of dry *tert*-butyl alcohol (Note 1), and is contained in a 1-l. flask fitted with a reflux condenser that is protected from moisture with a calcium chloride tube. The resultant thick suspension is heated at gentle reflux for 3–4 hours on a steam bath, then cooled, diluted with about 1 l. of water, and acidified to Congo red with 5N sulfuric acid. The mixture, containing the precipitated liquid dodecenoic acids, is extracted with two 100-ml. portions of hexane (or a comparable petroleum ether fraction), and the combined hexane solutions are washed with water and dried by filtration through anhydrous sodium sulfate. The hexane is removed by flash distillation, and the residual acid is fractionally distilled through a 2-ft. Podbielniak-type column (Notes 4 and 5). After a small fore-run, the main fraction of dodecenoic acids is collected over a 3° range at about 166–169°/3 mm. The yield is 14–15 g. (47–50%), n_D^{25} ca. 1.4610 (Note 6).

The distilled mixture of dodecenoic acids is dissolved in 150 ml. of commercial absolute ethanol containing 1.3 ml. of concentrated sulfuric acid and allowed to stand in a stoppered flask for 2 hours at 20°. The solution is diluted with 600 ml. of water and extracted with two 150-ml. portions of 60–68° petroleum ether. The extracts are washed with water and percolated through a Kies extraction apparatus (Note 7) consisting of three stages containing, respectively, 9.9 g. of 85% potassium hydroxide (0.15 mole) in 250 ml. of 20% ethanol, 2.5 g. of 85% potassium hydroxide (0.038 mole) in 125 ml. of 20% ethanol, and 125 ml. of water. An additional 250 ml. of petroleum ether is then passed through the extraction apparatus. The three aqueous layers are combined, acidified to Congo red with 5N sulfuric acid, and extracted with petroleum ether. The combined organic layers are washed with water and dried over anhydrous sodium sulfate. The solvent is removed by flash distillation and the residue distilled in a modified Claisen flask. The yield of colorless 2-dodecenoic acid, b.p. 155–158°/3 mm., 127–130°/0.15 mm., is 8–10 g. (27–34%), n_D^{25} 1.4629, λ_{max} 210 mμ (ϵ 13,650) in hexane, m.p. 13–18°.

2. Notes

1. The precautions for handling potassium and the procedure for preparing anhydrous potassium *tert*-butoxide have already been described (p. 134).

2. The submitters employed a sample of dodecanoic acid, m.p. 42.5–43°, obtained by fractional distillation of commercial material. The checkers used Eastman Kodak Company yellow label grade dodecanoic acid, m.p. 44.5–45°. If a product free of homologous material is desired, purified dodecanoic acid should be used as starting material.

3. The bromine was dried with phosphorus pentoxide and filtered into the addition funnel through a plug of glass wool.

4. A simplified Podbielniak column [2] was employed. Other columns of comparable efficiency should be suitable.

5. This distillation is of importance; if omitted, the final 2-dodecenoic acid is difficult to purify.

6. The mixture of dodecenoic acids exhibited λ_{max} at 210 mμ (ϵ 11,980) in hexane. From the extinction coefficient (13,650) for the pure 2-isomer and that (about 1000) for the 3-isomer, it is calculated that this mixture contains about 13% of the latter. A small amount of dodecanoic acid also appears to be present.

7. The Kies extraction apparatus [3] is useful in minimizing emulsion formation. The checkers performed the countercurrent extractions successfully in separatory funnels. The solutions must be mixed by mild rocking of the funnels; otherwise serious emulsions will be produced.

3. Methods of Preparation

Higher-molecular-weight normal 2-alkenoic acids have been prepared in poor yields by the Doebner condensation of aldehydes with malonic acid,[4–6] and by the Reformatsky reaction of aldehydes with ethyl bromoacetate followed by dehydration.[7] The α-iodo acid, prepared from the bromo acid, has been dehydrohalogenated with potassium hydroxide in ethanol,[8] but large quantities of the α-hydroxy acid are formed as a by-product which is difficult to separate in some instances. It has been reported that 2-dodecenoic acid is formed by the treatment of 3-tridecen-2-one with sodium hypobromite.[9] The present procedure is an adaptation of a published procedure.[4]

[1] University of California, Berkeley, California.

[2] Cason and Rapoport, *Laboratory Text in Organic Chemistry*, 2nd ed., p. 293, Prentice-Hall, Englewood Cliffs, New Jersey, 1962.

[3] Kies and Davis, *J. Biol. Chem.*, 189, 637 (1951).

[4] Cason, Allinger, and Sumrell, *J. Org. Chem.*, 18, 850 (1953).

[5] Lauer, Gensler, and Miller, *J. Am. Chem. Soc.*, 63, 1153 (1941).

[6] Zaar, *Ber. Schimmel and Co., Akt.-Ges.*, 299 (1929) [*C. A.*, 24, 2107 (1930)].

[7] Cason and Sumrell, *J. Org. Chem.*, **16**, 1181 (1951).

[8] Myers, *J. Am. Chem. Soc.*, **73**, 2100 (1951); Sweet and Estes, *J. Org. Chem.*, **21**, 1426 (1956).

[9] Kologrivova and Belov, *Zhur. Obshchei Khim.*, **28**, 1269 (1958) [*C. A.*, **52**, 19929 (1958)].

ETHANEDITHIOL *

(1,2-Ethanedithiol)

$$BrCH_2CH_2Br + 2NH_2CSNH_2 \rightarrow$$

$$
\begin{array}{c}
CH_2-S-C\overset{NH}{\underset{NH_2}{\diagdown}} \\
\mid \\
CH_2-S-C\overset{NH}{\underset{NH_2}{\diagdown}}
\end{array} \cdot 2HBr
$$

$$
\begin{array}{c}
CH_2-S-C\overset{NH}{\underset{NH_2}{\diagdown}} \\
\mid \qquad\qquad \cdot 2HBr + 8KOH \rightarrow \\
CH_2-S-C\overset{NH}{\underset{NH_2}{\diagdown}}
\end{array}
$$

$$KSCH_2CH_2SK + 4NH_3 + 2K_2CO_3 + 2KBr + 2H_2O$$

$$KSCH_2CH_2SK + H_2SO_4 \rightarrow HSCH_2CH_2SH + K_2SO_4$$

Submitted by A. JOHN SPEZIALE.[1]
Checked by R. S. SCHREIBER and R. W. KRATZ.

1. Procedure

Caution! This preparation requires the use of a good hood.

In a 5-l. round-bottomed flask fitted with an efficient reflux condenser are placed 2750 ml. of 95% ethanol and 609 g. (8.0 moles) of thiourea. The mixture is brought to the reflux temperature on a steam bath, and the refluxing solution is almost clear.

The steam is turned off, and 751.5 g. (4.0 moles) of ethylene dibromide is added in one portion. Within 5 minutes a vigorous reaction (Note 1) ensues and ethylene diisothiuronium bromide separates from solution. The exothermic reaction is allowed to continue to completion without further application of heat. The isothiuronium salt is collected by filtration and dried. The salt melts with decomposition at 225–227° (Note 2) and weighs 1104 g. (81%).

Concentration of the filtrate to a volume of about 250 ml. and recrystallization from 95% ethanol of the crude isothiuronium salt which separates gives an additional 130 g. of material. The total yield of the salt is 1234 g. (90%).

A mixture of 255 g. (0.75 mole) of ethylene diisothiuronium bromide and 640 g. (9.7 moles) of 85% potassium hydroxide in 1360 ml. of water is placed in a 5-l. round-bottomed three-necked flask and boiled under reflux for 5 hours. Ammonia is evolved during the reflux period. The flask is then equipped with a separatory funnel, a gas-inlet tube, and a condenser set for steam distillation (Note 3). Nitrogen is admitted through the inlet tube, and a cooled solution of 415 ml. of sulfuric acid in 760 ml. of water is added dropwise (Note 4). The addition is continued until the reaction mixture becomes acid to Congo red paper, and then a 20% excess of acid is added. Approximately 725–850 ml. of the acid solution is required. The heat of neutralization is sufficient to distil part of the dithiol. At the end of the addition of the acid, the passage of nitrogen is discontinued and steam is admitted through the inlet tube. The steam distillation is continued until about 3 l. of distillate is collected. The oil is separated from the water in the distillate, which is then extracted with two 500-ml. portions of ether. The ether solution and the oil are dried separately over calcium chloride. After evaporation of the solvent, the residue is added to the oil and the crude product is fractionated through a 10-in. Vigreux column under reduced pressure in an atmosphere of nitrogen. The ethanedithiol boils at 63°/46 mm. (Note 5) and weighs 39–44 g. (55–62%); n_D^{20} 1.5589.

2. Notes

1. The reaction may be so vigorous that external cooling is re-

quired. A cloth wet with ice water and applied to the flask is sufficient to control the reaction.

2. The melting point seems to depend upon the rate of heating. Use of a Fisher-Johns melting-point apparatus gives a value of 240–242°.

3. At this point the reaction *must* be carried out in a good hood. The vapors of ethanedithiol may cause severe headache and nausea.

4. The alkaline solution should be at room temperature before the acidification is begun.

5. In one run the checkers observed a boiling point of 69°/46 mm.

3. Methods of Preparation

Ethanedithiol has been prepared from ethylene dichloride [2] or ethylene dibromide [3] and alcoholic potassium hydrosulfide; from ethylene dibromide and alcoholic sodium hydrosulfide; [4] from ethylene dichloride [5] or ethylene dibromide [6] and alcoholic sodium hydrosulfide under pressure; from ethylene dibromide and thiourea; [7] and by the catalytic hydrogenation with cobalt trisulfide [8] or nickel-on-kieselguhr [9] of the mixture resulting from the reaction of ethylene and sulfur. The present method is a modification of one described by Mathias.[7]

[1] Organic Chemicals Division, Monsanto Chemical Co., St. Louis, Missouri.

[2] Löwig and Weidmann, *Ann.*, **36**, 321 (1840).

[3] Fasbender, *Ber.*, **20**, 460 (1887).

[4] Meyer, *Ber.*, **19**, 3259 (1886); Biswell, U. S. pat. 2,436,137 [*C. A.*, **42**, 3430 (1948)].

[5] Tucker and Reid, *J. Am. Chem. Soc.*, **55**, 775 (1933).

[6] Simpson, *Can. J. Research*, **25B**, 20 (1947).

[7] Mathias, *Bol. fac. filosof., ciênc. e letras, Univ. São Paulo*, **14**, Quim. No. 1, 75 (1942) [*C. A.*, **40**, 2792 (1946)].

[8] Signaigo, U. S. pat. 2,402,456 [*C. A.*, **40**, 5767 (1946)].

[9] Lazier, Signaigo, and Werntz, U. S. pat. 2,402,643 [*C. A.*, **40**, 5764 (1946)].

ETHOXYACETYLENE *

(Ether, ethyl ethynyl)

$$ClCH_2CH(OC_2H_5)_2 + 3NaNH_2 \xrightarrow{\text{liq. } NH_3}$$
$$NaCl + C_2H_5ONa + Nac\equiv COC_2H_5 + 3NH_3$$
$$Nac\equiv COC_2H_5 + H_2O \rightarrow NaOH + HC\equiv COC_2H_5$$

Submitted by E. R. H. JONES, GEOFFREY EGLINTON, M. C. WHITING, and
B. L. SHAW.[1]
Checked by ARTHUR C. COPE and RONALD M. PIKE.

1. Procedure

Caution! *This preparation should be conducted in a hood to avoid exposure to ammonia, and thorough shielding of the apparatus is recommended.*

A solution of sodium amide in liquid ammonia is prepared according to a procedure previously described (Note 1) in a 1-l. three-necked flask (Note 2) equipped with a cold-finger condenser (cooled with Dry Ice) attached through a soda-lime tower to a gas absorption trap [2] and an inlet tube. Anhydrous liquid ammonia (500 ml.) is introduced from a commercial cylinder through the inlet tube, and 0.5 g. of hydrated ferric nitrate is added, followed by 38 g. (1.65 g. atoms) of clean, freshly cut sodium (Note 3). The inlet tube is replaced with a 100-ml. dropping funnel, and the mixture is agitated manually (Note 2) until all the sodium is converted into sodium amide, after which 76.5 g. (0.50 mole) of diethylchloroacetal (Note 4) is added over a period of 15–20 minutes. The mixture is swirled for an additional period of 15 minutes, after which the ammonia is evaporated in a stream of pure nitrogen. The flask is cooled to −70° in a Dry Ice-trichloroethylene bath (Note 5), and 325 ml. of a saturated solution of sodium chloride which has been cooled to −20° is added all at once and as rapidly as possible with vigorous agitation (Note 6). The flask is then fitted with a still head connected to a trap cooled to −70° with Dry Ice, and the contents of the flask are slowly heated to 100° on a steam bath (Note 7). The condensate is allowed to warm to 0°, after which the trap is again cooled to $-70°$ and the mixture is neutralized by the dropwise addition of a saturated aqueous solution of sodium dihydrogen phosphate. The aqueous layer is frozen by cooling with Dry Ice, and the supernatant liquid is decanted and dried over about 4 g. of anhydrous calcium chloride. The drying agent is removed by filtration, and the filtrate is distilled (Note 8) through a column containing a 20-cm. section packed with glass helices at partial reflux, yielding 20–21.2 g. (57–60%) of ethoxyacetylene, b.p. 49–51°/749 mm., n_D^{25} 1.3790 (Notes 9, 10, 11, and 12).

2. Notes

1. One of the procedures for converting sodium to sodium amide described in *Organic Syntheses* is used (p.763).[3]
2. The flask is clamped to the free end of a long (30-in.) Duraluminum rod which is fixed at the other (top) end to a rigid frame. The rod is clamped in a vertical position, and by moving the free end the contents of the flask can be swirled very conveniently.

The submitters used a 1-l. round-bottomed flask mounted in the same way, without provision for cooling or a condenser cooled with Dry Ice, and were successful in preparing sodium amide and adding diethylchloroacetal before excessive loss of liquid ammonia occurred. Conversion of sodium to sodium amide sometimes requires a considerable period of time, in which case use of the apparatus specified avoids difficulty and the necessity of adding more liquid ammonia.

3. More liquid ammonia should be added through the inlet tube if vaporization reduces the liquid volume to less than 300 ml.

4. Diethylchloroacetal (chloroacetaldehyde diethylacetal) supplied by Eastman Kodak Company is satisfactory.

5. *Important! This sodium derivative is extremely pyrophoric, and at this point and during the addition of the saturated sodium chloride solution it is essential that the contents of the flask be kept out of contact with air.* When the flask is immersed in the Dry Ice bath the nitrogen stream is increased to counteract the diminution in pressure due to rapid cooling.

6. The checkers found that a dropping funnel equipped with a pressure-equalizing tube provided the most satisfactory mode of addition.

7. It is advisable to allow the mixture to warm to room temperature gradually before applying heat.

8. The distillation should be conducted in a hood since the product has lachrymatory properties.

9. Others have reported b.p. $28°/300$ mm., n_D^{20} 1.3812,[4] and b.p. $48-50°/760$ mm.[5]

10. The submitters have used a similar procedure for the preparation of methoxyacetylene, b.p. $22.5-23.5°$, n_D^{16} 1.3693, from commercial dimethylchloroacetal in 60% yield. *Caution! Minor explosions have been reported to occur during the addition of saturated sodium chloride solution to the sodium derivative of methoxyacetylene. They may have resulted from ignition of the methoxyacetylene by particles of sodium which adhered to the walls of the flask. It is recommended that all apparatus be shielded during the preparation of methoxyacetylene.*

11. Ethoxyacetylene has proved to be a useful reagent for the conversion of ketones to acetylenic carbinols:

$$\begin{array}{c} \diagdown \\ \diagup \end{array} C{=}O \rightarrow \begin{array}{c} \diagdown \\ \diagup \end{array} \underset{\underset{OH}{|}}{C}{-}C{\equiv}COC_2H_5$$

which can be converted to α,β-unsaturated aldehydes, esters, and acids:[6]

$$\begin{array}{c} \diagdown \\ \diagup \end{array} \underset{\underset{OH}{|}}{C}{-}C{\equiv}COC_2H_5 \rightarrow$$

$$\begin{array}{c} \diagdown \\ \diagup \end{array} C{=}CHCHO, \quad \begin{array}{c} \diagdown \\ \diagup \end{array} C{=}CHCO_2C_2H_5, \quad \begin{array}{c} \diagdown \\ \diagup \end{array} C{=}CHCO_2H$$

It has been used in the synthesis of vitamin A aldehyde [7] and in the preparation of an intermediate in one of the total syntheses of cortisone.[8]

12. It has been reported [9] that the following modified procedure eliminates the dangers associated with the preparation of ethoxyacetylene.

The initial operations were the same as described above except that 46 g. (2 g. atoms) of sodium was employed (per 76.5 g. of chloroacetal), and the reaction mixture was stirred mechanically for 2 hours after the addition of the diethylchloroacetal. The dropping funnel and stirrer were removed and the flask was closed with a stopper provided with a wide plastic tube that came within 1 mm. of the bottom. The other end was placed near the bottom of an open 2-l. flask which contained 1 kg. of finely crushed ice.

When the stopper was fitted, the contents of the reaction flask were transferred to the ice by the ammonia pressure. The walls of the flask were rinsed with a little liquid ammonia and the washings also were forced onto the ice. The cold hydrolysis mixture was shaken repeatedly with 50-ml. portions of high-boiling petroleum (b.p. > 160°). The combined extracts were washed with ice water and dried over anhydrous sodium sulfate. Distillation afforded 21–23 g. (60–66%) of material which boiled at 49–55°/760 mm., n_D^{20} 1.3820.

3. Methods of Preparation

Ethoxyacetylene has been prepared from β-bromovinyl ethyl ether [4] or β-chlorovinyl ethyl ether [10] and potassium hydroxide; from α,β-dibromoethyl ethyl ether and potassium hydroxide; [5] and from diethyl chloroacetal, diethyl bromoacetal, α,β-dichloroethyl ethyl ether, or α,β-dibromoethyl ethyl ether and sodium amide. [11]

[1] University of Manchester, Manchester, England.

[2] Org. Syntheses Coll. Vol. 2, 4 (1943).

[3] Org. Syntheses Coll. Vol. 3, 219 (1955).

[4] Jacobs, Cramer, and Hanson, J. Am. Chem. Soc., 64, 223 (1942).

[5] Favorskii and Shostakovskii, J. Gen. Chem. U.S.S.R., 13, 1 (1943) [C. A., 38, 330 (1944)].

[6] Heilbron, Jones, Julia, and Weedon, J. Chem. Soc., 1949, 1823.

[7] van Dorp and Arens, Nature, 160, 189 (1947).

[8] Sarett, Arth, Lukes, Beyler, Poos, Johns, and Constantin, J. Am. Chem. Soc., 74, 4974 (1952).

[9] Arens and Brandsma, Private communication.

[10] Arens, Rec. trav. chim., 74, 271 (1955); van Dorp, Arens, and Stephenson, Rec. trav. chim., 70, 289 (1951).

[11] Eglinton, Jones, Shaw, and Whiting, J. Chem. Soc., 1954, 1860.

ETHYL α-ACETYL-β-(2,3-DIMETHOXYPHENYL)-PROPIONATE

(Hydrocinnamic acid, α-acetyl-2,3-dimethoxy-, ethyl ester)

Submitted by E. C. Horning, J. Koo, M. S. Fish, and G. N. Walker.[1]
Checked by T. L. Cairns and Charles W. Todd.

1. Procedure

A. *Ethyl α-acetyl-β-(2,3-dimethoxyphenyl)acrylate.* In a 1-l. round-bottomed flask fitted with a water-benzene separator [2] and a reflux condenser are placed 183 g. (1.1 moles) of 2,3-dimethoxybenzaldehyde (Note 1) and 130 g. (1.0 mole) of ethyl acetoacetate. The water-benzene separator is filled with benzene, an additional 70 ml. of benzene is added to the mixture, and the

2,3-dimethoxybenzaldehyde is brought into solution by warming. Piperidine (4 ml.) and glacial acetic acid (12 ml.) are added, and the mixture is heated under reflux for 2–3 hours (Note 2). The mixture is cooled, poured into a separatory funnel with 800 ml. of ether, and washed successively with 200-ml. portions of 5% hydrochloric acid, 5% sodium bicarbonate solution, and 5% acetic acid, and twice with water. The extract is dried over anhydrous magnesium sulfate (about 250 g.). After filtration, the ether and benzene are distilled under atmospheric pressure, and the residue is distilled under reduced pressure. The yield of viscous, yellow oil collected at 186–190°/2 mm. (Note 3) is 180–199 g. (64–72%), n_D^{25} 1.5507–1.5508.

B. *Ethyl α-acetyl-β-(2,3-dimethoxyphenyl)propionate.* The product obtained in Part A is divided into two approximately equal portions, and to each are added 125 ml. of ethyl acetate and 5 g. of 5% palladium-on-carbon catalyst (Note 4). Each solution is shaken with hydrogen at pressures between 20 and 40 lb. in a low-pressure apparatus. About 45 minutes is usually needed for complete reduction. The two solutions are then combined, and the catalyst is removed by filtration and washed with 20 ml. of ethyl acetate. The ethyl acetate is distilled at atmospheric pressure, and distillation of the residue under reduced pressure yields 158–176 g. (56–63% over-all) of a colorless product collected at 175–177°/3 mm., n_D^{25} 1.5042–1.5044 (Notes 3 and 5).

2. Notes

1. A practical grade of 2,3-dimethoxybenzaldehyde, although discolored, is satisfactory. It is best handled and weighed as a liquid. This is done by warming the container on a steam bath for several minutes.

2. The time of reflux is determined by the rate at which water separates from the reaction mixture. In general, the theoretical amount of water (18 ml.) is collected in the water separator after about 1 hour, but additional refluxing usually results in the separation of another 4–5 ml. Anhydrous materials are not necessary for the reaction.

3. Using a 6-in. Vigreux column with a wide side arm, the

checkers observed a boiling point of 166–168°/3 mm. for ethyl α-acetyl-β-(2,3-dimethoxyphenyl)propionate and 176–179°/ 2 mm. for the acrylate.

4. These quantities are convenient for hydrogenation in a Parr low-pressure apparatus; if a larger apparatus is available, it is unnecessary to divide the material. The palladium-on-carbon catalyst is prepared by Hartung's method,[3] using sufficient Norit to give a 5% catalyst. A few chips of Dry Ice should be placed in the bottle before the catalyst is added to provide an inert atmosphere. The checkers used a catalyst obtained from Baker and Company, Inc., 113 Astor Street, Newark, New Jersey.

5. The same procedure may be used with other aromatic aldehydes. The acrylate ester prepared from veratraldehyde is a crystalline compound melting at 82.5–84.5°. It may be purified by distillation under reduced pressure, b.p. 190–196°/0.8 mm. The yield is 53–63%. The hydrogenated product is obtained as a colorless oil, b.p. 180–183°/1.0 mm., in 38–44% over-all yield.

3. Methods of Preparation

Substituted α-benzylacetoacetic esters have usually been prepared by the general method of Leuchs,[4] in which ethyl acetoacetate is alkylated in ethanol solution with benzyl chloride and sodium ethoxide. The present method is based upon Cope's [5] procedure for the condensation of carbonyl compounds with ethyl acetoacetate, followed by catalytic reduction of the condensation product.

[1] University of Pennsylvania, Philadelphia, Pennsylvania.
[2] Org. Syntheses Coll. Vol. 3, 382 (1955).
[3] Org. Syntheses Coll. Vol. 3, 685 (1955), Method D.
[4] Leuchs, Ber., 44, 1510 (1911).
[5] Cope and Hofmann, J. Am. Chem. Soc., 63, 3456 (1941).

ETHYL AZODICARBOXYLATE *

(Formic acid, azodi-, diethyl ester)

$$H_2NNH_2 \cdot H_2O + 2ClCO_2C_2H_5 + Na_2CO_3 \rightarrow$$
$$C_2H_5O_2CNHNHCO_2C_2H_5 + 2NaCl + CO_2 + 2H_2O$$
$$C_2H_5O_2CNHNHCO_2C_2H_5 \xrightarrow[\text{HNO}_3]{\text{[O]}} C_2H_5O_2CN\!\!=\!\!NCO_2C_2H_5$$

Submitted by J. C. KAUER.[1]
Checked by WILLIAM G. DAUBEN and ROBERT L. CARGILL.

1. Procedure

A. *Ethyl hydrazodicarboxylate.* In a 3-l., three-necked flask equipped with a mechanical stirrer, two 500-ml. dropping funnels, and a thermometer (Note 1), is placed a solution of 75 g. (1.5 moles) of 100% hydrazine hydrate (or 88.5 g. of 85% hydrazine hydrate) in 750 ml. of 95% ethanol. The reaction flask is cooled in an ice bath. When the solution temperature drops to 10°, 326 g. (3 moles) of ethyl chloroformate is added dropwise with stirring at a rate sufficient to maintain the temperature between 15° and 20°. After exactly one-half of the ethyl chloroformate has been introduced, a solution of 159 g. (1.5 moles) of sodium carbonate in 750 ml. of water is added dropwise simultaneously with the remaining ethyl chloroformate. The addition of these two reactants is regulated so that the temperature does not rise above 20°. The addition of the chloroformate should be completed slightly in advance of the sodium carbonate, thus maintaining an excess of the chloroformate in the solution at all times. During the course of the addition of the reagents, a precipitate is formed.

After addition of the reactants is complete, the walls of the flask are washed down with 200 ml. of cold water, and the reaction mixture is allowed to stir for 30 minutes. The precipitate is then collected on a Büchner funnel, washed well with a total of 1 l. of cold water, and dried in a vacuum oven at 80°. There is obtained 215–225 g. (81–85%) of ethyl hydrazodicarboxylate which melts at 131–133°. It is sufficiently pure for the preparation of ethyl azodicarboxylate (Notes 2 and 3).

B. *Ethyl azodicarboxylate. Caution! Both ethyl and methyl azodicarboxylate are sensitive to heat. Thus, in a sealed capillary they explode violently when heated by a flame. Overheating is to be avoided, and distillations should be from a bath, not an electrically heated mantle. Distillations should be well shielded. Since copious fumes of nitrogen oxides are evolved during the oxidation by nitric acid, this operation and the subsequent work-up should be carried out in an efficient hood.*

A mixture of 200 g. of ethyl hydrazodicarboxylate in 125 ml. of 70% nitric acid is placed in a 1-l., three-necked flask equipped with a mechanical stirrer, gas-outlet tube, and thermometer. The flask is cooled in an ice bath, and, when the temperature of the solution reaches 5°, 220 ml. of ice-cold yellow fuming nitric acid (90–95% HNO_3, d_{15} = 1.49–1.50) is added. The reaction mixture is maintained at 0–5° for 2 hours with stirring and is then carefully poured on a stirred mixture of 500 g. of ice, 500 ml. of ice water, and 100 ml. of methylene chloride in a 2-l. beaker placed in a metal pan (Note 4). After the ice melts, the solution is transferred carefully to a 2-l. separatory funnel. The organic (lower) layer is removed, and the acid layer is extracted (Note 5) with three 100-ml. portions of methylene chloride. The combined organic layers are washed twice with 100-ml. portions of ice water and are then stirred mechanically for 10 minutes with 500 ml. of ice-cold 10% potassium bicarbonate solution. The layers are separated in a separatory funnel, and the organic layer is finally washed with 100 ml. of ice water and dried quickly with a small portion of anhydrous magnesium sulfate that is removed by filtration. The solution is dried overnight with a fresh portion of anhydrous magnesium sulfate. *Caution! The following distillations should be well shielded.* The methylene chloride is removed on a steam bath under reduced pressure, and the residue is rapidly distilled without fractionation under vacuum (1–5 mm.) from a flask immersed in an oil bath whose temperature is raised gradually from 75° to 130°. The crude distillate is then fractionally distilled under vacuum through a short column packed with glass helices, using an oil bath to heat the distillation flask. After a short fore-run, the main fraction is collected at 93–95°/5 mm. There is obtained 138–158 g. (70–80%) of ethyl azodicarboxylate which freezes at 6° (Notes 6, 7, and 8).

2. Notes

1. The thermometer and one of the funnels are fitted to a two-necked adapter so that, when the thermometer bulb is immersed in the solution, the range between 10° and 20° is easily visible.

2. Ethyl hydrazodicarboxylate may be purified by crystallization from dilute ethanol; m.p. 134–135°.

3. Methyl hydrazodicarboxylate, which is much more soluble in water than the ethyl ester, may be prepared by the following modification of the above procedure.

A solution of 100 g. (2 moles) of hydrazine hydrate in 500 ml. of 95% alcohol is treated as described above with a total of 378 g. (4 moles) of methyl chloroformate while maintaining the temperature below 20°. During the addition of the last half of the chloroformate, a warm (30–35°) solution of 212 g. (2 moles) of sodium carbonate in 800 ml. of water is added. The resulting slurry is stirred for 30 minutes, and the precipitate is filtered on a Büchner funnel, washed with 100 ml. of ice water, and air-dried. The filtrate is concentrated at reduced pressure (12–25 mm.) on a water bath to 700 ml. and is cooled in ice. The precipitate is filtered, washed with 100 ml. of ice water, and air-dried. The combined crops of crude methyl hydrazodicarboxylate are dried at 60° in a vacuum oven and are then stirred successively with two 800-ml. portions of warm acetone to separate inorganic impurities. The acetone solution is filtered, and evaporated at reduced pressure to yield 260 g. (88%) of methyl hydrazodicarboxylate (m.p. 127–131°) of sufficient purity for preparation of methyl azodicarboxylate.

A purer product (m.p. 131–132°) may be obtained by warming the acetone solution to 50° and adding to it 1.8 l. of hexane or heptane also warmed to 50°. The solution is seeded and allowed to cool slowly.

4. As a safety precaution, in case the beaker should break, it is placed in a metal pan.

5. The separatory funnel should be carefully vented at frequent intervals since large quantities of nitrogen oxides are liberated.

6. The submitter found that methyl azodicarboxylate can be prepared by a modification of this procedure. It is somewhat more water-soluble and much more susceptible to hydrolysis than the ethyl ester.

A mixture of 200 g. (1.35 moles) of methyl hydrazodicarboxylate, 400 ml. of chloroform (washed with water and dried over calcium chloride), and 250 ml. of concentrated nitric acid is cooled to 5°, and 350 ml. of ice-cold yellow fuming (90–95%) nitric acid is added. The solution is stirred for 2 hours at 0–5° and is poured onto 1 kg. of crushed ice. After most of the ice has melted, the organic layer is separated (Note 5) and is washed quickly with two 500-ml. portions of ice water containing some ice. It is then stirred for 2 minutes with a mixture of 300 ml. of ice-cold 10% potassium bicarbonate solution and 200 g. of crushed ice. The organic layer is separated, washed quickly with ice water, and dried and distilled as described for the ethyl ester. There is obtained 148 g. (75%) of methyl azodicarboxylate which boils at 90–91°/15 mm. and freezes at 10° (Note 7).

7. The submitter has found that methyl azodicarboxylate is often more reactive than the ethyl ester. Moreover, adducts prepared from the methyl ester are generally higher melting and lower boiling than those from the ethyl ester, and thus more easily purified.

A frequent annoying by-product of the reactions of azodicarboxylic esters is the corresponding hydrazodicarboxylate. In the case of the methyl ester this impurity is easily removed because of its moderate solubility in water and low solubility in such solvents as carbon tetrachloride.

3. Methods of Preparation

Ethyl hydrazodicarboxylate has been prepared by the reaction of ethyl chloroformate with hydrazine hydrate [2] or hydrazine sulfate in the presence of alkali.[3]

Ethyl azodicarboxylate has been prepared by oxidation of ethyl hydrazodicarboxylate with hypochlorous acid,[4] concentrated nitric acid,[3] and a mixture of concentrated and fuming nitric acid.[5,6]

¹ Contribution No. 633, from the Central Research Department, Experimental Station, E. I. du Pont de Nemours and Co., Wilmington, Delaware.

² Curtius and Heidenreich, *J. prakt. Chem.*, **52**, 476 (1895); Vogelesang, *Rec. trav. chim.*, **65**, 789 (1946).

³ Ingold and Weaver, *J. Chem. Soc.*, **127**, 381 (1925).

⁴ Rabjohn, *Org. Syntheses Coll. Vol.* **3**, 375 (1955).

⁵ Curtius and Heidenreich, *Ber.*, **27**, 774 (1894).

⁶ Diels and Fritzsche, *Ber.*, **44**, 3018 (1911).

ETHYL BENZOYLACETATE *

(Acetic acid, benzoyl-, ethyl ester)

$$CH_3COCH_2CO_2C_2H_5 + C_6H_5COCl + 2NaOH \rightarrow$$

$$(C_6H_5CO\overset{\overset{\displaystyle COCH_3}{|}}{C}CO_2C_2H_5)^- Na^+ + NaCl + 2H_2O$$

$$(C_6H_5CO\overset{\overset{\displaystyle COCH_3}{|}}{C}CO_2C_2H_5)^- Na^+ + NH_4Cl + H_2O \rightarrow$$

$$C_6H_5COCH_2CO_2C_2H_5 + NaCl + CH_3CO_2NH_4$$

Submitted by J. M. STRALEY and A. C. ADAMS (*Deceased*).¹
Checked by MAX TISHLER and M. A. KOZLOWSKI.

1. Procedure

In an open 3-l. three-necked flask, equipped with an efficient mechanical stirrer (Note 1) and two dropping funnels, are placed 500 ml. of water, 250 ml. of technical naphtha boiling at 95–110°, and 195 g. (1.5 moles) of freshly distilled ethyl acetoacetate. The mixture is cooled to 5° with a water-ice bath, and 65 ml. of 33% sodium hydroxide solution (33 g. sodium hydroxide in 100 g. solution) is added. As the temperature is maintained below 10° (Note 2) and the pH near 11 (Note 3), the mixture is stirred vigorously (Note 1), and there are added simultaneously from the two dropping funnels 230 g. (1.62 moles) of benzoyl chloride and 270 ml. of 33% sodium hydroxide solution. This addition should be made during about 2 hours. After addition is complete, the cooling bath is removed, and the mixture is allowed to come to room temperature. In order to ensure complete reaction, the mixture is finally brought to 35° during about 1 hour. The stirrer is then stopped, and the aqueous layer is separated and placed in a 2-l. Erlenmeyer flask (Note 4).

To the mixture is added 80 g. of technical ammonium chloride; then it is stirred slowly overnight. The specific gravity is brought to 1.13 by the addition of about 90 g. of sodium chloride, after which the mixture is transferred to a separatory funnel. About 10 ml. of benzene is used to rinse the flask and is added to the separatory funnel. The aqueous layer is withdrawn (Note 5), and the oil is washed three times with 100-ml. portions of cold water.

An additional 40 ml. of benzene is added (to accomplish drying on distillation), and the product is distilled under reduced pressure, using a short still head with no fractionating column (Note 6). The yield of ethyl benzoylacetate, b.p. 145–150°/12 mm., is 197–203 g. (68–71%) (Note 4).

2. Notes

1. Good stirring is essential. Slow stirring results in low yields.

2. Temperatures above 10° did not result in consistently good yields.

3. Lower pH did not give good yields. The pH was checked by means of filter paper, which had been dipped in an alcoholic solution of alizarin and then dried.

4. The naphtha layer may be used without further treatment for the next run. Yields of 76% have been obtained on such a second run without making allowance for recovered ethyl acetoacetate. Distillation of the naphtha layer together with the fore-run from the final distillation of ethyl benzoylacetate yields 11–14 g. of recovered ethyl acetoacetate and about 235 ml. of naphtha.

5. As high as 62 g. of benzoic acid has been recovered by acidification of the aqueous layer.

6. The chief impurity in the crude ester is a high-boiling material of unknown composition.

3. Methods of Preparation

The methods of preparation have been listed in two earlier volumes.[2,3] The present method, which is an adaptation of a process found in German documents,[4] is a shorter, more simple procedure which does not require use of dry solvent or metallic sodium.

[1] Research Laboratories, Tennessee Eastman Company, Kingsport, Tennessee.
[2] *Org. Syntheses Coll. Vol.* **2**, 266 (1943).
[3] *Org. Syntheses Coll. Vol.* **3**, 381 (1955).
[4] B.I.O.S., *Final Rept.* **1149**, 115.

ETHYL *tert*-BUTYL MALONATE

(Malonic acid, *tert*-butyl ethyl ester)

$$CH_2(CO_2C_2H_5)_2 + KOH \rightarrow C_2H_5O_2CCH_2CO_2K + C_2H_5OH$$

$$C_2H_5O_2CCH_2CO_2K + HCl \rightarrow C_2H_5O_2CCH_2CO_2H + KCl$$

$$C_2H_5O_2CCH_2CO_2H + CH_2{=}C(CH_3)_2 \underset{\longleftarrow}{\overset{H_2SO_4}{\longrightarrow}}$$

$$C_2H_5O_2CCH_2CO_2C(CH_3)_3$$

Submitted by R. E. STRUBE.[1]
Checked by WILLIAM S. JOHNSON and DUFF S. ALLEN, JR.

1. Procedure

A 2-l. three-necked flask, equipped with a sealed stirrer, a dropping funnel and a reflux condenser provided with a calcium chloride drying tube, is charged with 100 g. (0.625 mole) of diethyl malonate (Note 1) and 400 ml. of commercial absolute ethanol. Stirring is started, and a solution of 35 g. of potassium hydroxide pellets (Note 2) in 400 ml. of commercial absolute ethanol is added at room temperature during a period of 1 hour. A white crystalline precipitate forms during the addition, and, after all the hydroxide has been added, stirring is continued for an additional 2 hours. After the mixture has stood overnight, it is heated to boiling on the steam bath and filtered while hot with suction (Note 3). Precipitation of the potassium ethyl malonate is completed by cooling the filtrate in an ice bath. The salt is collected by suction filtration, washed with a small amount of ether, and dried under reduced pressure at room temperature. An additional amount of the potassium salt is obtained by concentrating the mother liquors on the steam bath to about 100–125 ml. The total yield is 80–87 g. (75–82%).

A 250-ml. three-necked flask provided with a stirrer, a dropping funnel, and a thermometer is charged with 80 g. (0.470 mole) of potassium ethyl malonate and 50 ml. of water. The mixture is cooled to 5° with an ice bath, and 40 ml. of concentrated hydrochloric acid is added over a 30-minute period while the temperature is maintained below 10°. The mixture is filtered with suction, and the precipitate of potassium chloride washed with 75 ml. of ether. The aqueous layer of the filtrate is separated and washed with three 50-ml. portions of ether. The combined ether solutions are dried over anhydrous magnesium sulfate; then most of the solvent is removed by distillation at atmospheric pressure, and the remainder under reduced pressure. Finally, the liquid residue of monoethyl malonate is dried at 50°/1 mm. for 1 hour. The yield is 58–62 g. (93–100%).

A 500-ml. Pyrex heavy-walled, narrow-mouthed pressure bottle is charged with 100 ml. of ether and 3.5 ml. of concentrated sulfuric acid. The solution is cooled with an ice bath to 5°, and 56 g. (0.42 mole) of monoethyl malonate and approximately 60 ml. (about 0.75 mole) of isobutylene (Note 4) are added. The bottle is immediately closed with a rubber stopper, which is clamped or wired in place, and is shaken mechanically at room temperature overnight (Note 5). The bottle is chilled in an ice-salt bath and then opened. The reaction mixture is poured into a 1-l. Erlenmeyer flask containing a cooled solution of 50 g. of sodium hydroxide in 200 ml. of water and 200 g. of ice. The mixture is swirled a few times and then transferred to a separatory funnel (Note 6). The layers are separated, and the aqueous portion is extracted with two 75-ml. portions of ether. The organic layers are combined and dried over anhydrous magnesium sulfate. The solution is concentrated in a 125-ml. round-bottomed flask (Note 7) and distilled at reduced pressure through a 10-cm. Vigreux column. The fraction distilling at 98–100°/22 mm. or 107–109°/24 mm. is collected. The yield is 42–47 g. (53–58%), n_D^{25} 1.4128, n_D^{23} 1.4142.

2. Notes

1. Diethyl malonate as supplied by the Eastman Kodak Company (white label grade) or by Abbott Laboratories may be used without further purification.

2. Potassium hydroxide (85% minimum assay) obtained from the Mallinckrodt Chemical Works is satisfactory.

3. A steam-heated Büchner or a warmed sintered glass funnel is recommended.

4. Technical grade isobutylene supplied by Matheson Company

was used. The isobutylene gas is liquefied by passage into a large test tube immersed in a Dry Ice-acetone bath.

5. For convenience, the reaction was carried out overnight. The reaction time may probably be shortened (compare the preparation of di-*tert*-butyl malonate, p. 261).

6. The mixture may be filtered, if necessary, to remove ice.

7. Since traces of acid will decompose the ester during the distillation, it is essential to wash the distillation apparatus carefully with a sodium hydroxide solution before rinsing and drying. The addition of some potassium carbonate or magnesium oxide before distillation is recommended (see Note 5 on p. 262).

3. Methods of Preparation

Ethyl *tert*-butyl malonate has been prepared by adding *tert*-butyl acetate and ethyl carbonate to sodium triphenylmethyl,[2] and from ethyl malonyl chloride and *tert*-butyl alcohol.[3] The present procedure is an adaptation of that for the preparation of di-*tert*-butyl malonate (p. 261).

[1] The Upjohn Company, Kalamazoo, Michigan.
[2] Hauser, Abramovitch, and Adams, *J. Am. Chem. Soc.*, **64**, 2714 (1942).
[3] Breslow, Baumgarten, and Hauser, *J. Am. Chem. Soc.*, **66**, 1287 (1944).

N-ETHYL-p-CHLOROANILINE

(Aniline, p-chloro-N-ethyl-)

$$p\text{-ClC}_6\text{H}_4\text{NH}_2 + (\text{C}_2\text{H}_5\text{O})_3\text{CH} \xrightarrow{\text{H}_2\text{SO}_4}$$

$$p\text{-ClC}_6\text{H}_4\underset{\overset{|}{\text{C}_2\text{H}_5}}{\text{NCHO}} + 2\text{C}_2\text{H}_5\text{OH}$$

$$p\text{-ClC}_6\text{H}_4\underset{\overset{|}{\text{C}_2\text{H}_5}}{\text{NCHO}} + \text{H}_2\text{O} \xrightarrow{\text{HCl}} p\text{-ClC}_6\text{H}_4\text{NHC}_2\text{H}_5 + \text{HCO}_2\text{H}$$

Submitted by ROYSTON M. ROBERTS and PAUL J. VOGT.[1]
Checked by JAMES CASON and MILTON FINGER.

1. Procedure

A. *N-Ethyl-p-chloroformanilide.* In a 300-ml. round-bottomed flask, equipped with a side tubulature just large enough to accommodate a thermometer, are placed 63.8 g. (0.50 mole) of p-chloroaniline and 111 g. (0.75 mole) of triethyl orthoformate (Note 1), and then 2.0 g. (0.02 mole) of concentrated sulfuric acid is added with mixing. The flask is attached to a 30-cm. column 2 cm. in diameter packed with glass helices (Note 2), which is surmounted by a distillation head equipped with a thermometer and condenser. A thermometer is connected through a slip joint made from a short section of rubber tubing to the side tubulature so that its bulb is in the reaction mixture; then the flask is heated in an oil bath. When the temperature of the oil bath reaches 115–120°, the reaction mixture begins to boil, and ethanol soon begins to distil at a vapor temperature of 78–80° at the top of the column. During the course of about 1 hour the bath temperature is raised to about 175°. This promotes a steady distillation of ethanol at a rate which begins to decrease after 30 minutes. The amount of ethanol that distils (70–75 ml.) is always in excess of the stoichiometric amount. Finally, the reaction mixture is kept in the oil bath at 175–180° for 30 minutes (Note 3); an additional small amount of volatile material distils during this time.

After the reaction mixture has cooled somewhat the flask is disconnected from the column, a Claisen head is attached, and the product is distilled at reduced pressure (Note 4). After a fore-run of about 20 g. (not readily condensed below 40 mm. pressure), the faintly yellow product is collected at 124–126°/3 mm., weight 73–79 g. (80–86%), n_D^{25} 1.5525–1.5540 (Note 4).

B. *N-Ethyl-p-chloroaniline.* In a 500-ml. round-bottomed flask are placed 70 g. (0.38 mole) of N-ethyl-p-chloroformanilide and 170 ml. of 10% hydrochloric acid. The mixture is heated under reflux for 1 hour, cooled, then neutralized, and finally made basic with 15% potassium hydroxide solution. The lower layer of N-ethyl-p-chloroaniline is separated, and the aqueous layer is saturated with potassium carbonate and extracted with two 200-ml. portions of ether. The ether extracts are combined with the bulk of the product, washed with two 100-ml. portions of water, and then dried over calcium chloride. After the ether has been removed by distillation, the residue is distilled at reduced pressure from a 125-ml. Claisen flask. N-Ethyl-p-chloroaniline is collected at 108–110°/5 mm. or 149–150°/40 mm., n_D^{25} 1.5650–1.5661, weight 52–55 g. (87–92%) (Note 5).

2. Notes

1. The checkers used, without purification, white label grades of p-chloroaniline and triethyl orthoformate from Eastman Organic Chemicals; the submitters used triethyl orthoformate from Kay-Fries Chemicals Inc., New York.

2. It is most convenient to make connections with standard taper joints. The checkers used with equal satisfaction a 50-cm. column randomly packed with short sections of glass tubing.

3. The submitters report that during this heating period the temperature of the reaction mixture may rise as high as 185–190° on account of an exothermic reaction; however, the checkers did not observe this temperature rise of the reaction mixture. The submitters also report that in some of the preparations mentioned in Note 5 the reaction is more exothermic and the temperature may rise as high as 244°, but this does not cause difficulty.

4. The submitters distilled the product through the 30-cm. packed column which was wrapped with a heating tape for this purpose. If this is done, there are collected about 20 g. of re-

covered triethyl orthoformate at 65–67°/40 mm. and about 2 g. of ethyl N-p-chlorophenylformimidate at 82–83°/40 mm., followed by the product, which has n_D^{25} 1.5559. The checkers obtained the same results when this distillation was carried out through a fractionating column; however, the yield and properties of N-ethyl-p-chloroaniline obtained from this material were the same as those of amine obtained from material distilled through a Claisen head.

5. This method is suitable for the mono-N-alkylation of other primary aromatic amines. Trimethyl and triethyl orthoformate are commercially available, and other alkyl orthoformates can be obtained readily from them by transesterification.[2] The following have been prepared in a similar manner by the submitters.[3]

Product	Yield, %	B.P., °C./mm.	n_D (t, °C)
N-Methylaniline	44	104–105/40	1.5701 (22)
N-Ethylaniline	66	92–93/16	—
N-Isoamylaniline	58	149–151/40	1.5212 (25)
N-Methyl-m-toluidine	67	120–121/40	1.5557 (25)
N-Ethyl-m-toluidine	69	125–127/40	1.5451 (23)
N-Methyl-p-chloroaniline	77	141–142/40	1.5799 (25)

3. Methods of Preparation

N-Ethyl-p-chloroaniline has been prepared by alkylation of p-chloroaniline with ethyl bromide [4,5] and by reduction of aceto-p-chloroanilide with lithium aluminum hydride.[6] The present procedure, which is based on the results of an investigation by Roberts and Vogt,[3] is a convenient general method for preparation of pure N-alkyl aromatic amines.

[1] University of Texas, Austin, Texas.

[2] Alexander and Busch, *J. Am. Chem. Soc.*, **74**, 554 (1952); Roberts, Higgins, and Noyes, *J. Am. Chem. Soc.*, **77**, 3801 (1955).

[3] Roberts and Vogt, *J. Am. Chem. Soc.*, **78**, 4778 (1956).

[4] Hofmann, *Ann.*, **74**, 143 (1850).

[5] Crowther, Mann, and Purdie, *J. Chem. Soc.*, **1943**, 58.

[6] Bory and Mentzer, *Bull. soc. chim. France*, **1953**, 814.

ETHYL CHLOROFLUOROACETATE

(Acetic acid, chlorofluoro-, ethyl ester)

$$\text{HCClFCF}_2\text{OC}_2\text{H}_5 + \text{H}_2\text{O} \xrightarrow{\text{H}_2\text{SO}_4} \text{HCClF}\overset{\overset{\displaystyle O}{\|}}{\text{C}}\text{OC}_2\text{H}_5 + 2\text{HF}$$

Submitted by BRUCE ENGLUND.[1]
Checked by R. S. SCHREIBER and BURRIS D. TIFFANY.

1. Procedure

Caution! Hydrogen fluoride vapors are highly corrosive and poisonous. An efficient hood should be used, and rubber gloves should be worn when dismantling the equipment.

A 2-l. three-necked round-bottomed flask is fitted with a mechanical stirrer and a 500-ml. dropping funnel (Note 1). The third neck is fitted with a thermometer and a length of rubber tubing to lead evolved hydrogen fluoride to the rear of the hood away from the operator. The flask is charged with 340 g. (2.09 moles) of crude 2-chloro-1,1,2-trifluoroethyl ethyl ether (p. 184) and cooled in an ice bath until the temperature of the halo ether is below 5°. From the dropping funnel, 228 ml. of 96% sulfuric acid (420 g., 4.1 moles) is added at such a rate that the temperature can be maintained at 5–15°. The addition requires 30–45 minutes, during which evolution of hydrogen fluoride begins. The reaction mixture is stirred at 10° for 2 hours, then carefully poured onto a mixture of 1 kg. of crushed ice and 500 ml. of water. The product, a nearly white oil, settles out as the lower layer. It is separated, washed until free of acid with three 25-ml. portions of saturated sodium bicarbonate solution (Note 2), then with four 25-ml. portions of water, and dried over 10 g. of Drierite. The weight of crude dried ester is 200–210 g. (68–71%). Fractional distillation gives 190–200 g. (65–68% yield) of pure ethyl chlorofluoroacetate, b.p. 129–130°, n_D^{25} 1.3925 (Note 3).

2. Notes

1. The hydrogen fluoride evolved etches the glass reaction vessel, but the same equipment may be used for 4–6 runs.

2. If the crude product is not washed free of acid considerable decomposition occurs during distillation with consequent reduction in yield.

3. The checkers found it necessary to use a moderately efficient (2 by 12 cm. helix-packed) column to effect satisfactory fractionation.

3. Methods of Preparation

Ethyl chlorofluoroacetate has been prepared by the action of sulfuric acid [2] or silica [3] on 2-chloro-1,1,2-trifluoroethyl ethyl ether, and by the reaction of sulfuryl chloride with ethyl fluoroacetate in the presence of benzoyl peroxide.[4] The procedure given is essentially that of Young and Tarrant.[2]

[1] Chemical Department, Experimental Station, E. I. du Pont de Nemours and Company, Wilmington, Delaware.

[2] Young and Tarrant, *J. Am. Chem. Soc.*, **71**, 2432 (1949).

[3] Hanford and Rigby, U. S. pat. 2,409,274 [*C. A.*, **41**, 982 (1947)].

[4] Bergmann, Moses, Neeman, Cohen, Kaluszyner, and Reuter, *J. Am. Chem. Soc.*, **79**, 4174 (1957).

ETHYL DIAZOACETATE *

(Acetic acid, diazo-, ethyl ester)

$$HCl \cdot NH_2CH_2CO_2C_2H_5 + NaNO_2 \rightarrow$$
$$N_2CHCO_2C_2H_5 + NaCl + 2H_2O$$

Submitted by N. E. Searle.[1]
Checked by Melvin S. Newman, G. F. Ottmann, and C. F. Grundmann.

1. Procedure

Diazoacetic esters are potentially explosive and therefore must be handled with caution. They are also toxic and prone to cause development of specific sensitivity. A well-ventilated hood should be used for the entire procedure.

A solution of 140 g. (1 mole) of ethyl glycinate hydrochloride [2] in 250 ml. of water is mixed with 600 ml. of methylene chloride in a 2-l. four-necked round-bottomed flask fitted with a stirrer, dropping funnel, thermometer, and nitrogen inlet tube, and cooled to −5° (Note 1). The flask is flushed with nitrogen and an ice-cold solution of 83 g. (1.2 moles) of sodium nitrite in 250 ml. of water is added with stirring. The temperature is lowered to −9° (Note 1), and 95 g. of 5% (by weight) sulfuric acid is added from the dropping funnel during a period of about 3 minutes (Note 2). The temperature may rise to a maximum of +1° with the cooling bath at −23° (Note 3). The reaction terminates within 10 minutes, when heat is no longer evolved.

The reaction mixture is transferred to an ice-cold 2-l. separatory funnel, and the yellow-green methylene chloride layer is run into 1 l. of cold 5% sodium bicarbonate solution. The aqueous layer is extracted once with 75 ml. of methylene chloride. The methylene chloride and sodium bicarbonate solutions are returned to the separatory funnel and shaken until no trace of acid remains, as shown by indicator paper (Note 4). The golden yellow organic layer is separated, transferred to a dry separatory funnel, and shaken for 5 minutes with 15 g. of granular anhydrous sodium sulfate. The dried ethyl diazoacetate solution is filtered through a cotton plug inserted in the separatory funnel stem, and the bulk of the solvent is distilled through an efficient column at a pressure of about 350 mm. (Note 5). The last traces of solvent are removed at a pressure of 20 mm. and a maximum pot temperature of 35° (Note 6). The yield is 90–100 g. (79–88%) of yellow oil, n_D^{25} 1.462. This product is pure enough for most synthetic work (Notes 7 and 8).

2. Notes

1. Lower temperatures may induce the solid methylene chloride dihydrate to separate and interfere with stirring. An acetone bath into which lumps of Dry Ice are introduced as required affords easily controlled cooling.

2. The stated amount of acid has been found sufficient to provide the strongly acidic medium which the reaction requires.

3. Higher reaction temperatures may result in reduced yields.

4. Traces of acid must be eliminated before the diazoacetate solution is concentrated.

5. Efficient solvent separation is indicated by absence of yellow color in the distillate. An 18-in. column packed with Berl saddles is satisfactory.

6. Higher temperatures should be avoided because of the explosive character of ethyl diazoacetate. The product should be placed in dark brown bottles and kept in a cool place. It should be used as soon as possible.

7. Distillation through a 7-in. column packed with glass helices gives a 65% over-all yield, b.p. 29–31°/5 mm. *Anal.* Calcd. for $C_4H_6N_2O_2$: N, 24.55. Found: N, 24.76. Heart-cut material has a refractive index of n_D^{25} 1.4616.[3] Principal loss is due to elimination of nitrogen with formation of high-boiling esters. Both the crude and the distilled products appear to function equally well as synthetic intermediates. *Distillation, even under reduced pressure, is dangerous, for the substance is explosive.*

8. The procedure has proved satisfactory for the preparation of the methyl, butyl, *n*-hexyl, 2-ethylhexyl, and decyl esters of diazoacetic acid.[4] *The methyl ester should be handled with particular caution since heat causes it to detonate with extreme violence.*

3. Methods of Preparation

Ethyl diazoacetate has been prepared from sodium nitrite and ethyl glycinate hydrochloride in the presence of diethyl ether.[5] The present procedure utilizes the unique ability of methylene chloride solvent to protect the diazoacetic ester from decomposition by aqueous mineral acid.[4] Hammond claims that the use of halogenated hydrocarbons, such as carbon tetrachloride, leads to a 95% yield of ethyl diazoacetate.[6]

[1] Contribution No. 360 from the Chemical Department, Experimental Station, E. I. du Pont de Nemours and Company, Wilmington, Delaware.

[2] *Org. Syntheses Coll. Vol.* **2**, 310 (1943).

[3] Data obtained by V. A. Engelhardt and H. E. Cupery.[1]

[4] Searle (to E. I. du Pont de Nemours and Company), U. S. pat. 2,490,714 [*C. A.*, **44**, 3519 (1950)].

[5] *Org. Syntheses Coll. Vol.* **3**, 392 (1955).

[6] Hammond (to National Distillers Products Corp.), U. S. pats. 2,691,649 and 2,691,650 [*C. A.*, **49**, 11690 (1955)].

ETHYL DIETHOXYACETATE *

(Glyoxylic acid, ethyl ester, diethyl acetal)

$Cl_2CHCO_2H + 3CH_3CH_2ONa \rightarrow$
$(CH_3CH_2O)_2CHCO_2Na + 2NaCl + 3CH_3CH_2OH$

$(CH_3CH_2O)_2CHCO_2Na + CH_3CH_2OH + HCl \rightarrow$
$(CH_3CH_2O)_2CHCO_2CH_2CH_3 + NaCl + H_2O$

Submitted by ROBERT BRUCE MOFFETT.[1]
Checked by CHARLES C. PRICE and CHARLES E. SCOTT.

1. Procedure

A 2-l. three-necked flask (Note 1), fitted with a sealed stirrer (Note 2) and an efficient reflux condenser (protected by a calcium chloride tube), is surrounded by a water bath. Sodium ethoxide is prepared in this flask by adding 31 g. (1.35 g. atoms) of sodium in portions to 450 ml. of absolute ethanol (Note 3). When practically all the sodium has dissolved, 50 g. (0.39 mole) of dichloroacetic acid is added with stirring at such a rate that the solvent refluxes smoothly. About 20 minutes is required. Sodium chloride soon begins to separate, and the solution becomes a yellow-orange color. After the initial reaction subsides, the mixture is heated under reflux with stirring for 3.5 hours.

The water is removed from the water bath and replaced by an ice-salt mixture. A thermometer is placed in the flask with the bulb below the surface of the liquid, and the mixture is cooled below 0°. Then a solution of about 27 g. (0.75 mole) of hydrogen chloride in 200 ml. of absolute ethanol (Note 4) is slowly added (Note 5) during about 40 minutes with stirring and cooling at such a rate that the temperature does not rise above 10°. The mixture is allowed to come to room temperature, and the stirring is continued for 3 hours. It is then allowed to stand overnight. The reaction mixture is again cooled to 0°, and the excess acid is neutralized to approximately pH 7 by slowly adding sodium ethoxide solution (Note 6). The mixture is tested from time to time during the addition of the alkali by placing a drop on moistened pH test paper. About 75 ml. of sodium ethoxide solution is required. The mixture is then filtered through a large Büchner funnel (Note 7), and the precipitate is extracted thoroughly with ether which is added to the alcoholic filtrate. The solid is discarded.

Most of the solvent is removed by distillation through a Vigreux column under reduced pressure (about 40 mm.) and a pot temperature less than 40°. The product is then transferred to a smaller flask (Note 8). The remainder of the solvent is removed through the Vigreux column at a pressure of about 15 mm. and a boiling point up to about 40°. A Dry Ice-cooled receiver (Note 9) is then attached, and the pressure is lowered by means of a high-vacuum pump. The crude product is distilled until no more material comes over while the flask is heated on a steam bath. This crude product, b.p. 87–88°/17 mm., 69–70°/10 mm., n_D^{25} 1.4073–1.4076, may be redistilled, a pinch of calcium carbonate being added, through an efficient column (Note 10) at a pressure of about 12 mm. The fore-run (boiling point up to about 60°/12 mm.) is discarded. The fraction, b.p. 60–81°/12 mm., is saved for redistillation with a subsequent run. The main fraction is obtained at a boiling range of 81–83°/12 mm. The yield of ethyl diethoxyacetate is 31–34 g. (45–50%) of colorless liquid, n_D^{25} 1.4075.

2. Notes

1. The submitter reports equally satisfactory results when the reaction is carried out in a 12-l. flask on ten times the scale described here.

2. A Hershberg stirrer is excellent for the purpose and may be connected by a mercury seal or a rubber seal [2] lubricated with glycerol (p. 546).

3. The absolute ethanol is dried with sodium and diethyl phthalate.[3]

4. The alcoholic hydrogen chloride can be prepared by passing hydrogen chloride from a cylinder (use safety trap) into 200 ml. of absolute ethanol cooled by an ice bath. From time to time a sample may be withdrawn and titrated with standard alkali to determine the concentration. The exact amount is not critical, but a considerable excess of hydrogen chloride must be used. If several runs are to be made it is convenient to prepare a large quantity of alcoholic hydrogen chloride at one time.

5. Precautions should be taken to prevent absorption of atmospheric moisture by the hydrogen chloride solution. The dropping funnel should be closed or protected by a calcium chloride tube.

6. Enough sodium ethoxide solution for four runs can be made in a 500-ml. flask by adding 18.4 g. of sodium portionwise to 300 ml. of absolute ethanol.

7. If difficulty in the filtration is encountered a filter aid may be used.

8. At this point it is convenient to combine several runs for distillation.

9. The Dry Ice-cooled receiver is conveniently constructed from a two-necked round-bottomed flask immersed up to the necks in a Dry Ice-ethanol mixture.

10. The submitter used a column packed with 12 in. of ⅛-in. glass helices and fitted with a variable reflux head. The checkers used a 12-in. Vigreux column.

3. Methods of Preparation

Ethyl diethoxyacetate has been prepared from dichloroacetic acid by the action of sodium ethoxide followed by esterification of the intermediate diethoxyacetic acid. This esterification has been carried out with ethyl iodide on the sodium salt or on the silver salt.[4-6] It has been more conveniently done with ethanol and acid.[7-9] Poorer yields are reported when the dichloroacetic acid is first esterified and then treated with sodium ethoxide.[10] Ethyl diethoxyacetate also has been prepared by the reaction of ethanol with the hemiacetal of ethyl glyoxylate in the presence of hydrogen chloride,[11] and by the treatment of 1,2-diethoxy-1,1,2-trichloroethane or 1,2-diethoxy-1-chloroethane with ethanol in the presence of pyridine.[12]

[1] The Upjohn Company, Kalamazoo, Michigan.
[2] Org. Syntheses Coll. Vol. 3, 368 (1955).
[3] Org. Syntheses Coll. Vol. 2, 155 (1943).
[4] Schreiber, Z. Chem., 1870, 167; Jahresber. Fortschr. Chem., 1870, 642.
[5] Johnson and Cretcher, Jr., J. Biol. Chem., 26, 106 (1916).
[6] Rugeley and Johnson, J. Am. Chem. Soc., 47, 2997 (1925).
[7] Wohl and Lange, Ber., 41, 3612 (1908).
[8] Blaise and Picard, Bull. soc. chim., [4] 11, 539 (1912).
[9] Johnson and Cretcher, Jr., J. Am. Chem. Soc., 37, 2147 (1915).

[10] Cope, *J. Am. Chem. Soc.*, **58**, 570 (1936).
[11] Korte, Paulus, and Störiko, *Ann.*, **619**, 63 (1958).
[12] Baganz, Domaschke, and Krüger, *Chem. Ber.*, **92**, 3167 (1959).

ETHYL ENANTHYLSUCCINATE

(Succinic acid, heptanoyl-, diethyl ester)

$$CH_3(CH_2)_5CHO + \begin{array}{c} CHCO_2C_2H_5 \\ \parallel \\ CHCO_2C_2H_5 \end{array} \xrightarrow{(C_6H_5CO_2)_2}$$

$$\begin{array}{c} O \\ \parallel \\ CH_3(CH_2)_5CCHCO_2C_2H_5 \\ | \\ CH_2CO_2C_2H_5 \end{array}$$

Submitted by TRACY M. PATRICK, JR., and FLOYD B. ERICKSON.[1]
Checked by T. L. CAIRNS and R. D. CRAMER.

1. Procedure

A mixture of 228 g. (2.0 moles) of enanthaldehyde (Note 1) and 172 g. (1.0 mole) of ethyl maleate (Note 2) is placed in a jacketed flask (Fig. 9) (Note 3) with condensers attached to the flask and the jacket openings. The jacket is charged with trichloroethylene (Note 4), which is heated to reflux, and, when the reaction mixture reaches 84–85°, 0.5 g. of benzoyl peroxide is added to the mixture through the condenser. After 3–8 hours an additional 0.5-g. portion of benzoyl peroxide is added, and heating is continued for a total of 18–24 hours (Note 5).

The reaction mixture is distilled through a short (8- to 12-in.) Vigreux column to give 108–111 g. of recovered enanthaldehyde, b.p. 64–65°/38 mm.; 6–16 g. of an intermediate fraction, b.p. 82–128°/1 mm.; and 202–216 g. (71–76% yield) of ethyl enanthylsuccinate, b.p. 119–122°/0.7 mm., n_D^{25} 1.4392–1.4398 (Notes 6, 7, and 8).

2. Notes

1. Good-quality enanthaldehyde such as the white label grade supplied by Eastman Kodak Company was distilled before use.

It is best stored in a brown bottle under nitrogen for protection against oxidation.

2. Ethyl maleate from Commercial Solvents Corporation, Terre Haute, Indiana, was distilled before use. Ethyl fumarate can be used also, but the yield of product is lower since a much greater proportion of high-boiling compounds is obtained. Aldehydes do not undergo free-radical addition to maleic anhydride under these conditions.

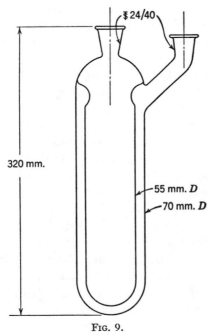

FIG. 9.

3. Although this piece of apparatus is not essential, it provides a convenient means of controlling the temperature for a mildly exothermic reaction which must be carried out overnight without close supervision. The submitters carried out one experiment in an ordinary apparatus heated by a mantle. The poorer temperature control (70–91°) resulted in a yield of only 44%.

4. Any other stable liquid having a boiling point in the neighborhood of 80–90° could be substituted for trichloroethylene.

5. The reaction time is not critical. For the sake of convenience the reaction can be started early in the day, the second

portion of peroxide added before leaving in the evening, and heating discontinued upon arrival the following morning.

6. The submitters collected material boiling in the range 128–150°/1 mm. and found this product to be satisfactory for the preparation of γ-oxocapric acid (Note 7).

7. The submitters state that ethyl enanthylsuccinate can be hydrolyzed and decarboxylated to γ-oxocapric acid. The reaction is carried out by heating 57 g. of the keto ester with a solution of 140 ml. of concentrated sulfuric acid in 250 ml. of water. The mixture is stirred while the ethanol is removed gradually by distillation over a period of 3 hours. The acid is taken up in benzene, extracted from the benzene with aqueous alkali, and liberated from the alkaline solution by acidification with concentrated hydrochloric acid. After recrystallization from 50% ethanol about 29 g. (78% yield) of γ-oxocapric acid is obtained as colorless crystals, m.p. 69–70°.

The submitters also state that γ-caprilactone can be prepared by hydrogenation of 50 g. of γ-oxocapric acid in 150 ml. of ethanol over 5 g. of Raney nickel at 135° and 1000 p.s.i. Filtration and distillation gives about 38 g. (83% yield) of γ-caprilactone, b.p. 109–110°/2.5 mm., n_D^{25} 1.4470.

8. A number of other acylsuccinic esters can be prepared in a similar fashion from the appropriate saturated aliphatic aldehydes and maleic esters.[2,3] The procedure is equally adaptable to the preparation of α-acyltricarballylic esters from aldehydes and aconitates.[2] Good temperature control (80–90°) is most important for success in these reactions.

3. Methods of Preparation

Ethyl enanthylsuccinate has been prepared by the free-radical addition of enanthaldehyde to ethyl maleate.[2]

[1] Monsanto Chemical Company, Dayton, Ohio.
[2] Patrick, *J. Org. Chem.*, **17**, 1009 (1952).
[3] Ladd, U. S. pat. 2,577,133 [*C. A.*, **46**, 6147 (1952)].

ETHYLENIMINE *

$$\underset{\substack{|\\ \overset{-}{O}SO_3^-}}{CH_2CH_2\overset{+}{N}H_3} \xrightarrow{\text{NaOH}} \underset{NH}{CH_2 \diagdown \diagup CH_2}$$

Submitted by C. F. H. ALLEN, F. W. SPANGLER, and E. R. WEBSTER.[1]
Checked by R. S. SCHREIBER, A. C. OTT, and M. F. MURRAY.

1. Procedure

Caution! This preparation should be carried out in a good hood, and it is advisable to use rubber gloves.

In a 5-l. flask surmounted by a water-cooled still head connected to a 30-in. spiral condenser set for downward distillation and connected to a well-cooled receiver (Note 1), 564 g. (4 moles) of β-aminoethylsulfuric acid (Note 2) is mixed with 1760 g. (1230 ml.) of 40% sodium hydroxide solution (704 g. of sodium hydroxide in 1056 ml. of water). The mixture is heated with a free flame until it just begins to boil. At this point external heating is discontinued (Note 3). The reaction that begins at the boiling point keeps the mixture boiling for several minutes. When this initial reaction has subsided, heating is resumed and about 500 ml. of distillate is collected as quickly as possible in the well-cooled receiver. To the chilled distillate 450–500 g. of potassium hydroxide pellets is added gradually, whereupon the imine separates as an upper layer. The organic layers from four such 4-mole runs are combined and left overnight in a refrigerator over about 400 g. of potassium hydroxide pellets. The aqueous layers are combined and distilled through a wrapped 10-in. Vigreux column attached to a 30-in. spiral condenser. The distillate boiling at 50–100° is chilled thoroughly, and 200–250 g. of potassium hydroxide pellets is added gradually. The upper layer of crude ethylenimine is separated and combined with the larger portion of base (Notes 4 and 5).

If an aqueous layer appears during drying of the combined organic layers, the upper layer (about 575–600 g.) is again separated, 200 g. of potassium hydroxide pellets is added, and the whole is distilled through the same apparatus as that used for distilling the aqueous portion. If no layer appears, the base is decanted from the hydroxide and distilled from a fresh 200-g. portion of potassium hydroxide. The fraction boiling at 50–100° (about 350 g.) is collected and dried over 100 g. of potassium hydroxide pellets.

The crude ethylenimine is separated and dried over fresh 100-g. portions of potassium hydroxide until an aqueous layer no longer appears (Note 6). It is then decanted from the drying agent and redistilled from 100 g. of potassium hydroxide.

The yield of ethylenimine (b.p. 56–58°) is 235–250 g. (34–37%). A stick of sodium hydroxide is added to act as a preservative, and the material is best stored in sealed bottles in a refrigerator (Notes 7, 8, and 9).

2. Notes

1. Cooling the receiver in a freezing mixture will cut the loss of the distillate to a minimum.

2. β-Aminoethylsulfuric acid of excellent quality is available from the B. F. Goodrich Company.

3. It is well to have an ice bath available to control the exothermic reaction, which may become quite violent.

4. The use of an efficient distilling column is recommended because the crude base contains higher-boiling by-products. One of these is the dimer, N-β-aminoethylethylenimine; b.p. 126–127.5°.

5. It has been suggested that the portion of ethylenimine, boiling at 50–100°, might be collected directly on distillation without separating the organic layer from the aqueous potassium hydroxide layer. This is not advisable, because heating ethylenimine in the presence of a base appears to increase polymerization. The quantity of the organic base contained in the concentrated aqueous solution of potassium hydroxide is sufficient, however, to warrant this distillation of the aqueous layer.

6. If the original separation is done carefully and if sufficient potassium hydroxide is used, an aqueous layer will separate during the first drying only. Should this not be the case, it may be

worth while to combine all aqueous portions obtained and redistil them to obtain any material boiling at 50–100°.

7. Yields of 26.5% and 32% of ethylenimine have been reported.[2,3]

8. Ethylenimine is strongly caustic and burns the skin. Inhalation of the vapor causes acute inflammation of the eyes, nose, and throat, with symptoms resembling those of bronchitis. After two or three days, the irritation subsides and the tissues return to normal, without suffering any apparent permanent injury. Continued exposure to the vapor may cause an individual to acquire an extreme sensitivity to it. Ethylenimine is also very inflammable and polymerizes with explosive violence under certain conditions.[4,5]

9. Redistillation over fresh potassium hydroxide of the residue from this final distillation gives an additional 10–15 g. of ethylenimine, boiling at 56–58°. This redistillation is advisable when the residues from three to four 16-mole batches are combined.

3. Methods of Preparation

Ethylenimine has been prepared from β-bromoethylamine hydrobromide by reaction with silver oxide,[6] potassium hydroxide,[7] or sodium methoxide;[8] from β-chloroethylamine hydrochloride by reaction with sodium methoxide[8] or sodium hydroxide;[9] from β-aminoethylsulfuric acid by reaction with sodium hydroxide;[2–4,10,11] and by heating oxazolidone, or substances yielding it, to 100–300°.[12]

[1] Eastman Kodak Company, Rochester, New York.

[2] Wenker, *J. Am. Chem. Soc.*, **57**, 2328 (1935); Leighton, Perkins, and Renquist, *J. Am. Chem. Soc.*, **69**, 1540 (1947).

[3] Jones, Langsjoen, Neumann, and Zomlefer, *J. Org. Chem.*, **9**, 125 (1944).

[4] Mills and Bogert, *J. Am. Chem. Soc.*, **62**, 1177 (1940).

[5] Pingree, *Am. Dyestuff Reptr.*, **35**, 124 (1946).

[4] Gabriel, *Ber.*, **21**, 1049 (1888).

[7] Gabriel, *Ber.*, **21**, 2665 (1888); Gabriel and Stelzner, *Ber.*, **28**, 2929 (1895).

[8] Knorr and Meyer, *Ber.*, **38**, 3130 (1905).

[9] U. S. pat. 2,212,146 [*C. A.*, **35**, 463 (1941)].

[10] Brit. pat. 460,888 [*C. A.*, **31**, 4676 (1937)].

[11] Reeves, Drake, and Hoffpauir, *J. Am. Chem. Soc.*, **73**, 3522 (1951).

[12] Sundén (to Stockholms Superfosfat Fabriks A/B), Swed. pat. 148,559 [*C. A.*, **50**, 2679 (1956)].

2-ETHYLHEXANONITRILE

(Hexanenitrile, 2-ethyl-)

$$CH_3(CH_2)_3CH(C_2H_5)CONH_2 + SOCl_2 \rightarrow$$
$$CH_3(CH_2)_3CH(C_2H_5)CN + SO_2 + 2HCl$$

Submitted by John A. Krynitsky and Homer W. Carhart.[1]
Checked by H. R. Snyder and Richard S. Colgrove.

1. Procedure

In a 1-l. round-bottomed flask, bearing an efficient reflux condenser (Note 1), are placed 286 g. (2 moles) of 2-ethylhexanamide (Note 2), 300 ml. of dry benzene (Note 3), and 357 g. (218 ml., 3 moles) of thionyl chloride (Note 4). The flask is placed in a water bath, which is heated quickly to 75–80° and maintained at that temperature for 4.5 hours (Note 5). The reaction mixture is transferred to a 1.5-l. beaker and cooled in an ice bath. A mixture of 100 g. of crushed ice and 100 ml. of water is added to decompose the excess thionyl chloride. Cold 50% potassium hydroxide solution is added in small portions, with stirring, until the mixture is alkaline to litmus (Note 6). The mixture is transferred to a separatory funnel, and the layers are separated. The aqueous portion is extracted with 100 ml. of benzene. The benzene solutions are combined and washed once with 150 ml. of 1% sodium carbonate solution and twice with 150-ml. portions of water (Note 7). The mixture is distilled from a modified Claisen flask, the bulk of the solvent being removed at atmospheric pressure. The yield of nitrile is 215–236 g. (86–94%); b.p. 118–120°/100 mm. (Notes 8, 9, and 10).

2. Notes

1. Unless the temperature of the water supplied to the condenser is below 20°, a larger amount of thionyl chloride may be required. The condenser should be attached to a gas trap.

2. The 2-ethylhexanamide was prepared in 86–88% yield from technical 2-ethylhexanoic acid (Carbide and Carbon Chemicals Corporation) by a method similar to that described previ-

ously,[2] except that the crude amide was filtered directly from the reaction mixture, washed well with water, and dried. The resulting product, which melted at 99–101°, was used without further purification. If the pure amide is desired, this product may be recrystallized (with 83–90% recovery) from 50% ethanol. For 100 g. of amide, 2 l. of 50% ethanol is used, and the hot solution is decolorized with charcoal. The product thus obtained is in the form of white needles which melt at 102–103°.

3. Benzene dried over sodium was used.

4. The submitters used Eastman Kodak Company white label grade thionyl chloride. The checkers purified commercial thionyl chloride (Hooker Electrochemical Company, refined grade) by the method of Cottle.[3]

5. It is advisable to carry out the reaction in a hood, as hydrogen chloride and sulfur dioxide are evolved. The evolution of gases stops just before the end of the specified heating period.

6. Approximately 150–200 ml. of the potassium hydroxide solution is required.

7. Drying of the solution is unnecessary, since the water present is removed in the next step by azeotropic distillation with benzene.

8. The checkers collected the product at 70.5–72°/10 mm.; n_D^{25} 1.4145.

9. Benzonitrile also can be prepared by this method in comparable yields, but a longer reaction time (7 hours) is required. The method was found to be unsatisfactory for the preparation of the nitriles of azelaic and phthalic acids from the corresponding diamides.

10. The submitters prepared palmitonitrile in more than 90% yield by heating the amide with a benzene solution of thionyl chloride for 6 hours followed directly by distillation. The nitrile so obtained was contaminated with a product having a strong sulfurous odor which could be removed by washing with aqueous mercuric acetate. By the described procedure, palmitonitrile free from objectionable odor was prepared in 80% yield. Troublesome emulsions were formed during the neutralization and washing steps; the addition of a small amount of ethanol aided in breaking the emulsions.

3. Methods of Preparation

Racemic 2-ethylhexanonitrile has been obtained only by the action of phosphorus pentachloride on 2-ethylhexanaldoxime;[4] the levorotatory form has been prepared from the active amide by the method described.[5] Other amides have been converted to nitriles by dehydration with thionyl chloride.[5, 6]

[1] Naval Research Laboratory, Washington, D. C.
[2] *Org. Syntheses Coll. Vol.* **3**, 490 (1955).
[3] Cottle, *J. Am. Chem. Soc.*, **68**, 1380 (1946).
[4] von Braun and Manz, *Ber.*, **67**, 1696 (1934).
[5] Levene and Kuna, *J. Biol. Chem.*, **140**, 263 (1941).
[6] Michaelis and Siebert, *Ann.*, **274**, 312 (1893).

ETHYL ISOCYANIDE

$$C_2H_5I + AgCN \rightarrow C_2H_5NC \cdot AgI$$
$$C_2H_5NC \cdot AgI + 2KCN \rightarrow C_2H_5NC + KAg(CN)_2 + KI$$

Submitted by H. L. JACKSON and B. C. McKUSICK.[1]
Checked by N. J. LEONARD and L. E. COLEMAN, JR.

1. Procedure

Caution! This preparation should be carried out in a well-ventilated hood because ethyl isocyanide has a vile odor. Since ethyl isocyanide has been known to explode,[2] all operations in which it is heated, including its distillation, should be carried out behind a shield of safety glass.

Silver cyanide (454 g., 3.40 moles) is added with stirring to 530 g. (3.40 moles) of ethyl iodide in a 3-l. three-necked round-bottomed flask, equipped with a reflux condenser and a sealed Hershberg stirrer. The third neck of the flask is closed with a stopper. The lower third of the flask is immersed in a steam bath, and the mixture is stirred vigorously until it turns to a viscous, homogeneous, brown liquid (1.7–2.3 hours). Stirring is interrupted, the steam bath is removed, and the stirrer is raised to a position just above the liquid (Note 1). Water (300 ml.) is added through the condenser to avoid loss of product during the addition. Potassium cyanide (610 g., 9.37 moles) and 260 ml. of water are then added through the third neck of the flask, and the mixture is stirred for about 10 minutes, during which the heavy brown liquid below the aqueous solution disappears and a brown layer of ethyl isocyanide appears above the aqueous solution. Stirring is discontinued, the reflux condenser is replaced by one arranged for distillation, and a thermometer extending into the aqueous layer is placed in the third neck. A receiver immersed in an ice bath is attached to the condenser, and the reaction mixture is heated by means of an electric heating mantle, causing a mixture of oil and water to distil. When the distillate contains almost no oil (Note 2), the distillation is discontinued (Note 3). The receiver contains about 200 ml. of crude ethyl isocyanide and 50 ml. of water. Sodium chloride (7 g.) is dissolved in the aqueous layer, and the ice-cold mixture is poured into a separatory funnel. The aqueous layer is separated and discarded. The ethyl isocyanide is washed with two 50-ml. portions of ice-cold saturated aqueous sodium chloride solution and is dried overnight with 10 g. of anhydrous magnesium sulfate. The decanted material is distilled through a 5- to 10-plate column (Note 4), giving 88–102 g. (47–55% yield) of ethyl isocyanide, b.p. 77–79°/760 mm., n_D^{20} 1.3632. The fore-run, b.p. 63–77°/760 mm., amounts to 15–46 g. and contains 10–15% of ethyl iodide and possibly a few drops of water. Additional pure ethyl isocyanide can be obtained by drying and redistilling this fore-run.

2. Notes

1. Sometimes the liquid crystallizes to a dense solid which immobilizes the stirrer if it has not been raised. The crystallization has no effect on subsequent steps except to necessitate a longer period of stirring after the potassium cyanide is added.

2. The temperature of the residual mixture is 115–120° at this point.

3. The aqueous cyanide solution, which is very toxic, can be disposed of by flushing it down the drain with a large volume of water.

4. The submitters used an 18-in. spinning-band column (inside

diameter 10 mm.).   The checkers employed a 12-in. helix-packed column.

3. Methods of Preparation

Ethyl isocyanide has been obtained by treating ethylamine and chloroform with potassium hydroxide,[3] by pyrolyzing the complex between ethyl isothiocyanate and triethylphosphine,[4] by heating cyanocobaltic (III) acid with ethanol,[5] by passing ethylene and hydrogen cyanide through an electric discharge,[6] or by heating ethyl iodide with silver cyanide [7-9] or other metal cyanides.[8, 10] The present procedure differs from earlier ones using silver cyanide mainly in that a stirrer is employed.   By this modification more than half of the silver cyanide can be converted to ethyl isocyanide, contrary to the belief of the earlier workers, who thought that this was precluded by formation of a complex, $C_2H_5NC \cdot AgCN$.

[1] Chemical Department, Experimental Station, E. I. du Pont de Nemours and Company, Wilmington, Delaware.

[2] Lemoult, *Compt. rend.*, **143**, 902 (1906).

[3] Hofmann, *Ann.*, **146**, 109 (1868).

[4] Hofmann, *Ber.*, **3**, 766 (1870).

[5] Hölzl, Meier-Mohar, and Viditz, *Monatsh.*, **53–54**, 237 (1929).

[6] Francesconi and Ciurlo, *Gazz. chim. ital.*, **53**, 327 (1923).

[7] Gautier, *Ann. chim. et phys.*, [4] **17**, 233 (1869).

[8] Guillemard, *Ann. chim. et phys.*, [8] **14**, 363 (1908).

[9] Lowry and Henderson, *Proc. Roy. Soc.* (*London*), **A136**, 487 (1932).

[10] Guillemard, *Bull. soc. chim. France*, [4] **1**, 530 (1907).

β-ETHYL-β-METHYLGLUTARIC ACID *

(Glutaric acid, 3-ethyl-3-methyl-)

$$CH_3COC_2H_5 + 2NCCH_2CO_2C_2H_5 + 2NH_3 \xrightarrow{CH_3CO_2^-NH_4^+}$$

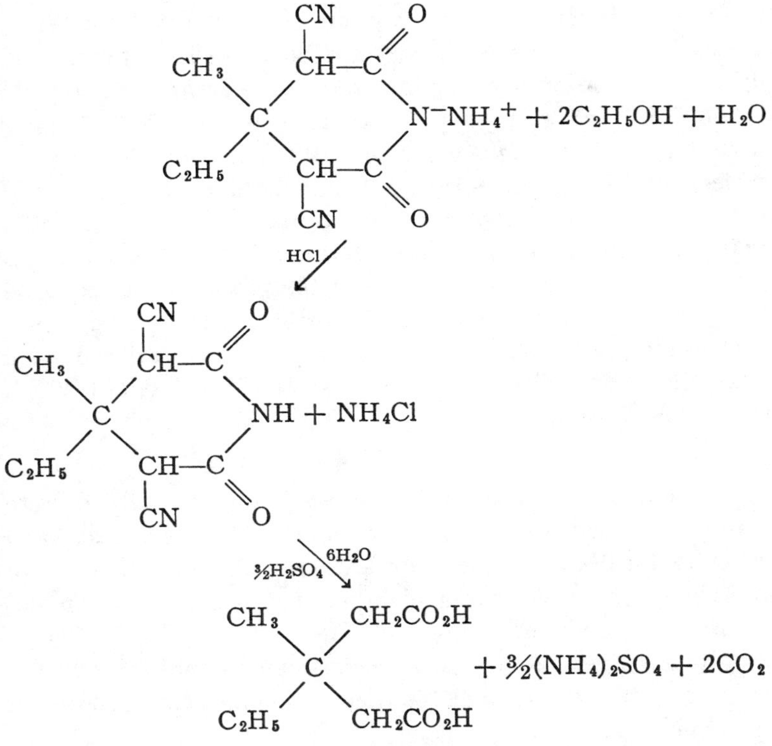

$+ 2C_2H_5OH + H_2O$

$+ NH_4Cl$

$+ \frac{3}{2}(NH_4)_2SO_4 + 2CO_2$

Submitted by H. H. FARMER and NORMAN RABJOHN.[1]
Checked by MELVIN S. NEWMAN and ROBERT HARPER.

1. Procedure

A. *α,α′-Dicyano-β-ethyl-β-methylglutarimide.* In a 2-l. round-bottomed flask are placed 452 g. (4.0 moles) of ethyl cyanoacetate, 144 g. (2.0 moles) of methyl ethyl ketone, 2 g. of ammonium acetate, and 800 ml. of 95% ethanol which contains 80 g. (4.7 moles) of anhydrous ammonia (Notes 1 and 2). The flask is stoppered and placed in a refrigerator.

After about 12 hours, the ammonium salt of the imide which has precipitated is removed by filtration (Note 3), washed on a Büchner funnel with about 200 ml. of ether and air-dried. It is dissolved in the minimum amount (about 800 ml.) of boiling water, and the solution is made acid to Congo red paper with concentrated hydrochloric acid. The free imide precipitates immediately and forms a white slurry which is cooled in an ice bath. The imide is collected on a Büchner funnel and dried at 100° in an oven, or in a vacuum desiccator. The yield is 266–287 g. (65–70%), m.p. 187–191° (Note 4).

B. *β-Ethyl-β-methylglutaric acid.* A mixture of 280 g. (1.36 moles) of α,α'-dicyano-β-ethyl-β-methylglutarimide (Note 5) and 1.35 l. of 65% (by weight) sulfuric acid is heated under reflux in a 5-l. round-bottomed flask for 8–10 hours (Note 6). The mixture is allowed to cool, and the precipitate is removed by filtration on a sintered glass (or other suitable type) filter. The crude β-ethyl-β-methylglutaric acid is recrystallized from water (Note 7). The yield of product is 174–191 g. (73–80%), m.p. 78–80°.

2. Notes

1. The ethyl cyanoacetate and methyl ethyl ketone were Eastman Kodak Company white label grade chemicals and were used without further purification. Commercial absolute ethanol was found to give a slightly better yield of the ammonium salt of the imide.

2. It is convenient to pass gaseous ammonia into the ethanol. It is advisable to carry out the preparation and manipulation of the ammonia solution in a hood.

3. The mother liquor is returned to the refrigerator and a second and a third crop of crystals may be collected after 24 and 48 hours. The first crop of crystals usually comprises 95% of the total yield.

4. In taking the melting point of this compound there was still some solid remaining at 200°. The crude imide is satisfactory for conversion to β-ethyl-β-methylglutaric acid and need not be dried before hydrolysis.

5. Although the ammonium salt of the imide may be hydrolyzed to the acid, the free imide appears to give better results.

6. The large flask is used because considerable foaming occurs during the first 2–3 hours of reaction.

7. A ratio of 3 ml. of water to 1 g. of acid gives satisfactory results. The use of activated carbon during the recrystallization is recommended.

3. Methods of Preparation

α,α'-Dicyano-β-ethyl-β-methylglutarimide apparently has been prepared only from the condensation of methyl ethyl ketone, ethyl cyanoacetate, and ammonia.[2-6]

β-Ethyl-β-methylglutaric acid has been prepared by the acid hydrolysis of α,α'-dicyano-β-ethyl-β-methylglutarimide,[2-6] 3-cyano-4-ethyl-6-imino-2-keto-4-methylpiperidine-5-carboxamide or the diimide of β-ethyl-β-methylpropane-α,α,α',α'-tetracarboxylic acid;[7] by the oxidation of β-ethyl-β-methyl-δ-valerolactone with chromic acid;[8] and by the reaction of sodium hypobromite on 1,4-dimethyl-1-ethyl-3,5-cyclohexanedione.[9]

The present procedure is essentially that of Guareschi[2] as detailed by Vogel.[4]

[1] University of Missouri, Columbia, Missouri.

[2] Guareschi, *Atti. accad. sci. Torino*, **36**, 443 (1900–1901) [*Chem. Zentr.*, **1901**, I, 821].

[3] Kon and Thorpe, *J. Chem. Soc.*, **115**, 686 (1919).

[4] Vogel, *J. Chem. Soc.*, **1934**, 1758.

[5] Benica and Wilson, *J. Am. Pharm. Assoc.*, **39**, 451 (1950).

[6] Lukeš and Ferles, *Collection Czechoslov. Chem. Communs.*, **16**, 252; *Chem. listy*, **45**, 386 (1951) [*C. A.*, **47**, 5870 (1953)].

[7] Thole and Thorpe, *J. Chem. Soc.*, **99**, 422 (1911).

[8] Sircar, *J. Chem. Soc.*, **1928**, 898.

[9] Becker and Thorpe, *J. Chem. Soc.*, **121**, 1303 (1922).

4-ETHYL-2-METHYL-2-OCTENOIC ACID

(2-Octenoic acid, 4-ethyl-2-methyl-)

$$n\text{-}C_4H_9\underset{\underset{C_2H_5}{|}}{CH}CHO + Br\underset{\underset{CH_3}{|}}{CH}CO_2C_2H_5 + Zn \rightarrow C_4H_9\underset{\underset{C_2H_5}{|}}{CH}\overset{\overset{OZnBr}{|}}{CH}\underset{\underset{CH_3}{|}}{CH}CO_2C_2H_5$$

$$2C_4H_9\underset{\underset{C_2H_5}{|}}{CH}\overset{\overset{OZnBr}{|}}{CH}\underset{\underset{CH_3}{|}}{CH}CO_2C_2H_5 + H_2SO_4 \rightarrow$$

$$2C_4H_9\underset{\underset{C_2H_5}{|}}{CH}\overset{\overset{OH}{|}}{CH}\underset{\underset{CH_3}{|}}{CH}CO_2C_2H_5 + ZnBr_2 + ZnSO_4$$

$$3C_4H_9\underset{\underset{C_2H_5}{|}}{CH}\overset{\overset{OH}{|}}{CH}\underset{\underset{CH_3}{|}}{CH}CO_2C_2H_5 + POCl_3 + 3\;\text{[pyridine]} \rightarrow$$

$$3C_4H_9\underset{\underset{C_2H_5}{|}}{CH}CH{=}\underset{\underset{CH_3}{|}}{C}CO_2C_2H_5 \;(\text{and}\; C_4H_9\underset{\underset{C_2H_5}{|}}{C}{=}CH\underset{\underset{CH_3}{|}}{CH}CO_2C_2H_5)$$

$$+ H_3PO_4 + 3\;\text{[pyridine]}\cdot HCl$$

$$C_4H_9\underset{\underset{C_2H_5}{|}}{C}{=}CH\underset{\underset{CH_3}{|}}{CH}CO_2C_2H_5 \xrightarrow[\text{Heat}]{H_2SO_4} C_4H_9\underset{\underset{O}{\|}}{\underset{O}{\overset{\overset{C_2H_5}{|}}{\big\langle}}}CH_3$$

$$C_4H_9\underset{\underset{C_2H_5}{|}}{CH}CH{=}\underset{\underset{CH_3}{|}}{C}CO_2C_2H_5 \xrightarrow[(2)\ HCl]{(1)\ KOH}$$

$$C_4H_9\underset{\underset{C_2H_5}{|}}{CH}CH{=}\underset{\underset{CH_3}{|}}{C}CO_2H + C_2H_5OH + KCl$$

Submitted by KENNETH L. RINEHART, JR., and EDWARD G. PERKINS.[1]
Checked by MELVIN S. NEWMAN and JOSEPH H. MANHART.

1. Procedure

A 3-l. three-necked flask is equipped with a mercury-sealed mechanical stirrer prepared from tantalum wire, a condenser arranged for distillation, and a 500-ml. pressure-equalizing dropping funnel. The flask is heated on a steam cone, and a slow stream of nitrogen is introduced from a cylinder through a line connected to the top of the dropping funnel. To the flask are added 98.1 g. (1.50 g. atoms) (Note 1) of freshly sandpapered zinc foil which has been cut into narrow strips and rolled loosely, and 750 ml. of thiophene-free benzene previously dried over sodium. To dry the apparatus and contents, 175–200 ml. of benzene is slowly distilled with stirring. Heating is interrupted, and the condenser is quickly arranged for reflux. A U-tube just closed with mercury is attached to the top of the condenser, and nitrogen flow is adjusted so that it bubbles slowly through the mercury. The benzene is heated once again to reflux, and a solution of 64.1 g. (0.50 mole) of 2-ethylhexanal (Note 2) and 271.5 g. (1.50 moles) of ethyl α-bromopropionate (*Caution! Note 3*) in 500 ml. of dried benzene is placed in the dropping funnel. The first 50 ml. of the solution is added to the flask at once. Usually reaction begins immediately, as evidenced by darkening of the zinc surface and clouding of the solution, but in some cases as much as 15 minutes elapses before the start of reaction. When the reaction has started, the remainder of the aldehyde-bromo ester solution is added during 1 hour, with stirring, as the solution is maintained under reflux. After addition is complete the mixture is heated for an additional 2 hours under reflux, and then cooled to room temperature.

The nitrogen line is removed, and to the solution is added 750 ml. of 12N sulfuric acid; the resulting mixture is stirred vigorously for 1 hour, and then decanted into a 3-l. separatory funnel. After the two phases have separated, the lower aqueous layer is drawn off into a 3-l. separatory funnel containing 1 l. of water which has been used to wash the reaction flask, and the diluted mixture is extracted twice with 350-ml. portions of benzene which have also been employed to rinse the reaction flask. The original organic layer and the combined benzene extracts are kept separate and are washed successively with 500-ml. portions of water, saturated sodium bicarbonate solution, and again with water. The two organic portions are now combined, allowed to stand over anhydrous sodium sulfate until clear, then transferred to a 2-l. distilling flask, from which solvent is distilled at atmospheric pressure, last traces under aspirator pressure. The residue remaining in the flask weighs 150–165 g. (Note 4) and is dehydrated without further purification.

To the residue which contains the crude hydroxy ester there is added 710 g. of pyridine (commercial reagent, c.p. grade), and the mixture is cooled to about 5° in an ice bath. To the cooled solution is added slowly with vigorous swirling 155 g. of phosphorus oxychloride (commercial reagent, c.p.); white crystals form almost immediately. The mixture is allowed to stand 8 hours at room temperature and is finally heated for 1.5 hours on the steam bath (Note 5). It is then cooled to room temperature and decanted into a 5-l. separatory funnel containing 1.25 kg. of cracked ice. Crystals that remain in the flask are decomposed with an additional 125 g. of ice. The flask is rinsed with 1.5 l. of water, and then with 400 ml. of hexane. These washes are added to the material in the separatory funnel. After the two layers have been shaken together thoroughly and then separated, the aqueous phase is extracted with two additional 400-ml. portions of fresh hexane. The three hexane extracts are not combined, but are washed in turn with two 500-ml. portions of $2N$ hydrochloric acid to remove pyridine. Excess hydrochloric acid is removed by washing with three 200-ml. portions of water to pH 4. The clear amber-colored solution is dried over anhydrous sodium sulfate, and solvent is removed by distillation as before. The residue, which weighs 120–135 g. and consists mainly of ethyl 4-ethyl-2-methyl-2-(and-3-)octenoates, is not distilled but is heated with sulfuric acid to convert the Δ^3-isomer to γ-lactone (Notes 6, 7, and 8).

To the residue is added 600 ml. of ethylene glycol, followed by 40 ml. of concentrated sulfuric acid. The resulting mixture is heated under vigorous reflux for 20 hours, and then cooled and transferred to a 2-l. stainless-steel or copper flask. To this material is added a solution prepared from 325 g. of potassium hydroxide, 300 ml. of water, and 300 ml. of 95% ethanol. The resultant solution is heated under reflux for 1 hour, and then cooled and transferred to a 5-l. separatory funnel, where it is diluted with 3 l. of water and acidified with 600 ml. of concentrated hydrochloric acid. The organic liquid which separates is diluted with 300 ml. of hexane and separated from the aqueous layer, which is extracted three additional times with 300-ml. portions of hexane. The hexane extracts are washed to pH

4 with three 300-ml. portions of water, and then combined and dried over anhydrous sodium sulfate. For removal of solvent by flash distillation, a 150-ml. round-bottomed flask, equipped with a ground joint attached to a distillation head and side tubulature attached to a dropping funnel, is heated by an oil bath whose temperature is maintained at 130–140°. After all the solution has been added at about the rate of distillation, the dropping funnel is replaced by a capillary, and the last of the solvent is removed at reduced pressure furnished by a water pump. The flask is finally attached to an efficient fractionating column (Note 9), and the residue of mixed α,β-unsaturated acid and γ-lactone is distilled carefully at reduced pressure. After 1–5 g. of fore-run, there are obtained 25–28.5 g. (27–31%) of 4-ethyl-4-hydroxy-2-methyloctanoic acid, γ-lactone, b.p. 118–121°/ 5.0 mm., n_D^{25} 1.4469–1.4473 (Note 10), 2–3 g. of intermediate, and 27.5–32 g. (30–35%) of 4-ethyl-2-methyl-2-octenoic acid, b.p. 140– 143°/5.0 mm., n_D^{25} 1.4613–1.4625 (Note 11).

2. Notes

1. The yield in this reaction is improved by an excess of zinc and bromo ester relative to aldehyde. The present ratio of zinc:bromo ester:aldehyde (3:3:1) gives 87% of intermediate β-hydroxy ester; when the ratio is reduced to 2:3:1, the yield is lowered to about 68%.

2. Commercially available 2-ethylhexanal (Eastman practical grade) is purified by fractional distillation; b.p. 163–163.2°/atm., n_D^{25} 1.4133. Other aldehydes are conveniently prepared by the Rosenmund reduction.[2] If the aldehyde is relatively unstable toward autoxidation,[3] a catalytic amount (0.5–1.0 g.) of hydroquinone is added with the aldehyde-bromo ester solution.

3. Ethyl α-bromopropionate is available commercially (Sapon Laboratories, n_D^{25} 1.4452) and is employed without purification. *Bromo esters are severe lachrymators, and operations that involve transferring these compounds from one vessel to another should be conducted in a well-ventilated hood.*

4. If ethyl 4-ethyl-2-methyl-3-hydroxyoctanoate is isolated by distillation of this residue, the yield is about 100 g. (87%), b.p. 122–124°/ 4.9 mm., n_D^{25} 1.4415.

5. During heating, the mixture becomes dark brown; however, most of the color is removed by subsequent washing with hydrochloric acid.

If terminal heating is omitted, the yield in the dehydration step is reduced by approximately 15%.

6. If no separation of isomers is required, as when the mixture is to be hydrogenated, the mixed esters may be obtained by distillation; yield 75–90 g. (71–85%, based on starting aldehyde). It is extremely difficult to separate the Δ^2- from the Δ^3-ester by fractional distillation, as the two boil only 7° apart; however, by careful fractionation and refractionation through an 0.8 × 125-cm. simple Podbielniak column with partial-reflux head,[4] it is possible to obtain [3] pure ethyl 4-ethyl-2-methyl-2-octenoate, b.p. 102–103°/4.8 mm., n_D^{25} 1.4478, and a nearly pure sample of ethyl 4-ethyl-2-methyl-3-octenoate, b.p. 94–95°/4.8 mm., n_D^{25} 1.4393.

7. When α,γ-dialkyl-β,γ-unsaturated esters and acids are heated with acid, they are slowly converted to γ-lactones. The corresponding α,β-unsaturated isomers are recovered unchanged.[3] Treating a mixture of the two isomeric esters or acids with sulfuric acid in refluxing glycol thus destroys the unconjugated isomer, while leaving the conjugated compound intact. Whereas the isomeric esters and acids have very similar boiling points (cf. Note 6), the γ-lactone boils 20–25° lower than the Δ^2-acid and thus may be separated easily from the acid by fractional distillation. The acid may also be extracted from the lactone with sodium carbonate, or its barium salt may be precipitated by methanolic barium hydroxide.

Lactonization as a means of obtaining Δ^2-acid free from Δ^3-isomer is useful for unsaturated acids or esters having an α-alkyl substituent. Conjugated acids or esters without an α-alkyl substituent undergo acid-catalyzed isomerization to the Δ^3-isomer and subsequent lactonization with loss in yield of the Δ^2-compound.[5] In the latter case, use may be made of the differential rate either of bromine addition [6] or of esterification [3] for the conjugated and unconjugated compounds as a means of obtaining pure conjugated acid.

8. In the present case, the proportion of Δ^2-isomer in the original dehydration mixture may be estimated to be 42% of the total unsaturated esters, while an equilibrium mixture contains about 67% of the conjugated compound.[3] Thus the amount of conjugated isomer in the mixture may be considerably increased by base-catalyzed isomerization of the unsaturated esters. For other unsaturated esters, both the composition of the dehydration mixture [3,7] and the equilibrium ratio of the two isomers [3,8] vary, depending on the position and

nature of alkyl substituents on the chain, and equilibration is not always desirable. For suitable compounds the following procedure is advantageous; scrupulous protection against moisture is essential.

A 2-l. round-bottomed flask is fitted with a coil condenser (cooling water *inside* the coil) having a large free space in the center and is protected from atmospheric moisture by a calcium chloride tube. The apparatus is dried thoroughly with a Bunsen burner, and 1.5 l. of commercial ethylene glycol is introduced into the flask, together with 50 ml. of diethyl phthalate. The glycol is heated (with salt bath or electric mantle) to a temperature slightly under reflux, the calcium chloride tube is removed, and 50 g. of sodium is added cautiously, in suitable pieces, through the condenser. It is necessary to wait after the addition of each piece, for sodium melts at these temperatures and dissolves in glycol exothermically with vigorous evolution of hydrogen. After all the sodium has been added, the solution is heated for 1 hour under reflux. The condenser is arranged for distillation into a graduated thoroughly dried pressure-equalizing dropping funnel, and, as before, the system is protected by a calcium chloride tube. Heating is resumed, and the first 300 ml. of distilled glycol, which contains any remaining water, is discarded. The separatory funnel is replaced by a 1-l. stainless-steel flask with standard taper joint, which has been thoroughly dried with a burner. In this flask is collected the next 750 ml. of distilled glycol. The condenser and a calcium chloride tube are transferred to the steel flask, the contents are heated to a temperature slightly under boiling, and 57.5 g. of sodium is added in large pieces as before. The residue of unsaturated esters is introduced from a large pipet, a boiling chip is added, and the mixture is heated for 20 hours under reflux. It is then cooled to room temperature and decanted cautiously into a 2-l. round-bottomed flask containing a mixture of 210 ml. of glycol and 70 ml. of concentrated sulfuric acid. To the flask is added an additional 70 ml. of concentrated sulfuric acid, and after the mixture has been heated under reflux for 12 hours it is worked up as described in the final paragraph of the main procedure. Upon fractional distillation of the products, there are obtained 17.5–23.9 g. (19–26%) of 4-ethyl-4-hydroxy-2-methyloctanoic acid, γ-lactone, and 34.1–39.6 g. (38–43%) of 4-ethyl-2-methyl-2-octenoic acid. In this instance, therefore, the yield of Δ^2-acid is increased approximately one-third by isomerizing the dehydration products before converting the Δ^3-acid to lactone.

9. An 0.8 × 125-cm. simple Podbielniak column [4] with partial reflux head is a suitable type.

10. An analytical sample of 4-ethyl-4-hydroxy-2-methyloctanoic acid, γ-lactone, has b.p. 115–117°/4.3 mm., n_D^{25} 1.4462.

11. An analytical sample of 4-ethyl-2-methyl-2-octenoic acid has b.p. 141–142°/4.6 mm., n_D^{25} 1.4628.

3. Methods of Preparation

The procedure employed has been previously described by Cason and Rinehart [3] and is a modification of the standard Reformatsky procedure.[9, 10] The Reformatsky reaction, which has been reviewed elsewhere,[9] has been widely employed with ketones, somewhat less frequently with aldehydes, and very seldom with α-alkyl aliphatic aldehydes.

4-Ethyl-2-methyl-2-octenoic acid has been prepared only by this method. An alternative synthesis of α-alkyl-α,β-unsaturated acids proceeds via α-bromination of the saturated acid, followed by dehydrohalogenation with quinoline at elevated temperatures.[11] The present method is especially well adapted to preparation of α,γ-dialkyl-α,β-unsaturated acids.

[1] University of Illinois, Urbana, Illinois.

[2] *Org. Syntheses Coll. Vol.* **3**, 627 (1955).

[3] Cason and Rinehart, *J. Org. Chem.*, **20**, 1591 (1955).

[4] Cason and Rapoport, *Laboratory Text in Organic Chemistry*, 2nd ed., p. 293, Prentice-Hall, Englewood Cliffs, New Jersey, 1962.

[5] Linstead, *J. Chem. Soc.*, **1932**, 115.

[6] Linstead, *J. Chem. Soc.*, **1927**, 355.

[7] Kon and Nargund, *J. Chem. Soc.*, **1932**, 2461.

[8] For a review of references, cf. Adkins, in Gilman's *Organic Chemistry*, 2nd ed., p. 1042, John Wiley & Sons, New York, 1943.

[9] Shriner, *Org. Reactions*, **1**, 2 (1942).

[10] Reference 4, p. 330.

[11] Cason, Allinger, and Allen, *J. Org. Chem.*, **18**, 857 (1953).

5-ETHYL-2-METHYLPYRIDINE [*]

(2-Picoline, 5-ethyl-)

$$4 \quad \begin{matrix} & O & \\ & / \ \backslash & \\ CH_3CH & & CHCH_3 \\ | & & | \\ O & & O \\ \backslash & & / \\ & CH & \\ & | & \\ & CH_3 & \end{matrix} \quad + 3NH_4OH \quad \xrightarrow{CH_3CO_2NH_4}$$

$$3 \quad \begin{matrix} C_2H_5 \\ \end{matrix} \bigotimes_{N}^{CH_3} + 15H_2O$$

Submitted by ROBERT L. FRANK, FREDERICK J. PILGRIM, and
EDWARD F. RIENER.[1,2]
Checked by R. S. SCHREIBER and T. L. ALDERSON.

1. Procedure

Two hundred and sixty-seven grams (296 ml., 4.38 moles) of
28% aqueous ammonium hydroxide, 207.5 g. (209 ml., 1.57 moles)
of paraldehyde, and 5.0 g. (0.065 mole) of ammonium acetate are
heated to 230° with continuous agitation in a 2-l. steel reaction
vessel (Note 1), and the temperature is maintained at 230° for 1
hour (Note 2). The autoclave is then allowed to cool, and the
two layers of the reaction mixture are separated (Note 3). To
the non-aqueous layer is added 60 ml. of chloroform, causing
separation of water which is combined with the aqueous layer.
The aqueous layer is extracted with three 50-ml. portions of
chloroform, and the extracts are combined with the main portion
of the chloroform solution. After removal of the chloroform by
distillation at atmospheric pressure, fractional distillation under
reduced pressure through a 30-cm. Fenske-type column [3] gives a
fore-run of water, paraldehyde, and α-picoline, b.p. 40–60°/17

mm., followed by 72–76 g. (50–53%) of 5-ethyl-2-methylpyridine, b.p. 65–66°/17 mm.; n_D^{20} 1.4971 (Note 4).

2. Notes

1. A steel reaction vessel of the type used for high-pressure catalytic hydrogenations is satisfactory. The pressure of the reaction mixture ranges from 800 to 3000 lb. A larger volume of reactants should not be used in a 2-l. reaction vessel.

2. The reaction is exothermic and in some reaction vessels may cause the temperature to rise above 230° for a short period. This has no apparent effect on the yield of product. The temperature measured is that of a thermocouple inserted in a well in the cover of the autoclave and corresponds to about 250° if the thermocouple is in the wall of the autoclave.

3. The mixture contains a small amount of solid material, apparently due to slight corrosion of the steel reaction vessel. If the solid causes the formation of an emulsion, it can be removed by filtration.

4. The yield may be increased to 60–70% by use of an 8:1 molar ratio of ammonium hydroxide to paraldehyde, but this is generally inconvenient because of the greatly increased volume of the reaction mixture.

3. Methods of Preparation

5-Ethyl-2-methylpyridine (also known as "aldehyde-collidine") has been prepared by heating aldehyde-ammonia; [4] aldehyde-ammonia and acetaldehyde [5-7] or paraldehyde; [7-9] aldol-ammonia and ammonia; [10] paraldehyde and ammonia; [11-13] acetamide, [14] or acetamide and phosphorus pentoxide; [15] ethylene glycol and ammonium chloride; [16] ethylidene chloride [17, 18] or bromide [19] and ammonia; ethylidene chloride and acetamide, ethylamine, or n-amylamine; [16] crotonic acid and a calcium chloride-ammonia complex; [20] and by passage of acetylene [21] or acetaldehyde [22] and ammonia over alumina and other catalysts.

A study has been made of catalysts for the present reaction, [23] and a mechanism for the synthesis of pyridine and its derivatives by the Beyer-Chichibabin method has been published. [24]

[1] University of Illinois, Urbana, Illinois.
[2] Work done under contract with the Office of Rubber Reserve.
[3] Fenske, Tongberg, and Quiggle, *Ind. Eng. Chem.*, **26**, 1169 (1934).
[4] Ador and Baeyer, *Ann.*, **155**, 297 (1870).
[5] Dürkopf and Schlaugk, *Ber.*, **21**, 294 (1888).
[6] Dürkopf, *Ber.*, **20**, 444 (1887).
[7] Tschitschibabin and Oparina, *J. prakt. Chem.*, [2] **107**, 138 (1924).
[8] Plath, *Ber.*, **21**, 3086 (1888).
[9] Ladenburg, *Ann.*, **247**, 42 (1888).
[10] Wurtz, *Ber.*, **8**, 1196 (1875).
[11] Farbwerke vorm. Meister, Lucius and Brüning, Brit. pat. 146,869 [*C. A.*, **14**, 3675 (1920)]; Austrian pat. 81,299 [*Chem. Zentr.*, **92** II, 35 (1921)]; French pat. 521,891 [*Chem. Zentr.*, **92** IV, 805 (1921)].
[12] Graf and Langer, *J. prakt. Chem.*, **150**, 153 (1938); Frank and Seven, *J. Am. Chem. Soc.*, **71**, 2629 (1949).
[13] Mahan (to Phillips Petroleum Co.), U. S. pat. 2,877,228 [*C. A.*, **53**, 13182 (1959)]; Farberov, Ustavshchikov, Kut'in, Vernova, and Yarosh, *Izvest. Vysshykh Ucheb. Zavedeniĭ, Khim. i Khim. Tekhnol.*, **1958**, No. 5, 92 [*C. A.*, **53**, 11364 (1959)]; Kudo, *Repts. Statist. Appl. Research, Union Japan. Scientists and Engrs.*, **6**, No. 1, 13 (1959) [*C. A.*, **53**, 21934 (1959)]; Frank, Blegen, Dearborn, Myers, and Woodward, *J. Am. Chem. Soc.*, **68**, 1368 (1946).
[14] Pictet and Stehelin, *Compt. rend.*, **162**, 877 (1916).
[15] Hesekiel, *Ber.*, **18**, 3095 (1885).
[16] Hofmann, *Ber.*, **17**, 1905 (1884).
[17] Kraemer, *Ber.*, **3**, 262 (1870).
[18] Dürkopf, *Ber.*, **18**, 920 (1885).
[19] Tawildarow, *Ann.*, **176**, 15 (1875).
[20] Fichter and Labhardt, *Ber.*, **42**, 4714 (1909).
[21] Tschitschibabin and Moschkin, *J. Russ. Phys. Chem. Soc.*, **54**, 611 (1922–1923); *J. prakt. Chem.*, [2] **107**, 109 (1924); Murahashi and Otuka, *Mem. Inst. Sci. Ind. Research, Osaka Univ.*, **7**, 121 (1950) [*C. A.*, **45**, 9052 (1951)].
[22] Tschitschibabin, Moschkin, and Tjaschelowa, *J. prakt. Chem.*, [2] **107**, 132 (1924).
[23] Arai, Osuka, Tanabe, Teramoto, and Ichikizaki, *J. Chem. Soc. Japan, Ind. Chem. Sect.*, **57**, 495 (1954) [*C. A.*, **49**, 15892 (1955)].
[24] Herzenberg and Boccato, *Chim. & ind. (Paris)*, **80**, 248 (1958).

ETHYL α-NITROBUTYRATE

(Butyric acid, 2-nitro-, ethyl ester)

$$CH_3CH_2CHCO_2C_2H_5 + NaNO_2 \xrightarrow[\text{Phloroglucinol}]{\text{DMF}}$$
$$|$$
$$Br$$

$$CH_3CH_2CHCO_2C_2H_5 + NaBr$$
$$|$$
$$NO_2$$

Submitted by NATHAN KORNBLUM and ROBERT K. BLACKWOOD.[1]
Checked by JAMES CASON, JOANNE FACAROS, and WILLIAM G. DAUBEN.

1. Procedure

Ethyl α-bromobutyrate (58.5 g., 0.30 mole) (Note 1) is poured into a stirred mixture of 600 ml. of N,N-dimethylformamide (DMF) (Note 1), 36 g. of sodium nitrite (0.52 mole) (Note 1), and 40 g. of anhydrous phloroglucinol (0.32 mole) (Note 2) contained in a 1-l. three-necked flask equipped with a sealed stirrer. The flask is closed, except for a tube containing calcium chloride, and immersed in a water bath maintained at room temperature (Note 3). Stirring is continued for 2.5 hours (Note 4); then the reaction mixture is poured into 1.2 l. of ice water layered over with 300 ml. of ether (Note 5). After separation of the upper layer, the aqueous phase is extracted with four 100-ml. portions of ether. The combined extracts are washed with four 100-ml. portions of water and then dried over anhydrous magnesium sulfate. The magnesium sulfate is removed by suction filtration and washed with four 25-ml. portions of ether which are combined with the filtered extract.

The ether is distilled through a small column (Note 6), under reduced pressure, from a 1-l. flask which is heated by a bath whose temperature is gradually raised to about 60°. The residual yellow liquid is transferred, with the aid of a little anhydrous ether, to a 100-ml. flask, and the remaining solvent is distilled through the column under reduced pressure. Rectification of the residue yields 2–3 g. of fore-run boiling in the range 33–71°/1 mm. which is followed by 33–36 g. (68–75%) of colorless ethyl α-nitrobutyrate, b.p. 71°/1 mm., n_D^{20} 1.4233 (Notes 7, 8, and 9).

2. Notes

1. The ethyl α-bromobutyrate employed was redistilled Eastman Kodak white label grade material, b.p. 64°/15 mm., n_D^{20} 1.4479. Technical DMF (du Pont) was used, and the sodium nitrite was an analytical grade.

Subsequent to the checking of this preparation, the submitters reported that DMSO (dimethyl sulfoxide) may be a somewhat better solvent for this preparation than is DMF. Since sodium nitrite is more soluble in DMSO, only 250 ml. of this solvent is required for the preparation. The more concentrated solution permits a reduction in reaction time to about 1.5 hours.

2. Ringwood Chemical Corporation technical grade phloroglucinol dihydrate was rendered anhydrous by heating for 3 hours at 110°. By reacting rapidly with any ethyl α-nitritobutyrate formed, it prevents nitrosation of the α-nitro ester. In the absence of phloroglucinol all the ethyl α-nitrobutyrate is destroyed.[2,3]

3. The reaction mixture becomes homogeneous and turns deep red-brown shortly after the addition of the α-bromo ester. The deep color is, presumably, due to nitrosated phloroglucinol; however, this in no way interferes with subsequent isolation of product.

4. Two hours allows more than sufficient time for complete reaction. Since the yield is not critically dependent on this factor, no attempt was made to establish the minimum reaction time. Even after a 17-hour reaction time there is no decrease in yield.

5. Approximately 200 ml. of this ether is required to saturate the aqueous DMF layer.

6. A 60 × 1 cm. externally heated column, packed with ⅛-in. glass helices and equipped with a total reflux variable take-off head, was used by the submitters.

7. Toward the end of the rectification the jacket of the column is heated to 90–95° in order to obtain the last few grams of product.

8. The ethyl α-nitrobutyrate dissolves rapidly in 10% aqueous sodium hydroxide and dissolves completely in saturated aqueous sodium carbonate on shaking for 2–3 minutes.

9. This procedure has been applied successfully to the synthesis of other α-nitro esters from α-bromo esters,[3] as listed below; ethyl bromoacetate is exceptional in that it fails to give ethyl nitroacetate.

SYNTHESIS OF α-NITRO ESTERS FROM α-BROMO ESTERS

α-Nitro Ester	Reaction Time, hr.	Yield, %
Ethyl α-nitropropionate	2	62
Ethyl α-nitrocaproate	5	74
Ethyl α-nitroisobutyrate [a]	44	78
Ethyl α-nitroisovalerate	150	67
Ethyl α-phenyl-α-nitroacetate	2.5	70

[a] No phloroglucinol employed.

3. Methods of Preparation

Ethyl α-nitrobutyrate may be prepared in 75% yield by the reaction of silver nitrite with ethyl α-iodobutyrate,[4] in 82% yield by the reaction of sodium nitrite with ethyl α-bromobutyrate in dimethyl sulfoxide,[5] and in 51% and 46% yields from the appropriately substituted malonic or acetoacetic ester by nitration with acetone cyanohydrin nitrate, followed by cleavage of the nitrated esters by means of sodium hydride.[6] It also has been prepared in 18% yield by direct nitration and subsequent decarboxylation of diethyl ethylmalonate.[7] The present method offers the advantage of using sodium nitrite.

[1] Purdue University, West Lafayette, Indiana. This research was supported, in part, by grants from the Explosives Department of E. I. du Pont de Nemours & Company and, in part, by the United States Air Force under contract No. AF 18 (600)-310 monitored by the Office of Scientific Research, Air Research and Development Command.

[2] Kornblum, Blackwood, and Mooberry, *J. Am. Chem. Soc.*, **78**, 1501 (1956).

[3] Kornblum, Blackwood, and Powers, *J. Am. Chem. Soc.*, **79**, 2507 (1957).

[4] Kornblum, Chalmers, and Daniels, *J. Am. Chem. Soc.*, **77**, 6654 (1955).

[5] Kornblum and Powers (to Purdue Research Foundation), U. S. pat. 2,816,909 [*C. A.*, **52**, 11896 (1958)].

[6] Emmons and Freeman, *J. Am. Chem. Soc.*, **77**, 4391 (1955).

[7] Kornblum and Eicher, *J. Am. Chem. Soc.*, **78**, 1494 (1956); Ulpiani, *Atti. reale accad. Lincei*, [5] **13**, II, 346 (1904).

ETHYL ORTHOCARBONATE *

(Orthocarbonic acid, tetraethyl ester)

$$CCl_3NO_2 + 4C_2H_5ONa \rightarrow C(OC_2H_5)_4 + 3NaCl + NaNO_2$$

Submitted by JOHN D. ROBERTS and ROBERT E. McMAHON.[1]
Checked by WILLIAM S. JOHNSON and WILLIAM E. LOEB.

1. Procedure

Caution! This preparation should be conducted in a hood to avoid exposure to chloropicrin.

A solution of sodium ethoxide is prepared under nitrogen from 70 g. (3.04 g. atoms) of sodium and 2 l. of absolute ethanol (Note 1) in a 3-l. three-necked flask which is equipped with mechanical stirrer, efficient reflux condenser, dropping funnel, and a thermometer which dips below the level of the liquid in the flask. Chloropicrin (100 g., 0.61 mole) (Note 2) is placed in the dropping funnel, and the stirred solution is heated to 58–60° with a water bath. The chloropicrin is added at a rate of 30–35 drops per minute until the reaction becomes self-sustaining (about 20 minutes), at which point the water bath is removed and the balance of the chloropicrin is added at a rate sufficient to maintain the temperature at 58–60° (Note 3). When the addition, which requires nearly 2 hours, is complete, the stirrer is stopped and the mixture is allowed to stand overnight.

The flask is connected to a 2 by 50 cm. Vigreux column equipped with a total-reflux partial take-off head, and all but about 400 ml. of the ethanol is removed at 200 mm. pressure with a reflux ratio greater than 5:1 (Note 4).

The residue is cooled, diluted with 1.2 l. of water, and transferred to a 2-l. separatory funnel. The organic layer is separated, washed with 200 ml. of saturated salt solution, and dried over anhydrous magnesium sulfate. The aqueous layer is extracted with a total of 800 ml. of ether used in several small portions. The ethereal extracts are combined, washed first with 500 ml. of water then with 500 ml. of saturated salt solu-

tion, and finally dried over anhydrous magnesium sulfate. The
ether is removed through a 1.8 by 25 cm. glass-helix-packed
fractionating column with a total-reflux partial take-off head.
The residue is combined with the balance of the crude product
and distilled through the fractionating column at atmospheric
pressure. The yield of ethyl orthocarbonate is 54–57.5 g.
(46–49%); b.p. 158–161°; n_D^{25} 1.3905–1.3908.

2. Notes

1. The absolute ethanol was a good commercial grade and contained less than 0.1% of water according to the paraffin-oil test.[2]
2. Chloropicrin is a skin irritant and a lachrymator. No difficulty was experienced when the preliminary steps were carried
out in a good hood.
3. Care should be taken to regulate the temperature and rate
of addition of chloropicrin as specified in order to avoid accumulation of unreacted chloropicrin in the reaction mixture during
the induction period; otherwise the reaction, which is strongly
exothermic, may get out of control.
4. A water bath should be used as a heat source to avoid overheating, which leads to lowered yields. This distillation should
be carried out carefully to prevent loss of product by co-distilla-
tion with the ethanol.

3. Methods of Preparation

The above procedure is essentially that of Tieckelmann and
Post.[3] Ethyl orthocarbonate has been prepared by the reaction
of chloropicrin and sodium ethoxide by Bassett [4] and Röse.[5]
Thiocarbonyl perchloride has been reported [3, 6] to react with
sodium ethoxide to give good yields of ethyl orthocarbonate.

[1] Massachusetts Institute of Technology, Cambridge, Massachusetts.
[2] Robertson, *Laboratory Practice of Organic Chemistry*, p. 177, The Macmillan
Company, New York, 1943.
[3] Tieckelmann and Post, *J. Org. Chem.*, **13**, 265 (1948).
[4] Bassett, *Ann.*, **132**, 54 (1864).
[5] Röse, *Ann.*, **205**, 249 (1880).
[6] Connolly and Dyson, *J. Chem. Soc.*, **1937**, 827.

ETHYL β,β-PENTAMETHYLENEGLYCIDATE *

(1-Oxaspiro[2.5]octane-2-carboxylic acid, ethyl ester)

$$+ \text{ClCH}_2\text{CO}_2\text{C}_2\text{H}_5 + \text{KOC(CH}_3)_3 \rightarrow$$

$$+ \text{(CH}_3)_3\text{COH} + \text{KCl}$$

Submitted by RICHARD H. HUNT, LELAND J. CHINN, and WILLIAM S. JOHNSON.[1]
Checked by N. J. LEONARD and F. P. HAUCK, JR.

1. Procedure

The reaction is conducted in a 500-ml. round-bottomed three-necked flask to which are attached (ground-glass joints) a rubber slip-sleeve-sealed wire stirrer, a thermometer, and a pressure-equalized dropping funnel. The top of the dropping funnel is connected to a system for exhausting and filling with nitrogen (p. 133). The apparatus is flame-dried at reduced pressure, and the flask is charged with 14.50 g. (0.148 mole) of freshly distilled cyclohexanone and 18.15 g. (0.148 mole) of freshly distilled ethyl chloroacetate. A solution of 6.0 g. (0.153 g. atom) of potassium in 125 ml. of dry *tert*-butyl alcohol (Notes 1 and 2) is introduced into the dropping funnel, and the system is exhausted and filled with nitrogen. The flask is cooled with an ice bath, stirring is commenced, and the solution of potassium *tert*-butoxide is added from the dropping funnel over a period of about 1.5 hours, the temperature of the reaction mixture being maintained at 10–15°. After the addition is complete, the mixture is stirred for an additional 1–1.5 hours at about 10°. Most of the *tert*-butyl alcohol is removed by distillation from the reaction flask at reduced pressure (water aspirator) and a bath temperature of 100°.

The oily residue is taken up in ether. The ether solution is washed with water, then with saturated aqueous sodium chloride solution, and is finally dried over anhydrous sodium sulfate. The residue obtained on evaporation of the ether is distilled through a 6-in. Vigreux column to give 22.5–26.0 g. (83–95% yield) of colorless glycidic ester, b.p. 134–137°/21 mm., 147–152°/30 mm., n_D^{25} 1.4568–1.4577 (Note 3).

2. Notes

1. The preparation of potassium *tert*-butoxide is carried out according to a procedure already described (p. 132). Particular attention should be paid to the precautions in handling potassium.

2. The *tert*-butyl alcohol may be dried over sodium (p.134). Scrupulously dry *tert*-butyl alcohol may be prepared by distilling alcohol thus treated from calcium hydride (about 1 g./4 l.), obtainable from Metal Hydrides, Inc.

3. This material is of satisfactory quality, as shown by its conversion to solid derivatives in good yield.[2]

3. Methods of Preparation

Ethyl β,β-pentamethyleneglycidate has been prepared in 65% yield by the condensation of cyclohexanone with ethyl chloroacetate in the presence of sodium ethoxide,[3, 4] and in 50% yield in the presence of sodium in xylene.[4, 5] The present procedure employs potassium *tert*-butoxide as the condensing agent.[2]

[1] University of Wisconsin, Madison, Wisconsin.

[2] Johnson, Belew, Chinn, and Hunt, *J. Am. Chem. Soc.*, **75**, 4995 (1953).

[3] Darzens and Lefebure, *Compt. rend.*, **142**, 714 (1906); Rodinov and Kiseleva, *Izvest. Akad. Nauk S.S.S.R., Otdel. Khim. Nauk*, **1952**, 278 [*C. A.*, **47**, 3300 (1953)]; *Akad. Nauk S.S.S.R., Inst. Org. Khim., Sintezy Org. Soedinenii, Sbornik*, **2**, 57 (1952) [*C. A.*, **48**, 570 (1954)].

[4] Chiurdoglu, Mathieu, Baudet, Delsemme, Planchon, and Tullen, *Bull. soc. chim. Belges*, **65**, 664 (1956).

[5] Lunt and Sondheimer, *J. Chem. Soc.*, **1950**, 2957; Martynov, *Zhur. Obshchei Khim.*, **23**, 2006 (1953) [*C. A.*, **49**, 3124 (1955)].

ETHYL PHENYLCYANOACETATE *

(Acetic acid, cyanophenyl-, ethyl ester)

$$C_6H_5CH_2CN + CO(OC_2H_5)_2 + NaOC_2H_5 \rightarrow$$

$$\overset{\displaystyle CN}{\underset{\displaystyle |}{Na^+[C_6H_5CCO_2C_2H_5]^-}} + 2C_2H_5OH$$

$$\overset{\displaystyle CN}{\underset{\displaystyle |}{Na^+[C_6H_5CCO_2C_2H_5]^-}} + CH_3CO_2H \rightarrow$$

$$\overset{\displaystyle CN}{\underset{\displaystyle |}{C_6H_5CHCO_2C_2H_5}} + CH_3CO_2Na$$

Submitted by E. C. HORNING and A. F. FINELLI.[1]
Checked by WILLIAM S. JOHNSON and H. WYNBERG.

1. Procedure

Sodium ethoxide is prepared from 12.0 g. (0.52 g. atom) of sodium and 300 ml. of anhydrous ethanol in a 1-l. three-necked round-bottomed flask fitted with a reflux condenser carrying a calcium chloride tube. After the sodium has dissolved completely, the condenser is arranged for distillation under reduced pressure and the excess ethanol is removed by heating the flask on a steam bath while the system is maintained at the pressure obtained with an ordinary aspirator (Note 1).

As rapidly as possible, after removal of the ethanol, the flask is fitted with a rubber-sealed stirrer, a dropping funnel, a distilling head containing a thermometer, and a condenser arranged for distillation into a flask protected by a calcium chloride tube. There are then added 300 ml. (292 g., 2.5 moles) of dry diethyl carbonate, 80 ml. of dry toluene, and 58.5 g. (0.50 mole) of phenylacetonitrile (Note 2). The flask is heated, with good stirring, and the cake of sodium ethoxide soon dissolves. When distillation has started, dry toluene is added dropwise at about the same rate that the distillate is collected. Approximately 200–250 ml. of toluene should be added in a period of 2 hours (Note 3) while stirring and distillation are continued.

The mixture is cooled and transferred to a 1-l. beaker. After addition of 300 ml. of cold water, the aqueous phase is acidified with 35–40 ml. of acetic acid. The layers are separated and the water solution is extracted with three 75-ml. portions of ether. The organic solutions are combined, washed with 100 ml. of water, and dried over anhydrous magnesium sulfate. The low-boiling solvents are removed by distillation at atmospheric pressure, and the residue is distilled under reduced pressure through a short (15-cm.) Vigreux column. After a 1–5 g. forerun, the product is collected at 125–135°/3–5 mm. (Note 4). The yield is 66–74 g. (70–78%).

2. Notes

1. The success of this procedure is dependent upon the quality of the sodium ethoxide. The ethanol should be dried before use,[2] and the sodium ethoxide should not be heated to a temperature higher than 90–100°. The dry material can be transferred, but in this case it is advisable to prepare it in the flask in which it is to be used.

2. Commercial phenylacetonitrile should be distilled before use. The diethyl carbonate and toluene are dried by distillation.

3. Any ethanol remaining in the sodium ethoxide, together with the ethanol produced during the reaction, is removed during this period. The progress of the carbethoxylation reaction can be followed by temperature readings. During the first half of the heating period distillation usually occurs at a vapor temperature of 80–85°, but as the reaction nears completion and the ethanol is removed, the temperature rises to 110–115°. Near the end of the period, the sodium salt of ethyl phenylcyanoacetate appears as a precipitate.

4. Other observed boiling points are 129–131°/3 mm., 145–150°/7–8 mm. The product is a colorless liquid, n_D^{25} 1.5012-1.5019.

3. Methods of Preparation

This procedure is a modification of the method of Wallingford, Jones, and Homeyer.[3] The carbethoxylation of phenylacetonitrile is the only method of preparative value for this compound.

[1] University of Pennsylvania, Philadelphia, Pennsylvania.
[2] *Org. Syn. Coll. Vol.* 2, 1955 (1943).
[3] Wallingford, Jones, and Homeyer, *J. Am. Chem. Soc.*, **64**, 576 (1942); Testa, Fontanella, Christiani, and Fava, *Ann.*, **614**, 158 (1958).

ETHYL (1-PHENYLETHYLIDENE)CYANOACETATE

(Cinnamic acid, α-cyano-β-methyl-, ethyl ester)

$$C_6H_5COCH_3 + CH_2(CN)CO_2C_2H_5 \xrightarrow[CH_3CO_2H]{CH_3CO_2NH_4}$$

$$CH_3C{=}C(CN)CO_2C_2H_5 + H_2O$$
$$\underset{\displaystyle C_6H_5}{|}$$

Submitted by S. M. McElvain and David H. Clemens.[1]
Checked by W. E. Parham, Perry W. Kirklin, Jr., and Wayland E. Noland.

1. Procedure

In a 1-l. three-necked round-bottomed flask fitted with a Hershberg stirrer and a constant water separator (Note 1) surmounted by a reflux condenser are placed 120 g. (1 mole) of acetophenone, 113 g. (1 mole) of ethyl cyanoacetate (Note 2), 15.4 g. (0.2 mole) of ammonium acetate, 48.0 g. (0.8 mole) of glacial acetic acid, and 200 ml. of benzene. The reaction mixture is stirred and heated under reflux for 9 hours during which time 28–33 ml. of lower layer is collected in the water separator (Note 3). To the cooled reaction mixture is added 100 ml. of benzene, and the whole is extracted with three 100-ml. portions of water. The combined aqueous layers are extracted with 30 ml. of benzene, which is then added to the organic layer from the previous extraction. Anhydrous magnesium sulfate (15 g.) is added, and, after swirling occasionally for 10 minutes, the mixture is filtered by suction and the magnesium sulfate washed with two 25-ml. portions of benzene. The benzene is removed by distillation at reduced pressure and the residual oil distilled rapidly through a 15-cm. column. The yield of ester is 113–125 g. (52–58%), b.p. 135–160° (0.35 mm.) (Note 4).

2. Notes

1. A typical water separator has been described by Cope et al.[2]

2. Eastman Kodak white label grade acetophenone and ethyl cyanoacetate are used without further purification. The checkers used Matheson, Coleman, and Bell acetophenone and ethyl cyanoacetate without further purification.

3. The checkers used ammonium acetate which was slightly moist; consequently 33.5–34.5 ml. of lower layer was collected.

4. The checkers report the refractive index of the product to be $n_D^{25.1}$ 1.5468–1.5469.

3. Methods of Preparation

The above procedure is essentially that described by Cope et al.[2] Ethyl (1-phenylethylidene)cyanoacetate has been prepared also by condensing acetophenone with ethyl cyanoacetate in the presence of zinc chloride and aniline,[3] and other catalysts.[4] Additional aralkylidenecyano esters have been prepared by the present procedure.[5]

[1] University of Wisconsin, Madison, Wisconsin.
[2] Cope, Hofmann, Wyckoff, and Hardenbergh, J. Am. Chem. Soc., **63**, 3452 (1941).
[3] Scheiber and Meisel, Ber., **48**, 238 (1915).
[4] Cragoe, Robb, and Sprague, J. Org. Chem., **15**, 381 (1950).
[5] McElvain and Clemens, J. Am. Chem. Soc., **80**, 3915 (1958).

ETHYL N-PHENYLFORMIMIDATE

(Formimidic acid, N-phenyl-, ethyl ester)

$$C_6H_5NH_2 + (C_2H_5O)_3CH \xrightarrow{H^+} C_6H_5N{=}CHOC_2H_5 + 2C_2H_5OH$$

Submitted by ROYSTON M. ROBERTS and PAUL J. VOGT.[1]
Checked by T. L. CAIRNS and J. J. DRYSDALE.

1. Procedure

A 500-ml. flask is equipped with a capillary through a side opening, and 94 g. (1.01 moles) of aniline and 1 ml. of concentrated hydrochloric acid are added. A 12-in. glass-helix-packed column is attached (Note 1), and the water introduced with the acid is removed by boiling; about 1 ml. of aniline is collected after the water has distilled. The flask and its contents are then cooled to room temperature, and 222 g. (1.50 moles) of ethyl orthoformate is added. The column is reattached, and ethanol (Note 2) is distilled as it is produced; the theoretical amount (92 g., 116 ml.) is obtained in about 2.25 hours.

The reaction mixture is allowed to cool slightly, and the pressure is lowered to 40 mm. (Note 3). The excess ethyl orthoformate is distilled at 65°/40 mm. After a small intermediate fraction of about 4 g., b.p. 65–117°/40 mm., the product distils at 117–118°/40 mm. (b.p. 87–88°/10 mm.; n_D^{25} 1.5248); the yield is 118–127 g. (78–84%). The residue amounts to about 14 g. and is mainly N,N'-diphenylformamidine (Note 4).

2. Notes

1. A Vigreux column may also be used since it is not difficult to separate the ethanol from ethyl orthoformate, the next most volatile component present. A total reflux, partial take-off head was used. Heat was supplied by an electric mantle; the column was heated with a glass-covered heating tape during the distillation of excess ethyl orthoformate and product.

2. A small amount (5–10 ml.) of lower-boiling material usually comes over before the ethanol; this is probably ethyl formate, produced by hydrolysis of the ethyl orthoformate.

3. A pressure regulator may conveniently be used in conjunction with a water aspirator.

4. If several runs are to be made, the residue may be saved and used as starting material, since the reaction proceeds via the initial formation of N,N'-diphenylformamidine and its subsequent reaction with ethyl orthoformate.[2,3]

3. Methods of Preparation

Ethyl N-phenylformimidate has been prepared from silver formanilide and ethyl iodide,[4] and from aniline and ethyl orthoformate.[3] This method incorporates the discovery[2] of the necessity of acid catalysis for satisfactory yields by the latter process.

[1] University of Texas, Austin, Texas.
[2] Roberts, *J. Am. Chem. Soc.*, **71**, 3848 (1949).
[3] Claisen, *Ann.*, **287**, 363 (1895); Roberts and Vogt, *J. Am. Chem. Soc.*, **78**, 4778 (1956).
[4] Comstock and Clapp, *Am. Chem. J.*, **13**, 527 (1891).

ETHYL α-(1-PYRROLIDYL)PROPIONATE

(1-Pyrrolidineacetic acid, α-methyl-, ethyl ester)

$$2 \boxed{}NH + BrCHCO_2C_2H_5 \rightarrow$$
$$\underset{CH_3}{|}$$

$$\boxed{}N-CHCO_2C_2H_5 + \boxed{}NH \cdot HBr$$
$$\underset{CH_3}{|}$$

Submitted by ROBERT BRUCE MOFFETT.[1]
Checked by N. J. LEONARD and S. GELFAND.

1. Procedure

A 1-l. three-necked round-bottomed flask is placed on a steam bath and fitted with a stirrer, reflux condenser, and dropping funnel. A solution of 181 g. (1 mole) of ethyl α-bromopropionate in 200 ml. of benzene is placed in the flask (Note 1), and 148 g. (2.1 moles) of pyrrolidine (Note 2) is added slowly with stirring at such a rate that the solvent refluxes gently. When the addition is complete (about 1 hour is required), the mixture is heated under reflux for 1 hour. After being cooled, the mixture is poured into about 500 ml. of ice water and acidified with dilute hydrochloric acid. The aqueous layer is separated, washed once with ether, and made strongly basic with cold 40% sodium hydroxide solution. The basic ester is extracted with four 200-ml. portions of ether. The ether extracts are combined, washed with 100 ml. of water, and dried over anhydrous potassium carbonate. The drying agent is removed by filtration and the ether by distillation. The residue is distilled under reduced pressure through a short fractionating column; b.p. $84°/12$ mm. $(95-96°/19$ mm., $99.5-100.5°/23$ mm., $104-105°/30$ mm.$)$; n_D^{20} 1.4478, n_D^{25} 1.4450; d_4^{25} 0.9724. The yield is 137-156 g. $(80-91\%)$.

2. Notes

1. The yield of product is lowered appreciably if the solution is preheated before addition of the pyrrolidine.
2. Pyrrolidine is obtainable from E. I. du Pont de Nemours and Company, Electrochemicals Division, Niagara Falls, New York.

3. Methods of Preparation

Ethyl α-(1-pyrrolidyl)propionate has been prepared by the reaction of pyrrolidine with ethyl α-bromopropionate.[2]

[1] The Upjohn Company, Kalamazoo, Michigan.
[2] Moffett, *J. Org. Chem.*, 14, 862 (1949).

ETHYL PYRUVATE *

(Pyruvic acid, ethyl ester)

$$CH_3CHOHCO_2C_2H_5 \xrightarrow{\text{KMnO}_4} CH_3COCO_2C_2H_5$$

Submitted by J. W. CORNFORTH.[1]
Checked by CHARLES C. PRICE, KENNETH N. CAMPBELL, and JOHN WARNKE.

1. Procedure

In a 1-l. round-bottomed flask fitted with a thermometer and a mechanical stirrer are placed 130 ml. of saturated aqueous magnesium sulfate solution, 500 ml. of light petroleum ether (Note 1), 50 g. (0.42 mole) of ethyl lactate (Note 2), and 20 g. (0.13 mole) of sodium dihydrogen phosphate dihydrate. The stirrer is started (Note 3), the temperature is brought to $15°$ by means of an ice-water bath, and 55 g. (0.35 mole) of powdered potassium permanganate is added during 25-30 minutes. Stirring is continued until the oxidation is complete (Note 4), the temperature being kept near $15°$ throughout the process. The petroleum ether solution is decanted and the sludge stirred with three 50-ml. portions of light petroleum ether. The combined petroleum ether extracts are evaporated on a steam bath under a short fractionating column (Note 5). The residual oil is shaken thoroughly with two 10-ml. portions of a saturated aqueous calcium chloride solution (Note 6) and then distilled under reduced pressure. Almost the whole product boils at 56–57°/20 mm. The yield is 25–27 g. (51–54%) of nearly pure ethyl pyruvate, n_D^{20} 1.4053. This product compares favorably with material prepared by esterification of pyruvic acid (Note 7). Further purification may be effected through the sodium bisulfite compound (Note 8).

2. Notes

1. Petroleum ether, b.p. 40–60°, was washed with concentrated sulfuric acid before use. The checkers used the hexane fraction of petroleum.

2. The ethyl lactate should be of good quality, as its impurities tend to appear in the final product. The submitter used a good commercial grade supplied by British Industrial Solvents, Ltd. Its specification included an ester content of not less than 99% (calculated as ethyl lactate).

The commercial 99% ethyl lactate available to the checkers did not give satisfactory results. It was purified by distillation through a fractionating column 8 by ¾ in., packed with glass beads. The portion having the following properties was used: b.p. 154–155°, n_D^{20} 1.4125, d_4^{20} 1.0302.

3. The thick lower layer is stirred continuously and not too fast. Vigorous agitation of the upper layer is not advisable. A short Hershberg wire stirrer was used.

4. The oxidation requires about 2.5 hours. Unreduced permanganate is easily detected by spotting on filter paper. If a cake of manganese dioxide is formed beyond the compass of the stirrer, it should be pushed down. It is rarely necessary to do this more than once.

5. The distillate of petroleum ether, which contains ethanol and some ethyl pyruvate, can be recovered for another run by shaking with a little concentrated sulfuric acid.

6. This treatment removes unoxidized ethyl lactate. Each

shaking should be continued for 5 minutes. It is convenient to separate the layers by centrifuging. Droplets of calcium chloride solution should not be present in the oil when it is to be distilled, or some polymerization will occur.

7. No satisfactory criterion of purity for ethyl pyruvate is available in the literature. The submitter used a method of assay which was devised [2] for the estimation of aldehydes. One hundred and sixteen milligrams of the ester is weighed in a 100-ml. conical flask, dissolved in 5 ml. of water, and treated with 0.3 ml. of saturated sodium bisulfite solution. After 1–2 minutes, a little starch solution is added, the mixture is chilled, and 0.1N iodine solution is run in as rapidly as possible until the blue color is stable for a few seconds (about 12 ml. is required). Six milliliters of saturated sodium bicarbonate solution is added, and titration with iodine solution is carried out in the ordinary way. The end point is stable for 1 minute or more. The theoretical volume of 0.1N iodine required for pure ethyl pyruvate in the second stage is 20 ml. Thus if n ml. is required the estimated purity is 5n%. The results are perfectly consistent but may be slightly lower than the true values owing to dissociation of the bisulfite complex. The results from four different samples are as tabulated.

Method of Preparation	Estimated Ethyl Pyruvate Content, %
(1) Oxidation of ethyl lactate	95
(2) Esterification of once-distilled pyruvic acid	93–94
(3) Sample (2) twice redistilled; fraction b.p. 147–148° taken	96
(4) Sample (1) purified through bisulfite complex (Note 8)	98.5

8. The bisulfite compound is best made in small batches. The ester (2.2 ml.) in a large test tube is underlaid with 3.6 ml. of saturated sodium bisulfite solution. The tube is chilled in a freezing mixture, and the layers are shaken together. Crystallization occurs rapidly, especially if seed crystals are present. After 3 minutes, 10 ml. of ethanol is added and the crystalline product is washed on a filter with ethanol and ether. The yield is 3.0 g. Sixteen grams of the bisulfite complex is mixed with 32 ml. of saturated magnesium sulfate solution, and 5 ml. of 40% formaldehyde is added. After shaking, the oil is separated and the

aqueous layer extracted with a little ether, which is added to the oil. After drying with magnesium sulfate the product is distilled at low pressure and affords 5.5 g. of ethyl pyruvate, b.p. 56°/20 mm. On redistillation, the purified ester boils at 147.5°/750 mm., n_D^{20} 1.4052, f.p. around −50°.

3. Methods of Preparation

Ethyl pyruvate can be prepared by esterification of pyruvic acid [3, 4] or by catalytic oxidation of ethyl lactate with air or oxygen.[5, 6] A process has been patented for the oxidation of ethyl lactate by acidified permanganate in dilute aqueous solution.[7] Ethyl pyruvate also has been obtained by the treatment of pyruvaldehyde diethyl acetal with N-bromosuccinimide,[8] by the reaction of pyruvyl chloride with ketene,[9] and by the reaction of pyruvic acid with diethyl pyrocarbonate.[10] Of minor interest are the preparations by pyrolysis of ethyl α-triphenylmethoxy-propionate [11] and by the action of diethylamine on ethyl *meso*-α,α'-dibromoadipate.[12]

[1] National Institute for Medical Research, London, England.
[2] Clausen, *J. Biol. Chem.*, **52**, 263 (1922).
[3] Archer and Pratt, *J. Am. Chem. Soc.*, **66**, 1656 (1944).
[4] Simon, *Bull. soc. chim. France*, [3] **13**, 474 (1895).
[5] C. H. Boehringer Sohn, Ger. pat. 447,838 [*Frdl.*, **15**, 382 (1928)].
[6] Kulka, *Can. J. Research*, **24B**, 221 (1946).
[7] Byk-Guldenwerke Chem. Fab. A. G., Ger. pat. 526,366 [*C. A.*, **25**, 4285 (1931)].
[8] Wright, *J. Am. Chem. Soc.*, **77**, 4883 (1955).
[9] Beránek, Smrt, and Šorm, *Chem. listy*, **48**, 679 (1954) [*C. A.*, **49**, 9545 (1955)].
[10] Thoma and Rinke, *Ann.*, **624**, 30 (1959).
[11] Hurd and Filachione, *J. Am. Chem. Soc.*, **59**, 1949 (1937).
[12] von Braun, Leistner, and Münch, *Ber.*, **59**, 1953 (1926).

1,1′-ETHYNYLENE-*bis*-CYCLOHEXANOL*

(Cyclohexanol, 1,1′-ethynylenedi-)

Submitted by G. FORREST WOODS and LOUIS H. SCHWARTZMAN.[1]
Checked by RICHARD T. ARNOLD and STUART W. FENTON.

1. Procedure

In a 2-l. three-necked flask (Note 1), fitted with a dropping funnel, condenser (equipped with a Drierite tube), and efficient stirrer driven by a powerful motor, is placed a mixture of 600 ml. of benzene, 56 g. (0.85 mole) of 85% potassium hydroxide (Note 2), and 76.4 g. of powdered calcium carbide (Note 3). While this mixture is being stirred vigorously, 85 g. (0.87 mole) of cyclohexanone is added over a period of 0.5–1 hour. The mixture is dark gray and will become warm, but no external cooling is necessary. Stirring is continued, and within 24 hours the contents congeal (Note 4). This semisolid is allowed to stand for an additional 4 days (Note 5).

The flask is immersed in an ice bath, and a solution containing 200 ml. of concentrated hydrochloric acid and 200 ml. of water is added cautiously (Note 6) over a period of 4–6 hours. The dark solid is separated by filtration with the aid of a large Büchner funnel. This impure product is air-dried and digested with 900 ml. of boiling carbon tetrachloride, and the insoluble portion is collected on a Büchner funnel and subsequently extracted with 100 ml. of hot acetone and again filtered. When the filtrates are kept overnight in a refrigerator, 47.3–50.3 g. (49–52%) of a colorless crystalline product separates; m.p. 106.5–109°. Partial evaporation of the combined filtrates, followed by effective cooling, gives an additional 7.8–12.9 g. (8–13%) of 1,1'-ethynylene-*bis*-cyclohexanol; m.p. 100–109° (Note 7).

2. Notes

1. The submitters employed a 12-l. resin flask equipped with Lightning Stirrer whose shaft and blades were of stainless steel when using 10 times the quantity of starting materials reported here.

Any evolution of acetylene is best accommodated by a rubber tube which leads outside or to a good hood.

2. Either pellets or flakes of potassium hydroxide are powdered in a ball mill.

3. Technical calcium carbide of approximately 100 mesh was obtained from the Union Carbide and Carbon Corporation, New York, New York. The amount of calcium carbide used is in excess and based on an activity of 75%.

4. Stirring is continued until the mass has set solidly; it is desirable for a channel to exist around the stirrer shaft to facilitate the subsequent decomposition.

5. Decreased yields result from shorter periods of standing.

6. It is advisable to bore several holes into the solid mass by means of a stirring rod in order to permit better contact with the acid. Initial addition of the acid should be slow, and a total of 4–6 hours should be allowed for this operation, since considerable heat and acetylene are evolved.

7. The product can be recrystallized from carbon tetrachloride or acetone, or sublimed at reduced pressure, to yield a product melting at 109–111°.

3. Methods of Preparation

This method is based on the procedure of Kazarin.[2] The same substance has been prepared by the reaction of the dimagnesium halide [3] or dilithium derivative [4] of acetylene with cyclohexanone, and also by the reaction of cyclohexanone with acetylene in the presence of potassium *tert*-butoxide followed by the preparation of the Grignard reagent of this compound and reaction again with cyclohexanone.[5]

It has been claimed that the yield of the present reaction is improved when it is carried out in the presence of an acetal or ethylene glycol dialkyl ether.[6] Petrov et al.[7] have reported that they obtained almost a 100% yield of the glycol when cyclohexanone was treated with acetylene and double the amount of potassium hydroxide.

[1] University of Maryland, College Park, Maryland.
[2] Kazarin, *J. Gen. Chem. U.S.S.R.*, **4**, 1347 (1934).
[3] Dupont, *Ann. chim.*, [8] **30**, 485 (1913).
[4] Viehe, Franchimont, and Valange, *Chem. Ber.*, **92**, 3064 (1959).
[5] Pinkney, Nesty, Wiley, and Marvel, *J. Am. Chem. Soc.*, **58**, 972 (1936).
[6] Bergmann, Sulzbacher, and Herman, *J. Appl. Chem. (London)*, **3**, 39, (1953).
[7] Petrov, Mitrofanova, and Lesyuchevskaya, *Doklady Akad. Nauk S.S.S.R.*, **68**, 83 (1949) [*C. A.*, **44**, 1903 (1950)].

FERROCENE *

(Iron, dicyclopentadienyl-)

I. METALLIC SODIUM METHOD

Submitted by G. Wilkinson.[1]
Checked by N. J. Leonard, Kenneth L. Rinehart, Jr., Donald J. Casey, and Sung Moon.

1. Procedure

In a 250-ml. three-necked flask, fitted with a mechanical stirrer, a reflux condenser, and an inlet for admission of nitrogen, is placed 100 ml. of tetrahydrofuran (Note 1). With stirring, 27.1 g. (0.166 mole) of anhydrous ferric chloride is added in portions, followed by 4.7 g. (0.084 g. atom) of iron powder (Note 2). The mixture is heated with stirring under nitrogen at the reflux temperature for 4.5 hours, giving a gray powder with a brown supernatant liquid.

During this time, a second system is assembled, consisting of a 500-ml. three-necked flask fitted with a mechanical stirrer, a reflux condenser topped with a calcium chloride-filled drying tube attached to a xylene-filled bubbler, and a pressure-equalizing dropping funnel through which a slow stream of nitrogen is passed into the flask. In the flask are placed 200 ml. of sodium-dried xylene and 11.5 g. (0.5 g. atom) of sodium. The mixture is heated to boiling, and the sodium is finely dispersed by rapid stirring (Note 3). Stirring is continued while the mixture is allowed to cool in a nitrogen atmosphere. The cooled mixture is allowed to settle, and the bulk of the xylene is siphoned. Tetrahydrofuran (200 ml.) (Note 1) is added through the separatory funnel, and to the stirred mixture, cooled in ice, is added 42 ml. (0.5 mole) of cyclopentadiene (Note 4) in portions during 1 hour (Note 5). Stirring is continued for 2–3 hours in the cold, after which only a small amount of sodium remains unreacted.

The cooled contents of the 250-ml. flask containing ferrous chloride (Note 6) are added to the cold sodium cyclopentadienide solution while passing a stream of nitrogen through both flasks. The combined mixture is stirred for 1.25 hours at a temperature just below reflux. Solvent is removed by distillation, and the ferrocene is extracted from the residue with several portions of refluxing petroleum ether (b.p. 40–60°). The product is obtained by evaporation of the petroleum ether solution. Ferrocene may be purified by recrystallization from pentane or cyclohexane (hexane, benzene, and methanol have also been used) or by sublimation. The yield is 31–34 g. (67–73%) (Note 7), m.p. 173–174°.

2. Notes

1. Tetrahydrofuran may be purified by refluxing over solid potassium hydroxide, followed by distillation from lithium alu-

minum hydride. Tetrahydrofuran may be replaced by ethylene glycol dimethyl ether (dimethoxyethane). The submitter has indicated that either solvent may be freed conveniently from water, alcohols, and moderate amounts of peroxides by passing the commercial solvent through a column (2 in. diameter × 2–3 ft. length) of Linde Air Products "Molecular Sieves" (type 13X $\frac{1}{16}$-in. pellets), at a rate of approximately 100 ml. per minute.

2. The quality of the iron used in preparing the ferrous chloride has a marked effect on the yield of ferrocene. The checkers employed Rascher and Betzold (730 N. Franklin, Chicago, Ill.) 300-mesh iron powder, reduced by hydrogen. When 40-mesh iron filings were used, the yield of ferrocene was much lower (ca. 33%).

3. The checkers employed a "Mixmaster"-type motor and a Hershberg stirrer made from tantalum wire.

4. Cyclopentadiene, b.p. 40°, is obtained by heating commercial 85% dicyclopentadiene (e.g., from Matheson, Coleman and Bell Company, Norwood, Ohio) under a short column ($\frac{3}{4}$ in. diameter × 8–12 in. length) filled with glass helices. The distilled cyclopentadiene is collected in a receiver which is maintained at Dry Ice temperature until the cyclopentadiene is used. Methylcyclopentadiene and other substituted cyclopentadienes such as indene may also be employed for the synthesis of the correspondingly substituted ferrocenes. In these cases, the reaction of the hydrocarbon with sodium is much slower than with cyclopentadiene, and refluxing for several hours is required to complete the reaction.

5. Under the best conditions, sodium cyclopentadienide gives pale yellow or orange solutions. Traces of air lead to red or purple solutions, as does insufficiently purified solvent, without, however, lowering the reaction yield appreciably. If 1,2-dimethoxyethane is used, in which sodium cyclopentadienide is less soluble than in tetrahydrofuran, white crystals may be obtained at this point.

6. Ferrous chloride may be substituted by ferric chloride directly, with a corresponding reduction in yield, since the sodium cyclopentadienide solution will reduce ferric chloride.

7. The submitter reported yields up to 90% by this method.

II. DIETHYLAMINE METHOD

$$\tfrac{2}{3}FeCl_3 \; + \; \tfrac{1}{3}Fe \; \longrightarrow \; FeCl_2$$

$$FeCl_2 \; + \; 2 \bigg/\!\!\!\bigg\backslash \; + \; 2(C_2H_5)_2NH \; \longrightarrow \; Fe \; + \; 2(C_2H_5)_2NH \cdot HCl$$

Submitted by G. WILKINSON.[1]
Checked by N. J. LEONARD, KENNETH L. RINEHART, JR., and PETER WOO.

1. Procedure

The conditions given in the preceding preparation are used for obtaining a suspension of 0.25 mole of ferrous chloride in 100 ml. of tetrahydrofuran (Note 1), contained in a 250-ml. flask. The tetrahydrofuran is then removed under reduced pressure until the residue is almost dry. The flask is cooled in an ice bath, and to the residue is added a mixture of 42 ml. (0.5 mole) of cyclopentadiene and approximately 100 ml. (about 1 mole) of diethylamine. The mixture is stirred vigorously at room temperature for 6–8 hours or, conveniently, overnight. The excess amine is removed under reduced pressure, and the residue is extracted repeatedly with refluxing petroleum ether. The extract is filtered hot, and the solvent is evaporated to leave ferrocene. The product is purified by recrystallization from pentane or cyclohexane or by sublimation. The yield is 34–39 g. (73–84%), m.p. 173–174°.

2. Note

1. All precautions with regard to the purification of tetrahydrofuran, the quality of the iron powder, the rapid stirring, the maintenance of a nitrogen atmosphere, and the handling of cyclopentadiene, described in the preceding preparation, are followed.

3. Methods of Preparation

The methods of preparation of ferrocene have been reviewed by Pauson[2] and by Fischer.[3] Ferrocene has been made by the reaction of ferric chloride with cyclopentadienylmagnesium bromide,[4] by the direct thermal reaction of cyclopentadiene with iron metal,[5] by the direct interaction of cyclopentadiene with iron carbonyl,[6] by the reaction of ferrous oxide and cyclopentadiene in the presence of chromic oxide,[7] by the reaction of ferrous chloride with sodium cyclopentadienide in liquid ammonia,[8] and from cyclopentadiene and ferrous acetylacetone-dipyridine complex.[9] Method I is based on that developed by Wilkinson and his co-workers for ferrocene and many analogous compounds.[10]

Although not so generally applicable for the preparation of dicyclopentadienyl metal compounds as the sodium cyclopentadienide procedure, Method II represents the simplest preparation of ferrocene. The amine procedure also may be employed for dicyclopentadienylnickel (about 80% yield), using nickel bromide obtained by the action of bromine on nickel metal powder and 1,2-dimethoxyethane as the solvent. The method of preparation given here is a modified version [10d] of that originally described,[11] and it has been studied by others.[12, 13]

[1] Harvard University, Cambridge, Massachusetts; Imperial College of Science and Technology, London, England.

[2] Pauson, *Quart. Revs. (London)*, **9**, 391 (1955).

[3] Fischer, *Angew. Chem.*, **67**, 475 (1955).

[4] Kealy and Pauson, *Nature*, **168**, 1039 (1951).

[5] Miller, Tebboth, and Tremaine, *J. Chem. Soc.*, **1952**, 632.

[6] Wilkinson, *J. Am. Chem. Soc.*, **76**, 209 (1954); Piper, Cotton, and Wilkinson, *J. Inorg. Nuclear Chem.*, **1**, 165 (1955); Hallam, Mills, and Pauson, *J. Inorg. Nuclear Chem.*, **1**, 313 (1955).

[7] Hogan and Gardner (to Phillips Petroleum Co.), U. S. pat. 2,898,360 [*C. A.*, **54**, 569 (1960)].

[8] Fischer and Jira, *Z. Naturforsch.*, 8b, 217 (1953); see also Weinmayr, *J. Am. Chem. Soc.*, **77**, 3012 (1955); Ziegler, Froitzheim-Kühlhorn, and Hafner, *Chem. Ber.*, **89**, 434 (1956).

[9] Wilkinson, Pauson, and Cotton, *J. Am. Chem. Soc.*, **76**, 1970 (1954).

[10] Wilkinson and Birmingham, *J. Am. Chem. Soc.*, **76**, 4281 (1954); Wilkinson and Cotton, *Chem. & Ind. (London)*, **1954**, 307; Birmingham, Fischer, and Wilkinson, *Naturwiss.*, **42**, 96 (1955); Wilkinson, Cotton, and Birmingham, *J. Inorg. Nuclear Chem.*, **2**, 95 (1956); Cotton and Wilkinson, *Z. Naturforsch.*, 9b, 417 (1954); Wilkinson, *J. Am. Chem. Soc.*, **74**, 6146 (1952); Wilkinson, *J. Am. Chem. Soc.*, **74**, 6148 (1952); Cotton, Whipple, and Wilkinson, *J. Am. Chem. Soc.*, **75**, 3586 (1953); Wilkinson and Birmingham, *J. Am. Chem. Soc.*, **76**, 6210 (1954); Pauson and Wilkinson, *J. Am. Chem. Soc.*, **76**, 2024 (1954).

[11] Birmingham, Seyferth, and Wilkinson, *J. Am. Chem. Soc.*, **76**, 4179 (1954).

[12] Pruett and Morehouse, *Advances in Chem. Ser.*, **23**, 368 (1959).
[13] Titov, Lisitsyna, and Shemtova, *Doklady Akad. Nauk S.S.S.R.*, **130**, 340 (1960)
[*C. A.*, **54**, 10986 (1960)].

FLAVONE *

Method 1

$$o\text{-}C_6H_4(OH)COCH_3 \xrightarrow[C_5H_5N]{C_6H_5COCl} o\text{-}C_6H_4(OCOC_6H_5)COCH_3$$

Submitted by T. S. WHEELER.[1]
Checked by R. L. SHRINER and DONALD A. SCOTT.

1. Procedure

A. *o-Benzoyloxyacetophenone.* In a 100-ml. conical flask fitted with a calcium chloride drying tube are placed 13.6 g. (12 ml., 0.1 mole) of *o*-hydroxyacetophenone,[2] 21.1 g. (17.4 ml., 0.15 mole) of benzoyl chloride, and 20 ml. of pyridine (Note 1). The temperature of the reaction mixture rises spontaneously, and when no further heat is evolved (about 15 minutes) the mixture is poured with good stirring into 600 ml. of 3% hydrochloric acid containing 200 g. of crushed ice. The product is collected on a Büchner funnel and washed with 20 ml. of methanol, then with 20 ml. of water. The product is sucked as dry as possible and air-dried at room temperature. The yield of dry crude product melting at 81–87° is 22–23 g. It is recrystallized from 25 ml. of methanol, and the *o*-benzoyloxyacetophenone is obtained as white crystals; yield 19–20 g. (79–83%); m.p. 87–88°.

B. *Flavone.* In a 500-ml. round-bottomed three-necked flask, equipped with a mercury-sealed mechanical stirrer, a thermometer, and an air condenser closed with a calcium chloride drying tube in the second neck, are placed 20 g. (0.083 mole) of *o*-benzoyloxyacetophenone and 200 ml. of freshly distilled anhydrous glycerol (Note 2). A stream of nitrogen, dried by passage through a wash bottle containing sulfuric acid, is introduced through the third neck. The mixture is heated and maintained at 260° for two hours while being stirred continuously. The contents are cooled below 100° and then poured into 2 l. of water which is rendered slightly alkaline with aqueous sodium hydroxide. The mixture is stirred for 15 minutes, cooled, and kept at 0° (in a refrigerator) for 48 hours. The tan-colored crystals of flavone are collected on a filter and dried at 50°. The yield of crude product amounts to about 10 g.; m.p. 90–93°. The crude material is dissolved in 400 ml. of hot ligroin (b.p. 60–70°). Repeated partial evaporation of the solvent in stages, each followed by cooling, gives successive crops of flavone as white needles. The yield of pure flavone amounts to 8–9 g. (43–48%); m.p. 96–97°.

Method 2

$$o\text{-}C_6H_4(OH)COCH_3 \xrightarrow[\text{C}_5\text{H}_5\text{N}]{\text{C}_6\text{H}_5\text{COCl}} o\text{-}C_6H_4(OCOC_6H_5)COCH_3$$

$$o\text{-}C_6H_4(OCOC_6H_5)COCH_3 \xrightarrow[\text{C}_5\text{H}_5\text{N}]{\text{KOH}} o\text{-}C_6H_4(OH)COCH_2COC_6H_5$$

A. *o-Benzoyloxyacetophenone.* This is prepared as in Method 1.

B. *o-Hydroxydibenzoylmethane.* A solution of 20 g. (0.083 mole) of *o*-benzoyloxyacetophenone in 75 ml. of pyridine (Note 1) is prepared in a 300-ml. beaker and warmed to 50°. To the solution is added 7 g. of hot pulverized 85% potassium hydroxide (Note 3), and the mixture is mechanically stirred for 15 minutes, during which time a copious precipitate of the yellow potassium salt of *o*-hydroxydibenzoylmethane forms (Note 4). The mixture is cooled to room temperature and acidified with 100 ml. of 10% acetic acid. The diketone separates as a light-yellow precipitate which is collected on a filter and sucked dry (Note 5). The yield of crude *o*-hydroxydibenzoylmethane is 16–17 g. (80–85%); m.p. 117–120°.

C. *Flavone.* To a solution of 16.6 g. (0.069 mole) of the crude diketone in 90 ml. of glacial acetic acid, contained in a 250-ml.

conical flask, is added, with shaking, 3.5 ml. of concentrated sulfuric acid. The mixture is heated under a reflux condenser on a steam bath for 1 hour with occasional shaking and is then poured onto 500 g. of crushed ice with vigorous stirring. After the ice has melted, the crude flavone is collected on a filter, washed with water (about 1 l.) until free from acid, and finally dried at 50°. The yield of product is 14.5–15 g. (94–97%); m.p. 95–97°. The over-all yield of flavone based on *o*-hydroxy-acetophenone is 59–68%. The product may be recrystallized from ligroin as in Method 1.

2. Notes

1. Commercial pyridine is dried over solid sodium hydroxide and distilled through a fractionating column.

2. Glycerol is twice distilled under reduced pressure and used immediately in the reaction.

3. The potassium hydroxide is pulverized rapidly in a mortar previously heated at 100°.

4. The mixture usually becomes so thick and pasty that hand stirring is necessary.

5. *o*-Hydroxydibenzoylmethane can be crystallized from 95% ethanol and forms crystals melting at 120°, which give a strong enol reaction with ferric chloride. Crystallization is not necessary here.

3. Methods of Preparation

o-Benzoyloxyacetophenone has been prepared by the action of benzoyl chloride on a pyridine solution of *o*-hydroxyacetophenone.[3] The rearrangement of *o*-benzoyloxyacetophenone to *o*-hydroxydibenzoylmethane by alkali has been described.[4, 5] The latter diketone has been made by the base-catalyzed condensation of ethyl benzoate with *o*-hydroxyacetophenone.[6] The cyclization of *o*-hydroxydibenzoylmethane described in Method 2 is based on the work of Doyle, Gogan, Gowan, Keane, and Wheeler.[5] Cyclization has also been effected by use of glacial acetic acid containing hydrogen chloride or sodium acetate.[4]

Other methods of preparing flavone include: the action of

ethanolic alkali on 2'-acetoxy-α,β-dibromochalcone; [7] Claisen condensation of ethyl o-ethoxybenzoate and acetophenone, and cyclization of the resulting 1,3-diketone with hydriodic acid; [8] and treatment of 3-bromoflavanone with potassium hydroxide in ethanol.[9] Flavone has also been prepared from ethyl phenyl-propiolate by condensation with sodium phenoxide and subsequent cyclization with phosphorus pentachloride in benzene; [10] by fusing o-hydroxyacetophenone with benzoic anhydride and sodium benzoate; [11] by the dehydrogenation of 2'-hydroxychalcone with selenium dioxide; [12] by the action of alkali on flavylium chloride; [13] by the acid hydrolysis of 3-benzoyl-4-hydroxy-coumarin; [14] by the condensation of benzamide with 2-hydroxy-acetophenone; [15] and by heating β-morpholino-2-chlorochalcone which cyclizes to 4-morpholinoflavylium chloride, and which in turn may be hydrolyzed to flavone.[16]

Method 1 is a new procedure for the direct production of flavone from o-benzoyloxyacetophenone and has been successfully applied to the synthesis of other flavones. Method 2, which involves the Baker-Venkataraman transformation, is recommended because of its high over-all yield and the reproducibility of the results. Mozingo and Adkins' method [6] is satisfactory, but the yield of o-hydroxydibenzoylmethane is variable.

[1] University College, Dublin, Ireland.

[2] *Org. Syntheses Coll. Vol.* **2**, 545 (1943); *Org. Syntheses Coll. Vol.* **3**, 389 (1955).

[3] Freudenberg and Orthner, *Ber.*, **55**, 1748 (1922); Baker, *J. Chem. Soc.*, **1933**, 1386.

[4] Baker, *J. Chem. Soc.*, **1933**, 1386; Bhalla, Mahal, and Venkataraman, *J. Chem. Soc.*, **1935**, 868; Virkar and Wheeler, *J. Chem. Soc.*, **1939**, 1681.

[5] Doyle, Gogan, Gowan, Keane, and Wheeler, *Sci. Proc. Roy. Dublin Soc.*, **24**, 291 (1948).

[6] Mozingo and Adkins, *J. Am. Chem. Soc.*, **60**, 672 (1938).

[7] Feuerstein and Kostanecki, *Ber.*, **31**, 1757 (1898).

[8] Kostanecki and Tambor, *Ber.*, **33**, 330 (1900).

[9] Kostanecki and Szabranski, *Ber.*, **37**, 2634 (1904); Nakagawa and Tsukahima, *Nippon Kagaku Zasshi*, **75**, 485 (1954) [*C. A.*, **51**, 11339 (1957)].

[10] Ruhemann, *Ber.*, **46**, 2188 (1913); Bogert and Marcus, *J. Am. Chem. Soc.*, **41**, 87 (1919).

[11] Chadha and Venkataraman, *J. Chem. Soc.*, **1933**, 1073.

[12] Mahal, Rai, and Venkataraman, *J. Chem. Soc.*, **1935**, 866.

[13] Hill and Melhuish, *J. Chem. Soc.*, **1935**, 1165.

[14] Veres and Horak, *Chem. listy*, **48**, 1644 (1954), *Collection Czechoslov. Chem. Communs.*, **20**, 371 (1955) [*C. A.*, **49**, 14756 (1955)].

[15] Gowan, O'Connor, and Wheeler, *Chem. & Ind. (London)*, **1954**, 1201.

[16] Southwick and Kirchner, *J. Am. Chem. Soc.*, **79**, 689 (1957).

9-FLUORENECARBOXYLIC ACID*

(Fluorene-9-carboxylic acid)

$$(C_6H_5)_2CCO_2H \xrightarrow{AlCl_3} \quad CHCO_2H$$
$$\underset{OH}{|}$$

Submitted by Henry J. Richter.[1]
Checked by Richard T. Arnold, B. C. W. Hummel, and W. J. Wolf.

1. Procedure

Caution! Because of the evolution of considerable amounts of hydrogen chloride, this preparation must be conducted in a good hood or the apparatus must be attached to a gas trap.

A mixture of 45.6 g. (0.2 mole) of benzilic acid (Note 1) in 700 ml. of anhydrous thiophene-free benzene, contained in a 2-l. three-necked flask fitted with a reflux condenser (attached to a calcium chloride drying tube) and a motor-driven sealed stirrer, is cooled in an ice bath until a crystalline mass results. To the stirred mixture is added, in one portion, 80 g. (0.6 mole) of anhydrous aluminum chloride. The stirred mixture is heated until refluxing begins and is maintained at this temperature for 3 hours. During this period much hydrogen chloride is evolved, and the initially yellow solution soon becomes deep red. The solution is cooled and decomposed by the cautious addition of small pieces of ice, and then 400 ml. of water is added cautiously, followed by 200 ml. of concentrated hydrochloric acid. The benzene is removed by steam distillation, and the product is separated by filtration from the hot mixture. The lumps of product are crushed and extracted with 400 ml. of boiling 10% sodium carbonate solution. The mixture is filtered, and the extraction is repeated on the undissolved residue with an additional 200 ml. of hot 10% sodium carbonate solution. The basic filtrates are combined, 3–4 g. of Norit is added, and the mixture is heated to boiling. The Norit is separated by filtration, and the cooled solution is strongly acidified with cold concentrated hydrochloric acid (Note 2). The solid is collected on a Büchner funnel, washed with two 100-ml. portions of water, and dried (Note 3). The 9-fluorenecarboxylic acid so obtained weighs 39–41 g. (93–97%) and melts at 215–222°.

This product can be further purified by stirring it for several minutes with 200 ml. of benzene at 45°. The insoluble portion is collected on a Büchner funnel and washed first with 40 ml. of cold benzene and then with 40 ml. of petroleum ether (b.p. 28–38°). There is thus obtained 30–34 g. (71–81%) of almost colorless 9-fluorenecarboxylic acid melting at 219–222° with some previous sintering (Note 4).

2. Notes

1. The Distillation Products Industries grade melting at 150–151° is satisfactory.

2. Good results are usually obtained if the temperature during neutralization is not allowed to exceed 15°.

3. If the occluded hydrochloric acid and aluminum salts are effectively removed during the washing operation, this product can be dried in a steam oven without discoloration.

4. The melting points reported for this compound range from 210° to 230° [2-6] and appear to be a function of the rate of heating. The product obtained above, showing a neutralization equivalent of 215–218 (calculated 210), has proved very satisfactory. The acid may be further purified by crystallization from 50% ethanol using 5–6 ml./g. There is then obtained 60–70% of acid melting at 221–223° with some previous sintering. All melting points are uncorrected.

3. Methods of Preparation

The procedure described is essentially that of Arnold, Parham, and Dobson [2] based on the reaction reported by Vorlander and Pritzsche.[3] This acid has also been prepared from ethyl trichloroacetate in benzene with aluminum chloride,[4] from fluorene by metalation with n-butyllithium,[5,6] sodium triphenylmethyl,[7] lithium triphenylgermanium,[8] lithium in tetrahydrofuran,[9] phenyllithium,[6] o-tolyllithium,[6] mesityllithium,[6] potassium hydroxide,[10] or ethylsodium-diethylzinc complex followed by carbonation,[11] from diphenyleneketene and water,[12] and from 9-fluorenylmagnesium bromide and carbon dioxide.[13]

[1] Hamline University, St. Paul, Minnesota.
[2] Arnold, Parham, and Dodson, *J. Am. Chem. Soc.*, **71**, 2439 (1949).
[3] Vorlander and Pritzsche, *Ber.*, **46**, 1793 (1913).
[4] Delacre, *Bull. soc. chim. France*, [3] **27**, 875 (1902).
[5] Burtner and Cusic, *J. Am. Chem. Soc.*, **65**, 262 (1943).
[6] Morton, *Iowa State Coll. J. Sci.*, **28**, 367 (1954) [*C. A.*, **49**, 8162 (1955)].
[7] Schlenk and Bergmann, *Ann.*, **463**, 98 (1928).
[8] Gilman and Gerow, *J. Org. Chem.*, **23**, 1582 (1958).
[9] Gilman and Gorsich, *J. Org. Chem.*, **23**, 550 (1958).
[10] Morton, Claff, and Kagen, *J. Am. Chem. Soc.*, **76**, 4556 (1954).
[11] Ludwig and Schulze, *J. Org. Chem.*, **24**, 1573 (1959).
[12] Staudinger, *Ber.*, **39**, 3064 (1906).
[13] Grignard and Courtot, *Compt. rend.*, **152**, 1493 (1911).

5-FORMYL-4-PHENANTHROIC ACID

(4-Phenanthrenecarboxylic acid, 5-formyl-)

$$+ \ O_3 \longrightarrow$$

$$CO_2H \quad CHO$$

Submitted by R. E. Dessy and M. S. Newman.[1]
Checked by J. D. Roberts and D. I. Schuster.

1. Procedure

A solution of 25 g. (0.125 mole) of pyrene (Note 1) in 100 ml. of dimethylformamide (Notes 2 and 3) is treated with a 50% excess of ozone (Note 4). The solution of the ozonide is added at a moderate rate, with stirring, to 500 ml. of 1% aqueous acetic acid. The suspension is allowed to stand overnight (Note 5), and the resulting solid is collected by filtration and washed with water.

The moist solid is suspended in 200 ml. of 10% aqueous potassium hydroxide solution, and the suspension is boiled for 5 minutes. The hot solution is filtered, and the remaining solid is again extracted, using 100 ml. of potassium hydroxide solution.

To the dark-brown combined filtrates is added 100 ml. of potassium hypochlorite solution (Note 6), and the resulting solution is permitted to stand overnight. The mixture is then heated on a steam bath for 4 hours. The resulting orange solution is filtered while hot, and 100 ml. of 35% sodium hydroxide solution is added. The solution is cooled to 5°, the resulting solid is collected by filtration and washed with a small amount of saturated sodium chloride solution.

The moist sodium salt is digested with 50 ml. of cold $6N$ hydrochloric acid, and after several hours the mixture is filtered and the resulting solid acid dried.

The crude acid is dissolved in 100 ml. of boiling dimethylformamide, and 100 ml. of hot glacial acetic acid is added. Water is added to the hot solution until it becomes cloudy, and then just enough dimethylformamide is added to render the solution clear again. It is cooled to 5°, and the resulting acid collected by filtration and washed with glacial acetic acid. Upon drying, 10–11.5 g. (32–38%) of 5-formyl-4-phenanthroic acid (Note 7), melting at 272–276° (Note 8), is obtained.

2. Notes

1. Technical pyrene, Reilly Tar and Chemical Corp., was employed. Purification did not improve the over-all yield, and purer pyrene is not sufficiently soluble in dimethylformamide.

2. Freshly distilled dimethylformamide should be employed. The yields with the technical grade solvent were very erratic.

3. Complete solution is attained by heating the mixture for 5 minutes on a steam bath.

4. An ozonizer similar to that described by Henne and Perilstein [2] was employed. At an oxygen flow rate of 30 l./hr. it produced about 30 millimoles of O_3 per hour (3% conversion). Under these conditions the ozonization of 25 g. of pyrene requires about 6 hours.

5. Filtration of the hydrolyzate immediately after decomposition is difficult because of the fine nature of the solid. Upon standing, coagulation takes place to yield a granular brown solid.

6. The potassium hypochlorite solution was prepared [3] from the calcium hypochlorite sold by Mathieson Chemical Corporation under the trade name HTH. If the HTH reagent used is not fresh, it is found that subsequent heating of the filtrate with the potassium hypochlorite solution does not result in an orange solution. The solution remains dark brown, and the product is distinctly brown. The yield is not affected.

7. Unpublished experiments indicate that 5-formyl-4-phenanthroic acid exists mainly in the cyclic hydroxylactone form.

8. The checkers found that the melting point depends on the rate of heating. A reproducible melting point was obtained if the sample was placed in the bath at 270° and the temperature raised at the rate of two degrees per minute.

3. Methods of Preparation

The only reported method of preparation of 5-formyl-4-phenanthroic acid is by the reaction described here.[4]

[1] Ohio State University, Columbus, Ohio.
[2] Henne and Perilstein, *J. Am. Chem. Soc.*, **65**, 2183 (1943).
[3] *Org. Syntheses Coll. Vol.* **2**, 429 (1943).
[4] Vollmann, Becker, Corell, and Streeck, *Ann.*, **531**, 65 (1937).

FUMARONITRILE *

$$C_2H_5O_2CCH=CHCO_2C_2H_5 \xrightarrow[NH_4Cl]{NH_4OH} H_2NCOCH=CHCONH_2$$

$$H_2NCOCH=CHCONH_2 \xrightarrow{P_2O_5} NCCH=CHCN$$

Submitted by DAVID T. MOWRY and JOHN MANN BUTLER.[1]
Checked by CHARLES C. PRICE and RICHARD D. GILBERT.

1. Procedure

Caution! Fumaronitrile is both a vesicant and a lachrymator.

A. *Fumaramide.* A mixture of 516 g. (3.0 moles) of diethyl fumarate (Notes 1 and 2), 600 ml. of concentrated ammonium hydroxide (28%, sp. gr. 0.90, 9.0 moles), and 60 g. of ammonium chloride is placed in a 2-l. flask equipped with a stirrer and a thermometer. The reaction mixture is stirred at 25–30° with slight cooling for 7 hours (Note 3). The thick slurry of fumaramide is then filtered with suction, reslurried with 1 l. of water, filtered, and washed with about 50 ml. of ethanol. The white crystalline product is then dried in air or in an oven below 75° (Note 4). The yield is 270–300 g. (80–88%).

B. *Fumaronitrile.* The dry, finely powdered amide (Note 5) (228 g., 2.0 moles) and 613 g. (4.3 moles) of phosphorus pentoxide are placed in a 3-l. flask and thoroughly mixed by shaking. The flask is connected to a 1-l. suction flask receiver by means of a short 17-mm. i.d. 60° elbow extending about 15 cm. into the receiver. The receiver is cooled by immersion in an ice bath or by cold running water. The system is evacuated to 15–30 mm. by means of a water aspirator. The flask is then heated with one or two burners, using large soft flames. Heating should be started at the side and moved toward the bottom as the reaction progresses (Note 6). The reaction mass froths and blackens, and the product distils and sublimes into the receiver. The elbow leading to the receiver must be heated occasionally to melt condensed fumaronitrile. Heating is continued until no more fumaronitrile distils from the reaction flask (1.5 to 2 hours) (Note 7). The product, usually white and sufficiently pure (m.p. 93–95°) for most purposes, is obtained in a yield of 125–132 g. (80–85%). The product is recrystallized conveniently (in a fume hood) by dissolving it in 150 ml. of hot benzene and decanting or filtering the solution into 500 ml. of hexane or petroleum ether (Notes 8 and 9). The yield of long, glistening prisms, m.p. 96°, amounts to 117–125 g. (75–80%).

2. Notes

1. Diethyl fumarate liquid and vapors often cause reddening and itching of the skin, which usually disappear after a few hours.

2. Dimethyl fumarate gives equally good results but is less convenient to handle because it is a solid.

3. Additional ammonium hydroxide may be added if the slurry becomes too thick.

4. Temperatures higher than 75° often cause yellowing of the

amide and excessive foaming during the dehydration with phosphorus pentoxide.

5. The amide is considered sufficiently dry if no heat is evolved when a sample is shaken vigorously in a test tube with phosphorus pentoxide powder.

6. On somewhat smaller runs it may be more convenient to effect heating by the use of a wax or Wood's metal bath or an electric mantle. A bath or mantle temperature of 200° is sufficient for optimum yield. The reaction may froth vigorously if heated too rapidly or if the fumaramide is impure.

7. The cooled reaction flasks are easily cleaned by first soaking overnight in water and then rinsing with dilute sodium hydroxide solution.

8. *Caution!* This should be done in a hood. Fumaronitrile vapors and dust are irritating to the mucous membranes and are both vesicatory and lachrymatory. In the event of skin contact, the area should be washed promptly and thoroughly with soap and water to avoid irritation and blistering.

9. The principal impurities are a benzene-insoluble brown tar and hexane-soluble ethyl β-cyanoacrylate.

3. Methods of Preparation

The preparation described is based on the method of deWolf and van de Straete.[2] Fumaronitrile also has been prepared by the reaction of diiodoethylene with cuprous cyanide [3] and by hydrolyzing the product from the reaction of hydrogen cyanide with chloroacrylonitrile.[4] Pace claims that the yield of nitrile is improved when fumaramide is heated with phosphorus pentoxide in the presence of metal powders such as magnesium, cadmium, zinc, iron, and aluminum.[5]

[1] Monsanto Chemical Company, Dayton, Ohio.

[2] deWolf and van de Straete, *Bull. classe sci., Acad. roy. Belg.*, **21**, 216 (1935) [*C. A.*, **29**, 3985 (1935)].

[3] Jennen, *Bull. classe sci., Acad. roy. Belg.*, **22**, 1169 (1936) [*C. A.*, **31**, 1010 (1937)]; Gaade, *Rec. trav. chim.*, **65**, 823 (1946).

[4] Mowry and Yanko (to Monsanto Chemical Co.), U. S. pat. 2,471,767 [*C. A.*, **43**, 7498 (1949)].

[5] Pace (to Wingfoot Corp.), U. S. pat. 2,438,019 [*C. A.*, **42**, 4606 (1948)].

FURFURAL DIACETATE

(2-Furanmethanediol, diacetate)

$$\text{(furanyl)}C{=}O + (CH_3CO)_2O \xrightarrow{H^+} \text{(furanyl)}C(OCOCH_3)_2$$

Submitted by R. T. BERTZ.[1]
Checked by JAMES CASON, WILLIAM G. DAUBEN, W. B. FEARING, and B. P. SUMMERER.

1. Procedure

In a 300-ml. Claisen flask, whose side neck is elongated by a 10-cm. indented section, 102 g. (1 mole) of acetic anhydride and 0.1 ml. (Note 1) of concentrated sulfuric acid are mixed by hand swirling. The mixture is cooled to 10° by swirling in an ice bath, then there is added, during about 10 minutes, 96 g. (1 mole) of recently distilled furfural (Note 2). The temperature is maintained at 10–20°. After addition is complete and the contents of the flask have been well mixed by swirling, the cooling bath is removed and the reaction allowed to warm up spontaneously. A maximum temperature of about 35° is usually reached in about 5 minutes. After the temperature has dropped to that of the room (20–30 minutes), 0.4 g. (Note 1) of anhydrous sodium acetate is added, the flask is fitted for distillation at reduced pressure, and the mixture is distilled from an oil bath. A fore-run, weighing 50–70 g. and consisting principally of a mixture of acetic anhydride, furfural, and furfural diacetate (Note 3), is collected at 50–140°/20 mm. (Note 4). The product, collected at 140–142°/20 mm., weighs 129–139 g. (65–70%) and melts at 52–53° (Notes 4 and 5).

2. Notes

1. In order to freeze the equilibrium during distillation, it is imperative that the sulfuric acid catalyst be previously neutralized by sodium acetate; hence it is important that the small

quantities of sulfuric acid and sodium acetate be measured carefully.

2. The furfural used by the checkers was collected at 158–163°. In several runs made with technical grade furfural, which had been stored several years and was black and opaque, a satisfactory product was obtained but yields were 45–53%.

3. This fore-run turns black on standing, but if stored no more than a few days it may be assumed to be an equimolar mixture of furfural and acetic anhydride and may be used as starting material for subsequent runs. If no additional runs are to be made, the yield may be increased by 5–10% by redistilling the fore-run.

4. If collection of the product is started too soon it rapidly darkens and becomes partly liquid on standing. It is advisable to begin collecting the product only after a small sample of distillate, collected separately, sets to a crystalline mass on cooling. Products collected in this manner darkened slightly but remained solid after storage for 2 months. Original distillation of the reaction mixture through a 50-cm. column appeared not to improve the stability to storage, but a sample redistilled in a Claisen flask remained nearly white after storage for 6 months. All stored samples have a strong odor of acetic acid.

5. The submitter has carried out the preparation similarly on a 10-mole scale.

3. Methods of Preparation

Furfural diacetate has been prepared from furfural and acetic anhydride in the presence of various catalysts, including cation-exchange resins.[2-5] It has been claimed that the furfural diacetate preparation of Knoevenagel is improved by the addition of water to the reaction product, followed by ether extraction.[6] The present method is adapted to recovery and reuse of unreacted starting materials.

[1] Jacob V. Heemskerklaan 16, Katwyk aan Zee, Holland.
[2] Knoevenagel, *Ann.*, **402**, 119 (1913).
[3] Scheibler, Sotscheck, and Friese, *Ber.*, **57**, 1445 (1924).
[4] Gilman and Wright, *Rec. trav. chim.*, **50**, 833 (1931).

[5] Yamada, Chibata, and Tsurni, Jap. pat. 3917 (1955) [C. A., 51, 16542 (1957)]; *Pharm. Bull. (Japan)*, 2, 59 (1954) [C. A., 50, 214 (1956)].

[6] Ivanov and Fabrikant, *Compt. rend. acad. bulgare sci.*, 7, No. 1, 21 (1954) [C. A., 49, 6214 (1955)].

2-FURFURYL MERCAPTAN *

(2-Furanmethanethiol)

$$\underset{O}{\square}CH_2OH + H_2N-\overset{\overset{S}{\|}}{C}-NH_2 \xrightarrow{HCl}$$

$$\underset{O}{\square}CH_2-S-\overset{\overset{NH}{\|}}{C}-NH_2 \cdot HCl$$

$$\underset{O}{\square}CH_2-S-\overset{\overset{NH}{\|}}{C}-NH_2 \cdot HCl \xrightarrow{NaOH} \underset{O}{\square}CH_2SH + (H_2NCN)_n$$

Submitted by HELMER KOFOD.[1]
Checked by RICHARD T. ARNOLD and ERICH MARCUS.

1. Procedure

Caution! The following operations should be carried out in an effective hood (Note 1).

In a 3-l. round-bottomed flask are placed 380 g. (5 moles) of thiourea (Note 2), 500 ml. of water, and 400 ml. of concentrated hydrochloric acid (12.5N) (Note 3). The solid is dissolved by gentle heating, and the solution is cooled to 30°. Furfuryl alcohol (490 g., 434 ml., 5 moles) (Note 4) is added to the reaction mixture. The reaction, which usually commences spontaneously within a few minutes (Note 5), is strongly exothermic and should be controlled by suitable cooling with tap water so as to hold the temperature near 60° (Note 6). When the reaction subsides, cooling is discontinued and the clear, dark green solution is allowed to stand at room temperature for 12 hours.

A solution of 225 g. of sodium hydroxide (Note 3) in 250 ml. of water is poured into the reaction mixture. A heavy brown oil separates, consisting of S-2-furfurylisothiourea, which has already partially decomposed to 2-furfuryl mercaptan. The flask is quickly fitted with a steam-inlet tube and condenser. Steam distillation is continued as long as the distillate contains oily drops. The mercaptan is separated from the aqueous phase by means of a separatory funnel (Notes 7 and 8). The product is dried with calcium chloride; yield 313–340 g. (55–60%). The 2-furfuryl mercaptan so obtained is of a high degree of purity (Note 9) but can be distilled without decomposition in a nitrogen atmosphere; b.p. 160°/759 mm., 84°/65 mm., n_D^{20} 1.533.

2. Notes

1. The odor of the mercaptan is extremely disagreeable, and the substance in high concentration causes headache. An effective hood is absolutely essential.

2. The checkers employed practical grade thiourea obtained from Matheson, Coleman and Bell, East Rutherford, New Jersey.

3. It is desirable to determine the exact concentration of the hydrochloric acid and the composition of the sodium hydroxide by titration.

4. The checkers employed practical grade furfuryl alcohol obtained from the Eastman Kodak Company.

5. If the reaction does not start, the flask is heated gently until a spontaneous temperature rise sets in.

6. Temperatures above 60° and particularly supplementary refluxing are to be avoided since under these conditions the sensitive furan ring is attacked.

7. The mercaptan is almost insoluble in water, and the aqueous phase contains too little product to justify extraction.

8. It is convenient to use a 2-l. separatory funnel as a receiver during the steam distillation.

9. The checkers found the undistilled product to be essentially pure; n_D^{25} 1.5285. The distilled material was obtained with only mechanical losses; b.p. 84°/65 mm., n_D^{25} 1.5280.

3. Methods of Preparation

Furfuryl mercaptan cannot be prepared according to the classical method using furfuryl chloride and potassium sulfide.[2]

It has been prepared by reduction of 2-furfuryl disulfide, obtained from furfural and ammonium hydrosulfide.[3] The mercaptan has also been obtained in 33% yield [2] by the reaction of furfuryl chloride with thiourea and subsequent decomposition of the intermediate S-2-furfurylisothiourea according to the general method described in *Organic Syntheses*.[4] In the present method, which has been published previously, the use of the very unstable and difficultly available furfuryl halides is avoided.[5]

The formation of mercaptans directly from alcohols may be applied to the preparation of a large number of mercaptans, but usually much longer reaction periods or higher temperatures and higher concentrations of hydrogen halides are required.[6] Under such conditions the furan ring is destroyed.

[1] Danmarks farmaceutiske Højskole, Copenhagen, Denmark.
[2] Kirner and Richter, *J. Am. Chem. Soc.*, **51**, 3131 (1929).
[3] Staudinger and Reichstein, Can. pat. 283,765 [*C. A.*, **22**, 4537 (1928)].
[4] *Org. Syntheses Coll. Vol.* **3**, 363 (1955).
[5] Kofod, *Acta Chem. Scand.*, **7**, 1302 (1953).
[6] Frank and Smith, *J. Am. Chem. Soc.*, **68**, 2103 (1946).

2-FUROIC ACID *

$$2 \underset{O}{\boxed{}}\text{CHO} + O_2 \xrightarrow[\text{Cu}_2\text{O/Ag}_2\text{O}]{\text{NaOH}} 2 \underset{O}{\boxed{}}\text{CO}_2\text{H}$$

Submitted by R. J. Harrisson and M. Moyle.[1]
Checked by James Cason and W. N. Baxter.

1. Procedure

A 1-l. flask (Note 1) is fitted with a condenser, an efficient stirrer, two dropping funnels, a thermometer extending well into the flask, and a delivery tube which extends far enough to be below the surface of the stirred liquid.

To the flask is added 250 ml. of a 2.5% solution of sodium hydroxide and a cuprous oxide-silver oxide catalyst (Note 2). In the two dropping funnels are placed, respectively, 96 g. (1.0 mole) of furfural (Note 3) and a solution of 40 g. (1.0 mole) of sodium hydroxide in 100 ml. of water. The contents of the flask are heated to about 55°, vigorous stirring is started, and a rapid stream of oxygen is bubbled through as the contents of the two funnels are added simultaneously (Note 4) to the reaction mixture at such a rate (20–25 min.) as to maintain the temperature at 50–55° (Note 5) without external heating. After the additions have been completed, vigorous stirring and passage of oxygen are continued until the temperature drops below 40° (15–30 min., depending on the rate of oxygen flow).

The catalyst is separated by filtration, and the aqueous solution is extracted with three 30-ml. portions of ether (Note 6), acidified with 30% sulfuric acid, boiled for 45 min. (Note 7) with 6–7 g. of carbon, then filtered hot. The filtrate is cooled to 0° and allowed to stand at this temperature for 1 hour or longer. The 2-furoic acid, which separates as pale pink needles, m.p. 130–132°, is collected by suction filtration and washed with small portions of ice water (Note 8). The yield is 96–101 g. (86–90%).

2. Notes

1. Since oxygen is passed through the solution at 55°, it is best to operate under a reflux condenser. Ground joint fittings are not essential but convenient. The checkers used a 4-necked flask with ground joints; the thermometer was hung through the condenser, and one neck was arranged for attachment of the gas delivery tube and one of the separatory funnels.

2. The catalyst used is commercial cuprous oxide (9.6 g., 10% of the weight of the furfural) and commercial silver nitrate (0.5 g., 0.5% of the weight of the furfural) in 15 ml. of water. The cuprous oxide is suspended in the rapidly stirred 2.5% sodium hydroxide solution, and the solution of silver nitrate is added to give a dark brown suspension of cuprous oxide and silver oxide which is used directly. So long as oxygen is being passed through the reaction mixture, the life of the catalyst appears to be unlimited.

3. A sample of commercial furfural is purified by simple distillation or steam distillation immediately before use. The submitters have carried out this reaction on an 8-mole scale following the same procedure except that a total reaction time of about 4 hours was required.

4. The simultaneous addition of furfural and sodium hydroxide maintains the concentration of sodium hydroxide at about 2.5%, at which concentration only a small percentage of furfuryl alcohol is formed in the accompanying Cannizzaro reaction.

5. Below 50°, the reaction is too slow; above 55°, the reaction becomes violent and cooling becomes necessary. Frothing may be checked by addition of small amounts of benzene.

6. Ether extraction at this stage removes furfuryl alcohol (about 12 g.). If this is not removed, the reaction mixture becomes resinous on acidification and heating.

7. During the period of boiling in acid solution a small amount of polymeric material forms, and this is separated when the charcoal is removed by filtration.

8. The solubility of 2-furoic acid in water is 2.8 g. per 100 ml. at 0°; however, its solubility in the salt solution resulting from the reaction is quite low. Usually it is not possible to obtain further material from the filtrate by salting out with sodium chloride; however, in case a lowered yield is obtained in the first crop it may be profitable to investigate the possibility of obtaining additional material from the filtrate.

3. Methods of Preparation

2-Furoic acid has been made by oxidation of lactose followed by pyrolysis, by the oxidation of 2-acetylfuran, 2-methylfuran, or furfuryl alcohol using potassium ferricyanide in alkaline medium, and by other methods already listed.[2] In addition, furfural has been oxidized to furoic acid by air in the presence of sodium hydroxide and silver nitrate,[3] by hydrogen peroxide in the presence of pyridine or picoline,[4] by sodium or potassium hypochlorite,[5] by means of sodium hydroxide in methanol or sodium hydroxide in methanol followed by hydrogen peroxide,[6] and by sodium hydroxide in the presence of a copper oxide-silver oxide-carbon catalyst.[7]

[1] University of Sydney, Sydney, Australia.

[2] *Org. Syntheses Coll. Vol.* 1, 276 (1941).

[3] Taniyama, *Toho-Reiyon Kenkyû Hôkoku*, 2, 51 (1955) [*C. A.*, 53, 4247 (1959)].

[4] Baba, *Kagaku Kenkyusho Hôkoku*, 33, 168 (1957) [*C. A.*, 52, 7267 (1958)]; Physical and Chemical Researches, Inc. (by Midorikawa and Baba), Jap. pat. 6113 (1959) [*C. A.*, 54, 1544 (1961)].

[5] Farbenfabriken Bayer A.-G., Ger. pat. 908,023 [C. A., 52, 10201 (1958)]; Arita and Odawara (to Asahi Chemical Industries Co.), Jap. pat. 1130 (1950) [C. A., 47, 2214 (1953)].

[6] Kuwada et al. (to Nippon Petroleum Oil Co.), Jap. pat. 3730 (1951) [C. A., 47, 8096 (1953)].

[7] Terai et al. (to Noguchi Research Institute, Inc.), Jap. pat. 1111 (1951) [C. A., 47, 3883 (1953)].

GLUTARIC ACID* AND GLUTARIMIDE

1. FROM γ-BUTYROLACTONE

$$
\begin{array}{c}
CH_2\!\!-\!\!CH_2 \\
| \qquad\quad | \\
O \qquad CH_2 \\
\diagdown\; C\; \diagup \\
\parallel \\
O
\end{array}
+ \text{ KCN} \xrightarrow{190\text{--}195^\circ}
\begin{array}{c}
CO_2K \\
| \\
(CH_2)_3 \\
| \\
CN
\end{array}
\xrightarrow{\text{Dilute HCl}}
\begin{array}{c}
CO_2H \\
| \\
(CH_2)_3 \\
| \\
CONH_2
\end{array}
$$

$$
\begin{array}{c}
CO_2H \\
| \\
(CH_2)_3 \\
| \\
CONH_2
\end{array}
\xrightarrow[\text{Heat}]{H_2O,\ H^+} HO_2C\!\!-\!\!(CH_2)_3\!\!-\!\!CO_2H + NH_4^+
$$

$$
\xrightarrow{\text{Heat}}
\begin{array}{c}
O \\
\diagup\!\!\!\!\diagdown \\
CH_2\!\!-\!\!C \\
\diagup \qquad\qquad \diagdown \\
CH_2 \qquad\qquad NH \\
\diagdown \qquad\qquad \diagup \\
CH_2\!\!-\!\!C \\
\diagdown\!\!\!\!\diagup \\
O
\end{array}
+ H_2O
$$

Submitted by G. Paris, L. Berlinguet, and R. Gaudry.[1]
Checked by James Cason and Edwin R. Harris.

1. Procedure

Caution! This preparation should be carried out in a good hood since poisonous hydrogen cyanide may be evolved.

In a 500-ml. three-necked flask fitted with a sealed mechanical stirrer and a reflux condenser are placed 86 g. (1 mole) of γ-butyrolactone (Note 1) and 72 g. (1.1 moles) of potassium cyanide (Note 2). As the contents of the flask are stirred, the mixture is heated in an oil bath for 2 hours at a temperature of 190–195° (Note 3). There is an initial vigorous reaction which soon subsides. After the completion of the heating period the mixture is cooled to about 100°, and the potassium salt of the cyano acid is dissolved in about 200 ml. of hot water. The warm solution is cautiously acidified to Congo red by the addition of about 90 ml. of concentrated hydrochloric acid. The resultant solution, which contains glutaric acid monoamide and potassium chloride, is used to prepare glutaric acid or glutarimide.

A. *Glutaric acid.* To the solution of monoamide is added 200 ml. of concentrated hydrochloric acid, and the mixture is heated under reflux in the hood for 1 hour. The reaction mixture is evaporated to dryness under reduced pressure, and the residue is dried by brief heating on a steam bath at reduced pressure. The residual crystalline solid is broken up, ground in a mortar, and extracted with four 200-ml. portions of boiling chloroform. The combined hot extracts are filtered by gravity through a fluted paper on a heated funnel and then concentrated to about 400 ml. After the solution has been cooled in water to effect crystallization, the glutaric acid is collected by suction filtration, washed with cold chloroform, and dried. The yield of slightly discolored glutaric acid, suitable for many purposes, is 105–110 g. (79.5–83.5%), m.p. 95–97°.

If a pure grade of glutaric acid is desired, it is decolorized by boiling for about 1 hour with 10 g. of charcoal in water solution. The charcoal is removed by filtration (Note 4), the water is evaporated under reduced pressure, and the dry residue is recrystallized from chloroform. The yield of white glutaric acid, m.p. 98–99°, is 94–99 g. (71–75%).

B. *Glutarimide.* The solution containing the monoamide is extracted with six 50-ml. portions of ether. The ether solution is dried over anhydrous sodium sulfate (or by filtering by gravity through a layer of the drying agent), and then the ether is evaporated by heating on a steam bath; the last portion is removed at reduced pressure. The oily residue of glutaric acid monoamide is placed in a 300-ml. round-bottomed flask which is fitted with a bent tube attached to a short condenser, and the flask is immersed in a bath (Note 5) held at 220–225°. Heating is continued until water no longer distils (3–4 hours). The cooled glutarimide is dissolved in about 200 ml. of water, and the solution is boiled for about 30 minutes with about 2 g. of charcoal. The charcoal is removed by filtration, water is removed by distillation at reduced pressure, and the dry residue is crystallized from about 125 ml. of 95% ethanol, with final cooling in an ice bath. The yield of glittering white crystals of glutarimide, m.p. 152–154°, is 65.5–73.5 g. (58–65%) (Note 6).

2. Notes

1. Butyrolactone from Eastern Chemical Corporation, 34 Spring Street, Newark 2, New Jersey, was used without purification.

2. Satisfactory results were obtained with potassium cyanide, 96–98% purity, from General Chemical Company or with material indicated as of 95% minimum purity, from Merck & Co., Inc. If potassium cyanide pellets are employed, they should be pulverized before use.

3. Since the reaction mixture is acidified after the heating period, it is most convenient to carry out the reaction in a forced-draft hood in order to provide protection against hydrogen cyanide. If higher temperatures than those specified are used, the reaction may get out of control during the initial vigorous reaction.

4. If filtration by gravity through a fluted paper fails to remove the last traces of charcoal, the filtrate should be refiltered by suction through a thin mat of filter aid such as Supercel.

5. The submitters used an oil bath. The checkers used a salt bath consisting of an equimolar mixture of potassium nitrate and sodium nitrite (Heat Transfer Salt). A salt bath should be handled only with proper precautions, which include wearing goggles and gloves and supporting the bath on a stand bolted to the bench.

6. This material is suitable for preparation of N-bromoglutarimide, as follows. In a 1-l. beaker, provided with a mechanical stirrer, 65 g. of potassium hydroxide is dissolved in 200 ml. of water. The vigorously stirred solution is cooled to about −5°, and as the temperature is maintained below 0° there is added gradually 113 g. (1 mole) of glutarimide and cracked ice. To this mixture is added in one portion 160 g. (1 mole) of bromine; then stirring is continued for 1 minute. The mixture is filtered by suction, and the precipitate is dissolved in hot water. On cooling, there crystallizes about 94 g. (49%) of N-bromoglutarimide which melts at about 165°. This product is usually suitable for use as a brominating agent. Pure N-bromoglutarimide, m.p. 180–185°, is obtained only after several recrystallizations from water.

II. FROM DIHYDROPYRAN

$$\text{(dihydropyran)} + H_2O \xrightarrow{H^+} OCHCH_2CH_2CH_2CH_2OH$$

$$OCHCH_2CH_2CH_2CH_2OH + 3[O] \xrightarrow{HNO_3}$$
$$HO_2CCH_2CH_2CH_2CO_2H + H_2O$$

Submitted by J. ENGLISH, JR., and J. E. DAYAN.[2]
Checked by ARTHUR C. COPE and MARK R. KINTER.

1. Procedure

In a 1-l. round-bottomed flask equipped with a reflux condenser are placed 400 ml. of 0.2N nitric acid (5 ml. of concentrated nitric acid, sp. gr. 1.42, and 395 ml. of water) and 168.3 g. (2 moles) of dihydropyran (Note 1). The mixture is heated on a steam bath or a boiling water bath; the yellowish upper layer dissolves suddenly after 25 to 45 minutes of heating. The flask is swirled to aid the dissolution at this time, and the period of heating is extended for an additional 5 to 10 minutes.

While the hydrolysis of the dihydropyran is taking place, 800 g. (575 ml., 9.25 moles) of concentrated nitric acid (sp. gr. 1.42) is placed in a 2-l. three-necked flask and cooled in an ice-salt bath in a well-ventilated hood. The flask should be equipped with an efficient stirrer, separatory funnel, reflux condenser, and a thermometer. When the temperature of the solution reaches 0°, 5.75 g. of sodium nitrite is added and stirring is continued until most of it has dissolved; the solution becomes yellow.

The solution obtained by the hydrolysis of dihydropyran is placed in the separatory funnel, and about 10 ml. is added to the nitric acid solution at a temperature below 0°. After the evolution of brown nitrogen dioxide fumes begins (in about 10 minutes), the addition is continued at a rate that allows the temperature to be held below 10° (Note 2). The addition requires about 3 hours (Notes 3 and 4). When the addition is completed, the blue-green solution is stirred for an additional 1.5 hours as brown fumes continue to be evolved. The cooling bath then is removed and stirring is continued as the temperature is allowed to rise slowly to 25–30° (Note 5). As the reaction nears completion, the color changes from blue-green to green to light yellow. Appearance of the light yellow color indicates the end of the reaction and normally requires 2 to 3 hours after the addition is completed. The volume of the solution then is reduced either by evaporation on a steam bath or by distillation under reduced pressure.

In the latter method, after removal of all the water, an additional 100 ml. of water is added and the distillation is repeated to remove the remaining nitric acid. The solid residue remaining from either method of removing water from the reaction mixture is recrystallized from a mixture of 100 ml. of ether and 1 l. of benzene. Insoluble sodium nitrate and succinic acid are removed by filtration of the hot solution. Upon cooling, 185–198 g. (70–75%) of glutaric acid is obtained in the first crop, m.p. 89.5–91.5° (Note 6). On concentrating the benzene solution to 200 ml. and cooling, an additional 18–23 g. of crude glutaric acid can be obtained (Note 7).

2. Notes

1. Dihydropyran is available from the Electrochemicals Department, E. I. du Pont de Nemours and Company, or can be prepared by the dehydration of tetrahydrofurfuryl alcohol.[3]

2. More efficient cooling may shorten the time required for the addition.

3. As the hydrolyzed dihydropyran solution cools, it may become cloudy and δ-hydroxyvaleraldehyde may separate as a red liquid. The separation can be avoided by continuing to heat most of the solution while a small part is left in the separatory funnel, but it is not essential, for the yield is not affected by the separation of phases at this point. Oxidation of the pure aldehyde in a similar manner is stated to give a 90% yield of glutaric acid.[4]

4. The separatory funnel is removed after the addition is completed to facilitate the removal of nitrogen oxides.

5. Care should be taken that the temperature does not rise above room temperature; if this occurs much succinic acid is pro-

duced. If the temperature rises too high the cooling bath should be replaced again until the temperature does not exceed 25–30° on its removal.

6. The glutaric acid obtained has a neutralization equivalent of 66.3 (theory 66.1) and is suitable for most synthetic work.

7. This material melts at 80–82° and can be purified by recrystallization. In a series of preparations it is advantageous to save these residues and combine them for recrystallization.

3. Methods of Preparation

Methods of preparation of glutaric acid are given in an earlier procedure.[5] In addition to these, glutaric acid has been obtained by the oxidation of glutaraldehyde in the presence of catalysts,[6] by the ozonization of cyclopentene followed by permanganate cleavage of the ozonide,[7] by the nitric acid oxidation of 2-cyanocyclopentanone,[8] by the oxidation of pentamethylene glycol with nitrogen tetroxide,[9] and by the condensation of acrylonitrile with ethyl malonate, followed by acid hydrolysis of the mono- and di-adduct.[10]

Glutarimide has been prepared from glutaric acid and sulfamide [11] or formamide,[12] by distillation of ammonium glutarate,[13] by hydrolysis of pentanedinitrile with acetic acid,[14] and by oxidation of piperidine with hydrogen peroxide.[15]

Method I, based on a published procedure,[16] offers a more convenient synthesis of glutaric acid and its imide and may be readily adapted to a large scale.

[1] Université Laval, Quebec City, Canada.

[2] Yale University, New Haven, Connecticut.

[3] Schniepp and Geller, *J. Am. Chem. Soc.*, **68**, 1646 (1946); *Org. Syntheses Coll. Vol.* **3**, 276 (1955).

[4] U. S. pat. 2,389,950 [*C. A.*, **40**, 1539 (1940)].

[5] *Org. Syntheses Coll. Vol.* **1**, 289 (1941).

[6] Guest, Stansbury, and Lykins, Brit. pat. 767,416 [*C. A.*, **51**, 12967 (1957)].

[7] N. v. de Bataafsche Petroleum Maatschappy, Brit. pat. 772,410 [*C. A.*, **51**, 12970 (1957)].

[8] Badische Anilin- & Soda-Fabrik Akt.-Ges., Ger. pat. 887,943 [*C. A.*, **52**, 13784 (1958)].

[9] Langenbeck and Richter, *Chem. Ber.*, **89**, 202 (1956).

[10] Hesse and Bücking, *Ann.*, **563**, 31 (1949).

[11] Kirsanov and Zolotov, *Zhur. Obshcheĭ Khim.*, **20**, 1145 (1950).

[12] Sugasawa and Shigehara, *J. Pharm. Soc. Japan*, **62**, 531 (1942).

[13] Bernheimer, *Gazz. chim. ital.*, **12**, 281 (1882).

[14] Seldner, *Am. Chem. J.*, **17**, 532 (1895).

[15] Wolffenstein, *Ber.*, **25**, 2777 (1892).

[16] Paris, Gaudry, and Berlinguet, *Can. J. Chem.*, **33**, 1724 (1955); Reppe et al., *Ann.*, **596**, 158 (1955).

GUANYLTHIOUREA *

(Urea, 1-amidino-2-thio-)

$$NH_2C(NH)NHCN + H_2S \rightarrow NH_2C(NH)NHCSNH_2$$

Submitted by FREDERICK KURZER.[1]
Checked by JOHN C. SHEEHAN, GEORGE BUCHI, and DAVID KNUTSON.

1. Procedure

A 500-ml. three-necked round-bottomed flask, supported in a water bath equipped with a thermostat, is fitted with a vertical air-condenser and, on the side necks, with a thermometer and a gas delivery tube, both of which nearly reach the bottom of the vessel. The flask contains a solution of 42 g. (0.5 mole) of N-cyanoguanidine (dicyandiamide) (Note 1) in 200 ml. of water (Note 2). This is kept at 75° during 12 hours, and at 65–70° during an additional 25–30 hours (Note 3) while a fairly slow stream (Note 4) of hydrogen sulfide is passed through (Note 5). The resulting deep yellow liquid is allowed to cool to 45° while passage of hydrogen sulfide is continued; it is made strongly alkaline with 15 ml. of 40% aqueous sodium hydroxide and is freed from suspended impurities (consisting usually of finely divided black particles) by rapid filtration under reduced pressure (Note 6). The clear, yellow filtrate is allowed to cool slowly to room temperature, when lustrous prismatic leaflets separate gradually. Crystallization is completed by storing the flask and contents at 0° for 24 hours. The crystalline mass is broken up with a spatula, collected by filtration, and washed with three 20-ml. portions of ice water, the second and third portions of washing liquid being collected separately and discarded (Note 7). The yield of crude guanylthiourea, forming large, nearly colorless prisms, m.p. 160–164° (dec., with previous sintering at 154–158° or somewhat lower) (Note 8), varies between 29.5 and 32.5 g. (50–55%).

The material is purified as follows: The large crystals are powdered (Note 9) and boiled with successive portions of methanol until solution is complete, the saturated extracts being decanted and collected; approximately 8–9 ml. of methanol per gram of crude solid is required. The combined hot solution is filtered, if necessary, with suction, is allowed to cool slowly to room temperature, and is then stored at 0° overnight. Almost pure guanylthiourea, forming colorless prisms, m.p. 170–172° (dec., with previous sintering in the range 166–170°), is collected by filtration and washed with a little methanol. The recovery, per crystallization, varies from 65% to 70%. Most of the material contained in the mother liquors may be obtained by evaporation to small volume (Note 10) and forms one or two crops of satisfactory quality (melting point ranging between 166° and 170°).

2. Notes

1. A good reagent grade of N-cyanoguanidine is satisfactory in this synthesis.

2. The smallest possible volume of water is employed to ensure direct crystallization of the product when the solution is finally allowed to cool. Smaller volumes than the stated amount of water, however, tend to block the delivery tube with separated crystalline material, particularly in the beginning stages of the reaction, when much of the less-soluble reactant is still present.

3. The total time of heating may be made up of several shorter periods. On being cooled overnight the reaction mixture deposits a thick crust of crystalline solid on the walls of the flask. After 8 hours, large, spikelike prisms of unchanged N-cyanoguanidine predominate, but later the separated mass consists of massive rhombic prisms of guanylthiourea, with small needles of 2,4-dithiobiuret filling the spaces between the prisms. Solution is once again rapidly effected when the temperature is raised to 70°, with passage of hydrogen sulfide.

4. A rate of 3–4 bubbles of hydrogen sulfide per second is satisfactory. Toward the end of the time of reaction, a slower stream suffices.

5. Since hydrogen sulfide is only slowly absorbed, most of the gas escapes through the air condenser. The reaction must therefore be performed in an efficient hood.

6. A preheated Büchner flask and filter of 5- to 7-cm. diameter carrying a double layer of filter paper are employed.

7. 2,4-Dithiobiuret may be isolated from the alkaline filtrate as follows: The stirred alkaline filtrate is acidified to Congo red with 25–30 ml. of concentrated hydrochloric acid. A pale yellow powdery solid is rapidly precipitated, with evolution of hydrogen sulfide; after storage at 0°, the solid (dry weight 7.5–9.0 g.) is collected by filtration, washed three times with 30-ml. portions of ice water, and pressed semidry. This material is added to boiling water (100 ml.) and stirred at 95–100° for 2 minutes, and the undissolved yellow powdery impurities (1.0–1.5 g., consisting largely of sulfur) are quickly removed by filtration with suction through a preheated funnel. The clear, pale yellow filtrate is rapidly filled with crystals, which are collected after storage at 0° for 12 hours. The yield of crude 2,4-dithiobiuret, forming small yellow needles, m.p. 178–180° (dec., previously sintering at 172–175°) is 5.5–7.0 g. (8–10%) (Note 11). Two crystallizations from boiling water (12 ml. per gram, recovery per crystallization: 80–90%) (Note 12), one with addition of carbon, affords nearly white lustrous needles of 2,4-dithiobiuret, m.p. 180–182° (dec., previously sintering slightly at 178–180°) (Note 13).

8. Colorless guanylthiourea melts to a pale yellow liquid, which resolidifies to a yellow crystalline mass on cooling; partial rearrangement to guanidine thiocyanate occurs during the fusion.

9. When ground in a mortar, the crude guanylthiourea evolves small quantities of occluded hydrogen sulfide.

10. The methanolic filtrates are rapidly distilled to approximately half volume at low temperature in a good vacuum. The solution is likely to froth considerably, and the use of a relatively large distilling flask is recommended. Small quantities of solid separating on the walls of the flask are redissolved by heating the residual liquid once again on a steam bath.

11. The yield of dithiobiuret increases on lengthening the time of reaction. After 100 hours' passage of hydrogen sulfide the yields of guanylthiourea and dithiobiuret were 25–35% and 15–18%, respectively.

12. Crystallization is carried out by adding the solid to the appropriate volume of boiling water, stirring at 95–100° for about 1 minute, and removing small quantities of suspended yellow im-

purities by filtration under reduced pressure using carefully pre-heated apparatus. When cooled slowly, the clear filtrate deposits the product in the form of large, lustrous needles.

13. On analysis, this material gives carbon, hydrogen, nitrogen, and sulfur percentages in excellent agreement with the calculated values. Two additional crystallizations from ethanol-water (10 ml. each, per gram) raise the melting point to 183–185° (dec.). These decomposition temperatures are somewhat influenced by the rate of heating; the quoted values are observed when the specimen is inserted at 160° and the bath temperature is raised at the approximate rate of 8° per minute.

3. Methods of Preparation

Guanylthiourea has been prepared by the prolonged interaction of saturated aqueous hydrogen sulfide with N-cyanoguanidine [2,3] or an amidinourea salt [4] at 60–80°. It is formed in small quantities when thiourea is heated with thiophosgene or phosphorus pentachloride at 100–110°.[5] Guanylthiourea also results from the acid hydrolysis of 4,6-diamino-2-thio-1,3,5-thiadiazine.[6,7] In all these syntheses the product is collected as a sparingly soluble salt, the base being subsequently isolated by comparatively laborious methods, e.g. from the oxalate [2] or the phosphate.[7] The present procedure,[8] based on Bamberger's method,[2] allows the direct isolation of the base. Small quantities of 2,4-dithiobiuret, which are also formed in this reaction as a by-product, are readily separated.

[1] Royal Free Hospital School of Medicine, University of London, England.

[2] Bamberger, *Ber.*, **16**, 1460 (1883).

[3] Slotta and Tschesche, *Ber.*, **62**, 1402 (1929).

[4] Bamberger, *Ber.*, **16**, 1461 (1883).

[5] Rathke, *Ber.*, **11**, 962 (1878).

[6] Thurston and Sperry, U. S. pat. 2,364,594 [*C. A.*, **39**, 4630 (1945)].

[7] Birtwell, Curd, Hendry, and Rose, *J. Chem. Soc.*, **1948**, 1653.

[8] Kurzer, *J. Chem. Soc.*, **1955**, 1.

D-GULONIC-γ-LACTONE *

(D-Gulonic acid, γ-lactone)

$$
\begin{array}{ccc}
\begin{array}{c}
\text{CHO} \\
| \\
\text{HCOH} \\
| \\
\text{HOCH} \\
| \\
\text{HCOH} \\
| \\
\text{CH}_2\text{OH}
\end{array}
&
+ \; \text{NaCN} + \text{NH}_4\text{Cl} \; \xrightarrow[\text{Lactonization}]{\text{Hydrolysis}}
&
\begin{array}{c}
\overline{}\text{CO} \\
| \\
\text{HCOH} \\
| \\
\text{HCOH} \\
| \\
\underline{}\text{OCH} \\
| \\
\text{HCOH} \\
| \\
\text{CH}_2\text{OH}
\end{array}
\end{array}
$$

Submitted by J. V. KARABINOS.[1]
Checked by R. T. ARNOLD, FRED SMITH, and BERTHA LEWIS.

1. Procedure

In a 500-ml. glass-stoppered Erlenmeyer flask, 30 g. (0.2 mole) of D-xylose and 10.7 g. (0.2 mole) of ammonium chloride are dissolved in 100 ml. of distilled water. Cracked ice (100 g.) is added to this mixture, followed by 10 g. (0.2 mole) of sodium cyanide, and the solution is maintained at 0–5° for 48 hours. Powdered barium hydroxide octahydrate (63 g., 0.2 mole) is added along with 100 ml. of water to the cyanohydrin mixture (Note 1), which is heated on a steam bath for 2 hours with occasional stirring. The basic barium gulonate (Note 2), which is allowed to separate overnight at 5°, is collected by filtration and washed with cold water (0°) until the washings are chloride-free. Excessive washing of the barium salt is to be avoided because of its solubility. The barium salt is suspended in 200 ml. of water, and the barium ion is precipitated quantitatively by sulfate ion (Note 3). After removal of the barium sulfate by suction filtration, the filtrate and washings are concentrated to a colorless syrup on a steam bath in a stream of dry air (Note 4). The resultant syrup is dissolved in 50 ml. of hot ethylene glycol monomethyl ether (methyl Cellosolve), sufficient ethyl acetate is added to incipient turbidity, and the solution is seeded with D-gulonic-γ-lactone (Note 5). The lactone, which is allowed to crystallize overnight, is collected by suction filtration, washed with ethanol and dried in a vacuum oven at 60°. The D-gulonic-γ-lactone (Note 6) has a melting point of 181–183° which is unchanged by recrystallization from aqueous ethanol. The yield is 10.7–11.6 g. (30–33%) (Note 7).

2. Notes

1. The barium hydroxide serves to hydrolyze any unchanged nitriles as well as to precipitate the aldonic acid.

2. The barium gulonate is undoubtedly contaminated with some epimeric idonate. The lactone of the latter substance is removed by recrystallization of the gulonic lactone from methyl Cellosolve.

3. It is convenient to titrate the suspended barium salt with $18N$ sulfuric acid (approx. 12–14 ml.) to a pH of 1.5 using a pH meter. After removal of the barium sulfate the slight excess of sulfate ion may be precipitated using barium chloride solution. The end point is taken when several drops of filtrate show no turbidity either upon addition of sulfuric acid or barium chloride solution.

4. Concentration in this manner allows sufficient time for the gulonic acid to be converted to the lactone in the presence of a trace of hydrochloric acid. The checkers observed also that an easily crystallized lactone was always obtained if concentration under reduced pressure was employed.

5. Crystallization is speeded considerably by seeding.

6. A small amount of less pure lactone may be obtained by evaporation of the mother liquor to a syrup and repetition of the methyl Cellosolve-ethyl acetate crystallization.

7. The submitter has reported yields up to 39% using the above procedure.

3. Methods of Preparation

The present method is adapted from that of Fischer [2] employing recently developed modifications of the cyanohydrin synthesis.[3, 4]

[1] Blockson Chemical Company, Joliet, Illinois.

[2] Fischer and Stahel, *Ber.*, **24**, 528 (1891).

[3] Karabinos, Hann, and Hudson, *J. Am. Chem. Soc.*, **75**, 4320 (1953).

[4] Isbell, Karabinos, Frush, Holt, Schwebel, and Galkowski, *J. Research Natl. Bur. Standards*, **48**, 163 (1952).

HEMIMELLITENE *

(Benzene, 1,2,3-trimethyl-)

$$CH_2\overset{\oplus}{N}(CH_3)_3\overset{\ominus}{I}$$

Submitted by W. R. Brasen and C. R. Hauser.[1]

Checked by William S. Johnson, Donald W. Stoutamire, and A. L. Wilds.

1. Procedure

A 3-l. round-bottomed three-necked flask, fitted with a reflux condenser and a sealed stirrer, is charged with 2 l. of hot water and 100 g. (0.33 mole) of 2,3-dimethylbenzyltrimethylammonium iodide (Note 1). The stirred suspension is heated on the steam bath, and 2760 g. of 5% sodium amalgam (Note 2) is added in 200- to 250-g. portions over a period of 45 minutes. Stirring and heating are continued for 24 hours; then the mixture is steam-distilled until no more oily material comes over. The distillate (1–1.5 l.) is extracted with three 50-ml. portions of ether. The combined extracts are washed with 50 ml. of 10% hydrochloric acid and 50 ml. of saturated sodium chloride solution, and dried over anhydrous calcium chloride. After removal of the ether by distillation, the residue is distilled from a Claisen flask, giving 33.5–35.5 g. (85–90% yield) of colorless hydrocarbon, b.p. 171–174°. On redistillation 85–90% of the material distils at 173–174°, n_D^{25} 1.5116–1.5120.

2. Notes

1. The methiodide is prepared from 2,3-dimethylbenzyldimethylamine according to the procedure for producing the lower homolog (p. 587, Notes 5 and 8).

2. The sodium amalgam is prepared in a 1-l. filter flask fitted with a two-hole rubber stopper carrying a dropping funnel and an outlet tube. The flask is charged with 138 g. (6 g. atoms) of sodium and 300 ml. of mineral oil. With a stream of nitrogen passing through the side arm of the flask, the flask is heated on a hot plate covered with an asbestos pad until the sodium melts, and then 2622 g. of mercury is added rapidly from the dropping funnel over a 1–2 minute period with swirling. The whole operation is carried out in a large pan to catch material in case of breakage. A vigorous exothermic reaction occurs, and the hands must be adequately protected with several layers of cloth or heavy gloves as the temperature approaches 400°. If the addition is rapid enough, the amalgam will be a liquid; otherwise the material will be partially solid and must be heated vigorously to produce a homogeneous melt. The mineral oil is decanted from the molten amalgam, which is poured into a shallow metal pan while still warm and is cut or broken (with a hammer) into small pieces as it solidifies. It is finally washed with petroleum ether and stored in a tightly stoppered bottle or under petroleum ether.

3. Methods of Preparation

Hemimellitine has been prepared by the action of sodium on 2,3-dimethyliodobenzene and methyl iodide,[2] by the reduction of the chloromethylation product of o-xylene,[3] by the catalytic hydrogenolysis of 2,3-dimethylbenzyl alcohol,[4] and by treatment of 1,5,5,6-tetramethyl-1,3-cyclohexadiene with a catalyst prepared from sodium and o-chlorotoluene.[5] The present procedure is based on the method of Kantor and Hauser.[6]

[1] Duke University, Durham, North Carolina. Work supported by the Office of Ordnance Research.
[2] von Auwers, Ann., 419, 116 (1919).
[3] von Braun and Nelles, Ber., 67, 1094 (1934).
[4] Smith and Spillane, J. Am. Chem. Soc., 62, 2639 (1940).
[5] Pines and Eschinazi, J. Am. Chem. Soc., 78, 5950 (1956).
[6] Kantor and Hauser, J. Am. Chem. Soc., 73, 4122 (1951).

HENDECANEDIOIC ACID

(Undecanedioic acid)

$$HO_2C(CH_2)_4\overset{\overset{\textstyle O}{\|}}{C}(CH_2)_4CO_2H \xrightarrow[\text{KOH}]{H_2NNH_2} HO_2C(CH_2)_9CO_2H$$

Submitted by Lois J. Durham, Donald J. McLeod, and James Cason.[1]
Checked by N. J. Leonard, D. H. Dybvig, and K. L. Rinehart, Jr.

1. Procedure

A 500-ml. round-bottomed flask is attached by a well-lubricated ground-glass joint to a reflux condenser with a side take-off having a stopcock which may be opened to permit distillation. In the flask are placed 170 ml. of commercial diethylene glycol and 30 g. (0.46 mole) of potassium hydroxide (U.S.P., or reagent grade, 85%). This mixture is heated carefully (*Caution! Note 1*) until the potassium hydroxide begins to melt and go into solution; then the heat is removed intermittently until the exothermic dissolution is completed. After the solution has been cooled to 80–100°, the condenser is removed and there are added to the flask 35 g. (0.152 mole) of 6-ketohendecanedioic acid (p. 555) and 22 ml. (22.4 g., 0.38 mole) of commercial 85% hydrazine hydrate. The condenser is immediately replaced, and the mixture is warmed cautiously until any exothermic reaction is complete and then heated under reflux for 1 hour.

A thermometer is suspended through the condenser by copper wire so that the bulb is in the heated liquid, the stopcock of the take-off attachment is opened, and the mixture is distilled sufficiently slowly so that the froth does not rise out of the flask. When the liquid temperature has reached 205–210° (after about 30 ml. of distillate has been collected), the stopcock in the take-off is closed, the thermometer is removed, and the mixture is heated under reflux for 3 hours. If the temperature is checked during this heating period, it is usually found to be in the range 190–200°.

At the end of the heating period the reaction mixture is cooled to about 100–110° (at lower temperatures a gelatinous precipitate separates) and is then poured into 150 ml. of water contained in a 1-l. Erlenmeyer flask. An additional 100 ml. of water is used to rinse the reaction flask. The diluted mixture is acidified to Congo red by slow addition of 6N hydrochloric acid as the mixture is stirred vigorously to ensure conversion of any precipitated potassium salt to the free acid. The mixture is then cooled by tap water for at least 30 minutes (Note 2). The white precipitate is collected by suction filtration, transferred to a beaker, and is heated with about 250 ml. of water until the solid has melted (Note 3). As the mixture is cooled, it is stirred vigorously by hand until the oil has resolidified. After the mixture has been cooled again, the precipitated acid is collected by suction filtration, washed with water, and dried (Note 4). The yield of hendecanedioic acid, m.p. 110.5–112° (Note 5), is 28.5–30.6 g. (87–93%).

2. Notes

1. When the temperature becomes high enough for the potassium hydroxide to melt under the diethylene glycol, solution occurs rapidly with evolution of heat sufficient to drive material out of the top of the condenser if the source of external heat is not removed immediately. For this reason it is wise not to add the compound being reduced until after solution has been accomplished; moreover, in the case of keto acids there is some evidence that yields are lowered somewhat if the acid is added before solution of the potassium hydroxide. Since diethylene glycol does not present a serious fire hazard, this initial heating is probably best done with a small flame which may be easily removed quickly. Subsequent heating with an electric mantle is recommended.

2. Somewhat higher yields appear to be obtained when the mixture is allowed to stand overnight.

3. Remelting the solid acid over water removes occluded impurities, including salt and diethylene glycol. Moreover, acid which has crystallized in lumps is dried faster than the initial fine precipitate.

4. The submitters suggest recrystallization from benzene as a means of purification.

5. The acid is probably polymorphic,[2] since the melting point

varies somewhat with the rate of heating and a solidified melt remelts more sharply than do the original crystals.

3. Methods of Preparation

Hendecanedioic acid has been prepared by hydrolysis of the corresponding dinitrile, obtained from 1,9-dibromononane or 1,9-diiodononane; [3-5] by oxidation of 11-hydroxyhendecanoic acid; [6,7] by the Arndt-Eistert synthesis from 9-carbethoxynonanoyl [8] and nonanedioyl chloride; [9] by the Willgerodt reaction on undecylenic acid; [10] and by treatment of (5,5'-dicarboxy-2,2'-dithienyl)-methane with Raney nickel catalyst and alkali.[11] The present method, previously described briefly,[12,13] appears to represent the most convenient preparation of dibasic acids having an odd number of carbon atoms greater than ten. It has been applied to several other dibasic acids.[14]

[1] University of California, Berkeley, California.

[2] Dupré la Tour, *Ann. phys.*, **18**, 199 (1932).

[3] von Braun and Danziger, *Ber.*, **45**, 1970 (1912).

[4] Chuit, *Helv. Chim. Acta*, **9**, 264 (1926).

[5] Arosenius, Ställberg, Stenhagen, and Tägtström-Eketorp, *Arkiv Kemi Mineral. Geol.*, **26A**, No. 19 (1948) [*C. A.*, **44**, 3883 (1950)].

[6] Walker and Lumsden, *J. Chem. Soc.*, **79**, 1191 (1901).

[7] Verkade, Hartman, and Coops, *Rec. trav. chim.*, **45**, 373 (1926).

[8] Kawasaki, *J. Pharm. Soc. Japan*, **70**, 485 (1950) [*C. A.*, **45**, 5624 (1951)].

[9] Canonica and Bacchetti, *Atti accad. nazl. Lincei, Rend. Classe sci. fis. mat. e nat.*, **10**, 479 (1951) [*C. A.*, **48**, 6377 (1954)].

[10] Pattison and Carmack, *J. Am. Chem. Soc.*, **68**, 2033 (1946).

[11] Buu-Hoï, Sy, and Xuong, *Compt. rend.*, **240**, 442 (1955).

[12] Cason, Taylor, and Williams, *J. Org. Chem.*, **16**, 1187 (1951).

[13] Canonica and Bacchetti, *Atti accad. nazl. Lincei Rend. Classe sci. fis. mat. e nat.*, **15**, 278 (1953) [*C. A.*, **49**, 8121 (1955)].

[14] Blomquist, Johnson, Diuguid, Shillington, and Spencer, *J. Am. Chem. Soc.*, **74**, 4203 (1952).

n-HEPTAMIDE

(Heptanamide)

$$2n\text{-}C_6H_{13}CO_2H + H_2NCONH_2 \rightarrow 2n\text{-}C_6H_{13}CONH_2 + CO_2 + H_2O$$

Submitted by J. L. GUTHRIE and NORMAN RABJOHN.[1]
Checked by WILLIAM S. JOHNSON and DUANE ZINKEL.

1. Procedure

In a 1-l. round-bottomed flask, fitted with a thermometer extending nearly to the bottom, are placed 60 g. (1 mole) of urea and 69 g. (0.5 mole) of 95% *n*-heptanoic acid (Note 1). A condenser (Note 2) is attached to the flask, and the mixture is heated by means of an electric mantle. When the temperature reaches 140°, the urea is in solution, and a rather vigorous evolution of gas occurs which continues for several minutes. The temperature is maintained at 170–180° for 4 hours (Note 3), and then the mixture is allowed to cool.

As soon as the temperature drops to 110–120°, 400 ml. of 5% sodium carbonate solution is added carefully through the condenser, and the mixture is shaken vigorously (Note 4). The mixture is cooled in an ice bath, and the product is collected on a Büchner funnel. The solid, when dry, is slightly colored, and weighs 57–64 g., m.p. 85–91°.

The crude material is boiled for a few minutes with 200 ml. of 95% ethanol and a small amount of decolorizing carbon (Note 5). The mixture is filtered by gravity, and 800 ml. of water is added to the filtrate. The resulting slurry is cooled in an ice-salt bath and the solid is collected by filtration on a Büchner funnel. The product, which is almost colorless is air-dried. It weighs 44–48 g. (68–74%) and melts at 91–94° (Note 6). Evaporation of the filtrate under reduced pressure and reprecipitation of the residue from 20 ml. of 95% ethanol and 80 ml. of water affords an additional 3–4 g. (5–6%) of material which melts at 90–93°.

2. Notes

1. Eastman Kodak Company, yellow label brand (95%), *n*-heptanoic acid was used.
2. A condenser should be chosen which has an inside diameter of at

least 1.5 cm.; otherwise frequent loosening of the sublimate is required to prevent clogging. A 3-ft., air-cooled tube with an internal diameter of about 2.5 cm. serves as a satisfactory condenser.

3. Temperatures below 170° lead to slightly lower yields, and temperatures above 180° cause excessive sublimation of urea. Although the reaction is nearly complete after 2 hours, the yield appears to be improved by additional heating.

4. Failure to make the mixture basic leads to the formation of a greasy, colored product.

5. A small amount of solid does not dissolve in the alcohol.

6. Recrystallization from dilute ethanol affords colorless material, m.p. 94–95°.

3. Methods of Preparation

Heptamide has been prepared by heating heptanoic acid with ammonia in a sealed tube [2] at 230°, by treating heptanoic anhydride with ammonia,[3] by passing ammonia through heptanoic acid [4] at 125–190°, by the rearrangement of heptaldehyde oxime in the presence of Raney nickel in a quartz tube at 150° for 5 minutes,[5] by the Willgerodt reaction with 2-, 3-, or 4-heptanone or heptanal,[6,7] and by the action of ammonia on heptanoyl chloride.[8]

The procedure described is based on the method of E. Cherbuliez and F. Landolt,[9] by which formic and acetic acids were converted into the corresponding amides.

[1] University of Missouri, Columbia, Missouri.
[2] Hofmann, Ber., 15, 979 (1882).
[3] Chiozza and Malerba, Ann., 91, 103 (1854).
[4] Mitchell and Reid, J. Am. Chem. Soc., 53, 1879 (1931).
[5] Paul, Bull. soc. chim., [5]4, 1115 (1937).
[6] Cavalieri, Pattison, and Carmack, J. Am. Chem. Soc., 67, 1783 (1945).
[7] King and McMillan, J. Am. Chem. Soc., 68, 1369 (1946).
[8] Philbrook, J. Org. Chem., 19, 623 (1954).
[9] Cherbuliez and Landolt, Helv. Chim. Acta, 29, 1438 (1946).

3-*n*-HEPTYL-5-CYANOCYTOSINE

[2(1H)-Pyrimidinone, 5-cyano-3-heptyl-4(3)-imino-]

$$n\text{-}C_7H_{15}NH_2 + HCl + NaCNO \rightarrow n\text{-}C_7H_{15}NHCONH_2 + NaCl$$

$$C_7H_{15}NHCONH_2 + HC(OC_2H_5)_3 + H_2C(CN)_2 \rightarrow$$

$$C_7H_{15}NHCONHCH{=}C(CN)_2 + 3C_2H_5OH$$

$$C_7H_{15}NHCONHCH{=}C(CN)_2 \xrightarrow[\text{(2) } CH_3CO_2H]{\text{(1) } NaOCH_3}$$

Submitted by BARBARA B. KEHM and CALVERT W. WHITEHEAD.[1]
Checked by M. S. NEWMAN and K. G. IHRMAN.

1. Procedure

A. *N-n-Heptylurea.* To a mixture of 24.1 g. (0.21 mole) of *n*-heptylamine (Note 1), 35 g. of cracked ice, and 150 ml. of ice-cold water is added 38 ml. of 5*N* hydrochloric acid (Note 2) with stirring. The mixture is heated on the steam bath at 70–80°, and 14.3 g. (0.22 mole) of sodium cyanate is added portion-wise. After 2–4 hours of continued heating, two layers separate. The product crystallizes on standing overnight at room temperature. It is collected on a Büchner funnel, washed with 100 ml. of cold water, and drained as dry as practical by suction. This solid is dissolved in 125 ml. of boiling ethyl acetate, and the resulting solution is cooled to room temperature. The white crystalline N-*n*-heptylurea is filtered and dried on a porcelain plate at room temperature. A yield of 28.5–29.5 g. (86–88%) of product melting at 110–111° is obtained (Note 3).

B. *3-n-Heptylureidomethylenemalononitrile.* In a 250-ml. round-bottomed flask fitted with a heating mantle and a reflux condenser are placed 28.5 g. (0.18 mole) of N-*n*-heptylurea, 11.9 g. (0.18 mole)

of malononitrile (Note 4), and 26.7 g. (0.18 mole) of triethyl orthoformate (Note 4). The mixture is heated under reflux for 2 hours and
then cooled in ice. The solid product is collected by suction filtration
on a Büchner funnel. The filtrate is concentrated on the steam bath
to incipient crystallization, cooled, and filtered. The two lots of tan
solid, 41–42 g., thus obtained are dissolved in 75 ml. of 75% ethyl
alcohol in a 250-ml. beaker, 2 g. of decolorizing carbon is added, and
the mixture is boiled for 2–3 minutes with constant stirring (necessary
to avoid vigorous bumping). The hot solution is filtered by gravity
into a 250-ml. Erlenmeyer flask through fluted filter paper. The
flask is stoppered and cooled in the refrigerator for 4 hours. The solid
product is collected by suction filtration on a Büchner funnel and
washed four times with 10-ml. portions of distilled water. The white
crystalline 3-*n*-heptylureidomethylenemalononitrile is dried at 50° in a
vacuum oven. It melts at 130–132° and amounts to 33.8–34.8 g.
(80–83%) (Note 5).

C. *3-n-Heptyl-5-cyanocytosine.* In a 250-ml. Erlenmeyer flask are
placed 33.8 g. (0.145 mole) of 3-*n*-heptylureidomethylenemalononitrile
and 70 ml. of methanol; then 8.5 g. (0.16 mole) of sodium methoxide
(Note 6) is added carefully in small portions (Note 7). The resulting
solution is allowed to stand at room temperature for 3 days in the
stoppered flask. The contents of the flask are dissolved in 300 ml. of
cold water in an 800-ml. beaker, and the solution is stirred as 11 ml. of
glacial acetic acid is added. The precipitated solid is collected by
suction filtration on a Büchner funnel and washed with three 40-ml.
portions of distilled water. The undried product is dissolved in 600
ml. of hot ethyl alcohol; then the solution is filtered into a 1-l. flask by
gravity through a fluted filter paper, concentrated on the steam bath
to 200 ml., and cooled in the refrigerator for 4 hours. The 3-*n*-heptyl-
5-cyanocytosine crystallizes in white needles, melts at 192–197° (Note
8), and amounts to 29.7–31.1 g. (88–92%) (Note 9).

Recrystallization of 20 g. of this product from 230 ml. of hot ethyl
alcohol affords 17.8 g. of fine colorless needles, m.p. 199.5–202.5°.

2. Notes

1. *n*-Heptylamine is available from Sapon Laboratories, P. O. Box
599, Lynbrook, New York, and from Distillation Products Industries,
Rochester 3, New York.

2. If excess acid is present, considerable foaming may occur on addition of sodium cyanate.

3. The melting point for N-n-heptylurea given in the literature [2] is 110–111°.

4. Commercially available reagents were employed.

5. The melting point for n-heptylureidomethylenemalononitrile is reported [3] to be 130–132°. An additional 1–1.5 g. may be obtained by evaporating the filtrate to a volume of 100 ml.

6. Sodium methoxide is available from Mathieson Chemical Corporation, Niagara Falls, New York. Alternatively, a solution prepared by dissolving 3.6 g. of sodium in 70 ml. of methanol may be used.

7. Upon addition of sodium methoxide to the methanol solution considerable heat is evolved.

8. The melting point of 3-n-heptyl-5-cyanocytosine is given in the literature [4] as 200°.

9. The submitters obtained comparable yields in all steps using 144.2 g. (1.25 moles) of heptylamine and correspondingly larger amounts of all reagents.

3. Methods of Preparation

N-n-Heptylurea has been prepared by the action of nitrourea on n-heptylamine.[5]

3-n-Heptyl-5-cyanocytosine has been prepared by a variation of this procedure by Whitehead.[4]

The procedure described here has given equally good yields of other 3-alkyl-5-cyanocytosines and 3-cycloalkyl-5-cyanocytosines; however, it does not yield 3-aryl-5-cyanocytosines from arylureas.

[1] Eli Lilly & Company, Indianapolis, Indiana.
[2] Forselles and Wahlforss, *Ber.*, **25** Referate, 636 (1892).
[3] Whitehead, *J. Am. Chem. Soc.*, **75**, 674 (1953).
[4] Whitehead and Traverso, *J. Am. Chem. Soc.*, **77**, 5871 (1955).
[5] Tseng and Ho, *J. Chinese Chem. Soc.*, **4**, 335 (1936) [*C. A.*, **31**, 1011 (1937)].

HEXAHYDRO-1,3,5-TRIPROPIONYL-*s*-TRIAZINE

(*s*-Triazine, hexahydro-1,3,5-tripropionyl)

$$3C_2H_5CN + 3CH_2O \xrightarrow{H_2SO_4} C_2H_5CO-N \begin{array}{c} COC_2H_5 \\ | \\ N \\ \diagup \quad \diagdown \\ CH_2 \quad CH_2 \\ | \qquad | \\ \diagdown \quad \diagup \\ CH_2 \end{array} N-COC_2H_5$$

Submitted by W. O. TEETERS and M. A. GRADSTEN.[1]
Checked by ARTHUR C. COPE and MARK R. KINTER.

1. Procedure

In a 3-l. three-necked flask equipped with a mechanical stirrer, a dropping funnel, a reflux condenser, and a thermometer are placed 110 g. (138 ml., 2 moles) of propionitrile (Note 1) and 8 g. (4.35 ml.) of concentrated sulfuric acid (sp. gr. 1.84). To the stirred mixture, which is heated to 90°, is added gradually a solution of 60 g. of trioxane (equivalent to 2 moles of formaldehyde) in 110 g. (138 ml., 2 moles) of propionitrile. During the addition the temperature of the reaction mixture is kept between 95° and 105° (Note 2). When the addition is complete (30 to 60 minutes) (Note 3) the reaction mixture is heated under reflux for an additional 3 hours, the internal temperature being kept at approximately 105°. The mixture is then allowed to cool to room temperature; during the cooling period the product crystallizes. The light-brown crystals are collected by filtration with suction on a Büchner funnel (Note 4). The solid (160 to 170 g.) is washed three times with 100-ml. portions of ether and air-dried. The crude yellow product (130 to 135 g., melting point between 164° and 170°) is recrystallized from 160 ml. of 90% ethanol. The product is collected on a Büchner funnel, washed on the funnel with 100 ml. of ether, and air-dried. In this way 105 to 115 g. (62 to 68%) of white crystals of hexahydro-1,3,5-tripropionyl-s-triazine melting at 170–172° is obtained (Note 5). Recrystallization from 95% ethanol (2.4 ml. per g.) results in 91% recovery of an analytically pure product, m.p. 173.2–174.1° (cor.).

2. Notes

1. A practical grade of propionitrile obtained from the Eastman Kodak Company was used.

2. During the addition external heating may be discontinued and the internal temperature regulated by the rate of addition.

3. The reaction mixture turns from light yellow to a reddish brown color during the addition.

4. An appreciable amount of propionitrile (100 to 110 g.) can be recovered from the filtrate. If subsequent batches of hexahydro-1,3,5-tripropionyl-s-triazine are being prepared the filtrate can be added to the next charge without isolating the propionitrile. To counteract the accumulation of impurities, activated carbon may be used in the recrystallization of the reaction product.

5. Another 35 to 50 g. of a less pure product (m.p. 160–165°) may be isolated by concentrating the mother liquor.

3. Methods of Preparation

A hexahydro-1,3,5-triacyl-s-triazine was first prepared by Duden and Scharff [2] from ammonium chloride, formalin, and benzoyl chloride or from hexamethylenetetramine and benzoyl chloride. Procedures similar to the one described [3,4] also have been used for the preparation of hexahydro-1,3,5-triacetyl-, tri(β-chloropropionyl)-, triacrylyl-, trimethacrylyl-, and tribenzoyl-s-triazine. Several of these compounds also have been prepared by Wegler and Ballauf [5] from the corresponding nitriles and paraformaldehyde in the presence of acetic anhydride and sulfuric acid.

[1] Sun Chemical Corp., New York, New York.
[2] Duden and Scharff, *Ann.*, **288**, 247 (1895).
[3] Gradsten and Pollock, *J. Am. Chem. Soc.*, **70**, 3079 (1948).
[4] Emmons, Rolewicz, Cannon, and Ross, *J. Am. Chem. Soc.*, **74**, 5524 (1952).
[5] Wegler and Ballauf, *Chem. Ber.*, **81**, 527 (1948).

HEXAMETHYLBENZENE *

(Benzene, hexamethyl-)

$$\text{OH} \quad \xrightarrow[\text{Al}_2\text{O}_3]{\text{CH}_3\text{OH}} \quad \text{CH}_3 \text{ ... } \text{CH}_3$$

Submitted by N. M. Cullinane, S. J. Chard, and C. W. C. Dawkins.[1]
Checked by T. L. Cairns and D. C. England.

1. Procedure

A solution of 100 g. (1.06 moles) of phenol in 1 l. of methanol is allowed to drop at a rate of 110 ml. per hour (Note 1) over an activated alumina catalyst (Note 2) heated to 530° (Note 3). The exit from the hot tube is attached to a receiver arranged to lead by-product gases to an efficient hood (Note 4). After addition of the methanol solution is finished, the pale yellow product is transferred to a Büchner funnel and washed with methanol. The yield of crude product melting at 135–145° is 112–115 g. (65–67%). Recrystallization from ethanol (50 g. in 650 ml.) gives 85% recovery or from benzene (50 g. in 130 ml.) gives 60% recovery of colorless hexamethylbenzene, m.p. 165–166°.

2. Notes

1. The submitters used a rate of 250 ml. per hour with a 2-in.-diameter tube, 16 in. long, packed with 300 g. of alumina, and a temperature in the catalyst bed about 370–380°. The checkers used 34 g. of alumina packed in a ⅞-in.-diameter tube 13 in. long.

2. The submitters used 4- to 8-mesh alumina from Peter Spence and Sons, Widnes, Lancashire, England. The checkers used 8- to 14-mesh Alorco H-41 obtained from the Aluminum Company of America, 1200 Alcoa Building, Pittsburgh 19, Pennsylvania.

3. Automatically controlling the outside of the catalyst tube to 370–400° gives a hot spot in the catalyst bed of 530° at the rate specified in equipment used by the checkers.

4. Gases formed in the reaction included carbon monoxide, methane, and hydrogen. The exact equation for the reaction is not known.

3. Methods of Preparation

Hexamethylbenzene has been prepared by passing the mixed vapors of acetone and methanol over alumina at 400°.[2] Briner, Plüss, and Paillard [3] have obtained it by passing different phenols mixed with methanol in an atmosphere of dry carbon dioxide over alumina at 410–440°. Hexamethylbenzene also has been prepared by the trimerization of dimethylacetylene in the presence of triethylchromium,[4] or a mixture of triisobutylaluminum and titanium tetrachloride.[5] The present method is based upon that of Cullinane and Chard.[6]

[1] University College, Cardiff, Wales.
[2] Reckleben and Scheiber, Ber., 46, 2363 (1913).
[3] Briner, Plüss, and Paillard, Helv. Chim. Acta, 7, 1046 (1924).
[4] Zeiss and Herwig, J. Am. Chem. Soc., 80, 2913 (1958).
[5] Franzus, Canterino, and Wickliffe, J. Am. Chem. Soc., 81, 1514 (1959).
[6] Cullinane and Chard, J. Chem. Soc., 1945, 821.

HEXAMETHYLENE DIISOCYANATE *
(Isocyanic acid, hexamethylene ester)

$$NH_2(CH_2)_6NH_2 + 2HCl \rightarrow NH_2(CH_2)_6NH_2 \cdot 2HCl$$
$$NH_2(CH_2)_6NH_2 \cdot 2HCl + 2COCl_2 \rightarrow OCN(CH_2)_6NCO + 6HCl$$

Submitted by MARK W. FARLOW.[1]
Checked by R. L. SHRINER and ROBERT C. JOHNSON.

1. Procedure

Caution! Hexamethylenediamine, hexamethylene diisocyanate, and phosgene are highly toxic. Exposure to vapors or solutions containing these materials should be avoided. All operations should be conducted in a hood.

A. *Hexamethylenediammonium chloride.* To a solution of 116 g. (1.0 mole) of hexamethylenediamine (Note 1) in 145 ml. of methanol in a 1-l. beaker is added slowly from a dropping funnel 175 ml. of concentrated hydrochloric acid (sp. gr. 1.19). The mixture is well stirred during the addition and cooled externally to keep the contents below 30°. The hexamethylenediammonium chloride is then precipitated by adding the solution slowly with stirring to approximately 2 l. of acetone. The precipitate is collected on a Büchner funnel, washed with 100 ml. of cold acetone, and dried in a vacuum oven at 75° for 12–18 hours (Note 2). The yield of dry product amounts to 170–187 g. (90–99%), m.p. 243–246° (Note 3).

B. *Hexamethylene diisocyanate.* A suspension of 94.5 g. (0.50 mole) of finely powdered hexamethylenediammonium chloride in 500 ml. of anhydrous redistilled amylbenzene (or tetralin) (Note 4) is prepared in a 1-l. three-necked flask fitted with an efficient mechanical stirrer (Note 5), a water-cooled reflux condenser, a thermometer, and a phosgene inlet tube (Note 6) extending well below the surface of the suspension. Stirring is started, the mixture is heated to 180–185° (Note 7), and gaseous chlorine-free phosgene (Note 8) is delivered to the mixture at a rate of about 33 g. (0.33 mole) per hour. Hydrogen chloride and excess phosgene escape through the condenser. The temperature is carefully maintained between 180° and 185°; after 8–15 hours (Note 9) solution of the hexamethylenediammonium chloride is essentially complete and hydrogen chloride is no longer evolved (Note 10). The reaction mixture is then filtered through a suction filter, and the filtrate is distilled at reduced pressure through a fractionating column, giving amylbenzene, b.p. 65–75°/10 mm. (Note 11), and 70–80 g. (84–95%) of hexamethylene diisocyanate boiling at 120–125°/10 mm. (92–96°/1 mm., 108–111°/5 mm.); n_D^{20} 1.4585; d_4^{20} 1.0528 (Note 12).

2. Notes

1. Hexamethylenediamine may be obtained from the Poly-chemicals Department, E. I. du Pont de Nemours and Company, Inc., Wilmington, Delaware. Usually the commercial product is a 70% aqueous solution from which the water may be distilled

at atmospheric pressure. The residue is suitable for preparing the dihydrochloride.

2. Thorough drying of the hydrochloride is essential to the success of the next step. Overnight drying in a vacuum oven at 70–100° is effective. The dry salt is not appreciably hygroscopic, but it should be preserved in a well-stoppered bottle until used.

3. This salt shows a marked tendency to sublime at and above 200°.

4. It is preferable to add the hexamethylenediaommnium chloride in three portions. This modification facilitates stirring and prevents clumping of the solid.[2]

Commercial amylbenzene is dried by distillation, and the fraction boiling at 184–194° is collected. This solvent has been withdrawn from the market (1950) and hence tetralin may be substituted. Commercial tetralin is washed with ferrous sulfate solution and distilled; the fraction boiling at 202–206° is used. o-Dichlorobenzene also has been recommended as a solvent.[3] When the latter was employed,[2] it was found advisable to use recovered material, since it contained some component that acted as a solvent for the hydrochloride. The phosgene was absorbed more rapidly, and the reaction was completed in only 9 hours.

5. Efficient agitation of the gas-liquid-solid reaction mixture is conducive to a high rate of reaction. The use of a reaction flask modified with creases [4] has given good results.

6. The end of the inlet tube should have a very coarse fritted glass disk in order to promote rapid reaction. A tube with a bulb in which many fine holes have been blown may also be used, but the reaction time is longer. If the inlet tube becomes clogged, it may be cleaned quickly by removing it from the reaction flask and dipping in warm cresol, which dissolves any polyhexamethylene urea that may form. The tube is then rinsed with amylbenzene or tetralin and replaced in the reaction flask.

7. The temperature of the reaction mixture should be maintained between 180° and 185° in order to obtain as rapid a reaction as possible. Higher temperatures lead to the formation of polyhexamethylene urea. A run carried out at the boiling point of tetralin (206°) gave an 84% yield of polymer.

8. Phosgene is available from the Niagara Chlorine Products

Co., Inc., Lockport, New York, or the Matheson Co., Inc., Rutherford, New Jersey. When phosgene containing small amounts of chlorine is used, the reaction appears to proceed normally but the product and recovered solvent are contaminated with chlorine-containing impurities. Chlorine in phosgene can be detected by bubbling a stream of the gas rapidly through clean mercury. Chlorine reacts with and discolors the mercury, whereas pure phosgene leaves the mercury unchanged. If chlorine is present, it may be removed by bubbling the phosgene through two wash bottles containing cottonseed oil.

9. The time required for complete reaction is dependent on the reaction temperature, on the design of the phosgene inlet tube, on the efficiency of agitation, and on the rate of phosgene addition. It is important that the reaction be continued until practically all the hexamethylenediammonium chloride has disappeared. If unreacted amine salt is present, it has a tendency to sublime with the diisocyanate during distillation.

10. When moist air is blown through a glass tube held at the end of the condenser, across the current of phosgene containing hydrogen chloride, the fogging typical of hydrogen chloride gas in a moist atmosphere is produced. Pure phosgene gives no visible effect under similar conditions.

11. The recovered solvent is suitable for succeeding preparations. If tetralin is used as the reaction medium, it is recovered as the low-boiling fraction, b.p. 60–70°/8 mm.

12. If the product contains chlorine as indicated by the alcoholic silver nitrate test, it may be purified by adding a small amount of anhydrous calcium oxide (0.5 g. per 50 g. of product) and redistilling under reduced pressure.

3. Methods of Preparation

Hexamethylene diisocyanate has been prepared by the action of phosgene on hexamethylenediammonium chloride [5] or on hexamethylenediammonium carbonate.[6] Metal chlorides such as those of cobalt, iron, mercury, or zinc have been stated to promote the reaction.[3] It has been claimed [7] that the vapor-phase reaction of an hexamethylenediamine-toluene solution with phosgene at 280–300° affords an 80% yield of the diisocyanate.

[1] E. I. du Pont de Nemours and Company, Wilmington, Delaware.

[2] Allen, Private communication.

[3] Burgoine and New, Imperial Chemical Industries, Ltd., Brit. pat. 574,222 [*C. A.*, **42**, 7788 (1948)].

[4] Morton, Darling, and Davidson, *Ind. Eng. Chem., Anal. Ed.*, **14**, 734 (1942).

[5] Farlow, E. I. du Pont de Nemours and Company, U. S. pat. 2,374,340 [*C. A.*, **39**, 3555 (1945)].

[6] Smith, P. B. 7416, *Synthetic Fiber Developments in Germany*, p. 47, Textile Research Institute, Inc., New York, 1946.

[7] Mashio and Nomachi, *J. Chem. Soc. Japan, Ind. Chem. Sect.*, **56**, 289 (1953) [*C. A.*, **48**, 10634 (1954)].

n-HEXYL FLUORIDE

(Hexane, 1-fluoro-)

$$CH_3(CH_2)_4CH_2Br + KF \xrightarrow{\text{Ethylene glycol}} CH_3(CH_2)_4CH_2F + KBr$$

Submitted by ARTHUR I. VOGEL, JAMES LEICESTER, and WILLIAM A. T. MACEY.[1]
Checked by T. L. CAIRNS and C. W. TULLOCK.

1. Procedure

Warning! The 1-fluoroalkanes are the most toxic of the fluoro compounds. This fact, together with their relatively high volatility, emphasizes the need for care in handling these materials. This preparation should be carried out in a good hood and the operator should wear rubber gloves (Note 1).

In a thoroughly dry 500-ml. three-necked round-bottomed flask, equipped with a mercury-sealed stirrer, a 100-ml. dropping funnel and a short fractionating column (Note 2), is placed a mixture of 116 g. (2.0 moles) of anhydrous finely powdered potassium fluoride (Note 3) and 200 g. of dry ethylene glycol (Note 4). The fractionating column carries a thermometer and is connected to a downward double-surface condenser with a filter flask as receiver. The round flask is heated at a bath temperature of 160–170°, and 165 g. (1.0 mole) of *n*-hexyl bromide, b.p. 154–156° (Note 5), is added dropwise during 5 hours; liquid passes over intermittently at 60–90° (temperature at the top of the fractionating column). The bath temperature is allowed to fall to 110–120°, and a slow stream of air is drawn through the apparatus by attaching the side arm of the filter flask to a water pump and replacing the dropping funnel by a narrow-bore tube dipping just below the surface of the liquid; stirring is maintained during this operation. It is advisable to interpose a U-tube cooled in ice between the water pump and the receiver in order to recover any uncondensed liquid. The combined distillates are then distilled at atmospheric pressure through an efficient fractionating column; after a small fore-run (up to 10 g.) of 1-hexene, the crude n-hexyl fluoride is collected at 89–92° (46–48 g.). The crude product is purified by cooling in ice and adding 1-ml. portions of a solution containing 9.0 g. of bromine and 6.0 g. of potassium bromide in 50 ml. of water until the organic layer acquires an orange color; after each addition the mixture is shaken vigorously for a minute or so. The volume of bromine-potassium bromide solution required is usually less than 5 ml. The aqueous layer is separated, the organic layer is washed with saturated aqueous potassium bromide solution until colorless and finally with water. The liquid is dried with anhydrous magnesium sulfate and distilled through an efficient fractionating column; the n-hexyl fluoride is collected at 91–92°. This procedure yields 42–47 g. (40–45% over-all yield based on the bromide employed) of a water-white product, n_D^{20} 1.375, n_D^{25} 1.372–1.373, d_4^{20} 0.8011. It has been kept for 1 year without change of physical properties and therefore appears to be stable (Note 6).

2. Notes

1. Data giving lethal doses of the 1-fluoroalkanes in mice may be found in Pattison's *Toxic Aliphatic Fluorine Compounds*, Elsevier Publishing Company, 1959, and in Pattison, *J. Am. Chem. Soc.*, **79**, 2311 (1957). It has been estimated that 0.1–0.2 g. of n-hexyl fluoride would be a fatal dose for a human being.[2]

2. Any fractionating column of moderate efficiency is satisfactory. The submitters employed a 20-cm. Dufton column containing a spiral 10 cm. in length, 2 cm. in diameter, with 8 turns of the helix. A 20–25 cm. Vigreux column may also be used.

3. Pure laboratory grade anhydrous potassium fluoride is

finely ground and kept for 48 hours in an oven at 180–210°; it is stored in a desiccator. Before use, the powdered salt is dried for 3 hours at 180° and ground again in a warm (50°) glass mortar.

4. Laboratory grade ethylene glycol is redistilled under diminished pressure, and the fraction boiling at 85–90°/7 mm. is used as the solvent for the potassium fluoride.

5. The *n*-hexyl bromide may be prepared from redistilled *n*-hexyl alcohol by the red phosphorus-bromine procedure or may be purchased from Eastman Kodak Company.

6. The procedure described has been employed by the submitters for the preparation of the alkyl fluorides listed below. Due regard must be paid to the boiling point of the alkyl bromide; as a general rule the bath temperature is maintained at about the boiling point of the alkyl bromide, with a minimum value (for *n*-amyl fluoride) of about 140–150°. For bromides of higher boiling point, the bath temperature is held at about 190°. In all cases, after the alkyl bromide has been added, the bath is allowed to cool 10–20° below the original reaction temperature, and a slow stream of air is drawn through the apparatus; the alkyl fluoride and ethylene glycol which pass over are collected, and the latter is removed by washing with water.

Alkyl fluoride	B.P.	d_4^{20}	n_D^{20}	$n_D^{/25}$	Yield, %
n-Amyl	63.5–65°	0.7917	1.3597	1.3562	27
n-Heptyl	119–121°	0.8060	1.3861	1.3833	35
n-Octyl	144–146°	0.8137	1.3955	1.3927	34
n-Nonyl	166–169°	0.8159	1.4033	1.4002	46
n-Decyl	186–188°	0.8197	1.4095	1.4068	37
n-Undecyl	70–71.5°/3 mm.	0.8239	1.4151	1.4122	45
n-Dodecyl	93–95°/3 mm.	0.8257	1.4192	1.4162	34
n-Tetradecyl	119–121°/3 mm., m.p. 8°	0.8277	1.4266	1.4236	43
n-Hexadecyl	150–152°/2 mm., m.p. 19°	0.8313	1.4322	1.4295	27

3. Methods of Preparation

Alkyl fluorides have been prepared by reaction between elementary fluorine and the paraffins,[3] by the addition of hydrogen fluoride to olefins,[4] by the reaction of alkyl halides with mercurous fluoride,[5] with mercuric fluoride,[6] with silver fluoride,[7] or with potassium fluoride under pressure,[8] and by the reaction of potassium fluoride with n-hexyl methanesulfonate[9] or n-hexyl p-toluenesulfonate.[10] The procedure used is based on that of Hoffmann[11] involving interaction at atmospheric pressure of anhydrous potassium fluoride with an alkyl halide in the presence of ethylene glycol as a solvent for the inorganic halide; a small amount of olefin accompanies the alkyl fluoride produced and is readily removed by treatment with bromine-potassium bromide solution.

The reaction of n-hexyl bromide with potassium fluoride in several glycols has been studied,[12] and the methods for the preparation of alkyl monofluorides have been reviewed.[13]

[1] Woolwich Polytechnic, London, S.E. 18, England.

[2] Sauer, Private communication.

[3] Hadley and Bigelow, J. Am. Chem. Soc., 62, 3302 (1940).

[4] Grosse and Linn, J. Org. Chem., 3, 26 (1938); Grosse, Wackher, and Linn, J. Phys. Chem., 44, 275 (1940).

[5] Swarts, Bull. classe sci. Acad. roy. Belg., 7, 438 (1921); Bull. soc. chim. Belg., 46, 10 (1937); Desreux, Bull. classe sci. Acad. roy. Belg., 20, 457 (1934); Bull. soc. chim. Belg., 44, 1 (1935); Henne and Renoll, J. Am. Chem. Soc., 60, 1060 (1938); Vogel, J. Chem. Soc., 1948, 649.

[6] Henne and Midgley, J. Am. Chem. Soc., 58, 884 (1936); see also Henne in Adams, Org. Reactions, 2, 57 (1944). Henne and Renoll, J. Am. Chem. Soc., 58, 887 (1936).

[7] Swarts, Bull. soc. chim. Belg., 30, 302 (1921).

[8] Gryszkiewicz-Trochimowski, Sporzynski, and Wnuk, Rec. trav. chim., 66, 413 (1947).

[9] Pattison and Millington, Can. J. Chem., 34, 757 (1956).

[10] Bergmann and Shahak, Chem. & Ind. (London), 1958, 157.

[11] Hoffmann, J. Am. Chem. Soc., 70, 2596 (1948); J. Org. Chem., 14, 105 (1949); 15, 425 (1950).

[12] Kitano and Fukui, Kôgyô Kagaku Zasshi, 60, 272 (1957) [C. A., 53, 7969 (1959)].

[13] Pattison, Nature, 174, 740 (1954).

4-HYDROXY-1-BUTANESULFONIC ACID SULTONE

(1-Butanesulfonic acid, 4-hydroxy-, δ-sultone)

$$O \underset{(CH_2)_4Cl}{\overset{(CH_2)_4Cl}{\big\langle}} + 2Na_2SO_3 \longrightarrow O \underset{(CH_2)_4SO_3Na}{\overset{(CH_2)_4SO_3Na}{\big\langle}} + 2NaCl$$

$$O \underset{(CH_2)_4SO_3Na}{\overset{(CH_2)_4SO_3Na}{\big\langle}} \xrightarrow[\text{2. Heat}]{\text{1. H}^+ \text{ (resin)}} 2 \underset{CH_2—SO_2}{\overset{CH_2—CH_2}{\big\langle CH_2 \quad\quad O}}$$

Submitted by A. O. SNODDY.[1]
Checked by T. L. CAIRNS and W. R. BRASEN.

1. Procedure

A mixture of 132.5 g. (1.05 moles) of sodium sulfite, 99.5 g. (0.5 mole) of bis-4-chlorobutyl ether (Note 1), and 450 ml. of water is placed in a creased 1-l. three-necked flask fitted with an efficient sealed stirrer and a reflux condenser. The third neck of the flask is closed with a stopper, and the mixture is heated and stirred vigorously under reflux until the ether has dissolved (Note 2). At the end of this time, heating is discontinued, and 60 ml. of concentrated c.p. hydrochloric acid is cautiously added to the solution. The mixture is then boiled with stirring until sulfur dioxide is no longer evolved. Solid barium chloride dihydrate (or a 10% aqueous solution of this salt) is added to the hot solution (Note 3) until all sulfate has been precipitated; then the barium sulfate is removed by suction filtration through a layer of filter aid.

The filtered solution is diluted to 2 l. with water and passed through a column of ion-exchange resin (Notes 4 and 5). A test portion of the effluent should yield no ash on evaporation and ignition of the residue. The column is washed with water until the effluent is no longer acid to litmus, and the washings are added to the original eluent. The total eluent is evaporated (Note 6) at reduced pressure (water aspirator) until the volume is about 250 ml.; then it is transferred to a 500-ml. pot which is equipped for vacuum distillation and attached to a short Vigreux column. After most of the remaining water and hydrochloric acid have been removed at the pressure obtainable with an aspirator, an oil pump is attached (Note 7), and the heating bath is cautiously raised to a temperature in the range 130–150°. The distillate is allowed to stand in a separatory funnel until the layers have separated (several hours may be required), then the crude sultone is withdrawn and distilled (Note 8) at reduced pressure. The yield of sultone, b.p. 134–136°/4 mm., is 99–109 g. (72–80%), n_D^{25} 1.4619–1.4625, d^{25} 1.3347, m.p. 12.5–14.5°.

2. Notes

1. Material supplied by Matheson, Coleman and Bell was used without purification.

2. This reaction may be carried out under pressure in a rocking autoclave at 180° in about 8 hours. When it is carried out under reflux with vigorous stirring in a creased flask it is complete in about 20 hours, whereas 50–60 hours is required if an ordinary flask is used.

3. Any sulfate ions resulting from oxidation of sulfite should be removed; otherwise they will be converted to sulfuric acid in the subsequent procedure and destroy sultone. Ordinarily, 12–13 g. of barium chloride dihydrate is required. The end point of the addition is conveniently determined with tetrahydroquinone indicator used in spot tests on filter paper [cf. *Ind. Eng. Chem., Anal. Ed.*, **9**, 331 (1937)].

4. Rohm & Haas Amberlite IR-120, which has been developed with $3N$ hydrochloric acid and then washed free from chlorides with water, is used in the form of a column 6 cm. in diameter and 55 cm. in length. Such a column contains approximately 1.1 kg. of resin (50% moisture) and is equivalent to about 2.2 moles of hydrogen chloride.

5. The checkers found that the procedure of Helberger and Lantermann,[2] which avoids the use of an ion-exchange resin, is also satisfactory in case it is regarded as inconvenient to set up the resin column. According to this procedure, after completion of the reaction with sodium sulfite, anhydrous hydrogen chloride is passed into the hot solution to liberate sulfur dioxide. After removal of sulfate as in the described procedure, the undiluted aqueous solution is saturated with hydrogen chloride gas at a temperature below 25°. The precipitated sodium chloride is removed by suction filtration, the filter cake is washed with two 50-ml. portions of $12N$ hydrochloric acid, and then

the combined filtrate and washings are worked up as described for the eluent from the resin column.

6. The checkers found that satisfactory results may also be obtained if the water is evaporated by heating the solution on a steam bath as air is aspirated through it.

7. At the bath temperature specified, and low pressure, dehydration and distillation occur; if a pressure of about 4 mm. is maintained, the vapor temperature is in the range 132–138°. The temperature of the bath should be raised slowly and with caution, or else the contents of the pot may froth through the column into the distillate. The receiver should be cooled in a Dry Ice-acetone bath to prevent vapors from reaching the oil pump.

8. Water is nearly insoluble in the sultone. The small amount of water and a very volatile impurity which are present in the crude sultone distil rapidly before the sultone is collected. An efficient cold trap should be used to protect the pump from the volatile materials.

3. Methods of Preparation

4-Hydroxy-1-butanesulfonic acid sultone has been made through the chlorosulfonation of 1-chlorobutane,[3] from 4-chlorobutyl acetate [2,4] which is prepared through the reaction of tetrahydrofuran and acetyl chloride,[2] from 4-chlorobutanol,[4] and from bis-4-chlorobutyl ether.[2] 4-Hydroxy-1-butanesulfonic acid sultone also has been obtained by treating sodium 4-hydroxybutane-1-sulfonate with cation-exchange resins,[5] by heating methoxybutanesulfonic acid under reduced pressure at 150–180°,[6] by concentrating an alcohol solution of 4-hydroxybutanesulfonic acid under reduced pressure,[7] and by heating 4-chlorobutanesulfonic acid with copper oxide and steam.[8]

Both 4-chlorobutanol and bis-4-chlorobutyl ether can be prepared from tetrahydrofuran (p. 266). The procedure described is based on the method of Helberger and Lantermann.[2]

[1] The Procter & Gamble Company, Miami Valley Laboratories, Cincinnati 31, Ohio.

[2] Helberger and Lantermann, *Ann.*, **586**, 158 (1954).

[3] Helberger, Manecke, and Fischer, *Ann.*, **562**, 23 (1949); Helberger, Manecke, and Heyden, *Ann.*, **565**, 22 (1949).

[4] Truce and Hoerger, *J. Am. Chem. Soc.*, **76**, 5357 (1954).

[5] Böhme Fettchemie G.m.b.H., Brit. pat. 774,563 [*C. A.*, **51**, 16519 (1957)].

[6] Böhme Fettchemie G.m.b.H., Ger. pat. 902,615 [*C. A.*, **50**, 9443 (1956)].

[7] Willems, *Bull. soc. chim. Belges*, **64**, 747 (1955).

[8] Henkel & Cie. G.m.b.H., Ger. pat. 860,637 [*C. A.*, **48**, 1412 (1954)].

6-HYDROXYNICOTINIC ACID *

(Nicotinic acid, 6-hydroxy-)

Submitted by J. H. Boyer and W. Schoen.[1]
Checked by T. L. Cairns and W. J. Linn.

1. Procedure

A. *Methyl coumalate.* In a 500-ml. round-bottomed flask provided with a thermometer is placed 139 ml. of concentrated sulfuric acid. To the acid is added, with swirling, 50 g. (0.36 mole) of pulverized coumalic acid (p. 201) in small portions. The reaction is slightly exothermic, and the mixture is maintained between 20° and 30° by occasional immersion of the flask into an ice bath. Methanol (70 ml.) is then added in small portions with frequent swirling, and the temperature is held between 25° and 35°. The mixture is heated on a steam bath for 1 hour, cooled to about 40°, and poured slowly with stirring into 800 ml. of water in a 2-l. beaker while the temperature is maintained below 40° by an ice bath (Note 1). Anhydrous sodium carbonate is added in small portions with stirring until the mixture is slightly alkaline (Note 2). The precipitated ester is freed of inorganic salts by slurrying four times with 100-ml. portions of cold water, filtered, and air-dried overnight. The yield of methyl coumalate, m.p. 68–70°, is 17.5–24.5 g. (32–45%). This crude product is used for the preparation of 6-hydroxynicotinic acid.

B. *6-Hydroxynicotinic acid.* In a 500-ml. beaker provided with a thermometer, magnetic stirring, and external cooling is placed 117 ml. of 14% ammonium hydroxide. With stirring, 45 g. (0.29 mole) of methyl coumalate is added over a period of

10 minutes, during which time the solution is kept below 20°. Stirring is continued for an additional 45 minutes at about 20° (Note 3). A solution of 600 ml. of approximately 17% aqueous sodium hydroxide is placed in a 2-l. beaker and heated almost to the boiling point. At the end of the 45-minute period, the ammoniacal solution is added to the hot sodium hydroxide solution, and the mixture is heated rapidly to the boiling point. After it has boiled vigorously for 5 minutes, the stirred solution is cooled in an ice bath to room temperature. With the temperature held below 30°, concentrated hydrochloric acid is added with stirring until the solution is strongly acid (Note 4). The heavy, yellow, microcrystalline solid which separates after stirring and cooling for about an hour is collected on a Büchner funnel, washed twice with water, and dried at 80°. The yield of bright yellow 6-hydroxynicotinic acid, m.p. 299–300° (dec., uncor.), is 29–37 g. (72–91%) (Note 5).

2. Notes

1. A turbid, brown solution containing a small amount of fine precipitate is obtained.
2. A small amount of curdy, brown precipitate is obtained at first. About 220 g. of anhydrous sodium carbonate is required.
3. Most of the ester dissolves; a turbid, red solution is formed.
4. About 250 ml. of acid is required.
5. The product is sufficiently pure for further synthetic work; a purer product may be obtained by recrystallization from 50% aqueous acetic acid.

3. Methods of Preparation

The procedure for preparing methyl coumalate is based on a method described by von Pechmann.[2] Methyl coumalate has also been prepared by direct esterification of the reaction mixture from malic acid and fuming sulfuric acid[3] and from coumalyl chloride and methanol.[4]

The procedure for preparing 6-hydroxynicotinic acid is also based on a method described by von Pechmann.[5] 6-Hydroxynicotinic acid has also been prepared by decarboxylation of 6-hydroxy-2,3-pyridinedicarboxylic acid;[6, 7] by heating 6-hydrazinonicotinic acid or its hydrazide with hydrochloric acid;[8] by the action of carbon dioxide on the sodium salt of α-pyridone at 180–200° and 20 atm.;[9] by heating the nitrile of 6-chloronicotinic acid with alcoholic sodium hydroxide or hydrochloric acid;[10] from 6-aminonicotinic acid;[11, 12] by the prolonged action of concentrated ammonium hydroxide on methyl coumalate;[3] and by carbonating 2-hydroxypyridine in the presence of potassium carbonate.[13]

[1] Tulane University, New Orleans, Louisiana.
[2] von Pechmann, *Ann.*, **264**, 279 (1891).
[3] Caldwell, Tyson, and Lauer, *J. Am. Chem. Soc.*, **66**, 1479 (1944).
[4] Wiley and Knabeschuh, *J. Am. Chem. Soc.*, **77**, 1615 (1955).
[5] von Pechmann and Welsh, *Ber.*, **17**, 2391 (1884).
[6] Königs and Geigy, *Ber.*, **17**, 589 (1884).
[7] Diamant, *Monatsh.*, **16**, 767 (1895).
[8] Marckwald and Rudzik, *Ber.*, **36**, 1114 (1903).
[9] Tschitschibabin and Kirssanow, *Ber.*, **57B**, 1162 (1924).
[10] Räth (to Schering-Kahlbaum A.-G.), Ger. pat. 447,303 [*Frdl.*, **15**, 1487 (1925–1927)].
[11] Marckwald, *Ber.*, **27**, 1323 (1894).
[12] Räth and Prange, *Ann.*, **467**, 9 (1928).
[13] Baine, Adamson, Barton, Fitch, Swayampati, and Jeskey, *J. Org. Chem.*, **19**, 510 (1954).

3-HYDROXYTETRAHYDROFURAN

(Furan, 3-hydroxy-1,2,3,4-tetrahydro-)

$$HOCH_2CH_2CHOHCH_2OH \xrightarrow{H^+} $$

Submitted by HANS WYNBERG and A. BANTJES.[1]
Checked by JOHN C. SHEEHAN and GREGORY L. BOSHART.

1. Procedure

A 500-ml. flask is charged with 318 g. (3 moles) of 1,2,4-trihydroxybutane (Note 1) and 3 g. of *p*-toluenesulfonic acid monohydrate. A few Carborundum boiling chips are added, the flask is equipped with a 30.5-cm. Vigreux column, condenser, and receiver arranged for vacuum distillation, and the contents are heated, with swirling, to dissolve the acid (Note 2). The flask is then heated in a bath held at 180–220° so that 300–306 g. of distillate, b.p. 85–87°/22 mm., is collected over a period of 2–2.5 hours (Note 3). The colorless liquid obtained is refractionated, the same apparatus being used, and two fractions are collected: the first, 50–60 g., b.p. 42–44°/24 mm., n_D^{25} 1.3343, is mainly water. After a negligible intermediate fraction, 215–231 g. (81–88%) of pure 3-hydroxytetrahydrofuran, b.p. 93–95°/26 mm., n_D^{25} 1.4497, $d_4^{20} = 1.095$, is collected (Note 4).

2. Notes

1. Supplied by the General Aniline and Film Corporation.

2. Considerable darkening occurs even when the acid is well dispersed. The yield appears not to be affected.

3. Other temperatures are: b.p. 75–77°/16 mm.; 90–92°/28 mm. This first distillate contains 14% (±3%) of water as determined by interpolation of the refractive indices.

4. Calcd. $M_D = 21.64$. Found: 21.72. As obtained by this single fractionation the submitters found the alcohol to be analytically pure: Calcd. for $C_4H_8O_2$: C, 54.53; H, 9.14. Found: C, 54.74; H, 9.32. Others have reported: b.p. 50°/1 mm., n_D^{18} 1.4486, $d_4^{20} = 1.090$,[2] and b.p. 81°/13 mm., $d^{18} = 1.07$, n_D^{18} 1.4478.[3]

3. Methods of Preparation

3-Hydroxytetrahydrofuran has been obtained during the preparation of 1,2,4-trihydroxybutane,[3] by hydrolysis of 4-chloromethyl-1,3-dioxane [2] and by acid catalyzed dehydration of 1,2,4-trihydroxybutane.[4] The present procedure is similar to that described by Reppe.[4]

[1] Tulane University, New Orleans, Louisiana.

[2] Price and Krishnamurti, J. Am. Chem. Soc., 72, 5335 (1950).

[3] Pariselle, Ann. chim. (Paris), [8]24, 315 (1911).

[4] Reppe, Ann., 596, 1 (1955), see p. 112; DBP 841 592 (1942), BASF (H. Krzikalla, E. Woldan).

INDAZOLE *

(1H-Indazole)

Submitted by C. Ainsworth.[1]
Checked by Max Tishler, George Gal, and G. A. Stein.

1. Procedure

A. *2-Hydroxymethylenecyclohexanone, Method 1.* A mixture of 23 g. (1 g. atom) of sodium metal cut in approximately 1-cm. cubes, 2 l. of dry ether, 98 g. (103 ml., 1 mole) of redistilled cyclohexanone, and 110 g. (120 ml., 1.5 moles) of ethyl formate is placed in a 5-l. three-necked flask equipped with a stirrer, stopper, and vent tube. The reaction is initiated by the addition of 5 ml. of ethyl alcohol to the stirred mixture, which is then placed in a cold water bath. Stirring is continued for 6 hours. After standing overnight, 25 ml. of ethyl alcohol is added, and the mixture is stirred for an additional hour. After the addition of 200 ml. of water, the mixture is shaken in a 3-l. separatory funnel. The ether layer is washed with 50 ml. of water, and the combined aqueous extracts are washed with 100 ml. of ether. The aqueous layer is acidified with 165 ml. of 6N hydrochloric acid, and the mixture is extracted twice with 300 ml. of ether. The ether solution is washed with 25 ml. of saturated sodium chloride solution and then is dried by the addition of approximately 30 g. of anhydrous magnesium sulfate powder. The drying agent is removed by suction filtration, and the ether is evaporated on the steam bath. The residue is distilled under reduced pressure using a 6-inch Vigreux column. After a small fore-run there is obtained 88–94 g. (70–74%) of 2-hydroxymethylenecyclohexanone, b.p. 70–72°/5 mm., n_D^{25} 1.5110 (Note 1).

2-Hydroxymethylenecyclohexanone, Method 2. A mixture of 50 g. (1 mole) of 48% sodium hydride dispersed in mineral oil (Note 2), 2 l. of dry ether, and 5 ml. of ethyl alcohol is placed in a 5-l. three-necked flask equipped with a stirrer, dropping funnel, and vent tube. The reaction vessel is cooled by means of a cold water bath, and a solution of 98 g. (103 ml., 1 mole) of redistilled cyclohexanone and 110 g. (120 ml., 1.5 moles) of ethyl formate is added dropwise during 1 hour. Stirring is continued for 6 hours (Note 3), and the solution is allowed to stand overnight. After the addition of 20 ml. of ethyl alcohol, the mixture is stirred for 1 hour. Water (200 ml.) is added to the flask with stirring, the mixture is shaken in a 3-l. separatory funnel and the organic layer separated. The product is isolated according to the procedure described in Method 1. The yield of 2-hydroxymethylenecyclohexanone is the same by both methods.

B. *4,5,6,7-Tetrahydroindazole.* A solution of 63 g. (0.5 mole) of 2-hydroxymethylenecyclohexanone and 500 ml. of methyl alcohol contained in a 2-l. beaker is treated with 25 ml. (0.5 mole) of hydrazine hydrate in small portions (Note 4). After standing for 30 minutes the mixture is concentrated by warming under reduced pressure on the steam bath. To aid in removing the water, 100 ml. of ethyl alcohol is added, and again the mixture is concentrated by heating under reduced pressure. The residue is dissolved in about 100 ml. of hot petroleum ether (Note 5), and after cooling in an ice bath for 1 hour the solid that separates is collected by suction filtration and washed with a small amount of cold petroleum ether. The crude 4,5,6,7-tetrahydroindazole, m.p. 79–80°, weighs 58–60 g. (95–98%) and is sufficiently pure to be used in the next step (Notes 6 and 7).

C. *Indazole.* A mixture of 50 g. (0.41 mole) of 4,5,6,7-tetrahydroindazole, 35 g. of 5% palladium on carbon (Note 8) and 1 l. of dry decalin is placed in a 3-l. round-bottomed flask and heated under reflux for 24 hours. The hot mixture is filtered by suction, using a previously heated Büchner funnel, and the filtrate is allowed to cool. After standing overnight, the indazole that separates is collected by suction filtration, washed with 100 ml.

of petroleum ether, and air-dried. The solid is dissolved in 750 ml. of hot benzene, and about 4 g. of decolorizing carbon is added. The mixture is filtered through a fluted filter paper and then is refiltered, using the same filter paper. Two liters of warm petroleum ether is added, and the solution is cooled in an ice bath for 2 hours. The indazole (m.p. 146–147°) is collected by suction filtration and after air drying weighs 24–25 g. (50–52%).

2. Notes

1. 2-Hydroxymethylenecyclohexanone begins to polymerize after standing at room temperature for several days.

2. Available from Metal Hydrides Inc., Beverly, Massachusetts.

3. After an hour or so, the ether comes to a boil, and the reaction mixture is cooled in a cold water bath.

4. The reaction is exothermic but is contained in the beaker.

5. The petroleum ether fraction used was Skellysolve B (boiling range 60–70°).

6. 4,5,6,7-Tetrahydroindazole can be distilled, b.p. 135–140°/5 mm. It is recrystallized from petroleum ether and melts at 84°.

7. The petroleum ether purification step may be eliminated with equally satisfactory results. After dehydration with ethyl alcohol, the residue is dried under reduced pressure to constant weight and used directly for the next step.

8. The palladium catalyst is prepared according to *Organic Syntheses Coll. Vol.* **3**, 686 (1955). It can be reused for the dehydrogenation of 4,5,6,7-tetrahydroindazole to indazole. The yield is somewhat better with used catalyst than with fresh catalyst.

3. Methods of Preparation

2-Hydroxymethylenecyclohexanone has been prepared by the reaction of cyclohexanone and alkyl formates.[2–4]

4,5,6,7-Tetrahydroindazole has been prepared by the hydrolysis of 1-carbamyl-4,5,6,7-tetrahydroindazole.[2,3] It was first prepared in the Lilly Research Laboratories by Dr. N. Easton by the reaction of 2-hydroxymethylenecyclohexanone and hydrazine hydrate.

Indazole has been prepared according to the method reported in *Organic Syntheses*.[5] Recently, indazole has been prepared by the hydrolysis or reduction of 3-cyanoindazole,[6] by heating 1-*o*-tolyl-3,3-dimethyltriazine,[7] by the coupling of N-nitroso-*o*-benzo-(or aceto-)toluidide,[8-10] and by the decomposition of *cis*-2-stilbenediazonium fluoroborate.[11] The present method employs milder reaction conditions.[12]

[1] The Lilly Research Laboratories, Indianapolis 6, Indiana.
[2] Wallach, Steindorff, and Grimmer, *Ann.*, **329**, 109 (1903).
[3] von Auwers, Buschmann, and Heidenreich, *Ann.*, **435**, 277 (1924).
[4] Plattner, Treadwell, and Scholz, *Helv. Chim. Acta*, **28**, 771 (1945).
[5] *Org. Syntheses Coll. Vol.* **3**, 475 (1955).
[6] Rousseau and Lindwall, *J. Am. Chem. Soc.*, **72**, 3047 (1950).
[7] Cook, Dickson, Jack, Loudon, McKeown, MacMillan, and Williamson, *J. Chem. Soc.*, **1950**, 139.
[8] Huisgen and Nakaten, *Ann.*, **573**, 181 (1951).
[9] Huisgen and Nakaten, *Ann.*, **586**, 84 (1954).
[10] Rondestvedt and Blanchard, *J. Am. Chem. Soc.*, **77**, 1769 (1955).
[11] DeTar and Chu, *J. Am. Chem. Soc.*, **76**, 1686 (1954).
[12] Ainsworth, *J. Am. Chem. Soc.*, **79**, 5242 (1957).

INDOLE-3-ALDEHYDE *

(Indole-3-carboxaldehyde)

$$\text{indole} + \text{HCON(CH}_3)_2 \xrightarrow[\text{2. H}_2\text{O}]{\text{1. POCl}_3} \text{indole-3-CHO} + \text{(CH}_3)_2\text{NH}$$

Submitted by Philip N. James and H. R. Snyder.[1]
Checked by Virgil Boekelheide and Richard N. Knowles.

1. Procedure

In a 1-l. round-bottomed, three-necked flask fitted with an

efficient mechanical stirrer, a drying tube containing Drierite, and a 125-ml. dropping funnel is placed 288 ml. (274 g., 3.74 moles) of freshly distilled dimethylformamide (Note 1). The flask and its contents are cooled in an ice-salt bath for about 0.5 hour, and 86 ml. (144 g., 0.94 mole) of freshly distilled phosphorus oxychloride (Note 2) is subsequently added with stirring to the dimethylformamide over a period of 0.5 hour. The pinkish color of the formylation complex may be observed during this step. The 125-ml. dropping funnel is replaced with a 200-ml. dropping funnel, and a solution of 100 g. (0.85 mole) of indole (Note 3) in 100 ml. (95 g., 1.3 moles) of dimethylformamide is added to the yellow solution over a period of 1 hour during which time the temperature should not rise above 10°. Once the solution is well mixed, the dropping funnel is replaced with a thermometer, and the temperature of the viscous solution is brought to 35°. The syrup is stirred efficiently at this temperature for 1 hour, or for 15 minutes longer than is necessary for the clear yellow solution to become an opaque, canary-yellow paste (Note 4). At the end of the reaction period, 300 g. of crushed ice is added to the paste (Note 5) with careful stirring, producing a clear, cherry-red aqueous solution.

This solution is transferred with 100 ml. of water to a 3-l. three-necked flask containing 200 g. of crushed ice and fitted with an efficient mechanical stirrer and a separatory funnel containing a solution of 375 g. (9.4 moles) of sodium hydroxide in 1 l. of water. The aqueous base is added dropwise with stirring until about one-third of it has been added (Note 6). The remaining two-thirds is added rapidly with efficient stirring (Note 7), and the resulting suspension is heated rapidly to the boiling point and allowed to cool to room temperature, after which it is placed in a refrigerator overnight. The precipitate is collected on a filter and resuspended in 1 l. of water. Most of the inorganic material dissolves, and the product is then collected on a filter, washed with three 300-ml. portions of water and air-dried, yielding about 120 g. (97%) of indole-3-aldehyde, m.p. 196–197°. The indole-3-aldehyde resulting from this procedure is sufficiently pure for most purposes, but it may be recrystallized from ethanol if desired (Note 8).

2. Notes

1. Freshly distilled Merck reagent grade or du Pont technical grade, dimethylformamide, b.p. 151–153°, was used.

2. Mallinckrodt analytical reagent grade phosphorus oxychloride was freshly distilled, b.p. 106–108°.

3. Dow Chemical Company indole was employed. It was recrystallized once (150 g. per 1.8 l.) from 60–90° petroleum ether, m.p. 52–53°.

4. The precipitation described here did not occur in all runs, but no appreciable effect on the yield or purity of the final product was noticed if the stirring and heating of the greenish yellow solution were continued for at least one hour.

5. Reaction between the non-aqueous paste and water (or ice) is exothermic, so it is sometimes helpful to cool the paste in an ice bath before adding the ice. In any case, no trouble should be encountered provided the 300 g. of ice is added at once.

6. The point at which rapid addition should begin is easily recognized by the disappearance of the red color of the solution and the appearance of a greenish blue or greenish yellow color.

7. Near the end of the addition, the entire contents of the flask may set up solid, stopping the stirrer. The use of a powerful stirrer at this point is desirable, for by the addition of about 100 ml. of water with rapid stirring, the cake is returned to the condition of a thick slurry. During the heating period which follows, the setting-up may again occur, but rapid and efficient stirring is usually sufficient to break up the cake. By the time the temperature has reached the boiling point, a clear yellow-orange solution should be obtained.

There is considerable evolution of dimethylamine during the heating period, especially near the boiling point.

8. About 8.5 ml. of 95% ethanol is required per gram of aldehyde. The recovery of aldehyde in this recrystallization is seldom better than 85%, and the melting point is raised only 1–2°. Concentration of mother liquors to about 15% of their original volume yields another 12–13% of aldehyde which is nearly as pure as the first crop.

3. Methods of Preparation

Indole-3-aldehyde may be prepared by direct formylation of indole with dimethylformamide [2,3] or N-methylformanilide [4] using phosphorus oxychloride as a catalyst, by the Reimer-Tiemann reaction,[5,6] by a modified Gattermann reaction on 2-carbethoxyindole,[6] by formylation of the potassium salt of indole with carbon monoxide under vigorous conditions of heat and pressure,[3] by the Grignard reaction,[7] by hydrolysis and decarboxylation of the anil of 3-indolylglyoxylic acid,[8] by a modified Sommelet reaction on gramine [9] and on indole itself,[10] and by oxidation and hydrolysis of N-skatyl-N-phenylhydroxylamine.[11]   The method described above is essentially that of Smith.[2]   It is far superior to other methods reported for the preparation of indole-3-aldehyde because it is extremely simple and convenient, the yield of aldehyde is nearly quantitative, and the product is obtained in a state of high purity.   Two other examples of the use of the dimethylformamide procedure are described in *Organic Syntheses* (pp. 331, 831).

[1] University of Illinois, Urbana, Illinois.

[2] Smith, *J. Chem. Soc.*, **1954**, 3842.

[3] Tyson and Shaw, *J. Am. Chem. Soc.*, **74**, 2273 (1952).

[4] Shabica, Howe, Ziegler, and Tishler, *J. Am. Chem. Soc.*, **68**, 1156 (1946).

[5] Ellinger, *Ber.*, **39**, 2515 (1906); Ellinger and Flamand, *Z. physiol. Chem.*, **55**, 8 (1908).

[6] Boyd and Robson, *Biochem. J.*, **29**, 555 (1935).

[7] Dow Chemical Company, British Pat. 618,638 (Feb. 24, 1949) [*C.A.*, **43**, 5806 (1949)].

[8] Elks, Elliott, and Hems, *J. Chem. Soc.*, **1944**, 629.

[9] Snyder, Swaminathan, and Sims, *J. Am. Chem. Soc.*, **74**, 5110 (1952).

[10] Swaminathan and Ranganathan, *Chem. & Ind.* (*London*), **1955**, 1774.

[11] Thesing, *Chem. Ber.*, **87**, 507 (1954); Thesing, Müller, and Michel, *Chem. Ber.*, **88**, 1027 (1955).

IODOCYCLOHEXANE *

(Cyclohexane, iodo-)

$$\underset{\substack{H_2C \\ | \\ H_2C}}{\overset{\substack{H \\ C}}{\diagup}} \overset{\substack{CH \\ \| \\ CH_2}}{\diagdown} + KI + H_3PO_4 \rightarrow \underset{\substack{H_2C \\ | \\ H_2C}}{\overset{CHI}{\diagup}} \overset{\substack{CH_2 \\ | \\ CH_2}}{\diagdown} + KH_2PO_4$$

Submitted by HERMAN STONE and HAROLD SHECHTER.[1]
Checked by T. L. CAIRNS and V. A. ENGELHARDT.

1. Procedure

Forty-one grams (0.5 mole) of cyclohexene (Note 1) is added to a mixture of 250 g. (1.5 moles) of potassium iodide in 221 g. (2.14 moles) of 95% orthophosphoric acid (Notes 2, 3, and 4) contained in a 1-l. three-necked flask equipped with a reflux condenser, a sealed mechanical stirrer, and a thermometer. The mixture is stirred and heated at 80° for 3 hours, after which it is allowed to cool and treated with 150 ml. of water and 250 ml. of ether with continued stirring (Notes 5 and 6). The ether extract is separated, decolorized with 50 ml. of 10% aqueous sodium thiosulfate solution, washed with 50 ml. of saturated sodium chloride solution, and dried with anhydrous sodium sulfate (50 g.). The ether is evaporated on a steam bath, and the product is distilled from a modified Claisen flask under reduced pressure. The portion boiling at 48–49.5°/4 mm. is collected. The yield of iodocyclohexane is 93–95 g. (88–90%), n_D^{20} 1.551, d_4^{20} 1.625.

2. Notes

1. Cyclohexene was obtained from Eastman Kodak Company.
2. The 95% orthophosphoric acid is prepared by adding 174 g. (102 ml., 1.5 moles) of 85% phosphoric acid with stirring to 47 g.

of phosphoric anhydride. The solution should be cooled to room temperature before the addition of potassium iodide; otherwise evolution of hydrogen iodide and formation of iodine will take place. After the cyclohexene has been added, the mixture can be heated as desired, since the hydrogen iodide reacts as rapidly as it is generated.

3. Although 95% orthophosphoric acid is recommended for this method, commercial phosphoric acid (85%) may be used, but the reaction proceeds more slowly and the yield is lower.

4. This procedure has been used for the conversion of other olefins to iodides in excellent yield. Yields of 2-iodohexane and 2,3-dimethyl-2-iodobutane from 1-hexene and 2,3-dimethyl-2-butene were 94.5 and 91.4%, respectively.

5. Excess potassium iodide can be recovered by filtering the acid layer, after adding sufficient water to dissolve precipitated inorganic phosphates.

6. If the acid layer has an iodine color, another extraction with 100 ml. of ether is recommended.

3. Methods of Preparation

Iodocyclohexane has been prepared by the action of phosphorus and iodine on cyclohexanol,[2] and from hydrogen iodide and cyclohexanol,[3] chlorocyclohexane,[4] or cyclohexyl ether.[5] It has also been prepared by reaction of potassium iodide and chlorocyclohexane,[6] by the reaction of iodine with cyclohexyldiphenyl phosphite,[7] and by the condensation of triphenyl phosphite, methyl iodide, and cyclohexanol.[8]

[1] Ohio State University, Columbus, Ohio.
[2] Freundler and Damon, *Compt. rend.*, **141**, 593 (1905).
[3] Baeyer, *Ann.*, **278**, 107 (1894).
[4] Markownikoff, *Ann.*, **302**, 12 (1898).
[5] Lacourt, *Bull. soc. chim. Belges*, **36**, 353 (1927).
[6] Conant and Hussey, *J. Am. Chem. Soc.*, **47**, 476 (1925).
[7] Forsman and Lipkin, *J. Am. Chem. Soc.*, **75**, 3145 (1953).
[8] Rydon and Landauer (to National Research Development Corp.), Brit. pat. 695,468 [*C. A.*, **48**, 10047 (1954)]; Landauer and Rydon, *J. Chem. Soc.*, **1953**, 2224.

2-IODOTHIOPHENE *

(Thiophene, 2-iodo-)

$$2 \; \underset{\underset{S}{\overset{HC-CH}{\Big|\Big|}}}{\overset{HC-CH}{\Big|\Big|}} + I_2 \xrightarrow{HNO_3} 2 \; \underset{\underset{S}{\overset{HC-CH}{\Big|\Big|}}}{\overset{HC-CH}{\Big|\Big|}} + H_2O$$

Submitted by HENRY Y. LEW and C. R. NOLLER.[1]
Checked by CLIFF S. HAMILTON and FRANK A. BOWER.

1. Procedure

In a 200-ml. three-necked flask fitted with a mechanical stirrer (Note 1), a reflux condenser, and a separatory funnel, and set up in a hood, are placed 38 g. (0.15 mole) of iodine and 42 g. (39 ml., 0.50 mole) of thiophene (Note 2). A solution of 28 ml. (0.44 mole) of nitric acid (sp. gr. 1.42) diluted with an equal volume of water is placed in the separatory funnel, from which approximately one-fourth of the nitric acid is added slowly and with vigorous stirring. Slight heating may be necessary to start the reaction, but once initiated it proceeds vigorously with the evolution of brown oxides of nitrogen. Cooling with an ice bath may be necessary to control the reaction. After the evolution of gases has subsided, the remaining nitric acid is added dropwise, and the reaction proceeds smoothly at room temperature with continual evolution of oxides of nitrogen. After all the nitric acid has been added, the solution is heated under reflux on a water bath for 30 minutes.

The reaction mixture is allowed to stand, and the red organic layer is separated, mixed with 40 ml. of 10% sodium hydroxide solution, and steam-distilled (Note 3). The yellow oil is separated, dried over anhydrous calcium chloride, and distilled at reduced pressure from a modified Claisen flask. The yield is 43–45 g. (68–72%); b.p. 89–93°/36 mm.; n_D^{25} 1.6465.

2. Notes

1. The seal for the mechanical stirrer used (Fig. 10) is made from two one-hole rubber stoppers and a piece of glass tubing with glycerin as a seal and lubricant. According to the submitters, it is better than a mercury seal, not only for reactions where halogens, halogen acids, or compounds that react with mercury are present, but also for practically any other reaction since the handling of mercury always requires a considerable amount of care.

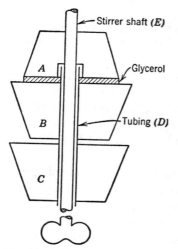

FIG. 10. Seal for a mechanical stirrer made from rubber stoppers and glass tubing using glycerol as a seal and lubricant.

The glass tubing D, about 10 cm. in length, is just large enough in diameter to permit the stirrer E to rotate freely. The smaller of the two one-hole rubber stoppers, A, is bored to a depth of about 7 mm. with a cork-borer whose diameter is approximately 3 mm. larger than that of tubing D, and then this section is cut out with a pair of scissors. Rubber stopper B is 10–15 mm. larger in diameter than A. When assembled, tubing D fits tightly in rubber stopper B, and protrudes out at the top about 4 mm., and rubber stopper A fits tightly about stirrer E and rests on top of rubber stopper B. Glycerol is used as a seal and as a lubricant between the contact surfaces of rubber stoppers A and B, and the portion of tubing D protruding out of rubber stopper B acts as a wall in preventing glycerol from flowing down inside the tubing. When the stirrer motor is on, rubber stopper A rotates with stirrer E. Rubber stopper C fits tightly over tubing D and in the mouth of the reaction flask.

2. When the preparation is carried out in larger quantities, evolution of heat accompanies the mixing of iodine and thiophene, and unless the mixture is stirred well, it will solidify into a

hard mass.   By keeping the quantities reasonably small, this tendency to solidify is reduced to a minimum.

3. Toward the end of the distillation a small amount (0.5–2.0 g.) of 2-iodo-5-nitrothiophene is collected in the condenser and the receiver.   The amount formed increases with increase in the temperature of the reaction mixture.

3. Method of Preparation

2-Iodothiophene has been prepared by the action of iodine and mercuric oxide on thiophene.[2]   The present method has been published.[3]

[1] Stanford University, Stanford, California.
[2] Meyer and Kreis, *Ber.*, **17**, 1558 (1884); Thyssen, *J. prakt. Chem.*, [2] **65**, 5 (1902); *Org. Syntheses Coll. Vol.* **2**, 357 (1943).
[3] Lew and Noller, *J. Am. Chem. Soc.*, **72**, 5715 (1950).

4-IODOVERATROLE

(Veratrole, 4-iodo-)

$$\text{(OCH}_3\text{, OCH}_3\text{ benzene)} + I_2 + CF_3CO_2Ag \rightarrow$$

$$\text{(OCH}_3\text{, OCH}_3\text{, I benzene)} + CF_3CO_2H + AgI$$

Submitted by DONALD E. JANSSEN and C. V. WILSON.[1]
Checked by MAX TISHLER and GEORGE PURDUE.

1. Procedure

A. *Silver trifluoroacetate.*   To a suspension of 187 g. (0.81 mole) of silver oxide (Note 1) in 200 ml. of water is added 177 g. (1.55 moles) of trifluoroacetic acid (Note 2).   The resulting solution is filtered, and the filtrate is evaporated to dryness under reduced pressure.   The dry silver trifluoroacetate thus obtained is purified by placing it in a Soxhlet thimble and extracting with ether, or by dissolving the salt in 1.2 l. of ether, filtering through a thin layer of activated carbon, and evaporating the filtered ether solution to dryness. The yield of colorless crystalline salt obtained after removal of the ether is 300 g. (88%).

B. *4-Iodoveratrole.* In a 3-l. three-necked round-bottomed flask, fitted with a sealed stirrer, a dropping funnel, and a reflux condenser protected with a drying tube, is placed 110 g. (0.5 mole) of dry silver trifluoroacetate (Note 3). The flask is flamed to remove all moisture, and 69 g. (0.5 mole) of dry veratrole is added (Note 2). To the stirred suspension a solution of 127 g. (0.5 mole) of iodine in 1.6 l. of chloroform is added through the dropping funnel over a period of 2 hours. After stirring an additional hour, the mixture is filtered and the precipitated silver iodide is washed with 100 ml. of chloroform. The solvent is removed from the filtrate and washings under vacuum, and the residue is distilled through an 8-in. Vigreux column. The fraction boiling at 152–155°/15 mm. weighs 112–120 g. and constitutes a yield of 85–91% (Notes 4 and 5).

2. Notes

1. The silver oxide was prepared by adding, with manual stirring, 66 g. of 98% sodium hydroxide (1.62 moles) in 2 l. of water to a solution of 274 g. (1.62 moles) of silver nitrate in 500 ml. of water. The precipitate was collected by filtration and washed with water until free from alkali. The wet cake can be dried or preferably used moist for reaction with trifluoroacetic acid.

2. The trifluoroacetic acid and veratrole used were Eastman Kodak Company white label grade.

3. Commercially available silver acetate may be used in place of the silver trifluoroacetate, but the yield is somewhat lower (75–80%).

4. The product, n_D^{25} 1.612, solidifies on chilling. Recrystallization from ethanol gives solid of melting point 34–35°.

5. Iodination in the presence of mercuric oxide [2-6] gives yields of about 40–55%, and even after careful distillation the product is contaminated with mercury salts.

3. Methods of Preparation

4-Iodoveratrole has been prepared by iodination of veratrole in the presence of mercuric oxide [2-6] and by methylation of 4-iodoguaiacol with methyl iodide in alcoholic sodium ethoxide solution.[7]

[1] Eastman Kodak Company, Rochester 17, New York.
[2] Ritchie, *J. Proc. Roy. Soc. N. S. Wales*, **78**, 134 (1945) [*C. A.*, **40**, 876 (1946)].
[3] Jurd, *Australian J. Sci. Research*, **2A**, 246 (1949) [*C. A.*, **45**, 2887 (1951)].
[4] Seer and Karl, *Monatsh.*, **34**, 647 (1913).
[5] Bruce and Sutcliffe, *J. Chem. Soc.*, **1955**, 4435.
[6] Gutzke, Fox, Ciereszko, and Wender, *J. Org. Chem.*, **22**, 1271 (1957).
[7] Tassilly and Leroide, *Compt. rend.*, **144**, 757 (1907); *Bull. soc. chim. France*, [4] **1**, 932 (1907) [*C. A.*, **1**, 1848 (1907)].

ISODEHYDROACETIC ACID * AND ETHYL ISODEHYDROACETATE *

(Coumalic acid, 4,6-dimethyl-, and coumalic acid, 4,6-dimethyl-, ethyl ester)

$$2CH_3COCH_2CO_2C_2H_5 \xrightarrow{H_2SO_4} \underset{H_3C}{\overset{HO_2C}{\bigvee}} \underset{O \quad O}{\overset{CH_3}{}} + 2C_2H_5OH$$

$$2CH_3COCH_2CO_2C_2H_5 \xrightarrow{H_2SO_4} \underset{H_3C}{\overset{C_2H_5O_2C}{\bigvee}} \underset{O \quad O}{\overset{CH_3}{}} + C_2H_5OH$$

(Note 1)

Submitted by NEWTON R. SMITH and RICHARD H. WILEY.[1,2]
Checked by R. S. SCHREIBER and H. H. FALL.

1. Procedure

In a 2-l. three-necked flask fitted with a thermometer, a stirrer, and a dropping funnel is placed 900 ml. of concentrated sulfuric acid. To the acid, cooled in an ice bath, is added, with stirring, 635 ml. (650 g., 5 moles) of ethyl acetoacetate (Note 2) at such a rate that the temperature remains between 10° and 15°. When all the ester has been added, the flask is stoppered with a calcium chloride drying tube and allowed to stand at room temperature. After 5–6 days (Note 3) the reaction mixture is poured onto 2 kg. of crushed ice while being stirred vigorously with a wooden paddle. The solid is collected on a large Büchner funnel, washed with two 200-ml. portions of cold water, and sucked as dry as possible.

The filtrate, *ca.* 4 l. in volume, is extracted with three 1.5-l. portions of ether (Note 4), and the sulfuric acid mother liquors are discarded. The ether extracts are combined and used to dissolve the solid mixture of acid and ester. If necessary, more ether can be added to assist in dissolving the solid. The ether solution is washed with 50 ml. of cold water and extracted with approximately ten 100-ml. portions of saturated sodium carbonate solution until all the isodehydroacetic acid has been removed (Note 5). The combined sodium carbonate extracts are acidified with an excess of concentrated hydrochloric acid, and the finely divided acid which precipitates is redissolved by heating to the boiling point. The hot solution is filtered with the aid of suction and is cooled in an ice bath; the solid is collected on a filter. The crude isodehydroacetic acid is dissolved in 400 ml. of hot water, and this solution is treated with decolorizing carbon, filtered, and cooled slowly to effect crystallization. The yield of isodehydroacetic acid is 91–115 g. (22–27%); m.p. 154–155°.

The ether extract, which contains ethyl isodehydroacetate, is dried for 24 hours over anhydrous sodium sulfate, and the ether is then removed by heating on a water bath. The residue is distilled from a 250-ml. Claisen flask, under reduced pressure, to give 130–175 g. (27–36%) of ethyl isodehydroacetate; b.p. 185–192°/35 mm.; m.p. 18–20° (Notes 6 and 7).

2. Notes

1. Two equations are written since isodehydroacetic acid is not esterified rapidly and its ester is not hydrolyzed rapidly under the conditions used here.

2. The checkers used Eastman Kodak Company white label grade ethyl acetoacetate as received.

3. If the reaction mixture is allowed to stand only 24 hours, the yields drop about 10% for both the acid and its ester.

4. Occasionally a large amount of carbon dioxide is present in the solution, and caution should be exercised in the extraction with ether.

5. The checkers found it convenient to test the pH of the extract with Hydrion papers. When the pH of the extract became the same as that of the saturated sodium carbonate solution, the color of the extract changed from green to orange. Thereafter, extraction was continued until neutralization of small aliquots with concentrated hydrochloric acid produced no solid.

6. The fore-run of 30–40 g. consists mostly of a mixture of 4,6-dimethyl-1,2-pyrone and ethyl isodehydroacetate.

7. The ester may be hydrolyzed to the acid by heating on a steam bath with 5 times its weight of concentrated sulfuric acid for 5–8 hours. The yield of acid is 40–50% with about 30% of the original ester recovered.

3. Methods of Preparation

Isodehydroacetic acid has been prepared by the action of sulfuric acid on acetoacetic ester.[3-5] The ethyl ester has been prepared by the action of dry hydrogen chloride on acetoacetic ester [6, 7] and by the sodium-catalyzed condensation of ethyl β-chloroisocrotonate with ethyl acetoacetate.[3] The methyl ester of isodehydroacetic acid has been prepared by the thermal rearrangement of pyrazolines.[8]

[1] University of Louisville, Louisville, Kentucky.

[2] The submitters wish to thank the Research Corporation for a grant under which this work was done.

[3] Anshutz, Bendix, and Kerp, Ann., **259**, 148 (1890).

[4] Hantzsch, Ann., **222**, 1 (1884).

[5] Wiley and Smith, J. Am. Chem. Soc., **73**, 3531 (1951).

[6] Duisberg, Ber., **15**, 1387 (1882).

[7] Goss, Ingold, and Thorpe, J. Chem. Soc., **123**, 348 (1923).

[8] Büchner and Schröder, Ber., **35**, 782 (1902).

ISOPHORONE OXIDE *

(Cyclohexanone, 2,3-epoxy-3,5,5-trimethyl-)

$$(CH_3)_2 \text{—cyclohexenone} + H_2O_2 \xrightarrow{NaOH} (CH_3)_2 \text{—epoxide} + H_2O$$

Submitted by RICHARD L. WASSON and HERBERT O. HOUSE.[1]
Checked by JAMES CASON and RALPH J. FESSENDEN.

1. Procedure

In a 1-l. three-necked flask, equipped with a dropping funnel, a mechanical stirrer, and a thermometer, is placed a solution of 55.2 g. (0.4 mole) of isophorone (Note 1) and 115 ml. (1.2 moles) of 30% aqueous hydrogen peroxide (*Caution! avoid contact with skin*) in 400 ml. of methanol. After the contents of the flask have been cooled to 15° by means of an ice bath, 33 ml. (0.2 mole) of 6N aqueous sodium hydroxide is added, dropwise and with stirring, over a period of 1 hour. During the addition the temperature of the reaction mixture is maintained at 15–20° with a bath of cold water (Note 2). After the addition is complete, the resulting mixture is stirred for 3 hours as the temperature of the reaction mixture is maintained at 20–25° (Notes 3 and 4). The reaction mixture is then poured into 500 ml. of water, and the resulting mixture is extracted with two 400-ml. portions of ether. The combined extracts are washed with water and dried over anhydrous magnesium sulfate. After the bulk of the ether has been removed by distillation (or flash distillation) through a 30-cm. Vigreux column (Note 5) at atmospheric pressure, the residual liquid is distilled through the Vigreux column under reduced pressure. The yield of isophorone oxide (Note 4) is 43–44.5 g. (70–72%), b.p. 70–73°/5 mm., n_D^{25} 1.4500–1.4510.

2. Notes

1. A technical grade of isophorone, b.p. 80–84°/9 mm., n_D^{25} 1.4755, purchased from Eastman Kodak Company, was employed for this preparation.

2. If the temperature of the reaction mixture is less than 15°, the reaction does not begin. When the resulting mixture is subsequently warmed to room temperature the exothermic reaction which results is difficult to control (Note 3).

3. If the temperature of the reaction mixture is allowed to rise above 30°, the yield of isophorone oxide is diminished.

4. If desired, the course of the reaction may be followed by means of the optical density of the reaction mixture at 235 mμ. The ultraviolet spectrum of isophorone has a maximum at 235 mμ (ϵ 13,300); the ultraviolet spectrum of isophorone oxide has a maximum at 292 mμ (ϵ 43). A total reaction time of 4 hours under the conditions specified was found to be ample for the complete conversion of isophorone to its oxide. If the conversion is not complete, the product cannot be separated from the unchanged isophorone without recourse to precise fractional distillation. The absence of isophorone from the final product may be verified by examination of the spectrum at 235 mμ.

5. When the checkers distilled the product through a simple type of Podbielniak column of 65-cm. length, with heated jacket and partial reflux head, the boiling range was 1°, 74–75°/6 mm., but the yields and index of refraction were the same as those reported by the submitters.

3. Method of Preparation

Isophorone oxide has been prepared by the epoxidation of isophorone with alkaline hydrogen peroxide.[2,3]

[1] Massachusetts Institute of Technology, Cambridge 39, Massachusetts.
[2] Treibs, *Ber.*, **66**, 1483 (1933).
[3] House and Wasson, *J. Am. Chem. Soc.*, **79**, 1488 (1957).

ITACONYL CHLORIDE

$$\begin{matrix} CH_2{=}CCO_2H \\ | \\ CH_2CO_2H \end{matrix} + 2PCl_5 \rightarrow \begin{matrix} CH_2{=}CCOCl \\ | \\ CH_2COCl \end{matrix} + 2POCl_3 + 2HCl$$

Submitted by Henry Feuer and Stanley M. Pier.[1]
Checked by William S. Johnson and Ernest F. Silversmith.

1. Procedure

In a 500-ml. round-bottomed flask fitted with a reflux condenser and a drying tube leading to a gas-absorption trap [2] are placed 234 g. (1.1 mole) of phosphorus pentachloride (Note 1) and 65 g. (0.5 mole) of itaconic acid (Note 2). The reagents are mixed by shaking the flask; after a few minutes a vigorous reaction commences, resulting in partial liquefaction of the mixture and copious evolution of hydrogen chloride. When the initial reaction subsides, the mixture is gently heated to cause reflux of phosphorus oxychloride until all the solid dissolves; then heating is continued for an additional 15 minutes (Note 3). The reflux condenser is replaced by a 12-in. Vigreux column, and the phosphorus oxychloride is removed by distillation at reduced pressure provided by a water aspirator (Note 4), the major portion coming over at about 45°/85 mm. When all the phosphorus oxychloride has been removed, the pressure is reduced (vacuum pump) and the material boiling at 70–75°/2 mm. is collected. Liquid boiling in this range weighs 50–55 g., representing a yield of 60–66%. This material, n_D^{20} 1.4915, n_D^{25} 1.4900, is pure enough for most purposes, but it may be further refined by distillation through a packed column, yielding 47–53 g. of a water-white liquid, n_D^{20} 1.4919, boiling at 71–72°/2 mm.

2. Notes

1. The slight molar excess of phosphorus pentachloride has been found to increase the yield of product. It is best to use apparatus with ground-glass joints.

2. Chas. Pfizer and Company technical grade itaconic acid was employed without purification.[3]

3. Heating for a longer period results in a rather sudden change in color from pale yellow to deep orange or red, and a decrease in yield.

4. Considerable dissolved hydrogen chloride is liberated at this point and passes into the water aspirator. A mechanical vacuum pump should not be used at this stage because it would be damaged by corrosion.

3. Methods of Preparation

Itaconyl chloride has been prepared previously only by the reaction of itaconic anhydride with phosphorus pentachloride.[4]

[1] Purdue University, Lafayette, Indiana.
[2] Org. Syntheses Coll. Vol. 2, 4 (1943).
[3] See also Org. Syntheses Coll. Vol. 2, 369 (1943).
[4] Petri, Ber., 14, 1635 (1881).

6-KETOHENDECANEDIOIC ACID

(Undecanedioic acid, 6-oxo-)

$$2CH_3O_2C(CH_2)_4COCl \xrightarrow[\substack{(2)\ KOH \\ (3)\ HCl}]{(1)\ N(C_2H_5)_3} HO_2C(CH_2)_4\overset{\overset{\textstyle O}{\|}}{C}(CH_2)_4CO_2H$$

Submitted by Lois J. Durham, Donald J. McLeod, and James Cason.[1]
Checked by N. J. Leonard, D. H. Dybvig, and K. L. Rinehart, Jr.

1. Procedure

In a 1-l. three-necked flask fitted with a sealed mechanical stirrer, a 125-ml. dropping funnel, a thermometer, and a drying tube filled with calcium chloride, are placed 500 ml. of dry benzene (Note 1) and 89.3 g. (0.5 mole) of δ-carbomethoxyvaleryl chloride (Note 2). The thermometer is adjusted to extend into the stirred liquid but not into the path of the stirrer. The mixture is cooled, with stirring, to 3–5° in an ice bath, then 50.6 g.

(0.5 mole) of triethylamine (Note 3) is added as rapidly as is consistent with keeping the temperature of the reaction mixture below 25° (3–5 minutes). When the mildly exothermic reaction has subsided, the ice bath is removed and a warm water bath is used to raise the temperature of the reaction mixture to 33–35° during 10–15 minutes. A heavy white precipitate of triethylamine hydrochloride separates. After the reaction mixture has been warmed to about 35°, the water bath is removed and stirring is continued without heating for 30 minutes.

The reaction mixture is filtered (Note 4) with suction, and the amine salt is washed with about 200 ml. of benzene. The filtrate and washings are combined and transferred to a 1-l. round-bottomed flask, benzene is removed at reduced pressure, and to the residue is added 500 ml. of $2N$ aqueous potassium hydroxide. This mixture is heated under reflux for 4 hours, by which time the solution should become completely homogeneous. The cooled solution is extracted with three 100-ml. portions of ether, then acidified to Congo red with concentrated hydrochloric acid (approximately 95 ml.). After the solution has been cooled in ice for at least 1 hour, the precipitated white solid is collected by suction filtration, washed with water, and recrystallized from a minimal amount of hot water (105–125 ml. required at about 90°). The yield of colorless 6-ketohendecanedioic acid, m.p. 108–109° (Note 5) is 35–37 g. (60–64%).

2. Notes

1. A quantity of thiophene-free benzene is conveniently dried by distilling about one-fourth of it, then cooling the residue with protection from moisture by use of a calcium chloride tube.

2. This ester acid chloride is prepared by allowing 100 g. (0.63 mole) of redistilled commercial methyl hydrogen adipate (b.p. 155–156°/7 mm., 172–173°/13 mm.) to stand overnight at room temperature with 150 g. (1.25 moles) of thionyl chloride. A Claisen head is attached, and the thionyl chloride is removed at aspirator pressure on a steam bath. A pump is attached, and the ester acid chloride is distilled; the yield is at least 94 g. (84%), b.p. 114–115°/1 mm.

3. Triethylamine purified by drying over sodium hydroxide pellets and distilling from α-naphthyl isocyanate was found to give no better results than amine which had been distilled through a half-meter Vigreux column and collected over the range 89.5–90°.

4. Frequently the flocculent precipitate of triethylamine hydrochloride is filtered with some difficulty; accordingly, a sufficiently large Büchner funnel should be used for the filtration.

5. Titration of this acid gives an equivalent weight in the range 115–116 (theory, 115). The highest melting point recorded for this acid is 111°.[2]

3. Methods of Preparation

6-Ketohendecanedioic acid has been prepared by the reactions described,[3,4] by the dialkylation of diethyl acetonedicarboxylate with ethyl γ-iodobutyrate in the presence of sodium ethoxide followed by hydrolysis and decarboxylation,[2,5] and by the permanganate oxidation of 6-(1'-cyclohexenyl)-1-hexene,[6] by the reaction of 1-morpholinocyclohexene with glutaryl chloride in the presence of triethylamine,[7] and by hydrolysis of the reaction product obtained by treating bromomagnesium cyclopentanone-2-carboxylate with adipyl chloride.[8] The present method is a simplification of the procedure originally described by Sauer.[3] This method is practical for the preparation of symmetrical keto dibasic acids and esters.[9]

[1] University of California, Berkeley, California.

[2] English, *J. Am. Chem. Soc.*, **63**, 941 (1941).

[3] Sauer, *J. Am. Chem. Soc.*, **69**, 2444 (1947).

[4] Cason, Taylor, and Williams, *J. Org. Chem.*, **16**, 1187 (1951).

[5] Leonard and Goode, *J. Am. Chem. Soc.*, **72**, 5404 (1950).

[6] Kreuchunas, *J. Am. Chem. Soc.*, **75**, 4278 (1953).

[7] Hünig and Lücke, *Chem. Ber.*, **92**, 652 (1959).

[8] Plešek, *Chem. listy*, **49**, 1840 (1955) [*C. A.*, **50**, 9294 (1956)].

[9] Blomquist, Johnson, Diuguid, Shillington, and Spencer, *J. Am. Chem. Soc.*, **74**, 4203 (1952).

β-KETOISOÖCTALDEHYDE DIMETHYL ACETAL

(Heptanal, 6-methyl-3-oxo-, 1-dimethyl acetal)

$$(CH_3)_2CHCH_2CH_2COCH{=}CHCl + 2CH_3OH + NaOH \rightarrow$$
$$(CH_3)_2CHCH_2CH_2COCH_2CH(OCH_3)_2 + NaCl + H_2O$$

Submitted by CHARLES C. PRICE and JOSEPH A. PAPPALARDO.[1]
Checked by R. S. SCHREIBER, WM. BRADLEY REID, JR., and R. W. JACKSON.

1. Procedure

A dry 1-l. three-necked flask fitted with a mercury-sealed stirrer, a calcium chloride drying tube, and a 500-ml. dropping funnel, protected by a calcium chloride drying tube, is surrounded by an ice-salt mixture at −11°. Anhydrous methanol (130 ml.) (Notes 1 and 2) and 161 g. (1 mole) of β-chlorovinyl isoamyl ketone (p. 186) are poured into the flask (Note 3). A solution of 43 g. (1.04 moles) of sodium hydroxide (97%) and 350 ml. of absolute methanol (Note 4) is added dropwise with stirring over a period of 2 hours, during which time the bath temperature is kept between −11° and −8°.

The reaction mixture is poured, with stirring, into 1 kg. of a saturated sodium chloride solution (Note 5). The mixture is extracted with four 100-ml. portions of low-boiling petroleum ether. The extracts are combined and dried over anhydrous potassium carbonate. The liquid is decanted, the potassium carbonate is washed with 25 ml. of low-boiling petroleum ether, and the solution is added to the main fraction. A pinch of anhydrous potassium carbonate is added to the petroleum ether solution, and the solution is then distilled from a Claisen flask to give a colorless liquid; yield 151–169 g. (80–90%); b.p. 122–125°/25 mm.; n_D^{25} 1.4260; d_4^{25} 0.932 (Note 6).

2. Notes

1. Absolute methanol (C.P. grade) was used.
2. The checkers ran this preparation at one-third the scale described here and obtained equivalent results.

3. When this procedure is used to make the dimethyl acetal of β-ketobutyraldehyde, the methanol and β-chlorovinyl methyl ketone must be mixed quickly and cooled well. If these two liquids are placed in the same flask without immediate cooling and mixing, the ketone may decompose rapidly with the evolution of heat and large amounts of hydrogen chloride.

4. It takes about 1 hour of shaking to dissolve the sodium hydroxide pellets.

5. If this reaction is used to prepare the water-soluble dimethyl acetal of β-ketobutyraldehyde, the product of the reaction is not poured into the saturated sodium chloride solution. Instead, the methanol solution is filtered from the sodium chloride and distilled.

6. The product darkens a little on standing but undergoes no change in refractive index.

3. Methods of Preparation

The preparation described [2] has been used by Nelles [3] to make a variety of β-ketoacetals.

[1] University of Notre Dame, Notre Dame, Indiana.
[2] Price and Pappalardo, *J. Am. Chem. Soc.*, **72**, 2613 (1950).
[3] Nelles, U. S. pat. 2,091,373 [*C. A.*, **31**, 7444 (1937)]; Brit. pat. 466,890 [*C. A.*, **31**, 7886 (1937)].

LAURONE *

(12-Tricosanone)

$$2C_{11}H_{23}COCl + 2(C_2H_5)_3N \rightarrow$$

$$\underset{\underset{C_{10}H_{21}}{|}}{C_{11}H_{23}COC}{=}C{=}O + 2(C_2H_5)_3N \cdot HCl$$

$$\underset{\underset{C_{10}H_{21}}{|}}{C_{11}H_{23}COC}{=}C{=}O + H_2O \rightarrow C_{11}H_{23}COC_{11}H_{23} + CO_2$$

Submitted by J. C. SAUER.[1]
Checked by WILLIAM S. JOHNSON and H. C. DEHM.

1. Procedure

Into a 3-l. three-necked round-bottomed flask fitted with a mechanical stirrer, dropping funnel, and reflux condenser provided with a calcium chloride drying tube is placed 1260 ml. (approximately 900 g.) of anhydrous ether. Stirring is commenced, and 153.0 g. (0.7 mole) of lauroyl chloride (Note 1) is added rapidly through one of the flask openings. The solution is cooled in ice water, and 70.7 g. (0.7 mole) of triethylamine (Note 2) is added over a period of 10 minutes through the dropping funnel in a fine stream. Stirring is discontinued after 1 hour, and the mixture is allowed to come to room temperature. After 12 to 24 hours, the mixture of decylketene dimer (Note 3) and triethylamine hydrochloride is extracted once with 125 ml. of an aqueous 2% sulfuric acid solution to remove the amine salt.

Procedure A. The wet ether layer is transferred to a 3-l. distillation flask and distilled to remove most of the solvent. The warm oily residue is transferred to a 1-l. beaker and mixed with 500 ml. of 2% potassium hydroxide solution. The mixture is heated on a steam bath for 1 hour with occasional stirring and is then chilled in ice water. The waxy cake which settles out on top of the aqueous suspension is skimmed from the surface and dissolved in a mixture of 400 ml. each of acetone and methanol.

The hot solution is filtered through a steam-jacketed funnel and cooled in ice water, and the precipitate is collected on a Büchner funnel with suction. The product is washed on the funnel with cold methanol; after air drying overnight it amounts to 55–65 g. (46–55%), m.p. 62–64°.

Procedure B. The following alternative isolation procedure yields a somewhat purer product. The wet ether layer which has been washed with dilute sulfuric acid to remove amine salt is transferred to a 3-l. distillation flask, 150 ml. of 2% sulfuric acid is added, and the mixture is distilled until nearly all the ether is removed. The hot, oily layer is separated in a separatory funnel and distilled (Note 4). The yield of the fraction distilling at 215–230°/3 mm. is 64–75 g. (54–63%). After recrystallization from 750 ml. of acetone, the laurone weighs 55–65 g. (46–55%), m.p. 68–69° (Note 5).

2. Notes

1. A commercial lauric acid, such as that available from Armour and Company, was converted into the acid chloride by reaction with thionyl chloride. The checkers employed 1 kg. of thionyl chloride for 1201 g. of acid. The product was distilled through a 12-in. Vigreux column, giving 1145 g. (87%) of colorless acid chloride, f.p. −15° to −18°.

2. Triethylamine was purified by the following procedure: fractional distillation, addition of about 2% phenyl isocyanate to the distillate, and redistillation.

3. If desired, decylketene dimer can be isolated at this point by filtering the reaction mixture and concentrating the filtrate. The mixture should be handled at all times under anhydrous conditions. The filtration should be carried out by the inverted filtration method.[2] Difficulties are usually encountered in the filtration step since the amine salt frequently separates as a gel. Seeding the ether solution of lauroyl chloride with triethylamine hydrochloride usually aids in preventing this gel formation. It is necessary to rinse the amine salt several times with ether to extract the dimer, which is usually contaminated with traces of triethylamine hydrochloride.

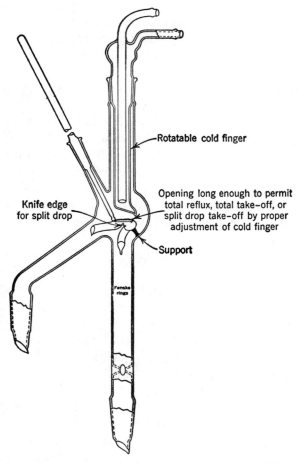

Rotatable cold finger

Knife edge
for split drop

Opening long enough to permit
total reflux, total take-off, or
split drop take-off by proper
adjustment of cold finger

Support

Fenske
rings

FIG. 11.

4. The electrically heated fractionating column used by the submitter for distilling laurone is pictured in part in Fig. 11. This still, with a column length of 8 in., was designed by Dr. H. J. Sampson of the Rayon Department of E. I. du Pont de Nemours and Company, Inc., Waynesboro, Virginia.

5. Other acid chlorides of the type RCH_2COCl can be similarly dehydrochlorinated. For example, caproyl chloride (1.2 moles) was converted to di-n-amyl ketone, b.p. 98–102°/15 mm., in 60–71.5% yield. In this case, it was found preferable to remove

the amine salt from the reaction mixture by washing with 2% sulfuric acid. The butylketene dimer was then extracted from the reaction mixture by washing with 5% sodium hydroxide solution; the alkaline solution was acidified with sulfuric acid and steam-distilled. The oily layer in the distillate was separated and fractionated.

3. Methods of Preparation

Laurone has been prepared by hydrating and decarboxylating decylketene dimer.[3] It has also been prepared by distilling calcium laurate;[4] by heating lauric acid with phosphorus pentoxide;[5] by heating barium laurate under reduced pressure;[6] by the ester condensation of ethyl laurate with sodium ethoxide[7] or of methyl laurate with sodium hydride[8] or diisopropylaminomagnesium bromide[9] followed by ketonic hydrolysis; by catalytic ketonization of lauric acid over a chromate catalyst;[10] or by passing lauric acid over thorium dioxide at 400°.[11]

[1] E. I. du Pont de Nemours and Company, Wilmington, Delaware.

[2] *Org. Syntheses Coll. Vol.* **2**, 610 (1943).

[3] Sauer, *J. Am. Chem. Soc.*, **69**, 2444 (1947); Piekarski, *J. recherches centre natl. recherche sci., Labs. Bellevue* (*Paris*), No. **40**, 197 (1958) [*C. A.*, **52**, 19922 (1958)].

[4] Overbeck, *Ann.*, **84**, 289 (1852).

[5] Kipping, *J. Chem. Soc.*, **57**, 980 (1890).

[6] Krafft, *Ber.*, **15**, 1711 (1882).

[7] Strating, Backer, Lolkema, and Benninga, *Rec. trav. chim.*, **55**, 903 (1936).

[8] Hansley, U. S. pat. 2,158,071 [*C. A.*, **33**, 6342 (1939)].

[9] Frostick and Hauser, *J. Am. Chem. Soc.*, **71**, 1350 (1949).

[10] Wortz, U. S. pat. 2,108,156 [*C. A.*, **32**, 2542 (1938)].

[11] Pickard and Kenyon, *J. Chem. Soc.*, **99**, 57 (1911).

LAURYLMETHYLAMINE

(Dodecylamine, N-methyl-)

$$CH_3(CH_2)_{10}CONHCH_3 \xrightarrow{\text{LiAlH}_4} CH_3(CH_2)_{10}CH_2NHCH_3$$

Submitted by C. V. Wilson and J. F. Stenberg.[1]
Checked by N. J. Leonard and C. W. Schimelpfenig.

1. Procedure

In a 5-l. three-necked flask, fitted with a ball-joint sealed stirrer and a Soxhlet extractor (70 mm. internal diameter $\times$ 300 mm. length of body) carrying a large-capacity condenser, are placed 1.8 l. of dry ether (Note 1) and 38 g. (1 mole) of finely divided lithium aluminum hydride (Note 2). In the Soxhlet cup is placed 160 g. (0.75 mole) (Note 3) of N-methyllauramide (m.p. 67–69°) (Note 4). The mixture is heated under gentle reflux with efficient stirring over a 5-hour period and then stirred overnight at room temperature (Note 5); the N-methyllauramide is extracted from the cup during the first 3 hours.

The excess lithium aluminum hydride and the metallic complexes are decomposed by the careful addition of 82 ml. of distilled water, from a dropping funnel, to the well-stirred mixture. The reaction mixture is stirred for an additional 30 minutes, filtered with suction, and the solid is washed with several 100-ml. portions of ether. After the ether is removed from the filtrates, the residual oil is distilled under reduced pressure. The yield of laurylmethylamine, a colorless liquid boiling at 110–115°/1.2–1.5 mm., is 121–142 g. (81–95%) (Note 6).

2. Notes

1. It is preferable to use ether subjected to final drying by distillation from lithium aluminum hydride.
2. The yield depends upon the use of high-quality, fresh lithium aluminum hydride.
3. If a Soxhlet extractor having a smaller capacity is employed, the cup will have to be recharged during the course of the reaction.

4. N-Methyllauramide, N-methylmyristamide, and N-methylpelargonamide can be prepared in 95–98% yield by adaptation of the method used by Roe, Scanlan, and Swern [2] for the preparation of amides of oleic and 9,10-dihydroxystearic acids.

5. Stirring overnight is a matter of convenience. In the preparation of methylnonylamine, refluxing for an hour after the addition of the amide was found by the submitters to be sufficient.

6. Methylnonylamine and methylmyristylamine were prepared by the submitters in 89–92% yield using the same procedure; with methylmyristylamine a longer reflux period was required, owing to the lower solubility of the amide.

3. Methods of Preparation

Laurylmethylamine has been prepared by the reaction of lauryl alcohol with methylamine under pressure in the presence of catalysts at high temperature,[3] by heating lauryl chloride with methylamine in alcoholic or aqueous medium under pressure,[4-6] by the reaction of lauryl halides with aqueous methylamine,[7] by the hydrolysis of o-carboxy-N-methyl-N-laurylbenzenesulfonamide,[8] and by the pyrolysis of β-cyanoethyllaurylmethylamine.[9] Cetylmethylamine has been prepared by the catalytic debenzylation of benzylcetylmethylamine.[10]

[1] Eastman Kodak Company, Rochester 17, New York.
[2] Roe, Scanlan, and Swern, J. Am. Chem. Soc., 71, 2215 (1949).
[3] I. G. Farbenindustrie A.-G., Fr. pat. 779,913 [C. A., 29, 5458 (1935)].
[4] Westphal and Jerchel, Ber., 73B, 1002 (1940).
[5] Ralston, Reck, Harwood, and DuBrow, J. Org. Chem., 13, 186 (1948).
[6] Zerweck and Gofferjé (I. G. Farbenindustrie A.-G.), Ger. pat. 657,358 [C. A., 32, 4175 (1938)].
[7] I. G. Farbenindustrie A.-G., Fr. pat. 784,599 [C. A., 30, 107 (1936)].
[8] Abe, J. Pharm. Soc. Japan, 75, 159 (1955) [C. A., 50, 1779 (1956)].
[9] DuBrow and Harwood, J. Org. Chem., 17, 1043 (1952).
[10] Birkofer, Ber., 75B, 429 (1942).

2-MERCAPTO-4-AMINO-5-CARBETHOXYPYRIMIDINE

(5-Pyrimidinecarboxylic acid, 4-amino-2-mercapto-, ethyl ester)

AND 2-MERCAPTO-4-HYDROXY-5-CYANOPYRIMIDINE

(5-Pyrimidinecarbonitrile, 4-hydroxy-2-mercapto-)

$$C_2H_5OCH{=}C(CN)CO_2C_2H_5 + CS(NH_2)_2 \xrightarrow{C_2H_5ONa}$$

$$+$$

Submitted by T. L. V. ULBRICHT, TAKUO OKUDA, and CHARLES C. PRICE.[1]
Checked by B. C. McKUSICK and STEPHEN PROSKOW.

1. Procedure

A. *2-Mercapto-4-amino-5-carbethoxypyrimidine.* A 5-l. three-necked, round-bottomed flask mounted in a heating mantle is fitted with a 250-ml. dropping funnel, an efficient, sealed, mechanical stirrer, and a reflux condenser connected to a calcium chloride drying tube. Absolute ethanol (625 ml.) is placed in the flask, the stirrer is started, and 23 g. (1 g. atom) of freshly cut sodium is added in portions. After the sodium has dissolved, 76.1 g. (1 mole) of thiourea is added to the warm, stirred solution in one portion. When the bulk of the thiourea has dissolved, 169 g. (1 mole) of liquefied ethyl ethoxymethylenecyanoacetate is added from the dropping funnel to the stirred mixture over a period of 2 hours (Note 1). This rate of addition keeps the reaction mixture warm. The solution is then stirred and gently refluxed for 6 hours. The sodium salt of the carbethoxypyrimidine may precipitate during the course of the reaction.

The reaction mixture is cooled to 50–60°, and 1.75 l. of water is added, followed by 65 ml. of acetic acid to make the mixture distinctly acidic. The resulting suspension is stirred and boiled for 5 minutes in order to effect complete decomposition of the sodium salt.

The mixture is cooled to 25°, and the crystalline 2-mercapto-4-amino-5-carbethoxypyrimidine is collected on a 10-cm. Büchner funnel and washed successively with five 50-ml. portions of water, 50 ml. of acetone, and 50 ml. of ether (Note 2). The carbethoxypyrimidine weighs 152–159 g. (76–80%) and melts with decomposition at 259–260° (Note 3) after being dried for 5 hours at 110° and atmospheric pressure. It is in the form of a cream-colored powder that is sufficiently pure for synthetic purposes. Pure carbethoxypyrimidine can be obtained by recrystallizing the crude product once from 50% acetic acid, using 170 ml. per gram of pyrimidine.

B. *2-Mercapto-4-hydroxy-5-cyanopyrimidine.* The aqueous filtrate from which the crude 2-mercapto-4-amino-5-carbethoxypyrimidine separated is cooled overnight at 0°, and the cyanopyrimidine that precipitates is collected on a suction filter. The crude product is recrystallized from about 200 ml. of 10% acetic acid with 1 g. of decolorizing charcoal added. Two additional recrystallizations done similarly give the pure cyanopyrimidine as faintly yellow crystals, m.p. 265–272° (dec.) (Note 3). The yield is 10–18 g. (7–12%).

2. Notes

1. Ethyl ethoxymethylenecyanoacetate can be prepared in the laboratory from ethyl cyanoacetate and ethyl orthoformate according to the directions of de Bellemont.[2] The submitters and checkers used a commercial product, m.p. 45–50°, obtained from Kay-Fries, Inc., New York. The liquefied product is weighed and poured into the dropping funnel. An infrared heating lamp is used to keep it liquid during the addition.

2. For complete removal of a yellow impurity, the product should be stirred well with each portion of water before filtration. If the solid is not washed with organic solvents, drying of the caked product will be slow.

3. The decomposition point is greatly dependent on the rate of heating. The checkers found that the carbethoxypyrimidine heated on a Fisher-Johns melting-point block at a rate of 4° per minute decomposed at 280–285°. Under the same conditions, the cyanopyrimidine decomposed at 285–289°. Both products

started to darken around 260°. In the infrared, the carbethoxy-pyrimidine has a strong band at 5.88 μ and no absorption in the 4.4 μ range, whereas the cyanopyrimidine has a strong band at 4.48 μ and no absorption at 5.88 μ.

3. Methods of Preparation

The described procedure is based on the methods of Johnson and Ambler [3] and Anderson et al.,[4] as modified by Ulbricht and Price.[5] This procedure is illustrative of a general method of preparing pyrimidines, wherein one condenses thiourea, guanidine, or an amidine with alkoxymethylenemalonic esters, alkoxymethylenecyanoacetic esters, or alkoxymethylenemalononitrile. Kenner and Todd recently reviewed the synthesis of pyrimidines.[6]

2-Mercapto-4-amino-5-carbethoxypyrimidine has been converted to 2-methylmercapto-4-amino-5-hydroxymethylpyrimidine,[5] an antimetabolite possessing antitumor activity,[7] by methylation of the mercapto group followed by reduction of the ester group to a hydroxymethyl group with lithium aluminum hydride.[5]

[1] University of Pennsylvania, Philadelphia 4, Pennsylvania; supported in part by U.S.P.H.S. Grant No. CY–2189.

[2] de Bellemont, *Bull. soc. chim. France*, [3] **25**, 18 (1901).

[3] Johnson and Ambler, *J. Am. Chem. Soc.*, **33**, 978 (1911).

[4] Anderson, Halverstadt, Miller, and Roblin, *J. Am. Chem. Soc.*, **67**, 2197 (1945).

[5] Ulbricht and Price, *J. Org. Chem.*, **21**, 567 (1956).

[6] Kenner and Todd, "Pyrimidine and Its Derivatives" in Elderfield, *Heterocyclic Compounds*, Vol. 6, pp. 234–323, John Wiley & Sons, New York, 1957.

[7] Okuda and Price, *J. Org. Chem.*, **23**, 1738 (1958).

2-MERCAPTOBENZIMIDAZOLE *

(2-Benzimidazolethiol)

$$\text{(structure)} \quad \begin{array}{c} NH_2 \\ NH_2 \end{array} + C_2H_5OCS_2K \rightarrow$$

$$\text{(structure)} \quad C-SK + C_2H_5OH + H_2S$$

$$\text{(structure)} \quad C-SK + CH_3CO_2H \rightarrow$$

$$\text{(structure)} \quad C-SH + CH_3CO_2K$$

Submitted by J. A. Van Allan and B. D. Deacon.[1]
Checked by Cliff S. Hamilton and Yao-Hua Wu.

1. Procedure

A mixture of 32.4 g. (0.3 mole) of *o*-phenylenediamine,[2] 52.8 g. (0.33 mole) of potassium ethyl xanthate (Note 1), 300 ml. of 95% ethanol, and 45 ml. of water in a 1-l. flask is heated under reflux for 3 hours.   Norit (12 g.) is then added cautiously, and after the mixture has been heated at the reflux temperature for 10 minutes the Norit is removed by filtration.   The filtrate is heated to 60–70°, 300 ml. of warm tap water (60–70°) is added, and then 25 ml. of acetic acid in 50 ml. of water is added with good stirring. The product separates as glistening white crystals, and the mixture is placed in a refrigerator for 3 hours to complete the crystallization. The product is collected on a Büchner funnel and dried overnight at 40°. The yield is 37.8–39 g. (84–86.5%) of 2-mercaptobenzimidazole melting at 303–304° (cor.) (Notes 2 and 3).

2. Notes

1. The potassium ethyl xanthate may be replaced by 19 g. of potassium hydroxide and 26 g. (21 ml., 0.34 mole) of carbon disulfide. The yield and quality of the product are the same.

2. The quality of the product is excellent. Recrystallization may be effected from 95% ethanol as a solvent. Recovery is about 90%; there is no change in the melting point.

3. 2-Mercaptobenzoxazole, m.p. 193–195°, can be prepared in 80% yield by a similar procedure, using o-aminophenol in place of o-phenylenediamine.

3. Methods of Preparation

2-Mercaptobenzimidazole has also been prepared from o-phenylenediamine by heating the thiocyanate to 120–130°; [3] by heating with aqueous potassium thiocyanate (in which case 2-aminophenylthiourea is a by-product); [4] by the action of thiophosgene in chloroform; [5] by heating with carbon disulfide in alcohol [6] or in water; [7] and by heating the hydrochloride with thiourea to 170–180°. [8]

[1] Eastman Kodak Company, Rochester, New York.

[2] *Org. Syntheses Coll. Vol.* **2**, 501 (1943).

[3] Lellmann, *Ann.*, **221**, 9 (1883).

[4] Frerichs and Hupka, *Arch. Pharm.*, **241**, 165 (1903).

[5] Billeter and Steiner, *Ber.*, **20**, 231 (1887).

[6] Gucci, *Gazz. chim. ital.*, [1] **23**, 295 (1893).

[7] Kawaoka, *J. Soc. Chem. Ind.*, *Japan* **43**, Suppl. binding 223 (1940) [*C. A.*, **35**, 2368 (1941)].

[8] Kym, *J. prakt. Chem.*, [2] **75**, 324 (1907).

METHANESULFONYL CHLORIDE

$$CH_3SO_3H + SOCl_2 \rightarrow CH_3SO_2Cl + SO_2 + HCl$$

Submitted by PETER J. HEARST and C. R. NOLLER.[1]
Checked by R. S. SCHREIBER, B. D. ASPERGREN, and R. V. HEINZELMANN.

1. Procedure

In a 1-l. three-necked flask fitted with a mechanical stirrer, a reflux condenser, a thermometer, and a separatory funnel (Note 1), and set up in a hood, is placed 152 g. (105 ml., 1.5 moles) of methanesulfonic acid (Note 2). The acid is heated to 95° on a steam bath, and 238 g. (146 ml., 2.0 moles) of thionyl chloride (Note 3) is added over a period of 4 hours. The temperature is kept at 95° throughout the addition and for 3.5 hours after it is completed.

The product is transferred to a modified Claisen flask (Note 4) and distilled under reduced pressure, heat being supplied by an oil bath (Note 5). Most of the thionyl chloride distils at room temperature. The yield of almost colorless product distilling at 64–66°/20 mm. (Note 6) is 122–143 g. (71–83%); n_D^{23} 1.451.

2. Notes

1. The checkers recommend the use of silicone grease on all glass joints.

2. The methanesulfonic acid is a commercial product supplied by the Standard Oil Company of Indiana and reported to be 95% pure and to contain 2% water.

3. Eastman Kodak Company thionyl chloride (b.p. 75–76°) was used without further purification.

4. The checkers used a Claisen head with an attached 10-cm. Vigreux column.

5. A free flame should be avoided, because local superheating causes charring and decomposition. The fumes from the decomposition cause the product (which normally is colorless) to darken. The bath temperature should not exceed 115° at the end of the distillation.

6. The checkers observed a boiling point of 61–62°/18 mm.

3. Methods of Preparation

Methanesulfonyl chloride has been prepared by the chlorination of methyl thiocyanate,[2] S-methylthiourethan,[3] sodium methylthiosulfate,[4] S-methylisothiuronium sulfate [5] or S-methylisothiuronium p-toluenesulfonate,[6] and dimethyl disulfide in the presence of water; [7] from sodium methanesulfonate by the action of phosphorus pentachloride,[8,9] phosphorus oxychloride,[9] or benzotrichloride; [10] from methanesulfonic acid by the action of phosphorus pentachloride [11] or thionyl chloride; [12] or by the reaction of methylmagnesium iodide with sulfuryl chloride.[13]

[1] Stanford University, Stanford, California.

[2] Johnson and Douglass, *J. Am. Chem. Soc.*, **61**, 2548 (1939); Johnson, U. S. pat. 2,174,856 [*C. A.*, **34**, 778 (1940)]; Kostova, *Zhur. Obshcheĭ Khim.* (J. Gen. Chem.), **18**, 729 (1948) [*C. A.*, **43**, 120 (1949)].

[3] Battegay and Krebs, *Compt. rend.*, **206**, 1262 (1938).

[4] Douglass and Johnson, *J. Am. Chem. Soc.*, **60**, 1486 (1938).

[5] Johnson and Sprague, *J. Am. Chem. Soc.*, **58**, 1348 (1936).

[6] Klamann and Drahowzal, *Monatsh.*, **83**, 463 (1952).

[7] Dijkstra and Backer, *Rec. trav. chim.*, **73**, 569 (1954).

[8] Billeter, *Ber.*, **38**, 2019 (1905); Helferich and Gnüchtel, *Ber.*, **71**, 712 (1938).

[9] Dutt, *J. Chem. Soc.*, **125**, 1463 (1924).

[10] I. G. Farbenind., A.-G., Ger. pat. 574,836 [*C. A.*, **27**, 4543 (1933)].

[11] Carius, *Ann.*, **114**, 142 (1860); McGowan, *J. prakt. Chem.*, [2] **30**, 281 (1884).

[12] Noller and Hearst, *J. Am. Chem. Soc.*, **70**, 3955 (1948).

[13] Cherbuliez and Schnauder, *Helv. Chim. Acta*, **6**, 256 (1923).

o-METHOXYPHENYLACETONE

[2-Propanone, 1-(*o*-methoxyphenyl)-]

$$
\underset{OCH_3}{\overset{CHO}{\bigcirc}} + C_2H_5NO_2 \xrightarrow{C_4H_9NH_2} \underset{OCH_3}{\overset{CH=\overset{\overset{\displaystyle CH_3}{|}}{C}-NO_2}{\bigcirc}}
$$

$$
\underset{OCH_3}{\overset{CH=\overset{\overset{\displaystyle CH_3}{|}}{C}-NO_2}{\bigcirc}} \xrightarrow[H_2O]{Fe/HCl} \underset{OCH_3}{\overset{CH_2-\overset{\overset{\displaystyle CH_3}{|}}{C}=O}{\bigcirc}}
$$

Submitted by R. V. HEINZELMAN.[1]
Checked by M. TISHLER and H. L. SLATES.

1. Procedure

A. *1-(o-Methoxyphenyl)-2-nitro-1-propene.* A 1-l. round-bottomed flask is fitted with an electric heating mantle, a modified Dean and Stark water separator (Note 1), and a reflux condenser. To the flask are added in this order 200 ml. of reagent-grade toluene, 136 g. (1.0 mole) of *o*-methoxybenzaldehyde (Note 2), 90 g. (1.1 moles) of commercial nitroethane (Note 3), and 20 ml. of *n*-butylamine (Note 4). The solution is heated to produce a rapid reflux until the separation of water ceases (Note 5). The toluene solution is used directly in the next step (Note 6).

B. *o-Methoxyphenylacetone.* A 3-l. three-necked round-bottomed flask is equipped with an electric heating mantle, two reflux condensers, a dropping funnel, and a high-speed whip stirrer (Note 7). The toluene solution from Part A is placed in the flask, and 500 ml. of water, 200 g. of powdered iron (Note 8), and 4 g. of ferric chloride are added. With vigorous agitation the suspension is heated to about 75°, and 360 ml. of concentrated hydrochloric acid is added over a 2-hour period (Note 9); heating and stirring are continued for an additional 30 minutes.

The suspension is transferred to a 5-l. three-necked round-bottomed flask and subjected to steam distillation until 7–10 l. are collected (Note 10). The toluene layer is removed, and the aqueous layer is extracted with 1 l. of fresh toluene. The combined toluene layers are agitated for 30 minutes with a solution of 26 g. of sodium bisulfite in 500 ml. of water (Note 11). The toluene layer is washed with water, and the solvent is removed at water-pump pressure on the steam bath. The resulting orange liquid weighs 107–120 g. (65–73%); n_D^{20} 1.5250–1.5270, and is sufficiently pure for most uses. It is purified by distillation through a 12-in. Vigreux column, and the fraction boiling at 128–130°/14 mm. is collected. The yield is 102–117 g. (63–71%, based on the methoxybenzaldehyde used), n_D^{20} 1.5250–1.5260 (Notes 12, 13).

2. Notes

1. The model manufactured by the Corning Glass Works, Corning, New York, and listed as No. 3622 was used. See also *Org. Syntheses Coll. Vol.* **3**, 382 (1955).

2. *o*-Methoxybenzaldehyde is available from Eastman Kodak Company, Rochester, New York, and from the Matheson Company, East Rutherford, New Jersey. It may also be prepared according to Baeyer and Villiger.[2]

3. The nitroethane was obtained from Commercial Solvents Corporation, Terre Haute, Indiana, as a 90% pure product. The amount used is a 10% excess based on its nitroethane content. The chief contaminant is 2-nitropropane, which does not interfere in the reaction.

4. It is desirable to swirl the flask after each addition to prevent the formation of layers.

5. Half of the water is collected in about an hour, and the theoretical amount (18 ml.) in about 5 hours. Water removal usually ceases at about 105% of theory. Insufficient reflux rate causes incomplete water removal or an unduly prolonged reaction time.

6. The pure nitroölefin can be obtained by removing the toluene on the steam bath at water-pump pressure and recrystallizing the resulting oil from ethanol or petroleum ether. With petroleum ether, particularly, the volume should be great enough that the product remains in solution until the solution temperature is sufficiently low to prevent oiling out. Addition of seed crystals will encourage crystallization. Alternatively, the yellow oil may be distilled at reduced pressure. The nitroölefin boils at

135–138°/1 mm. and crystallizes in the receiver when seeded. The yield is 150–175 g. (80–90%). The yellow crystals melt at 51–52° when pure. Although no difficulty has been experienced with this compound, the usual safety precautions should be observed when distilling an unsaturated nitro compound. The material is somewhat lachrymatory and irritates the skin.

7. Rapid agitation is necessary to keep the iron in suspension and to mix the two liquid layers.

8. A 40-mesh grade was used, but material up to 100 mesh has been used successfully. However, with the finer material the reaction is somewhat more vigorous.

9. The reaction mixture should reflux vigorously. When the addition time is increased to 6 hours, the yield is not appreciably changed. The iron-acid ratio appears to be important; however, doubling the amounts of both ingredients produces no change in yield.

10. The 5-l. flask is either heated with an electric mantle or placed on a steam bath to prevent condensation of steam and increase in volume of the suspension. The steam distillation must be continued beyond the point at which the distillate becomes clear.

11. The bisulfite treatment removes any aldehydic material present at this point. Since this ketone is quite inert to sodium bisulfite, the yield is not lowered by this procedure.

12. The pure ketone has n_D^{20} 1.5240 and a boiling point of 128–130°/14 mm., 150°/30 mm. Insufficient removal of o-methoxybenzaldehyde will cause the refractive index to be high to the extent of about 0.0003 for each per cent present. The use of distilled nitroölefin eliminates the aldehyde, but this advantage is offset by the distillation hazard and slightly lower over-all yields.

13. This procedure is quite general for other aromatic aldehydes. An excess of bisulfite must be avoided in the washing step, since many phenyl-substituted acetones react appreciably with it.

3. Methods of Preparation

The present procedure is that described by the submitter.[3] It is an improved modification of that described by Hoover and

Hass[4] for the corresponding *para* isomer. *o*-Methoxyphenyl-acetone also has been prepared from the glycidic ester,[5] by the hydrogenation of *o*-methoxybenzalacetone,[6] by treating *o*-methoxyphenylacetyl chloride with methylzinc iodide,[7] by the hydrolysis of *o*-methoxy-α-acetylbenzyl cyanide,[8] by the oxidation of 1-(*o*-methoxyphenyl)propene with lead oxide in acetic acid,[9] and by the condensation of *o*-methoxybenzoyl chloride with diethyl ethoxymagnesiummmalonate, followed by hydrolysis of the substituted malonic ester.[10]   The methods of synthesis of *o*-methoxyphenylacetone have been studied.[11]

[1] The Upjohn Company, Kalamazoo, Michigan.
[2] Baeyer and Villiger, *Ber.*, **35**, 3023 (1902); Spath, *Monatsh.*, **34**, 1995 (1917).
[3] Heinzelman, *J. Am. Chem. Soc.*, **75**, 921 (1953); U. S. pats. 2,557,051 (1951) and 2,601,282 (1952).
[4] Hoover and Hass, *J. Org. Chem.*, **12**, 501 (1947).
[5] Wolf, Ger. pat. 752,328 (1950) [*C. A.*, **45**, 1626 (1951).]
[6] Chen and Barthel, *J. Am. Chem. Soc.*, **75**, 4287 (1953).
[7] Imai and Abe (to Sumitomo Chemical Industries Co.), Jap. pat. 4572 (1952) [*C. A.*, **48**, 8261 (1954)].
[8] Horii, Satoda, and Inoi (to Nippon Drug Co.), Jap. pat. 7166 (1956) [*C. A.*, **52**, 9210 (1958)].
[9] Tanaka and Seki, *Yakugaku Zasshi*, **77**, 310 (1957) [*C. A.*, **51**, 11278 (1957)].
[10] Petropoulos and Tarbell, *J. Am. Chem. Soc.*, **74**, 1249 (1952).
[11] Horii, Tsuji, and Inoi, *Yakugaku Zasshi*, **77**, 252 (1957) [*C. A.*, **51**, 8671 (1957)].

p-METHOXYPHENYLACETONITRILE *

(Acetonitrile, *p*-methoxyphenyl-)

$$CH_3O\langle\bigcirc\rangle CH_2OH + HCl \rightarrow CH_3O\langle\bigcirc\rangle CH_2Cl + H_2O$$

$$CH_3O\langle\bigcirc\rangle CH_2Cl \xrightarrow[\text{NaI, CH}_3\text{COCH}_3]{\text{NaCN}} CH_3O\langle\bigcirc\rangle CH_2CN$$

Submitted by KURT RORIG, J. DERLAND JOHNSTON, ROBERT W. HAMILTON, and THOMAS J. TELINSKI.[1]
Checked by WILLIAM S. JOHNSON, STANLEY SELTZER, and PETER YATES.

1. Procedure

In a 1-l. flask fitted with a paddle-blade stirrer are placed 138 g. (1 mole) of anisyl alcohol (Note 1) and 248 ml. of concentrated hydrochloric acid. After stirring vigorously for 15 minutes the contents of the flask are transferred to a separatory funnel. The lower layer (anisyl chloride) is separated, dried over 20 g. of granular calcium chloride for about 30 minutes, and filtered to remove the drying agent.

In a 2-l. three-necked round-bottomed flask, fitted with an efficient sealed stirrer and a reflux condenser capped by a drying tube, are placed the dried anisyl chloride (Notes 2 and 3), 73.6 g. (1.5 moles) of finely powdered sodium cyanide, 10 g. of sodium iodide, and 500 ml. of dry acetone (Note 4). The heterogeneous reaction mixture is heated under reflux with vigorous stirring for 16–20 hours, then cooled and filtered with suction. The solid on the filter is washed with 200 ml. of acetone and discarded (Note 5). The combined filtrates are distilled to remove the acetone. The residual oil is taken up in 300 ml. of benzene and washed with three 100-ml. portions of hot water. The benzene solution is dried over anhydrous sodium sulfate for about 15 minutes, and the solvent is removed by distillation at the reduced pressure of the water aspirator (Note 6). The residual *p*-methoxyphenyl-acetonitrile is purified by distillation under reduced pressure through an 8-in. Vigreux column; b.p. 94–97°/0.3 mm.; n_D^{25} 1.5285–1.5291. The yield is 109–119 g., or 74–81% based on anisyl alcohol (Notes 7 and 8).

2. Notes

1. Givaudan-Delawanna (330 W. 42nd Street, New York 18, N. Y.) "Anisic Alcohol" of 97% minimum purity was used.

2. The crude anisyl chloride is unstable. It should be used the same day it is made.

3. This step should be performed in a well-ventilated hood.

4. The acetone is dried over about one-quarter its volume of granular calcium chloride for one day. The dried acetone is then filtered and distilled.

5. This residue should be discarded with due regard for the unused sodium cyanide it contains.

6. The undistilled *p*-methoxyphenylacetonitrile weighs 125–139 g. (85–95%) and has a refractive index close to that of the distilled product. It can be used for many purposes, such as

condensation with aromatic aldehydes to yield α-p-methoxy-phenylcinnamonitriles, without further purification.

7. The submitters have carried out this preparation on five times the scale described here with comparable yields.

8. This method is particularly applicable to the more reactive benzyl halides which are easily hydrolyzed in the aqueous media usually employed for the metathetical reaction with alkali cyanides. For example, anisyl chloride treated with sodium cyanide in aqueous dioxane gives, as a by-product, 5–10% of anisyl alcohol as determined by infrared analysis. The use of anhydrous acetone not only prevents hydrolysis to the alcohol but also decreases the formation of isonitriles. This method was also applied successfully by the submitters to the preparation of p-chlorophenylacetonitrile in 74% yield.

3. Methods of Preparation

This method is an adaptation of that of Dengel.[2] p-Methoxy-phenylacetonitrile can also be prepared by the metathetical reaction of anisyl chloride with alkali cyanides in a variety of aqueous solvent mixtures;[3-11] by the nitration of phenylacetonitrile, followed by reduction, diazotization, hydrolysis, and methylation;[12, 13] by the reduction of α-benzoxy-p-methoxy-phenylacetonitrile (prepared from anisaldehyde, sodium cyanide, and benzoyl chloride);[14] by the reaction of acetic anhydride with the oxime of p-methoxyphenylpyruvic acid;[15] and through the condensation of p-methoxybenzaldehyde with rhodanine.[16]

[1] G. D. Searle and Company, Chicago 80, Illinois.

[2] Dengel, German pat. application (DBP, Anm. K2355, March 30, 1950; Knoll-A.G.) as reported by Müller, *Methoden der Organischen Chemie* (Houben-Weyl), Vol. 8, p. 294, Georg Thieme Verlag, Stuttgart, 1952.

[3] Shriner and Hull, *J. Org. Chem.*, **10**, 230 (1945).

[4] Métayer, *Ann. chim. Paris*, [12] **4**, 210 (1949).

[5] Van Heyningen, *J. Am. Chem. Soc.*, **74**, 4862 (1952).

[6] Dankova, Evdokimova, Stepanov, and Preobrazhenskiĭ, *Zhur. Obshcheĭ Khim.*, **18**, 1724 (1948) [*C. A.*, **43**, 2606 (1949)].

[7] Lapiné, *Bull. soc. chim. France*, [5] **6**, 390 (1939).

[8] Lee, Ziering, Berger, and Heineman, *Jubilee Vol. Emil Barell*, **1946**, 280 [*C. A.*, **41**, 6252 (1947)].

[9] Livshits, Bazilevskaya, Bainova, Dobrovinskaya, and Preobrazhenskiĭ, *Zhur. Obshcheĭ Khim.*, **17**, 1675 (1947) [*C. A.*, **42**, 2606 (1948)].

[10] Cagniant, *Ann. chim. (Paris)*, **7**, 442 (1952).
[11] Burckhalter, Jackson, Sam, and Meyer, *J. Am. Chem. Soc.*, **76**, 4112 (1954).
[12] Pschorr, Wolfes, and Buckow, *Ber.*, **33**, 171 (1900).
[13] Silverman and Bogert, *J. Org. Chem.*, **11**, 43 (1946).
[14] Campbell and McKail, *J. Chem. Soc.*, **1948**, 1255; Wawzonek and Fredrickson, *J. Electrochem. Soc.*, **106**, 325 (1959).
[15] Baker and Eastwood, *J. Chem. Soc.*, **1929**, 2902; Seshadri and Varadarajan, *Proc. Indian Acad. Sci.*, **37A**, 145 (1953).
[16] Yoder, Cheng, and Burroughs, *Proc. Iowa Acad. Sci.*, **61**, 271 (1954) [*C. A.*, **49**, 13236 (1955)].

METHYL *p*-ACETYLBENZOATE

(Benzoic acid, *p*-acetyl-, methyl ester)

Submitted by WILLIAM S. EMERSON and GEORGE F. DEEBEL.[1]
Checked by ARTHUR C. COPE and WARREN N. BAXTER.

1. Procedure

A 200-ml. three-necked flask is equipped with a thermometer, a mechanical stirrer, a gas inlet tube extending as far into the flask as the stirrer permits, and a trap (designed to return the heavier layer of the condensate) (Note 1), which is attached to a bulb and a coil-type reflux condenser connected in series. Methyl *p*-ethylbenzoate (98 g., 0.6 mole) (Note 2) and a mixture of 1 g. of chromium oxide and 4 g. of calcium carbonate, prepared by grinding the solids together in a mortar, are added. An air line with a bleed control valve (Note 3) is connected from the top of the condenser to a water aspirator. The flask is heated with a hemispherical electric mantle (Note 4).

The mixture is stirred vigorously at a temperature of 140–150°, and air is drawn through it (Note 5) for 24 hours while the water formed is collected in the trap (Note 6). At the end of this period, the mixture is cooled, diluted with 100 ml. of benzene (Note 7), and filtered with suction to remove the catalyst (Note 8). The catalyst is washed with 10 ml. of benzene on the funnel, and the washings are combined with the filtrate. The benzene is distilled from the clear, yellow solution (conveniently at a pressure of 100 mm. or lower to avoid superheating and partial decomposition of the product) through a fractionating column (Note 9). After removal of the benzene, the pressure is reduced and the residue is fractionally distilled. There is obtained 29–32 g. of unreacted methyl p-ethylbenzoate (b.p. 118–121°/16 mm.), and 43–45 g. (40–42% conversion and 60% yield) of methyl p-acetylbenzoate; b.p. 149–150°/7 mm. (Notes 10 and 11). The high-boiling residue weighs 13–18 g. The crude methyl p-acetylbenzoate can be purified by crystallization from the minimum amount of 1:1 benzene-commercial hexane required for solution with a recovery of 82–85%, collected in two crops; m.p. 92–95°.

2. Notes

1. A Dean and Stark trap [2] modified to provide for continuous return of the heavier organic layer is most convenient. Such a trap can be constructed like a small Wehrli extractor.[3]

2. Methyl p-ethylbenzoate was prepared from ethylbenzene (1 kg.), which was converted to p-ethylacetophenone by adding aluminum chloride (792 g.), followed by acetyl chloride (462 g.), which was added with stirring at 0–5° over a period of 3 hours. The mixture was stirred for 1 hour longer, allowed to stand overnight, and washed with iced dilute hydrochloric acid. The ethylbenzene was removed by distillation, and the crude product was oxidized to p-ethylbenzoic acid with alkaline hypochlorite.[4] Recrystallization of the crude acid from 95% ethanol yielded 350 g. (41%) of the pure acid; m.p. 110–112°. Esterification with methanol in the presence of hydrogen chloride [4] yielded methyl p-ethylbenzoate (77–79%).

3. A T-tube in the connection between the top of the condenser and the water aspirator is attached to a Bunsen burner. The needle valve of the burner serves as a bleed control valve, regulating the amount of air drawn through the reaction mixture.

4. An oil bath can be used.

5. The air stream should be introduced as rapidly as possible, preferably at a rate which will just permit return of the liquid from the coil condenser. It is equally satisfactory to force air

through the mixture from a cylinder or compressed-air line if the air is clean and free from oil. Use of a rapid stream of air and vigorous stirring prevents the catalyst from settling.

6. Removal of the water should be as complete as possible, in order to prevent coagulation of the catalyst, which retards the oxidation.

7. The benzene serves to keep the methyl *p*-acetylbenzoate in solution. If crystallization occurs, the mixture should be warmed to dissolve the ester before filtration.

8. The catalyst should be removed completely, because distillation of the product in the presence of small amounts of the catalyst results in some decomposition with a corresponding lowering of the yield. Use of a filter aid such as Super-Cel aids in separating the catalyst.

9. The submitters used a 120-cm. Vigreux column in distilling the benzene and a 60-cm. Vigreux column in fractionating the product. The checkers used a Widmer column with a 17-cm. spiral for the entire distillation.

10. Cooling water should not be circulated through the condenser during distillation of the product, because the methyl *p*-acetylbenzoate tends to crystallize in the condenser.

11. The submitters have obtained similar yields (40–54% conversion and 60–66% yield) in oxidations on 10 times this scale.

3. Methods of Preparation

Methyl *p*-acetylbenzoate has been prepared by the esterification of *p*-acetylbenzoic acid with methanol in the presence of hydrogen chloride,[5] by the hydrogenation of methyl *p*-trichloroacetylbenzoate in the presence of a palladium on calcium carbonate catalyst,[6] and by the air oxidation of methyl *p*-ethylbenzoate.[4]

[1] Monsanto Chemical Company, Dayton, Ohio.
[2] Dean and Stark, *Ind. Eng. Chem.*, **12**, 486 (1920).
[3] Wehrli, *Helv. Chim. Acta*, **20**, 927 (1937).
[4] Emerson, Heyd, Lucas, Chapin, Owens, and Shortridge, *J. Am. Chem. Soc.*, **68**, 674 (1946); Monsanto Chemical Co., Brit. pat. 636,196 [*C. A.*, **44**, 8951 (1950)].
[5] Meyer, *Ann.*, **219**, 234 (1883); Bergmann and Blum, *J. Org. Chem.*, **24**, 549 (1959); Sergeev and Sladkov., *Zhur. Obshcheĭ Khim.*, **27**, 817 (1957) [*C. A.*, **51**, 16348 (1957)].
[6] Feist, *Ber.*, **67**, 938 (1934).

o-METHYLBENZYL ALCOHOL*

(Benzyl alcohol, o-methyl-)

$$\text{(structure: benzene ring with } CH_2N(CH_3)_2 \text{)} \quad CH_3 + C_2H_5Br \rightarrow o\text{-}CH_3C_6H_4CH_2\overset{\oplus}{N}(CH_3)_2\overset{\ominus}{Br} \underset{CH_3CO_2H}{\overset{NaOCOCH_3}{\longrightarrow}}$$
$$\underset{C_2H_5}{|}$$

$$o\text{-}CH_3C_6H_4CH_2OCOCH_3 \overset{NaOH}{\longrightarrow} o\text{-}CH_3C_6H_4CH_2OH$$

Submitted by W. R. BRASEN and C. R. HAUSER.[1]
Checked by WILLIAM S. JOHNSON, DONALD W. STOUTAMIRE, and A. L. WILDS.

1. Procedure

A. *o-Methylbenzyl acetate.* A solution of 29.8 g. (0.20 mole) of 2-methylbenzyldimethylamine (p. 585) and 32.7 g. (0.30 mole) of ethyl bromide in 40 ml. of absolute ethanol is placed in a 500-ml. round-bottomed flask fitted with a reflux condenser capped with a calcium chloride drying tube. The solution is heated under reflux on the steam bath for 1 hour; then an additional 10.8 g. (0.10 mole) of ethyl bromide is added and the heating continued for an additional 3 hours. The solvent and residual ethyl bromide are removed at reduced pressure (water aspirator) while the flask is heated in a water bath kept at about 60° (Note 1). The oily residue is treated with about 300 ml. of absolute ether, and on scratching crystallization is induced. The product is collected on a Büchner funnel, washed with two 50-ml. portions of anhydrous ether, and dried in a vacuum desiccator. The yield of colorless 2-methylbenzylethyldimethylammonium bromide is 47.5–49.0 g. (92–95%) (Note 2). It is hygroscopic and should therefore not be exposed to moist air.

In a 500-ml. round-bottomed flask, fitted with a reflux condenser capped with a calcium chloride drying tube, are placed 38.7 g. (0.15 mole) of the quaternary ammonium bromide (Note 3) described above, 24.6 g. (0.3 mole) of fused sodium acetate, and 100 ml. of glacial acetic acid. The mixture is boiled under reflux for 24 hours (Note 4) and then allowed to cool. It is transferred to a large beaker (Note 5), 250 ml. of water is added, and the acid is partially neutralized by the addition of 84 g. of solid sodium bicarbonate. The mixture is extracted with three 75-ml. portions of ether, and the combined ether solutions are washed with two or more 50-ml. portions of saturated sodium bicarbonate solution until all the acetic acid has been removed. The ether layer is then washed with 50 ml. of saturated sodium chloride solution and dried over anhydrous sodium sulfate. The ether is removed by distillation, and the residue is distilled under reduced pressure. The yield of colorless liquid acetate, b.p. 119–121°/15 mm. or 129–131°/31 mm., is 21.6–22.4 g. (88–91%), n_D^{25} 1.5041–1.5045 (Note 6).

B. *o-Methylbenzyl alcohol.* A solution of 5 g. (0.12 mole) of sodium hydroxide in 50 ml. of water is added to a solution of 16.4 g. (0.1 mole) of 2-methylbenzyl acetate (prepared as described above, part A) in 50 ml. of methanol contained in a 250-ml. round-bottomed flask fitted with a reflux condenser. The mixture is boiled under reflux for 2 hours, cooled, diluted with 50 ml. of water, and extracted with three 75-ml. portions of ether. The combined ether solutions are washed with 50 ml. of water and 50 ml. of saturated sodium chloride solution and dried over anhydrous sodium sulfate. The solvent is removed by distillation, finally at reduced pressure to remove the last traces of methyl alcohol, and the residue is dissolved in 50 ml. of boiling 30–60° petroleum ether. The colorless crystals obtained on cooling, finally in the ice bath, are collected by suction filtration, washed with a few milliliters of cold petroleum ether, and air-dried. Concentration of the mother liquors to 6–7 ml. and cooling gives an additional crop. The total yield of product melting between 33–34° and 35–36° is 11.6–11.8 g. (95–97%) (Note 7).

2. Notes

1. A capillary ebullition tube is used to prevent bumping.

2. 2,3-Dimethylbenzylethyldimethylammonium bromide can be prepared similarly in comparable yield. The salts are suitable for use in the displacement reaction without purification.

3. The ethobromides are preferable to methiodides in this re-

action, because the former salts are more soluble in glacial acetic acid and do not liberate halogen as do the iodides.

4. The sodium acetate dissolves as the mixture reaches reflux temperature, but a small amount of solid (perhaps sodium bromide) remains undissolved throughout the heating.

5. Considerable foaming may occur during the neutralization, and material may be lost if the process is carried out in a small narrow-mouthed vessel.

6. The submitters have used the same procedure for the preparation of 2,3-dimethylbenzyl acetate from the corresponding quaternary salt (Note 2). The yield of material b.p. 127–129°/9 mm. was 94%.

7. Hydrolysis of 2,3-dimethylbenzyl acetate (Note 6) by this procedure gave 2,3-dimethylbenzyl alcohol, m.p. 65–66°, in 96% yield.

3. Methods of Preparation

The vicinal methylbenzyl alcohols have been prepared in general by the abnormal reaction of the appropriately substituted Grignard reagent with formaldehyde.[2] The present method is preferred over the former because it yields purer products than the Grignard approach.

o-Methylbenzyl alcohol also has been prepared by the lithium aluminum hydride reduction of o-toluic acid [3] or ethyl o-toluate,[4] and by the hydrolysis of o-xylyl bromide.[4]

[1] Duke University, Durham, North Carolina. Work supported by the Office of Ordnance Research.

[2] Reichstein, Cohen, Ruth, and Meldahl, *Helv. Chim. Acta*, **19**, 412 (1936); Siegel, Boyer, and Jay, *J. Am. Chem. Soc.*, **73**, 3237 (1951).

[3] Daub and Castle, *J. Org. Chem.*, **19**, 1571 (1954).

[4] Dev, *J. Indian Chem. Soc.*, **32**, 403 (1955).

2-METHYLBENZYLDIMETHYLAMINE

(Benzylamine, N,N,o-trimethyl-)

$$C_6H_5CH_2N(CH_3)_2 + CH_3I \rightarrow C_6H_5CH_2\overset{\oplus}{N}(CH_3)_3\overset{\ominus}{I}$$

Submitted by W. R. BRASEN and C. R. HAUSER.[1]
Checked by WILLIAM S. JOHNSON, MARY E. MILLS, and A. L. WILDS.

1. Procedure

A. *Benzyltrimethylammonium iodide.* A solution of 135 g. (1 mole) of N,N-dimethylbenzylamine (Note 1) in 200 ml. of commercial absolute ethanol (Note 2) is placed in a three-necked flask fitted with a 125-ml. dropping funnel, a reflux condenser, and a rubber slip-sleeve-sealed stirrer. The openings of the dropping funnel and condenser are protected from atmospheric moisture with drying tubes. While the solution is stirred rapidly, 190 g. (1.34 moles) of methyl iodide contained in the dropping funnel is added slowly at first, then at such a rate as to cause gentle refluxing of the solution. After the addition is complete (30 minutes), the solution is boiled under reflux on the steam bath for 30 minutes more and transferred to a 2-l. Erlenmeyer flask with the aid of an additional 25-ml. of absolute ethanol for rinsing. On cooling to room temperature a large portion of the methiodide crystallizes, and the remaining material is then precipitated by the addition of 1 l. of anhydrous ether with stirring (Note 3). The product is separated by suction filtration, washed with two 100-ml. portions of anhydrous ether (Note 4), and dried in air at room temperature. The yield of benzyltrimethylammonium iodide, m.p. 178–179° (dec.), is 260–274 g. (94–99%), and is pure enough for most purposes (Note 5).

B. *2-Methylbenzyldimethylamine.* (*Caution! This preparation should be conducted in a good hood to avoid exposure to ammonia.*) To 800 ml. of liquid ammonia contained in a 2-l. three-necked flask fitted with a rubber slip-sleeve-sealed wire stirrer and an air-cooled reflux condenser, sodium is added in small pieces until the blue color persists (Note 6). At this point 0.5 g. of granulated ferric nitrate is added, and then 27.8 g. (1.2 g. atoms) of sodium (cut into approximately 0.5-g. pieces), is introduced at such a rate that stirring is not hindered. After all the sodium has been added (about 15 minutes), the mixture is stirred until the blue color disappears and the grayish-black suspension of sodium amide remains (15–20 minutes). Stirring is discontinued, and the mixture is swirled to wash down the mirror of sodium which forms on the upper walls of the flask (Note 7).

A 500-ml. Erlenmeyer flask is charged with 277 g. (1 mole) of benzyltrimethylammonium iodide (prepared as described in Part A) and is connected with a short section of large-diameter rubber tubing to the third neck of the flask containing the sodium amide. Stirring is started, and the salt is shaken in at a steady rate and as rapidly as possible without serious loss of material through the condenser. This operation requires about 10–15 minutes (Note 8). The greenish-violet color, which is first produced by addition of the salt, persists for about 15 minutes after all the material has been added. During this and the subsequent reaction period, more ammonia is added as necessary to maintain the original volume. The mixture is stirred for an additional 2 hours, and then 27 g. (0.5 mole) of ammonium chloride is added cautiously to destroy excess sodium amide.

The flask is fitted with a dropping funnel, and enough (about 100 ml.) water is added, dropwise at first, to bring all the solid material into solution. The mixture is stirred until it reaches room temperature, 70 ml. of ether is added, and the organic layer is separated. The water layer is extracted with two 70-ml. portions of ether, and the combined ether solutions are washed with two 50-ml. portions of saturated salt solution and dried over anhydrous potassium carbonate. After filtration and removal of the ether by distillation, the amine is distilled through a 10-cm. Vigreux column. The fraction, b.p. 72–73°/9 mm., 97–99°/13 mm., or 197–198°/atm., amounts to 134–141.5 g. (90–95% yield), n_D^{20} 1.5050–1.5060 (Note 8). On redistillation all but about 5% of the material is recovered, b.p. 78–79°/12 mm., n_D^{20} 1.5049–1.5052.

2. Notes

1. N,N-Dimethylbenzylamine supplied by the Rohm and Haas Company was used without purification.

2. Methanol may be substituted in the preparation of this as well as other quaternary salts, which, however, are generally more soluble in this medium and therefore require a larger volume of ether for complete precipitation.

3. This isolation procedure produces a granular product which is more desirable for use in the rearrangement reaction than the fluffy material obtained on rapid cooling.

4. The ether should be free of peroxides; otherwise iodine will be liberated from the salt and will color the product.

5. This procedure is suitable for the preparation of the following quaternary salts with changes only in the proportion of ethanol employed: 2-methylbenzyltrimethylammonium iodide, using 350 ml. of ethanol per mole of amine (yield 98–99%), and 2,3-dimethylbenzyltrimethylammonium iodide, using 500 ml. of ethanol per mole of amine (yield 98–99%).

6. Only about 0.05 g. of sodium is required unless the ammonia is quite wet.

7. The complete removal of sodium is indicated by the lack of formation of blue coloration when the solution is swirled over the upper walls of the flask.

8. Essentially the same procedure may be used for the preparation of 2,3-dimethylbenzyldimethylamine from 2-methylbenzyltrimethylammonium iodide except that the time of addition of the latter to the sodium amide is increased to 75 minutes. The checkers found it necessary also to use a larger volume of ammonia. Thus, on a 0.3-mole scale, 1 l. of ammonia was employed, and a total of 600 ml. more ammonia was added during the reaction to maintain the volume at 800–1000 ml.

Owing to the formation of neutral side products, the reaction mixture is extracted with excess $4N$ hydrochloric acid, and the mixture of amines is liberated from the acid solution by neutralization with sodium hydroxide. The amine fraction is taken up in ether and dried over anhydrous sodium sulfate or solid sodium hydroxide. After removal of the ether, the 2,3-dimethylbenzyldimethylamine distils at 101–102°/15 mm., 107–109°/19 mm., n_D^{23} 1.5100–1.5102. The yield is 60–70%.

3. Methods of Preparation

This procedure is based on the method of Kantor and Hauser.[2] 2-Methylbenzyldimethylamine has also been prepared from o-xylyl bromide and hexamethylenetetramine.[2]

[1] Duke University, Durham, North Carolina. Work supported by the Office of Ordnance Research.

[2] Kantor and Hauser, *J. Am. Chem. Soc.*, **73**, 4122 (1951).

O-METHYLCAPROLACTIM

(2H-Azepine, 7-methoxy-3,4,5,6-tetrahydro-)

$$2 \; \overline{[-(CH_2)_5CONH-]} + 2(CH_3)_2SO_4 \rightarrow$$

$$2 \left[\begin{array}{c} OCH_3 \\ | \\ \overline{-(CH_2)_5C{=}NH-} \end{array} \right]^{+} 2OSO_3CH_3^{-}$$

$$2 \left[\begin{array}{c} OCH_3 \\ | \\ \overline{-(CH_2)_5C{=}NH-} \end{array} \right]^{+} 2OSO_3CH_3^{-} + K_2CO_3 \rightarrow$$

$$2 \begin{array}{c} OCH_3 \\ | \\ \overline{-(CH_2)_5C{=}N-} \end{array} + K(OCH_3)SO_3 + CO_2 + H_2O$$

Submitted by RICHARD E. BENSON and THEODORE L. CAIRNS.[1]
Checked by CHARLES C. PRICE, KENNETH N. CAMPBELL, and RICHARD McBRIDE.

1. Procedure

In a 5-l. three-necked flask equipped with a reflux condenser, a sealed mechanical stirrer, and a 1-l. dropping funnel are placed 678 g. (6.0 moles) of ε-caprolactam [2] and 2 l. of benzene (Note 1). The mixture is heated on a steam bath to reflux temperature, during which time all the solid dissolves. At this point 569 ml. (757 g., 6.0 moles) of dimethyl sulfate (Note 1) is added with stirring in a thin stream through the dropping funnel. The rate is about 4 ml. per minute, and the addition requires 2.5 hours. The stirring and heating are continued for an additional 2 hours, during which time two separate phases appear in the reaction mixture. The stirring is then discontinued and the mixture is heated under reflux for an additional 14 hours.

The mixture is cooled to room temperature, and 600 ml. of 50% potassium carbonate is added slowly through the dropping funnel with stirring (Note 2) to the reaction mixture over a period of 30 minutes. After the vigorous evolution of carbon dioxide has subsided, the mixture is stirred (Note 2) slowly for 30 minutes. The potassium methyl sulfate (Note 3) present is removed by filtration, the solid filter cake is washed with two 100-ml. portions of ether, and the washings are combined with the original filtrate. The filtrate is transferred to a 4-l. separatory funnel, the aqueous layer withdrawn, and the organic layer transferred to a 3-l. round-bottomed flask. The ether and benzene are removed by distillation at slightly reduced pressure (200–600 mm.) and the product distilled through an 8-in. Vigreux column. After a fore-run of benzene and O-methylcaprolactim, the fraction boiling at 65–67°/24 mm., n_D^{25} 1.4610, d_4^{25} 0.9598, is collected. The yield is 450–473 g. (59–62%). An additional quantity of the imino ether can be recovered from the fore-run by distillation through an efficient column, making the total yield 463–517 g. (61–68%) (Note 4).

2. Notes

1. Commercial grade reagents were used throughout. The ϵ-caprolactam was obtained from the Explosives Department, E. I. du Pont de Nemours and Company, Inc., Wilmington, Delaware.

2. Vigorous stirring at this point should be avoided or otherwise an emulsion will be formed that is difficult to break.

3. The solid that separated was not identified, but it was presumed to be potassium methyl sulfate.

4. The corresponding O-ethyl derivative (b.p. 81–82°/26 mm.) can be prepared in 52% yield by a similar procedure.

3. Methods of Preparation

O-Methylcaprolactim has been prepared by the reaction of

cyclohexanone oxime, p-toluenesulfonyl chloride, and methanol;[3] and by the procedure[4] described above, which is a modification of the method given in the patent literature.[5]

[1] E. I. du Pont de Nemours and Company, Wilmington, Delaware.
[2] *Org. Syntheses Coll. Vol.* **2**, 371 (1943).
[3] Schmidt and Zutavern, Ger. pat. 532,969 [*Frdl.*, **18**, 3050 (1931)].
[4] Benson and Cairns, *J. Am. Chem. Soc.*, **70**, 2115 (1948).
[5] Schlack, U. S. pat. 2,356,622 [*C. A.*, **39**, 1420 (1945)].

3-METHYLCOUMARONE

(Benzofuran, 3-methyl-)

$$CH_3COCH_2CO_2C_2H_5 + SO_2Cl_2 \rightarrow$$
$$CH_3COCHClCO_2C_2H_5 + HCl + SO_2$$

Submitted by WERNER R. BOEHME.[1]
Checked by JAMES CASON and KENNETH L. RINEHART.

1. Procedure

A. *Ethyl 3-methylcoumarilate.* Dry sodium phenolate (116 g.,

1 mole) (Note 1) and 1 l. of dry thiophene-free benzene (Note 2) are placed in a 2-l. three-necked flask fitted with mechanical stirrer, dropping funnel, and reflux condenser with drying tube. The suspension is heated to the boiling point on the steam bath, heating is moderated, and 165 g. (1 mole) of ethyl α-chloroacetoacetate (Note 3) is added with stirring through the dropping funnel at such a rate as to maintain gentle refluxing (Note 4). Refluxing and stirring are continued on the steam bath for 4 hours after the addition of the chloroester has been completed. The light brown suspension is cooled to room temperature, extracted with two 500-ml. portions of water, and dried superficially by filtration through a layer of anhydrous magnesium sulfate. The solvent is removed by distillation on the steam bath under water-aspirator vacuum. The brown oil remaining consists of crude ethyl α-phenoxyacetoacetate and weighs 188–200 g. (85–90%).

Concentrated sulfuric acid (195 ml.) is placed in a 2-l. three-necked flask immersed in an ice-salt bath and fitted with a dropping funnel, a thermometer, and a mechanical stirrer. Ethyl α-phenoxyacetoacetate (195 g.) is added with stirring through the dropping funnel while the temperature within the flask is maintained below 5°. About 1 hour is required for the addition. The mixture, which solidifies soon after all the ester has been added, is allowed to stand in the ice bath for 1 hour longer. Ice (500 g.) and water (500 ml.) are added with stirring and external cooling. The mixture is then extracted with two 250-ml. portions of benzene. The combined extracts are washed with 100 ml. of water, then with 100 ml. of saturated aqueous sodium bicarbonate solution, and finally are dried superficially by filtration through a layer of anhydrous magnesium sulfate. The solvent is distilled from the dried extracts, and the residue is fractionated under reduced pressure from a 250-ml. Claisen flask fitted with a short air condenser. The pale-yellow oil boils at 162–167°/16 mm. and solidifies on cooling. The product is triturated with a little petroleum ether (b.p. 35–60°) and dried at room temperature. The almost colorless rhombic plates, melting at 49–51°, weigh 60–75 g. (34–42%).

B. *3-Methylcoumarilic acid.* Ethyl 3-methylcoumarilate (70 g.) and 500 ml. of 10% aqueous potassium hydroxide solution are refluxed for 1 hour. The clear yellowish solution is acidified while hot (Note 5) with a slight excess of concentrated hydrochloric acid to precipitate the 3-methylcoumarilic acid. The suspension is cooled to room temperature, and the colorless solid is filtered with suction. The filter cake is resuspended in 500 ml. of cold water, stirred vigorously for several minutes, and filtered again with suction. The colorless powder, after being dried in a desiccator under reduced pressure, weighs 54–57 g. (90–95%) and melts at 192–193°.

C. *3-Methylcoumarone.* Dry 3-methylcoumarilic acid (50 g.) is distilled from a 250-ml. Claisen flask fitted with a long air condenser and immersed in a Wood's metal bath heated slowly to 280°. Carbon dioxide is evolved, and a cloudy liquid distils at 190–220°. The crude product is purified by redistillation through a Vigreux column (Note 6). The clear colorless distillate, weighing 31.5–33 g. (84–88%), boils at 195–197°, n_D^{25} 1.5520.

2. Notes

1. Sodium phenolate may be prepared *in situ* by evaporating molar equivalents of phenol and sodium hydroxide solution in the reaction flask on the steam bath under reduced pressure and drying the residue by heating the flask for several hours longer on the steam bath under reduced pressure. The solid cake of dry sodium phenolate breaks up in the succeeding step of the synthesis.

2. Benzene is conveniently dried by slowly distilling about 20% of it and cooling the residue with protection from atmospheric moisture by use of a calcium chloride tube.

3. Ethyl α-chloroacetoacetate is prepared by the general method of Allihn.[2] *Caution! Since the substance is a severe lachrymator and gases evolved during the preparation are difficult to absorb, the entire preparation should be carried out in a hood.* In a 1-l. three-necked flask fitted with dropping funnel, a thermometer, and a mechanical stirrer, and connected to a gas-absorption trap,[3] is placed 260 g. (2 moles) of technical grade ethyl acetoacetate. Sulfuryl chloride (270 g., 2 moles) is added slowly with stirring and external cooling, the temperature being maintained between 0° and 5°. About 3 hours is required for the addition. The solution is allowed to stand overnight at room temper-

ature, and hydrogen chloride and sulfur dioxide are removed at 40–50° under water-aspirator vacuum. The residual dark-amber liquid is distilled through a short Vigreux column at reduced pressure. After a small fore-run, the ethyl α-chloroacetoacetate distils at 85–89°/17 mm. The yield of colorless liquid is 308–321 g. (93–97%). The checkers, using a 12-cm. unheated Vigreux column, obtained this yield only after redistilling the combined fore-run and after-run. Ethyl α-chloroacetoacetate is currently available from Distillation Products Industries, Rochester, New York.

4. The time of addition is 20–30 minutes. Some external heating is usually necessary to maintain reflux during the addition.

5. If the solution is allowed to cool much below 70° the potassium salt of 3-methylcoumarilic acid crystallizes in the form of colorless needles.

6. The checkers used an 18-in. Podbielniak column with simple wire spiral, heated jacket, and partial-reflux head.

3. Methods of Preparation

3-Methylcoumarone has been prepared by the cyclization of ethyl α-phenoxyacetoacetate followed by hydrolysis and decarboxylation of the resulting ethyl 3-methylcoumarilate,[4] by debromination and rearrangement of 3,4-dibromo-4-methylcoumarin to 3-methylcoumarilic acid followed by decarboxylation,[4,5] by cyclization of phenoxyacetone with concentrated sulfuric acid,[6] and by treatment of 3-coumaranone with methylmagnesium iodide followed by dehydration of the resulting carbinol.[7]

The procedure described is a modification of the method of Hantzsch.[4]

[1] The National Drug Company, Philadelphia, Pennsylvania.
[2] Allihn, *Ber.*, **11**, 567 (1878).
[3] *Org. Syntheses Coll. Vol.* **2**, 4 (1943).
[4] Hantzsch, *Ber.*, **19**, 1290 (1886).
[5] Peters and Simonis, *Ber.*, **41**, 832 (1908).
[6] Stoermer, *Ber.*, **28**, 1254 (1895).
[7] Stoermer and Barthelmes, *Ber.*, **48**, 67 (1915).

METHYL CYCLOPENTANECARBOXYLATE

(Cyclopentanecarboxylic acid, methyl ester)

$$+ \; CH_3ONa \; \rightarrow \qquad \qquad + \; NaCl$$

Submitted by D. W. GOHEEN and W. R. VAUGHAN.[1]
Checked by N. J. LEONARD, M. J. KONZ, W. H. PITTMAN, and
K. L. RINEHART, JR.

1. Procedure

A dry 1-l. three-necked, round-bottomed flask is equipped with an efficient stirrer (Note 1), a spiral reflux condenser, and a dropping funnel, and all openings are protected by calcium chloride drying tubes. A suspension of 58 g. (1.07 moles) of sodium methoxide (Notes 2 and 3) in 330 ml. of anhydrous ether (Note 4) is added, and stirring is begun. To the stirred suspension is added dropwise a solution of 133 g. (1 mole) of 2-chlorocyclohexanone [2] (Notes 5 and 6) diluted with 30 ml. of dry ether. The exothermic reaction is regulated by the rate of addition of the chloroketone; about 40 minutes is required for the addition. After the addition of the chloroketone is complete, the mixture is stirred and heated under reflux for 2 hours (Note 7) and is then cooled. Water is added until the salts are dissolved (Note 8). The ether layer is separated, and the aqueous layer is saturated with sodium chloride. After extraction of the aqueous layer with two 50-ml. portions of ether, the ethereal solutions are combined and washed successively with 100-ml. portions of 5% hydrochloric acid, 5% aqueous sodium bicarbonate solution, and saturated sodium chloride solution. The ether solution is dried over magnesium sulfate, and the magnesium sulfate is removed by filtration and washed with ether. Removal of the ether by distillation at atmospheric pressure leaves the crude ester, which is distilled, with fractionation (Note 9), at 70–73°/48 mm., n_D^{25} 1.4341. The yield of methyl cyclopentanecarboxylate is 72–78 g. (56–61%) (Note 10).

2. Notes

1. A mercury seal and a Hershberg stirrer made from tantalum wire are suitable.

2. Commercial (Matheson Co., Inc.) sodium methoxide is most convenient. The reaction can be run using sodium methoxide prepared from sodium and methanol, but this procedure is more tedious since it requires the removal of a considerable amount of methanol.

3. A slight excess of sodium methoxide should always be used. When an equivalent amount is employed, slightly lower yields are obtained.

4. Commercial anhydrous analytical reagent ether, from sealed cans, was employed by the checkers without further drying.

5. The 2-chlorocyclohexanone [2] employed by the checkers had b.p. 98–99°/14.5 mm., n_D^{25} 1.4826. Purity is critical in determining the yield of methyl cyclopentanecarboxylate.

6. Chlorocyclohexanone is added to the sodium methoxide, since the reverse mode of addition results in lower yields through increased formation of high-boiling condensation products.

7. The submitters report that equivalent yields are obtained when the mixture is allowed to stand overnight after the addition of the chloroketone is complete.

8. The checkers found that approximately 175 ml. of water is required.

9. The checkers employed a Podbielniak column, 0.8 x 125 cm., with tantalum wire spiral and partial reflux head.[3]

10. The residue of higher-boiling material arises from the condensation of both the starting material and the product under the influence of sodium methoxide.

3. Methods of Preparation

Methyl cyclopentanecarboxylate has been prepared by the Favorskiĭ rearrangement of 2-chlorocyclohexanone with sodium methoxide.[4] Other alkyl esters of cyclopentanecarboxylic acid have been prepared by employing the corresponding alkoxides with 2-chlorocyclohexanone.[4-6] The Favorskiĭ reaction has been reviewed elsewhere.[7]

The methyl ester has also been obtained by esterification of cyclopentanecarboxylic acid.[8] The acid, in turn, has been prepared by the Favorskiĭ rearrangement,[6, 7, 9-11] by the reaction of cyclopentyl Grignard reagent with carbon dioxide,[12] by the carbonylation of cyclopentyl alcohol with nickel carbonyl [13] or with formic acid in the presence of sulfuric acid,[14] by the hydrogenation of cyclopentene-1-carboxylic acid prepared from ethyl cyclopentanone-2-carboxylate [15] or from cyclopentanone cyanohydrin,[16] and from cyclopentene and carbon monoxide in the presence of sulfuric acid.[17]

[1] University of Michigan, Ann Arbor, Michigan.

[2] *Org. Syntheses Coll. Vol.* **3**, 188 (1955).

[3] Cason and Rapoport, *Laboratory Text in Organic Chemistry*, 2nd ed., p. 293, Prentice-Hall, Englewood Cliffs, N. J., 1962.

[4] Borowitz, Ph.D. Thesis, Columbia University, 1956.

[5] Loftfield, *J. Am. Chem. Soc.*, **72**, 632 (1950); **73**, 4707 (1951).

[6] Mousseron, Jacquier, and Fontaine, *Bull. soc. chim. France*, **1952**, 767.

[7] Jacquier, *Bull. soc. chim. France*, **1950**, D35.

[8] Kohlrausch and Skrabal, *Monatsh.*, **70**, 44 (1937).

[9] Jackman, Bergman, and Archer, *J. Am. Chem. Soc.*, **70**, 497 (1948).

[10] Favorskiĭ and Boshovskiĭ, *J. Russ. Phys.-Chem. Soc.*, **46**, 1097 (1914); **50**, 582 (1920); *Bull. soc. chim. France*, [4] **18**, 615 (1915).

[11] Mousseron and Jacquier, *Compt. rend.*, **229**, 374 (1949); *Bull. soc. chim. France*, **1950**, 698.

[12] Kharasch and Reinmuth, *Grignard Reactions of Nonmetallic Substances*, p. 918, Prentice-Hall, Englewood Cliffs, N. J, 1954.

[13] Adkins and Rosenthal, *J. Am. Chem. Soc.*, **72**, 4550 (1950).

[14] Koch and Haaf, *Angew. Chem.*, **70**, 311 (1958).

[15] Skraup and Binder, *Ber.*, **62**, 1127 (1929).

[16] Farquharson and Sastri, *Trans. Faraday Soc.*, **33**, 1474 (1937).

[17] Puzitskiĭ, Eĭdus, Ryabova. and Guseva, *Doklady Akad. Nauk S.S.S.R.*, **128**, 555 (1959) [*C. A.*, **54**, 7584 (1960)].

METHYL CYCLOPROPYL KETONE*

(Ketone, cyclopropyl methyl)

$$CH_3\overset{O}{\underset{\displaystyle}{C}}-\underset{\underset{\displaystyle CH_2-CH_2}{|}}{CH}\overset{\overset{\displaystyle O}{\|}}{\underset{\displaystyle}{C}}O + HCl \rightarrow CH_3\overset{O}{\underset{\displaystyle}{C}}CH_2CH_2CH_2Cl + CO_2$$

$$CH_3\overset{O}{\underset{\displaystyle}{C}}CH_2CH_2CH_2Cl + NaOH \rightarrow$$

$$CH_3\overset{O}{\underset{\displaystyle}{C}}-\underset{\displaystyle}{CH}\overset{\overset{\displaystyle CH_2}{\diagdown}}{\diagup}CH_2 + NaCl + H_2O$$

Submitted by GEORGE W. CANNON, RAY C. ELLIS, and JOSEPH R. LEAL.[1]
Checked by R. S. SCHREIBER, WM. BRADLEY REID, JR., and
R. D. BIRKENMEYER.

1. Procedure

A. *5-Chloro-2-pentanone.* A mixture of 450 ml. of concentrated hydrochloric acid, 525 ml. of water, and 384 g. (3 moles) of α-acetyl-γ-butyrolactone (Note 1) and a boiling chip are placed in a 2-l. distilling flask fitted with a 90-cm. bulb-type condenser and a receiver immersed in an ice water bath (Note 2). Carbon dioxide is evolved immediately. Heating of the reaction mixture is begun at once, and the temperature is raised at such a rate that the reaction mixture does not foam into the condenser. In about 10 minutes the color changes from yellow to orange to black, the effervescence begins to subside, and distillation commences. The distillation is continued as rapidly as possible (Note 3). After 900 ml. of distillate has been collected, 450 ml. of water is added to the distilling flask and another 300 ml. of distillate is collected.

The yellow organic layer in the distillate is separated (Note 4), and the aqueous layer is extracted with three 150-ml. portions of ether. The ether extracts are combined with the organic layer and dried for 1 hour over 25 g. of calcium chloride. A saturated calcium chloride layer forms in the bottom. The ether solution is decanted and dried with an additional 25 g. of calcium chloride. The ether is removed by distillation through a 30-cm. column packed with glass helices and fitted with a total condensation, variable take-off head. The residual crude 5-chloro-2-pentanone weighs 287–325 g. (79–90%) (Note 5). When 290 g. of this material is fractionated through a wrapped 12-in. Vigreux column, the major portion boils at 70–72°/20 mm., n_D^{25} 1.4371, and weighs 258–264 g. (89–91%).

B. *Methyl cyclopropyl ketone.* A 2-l. three-necked flask is fitted with a sweep-type stirrer made from ¼-in. iron rod (Note 6), a reflux condenser, and a 500-ml. dropping funnel. In the flask is placed a solution of 180 g. (4.5 moles) of sodium hydroxide pellets in 180 ml. of water. To this solution is added over a period of 15–20 minutes 361.5 g. (342 ml., approximately 3 moles) of the crude 5-chloro-2-pentanone (Note 7). If the reaction mixture does not begin to boil during the addition, boiling is initiated by slight heating of the flask and is continued for 1 hour. Three hundred and seventy milliliters of water is then added slowly to the reaction mixture over a 20-minute period, and the mixture is heated under reflux for an additional hour.

The condenser is arranged for distillation, and a water-ketone mixture is distilled until all the organic layer is removed from the reaction mixture. The aqueous layer of the distillate is saturated with potassium carbonate, and the upper layer of methyl cyclopropyl ketone is separated. The aqueous layer is extracted with two 150-ml. portions of ether. The ether extracts and the ketone layer are combined and dried over 25 g. of calcium chloride for 1 hour. The ether solution is decanted and dried with an additional 25 g. of calcium chloride. The dried ether solution is fractionated through the 30-cm. column described in Part A (Note 8). The yield of methyl cyclopropyl ketone, b.p. 110–112°, n_D^{25} 1.4226, is 193–210 g. (77–83%).

2. Notes

1. Available from U. S. Industrial Chemicals, Inc. Its preparation has been described by several workers.[2–4]

2. Efficient condensation is important; otherwise some of the product is swept out by the carbon dioxide, and the yield of the chloride is decreased.

3. Any delay in distilling the chloride results in a decrease in yield. If the reaction mixture is allowed to stand overnight before removal of the chloride, the yield is less than 50 %.

4. The chloride is usually the bottom layer. However, occasionally some of it also will be found on top. If so, the addition of 50–100 ml. of ether will cause all the chloride to be in the upper layer and will facilitate separation.

5. Corresponding runs beginning with 128 g. (1 mole) of α-acetyl-γ-butyrolactone gave 107–112 g. (89–93%) of crude 5-chloro-2-pentanone. The checkers consistently obtained yields of 79–81% using U. S. Industrial Chemicals, Inc., lactone.

6. A Hershberg stirrer constructed with a ¼-in. iron or stainless-steel shaft is also satisfactory. Glass stirrers are not recommended because they may break.

7. The use of distilled chloride does not result in better over-all yields.

8. A fractionating column is necessary for the separation of the ether-ketone solution. With an ordinary distilling flask, a ketone-ether mixture, b.p. 41°, is obtained with a resultant decrease in the yield of pure ketone. A well-insulated or preferably heated column is necessary for good fractionation.

3. Methods of Preparation

Methyl cyclopropyl ketone has been prepared from ethyl acetoacetate and ethylene bromide,[5] and by the action of methylmagnesium bromide on cyclopropyl cyanide.[6, 7] The procedure described for its preparation from 5-chloro-2-pentanone is similar to that of Zelinsky and Dengin [8] and has been employed by Smith and Rogier.[9] Methyl cyclopropyl ketone also has been obtained by heating trimethyl-γ-acetopropylammonium bromide [10] or 5-bromo-2-pentanone with potassium hydroxide.[11]

5-Chloro-2-pentanone has been prepared by a number of methods,[12] including treatment of acetopropyl alcohol with gaseous hydrogen chloride.[13] The procedure given is essentially that of

Boon [14] and of Forman.[15]   A similar procedure has been used for the preparation of the corresponding bromo- and iodoketones.[14]

[1] University of Massachusetts, Amherst, Massachusetts.
[2] Knunyantz, Chelintzev, and Osetrova, *Compt. rend. acad. sci. U.R.S.S.*, [N.S.] **1**, 312 (1934) [*C. A.*, **28**, 4382 (1934)].
[3] Matukawa et al., Japan. pat. 134,284 [*C. A.*, **35**, 7421 (1941)].
[4] Johnson, U. S. pat. 2,443,827 [*C. A.*, **43**, 678 (1949)].
[5] Freer and Perkin, *J. Chem. Soc.*, **51**, 820 (1887).
[6] Bruylants, *Rec. trav. chim.*, **28**, 180 (1909) [*C. A.*, **3**, 2700 (1909)].
[7] Bruylants, *Bull. soc. chim. Belges*, **36**, 519 (1927) [*C. A.*, **22**, 582 (1928)].
[8] Zelinsky and Dengin, *Ber.*, **55B**, 3360 (1922).
[9] Smith and Rogier, *J. Am. Chem. Soc.*, **73**, 4049 (1951).
[10] Slobodin and Selezneva, *Zhur. Obshcheĭ Khim.*, **23**, 886 (1953) [*C. A.*, **48**, 4449 (1954)].
[11] Favorskaya and Shcherbinskaya, *Zhur. Obshcheĭ Khim.*, **23**, 1485 (1953) [*C. A.*, **48**, 11358 (1954)].
[12] For a summary with references see Huntress, *Organic Chlorine Compounds*, p. 1274, John Wiley & Sons, New York, New York, 1948.
[13] Meshcheryakov and Glukhovtsev, *Izvest. Akad. Nauk S.S.S.R., Otdel. Khim. Nauk*, **1959**, 1490 [*C. A.*, **54**, 1346 (1960)].
[14] Boon, U. S. pat. 2,370,392 [*C. A.*, **39**, 4090 (1945)].
[15] Forman, U. S. pat. 2,397,134 [*C. A.*, **40**, 4394 (1946)].

2-METHYL-2,5-DECANEDIOL

(2,5-Decanediol, 2-methyl-)

$$CH_3(CH_2)_4\underset{\underset{O}{|\rule{0pt}{0pt}}}{CH}(CH_2)_2CO + 2CH_3MgBr \rightarrow$$

$$\left[CH_3(CH_2)_4\underset{\underset{OMgBr}{|}}{CH}(CH_2)_2\underset{\underset{CH_3}{|}}{\overset{\overset{CH_3}{|}}{C}}OMgBr \right] \xrightarrow[\text{HCl}]{\text{H}_2\text{O}}$$

$$CH_3(CH_2)_4CHOH(CH_2)_2\underset{\underset{OH}{|}}{C}(CH_3)_2$$

$$+ \ MgBr_2, MgCl_2$$

Submitted by J. COLONGE and R. MAREY.[1]
Checked by V. BOEKELHEIDE and H. KAEMPFEN.

1. Procedure

A 2-l. flask containing 1.0 l. of anhydrous ether (Note 1) is fitted with a stopper bearing an inlet tube dipping below the surface of the ether and an outlet tube protected by a calcium chloride drying tube. After the ether has been cooled thoroughly in an ice-salt bath, the flask is placed on a balance and cold methyl bromide (Note 2) is introduced through the inlet tube until the gain in weight is 200 g. (2.1 moles).

In a 3-l. three-necked flask, equipped with a sealed mechanical stirrer, reflux condenser, and a pressure-equalizing separatory funnel (Note 3), are placed 48 g. (2 g. atoms) of magnesium turnings, 500 ml. of anhydrous ether, and a small crystal of iodine. The cold methyl bromide solution is transferred to the separatory funnel and slowly added, with stirring. The reaction starts spontaneously, and the remainder of the methyl bromide is added at a rate such that the solution boils gently under reflux. Generally, the addition is complete at the end of 1–2 hours and all the magnesium should be dissolved. After the stirred solution of methylmagnesium bromide is well cooled by using an ice bath, a solution of 78.0 g. (0.5 mole) of γ-nonanoic lactone (Note

4) in 100 ml. of dry ether is added slowly over a period of 30 minutes. When the addition is complete, the mixture is placed on a steam bath and boiled under reflux for 2 hours. Then the condenser is arranged for downward distillation (Note 5), and the ether is removed.

To the thick, syrupy residue is added 200 ml. of benzene, and, after the solution is cooled in an ice bath and the condenser is set for reflux, 350 ml. of water is slowly added through the separatory funnel, with stirring. This is followed by the cautious addition of 325 ml. of a 20% solution of hydrochloric acid, and stirring is continued until all the precipitate dissolves. The organic layer is then separated, and the aqueous layer and flask are washed with 50 ml. of benzene. The combined benzene extracts are washed successively with water, a 5% solution of sodium carbonate, and again with water. Concentration of the benzene solution gives 88.5 g. of an oily residue. Careful fractional distillation (Note 6) of this residue gives, after a fore-run, 53.0 g. (57%) of the pure 2-methyl-2,5-decanediol boiling at 65–69°/2 mm., n_D^{25} 1.4420.

2. Notes

1. Commercial anhydrous ether should be dried over sodium or sodium hydride before use.

2. Commercial methyl bromide (Eastman Kodak Company, Rochester, New York) was used without purification.

3. The separatory funnel is fitted to an adapter tube extending to the bottom of the flask so that the methyl bromide solution is introduced below the surface of the mixture. A drying tube is placed in the condenser outlet.

4. Commercial γ-nonanoic lactone (Aldrich Chemical Co., Milwaukee, Wisconsin) was purified by distillation prior to use. The refractive index of the pure lactone is n_D^{25} 1.4449.

5. As the removal of ether proceeds, the viscous solution becomes difficult to stir and stirring may be stopped without harm.

6. The checkers found that an ordinary Vigreux column was ineffective in separating lower-boiling impurities. An efficient fractionating column 1 m. in length and of 5 mm. I.D. gave excellent results. The infrared spectrum of the product gave no evidence of impurities.

3. Methods of Preparation

The preparation of 2-methyl-2,5-decanediol has not been described elsewhere in the literature.

[1] École de Chimie Industrielle de Lyon and Établissement Descollonges Frères (Lyon).

N-METHYL-2,3-DIMETHOXYBENZYLAMINE

(Benzylamine, 2,3-dimethoxy-N-methyl-)

$$\text{C}_6\text{H}_3(\text{CHO})(\text{OCH}_3)_2 + \text{CH}_3\text{NH}_2 + \text{H}_2 \xrightarrow{\text{Ni}} \text{C}_6\text{H}_3(\text{CH}_2\text{NHCH}_3)(\text{OCH}_3)_2 + \text{H}_2\text{O}$$

Submitted by DON M. BALCOM and C. R. NOLLER.[1]
Checked by HOMER ADKINS and WALTER W. GILBERT.

1. Procedure

To a solution of 41.6 g. (0.25 mole) of purified 2,3-dimethoxybenzaldehyde (Note 1) in 150 ml. of 95% ethanol is added 23.4 g. (0.75 mole) of methylamine in 50 to 75 ml. of water (Note 2). The mixture is heated to boiling and placed in a 500-ml. heavy-walled bottle (Note 3), and 6 g. of Raney nickel catalyst is added (Note 4). The bottle is connected to a low-pressure hydrogenation apparatus,[2] the system is flushed with hydrogen, and the mixture is shaken with hydrogen at an initial pressure of 45 lb. until 0.25 mole of hydrogen is absorbed and the absorption ceases (Note 5).

After the completion of the reduction the catalyst is removed by filtration and the alcohol is evaporated on the steam bath. To the resulting syrupy solution is added about 85 ml. of 3N hydrochloric acid (sufficient to make the solution acid to Congo red paper). The solution is extracted with three 50-ml. portions of ether. The aqueous layer is made alkaline with 50 ml. of 6N sodium hydroxide, and the amine layer is separated. The aqueous layer is extracted with three 50-ml. portions of ether, and the extracts are combined with the amine layer. The ether solution is dried overnight with solid potassium hydroxide and is filtered to remove suspended matter; the flask is rinsed with 25 ml. of ether. The ether is removed on a steam bath, and the amine is distilled under reduced pressure from a modified Claisen flask having a 15-cm. indented column. The N-methyl-2,3-dimethoxybenzylamine distils at 120–124°/8 mm. and is obtained in a yield of 39–42 g. (86–93%) (Notes 6 and 7).

2. Notes

1. The 2,3-dimethoxybenzaldehyde employed was a dark-colored product supplied by the Monsanto Chemical Company or Eastman Kodak Company. It was purified by distilling at reduced pressure and crystallizing from methanol, after which it was colorless and melted at 52–54°.

2. The submitters used 72 g. of a 33% water solution of methylamine; the checkers used 100 g. of a 23% solution.

3. The submitters used a pint hydrogenation bottle wound with 30 ft. of No. 24 Nichrome wire insulated with asbestos paper. Before the reaction the current was adjusted by means of a variable transformer so that the solution was maintained at approximately 70° during the reaction. The checkers used a 500-ml. centrifuge bottle without provision for heating.

4. The submitters used Raney nickel catalyst prepared as described by Mozingo.[3] The checkers used W-6 Raney nickel catalyst.[4]

5. The time required depends upon the activity of the catalyst and the temperature of reaction. The submitters reported that 90–95% of the calculated amount of hydrogen was taken up in 90 minutes at 70° whereas 20 hours was required at room temperature. The checkers found the reaction to be complete after 36 to 41 minutes, the temperature dropping during the period from 70° to 30° since no provision was made for heating. When the whole hydrogenation was carried out at room temperature the period of reaction was 90 minutes with the W-6 Raney nickel catalyst.

6. The submitters state that with a 1:1 mole ratio of aldehyde to amine the yield dropped to 75%, presumably because more of the aldehyde was reduced.

7. The submitters report that similar yields of methylaryl-amines were obtained from benzaldehyde, anisaldehyde, vera-traldehyde, and piperonal.

3. Methods of Preparation

N-Methyl-2,3-dimethoxybenzylamine has been prepared by the reaction of 2,3-dimethoxybenzyl chloride with methylamine [5] and by the catalytic reduction of a mixture of 2,3-dimethoxybenzal-dehyde and methylamine using Adams' platinum catalyst.[6]

[1] Stanford University, Stanford, California.
[2] *Org. Syntheses Coll. Vol.* **1**, 61 (1941).
[3] *Org. Syntheses Coll. Vol.* **3**, 181 (1955).
[4] Adkins and Billica, *J. Am. Chem. Soc.*, **70**, 695 (1948); *Org. Syntheses Coll. Vol.* **3**, 176 (1955).
[5] Douetteau, *Bull. soc. chim. France*, [4] **11**, 655 (1912); Ishii and Mariani, *Anales asoc. quím. argentina*, **33**, 167 (1945) [*C. A.*, **41**, 2399 (1947)].
[6] Sapp, Dissertation, Stanford University, 1940.

N-METHYL-1,2-DIPHENYLETHYLAMINE * AND HYDROCHLORIDE

(Ethylamine, 1,2-diphenyl-N-methyl-)

$$C_6H_5CHO + CH_3NH_2 \rightarrow C_6H_5CH{=}NCH_3 + H_2O$$

$$C_6H_5CH{=}NCH_3 + C_6H_5CH_2MgCl \rightarrow C_6H_5\underset{\underset{CH_3NMgCl}{|}}{C}HCH_2C_6H_5$$

$$C_6H_5\underset{\underset{CH_3NMgCl}{|}}{C}HCH_2C_6H_5 + H_2O \rightarrow C_6H_5\underset{\underset{CH_3NH}{|}}{C}HCH_2C_6H_5 + Mg(OH)Cl$$

$$C_6H_5\underset{\underset{CH_3NH}{|}}{C}HCH_2C_6H_5 + HCl \rightarrow C_6H_5\underset{\underset{CH_3NH \cdot HCl}{|}}{C}HCH_2C_6H_5$$

Submitted by Robert Bruce Moffett.[1]
Checked by N. J. Leonard and L. A. Miller.

1. Procedure

A. *N-Benzylidenemethylamine.* A solution of 31.9 g. (0.3 mole) of benzaldehyde in 80 ml. of benzene contained in a 300-ml. round-bottomed flask is cooled to approximately 10°. To this is added a solution of 14 g. (0.45 mole) of anhydrous methylamine in 50 ml. of benzene (Note 1). On standing, the solution becomes warm and turns milky. After 1 hour the flask is connected to a Dean-Stark water separator ^{2} which is attached to a reflux condenser, and the solvent is caused to reflux until no more water separates (Note 2). The water separator is then replaced by an 8-in. Vigreux column (Note 3), and the solution is distilled under reduced pressure. After removal of the solvent, the product distils at 92–93°/34 mm. The yield is 31–34 g. (87–95%) of colorless liquid, n_D^{25} 1.5497, n_D^{20} 1.5528.

B. *N-Methyl-1,2-diphenylethylamine.* Benzylmagnesium chloride is prepared in a 1-l. three-necked flask as described previously,[3] using 19.5 g. (0.8 g. atom) of magnesium, 92 ml. (102 g., 0.8 mole) of benzyl chloride, and 300 ml. of anhydrous ether. From the dropping funnel a solution of 24.0 g. (0.2 mole) of N-benzylidenemethylamine in 50 ml. of anhydrous ether or benzene (Note 3) is added slowly to the Grignard reagent with stirring. After being stirred at reflux temperature for 2 hours, the mixture is cooled and poured slowly into a mixture of ice and 200 ml. of concentrated hydrochloric acid. The layers are separated, the ether layer is extracted with 100 ml. of water, and the ether solution is discarded. The aqueous layer and water extract are combined, washed with 100 ml. of ether, and made strongly basic with about 600 ml. of 20% aqueous sodium hydroxide solution. The aqueous suspension of magnesium hydroxide is extracted with 800 ml. of ether in a continuous extractor for 48 hours. The ether extract is washed with about 150 ml. of water and dried over anhydrous potassium carbonate. The solution is filtered from the drying agent, the solvent is removed by distillation on a steam bath, and the residue is distilled from a Claisen flask under reduced pressure. The product distils at 83–90°/0.04 mm., 90–93°/0.2 mm., 94–97°/0.3 mm. The yield is 38.4–40.5 g. (91–96%) of colorless liquid, n_D^{25} 1.5640, n_D^{20} 1.5667.

C. *N-Methyl-1,2-diphenylethylamine hydrochloride.* Hydrogen chloride gas (Note 4) is passed into a stirred solution of 30 g. (0.14 mole) of N-methyl-1,2-diphenylethylamine in 500 ml. of anhydrous ether until saturated or until a drop of the ether on moistened pH test paper indicates that it is strongly acid. The hydrochloride separates as a colorless crystalline precipitate. It is collected on a suction filter, washed with ether, and dried. The yield is 34.2–35.1 g. (97–100%), and the product is practically pure, m.p. 184–186°. If desired it can be recrystallized by dissolution in a little methanol followed by addition of absolute ether.

2. Notes

1. Methylamine is most conveniently obtained commercially in cylinders. However, it can be generated by adding 50% aqueous sodium hydroxide solution dropwise to a flask containing the hydrochloride, and allowing the amine to distil. It can also be generated by allowing an aqueous solution of methylamine to drop into a flask containing solid sodium or potassium hydroxide. The methylamine is distilled directly below the surface of a weighed quantity of benzene kept just above its freezing point. The resulting solution is reweighed to determine the concentration, or an aliquot can be titrated with standard acid.

2. The collection of the theoretical amount of water (about 5.4 ml.) requires approximately 3 hours.

3. For the preparation of N-methyl-1,2-diphenylethylamine it is not absolutely necessary to distil the N-benzylidenemethylamine. The dried benzene solution can be used directly.

4. Hydrogen chloride gas is most conveniently obtained in a cylinder which should be connected to the outlet tube through a safety trap. It can be generated if desired.[4]

3. Methods of Preparation

The only practical method for preparing N-benzylidenemethylamine is by the reaction of benzaldehyde with methylamine.[5-7]

N-Methyl-1,2-diphenylethylamine has been prepared in 8% yield by the Leuckart reaction from deoxybenzoin and methylammonium formate [8] and by the present method.[9]

[1] The Upjohn Company, Kalamazoo, Michigan.

[2] *Org. Syntheses Coll. Vol.* **3**, 382 (1955).

[3] *Org. Syntheses Coll. Vol.* **1**, 471 (1941).

[4] *Org. Syntheses Coll. Vol.* 1, 293, 534 (1941).

[5] Zaunschirm, *Ann.*, **245**, 279 (1888).

[6] Kindler, *Ann.*, **431**, 187 (1923).

[7] Campbell, Helbing, Florkowski, and Campbell, *J. Am. Chem. Soc.*, **70**, 3868 (1948).

[8] Goodson, Wiegand, and Splitter, *J. Am. Chem. Soc.*, **68**, 2174 (1946).

[9] Moffett and Hoehn, *J. Am. Chem. Soc.*, **69**, 1792 (1947).

trans-2-METHYL-2-DODECENOIC ACID

(*trans*-2-Dodecenoic acid, 2-methyl-)

$$CH_3(CH_2)_8CH_2\overset{\underset{\displaystyle CH_3}{|}}{C}HCO_2H \xrightarrow[PBr_3]{Br_2} CH_3(CH_2)_8CH_2\overset{\underset{\displaystyle Br}{|}}{\underset{\displaystyle CH_3}{C}}COBr \xrightarrow{CH_3OH}$$

$$CH_3(CH_2)_8CH_2\overset{\underset{\displaystyle Br}{|}}{\underset{\displaystyle CH_3}{C}}CO_2CH_3 \xrightarrow{C_9H_7N}$$

$$CH_3(CH_2)_8CH{=}\overset{\underset{\displaystyle}{|}}{\underset{\displaystyle CH_3}{C}}CO_2CH_3 \xrightarrow[(2)H^+]{(1)KOH} CH_3(CH_2)_8CH{=}\overset{\underset{\displaystyle}{|}}{\underset{\displaystyle CH_3}{C}}CO_2H$$

Submitted by C. FREEMAN ALLEN and MAX J. KALM.[1]
Checked by WILLIAM S. JOHNSON and H. W. WHITLOCK, JR.

1. Procedure

Caution! The bromination and dehydrobromination steps should be carried out in a hood.

Thirty grams (0.14 mole) of 2-methyldodecanoic acid is brominated exactly as described in the preparation of 2-methylenedodecanoic acid (p. 616). The crude product, after the 18-hour heating period, is allowed to cool; then 56 ml. (1.4 moles) of commercial absolute methanol is added at such a rate that the exothermic reaction is kept under control (Note 1). The resultant pale-orange two-phase mixture is heated under reflux with stirring for 15 minutes, cooled, and diluted with 150 ml. of water containing about 2 g. of sodium sulfite. The bromo ester is extracted with two portions (75 ml. and 25 ml.) of petroleum ether

(Note 2). The extracts are combined, washed with water, and dried over anhydrous sodium sulfate. The solvent is removed by flash distillation of the filtered solution from a 250-ml. flask heated on a steam bath, and the last traces of solvent are eliminated by reducing the pressure with a water aspirator. The crude bromo ester remaining amounts to 41.5–42.5 g. and is thermally unstable to distillation at reduced pressure.

The crude bromo ester is mixed with 82.5 ml. (0.70 mole) of pure quinoline (Note 3) in a 250-ml. round-bottomed flask equipped with an air condenser, and the mixture is heated for 3 hours with an oil bath maintained at 160–170°. The black mixture is cooled, treated with 150 ml. of 20% hydrochloric acid, then shaken thoroughly with 200 ml. of petroleum ether (Note 2) until most of the tarry material has dissolved (Note 4). The aqueous phase is separated and washed with an additional 200 ml. of petroleum ether, and the combined organic extracts are washed with 10% hydrochloric acid and then with water. This washing cycle is repeated until the washes are colorless (two acid washes usually suffice); finally, the petroleum ether solution is washed once more with water. The organic layer is dried over anhydrous sodium sulfate, the solvent is flash-distilled as described above, and the residual liquid ester is distilled through a 61-cm. Podbielniak-type column (Note 5). The colorless unsaturated ester (Note 6) distils at 153–154°/14.5 mm. after a small fore-run. The yield is 22–27 g. (70–85.5% based on 2-methyldodecanoic acid), n_D^{25} 1.4520–31, λ_{max} 214 mμ, ϵ 12,300, in hexane (Note 7), λ_{max} 217 mμ, ϵ 12,800, in 95% ethanol.

The ester is hydrolyzed by heating under reflux for 1.5 hours with 50 ml. of 95% ethanol and 4.4 g. of 85% potassium hydroxide (0.066 mole) for each 10 g. (0.044 mole) of ester. Two-thirds of the ethanol is removed by distillation; then the residue is diluted with five volumes of water and acidified to Congo red with 5N sulfuric acid. The organic acid is extracted with two 150-ml. portions of petroleum ether (Note 2), washed with water, and dried over anhydrous sodium sulfate. The petroleum ether is removed from the extracts by flash distillation as described above, and the residual acid is distilled at reduced pressure through a 61-cm. Podbielniak-type column (Note 5). The 2-methyl-2-dodecenoic acid, distilling at 166–168°/3 mm., is obtained in 68–83%

over-all yield (20–24.5 g. from 22–27 g. of ester) (Notes 8 and 9), m.p. 28.5–32° to 29.5–32.4°, λ_{max} 218 mμ, ϵ 12,900, in hexane (Note 7), λ_{max} 216–217 mμ, ϵ 12,800, in 95% ethanol.

2. Notes

1. About 20 minutes is required. It may be desirable to cool the flask during the addition.

2. Commercial hexane, b.p. 65–68°, from petroleum fractionation is satisfactory.

3. Eastman Kodak Company white label grade *synthetic* quinoline was used. The use of coal tar quinoline introduces non-extractable aromatic impurities which contaminate the product.

4. The tarry material apparently consists of quinoline salts and some insoluble polymers.

5. A simplified Podbielniak column was employed.[2] Other columns of comparable efficiency should be suitable.

6. This methyl *trans*-2-methyl-2-dodecenoate is contaminated with 10–15% of methyl 2-methylenedodecanoate. Little, if any, *cis* isomer, which boils at the same temperature as the methylene isomer, is present. The methylene ester, which boils less than 10° below the desired isomer, can be separated by careful fractionation through an efficient column such as the 1.5-m. simple Podbielniak-type column.[2] The esters are more easily fractionated than are the higher-boiling acids.

A significant fraction of the α,β-unsaturated ester is transformed into an unidentified material of similar molecular weight when stored for several weeks in contact with air. The contaminant cannot be separated by ordinary distillation. The acid is much more stable to storage.

7. Optically pure hexane is preferred to ethanol as a solvent for absorption measurements below about 220 mμ. Commercial hexane from petroleum fractionation can usually be rendered optically pure by stirring overnight twice with 15% fuming sulfuric acid (1 lb. of the acid to about 3 l. of hexane), followed by a wash with 5% aqueous sodium hydroxide and distillation from sodium hydroxide pellets.

8. The carbon-carbon double bond is not isomerized to a detectable extent during saponification of the ester or distillation

of the acid; thus the 2-methyl-2-dodecenoic acid will have the same isomeric composition as the sample of ester from which it was obtained, and it consists entirely of α,β-unsaturated isomers.

9. *trans*-2-Methyl-2-dodecenoic acid, freed from isomeric impurities by fractionation through a 1.5-m. Podbielniak-type column, distilled at 146–147°/1.4 mm., λ_{max} 217 mμ, ϵ 14,500, in hexane. Since several consecutive fractions showed these properties, they are believed to represent the properties of the pure isomer. The quantity of such material recovered is entirely dependent on the efficiency of the column and the distillation procedure. Pure samples of solid acids may be readily secured by crystallization. Those prepared [3] by this procedure include: 2-methyl-2-eicosenoic acid, m.p. 66.3–67.6°, λ_{max} 217 mμ, ϵ 13,490, in 54% yield; 2-methyl-2-hexacosenoic acid, m.p. 85.4–86.2°, λ_{max} 217 mμ, ϵ 14,000, in 20% yield; 2,4-dimethyl-2-pentacosenoic acid, m.p. 69.5–70.3°, λ_{max} 218 mμ, ϵ 14,550, in 19% yield. The lowered yields arise from difficulties in purification of the higher molecular weight isomers.

3. Methods of Preparation

2-Methyl-2-dodecenoic acid has been prepared by bromination of methyl 2-methyldodecanoate with N-bromosuccinimide, followed by dehydrobromination with quinoline and saponification of the ester.[4] The present procedure is an adaptation of the method of Cason, Allinger, and Williams.[4]

[1] University of California, Berkeley, California.
[2] Cason and Rapoport, *Laboratory Text in Organic Chemistry*, 2nd ed., p. 293, Prentice-Hall, Inc., Englewood Cliffs, New Jersey, 1962.
[3] Cason and Kalm, *J. Org. Chem.*, **19**, 1836 (1954).
[4] Cason, Allinger, and Williams, *J. Org. Chem.*, **18**, 842 (1953).

METHYLENECYCLOHEXANE
(Cyclohexane, methylene-)

AND N,N-DIMETHYLHYDROXYLAMINE
HYDROCHLORIDE
(Hydroxylamine, N,N-dimethyl-, hydrochloride)

$$\text{Cyclohexyl-}CH_2\text{---}N(CH_3)_2 \xrightarrow{H_2O_2} \text{Cyclohexyl-}CH_2\text{---}\overset{\oplus}{N}(CH_3)_2 \ \ O^{\ominus}$$

$$\text{Cyclohexyl-}CH_2\text{---}\overset{\oplus}{N}(CH_3)_2 \ \ O^{\ominus} \xrightarrow{\text{Heat}} \text{Cyclohexylidene-}CH_2 + (CH_3)_2NOH$$

$$(CH_3)_2NOH + HCl \longrightarrow (CH_3)_2NOH \cdot HCl$$

Submitted by Arthur C. Cope and Engelbert Ciganek.[1]
Checked by William E. Parham and Robert Koncos.

1. Procedure

In a carefully cleaned 500-ml. Erlenmeyer flask, covered with a watch glass, are placed 49.4 g. (0.35 mole) of N,N-dimethyl-cyclohexylmethylamine (Note 1), 39.5 g. (0.35 mole) of 30% hydrogen peroxide, and 45 ml. of methanol. The homogeneous solution is allowed to stand at room temperature for 36 hours. After 2 and 5 hours hydrogen peroxide (39.5-g. portions each time) is added (Notes 2, 3). The excess hydrogen peroxide is destroyed by stirring the mixture with a small amount of platinum black (Note 4) until the evolution of oxygen ceases. The solution is filtered into a 500-ml. round-bottomed flask and concentrated at a bath temperature of 50–60° (Note 5), a water aspirator being used initially and an oil pump finally, until the amine oxide hydrate solidifies. A Teflon-covered stirring bar is introduced into the flask, which is then connected by a 20-cm. column to a trap (reversed to avoid plugging) cooled in Dry Ice-acetone. The flask is heated in an oil bath to 90–100°, and the apparatus is evacuated to a pressure of ca. 10 mm. with stirring of the liquefied amine oxide hydrate. When the content of the flask resolidifies, the temperature of the oil bath is raised to 160°. The amine oxide decomposes completely within about 2 hours at this temperature. Water (100 ml.) is added to the contents of the trap. The olefin layer is removed with a pipette and washed with two 5-ml. portions of water, two 5-ml. portions of ice-cold 10% hydrochloric acid (Notes 6, 7), and one 5-ml. portion of 5% sodium bicarbonate solution. The olefin is cooled in a Dry Ice-acetone bath and filtered through glass wool (Note 8). Distillation over a small piece of sodium through a semimicro column [2] yields 26.6–29.6 g. (79–88%) of methylenecyclohexane, b.p. 100–102° (Note 9), n_D^{25} 1.4474 (Note 10).

The aqueous layer is combined with the two neutral aqueous extracts and acidified by addition of 45 ml. of concentrated hydrochloric acid. The solution is concentrated under reduced pressure at 60–70° until no more distillate comes over. The residue, which solidifies on cooling, is dried in a vacuum desiccator over potassium hydroxide pellets to yield 30.7–32.7 g. (90–96%) of crude N,N-dimethylhydroxylamine hydrochloride, m.p. 103–106° (sealed tube). Crystallization from 40 ml. of isopropyl alcohol gives 26.6–30.7 g. (78–90%) of the pure hydrochloride, m.p. 106–108° (sealed tube).

2. Notes

1. The preparation of N,N-dimethylcyclohexylmethylamine is described on p.339 .

2. Many amines are oxidized much more rapidly than the one used in this preparation, and it is often necessary to cool such reaction mixtures in order to avoid decomposition of the amine oxide or a vigorous exothermic reaction.

3. The completion of the oxidation should be tested by adding 1 drop of an alcoholic phenolphthalein solution and 3 drops of water to 1 drop of the oxidation mixture on a porcelain spot plate. Amine oxides give no color with phenolphthalein.

4. Prepared by the procedure of Feulgen, *Ber.*, **54**, 360 (1921), and added as an aqueous suspension.

5. Some amine oxides decompose at slightly elevated temperatures. In these cases, removal of the solvents should be carried

out at room temperature. It is convenient to use a rotary evaporator for removal of the solvents.

6. Methylenecyclohexane does not rearrange to 1-methylcyclohexene under these conditions. In preparations of those olefins which are more sensitive to acid, washing with acid should be omitted.

7. By making the acid extracts strongly alkaline, extracting the basic material with ether, and distilling the ether extracts, 1.0–2.5 g. (2–5%) of N,N-dimethylcyclohexylmethylamine containing a small amount of a higher-boiling basic compound of unknown structure may be recovered.

8. The material obtained in this manner is of high purity before distillation; it has the same refractive index as the distilled methylenecyclohexane, and no impurities could be detected by gas chromatographic analysis on two different columns.

9. Most of the material boils at 101–102°; the small fore-run has the same refractive index as the main fraction.

10. 1-Methylcyclohexene is completely absent, as shown by gas chromatography on a column packed with 30% by weight of a 52% solution of silver nitrate in tetraethylene glycol (Dow Chemical Company) on 48–100 mesh "firebrick" at 60°. It is estimated that the presence of less than 0.01% of this isomer could have been detected.

3. Methods of Preparation

The present preparation of methylenecyclohexane is an example of an amine oxide pyrolysis. This route from amines to olefins in many cases yields pure olefins where the alternative method, the Hofmann exhaustive methylation reaction, is accompanied by some rearrangement to more stable isomeric olefins.

Methylenecyclohexane has been prepared by treatment of cyclohexylmethyl iodide with alcoholic potassium hydroxide solution,[3] by thermal decarboxylation of cyclohexylideneacetic acid and of cyclohexane-1-acetic acid, [4–6] and of cyclohexane-1,1-diacetic acid; [7] by pyrolysis of the xanthate,[8] the acetate,[9–13] and the stearate [14] of cyclohexanemethanol; by the action of triphenylphosphine-methylene on cyclohexanone; [15] by the pyrolysis of N,N-dimethylcyclohexylmethylamine methohydroxide [16, 17] and of N,N-dimethylcyclohexylmethylamine N-oxide; [16, 17] and by treatment of benzylhexahydrobenzyldimethylammonium bromide with sodium amide in liquid ammonia.[18]

N,N-Dimethylhydroxylamine has been prepared by the action of methylmagnesium iodide on ethyl nitrate.[19]

[1] Massachusetts Institute of Technology, Cambridge 39, Massachusetts. Supported by the Office of Ordnance Research, U. S. Army, under Contract No. DA–19–020–ORD–4542.

[2] Gould, Holzman, and Niemann, *Anal. Chem.*, **20**, 361 (1948).

[3] Faworsky and Borgmann, *Ber.*, **40**, 4863 (1907).

[4] Wallach and Isaac, *Ann.*, **347**, 328 (1906); Wallach, *ibid.*, **365**, 255 (1909).

[5] Linstead, *J. Chem. Soc.*, **1930**, 1603.

[6] Šorm and Beránek, *Chem. listy*, **47**, 708 (1953) [*C. A.*, **49**, 194 (1955)].

[7] Vogel, *J. Chem. Soc.*, **1933**, 1028.

[8] Aleksandrovich, *J. Gen. Chem. U.S.S.R.*, **3**, 48 (1933) [*C. A.*, **28**, 2337 (1934)].

[9] Arnold and Dowdall, *J. Am. Chem. Soc.*, **70**, 2590 (1948).

[10] van der Bij and Kooyman, *Rec. trav. chim.*, **71**, 837 (1952).

[11] Levina and Mezentsova, *Uchenye Zapiski Moskov. Gosudarst. Univ. im. M. V. Lomonosova*, **No. 132**, *Org. Khim.*, **7**, 241 (1950) [*C. A.*, **49**, 3847 (1955)].

[12] Bailey, Hewitt, and King, *J. Am. Chem. Soc.*, **77**, 357 (1955).

[13] Nevitt and Hammond, *J. Am. Chem. Soc.*, **76**, 4124 (1954).

[14] Waldmann and Schubert, *Chem. Ber.*, **84**, 139 (1951).

[15] Wittig and Schöllkopf, *Chem. Ber.*, **87**, 1318 (1954).

[16] Cope, Bumgardner, and Schweizer, *J. Am. Chem. Soc.*, **79**, 4729 (1957).

[17] Baumgarten, Bower, and Okamoto, *J. Am. Chem. Soc.*, **79**, 3145 (1957).

[18] Bumgardner, *Chem. & Ind. (London)*, **1958**, 1555.

[19] Hepworth, *J. Chem. Soc.*, **119**, 251 (1921).

2-METHYLENEDODECANOIC ACID

(Dodecanoic acid, 2-methylene-)

$$CH_3(CH_2)_9\overset{\overset{\displaystyle CH_3}{|}}{C}HCO_2H \xrightarrow[PBr_3]{Br_2} CH_3(CH_2)_9\overset{\overset{\displaystyle CH_3}{|}}{\underset{\underset{\displaystyle Br}{|}}{C}}COBr \xrightarrow{KOC(CH_3)_3}$$

$$CH_3(CH_2)_9\overset{\overset{\displaystyle CH_2}{\|}}{C}CO_2C(CH_3)_3 \xrightarrow[(2)\ H_2SO_4]{(1)\ NaOH} CH_3(CH_2)_9\overset{\overset{\displaystyle CH_2}{\|}}{C}CO_2H$$

Submitted by C. FREEMAN ALLEN and MAX J. KALM.[1]
Checked by WILLIAM S. JOHNSON and KENNETH L. WILLIAMSON.

1. Procedure

Caution! The bromination step should be carried out in a hood, and appropriate precautions should be employed in handling potassium (Note 1).

A 250-ml. three-necked flask fitted (glass joints) with a sealed mechanical glass stirrer, an addition funnel, and a reflux condenser capped with a calcium chloride drying tube is charged with 30.0 g. (0.140 mole) of 2-methyldodecanoic acid (Note 2) and 13.7 ml. (0.144 mole) of phosphorus tribromide (Note 3). Stirring is commenced, and 14.6 ml. (0.284 mole) of dry bromine (Note 4) is introduced slowly from the addition funnel until the reaction mixture retains a deep bromine coloration. The addition requires about 10 minutes to this stage (Note 5). The remainder of the bromine is then added all at once, and the flask heated in a bath maintained at 85–90° (Note 6) for 1.5 hours. An additional 3.6 ml. (0.07 mole) of bromine is then added and the heating at 85–90° continued for 18 hours. The mixture is cooled to room temperature and poured into a 1-l. separatory funnel containing about 150 ml. of water and 200 g. of cracked ice. The transfer is completed with the addition of 150 ml. of benzene, and the separatory funnel is shaken vigorously for about 10 minutes, during which time most of the ice melts and the originally denser organic phase becomes the upper phase.

The aqueous layer is separated and washed with 100 ml. of benzene, while the organic layer is vigorously shaken with another 200-ml. portion of ice water. This ice-water wash is also shaken with the 100-ml. portion of benzene and is then discarded. The combined benzene extracts are filtered through anhydrous sodium sulfate to remove suspended water; then the benzene and residual bromine are removed under reduced pressure (water aspirator) at a bath temperature of 70° or below (Note 7). The crude bromoacyl bromide is added slowly, at room temperature, to a solution of 13.7 g. (0.35 g. atom) of potassium metal (Note 1) in 300 ml. of dry *tert*-butyl alcohol (Note 1) contained in a well-dried 1-l. flask fitted with a reflux condenser that is protected from moisture with a calcium chloride drying tube. The resultant suspension is heated at reflux for 1 hour, cooled, and diluted with three volumes of water. The mixture, containing insoluble *tert*-butyl 2-methylenedodecanoate, is extracted with two 100-ml. portions of low-boiling petroleum ether (Notes 8 and 9). The combined petroleum ether extracts are washed with water and filtered through anhydrous sodium carbonate into a distilling flask from which the solvent is flash-distilled. The residual *tert*-butyl 2-methylenedodecanoate is then distilled through a 61-cm. Podbielniak-type column (Note 10) at 129–130°/3.0 mm.; the yield is 18.5–21 g. of semipurified ester, n_D^{25} 1.4405–1.4413 (Notes 11 and 12).

The *tert*-butyl 2-methylenedodecanoate is hydrolyzed by heating under reflux with ethanolic potassium hydroxide for 6 hours; 40 ml. of 95% ethanol and 3.7 g. (0.056 mole) of 85% potassium hydroxide are used for each 10 g. (0.037 mole) of the ester. The hydrolysis mixture is cooled, diluted with three volumes of water, and extracted with two 100-ml. portions of petroleum ether which are discarded. The aqueous phase is acidified to Congo red with 5N sulfuric acid, and the 2-methylenedodecanoic acid is extracted with two 150-ml. portions of low-boiling petroleum ether (Note 8). These extracts are combined, washed with three 100-ml. portions of water to ensure complete removal of *tert*-butyl alcohol, and dried by filtration through anhydrous sodium sulfate. The solvent is removed by flash distillation, and the acid is subjected to rapid distillation from a Claisen flask (Note 13). 2-Methylenedodecanoic acid containing less than 5% of

2-methyl-2-dodecenoic acid is collected at 149–151°/1.7 mm., m.p. about 32°, λ_{max} 209 mμ, ϵ 7800, in hexane (Note 14). The yield is 10.5–12 g. (35–40% from 2-methyldodecanoic acid), not including the acid obtained on acidification of the dehydrohalogenation mixture (Note 9).

2. Notes

1. The precautions for handling potassium and the procedure for preparing anhydrous potassium *tert*-butoxide have already been described (p.132).

2. 2-Methyldodecanoic acid was prepared as follows according to the method of Cason, Allinger, and Williams.[2] A three-necked flask fitted with a sealed mechanical stirrer, dropping funnel, and efficient reflux condenser is charged with 1.4 l. of absolute ethanol; then, while stirring, 48.3 g. (2.1 g. atoms) of sodium is added gradually in small pieces. The dropping funnel is charged with 383 g. (2.2 moles) of diethyl methylmalonate (Matheson, Coleman and Bell) which is added to the sodium ethoxide solution over a period of about 20 minutes; then the mixture is boiled under reflux for 5 minutes. The dropping funnel is charged with 442 g. (2.0 moles) of *n*-decyl bromide (Eastman Kodak Company, white label brand) which is added to the mixture as rapidly as is allowed by the exothermic reaction. After the addition is complete (about 20 minutes), the mixture is boiled under reflux for 2 hours, then neutralized with a few drops of glacial acetic acid. About two-thirds of the alcohol is removed by distillation, and 2 l. of water is added to the residue. The organic phase is separated, and the aqueous phase extracted with three 250-ml. portions of benzene. The organic phase and extracts are combined, washed with water, and dried over anhydrous sodium sulfate. The residue obtained upon evaporation of the solvent is treated with a solution of 447.5 g. of 85% potassium hydroxide pellets in 3.5 l. of 95% ethanol and the mixture heated at reflux, with stirring, for 4 hours. About two-thirds of the solvent is removed by distillation, 3 l. of water is added, followed by sufficient (about 2 l.) 6N sulfuric acid (cooling is necessary) to bring the pH of the solution to 1–2. The organic phase is separated, and the aqueous phase (contain-

ing some precipitated sulfates) is extracted with two portions of ether. The organic phase and extracts are combined, washed with water, then with saturated sodium chloride solution, and finally dried over anhydrous sodium sulfate. The residue obtained upon evaporation of the ether is heated at 180–190°, at which temperature decarboxylation occurs smoothly over a 20-minute period. The material is then distilled from a modified Claisen flask. The yield of product, b.p. 159–161°/4.4 mm., is 262–318 g. (61–74%), n_D^{25} 1.4404–1.4408.

3. Phosphorus tribromide (Eastman Kodak Company, white label brand) was freshly distilled before use. The full molar equivalent accelerates the desired α-bromination.

4. Bromine is conveniently dried over phosphorus pentoxide, then filtered into the addition funnel through a plug of glass wool.

5. The first mole equivalent of bromine reacts with the phosphorus tribromide to form the solid pentabromide which, in turn, is rapidly consumed in the formation of the acyl bromide.

6. The temperature is critical. At lower temperatures the reaction is very slow, and at higher temperatures partial dehydrobromination and subsequent allylic bromination result in contamination of the product with dienoic acid, as evidenced by the characteristic intense absorption at 275 mμ in hexane.

7. Higher temperatures at this point tend to promote dehydrobromination under the acidic conditions with the formation of 2-methyl-2-dodecenoic acid.

8. Commercial hexane from petroleum fraction (b.p. 65–68°) is satisfactory.

9. Acidification of the alkaline aqueous layer to Congo red with 5N sulfuric acid liberates 2–4 g. of crude 2-methylenedodecanoic acid which can be isolated by extraction with petroleum ether as described below. Distillation of such material gave a product which did not crystallize readily.

10. A simplified Podbielniak column [3] was employed. Other columns of comparable efficiency should be suitable.

11. If the distillation is carried out in a Claisen flask, the final product is impure, apparently contaminated with about 10% of 2-methyl-2-dodecenoic acid.

12. Pure *tert*-butyl 2-methylenedodecanoate may be isolated by refractionation of this product through an efficient column.

The yield in such an experiment was 15–17 g. of material with n_D^{25} 1.4400.

13. Fractionation through a column results in some isomerization.

14. 2-Methylenedodecanoic acid, free from any isomeric impurities, can be obtained by recrystallization from acetone. The pure acid is recovered in over 70% yield, m.p. 33.3–34.2°, λ_{max} 210 mμ, ϵ 7500, in hexane. It is very difficult to obtain pure methylene acid from Claisen-distilled *tert*-butyl ester (cf. Note 11).

3. Methods of Preparation

This preparation is based on the method of Cason, Allinger, and Williams.[2]

[1] University of California, Berkeley, California.

[2] Cason, Allinger, and Williams, *J. Org. Chem.*, **18**, 842 (1953).

[3] Cason and Rapoport, *Laboratory Text in Organic Chemistry*, 2nd ed., p. 293, Prentice-Hall, Inc., Englewood Cliffs, New Jersey, 1962.

1-METHYL-3-ETHYLOXINDOLE *

(Oxindole, 3-ethyl-1-methyl)

$$CH_3CH_2CHBrCO_2H + SOCl_2 \rightarrow$$
$$CH_3CH_2CHBrCOCl + HCl + SO_2$$
$$CH_3CH_2CHBrCOCl + 2C_6H_5NHCH_3 \rightarrow$$
$$C_6H_5N(CH_3)COCHBrCH_2CH_3 + C_6H_5NHCH_3 \cdot HCl$$

Submitted by M. W. Rutenberg and E. C. Horning.[1]
Checked by William S. Johnson and C. A. Erickson.

1. Procedure

In a 500-ml. round-bottomed flask fitted with a calcium chloride drying tube are placed 226 g. (1.35 moles) of α-bromo-*n*-butyric acid (Note 1) and 284 g. (175 ml., 2.39 moles) of thionyl chloride (Note 2). A small piece of porous plate is added, and the reaction mixture is allowed to stand at room temperature for 48 hours (Note 3). The excess thionyl chloride is removed by distillation, and the acid chloride is collected at 147–153° (Note 4). The yield of colorless product is 168–197 g. (67–78%).

In a 1-l. three-necked flask fitted with a Hershberg wire stirrer, a reflux condenser equipped with a calcium chloride drying tube, and a dropping funnel, are placed 237 g. (2.21 moles) of methylaniline (Note 5) and 300 ml. of dry benzene (Note 6). Stirring is started, and the reaction mixture is cooled in an ice-water bath during the dropwise addition of 197 g. (1.06 moles) of α-bromo-*n*-butyryl chloride diluted with approximately 40 ml. of dry benzene. The addition requires approximately 1 hour, and the reaction mixture becomes thick owing to the separation of methylaniline hydrochloride. The mixture is stirred for an additional 30 minutes and is then set aside protected by a calcium chloride tube for approximately 12 hours. The colorless methylaniline hydrochloride is removed by filtration with suction and washed with two 25-ml. portions of benzene. The combined filtrate and washings are washed with three 100-ml. portions of 5% hydrochloric acid to remove excess methylaniline, and then with two 100-ml. portions of water (Note 7). The benzene layer is dried over anhydrous magnesium sulfate. The drying agent is removed by filtration and the solvent is removed by distillation (the last traces of solvent are removed with an aspirator) to give N-methyl-α-bromo-*n*-butyranilide (Note 8).

The crude N-methyl-α-bromo-*n*-butyranilide is placed in a 500-ml. three-necked round-bottomed flask equipped with a Hershberg wire stirrer and a reflux condenser fitted with a calcium chloride drying tube. To the stirred liquid, cooled in an ice-water bath, is added 281 g. (2.1 moles) of aluminum chloride (reagent grade, powdered) in portions over a period of about 30 minutes (Note 9). A thermometer is then introduced, and the cooling bath is replaced by a source of heat (Note 10). The reaction commences at about 80° with the evolution of hydrogen bromide (Note 11) and becomes very vigorous at 95–105°. At this point, no external heat need be applied, because the heat of reaction carries the temperature to 110–115°. When the reaction slows down, the temperature is raised to 160–170°, and the dark mixture is then allowed to cool to about 80–90°. The reaction mixture is poured *cautiously* into a 3-l. beaker about one-fourth full of cracked ice. Additional ice is added as required. The last traces of product are removed from the flask with the aid of hydrochloric acid and ice. Concentrated hydrochloric acid (75 ml.) is added to aid the decomposition of the aluminum chloride complex. The brown oil is separated after adding 75 ml. of ether, and the aqueous layer is extracted with two 100-ml. portions of ether. The combined organic layers are washed with two 75-ml. portions of 5% hydrochloric acid, two 100-ml. portions of water, two 75-ml. portions of saturated sodium bicarbonate solution, and two 100-ml. portions of water. After drying over anhydrous magnesium sulfate the ether is distilled, and the residue is distilled under reduced pressure. The main fraction is collected at 103–107°/0.5 mm. The yield is 126–131 g. (68–71% based on the acid chloride) of pale yellow 1-methyl-3-ethyloxindole, n_D^{25} 1.5569–1.5580.

2. Notes

1. The α-bromo-n-butyric acid obtained from the Eastman Kodak Company, which had a boiling range of 2°, proved to be satisfactory.

2. Commercial thionyl chloride was distilled before use.

3. The reaction mixture is allowed to stand under an efficient hood during this period. A yield of 166 g. of product may be obtained by refluxing the mixture for 2.5 hours in a 1-l. flask equipped with condenser and drying tube.

4. The submitters used apparatus fitted with ground-glass joints for this and subsequent operations.

5. Redistilled N-methylaniline was used, b.p. 195–196°.

6. The benzene was dried by distillation.

7. Sodium chloride was added to the hydrochloric acid solution and to the wash water to reduce emulsion formation.

8. The crude N-methyl-α-bromo-n-butyranilide may be used directly for the ring closure or, if desired, may be purified by distillation (b.p. 117–118°/0.4 mm., 125–127°/0.8 mm., 175–184°/24 mm.).

9. The aluminum chloride is introduced into the reaction mixture without exposure to the atmosphere.[2] The mixture becomes quite viscous, and a powerful stirrer is needed.

10. An electric heating jacket was found to be most satisfactory.

11. The reaction should be carried out under an efficient hood with provision for the absorption of evolved hydrogen bromide.

3. Methods of Preparation

This method is a general one for the preparation of oxindoles and 1-methoxindoles, and is based on the procedure of Stollé[3] as developed by Julian and Pikl.[4]

1-Methyl-3-ethyloxindole has been prepared previously by methylation of 3-ethyloxindole with methyl iodide.[5] It has also been made from 1-methyloxindole by acylation with ethyl acetate in the presence of sodium ethoxide, followed by hydrogenation over a palladium catalyst.[6]

[1] University of Pennsylvania, Philadelphia 4, Pennsylvania.

[2] Fieser, *Experiments in Organic Chemistry*, 3rd ed., p. 265, D. C. Heath and Company, Boston, 1955.

[3] Stollé, *J. prakt. Chem.*, [2] **128**, 1 (1930).

[4] Julian and Pikl, *J. Am. Chem. Soc.*, **57**, 563 (1935). Cf. Porter, Robinson, and Wyler, *J. Chem. Soc.*, **1941**, 620.

[5] Brunner, *Monatsh.*, **18**, 545 (1897).

[6] Julian, Pikl, and Wantz, *J. Am. Chem. Soc.*, **57**, 2026 (1935).

9-METHYLFLUORENE

(Fluorene, 9-methyl-)

$$\xrightarrow[\text{CH}_3\text{ONa}]{\text{CH}_3\text{OH}}$$

H   H   CH$_3$ H

Submitted by KURT L. SCHOEN and E. I. BECKER.[1]
Checked by WILLIAM S. JOHNSON and V. B. HAARSTAD.

1. Procedure

An 850-ml. steel bomb is charged with a solution of 23 g. (1.0 g.

atom) of sodium in 450 ml. of absolute methanol and 113 g. (0.68 mole) of fluorene (Note 1). The vessel is then closed, heated to 220° (Note 2), and rocked for 16 hours (Note 3). The reaction vessel is allowed to cool, and the contents are transferred to a 2-l. beaker with the aid of small volumes of benzene and then water to complete the transfer. The reaction mixture is diluted with an equal volume of water, neutralized with concentrated hydrochloric acid, and extracted with three 150-ml. portions of benzene. The combined benzene extracts are washed with three 200-ml. portions of water, and the solvent is removed by distillation at atmospheric pressure. The residue is recrystallized from methanol (1 l. per 100 g. of solute) to give 96–106 g. (78–86% yield) (Note 4) of colorless 9-methylfluorene, m.p. 44–45° (Note 5).

2. Notes

1. A commercial grade of fluorene was purified by crystallization from methanol until the m.p. was 113–114°.

2. In the checkers' experience the temperature must not be below 220° or a diminution in yield will result.

3. Without rocking, the crude product is colored and the yield is slightly lower.

4. This is a total yield of material obtained in 2–3 crops. In a typical run the first crop amounted to 90 g., m.p. 44–45°, and the second, obtained on concentrating and cooling the mother liquor, amounted to 16 g., m.p. 44–45°.

The crude product may alternatively be purified by rapid distillation at reduced pressure to give 114–116 g., b.p. 95–100°/1 mm. Redistillation affords 102–105 g. (83–86% yield), b.p. 96–98°/0.6 mm., of colorless 9-methylfluorene which solidifies.

5. The submitters state that the procedure is general and has been carried out with normal alcohols from C_1 to C_7. In an analogous procedure 10 g. of fluorene was treated with 40 ml. of alcohol and 2.3 g. of sodium (in a Carius tube) to give 52–84% of redistilled 9-alkylfluorene. 9-n-Octadecylfluorene was prepared from 13.3 g. of fluorene, 16.2 g. of n-octadecyl alcohol, and 2 g. of sodium. In this case the reaction was carried out in a flask (equipped with a condenser) that was heated in an oil bath

for 16 hours at 210°. The solid product was purified by crystallization.

Product	B.P./1 mm.	n_D^{25}	d_4^{25}	Yield, %
9-Ethylfluorene	123–124°	1.6180	1.0508	84
9-n-Propylfluorene	126–128°	1.6050	1.0326	72
9-n-Butylfluorene	140°	1.5956	1.0197	78
9-n-Pentylfluorene	144–146°	1.5929	1.0153	66
9-n-Hexylfluorene	156–158°	1.5757	0.9900	68
9-n-Heptylfluorene	163–165°	1.5717	0.9827	58
9-n-Octadecylfluorene	65–66.4° (m.p.)	—	—	92

3. Methods of Preparation

Generally, fluorene has been alkylated in the 9-position by reaction of 9-acyl- or ester-substituted fluorenes with sodium alkoxide and an alkyl halide followed by removal of the activating group, by treating a 9-fluorenyl organometallic compound with an alkyl halide, by reduction of a 9-fluorenylidene derivative, by hydrogenolysis of a 9-alkyl-9-hydroxyfluorene, by hydrogenolysis of a 9-halogen-9-alkylfluorene, and by cyclization of a diphenylalkylcarbinol with phosphorus pentoxide.[2]   The present procedure is based on the method of Shoen and Becker.[3]

9-Methylfluorene has been prepared by cleavage of ethyl 9-methyl-9-fluorenylglyoxylate,[4] by the decarboxylation of 9-methylfluorene-9-carboxylic acid,[4] by the decarboxylation of 9-fluorenylacetic acid,[5] by the cleavage of 9-methyl-9-acetylfluorene with alcoholic potassium hydroxide or soda-lime,[6] by the reduction of 9-methyl-9-fluorenol with hydriodic acid in acetic acid,[7] by the reaction of 9-fluorenyllithium [8] or -sodium [9] with methyl iodide or methyl sulfate,[9] by the cyclization of diphenylmethylcarbinol over platinum-on-carbon at 300°,[10] by the reaction of ethyl 9-methoxymethyl-9-fluorenylcarboxylate,[11] by the diazotization and heating of 2-ethyl-2-aminobiphenyl,[12] by the dehydration and then reduction of 9-methyl-9-fluorenol,[13] by the thermal decomposition of tetramethylammonium 9-fluorenide,[14] and by recovery from coal tar.[15]

[1] Polytechnic Institute of Brooklyn, Brooklyn, New York.

² Josephy and Radt, *Elsevier's Encyclopedia of Organic Chemistry*, Vol. 13, Series III, pp. 29 ff., Elsevier Publishing Company, Inc., New York, 1946.
³ Shoen and Becker, *J. Am. Chem. Soc.*, **77**, 6030 (1955).
⁴ Wislicenus and Mocker, *Ber.*, **46**, 2772 (1913).
⁵ Mayer, *Ber.*, **46**, 2579 (1913).
⁶ Meerwein, *Ann.*, **396**, 242 (1913).
⁷ Wanscheidt and Moldavski, *Ber.*, **64**, 917 (1931).
⁸ Blum-Bergmann, *Ann.*, **484**, 26 (1930).
⁹ Greenhow, White, and McNeil, *J. Chem. Soc.*, **1951**, 2848.
¹⁰ Zelinsky and Gawerdowskaja, *Ber.*, **61**, 1049 (1928).
¹¹ Pinck and Hilbert, *J. Am. Chem. Soc.*, **69**, 723 (1947).
¹² Mascarelli and Longo, *Gazz. chim. ital.*, **71**, 397 (1941).
¹³ Badger, *J. Chem. Soc.*, **1941**, 535.
¹⁴ Wittig, Heintzeler, and Wetterling, *Ann.*, **557**, 201 (1947).
¹⁵ I. G. Farbenind, A.-G. (Pier and Schoenemann), Ger. pat. 659,878 (May 12, 1938) [*C.A.*, **32**, 6844 (1938)].

5-METHYLFURFURYLDIMETHYLAMINE

(Furfurylamine, N,N,5-trimethyl-)

$$CH_3 \underset{O}{\underset{}{\boxed{}}} + CH_2O + HN(CH_3)_2 \rightarrow$$

$$CH_3 \underset{O}{\underset{}{\boxed{}}} CH_2N(CH_3)_2 + H_2O$$

Submitted by ERNEST L. ELIEL and MILTON T. FISK.[1]
Checked by JAMES CASON, MARY S. NAKATA, and WILLIAM G. DAUBEN.

1. Procedure

To 200 ml. of glacial acetic acid in a 1-l. round-bottomed flask is added slowly, with cooling in an ice bath, 151 ml. (54 g. of dimethylamine, 1.2 moles) of 40% aqueous dimethylamine solution, followed by 90 ml. (36 g. of formaldehyde, 1.2 mole) of 37% aqueous formaldehyde (formalin) solution. The flask is removed from the ice bath and equipped with a reflux condenser, through which 82 g. (90 ml., 1 mole) of 2-methylfuran (Note 1) is added all at once. Upon gentle swirling of the flask, an exothermic reaction may set in spontaneously; if it does not, the flask is heated on a steam bath until reaction commences. In any event, the spontaneous reaction is allowed to proceed without further external heating. When it ceases, the reaction mixture is heated on a steam bath for another 20 minutes, cooled, and without delay poured into a cold solution of 250 g. of sodium hydroxide (Note 2) in 800 ml. of water.

The reaction mixture is steam-distilled until the distillate is only faintly alkaline (Note 3). To the distillate is added sodium hydroxide (Note 2) to the extent of 10 g. for each 100 ml. of distillate. The strongly alkaline solution is cooled and extracted with two 300-ml. portions of ether. The combined ether layers are dried over 25 g. of solid potassium hydroxide (Note 4), decanted, and concentrated. The residue is distilled under reduced pressure; the yield of 5-methylfurfuryldimethylamine boiling at 62–63°/13 mm. is 96–106 g. (69–76%), n_D^{25} 1.4616–1.4620.

2. Notes

1. The submitters used du Pont 2-methylfuran. Since the stabilizer contained in this material inhibits the reaction, it should be removed before use. The liquid is stored over solid potassium hydroxide (10 g. for each 100 ml. of 2-methylfuran) for 24 hours, decanted, and stored over the same amount of fresh potassium hydroxide at least overnight. A more rapid but less convenient method of removing the stabilizer is to extract the liquid with 10% aqueous potassium hydroxide until the extracts are only faintly colored.

2. The submitters used commercial sodium hydroxide flakes.

3. The steam distillate amounts to 2–3 l.

4. If a large aqueous phase appears, the amine should be decanted and dried further over a fresh portion of potassium hydroxide pellets.

3. Methods of Preparation

The first preparation of 5-methylfurfurylamines by the Mannich reaction was by Holdren and Hixon.[2] The present modification has been published [3] without the steam-distillation step, which facilitates separation of the product.

[1] University of Notre Dame, Notre Dame, Indiana.
[2] Holdren and Hixon, J. Am. Chem. Soc., 68, 1198 (1946).
[3] Eliel and Peckham, J. Am. Chem. Soc., 72, 1210 (1950).

3-METHYL-2-FUROIC ACID
(2-Furoic acid, 3-methyl-)

AND 3-METHYLFURAN
(Furan, 3-methyl-)

Submitted by D. M. BURNESS.[1]
Checked by JAMES CASON and ROBERT B. HUTCHISON.

1. Procedure

A. *3-Methyl-2-furoic acid.* A mixture of 35 g. (0.25 mole) of methyl 3-methyl-2-furoate (p. 649) and 80 ml. of aqueous 20% sodium hydroxide is heated under reflux for 2 hours. The solution is cooled, acidified with about 50 ml. of concentrated hydrochloric acid (sp. gr. 1.18), stirred vigorously for a few minutes to ensure freeing of the acid from its salt, then cooled to room temperature before the product is collected by suction filtration. The product is washed with about 25 ml. of water used in two portions, drained well on the funnel, then dried. The yield of essentially pure 3-methyl-2-furoic acid is 28.5–29.5 g. (90–93%), m.p. 134–135° (Note 1).

B. *3-Methylfuran.* A mixture of 25 g. of 3-methyl-2-furoic acid, 50 g. of quinoline (Note 2), and 4.5 g. of copper powder is placed in a 125-ml. round-bottomed flask attached by a ground joint to a 30-cm. simple Vigreux column which delivers to a water-cooled condenser. The condenser is connected to a small distilling flask which serves as a receiver, with the tip of the condenser extending to the edge of the bulb of the flask. The receiver is cooled in an ice-salt bath (Note 3). The round-bottomed flask is heated by means of an electric mantle or liquid bath.

When the quinoline is heated to boiling by raising the bath temperature to about 250°, carbon dioxide is evolved at a moderate rate; the reaction is usually completed in 2–3 hours. Near the end of the reaction, heat is increased to about 265°, and the last distillate is collected until the temperature at the top of the column begins to rise rapidly above 65°. The contents of the receiver are decanted from any ice present and dried over about 1.5 g. of anhydrous magnesium sulfate, followed by Drierite, in a tightly closed flask. Redistillation yields 13.5–14.5 g. (83–89%) of colorless 3-methylfuran, b.p. 65.5–66°, n_D^{25} 1.4295–1.4315 (Notes 4 and 5).

2. Notes

1. In a run thirty times the size described, the submitter obtained a yield of 85%.
2. The quinoline should be dried by distillation from anhydrous barium oxide.
3. For larger runs, the side arm of the distilling flask should be attached to a cold trap immersed in an ice-salt bath, for about 10% of the product is likely to pass through the first receiver.
4. The product turns yellow on standing, even overnight. It can be stabilized with 0.1% hydroquinone or similar material.[2]
5. The submitter obtained similar results in runs about twenty times the size described. Over-all yields for the four steps starting with 4,4-dimethoxy-2-butanone (p. 649) were consistently in the range 50–55%.

3. Methods of Preparation

3-Methyl-2-furoic acid has been prepared by the oxidation of 3-methyl-2-furaldehyde [3] and by the degradation of 3-methyl-2-isovalerylfuran (Elsholtzia ketone).[4] 3-Methylfuran has been prepared by the present method [5] and more recently by a three-step method starting with methallyl chloride and ethyl orthoformate.[6] Circuitous routes from citric acid [7] and malic acid [8] have also been used.

[1] Eastman Kodak Co., Rochester, New York.
[2] Cass, U. S. pat. 2,489,265 (1949, to du Pont) [C. A., 44, 1543 (1950)].
[3] Reichstein, Zschokke and Goerg, Helv. Chim. Acta, 14, 1277 (1931).

[4] Asahina and Murayama, *Arch. Pharm.*, **252**, 442 (1914); Asahina, *Acta phytochim. (Japan)*, **2**, 12 (1924) [*Chem. Zentr.*, **1924**, II, 1694].

[5] Burness, *J. Org. Chem.*, **21**, 102 (1956); U. S. pat. 2,772,295 [*C. A.*, **51**, 7424 (1957)].

[6] Cornforth, *J. Chem. Soc.*, **1958**, 1310.

[7] Rinkes, *Rec. trav. chim.*, **50**, 1127 (1931); Reichstein and Zschokke, *Helv. Chim. Acta*, **14**, 1270 (1931).

[8] Gilman and Burtner, *J. Am. Chem. Soc.*, **55**, 2903 (1933).

β-METHYLGLUTARIC ANHYDRIDE *

(Glutaric anhydride, 3-methyl-)

$$CH_2(CO_2C_2H_5)_2 + CH_3CH{=}CHCO_2CH_3 \xrightarrow{NaOC_2H_5}$$

$$\left[\begin{array}{c} CH_2CO_2C_2H_5 \\ | \\ CH_3CH \\ | \\ CH(CO_2C_2H_5)_2 \end{array} \right]$$
(as sodio derivative)

$$\left[\begin{array}{c} CH_2CO_2C_2H_5 \\ | \\ CH_3CH \\ | \\ CH(CO_2C_2H_5)_2 \end{array} \right] \xrightarrow[\text{(2) } -CO_2]{\text{(1) } H_2O,\ H^+} [CH_3CH(CH_2CO_2H)_2]$$
(as sodio derivative)

$$[CH_3CH(CH_2CO_2H)_2] \xrightarrow[\text{anhydride}]{\text{Acetic}} \begin{array}{c} CH_2{-}C \\ / \quad \quad \backslash \\ CH_3CH \quad \quad O \\ \backslash \quad \quad / \\ CH_2{-}C \end{array} \begin{array}{c} O \\ \\ \\ O \end{array}$$

Submitted by JAMES CASON.[1]
Checked by B. C. McKUSICK, R. D. SMITH, and W. R. HATCHARD.

1. Procedure

A. *Triethyl 2-methyl-1,1,3-propanetricarboxylate* (not isolated). A 1-l. three-necked flask is fitted with a mechanical stirrer, a reflux condenser protected by a calcium chloride tube, and a 250-ml. dropping funnel. All parts of the apparatus should be scrupulously dry. Three hundred milliliters of anhydrous ethanol

(Note 1) is placed in the flask, and 14.1 g. (0.61 g. atom) of clean sodium, cut in pieces as large as will easily pass through a neck of the flask, is added rapidly. The neck of the flask is immediately closed, and the mixture is stirred until all the sodium has dissolved; the cooling bath is removed, and heat is applied if the reaction becomes sluggish at the end.

After all the sodium has dissolved, a mixture of 115 g. (0.72 mole) of diethyl malonate (Note 2) and 60 g. (0.60 mole) of methyl crotonate (Note 2) is added from the separatory funnel. This addition is made as rapidly as is consistent with keeping the exothermic reaction under control. After the exothermic reaction has subsided, the mixture is heated for 1 hour under reflux with stirring. An oil bath is recommended for heating. At the end of the heating period the condenser is changed to a position for distilling, and the temperature of the oil bath is raised sufficiently to distil alcohol fairly rapidly from the stirred mixture. Distillation is continued until most of the alcohol has been distilled. There is left a residue of the sodio derivative of triethyl 2-methyl-1,1,3-propanetricarboxylate (Note 3). If water is added to obtain the free ester, considerable heat is generated and cooling must be adequate to prevent partial hydrolysis of the ester by the alkali liberated. For present purposes there is no advantage in attempting isolation of the ester, and the residue is processed as described below.

B. *β-Methylglutaric anhydride.* The residue described above is cooled in an ice bath during the successive addition of 200 ml. of water and 450 ml. of concentrated hydrochloric acid. The resultant mixture is heated under reflux, with stirring, for 8 hours (Note 4). The condenser is again set for distillation, the bath temperature is raised, and, with continued stirring, water and alcohol are distilled. The bath is finally heated to 180–190° until gas evolution ceases (usually about 1 hour).

The stirrer is removed, 125 ml. of technical acetic anhydride is added to the residue, and, after thorough mixing, the mixture is heated on a steam bath for 1 hour. The condenser protected by a calcium chloride tube is left attached, and the other necks of the flask are closed. At the end of the heating period salt is removed from the cooled reaction mixture by filtering it with suction, using a filter aid mat, into a 250-ml. Claisen flask. The reaction vessel and filter are washed with a few milliliters of acetic acid.

The combined filtrate and washings are distilled at reduced pressure, using a water pump, until all acetic acid and acetic anhydride have been removed. An oil pump is then connected, and the distillation is continued. β-Methylglutaric anhydride is collected at 118–122°/3.5 mm. The yield of semisolid anhydride (Note 5) is 46–58 g. (60–76%) (Note 6).

2. Notes

1. Commercial absolute ethanol from a freshly opened bottle is often satisfactory; otherwise it can be dried by treatment with sodium,[2] sodium ethoxide and diethyl phthalate,[3] magnesium methoxide,[4] or aluminum tert-butoxide.[5]

2. Because commercial diethyl malonate is likely to contain small amounts of water and acid, it should be distilled from a Claisen flask at reduced pressure before use. The material used should be collected over a two- or three-degree range. The boiling point of malonic ester is 98°/20 mm.

Commercial methyl crotonate is rather impure and should be distilled at atmospheric pressure through a simple Vigreux or packed column which is 40–60 cm. in length. The fraction used should be collected over a two-degree range, and 70–85% of material boiling in such a range may usually be obtained from the commercial product. The boiling point is 117–118°.

3. Because of transesterification, the ester is principally the triethyl ester.

4. Sometimes the mixture becomes homogeneous after 3–5 hours, and heating can be stopped. At other times oily material remains even after 8 hours, but nothing is gained by further heating.

5. Pure β-methylglutaric anhydride melts at 46°.[6] The β-methylglutaric anhydride obtained in this preparation varies in its appearance at 25° from an almost completely crystalline mass to a mixture of about one-third solid and two-thirds liquid. However, the submitter has found that product of either appearance can be converted to methyl hydrogen β-methylglutarate in 80–85% yield. Further purification is troublesome, and the

product of the present procedure is pure enough for most purposes.
6. The submitter has obtained yields of 85–90%.

3. Methods of Preparation

β-Methylglutaric anhydride is obtained from the acid, which
has been prepared by condensation of acetaldehyde with cyanoacetamide [7,8] or by the oxidation of 3-methyl-1,3-cyclohexane-
dione with periodic acid.[9] The present method, which is a simplification of that published by Ställberg-Stenhagen,[10] gives a
higher yield and is much better adapted to the preparation of
large quantities.

[1] University of California, Berkeley, California.
[2] Org. Syntheses Coll. Vol. 1, 259 (1941).
[3] Org. Syntheses Coll. Vol. 2, 155 (1943).
[4] Org. Syntheses Coll. Vol. 1, 249 (1941).
[5] Org. Syntheses Coll. Vol. 3, 672 (1955).
[6] Darbishire and Thorpe, J. Chem. Soc., 87, 1717 (1905).
[7] Day and Thorpe, J. Chem. Soc., 117, 1465 (1920).
[8] Org. Syntheses Coll. Vol. 3, 591 (1955).
[9] Wolfrom and Bobbitt, J. Am. Chem. Soc., 78, 2489 (1956).
[10] Ställberg-Stenhagen, Arkiv Kemi Mineral. Geol., 25A, No. 10 (1947) [C. A.,
42, 5851 (1948)].

METHYLGLYOXAL-ω-PHENYLHYDRAZONE

(Pyruvaldehyde, 1-phenylhydrazone)

$$CH_3COCH_2CO_2C_2H_5 + KOH \rightarrow CH_3COCH_2CO_2K + C_2H_5OH$$

$$CH_3COCH_2CO_2K + HCl \rightarrow CH_3COCH_2CO_2H + KCl$$

$$CH_3COCH_2CO_2H + C_6H_5\overset{\oplus}{N_2}\ \overset{\ominus}{Cl} \rightarrow$$
$$CH_3COCH{=}NNHC_6H_5 + CO_2 + HCl$$

Submitted by GEORGE A. REYNOLDS and J. A. VANALLAN.[1]
Checked by T. L. CAIRNS and C. T. HANDY.

1. Procedure

In a 4-l. beaker equipped with a mechanical stirrer is placed
a solution of 35 g. (0.53 mole) of 85% potassium hydroxide in

1120 ml. of water. To the solution is added, with stirring, 65 g. (64 ml., 0.5 mole) of ethyl acetoacetate (Note 1). The mixture is allowed to stand at room temperature for 24 hours.

Forty-seven grams (48 ml., 0.5 mole) of aniline is dissolved in 200 ml. of aqueous hydrochloric acid (prepared from equal volumes of concentrated acid and water) in a 2-l. beaker. The beaker is equipped with a mechanical stirrer and immersed in an ice-salt bath. After the solution has cooled to ±5°, 36 g. (0.52 mole) of sodium nitrite dissolved in 1 l. of water is added slowly, with stirring, from a separatory funnel. The tip of the stem of the separatory funnel should dip well below the surface of the liquid. The rate of addition is adjusted to maintain the temperature below 10°. A drop of the reaction mixture is tested from time to time with starch-iodide paper (Note 2). The sodium nitrite solution is added until nitrous acid persists in the solution during a 5-minute interval.

The solution of potassium acetoacetate is cooled to 0°, and 45 ml. of concentrated hydrochloric acid in 150 ml. of ice water is added slowly with stirring (Note 3). The diazonium salt solution is then added over a period of 20 minutes, and the mixture is made basic by the addition of 82 g. of sodium acetate dissolved in 300 ml. of water (Note 4). The temperature of the reaction mixture is raised slowly to 50° and maintained at this value for 2 hours; the solid that separates is collected on a filter and dried. The yield of crude product is 72–77 g. (89–95%). Purification can be effected by recrystallization from 200 ml. of toluene. The purified product weighs 59–66 g. (73–82%) (Notes 5 and 6); m.p. 148–150°.

2. Notes

1. Commercial ethyl acetoacetate was used.
2. The test is made by diluting the test drop on a watch glass with about 1 ml. of water and then placing a drop of this solution on the starch-iodide paper.
3. The solution is neutralized slowly in order to keep it cold so that the acetoacetic acid will not be decomposed.
4. The reaction proceeds much more rapidly in basic solution.
5. An additional 4–6 g. of product separates slowly from the filtrate.

6. This general procedure is effective for the preparation of many types of phenylhydrazones. For example, a substituted diazo compound can be employed.[2] Alkylated acetoacetic esters [3] and ethyl benzoylacetate [4] may be used. For the higher homologs, the α-formyl derivatives of ketones may be used in place of ethyl acetoacetate.[5,6] Ethyl pyridylacetates may also be substituted for ethyl acetoacetate.[7] The products in these cases are the phenylhydrazones of 2-acylpyridines.

3. Methods of Preparation

The procedure described is essentially that of Japp and Klingemann.[3] Methylglyoxal-ω-phenylhydrazone may also be prepared by heating phenylazoacetoacetic acid at 170–180° [8,9] or by warming ethyl phenylazoacetoacetate with a solution of sodium hydroxide in dilute ethanol. [8,9]

[1] Eastman Kodak Company, Rochester, New York.
[2] Stierlin, *Ber.*, **21**, 2124 (1888).
[3] Japp and Klingemann, *Ann.*, **247**, 218 (1888).
[4] Bamberger and Schmidt, *Ber.*, **34**, 2009 (1901).
[5] Benary, *Ber.*, **59**, 2198 (1926).
[6] Benary, Meyer, and Charisius, *Ber.*, **59**, 108, 600 (1926).
[7] Frank and Phillips, *J. Am. Chem. Soc.*, **71**, 2804 (1949).
[8] Richter and Munzer, *Ber.*, **17**, 1928 (1884).
[9] Japp and Klingemann, *Ann.*, **247**, 198 (1888).

METHYL HYDROGEN HENDECANEDIOATE

(Undecanedioic acid, methyl ester)

$$HO_2C(CH_2)_9CO_2H + 2CH_3OH \xrightarrow{H^+}$$
$$CH_3O_2C(CH_2)_9CO_2CH_3 + 2H_2O$$

$$CH_3O_2C(CH_2)_9CO_2CH_3 \xrightarrow{Ba(OH)_2} \xrightarrow{H^+} CH_3O_2C(CH_2)_9CO_2H$$

Submitted by Lois J. Durham, Donald J. McLeod, and James Cason.[1]
Checked by N. J. Leonard, D. H. Dybvig, and K. L. Rinehart, Jr.

1. Procedure

Dimethyl hendecanedioate is prepared by heating 23 g. (0.106

mole) of hendecanedioic acid (p. 510) under reflux for 2 hours with a solution of 8 ml. of concentrated sulfuric acid in 80 ml. of methanol. After the reaction mixture has been diluted with 3 volumes of water, it is extracted with two 75-ml. portions of benzene. The benzene extracts are washed successively with 250-ml. portions of water, 5% aqueous sodium carbonate solution, and water. After the benzene has been removed under reduced pressure, the residue (Note 1) is transferred to a 250-ml. Erlenmeyer flask containing 127 ml. of a $0.915N$ solution of barium hydroxide (0.058 mole) in commercial anhydrous methanol (Note 2). The flask is immediately closed with a soda-lime tube and swirled to mix the solution. The barium salt of the half ester begins to precipitate after about 2 minutes.

After the flask has been allowed to stand at room temperature (20–25°) for at least 17 hours (Note 3), the barium salt is collected by suction filtration and washed with about 20 ml. of methanol (Note 4). The moist barium salt is shaken for a few minutes in a separatory funnel with a mixture of 100 ml. of $4N$ hydrochloric acid and 100 ml. of ether. The aqueous layer, together with any precipitated barium chloride, is removed and extracted again with 100 ml. of ether. The two ether extracts are combined and washed with three 100-ml. portions of water, the solvent is removed, and the residue (Note 5) is distilled through a half-meter column (Note 6). There is essentially no fore-run (see Note 4). The pure half ester is collected at 165–168°/2 mm., weight 14.6–15.7 g. (60–64%), m.p. 44–46° (Note 7).

2. Notes

1. The submitters state that the residue may be distilled to give dimethyl hendecanedioate in 98% yield (25.5 g.).

2. An approximately $1.0N$ solution of *anhydrous* barium hydroxide in methanol is prepared, and the exact normality is determined by titration. The checkers employed 143 ml. of an $0.814N$ solution of barium hydroxide in anhydrous methanol. This procedure is not suitable for making ethyl esters on account of the low solubility of barium hydroxide in ethanol.

3. Periodic titration of aliquots of the reaction mixture has

shown that after 16 hours about 95% of the original barium hydroxide has reacted.

4. If an insufficient amount of methanol is employed to wash the precipitate, a considerable quantity of recovered diester, b.p. ca. 145°/2 mm., is obtained as a fore-run in the fractional distillation of the product. The diester may be recovered by dilution of the filtrate with water and extraction.

5. The crude product obtained directly from the barium salt is 90–95% half ester, and the remainder is diacid. This material may be used directly in reactions where small amounts of diacid are not objectionable; however, distillation is necessary in order to obtain a pure sample of half ester. This distillation is simplified because there is no lower-boiling diester and only a small amount of higher-boiling diacid.

6. The submitters used a simple type of Podbielniak column.[2] The checkers used a 122-cm. column of similar design.

7. Titration of this half ester gives an equivalent weight, within experimental error, of the calculated value of 230.3. The yield in this preparation may be regarded as close to theoretical on the basis of recovery and re-use of diacid and diester.

3. Methods of Preparation

Methyl hydrogen hendecanedioate has been reported as a by-product in the ozonolysis of methyl 11-dodecenoate,[3] but the only preparative procedure reported is that presently described.[4] Half esters have usually been prepared by partial esterification [5] and direct fractional distillation of the three products of the reaction; however, some modification [6] of this procedure is required for the higher-boiling half esters. The present method [4,7] is considerably less laborious than the partial esterification procedure and is a particular advantage for higher-boiling esters where a prolonged fractional distillation at high temperatures permits disproportionation of the half ester. This method is not satisfactory for lower-molecular-weight half esters, for their salts are too soluble in methanol. Sebacic acid gives satisfactory results by this method; azelaic acid, poor results; and lower-molecular-weight dibasic acids fail to give significant amounts of half ester.

[1] University of California, Berkeley, California.
[2] Cason and Rapoport, *Laboratory Text in Organic Chemistry*, 2nd ed., p. 293, Prentice-Hall, Inc., Englewood Cliffs, New Jersey, 1962.
[3] Lycan and Adams, *J. Am. Chem. Soc.*, **51**, 625 (1929).
[4] Cason, Taylor, and Williams, *J. Org. Chem.*, **16**, 1187 (1951); Cason and McLeod, *J. Org. Chem.*, **23**, 1497 (1958).
[5] *Org. Syntheses Coll. Vol.* **2**, 276 (1943).
[6] Jones, *J. Am. Chem. Soc.*, **69**, 2350 (1947).
[7] Signer and Sprecher, *Helv. Chim. Acta*, **30**, 1001 (1947).

4-METHYL-6-HYDROXYPYRIMIDINE

(4-Pyrimidinol, 6-methyl-)

$$S{=}C(NH_2)_2 + CH_3COCH_2CO_2C_2H_5 \xrightarrow{\text{NaOCH}_3}$$

$$\xrightarrow{\text{Ni—H}_2}$$

Submitted by H. M. Foster and H. R. Snyder.[1]
Checked by R. T. Arnold and P. E. Throckmorton.

1. Procedure

A. *2-Thio-6-methyluracil.* In a 2-l. flask are placed 76 g. (1 mole) of thiourea, 130 g. (1 mole) of commercial ethyl acetoacetate, 120 g. of commercial sodium methoxide, and 900 ml. of methanol. The reaction mixture is heated gently on the steam bath and is permitted to evaporate to dryness in a hood over a period of about 8 hours. The residue is dissolved in 1 l. of hot water; the solution is treated with a few grams of activated carbon and is filtered. The hot filtrate is carefully treated (Note 1) with 120 ml. of glacial acetic acid. The thiouracil precipitates rapidly and is collected on a 4-in. Büchner funnel. The still wet solid filter cake is suspended in a boiling solution of 1 l. of water and 20 ml. of glacial acetic acid. The slurry is stirred and mixed thoroughly to break up lumps and is then refrigerated.

The product is collected on a 4-in. Büchner funnel and is washed with about 200 ml. of cold water in four portions. The solid is permitted to drain (with suction) for several hours and is then transferred to an oven at 70° for more complete drying. The yield of oven-dried 2-thio-6-methyluracil is 98–119 g. (69–84%). This material is sufficiently pure for the desulfurization reaction (Note 2).

B. *4-Methyl-6-hydroxypyrimidine.* To a hot solution of 10 g. (0.07 mole) of 2-thio-6-methyluracil in 200 ml. of distilled water and 20 ml. of concentrated aqueous ammonia in a 500-ml. round-bottomed flask is added 45 g. (wet paste) of Raney nickel catalyst (Note 3). About 30 ml. of distilled water is used to wash all the nickel catalyst into the reaction flask. The mixture is heated under reflux in a hood for about 1.5 hours. The catalyst is permitted to settle, and the clear solution is decanted and filtered by gravity. The catalyst is washed with two 75-ml. portions of hot water and is discarded (Note 4). The combined filtrate and washings (Note 5) are evaporated to dryness on a steam bath. The residue is placed in an oven at 70° to complete the drying process (Note 6). The yield of crude pyrimidine, m.p. 136–142°, is 7.0–7.2 g. (90–93%).

The crude product is best purified by sublimation under reduced pressure (100–110°/1 mm.) (recovery 90–95%). Purification can also be effected by recrystallization from acetone (recovery 80–90%), ethyl acetate (recovery 70–80%), or ethanol (recovery 60–70%). The purified 4-methyl-6-hydroxypyrimidine melts at 148–149°.

2. Notes

1. The hot solution tends to foam and froth badly when the acetic acid is added, and if care is not taken mechanical loss of product may result.

2. 2-Thio-6-methyluracil does not possess a clearly defined melting point but shows marked decomposition above 280°.[2]

3. The activity of the Raney nickel catalyst greatly affects the yield of the desulfurized pyrimidine. A catalyst described by Brown[3] gave very satisfactory results. A Raney nickel C described by Hurd and Rudner[4] is perhaps a more reactive catalyst; however, the yield of desulfurized pyrimidine was not sufficiently better to warrant its use.

4. Care must be taken not to allow the nickel to dry too completely lest it ignite.

5. The filtrate and washings should be clear and colorless. If the solution is blue or green (indicative of the presence of dissolved nickel) the solution should be treated with hydrogen sulfide, or better with dimethylglyoxime and ammonia, to precipitate the nickel.

6. The 4-methyl-6-hydroxypyrimidine is surprisingly volatile, and loss of product may occur if the material is heated on the steam bath for an appreciable period of time.

3. Methods of Preparation

4-Methyl-6-hydroxypyrimidine can be prepared by heating 2,6-dichloro-4-methylpyrimidine with red phosphorus and hydriodic acid [5] and by treating 2-thio-6-methyluracil with hydrogen peroxide.[6] The present synthesis is modeled after the work of Brown,[3] who has described the desulfurization of several thiopyrimidines.

The procedure for the synthesis of 2-thio-6-methyluracil is a modification of the method described by Wheeler and Merriam [7] for the preparation of 2-methylthio-6-methyluracil.

[1] University of Illinois, Urbana, Illinois.
[2] List, *Ann.*, **236**, 6 (1886).
[3] Brown, *J. Soc. Chem. Ind.* (*London*), **69**, 355 (1950).
[4] Hurd and Rudner, *J. Am. Chem. Soc.*, **73**, 5158 (1951).
[5] Gabriel and Colman, *Ber.*, **32**, 2931 (1899).
[6] Williams, Ruehle, and Finkelstein, *J. Am. Chem. Soc.*, **59**, 526 (1937).
[7] Wheeler and Merriam, *Am. Chem. J.*, **29**, 486 (1903).

1-METHYLISOQUINOLINE *
(Isoquinoline, 1-methyl-)

$$\text{[isoquinoline]} + C_6H_5COCl + KCN \rightarrow$$

$$\text{[1-cyano-2-benzoyl-1,2-dihydroisoquinoline]} \quad N{-}COC_6H_5 + KCl$$

$$\text{[Reissert compound]} \quad N{-}COC_6H_5 + C_6H_5Li \xrightarrow[\text{dioxane}]{\overset{CH_3I,}{\text{ether}}}$$

$$\quad N{-}COC_6H_5 + C_6H_6 + LiI$$
$$\overset{|}{CH_3} \quad CN$$

$$\quad N{-}COC_6H_5 + 2KOH \xrightarrow[C_2H_5OH]{H_2O}$$
$$\overset{|}{CH_3} \quad CN$$

$$\quad N + KCN + C_6H_5CO_2K$$
$$\overset{|}{CH_3}$$

Submitted by J. WEINSTOCK and V. BOEKELHEIDE.[1]
Checked by N. J. LEONARD, TERRY W. MILLIGAN, and WILLIAM R. SHERMAN.

1. Procedure

Caution! All the operations should be carried out in a well-ventilated hood because of the toxic nature of hydrogen cyanide and the cyanide solutions.

A. *1-Cyano-2-benzoyl-1,2-dihydroisoquinoline* [2] (*Reissert's compound* [3]) (*Note 1*). In a 5-l. three-necked flask equipped with a Hershberg stirrer, a dropping funnel, and a condenser is placed a solution of 391 g. (6.0 moles) of potassium cyanide in 2.5 l. of water and 258 g. (2.0 moles) of isoquinoline (freshly distilled from zinc dust). The mixture is maintained below 25° by immersion in an ice bath (Note 2). The stirrer is started, and, when the isoquinoline has formed an emulsion with the aqueous solution, 562 g. (4.0 moles) of benzoyl chloride is added over 3 hours. The stirring is continued another hour or until the Reissert's compound has separated as small, hard, tan spheres. The reaction mixture is cooled further in the ice bath, and the product is collected on a large Büchner funnel. It is washed on the funnel with successive portions of 400 ml. of water, 400 ml. of 3N hydrochloric acid, and 500 ml. of water. The product is then recrystallized from 2–3 l. of commercial absolute ethanol using 2.5 g. of activated carbon to effect partial decolorization of the solution. The hot filtration to remove the carbon is done by means of a heated Büchner funnel. The filtrate is cooled in an ice bath and filtered when cold. Long standing in the presence of the supernatant liquid causes the adsorption of dark material by the product. The cream-colored crystals which separate are collected on a Büchner funnel, washed with 100 ml. of cold 95% ethanol, and dried in air overnight. The yield of dry 1-cyano-2-benzoyl-1,2-dihydroisoquinoline, m.p. 125–127°, sufficiently pure for use in the next step, is 303–400 g. (58–77%).

B. *2-Benzoyl-1-cyano-1-methyl-1,2-dihydroisoquinoline.* A 3-l. round-bottomed flask, a 500-ml. dropping funnel, a condenser, and a stirrer are dried in an oven and then arranged so that a nitrogen atmosphere can be maintained in the flask with the use of a mercury bubbler. The apparatus is flushed with dry nitrogen for 1 hour, and 83.5 g. (0.32 mole) of 1-cyano-2-benzoyl-1,2-dihydroisoquinoline, 350 ml. of dry dioxane (Note 3), and 100 ml. of anhydrous ether are added. The stirrer is started, and, when the solid is dissolved completely, the flask is immersed in an ice-salt bath at −10°. Then 450 ml. of a 0.78N ether solution of phenyllithium (0.35 mole) (Note 4) is added dropwise, with stirring, during 30 minutes. The reaction mixture turns a deep red, and as the addition is continued a red solid separates. Ten minutes after the addition is complete 56.2 g. (0.40 mole) of methyl iodide is added, and the reaction mixture is stirred in the cold for 2 hours, then overnight at room temperature. The reaction mixture is transferred to a separatory funnel and washed with three 50-ml. portions of water. The organic solution is filtered, and the solvent is removed under reduced pressure. If the residue does not crystallize immediately upon evaporation of the solvent, crystallization may be induced by scratching the sides of the flask and cooling. The crystals are transferred to a Büchner funnel, washed on the funnel with 50 ml. of cold 95% ethanol, and dried. The yield of dry 2-benzoyl-1-cyano-1-methyl-1,2-dihydroisoquinoline in the form of cream-colored crystals, m.p. 120–122°, is 62–63 g. (71–72%) (Note 5).

C. *1-Methylisoquinoline*. In a 500-ml. round-bottomed flask equipped with a reflux condenser are placed 62.2 g. (0.227 mole) of 2-benzoyl-1-cyano-1-methyl-1,2-dihydroisoquinoline, 50 ml. of 95% ethanol, and a solution of 37.6 g. (0.57 mole) of 85% potassium hydroxide in 100 ml. of water. The mixture is heated under reflux for 1.5 hours, during which time the solid dissolves and the solution becomes homogeneous. After the solution has cooled, it is extracted with four 75-ml. portions of ether. The combined ethereal extracts are washed with two 25-ml. portions of water and dried with anhydrous magnesium sulfate. After removal of the drying agent by filtration and of the solvent by concentration under vacuum, the residue is distilled under reduced pressure to give 24–26 g. (74–80%) of colorless 1-methylisoquinoline, b.p. 81°/1 mm.; n_D^{25} 1.6102, n_D^{20} 1.6119 (Note 6).

2. Notes

1. This is essentially the procedure of Padbury and Lindwall.[2]

2. If the mixture is not kept cold during the addition of the benzoyl chloride, the product is likely to be highly colored.

3. The dioxane is dried as described by Fieser.[4] It is kept in a glass-stoppered bottle sealed with wax. The ether is commercial anhydrous ether dried over sodium wire.

4. A convenient preparation of phenyllithium is described by Wittig.[5] The ethereal solution may be titrated by adding an aliquot to water and titrating to the methyl orange end point with standardized hydrochloric acid.

5. Another 10 g. (11%) of impure 2-benzoyl-1-cyano-1-methyl-1,2-dihydroisoquinoline may be obtained by adding water to the cold mother liquor until it becomes turbid. The mixture is heated until the turbidity disappears, and the resulting solution is cooled slowly, then refrigerated. The additional product obtained by this method is contaminated with Reissert's compound and should not be used in the next step if pure 1-methylisoquinoline is desired.

6. This method may be used to prepare other 1-substituted isoquinolines. The submitters have prepared 1-benzylisoquinoline and 1-butylisoquinoline by this method.[6]

3. Methods of Preparation

1-Methylisoquinoline has been prepared by the catalytic dehydrogenation of 1-methyl-3,4-dihydroisoquinoline prepared by the Bischler-Napieralski reaction, which involves treating β-phenylethylacetamide with a strong dehydrating reagent at elevated temperatures.[7-12] 1-Methyl-3,4-dihydroisoquinoline has been prepared from β-phenylethylacetamide by using polyphosphoric acid, and the same reagent produced some 1-methylisoquinoline from N-acetyl-dl-phenylalanine.[13] 1-Methylisoquinoline has also been prepared from β-phenyl-β-hydroxyethylacetamide by using phosphorus pentoxide,[14] and by cyclization of the Schiff base obtained from acetophenone and aminoacetal [15] or from α-phenylethylamine and glyoxalsemiacetal.[16]

[1] University of Rochester, Rochester, New York.
[2] Padbury and Lindwall, *J. Am. Chem. Soc.*, **67**, 1268 (1945.)
[3] Reissert, *Ber.*, **38**, 3427 (1905).
[4] Fieser, *Experiments in Organic Chemistry*, 3rd ed., p. 284, D. C. Heath and Co., Boston, 1955.
[5] Wittig, *Newer Methods of Preparative Organic Chemistry*, p. 576, Interscience Publishers, Inc., New York, 1948; see also *Org. Syntheses Coll. Vol.* **3**, 757 (1955).
[6] Boekelheide and Weinstock, *J. Am. Chem. Soc.*, **74**, 660 (1952).
[7] Whaley and Hartung, *J. Org. Chem.*, **14**, 650 (1949).
[8] Leonard and Boyer, *J. Am. Chem. Soc.*, **72**, 2980 (1950).
[9] Späth, Berger, and Kuntara, *Ber.*, **63B**, 134 (1930).
[10] Barrows and Lindwall, *J. Am. Chem. Soc.*, **64**, 2430 (1942).
[11] Dey and Ramanathan, *Proc. Natl. Inst. Sci. India*, **9A**, 193 (1943).
[12] Pictet and Kay, *Ber.*, **42**, 1973 (1909); Sugasawa and Tachikawa, *Tetrahedron*, **4**, 205 (1958).

[13] Snyder and Werber, *J. Am. Chem. Soc.*, **72**, 2962 (1950).
[14] Mills and Smith, *J. Chem. Soc.*, **121**, 2724 (1922); Pictet and Gams, *Ber.*, **43**, 2384 (1910); Ukita, Nakazawa, and Tamura, *Japan. J. Exptl. Med.*, **21**, 259 (1951) [*C. A.*, **47**, 12392 (1953)].
[15] Pomeranz, *Monatsh.*, **15**, 299 (1894).
[16] Schlittler and Müller, *Helv. Chim. Acta*, **31**, 914 (1948).

METHYLISOUREA HYDROCHLORIDE

(Pseudourea, 2-methyl-, hydrochloride)

$$CaNCN + 2CH_3CO_2H \rightarrow NH_2CN + Ca(O_2CCH_3)_2$$
$$NH_2CN + CH_3OH + HCl \rightarrow NH_2C(OCH_3)NH \cdot HCl$$

Submitted by FREDERICK KURZER and ALEXANDER LAWSON.[1]
Checked by WILLIAM S. JOHNSON and WILLIAM T. TSATSOS.

1. Procedure

A. *Cyanamide.* In a large mortar are placed 57 g. (54 ml., 0.75 mole) (Note 1) of glacial acetic acid and 135 ml. of water (Note 2). To this solution, 40 g. (0.5 mole) of calcium cyanamide (Note 3) is added slowly (Note 4) with good stirring and grinding. As the introduction of the calcium cyanamide proceeds, small quantities of acetylene are evolved, while the initially thin cream gradually turns into a thick, dark-gray to black paste. The mixture must remain acidic to litmus throughout the addition. After being dried at 40–50° for 12–18 hours in a vacuum oven, at a pressure of 30 mm. or less, the material is obtained as a pale-gray, dry powder. This is extracted exhaustively in a Soxhlet apparatus, for 2- to 3-hour periods, with two successive 400-ml. portions of ether (Note 5) containing a few drops of dilute acetic acid. The ethereal extracts (Note 6) are each dried over 30-g. portions of anhydrous sodium sulfate and combined, and the solvent is removed under reduced pressure (Note 7). The colorless, viscous, oily residue of cyanamide is suitable for use in the next stage. The yield of cyanamide is 10.5–15.8 g. (50–75% calculated on the basis of the formula CaNCN) (Notes 8 and 9).

B. *Methylisourea hydrochloride.* The crude cyanamide obtained as described above (Part A) is taken up in 100 ml. of anhydrous methanol, and the clear solution is decanted from a trace of insoluble oily material if necessary. Anhydrous hydrogen chloride is passed into the clear colorless liquid until an increase in weight amounting to 1 g. of hydrogen chloride per gram of crude cyanamide (1.15 moles) is attained. During the addition of the hydrogen chloride, the cyanamide solution is maintained at room temperature by external ice-cooling. The resulting clear liquid is set aside for 3–4 days, and the methanol is removed by distillation under reduced pressure. The residual colorless crystalline solid is dried in a vacuum desiccator containing potassium hydroxide and phosphorus pentoxide; it consists of methylisourea hydrochloride. The yield is 1.8–2.1 g. per gram of cyanamide (69–80%) (Note 10).

The material may be crystallized from boiling methanol (1 ml. per gram of crude material) and forms lustrous, colorless, thick, prismatic needles, which melt with decomposition (Notes 11 and 12). They are separated by suction filtration at 0° (Note 13) and quickly rinsed with a very little ice-cold methanol. The product is hygroscopic and is quickly pressed between filter paper, then dried in a vacuum desiccator over phosphorus pentoxide. The filtrates yield a further crop on partial evaporation under reduced pressure, the total recovery of recrystallized material being 85–90% (Note 14).

2. Notes

1. The exact amount of acetic acid required by a particular sample of calcium cyanamide is first determined volumetrically as follows: A weighed sample of calcium cyanamide (approximately 1 g.) is suspended in about 50 ml. of distilled water and titrated with standard hydrochloric acid (preferably of approximately normal strength), using phenolphthalein as indicator. Acid is added until the pink color of the indicator does not reappear within 2–3 minutes. From the results of the titration the amount of acetic acid required is calculated by proportion, a 10% excess being allowed to ensure that the reaction mixture remains acid throughout the experiment.

2. In order to facilitate drying at a later stage, as little water is

used as will produce a paste that can still be effectively mixed.

3. Commercial calcium cyanamide (nitrolime), containing carbon and small quantities of calcium carbide, is suitable for this preparation.

4. The addition of the calcium cyanamide to the acid should be slow enough to ensure thorough mixing and to prevent the reaction mixture from becoming hot.

5. The checkers found it desirable to saturate the ether (by shaking) with water before extraction. The Soxhlet extraction should not be prolonged for more than 3–4 hours. If a longer extraction period is employed, a fresh portion of ether should be used so that the extracted material will not be subjected to heat for too long a period.

6. Cyanamide may be kept unchanged at $0°$ in ethereal solution in the presence of traces of acetic acid. The ethereal extracts from several runs of calcium cyanamide may therefore be combined and worked up collectively.

7. Distillation is best carried out from a previously weighed small flask, and the weighed residue of cyanamide is immediately dissolved in methanol for the next stage.

8. The yields of cyanamide from commercial calcium cyanamide vary from sample to sample but do not fluctuate greatly for one particular specimen. The present procedure was found to be satisfactory for preparing 15- to 30-g. batches of cyanamide.

9. If desired, the cyanamide may be crystallized from a mixture of ether and benzene; however, the crystallization is difficult because of the tendency of the material to oil.

10. This material is satisfactory for most synthetic purposes without further purification.

11. Methylisourea hydrochloride decomposes on heating with evolution of methyl chloride. The decomposition temperature depends on the rate of heating, but reproducible values are obtainable if the rate of heating is controlled. Samples of pure (98–99%) (Note 12) methylisourea hydrochloride, introduced into the melting-point tube without undue previous exposure to atmospheric moisture, placed in the melting-point bath at $60°$, and heated at the rate of $12°$ per minute, sinter at $118–119°$ and decompose at $122–124°$ (the mass moving rapidly up in the melting-point tube).

12. The purity of the crystallized product, determined volumetrically by Volhard's method, exceeds 98%. In this procedure, 10 ml. of a 1% solution of methylisourea hydrochloride is acidified with a few drops of nitric acid and treated with 20 ml. of 0.1N silver nitrate. After removal of the silver chloride by filtration, the excess of the silver nitrate is estimated with 0.1N thiocyanate solution, using ferric alum as indicator. Alternatively, 10-ml. portions of 0.1N silver nitrate, acidified with nitric acid, may be titrated directly with the 1% methylisourea hydrochloride solution in the presence of tartrazine.

Owing to the presence of small quantities of free hydrochloric acid in the crude product, the above procedures are applicable to recrystallized specimens only.

13. It has been suggested [2] that it is best to cool the solution to −10°, cool the wash solvent to −15° in ice-salt, and to put some of the cold solvent through the funnel before filtration (the addition of a few chips of Dry Ice to this solvent on the funnel is helpful).

14. Methylisourea hydrochloride should be kept in a desiccator, even for brief storage, and especially for extended periods of time.

3. Methods of Preparation

Methylisourea hydrochloride has been prepared by the action of hydrogen chloride on a suspension of silver cyanamide [3] or a solution of cyanamide [4,5] in methanol,[4] and by the action of dimethyl sulfate on urea.[6] The free base is obtained by treating the salt with powdered potassium hydroxide in a water-ether mixture [4] or with sodium methoxide in methanol.[7]

An alternative laboratory preparation of cyanamide and a selection of references to the literature have appeared in *Inorganic Syntheses*.[8] The present method is that of Werner.[9]

[1] School of Medicine, University of London, London, England.

[2] Cason, Private communication.

[3] Stieglitz and McKee, *Ber.*, **33**, 810 (1900); McKee, *Am. Chem. J.*, **26**, 244 (1901).

[4] Stieglitz and McKee, *Ber.*, **33**, 1517 (1900); McKee, *Am. Chem. J.*, **26**, 245 (1901).

[5] Basterfield and Powell, *Can. J. Research*, **1**, 261 (1929); see also Cox and Raymond, *J. Am. Chem. Soc.*, **63**, 300 (1941).

[6] Werner, *J. Chem. Soc.*, **1914**, 927; Janus, *J. Chem. Soc.*, **1955**, 3551.

[7] Kapfhammer and Müller, *Z. physiol. Chem.*, **225**, 7 (1934); Rodionov and Urbanskaya, *Zhur. Obshchei Khim.* (*J. Gen. Chem. U.S.S.R.*), **18**, 2023 (1948) [*C. A.*, **43**, 3793 (1949)].

[8] Pinck and Salisbury, *Inorg. Syntheses*, **3**, 39 (1950).

[9] Werner, *J. Chem. Soc.*, **1916**, 1325; Werner, *The Chemistry of Urea*, Longmans, London, 1923, p. 184.

METHYL 3-METHYL-2-FUROATE

(2-Furoic acid, 3-methyl-, methyl ester)

$$(CH_3O)_2CHCH_2COCH_3 + ClCH_2CO_2CH_3 \xrightarrow{NaOCH_3}$$

$$(CH_3O)_2CHCH_2\overset{\overset{\displaystyle CH_3}{|}}{C}\underset{O}{\diagdown\diagup}CHCO_2CH_3 + NaCl$$

$$(CH_3O)_2CHCH_2\overset{\overset{\displaystyle CH_3}{|}}{C}\underset{O}{\diagdown\diagup}CHCO_2CH_3 \xrightarrow{160°}$$

$$+ 2CH_3OH$$

Submitted by D. M. BURNESS.[1]
Checked by JAMES CASON and ROBERT B. HUTCHISON.

1. Procedure

A. *Methyl 5,5-dimethoxy-3-methyl-2,3-epoxypentanoate.* A 2-l. three-necked flask is equipped with a sealed centrifugal stirrer (Note 1), a thermometer inserted through an adapter with a side arm connected to a source of dry nitrogen, and a 250-ml. Erlenmeyer addition flask.[2] The apparatus is dried with a free flame in a slow stream of nitrogen; from this point the reaction is conducted in an atmosphere of nitrogen (Note 2).

A mixture of 132 g. (1.0 mole) of 4,4-dimethoxy-2-butanone (Note 3), 174 g. (1.6 moles) of methyl chloroacetate (Note 3), and 800 ml. of dry ether is placed in the reaction flask, then 86 g.

(1.6 moles) of sodium methoxide (Note 4) is placed in the addition flask. The solution is cooled in an ice-salt bath to $-10°$, then the sodium methoxide is added gradually at a rate such that a temperature below $-5°$ can be maintained (about 2 hours). The mixture is stirred for an additional 2 hours (Note 5) and then allowed to come to room temperature overnight. It is cooled again to $0°$ and made slightly acidic by the addition of a solution of 10 ml. of glacial acetic acid in 150 ml. of water. The ether is decanted, and the residual slurry is extracted with three 100-ml. portions of ether. The combined ether solutions are washed in a separatory funnel with 50 ml. of saturated sodium chloride solution to which is added 1-g. portions of sodium bicarbonate until the washings are no longer acidic. After each bicarbonate addition, the mixture is shaken for at least 1 minute before a test for acidity is made. Finally, the ether phase is washed with saturated sodium chloride solution, then dried over 20–25 g. of anhydrous magnesium sulfate. Distillation of the solvent leaves a nearly quantitative yield of crude glycidic ester (Note 6).

B. *Methyl 3-methyl-2-furoate.* The crude glycidic ester prepared as described above is placed in a 300-ml. flask which is attached to a 12-cm. column filled with $\frac{3}{16}$-inch glass helices (or a 50-cm. simple Vigreux column) and heated in a liquid bath. When the pot temperature reaches about 160°, or before, methanol begins to distil. Heating is continued until the distillation of methanol essentially ceases and about the theoretical amount (64 g.) has been collected. After the heating bath has been allowed to cool, the product is distilled at reduced pressure; b.p. 72–78°/8 mm., yield 91–98 g. (65–70%) (Note 7). The ester solidifies in the receiver as an essentially pure compound, m.p. 34.5–36.5° (Note 8).

2. Notes

1. A stirring assembly which makes use of a lubricated ball-joint seal [3] is convenient. The checkers used a Hershberg stirrer rather than a centrifugal stirrer.

2. Maintenance of a low positive pressure of nitrogen on the system is accomplished by insertion of a T-tube in the nitrogen line for attachment of a U-tube whose bend is just closed with mineral oil.

3. The 4,4-dimethoxy-2-butanone [4] may be obtained from Aldrich Chemical Co., Milwaukee, Wisconsin, under the name of 3-ketobutyraldehyde dimethyl acetal. This and the methyl chloroacetate are preferably dried over Drierite and distilled before use. The pure acetal has b.p. 55–56°/8 mm., n_D^{25} 1.4119. The presence of 4-methoxy-3-buten-2-one, which raises the index of refraction, can be tolerated as an impurity, for it leads to the same reaction product.[5] Commercial methyl chloroacetate usually contains considerable low-boiling material which is best separated by distillation through a 50-cm. simple Vigreux column. The chloroacetate is collected at 131–132°.

4. The submitter reports that the commercial 95% "Sodium Methylate" from Mathieson Chemical Corp. is satisfactory, provided that either fresh material or material which has been opened previously only under dry nitrogen is used. The checkers experienced such erratic results with commercial sodium methoxide (even previously unopened bottles) that freshly prepared material was used. For this purpose, 37 g. of clean sodium, cut in 1- to 3-g. pieces, was added portionwise to 800 ml. of stirred anhydrous methanol contained in a 2-l. three-necked flask equipped with a condenser. After the sodium had dissolved, the methanol was removed by distillation at reduced pressure, and the residual white sodium methoxide was dried by heating at 150° under aspirator vacuum.

5. The stream of nitrogen may be discontinued at this point if the outlet tube from the flask is closed with a Drierite tube.

6. The submitter reports that the residual glycidic ester was distilled through a 15-cm. Vigreux column to yield 185–195 g. of crude product, b.p. 113–122°/8 mm. Redistillation through a 25-cm. column packed with $\frac{3}{16}$-inch glass helices was reported to give 157–164 g. (77–80%) of product; b.p. 93°/0.7 mm. to 89°/1 mm.; n_D^{25} 1.4405–1.4419. The drop in boiling point was attributed to decomposition during distillation to yield methanol and methyl 3-methyl-2-furoate. The checkers found that in most runs the product obtained from the first distillation consisted largely of the furoate.

The submitter has prepared methyl 5,5-dimethoxy-3-phenyl-2,3-epoxypentanoate by essentially the same procedure as here described.

7. In a run 15 times this size, a 71% yield was obtained by the submitter.

8. Recrystallization from ethanol raises the melting point to 36.5–37°.

3. Methods of Preparation

Methyl 5,5-dimethoxy-3-methyl-2,3-epoxypentanoate has been prepared only by the procedure described or in like manner from 4-methoxy-3-buten-2-one.[5]

Methyl 3-methyl-2-furoate has been prepared previously, presumably from the acid.[6]

[1] Eastman Kodak Co., Rochester, New York.

[2] *Org. Syntheses Coll. Vol.* **3**, 550 (1955).

[3] *Organic Chemical Bulletin*, **24**, No. 3, Eastman Kodak Co., Rochester, New York, 1952.

[4] Burness, U. S. pat. 2,760,985 [*C. A.*, **51**, 2854 (1957)].

[5] Burness, *J. Org. Chem.*, **21**, 102 (1956); U. S. pat. 2,772,295 [*C. A.*, **51**, 7424 (1957)].

[6] Asahina, *Acta phytochim.* (*Japan*), **2**, 12 (1924) [*Chem. Zentr.*, **1924**, II, 1694].

METHYL γ-METHYL-γ-NITROVALERATE

(Valeric acid, 4-methyl-, 4-nitro-, methyl ester)

$$(CH_3)_2CHNO_2 + CH_2{=}CH{-}CO_2CH_3 \xrightarrow{\text{Triton B}}$$

$$(CH_3)_2CCH_2CH_2CO_2CH_3$$
$$\underset{NO_2}{|}$$

Submitted by R. B. Moffett.[1]
Checked by N. J. Leonard and B. L. Ryder.

1. Procedure

A 500-ml. three-necked flask is fitted with a stirrer, a dropping funnel, and a thermometer placed so that the bulb is near the bottom of the flask. In the flask are placed 89 g. (1 mole) of 2-nitropropane (Note 1), 50 ml. of dioxane, and 10 ml. of a 40% aqueous solution of benzyltrimethylammonium hydroxide (Triton B) (Note 2), and the contents of the flask are warmed to 70°. Eighty-six grams (1 mole) of redistilled methyl acrylate

(Note 3) is added, with stirring, during 15 minutes. The temperature rises to about 100° during the addition and then drops to about 85°. The mixture is stirred and heated on a steam bath for 4 hours. After cooling, the contents of the flask are acidified with dilute hydrochloric acid and extracted with ether. The ether layer is washed twice with water, then with approximately 50 ml. of 0.1% sodium bicarbonate solution, and finally again with water. After the ethereal solution has been dried over anhydrous sodium sulfate, the drying agent is separated, the solvent is removed by distillation, and the product is distilled through a short fractionating column. A nearly colorless oil is obtained in a yield of 140–151 g. (80–86%); b.p. 79°/1 mm.; n_D^{20} 1.4408.

2. Notes

1. 2-Nitropropane from Commercial Solvents Corporation, Terre Haute, Indiana, was redistilled before use.

2. Available from Commercial Solvents Corporation or Rohm and Haas Company, Philadelphia, Pennsylvania.

3. Although the submitter knows of no case of an explosion with this type of nitro compound, it is recommended that adequate safety shields be employed during both the reaction and the distillation.

3. Methods of Preparation

Methyl γ-methyl-γ-nitrovalerate has been prepared by the Michael-type condensation of 2-nitropropane with methyl acrylate in the presence of a quaternary ammonium hydroxide [2] or triethylamine.[3]

[1] The Upjohn Company, Kalamazoo, Michigan.
[2] Bruson, U. S. pat. 2,342,119 and 2,390,918 [C. A., **38**, 4619 (1944)].
[3] Kloetzel, J. Am. Chem. Soc., **70**, 3571 (1948).

654 ORGANIC SYNTHESES

3-METHYL-4-NITROPYRIDINE-1-OXIDE

(3-Picoline, 4-nitro-, 1-oxide)

$$\underset{\underset{O}{\downarrow}}{\overset{\displaystyle \bigcirc\!\!\!\!\!\diagup^{CH_3}}{N}} + HNO_3 \xrightarrow{H_2SO_4} \underset{\underset{O}{\downarrow}}{\overset{\displaystyle \overset{NO_2}{\bigcirc\!\!\!\!\!\diagup^{CH_3}}}{N}} + H_2O$$

Submitted

Submitted by E. C. TAYLOR, JR.,[1] and ALDO J. CROVETTI.[2]
Checked by MAX TISHLER and HENRY B. LANGE.

1. Procedure

One hundred and eighty grams (1.65 moles) of liquefied 3-methylpyridine-1-oxide (Note 1) is added to 630 ml. of cold (0–5°) sulfuric acid (sp. gr. 1.84) contained in a 3-l. round-bottomed flask immersed in an ice-salt bath. The resulting mixture is cooled to about 10°, and 495 ml. of fuming yellow nitric acid (sp. gr. 1.50) is added in 50-ml. portions with shaking. An efficient spiral condenser (52 × 4 cm.) is attached to the flask, and the latter is placed in an oil bath. The temperature is slowly raised to 95–100° during 25–30 minutes, at which time gas evolution begins. After about 5 minutes the rate of gas evolution increases, and the oil bath is removed. A spontaneous and vigorous reaction commences which must be controlled by the application of an ice-water bath (Note 2). After the vigorous reaction has subsided to a moderate rate (about 5 minutes) the ice-water bath is removed, and the reaction is allowed to proceed for an additional 5–10 minutes. The oil bath is then replaced, and heating is continued at 100–105° for 2 hours.

The reaction mixture is cooled to 10° and poured onto 2 kg. of crushed ice contained in a 4-l. beaker. Addition of 1.36 kg. of sodium carbonate monohydrate (*Hood!*) (Note 3) in small portions with stirring causes the separation of the yellow crystalline product along with sodium sulfate. The mixture is then allowed to stand for 3 hours to expel nitrogen oxides. The yellow solid is collected by suction filtration, thoroughly washed with water, and rendered as dry as possible on the filter. The filtrates (about 4 l.) are transferred to a separatory funnel. The collected solid is extracted twice with 400–500 ml. portions of boiling chloroform, the combined extracts are used to extract the aqueous filtrates contained in the separatory funnel, and the extraction is repeated with several fresh 500-ml. portions of chloroform. The combined chloroform extracts are then given preliminary drying over anhydrous sodium sulfate and evaporated to dryness by distillation under reduced pressure. The residue is transferred to a 2-l. Erlenmeyer flask and dissolved in 1.5 l. of boiling acetone. The acetone solution is concentrated on a steam bath to 800–900 ml. (crystallization begins when the volume is about 1 l.) and then cooled at 5° for 6–8 hours. The product is filtered by suction, the filtrates are removed and saved, and the collected solid is washed with ether and dried. The yield is 162–173 g. (64–68%), m.p. 137–138°. The acetone filtrates mentioned above are boiled down to 150 ml. and chilled in an ice bath, and the crude product so obtained (m.p. 131–135°) is recrystallized from acetone to give an additional 13.5–16.5 g., m.p. 136–138°. The total yield is 178–187 g. (70–73%).

2. Notes

1. Freshly distilled 3-methylpyridine-1-oxide (b.p. 101–103°/ 0.7–0.8 mm.) will remain in a supercooled liquid state for several hours before solidifying. The highly hygroscopic solid may be melted on a steam bath in a tightly closed, previously weighed flask, and the melt poured slowly into the sulfuric acid. A large amount of heat is liberated in the mixing.

3-Methylpyridine-1-oxide (3-picoline-1-oxide) may be prepared by a method similar to that employed for pyridine-1-oxide (p.828). To a mixture of 600–610 ml. of glacial acetic acid and 200 g. (2.15 moles) of freshly distilled 3-methylpyridine (b.p. 141–143°) contained in a 2-l. round-bottomed flask is added, with shaking, 318 ml. (2.76 moles) of cold (5°) 30% hydrogen peroxide. The mixture is heated in an oil bath for 24 hours, with the internal temperature adjusted to 70 ± 5°. The excess acetic acid and water are removed under reduced pressure (30 mm.), and, after 500 ml. of distillate has been collected, the residue is diluted with 200 ml. of water and concentrated again, with the collection of 200 ml. of distillate. The residual mixture is cooled to 0–5° in an ice-salt bath, and 500 ml. of cold (0–5°) 40% aqueous sodium hydroxide solution is added slowly with shaking. The strongly alkaline solution is extracted with 2 l. of chloroform, and the extracts are given preliminary drying over anhydrous sodium carbonate. The extracts are filtered and concentrated by distillation under reduced pressure. The product is distilled under vacuum, b.p. 84–85°/0.3 mm., and the yield of 3-methylpyridine-1-oxide is 175–180 g. (73–77%).

2. Vigorous refluxing with evolution of nitrogen oxides occurs. Serious flooding of the condenser may result if no cooling is applied.

3. During the addition of sodium carbonate, large volumes of nitrogen oxides are evolved. In experiments where smaller quantities of sodium carbonate were used, lower yields (ca. 62%) were obtained.

3. Methods of Preparation

3-Methylpyridine-1-oxide has been prepared by the oxidation of 3-methylpyridine with hydrogen peroxide in glacial acetic acid,[3, 4] with 40% peracetic acid and sodium acetate,[5] and with perbenzoic acid in benzene.[6]

3-Methyl-4-nitropyridine-1-oxide has been prepared by the nitration of 3-methylpyridine-1-oxide hydrochloride with a mixture of concentrated sulfuric acid and potassium nitrate,[5] and by the nitration of 3-methylpyridine-1-oxide with a mixture of concentrated sulfuric acid and fuming nitric acid.[7] The preparation of this compound has been mentioned briefly by Talikowa.[8]

[1] Princeton University, Princeton, New Jersey.
[2] University of Illinois, Urbana, Illinois.
[3] Ochiai, Ikehara, Kato, and Ikekawa, *J. Pharm. Soc. Japan*, **71**, 1385 (1951) [*C. A.*, **46**, 7101 (1952)].
[4] Boekelheide and Linn, *J. Am. Chem. Soc.*, **76**, 1286 (1954).
[5] Herz and Tsai, *J. Am. Chem. Soc.*, **76**, 4184 (1954).
[6] Matsumura, *J. Chem. Soc. Japan*, **74**, 446 (1953) [*C. A.*, **48**, 6442 (1954)].
[7] Itai and Ogura, *J. Pharm. Soc. Japan*, **75**, 292 (1955) [*C. A.*, **50**, 1808 (1956)].
[8] Talikowa, *Wiadomości Chem.*, **7**, 169 (1953) [*C. A.*, **48**, 1337 (1954)].

3-METHYLOXINDOLE

(Oxindole, 3-methyl-)

$$\text{NHNH}_2 + (\text{CH}_3\text{CH}_2\text{CO})_2\text{O} \rightarrow$$

$$\text{NHNHCOCH}_2\text{CH}_3 + \text{CH}_3\text{CH}_2\text{CO}_2\text{H}$$

Submitted by ABRAHAM S. ENDLER and ERNEST I. BECKER.[1]
Checked by JAMES CASON and WARREN N. BAXTER.

1. Procedure

A. *β-Propionylphenylhydrazine.* To 130 g. (1.0 mole) of propionic anhydride (Note 1), contained in a 500-ml. wide-mouthed Erlenmeyer flask which is cooled in an ice bath, there is added slowly with swirling 108 g. (1.0 mole) of phenylhydrazine (Note 1) at such a rate that the maximum temperature does not exceed 60°. After addition is complete, the flask is corked (not so tightly as to give a vacuum which may collapse the flask) and allowed to stand 72 hours at room temperature. At the end of this period, the resultant solid cake is broken up and slurried with 100 ml. of toluene. The suspension is cooled to 5° in an ice bath; then the product is collected by suction filtration and washed with 200 ml. of a mixture of equal parts of cyclohexane and toluene, precooled to 5°. The yield of vacuum-dried, almost white crystals, m.p. 158–159°, is 140–145 g. (85–88%).

B. *3-Methyloxindole.* A mixture of 33 g. (0.20 mole) (Note 2) of β-propionylphenylhydrazine and 14 g. (0.33 mole) of freshly ground commercial calcium hydride (Note 3) is placed in a 500-ml. round-bottomed flask, which is equipped with a 15-cm. air condenser of

2.5 cm. diameter. The flask is immersed to two-thirds its depth in an oil bath, in a forced-draft hood, and the bath is heated, cautiously as 190° is approached. In the range 190–215°, a very vigorous exothermic reaction sets in (sometimes after a brief delay), and considerable gas and vapor is expelled through the air condenser. The vigorous reaction abates in 5 minutes or less, and the oil bath is raised to about 230°. After heating at this temperature has been continued for 30 minutes, the flask is removed from the bath and cooled to room temperature.

A mixture of 50 ml. of methanol and 20 ml. of water is cautiously added to the cooled reaction mixture, and this is followed by slow addition of concentrated hydrochloric acid (Note 4) until the pH is brought to 1–2 (Hydrion test paper). When effervescence has stopped, 50 ml. of water is added, and the contents of the flask are boiled gently for 1 hour. Additional concentrated hydrochloric acid is added as needed to maintain the specified pH (Note 5). At this point, all solid material should have been decomposed. The mixture is transferred to a 600-ml. beaker, using 10 ml. of methanol to wash out the flask, sufficient sodium hydroxide is added (Note 6) to bring the pH to about 3 (Hydrion test paper or more accurately with methyl orange indicator), and water is added to bring the volume to about 300 ml. The mixture is stirred in an ice bath until the oily layer solidifies, and the crude crystalline material is collected by suction filtration at about 15° and washed with 100 ml. of water. After vacuum drying at 70°, the yield of crude product is 16–19 g., but this material should be purified before further use.

The crude product is distilled in a short-path distillation apparatus composed of a 125-ml. Claisen flask connected by a ground-glass joint to a receiver which is a 300-ml. round-bottomed flask with side tubulature for evacuation. A 50-g. portion (Note 7) of crude 3-methyloxindole is placed in the distillation flask, which is completely immersed in an oil bath. The product is collected at approximately 132°/1.5 mm. The solid distillate is dissolved in 75 ml. of hot methanol, then 25 ml. of hot water is added. After crystallization has been continued for 24 hours at about 20°, there is obtained 35–37 g. (41–44% from β-propionylphenylhydrazine) of light-yellow crystals, m.p. 122.5–123.5° (softening at 121.5°) (Note 8). Five grams of less pure material is recoverable from the mother liquor and may be distilled with a succeeding batch.

2. Notes

1. Phenylhydrazine and propionic anhydride from Fisher Scientific Company were used without purification by the submitters. The checkers distilled the propionic anhydride before use, since old samples contained considerable propionic acid.

2. The ring closure is so highly exothermic that runs no larger than that described are recommended. Several lots may be combined for distillation, as suggested in the description.

3. The checkers used 40-mesh calcium hydride, from Metal Hydrides, Inc. The submitters report that there may also be used freshly ignited lime (22 g., 0.39 mole), but with a reduction of 10–20% in the yield.

4. The amount of concentrated hydrochloric acid is usually 40 ml. or more, depending on the quality of the calcium hydride.

5. This procedure is for the purpose of hydrolyzing unchanged starting material.

6. The final pH is brought to about 3 in order to minimize the solubility of 3-methyloxindole in strong acid without precipitating the phenylhydrazine which is present at this point.

7. The checkers obtained the same yields when the product of a single ring closure was distilled and crystallized.

8. After two crystallizations from butanol and two from toluene, the product may be obtained in about 25% recovery as white crystals of m.p. 123.8–124.6°.

3. Methods of Preparation

3-Methyloxindole has been prepared by the reduction of α-(2-nitrophenyl)propionic acid,[2] by heating β-propionylhydrazine with lime[3] or with sodium alkoxides,[4] by the reduction of the benzoyl derivative of oxindole-3-aldehyde[5] or 3-(methylaminomethylene)-oxindole,[6] and by the oxidation of skatole with potassium persulfate.[7]

[1] Polytechnic Institute of Brooklyn, Brooklyn, New York.

[2] Trinius, Ann., **227**, 274 (1885).

[3] Brunner, Monatsh., **18**, 533 (1897).

[4] C. F. Boehringer and Söhne (Waldhof b. Mannheim), Ger. pat. 218,727, Kl. 12 p., Jan. 10, 1910 [Frdl., **9**, 968 (1908–1910)].

[5] L. Horner, Ann., **548**, 134 (1941).

[6] Wenkert, Udelhofen, and Bhattacharyya, J. Am. Chem. Soc., **81**, 3763 (1959).

[7] Dalgliesh and Kelly, J. Chem. Soc., **1958**, 3726.

3-METHYL-1,5-PENTANEDIOL *

(1,5-Pentanediol, 3-methyl-)

$$\begin{array}{c} CH_3 \\ | \\ CH \\ HC \diagup \quad \diagdown CH_2 \\ HC \diagdown \quad \diagup CHOCH_3 \\ O \end{array} \xrightarrow[HCl]{H_2O} \begin{array}{c} CH_3 \\ | \\ OHCCH_2CHCH_2CHO \end{array}$$

$$\begin{array}{c} CH_3 \\ | \\ OHCCH_2CHCH_2CHO \end{array} \xrightarrow[Raney\ Ni]{H_2} \begin{array}{c} CH_3 \\ | \\ HO(CH_2)_2CH(CH_2)_2OH \end{array}$$

Submitted by RAYMOND I. LONGLEY, JR., and WILLIAM S. EMERSON.[1]
Checked by T. L. CAIRNS and JOHN F. HARRIS, JR.

1. Procedure

In a 2-l. three-necked flask equipped with a stirrer and thermometer are placed 336 g. (2.62 moles) of 3,4-dihydro-2-methoxy-4-methyl-2H-pyran (p. 311), 630 ml. of water, and 24 ml. of concentrated hydrochloric acid (sp. gr. 1.19). The mixture is stirred for 2 hours, during which the temperature may reach 50° but should not be permitted to rise higher. Solid sodium bicarbonate is then added until the solution is neutral to pH indicator paper (Note 1). The entire reaction mixture weighing about 1 kg. together with 39 g. of Raney nickel [2] is introduced into a 3-l. stainless-steel rocking hydrogenation autoclave. A hydrogen pressure of at least 1625 p.s.i. (Note 2) is applied, and the autoclave is heated to 125° and held there with shaking for 4 hours. The mixture is allowed to cool overnight, and the catalyst is separated either by suction filtration through Filter-Cel or by centrifugation. The solution is distilled through a 12-in. Vigreux column. After the methanol and water are separated, the 3-methyl-1,5-pentanediol distils at 139–146°/17 mm., 149–150°/25 mm. The yield is 251–256 g. (81–83%), n_D^{25} 1.4512 –1.4521.

2. Notes

1. The submitters report that β-methylglutaraldehyde may be isolated at this point from an analogous hydrolysis. The hydrolysis is carried out with 196 g. of 3,4-dihydro-2-methoxy-4-methyl-2H-pyran in 650 ml. of water and 15 ml. of concentrated hydrochloric acid for 3 hours. After neutralization with sodium bicarbonate, the solution is saturated with sodium chloride and extracted continuously with ether for 20 hours. The ether is removed by distillation, and the product is dried thoroughly by azeotropic distillation using a benzene-hexane mixture. Distillation affords β-methylglutaraldehyde, b.p. 85–86°/15 mm., n_D^{25} 1.4307–1.4351. Yields up to 90% have been secured. The aldehyde polymerizes on standing but is stable as a 50% solution in water or ether. The monomer may be recovered by careful destructive distillation of the polymer.

2. The initial hydrogen pressure should be high enough that it does not fall below 1000 p.s.i. during the shaking period.

3. Methods of Preparation

3-Methyl-1,5-pentanediol has been prepared by the hydrogenation of ethyl β-methylglutarate;[3] by the hydrogenation of ethyl β-methyl-α,γ-dicarbethoxyglutarate;[4] by the hydrogenation of β-methylglutaraldehyde;[5] and by heating 3,4-dihydro-2-methoxy-4-methyl-2H-pyran with water, hydrogen, and copper-chromium oxide, nickel on kieselguhr,[5] or Raney nickel catalyst.[6]

[1] Central Research Department, Monsanto Chemical Company, Dayton 7, Ohio.

[2] *Org. Syntheses Coll. Vol.* **3**, 181 (1955).

[3] Paden and Adkins, *J. Am. Chem. Soc.*, **58**, 2487 (1936).

[4] Wojcik and Adkins, *J. Am. Chem. Soc.*, **55**, 4939 (1933).

[5] Longley, Emerson, and Shafer, *J. Am. Chem. Soc.*, **74**, 2012 (1952); Ashley, Collins, Davis, and Sirett, *J. Chem. Soc.*, **1958**, 3298.

[6] Smith, U. S. pat. 2,546,019 [*C. A.*, **45**, 7589 (1951)].

β-METHYL-β-PHENYL-α,α'-DICYANOGLUTARIMIDE

(Glutarimide, 2,4-dicyano-3-methyl-3-phenyl-)

$C_6H_5(CH_3)C=C(CN)CO_2C_2H_5$

$+ CH_2(CN)CONH_2 + NaOC_2H_5 \rightarrow$

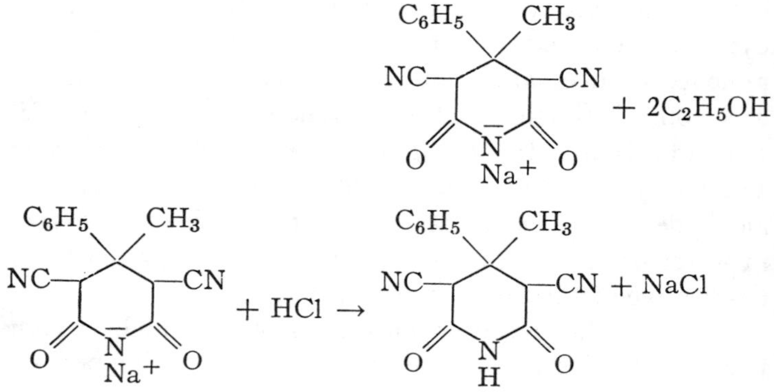

$+ 2C_2H_5OH$

$+ HCl \rightarrow$

$+ NaCl$

Submitted by S. M. McElvain and David H. Clemens.[1]
Checked by W. E. Parham, Perry W. Kirklin, Jr., and Wayland E. Noland.

1. Procedure

In a 2-l. Erlenmeyer flask fitted with a reflux condenser and arranged for magnetic stirring are placed 400 ml. of absolute ethanol (Note 1) and 11.5 g. (0.5 g. atom) of sodium added in small portions. After the sodium has reacted (Note 2), the clear solution is cooled to room temperature, and 42.0 g. (0.5 mole) of finely powdered cyanoacetamide (Note 3) is added with stirring over a period of 1 minute. Immediately thereafter 107.6 g. (0.5 mole) of ethyl (1-phenylethylidene)cyanoacetate (p. 463) is added. After about 20 minutes, the mixture becomes homogeneous and is allowed to stand at room temperature for 2 hours. Water (650 ml.) is added, followed by 100 ml. of concentrated hydrochloric acid in 1 portion. The resulting suspension is stirred thoroughly with a glass rod and placed in a refrigerator overnight. The product is then filtered by suction. The filter cake is sucked as dry as possible using a rubber dam, stirred to a paste with a mixture of 150 ml. of water and 50 ml. of 95% ethanol, and sucked dry. This process is repeated using 200 ml. of water, and the product is dried to constant weight in an oven at 45°. The yield is 114–116 g. (90–92%) of the dicyanoglutarimide, m.p. 274–278° (dec.) (Note 4).

2. Notes

1. Commercially available absolute ethanol is used without further drying.

2. The checkers report that the sodium ethoxide solution should be used promptly in order to avoid the formation of colored impurities.

3. Eastman Kodak white label grade is used after grinding in a mortar.

4. Recrystallization from absolute ethanol gives glistening plates melting at 286–287° (dec.).

3. Methods of Preparation

β-Methyl-β-phenyl-α,α'-dicyanoglutarimide has been prepared in low yield by the Guareschi condensation of acetophenone, ethyl cyanoacetate, and ammonia.[2]

[1] University of Wisconsin, Madison, Wisconsin.

[2] Phalnikar and Nargund, *J. Univ. Bombay*, **6**, Pt. II, 102 (1937) [*C.A.*, **32**, 3763 (1938)].

β-METHYL-β-PHENYLGLUTARIC ACID

(Glutaric acid, 3-methyl-3-phenyl-)

$$C_6H_5 \quad CH_3$$

$$NC \underset{\underset{H}{N}}{\overset{O}{\bigcirc}} CN \quad \xrightarrow[CH_3CO_2H]{\substack{H_2O \\ H_2SO_4}}$$

$$HO_2CCH_2\underset{C_6H_5}{\overset{CH_3}{C}}CH_2CO_2H + 3NH_4HSO_4 + 2CO_2$$

Submitted by S. M. McElvain and David H. Clemens.[1]
Checked by W. E. Parham, Perry W. Kirklin, Jr., and
Wayland E. Noland.

1. Procedure

In a 3-l. round-bottomed flask fitted with a small glass paddle
stirrer and a reflux condenser are placed 101 g. (0.4 mole) of
β-methyl-β-phenyl-α,α'-dicyanoglutarimide (p. 662), and a mixture of 500 ml. of water, 500 g. of concentrated sulfuric acid, and
400 ml. of glacial acetic acid. Without starting the stirrer (Note
1), the mixture is heated under reflux for 2 hours. Then the stirrer ˙s cautiously started, and reflux is continued for a total of 80
ʰ ɹrs. The reaction mixture is transferred to a 6-l. Erlenmeyer
ʌsk, 3 l. of water is added and the mixture cooled in a refrigera-
ʟor overnight. The precipitated acid is filtered by suction (Note
2), washed with 100 ml. of water, and air-dried to constant
weight. The crude, dry product is swirled for 5 minutes with 200
ml. of benzene, filtered by suction, washed with two 100-ml. portions of benzene, and again air-dried. The yield of acid, m.p.
136–140°, is 64.5–68 g. (72.5–76.5%) (Notes 3 and 4).

2. Notes

1. Use of the stirrer during the first 2 hours of the hydrolysis
results in excessive foaming.

2. In some runs a small amount of dark tar adhered to the side of the flask. This material was not isolated in the filtration.

3. The product is pure enough for most purposes, but it may be further purified by recrystallization from water (100 ml. for 20 g. of acid) to give material melting at 140–142° in 95% yield.

4. If the alkyl group of a β-alkyl-β-phenyl-α,α'-dicyanoglutarimide is larger than the methyl group, the hydrolysis to the corresponding glutaric acid should be modified as described by McElvain and Clemens, *J. Am. Chem. Soc.*, **80**, 3915 (1958).

3. Methods of Preparation

β-Methyl-β-phenylglutaric acid has been prepared by the hydrolysis of β-methyl-β-phenyl-α,α'-dicyanoglutarimide with sulfuric acid, and also by acid hydrolysis of the condensation product of 1,1-dichloroethylbenzene and ethyl sodiomalonate.[2]

[1] University of Wisconsin, Madison, Wisconsin.

[2] Phalnikar and Nargund, *J. Univ. Bombay*, **6**, Pt. II, 102 (1937) [*C.A.*, **32**, 3763 (1938)].

1-METHYL-3-PHENYLINDANE

(Indan, 1-methyl-3-phenyl-)

Submitted by MILTON J. ROSEN.[1]
Checked by R. T. ARNOLD and WILLIAM K. WITSIEPE.

1. Procedure

In a 500-ml. three-necked round-bottomed flask fitted with a mechanical stirrer and a reflux condenser are placed 50 g. (0.48 mole) of styrene (Note 1) and a previously cooled solution of 100 ml. of concentrated sulfuric acid in 150 ml. of water. The mixture is stirred vigorously (Note 2) and heated under reflux in an oil bath for approximately 4 hours.

Without interrupting stirring or heating, 50 ml. of concentrated sulfuric acid is added slowly through the condenser, and the mixture is stirred and heated for an additional 12 hours (Note 3).

The reaction mixture is cooled, cautiously poured into 250 ml. of cold water with stirring, and allowed to separate into layers. The upper hydrocarbon layer is removed, and the lower layer is extracted with three 50-ml. portions of ether. The combined ether extracts and hydrocarbon layer are washed successively with about 30 ml. each of a saturated solution of sodium bicarbonate, water, and a saturated solution of calcium chloride, and then dried over anhydrous calcium chloride. The ether is removed by distillation, and the product is distilled under reduced pressure. The yield of 1-methyl-3-phenylindane, b.p. 168–169°/16 mm. (Note 4), n_D^{20} 1.5811 ± 0.0005, is 38.5–40.5 g. (77–81%) (Note 5).

2. Notes

1. Commercial styrene, distilled from a water bath at about 80–100 mm. pressure just before use, is employed.

2. It is essential to use an efficient stirring device, capable of forming a dispersion of the styrene in the acid layer.

3. The sulfuric acid is added in two portions in order to minimize higher-polymer formation.[2]

4. The product also distils at 150–151°/6.5 mm.

5. The submitter states that α-methylstyrene can be converted to 1,1,3-trimethyl-3-phenylindane, b.p. 154–155°/8 mm., m.p. 50.4–52.1°, by the same general procedure. The yield is 86–89% of the theoretical amount. The 1,1,3-trimethyl-3-phenylindane may be purified further by one recrystallization from three times its weight of 80% isopropyl alcohol. The yield of purified product, m.p. 51.8–52.3°, is 80–83% (based on the original weight of monomer used).

3. Methods of Preparation

1-Methyl-3-phenylindane has been prepared by the treatment

of dimeric styrene [3] with aqueous sulfuric acid, and by the hydrogenation of 1-methyl-3-phenyl-Δ^2-indene.[4]

[1] Brooklyn College, Brooklyn, New York.
[2] Rosen, *J. Org. Chem.*, **18**, 1701 (1953).
[3] Spoerri and Rosen, *J. Am. Chem. Soc.*, **72**, 4918 (1950).
[4] Müller and Körmendy, *J. Org. Chem.*, **18**, 1237 (1953).

METHYL 2-THIENYL SULFIDE

[Thiophene, 2-(methylthio)-]

Submitted by J. CYMERMAN-CRAIG and J. W. LODER.[1]
Checked by CHARLES C. PRICE and E. A. DUDLEY.

1. Procedure

In a 1-l. three-necked flask fitted with a liquid-sealed mechanical stirrer, a reflux condenser, and a dropping funnel are placed 8 g. (0.33 g. atom) of magnesium turnings and 600 ml. of absolute ether. There is placed in the dropping funnel 70 g. (0.33 mole) of 2-iodothiophene (p. 545), the stirrer is started, and about 10 ml. of the iodothiophene is added. The reaction generally begins within a few minutes (Note 1), and the iodothiophene is then added dropwise at such a rate that moderate refluxing occurs. When the addition is complete, the mixture is refluxed gently until only a small residue of unreacted magnesium remains. The solution is then cooled in an ice bath, the dropping funnel is removed, and 10.7 g. (0.33 g. atom) of finely powdered sulfur (Note 2) is added (Note 3), the funnel is replaced, and the mixture is refluxed (Note 4) for 45 minutes. The solution is again cooled in an ice bath, and 22.6 ml. (0.36 mole) of methyl iodide is added dropwise from the funnel, and the stirring is then discontinued (Note 5). The reaction mixture is refluxed 10 hours. It is then cooled, and an aqueous solution of ammonium chloride is run in with vigorous stirring (Note 6). The liquid is transferred to a separatory funnel, the lower aqueous layer is run off, and the ethereal solution is washed three times with a 2% solution of potassium hydroxide, then with water, and finally is dried over anhydrous sodium sulfate. The ether is removed by distillation at ordinary pressure, and the residual dark liquid is distilled under reduced pressure. The yield of colorless methyl 2-thienyl sulfide is 23–26 g. (53–60%), b.p. 82–86°/22 mm., n_D^{25} 1.5978 (Notes 7 and 8).

2. Notes

1. If the reaction does not start, it may be assisted by the addition of a small amount of methylmagnesium iodide in ethereal solution.

2. The sulfur was distilled and then ground in a mortar before use.

3. The mechanical stirring is continued. The sulfur dissolves during the refluxing period to give a clear yellow solution. A sludge which adheres to the bottom of the flask may form, but the yield is unaltered.

4. On reheating, a vigorous reaction with the sulfur occurs. A means of cooling the reaction flask should be at hand to ensure control of the reaction.

5. The solution may be left overnight at this stage, sealed under nitrogen.

6. The decomposition of the unreacted Grignard reagent is best carried out in a hood to remove the strong odor of thiols.

7. Considerable decomposition occurs if distillation is attempted at atmospheric pressure. The product is sometimes pale yellow, and darkens slightly on standing.

8. A dark oil remains which decomposes at this pressure when strongly heated.

3. Methods of Preparation

Methyl 2-thienyl sulfide has been prepared by the action of phosphorus trisulfide on dimethyl succinate [2] and by the action of methyl iodide on the sodium salt of 2-thiophenethiol,[3] both of which methods are of little preparative value. The procedure described above is that of Cymerman-Craig and Loder.[4] 2-Bromothiophene has been used in the latter method in place of the iodine derivative.[5]

[1] University of Sydney, Sydney, Australia.
[2] Steinkopf and Leonhardt, *Ann.*, **495**, 166 (1932).
[3] Meyer and Neure, *Ber.*, **20**, 1756 (1887).
[4] Cymerman-Craig and Loder, *J. Chem. Soc.*, **1954**, 237.
[5] Gol'dfarb, Kalik, and Kirmalova, *Zhur. Obshchei Khim.*, **29**, 2034 (1959) [*C. A.*, **54**, 8775 (1960)].

METHYL β-THIODIPROPIONATE

(Propionic acid, 3,3'-thiodi-, dimethyl ester)

$$2CH_2{=}CHCO_2CH_3 + H_2S \xrightarrow{CH_3CO_2Na} S(CH_2CH_2CO_2CH_3)_2$$

Submitted by EDWARD A. FEHNEL and MARVIN CARMACK.[1]
Checked by ARTHUR C. COPE and JAMES J. RYAN.

1. Procedure

Caution! This preparation should be conducted in a hood to avoid exposure to poisonous hydrogen sulfide.

A mixture of 150 g. (1.74 moles) of methyl acrylate (Note 1), 100 g. (0.73 mole) of sodium acetate trihydrate, and 800 ml. of 95% ethanol (Note 2) is placed in a 2-l. two-necked flask fitted with an efficient reflux condenser and a sintered-glass bubbler tube which reaches almost to the bottom of the flask. The mixture is heated on the steam bath until all the solid is dissolved and the solution is refluxing gently. A steady stream of hydrogen sulfide gas (Note 3) is introduced into the boiling solution through the bubbler tube while heating is continued for a period of 25 hours. The gas flow is then stopped, the condenser is changed for distillation, and the solvent, along with some unreacted methyl acrylate, is distilled from the mixture on the steam bath. About 200 ml. of ether and 400 ml. of water are added to the residue in the flask, and after thorough agitation the layers are separated. The aqueous layer is washed with four 50-ml. portions of ether, and the washings are added to the original ether layer. The combined ether extracts are dried over anhydrous sodium sulfate, the ether is removed by distillation on the steam bath, and the residue is distilled under reduced pressure. Methyl β-thiodipropionate is obtained as a colorless oil, b.p. 162–164°/18 mm., 138–139°/6 mm.; n_D^{25} 1.4713. The yield is 128–145 g. (71–81%).

2. Notes

1. A good grade of commercial methyl acrylate containing hydroquinone is entirely satisfactory. Material of doubtful quality may be redistilled (into a receiver containing hydroquinone), b.p. 78–81°.

2. Either methanol or 95% ethanol may be used as the solvent. No ester interchange was observed to occur under the conditions employed.

3. Commercial tank hydrogen sulfide was used. The flow was regulated by passing the gas through a gas-washing bottle containing a little water. A rate of about 3–5 bubbles per second was maintained during the reaction.

3. Methods of Preparation

β-Thiodipropionic acid esters have been prepared by the addition of hydrogen sulfide to the corresponding acrylic esters in the presence of basic catalysts with [2] or without [3] solvents. The ethyl ester has also been prepared by the treatment of ethyl β-chloropropionate with sodium sulfide.[4]

[1] University of Pennsylvania, Philadelphia, Pennsylvania.

[2] I. G. Farbenind. A.-G., Fr. pat. 797,606 [*Chem. Zentr.*, **107**, II, 1062 (1936)] [*C. A.*, **30**, 8244 (1936)]; cf. also Ger. pat. 669,961 [*C. A.*, **33**, 5415 (1939)].

[3] Gershbein and Hurd, *J. Am. Chem. Soc.*, **69**, 241 (1947).

[4] Arndt and Bekir, *Ber.*, **63**, 2393 (1930).

3-METHYLTHIOPHENE *

(Thiophene, 3-methyl-)

$$\underset{\substack{|\\CO_2Na}}{CH_2}\text{---}\underset{\substack{|\\CO_2Na}}{CHCH_3} \xrightarrow{P_4S_7} \underset{\substack{\|\\HC}}{HC}\text{---}\underset{\substack{\|\\CH}}{CCH_3}$$
$$\diagdown S \diagup$$

Submitted by R. F. Feldkamp and B. F. Tullar.[1]
Checked by Cliff S. Hamilton, Y. H. Wu, and William J. Raich.

1. Procedure

Caution! This preparation should be conducted in a well-ventilated hood to avoid exposure to hydrogen sulfide.

A 1-l. three-necked round-bottomed flask is fitted with a ground-glass-sealed stirrer, an immersed thermometer, a gas inlet tube, an addition funnel (Note 1), and a distilling head, wrapped with asbestos cloth, connected to a condenser arranged for distillation. The flask is charged with 150 ml. of mineral oil (Note 2), and the system is thoroughly swept out with a slow stream of carbon dioxide admitted through the gas inlet tube while the flask is heated with an electric heating mantle. When the temperature of the oil reaches 240–250°, a slurry of 90 g. (0.51 mole) of powdered anhydrous disodium methylsuccinate (Note 3) and 100 g. (0.287 mole) of phosphorus heptasulfide (Note 4) in 250 ml. of mineral oil is placed in the addition funnel. With efficient stirring and a slow continuous stream of carbon dioxide passing through the system, the slurry is added to the hot mineral oil at such a rate as to effect fairly rapid distillation of 3-methylthiophene accompanied by considerable gas evolution (mostly hydrogen sulfide). During the addition, which requires about 1 hour, the temperature is maintained at 240–250° (Note 5). The temperature is then raised to 275° and stirring continued in the inert atmosphere for an additional hour or until distillation ceases. The total distillate, amounting to 33–38 ml., is washed with two 50-ml. portions of 5% sodium hydroxide solution and finally with 50 ml. of water (Note 6). The crude 3-methyl-thiophene is then distilled (Note 7). A small fore-run is discarded, and the fraction boiling between 112° and 115° is collected. The yield is 26–30 g. (52–60%), n_D^{25} 1.5170 ± 0.0005 (Notes 8, 9, and 10).

2. Notes

1. A gravity funnel fitted with a glass rod of suitable diameter is satisfactory for regulating the addition of the slurry.

2. Dowtherm A may also be used as a solvent for this reaction. However, because the boiling point (about 265°) is close to the reaction temperature, considerable quantities of Dowtherm distil along with the 3-methylthiophene. The total distillate amounts to 60–75 ml., which after washing and fractionally distilling gives the same yield as with mineral oil. Dowtherm gives a more fluid slurry and final residue and can be easily recovered by distillation at reduced pressure.

3. Disodium methylsuccinate was made by hydrogenating a concentrated solution of itaconic acid supplied by Chas. Pfizer and Company in aqueous sodium hydroxide (pH 8.7) over Raney nickel catalyst at 50 p.s.i. and 80–100°. After the catalyst was removed by filtration, the product was isolated by evaporation of the water and the residue was dried in a vacuum oven at 70–80°.

4. Phosphorus heptasulfide was obtained from the Oldbury Electrochemical Company. It has been shown [2] that the "phosphorus trisulfide" used by earlier workers for such fusions was actually somewhat impure phosphorus heptasulfide.

5. The rate of addition of the slurry and the reaction temperature should be carefully controlled. About two-thirds of the crude 3-methylthiophene distillate is collected during the addition of the slurry. Further heating at 260–275° gives the remainder of the material.

6. An emulsion may be obtained on further washing with water. A small amount of sodium chloride (2–3 g.) dissolved in the wash water assists in breaking such emulsions.

7. Rather violent foaming may occur during this distillation but is easily controlled in a 250-ml. flask. The distilling head and

flask may be wrapped in asbestos cloth in order to increase the speed of distillation.

8. When fresh mineral oil was used (first run), a yield of 52–54% was obtained. However, when the recovered mineral oil (Note 9) was used, the yield increased to 60%.

9. The reaction residue is allowed to cool to room temperature and is filtered by suction. The dark mineral oil filtrate may be reused in the process. The malodorous filter cake is not pyrophoric and is almost completely soluble in water or dilute alkali with liberation of hydrogen sulfide.

10. The submitters obtained comparable yields on twice the scale described. They also applied the same procedure to disodium succinate to give a 25% yield of thiophene comparable to the yield obtained by dry fusion.[3]

3. Methods of Preparation

3-Methylthiophene has been prepared by the dry fusion of a salt of methylsuccinic acid and phosphorus "trisulfide."[4] This reaction was later investigated quite completely in respect to ratio of reactants, rate of heating, carbon dioxide atmosphere, and dilution of reactants with sand.[5] An excellent technical method for preparing methylthiophenes has been described which involves a vapor-phase reaction of preheated sulfur with pentanes.[6] 3-Methylthiophene has also been prepared by adding 50% crude isoprene (amylenes) to molten sulfur at 350°,[7] and by passing a mixture of 2-methyl-2-butene and sulfur dioxide over a chromium oxide-aluminum oxide catalyst at 450°.[8]

[1] Sterling-Winthrop Research Institute, Rensselaer, New York.

[2] Pernert and Brown, *Chem. Eng. News*, **27**, 2143 (1949).

[3] *Org. Syntheses Coll. Vol.* **2**, 578 (1943).

[4] Volhard and Erdmann, *Ber.*, **18**, 454 (1885).

[5] Linstead, Noble, and Wright, *J. Chem. Soc.*, **1937**, 915.

[6] Rasmussen and Hansford, U. S. pat. 2,450,686 [*C. A.*, **43**, 1067 (1949)].

[7] Shepard, Henne, and Midgley, Jr., *J. Am. Chem. Soc.*, **56**, 1355 (1934).

[8] Yur'ev and Khmel'nitskiĭ, *Doklady Akad. Nauk S.S.S.R.*, **92**, 101 (1953) [*C. A.*, **48**, 10725 (1954)].

METHYL p-TOLYL SULFONE

(Sulfone, methyl p-tolyl)

$$p\text{-}CH_3C_6H_4SO_2Cl + Na_2SO_3 + 2NaHCO_3 \rightarrow$$
$$p\text{-}CH_3C_6H_4SO_2Na + H_2O + NaCl + Na_2SO_4 + 2CO_2$$
$$2p\text{-}CH_3C_6H_4SO_2Na + (CH_3)_2SO_4 \rightarrow$$
$$2p\text{-}CH_3C_6H_4SO_2CH_3 + Na_2SO_4$$

Submitted by L. Field and R. D. Clark.[1]
Checked by John C. Sheehan and M. Gertrude Howell.

1. Procedure

In a 4-l. beaker (Note 1) provided with a mechanical stirrer and thermometer are placed 600 g. (4.76 moles) of anhydrous sodium sulfite, 420 g. (5.0 moles) of sodium bicarbonate, and 2.4 l. of water. The mixture is heated on a hot plate at 70–80° and is maintained at this temperature by switching the hot plate off occasionally, while 484 g. (2.54 moles) of p-toluenesulfonyl chloride (Note 2) is added in portions of 5–10 g., with stirring, during 3 hours. When addition is complete, the mixture is heated and stirred at 70–80° for 1 hour (Note 3). The mixture is then removed from the hot plate and allowed to stand for 4 to (preferably) 10 hours.

The solid sodium p-toluenesulfinate which separates is collected by filtration and mixed with 400 g. (4.76 moles) of sodium bicarbonate and 490 g. (370 ml., 3.88 moles) of dimethyl sulfate (Note 4) in a 3-l. three-necked round-bottomed flask equipped with a mechanical stirrer, a reflux condenser, and a 1-l. separatory funnel containing 925 ml. of water. Water (75–100 ml.) is added from the separatory funnel to make the mixture fluid enough for stirring. The remainder of the water is then added dropwise, with stirring, during 3 hours. The mixture is then heated under reflux for 20 hours.

After the mixture is cooled to 75°, 200 ml. of benzene is added (Note 5). The mixture is stirred briefly, and the liquid is decanted from the solid into a 5-l. separatory funnel. The aqueous layer is separated and extracted again with 200 ml. of benzene.

The aqueous layer is then returned to the separatory funnel, and the solid in the reaction flask is washed in with it by means of 2 l. of water. The mixture is shaken with 200-ml. portions of benzene until all solid has dissolved (usually three portions of benzene suffice). All the benzene extracts are combined and dried with 20 g. of anhydrous calcium chloride. The drying agent is removed by filtration and washed with two 20-ml. portions of benzene. Benzene is removed from the filtrate by distillation under reduced pressure (Note 6), and the solid which separates is dried further at about 10 mm. and room temperature to constant weight. The yield (Note 2) of methyl *p*-tolyl sulfone is 298–317 g. (69–73%), m.p. 83–87.5°. Further purification is generally unnecessary, but, if desired, the product may be recrystallized from carbon tetrachloride or ethanol-water (1:1). The submitters state that the method may be extended to the preparation of methyl phenyl sulfone and, presumably, of methyl aryl sulfones generally (Note 7).

2. Notes

1. A porcelainized metal bucket is a convenient alternative.

2. The submitters used Eastman Kodak Company practical grade; although the solid is somewhat oily, the m.p. is 66–68°. The checkers used *p*-toluenesulfonyl chloride purchased from Matheson, Coleman and Bell and obtained a yield of 78–82%.

3. The volume at the end of the heating period should not exceed 2.4 l. If, after 1 hour of heating, the volume exceeds 2.4 l., the mixture is heated longer.

4. Eastman Kodak Company practical grade was used, b.p. 69–70°/10 mm. Dimethyl sulfate is toxic and must be handled with caution. This part of the preparation should be run in a hood with provision for containing the contents should breakage occur. It is unlikely that any dimethyl sulfate survives the 20-hour reflux period however, and the submitters reported that no difficulty whatever was encountered in handling the mixture after this point without special precautions; nevertheless, they recommend that the possible presence of dimethyl sulfate be borne in mind. Ammonia is a specific antidote for dimethyl sulfate and should be at hand to destroy any accidentally spilled.

A solution of a detergent in dilute ammonia water may be used to clean glassware used in transfers. The hazards associated with dimethyl sulfate are described by Sax.[2]

5. Extraction with benzene improves the yield somewhat but offers the more important advantage of permitting effective drying of the sulfone when it is to be used in metalation reactions. If this advantage is not sought, the reaction mixture simply can be allowed to cool to room temperature and to stand until crystallization is complete (2–3 hours). The solid is then collected by filtration and washed with water (about six 200-ml. portions) until the washings give no precipitate with barium chloride solution. The yield is 272 g. (63%); it can be increased by 12.6 g. (66%) by extraction of the mother liquor and wash water. The sulfone thus obtained contains only 0.25% of benzene-insoluble material and has m.p. 86.5–87.5°.

6. If the benzene is removed at a temperature not exceeding 50°, the sulfone is obtained as well-formed crystals.

7. According to the submitters, methyl phenyl sulfone[3] is obtained similarly from benzenesulfonyl chloride in 66–69% yields, m.p. 86–88°.

3. Methods of Preparation

Methyl p-tolyl sulfone has been prepared by oxidation of methyl p-tolyl sulfide with hydrogen peroxide[4,5] or ruthenium tetroxide,[6] by alkylation of sodium p-toluenesulfinate with methyl iodide[7,8] or with methyl potassium sulfate,[9] by decarboxylation of p-tolylsulfonylacetic acid,[7] by thermal decomposition of tetramethylammonium p-toluenesulfinate,[10] by reaction of cis-bis-(p-tolylsulfonyl)ethene with sodium hydroxide (low yield),[11] by the reaction of methanesulfonyl chloride with toluene in the presence of aluminum chloride (mixture of isomers),[12] by reaction of alkali with 3-p-tolylsulfonyl-7-hydroxynaphtho-α-pyrone,[13] by heating allyl p-tolyl sulfone with sodium hydroxide;[14] and by the decomposition of sodium p-toluenesulfonyl-acetate.[15]

The method described here is that of Field and Clark.[3] It involves preparation of sodium p-toluenesulfinate by the procedure of Oxley et al.[8] and alkylation by modification of a method used by Baldwin and Robinson [16] for the preparation of methyl phenyl sulfone.

[1] Vanderbilt University, Nashville, Tennessee.
[2] Sax, *Handbook of Dangerous Materials*, p. 147, Reinhold Publishing Corporation, New York, 1951.
[3] Field and Clark, *J. Org. Chem.*, **22**, 1129 (1957).
[4] Zinckc and Frohneberg, *Ber.*, **43**, 837 (1910).
[5] Gilman and Beaber, *J. Am. Chem. Soc.*, **47**, 1449 (1925).
[6] Djerassi and Engle, *J. Am. Chem. Soc.*, **75**, 3838 (1953).
[7] Otto, *Ber.*, 18, 154 (1885).
[8] Oxley, Partridge, Robson, and Short, *J. Chem. Soc.*, **1946**, 763.
[9] Otto, *Ann.*, **284**, 300 (1895).
[10] Meyer, *Chem. Zentr.*, **80**, 1800 (1909).
[11] Truce and McManimie, *J. Am. Chem. Soc.*, **76**, 5745 (1954).
[12] Truce and Vriesen, *J. Am. Chem. Soc.*, **75**, 5032 (1953).
[13] Tröger and Dunkel, *J. prakt. Chem.*, **104**, 311 (1922).
[14] Backer, Strating, and Drenth, *Rec. trav. chim.*, **70**, 365 (1951).
[15] O'Connor and Verhoek, *J. Am. Chem. Soc.*, **80**, 288 (1958).
[16] Baldwin and Robinson, *J. Chem. Soc.*, **1932**, 1445.

β-METHYL-δ-VALEROLACTONE *

(Valeric acid, 5-hydroxy-3-methyl-, δ-lactone)

$$HO(CH_2)_2\overset{\displaystyle CH_3}{\underset{\displaystyle |}{CH}}(CH_2)_2OH \xrightarrow[\text{chromite}]{\text{Copper}}$$

Submitted by RAYMOND I. LONGLEY, JR., and WILLIAM S. EMERSON.[1]
Checked by T. L. CAIRNS and W. W. GILBERT.

1. Procedure

A 1-l. three-necked flask fitted with an efficient stirrer, a thermometer, and a reflux condenser attached to a device for measuring gas evolution (Note 1) is charged with 197 g. (1.67 moles) of 3-methyl-1,5-pentanediol (p. 660) and 10 g. of copper chromite (Note 2). The mixture is heated rapidly to 200° (Note 3) with good stirring and is held at 195–205° for 1.5–3.0 hours, during which time 3.1 cu. ft. of hydrogen is evolved (Note 4). The product is distilled directly from the flask with stirring through a 2 by 120 cm. Vigreux column (Note 5). The yield of β-methyl-δ-valerolactone is 172–180 g. (90–95%), b.p. 110–111°/ 15 mm., n_D^{25} 1.4495.

2. Notes

1. A standard wet test meter may be used.

2. Copper chromite is prepared according to *Organic Syntheses* [2] and washed with sodium bicarbonate solution. The glycol is slurried with sodium bicarbonate and filtered before use.

3. At this point gas evolution becomes so rapid that the temperature tends to drop slightly.

4. If gas evolution subsides more catalyst may be added.

5. The column is substituted for the reflux condenser in the same set-up. Stirring during distillation prevents serious bumping.

3. Methods of Preparation

β-Methyl-δ-valerolactone has been prepared by heating 3-methyl-1,5-pentanediol with copper chromite in the liquid phase,[3] by passing the vapors of 3-methyl-1,5-pentanediol over copper on pumice,[3] by heating 2-methoxy-4-methyl-3,4-dihydro-2H-pyran with water and copper chromite,[3] by treating 3-methylglutaraldehyde with aqueous alkali,[3] and by reducing β-methylglutaric anhydride with sodium and ethanol.[4] The present method was first developed by Kyrides and Zienty.[5]

[1] Monsanto Chemical Company, Dayton 7, Ohio.

[2] *Org. Syntheses Coll. Vol.* **2**, 142 (1943).

[3] Longley, Emerson, and Shafer, *J. Am. Chem. Soc.*, **74**, 2012 (1952); Emerson, Longley, and Shafer (to Monsanto Chemical Co.), U. S. pat. 2,680,118 [*C. A.*, **49**, 6315 (1955)].

[4] Sircar, *J. Chem. Soc.*, **1928**, 898.

[5] Kyrides and Zienty, *J. Am. Chem. Soc.*, **68**, 1385 (1946).

MONOBENZALPENTAERYTHRITOL

(m-Dioxane-5,5-dimethanol, 2-phenyl-)

$$C_6H_5CHO + C(CH_2OH)_4 \xrightarrow{H^+}$$

$$C_6H_5CH \underset{O—CH_2}{\overset{O—CH_2}{\diagup}} C \underset{CH_2OH}{\overset{CH_2OH}{\diagdown}} + H_2O$$

Submitted by C. H. Issidorides and R. Gulen.[1]
Checked by M. S. Newman and Arlen B. Mekler.

1. Procedure

In an open 3-l. three-necked flask are placed 180 g. (1.32 moles) of pentaerythritol (Note 1) and 1.3 l. of water. The flask is fitted with an efficient mechanical stirrer and a graduated dropping funnel containing 147 g. (1.38 moles) of benzaldehyde (Note 2). The mixture in the flask is heated until all the solid dissolves and is then allowed to cool undisturbed (Note 3).

When the solution has cooled to room temperature, stirring is started and 6.6 ml. of concentrated hydrochloric acid is added through the open neck of the flask, followed by 30 ml. of benzaldehyde from the dropping funnel. When the precipitate of monobenzalpentaerythritol starts forming, dropwise addition of benzaldehyde is begun (Note 4). After the addition of benzaldehyde is completed, the mixture is stirred for an additional 3 hours (Note 5). The precipitate is collected (Note 6) on a Büchner funnel and washed with ice-cold water which has been made slightly alkaline by addition of sodium carbonate. The solid is transferred to a 3-l. round-bottomed flask, 1 l. of water (slightly alkaline with sodium carbonate) is added, and the mixture is heated to 100° (Note 7). After about 10 minutes at this temperature the hot mixture is filtered quickly through a fluted filter paper (Note 8). The solid remaining on the filter paper is washed with 50 ml. of hot water (made slightly alkaline with sodium carbonate) (Note 9). The combined aqueous filtrates are cooled in an ice bath for several hours, and the crystals are collected on a

Büchner funnel and dried. The dry product is heated under reflux for 15 minutes in an Erlenmeyer flask with 200 ml. of toluene, and the hot mixture is allowed to cool to room temperature, with continuous agitation (stirring rod) to prevent formation of hard lumps. Finally, the mixture is cooled in an ice bath for 5 hours, and the solid product is collected on a Büchner funnel and dried (Note 10). The yield of monobenzalpentaerythritol melting at 134–135° is 215–227 g. (73–77%).

2. Notes

1. Eastman Kodak Company white label grade pentaerythritol was used. The checkers used a commercial sample obtained from the Heyden Chemical Corporation.

2. The third neck may be left open and used later for addition of hydrochloric acid and for introduction of the thermometer.

3. The temperature of the solution should not be allowed to go below 25°; otherwise pentaerythritol will precipitate.

4. The addition should take about 2.5 hours. The temperature of the mixture should be kept at 25–29°.

5. If stirring is continued for a substantially longer period, the yield of monobenzalpentaerythritol is somewhat decreased.

6. The product should be collected immediately, as losses result if there is delay at this point.

7. The mixture should be stirred continuously during the heating. The use of a mechanical stirrer is recommended.

8. Use of a steam-heated funnel is recommended.

9. The solid remaining finally on the filter paper may be recrystallized from 1-butanol to give 1–2 g. of dibenzalpentaerythritol melting at 159–160°.

10. The product and the toluene in the Erlenmeyer flask form a solid mass which is difficult to remove. The operation is made easier by adding 70 ml. of ice-cold toluene to the flask and stirring, before transferring the product to the Büchner funnel. One or two additional 20-ml. portions of cold toluene may be used to remove the product completely.

3. Methods of Preparation

The procedure described is based on the method of E. Bogra-chov.[2]

[1] American University of Beirut, Beirut, Lebanon.
[2] E. Bograchov, *J. Am. Chem. Soc.*, **72**, 2268 (1950).

MONOBROMOPENTAERYTHRITOL

[1,3-Propanediol, 2-(bromomethyl)-2-(hydroxymethyl)-]

$$C(CH_2OH)_4 + HBr \rightarrow BrCH_2C(CH_2OH)_3 + H_2O$$

Submitted by S. Wawzonek, A. Matar, and C. H. Issidorides.[1]
Checked by Charles C. Price and G. Venkat Rao.

1. Procedure

In a 3-l. two-necked flask (Note 1) fitted with a dropping funnel and a reflux condenser are placed 200 g. (1.47 moles) of penta-erythritol, 1.5 l. of glacial acetic acid, and 17 ml. of 48% hydro-bromic acid (Note 2). After a reflux period of 1.5 hours, 170 ml. of 48% hydrobromic acid is added and the solution is heated under reflux for an additional 3 hours. At the end of this time 96 ml. of 48% hydrobromic acid is added and the heating under reflux is continued for 3 hours. The solution is distilled under reduced pressure to remove as much of the acetic acid and the water as possible, first on a steam bath and finally for 15 minutes in an oil bath at 140–150°, as the pressure is reduced to 10 mm. The viscous residue is transferred to a 2-l. flask and treated with 750 ml. of 98% ethanol and 50 ml. of 48% hydrobromic acid. The flask is provided with an efficient fractionating column (Note 3), and the solution is fractionated slowly until about 500 ml. of distillate is collected. Then a second 750-ml. portion of ethanol is added, and the fractionation is continued slowly until 750 ml. more distillate is collected (Note 4). Finally, the flask is fitted with a Claisen head and a condenser set for downward distillation, and the remaining alcohol is removed as completely as possible under reduced pressure.

Benzene (500 ml.) is added to the residue and distilled at atmospheric pressure. The last traces of benzene are removed by heating for 15 minutes in an oil bath at 150°, as the pressure is reduced to 8 mm. The same procedure is repeated, using a second 500-ml. portion of benzene (Note 5). The viscous residue is then heated under reflux for several hours with 500 ml. of dry ether, with frequent shaking, until it becomes white and granular (Note 6). After cooling thoroughly, the ether is decanted, and the solid is washed twice with two 200-ml. portions of dry ether. The solid is powdered thoroughly and dried in a vacuum desiccator. The dry solid is then extracted exhaustively in a Soxhlet extractor with 600 ml. of dry ether (Note 7). The ether extract is cooled overnight in an ice bath, and the precipitated monobromopentaerythritol is collected by filtration and washed with two 200-ml. portions of cold, dry ether. The yield of crude product melting at 72–73° is 145–160 g. (49–54% of the theoretical). One recrystallization from a mixture of 3 parts of chloroform and 2 parts of ethyl acetate by volume raises the melting point to 75–76°, recovery 75–85%.

2. Notes

1. For best results the flask should have standard-taper, ground-glass fittings.

2. Eastman Kodak Company white label grade pentaerythritol and 48% hydrobromic acid were used.

3. A 40-cm. column packed with glass beads is satisfactory.

4. The fractionation should be carried out slowly to ensure complete alcoholysis of the bromoacetate. The boiling point during the collection of the first 500 ml. of distillate remains constant at around 72°, corresponding to the ethanol-ethyl acetate azeotrope.

5. The purpose of this operation is to remove completely the water present in the product. Toluene may be substituted for benzene.

6. If the product has a tendency to form a hard mass, it is advisable to break up the solid with a stirring rod.

7. The extraction is very slow and requires several hours for completion, depending upon the rate of refluxing of the ether.

Usually crystals of the monobromopentaerythritol begin to deposit on the walls of the extraction flask after the first hour. At the end of the extraction 30–35 g. of unchanged pentaerythritol remains in the extraction thimble. Dibromopentaerythritol, formed as a side product, is present in the ether washings.

3. Methods of Preparation

Monobromopentaerythritol has been prepared by the action of 66% hydrobromic acid on pentaerythritol in glacial acetic acid [2] and by the action of 66% hydrobromic acid on pentaerythritol [3] at 120°. The procedure described is a modification of the method of Beyaert and Hansens.[3]

[1] State University of Iowa, Iowa City, Iowa.
[2] Beyaert and Hansens, *Natuurw. Tijdschr.* (*Ghent*), **22**, 249 (1940) [*C. A.*, **37**, 5373 (1943)].
[3] Barbiere and Matti, *Bull. soc. chim. France* [5]**5**, 1565 (1938).

MONOVINYLACETYLENE

(1-Buten-3-yne)

$$Cl-CH_2CH=\overset{\displaystyle Cl}{\overset{\displaystyle |}{C}}-CH_3 \xrightarrow[\text{HOCH}_2\text{CH}_2\text{OH}]{\text{KOH}} H_2C=CH-C\equiv CH$$

Submitted by G. F. HENNION, CHARLES C. PRICE, and THOMAS F. McKEON, JR.[1]
Checked by MAX TISHLER and JOHN E. ALLEGRETTI.

1. Procedure

A 2-l. three-necked flask, heated by a Carbowax bath, is equipped with a motor-driven Trubore stirrer, Teflon paddle, Trubore bearing (Note 1), a Friedrichs condenser, and a 250-ml. dropping funnel. The dropping funnel is connected to the flask by a 24/40 ground-glass joint with a side arm made of 7-mm. tubing. The side arm is connected through a calcium chloride drying tower and a bubbler to a nitrogen tank.

The top of the condenser is connected to a horizontal tube (ca. 2 x 25 cm.) partially filled with granular anhydrous calcium chloride. The horizontal tube is then connected to a 100-ml.

graduated cylinder immersed in an acetone-Dry Ice mixture for collection of the product. The cylinder is also equipped with an escape tube protected by a calcium chloride drying tube. Four hundred grams of powdered technical potassium hydroxide flakes (Note 2) is placed in the flask, and 500 ml. of ethylene glycol is added. This mixture is stirred vigorously while adding 100 ml. of *n*-butyl Cellosolve (Note 3). The system is swept with a rapid stream of nitrogen for 15–20 minutes while the temperature of the oil bath is raised to 165–170°. The flow of nitrogen is then reduced to a rate just sufficient to maintain an atmosphere of nitrogen in the system.

One hundred and twenty-five grams of 1,3-dichloro-2-butene (1.0 mole) (Note 4) is added at a rate of about 3 drops per second. Addition should be complete in 0.75–1 hour (Note 5). During the addition the temperature of the oil bath is maintained at 165–170° (Note 6), and the reaction mixture is stirred vigorously. Heating is continued for 1 hour after the addition of dichlorobutene is complete.

The yield of crude product obtained is 31.2 g. (39 ml., 60%) (Notes 7 and 8). The crude product may be purified by distillation through a low-temperature column (Note 9) to yield 22.4–24.8 g. (28–31 ml., 43–48%) of monovinylacetylene, b.p. 0–6° (Note 10) The storage of monovinylacetylene in the presence of oxygen has been reported to lead to explosive compounds. It is therefore suggested that the product be stored under an inert atmosphere.

2. Notes

1. An oil-sealed stirrer was found unsuitable because the pressure drop through the system was greater than the pressure drop across the oil-sealed stirrer. The Trubore stirrer and bearing were purchased from Ace Glass Inc., Vineland, New Jersey.

2. The potassium hydroxide was weighed as flakes, then ground rapidly with a mortar and pestle to roughly the consistency of granulated sugar, and added as soon as possible to the reaction flask to minimize moisture uptake. The flakes may be used directly without grinding, the only difference being that the mixture of potassium hydroxide, glycol, and Cellosolve should be

heated for a slightly longer time before any dichlorobutene is added in order to allow as complete a dispersion as possible.

3. The butyl Cellosolve is added to control the foaming during the reaction.

4. The dichlorobutene used was Eastman Kodak Company technical grade. Distillation before use was not observed to increase the yields.

5. The rate of addition is very important. If the time of addition of 1 mole of dichlorobutene goes much beyond 1 hour, the yield will decrease noticeably.

6. This is apparently an optimum temperature. Lower temperatures lead to lower yields, while higher temperatures do not change the yield appreciably.

7. The calculations of yields are based on $d^{-80} = 0.8$ extrapolated from data at higher temperature.[2]

8. The preparation also has been run on a 2-mole scale, using double the quantities specified in a 3-l. flask. The yields obtained were comparable to those obtained on the 1-mole scale.

9. The low-temperature column shown in Fig. 12 was 75 cm. long, made of 7-mm. glass tubing. The entire column was surrounded by an evacuated, silvered jacket. The column, for the lower 60 cm., was a tightly coiled spiral (3.8 cm. diameter). The remaining length was straight and was surrounded by a coolant cup in which liquid of any temperature could be placed to control the refluxing temperature. For the purpose, a calcium chloride-water-Dry Ice mixture was used to keep the temperature of the cup at approximately −5 to 0°. The vapor from the column was condensed and collected in a graduated cylinder immersed in an acetone-Dry Ice bath (Note 10). The major features of infrared spectra of the vapor from redistilled product are summarized in Table I (Note 11).

10. Warming the still pot with 60° water was necessary to distil all the volatile gases toward the end of the distillation.

11. The spectrum of the fractionated monovinylacetylene was comparable, in band peaks and intensities, with previously reported spectra. A previously reported band at 5.8 μ, however, was not found in any of the samples, suggesting that an impurity was present in the sample previously reported.

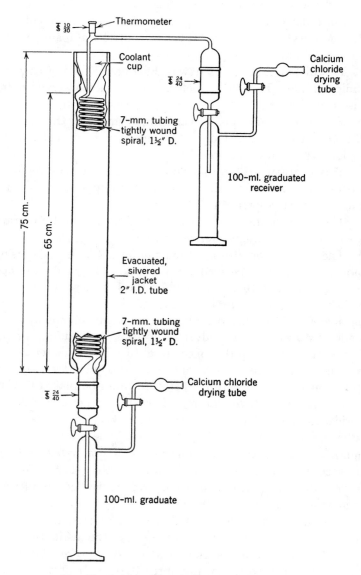

Fᴵɢ. 12. Low-temperature distillation column.

TABLE I

THE MAJOR BANDS FOR THE INFRARED SPECTRUM OF MONOVINYLACETYLENE GAS

(350 mm. pressure, 5-cm. cell)

Wavelength, μ	Absorption, %	Wavelength, μ	Absorption, %
3.02	93	6.20	87
3.22	62	7.0–7.15	33
3.30	67	7.9–8.1	98
4.73	9	9.17	49
5.13	19	9.37	18
5.41	59	10.2–11.2	99
5.46	52	13.67	27
5.80	—		

3. Methods of Preparation

Monovinylacetylene has been prepared by the decomposition of a diquaternary ammonium base,[3] by the dimerization of acetylene,[4] by the dehydrohalogenation of dihalobutenes in liquid ammonia,[5] and by heating dimethylvinylethynylcarbinol with potassium hydroxide.[6] The procedure described has been published.[7]

[1] University of Notre Dame, Notre Dame, Indiana.

[2] Kuchinskaya and Anitina, *Khim. Referat. Zhur.*, **2**, No. 5, 65 (1939) [*C. A.*, **34**, 2783 (1940)].

[3] Willstätter and Wirth, *Ber.*, **46**, 535 (1913); Slobodin, *Zhur. Obshcheĭ Khim.*, **27**, 2473 (1957) [*C. A.*, **52**, 7119 (1958)].

[4] Nieuwland, Calcott, Downing, and Carter, *J. Am. Chem. Soc.*, **53**, 4197 (1931); Apotheker (to E. I. du Pont de Nemours and Co.), U. S. pat. 2,875,258 [*C. A.*, **53**, 12172 (1959)]; Komada, *J. Chem. Soc. Japan*, **63**, 949, 955, 963, 970 (1942) [*C. A.*, **41**, 3743 (1947)]; Iguchi and Kanno, *J. Soc. Chem. Ind. Japan*, **45**, Suppl. binding, 9–10 (1942) [*C. A.*, **44**, 8313 (1950)]; Sugino, Aiya, and Ariga, *J. Soc. Chem. Ind. Japan*, **46**, 573 (1943); **47**, 199 (1944) [*C. A.*, **42**, 6310 (1948)]; Tichý, *Chem. Průmysl*, **5**, 493 (1955) [*C. A.*, **50**, 10439 (1956)]; Apotheker (to E. I. du Pont de Nemours and Co.), U. S. pat. 2,924,631 [*C. A.*, **54**, 11989 (1960)].

[5] Croxall and Van Hook (to Rohm and Haas Co.), U. S. pat. 2,623,077; *J. Am. Chem. Soc.*, **76**, 1700 (1954).

[6] Sokolov, Litvinenko, and Isin, *Izvest. Akad. Nauk Kazakh. S.S.R., Ser. Khim.*, **1959**, No. 2, 68 [*C. A.*, **53**, 19840 (1959)].

[7] Hennion, Price, and McKeon, *J. Am. Chem. Soc.*, **76**, 5160 (1954).

MUCOBROMIC ACID

(Acrylic acid, 2,3-dibromo-3-formyl-)

$$\begin{array}{c} HC\text{---}CH \\ \| \qquad \| \\ HC \qquad C\text{---}CHO \\ \diagdown \diagup \\ O \end{array} + 5Br_2 + 3H_2O \rightarrow$$

$$\begin{array}{c} BrC\text{---}CHO \\ \| \\ BrC\text{---}CO_2H \end{array} + CO_2 + 8HBr$$

Submitted by G. A. TAYLOR.[1]
Checked by B. C. McKUSICK, E. L. MARTIN, and W. R. BRASEN.

1. Procedure

A mixture of 50 g. (45 ml., 0.52 mole) of freshly distilled furfural and 500 ml. of water is stirred vigorously in a 2-l. three-necked round-bottomed flask equipped with a dropping funnel and a thermometer that dips into the liquid. The flask is immersed in an ice bath, and 450 g. (144 ml., 2.81 moles) of bromine is added, while the temperature of the reaction mixture is kept below 5° (Note 1). After the addition is complete, the thermometer is replaced by a reflux condenser, and the mixture is stirred and boiled for 30 minutes. The reflux condenser is replaced by a still head and condenser, and excess bromine is removed by distilling the liquid until the distillate is almost colorless (Note 2).

The reaction mixture is evaporated to dryness under reduced pressure at a water pump on a steam bath, using a trap cooled in ice and salt to condense the hydrobromic acid (Note 3). The solid residue is cooled in an ice bath and triturated with 30–50 ml. of ice water. A few grams of sodium bisulfite, dissolved in water, is added to discharge a slight yellow discoloration. The cold mixture is filtered with suction to separate crude mucobromic acid, which is washed with two small portions of ice water. The crude mucobromic acid weighs 125–132 g. (93–99%). It is dissolved in about 110 ml. of boiling water, 2–5 g. of decolorizing carbon is added, the hot mixture is stirred for 10 minutes and filtered, and the filtrate is cooled to 0–5°. Colorless crystals of mucobromic acid separate from the filtrate; weight 100–112 g. (75–83%); m.p. 124–125°.

2. Notes

1. If the temperature is allowed to rise much above 10°, the yield is considerably reduced. Without cooling, the mixture becomes quite hot, the yield is decreased by half, and tarry material is formed.

2. Mucobromic acid can be obtained in about 63% yield (85 g.) by adding 5 g. of decolorizing carbon at this point, stirring the mixture at the boil for 10 minutes, filtering it hot, and cooling the filtrate to 0°. The crude mucobromic acid that crystallizes, weight about 105 g., is recrystallized from 120 ml. of water.

3. It is important to get rid of all the hydrobromic acid, for its presence increases the amount of mucobromic acid lost in the trituration step.

3. Methods of Preparation

This preparation is adapted from that described by Simonis.[2] It closely follows the practical details given in *Organic Syntheses*[3] for the preparation of mucobromic acid from the relatively expensive furoic acid.

[1] Department of Biochemistry and Dyson Perrins Laboratory, Oxford University, Oxford, England.

[2] Simonis, *Ber.*, **32**, 2085 (1899).

[3] *Org. Syntheses Coll. Vol.* **3**, 621 (1955).

1-NAPHTHALDEHYDE *

$$C_{10}H_7CH_2Cl + (CH_2)_6N_4 \rightarrow [C_{10}H_7CH_2 \cdot C_6H_{12}N_4]^+Cl^-$$

$$[C_{10}H_7CH_2 \cdot C_6H_{12}N_4]^+Cl^- + 6H_2O \rightarrow$$

$$C_{10}H_7CHO + CH_3NH_2 + 2NH_3 + 5HCHO + NH_4Cl$$

Submitted by S. J. ANGYAL, J. R. TETAZ, and J. G. WILSON.[1]
Checked by R. S. SCHREIBER and PAUL E. MARLATT.

1. Procedure

Caution! *Precautions should be taken to avoid contact with* *1-chloromethylnaphthalene, which is a lachrymator and a vesicant, and with the aldehyde, which seems to possess the same properties to a lesser degree.*

In a 1-l. flask fitted with a reflux condenser are placed 106 g. (0.6 mole) of 1-chloromethylnaphthalene[2] (Note 1), 168 g. (1.2 moles) of hexamethylenetetramine, 250 ml. of glacial acetic acid, and 250 ml. of water. This mixture is heated under reflux for 2 hours. In about 15 minutes the solution becomes homogeneous, and then an oil starts to separate. After the reflux period, 200 ml. of concentrated hydrochloric acid is added and refluxing is continued for an additional 15 minutes (Note 2). After cooling, the mixture is extracted with 300 ml. of ether; the ether layer is washed three times with 100-ml. portions of water, then with 100 ml. of 10% sodium carbonate solution (Note 3), and again with 100 ml. of water. The ether extract is dried with about 15 g. of anhydrous sodium sulfate and filtered, and the ether is removed by distillation. The residual liquid is distilled under reduced pressure, the distillate being collected at 105–107°/0.2 mm. or 160–162°/18 mm. (Note 4). The yield of colorless 1-naphthaldehyde freezing between 0.0° and 2.5° (Note 5) is 70–77 g. (75–82%).

2. Notes

1. The chloromethylnaphthalene used melted at 24–26°. Material with a lower melting point can be used, but the yield is correspondingly smaller; e.g., a sample having a melting point of

15–18° gave a 73% yield of slightly impure 1-naphthaldehyde. The checkers found that crude chloromethylnaphthalene obtained from the preparation in *Organic Syntheses* [2] could be used with good results. Naphthalene, paraformaldehyde, hydrochloric acid, and phosphoric acid are heated under reflux according to the procedure described. After the crude product is washed with water, 10% potassium carbonate, and water, it is dissolved directly in 500 ml. of glacial acetic acid, diluted with 500 ml. of water, and treated with hexamethylenetetramine by the procedure described above. The over-all yield of almost colorless 1-naphthaldehyde is 162 g., b.p. 162–164°/18 mm.; n_D^{25} 1.6503 (52% yield based on naphthalene).

In this variation of the preparation, it is best to use a wide-bore tube as a condenser to remove the unreacted naphthalene. After the naphthalene has been distilled, the wide-bore tube is replaced with an ordinary condenser and the naphthaldehyde is distilled in the usual manner.

2. The various amines and aldehydes present combine to form Schiff's bases. If these are not hydrolyzed by a strong acid, they will contaminate the final product.

3. Care should be exercised when washing the solution with sodium carbonate because some carbon dioxide is evolved.

4. The brown distillation residue contains some methylene-α-naphthylmethylamine.

5. The melting point of 1-naphthaldehyde given by Stephen [3] (33–34°) is apparently incorrect. A sample that was purified through the bisulfite addition compound and redistilled had a freezing point of 2.5°.

In no instance could the checkers obtain a completely colorless product even though it was redistilled several times with ordinary laboratory distilling apparatus.

3. Methods of Preparation

1-Naphthaldehyde has been prepared from calcium α-naphthoate by distillation with calcium formate; [4] from α-naphthylcarbinol by oxidation with chromic acid, [5, 6] N-bromosuccinimide, [7] or N-chlorosuccinimide; [8] from α-naphthylglyoxylic acid by heating with aniline and hydrolyzing the anil; [9] and from α-naphthylmagnesium bromide and ethoxymethyleneaniline [10, 11] or ethyl orthoformate.[12, 13] This Grignard reagent also has been converted to the dithioacid with carbon disulfide and the acid taken to 1-naphthaldehyde (through the semicarbazone).[14]

1-Naphthaldehyde has been made from α-naphthonitrile by reduction with stannous chloride,[3, 15, 16] sodium triethoxyaluminohydride,[17] lithium triethoxyaluminohydride,[18] or diisobutylaluminum hydride; [19] from naphthalene by the action of aluminum chloride, hydrogen cyanide, and hydrochloric acid,[20] by treatment with dichloromethyl methyl ether in the presence of stannic chloride,[21] and by the reaction with carbon monoxide in the presence of boron trifluoride and hydrogen fluoride; [22] from α-naphthoyl chloride by reduction with lithium tri-*tert*-butoxyaluminohydride; [23] from α-naphthoic acid N,N-dimethylamide by reduction with lithium diethoxyaluminohydride; [24] from the 1,3-diphenyltetrahydroimidazole derivative of 1-naphthaldehyde by hydrolysis; [25] from the reaction of α-naphthylmethylpyridinium bromide with *p*-nitrosodimethylaniline and hydrolysis of the resulting nitrone; [26] by the decomposition of 1-(benzenesulfonhydrazidocarbonyl)naphthalene with sodium carbonate; [27] and by the oxidation of α-methylnaphthalene with selenium dioxide.[28]

1-Naphthaldehyde has been obtained by means of the Sommelet reaction [29] from α-chloro- or α-bromomethylnaphthalene and hexamethylenetetramine in aqueous alcohol [30–34] or glacial acetic acid.[35–37] This method has been improved in the present procedure by the use of 50% acetic acid as a solvent.

[1] University of Sydney, Sydney, Australia.
[2] *Org. Syntheses Coll. Vol.* **3**, 195 (1955).
[3] Stephen, *J. Chem. Soc.*, **127**, 1877 (1925).
[4] Lugli, *Gazz. chim. ital.*, **11**, 394 (1881).
[5] Bamberger and Lodter, *Ber.*, **21**, 259 (1888).
[6] Ziegler, *Ber.*, **54**, 739 (1921).
[7] Lecomte and Gault, *Compt. rend.*, **238**, 2538 (1954).
[8] Hebbelynck, *Ind. chim. belge*, **16**, 483 (1951) [*C. A.*, **46**, 10127 (1952)].
[9] Rousset, *Bull. soc. chim. France*, [3] **17**, 303 (1897).
[10] Monier-Williams, *J. Chem. Soc.*, **89**, 275 (1906).
[11] Gattermann, *Ann.*, **393**, 227 (1912).
[12] Bodroux, *Compt. rend.*, **138**, 701 (1904); *Bull. soc. chim. France*, [3] **31**, 587 (1904).
[13] Kikkoji, *Biochem. Z.*, **35**, 67 (1911).
[14] Wuyts, Berman, and Lacourt, *Bull. soc. chim. Belg.*, **40**, 665 (1931).
[15] King, l'Ecuyer, and Openshaw, *J. Chem. Soc.*, **1936**, 353.

[16] Stephen and Stephen, *J. Chem. Soc.*, **1956**, 4695.
[17] Hesse and Schrödel, *Angew. Chem.*, **68**, 438 (1956).
[18] Brown, Shoaf, and Garg, *Tetrahedron Letters*, **3**, 9 (1959).
[19] Zakharkin and Khorlina, *Doklady Akad. Nauk S.S.S.R.*, **116**, 422 [*C. A.*, **52**, 8040 (1958)].
[20] Kinkel, Ayling, and Beynon, *J. Chem. Soc.*, **1936**, 342.
[21] Rieche, Gross, and Höft, *Chem. Ber.*, **93**, 88 (1960).
[22] Gresham and Tabet (to E. I. du Pont de Nemours & Co.), U. S. pat. 2,485,237 [*C. A.*, **44**, 2027 (1950)].
[23] Brown and Rao, *J. Am. Chem. Soc.*, **80**, 5377 (1958).
[24] Brown and Tsukamoto, *J. Am. Chem. Soc.*, **81**, 502 (1959).
[25] Bestmann and Schulz, *Chem. Ber.*, **92**, 530 (1959).
[26] Kröhnke, Ger. pat. 755,943 [*C. A.*, **52**, 5451 (1958)].
[27] Brown, Carter, and Tomlinson, *J. Chem. Soc.*, **1958**, 1843.
[28] Jensen, Kjaer, and Linholt, *Acta Chem. Scand.*, **6**, 180 (1952).
[29] Sommelet, *Compt. rend.*, **157**, 852 (1913).
[30] Mayer and Sieglitz, *Ber.*, **55**, 1846 (1922).
[31] Anderson and Short, *J. Chem. Soc.*, **1933**, 485.
[32] Coles and Dodds, *J. Am. Chem. Soc.*, **60**, 853 (1938).
[33] Rupe and Brentano, *Helv. Chim. Acta*, **19**, 586 (1936).
[34] Ruggli and Preuss, *Helv. Chim. Acta*, **24**, 1350 (1941).
[35] Badger, *J. Chem. Soc.*, **1941**, 536.
[36] Price and Voong, *J. Org. Chem.*, **14**, 115 (1949).
[37] Angyal, Morris, Tetaz, and Wilson, *J. Chem. Soc.*, **1950**, 2141.

NAPHTHALENE-1,5-DISULFONYL CHLORIDE

(1,5-Naphthalenedisulfonyl chloride)

$$\text{(naphthalene with } SO_3Na \text{ and } NaO_3S\text{)} + 2PCl_5 \rightarrow \text{(naphthalene with } SO_2Cl \text{ and } ClO_2S\text{)} + 2POCl_3 + 2NaCl$$

Submitted by P. D. Caesar.[1]
Checked by R. T. Arnold and George P. Scott.

1. Procedure

In a 1-l. round-bottomed flask is placed a mixture of 204.3 g. (1 mole) of finely divided phosphorus pentachloride and 132.8 g. (0.4 mole) of the disodium salt of naphthalene-1,5-disulfonic acid which has been dried previously at 140° for 48 hours (Note 1). The flask is provided with an air condenser which is fitted at the top with a calcium chloride drying tube. It is then placed in an oil bath which is heated to 110°, and the mixture is maintained at that temperature for 1 hour. The condenser is removed for brief intervals now and then during the heating period, and the reactants are stirred by means of a glass rod. At the end of the heating period, the product is a thick paste.

The flask and contents are placed on a steam cone and heated for 2 hours under vacuum (furnished by a water aspirator) (Note 2) in order to remove the phosphorus oxychloride formed in this process as well as most of the unreacted phosphorus pentachloride. The dry cake is pulverized in a mortar and transferred to a 4-l. beaker. To this is added 750 ml. of distilled water and 2 l. of chloroform. The mixture is placed on a steam bath, heated to boiling, and stirred vigorously until nearly all the solid dissolves. By means of a separatory funnel the layers are separated while still hot. The chloroform solution is again heated to boiling and filtered through a large fluted filter into an Erlenmeyer flask.

After sufficient chloroform has been evaporated to give a solution volume approximating 250 ml., the solution is cooled in an ice bath and the crystalline product collected on a filter. By further concentration of the mother liquor, an additional quantity of naphthalene-1,5-disulfonyl chloride is obtained. A total yield of 85–115 g. (65–88%) of recrystallized material results; m.p. 181–183°.

2. Notes

1. The powdered solids should be thoroughly mixed before heating. This may be done by inserting a rubber stopper into the neck of the reaction flask and shaking vigorously for about 2 minutes. However, an appreciable pressure develops in the flask and care must be taken in removing the stopper.

2. The reaction flask should be connected to the receiver by a tube of large bore and equipped, preferably, with ground-glass fittings. Two traps between the receiver and the aspirator are desirable to assure no contact between the phosphorus chlorides and water.

3. Methods of Preparation

Naphthalene-1,5-disulfonyl chloride has been prepared by the reaction of naphthalene with chlorosulfonic acid; [2-7] however, the yields are generally poor and the conditions difficult to reproduce. It also has been obtained by treating disodium 1,5-naphthalene-disulfonate with chlorosulfonic acid.[8]   The present method has been published.[9]

[1] University of Illinois, Urbana, Illinois.
[2] Armstrong, *J. Chem. Soc.*, **24**, 173 (1871).
[3] Armstrong, *Ber.*, **15**, 204 (1882).
[4] Corbellini, *Giorn. chim. ind. applicata*, **9**, 118 (1927) [*C. A.*, **22**, 2938 (1928)].
[5] Pollak, Heimberg-Krauss, Katscher, and Lustig, *Monatsh.*, **55**, 358 (1930).
[6] General Tire & Rubber Co., Brit. pat. 802,654 [*C. A.*, **53**, 11330 (1959)].
[7] Marotta and Swisher (to Monsanto Chemical Co.), U. S. pat. 2,827,487 [*C. A.*, **52**, 16324 (1958)].
[8] Spryskov and Apar'eva, *Zhur. Obshchei Khim.* (*J. Gen. Chem.*), **19**, 1576 (1949) [*C. A.*, **44**, 1082 (1950)].
[9] Marvel and Caesar, *J. Am. Chem. Soc.*, **73**, 1097 (1951).

1,5-NAPHTHALENEDITHIOL

Submitted by P. D. CAESAR.[1]
Checked by RICHARD T. ARNOLD, W. E. PARHAM, and R. M. SCRIBNER.

1. Procedure

A 2-l. round-bottomed flask having standard-taper, ground-glass fittings is equipped with a bulb condenser (Note 1) and an efficient Hershberg stirrer.   To this are added with stirring 600 g. (2.0 moles) of 33% sulfuric acid, 20 g. (0.06 mole) of finely divided 1,5-naphthalenedisulfonyl chloride (p. 693), and 100 g. (1.5 g. atoms) of zinc dust amalgam (Note 2) at room temperature (Note 3).   The zinc dust amalgam is added directly after the disulfonyl chloride in the course of 2–5 minutes.   The mixture is heated to reflux, held there for about 6 hours, and allowed to cool overnight without agitation (Note 4).

The product is filtered, and the precipitate is extracted with a total of 1 l. of warm ether (Note 5). The combined ether extracts are evaporated to a volume of 50 ml., cooled, and filtered. The filtrate is further evaporated to a volume of about 10 ml., cooled, and again filtered. The precipitates melt at 119–121° and total 7.1–9.1 g. This represents a yield of 60–77%.

The 1,5-naphthalenedithiol can be further purified to a melting point of 120–121° by sublimation under high vacuum in a molecular still, followed by reprecipitation of the water-soluble disodium salt of the sublimate from excess hydrochloric acid. The pure compound obtained from 9.1 g. of product weighs 8.6 g. (73%).

2. Notes

1. The product collects in the condenser, and it may be necessary to clear the condenser with a glass rod.

2. A good grade of zinc and mercury(II) chloride should be used. The zinc dust amalgam is prepared by dissolving 20 g. of mercury(II) chloride in a solution of 10 ml. of concentrated hydrochloric acid in 300 ml. of distilled water, and adding quickly, with stirring, 100 g. of zinc dust (Schaar chemicals, 95% purity). After 10–15 minutes of stirring and crushing lumps, the mixture is filtered through a Büchner funnel, and the zinc dust amalgam is carefully washed with a total of 500 ml. of distilled water containing a trace of hydrochloric acid. The water is then removed by ethanol, the ethanol by ether, and most of the ether by air. It is advisable to remove the zinc dust amalgam from the Büchner funnel and add it to the reduction mixture before all the ether is removed to assure minimum contact with the air.

An occasional batch of zinc dust failed to effect the desired reduction, possibly because of excessive oxide deposition on the surface of the zinc. It is suggested, therefore, that the surface of the zinc dust be cleaned with dilute hydrochloric acid just before amalgamation.

3. This technique eliminates the long induction period at 0° and the violent foaming described in the preparation of thio-

phenol.[2] However, in a larger-scale operation it would be advisable to check the rate of addition of the zinc dust somewhat, lest an exception arise.

4. The overnight period was a matter of convenience and is not considered to be vital to the completeness of the reaction.

5. When unamalgamated zinc dust is used, a considerable proportion of yellow insoluble product is often noted at this point. Since the disulfide has been isolated as an intermediate in a similar reduction of *m*-chlorosulfonylbenzoic acid,[3] it is probable that this material is a mixture of disulfides of varying molecular weight.

3. Application

This method has been applied successfully to the preparation of phenoxybenzene-4,4'-dithiol (84% of the theoretical amount), diphenylmethane-4,4'-dithiol, and *m*-sulfhydrylbenzoic acid[3] (80%). It did not prove satisfactory for the preparation of higher-melting thiols of lower solubility, such as 2,7-naphthalenedithiol, 2,6-naphthalenedithiol, and 4,4'-biphenyldithiol. These were better prepared by the use of tin(II) chloride 2-hydrate in glacial acetic acid saturated with hydrogen chloride.[4]

4. Methods of Preparation

1,5-Naphthalenedithiol can be prepared by adding 1,5-naphthalenedisulfonyl chloride to an ethanol solution of tin(II) chloride 2-hydrate saturated with hydrogen chloride.[5] An 80% yield of the crude dithiol melting at 103° was previously reported using zinc dust and sulfuric acid.[6]

[1] University of Illinois, Urbana, Illinois.
[2] *Org. Syntheses Coll. Vol.* 1, 504 (1941).
[3] Tennyson, Private communication.
[4] Marvel and Caesar, *J. Am. Chem. Soc.*, **73**, 1097 (1951).
[5] Corbellini and Albenga, *Gazz. chim. ital.*, **61**, 111 (1931); Tilak, *Proc. Indian Acad. Sci.*, **33**, 71 (1951).
[6] Braun and Ebert, *Ber.*, **25**, 2735 (1892).

1,4-NAPHTHOQUINONE *

Submitted by E. A. BRAUDE and J. S. FAWCETT.[1]
Checked by CHARLES C. PRICE, R. S. SCHREIBER, R. D. BIRKENMEYER,
PAUL F. KIRK, and WILLIAM BRADLEY REID, JR.

1. Procedure

In a 2-l. three-necked flask, fitted with a mechanical stirrer, a 1-l. dropping funnel, and a thermometer, is placed a solution of 120 g. (1.2 moles) of pure chromium trioxide (Note 1) in 150 ml. of 80% aqueous acetic acid. The flask is surrounded by a freezing mixture of ice and salt, and, when the temperature of the contents of the flask has fallen to 0°, a solution of 64 g. (0.5 mole) of naphthalene in 600 ml. of glacial acetic acid is gradually added, with constant stirring, over a period of 2–3 hours. The internal temperature is maintained at about 10–15°. Stirring is continued overnight, during which time the reaction mixture and cooling bath gradually attain room temperature (Note 2). The dark green solution is then set aside for 3 days and occasionally stirred.

The crude naphthoquinone is precipitated by pouring the reaction mixture into 6 l. of water. The yellow precipitate is filtered, washed with 200 ml. of water, and dried in a desiccator. The product can be crystallized from 500 ml. of petroleum ether (b.p. 80–100°) (Note 3) and separates in the form of long yellow needles, m.p. 124–125°. The yield is 14–17 g. (18–22%) (Note 4).

2. Notes

1. An equivalent quantity of technical grade chromium trioxide, ground to a fine powder, may be employed.
2. It is necessary to continue vigorous stirring at this stage in

order to prevent local overheating and to keep the mixture from setting to a solid mass.

3. Crystallization from petroleum ether (b.p. 80–100°) is far more convenient than steam distillation as a method of purification, and a product of high purity is obtained after a single crystallization. The checkers used Skellysolve C (b.p. 88–115°) with consistent results. Crystallization from ether has also been employed.[2]

4. The yield is substantially unchanged on increasing the proportion of chromium trioxide used in the oxidation. The submitters report consistent yields of 25–28 g. (32–35%).

3. Methods of Preparation

The present method of preparation is adapted from Miller.[3] Although the yield is relatively low, the method is less costly and time-consuming than those starting from α-naphthol[2] or 1,4-benzoquinone.[4] Other methods that have been employed include the oxidation of naphthalene with hydrogen peroxide,[5] the oxidation of 1,4-naphthalenediamine[6] and naphthylaminesulfonic acid[7] and the oxidation of 4-amino-1-naphthol prepared by electrolytic reduction of 1-nitronaphthalene.[8]

[1] Imperial College of Science and Technology, London, England.

[2] Conant and Fieser, J. Am. Chem. Soc., 46, 1862 (1924); Fieser and Fieser, J. Am. Chem. Soc., 57, 491 (1935); Org. Syntheses Coll. Vol. 1, 383 (1946).

[3] Miller, J. Russ. Phys. Chem. Soc., 16, 414 (1884); cf. Japp and Miller, J. Chem. Soc., 39, 220 (1881).

[4] Fieser, J. Am. Chem. Soc., 70, 3165 (1948); Grinev and Terentév, Vestnik Moskov. Univ., Ser. Mat. Mekh. Astron. Fiz., Khim., 12, No. 6, 147 (1957) [C. A., 53, 3187 (1959)].

[5] Arnold and Larson, J. Org. Chem., 5, 250 (1940); Ibuki, J. Chem. Soc. Japan, Pure Chem. Sect., 70, 286 (1949) [C. A., 45, 4702 (1951)].

[6] Liebermann, Ann., 183, 242 (1876); Russig, J. prakt. Chem., 62, 31 (1900).

[7] Monnet, Reverdin, and Nolting, Ber., 12, 2306 (1879).

[8] Harman and Cason, J. Org. Chem., 17, 1058 (1952).

α-NAPHTHYL ISOTHIOCYANATE

(Isothiocyanic acid, 1-naphthyl ester)

$$\text{NH}\overset{\overset{\displaystyle S}{\|}}{\text{C}}\text{NH}_2 \rightarrow \text{NCS} + \text{NH}_3$$

Submitted by J. CYMERMAN-CRAIG, M. MOYLE, and R. A. WHITE.[1]
Checked by N. J. LEONARD and F. H. OWENS.

1. Procedure

In a 500-ml. round-bottomed flask fitted with a reflux condenser are placed 16.2 g. (0.08 mole) of dry α-naphthylthiourea (Note 1) and 180 ml. of redistilled chlorobenzene. The flask is heated at the reflux temperature by means of an electric heating mantle. Evolution of ammonia begins almost at once, and all of the solid dissolves after 30–45 minutes. The solution is maintained at reflux for 8 hours (Note 2) and then is evaporated on a steam bath at water-pump pressure to remove all of the chlorobenzene. The residue crystallizes on cooling and is extracted with four 30-ml. portions of boiling hexane (Note 3). Removal of solvent from the combined hexane extracts affords pale yellow crystals of α-naphthyl isothiocyanate, m.p. 58–59°. The yield is 12.7–13.0 g. (86–88%). Recrystallization from hexane (9 ml. of hexane for 1 g. of solute) gives colorless needles, melting point unchanged (Note 4).

2. Notes

1. α-Naphthylthiourea may be prepared by the method of Frank and Smith, *Org. Syntheses Coll. Vol.* **3**, 735 (1955), or by the method of de Clermont, *Ber.*, **9**, 446 (1876), and Bertram, *Ber.*, **25**, 48 (1892).

2. The checkers found that a 24-hour heating period increased the yield of α-naphthyl isothiocyanate to 95%.

3. A fifth extraction yields no further product. The residue

insoluble in hexane was found by the submitters to consist of 1.3–1.9 g. of a mixture of equal parts of di-α-naphthylthiourea and α-naphthylthiourea, m.p. 178–181°.

4. The method is generally applicable to the preparation of aryl isothiocyanates. Using this procedure, the submitters have prepared the following isothiocyanates, with the yields and times of refluxing indicated: phenyl, 44%, 8 hours; o-chlorophenyl, 46, 8; p-bromophenyl, 73, 8; p-biphenylyl, 49, 6; β-naphthyl, 70, 10; 9-phenanthryl, 70, 10; 1-pyrenyl, 72, 10.

3. Methods of Preparation

Aryl isothiocyanates can be prepared by the action of thiophosgene on the arylamine[2] (this reaction fails with naphthyl compounds),[3] by fission of a sym-diarylthiourea with acidic reagents[4] (this reaction involves the loss of half the amine used), and by the decomposition of an ammonium aryldithiocarbamate[5] (low yields are reported for naphthyl and other compounds).[5, 6] The procedure described here is that of Baxter, Cymerman-Craig, Moyle, and White.[7]

[1] University of Sydney, Sydney, Australia.

[2] Org. Syntheses Coll. Vol. 1, 165 (1941).

[3] Connolly and Dyson, J. Chem. Soc., 1935, 679.

[4] Werner, J. Chem. Soc., 59, 396 (1891).

[5] Org. Syntheses Coll. Vol. 1, 447 (1941).

[6] Dains, Brewster, and Olander, Univ. Kansas Sci. Bull., 13, 1 (1922) [C. A., 17, 543 (1923)].

[7] Baxter, Cymerman-Craig, Moyle, and White, Chem. & Ind. (London), 1954, 785; J. Chem. Soc., 1956, 659.

NEOPHYL CHLORIDE

[Benzene, (2-chloro-1,1-dimethylethyl)-]

$$CH_3-\underset{\underset{CH_2Cl}{|}}{C}=CH_2 \ + \ \bigcirc \ \xrightarrow{H_2SO_4} \ \bigcirc-\underset{\underset{CH_3}{|}}{\overset{\overset{CH_3}{|}}{C}}-CH_2Cl$$

Submitted by W. T. SMITH, JR., and J. T. SELLAS.[1]
Checked by N. J. LEONARD and R. O. KERR.

1. Procedure

In a 2-l. three-necked flask equipped with a mechanical stirrer, thermometer, and dropping funnel are placed 500 g. (570 ml., 6.4 moles) of benzene (Note 1) and 34.6 g. (18.8 ml.) of concentrated sulfuric acid (sp. gr. 1.84). The resultant mixture is brought to 20°. To this is added dropwise 201 g. (219 ml., 2.22 moles) of methallyl chloride (Note 2) over a period of 12 hours, during which time vigorous stirring is maintained and the temperature is kept at 20° with the aid of a water bath. The mixture, which becomes an amber color, is stirred for an additional 12 hours.

The reaction mixture is then transferred to a 1-l. separatory funnel, and the sulfuric acid layer is removed. The remaining benzene solution is then washed with four 200-ml. portions of distilled water (Note 3). In this step the amber color disappears and the liquid becomes colorless. The benzene solution is dried with anhydrous sodium sulfate and transferred to a 1-l. distilling flask. The benzene is removed by distillation under a pressure of about 45 mm. The liquid residue is poured into a 500-ml. flask and distilled through a 40-cm. Vigreux column under reduced pressure. The yield of neophyl chloride boiling at 97–98°/10 mm. is 262–275 g. (70–73%) (Notes 4 and 5); n_D^{20} 1.5250.

2. Notes

1. The benzene is purified by washing with three 80-ml. por-

tions of concentrated sulfuric acid and then drying with anhydrous sodium sulfate. It is used directly after removal of the sodium sulfate by filtration.

2. The methallyl chloride used was a redistilled commercial sample and had the following properties: b.p. 71–72°/760 mm., n_D^{20} 1.4274, d_4^{20} 0.918.

3. To ensure good results it is necessary to remove all of the sulfuric acid by washing the mixture thoroughly with water. The final washing should be neutral to litmus.

4. If the residue in the distilling flask is dissolved in ether, treated with activated carbon, and evaporated to dryness, the solid so obtained can be recrystallized from about 25 ml. of 95% ethanol to give 10 g. (2.7%) of p-di(chloro-*tert*-butyl)benzene; m.p. 54.5–55.0°.

5. In a run in which the temperature was kept at 10–15° during the addition of methallyl chloride, the methallyl chloride was added over a period of 1 hour. Stirring was continued for 1 hour, and the reaction mixture was worked up as described above to give a 53% yield.

3. Methods of Preparation

(Chloro-*tert*-butyl)benzene has been prepared by the direct chlorination of *tert*-butylbenzene in the presence of strong light;[2] by chlorination of *tert*-butylbenzene with sulfuryl chloride in the presence of benzoyl peroxide;[3] by the action of thionyl chloride on the corresponding alcohol;[4] and by the hydrogen fluoride-catalyzed alkylation of benzene with methallyl chloride.[5] The sulfuric acid-catalyzed alkylation described here is based on the procedure of Whitmore, Weisgerber, and Shabica.[6] A variation using a shorter reaction time has also been described.[7]

[1] State University of Iowa, Iowa City, Iowa.
[2] Truce, McBee, and Alfieri, *J. Am. Chem. Soc.*, 71, 752 (1949).
[3] Kharasch and Brown, *J. Am. Chem. Soc.*, 61, 2147 (1939).
[4] Haller and Ramart, *Compt. rend.*, 174, 1211 (1922).
[5] Calcott, Tinker, and Weinmayr, *J. Am. Chem. Soc.*, 61, 1010 (1939).
[6] Whitmore, Weisgerber, and Shabica, *J. Am. Chem. Soc.*, 65, 1469 (1943).
[7] Schmerling and Ipatieff, *J. Am. Chem. Soc.*, 67, 1862 (1945).

NICOTINAMIDE-1-OXIDE *

$$\text{pyridine-CONH}_2 \xrightarrow[\text{H}_2\text{O}_2]{\text{CH}_3\text{CO}_2\text{H}} \text{pyridine-N-oxide-CONH}_2 + \text{H}_2\text{O}$$

Submitted by E. C. Taylor, Jr.,[1] and Aldo J. Crovetti.[2]
Checked by Charles C. Price and Walter A. Schroeder.

1. Procedure

In a 2-l. round-bottomed flask with ground-glass joint (Note 1) are placed 100 g. (0.82 mole) of powdered nicotinamide (Note 2) and 1 l. of c.p. glacial acetic acid, and the mixture is warmed with occasional shaking on a steam bath until a clear solution is obtained. To this mixture is added 160 ml. (1.39 moles) of cold 30% hydrogen peroxide. An air condenser is attached to the reaction flask, and the mixture is heated on a steam bath for 3.5 hours.

The reaction mixture is then distilled under reduced pressure (80–100 mm.) (Note 3). After 600–700 ml. has distilled, the mixture is diluted with 150–200 ml. of distilled water and the distillation is continued. The product separates near the end of the distillation, causing somewhat vigorous bumping for a short period. When the bumping has almost ceased, the pressure is reduced to 20 mm. and the distillation continued almost to dryness (Note 4).

The major portion of wet solid is removed from the flask and transferred to a 1-l. Erlenmeyer flask. The remaining solid is washed out with a little distilled water, and the washings are transferred to the flask. The solid is dissolved in the smallest amount of boiling water required, the flask removed from the heat source, and 50 ml. of ethyl alcohol added (Note 5). The flask is allowed to cool slowly, and, after the major portion of the product has separated, the flask is cooled to 5° overnight. The solid is removed by filtration and washed with cold alcohol, then acetone, and finally ether. The white, crystalline, air-dried product weighs 82–93 g. (73–82%), m.p. 291–293° dec. (rapid heating); the compound starts to turn brown at about 280–285° (Note 6).

2. Notes

1. Since hydrogen peroxide attacks rubber stoppers, glass-jointed equipment is recommended.

2. U.S.P. Niacinamide (Mallinckrodt) was used.

3. For the distillation, the still head consisted of a Claisen-type adapter with a parallel side arm (24/40 standard taper joints), 21 cm. high and 10.5 cm. wide. An ordinary straight still head is attached to the parallel side arm. This large still head prevents any bumping solid from entering the condenser.

4. Distilling to complete dryness exposes the solid to prolonged heating and causes oxidation of the product, which is obtained colored and in lower yield.

5. The alcohol serves to retain the brown color upon recrystallization and to decompose any excess hydrogen peroxide.

6. Because of the questionable value of the melting point as a criterion of purity, the checkers analyzed two samples of the product: Calcd. for $C_6H_6N_2O$: C, 52.17; H, 4.34. Found (Sample 1): C, 52.34; H, 4.44. Found (Sample 2): C, 52.61, 52.63; H, 4.46, 4.53.

3. Methods of Preparation

The procedure given is essentially that described by Taylor and Crovetti.[3] Nicotinamide-1-oxide (m.p. 275–276° dec.) has also been prepared by the alkaline hydrolysis of nicotinonitrile-1-oxide [4] and by the action of ammonium hydroxide on methyl nicotinate-1-oxide.[5] The melting point of the product prepared by the latter synthesis is reported to be 282–284° dec.

[1] Princeton University, Princeton, New Jersey.

[2] University of Illinois, Urbana, Illinois.

[3] Taylor and Crovetti, *J. Org. Chem.*, **19**, 1633 (1954).

[4] Jujo, *J. Pharm. Soc. Japan*, **66**, 21 (1946) [*C. A.*, **45**, 6200 (1951)].

[5] Shimizu, Naito, Ohta, Yoshikawa, and Dohmori, *J. Pharm. Soc. Japan*, **72**, 1474 (1952) [*C. A.*, **47**, 8077 (1953)].

NICOTINONITRILE *

$$\text{(structure)} \xrightarrow{\text{P}_2\text{O}_5} \text{(structure)} + \text{H}_2\text{O}$$

Submitted by PEYTON C. TEAGUE and WILLIAM A. SHORT.[1]
Checked by ARTHUR C. COPE, JOHN C. SHEEHAN, and LOUIS A. COHEN.

1. Procedure

In a dry 1-l. round-bottomed flask are placed 100 g. (0.82 mole) of powdered nicotinamide and 100 g. (0.70 mole) of phosphorus pentoxide. The flask is stoppered and shaken to mix the two powders. It is then connected by means of a 10-mm. i.d. tube to an 80-cm. air condenser arranged for distillation. A 125-ml. Claisen flask immersed in an ice-salt bath is used as the receiver (Note 1). The pressure is reduced to 15–20 mm., and the mixture is heated with a large free flame of a high-temperature burner (such as a Fisher or Meker type). The flame is moved about freely to melt the material as rapidly as possible, and then the mixture is heated vigorously until nothing more comes over or until foam reaches the top of the flask (15–20 minutes). The apparatus is allowed to cool (Note 2), and the product is rinsed out of the tube and condenser with ether (Note 3). The ether solution is added to the distillate, the ether is distilled on a steam bath, and the product is distilled at atmospheric pressure using an air condenser. The yield of nicotinonitrile, boiling at 205–208° and melting at 50–51°, is 71–72 g. (83–84%).

2. Notes

1. To prevent possible clogging of the condenser by the solid nicotinonitrile, the end of the condenser should not be constricted and should not extend far into the receiver.

2. The residue left in the flask may be removed by carefully

adding water, allowing the mixture to stand overnight, and then washing repeatedly with water.

3. A small amount of material insoluble in ether but soluble in water remains in the condenser. The nicotinonitrile can be washed from the condenser more easily with acetone. If acetone is used, it should be removed by distillation under reduced pressure before the product is distilled.

3. Methods of Preparation

The method described is essentially that of La Forge.[2] Nicotinonitrile has also been prepared from nicotinic acid by heating with ammonium acetate and acetic acid,[3] from 3-pyridinesulfonic acid by fusion of the sodium salt with sodium cyanide,[4] from 3-bromopyridine and cuprous cyanide,[5] from nicotinamide and benzenesulfonyl or p-toluenesulfonyl chloride in pyridine,[6] from nicotinic acid and ammonia in the presence of a dehydrating catalyst,[7] from β-picoline and ammonia,[8] and from heating a mixture of nicotinic acid and lead thiocyanate.[9]

[1] University of South Carolina, Columbia, South Carolina.

[2] La Forge, J. Am. Chem. Soc., 50, 2477 (1928).

[3] Adkins, Wolff, Pavlic, and Hutchinson, J. Am. Chem. Soc., 66, 1293 (1944).

[4] McElvain and Goese, J. Am. Chem. Soc., 65, 2233 (1943).

[5] McElvain and Goese, J. Am. Chem. Soc., 63, 2283 (1941).

[6] Stephens, Bianco, and Pilgrim, J. Am. Chem. Soc., 77, 1701 (1955).

[7] Scudi, Maschetto, and Mayurnik (to Nepera Chemical Co., Inc.), U. S. pat. 2,680,742 [C. A., 49, 6316 (1955)].

[8] Hadley and Wood, Brit. pat. 777,746 [C. A., 51, 18011 (1957); Porter, Erchak, and Cosby (to Allied Chemical and Dye Corp.), U. S. pat. 2,510,605 [C. A., 45, 187 (1951)]; Mayurnik, Moschetto, Bloch, and Scudi, Ind. Eng. Chem., 44, 1630 (1952).

[9] Spasov and Golovinskiï, Compt. rend. acad. bulgare sci., 11, No. 4, 287 [C. A., 53, 18026 (1959)].

o-NITROACETOPHENONE *

(Acetophenone, 2′-nitro-)

$$CH_2(CO_2C_2H_5)_2 + Mg + C_2H_5OH \rightarrow$$
$$C_2H_5OMgCH(CO_2C_2H_5)_2 + H_2$$

$$C_2H_5OMgCH(CO_2C_2H_5)_2 + o\text{-}NO_2C_6H_4COCl \rightarrow$$
$$o\text{-}NO_2C_6H_4COCH(CO_2C_2H_5)_2 + MgClOC_2H_5$$

$$o\text{-}NO_2C_6H_4COCH(CO_2C_2H_5)_2 + 2H_2O \xrightarrow{H_2SO_4}$$
$$o\text{-}NO_2C_6H_4COCH_3 + 2C_2H_5OH + 2CO_2$$

Submitted by George A. Reynolds and Charles R. Hauser.[1]
Checked by Cliff S. Hamilton and Yao-Hua Wu.

1. Procedure

In a 500-ml. three-necked round-bottomed flask equipped with a mercury-sealed stirrer, a dropping funnel, and a reflux condenser (protected by a drying tube) is placed 5.4 g. (0.22 g. atom) of magnesium turnings. Five milliliters (0.085 mole) of absolute ethanol and 0.5 ml. of carbon tetrachloride are added. If the reaction does not start immediately, the flask is heated for a short time on a steam bath. After the reaction has proceeded for several minutes, 150 ml. of absolute ether is added cautiously with stirring. A solution of 35.2 g. (0.22 mole) of diethyl malonate (Note 1), 20 ml. (0.34 mole) of absolute ethanol, and 25 ml. of absolute ether is added with stirring at such a rate that rapid boiling is maintained; heat is supplied when necessary. The mixture is heated under reflux on a steam bath for 3 hours, at which time most of the magnesium has dissolved. To the gray solution is added 37 g. (0.2 mole) of o-nitrobenzoyl chloride (Note 2), dissolved in 50 ml. of ether, in a period of 15 minutes. Heating under reflux on the steam bath is continued throughout the addition of the o-nitrobenzoyl chloride and until the green solution becomes too viscous to stir. The reaction mixture is cooled and shaken with dilute sulfuric acid (25 g. of concentrated sulfuric acid in 200 ml. of water) until all the solid has dissolved (Note 3).

The ether phase is separated and the aqueous layer extracted with 75 ml. of ether. The ether extracts are combined and washed with water, and the solvent is removed by distillation. To the crude diethyl *o*-nitrobenzoylmalonate is added a solution of 60 ml. of glacial acetic acid, 7.6 ml. of concentrated sulfuric acid, and 40 ml. of water, and the mixture is heated under reflux for 4 hours (Note 4) or until no more carbon dioxide is evolved. The reaction mixture is chilled in an ice bath, made alkaline with 20% sodium hydroxide solution, and extracted with several portions of ether. The combined ethereal extracts are washed with water and dried with anhydrous sodium sulfate followed by Drierite, and the solvent is removed by distillation. On fractional distillation of the residue, 27.0–27.4 g. (82–83%) of light-yellow *o*-nitroacetophenone boiling at 158–159°/16 mm. is obtained (n_D^{25} 1.548, n_D^{20} 1.551, d_4^{25} 1.236) (Note 5).

2. Notes

1. The checkers found that the yield of final product was cut in half unless the commercial grade of diethyl malonate was purified by distillation.

2. The *o*-nitrobenzoyl chloride can be prepared from the commercially available acid and thionyl chloride. It has been reported [2] that, when a particularly pure sample of *o*-nitrobenzoyl chloride was used, no difficulty was encountered in stirring the reaction mixture, and that the precipitate formed in a granular state. As a consequence, the subsequent decomposition with sulfuric acid was simple and rapid. The final product boiled at 107°/0.5 mm., and melted at 26.8°; the supercooled liquid had n_D^{25} 1.5499.

3. The solution of the magnesium complex, which is difficult to decompose, is facilitated by mechanical shaking of the mixture for 30 minutes.

4. The decarboxylation is almost complete within 2 to 3 hours.

5. *p*-Nitroacetophenone may be prepared in 73% yield by a similar procedure. Various other methyl ketones have been prepared by this procedure.[3] It has been reported [2] that *o*-chloroacetophenone was obtained in 82% yield by this method; the product possessed the following physical properties; b.p. 70°/1.5 mm., n_D^{25} 1.5404; it did not solidify at −50°.

3. Methods of Preparation

This procedure is an adaptation of one described by Walker and Hauser.[3] Schofield and Swain [4] state that, in their opinion, for both convenience and economy, this method [3] is the best yet described for the preparation of o-nitroacetophenone.

o-Nitroacetophenone has also been prepared by the treatment of ethyl o-nitrobenzoylacetoacetate with sulfuric acid in ethanol,[5] by the direct nitration of acetophenone,[6, 7] by the reaction of o-nitrobenzaldehyde with diazomethane,[8] by the oxidation of o-nitroethylbenzene with potassium permanganate,[9] by the hydrolysis of o-nitroacetophenone oxime (obtained by the nitrosation of o-nitroethylbenzene with amyl nitrate),[10] and by the oxidation of o-nitrophenylmethylcarbinol with chromic acid.[10]

[1] Duke University, Durham, North Carolina.
[2] Ford-Moore, Private communication.
[3] Walker and Hauser, J. Am. Chem. Soc., 68, 1386 (1946).
[4] Schofield and Swain, J. Chem. Soc., 1948, 384.
[5] Kermack and Smith, J. Chem. Soc., 1929, 814.
[6] Morgan and Moss, J. Soc. Chem. Ind. (London), 42, 461 (1923).
[7] Elson, Gibson, and Johnson, J. Chem. Soc., 1930, 1128.
[8] Arndt, Z. angew. Chem., 40, 1099 (1927).
[9] Kochergin, Tilkova, Zasosov, and Grigorovskiĭ, Zhur. Priklad. Khim., 32, 1806 (1959) [C. A., 54, 4458 (1960)].
[10] Ford-Moore and Rydon, J. Chem. Soc., 1946, 679.

9-NITROANTHRACENE*

(Anthracene, 9-nitro-)

Submitted by CHARLES E. BRAUN, CLINTON D. COOK,
    CHARLES MERRITT, JR., and JOSEPH E. ROUSSEAU.[1]
Checked by WILLIAM S. JOHNSON, GEORGE N. SAUSEN, and PAUL R. SHAFER.

1. Procedure

Twenty grams (0.112 mole) of finely powdered anthracene (Note 1) is suspended in 80 ml. of glacial acetic acid (Note 2) in a 500-ml. three-necked round-bottomed flask fitted with a 150-ml. dropping funnel, a thermometer, and an efficient motor-driven stirrer. The flask is immersed in a water bath at 20–25°, and 8 ml. (0.126 mole) of concentrated nitric acid (70% by weight, sp. gr. 1.42), essentially free of oxides of nitrogen, is added slowly from the dropping funnel with vigorous stirring. The rate of addition is controlled so that the reaction temperature does not exceed 30°. About 15–20 minutes is required for this step.

After all the nitric acid has been added, the mixture is stirred until a clear solution is obtained (about 30 minutes), and stirring is then continued for an additional 30 minutes. The solution is filtered to remove any anthracene, and a mixture of 50 ml. (0.60 mole) of concentrated hydrochloric acid (37% by weight, sp. gr. 1.19) and 50 ml. of glacial acetic acid is added slowly to the filtrate with vigorous stirring. The pale-yellow precipitate of 9-nitro-10-chloro-9,10-dihydroanthracene which forms is separated by suction filtration on a sintered-glass funnel and is washed with two 25-ml. portions of glacial acetic acid and then with water until the washings are neutral. The product is removed from the funnel and triturated thoroughly with 60 ml. of warm (60–70°) 10% sodium hydroxide solution (Note 3). The crude orange nitroanthracene is separated from the warm slurry by suction filtration and is treated with four 40-ml. portions of 10% sodium hydroxide solution (Note 3). The product is finally washed thoroughly with warm water until the washings are neutral to litmus. This treatment requires about 1.5–2 l. of water. The crude 9-nitroanthracene is air-dried and recrystallized from glacial acetic acid (Notes 2 and 4). The yield of bright orange-yellow needles is 15–17 g. (60–68%), m.p. 145–146°.

2. Notes

1. Anthracene of good quality is required. The Eastman Kodak Company product, m.p. 215–217°, is satisfactory, or practical grade anthracene may be purified by codistillation with ethylene glycol.[2]

2. The checkers employed E. I. du Pont de Nemours and Company, Inc., C.P. acetic acid, which was further purified by distillation from potassium permanganate.

3. The checkers found it desirable to carry out the trituration by grinding the mixture in a mortar, because the nitrochloride has a tendency to form small, hard granules which otherwise may not come in contact with the alkali. The later treatments with alkali may be carried out satisfactorily in a beaker, or directly on the funnel, if thorough mixing is obtained. It is desirable to remove as much of the mother liquor as possible by suction from each of the alkali treatments.

4. For recrystallization, 10 ml. of glacial acetic acid is used for each gram of dried product. It is important that the dissolution be carried out rapidly; otherwise some decomposition may occur, producing anthraquinone as a contaminant. A satisfactory technique is to add the crude, dried, and crushed product rapidly in small portions to the total amount of boiling acetic acid. The solution should then be filtered through a steam-heated funnel.

3. Methods of Preparation

The procedure described is a modification of that of Dimroth.[3] 9-Nitroanthracene has also been prepared by nitration of anthracene with copper nitrate in glacial acetic acid and with diacetylorthonitric acid.[4] Other methods include direct nitration in acetic acid solution with nitric acid and acetic anhydride,[5] or by nitryl fluoride.[6]

[1] University of Vermont, Burlington, Vermont.
[2] Fieser, *Experiments in Organic Chemistry*, 2nd ed., p. 345, footnote 13, D. C. Heath and Company, Boston, Massachusetts, 1941.
[3] Dimroth, *Ber.*, **34**, 221 (1901).
[4] Braun, Cook, and Rousseau, unpublished work.
[5] Meisenheimer and Connerade, *Ann.*, **330**, 133 (1904).
[6] Hetherington and Robinson, *J. Chem. Soc.*, **1954**, 3512.

o- AND *p*-NITROBENZALDIACETATE

(Toluene-α,α-diol, *o*-nitro-, diacetate)

(Toluene-α,α-diol, *p*-nitro-, diacetate)

$$CH_3 \qquad\qquad CH(O_2CCH_3)_2$$

$$NO_2 \qquad\qquad NO_2$$

Submitted by Tamio Nishimura.[1]
Checked by T. L. Cairns and R. E. Foster.

1. Procedure

A. *p-Nitrobenzaldiacetate.* In a 2-l., three-necked flask provided with a mechanical stirrer, dropping funnel, and thermometer, surrounded by an ice-salt bath, are placed 400 ml. of acetic anhydride (Note 1) and 50 g. (0.36 mole) of *p*-nitrotoluene (Note 2). To this solution is added slowly, with stirring, 80 ml. of concentrated sulfuric acid. When the mixture has cooled to 0°, a solution of 100 g. (1.0 mole) of chromium trioxide (Note 3) in 450 ml. of acetic anhydride is added slowly, with stirring, at such a rate that the temperature does not exceed 10° (Note 4), and stirring is continued for 2 hours at 5–10° in an ice-water bath after the addition is complete. The contents of the flask are poured into two 3-l. beakers one-third filled with chipped ice, and water is added until the total volume is 5–6 l. The solid is separated by suction filtration and washed with water until the washings are colorless. The product is suspended in 300 ml. of 2% aqueous sodium carbonate solution and stirred. After thorough mixing, the solid is collected on a filter (Note 5), washed with water and finally with 20 ml. of ethanol. The product, after drying in a vacuum desiccator, weighs 60–61 g. (65–66% of the theoretical amount), m.p. 121–124° (Note 6).

B. *o-Nitrobenzaldiacetate.* *o*-Nitrotoluene (the fraction boiling at 217–219° of commercial product) is treated in a manner exactly similar to that for *p*-isomer, except that the reaction mixture is stirred mechanically for 3 hours at 5–10° after the addition of chromium trioxide is complete. Washing of the crude product with ethanol is omitted; instead, the crude product is heated under reflux with petroleum ether for 30 minutes. The product melts at 82–84° with preliminary softening. The yield is 33.6–34.5 g. (36–37%).

2. Notes

1. The industrial grade of acetic anhydride was used without further purification.

2. Commercial *p*-nitrotoluene (m.p. 53–54°) was used.

3. The chromium trioxide was of 97% purity. Cooling is necessary on dissolving chromium trioxide in acetic anhydride. *Caution! Addition of acetic anhydride to solid chromium trioxide has resulted in explosive decompositions. The trioxide should be added in small portions to the cooled anhydride.*

4. With a good ice-salt bath, the time required for the addition is 1.5–2.0 hours.

5. By acidification of the sodium carbonate washings, 2–4 g. of *p*-nitrobenzoic acid, m.p. 235–237°, is obtained.

6. A similar procedure may be used for the preparation of *p*-cyanobenzaldiacetate from *p*-tolunitrile. Information submitted by Rorig and Nicholson, of G. D. Searle and Company, indicates that the critical step in this preparation is to maintain the reaction temperature below 10° throughout the process.

Exposure of *p*-cyanobenzaldiacetate to excess chromic, acetic, and sulfuric acids causes a reduction in yield. During the oxidation care should be taken to prevent chromium trioxide from adhering to the walls of the flask above the reaction mixture and then dropping in large amounts into the solution.

3. Methods of Preparation

These have been reviewed in an earlier volume.[2] *p*-Nitrobenzaldehyde has also been prepared in 32% yield by heating *p*-nitrobenzyl chloride with potassium *tert*-butylperoxide.[3] The preparation of *o*-nitrobenzaldiacetate has been described in an earlier volume.[4]

o- and *p*-Nitrobenzaldiacetate also have been obtained from the corresponding aldehydes by reaction with acetic anhydride in the presence of phosphoric acid.[5]

[1] Kitasato Institute for Infectious Diseases, 138, Shiba-Shirokane-Sanko-cho, Minato-ku, Tokyo, Japan.

[2] *Org. Syntheses Coll. Vol.* 2, 441 (1943).

[3] Campbell and Coppinger, *J. Am.. Chem. Soc.*, **73**, 1788 (1951); U. S. pat. 2,628,256 [*C. A.*, 48, 724 (1954)].

[4] *Org. Syntheses Coll. Vol.* 3, 641 (1955).

[5] Davey and Gwilt, *J. Chem. Soc.*, **1955**, 1384.

m-NITROBENZAZIDE

(Benzoyl azide, *m*-nitro-)

$$m\text{-NO}_2\text{C}_6\text{H}_4\text{CO}_2\text{H} + \text{SOCl}_2 \rightarrow m\text{-NO}_2\text{C}_6\text{H}_4\text{COCl} + \text{SO}_2 + \text{HCl}$$

$$m\text{-NO}_2\text{C}_6\text{H}_4\text{COCl} + \text{NaN}_3 \rightarrow m\text{-NO}_2\text{C}_6\text{H}_4\text{CON}_3 + \text{NaCl}$$

Submitted by Jon Munch-Petersen.[1]
Checked by William S. Johnson and W. David Wood.

1. Procedure

A. *m-Nitrobenzoyl chloride.* In a 1-l. round-bottomed flask are placed 200 g. (1.2 moles) of crude *m*-nitrobenzoic acid [2] and 500 g. (300 ml., 4.2 moles) of thionyl choride (Note 1). The flask is fitted (ground-glass joint) with a reflux condenser carrying a calcium chloride drying tube leading to a gas-absorption trap [3] and is heated on a steam bath for 3 hours. The condenser is then set for downward distillation, and as much of the excess thionyl chloride as possible is distilled at the temperature of the steam bath. The residue is transferred to a 250-ml. Claisen flask and distilled at reduced pressure (water pump), b.p. 153–154°/12 mm. (Note 2). The yield is 200–217 g. (90–98%), m.p. 33°.

B. *m-Nitrobenzazide.* In a 2-l. round-bottomed flask fitted with an efficient mechanical stirrer is placed a solution of 78 g. (1.2 moles) of commercial sodium azide in 500 ml. of water (Note 3). The flask is surrounded by a water bath kept at 20–25°. The stirrer is started, and over a period of about 1 hour a solution of 185.5 g. (1 mole) of *m*-nitrobenzoyl chloride in 300 ml. of acetone (previously dried over anhydrous potassium carbonate) is added from a dropping funnel. *m*-Nitrobenzazide separates at once as a white precipitate. Stirring is continued for 30 minutes after the addition is complete; then 500 ml. of water is added and the reaction mixture stirred for an additional 30 minutes. The azide is separated on a suction filter, washed with water, and dried in the air. The yield of crude product, m.p. 68°, is 189 g. (98%) (Note 4). It may be recrystallized from a mixture of equal parts of benzene and ligroin (b.p. 100–140°), when the temperature is kept below 50° (Note 5). The product thus obtained consists of almost colorless crystals, m.p. 68–69° (Note 6), the recovery being 80–90% (Note 7).

2. Notes

1. Eastman Kodak Company white label grade thionyl chloride is satisfactory.

2. Since the product crystallizes readily, water cooling should be applied only at the receiver, not at the side arm.

3. The reaction should preferably be carried out in a hood, as hydrazoic acid may be liberated in small amounts. This compound, which is volatile, is highly toxic, and its inhalation may cause temporary headache and giddiness.

4. This product is sufficiently pure for general reagent use.

m-Nitrobenzazide is recommended [4-7] as a reagent for the characterization and estimation of aliphatic and aromatic hydroxyl compounds. It reacts to form nicely crystalline *m*-nitrophenylcarbamic esters,[5, 6, 8] in which the nitro group may be titrated with titanous chloride. With amines it forms substituted *m*-nitrophenylureas.[9, 10]

5. At·higher temperatures a Curtius rearrangement into the isocyanate may occur, nitrogen being liberated. An alternative procedure for recrystallization (preferred by the checkers) consists in dissolving the crude product in a small amount of benzene (if the solution is discolored it may be treated with decolorizing carbon) and adding an equal volume of ligroin. On seeding, the product crystallizes.

6. The melted compound decomposes with liberation of nitrogen.

7. Using the same procedure, *p*-nitrobenzazide, m.p. 71–72° (Note 6), may be prepared. The yield of crude product is 90%, and of recrystallized product 70%.

3. Methods of Preparation

m-Nitrobenzazide has been prepared by the action of nitrous acid on *m*-nitrobenzhydrazide, which is obtained by treating methyl *m*-nitrobenzoate with hydrazine hydrate.[5, 7] The procedure described here is mentioned by Naegeli and Tyabji [11] and is similar to that given for benzazide.[12]

[1] Technical University of Denmark, Copenhagen, Denmark.
[2] *Org. Syntheses Coll. Vol.* 1, 391 (1941).
[3] *Org. Syntheses* Coll. Vol. 2, 4 (1943).
[4] Veibel, *Anal. Chem.*, 23, 665 (1951).
[5] Sah and Woo, *Rec. trav. chim.*, 58, 1013 (1939).
[6] Veibel and Lillelund, *Dansk Tidsskr. Farm.*, 14, 236 (1940) [*C. A.*, 35, 2444 (1941)].
[7] Veibel, Lillelund, and Wangel, *Dansk Tidsskr. Farm.*, 17, 183 (1943) [*C. A.*, 39, 1608 (1945)].
[8] Hoeke, *Rec. trav. chim.*, 54, 505 (1935).
[9] Sah et al., *J. Chinese Chem. Soc.*, 13, 22 (1946) [*C. A.*, 42, 148 (1948)].
[10] Karrman, *Svensk Kem. Tidskr.*, 60, 61 (1948) [*C. A.*, 42, 5804 (1948)].
[11] Naegeli and Tyabji, *Helv. Chim. Acta*, 16, 361 (1933).
[12] Barret and Porter, *J. Am. Chem. Soc.*, 63, 3434 (1941).

m-NITROBIPHENYL*

(Biphenyl, 3-nitro-)

$$\text{(ring)}^{NH_2}_{NO_2} + NaNO_2 + 2HCl \rightarrow \text{(ring)}^{N_2Cl}_{NO_2} + NaCl + H_2O$$

$$\text{(ring)}^{N_2Cl}_{NO_2} + (CH_3)_2NH + Na_2CO_3 \rightarrow$$

$$\text{(ring)}^{N=NN(CH_3)_2}_{NO_2} + NaCl + NaHCO_3$$

$$\text{(ring)}^{N=NN(CH_3)_2}_{NO_2} + C_6H_6 + C_7H_7SO_3H \rightarrow$$

$$\text{(ring)}^{C_6H_5}_{NO_2} + N_2 + C_7H_7SO_3NH_2(CH_3)_2$$

Submitted by C. E. Kaslow and R. M. Summers.[1]
Checked by James Cason and B. H. Walker.

1. Procedure

A. *1-(m-Nitrophenyl)-3,3-dimethyltriazene.* To a 3-l. three-necked flask containing 276 g. (2 moles) of technical grade *m*-nitroaniline (Note 1) are added 250 ml. of concentrated hydrochloric acid and 500 ml. of hot water. The contents of the flask are heated to about 85° to dissolve the *m*-nitroaniline; then 550 ml. of concentrated hydrochloric acid is added, and the solution is cooled rapidly. A stirrer, a thermometer, and a long-stemmed dropping funnel are attached to the flask, and its contents are then cooled to −3 to −5° by means of a salt-ice bath. A solution of 144 g. (2.09 moles) of sodium nitrite in 350 ml. of water is added dropwise under the surface of the acid solution while it is being stirred. The rate of addition is regulated (Note 2) so that the temperature does not rise above 0°. The stirring is continued for 15–20 minutes after the sodium nitrite solution has been added; then a solution of 8–10 g. of urea in 25 ml. of water is added during about 15 minutes (foaming), and the stirring is discontinued. The diazonium salt solution must be kept cold while the next step is proceeding.

Two and one-half liters of water in a 3-gal. crock is stirred vigorously with a mechanical stirrer, and 870 g. (7 moles) of pulverized sodium carbonate monohydrate (Note 3) is added portionwise (Note 4). Crushed ice is added to the sodium carbonate suspension until the temperature is lowered to 10°; then 423 g. (2.35 moles) of 25% dimethylamine solution (Note 5) is added. The ice-cold *m*-nitrobenzenediazonium chloride solution is added over a period of 25–35 minutes under the surface of the vigorously stirred dimethylamine solution by means of a dropping funnel (Note 6) while the temperature is maintained at about 10° by the addition of ice. The solution is stirred for 15–20 minutes after the addition is complete. The crude yellow triazene is removed by filtration on a large Büchner funnel. The cake is washed twice by removal and thorough mixing with 2–2.5 l. of water. After the second washing the cake is pressed as dry as possible, then removed and dissolved in 1.8–2.0 l. of boiling 95% ethanol contained in a 4-l. flask under a reflux condenser. The triazene is allowed to crystallize as the ethanol is cooled in water, then removed by filtration, and washed with two 200-ml. portions of 95% ethanol. After drying in air at room temperature, the yield of 1-(*m*-nitrophenyl)-3,3-dimethyltriazene is 348–365 g. (89–94%), m.p. 100.8–101.5°.

B. *m-Nitrobiphenyl*. To a 5-l. three-necked flask equipped with a sealed mechanical stirrer, a dropping funnel, and a reflux condenser are added 116.4 g. (0.6 mole) of 1-(*m*-nitrophenyl)-3,3-dimethyltriazene and 2.5 l. of benzene. The benzene solution is heated to maintain refluxing and stirred vigorously while a solution of 148 g. (0.8 mole) of 94% toluenesulfonic acid (Note 7) in 750 ml. of benzene is added dropwise (Note 8) over a period of 4–4.5 hours. The refluxing is continued for 1–1.5 hours (Note 9) after the toluenesulfonic acid has been added. The solution is allowed to cool somewhat, 800 ml. of water is added cautiously with stirring through the separatory funnel, and then the water layer is removed. The benzene layer is extracted twice with

500-ml. portions of water, then three times with 500-ml. portions of 5% sodium hydroxide solution, and finally with a 500-ml. portion of water. The benzene solution is shaken with 30–40 g. of anhydrous calcium chloride to remove most of the water. The benzene is removed by distillation by dripping it into a 500-ml. Claisen flask which is heated sufficiently to maintain a rapid rate of distillation. After most of the benzene is removed, the residue is transferred to a 125-ml. Claisen flask. The low-boiling material is removed at a pressure of 20–30 mm. up to a bath temperature of 135–140°. The residue is then distilled at 0.1 mm. pressure. After a fore-run of nitrobenzene (2–5 ml.), the *m*-nitrobiphenyl is distilled (Note 10) at 115–118° at 0.1 mm. while the bath temperature is maintained at 155–160°. The yield of the crude yellow oil is 50–60 g. (42–50%) (Note 11). For purification, the substance is dissolved in 50 ml. of hot methanol. Upon cooling, two layers separate. Crystallization is induced by scratching the inner surface with a glass rod. When the crystalline mass is cold, the yellow solid is removed by filtration and washed with two 30-ml. portions of cold methanol. After drying in air, the yield of *m*-nitrobiphenyl is 40–50 g. (34–42%), m.p. 58.5–59.5°.

2. Notes

1. E. I. du Pont de Nemours and Company technical product, m.p. 111–112°, was used.
2. The time required for the addition varies from 1.5 to 2 hours.
3. There must be an excess of sodium carbonate to prevent the troublesome frothing caused by the liberation of carbon dioxide near the end of the coupling reaction.
4. The addition must be regulated so as to prevent the formation of large chunks of sodium carbonate decahydrate.
5. Obtained from Eastman Organic Chemicals Department, Distillation Products Industries, Rochester, New York.
6. Small portions are added to the separatory funnel while the main portion is kept in the ice bath.
7. Monsanto Chemical Company anhydrous toluenesulfonic acid, which is about 80% of the *para* isomer, was used.
8. Some toluenesulfonic acid hydrate which does not dissolve

in benzene must be pushed through the stem of the funnel by means of a wire or a thin glass rod.

9. The evolution of nitrogen gas may be followed by attaching a bubble counter containing kerosene to the top of the reflux condenser. Generally, the evolution of nitrogen is complete within an hour.

10. The *m*-nitrobiphenyl has a tendency to crystallize in the side arm of the Claisen flask. Arrangements must be made to keep the side arm sufficiently warm to prevent crystallization.

11. After the oil has congealed to a solid, the substance melts at 53–57°.

3. Methods of Preparation

The procedure described for the preparation of 1-(*m*-nitrophenyl)-3,3-dimethyltriazene is the method of Elks and Hey,[2] and the preparation of *m*-nitrobiphenyl is also a modification of their procedure. The other principal methods for the preparation of *m*-nitrobiphenyl are the decomposition of N-nitroso-*m*-nitroacetanilide in benzene [3] and the decomposition of alkaline *m*-nitrobenzenediazohydroxide in benzene.[4] Other methods that have been reported include the decomposition of potassium *m*-nitrobenzenediazotate in benzene with acetyl chloride,[5] the decomposition of *m*-nitrobenzoyl peroxide in boiling benzene,[6] the decomposition of benzenediazonium borofluoride in nitrobenzene [7] at 70°, and the reduction of 4-(3'-nitrophenyl)-benzenediazonium acid sulfate in boiling ethanol.[8]

[1] Indiana University, Bloomington, Indiana.

[2] Elks and Hey, *J. Chem. Soc.*, **1943**, 441.

[3] Bachmann and Hoffman, in Adams, *Organic Reactions*, Vol. 2, p. 249, John Wiley & Sons, 1944.

[4] Gomberg and Bachmann, *J. Am. Chem. Soc.*, **46**, 2339 (1924); Blakey and Scarborough, *J. Chem. Soc.*, **1927**, 3000; Elks, Haworth, and Hey, *J. Chem. Soc.*, **1940**, 1284.

[5] Jacobsen and Loeb, *Ber.*, **36**, 4082 (1903).

[6] Hey and Walker, *J. Chem. Soc.*, **1948**, 2213.

[7] Nesmeyanov and Makarova, *Bull. acad. sci. U.R.S.S., Classe sci. chim.*, **1947**, 213 [*C. A.*, **42**, 5441*a* (1948)].

[8] Fichter and Sulzberger, *Ber.*, **37**, 878 (1904).

o-NITROCINNAMALDEHYDE*

(Cinnamaldehyde, o-nitro-)

$$\text{C}_6\text{H}_5\text{CH}{=}\text{CHCHO} + \text{HNO}_3 \rightarrow \underset{\text{NO}_2}{\text{C}_6\text{H}_4}\text{CH}{=}\text{CHCHO} + \text{H}_2\text{O}$$

Submitted by ROBERT E. BUCKLES and M. PETER BELLIS.[1]
Checked by RICHARD T. ARNOLD and JAMES D. GROVES.

1. Procedure

A 1-l. three-necked round-bottomed flask, fitted with a dropping funnel and a mechanical stirrer, is cooled in an ice-salt mixture. To the flask are added 55.5 g. (50 ml., 0.42 mole) of freshly distilled cinnamaldehyde (Note 1) and 225 ml. of acetic anhydride. When the temperature of the solution has reached 0–5° a solution of 18 ml. of concentrated nitric acid (sp. gr. 1.42) in 50 ml. of glacial acetic acid is added slowly through the dropping funnel while the mixture is stirred. The time of addition is 3–4 hours, during which the temperature is kept below 5°. After the addition is complete, the mixture is allowed to warm slowly to room temperature. The reaction flask is then dismantled and stoppered, and the reaction mixture is allowed to stand 2 days.

At the end of this time, hydrochloric acid (20%) is added cautiously to the cooled solution until a precipitate begins to appear (Note 2). The addition of acid is then stopped, and the solution is allowed to cool in an ice bath or refrigerator until precipitation of the solid is completed. The light-yellow needles are collected on a Büchner funnel and dried in air. About 24 g. of o-nitrocinnamaldehyde, m.p. 125–127°, is obtained (Note 3). Additional product can be isolated by cautiously adding water to the mother liquor until precipitation is observed and then cooling the resultant mixture for several hours in an ice bath. Recrystallization from 95% ethanol gives 5–10 g. of o-nitrocinnamaldehyde, m.p. 126–127°. The total yield is 27–34 g. (36–46%).

2. Notes

1. As with other aromatic aldehydes, cinnamaldehyde is readily oxidized by air to its corresponding carboxylic acid. The latter must be separated just before the use of the aldehyde.

2. A good deal of heat is evolved when the hydrochloric acid is added to the reaction mixture, owing to the hydrolysis of acetic anhydride. The reaction mixture will become excessively hot unless it is cooled in an ice bath.

3. The product obtained is pure enough for most purposes. Recrystallization from 95% ethanol yields a nearly white product, m.p. 126–127.5°.

3. Methods of Preparation

o-Nitrocinnamaldehyde has been prepared by the condensation of o-nitrobenzaldehyde with acetaldehyde.[2,3] The direct nitration of cinnamaldehyde with potassium nitrate in sulfuric acid yields o-nitrocinnamaldehyde along with p-nitrocinnamaldehyde.[3] o-Nitrocinnamaldehyde has been obtained in 85% yield by the nitration of cinnamaldehyde diacetate.[4] The nitration of cinnamaldehyde in acetic anhydride yields o-nitrocinnamaldehyde as the only product,[5] and the procedure is the basis for the method given above.

[1] State University of Iowa, Iowa City, Iowa.
[2] Baeyer and Drewsen, Ber., 16, 2205 (1883).
[3] Diehl and Einhorn, Ber., 18, 2335 (1885).
[4] Davey and Gwilt, J. Chem. Soc., 1955, 1384.
[5] Mills and Evans, J. Chem. Soc., 117, 1035 (1920).

1-NITROÖCTANE

(Octane, 1-nitro-)

$$CH_3(CH_2)_7Br + AgNO_2 \rightarrow CH_3(CH_2)_7NO_2 + AgBr$$

Submitted by N. KORNBLUM and H. E. UNGNADE.[1]
Checked by JOHN C. SHEEHAN and M. GERTRUDE HOWELL.

1. Procedure

1-Bromoöctane (96.5 g., 0.5 mole) (Note 1) is added dropwise during 2 hours to a stirred suspension of silver nitrite (116 g., 0.75 mole) (Note 2) in 150 ml. of dry ether (Mallinckrodt anhydrous), contained in a 500-ml. three-necked round-bottomed flask equipped with dropping funnel, reflux condenser, and a sealed Hershberg-type stirrer (Trubore) and immersed in a 1-gal. Dewar flask filled with ice and water (Note 3). The mixture is stirred for 24 hours in an ice bath. The bath is removed, and stirring is continued at room temperature (26–28°) until the supernatant liquor gives a negative test for halides (approximately 40 hours, Notes 4 and 5).

The silver salts are removed by filtration, slurried with two 100-ml. portions of dry ether, and the ether washings are added to the ethereal solution of reaction products (Note 6). The combined ethereal solutions are distilled at atmospheric pressure through a 2 x 45 cm. column packed with 4-mm. Pyrex helices (Note 7). The residue remaining after removal of the ether is fractionated under reduced pressure. The yellow liquid distilling at 37°/3 mm. has n_D^{20} 1.4127–1.4129 and weighs 11.3 g. (14%); this is 1-octyl nitrite. It is followed by an interfraction of b.p. 38–70°/3 mm., n_D^{20} 1.4133–1.4320, yield 6.83 g., which contains some nitrite, some 1-octanol, and a little 1-nitroöctane. Finally, pure, colorless 1-nitroöctane distils at 66°/2 mm., n_D^{20} 1.4321–1.4323, yield 59.6–63.6 g. (75–80%) (Notes 8 and 9).

2. Notes

1. 1-Bromoöctane b.p. 50–51°/0.8 mm., n_D^{20} 1.4526, was employed.

2. Silver nitrite can be prepared as follows: silver nitrate (169.9 g., 1 mole) dissolved in 500 ml. of distilled water is added in small portions, with vigorous shaking, to a solution of 76 g. (1.1 mole) of sodium nitrite, dissolved in 250 ml. of distilled water contained in a 1-l. Erlenmeyer flask. (These operations are best carried out under a yellow safelight or, in any case, with minimum exposure to light.) Then the mixture is allowed to stand in the dark for 1 hour. The yellow precipitate is collected by filtration with suction, suspended in 250 ml. of distilled water, and again filtered. The washing is repeated twice, and the product is collected by filtration and dried to constant weight in a vacuum desiccator over potassium hydroxide pellets; yield 134 g. (87%). The silver nitrite drying process can be facilitated by washing the material with methanol.

3. It is preferable to carry out the entire reaction in a dark room equipped with a yellow safelight. The reaction mixture should be protected from moisture by means of drying tubes.

4. The stirrer is stopped, and the precipitate is allowed to settle. Unchanged alkyl halide is detected in the supernatant liquor by the Beilstein test (copper wire spiral) or by adding a few drops of the ethereal solution to alcoholic silver nitrate.

5. According to the submitters, with primary straight-chain bromides the time needed to reach a negative test for halide is 24 hours at 0° followed by 48 ± 12 hours at room temperature. When primary straight-chain iodides are employed, the reaction time is shorter: 24 hours at ice temperature followed by 36 ± 12 hours at room temperature.

6. It is more difficult to remove an alcohol from the corresponding nitroalkane than it is to separate the nitrite ester and the nitroalkane. Minimal exposure to a moist atmosphere is, therefore, desirable since anhydrous ether is hygroscopic and nitrite esters hydrolyze readily, especially if a little acid is present.

7. A column of high efficiency is undesirable because of the thermal instability of the nitrite, and a column of lesser efficiency cannot accomplish a complete separation of the products. The column used by the submitters was equipped with a total condensate, partial take-off head with small holdup.

8. 1-Nitroöctane is completely soluble in aqueous alkali. It is converted to a crystalline colorless sodium salt on shaking with

20% aqueous sodium hydroxide, and this salt dissolves completely on adding enough water to make the solution 10% aqueous base. That the nitro compound is free from nitrate is shown by the absence of the infrared absorption bands at 6.15, 7.85 and 11.6 μ characteristic of nitrate esters.

9. A recent study [2] has shown that this is a general reaction for primary straight-chain bromides and iodides; in contrast, primary chlorides fail to react. Primary bromides and iodides having branched chains also give excellent yields of nitro compounds, especially if the branching is not on the carbon *alpha* to the one holding the halogen (Table I).

TABLE I

YIELDS OF NITRO COMPOUNDS

Halide	RNO_2, %	Halide	RNO_2, %
n-Butyl Br	73	*n*-Octyl I	83
n-Butyl I	74	Isoamyl Br	72
n-Hexyl Br	76	Isoamyl I	78
n-Hexyl I	78	Isobutyl Br	18
n-Heptyl Br	79	Isobutyl I	59
n-Heptyl I	82	Neopentyl I	0
n-Octyl Br	80		

The reaction of silver nitrite with secondary halides gives yields of nitroparaffins in the vicinity of 15%, while with tertiary halides the yields are even lower (0–5%). There is no question that the reaction of silver nitrite with alkyl halides is useful only for the synthesis of primary nitroparaffins.

3. Methods of Preparation

The present procedure is based on a published paper.[2] 1-Nitroöctane has been prepared from 1-iodoöctane and silver nitrite,[3] from octane by boiling with nitric acid,[4, 5] from 1-nitrooctylene by catalytic hydrogenation,[6] from *n*-octaldehyde oxime and peroxytrifluoroacetic acid in acetonitrile,[7] and from 1-bromoöctane and sodium nitrite in dimethyl sulfoxide or dimethylformamide.[8] 1-Nitroöctane also has been obtained from *n*-octyl *p*-toluenesulfonate and sodium nitrite in 17% yield.[9]

[1] Purdue University, Lafayette, Indiana.
[2] Kornblum, Taub, and Ungnade, *J. Am. Chem. Soc.*, **76**, 3209 (1954).
[3] Eichler, *Ber.*, **12**, 1883 (1879).
[4] Worstall, *Am. Chem. J.*, **20**, 213 (1898); **21**, 228 (1899).
[5] Urbanski and Slon, *Roczniki Chem.*, **17**, 161 (1937) [*C. A.*, **31**, 6190 (1937)].
[6] de Mauny, *Bull. soc. chim. France*, **7**, 133 (1940) [*C. A.*, **34**, 5413 (1940)].
[7] Emmons and Pagano, *J. Am. Chem. Soc.*, **77**, 4557 (1955).
[8] Kornblum and Powers, *J. Org. Chem.*, **22**, 455 (1957).
[9] Drahowzal and Klamann, *Monatsh.*, **82**, 975 (1951).

1-(p-NITROPHENYL)-1,3-BUTADIENE

[1,3-Butadiene, 1-(p-nitrophenyl)-]

$$p\text{-}O_2NC_6H_4NH_2 \xrightarrow[\text{NaNO}_2]{\text{HCl}} p\text{-}O_2NC_6H_4N_2Cl$$

$$CH_2{=}CH{-}CH{=}CH_2 + p\text{-}O_2NC_6H_4N_2Cl \rightarrow$$
$$p\text{-}O_2NC_6H_4CH_2CH{=}CHCH_2Cl + N_2$$

$$p\text{-}O_2NC_6H_4CH_2CH{=}CHCH_2Cl + KOH \rightarrow$$
$$p\text{-}O_2NC_6H_4CH{=}CH{-}CH{=}CH_2 + KCl + H_2O$$

Submitted by Gus A. Ropp and Eugene C. Coyner.[1]
Checked by Arthur C. Cope and David J. Marshall.

1. Procedure

A. *1-(p-Nitrophenyl)-4-chloro-2-butene.* *p*-Nitroaniline hydrochloride is prepared by heating 138 g. (1.0 mole) of *p*-nitroaniline (Note 1) with 240 ml. of concentrated hydrochloric acid and 100 ml. of water on a steam bath for 15 minutes with occasional stirring. The mixture is cooled in an ice-salt bath and stirred rapidly in order to precipitate the hydrochloride as fine crystals. Cracked ice (100 g.) is added, and a solution of 70 g. of sodium nitrite is added dropwise with rapid mechanical stirring during a 1-hour period while the temperature of the reaction mixture is held between −4° and +4.5° by cooling with the ice-salt bath. The mixture is stirred for an additional period of 20 minutes and then is filtered through a chilled funnel into an ice-cooled filter flask. The filtrate is kept below 4° (Note 2) and is added through a dropping funnel during 90 minutes to a cold, vigorously stirred mixture composed of 1 l. of acetone, a solution of 80 g. of sodium acetate trihydrate in 100 ml. of water, a solution of 30 g. of cupric chloride in 50 ml. of water, and 130 ml. of liquid butadiene (Note 3). The reaction mixture is kept at $-3°$ to $+2°$ by means of an ice-salt bath while the diazonium salt solution is added. After the addition is completed the cooling bath is removed and the mixture is stirred for 16 hours. One liter of ether is added, and after several minutes' stirring the ethereal layer is separated, washed with four 1-l. portions of water, and dried over 20 g. of anhydrous magnesium sulfate. The solvent is removed by distillation at 15 mm. by heating on a steam bath, leaving a dark brown oily residue (187–199 g.) of crude 1-(*p*-nitrophenyl)-4-chloro-2-butene (Note 4).

B. *1-(p-Nitrophenyl)-1,3-butadiene.* The crude 1-(*p*-nitrophenyl)-4-chloro-2-butene obtained in Part A is dissolved in a mixture of 500 ml. of ligroin, b.p. 90–100°, and 500 ml. of benzene; 5 g. of decolorizing carbon is added, and the mixture is heated under reflux for 2 hours. After filtration to separate the decolorizing carbon the solvents are removed by distillation from a steam bath under reduced pressure, and the residual clear oil is dissolved in 400 ml. of methanol. A solution of 112 g. of potassium hydroxide in 600 ml. of methanol is added from a dropping funnel during 30 minutes while the mixture is stirred mechanically and kept at 15–30° by cooling with a bath of cold water. After being stirred for an additional period of 5 minutes the mixture, which contains some precipitated product, is poured into 1.2 l. of cold water. The crude product is collected on a filter, washed well with cold water, and air-dried. It is dissolved in 700 ml. of hot ligroin, b.p. 90–100°, and the solution is treated with 5 g. of decolorizing carbon, and filtered. On cooling, 1-(*p*-nitrophenyl)-1,3-butadiene separates as a yellow crystalline solid which is collected on a filter and dried in a desiccator. The yield of pure product, m.p. 77–79° (Note 5), is 100–108 g. (57–62% based on *p*-nitroaniline).

2. Notes

1. Either a pure grade of *p*-nitroaniline obtained from the Eastman Kodak Company or a technical grade purified by one recrystallization from ethanol was used, m.p. 147.5–148°.

2. The filtrate is kept in an ice-salt bath and transferred to the dropping funnel in small amounts in order to keep the temperature below 4°.

3. Butadiene from a commercial cylinder is passed through an 8-mm. glass tube leading to the bottom of a graduated cylinder cooled with Dry Ice and acetone, where it condenses and is measured as a liquid.

4. The submitters report that small samples of the crude product can be distilled in order to obtain pure 1-(*p*-nitrophenyl)-4-chloro-2-butene, b.p. 160–165°/1 mm.

5. Two recrystallizations from ligroin raise the melting point of the 1-(*p*-nitrophenyl)-1,3-butadiene to a constant value of 78.6–79.4°. The product can be kept for several weeks in a dark bottle at room temperature without evidence of decomposition.

3. Methods of Preparation

1-(*p*-Nitrophenyl)-1,3-butadiene has been prepared only by the method described,[2] which is an example of the Meerwein reaction (addition of diazonium salts to a carbon-carbon double bond with the elimination of nitrogen).[3]

[1] University of Tennessee, Knoxville, Tennessee.

[2] Coyner and Ropp, *J. Am. Chem. Soc.*, **70**, 2283 (1948); Dombrovskiĭ, *Doklady Akad. Nauk S.S.S.R.*, **111**, 827 (1956) [*C. A.*, **51**, 9507 (1957)].

[3] Müller, *Angew. Chem.*, **61**, 179 (1949).

trans-o-NITRO-α-PHENYLCINNAMIC ACID

[Acrylic acid, trans-3-(o-nitrophenyl)-2-phenyl-]

Submitted by DeLos F. DeTar.[1]
Checked by Charles C. Price and J. D. Berman.

1. Procedure

A mixture of 30.2 g. (0.20 mole) of o-nitrobenzaldehyde,[2] 40 g. (0.29 mole) of phenylacetic acid, 100 ml. (1.08 moles) of acetic anhydride, and 20 g. (0.20 mole) of triethylamine is refluxed for 15 minutes in a 500-ml. flask. The solution is cooled to 90°, and 100 ml. of cold water is added over a 5-minute period at a rate that maintains the temperature above 90° (Note 1). The solution is filtered at 95–100° and cooled to 20°. trans-o-Nitro-α-phenylcinnamic acid precipitates in the form of light-orange crystals. It is separated by filtration and washed with 60 ml. of 50% acetic acid and with water. The dried acid weighs 39–42 g. (72–78%) and melts at 195–198°, which corresponds to a purity of about 98% (Note 2). After recrystallization from 500 ml. of toluene, it is in the form of yellow prisms weighing 38–39 g. (71–72%) and melting at 197.8–198.3°.

2. Notes

1. If the temperature gets too high, more cold water may be added. If the water is added too rapidly at first, the temperature drops below 90° and the rate of hydrolysis becomes very slow. Heating such an incompletely hydrolyzed mixture above 90° may cause it to boil violently.

2. Approximate melting points of mixtures of the trans- and cis-o-nitro-α-phenylcinnamic acids are as follows:

Per Cent *cis*	Sinter Point	M.P.
33		134–185°
23	130°	136–187°
13	131°	165–192°
8	135°	184–196°
3.5	140°	192–198°
1.5	170°	195–198°
—	196°	197.8–198.3°

3. Methods of Preparation

trans-o-Nitro-α-phenylcinnamic acid has been prepared by the condensation of *o*-nitrobenzaldehyde with sodium phenylacetate in the presence of acetic anhydride with [3] or without [4] fused zinc chloride as a catalyst. It has also been prepared by the condensation of *o*-nitrobenzaldehyde with phenylacetic acid in the presence of acetic anhydride and triethylamine.[5]

[1] University of South Carolina, Columbia, South Carolina.
[2] *Org. Syntheses Coll. Vol.* **3**, 641 (1955).
[3] Pschorr, *Ber.*, **29**, 496 (1896).
[4] Oglialoro and Rosini, *Gazz. chim. ital.*, **20**, 396 (1890).
[5] Bakunin and Peccerillo, *Gazz. chim. ital.*, **65**, 1145 (1935).

m-NITROSTYRENE *

(Styrene, *m*-nitro-)

$$m\text{-}NO_2C_6H_4CHO + CH_2(CO_2H)_2 \xrightarrow{\text{Pyridine}}$$
$$m\text{-}NO_2C_6H_4CH{=}CHCO_2H + CO_2 + H_2O$$

$$m\text{-}NO_2C_6H_4CH{=}CHCO_2H \xrightarrow[\text{Quinoline}]{\text{Heat, Cu}}$$
$$m\text{-}NO_2C_6H_4CH{=}CH_2 + CO_2$$

Submitted by RICHARD H. WILEY and NEWTON R. SMITH.[1]
Checked by RICHARD T. ARNOLD, WILLIAM E. PARHAM, and DARWIN D. DAVIS.

1. Procedure

A. *m-Nitrocinnamic acid.* In a 1-l. round-bottomed flask fitted with a reflux condenser are placed 151 g. (1 mole) of *m*-nitrobenzaldehyde (Note 1), 115 g. (1.1 moles) of malonic acid, 250 ml. of 95% ethanol, and 25 ml. of pyridine. The mixture is heated on a steam bath under gentle reflux for 6–8 hours and cooled. The large masses of crystals are broken up with a spatula, and the reaction mixture is cooled in an ice bath. The solid is collected on a Büchner funnel, and the residue is washed with 100 ml. of cold ethanol and then with two 100-ml. portions of ether. The crude *m*-nitrocinnamic acid is suspended in 300 ml. of ethanol and digested on a steam plate for 2–3 hours. The mixture is cooled and filtered, and the solid is air-dried. The product, 144–155 g. (75–80%), is a light-yellow solid and melts at 200–201° (Note 2).

B. *m-Nitrostyrene.* In a 250-ml. two-necked flask equipped with a 250° thermometer and an air condenser are placed 30 g. (0.155 mole) of *m*-nitrocinnamic acid, 2 g. of copper powder, and 60 ml. of dry quinoline (Note 3). The flask is heated with a Bunsen burner to 185–195°, during which time a steady stream of carbon dioxide is evolved. After 2–3 hours (Note 4), the reaction mixture is cooled and poured into a mixture of 75 ml. of concentrated hydrochloric acid and 175 g. of ice. The *m*-nitrostyrene is isolated by steam distillation, approximately 1 l. of distillate being collected. The aqueous distillate is extracted with three 50-ml. portions of chloroform, which are combined and dried over anhydrous sodium sulfate. A 50-ml. Claisen flask, equipped with a dropping funnel, is heated on a steam bath, and the filtered chloroform extract is added dropwise. Heating is continued until all the solvent has been removed. The residue is then distilled at 3–5 mm. pressure. Following a small fore-run of less than 1 g. (Note 5), the *m*-nitrostyrene distils as a yellow liquid. The yield is 13–14 g. (56–60%), b.p. 90–96°/3.5 mm., n_D^{20} 1.5836, n_D^{27} 1.5800–1.5802 (Notes 6 and 7).

2. Notes

1. The checkers used technical *m*-nitrobenzaldehyde melting at 53–56°.

2. This procedure has been used by the submitters and others to prepare the following cinnamic acids from substituted benzaldehydes: *o*-nitrocinnamic acid (70%),[2] *p*-nitrocinnamic acid

(77%),[2] *m*-cyanocinnamic acid (71%),[3] *o*-chlorocinnamic acid (82%),[4] *m*-chlorocinnamic acid (53%),[4] *p*-chlorocinnamic acid (73%),[4] 2,4-dichlorocinnamic acid (70%),[4] 3,4-dichlorocinnamic acid (81%),[4] *m*-bromocinnamic acid (31%),[4] *p*-methoxycinnamic acid (60%),[4] and 3,4-dimethoxycinnamic acid (77%).[4]

The method constitutes a simple preparation of ethanol-insoluble cinnamic acids, of a high degree of purity when compared with the Perkin reaction [5] or the usual procedure for the Doebner reaction (p. 327), which uses a large excess of pyridine. A useful modification of this reaction is to warm the reactants on a steam plate in the absence of alcohol.[6,7]

3. Quinoline that has been purified by steam distillation of an acid solution should be used. Crude quinoline sometimes contains non-basic, high-boiling impurities such as nitrobenzene, which make the purification of *m*-nitrostyrene more difficult.

4. The checkers obtained a lower yield (9–11 g.) when a 2-hour heating period was used. The reaction must be carried out at 195° until the evolution of carbon dioxide has practically ceased (usually 2.75 hours). The checkers used Mallinckrodt copper powder.

5. The checkers obtained essentially no fore-run and only trace amounts of solid residue. The refractive index of the first 200 mg. was n_D^{27} 1.5800; of the center fractions, n_D^{27} 1.5802; and of the last 700 mg., n_D^{27} 1.5800.

6. This procedure has been used in the preparation of other nitrostyrenes in the following yields: *o*-nitrostyrene (40%),[2] *p*-nitrostyrene (41%),[2] and 3-nitro-4-hydroxystyrene (60%).[2] A better procedure for more volatile styrenes involves simultaneous decarboxylation and codistillation with quinoline from the reaction flask. This method has been used to prepare the following styrenes: *o*-chlorostyrene (50%),[4] *m*-chlorostyrene (65%),[4] *p*-chlorostyrene (51%),[4] *m*-bromostyrene (47%),[4] *o*-methoxystyrene (40%),[4] *p*-methoxystyrene (76%),[4] *m*-cyanostyrene (51%),[3] and *p*-formylstyrene (52%).[8]

7. Larger runs usually give smaller percentage yields.

3. Methods of Preparation

m-Nitrocinnamic acid has been prepared from *m*-nitrobenzaldehyde with sodium acetate and acetic anhydride,[5] and by the condensation of m-nitrobenzaldehyde with malonic acid in the presence of bases.[7,9]

m-Nitrostyrene has been prepared by boiling the sodium salt of β-bromo-β-(m-nitrophenyl)propionic acid;[10] by the dehydration of m-nitrophenylmethylcarbinol with phosphorus pentoxide,[11] potassium bisulfate,[12] or phosphoric acid[13]; by the decarboxylation of m-nitrocinnamic acid;[14] and by the condensation of m-nitrobenzaldehyde with acetic anhydride over bentonite at high temperatures.[15]

[1] University of Louisville, Louisville, Kentucky.

[2] Wiley and Smith, J. Am. Chem. Soc., 72, 5198 (1950).

[3] Wiley and Smith, J. Am. Chem. Soc., 70, 1560 (1948).

[4] Walling and Wolfstirn, J. Am. Chem. Soc., 69, 852 (1947).

[5] Schiff, Ber., 11, 1783 (1878); Tiemann and Oppermann, Ber., 13, 2060 (1880); Reich and Koehler, Ber., 46, 3732 (1913); Posner, J. prakt. Chem., [2] 82, 425 (1910); Böch, Lock, and Schmidt, Monatsh., 64, 408 (1934); Org. Syntheses Coll. Vol. 1, 398 (1941).

[6] Pandya, Ittyerah, and Pandya, J. Univ. Bombay, 10, pt. 3, 78 (1941); Pandya and Pandya, Proc. Indian Acad. Sci., 14A, 112 (1941).

[7] Dutt, Quart. J. Indian Chem. Soc., 1, 297 (1925); Kurien, Pandya, and Surange, Quart. J. Indian Chem. Soc., 11, 824 (1934).

[8] Wiley and Hobson, J. Am. Chem. Soc., 71, 2429 (1949).

[9] Knoevenagel, Ber., 31, 2610 (1898).

[10] Prausnitz, Ber., 17, 597 (1884).

[11] Marvel, Overberger, Allen, and Saunders, J. Am. Chem. Soc., 68, 736 (1946).

[12] Matsui, J. Soc. Chem. Ind. Japan, 45 (supplementary binding), 437 (1942).

[13] Arcus, J. Chem. Soc., 1958, 2428.

[14] Wiley and Smith, J. Am. Chem. Soc., 70, 2295 (1948); LoVecchio and Monforte, Atti soc. peloritana sci. fis. mat. e nat., 2, 111 (1955–56) [C. A., 51, 7328 (1957)].

[15] Levi and Nicholls, Ind. Eng. Chem., 50, 1005 (1958).

6-NITROVERATRALDEHYDE *

(Veratraldehyde, 6-nitro-)

$$CH_3O-C_6H_3(OCH_3)-CHO \xrightarrow{HNO_3} CH_3O-C_6H_2(OCH_3)(NO_2)-CHO$$

Submitted by CHARLES A. FETSCHER.[1]
Checked by RICHARD T. ARNOLD, W. E. PARHAM, and DONALD A. LEISTER.

1. Procedure

The product of this reaction is quite sensitive to light, and the entire procedure should be carried out in semidarkness (Note 1).

A wide-mouth 1-l. Erlenmeyer flask is supported inside a water bath of at least 2-l. capacity so that the bottom of the flask is not in contact with the bottom of the bath. The bath is filled with water at about 15° to cover at least half the height of the flask. The flask is fitted with a moderate-speed stainless-steel propel-ler-type stirrer, and 350 ml. of nitric acid (sp. gr. 1.4) at 20° is poured into it. Veratraldehyde,[2] 70 g. (0.42 mole) (Note 2), is crushed at least as fine as rice grains and is slowly added in small portions to the acid. The rate of addition should be such that it requires about 1 hour to add all the aldehyde. It is helpful, although not usually necessary, to add two or three ice cubes to the bath at the start of the nitration. The internal temperature is checked from time to time and should be held between 18° and 22°. The mixture is stirred for 10 minutes after the addition of the last of the aldehyde.

The mixture is then poured into 4 l. of vigorously agitated cold water (Note 3) in a suitable opaque container. From this point onward the protection of the product from light is extremely important. The stirring is continued for a few minutes; then the batch is filtered through a 24-cm. Büchner funnel. The container and funnel are kept covered with an opaque sheet of some kind except while the transfer is being made. The cake is sucked down well and then returned to the crock and reslurried a few minutes with 2 l. of cold water. It is then refiltered, pressed out well with a spatula, and drained as well as possible.

The filter cake at this point is 60% to 80% water, and the material is sensitive to heat and to light. The drying, therefore, is difficult and slow, and the exact procedure will depend upon the equipment available. One satisfactory method is to set the Büchner funnel containing the wet material in a large forced-draft oven for 8 hours at 50°. The material, still very wet, is then easily spread on a tray and is placed in a dark but ventilated storeroom, where it is allowed to air-dry for 48 hours or until the weight of the product is less than about 90 g. The product now contains from 10% to 20% of water and is best recrystallized without more thorough drying.

The material is dissolved in 2 l. of boiling 95% ethanol. It is not necessary to filter this solution for the first crystallization. Upon standing overnight, the solution is solid with precipitate. This structure is easily broken up and is filtered on a large Büchner funnel. The mother liquor is concentrated to about 700 ml. and allowed to cool. The second crop of solid is added to the first and dried in a vacuum oven at 50° overnight. The dry material weighs 65–70 g., corresponding to a yield of 73–79%, and melts at 129–131°. It is sufficiently pure for most purposes. One additional crystallization from 1 l. of 95% ethanol gives 55–60 g. of pure material, melting at 132–133°.

2. Notes

1. A sample of 6-nitroveratraldehyde of original melting point 133°, after 9 hours' exposure to the diffused light of the laboratory, showed a melting point of 88–95°. A small quantity of 6-nitrosoveratric acid, m.p. 185–190°, was isolated from the altered material.

2. The veratraldehyde should have a minimum melting point of 43°. Since there is no difficulty in controlling this reaction, larger batches would undoubtedly be as efficient. The phenomenal bulk of the product and the necessity of minimizing exposure to light make it impractical to handle appreciably more at one time. Smaller batches are perfectly feasible, although it is suggested that for very small amounts of aldehyde somewhat

more than the proportional quantity of nitric acid be used. It has been found satisfactory to use 100 ml. of acid with 15 g. of aldehyde.

3. If less water is used, the material is more difficult to break up and wash adequately, and the crude material is more heat-sensitive. Very vigorous agitation is desirable during this precipitation.

3. Methods of Preparation

6-Nitroveratraldehyde has always been prepared by direct nitration of the aldehyde. This preparation is a modification of that given by Salway.[3]

[1] Cluett, Peabody and Company, Troy, New York.

[2] *Org. Syntheses Coll. Vol.* **2**, 619 (1943).

[3] Salway, *J. Chem. Soc.*, **95**, 1163 (1909).

NORBORNYLENE *

(2-Norbornene)

$$+ \quad 2CH_2{=}CH_2 \longrightarrow 2$$

Submitted by J. MEINWALD and N. J. HUDAK.[1]
Checked by J. D. ROBERTS, C. M. SHARTS, and W. G. WOODS.

1. Procedure

A 1-l. steel bomb is charged with 200 g. (1.5 moles) of dicyclo-pentadiene (Note 1). The bomb is flushed with ethylene (Note 2) and then filled while shaking to an initial pressure of 800–900 p.s.i. at 25°. Shaking is continued as the bomb is slowly heated (Note 3) to 190–200° and maintained at this temperature for 7 hours (Note 4). At the end of this period, the reaction vessel is cooled and vented, and the crude product is transferred to a simple distillation apparatus (Note 5). A fraction boiling between 93° and 100° is collected, yield 162–202 g. (57–71%, based on dicyclopentadiene) (Note 6). The norbornylene may be redistilled with negligible losses to give a final product, b.p. 94–97°/740 mm., m.p. 44–44.5° (sealed capillary).

2. Notes

1. The dicyclopentadiene used by the submitters was supplied by the Enjay Company. No preliminary purification is required. Technical (85%) dicyclopentadiene has been found by the checkers to give 54–56% yields of norbornylene without preliminary purification.
2. C.P. grade ethylene was obtained from the Matheson Company.
3. To avoid complications due to the exothermic nature of this reaction,[2] a rate of heating of about 50° per hour was adopted (cf. Note 6).
4. Near 180°, the maximum pressure (about 2350 p.s.i.) is developed.
5. In spite of the low melting point of norbornylene, the product has a remarkable tendency to crystallize. Care should therefore be taken

to prevent premature solidification of the distillate.   A short-path, air-cooled assembly using rather wide-diameter tubing is convenient for this purpose.

6. The submitters report the same yields using a 3-l. bomb and 3.68 moles of dicyclopentadiene.   Larger-scale preparations may necessitate special control procedures.

3. Methods of Preparation

The procedure described above is essentially that of Thomas.[2]   Norbornylene has also been prepared by the additon of ethylene to monomeric cyclopentadiene [3] (p. 238), by dehydration of β-norborneol with phosphorus pentoxide,[4] and by dehydrohalogenation of norbornyl chloride or bromide using quinoline.[4, 5]

[1] Cornell University, Ithaca, New York.
[2] Thomas, *Ind. Eng. Chem.*, **36**, 310 (1944); Thomas and Universal Oil Products, U. S. pat. 2,340,908 [*C. A.*, **38**, 4273 (1944)].
[3] Joshel and Butz, *J. Am. Chem. Soc.*, **63**, 3350 (1941).
[4] Komppa and Beckmann, *Ann.*, **512**, 175 (1934).
[5] Alder and Rickert, *Ann.*, **543**, 10 (1940).

OLEOYL CHLORIDE *

$$n\text{-}C_{17}H_{33}CO_2H + SOCl_2 \rightarrow n\text{-}C_{17}H_{33}COCl + SO_2 + HCl$$

Submitted by C. F. H. ALLEN, J. R. BYERS, JR., and W. J. HUMPHLETT.[1]
Checked by MAX TISHLER and R. CONNELL.

1. Procedure

Seventy grams (0.25 mole) of oleic acid (Note 1) is placed in the dropping funnel H of the tangential apparatus (Fig. 13; Note 2).   The thionyl chloride distillation is started and regulated (Note 2); the upper part of the column should be filled with the vapor, and reflux should be constant and steady.   The acid is dropped in at the top of the column over a period of 35 minutes (120 g. per hour).   The product that collects in the receiver I contains about 25–27% of thionyl chloride (Note 2) if the heated lower leg K is employed.   The product in

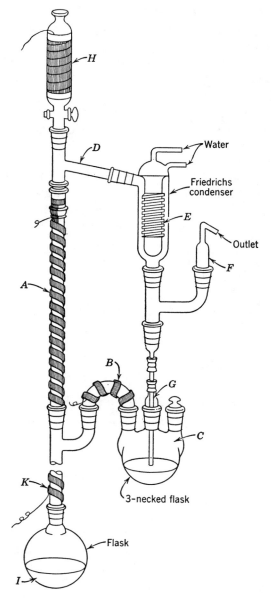

FIG. 13.

receiver I is heated on a steam bath under reduced pressure until the more volatile components are completely removed. The residue of crude acid chloride weighs 72–74 g. (97–99% yield). The infrared spectrum of this product shows no trace of a carboxyl band (Notes 1, 3, and 4).

The crude acid chloride will serve for most purposes. It can be distilled at very low pressures (b.p. 99–109°/25 μ) to yield a water-white product; n_D^{25} 1.4580–1.4613 (Note 5). Small amounts of oleoyl chloride may be distilled at higher pressures; b.p. 180–185°/1–2 mm. The infrared absorption curve of the oleic acid obtainable by hydrolysis is the same as that of the oleic acid used; thus no isomerization during the reaction is indicated.

2. Notes

1. The oleic acid used by the submitters had a freezing point of −2° to +13° and a boiling point of 182–187°/1–2 mm. It was a 90% middle cut from Emerson 233 as obtained from Emery Industries, Inc., Cincinnati, Ohio. The checkers found that the amount of color in the product was a function of the purity of the oleic acid. The product from distilled Emerson 233 was dark red-brown, whereas that from a purified grade of oleic acid was pale yellow.

2. The tangential apparatus [2] is built from stock pieces of glassware having 24/40 standard taper glass joints, and it is clamped to two ring stands. The reactor consists of a column 50 cm. long and 2 cm. inside diameter, packed with ⅛-in. glass helices of the type used for packing distillation columns, and wrapped with a 275-watt flexible heating tape A, 6 ft. long and ½ in. wide, with lead wires connected to a variable transformer such as the Powerstat or Variac, which is attached to a source of 110-volt power. Near the bottom of the column is a heated side arm connection B to a three-necked flask reservoir C. The use of the leg K, wrapped with a heating tape, removes much of the excess of thionyl chloride and is particularly advantageous with acid chlorides (such as oleoyl chloride) that are sensitive to heat. The crude effluent from the reactor ordinarily contains 40–50% of thionyl chloride, which is reduced one-half or more by this leg. It is not essential for use with chlorides insensitive to heat. Vapor supplied by boiling the liquid in the reservoir is forced to circulate via B through the column by sufficient heat input to the column and the lower leg

K; the excess escapes at the top through the side arm D which is connected to a Friedrichs condenser E. The bottom of the condenser is connected to a Y-shaped section with an outlet F for effluent gas and a return G which extends below the surface of the liquid in the reservoir. At the top of the column is a 250-ml. dropping funnel H for admitting reactants to the column. Since oleic acid is a liquid at room temperature, heating of this funnel is not required in the present preparation. At the bottom of the column is a 500-ml. flask I for receiving the product.

3. Other acid chlorides can be prepared similarly. (*a*) Palmitoyl chloride: Palmitic acid is a solid at room temperatures (m.p. 61–62°) and is not sufficiently soluble in palmitoyl chloride or thionyl chloride, and so these cannot be used as solvents. It is admitted to the reactor in the melted condition by warming the acid in the dropping funnel H. The quality of the chloride naturally depends on the homogeneity of the starting acid; palmitic acid usually contains a little stearic acid, which does not affect the melting point. Palmitoyl chloride, n_D^{25} 1.4489, can be obtained as a water-white product by distillation at very low pressures (b.p. 110°/15–24 μ), or with negligible decomposition at boiling points as high as 165°.

(*b*) Ricinoleoyl chloride: b.p. 205–210°/8 mm., 125–130°/25 μ; n_D^{25} 1.4759; this substance decomposes significantly at the higher distillation temperature.

(*c*) 2,4-Di-*tert*-amylphenoxyacetyl chloride: In this instance, the corresponding solid acid (20 g.) is dissolved in the acid chloride (80 g.) as a solvent and admitted at the rate of 70 g. of acid per hour. The crude acid chloride, after removal of the excess thionyl chloride, contains about 1% of unchanged acid (infrared determination). The yield of distilled acid chloride, b.p. 143–146°/2 mm., n_D^{25} 1.5062, is 85%.

4. The described procedure is not suitable for all acids. For instance, the acid chloride must not have a boiling point so near that of thionyl chloride that they are inseparable by distillation. Certain high-molecular-weight acids give dehydration products, presumably diketenes: e.g., behenic and dihydroxystearic acids.

5. The refractive index will vary with the purity of the oleic acid. Oleic acid purified by low-temperature crystallization and by conversion to the oleic acid-urea complex (95.3% oleic, 0.7% linoleic) yielded a product with n_D^{25} 1.4613.

3. Methods of Preparation

This modification of the continuous reactor (cf. benzoylacetanilide, p. 80) with countercurrent distillation is preferred for reactions in which a large amount of solvent or excess of one reactant is essential, but increase of total volume is undesirable. It is especially useful if the substances involved are heat-sensitive; with this apparatus the reactants are heated for only a few minutes at most. It is particularly applicable to the preparation of acid chlorides from carboxylic acids and thionyl chloride (cf. Notes 3 and 4). An indefinite amount of product can be prepared by replenishing the reactants as they are consumed.

Oleoyl chloride has been prepared by treatment of oleic acid with thionyl chloride,[3] phosphorus trichloride or pentachloride, and oxalyl chloride.[4] The highest yield (86%) reported was secured by use of oxalyl chloride in carbon tetrachloride, but the more economical phosphorus trichloride gave a yield of 60%. The standard procedures for obtaining aliphatic acid chlorides have been described many times without inclusion of details other than physical properties. Only references to the procedures useful in the laboratory are given.

[1] Eastman Kodak Company, Rochester, New York.
[2] Allen, Byers, Humphlett, and Reynolds, *J. Chem. Educ.*, **32**, 394 (1955).
[3] Verkade, *Rec. trav. chim.*, **62**, 393 (1943); Fierz-David and Kuster, *Helv. Chim. Acta*, **22**, 82 (1939).
[4] Bauer, *Oil & Soap*, **23**, 1 (1946).

PARABANIC ACID*

$$H_2NCONH_2 + \begin{matrix} CO_2C_2H_5 \\ | \\ CO_2C_2H_5 \end{matrix} + 2CH_3ONa \rightarrow$$

$$\begin{matrix} CO-CO \\ NaN \diagup \quad \diagdown NNa + 2CH_3OH + 2C_2H_5OH \\ \diagdown \quad \diagup \\ CO \end{matrix}$$

$$\begin{matrix} CO-CO \\ NaN \diagup \quad \diagdown NNa + 2HCl \rightarrow \\ \diagdown \quad \diagup \\ CO \end{matrix} \quad \begin{matrix} CO-CO \\ HN \diagup \quad \diagdown NH + 2NaCl \\ \diagdown \quad \diagup \\ CO \end{matrix}$$

Submitted by Joseph I. Murray.[1]
Checked by M. S. Newman and Tadamichi Fukunaga.

1. Procedure

Eight hundred fifty milliliters of absolute methanol (magnesium-dried) is distilled directly, through a condenser, into a 1-l. three-necked reaction flask equipped with a sealed mechanical stirrer and an efficient reflux condenser protected by a drying tube. After about 300 ml. of methanol has been distilled, the drying tube is removed from the reflux condenser, and small pieces of clean sodium are added to the stirred methanol at such a rate that the alcohol vapors do not escape from the condenser (Note 1). When a total of 23 g. (1 g. atom) of sodium has been added, the drying tube is replaced, the distillation of methanol is completed (Note 1), and the system is brought to a temperature of 20–25° (other temperatures are less favorable). The inlet condenser is removed, 30 g. (0.5 mole) of dry finely ground urea (Note 2) is quickly added to the reaction mixture, and the opening is then closed by attaching a dropping funnel. When the urea has dissolved completely, the addition of 70 g. (0.48 mole) of diethyl oxalate (Note 2) is begun (Note 3). A white precipitate forms immediately. Stirring is continued for 1 hour after the addition of the ester has been completed (Note 4). One hundred milliliters of concentrated hydrochloric acid is added dropwise to the mixture at a rate that causes little increase in the solution temperature, stirring is continued for a few minutes, and the mixture is filtered. The residue is washed twice with a small quantity of methanol, and the filtrate and washings are transferred to a 2-l. two-necked flask equipped with a sealed stirrer and a vacuum take-off. The alcohol is removed under the reduced pressure of a good water aspirator (Note 5). The practically dry solid residue is washed from the evaporator with water which is used as the recrystallization solvent (Note 6). When the bulk of the solid has dissolved, the solution is cooled slightly below the boiling point, 2 g. of activated carbon is added, and the solution is filtered through a hot Büchner funnel by use of moderate suction. The solution yields a white, crystalline precipitate of parabanic acid upon standing in a refrigerator overnight. The yield of purified, dried product melting at 241–243° with decomposition is 39–41.5 g. (71.5–76%).

2. Notes

1. The checkers used Mallinckrodt analytical reagent methanol without drying and distilling. The sodium was added to 300 ml. of the methanol, and the remainder of the methanol was added dropwise. The results obtained in this way were the same as those described by the submitter.

The hydrogen released during the formation of the sodium methoxide prevents moisture from entering the system. More rapid addition of the sodium and reduced loss of methanol vapor are realized if the reaction flask is surrounded by an ice bath during the addition of the metal. Calcium chloride should not be used in the drying tube because of possible clogging. Drierite ($CaSO_4$) is preferable.

2. The checkers used reagent grade urea (Baker) and diethyl oxalate (Matheson, Coleman and Bell), without purification.

3. Ester addition must be quite slow (2 drops per second or less) to prevent emulsion formation and extremely low yields.

4. At the end of 1 hour, the precipitated salt should be in a fine crystalline form which settles readily when the stirring is momentarily stopped. A viscous, creamy suspension which is poorly mixed by the stirring usually gives poor yields. Factors contributing to this condition have been found to be: insufficient amount of solvent initially, too rapid addition of ester, and improper temperature control.

5. Considerable frothing may be encountered during the first stages of the evaporation, particularly when a steam bath is used to hasten removal of the solvent. Rapid stirring helps to cut down frothing and lessens bumping.

6. The solid may go into solution rather slowly even in boiling water.

3. Methods of Preparation

Parabanic acid can be prepared by the condensation of urea with diethyl oxalate in an ethanolic solution of sodium ethoxide,[2] by reaction of urea with an ethereal solution of oxalyl chloride,[3] by oxidizing uric acid with an acid solution of perhydrol,[4] or by the action of hot, concentrated nitric acid on uric acid.[5] The present method gives better yields than the previously reported methods and is better adapted to larger-scale preparations.

[1] University of Buffalo, Buffalo, New York.
[2] Michael, *J. prakt. Chem.*, [2]**35**, 457 (1886).
[3] Biltz and Topp, *Ber.*, **46**, 1387 (1913).
[4] Biltz and Schiemann, *Ber.*, **59**, 721 (1926).
[5] Behrend and Asche, *Ann.*, **416**, 226 (1918).

1,4-PENTADIENE *

I. METHOD A

$$CH_3CO_2CH_2CH_2CH_2CH_2CH_2O_2CCH_3 \xrightarrow{575°}$$
$$CH_2{=}CHCH_2CH{=}CH_2 + 2CH_3CO_2H$$

Submitted by R. E. BENSON and B. C. McKUSICK.[1]
Checked by N. J. LEONARD and A. G. COOK.

1. Procedure

The apparatus (Fig. 14) is similar to that described on p. 795 and in a previous volume.[2] It consists of a Pyrex glass reaction tube, 90 cm. long by 45 mm. in outside diameter, mounted vertically in an electric furnace about 50 cm. long. Attached to the top of the tube are a graduated dropping funnel (Note 1), an inlet tube for nitrogen, and a thermocouple well extending to the bottom of the heated section and holding a movable thermocouple. The entire heated section, which begins 10 cm. from the top of the tube, is packed with Pyrex glass rings 10 mm. in outside diameter by about 10 mm. in length held in place by a plug of glass wool supported by indentations in the tube. The lower end of the tube is attached to a 1-l. round-bottomed flask immersed in an ice bath and having a side arm from which vapors pass successively through a trap immersed in an ice bath and a trap immersed in a bath of Dry Ice and acetone. Each trap is capable of holding about 200 ml. of liquid. The temperature of the hottest part of the tube, which is located near the middle of the heated section, is raised to 575° ± 10° while nitrogen (Note 2) is passed successively through a flowmeter and the tube at a rate

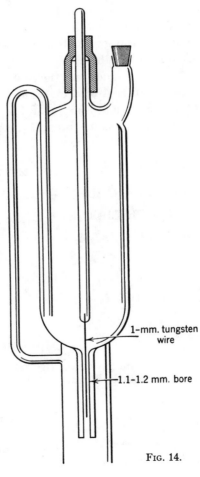

FIG. 14.

of 4–6 l./hr. Under these conditions (Note 3), 658 g. (645 ml., 3.5 moles) of 1,5-pentanediol diacetate (Note 4) is added to the tube over a period of 3.5 hours. The contents of the three receivers are combined and distilled at atmospheric pressure through a 15-cm. indented Claisen-type still head; the condenser is cooled with ice water, and the receiver is immersed in an ice bath. The fraction boiling at 25–55°, wt. 170–190 g., is redistilled through a 60-cm. column packed with glass helices or a column of similar efficiency to give 150–170 g. (63–71%) of 1,4-pentadiene, b.p. 26–27.5°/760 mm., n_D^{25} 1.3861–1.3871 (Note 5).

1-mm. tungsten wire

1.1–1.2 mm. bore

2. Notes

1. A Hershberg dropping funnel[3] modified by addition

of a pressure-equalizing arm (Fig. 14) makes it easy to add the diacetate at a constant rate.

2. A dry, oxygen-free grade of commercial nitrogen is used. Nitrogen can be omitted without diminishing the yield by more than a few per cent.

3. As the addition of the diacetate begins, the temperature of the hottest part of the tube (the location of which generally shifts lower at this time) decreases, necessitating an increase in current in the electric furnace.

4. 1,5-Pentanediol diacetate,[4, 5] b.p. 85–90°/0.9 mm., n_D^{25} 1.4253, is obtained in 92–94% yield by adding a 10% excess of acetic anhydride to 1,5-pentanediol [6] at 120–140°, heating the mixture under reflux for 2 hours, and distilling it at reduced pressure. The practical grade of 1,5-pentanediol sold by Eastman Kodak Company may be used.

5. The residue from the first distillation is a mixture of acetic acid, 4-penten-1-ol acetate, and 1,5-pentanediol diacetate. Another 15–35 g. (6–15%) of 1,4-pentadiene can be obtained by passing the residue through the pyrolysis tube under the conditions described above.

II. METHOD B

$$CH_3CHO + C_2H_5OH + HCl \rightarrow CH_3CH(Cl)OC_2H_5 + H_2O$$

$$CH_3CH(Cl)OC_2H_5 + Br_2 \rightarrow BrCH_2CH(Br)OC_2H_5 + HCl$$

$$BrCH_2CH(Br)OC_2H_5 + CH_2{=}CHCH_2MgBr \rightarrow$$

$$\begin{array}{c} BrCH_2CHOC_2H_5 + MgBr_2 \\ | \\ CH_2CH{=}CH_2 \end{array}$$

$$\begin{array}{c} BrCH_2CHOC_2H_5 + Zn \rightarrow \\ | \\ CH_2CH{=}CH_2 \end{array}$$

$$CH_2{=}CHCH_2CH{=}CH_2 + ZnBr(OC_2H_5)$$

Submitted by OLIVER GRUMMITT, E. P. BUDEWITZ, and C. C. CHUDD.[7]
Checked by WILLIAM S. JOHNSON, E. SAITO, and DONALD J. REIF.

1. Procedure

A. α-Chloroethyl ethyl ether. A mixture of 200 g. (201 ml.) of redistilled paraldehyde, b.p. 121–122.5° (equivalent to 4.54 moles of acetaldehyde), and 200 g. (254 ml., 4.34 moles) of absolute ethanol is placed in a 1-l. three-necked flask fitted with a mechanical stirrer and a gas inlet tube reaching to the bottom of the flask. The mixture is cooled to $-5°$ in a mixture of Dry Ice and acetone, and dry hydrogen chloride (Note 1) is passed into the stirred reaction mixture maintained at about $-5°$ until 200 g. (5.48 moles) has been absorbed. During this operation, which requires about 2 hours, the reaction mixture separates into two layers. The upper layer of crude α-chloroethyl ethyl ether is removed, and the dissolved hydrogen chloride is swept out by bubbling dry nitrogen gas through the mixture. The product is dried overnight with 25–50 g. of anhydrous calcium sulfate (Note 2). The yield of crude α-chloroethyl ethyl ether (Note 3) is 411–432 g. (87–92% based on ethanol). The product is not distilled, since it decomposes readily.

B. α,β-Dibromoethyl ethyl ether. Four hundred and twenty-five grams (3.92 moles) of the crude α-chloroethyl ethyl ether (part A) is placed in a dry 1-l. three-necked flask, fitted with a dropping funnel, mechanical stirrer, and an outlet tube leading to a gas absorption trap [8] to dispose of the hydrogen chloride evolved. The flask is cooled in an ice bath, and 625 g. (200 ml., 3.92 moles) of bromine is added in small portions from the dropping funnel with stirring. The reaction mixture is allowed to become almost colorless after each addition. When all of the bromine has been added (5–6 hours), a slow current of dry nitrogen is bubbled through the reaction mixture to sweep out the hydrogen chloride (Note 4). The product is dried overnight with 25–50 g. of anhydrous calcium sulfate and then distilled at reduced pressure (Note 5). The yield of colorless α,β-dibromoethyl ethyl ether is 599–663 g. (66–73% based on the α-chloroethyl ethyl ether), b.p. 70–75°/27 mm., n_D^{20} 1.5097–1.5102.

C. Allylmagnesium bromide (Note 6). In a dry 5-l. three-necked flask, equipped with a sealed stirrer, a reflux condenser and drying tube, a pressure-equalized dropping funnel, and a nitrogen inlet tube, are placed 195 g. (8.0 g. atoms) of dry magnesium turnings and 2.4 l. of anhydrous ether. The flask is cooled in an ice bath, a small crystal of iodine is added, and a solution of 400 g. (287 ml., 3.31 moles) of allyl bromide (redistilled, b.p. 69–71°) in an equal volume of anhydrous ether is added dropwise over a period of 17 hours. During the reaction a slow stream of dry, oxygen-free nitrogen is passed through the flask. After the addition is complete, the reaction mixture is stirred for 30 minutes. The Grignard reagent is decanted into a dry, graduated storage bottle; the residue in the flask is washed with 150–200 ml. of dry ether; and the wash liquid is added to the storage bottle. Samples of the clear supernatant solution are analyzed by acidimetric titration.[9] The yield of allylmagnesium bromide is 2.62–2.95 moles (79–89%).

D. *α-Allyl-β-bromoethyl ethyl ether*. The same apparatus is used as in the preparation of allylmagnesium bromide. The flask is charged with an amount of the Grignard solution (part C) equivalent to 2.78 moles of allylmagnesium bromide (or chloride) and cooled in an ice bath. A solution of 580 g. (2.5 moles) of $α,β$-dibromoethyl ethyl ether (part B) in an equal volume of anhydrous ether is added slowly with stirring over a period of 3–4 hours. The mixture is allowed to stand overnight and is then hydrolyzed with 75 ml. of 20% acetic acid followed by 500 ml. of water. The ether layer is separated, washed with four 100-ml. portions of 10% aqueous sodium bicarbonate solution followed by four 100-ml. portions of saturated aqueous sodium chloride solution, dried over 100 g. of anhydrous calcium sulfate, and distilled under reduced pressure. The yield of colorless α-allyl-β-bromoethyl ethyl ether is 370–396 g. (77–82% based on the $α,β$-dibromoethyl ethyl ether), b.p. 72–75°/21 mm., n_D^{20} 1.4600–1.4606.

E. *1,4-Pentadiene*. A 2-l. three-necked flask is equipped with a sealed stirrer and a 28-cm. reflux condenser, at the top of which is a 2.5 × 35 cm. Vigreux fractionating column attached to an efficient condenser arranged for distillation. Water at 35–40° is pumped through the reflux condenser, ice water is pumped through the downward condenser, and the receiver is ice-cooled and attached to a trap cooled by Dry Ice. The flask is charged with 380 g. (1.97 moles) of α-allyl-β-bromoethyl ethyl ether (part D) in 550 ml. of *n*-butyl alcohol; then 550 g. (8.4 g. atoms) of zinc dust and 2 g. of anhydrous zinc chloride are added. The mixture is stirred vigorously and heated gradually to the point where the pentadiene distils at a rate of about one drop every 2 seconds. The reflux condenser and column return most of the butyl alcohol to the flask. The reaction takes 5–6 hours for completion. The distillate is washed with five 100-ml. portions of ice water to remove most of the butyl alcohol and is then dried overnight with 15 g. of calcium chloride. The crude dry product is distilled through a 2.5 × 35 cm. Vigreux column to give 97–102 g. (72–76%) of colorless 1,4-pentadiene, b.p. 26–27°/740 mm., n_D^{20} 1.3887–1.3890.

2. Notes

1. Commercial hydrogen chloride from a cylinder is dried by passage through a train consisting of a wash bottle of concentrated sulfuric acid, a 25-cm. calcium chloride tube, and finally an empty safety trap.

2. This ether is a lachrymator, hydrolyzes rapidly in the presence of moisture, and resinifies readily at temperatures above 0°. It is advisable to store it in a Dry Ice chest during drying and until the next step is to be run. Hydrolysis at low temperatures appears to be negligible.

3. Analysis of the crude material for chlorine gave 31.74% (calcd. 32.69%).

4. It is necessary to remove the hydrogen chloride because it promotes decomposition of the dibromoether.

5. Distillation is required to remove aldehyde, alcohol, and water which would react with the Grignard reagent in the next step. α,β-Dibromoethyl ethyl ether is also a lachrymator.

6. Either allylmagnesium bromide or allylmagnesium chloride may be used. The former is more conveniently prepared in higher yield and in a more concentrated solution, but allyl chloride is considerably less expensive than allyl bromide. The submitters state that the chloride is prepared just as the bromide with the following exceptions: the temperature is maintained between −10 and −15° with a Dry Ice-acetone cooling bath; 76.5 g. (82 ml., 1 mole) of allyl chloride (b.p. 45–47°) dissolved in an equal volume of dry ether is used (the amounts of all other reagents are unaltered); the addition is carried out over a 12-hour period, then the cooling bath is removed, and the mixture (containing considerable solid) is stirred until it reaches room temperature, during which time most of the solid material dissolves.

The yield of allylmagnesium chloride as determined by acidimetric titration [9] is 0.67–0.69 mole (67–69%).

3. Methods of Preparation

1,4-Pentadiene has been prepared by the interaction of allyl bromide and vinyl bromide in the presence of magnesium; [10] by the pyrolysis of 1,5-pentanediol diacetate [4, 5] or 4-penten-1-ol acetate; [5, 11] by the reaction of α-allyl-β-bromoethyl ethyl ether with zinc; [12, 13] by the thermal decomposition of dimethyl(4-pentenyl)amine oxide; [14] and by the reaction of vinylmagnesium bromide with allyl bromide. [15]

Method I is based on the work of Schniepp and Geller [4] and of Paul and Tchelitcheff, [5] while Method II is essentially that of Shoemaker and Boord [12] with some modifications. [13]

[1] Chemical Department, Experimental Station, E. I. du Pont de Nemours and Company, Wilmington, Delaware.

[2] *Org. Syntheses Coll. Vol.* **3**, 30 (1955).

[3] *Org. Syntheses Coll. Vol.* **2**, 129 (1943).

[4] Schniepp and Geller, *J. Am. Chem. Soc.*, **67**, 54 (1945).

[5] Paul and Tchelitcheff, *Bull. soc. chim. France*, [5] **15**, 108 (1948).

[6] *Org. Syntheses Coll. Vol.* **3**, 693 (1955).

[7] Western Reserve University, Cleveland, Ohio.

[8] *Org. Syntheses Coll. Vol.* **2**, 4 (1943).

[9] Gilman, Zoellner, and Dickey, *J. Am. Chem. Soc.*, **51**, 1577 (1929).

[10] Kogerman, *J. Am. Chem. Soc.*, **52**, 5060 (1930).

[11] Paul and Normant, *Bull. soc. chim. France*, [5] **11**, 367 (1944).

[12] Shoemaker and Boord, *J. Am. Chem. Soc.*, **53**, 1505 (1931).

[13] Kistiakowsky, Ruhoff, Smith, and Vaughan, *J. Am. Chem. Soc.*, **58**, 146 (1936); Elsner and Wallsgrove, *J. Inst. Petrol.*, **35**, 259 (1949).

[14] Cope and Bumgardner, *J. Am. Chem. Soc.*, **79**, 960 (1957).

[15] Normant, *Compt. rend.*, **239**, 1811 (1954).

PENTAERYTHRITYL TETRABROMIDE *

[Propane, 1,3-dibromo-2,2-bis(bromomethyl)-]

$$C(CH_2OH)_4 + 4C_6H_5SO_2Cl \xrightarrow{C_5H_5N} C(CH_2OSO_2C_6H_5)_4 + 4HCl$$

$$C(CH_2OSO_2C_6H_5)_4 + 4NaBr \rightarrow C(CH_2Br)_4 + 4C_6H_5SO_3Na$$

Submitted by HERSHEL L. HERZOG.[1]
Checked by T. L. CAIRNS and D. W. WOODWARD.

1. Procedure

In a 5-l. three-necked round-bottomed flask equipped with a powerful mechanical stirrer (Note 1), a thermometer, and a 1-l. dropping funnel are placed 130 g. (0.96 mole) of technical grade pentaerythritol (Note 2) and 650 ml. of pyridine. The stirrer is started, and to the resulting suspension is added dropwise 750 g. (4.24 moles) of benzenesulfonyl chloride (Note 3) at such a rate that the temperature of the reaction does not rise above 30–35°. The addition requires about 2 hours. The resulting slurry is stirred at 40° for 1 hour after the addition is complete. The slurry is then added slowly (Note 4) to a vigorously stirred solution of 800 ml. of concentrated hydrochloric acid in 1 l. of water and 2 l. of methanol contained in a 9 by 15 in. battery jar. The resulting suspension of granular white pentaerythrityl benzene-sulfonate is cooled by addition of 500 g. of ice, filtered with suction, and washed with 5 l. of water and then with 1 l. of cold methanol in two portions.

The crude, slightly wet pentaerythrityl benzenesulfonate is added to 1 l. of diethylene glycol (Note 5) in a 4-l. Erlenmeyer flask equipped with a Hershberg stirrer. Then 600 g. (5.8 moles) of sodium bromide is added, and the mixture is heated in an oil bath at 140–150° with slow stirring (60–120 r.p.m.) overnight. The resulting orange mixture is allowed to cool to about 90°, 2 l. of ice water is added rapidly with stirring, and finally the mixture is cooled to 10° by direct addition of ice. The precipitate is filtered with suction, washed with 2 l. of water, and pressed dry. The yield is 315–323 g. of a crude tan crystalline solid, m.p. 147–149°. The solid is dissolved in 2 l. of boiling acetone and filtered by gravity on a steam-heated funnel. On cooling, the solution deposits colorless glistening plates, which are filtered with suction and washed with 100 ml. of cold 95% ethanol, yielding 150–160 g. of pentaerythrityl tetrabromide, m.p. 159–160° (Note 6). By repeated concentration and cooling of the mother liquor, an additional 90–100 g. of pentaerythrityl tetrabromide, m.p. 156.5–158° (Note 6), is obtained. The combined yield is 228–260 g. (68–78%) (Note 7).

2. Notes

1. A powerful stirrer is necessary to mix the reactants, particularly in the later stages of the reaction when the mixture is quite viscous. The submitter and checkers used a "Lightnin" stirrer.

2. Heyden Chemical Corporation technical grade pentaerythritol (Pentek) was found to be satisfactory. It contains about 90% pentaerythritol, the remainder being principally dipentaerythritol.

3. Eastman Kodak Company practical grade benzenesulfonyl chloride was used.

4. Crystallization is extremely slow at first and becomes satisfactory only when the mixture is well seeded. It is well to remove a small portion first and work it up in the hydrochloric acid solution with a spatula to induce crystallization. The mixture should be added slowly at first but more rapidly toward the end.

5. Eastman Kodak Company practical grade diethylene glycol was used.

6. The checkers observed melting points about 2° lower.

7. The yield is based on the assumption that 90% of the starting material is pentaerythritol.

3. Methods of Preparation

The procedure given was developed by Buchman, Herzog, and Fujimoto.[2] Pentaerythrityl bromide has also been prepared from phosphorus tribromide and pentaerythritol,[3] and by the action of hydrobromic acid on pentaerythrityl tetraacetate in acetic acid.[4]

[1] California Institute of Technology, Pasadena, California.
[2] Buchman, Herzog, and Fujimoto, unpublished results.
[3] *Org. Syntheses Coll. Vol.* **2**, 476 (1943).
[4] Perkin and Simonsen, *J. Chem. Soc.*, **87**, 860 (1905).

4-PENTYN-1-OL

$$\begin{matrix} CH_2\!\!-\!\!CH_2 \\ | \qquad | \\ CH_2 \quad CHCH_2Cl \\ \diagdown \!\! O \!\! \diagup \end{matrix} + 3NaNH_2 \xrightarrow{\text{Liq. NH}_3}$$

$$NaCl + 3NH_3 + NaC{\equiv}CCH_2CH_2CH_2ONa$$

$$NaC{\equiv}CCH_2CH_2CH_2ONa + 2NH_4Cl \rightarrow$$
$$2NaCl + 2NH_3 + HC{\equiv}CCH_2CH_2CH_2OH$$

Submitted by E. R. H. Jones, Geoffrey Eglinton, and M. C. Whiting.[1]
Checked by Arthur C. Cope and Ronald M. Pike.

1. Procedure

Caution! This preparation should be conducted in a hood to avoid exposure to ammonia.

A solution of sodium amide in liquid ammonia is prepared according to a procedure previously described (Note 1) in a 3-l. three-necked round-bottomed flask equipped with a cold-finger condenser (cooled with Dry Ice) attached through a soda-lime tower to a gas-absorption trap,[2] a mercury-sealed stirrer, and an inlet tube. Anhydrous liquid ammonia (1 l.) is introduced from a commercial cylinder through the inlet tube, and 1 g. of hydrated ferric nitrate is added, followed by 80.5 g. (3.5 g. atoms) of clean, freshly cut sodium (Notes 1 and 2). The inlet tube is replaced with a 250-ml. dropping funnel, and the mixture is stirred until all the sodium is converted into sodium amide, after which 120.5 g. (1 mole) of tetrahydrofurfuryl chloride[3] (Note 3) is added over a period of 25 to 30 minutes. The mixture is stirred for an additional period of 1 hour, after which 177 g. (3.3 moles) of solid ammonium chloride is added in portions at a rate that permits control of the exothermic reaction. The flask is allowed to stand overnight in the hood while the ammonia evaporates.

The residue is extracted thoroughly with ten 250-ml. portions of ether, which are decanted through a Büchner funnel (Note 4). The ether is distilled, and the residue is fractionated at a reflux ratio of about 5 to 1, through a column containing a 20-cm. section packed with glass helices yielding 63–71 g. (75–85%) of 4-pentyn-1-ol, b.p. 70–71°/29 mm., n_D^{25} 1.4443 (Note 5).

2. Notes

1. Procedures for converting sodium to sodium amide are given on p. 763 and in a previous volume.[4]

2. More liquid ammonia should be added through the inlet tube if vaporization reduces the liquid volume to less than 750 ml.

3. Freshly distilled tetrahydrofurfuryl alcohol should be used in the preparation of tetrahydrofurfuryl chloride according to the procedure of *Organic Syntheses*.[3]

4. Ether extraction of the solid must be thorough or the yield will be reduced. A large Soxhlet extractor may be used if desired.

5. Others have reported b.p. 154–155°, n_D^{19} 1.4432;[5] b.p. 154–155°, $n_D^{22.5}$ 1.4450.[6] A sample purified through the silver derivative had b.p. 77°/37 mm., n_D^{15} 1.4464. The α-naphthylurethan of 4-pentyn-1-ol crystallized as needles from 60–80° petroleum ether; m.p. 79–80°.

3. Methods of Preparation

4-Pentyn-1-ol has been prepared from 4-penten-1-ol [3] by bromination followed by dehydrobromination with alkali;[6] by the reaction of 3-bromodihydropyran or 3,4-dihydro-2H-pyran with n-butylsodium, n-butyllithium, or n-butylpotassium;[5,7] by the reaction of dihydropyran or 2-methylenetetrahydrofuran with n-amylsodium or n-butyllithium;[7] by the reduction of ethyl 4-pentynoate with lithium aluminum hydride;[8] and by the method used in this preparation.[9]

[1] Victoria University of Manchester, Manchester, England.
[2] *Org. Syntheses Coll. Vol.* **2**, 4 (1943).
[3] *Org. Syntheses Coll. Vol.* **3**, 698 (1955).
[4] *Org. Syntheses Coll. Vol.* **3**, 219 (1955).

⁵ Paul and Tchelitcheff, *Compt. rend.*, **230**, 1473 (1950); Paul, *Angew. Chem.*, **63**, 304 (1951); Paul, *Bull. soc. chim. France*, **18**, 109 (1951).

⁶ Lespieau, *Compt. rend.*, **194**, 287 (1932).

⁷ Paul and Tchelitcheff, *Bull. soc. chim. France*, **19**, 808 (1952).

⁸ Colonge and Gelin, *Bull. soc. chim. France*, **1954**, 799.

⁹ Eglinton, Jones, and Whiting, *J. Chem. Soc.*, **1952**, 2873.

PHENANTHRENEQUINONE *

$$\xrightarrow[\text{H}_2\text{SO}_4]{\text{CrO}_3}$$

Submitted by RAY WENDLAND and JOHN LALONDE.[1]
Checked by WILLIAM S. JOHNSON and SOL SHULMAN.

1. Procedure

In a 3-l. three-necked flask equipped with a reflux condenser, a sealed mechanical stirrer, and a 1-l. dropping funnel, are placed 100 g. (0.56 mole) of phenanthrene (Note 1), 210 g. (2.1 moles) of chromic acid (Note 2), and 1 l. of water. The stirrer is started, and 450 ml. of concentrated sulfuric acid is added from the dropping funnel into the suspension at such a rate that gentle boiling is induced (Note 3). After addition of the sulfuric acid is complete, a mixture of 210 g. (2.1 moles) of chromic acid and 500 ml. of water is added carefully to the reaction mixture from the dropping funnel (Note 4). The resulting mixture is boiled under reflux for 20 minutes.

After being cooled to room temperature the reaction mixture is poured into an equal volume of water and chilled to 10° in an ice bath. The crude precipitate is separated by suction filtration and thoroughly washed with cold water until the washings no longer show any chrome green color. The precipitate is triturated with three 300-ml. portions of boiling water and filtered to remove the diphenic acid formed in the reaction. The precipitate is then triturated with several (4–6) 300-ml. portions of hot 40% sodium bisulfite solution and again filtered (Note 5). The insoluble material is a mixture of anthraquinone and some resinous products derived from anthracene and other contaminants present in the starting material (Note 6). The sodium bisulfite filtrates are combined and cooled to 5° in an ice bath. The precipitate which separates is collected by suction filtration; then it is transferred to a 1-l. beaker and finely dispersed in 300 ml. of water. To this suspension is added, with good stirring, 500 ml. of a saturated solution of sodium carbonate. The deep orange phenanthrenequinone which is liberated is separated by suction filtration (Note 7), washed well with cold water, and dried on a porcelain plate. The yield of product, melting at 205–208° cor., is 52–56 g. (44–48%). Further purification may be accomplished by crystallization from 95% ethanol (100 ml./g.). The recovery of first-crop material is over 80%, m.p. 208.5–210° cor.

2. Notes

1. Reilly Tar and Chemical Company practical grade phenanthrene is crystallized from boiling toluene using Norit. One crystallization is sufficient to produce material melting at about 99.5°. The checkers employed Eastman Kodak Company technical grade (90%) phenanthrene.

2. Technical grade chromic acid (99.5% CrO_3) in flake form was used.

3. It is safe to add 100 ml. of acid all at once to start the oxidation, but as soon as the temperature rises to 70–75° the remainder must be added slowly in order to avoid violent boiling.

4. One-half (250 ml.) of the mixture is added carefully (the reaction may become vigorous at this stage), and 20–25 minutes is allowed for the oxidation to proceed spontaneously. The remaining mixture is then added slowly.

5. A heated Büchner funnel is preferably used to prevent clogging by crystallization.

6. The anthraquinone may be purified by extracting the insoluble material with a 150-ml. solution of potassium hydroxide and sodium hydrosulfite (approximately 10% by weight of each).

The resulting red solution is quickly filtered by suction, and hydrogen peroxide is added to the filtrate until a yellow precipitate appears. Dilute hydrochloric acid is added until the mixture is acidic to litmus, and the precipitate is collected by suction filtration, washed well with water, and air-dried. Five to seven grams of anthraquinone, m.p. 280–283°, may thus be obtained.

7. The filtrate should be treated with more sodium carbonate to test for completeness of precipitation.

3. Methods of Preparation

Phenanthrenequinone has been prepared by treatment of phenanthrene with chromic acid in acetic acid;[2] potassium dichromate in sulfuric acid;[3-5] hydrogen peroxide in acetic acid;[6,7] selenium dioxide above 250°;[8] iodic acid in acetic acid,[9] and chromyl chloride.[10] It also can be prepared from benzil with aluminum chloride at 120°[11] and from biphenyl-2,2'-dialdehyde with potassium cyanide.[12]

[1] North Dakota State University, Fargo, North Dakota.
[2] Graebe, Ann., 167, 131 (1873); Kato, Maezawa, and Hashimoto, Yûki Gôsei Kagaku Kyokaishi, 15, 402 (1957) [C. A., 51, 16392 (1957)].
[3] Anschütz and Schultz, Ann., 196, 32 (1879).
[4] Oyster and Adkins, J. Am. Chem. Soc., 43, 208 (1921).
[5] Underwood and Kochmann, J. Am. Chem. Soc., 46, 2069 (1924).
[6] Henderson and Boyd, J. Chem. Soc., 1910, 1659.
[7] Charrier and Moggi, Gazz. chim. ital., 57, 736 (1927).
[8] Postowsky and Lugowkin, Ber., 68, 852 (1935).
[9] Fuson and Tomboulian, J. Am. Chem. Soc., 79, 956 (1957).
[10] Wheeler, Can. J. Research, 36, 949 (1958).
[11] Scholl and Schwarzer, Ber., 55, 324 (1922).
[12] Mayer, Ber., 45, 1105 (1912).

PHENYLACETAMIDE *

(Acetamide, 2-phenyl-)

$$C_6H_5CH_2CN \xrightarrow[\text{HCl}]{\text{H}_2\text{O}} C_6H_5CH_2CONH_2$$

Submitted by WILHELM WENNER.[1]
Checked by WILLIAM S. JOHNSON and ROBERT E. IRELAND.

1. Procedure

In a 3-l. three-necked round-bottomed flask equipped with glass joints are placed 200 g. (1.71 moles) of benzyl cyanide (Note 1) and 800 ml. of 35% hydrochloric acid (Note 2). The flask is fitted with a reflux condenser, a thermometer, and an efficient mechanical stirrer (Note 3). At a bath temperature of about 40° (Note 4) the mixture is stirred vigorously. Within a period of 20–40 minutes the benzyl cyanide goes into solution (Note 3). During this time, the temperature of the reaction mixture rises about 10° above that of the bath. The homogeneous solution is kept in the bath with, or without, stirring for an additional 20–30 minutes (Note 5). The warm water in the bath is replaced by tap water at about 15–20°, and the thermometer is replaced by a dropping funnel from which 800 ml. of cold distilled water is added with stirring (Note 6). After the addition of about 100–150 ml., crystals begin to separate. When the total amount of water has been added, the mixture is cooled externally with ice water for about 30 minutes (Note 7). The cooled mixture is filtered by suction. Crude phenylacetamide remains on the filter and is washed with two 100-ml. portions of water. The crystals are then dried at 50–80°. The yield of crude phenylacetamide is 190–200 g. (82–86%). It is sufficiently pure for most purposes although it contains traces of phenylacetic acid. If pure phenylacetamide is desired, the crude, wet solid is stirred for about 30 minutes with 500 ml. of a 10% solution of sodium carbonate, collected on a suction funnel, washed with two 100-ml. portions of cold water, and dried. The yield of this product is 180–190 g. (78–82%); m.p. 154–155° (Notes 8 and 9).

2. Notes

1. The quality of the benzyl cyanide markedly affects the yields. Material prepared according to directions given previously [2] is satisfactory. Several commercially available grades were also found to be usable without distillation.

2. The hydrochloric acid must be of at least 30% strength.

3. Efficient stirring is of prime importance for satisfactory reaction, because intimate mixing of the heterogeneous mixture is necessary. The rate of dissolution of the nitrile depends on the efficiency of the stirring.

4. The reaction proceeds slowly at lower temperatures. Temperatures above 50° are not recommended because of the high volatility of hydrochloric acid.

5. This additional warming ensures complete reaction of some dissolved benzyl cyanide. The phenylacetamide is not readily hydrolyzed under these conditions.

6. The rate of addition is not critical.

7. If phenylacetic acid is desired, the suspension of phenylacetamide is refluxed with stirring and the phenylacetamide redissolves. After about 30 minutes, the mixture becomes turbid and the product begins to separate as an oil. After 6 hours the mixture is cooled, first with tap water and then by an ice-water bath. When the temperature has dropped to about 40–50°, the phenylacetic acid crystallizes. After cooling at 0° for about 4 hours (the acid is rather soluble in warm water), the mixture is filtered by suction. The crude, colorless phenylacetic acid is washed with two 100-ml. portions of cold water and dried in a desiccator. The yield of crude acid is 180–195 g. (77.5–84%). It melts at 66–70° and is sufficiently pure for most purposes. The mother liquor on extraction with two 150-ml. portions of benzene and evaporation yields an additional 3–5 g. of acid. To prepare the pure acid, vacuum distillation (as described by Adams and Thal [3]) is simpler and gives higher yields than recrystallization from ligroin.

8. Further purification is effected by recrystallization from 95% ethanol or benzene, yielding the pure compound of m.p. 156°.

9. The following arylacetamides have been prepared from the corresponding nitriles by the same method in the indicated

yields: [4] p-methylphenylacetamide (70%), p-isopropylphenylacetamide (90%), 1-naphthylacetamide (54%), 5,6,7,8-tetrahydro-2-naphthylacetamide (90%), p-methoxyphenylacetamide (76%), 3,4-dimethoxyphenylacetamide (82%), and 2,3-dimethoxyphenylacetamide (91%). Only in the cases of the alkoxy-substituted nitriles are the resulting amides soluble in the reaction mixture; the other nitriles do not dissolve completely at any time during the reaction.

3. Methods of Preparation

Phenylacetamide has been obtained by a wide variety of reactions: from benzyl cyanide with water at 250–260°; [5] from benzyl cyanide with water and cadmium oxide at 240°; [6] from benzyl cyanide with sulfuric acid; [7,8] by saturation of an acetone solution of benzyl cyanide with potassium hydrosulfide; [9] from benzyl cyanide with sodium peroxide; [10] by electrolytic reduction of benzyl cyanide in sodium hydroxide; [11] from ethyl phenylacetate with alcoholic [12] or aqueous [13] ammonia; from phenylacetic acid with ammonium acetate [14] or urea; [15] from diazoacetophenone with ammoniacal silver solution; [16] from phenylacetic acid imino ether hydrochloride and water; [17] from acetophenone with ammonium polysulfide; [18] from benzoic acid; [19] by heating the ammonium salt of phenylacetic acid; [20] and by heating cinnamic acid with a mixture of sulfur and ammonium hydroxide.[21]

The literature on the preparation of phenylacetic acid is reviewed in an earlier volume of this series.[3]

The present method is that of Wenner [22] and is applicable to other arylacetonitriles.[4]

[1] Hoffmann-La Roche, Inc., Nutley, New Jersey.
[2] *Org. Syntheses Coll. Vol.* **1**, 107 (1941).
[3] *Org. Syntheses Coll. Vol.* **1**, 436 (1941).
[4] Wenner, *J. Org. Chem.*, **15**, 548 (1950).
[5] Bernthsen, *Ann.*, **184**, 318 (1877).
[6] I. G. Farbenind., Ger. pat. 551,869 [*C. A.*, **26**, 4826 (1932)].
[7] Maxwell, *Ber.*, **12**, 1764 (1879).
[8] Purgotti, *Gazz. chim. ital.*, **20**, 173, 593 (1891).
[9] Weddige, *J. prakt. Chem.*, [2] **7**, 99 (1873).
[10] Deinert, *J. prakt. Chem.*, [2] **52**, 432 (1895).
[11] Ogura, *Mem. Coll. Sci., Kyoto Imp. Univ.*, **12A**, 339 (1929) [*C. A.*, **24**, 2060 (1930)].

[12] Fischer and Dilthey, *Ber.*, **35**, 856 (1902).

[13] Meyer, *Monatsh.*, **27**, 34 (1906).

[14] Kao and Ma, *J. Chem. Soc.*, **1930**, 2788; **1931**, 443.

[15] Das-Gupta, *J. Ind. Chem. Soc.*, **10**, 117 (1933).

[16] Wolff, *Ann.*, **394**, 43 (1912).

[17] Houben, *Ber.*, **59**, 2878 (1926).

[18] Willgerodt and Scholtz, *J. prakt. Chem.*, [2] **81**, 384 (1910); British Petroleum Ltd., Brit. pat. 772,443 [*C. A.*, **51**, 14811 (1957)]; Shchukina and Golombik, *Med. Prom. S.S.S.R.*, **11**, No. 7, 42 (1957) [*C. A.*, **52**, 10943 (1958)]; DeTar and Carmack, *J. Am. Chem. Soc.*, **68**, 2025 (1946); Carmack and DeTar, *J. Am. Chem. Soc.*, **68**, 2029 (1946); Pattison and Carmack, *J. Am. Chem. Soc.*, **68**, 2033 (1946).

[19] Arndt and Eistert, *Ber.*, **68**, 200 (1935).

[20] Menschutkin, *Ber.*, **31**, 1429 (1898).

[21] Davis and Carmack, *J. Org. Chem.*, **12**, 76 (1947).

[22] Wenner, U. S. pat. 2,489,348 [*C. A.*, **44**, 2559 (1950)].

PHENYLACETYLENE *

(Benzene, ethynyl-)

$$C_6H_5CHBrCH_2Br + 2NaNH_2 \xrightarrow{\text{liq. NH}_3}$$
$$C_6H_5C\equiv CH + 2NH_3 + 2NaBr$$

Submitted by Kenneth N. Campbell and Barbara K. Campbell.[1]
Checked by R. S. Schreiber and H. E. Cupery.

1. Procedure

Caution! Avoid contact with styrene dibromide, which is a skin irritant. This preparation should be conducted in a hood to avoid exposure to ammonia.

A 5-l. three-necked flask is equipped with a high-speed, motor-driven stirrer passing through a bushing in the center neck (Note 1). The side necks are equipped with rubber stoppers each carrying a short length of 8-mm. glass tubing, bent at right angles. A 10–12 in. length of stout, flexible iron wire is passed through one of these pieces of tubing. Two liters of liquid ammonia (Note 2) and 2 g. of ferric nitrate hydrate are placed in the flask. One hundred grams of sodium (4.35 g. atoms) is cut into rectangular pieces about 3 by ¾ by ¾ in. in size. One of the pieces of sodium is hooked onto the lower end of the iron wire and lowered into the liquid ammonia. Stirring is not necessary during this part of the reaction, but it is advisable. When the lump of sodium has reacted, the solution turns from blue to gray, and the remaining pieces of sodium are added in the same manner. The addition requires about 45 minutes (Note 3).

The stopper carrying the iron wire is removed, 2 g. of anil. is added, and then 528 g. (2 moles) of finely powdered, dry styrene dibromide is added gradually with vigorous stirring. The addition requires about 1 hour (Note 4). Stirring is continued for 2 hours (Note 5) after the addition has been completed, after which 600 ml. of concentrated ammonium hydroxide is added, followed by 1 l. of distilled water, and the mixture is allowed to stand until the frost on the outside of the flask is entirely melted.

The aqueous solution is then steam-distilled from the same flask (Note 6) until no more oil passes over. This usually requires about 6 hours, and 1.5–2 l. of distillate is collected. The phenylacetylene in the distillate is separated and washed several times with distilled water to remove ammonia (Note 7). The washed material is dried over anhydrous magnesium sulfate and distilled through an efficient column (Note 8) under reduced pressure. Almost the entire product distils at 73–74°/80 mm. The yield is 93–106 g. (45–52%); n_D^{20} 1.5465–1.5484.

2. Notes

1. A suitable stirrer has been described earlier.[2]

2. Additional liquid ammonia should be added from time to time. Liquid ammonia can be handled satisfactorily in fairly large amounts in an open flask, as the frost that quickly forms on the outside of the flask slows down evaporation.

3. This is an excellent method for making sodium amide for many purposes. If the sodium amide is to be used in another solvent, the solvent should be added to the liquid ammonia after the sodium amide is prepared; the ammonia is allowed to evaporate, and the last traces of ammonia are expelled by heating the flask on a steam bath.

4. The styrene dibromide must not be added too rapidly, or the heat of reaction may cause rapid boiling of the ammonia and possible loss of part of the mixture.

5. In one run an increase of this stirring period to 2.8 hours resulted in an 11% increase in yield.

6. Because the large amount of ammonia that comes over may entrain considerable phenylacetylene, a very efficient cooling system[3] is essential.

7. If acid is used to remove the ammonia, the product is likely to be dark colored.

8. The checkers used an 18-in. column packed with Berl saddles.

3. Methods of Preparation

Phenylacetylene has been prepared by treatment of β-bromostyrene with potassium hydroxide[4] and with sodium amide in liquid ammonia;[5] from styrene dibromide by treatment with sodium amide in liquid ammonia[6] or potassium hydroxide in methanol;[7] and by the reduction of phenylchloroacetylene.[8]

[1] University of Notre Dame, Notre Dame, Indiana.
[2] Org. Syntheses Coll. Vol. 1, 34 (1941).
[3] Org. Syntheses Coll. Vol. 2, 89 (1943).
[4] Org. Syntheses Coll. Vol. 1, 438 (1941).
[5] Vaughn, Vogt, and Nieuwland, J. Am. Chem. Soc., 56, 2120 (1934).
[6] Campbell and O'Connor, J. Am. Chem. Soc., 61, 2898 (1939).
[7] Fiesselmann and Sasse, Chem. Ber., 89, 1775 (1956).
[8] Viehe, Franchimont, and Valange, Chem. Ber., 92, 3064 (1959).

γ-PHENYLALLYLSUCCINIC ACID

(Succinic acid, cinnamyl-)

$$C_6H_5CH_2CH{=}CH_2 + \underset{CHCO}{\overset{CHCO}{\Big|}}\!\!\diagdown O \rightarrow C_6H_5CH{=}CHCH_2\underset{\diagdown}{\overset{CHCO}{\Big|}}\!\!\underset{CH_2CO}{\diagup}O$$

$$C_6H_5CH{=}CHCH_2\underset{CH_2CO}{\overset{CHCO}{\Big|}}\!\!\overset{}{\diagup}O + 2Na_2CO_3 + H_2O \rightarrow$$

$$C_6H_5CH{=}CHCH_2\underset{CH_2CO_2Na}{\overset{CHCO_2Na}{\Big|}} + 2NaHCO_3$$

$$C_6H_5CH{=}CHCH_2\underset{CH_2CO_2Na}{\overset{CHCO_2Na}{\Big|}} + 2H_2SO_4 \rightarrow$$

$$C_6H_5CH{=}CHCH_2\underset{CH_2CO_2H}{\overset{CHCO_2H}{\Big|}} + 2NaHSO_4$$

Submitted by CHRISTIAN S. RONDESTVEDT, JR.[1]
Checked by CHARLES C. PRICE and WM. J. BELANGER.

1. Procedure

A mixture of 35.4 g. (0.3 mole) of allylbenzene, 29.4 g. (0.3 mole) of maleic anhydride, and 50 ml. of o-dichlorobenzene (Note 1) in a 200-ml. round-bottomed flask is heated under reflux for 22 hours under an air condenser. While the orange mixture cools to 50°, the flask is equipped for vacuum distillation. At a bath temperature below 130°, the solvent and unreacted starting materials are removed by vacuum distillation with a water pump. The boiling range is 66–72°/23 mm. The viscous residue is poured while hot into a 125-ml. sausage flask (Note 2), and the transfer is completed with small amounts of acetone. After removal of the acetone by vacuum distillation with a water pump, the product is distilled, b.p. 199–206°/2 mm. (bath temperature 220–270°), to give 27–35 g. (42–54%) of a pale yellow liquid which solidifies readily (Note 3).

The product is melted in the receiver and poured into 100 ml. of benzene. An additional 25 ml. of hot benzene is used to rinse the receiver. The benzene solution is brought to boiling, filtered, and diluted with approximately 100 ml. of petroleum ether (60–75°) until faintly turbid. It is reheated to boiling, allowed to cool, and finally refrigerated for 4 hours. The white crystals are collected on a Büchner funnel, washed with two 25-ml. portions of cold 1:1 benzene-petroleum ether, pressed dry, and air-dried. The yield of anhydride melting at 103–105° is 24–31 g. (37–48%) (Note 4).

The anhydride is readily hydrolyzed by boiling a mixture of 21.6 g. (0.1 mole) of anhydride, 22.0 g. (0.207 mole) of anhydrous sodium carbonate, and 250 ml. of water for 2 hours on a hot plate. The pale yellow solution is cooled and extracted with 100 ml. of isopropyl ether (Note 5). The ether extract is washed with 50 ml. of water, and the combined water layers are acidified to Congo red by the slow addition of 10% sulfuric acid. The acid separates as an oil which quickly solidifies on cooling and stirring. It is collected on a Büchner funnel, washed with cold water, pressed dry on the funnel, and finally air-dried. The yield is 22 g. (94% based on the anhydride), m.p. 140–143°. The acid is conveniently recrystallized from acetonitrile (Note 6), using 5 ml. per gram of crude acid. The recovery of pure acid having a melting point of 142–143° is 85% (Note 7).

2. Notes

1. Allylbenzene can be prepared from phenylmagnesium bromide and allyl bromide.[2] The maleic anhydride used was Eastman Kodak Company white label grade. Slightly higher yields are obtained if it is freshly distilled at 25 mm. Commercial o-dichlorobenzene should be distilled before use.

2. A suitable flask is prepared from a 125-ml. distilling flask

by replacing the narrow side arm with a 150-mm. length of 10-mm. tubing. The side arm of a second flask is cut off to 25 mm., and the two flasks are connected by inserting the 10-mm. side arm into the bulb of the second flask.

A few boiling stones or sticks are added to the first flask containing the material to be distilled, a rubber stopper bearing a thermometer is inserted, and vacuum is applied to the shortened side arm of the receiver.

3. A Wood's metal bath is convenient as a high-temperature heat source.

4. Once-crystallized anhydride is sufficiently pure for conversion to the acid. A second recrystallization gives pure material, m.p. 106.0–106.5°.

5. Ethyl ether may be used. The aqueous layers must then be heated to boiling and cooled before acidification; otherwise the acid is slow to crystallize.

6. Acetonitrile is most convenient, but ethanol, aqueous acetic acid, or aqueous dioxane may be used.

7. The acid is partially dehydrated near its melting point. The reported melting point was observed by immersing the capillary at 140° and heating at 2° per minute.

3. Method of Preparation

γ-Phenylallylsuccinic anhydride and the derived acid have been prepared by heating maleic anhydride with excess allylbenzene in an autoclave at 170–175° for 24 hours.[3] The above procedure is more convenient since an autoclave is unnecessary.

[1] University of Michigan, Ann Arbor, Michigan.
[2] Hershberg, *Helv. Chim. Acta*, **17**, 351 (1934).
[3] Alder, Pascher, and Schmitz, *Ber.*, **76**, 27 (1943).

N-PHENYLBENZAMIDINE *

(Benzamidine, N-phenyl-)

$$C_6H_5CN + C_6H_5NH_2 \xrightarrow{AlCl_3} C_6H_5C-NHC_6H_5$$
$$\overset{\|}{NH}$$

Submitted by F. C. COOPER and M. W. PARTRIDGE.[1]
Checked by T. L. CAIRNS, R. E. BENSON, and V. J. WEBERS.

1. Procedure

Sixty-two grams (61 ml., 0.67 mole) of aniline (Note 1) is mixed with 68.5 g. (0.66 mole) of benzonitrile in a 250-ml., wide-mouthed flask, and, during about 20 minutes, 89 g. (0.67 mole, calculated as AlCl₃) of a freshly opened sample of powdered, anhydrous aluminum chloride is added in portions with thorough stirring (Note 2). The mixture is then heated at 200° for 30 minutes (Note 3), and, while still molten, is poured slowly into a thoroughly stirred mixture of 20 ml. of concentrated hydrochloric acid and 1.6 l. of water. After the addition of 20 g. of activated carbon, the suspension is stirred while being externally cooled in running water and is then filtered through a kieselguhr filter (Note 4). The filtrate is poured in a steady stream into a stirred solution of 220 g. of sodium hydroxide in 1.2 l. of water. The flocculent precipitate is collected on alkali-resistant paper in a 12-cm. Büchner funnel with the aid of suction, washed with water (Note 5), broken up thoroughly, and air-dried at room temperature to constant weight. The yield of white product, m.p. 111–115°, is 90–96 g. (69–74%). This material is sufficiently pure for most purposes. Recrystallization from benzene (60 ml. per 10 g. of amidine) yields 56–74 g. of white powder; concentration of the mother liquors raises the total yield of N-phenylbenz-amidine to 69–86 g. (53–66%), melting at 114–115.5°.

2. Notes

1. It is preferable to use aniline freshly redistilled from a small quantity of zinc dust.

2. The reaction is strongly exothermic. Although loss of reagents by volatilization is small, it is advisable to close the flask with a loose plug of cotton wool.

3. Below about 180° the mixture is too stiff to be stirred, but at 200° it is a mobile liquid.

4. A suitable filter is prepared by distributing a slurry of 10–15 g. of "Super-Cel" in water on a filter paper in a 12-cm. Büchner funnel and washing with water with the aid of suction until a clear filtrate is obtained.

5. Washing is best effected by vigorously stirring the cake with water until it is completely dispersed, collecting again, and draining well; three such washings are usually sufficient.

3. Methods of Preparation

This method is based on the procedure of Oxley, Partridge, and Short.[2] N-Phenylbenzamidine has also been prepared by heating aniline hydrochloride with benzonitrile or thiobenzamide [3] or by heating aniline benzenesulfonate with benzonitrile; [4] by the action of sodium or sodamide on a mixture of aniline and benzonitrile; [5] by treating phenylcyanamide with phenylmagnesium bromide; [6] by the interaction of aniline and benziminoethyl ether hydrochloride; [7] by the reaction between N-phenylbenzimidyl chloride and ammonia; [8] by the action of sodamide on benzylidene aniline; [7,9] by hydrogenating benzanilide oxime; [10] by treating benzophenone oxime benzenesulfonate with ammonia; [11] and by the reaction of 2-nitrobutyl benzimidate hydrochloride with aniline.[12]

[1] The University, Nottingham, England.

[2] Oxley, Partridge, and Short, *J. Chem. Soc.*, **1947**, 1112; Short and Partridge (Boots Pure Drug Company), Brit. pat. 598,453 [*C. A.*, **42**, 6854 (1948)]; U. S. pat. 2,450,386 [*C. A.*, **43**, 3456 (1949)].

[3] Bernthsen, *Ann.*, **184**, 348 (1877).

[4] Oxley and Short, *J. Chem. Soc.*, **1946**, 147.

[5] Lottermoser, *J. prakt. Chem.*, **54**, 116 (1896); Cooper and Partridge, *J. Chem. Soc.*, **1953**, 255.

[6] Busch and Hobein, *Ber.*, **40**, 4298 (1907).

[7] Lossen and Kobbert, *Ann.*, **265**, 138 (1891); Sugasawa and Ohara, *J. Pharm. Soc. Japan*, **72**, 1036 (1952) [*C. A.*, **47**, 7461 (1953)].

[8] Ghadiali and Shah, *J. Univ. Bombay*, **6**, 127 (1937) [*C. A.*, **32**, 3761 (1938)].

[9] Kirssanow and Iwastchenko, *Bull. soc. chim. France*, [5] **2**, 2118 (1935).

[10] Barber and Self (May and Baker Ltd.), U. S. pat. 2,375,611 [*C. A.*, **39**, 3544 (1945)].

[11] Oxley and Short, *J. Chem. Soc.*, **1948**, 1519.

[12] Cooper and Partridge, *J. Chem. Soc.*, **1952**, 5036.

trans-1-PHENYL-1,3-BUTADIENE

(1,3-Butadiene, 1-phenyl-, *trans*-)

$$C_6H_5CH{=}CHCHO + CH_3MgBr \rightarrow$$

$$C_6H_5CH{=}CHCH(OMgBr)CH_3$$

$$2\,C_6H_5CH{=}CHCH(OMgBr)CH_3 + H_2SO_4 \rightarrow$$

$$2\,C_6H_5CH{=}CHCH{=}CH_2 + MgSO_4 + MgBr_2 + 2H_2O$$

Submitted by OLIVER GRUMMITT and ERNEST I. BECKER.[1]
Checked by CHARLES C. PRICE and T. L. PATTON.

1. Procedure

In a 1-l. three-necked flask equipped with a mercury-sealed stirrer, a reflux condenser protected with a calcium chloride drying tube, a separatory funnel, a nitrogen inlet tube, and a thermometer is placed 0.515 mole of methylmagnesium bromide in 250–350 ml. of ether (a 1.5–2.0N solution). The mixture is cooled to a temperature below 10° by means of an ice-water bath, the stirrer is started, and a solution of 66.1 g. (0.50 mole) of cinnamaldehyde (Note 1) in 60 ml. of absolute ether is added, the rate of addition being controlled so that the temperature is kept below 10°. Throughout the addition, which takes about 1 hour, a slow stream of dry nitrogen is passed through the flask (Note 2).

The flask is detached from the condenser and stirrer, and its contents are transferred to a 500-ml. separatory funnel. The apparatus is then reassembled, without the nitrogen inlet tube or the drying tube, and 175 ml. of 30% sulfuric acid (by weight) is placed in the flask. Without cooling, but with efficient stirring with a Hershberg Nichrome wire stirrer at 1500–1700 r.p.m. (Note 3),[2] the ethereal solution of the cinnamaldehyde-methylmagnesium bromide adduct is added rapidly to the acid. The time for this addition (5–7 minutes) is limited by the efficiency of the condenser. Heat then is applied to maintain gentle reflux until the *total* time elapsed from the initiation of hydrolysis is 20 minutes. The contents of the flask are *immediately* transferred to a 1-l. separatory funnel, the lower aqueous layer is discarded, and the ether layer is washed successively with 50 ml. of water, a mixture of 50 ml. of 5% aqueous sodium hydroxide and 50 ml. of saturated ammonium chloride solution, and 50 ml. of water. Before each of the washings the air in the separatory funnel is displaced with nitrogen. When the second wash solution is added, 0.3 g. of phenyl-β-naphthylamine is dissolved in the ether layer. The washed solution is dried with 20 g. of anhydrous sodium sulfate for 30 minutes and then with 15 g. of anhydrous potassium carbonate for 12 hours.

The ethereal solution is filtered and concentrated by distillation from a steam bath to a residual volume of 80–100 ml. Some water separates at this time, and the mixture is cooled and then dried with about 15 g. of anhydrous potassium carbonate. The concentrated solution is filtered into a 125-ml. modified Claisen flask [3] and distilled under reduced pressure in a nitrogen atmosphere into a receiver containing 0.3 g. of phenyl-β-naphthylamine. In this manner 52–54 g. (80–83%) of crude *trans*-1-phenyl-1,3-butadiene is obtained, b.p. 81–85°/10–11 mm.; n_D^{25} 1.606–1.608, which may contain some methylstyrylcarbinol and water. This material is dried with 5 g. of anhydrous potassium carbonate, filtered, and distilled as before. The yield of *trans*-1-phenyl-1,3-butadiene is 47–49 g. (72–75%), b.p. 78–81°/8 mm.; n_D^{25} 1.607–1.608. This product is satisfactory for most purposes (Notes 4 and 5).

2. Notes

1. Cinnamaldehyde obtained from the Eastman Kodak Company was purified by washing a solution in an equal volume of

ether with aqueous sodium carbonate and then with water, dried, and distilled under nitrogen; b.p. 101–102°/2–3 mm.; n_D^{20} 1.6195.

2. The procedure may be altered at this point so that *trans*-methylstyrylcarbinol is obtained. It is necessary, however, to observe the precaution that all apparatus coming in contact with the *trans*-methylstyrylcarbinol be free from traces of acid.

The solution is stirred for 30 minutes after the addition is complete. Then 125 ml. of a saturated solution of ammonium chloride (about 28%), which has been neutralized to litmus with concentrated ammonium hydroxide, is added dropwise, the temperature being held at 5–10°. This addition takes from 1 to 1.5 hours. After decanting the ether layer, breaking up the precipitate and extracting it with two 60-ml. portions of absolute ether, and adding the extracts to the main solution, the solution is distilled from a steam bath until the residual volume is about 100 ml. The solution is transferred to a Claisen flask, and the residual ether is removed by evacuation with a water pump. After the discard of a small fore-run, the product is collected at 93–94°/1.5 mm.; yield, 65–67 g. (88–90%).

Upon cooling at 0–10° the *trans*-methylstyrylcarbinol forms a mass of white crystals melting at 30–34°. These may be purified by crystallization from petroleum ether (b.p. 30–35°) -methylene chloride (6:1). For each 30 g. of the carbinol, 350 ml. of the solvent mixture is used. The solution is cooled to −75° to −80° in Dry Ice and kept at that temperature for about 3 hours. The solution is filtered quickly by suction through a chilled funnel, washed with the cold solvent mixture, and dried in a vacuum desiccator. The yield of pure *trans*-methylstyrylcarbinol is 28.5 g., m.p. 33.5–34.5°; n_D^{35} 1.5598; d_4^{35} 0.9995.

3. The stirring must be vigorous in order to mix the ether and aqueous layers. This is absolutely essential for the production of reasonable yields. Slower stirring necessitates a longer time for the hydrolysis with consequent longer contact time between the 1-phenyl-1,3-butadiene and the sulfuric acid, which results in extensive polymerization of the product and corresponding decrease in yield.

4. Pure *trans*-1-phenyl-1,3-butadiene was obtained by distillation of the twice-distilled product under nitrogen through a 12-plate column of the total reflux-variable takeoff type [2] after add-

ing 0.5% of phenyl-β-naphthylamine. The packed section of the column was an 18-in. section of Pyrex tubing (10 mm. o.d.) filled with 4-mm. single-turn glass helices. Insulation was provided by means of a vacuum jacket, and heat losses were compensated by resistance wire wound on the jacket. About 50% of the sample taken was collected, b.p. 86°/11 mm.; n_D^{25} 1.6086–1.6090; d_4^{25} 0.9235–0.9239.

5. Present evidence indicates that 1-phenyl-1,3-butadiene [4] prepared by this method is the *trans* isomer.[5]

3. Methods of Preparation

1-Phenyl-1,3-butadiene has been prepared by the decarboxylation of allocinnamylideneacetic acid [6] and of cinnamylidenemalonic acid;[7] the dehydration of methylstyrylcarbinol from the Grignard addition of methylmagnesium halide to cinnamaldehyde;[8] the rearrangement and dehydration of the alcohol intermediate formed by the Grignard addition of phenylmagnesium bromide to crotonaldehyde;[9] the formation of methylstyrylcarbinol, its conversion to methylstyrylcarbinyl chloride, and dehydrohalogenation;[10] a modified Wurtz reaction in which benzyl chloride is coupled with allyl chloride by means of sodium in liquid ammonia;[11] the condensation of styrene and acetaldehyde and dehydration of the intermediate in the presence of sulfuric acid in acetic acid;[12] and by the dehydrochlorination of 4-chloro-1-phenyl-2-butene (prepared by the condensation of butadiene with benzenediazonium chloride).[13] Reference 12 describes the preparation of 1-phenyl-1,3-butadiene by pyrolysis of 1-phenyl-1,3-butyleneglycol diacetate and 2,6-dimethyl-4-phenyl-1,3-dioxane. The present method is a modification of the procedure of von der Heide.[8]

trans-Methylstyrylcarbinol has been prepared by several methods: the hydrolysis of the addition product formed from methylmagnesium halide and cinnamaldehyde in a variety of ways;[10, 14] hydrolysis of the addition compound formed from styrylmagnesium bromide and acetaldehyde;[4] hydrolysis and hydrogenation of the product formed in the Grignard reaction of phenylethynylmagnesium bromide and acetaldehyde;[15] the addition of hypobromous acid to 1-phenyl-1,3-butadiene followed by reduction with sodium amalgam in acetic acid;[16] the allylic rearrangement of 1-phenyl-1-acetoxy-3-butene to 1-phenyl-3-acetoxy-1-butene followed by saponification;[17] and the reduction of benzalacetone by means of aluminum isopropoxide.[18] The method employed here is essentially that of Kenyon, Partridge, and Phillips.[14]

[1] Western Reserve University, Cleveland, Ohio.

[2] *Org. Syntheses Coll. Vol.* **2**, 117 (1943).

[3] *Org. Syntheses Coll. Vol.* **1**, 130, Fig. 9*b* (1941).

[4] Wright, *J. Org. Chem.*, **1**, 457 (1936).

[5] Grummitt and Christoph, *J. Am. Chem. Soc.*, **71**, 4157 (1949).

[6] Liebermann and Riiber, *Ber.*, **33**, 2400 (1900); Doebner and Staudinger, *Ber.*, **36**, 4318 (1903).

[7] Liebermann and Riiber, *Ber.*, **35**, 2696 (1902); Riiber, *Ber.*, **37**, 2272 (1904); Doebner and Schmidt, *Ber.*, **40**, 148 (1907).

[8] von der Heide, *Ber.*, **37**, 2101 (1904); Klages, *Ber.*, **37**, 2301 (1904); von Auwers and Eisenlohr, *J. prakt. Chem.*, [2] **84**, 42 (1911); Muskat and Herrman, *J. Am. Chem. Soc.*, **53**, 252 (1931); Flood, Hladky, and Edgar, *Ind. Eng. Chem.*, **25**, 1234 (1933); Briegleb and Kambeitz, *Z. physik. Chem.*, **32B**, 305 (1936).

[9] Blumenfeld, *Ber.*, **74B**, 524 (1941).

[10] Klages, *Ber.*, **35**, 2649 (1902); Muskat and Herrman, *J. Am. Chem. Soc.*, **53**, 252 (1931).

[11] Kharasch, Nudenberg, and Fields, *J. Am. Chem. Soc.*, **66**, 1276 (1944).

[12] Emerson, *J. Org. Chem.*, **10**, 464 (1945).

[13] Dombrovskiĭ, *Doklady Akad. Nauk S.S.S.R.*, **111**, 827 (1956) [*C. A.*, **51**, 9507 (1957)]; Dombrovskiĭ and Terent'ev, *Zhur. Obshcheĭ Khim.*, **27**, 415 (1957) [*C. A.*, **51**, 15454 (1957)].

[14] Kenyon, Partridge, and Phillips, *J. Chem. Soc.*, **1936**, 85.

[15] Campbell, Campbell, and McGuire, *Proc. Indiana Acad. Sci.*, **50**, 87 (1940) [*C. A.*, **35**, 5872 (1941)].

[16] Ingold and Smith, *J. Chem. Soc.*, **1931**, 2752.

[17] Burton, *J. Chem. Soc.*, **1929**, 455.

[18] Lund, *Kem. Maanedsblad*, **17**, 169 (1936) [*Chem. Zentr.*, **108 I**, 3480 (1937)]. See also Wilds, *Organic Reactions*, Vol. 2, p. 214, John Wiley & Sons, New York, 1944.

α-PHENYL-α-CARBETHOXYGLUTARONITRILE

(Butyric acid, 2,4-dicyano-2-phenyl-, ethyl ester)

$$\underset{\underset{CO_2C_2H_5}{|}}{\overset{\overset{CN}{|}}{C_6H_5CH}} \quad + \ CH_2{=}CHCN \ \xrightarrow{KOH} \ \underset{\underset{CO_2C_2H_5}{|}}{\overset{\overset{CN}{|}}{C_6H_5CCH_2CH_2CN}}$$

Submitted by E. C. HORNING and A. F. FINELLI.[1]
Checked by WILLIAM S. JOHNSON and H. WYNBERG.

1. Procedure

In a 500-ml. three-necked flask equipped with a stirrer, a dropping funnel, and a thermometer is placed a solution of 57.0 g. (0.30 mole) of ethyl phenylcyanoacetate (p. 461) in 80 ml. of *tert*-butyl alcohol. The solution is heated to 40°, and with stirring the dropwise addition of a solution of 33.0 g. (0.62 mole) of acrylonitrile (Note 1) in 30 ml. of *tert*-butyl alcohol is started. After the addition of about 10–15 drops, 1.0 ml. of a 30% solution of potassium hydroxide in methanol is added, and the temperature is maintained at 40–45° by occasional external cooling while the remaining solution is added slowly. When about one-half of the acrylonitrile has been added, an additional 1.0 ml. of the potassium hydroxide solution is added to ensure the presence of a basic catalyst throughout the reaction. When the addition is completed (after about 30 minutes) and the temperature is no longer maintained above 40° by the exothermic reaction (another 30 minutes), the mixture is heated with a hot-water bath to keep the temperature at 40–45° for 1 hour.

The solution is diluted with 250 ml. of water and acidified with 30–40 ml. of 10% hydrochloric acid. The product is separated after the addition of 100 ml. of ether, and the aqueous solution is extracted with two 50-ml. portions of ether. The combined extracts are washed with 50 ml. of water and dried over anhydrous magnesium sulfate. The ether is distilled at atmospheric pressure, and the residue is distilled under reduced pressure through a short (15-cm.) Vigreux column. After a fore-run of a few grams, the product is collected at 157–167°/0.5–1 mm. (Note 2). The yield is 50–61 g. (69–83%).

2. Notes

1. The acrylonitrile should be distilled before use. Acrylonitrile vapors are toxic, and the distillation as well as the subsequent reaction should be carried out in a hood.

2. Other observed boiling points are 165–167°/1 mm., 195–200°/6 mm. The product is a colorless, viscous oil, n_D^{25} 1.5100–1.5103.

3. Methods of Preparation

α-Phenyl-α-carbethoxyglutaronitrile has been prepared by the reaction of ethyl α-cyanophenylacetate with β-chloropropionitrile in the presence of sodium amide.[2] The present procedure has been published,[3] and it follows the general method described by Bruson [4] for the cyanoethylation of arylacetonitriles.

[1] University of Pennsylvania, Philadelphia, Pennsylvania.

[2] Bergel, Morrison, and Rinderknecht (to Hoffmann-La Roche, Inc.), U. S. pat. 2,446,803 [C. A., **43**, 695 (1949)].

[3] Horning and Finelli, J. Am. Chem. Soc., **71**, 3204 (1949).

[4] Bruson and Riener, J. Am. Chem. Soc., **65**, 25 (1943).

α-PHENYLCINNAMIC ACID *

(Acrylic acid, 2,3-diphenyl-)

$$C_6H_5CHO + C_6H_5CH_2CO_2H \xrightarrow{(C_2H_5)_3N} \begin{array}{c} C_6H_5CH{=}CCO_2H \\ | \\ C_6H_5 \end{array}$$

Submitted by ROBERT E. BUCKLES and KEITH BREMER.[1]
Checked by T. L. CAIRNS and J. C. LORENZ.

1. Procedure

In a 500-ml. round-bottomed flask are placed 40.5 ml. (42.4 g., 0.40 mole) of freshly purified benzaldehyde (Note 1), 54.6 g. (0.40 mole) of phenylacetic acid,[2] 40 ml. of anhydrous triethylamine (Note 2), and 80 ml. of acetic anhydride (Note 3). The mixture is boiled gently under reflux for 5 hours. After the heating period is over, the 500-ml. flask containing the reaction mixture is incorporated into a steam-distillation apparatus (Note 4). The reaction mixture is distilled with steam until the distillate coming over is no longer cloudy, and then about 50 ml. more of the distillate is collected. The distillate can be discarded. The aqueous residue is cooled, and the solution is then separated from the solid by decantation. The solid is dissolved in 500 ml. of hot 95% ethanol, and 500 ml. of water, including the solution originally decanted from the crude solid, is added to the hot solution. The mixture is heated to boiling, and 2 g. of decolorizing carbon is added. The hot solution is filtered, and the filtrate is immediately acidified to Congo red with 6N hydrochloric acid. The solution is cooled, and the resulting crystals are removed from the mixture by filtration. The yield of crude α-phenylcinnamic acid (m.p. around 161–165°) is 60–67 g. The product is purified by crystallization from aqueous ethanol (Note 5). The over-all yield of purified product, m.p. 172–173°, is 48–53 g. (54–59%). The product is the isomer with the two phenyl groups *cis* to each other since decarboxylation yields *cis*-stilbene [3] (see p. 857).

2. Notes

1. Benzaldehyde, suitable for this synthesis, is purified in the following way. A 60-g. (58-ml.) sample is washed with two 20-ml. portions of 10% sodium carbonate and then with water. It is then dried over 5–10 g. of anhydrous magnesium sulfate. A few small crystals of hydroquinone or catechol are added with the drying agent. The dry benzaldehyde is decanted through a cotton plug into a Claisen flask; it is distilled under reduced pressure, preferably below 30 mm.

2. Sharples anhydrous grade triethylamine was used without further purification.

3. The acetic anhydride is carefully fractionated; the 137–139° fraction is collected.

4. A simple steam-distillation apparatus such as that given by Fieser [4] is entirely satisfactory. It is usually necessary to heat the distillation flask with a steam bath or a small flame in order to minimize the accumulation of excess water in the flask.

5. The submitters used 5 ml. of 95% ethanol and 5 ml. of water per gram of crude product for recrystallization. The checkers found use of 3:2 ethanol:water by volume more convenient.

3. Methods of Preparation

α-Phenylcinnamic acid has been prepared by the distillation of benzylmandelic acid,[5] by the condensation of phenylacetyl chloride [6] or phenylacetic acid [7] with benzaldehyde in the presence of triethylamine; by the reaction of sodium or potassium phenylacetate with benzaldehyde in acetic anhydride; [8-11] and by the treatment of ethyl α-bromo-α,β-diphenylpropionate with potassium hydroxide.[12] The most convenient synthesis appears to be that described above.[13-15]

[1] State University of Iowa, Iowa City, Iowa.
[2] Org. Syntheses Coll. Vol. 1, 436 (1941).
[3] Taylor and Crawford, J. Chem. Soc., 1934, 1130.
[4] Fieser, Experiments in Organic Chemistry, 3rd ed., p. 150, D. C. Heath and Company, Boston, 1955.
[5] Malkin and Robinson, J. Chem. Soc., 127, 376 (1925).
[6] Katoh, Science Repts. Tokyo Bunrika Daigaku, 2, 257 (1935).
[7] Ishikawa and Tukeuchi, Science Repts. Tokyo Bunrika Daigaku, A3, 231 (1939); Zimmerman and Ahramjian, J. Am. Chem. Soc., 81, 2086 (1959); Riemschneider and Kampfer, Monatsh., 90, 518 (1959).
[8] Oglialoro, Gazz. chim. ital., 8, 429 (1887).
[9] Bakunin, Gazz. chim. ital., 31 II, 77 (1901).
[10] Posner, J. prakt. Chem., [2] 82, 437 (1910).
[11] Johnson, in Adams, Organic Reactions, Vol. 1, p. 252, John Wiley & Sons, New York, 1942.
[12] D'yakonov and Vinogradova, Zhur. Obshcheĭ Khim., 23, 244 (1953) [C. A., 48, 3318 (1954)].
[13] Buckles and Hausman, J. Am. Chem. Soc., 70, 415 (1948).
[14] Buckles, J. Chem. Educ., 27, 210 (1950).
[15] Buckles, Bellis, and Coder, J. Am. Chem. Soc., 73, 4972 (1951).

2-PHENYLCYCLOHEPTANONE

(Cycloheptanone, 2-phenyl-)

$$C_6H_5CH_2NH_2 + ClCO_2C_2H_5 + NaOH \longrightarrow C_6H_5CH_2NHCO_2C_2H_5 + NaCl + H_2O$$

$$C_6H_5CH_2NHCO_2C_2H_5 + NaNO_2 + HNO_3 \longrightarrow C_6H_5CH_2N(NO)CO_2C_2H_5 + NaNO_3 + H_2O$$

Submitted by C. DAVID GUTSCHE and HERBERT E. JOHNSON.[1]
Checked by N. J. LEONARD and F. P. HAUCK, JR.

1. Procedure

A. *Ethyl N-benzylcarbamate.* A 12-l. three-necked flask fitted with a sturdy Hershberg-type stirrer and two 1-l. addition funnels is immersed in an ice bath and charged with 1 kg. (9.33 moles) of benzylamine, 500 ml. of ice water, and 1.5 kg. of chopped ice. To the stirred mixture 525 g. (4.83 moles) of ethyl chlorocarbonate is added dropwise while the temperature is maintained at 10–15° (1.0–1.5 hours) (Note 1). An additional 500 ml. of water and 1 kg. of chopped ice are then added to the flask, and a second 525-g. portion (4.83 moles) of ethyl chlorocarbonate is introduced. Simultaneously with this, an ice-cold solution of 400 g. (10 moles) of sodium hydroxide in 1.3 l. of water is added dropwise at such a rate that equal fractions of the ethyl chlorocarbonate and sodium hydroxide solutions are introduced over equal periods of time, the temperature being maintained throughout at 10–15° (2.5–3.0 hours). The reaction mixture is stirred for an additional 30 minutes and is then filtered through a Büchner funnel. The solid product is washed with copious amounts of cold water and is air-dried to yield 1.6 kg. (96%) of glistening white crystals, m.p. 45–47°.

B. *Ethyl N-nitroso-N-benzylcarbamate* (Note 2). In a 12-l. three-necked flask fitted with a thermometer, a 2-l. addition funnel (Note 3), and a gas outlet tube are placed a solution of 360 g. (2.0 moles) of ethyl N-benzylcarbamate in 2 l. of ether and a solution of 1.2 kg. (17.4 moles) of sodium nitrite in 2 l. of water. A stirrer is not used. The reaction mixture is cooled by means of a water bath to 20° and treated with a solution of 1 l. each of concentrated nitric acid and water, contained in the addition funnel. Enough of this solution is added to impart a permanent green color to the aqueous layer, and the remainder is then added over a period of 5 hours at such a rate as to keep the aqueous phase green (Note 4) and the temperature at 25–30°. The reaction mixture is allowed to stand an additional 30 minutes, and the layers are separated. The ether layer is washed with 200-ml. portions of 10% potassium carbonate solution (Note 5) until the evolution of gas ceases and is then dried over anhydrous potassium carbonate. The ether is removed under vacuum on a water bath kept below 50° (Note 6), a residue of 400–415 g. (95–99%) of a bright orange oil (Note 7) being left.

C. *2-Phenylcycloheptanone.* In a 2-l. three-necked flask fitted with a 500-ml. addition funnel, a sealed Hershberg stirrer, and a reflux condenser (Note 8) are placed 392 g. (4.0 moles) of freshly distilled cyclohexanone, 30 g. of finely powdered potassium carbonate, and 400 ml. of absolute methanol. To the stirred mixture is added 415 g. (2.0 moles) of ethyl N-nitroso-N-benzylcarbamate over a period of 1.5 hours during which time the reaction temperature is maintained at 25° by means of an ice-water bath. The dark red reaction mixture is then allowed to stand at room temperature until the evolution of nitrogen has ceased (24–28 hours) (Note 9). The solid material is removed by filtration, the lower-boiling materials are removed by evaporation under reduced pressure on the steam bath (Note 10), and the residue is distilled through an efficient column. A fore-run consisting of 30–60 g. of material is discarded or refractionated (Note 10), and the fraction with b.p. 94–96°/0.4 mm. (124–126°/2 mm., 136–138°/4 mm.) is collected. It amounts to 155–177 g. (41–47%) of 2-phenylcycloheptanone, n_D^{20} 1.5395–1.5398, which is pure enough for most purposes, but which may be purified further by recrystallization from petroleum ether (b.p. 30–60°) and obtained as colorless, very long needles; m.p. 21–23° (Note 11).

2. Notes

1. During this time ethyl N-benzylcarbamate begins to separate from solution as a white solid.

2. Although the benzyl nitroso compound appears to be a much less active vesicant than the methyl nitroso compound, it is, nevertheless, a wise precaution to wear heavy rubber gloves during the isolation of this product.

3. The stem of the addition funnel should reach to the bottom of the flask.

4. The color may appear yellow green, emerald green, or blue-green, depending upon the size of the run, the amount of nitric acid that has been added, and the room lighting.

5. Seven to nine portions of carbonate solution are sufficient if each portion is shaken very thoroughly with the ether solution. Caution should be observed because of pressure build-up in the separatory funnel!

6. Ethyl N-nitroso-N-benzylcarbamate is heat sensitive and, if the temperature is too high, *may detonate violently*. The submitters state that attempts to distil the nitroso compound under high vacuum have resulted in explosions.

7. The submitters state that the nitroso compound is stable at low temperature and can be stored in a refrigerator for several months or longer with no signs of deterioration.

8. The reflux condenser is an optional but convenient appendage for the third neck of the flask. To follow the evolution of nitrogen during the reaction, the exit from the condenser can be led either to a eudiometer tube (theoretical nitrogen evolution about 50 l. for the experiment described) or to a bubbler.

9. It is necessary to allow the reaction mixture to stand for a rather prolonged period, since about 40% of the nitrogen is evolved during this time.

10. The lower-boiling material includes methyl benzyl ether, which may be isolated, by careful fractionation through an efficient column, in about 25% yield, b.p. 74–77°/30 mm.

11. In a similar fashion the following 2-arylcycloheptanones have been prepared by the submitters:

	Yield, %	Melting Point, or Refractive Index at 25°
2-(o-Methylphenyl)cycloheptanone	29	1.5348
2-(p-Methylphenyl)cycloheptanone	26	57–58°
2-(o-Methoxyphenyl)cycloheptanone	7	1.5407
2-(m-Methoxyphenyl)cycloheptanone	42	1.5418
2-(p-Methoxyphenyl)cycloheptanone	20	58–59°

3. Methods of Preparation

Ethyl N-benzylcarbamate and its nitroso compound have been prepared by methods similar to those described for ethyl N-methylcarbamate and its nitroso compound.[2,3] 2-Phenylcycloheptanone has been prepared by the reaction of ethyl N-nitroso-N-benzylcarbamate [4] with cyclohexanone,[5] by the reaction of phenyldiazomethane with cyclohexanone,[6] by the reaction of ethyl N-nitroso-N-methylcarbamate with 2-phenylcyclohexanone,[5] and by the rearrangement of 1-phenyl-2-cyclohexylethylene oxide.[7]

[1] Washington University, St. Louis, Missouri.

[2] Org. Syntheses Coll. Vol. 2, 278 (1943).

[3] Org. Syntheses Coll. Vol. 2, 464 (1943).

[4] v. Pechmann, Ber., 31, 2640 (1898).

[5] Gutsche, J. Am. Chem. Soc., 71, 3513 (1949); Gutsche and Johnson, J. Am. Chem. Soc., 77, 109 (1955).

[6] Burger, Walter, Bennet, and Turnbull, Science, 112, 306 (1950); Gutsche and Jason, J. Am. Chem. Soc., 78, 1184 (1956).

[7] Tiffeneau, Weill, Gutmann, and Tchoubar, Compt. rend., 201, 277 (1935).

PHENYLDICHLOROPHOSPHINE

(Phosphonous dichloride, phenyl-)

$$\text{C}_6\text{H}_5 + PCl_3 + AlCl_3 \rightarrow \text{C}_6\text{H}_5\text{-}PCl_2\cdot AlCl_3 + HCl$$

$$\text{C}_6\text{H}_5\text{-}PCl_2\cdot AlCl_3 + POCl_3 \rightarrow \text{C}_6\text{H}_5\text{-}PCl_2 + AlCl_3\cdot POCl_3$$

Submitted by B. Buchner and L. B. Lockhart, Jr.[1]
Checked by Cliff S. Hamilton and P. J. Vanderhorst.

1. Procedure

In an all-glass apparatus consisting of a 1-l. three-necked flask equipped with a long-stem thermometer, a rubber-sealed mechanical stirrer, and a suitable condenser (Note 1) are placed 165 g. (1.2 moles) of phosphorus trichloride, 23.4 g. (0.3 mole) of benzene, and 53 g. (0.4 mole) of anhydrous aluminum chloride. The mixture is stirred continuously and heated (Note 2). As the temperature increases, the mixture becomes a homogeneous yellow solution and begins to reflux. After 2 hours, the reaction mixture is heated under reflux as vigorously as possible (Note 3). At the end of the third hour, the evolution of hydrogen chloride has almost ceased. The heat source is removed, and, while the mixture is still hot, 62 g. (0.4 mole) of phosphorus oxychloride is added gradually (Note 4) from a dropping funnel (Note 5). The granular precipitate of aluminum chloride-phosphorus oxychloride complex settles rapidly. After the apparatus is disassembled, 6–8 petroleum ether extractions of 100 ml. each are performed to remove phenyldichlorophosphine and the unreacted starting materials from the reaction flask. The residue is transferred to a Büchner funnel and washed with several small portions of petroleum ether, and the combined extracts and washings are concentrated under reduced pressure. Crude phenyldichlorophosphine is removed by distilling to dryness under reduced pressure and is purified by fractionating through a satisfactory column (Note 6). The product distils at 68–70°/1 mm. (90–92°/ 10 mm.), n_D^{25} 1.5962 (Note 7), and weighs 38.5–42 g. (72–78%).

2. Notes

1. The submitters and checkers used a Friedrichs condenser. The condenser outlet was connected to a gas absorption trap filled with sodium hydroxide solution to neutralize escaping acid vapors. A tube filled with Drierite was inserted between the condenser and trap to absorb moisture which might diffuse from the trap.

2. Slow heating is desirable to prevent too rapid evolution of hydrogen chloride.

3. Cold water, approximately 0°, is circulated by means of a water pump in order to increase the efficiency of the condenser.

4. The reaction between phosphorus oxychloride and aluminum chloride is exothermic.

5. The thermometer is replaced by a dropping funnel.

6. The submitters and checkers employed a 20-cm. column packed with glass helices.

7. The checkers obtained an average value of n_D^{24} 1.5919.

3. Methods of Preparation

Phenyldichlorophosphine has been prepared by the vapor-phase reaction of benzene and phosphorus trichloride over pumice in a hot tube [2] and by the action of diphenylmercury [3] or phenylzinc bromide [4] on phosphorus trichloride. The method described here is a Michaelis' modification of a Friedel-Crafts reaction.[5] It has been claimed [6] that pyridine is advantageous for the removal of aluminum chloride from the reaction mixture.

[1] Naval Research Laboratory, Washington, D. C.

[2] Michaelis, *Ber.*, **6**, 601 (1873).

[3] Michaelis, *Ann.*, **181**, 288 (1876).

[4] Weil, Prijs, and Erlenmeyer, *Helv. Chim. Acta*, **35**, 1412 (1952).

[5] Michaelis, *Ber.*, **12**, 1009 (1879); Buchner and Lockhart, *J. Am. Chem. Soc.*, **73**, 755 (1951).

[6] Gefter, *Zhur. Obshcheĭ Khim.*, **28**, 1338 (1958) [*C. A.*, **52**, 19999 (1958)].

4-PHENYL-*m*-DIOXANE *

(*m*-Dioxane, 4-phenyl-)

$$C_6H_5CH{=}CH_2 + 2CH_2O \xrightarrow{H_2SO_4} \begin{array}{c} C_6H_5CHCH_2CH_2 \\ | \qquad\qquad | \\ O \qquad\quad O \\ \diagdown \quad \diagup \\ CH_2 \end{array}$$

Submitted by R. L. Shriner and Philip R. Ruby.[1]
Checked by Richard T. Arnold, W. E. Parham, and John E. Franz.

1. Procedure

In a 2-l. round-bottomed flask, fitted with a reflux condenser and mechanical stirrer, are placed 675 g. (8.3 moles) of 37% formalin, 48 g. of sulfuric acid (sp. gr. 1.84), and 312 g. (3 moles) of styrene. The resulting mixture is gently refluxed and stirred for 7 hours. The mixture is cooled, and 500 ml. of benzene is stirred in. The layers are separated, and the aqueous layer is extracted with 500 ml. of benzene. The benzene solutions are combined and washed with two 750-ml. portions of water. The benzene is removed by distillation, and the residual liquid is fractionated under reduced pressure. At 2 mm. pressure a fore-run is collected separately, up to a temperature of 96° (Note 1); then the main fraction is collected at 96–103°/2 mm. The yield of 4-phenyl-*m*-dioxane amounts to 353–436 g. (72–88%): n_D^{20} 1.5300–1.5311; d_4^{20} 1.092–1.093 (Note 2).

2. Notes

1. The amount of fore-run and the yield depend on the efficiency of the fractionation. With a 7-cm. distilling head, a fore-run of 75 g. boiling at 84–96°/2 mm. was collected, whereas with a heated Vigreux column (2 cm. by 35 cm.) the fore-run amounted to only 11 g. and the higher yields were obtained. The fore-run may be refractionated to obtain additional product. The checkers used a 2 cm. by 20 cm. column packed with stainless-steel helices, and collected their product (72–75% yield) over a 1° boiling range (94–95°/2 mm., n_D^{20} 1.5300).

2. This modification of the Prins[2] reaction has been applied to other olefins.[3]   The aryl olefins give the best yields; see the tabulation.

Aryl Olefin	Yield of Substituted *m*-Dioxane, %
α-Methylstyrene	58
Propenylbenzene	66
Anethole	89
Isosafrole	84
1-(3′,4′-Dimethoxyphenyl)-1-propene	68
1-(*p*-Cumyl)-1-propene	96

3. Methods of Preparation

4-Phenyl-*m*-dioxane was obtained by Prins[2] by the reaction between styrene and formaldehyde in the presence of sulfuric acid.   The correct structure was pointed out by Fourneau, Benoit, and Firmenich.[4]   The above procedure is essentially that given by Shortridge[5] and by Beets[3] and mentioned in a patent.[6]   Methylphenylcarbinol has been substituted for styrene.[3]

[1] State University of Iowa, Iowa City, Iowa.

[2] Prins, *Chem. Weekblad*, **14**, 932 (1917); **16**, 1072, 1510 (1919); *Proc. Acad. Sci. Amsterdam*, **22**, 51 (1919).

[3] Beets, *Rec. trav. chim.*, **70**, 20 (1951); Beets and Van Essen, *Rec. trav. chim.*, **70**, 25 (1951); Drukker and Beets, *Rec. trav. chim.*, **70**, 29 (1951).

[4] Fourneau, Benoit, and Firmenich, *Bull. soc. chim. France*, **47**, 858 (1930).

[5] Shortridge, *J. Am. Chem. Soc.*, **70**, 873 (1948).

[6] Engel, U. S. pat. 2,417,548 [*C. A.*, **41**, 3493 (1947)].

o-PHENYLENE CARBONATE

(Carbonic acid, o-phenylene ester)

$$\text{C}_6\text{H}_4(\text{OH})_2 + \text{COCl}_2 + 2\text{NaOH} \rightarrow$$

$$\text{o-C}_6\text{H}_4(\text{O})_2\text{C}{=}\text{O} + 2\text{NaCl} + 2\text{H}_2\text{O}$$

Submitted by R. S. HANSLICK, W. F. BRUCE, and A. MASCITTI.[1]
Checked by ARTHUR C. COPE, HARRIS E. PETREE, and ELMER R. TRUMBULL.

1. Procedure

Caution! This preparation should be conducted in a hood to avoid exposure to toxic phosgene.

In a 5-l. three-necked flask, filled with nitrogen, 110 g. (1.0 mole) of recrystallized catechol (Note 1) is dissolved in 250 ml. of deaerated water (Note 2) containing 88 g. (2.2 moles) of sodium hydroxide. The flask is fitted with a gas inlet tube, a thermometer dipping into the liquid, and an efficient glass mechanical stirrer with a gas-tight rubber slip seal and is immersed in an ice-salt bath. A positive nitrogen pressure of about 1 cm. is maintained by attaching the inlet tube to a source of nitrogen through a line containing a T-tube dipping into mercury. A solution of 200–225 g. (2.0–2.3 moles) of commercial phosgene in 750 ml. of toluene is prepared at 0° by bubbling the gas into toluene in a tared flask (Note 3), and the solution is added to the flask in portions of about 50 ml. with good mechanical stirring over a period of 60 to 75 minutes. During the addition the temperature is maintained at 0–5° by periodic addition to the mixture of clean cracked ice, free from dirt and iron rust. After addition of the toluene solution of phosgene is completed, the mixture is stirred at 0–5° for 1 hour and then allowed to come to room temperature. The mixture is filtered with suction, and the solid is pressed on the funnel to remove as much water as possible. The aqueous portion of the filtrate is separated, and the solid on the funnel is added to the toluene in the filtrate and dissolved by warming. The warm toluene solution is filtered and distilled under reduced pressure (water aspirator) until the product begins to crystallize. The residue is warmed to redissolve the solid, and then chilled. The *o*-phenylene carbonate is collected on a suction filter and dried in a vacuum desiccator; the yield is 98–110 g., m.p. 119–120°.

Concentration of the filtrate yields a second crop of impure product, which is recrystallized from toluene and then melts at 119–120°. The combined yield of pure white *o*-phenylene carbonate from the first and second crops is 107–116 g. (79–85%).

2. Notes

1. Catechol obtained from the Koppers Company, Pittsburgh, Pennsylvania, was recrystallized from toluene.

2. Water that was deaerated by boiling was used, and an atmosphere of nitrogen essentially free from oxygen (such as the Seaford grade of the Air Reduction Company) was maintained, in order to prevent discoloration of the alkaline solution of catechol due to oxidation.

3. Phosgene from a commercial cylinder was used (Matheson Company or Ohio Chemical Company). For the preparation of a solution of phosgene in toluene see *Organic Syntheses.*[2]

3. Methods of Preparation

o-Phenylene carbonate has been prepared by the distillation of *o*-hydroxyphenyl ethyl carbonate[3] and by the reaction of catechol with phosgene.[3,4]

[1] Research Laboratory, Wyeth Institute, Philadelphia, Pennsylvania.
[2] *Org. Syntheses Coll. Vol.* **3**, 167 (1955).
[3] Einhorn and Lindenberg, *Ann.*, **300**, 141 (1898).
[4] Nachfolger, Ger. pat. 72,806 [*Chem. Zentr.*, **65 I**, 805 (1894)].

α-PHENYLGLUTARIC ANHYDRIDE *

(Glutaric anhydride, 2-phenyl-)

$$\underset{\underset{CO_2C_2H_5}{|}}{\overset{\overset{CN}{|}}{C_6H_5C}}CH_2CH_2CN \xrightarrow[CH_3CO_2H]{HCl, H_2O} \underset{\underset{CO_2H}{|}}{C_6H_5CH}CH_2CH_2CO_2H$$

$$\underset{\underset{CO_2H}{|}}{C_6H_5CH}CH_2CH_2CO_2H + (CH_3CO)_2O \longrightarrow$$

$$\underset{\underset{CO\text{——}O}{|\quad\quad|}}{C_6H_5CH}CH_2CH_2CO + 2CH_3CO_2H$$

Submitted by E. C. HORNING and A. F. FINELLI.[1]
Checked by WILLIAM S. JOHNSON and H. WYNBERG.

1. Procedure

In a 500-ml. flask equipped with a reflux condenser are placed 48.4 g. (0.20 mole) of α-phenyl-α-carbethoxyglutaronitrile (p. 776), 225 ml. of hydrochloric acid (sp. gr. 1.19), and 50 ml. of acetic acid. The mixture is heated under reflux for 10 hours. The solution is cooled, transferred to a 1-l. separatory funnel, and diluted with 300 ml. of water. The α-phenylglutaric acid is extracted with five 100-ml. portions of ether-ethyl acetate (1:1) (Note 1). The extracts are combined, washed once with saturated sodium chloride solution, and dried over anhydrous magnesium sulfate. The solvents are removed as completely as possible by distillation from a steam bath, and the residue is transferred to a 200-ml. flask. Acetic anhydride (50 ml.) is added, and the solution is heated under gentle reflux for 1 hour. The excess acetic anhydride is removed by distillation at atmospheric pressure, and the residue is distilled under reduced pressure through a short (15-cm.) Vigreux column with an air-cooled side arm. The product is collected at 178–188°/0.5–1 mm. (Note 2). The yield is 31.1–32.7 g. (82–86%); m.p. 90–94°.

This material may be recrystallized by dissolving the product in hot ethyl acetate (2 ml. per g. of the anhydride) and adding an equal volume of hexane or 60–68° petroleum ether slowly to the hot solution. The solution is allowed to cool, and when crystallization occurs (usually at 40–50°) an additional volume (2 ml. per g. of the anhydride) of hexane is added. The mixture is cooled, and the crystalline product is removed by filtration and washed with cold hexane (5 ml. per g. of the anhydride). The recovery of colorless material, m.p. 95–96°, is 90–92%.

2. Notes

1. The extraction may be facilitated by saturation of the aqueous layer with sodium chloride.

2. It is necessary to flame the column and side arm. The product obtained in this way is a light-yellow or cream-colored solid which need not be recrystallized unless a colorless sample is desired. If the final distillation is carried out too slowly or at pressures above 2 mm. considerable decomposition may occur, reducing the yield of the product.

3. Methods of Preparation

α-Phenylglutaric acid has been prepared by the hydrolysis and decarboxylation of diethyl α-phenyl-α-carbethoxyglutarate (prepared by the alkylation of diethyl phenylmalonate with ethyl β-iodopropionate) with hydrochloric acid.[2] The anhydride may be obtained from the acid by direct distillation under reduced pressure, although the use of acetic anhydride results in a purer product.

[1] University of Pennsylvania, Philadelphia, Pennsylvania.
[2] Fichter and Merckens, *Ber.*, **34**, 4175 (1901).

1-PHENYL-1-PENTEN-4-YN-3-OL

(1-Penten-4-yn-3-ol, 1-phenyl-)

$$HC\equiv CH + C_2H_5MgBr \xrightarrow{\text{Tetrahydrofuran}} HC\equiv CMgBr + C_2H_6$$

$$HC\equiv CMgBr + C_6H_5CH\!=\!CHCHO \xrightarrow[\text{NH}_4\text{Cl}]{\text{H}_2\text{O}}$$

$$HC\equiv CCHOHCH\!=\!CHC_6H_5$$

Submitted by Lars Skattebøl, E. R. H. Jones, and Mark C. Whiting.[1]
Checked by Melvin S. Newman and Raymond E. Dessy.

1. Procedure

A. *Ethynylmagnesium bromide.* A 500-ml. three-necked flask, equipped with a sealed mechanical stirrer, a reflux condenser, and a pressure-equalized dropping funnel, is arranged for carrying out a reaction in an atmosphere of nitrogen by fitting into the top of the condenser a T-tube attached to a low-pressure supply of nitrogen and to a mercury bubbler. For later use there is also prepared a dry 1-l. three-necked flask equipped with a sealed mechanical stirrer, a gas inlet tube which will dip below the surface of 200 ml. of liquid in the flask, a 500-ml. dropping funnel, and a gas outlet protected by a calcium chloride drying tube.

The 500-ml. flask is dried by warming with a soft flame as a slow stream of nitrogen is passed through the system. A solution of ethylmagnesium bromide is prepared in this flask from 12 g. (0.5 g. atom) of magnesium turnings, 60 g. (0.55 mole) of ethyl bromide (dried over calcium chloride), and 300 ml. of tetrahydrofuran (Notes 1 and 2).

After the preparation of ethylmagnesium bromide is completed, the separatory funnel is replaced by a bent tube which reaches the bottom of the flask and is bent at the outer end for downward delivery (Note 3). The warm (40–50°) solution is forced under nitrogen pressure (by carefully pinching off the tube to the mercury bubbler) into the 500-ml. separatory funnel which has been prepared for attachment to the 1-l. flask. After all the solution has passed into the separatory funnel, nitrogen flow is allowed to continue briefly in order to displace the air above the

solution, then the funnel is stoppered loosely and attached to the
1-l. flask. Two hundred milliliters of purified tetrahydrofuran is
now placed in the flask, acetylene (Note 4) is introduced through
the gas-inlet tube at the rate of 15–20 l. per hour, and the stirrer
is started. After 5 minutes, about 5 ml. of the solution of ethylmagnesium bromide is added in 1 portion. Almost at once, there
appears a froth of ethane which is easily distinguishable from the
larger bubbles of acetylene. When the frothing subsides, portionwise addition of the ethylmagnesium bromide solution is continued until the total solution has been added. This requires
about 3 hours, and the temperature of the reaction rises 5–10°
above room temperature. The solution of ethynylmagnesium
bromide is homogeneous at 30° (Note 5).

B. *1-Phenyl-1-penten-4-yn-3-ol.* The stirred solution of ethynylmagnesium bromide is cooled in ice water as there is added
dropwise during about 45 minutes a solution of 47.5 g. (0.36 mole)
of freshly distilled cinnamaldehyde (Note 6) in 50 ml. of purified
tetrahydrofuran. After addition is complete, stirring is continued overnight as the solution is allowed to warm to room
temperature. The brown homogeneous reaction mixture is added
carefully to 1.5 l. of cooled saturated ammonium chloride solution,
then the aqueous phase is extracted with three 250-ml. portions
of ether. The ether extracts are combined with the tetrahydrofuran solution and dried over anhydrous magnesium sulfate.
After evaporation of the solvent, the product is distilled (Note 7)
at a pressure of about 0.1 mm., with the heating bath at about
90°. The distillate, which solidifies on cooling, is crystallized
from petroleum ether (b.p. 40–60°) to yield 33–39 g. (58–69%)
of the unsaturated alcohol of m.p. 67–68° (Note 8).

2. Notes

1. If less solvent is used, the ethylmagnesium bromide may
crystallize on cooling.

2. If ether is substituted for tetrahydrofuran in this preparation, the acetylenic glycol is the sole product. The submitters
purified the tetrahydrofuran by shaking with potassium hydroxide pellets, heating under reflux with sodium metal, and
finally distilling. They report that heating under reflux with

sodium diphenylketyl gives a better sample of tetrahydrofuran, but that this does not improve the yield.

The checkers purified the solvent by shaking with potassium hydroxide, distilling from lithium aluminum hydride, then storing over sodium wire.

3. It is advisable to prepare the delivery tube prior to its need.

4. The submitters purified the acetylene by passing it through a trap cooled to −80°, then through concentrated sulfuric acid, and finally through soda-lime. The checkers purified the acetylene by passing it first through a tower of 10-mesh alumina, then through concentrated sulfuric acid.

5. No outside cooling is employed during the preparation of ethynylmagnesium bromide. If this solution of ethynylmagnesium bromide is cooled to 0°, a crystalline complex separates. If part of the solvent is evaporated, even at 40° under reduced pressure, there occurs disproportionation to acetylene and the bisbromomagnesium derivative.

6. Methyl ethyl ketone, crotonaldehyde, and acrolein react similarly with ethynylmagnesium bromide. The respective yields of acetylenic alcohols are 69%, 84%, and 40%.

7. The submitters used a still of the evaporative type, such as a Hickman still. The checkers used a similar still modified to include a magnetic stirring bar. This modification greatly decreases the time required for removal of the last traces of solvent, minimizes the danger of bumping just before evaporative distillation occurs, and increases the rate of distillation of the product.

8. The checkers report yields as high as 84% and believe that the use of magnetic stirring in the Hickman still is responsible for the higher yield.

3. Methods of Preparation

The preparation of ethynylmagnesium bromide in ether has been described;[2,3] however, the subsequent history of the compound has been controversial. The submitters have been unable to prepare ethynylcarbinols by the earlier procedures.

1-Phenyl-1-penten-4-yn-3-ol has been prepared in liquid ammonia from cinnamaldehyde and sodium acetylide in 2% yield,[4] and from the sodium bisulfite compound of cinnamaldehyde and sodium acetylide in 13.5% yield.[5] The present procedure has been published.[6]

[1] The Dyson Perrins Laboratory, Oxford, England.
[2] Salkind and Rosenfeld, *Ber.*, **57**, 1690 (1924).
[3] Grignard, Lapayre, and Tcheou, *Compt. rend.*, **187**, 517 (1928).
[4] Jones and McCombie, *J. Chem. Soc.*, **1942**, 733.
[5] Cymerman and Wilks, *J. Chem. Soc.*, **1950**, 1208.
[6] Jones, Skattebøl, and Whiting, *J. Chem. Soc.*, **1956**, 4765.

1-PHENYLPIPERIDINE

(Piperidine, 1-phenyl)

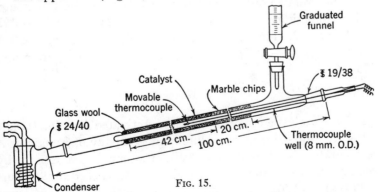

$$
\begin{array}{c} CH_2 \\ CH_2 \quad CH_2 \\ | \qquad | \\ CH_2 \quad CH_2 \\ \diagdown \diagup \\ O \end{array}
+ C_6H_5NH_2 \xrightarrow{Al_2O_3}
\begin{array}{c} CH_2 \\ CH_2 \quad CH_2 \\ | \qquad | \\ CH_2 \quad CH_2 \\ \diagdown \diagup \\ N \\ | \\ C_6H_5 \end{array}
+ H_2O
$$

Submitted by A. N. Bourns, H. W. Embleton, and Mary K. Hansuld.[1]
Checked by R. T. Arnold, William E. Parham, and Carl Serres.

1. Procedure

The apparatus (Fig. 15) consists of a reaction tube, 100 cm. long

Graduated funnel

$ 19/38

Catalyst

Marble chips

Movable thermocouple

Glass wool

$ 24/40

42 cm.

20 cm.

100 cm.

Thermocouple well (8 mm. O.D.)

Condenser

Fig. 15.

and 3 cm. in diameter, provided with an inlet tube to which a graduated dropping funnel is connected. The lower end of the reaction tube is fitted with a Friedrichs condenser and receiver. A thermocouple well, concentric with the reaction tube, is passed through the upper end by means of a standard ground-glass joint. The tube is packed with 200 ml. of 4- to 8-mesh alumina (Note 1) held in place by a plug of glass wool supported by indentations in the tube, followed by 100 ml. of marble chips which serve as a vaporizer and preheater for the reactants. The reaction tube is supported in an electrically heated furnace (Note 2) extending from the inlet tube to the joint connecting the condenser.

The reaction tube is heated to 330–340° (Note 3) and is swept out with a stream of nitrogen gas. A solution of 258 g. (3 mole) of tetrahydropyran (Note 4) and 559 g. (6 mole) of aniline is then placed in the graduated funnel and introduced into the reaction tube at a rate of 90–100 ml. per hour. The product, which is collected in a 1-l. flask containing 10 g. of sodium chloride, consists of a light-yellow oil and a lower aqueous layer. After the addition of the reactants is complete, a slow stream of nitrogen is passed through the reaction tube for 20–30 minutes to remove any product adsorbed on the catalyst. The lower aqueous layer is separated and discarded, and the upper organic layer is dried over sodium hydroxide pellets. The product is fractionated under reduced pressure using a Whitmore-Lux column [2] (Note 5) only to remove the small amount of unchanged tetrahydropyran and excess aniline. The column is permitted to drain, and the residue is distilled from an ordinary Claisen flask under reduced pressure. 1-Phenylpiperidine is obtained as a colorless or very light-yellow liquid, b.p. 123–126°/12.5 mm., 133–136°/21 mm., n_D^{25} 1.5603. The yield is 403–435 g. (83–90% based on tetrahydropyran) (Note 6).

2. Notes

1. The catalyst employed was Alcoa Activated Alumina (Grade F-1, 4- to 8-mesh) from the Aluminum Company of America. A fresh catalyst is brought to a condition of maximum activity by passing a slow stream of air through the catalyst bed for 94 hours at 390–405°. Without this pretreatment, yields are 5–10% lower than those reported here. The catalyst is reactivated after each run by passing air through it for 39 hours at 390–405°.

2. The furnace may be of the construction described in a previous volume.[3] It is desirable, although not essential, to provide separately controlled heating elements for each of the two packed zones of the reaction tube.

3. The temperature of the furnace is measured by a thermocouple which can be moved to various positions in the thermocouple well. The catalyst temperature should be maintained at 325–345°, although it may be as low as 320° at the ends of the catalyst zone, depending upon the construction of the furnace.

4. Eastman Kodak Company practical grade tetrahydropyran may be used without purification.

5. A column of equivalent efficiency may be employed.

6. 1-*m*-Tolylpiperidine (b.p. 141–143°/16 mm., n_D^{25} 1.5535) and 1-*p*-tolylpiperidine (b.p. 140–143°/15 mm., n_D^{25} 1.5509) may be prepared in 85–90% yield by a similar procedure. 1-*o*-Tolylpiperidine (b.p. 123–125°/15 mm., n_D^{25} 1.5391) is obtained in 60% yield under similar reaction conditions, but it is necessary to fractionate the product in order to obtain pure material.

3. Methods of Preparation

1-Phenylpiperidine has been prepared by warming aniline with 1,5-dibromopentane; [4,5] heating 5-anilino-1-bromopentane; [6] the dehydration of 5-anilino-1-pentanol over alumina; [7] the electrolytic reduction of N-phenylglutarimide; [8] the catalytic hydrogenation of 1-phenyl-3-hydroxypyridinium chloride; [9] the action of bromobenzene on piperidine in the presence of lithium [10] or sodium amide; [11] the reaction of fluorobenzene, 1-methylpiperidine, and phenyllithium; [12] the action of diphenylsulfone on piperidine in the presence of sodamide; [13] the diazotization and deamination of 1-(2-aminophenyl)piperidine [14] and of 1-(4-aminophenyl)piperidine; [15] the reduction of N-phenylglutarimide with lithium aluminum hydride; [16] the reaction of pentamethylene glycol dibenzenesulfonate with aniline; [17] the action of lithium piperidide on chlorobenzene; [18] and the present method.[19]

[1] Hamilton College, Hamilton, Ontario.
[2] Whitmore and Lux, *J. Am. Chem. Soc.*, **54**, 3451 (1932).
[3] *Org. Syntheses Coll. Vol.* **3**, 314 (1955).
[4] von Braun, *Ber.*, **37**, 3212 (1904).

[5] Paul, *Bull. soc. chim. France*, **53**, 1489 (1933).

[6] von Braun, *Ber.*, **40**, 3920 (1907).

[7] Scriabine, *Bull. soc. chim. France*, **1947**, 454; Société des usines chimiques Rhône-Poulenc, French pat. 880,986 [*C. A.*, **48**, 739 (1954)].

[8] Sakurai, *Bull. Chem. Soc. Japan*, **13**, 482 (1938).

[9] Koelsch and Carney, *J. Am. Chem. Soc.*, **72**, 2285 (1950).

[10] Horning and Bergstrom, *J. Am. Chem. Soc.*, **67**, 2110 (1945).

[11] Brotherton and Bunnett, *Chem. & Ind.* (*London*), **1957**, 80; Bunnett and Brotherton, *J. Org. Chem.*, **22**, 832 (1957).

[12] Wittig and Merkle, *Ber.*, **76B**, 109 (1943).

[13] Bradley, *J. Chem. Soc.*, **1938**, 458.

[14] Le Fevre, *J. Chem. Soc.*, **1932**, 1376.

[15] Lellmann and Geller, *Ber.*, **21**, 2279 (1888).

[16] Baddeley, Chadwick, and Taylor, *J. Chem. Soc.*, **1956**, 448.

[17] Reynolds and Kenyon, *J. Am. Chem. Soc.*, **72**, 1597 (1950).

[18] Kobrich, *Chem. Ber.*, **92**, 2985 (1959).

[19] Bourns, Embleton, and Hansuld, *Can. J. Chem.*, **30**, 1 (1952).

3-PHENYL-1-PROPANOL *

(1-Propanol, 3-phenyl-)

$$C_6H_5CHCH_2CH_2 \quad \xrightarrow[C_4H_9OH]{Na} \quad C_6H_5CH_2CH_2CH_2OH$$

Submitted by R. L. SHRINER and PHILIP R. RUBY.[1]
Checked by RICHARD T. ARNOLD, W. E. PARHAM, and HANS WYNBERG.

1. Procedure

In a 5-l. round-bottomed flask, equipped with two reflux condensers and a mechanical stirrer (Note 1), are placed 800 g. (925 ml.) of dry toluene and 168 g. (7.3 g. atoms) of sodium. The toluene is heated to boiling, the sodium is melted, and the stirrer is started. The source of external heat is removed, and a solution of 328 g. (2 moles) of 4-phenyl-*m*-dioxane (p. 786) in 311 g. (4.2 moles) of 1-butanol (Note 2) is added through the top of one of the condensers. The vapors should reflux about halfway up the condensers; about 30 to 60 minutes is used for the addition (Note 3). The mixture is cooled to room temperature, and a solution of 100 ml. of concentrated sulfuric acid in 800 ml. of water is added slowly with stirring. After the water layer is separated and discarded, 500 ml. of water is again added to the organic layer. Dilute sulfuric acid (5%) is added with shaking until the water layer is neutral to litmus paper. After the water layer is separated and discarded, the toluene and 1-butanol are removed from the organic layer by distillation. The remaining liquid is fractionated under reduced pressure (Note 4) to give 224–227 g. (82.2–83.4%) of 3-phenyl-1-propanol, b.p. 95–97°/0.4 mm. or 113–115°/3 mm., n_D^{20} 1.5268–1.5269, d_4^{20} 1.004–1.008.

2. Notes

1. The condenser should have a large bore to prevent flooding. A wide-sweep stirrer such as a Hershberg stirrer should be used, and the stirring motor must be capable of operating under heavy loads. The checkers suggest that the minimum size for the stirrer be 8-mm. glass rod.

2. The butanol should be freshly dried and distilled.

3. The addition must be as rapid as possible. An additional 100 to 400 ml. of toluene may have to be added to facilitate stirring.

4. A heated 35-cm. Vigreux column is recommended, but a Claisen flask can be used if care is taken. If a Claisen flask is used, the distillation must not be carried out too rapidly, particularly near the end, at which point some of the residue tends to codistil.

3. Methods of Preparation

Ethyl cinnamate has been reduced to 3-phenyl-1-propanol with sodium and ethanol,[2-5] hydrogen and copper chromite,[6] and sodium and ammonia.[7] The alcohol has also been prepared by reduction of the glyceride of cinnamic acid with sodium and amyl alcohol;[8] by reduction of cinnamic acid with lithium aluminum hydride;[9] by reduction of cinnamoyl chloride with sodium borohydride;[10] and by reduction of ethyl dihydrocinnamate with sodium and ethanol.[2,11] Cinnamaldehyde has been reduced to 3-phenyl-1-propanol with hydrogen and palladium,[12,13] platinum,[14-16] or nickel,[17-20] nickel in alkaline solution (no hydrogen),[21] lithium aluminum hydride,[22] electrolysis at a mercury[23] or lead[24] electrode, and with an unmentioned catalyst.[25] Reduction of cinnamyl alcohol to 3-phenyl-1-propanol has been effected by use of sodium and ethanol,[26] sodium amalgam and water,[27, 28] hydrogen and nickel [29] or palladium,[30] sodium and ammonia,[31] and lithium aluminum hydride.[32] Other syntheses have been brought about by reduction of ethyl α,β-epoxy-β-phenyldihydrocinnamate with sodium and amyl alcohol;[33] by reduction of ethyl benzoylacetate with hydrogen and copper chromite;[34] by reduction of acetonephenyllactic acid with hydrogen and copper chromite;[35] by reaction of ethyl alcohol with sodium benzylate;[36] by reaction of benzylmagnesium chloride with a mixture of ethylene chlorohydrin and ethylmagnesium chloride;[37] by reaction of trimethylene oxide with phenylmagnesium bromide;[38] by condensation of benzylmagnesium chloride with ethylene oxide;[39] and by hydrogenolysis of 1-phenyl-1,3-propanediol over nickel-on-kieselguhr.[40]

The reduction of 4-phenyl-m-dioxane to give 3-phenyl-1-propanol, as described here, is based on the procedure of Beets,[41] who used sodium and diisobutylcarbinol. Other substituted m-dioxanes may also be converted to substituted 3-aryl-1-propanols by the same procedure.[42] 3-Phenyl-1-propanol also has been obtained in 85% yield by the reduction of 4-phenyl-m-dioxane over copper chromite catalyst.[43]

[1] State University of Iowa, Iowa City, Iowa.

[2] Bouveault and Blanc, *Compt. rend.*, **137**, 328 (1903); *Bull. soc. chim. France*, [3] **31**, 1209 (1904).

[3] Shorygin, Bogacheva, and Shepeleva, *Khim. Referat. Zhur.*, **1940**, 114 [*C. A.*, **36**, 3793 (1942)].

[4] Clutterbuck and Cohen, *J. Chem. Soc.*, **123**, 2509 (1923).

[5] Krushevskii and Gildburg, Russ. pat. 31,008 [*C. A.*, **28**, 3425 (1934)].

[6] Adkins and Folkers, *J. Am. Chem. Soc.*, **53**, 1095 (1931); Wojcik and Adkins, *J. Am. Chem. Soc.*, **55**, 4939 (1933).

[7] Chablay, *Compt. rend.*, **156**, 1022 (1913).

[8] Darzens, *Compt. rend.*, **205**, 682 (1937).

[9] Nystrom and Brown, *J. Am. Chem. Soc.*, **69**, 2548 (1947).

[10] Chaikin and Brown, *J. Am. Chem. Soc.*, **71**, 122 (1949).

[11] Burrows and Turner, *J. Chem. Soc.*, **119**, 428 (1921).

[12] Straus and Grindel, *Ann.*, **439**, 307 (1924).

[13] Endoh, *Rec. trav. chim.*, **44**, 869 (1925).

[14] Skita, *Ber.*, **48**, 1692 (1915).

[15] Tuley and Adams, *J. Am. Chem. Soc.*, **47**, 3063 (1925).

[16] Vavon, *Compt. rend.*, **154**, 361 (1912); *Ann. chim. et phys.*, [9] **1**, 166 (1914).

[17] Adkins and Billica, *J. Am. Chem. Soc.*, **70**, 695 (1948).

[18] Braun and Kochendorfer, *Ber.*, **56**, 2175 (1923).

[19] Delepine and Hanegraaff, *Bull. soc. chim. France*, [5] **4**, 2087 (1937).

[20] Palfray, Sabetay, and Gauthier, *Compt. rend.*, **218**, 553 (1944).

[21] Papa, Schwenk, and Whitman, *J. Org. Chem.*, **7**, 587 (1942).

[22] Nystrom and Brown, *J. Am. Chem. Soc.*, **70**, 3738 (1948).

[23] Shima, *Mem. Coll. Sci. Kyoto Imp. Univ.*, **A12**, 70 (1929).

[24] Law, *J. Chem. Soc.*, **101**, 1030 (1912).

[25] Walter, Ger. pat. 296,507 [*C. A.*, **13**, 368 (1919)].

[26] Gray, *J. Chem. Soc.*, **127**, 1156 (1925).

[27] Hatton and Hodgkinson, *J. Chem. Soc.*, **39**, 319 (1881).

[28] Rugheimer, *Ann.*, **172**, 123 (1874).

[29] Tanaka, *Chem. Ztg.*, **48**, 25 (1924).

[30] Straus and Berkow, *Ann.*, **401**, 151 (1913).

[31] Chablay, *Compt. rend.*, **143**, 829 (1906); *Ann. chim. et phys.*, [9] 8, 191 (1914).

[32] Hochstein and Brown, *J. Am. Chem. Soc.*, **70**, 3484 (1948).

[33] Verley, *Bull. soc. chim. France*, [4] **35**, 488 (1924).

[34] Adkins and Billica, *J. Am. Chem. Soc.*, **70**, 3121 (1948).

[35] Oeda, *Bull. Chem. Soc. Japan*, **10**, 531 (1935).

[36] Guerbet, *Compt. rend.*, **146**, 300 (1908); *Bull. soc. chim. France*, [4] **3**, 503 (1908).

[37] Conant and Kirner, *J. Am. Chem. Soc.*, **46**, 241 (1924).

[38] Searles, *J. Am. Chem. Soc.*, **73**, 124 (1951).

[39] Ohara, Jap. pat. 4364 (1950) [*C. A.*, **47**, 3347 (1953)].

[40] Arnold (to E. I. du Pont de Nemours & Co.), U. S. pat. 2,555,912 [*C. A.*, **46**, 3074 (1952)].

[41] Beets, *Rec. trav. chim.*, **70**, 20 (1951).

[42] Beets and Van Essen, *Rec. trav. chim.*, **70**, 25 (1951); Drukker and Beets, *Rec. trav. chim.*, **70**, 29 (1951).

[43] Emerson, Heider, Longley, and Shafer, *J. Am. Chem. Soc.*, **72**, 5314 (1950).

PHENYLPROPARGYLALDEHYDE DIETHYL ACETAL*

(Propiolaldehyde, phenyl-, diethyl acetal)

$$C_6H_5C\equiv CH + HC(OC_2H_5)_3 \xrightarrow[\Delta]{ZnI_2}$$

$$C_6H_5C\equiv C-CH(OC_2H_5)_2 + C_2H_5OH$$

Submitted by B. W. Howk and J. C. Sauer.[1]
Checked by N. J. Leonard and S. W. Blum.

1. Procedure

Into a 300-ml. three-necked flask equipped with a nitrogen inlet, a thermometer, and a short fractionating column (Note 1) are charged 74.1 g. (0.50 mole) of triethyl orthoformate, 51.0 g. (0.50 mole) of phenylacetylene (Note 2), and 3.0 g. of zinc iodide (Note 3). Ethanol is slowly distilled from the reaction mixture, which must be heated to about 135° before refluxing in the still-head begins. A total of 29–35 ml. of distillate, b p. 65–88° (mostly 78°), is collected over a period of about 1 hour as the temperature of the reaction mixture gradually rises to 200° to 210° (Note 4). The reaction mixture is cooled to room temperature and filtered with suction. The flask and the small amount of precipitate on the filter paper are washed with 5 ml. of ether. The filtrate and ether washings are combined and distilled. After a small fore-run, phenylpropargylaldehyde diethyl acetal is collected at 99–100°/2 mm. The yield is 73–80 g. (72–78%), n_D^{25} 1.5153–1.5158. The synthesis is applicable to the preparation of other propargyl aldehyde acetals (Note 5).

2. Notes

1. The checkers found a 12-inch Vigreux column satisfactory.

2. The checkers purchased phenylacetylene (p. 763) from Gesellschaft für Teerverwertung mbH., Duisburg-Meiderich, Germany.

3. The submitters report that zinc nitrate appears to be equivalent to zinc iodide as a catalyst and that zinc chloride (commercial anhydrous grade) is satisfactory but requires 2–3 hours of heating and gives 64–70% yield.

4. Yields are lower under forcing conditions of prolonged heating.

5. The method has been applied by the submitters [2] to the preparation of cyclohexylmethylpropiolaldehyde diethyl acetal (54% yield) from cyclohexylmethylacetylene and triethyl orthoformate; of phenylethynyl n-butyl dimethyl ketal (40% yield) from phenylacetylene and trimethyl n-orthovalerate; and of phenylethynyl methyl diethyl ketal (34% yield) from phenylacetylene and triethyl orthoacetate. n-Butylpropiolaldehyde diethyl acetal was isolated in 32% yield by heating an equimolar mixture of 1-hexyne and triethyl orthoformate containing catalytic amounts of a zinc chloride-zinc iodide catalyst under autogenous pressure at 190° for 3 hours.

3. Methods of Preparation

The described method of preparing phenylpropargylaldehyde

diethyl acetal is that of Howk and Sauer.[2] The method for synthesizing phenylpropargylaldehyde diethyl acetal previously published in *Organic Syntheses* [3] involves three steps beginning with cinnamaldehyde; over-all yields are 49–62%. Other methods of preparative value are the interaction of the Grignard reagent of phenylacetylene with triethyl orthoformate, or the sodium derivative of phenylacetylene with either triethyl orthoformate or ethyl formate. These reactions are discussed critically by Raphael.[4] Phenylpropargylaldehyde diethyl acetal has also been made by the action of the phenyl Grignard reagent with the diethyl acetal of chloropropiolaldehyde.[5]

The acetylenic acetals are easily hydrolyzed to the corresponding aldehydes in high yields in the presence of dilute acids.[3,4] Acetylenic acetals have also been of value in the synthesis of α,β-unsaturated ethylenic acetals or aldehydes by partial catalytic hydrogenation of the triple bond.[4]

[1] Contribution No. 474 from the Central Research Department, Experimental Station, E. I. du Pont de Nemours & Co. (Inc.), Wilmington, Delaware.

[2] Howk and Sauer, *J. Am. Chem. Soc.*, **80**, 4607 (1958).

[3] *Org. Syntheses Coll. Vol.* **3**, 732 (1955).

[4] Raphael, *Acetylenic Compounds in Organic Synthesis*, pp. 68–75, Academic Press, New York, 1955.

[5] Zakharkin, *Doklady Akad. Nauk S.S.S.R.*, **105**, 985 (1955) [*C.A.*, **50**, 11237a (1956)].

PHENYLSUCCINIC ACID *

(Succinic acid, phenyl-)

$$C_6H_5CH=C(CO_2C_2H_5)_2 + KCN + 2H_2O \rightarrow$$
$$C_6H_5CHCH_2CO_2C_2H_5 + KHCO_3 + C_2H_5OH$$
$$\overset{|}{CN}$$

$$C_6H_5CHCH_2CO_2C_2H_5 + 3H_2O + HCl \rightarrow$$
$$\overset{|}{CN}$$

$$C_6H_5CHCH_2CO_2H + NH_4Cl + C_2H_5OH$$
$$\overset{|}{CO_2H}$$

Submitted by C. F. H. ALLEN and H. B. JOHNSON.[1]
Checked by H. R. SNYDER, D. J. MANN, and LEONARD E. MILLER.

1. Procedure

Caution! This preparation should be conducted in a well-ventilated hood to avoid exposure to hydrogen cyanide.

A. *Ethyl β-phenyl-β-cyanopropionate.* In a 5-l. round-bottomed three-necked flask suspended in an oil bath and fitted with a mechanical stirrer, a reflux condenser, and a 250-ml. dropping funnel is placed a solution of 200 g. (0.81 mole) of diethyl benzalmalonate [2] in 2 l. of absolute ethanol. The stirrer is started, and a solution of 56 g. (0.86 mole) of potassium cyanide in 100 ml. of water is added rapidly from the separatory funnel; a small amount of the potassium cyanide precipitates. The temperature of the oil bath is raised to 70° and maintained at 65–75° for 18 hours.

The mixture is cooled to 15°, and the precipitated potassium bicarbonate is collected on a Büchner funnel. The solid (weight 70–72 g.) is washed on the funnel with 100 ml. of 95% ethanol. The combined filtrate and wash liquor is transferred to a 5-l. round-bottomed flask and made slightly acid (*Caution!* Note 1) with dilute hydrochloric acid (about 15–20 ml. of the 10% acid is required). The solution is then concentrated under reduced pressure to a semi-solid residue (Note 1). The cooled residue is shaken with a mixture of 300 ml. of water and 500 ml. of ether. The material dissolves completely; the water layer is separated and washed with 200 ml. of ether. The ether solutions are combined, dried over 20 g. of calcium chloride, filtered into a 2-l. round-bottomed flask equipped with a glass joint, and concentrated by distillation (heating on a steam bath). The crude ethyl β-phenyl-β-cyanopropionate remains as a clear red oil weighing 130–140 g. It is sufficiently pure for use in the next step (Note 2).

B. *Phenylsuccinic acid.* To the crude ester obtained above is added 500 ml. of concentrated hydrochloric acid (sp. gr. 1.19). The flask is fitted to a condenser (Note 3), and the mixture is heated under reflux for 18 hours (Note 4). At the end of this time only a small amount of red oil remains (Note 5). The mixture is cooled, and the nearly solid cake which forms is broken up and collected on a glass filter cloth (Note 5). The crude tan-colored phenylsuccinic acid is washed with 300 ml. of cold water and dried at 60°. It then weighs 105–110 g. (67–70%) and melts at 163–164° (Notes 6 and 7).

2. Notes

1. Since hydrogen cyanide may be liberated during the acidification and the subsequent concentration, both operations should be carried out in a well-ventilated hood.

2. The pure ester can be obtained by distillation under reduced pressure (b.p. 161–164°/8 mm.).

3. Glass-jointed equipment is required in this step. The flask with a glass joint was used in the preceding operation only to avoid the necessity of transferring the product after the evaporation.

4. Hydrogen chloride is evolved during the first part of the refluxing; it may be disposed of by absorption in water in a gas trap.[3]

5. The red, oily impurity usually distributes itself as a film on the surface of the liquid and as a lump at the bottom of the flask. It solidifies on cooling and is most conveniently removed as a solid; the thin crust on the surface is lifted with a spatula, and the lump at the bottom of the flask is left undisturbed when the product is collected on the filter.

6. The pure acid can be obtained by recrystallization from water. For each 10 g. of acid about 300 ml. of water and 0.5 g. of Norit are required. The recovery is 85–90% of pure white acid melting at 165.5–166°.

7. The checkers found the product at this stage to be of sufficient purity for conversion to phenylsuccinic anhydride.

3. Methods of Preparation

Preparative methods for phenylsuccinic acid have been listed in an earlier volume.[4] More recent procedures have been based on the hydrolysis of ethyl α-phenyl-β-cyanopropionate,[5] tetraethyl 1-phenyl-1,1,2,2-ethanetetracarboxylate,[6] diethyl (α-phenyl-β-nitroethyl)malonate,[7] and ethyl [8] or methyl 2,3-dicyano-3-phenylpropionate.[9]

The procedure given above is more economical of time and materials than that previously published.[4] Applications of the present method, due originally to Bredt and Kallen,[10] have been published.[11, 12]

[1] Eastman Kodak Co., Rochester, New York.

[2] *Org. Syntheses Coll. Vol.* **3**, 377 (1955).

[3] *Org. Syntheses Coll. Vol.* **2**, 4 (1943).

[4] *Org. Syntheses Coll. Vol.* **1**, 451 (1941).

[5] Wideqvist, *Arkiv Kemi, Mineral. Geol.*, **26A**, No. 16 (1948) [*C. A.*, **43**, 2167 (1949)].

[6] Hsing and Li, *J. Am. Chem. Soc.*, **71**, 774 (1949).

[7] Perekalin and Sopova, *Zhur. Obshcheĭ Khim.*, **24**, 513 (1954) [*C. A.*, **49**, 6180 (1955)].

[8] Sen and Bagchi, *J. Org. Chem.*, **20**, 845 (1955).

[9] Miller and Long, *J. Am. Chem. Soc.*, **73**, 4895 (1951).

[10] Bredt and Kallen, *Ann.*, **293**, 344 (1896).

[11] Verkade and Hartman, *Rec. trav. chim.*, **52**, 945 (1933).

[12] Wideqvist, *Arkiv Kemi, Mineral. Geol.*, **14B**, No. 19, 6 pp. (1940) [*C. A.*, **35**, 3993 (1941)].

o-PHTHALALDEHYDE*

(Phthalaldehyde)

Submitted by J. C. Bill and D. S. Tarbell.[1]
Checked by Arthur C. Cope and Harris E. Petree.

1. Procedure

Caution! This preparation should be conducted in a hood, and rubber gloves should be worn, to avoid exposure to bromine as well as the by-product o-xylylene dibromide which is a lachrymator and skin irritant.

A. *α,α,α′,α′-Tetrabromo-o-xylene.* In a 2-l. three-necked flask equipped with an oil-lubricated Trubore stirrer, a dropping funnel, a thermometer extending nearly to the bottom of the flask, and a reflux condenser (Note 1) attached to a gas absorption trap[2] is placed 117 g. (1.1 moles) of dry *o*-xylene (Note 2). An ultraviolet lamp such as a General Electric R.S. Reflector Type 275-watt sun lamp is placed about 1 cm. from the flask so as to admit the maximum amount of light. The stirrer is started, and the *o*-xylene is heated to 120° with an electric heating mantle. A total of 700 g. (4.4 moles) of bromine (N.F. grade) is added in portions from the dropping funnel to the reaction flask at such a rate that the bromine color is removed as fast as it is added. After approximately one-half of the bromine has been added, the temperature is slowly increased to 175° for the remainder of the addition; the mixture becomes very dark toward the end of the reaction. The bromine can be added rapidly at first, but toward the end it must not be added at a rate exceeding 4–5 drops per minute in order to avoid loss of a visible amount of bromine with the evolved hydrogen bromide. After all the bromine has been added (10–14 hours), the mixture is illuminated and stirred at 170° for 1 hour. After removal of the stirrer, etc., the mixture is cooled and allowed to stand overnight to crystallize in the reaction flask exposed to the air.

The dark, solid tetrabromide is dissolved in 2 l. of hot chloroform (Note 3) and treated with 100 g. of 325-mesh Norit. The mixture is filtered with slight suction, the Norit is washed with hot chloroform, and the Norit treatment is repeated. The tan-colored filtrate from the second Norit treatment is concentrated to 250–300 ml. by distillation under reduced pressure and chilled to 0°. The solid product is collected on a cold Büchner funnel and washed with a small amount of cold chloroform. The filtrate is concentrated further and cooled to obtain a second crop of crystals, which is purified by recrystallization from chloroform. The yield of the tetrabromide obtained from the first crop (white) and second crop after recrystallization (light tan) is 344–370 g. (74–80%), m.p. 115–116°.

B. *o-Phthalaldehyde.* The $\alpha,\alpha,\alpha',\alpha'$-tetrabromo-*o*-xylene (344–370 g.) obtained as described above, part A, is placed in a 5-l. round-bottomed flask with 4 l. of 50% (by volume) ethanol and 275 g. of potassium oxalate. The mixture is heated under reflux for 50 hours (a clear yellow solution is formed after 25–30 hours). About 1750 ml. of the ethanol is then removed by distillation (which is stopped before the product begins to steam-distil), and 700 g. of disodium monohydrogen phosphate dodecahydrate ($Na_2HPO_4 \cdot 12H_2O$) is added to the aqueous residue. The mixture is steam-distilled rapidly (Note 4), using an efficient condenser, until 10–12 l. of distillate is collected and the distillate no longer gives a black color test for *o*-phthalaldehyde [3] when a portion is treated with concentrated ammonium hydroxide followed by glacial acetic acid. The distillate is then saturated with sodium sulfate at room temperature and divided into portions of approximately 4 l.; each portion is extracted first with 200 ml. and then with six 100-ml. portions of ethyl acetate. The combined ethyl acetate extracts are dried over anhydrous sodium sulfate, filtered, and concentrated under reduced pressure. The residue is crystallized from 90–100° ligroin, and a second crop is obtained by concentration of the mother liquor. The total yield of *o*-phthalaldehyde, m.p. 55.5–56°, is 87–94 g. (74–80% based on the tetrabromide, or 59–64% based on *o*-xylene).

2. Notes

1. Best results are obtained if an all-glass apparatus with ground-glass connections is used, as noted by Wawzonek and Karll.[4]

2. Pure commercial *o*-xylene (99% or higher purity) is dried by distillation until the distillate shows no further turbidity, and the residue is used.

3. The chloroform is allowed to stand over anhydrous calcium chloride overnight to remove water and ethanol.

4. The submitters state that the steam distillation is much more efficient if superheated (175–180°) steam is used, in which case special care must be taken to condense all the distillate.

3. Methods of Preparation

o-Phthalaldehyde has been made by the action of potassium hydroxide on *o*-(dichloromethyl)benzaldehyde,[5] by the hydrolysis of $\alpha,\alpha,\alpha',\alpha'$-tetrachloro-*o*-xylene,[6] by the hydrolysis of the $\alpha,\alpha,\alpha',\alpha'$-tetraacetate of *o*-phthalaldehyde,[3] by the oxidation of α,α'-dihydroxy-*o*-xylene with selenium dioxide [7] or N-chlorosuccinimide,[8] and by the reduction of phthalic acid N-methylanilide with diisobutylaluminum hydride.[9] The present method is essentially that of Thiele and Günther.[10, 11] Hydrolysis of the tetrabromide may also be carried out by treatment with fuming sulfuric acid followed by water.[12] For small-scale preparations of *o*-phthalaldehyde the reduction of N,N,N',N'-tetramethylphthalamide with lithium aluminum hydride is the method of preference.[13]

[1] University of Rochester, Rochester, New York.
[2] *Org. Syntheses Coll. Vol.* **2**, 4 (1943).
[3] Thiele and Winter, *Ann.*, **311**, 360 (1900).
[4] Wawzonek and Karll, *J. Am. Chem. Soc.*, **70**, 1666 (1948).
[5] Chaudhuri, *J. Am. Chem. Soc.*, **64**, 315 (1942).
[6] Colson and Gautier, *Ann. chim.* (*Paris*), **6**, 11, 28 (1887).
[7] Weygand, Kinkal, and Tietjen, *Chem. Ber.*, **83**, 394 (1950).

[8] Hebbelynck and Martin, *Bull. soc. chim. Belges*, **60**, 54 (1951) [*C. A.*, **46**, 7051 (1952)]; Hebbelynck, *Ind. chim. belge*, **16**, 483 (1951) [*C. A.*, **46**, 10127 (1952)].

[9] Zakharkin and Khorlina, *Izvest. Akad. Nauk S.S.S.R., Otdel. Khim. Nauk*, **1959**, 2146 [*C. A.*, **54**, 10932 (1960)].

[10] Thiele and Günther, *Ann.*, **347**, 107 (1906).

[11] Cope and Fenton, *J. Am. Chem. Soc.*, **73**, 1672 (1951).

[12] Weygand, Vogelbach, and Zimmermann, *Chem. Ber.*, **80**, 396 (1947).

[13] Weygand and Tietjen, *Chem. Ber.*, **84**, 625 (1951).

α-PHTHALIMIDO-o-TOLUIC ACID

(o-Toluic acid, α-phthalimido-)

Submitted by J. Bornstein, P. E. Drummond, and S. F. Bedell.[1]
Checked by John C. Sheehan and Y. L. Yeh.

1. Procedure

A 2-l. three-necked round-bottomed flask, fitted with a sealed stirrer and a reflux condenser carrying a drying tube, is charged with 100 g. (0.75 mole) of phthalide (Note 1), 150 g. (0.81 mole) of potassium phthalimide (Note 2), and 500 ml. of dimethylformamide (Note 3). The stirred suspension is heated under reflux by means of an electric mantle for 5 hours; the deep blue solution is then cooled to room temperature (Note 4). A solution of 300 ml. of glacial acetic acid in 500 ml. of water is added in one portion to the stirred reaction mixture, and the resulting yellow suspension, which becomes slightly warm, is stirred for an additional 30 minutes.

The precipitate is separated by suction filtration, pressed on the funnel, and washed successively with three 100-ml. portions of water and two 100-ml. portions of 95% ethanol. The product is transferred to a 1-l. Erlenmeyer flask, boiled for 10 minutes with 400 ml. of 60% ethanol with occasional stirring, filtered hot, washed twice with 50-ml. portions of 95% ethanol, and then dried in an oven at 90–100° for 6–12 hours. The crude α-phthalimido-*o*-toluic acid, which weighs 140–155 g., is divided into two equal portions, and each portion is dissolved in boiling propionic acid (Note 5). Each solution is treated with 1 tablespoon of Norit and filtered through an electrically heated gravity funnel. The filtrates are allowed to cool slowly to room temperature and are then refrigerated overnight. The crystals from the two portions are collected by suction filtration in one funnel and washed on the funnel with 400 ml. of 95% ethanol. The product is dried over potassium hydroxide in a vacuum desiccator. The yield of nearly white crystals of α-phthalimido-*o*-toluic acid is 126–141 g. (60–67% based on phthalide), m.p. 265.0–266.5°.

2. Notes

1. The phthalide was prepared according to *Organic Syntheses* [2] and was also purchased from Aldrich Chemical Company. The commercial product (200 g.) was recrystallized in 50-g. portions from 1.5 l. of water, the mother liquor from the first crop being employed for recrystallization of the subsequent portions. Each portion was treated with 2 tablespoons of Norit, filtered hot, allowed to cool to room temperature with occasional stirring, and then cooled to 5° before collecting the crystals which were washed on the funnel with small quantities of cold water. Final drying

was effected in a vacuum desiccator containing phosphorus pentoxide.

2. Eastman Kodak Company potassium phthalimide (200 g.) was digested with 450 ml. of boiling acetone for 15 minutes, filtered hot, washed on the funnel with 100 ml. of acetone, and dried at 100° for 6 hours.

3. The dimethylformamide was obtained from Eastman Kodak Company and was used without further purification.

4. The reaction mixture is most conveniently cooled by allowing it to stand at room temperature overnight. Occasionally the potassium salt of α-phthalimido-o-toluic acid precipitates at this point, but this does not interfere with the subsequent operations.

5. Approximately 1.33 l. of propionic acid is required for 78 g. of the crude α-phthalimido-o-toluic acid. Glacial acetic acid may be used as the solvent, but considerably larger volumes are required than when propionic acid is employed. This step should be carried out in a hood, since hot propionic acid vapors are very irritating.

3. Methods of Preparation

The present procedure is that described by the submitters.[3] α-Phthalimido-o-toluic acid has also been prepared by the acidolysis of the corresponding ethyl ester, obtained from the reaction of ethyl α-bromo-o-toluate with potassium phthalimide.[3]

[1] Boston College, Chestnut Hill, Massachusetts.

[2] *Org. Syntheses Coll. Vol.* 2, 526 (1943).

[3] Bornstein, Bedell, Drummond, and Kosloski, *J. Am. Chem. Soc.*, 78, 83 (1956).

PROPIOLALDEHYDE

$$HC\equiv C-CH_2OH \xrightarrow{CrO_3} HC\equiv C-CHO$$

Submitted by J. C. SAUER.[1]
Checked by JOHN C. SHEEHAN and E. R. GILMONT.

1. Procedure

Caution! Propiolaldehyde is a lachrymator.

A 3-l. three-necked round-bottomed flask is fitted with a thermometer, a graduated dropping funnel (Note 1), a stirrer (Note 2), a fine capillary tube for introducing nitrogen near the bottom of the flask, and an exit tube attached through a manometer to three traps set in series. In the flask are placed 360 ml. of 33% (by volume) propargyl alcohol [112.1 g. (120 ml., 2.0 moles)] (Note 3) and a cooled solution of 135 ml. of sulfuric acid and 200 ml. of water. The flask is cooled in an ice-salt mixture. While the contents of the flask are cooling, the first trap is cooled to about −15° with acetone and Dry Ice. The last two traps in the series are cooled to −78° with acetone and Dry Ice (Note 4). The pressure in the system is reduced to 40–60 mm., nitrogen is introduced through the capillary, and the mixture is stirred vigorously. A solution of 210 g. of commercial chromium trioxide (2.1 moles) in 400 ml. of water and 135 ml. of sulfuric acid is added dropwise in the course of about 3 hours while maintaining a reaction temperature of 2–10°. After the addition of the chromium trioxide, the ice bath is removed, and the flask is permitted to warm to room temperature while the pressure is gradually lowered to 14–20 mm. to remove the last of the aldehyde. The condensates of the three traps are combined (Note 5) and dried over anhydrous magnesium sulfate. The propiolaldehyde is distilled through a 16-in. column packed with platinum gauze. The fraction distilling at 54–57° weighs 38–44 g. (35–41%), n_D^{25} 1.4050 (Notes 6 and 7).

2. Notes

1. The end of the dropping funnel extends about 2 in. into

the flask from the opening and is drawn into a capillary. This is done to ensure the introduction of the chromic acid solution in the form of small droplets.

2. A "Trubore" stirring system with a 29/26 joint was used for stirring under vacuum.

3. Propargyl alcohol is available from the General Aniline and Film Corporation, Easton, Pennsylvania.

4. If the cooling bath for the first trap is lowered much below $-15°$, plugging of the trap is likely to occur. These traps are connected so that the vapor enters the larger, annular space, impinging on the cold wall before entering (and possibly plugging) the smaller inner tube.

5. The material in the first trap contains a considerable amount of water. In order to facilitate separation, the checkers saturated this mixture with sodium chloride. The upper layer, consisting of nearly pure propiolaldehyde, was combined with the contents of the second and third traps and dried over 5 g. of anhydrous magnesium sulfate. Distillation through a 12-in. vacuum-jacketed Vigreux column gave directly propiolaldehyde comparable in yield and quality with the final product described by the submitter in Note 6.

6. Considerable water is carried into the traps, and the distillate contains 3–10% water. The last of the water can be removed by drying the distillate a second time over magnesium sulfate and redistilling. In this way there is obtained 30–37 g. (28–34%) of propiolaldehyde distilling at 55–56°, n_D^{25} 1.4032–1.4034.

7. The material should be stored in glass-stoppered bottles, since contaminants from rubber stoppers may be sufficient to catalyze decomposition. A sample of propiolaldehyde underwent no noticeable change after four months' storage in a Dry Ice chest. However, this aldehyde undergoes extremely vigorous polymerization or decomposition in the presence of alkalies. For example, propiolaldehyde undergoes a change with almost explosive force in the presence of pyridine. Accordingly, *exceptional care should be used in the handling of propiolaldehyde.*

3. Methods of Preparation

The procedure described is that of Wille, Saffer, and Weisskopf.[2] Propiolaldehyde also has been prepared by the oxidation of propargyl alcohol using ammonium dichromate,[3] potassium dichromate,[4] or manganese dioxide in sulfuric acid.[5] In addition, it has been obtained by the electrolytic oxidation of propargyl alcohol[6] and by warming the dimethyl or diethyl acetal of propiolaldehyde with dilute sulfuric acid.[7]

[1] Contribution No. 364 from the Chemical Department, Experimental Station, E. I. du Pont de Nemours and Company, Wilmington, Delaware.

[2] Wille, Saffer, and Weisskopf, *Ann.*, **568**, 34 (1950).

[3] Quilico and Palazzo, *Proc. XI Intern. Congr. Pure and Appl. Chem.*, **2**, 253 (1947) [*C. A.*, **45**, 7107 (1951)].

[4] Maemoto (to Nissin Chemical Industries Co.), Jap. pat. 6159 (1951) [*C. A.*, **47**, 9997 (1953)].

[5] Copenhaver and Bigelow, *Acetylene and Carbon Monoxide Chemistry*, p. 124, Reinhold Publishing Corporation, New York, 1949.

[6] Wolff, *Chem. Ber.*, **87**, 668 (1954).

[7] Claisen, *Ber.*, **31**, 1022 (1898).

PSEUDOPELLETIERINE

$$\text{(structure)}\,OC_2H_5 + H_2O \longrightarrow OHC(CH_2)_3CHO + C_2H_5OH$$

$$\begin{array}{c}\text{CHO}\\ | \\ (CH_2)_3 \\ | \\ \text{CHO}\end{array} + CH_3NH_2 + \begin{array}{c}CH_2CO_2H\\ | \\ CO \\ | \\ CH_2CO_2H\end{array} \longrightarrow \left\langle CH_3\!\!-\!\!N \right\rangle\!\!=\!\!O + 2CO_2 + 2H_2O$$

Submitted by ARTHUR C. COPE, HUGH L. DRYDEN, JR., and CHARLES F. HOWELL.[1]
Checked by N. J. LEONARD, D. F. MORROW, and W. D. SMART.

1. Procedure

In a 3-l. round-bottomed flask equipped with a mechanical stirrer and flushed with a slow stream of nitrogen are placed 22 ml. (0.26 mole) of concentrated hydrochloric acid (sp. gr. 1.18), 165 ml. of "deoxygenated" water (Note 1), and 64 g. (0.5 mole) of 2-ethoxy-3,4-dihydro-2H-pyran (b.p. 62–65°/50 mm., n_D^{25} 1.4378) (Note 2). The mixture is stirred vigorously for 20 minutes and then allowed to stand for 1 hour.

To the resulting colorless solution of glutaraldehyde are added, in order, 350 ml. of water, 50 g. (0.74 mole) of commercial methylamine hydrochloride dissolved in 500 ml. of water, 83 g. (0.57 mole) of acetonedicarboxylic acid (Note 3) dissolved in 830 ml. of water, and a solution of 88 g. (0.25 mole) of disodium hydrogen phosphate dodeca-hydrate and 7.3 g. (0.18 mole) of sodium hydroxide dissolved in 200 ml. of water by heating. Carbon dioxide is evolved, and the pH of the solution, initially 2.5, increases to 4.5 after the mixture has been stirred under nitrogen for 24 hours. Concentrated hydrochloric acid (33 ml.) is added, and the solution is heated on the steam bath for 1 hour to complete the decarboxylation (Note 4). After the solution has been cooled to room temperature, 75 g. of sodium hydroxide in 100 ml. of water is added (Note 5), and the basic mixture is extracted with eight 250-ml. portions of methylene chloride (Note 6). The combined methylene chloride extracts are dried over sodium sulfate, concentrated to about 500 ml. (Note 7), and filtered through a layer of 400 g. of alumina (Note 8) packed in a 50-mm. column. The column is eluted with methylene chloride until about 1.5 l. of eluate has been collected. The eluate is concentrated under reduced pressure to yield crystalline but yellow pseudopelletierine. The solid is sublimed at 40° and 0.3 mm. to yield 47–55.5 g. (61–73%) of crude, nearly colorless pseudopelletierine (Note 9). The product is dissolved in 100 ml. of boiling pentane, 3 ml. of water is added, and the mixture is boiled until the aqueous layer disappears. After thorough chilling in a refrigerator, the crystals which separate are collected on a filter and washed well with ice-cold pentane. Evaporation of the combined filtrate and washings to 20 ml., followed by filtration and washing, yields a second crop of almost equally pure material. The combined pseudopelletierine hemihydrate weighs 47–55 g. and melts at 47–48.5°. Sublimation of the hemihydrate as described above removes the water of hydration and yields 44–52 g. (58–68%) of pure, colorless pseudopelletierine, m.p. 63–64° (sealed tube). Anhydrous material which has been prepared in this manner does not decompose on storage under dry conditions.

2. Notes

1. "Deoxygenated" water is prepared by passing a stream of nitrogen through ordinary distilled water and is used throughout the preparation until the condensation has been completed. The use of ordinary distilled water may lower the yield by no more than a few per cent.

2. This compound is prepared by the addition of ethyl vinyl ether to acrolein, under conditions similar to those described for a similar addition of methyl vinyl ether to crotonaldehyde (p. 311); see Longley and Emerson, *J. Am. Chem. Soc.*, **72**, 3079 (1950). Glutaraldehyde is available currently as a 30% aqueous solution from Carbide and Carbon Chemicals Company, 30 East 42nd Street, New York.

3. The preparation of acetonedicarboxylic acid is described in *Org. Syntheses Coll. Vol.* **1**, 10 (1941). The acid may also be obtained from Chas. Pfizer & Company, 630 Flushing Avenue, Brooklyn 6, New York.

4. Omission of the decarboxylation step decreases the yield of crude material to 57%. The temperature should reach about 80°.

5. The pH rises to about 12. A lower pH allows extraction of more of the dark-brown resin. The extraction must be performed promptly since the product can undergo self-condensation at this pH.

6. The employment of ether instead of methylene chloride requires the use of a continuous extractor for 2 days.

7. The checkers found that the purification of the pseudopelletierine could be simplified, at least in those preparations in which commercial acetonedicarboxylic acid was used. Thus, the crude product obtained by evaporation to dryness of the methylene chloride extracts can be sublimed directly. Two sublimations give pseudopelletierine of m.p. 62–64°, in 58–62% yield, comparable to the product obtained after the more extended purification procedure described in the text.

8. Chromatographic alumina (400 g.) is treated with 500 ml. of ethyl acetate at room temperature. After 48 hours, the alumina is collected on a filter and washed first with 1 l. of distilled water and then with 1 l. of methanol. After drying in air, the alumina is activated by heating at 120° for 3 hours at 50–100 mm.

9. Resublimation directly may serve for the final purification (see Note 7). Distillation is inconvenient because of the tendency of pseudopelletierine to crystallize in the condenser. The distilled product darkens rather rapidly even in the cold and melts at 47–53°. The once sublimed product also darkens slowly even when kept under dry nitrogen in a refrigerator. If anhydrous pseudopelletierine is exposed to moist air, the hemihydrate is formed, m.p. about 48°.

3. Methods of Preparation

Pseudopelletierine has been obtained from the bark of the pomegranate tree (*Punica granatum* L.).[2-5] The synthesis of the alkaloid from glutaraldehyde, methylamine, and calcium acetonedicarboxylate was first achieved by Menzies and Robinson.[6] The synthetic method subsequently was improved by Schöpf and Lehmann,[7] and others.[8, 9] The condensation of a dialdehyde with an amine and acetonedicarboxylic acid to form a heterobicyclic compound (an alkaloid or alkaloid analog, usually employing mild or so-called "physiological" conditions) is sometimes referred to as a Robinson-Schöpf synthesis.

The experimental procedure described is essentially one reported by Ziegler and Wilms[8] as subsequently modified,[9] except that the glutaraldehyde is prepared from 2-ethoxy-3,4-dihydro-2H-pyran instead of cyclopentene ozonide[8] or pyridine via dihydropyridine and glutaraldehyde dioxime.[9] Essentially these procedures have been reported briefly by other investigators.[10] Pseudopelletierine also has been prepared by the Dieckmann condensation of 1-methyl-2,6-piperidinediacetic acid diethyl ester with sodium.[11]

[1] Massachusetts Institute of Technology, Cambridge 39, Massachusetts.
[2] Tanret, *Compt. rend.*, **88**, 716 (1879); **90**, 696 (1880).
[3] Piccinini, *Gazz. chim. ital.*, **29**, II, 311 (1899).
[4] Hess and Eichel, *Ber.*, **50**, 1386 (1917); **52**, 1005 (1919).
[5] Chaze, *Compt. rend. soc. biol.*, **118**, 1065 (1935).
[6] Menzies and Robinson, *J. Chem. Soc.*, **125**, 2163 (1924).
[7] Schöpf and Lehmann, *Ann.*, **518**, 1 (1935).
[8] Ziegler and Wilms, *Ann.*, **567**, 31 (1950).
[9] Cope, Dryden, Overberger, and D'Addieco, *J. Am. Chem. Soc.*, **73**, 3416 (1951).
[10] Alder and Dortmann, *Ber.*, **86**, 1544 (1953).
[11] Putney and Soine, *J. Am. Pharm. Assoc.*, **44**, 17 (1955); Parker, Raphael, and Wilkinson, *J. Chem. Soc.*, **1959**, 2433.

PUTRESCINE DIHYDROCHLORIDE *

(1,4-Butanediamine dihydrochloride)

$$(CH_2)_4(CO_2C_2H_5)_2 \xrightarrow{N_2H_4}$$

$$(CH_2)_4(CONHNH_2)_2 \xrightarrow{HNO_2} (CH_2)_4(CON_3)_2$$

$$(CH_2)_4(CON_3)_2 \longrightarrow (CH_2)_4(NCO)_2 + 2N_2$$

$$(CH_2)_4(NCO)_2 \xrightarrow[H_2O]{HCl} (CH_2)_4(NH_2 \cdot HCl)_2$$

Submitted by PETER A. S. SMITH.[1]
Checked by JAMES CASON and WM. DEACETIS.

1. Procedure

A. *Adipyl hydrazide.* A solution of 120 ml. of 85% aqueous hydrazine hydrate (105 g., 2.0 moles) (Note 1) and 25 ml. of absolute ethanol is brought to a gentle boil in a 500-ml. three-necked flask provided by means of ground glass joints with a ball joint-sealed mechanical stirrer (Note 2), a reflux condenser, and a dropping funnel. One hundred and one grams (0.5 mole) of diethyl adipate [2] is added dropwise to the boiling stirred solution at such a rate that a separate liquid phase does not accumulate in the reaction mixture (Note 3). This operation requires 1–2 hours, at the end of which time the contents of the flask will have largely crystallized. The boiling is continued for 5 minutes after the completion of the addition, and the contents of the flask are then cooled to room temperature with running water.

The crystals are washed onto a Büchner funnel with the aid of about 100 ml. of absolute ethanol used in several portions. The precipitate is dried by suction, and the product is washed once with 25 ml. of ether. After drying in air (or at 100°) the adipyl hydrazide weighs 77–80 g. (88–92%) and is of good quality, with a melting point above 170° (Note 4). Concentration of the filtrate to a volume of about 25 ml. in an air stream on a steam bath yields an additional 2.5 g. of nearly pure material; total yield 91–95%.

B. *Putrescine dihydrochloride.* A 2-l. wide-mouthed Erlenmeyer flask containing 200 ml. of concentrated hydrochloric acid and 400 g. of cracked ice is clamped in an ice-salt bath and provided with an efficient stirrer (inefficient stirring may lower the yield). The stirring is started, and 80 g. (0.46 mole) of adipyl hydrazide is added all at once, followed by 500 ml. of ether (Note 5). While the temperature is maintained below 10° (Note 6), a solution of 80 g. (1.15 moles) of sodium nitrite in 150 ml. of water is added over about 30 minutes through a dropping funnel whose stem reaches below the bottom of the ether layer, but not into the path of the stirrer. *The operations in the next paragraph should be conducted without delay after the completion of the addition of the nitrite.*

The cold reaction mixture, which may be freed from suspended solid by rapid filtration through a cotton plug, is transferred to a 2-l. or 3-l. separatory funnel, and the aqueous layer is drawn off into the original reaction flask. The ether layer is poured into a 2-l. Erlenmeyer flask containing about 50 g. of anhydrous calcium chloride, and this flask is placed in the ice bath in which the diazotization was run. The cold aqueous layer is then extracted with two 100-ml. portions of chilled ether, which are combined with the first ether extract. After 5 minutes with occasional swirling, the ether solution of adipyl azide (Note 7) is sufficiently dry, and it is poured into a 2-l. round-bottomed flask containing 350–400 ml. of benzene (Note 8). The calcium chloride is rinsed with a 50-ml. portion of ether, which is added to the same flask.

The ether is distilled gently from a steam bath, preferably through a short fractionating column; nitrogen will be evolved at the same time. When the volume of the contents of the flask has reached about 400 ml., the flask is heated strongly on the steam bath for about 15 minutes to complete the decomposition of the azide (Note 9). The flask is removed from the steam bath, and 200 ml. of concentrated hydrochloric acid is added cautiously to the hot solution (Note 10). The flask is allowed to stand with occasional swirling until the carbon dioxide evolution has ceased (about 15 minutes). The mixture is heated strongly on a steam bath for about 15 minutes more, and the solvents are then distilled from a steam bath under aspirator vacuum (Note 11). When the contents of the flask have become a crystalline paste, the vacuum is temporarily disconnected, and the inner walls of the flask are washed down with about 50 ml. of ethanol. The vacuum is renewed cautiously, and the mixture is distilled to dryness (Note 12).

The residue of crystalline putrescine dihydrochloride is rinsed onto a Büchner funnel with the aid of 100–200 ml. of absolute ethanol used in several portions. The last portions of ethanol are used as wash liquid for the crystals. The crystals are finally pressed dry and washed with 25 ml. of ether. The air-dried product weighs 53–55 g. (72–74%) and melts above 275°. Analysis for chlorine indicates that the salt is anhydrous. Concentration of the filtrate to a volume of about 25 ml. yields an additional 1–2 g.; total yield 73–77%. The entire synthesis may be completed in one day.

2. Notes

1. Eastman Kodak Company practical grade was used.

2. Mechanical stirring is not strictly necessary; however, severe bumping is sometimes encountered if it is not used. Hydrazine attacks both cork and rubber. If a dropping funnel with ground glass joint is not available, an ordinary dropping funnel may be placed at the top of the reflux condenser.

3. If unreacted ester is allowed to accumulate, some secondary hydrazide may be formed, with consequent loss of yield in the next step.

4. The melting point of adipyl hydrazide reported in the literature [3] is 171°; however, values as high as 179° have been observed. The checkers observed values of 174–177° and 179–181° for

products of two runs, when the crystals were placed on a hot stage after the temperature had reached 160°.

5. Ethanol-free ether should be used. Dropwise addition of 50 ml. of concentrated sulfuric acid to 1 l. of ether, and distillation of the ether from a steam bath gives a satisfactory solvent.

6. To maintain this temperature at the rate of addition of the nitrite indicated will probably require the occasional addition of cracked ice to the reaction mixture. Only occasional checking of the temperature is necessary. If the temperature is kept below 0°, slightly better yields are obtained. The addition of Dry Ice directly to the ether layer accomplishes this easily. Temperatures above 10° cause a loss in yield, but apparently create no hazard.

7. Ethereal solutions of adipyl azide are quite safe, but the free azide is somewhat explosive and should not be isolated. If storage of an intermediate is desired, the azide should be converted to the urethane by the procedure given below. The urethane is quite stable to storage; also, the procedure *via* the urethane gives improved yields in some amine syntheses.

The dried ethereal solution of adipyl azide is added to a 2-l. flask containing 400 ml. of absolute ethanol, and the ether and some of the ethanol are distilled on a steam bath through a short fractionating column. When the volume has reached about 200 ml., the solution may be poured into a 500-ml. Erlenmeyer flask, and the remainder of the solvents removed on a steam bath by means of an air stream. The residual, waxy, crystalline cake of N,N'-dicarbethoxyputrescine weighs 84–91 g. (79–85%) and melts at 76–81°. It may be kept indefinitely.

If the urethane is to be hydrolyzed immediately, there is no need to isolate it. Instead, after distillation of most of the ether and alcohol from the flask in which the urethane is formed, 700 ml. of concentrated hydrochloric acid is added, and the mixture is heated under reflux for 4 hours. The solution is then distilled from a steam bath under aspirator vacuum (Note 11), and the putrescine dihydrochloride is isolated as described above. The total yield is 55–57 g. (74–77%).

The isolated urethane may be hydrolyzed by this procedure at any time. For the hydrolysis step, the yield is 89–91%.

8. Commercial benzene may be used.

9. The rearrangement of the azide is complete when nitrogen evolution ceases. This is usually concluded in the time indicated, but occasionally takes longer.

10. Carbon dioxide is evolved copiously at this point and may cause the solution to foam over if the acid is not added cautiously.

11. Alternatively, the solution may be poured into a 1-l. flask for this operation.

12. Solid putrescine dihydrochloride should not be heated unduly long at 100°, as it may turn pink.

3. Methods of Preparation

Putrescine dihydrochloride has been prepared by the Hofmann degradation of adipamide;[3-5] by the Curtius degradation of adipyl hydrazide through the urethane;[6] by the Curtius degradation of adipyl azide obtained from adipyl chloride and sodium azide;[7] by the Schmidt degradation of adipic acid with hydrogen azide;[8] by the reduction of succinonitrile,[9] succinaldoxime,[10, 11] or γ-phthalimidobutyronitrile[12, 13] with sodium; from N-benzoyl-γ-iodobutylamine;[14] and by the hydrolysis of 1,4-di-(N-phthalimido)butane.[15]

[1] University of Michigan, Ann Arbor, Michigan.

[2] *Org. Syntheses Coll. Vol.* 2, 264 (1943).

[3] von Braun and Lemke, *Ber.*, **55B**, 3529 (1922).

[4] Farbenfabriken F. Bayer and Company, Ger. pat. 232,072 [*Frdl.*, **10**, 106 (1910–1912)].

[5] Vogel, *J. Chem. Soc.*, **1929**, 1489.

[6] Curtius and Darmstaedter, *J. prakt. Chem.*, **91**, 11 (1915).

[7] Naegeli and Lendorff, *Helv. Chim. Acta*, **15**, 49 (1932).

[8] von Braun and Pinkernelle, *Ber.*, **67B**, 1059 (1934).

[9] Ladenburg, *Ber.*, **19**, 780 (1886).

[10] Ciamician and Zanetti, *Ber.*, **22**, 1970 (1889).

[11] Willstätter and Heubner, *Ber.*, **40**, 3871 (1907).

[12] Kanevskaya, *Zhur. Russ. Fiz. Khim. Obshchestva*, **59**, 646 (1927) [*Chem. Zentr.*, **1928**, I, 1026].

[13] Keil, *Ber.*, **59B**, 2816 (1926).

[14] Dudley and Thorpe, *Biochem. J.*, **19**, 845 (1925).

[15] Vassel (to International Minerals & Chemical Corp.), U. S. pat. 2,757,198 [*C. A.*, **51**, 2024 (1957)].

2,3-PYRAZINEDICARBOXYLIC ACID *

$$\begin{matrix} HC{=}O \\ | \\ HC{=}O \end{matrix} \cdot 2NaHSO_3 + \begin{matrix} H_2N \\ H_2N \end{matrix} \bigcirc \longrightarrow$$

$$\xrightarrow{KMnO_4}$$

$$\xrightarrow{HCl}$$

Submitted by REUBEN G. JONES and KEITH C. McLAUGHLIN.[1]
Checked by H. R. SNYDER and R. E. HECKERT.

1. Procedure

A. *Quinoxaline.* One hundred thirty-five grams (1.25 moles) of o-phenylenediamine [2] is dissolved in 2 l. of water, and the solution is heated to 70°. With stirring, a solution of 344 g. (1.29 moles) of glyoxal-sodium bisulfite [3] (Note 1) in 1.5 l. of hot water (about 80°) is added to the o-phenylenediamine solution. The mixture is allowed to stand for 15 minutes and then is cooled to about room temperature and 500 g. of sodium carbonate monohydrate (Note 2) is added. The quinoxaline separates as an oil or as a crystalline solid if the mixture is sufficiently cool. The mixture is extracted with three 300-ml. portions of ether. The combined extracts are dried over anhydrous magnesium sulfate or sodium sulfate, filtered, and concentrated on the steam bath. The residual liquid, consisting of almost pure quinoxaline, is distilled under reduced pressure, and the fraction boiling at 108–111°/12 mm. (m.p. 29–30°) is collected. It weighs 138–147 g. (85–90%) (Note 3).

B. *2,3-Pyrazinedicarboxylic acid.* A 12-l. three-necked flask is provided with an efficient mechanical stirrer, a reflux condenser, and a 1-l. dropping funnel. In the flask are placed 4 l. of hot (about 90°) water and 145 g. (1.12 moles) of quinoxaline. With rapid stirring a saturated aqueous solution of 1050 g. (6.6 moles) of potassium permanganate (Note 4) is added through the dropping funnel in a thin stream. The rate of addition of the permanganate solution is adjusted so that the reaction mixture boils gently. The addition requires about 1.5 hours.

The reaction mixture is cooled somewhat (Note 5) and filtered through a large Büchner funnel. The manganese dioxide cake is removed from the funnel and stirred to a smooth paste with 1 l. of fresh water. The slurry is filtered, and the washing is repeated. The total filtrate, about 10 l., is evaporated under reduced pressure on the steam bath to a volume of approximately 3 l. (Note 6). The solution is swirled or stirred gently while 550 ml. (6.6 moles) of 36% hydrochloric acid is cautiously added (Note 7). Evaporation under reduced pressure is then continued until a moist cake of solid potassium chloride and 2,3-pyrazinedicarboxylic acid remains in the flask (Note 8).

The moist cake is scraped from the flask and allowed to dry in a 16-in. porcelain evaporating dish until the odor of hydrochloric acid is faint (about 24 hours, but this time can be reduced to about 8 hours if a gentle stream of air is directed onto the solid). The solid material is returned to the dried flask and mixed thoroughly with about 200 ml. of water. Two liters of acetone is added, and the mixture is boiled under reflux for 15 minutes, then cooled to room temperature and filtered through a 6-in. Büchner funnel. The solid on the filter is returned to the flask, treated with 100 ml. of water, and extracted with 1 l. of boiling acetone as before (Note 9). The acetone filtrates are combined and distilled from a steam bath, finally under diminished pressure.

The solid in the flask is dissolved by refluxing with 2.5 l. of acetone. The mixture is cooled slightly, treated with 10 g. of decolorizing carbon, refluxed for an additional 5-minute period, and filtered hot. Evaporation of the pale yellow acetone filtrate leaves the acid as a light-tan crystalline solid. If the product still possesses an odor of hydrochloric acid it is dried in a vacuum desiccator over sodium hydroxide pellets for a few hours. Finally the product is dried in an oven at 100° for several hours

(Notes 10 and 11). The yield of material melting at 165–167° (dec.) is 140–145 g. (75–77%) (Note 12).

If a purer product is desired the material may be recrystallized (with about 17% loss) from 150 ml. of water; the hot solution is decolorized with carbon. The melting point after drying at 110° for several hours is then 183–185° (dec.) (Notes 11 and 13).

2. Notes

1. In the absence of sodium bisulfite, aqueous glyoxal solutions react with *o*-phenylenediamine to give only about 30% yields of quinoxaline together with large quantities of resinous by-products. Therefore, if an aqueous glyoxal solution is to be used in this preparation it should be mixed with a water solution of two molar equivalents of sodium bisulfite before it is added to the *o*-phenylenediamine solution.

2. Potassium carbonate or sodium hydroxide may also be used.

3. 2-Methylquinoxaline may be prepared in 88–92% yields from *o*-phenylenediamine and pyruvic aldehyde-sodium bisulfite by this same procedure.

4. The volume of this solution can be held to the minimum, about 3–5 l., by dissolving the permanganate in hot water (90–100°).

5. The only reason for cooling is to make the flask easier to handle during the filtration.

6. A 5-l. round-bottomed flask fitted with a separatory funnel for continuous addition of the solution during the concentration is convenient.

7. There is a vigorous evolution of carbon dioxide, and unless the acid is added slowly the contents of the flask may foam over.

8. Excess hydrochloric acid is present, and the 2,3-pyrazine-dicarboxylic acid tends to darken and decompose if it is heated too strongly or for too long a time.

9. The ease with which the dicarboxylic acid is removed from the potassium chloride depends upon the amount of water present. The potassium chloride should be set aside for an additional extraction if the yield of the crude product is low.

10. Drying at 100° converts any of the hydrated 2,3-pyrazine-dicarboxylic acids to the anhydrous form.

11. The product darkens somewhat on heating.

12. 2-Methyl-5,6-pyrazinedicarboxylic acid may be prepared in 70–75% yields from 2-methylquinoxaline by this same procedure. The crude acid (m.p. 155–160°, dec.) is somewhat unstable at elevated temperatures. It should not be heated above 100° for long periods of time. In order to obtain pure 2-methyl-5,6-pyrazinedicarboxylic acid (m.p. 175°, dec.) the crude product is best recrystallized from acetone.

13. The melting point is dependent on the rate of heating and the temperature of the bath at the time the sample is inserted. The figure given was obtained by insertion of the sample tube into a bath at 160°.

3. Methods of Preparation

Quinoxaline has been prepared by the reaction of glyoxal with o-phenylenediamine,[4] and 2-methylquinoxaline by the reaction of pyruvic aldehyde [5] or isonitrosoacetone [6] with o-phenylenediamine. 2,3-Pyrazinedicarboxylic acid has been prepared by the permanganate [7] or electrolytic [8] oxidation of quinoxaline. 2-Methyl-5,6-pyrazinecarboxylic acid has been obtained from 2-methylquinoxaline in the same way.[6,9] Also, the monopotassium salt of pyrazine-2,3-dicarboxylic acid has been prepared by the oxidation of quinoxaline with potassium permanganate.[10]

[1] Lilly Research Laboratories, Indianapolis, Indiana.

[2] Org. Syntheses Coll. Vol. 2, 501 (1943).

[3] Org. Syntheses Coll. Vol. 3, 438 (1955).

[4] Billman and Rendall, J. Am. Chem. Soc., 66, 540 (1944).

[5] Fischer and Taube, Ber., 57, 1502 (1924).

[6] Böttcher, Ber., 46, 3084 (1913).

[7] Gabriel and Sonn, Ber., 40, 4850 (1907); Sausville and Spoerri, J. Am. Chem. Soc., 63, 3153 (1941); Mager and Berends, Rec. trav. chim., 78, 109 (1959); Solomons and Spoerri, J. Am. Chem. Soc., 75, 679 (1953).

[8] Kimura, Yamada, Yoshizue, and Nagoya, Yakugaku Zasshi, 77, 891 (1957) [C. A., 52, 1181 (1958)].

[9] Leonard and Spoerri, J. Am. Chem. Soc., 68, 526 (1946).

[10] Rees and Dorf (to American Cyanamid Company), U. S. pat. 2,723,974 [C. A., 50, 10136 (1956)].

PYRIDINE-N-OXIDE *
(Pyridine-1-oxide)

$$\text{pyridine} + CH_3CO_3H \rightarrow \text{pyridine-N-oxide} \cdot CH_3CO_2H$$

$$\text{pyridine-N-oxide} \cdot CH_3CO_2H + HCl \rightarrow \text{pyridine-N-oxide} \cdot HCl + CH_3CO_2H$$

$$\text{pyridine-N-oxide} \cdot CH_3CO_2H \rightarrow \text{pyridine-N-oxide} + CH_3CO_2H$$

Submitted by HARRY S. MOSHER, LESLIE TURNER, and ALLAN CARLSMITH.[1]
Checked by N. J. LEONARD and E. D. SUTORIS.

1. Procedure

In a 1-l. three-necked flask equipped with a stirrer (Note 1), a thermometer, and a dropping funnel is placed 110 g. (1.39 moles) of pyridine. The pyridine is stirred, and 250 ml. (285 g., 1.50 moles) of 40% peracetic acid (Note 2) is added at such a rate that the temperature reaches 85° and is maintained there. After the addition, which requires 50–60 minutes, the mixture is stirred until the temperature drops to 40°.

A. *Pyridine-N-oxide hydrochloride.* The acetate is converted to the hydrochloride by bubbling a slight excess over the theoretical amount (51 g.) of gaseous hydrogen chloride into the reaction mixture by way of a 7-mm. gas inlet tube which replaces the dropping funnel in the reaction flask. The acetic acid and excess peracetic acid are removed by warming on the steam bath under vacuum (Note 3). The residual pyridine-N-oxide hydrochloride is purified by heating under reflux for 30 minutes with 300 ml. of isopropyl alcohol, cooling to room temperature, and filtering. The colorless crystals are washed with 50 ml. of isopropyl alcohol followed by 50 ml. of ether. The yield is 139–152 g. (76–83%) (Note 4), m.p. 179.5–181°.

B. *Pyridine-N-oxide.* The acetic acid solution is evaporated on the steam bath under the pressure of a water aspirator, and the residue (180–190 g.) is distilled at a pressure of 1 mm. or less in an apparatus suitable for collecting a solid distillate (Note 5). The vacuum pump must be protected with a Dry Ice trap capable of holding about 60 ml. of acetic acid, which distils as the pyridine-N-oxide acetate dissociates at low pressure. Heat is provided by an oil bath, the temperature of which is not allowed to rise above 130° (Note 6). The product is collected at 100–105°/1 mm. (95–98°/0.5 mm.). The yield is 103–110 g. (78–83%) of colorless solid, m.p. 65–66° (sealed capillary). The base is deliquescent and must be stoppered immediately.

2. Notes

1. A convenient seal for stirring under vacuum (see Note 3) is made by running an 8-mm. glass rod, with propeller or paddle stirrer at the end, through the outside member of an 18/9 spherical joint which is inserted into a suitable rubber stopper. The inner member of the 18/9 spherical joint is then slipped over the stirrer and held in place with a piece of rubber tubing. This rotating seal may then be lubricated with a drop of oil. Alternatively, one may use a Trubore stirring system.

2. Becco peracetic acid (40%) was used. The composition and properties of this commercial preparation are described fully in *Bulletin* 4 of the Buffalo Electro-Chemical Company, Buffalo, New York. The manufacturer's recommendations for storing and handling should be followed. Experiments using proportionate amounts of 10% or 20% peracetic acid in acetic acid were equally successful. The strength of the peracetic acid, as well as the progress of the reaction, can be determined iodimetrically.[2]

3. The vacuum evaporation proceeds much more smoothly and rapidly if the mixture is stirred mechanically during the process.

4. The submitters report that the same procedure is successful with four times the amounts given here. With the increased amounts, a water bath is used for cooling during the initial addition, which then requires about 45 minutes.

5. *Caution! Before distillation, absence of peroxide should be established by test with potassium iodide.*

The apparatus for distillation of solids in vacuum described in *Organic Syntheses* [3] is satisfactory, as is a combination of standard taper flasks, short column, and adaptors.

6. It is imperative that the pressure be maintained at 1 mm. or lower. Decomposition is usually extensive at higher pressures; however, the removal of the acetic acid may be initiated at 5–10 mm. pressure. The oil-bath temperature must not exceed 130° if decomposition is to be avoided. A fore-run of 15–20 g., b.p. 90–98°/0.5 mm., can be saved and redistilled in combination with similar cuts from successive runs. About 9–10 g. (7%) of additional crystalline pyridine-N-oxide is obtained per run in this manner.

3. Methods of Preparation

Pyridine-N-oxide has been prepared by oxidation of pyridine with perbenzoic acid,[4] with monoperphthalic acid,[5] with peracetic acid (hydrogen peroxide and acetic acid),[6,7] and with hydrogen peroxide and other carboxylic acids.[7]

[1] Stanford University, Stanford, California.

[2] Smit, *Rec. trav. chim.*, **49**, 691 (1930).

[3] *Org. Syntheses Coll. Vol.* **3**, 133 (1955).

[4] Meisenheimer, *Ber.*, **59**, 1848 (1926).

[5] Bobranski, Kochanska, and Kowalewska, *Ber.*, **71**, 2385 (1938).

[6] Ochiai, Ishikawa, and Zai-Ren, *J. Pharm. Soc. Japan*, **64**, 73 (1944) [*C. A.*, **45**, 8526h (1951)]; Hertog and Combé, *Rec. trav., chim.*, **70**, 581 (1951).

[7] Ochiai, Katada, and Hayashi, *J. Pharm. Soc. Japan*, **67**, 33 (1947) [*C. A.*, **45**, 9541i (1951)]; Ochiai, *J. Org. Chem.*, **18**, 534 (1953).

2-PYRROLEALDEHYDE *

(Pyrrole-2-carboxaldehyde)

$(CH_3)_2NCHO + POCl_3 \rightarrow [(CH_3)_2NCHO \cdot POCl_3]$

$+ CH_3CO_2NH_2(CH_3)_2 + 3CH_3CO_2H$

Submitted by ROBERT M. SILVERSTEIN, EDWARD E. RYSKIEWICZ, and
CONSTANCE WILLARD.[1]
Checked by MAX TISHLER and GEORGE PURDUE.

1. Procedure

In a 3-l. three-necked round-bottomed flask, fitted with a
sealed stirrer, a dropping funnel, and a reflux condenser, is placed
80 g. (1.1 moles) of dimethylformamide (Note 1). The flask is
immersed in an ice bath, and the internal temperature is maintained at 10–20°, while 169 g. (1.1 moles) of phosphorus oxychloride is added through the dropping funnel over a period of
15 minutes. An exothermic reaction occurs with the formation
of the phosphorus oxychloride-dimethylformamide complex.
The ice bath is removed, and the mixture is stirred for 15 minutes (Note 2).

The ice bath is replaced, and 250 ml. of ethylene dichloride is
added to the mixture. When the internal temperature has been
lowered to 5°, a solution of 67 g. (1.0 mole) of freshly distilled
pyrrole in 250 ml. of ethylene dichloride is added through a clean

dropping funnel to the stirred, cooled mixture over a period of 1 hour. After the addition is complete, the ice bath is replaced with a heating mantle, and the mixture is stirred at the reflux temperature for 15 minutes, during which time there is copious evolution of hydrogen chloride.

The mixture is then cooled to 25–30°, and to it is added through the dropping funnel a solution of 750 g. (5.5 moles) of sodium acetate trihydrate (Note 3) in about 1 l. of water, cautiously at first, then as rapidly as possible. The reaction mixture is again refluxed for 15 minutes, vigorous stirring being maintained all the while (Note 4).

The cooled mixture is transferred to a 3-l. separatory funnel, and the ethylene dichloride layer is removed. The aqueous phase is extracted three times with a total of about 500 ml. of ether. The ether and ethylene chloride solutions are combined and washed with three 100-ml. portions of saturated aqueous sodium carbonate solution, which is added cautiously at first to avoid too rapid evolution of carbon dioxide. The non-aqueous solution is then dried over anhydrous sodium carbonate, the solvents are distilled, and the remaining liquid is transferred to a Claisen flask and distilled from an oil bath under reduced pressure (Note 5). The aldehyde boils at 78° at 2 mm.; there is very little fore-run and very little residue. The yield of crude 2-pyrrolealdehyde is 85–90 g. (89–95%), as an almost water-white liquid which soon crystallizes. A sample dried on a clay plate melts at 35–40°. The crude product is purified by dissolving in boiling petroleum ether (b.p. 40–60°), in the ratio of 1 g. of crude 2-pyrrolealdehyde to 25 ml. of solvent, and cooling the solution slowly to room temperature, followed by refrigeration for a few hours. The pure aldehyde is obtained from the crude in approximately 85% recovery. The over-all yield from pyrrole is 78–79% of pure 2-pyrrolealdehyde, m.p. 44–45°.

2. Notes

1. The dimethylformamide is available as technical grade DMF from the Grasselli Chemicals Department of E. I. duPont de Nemours and Company, Wilmington, Delaware.

2. If the ice bath is not removed, the mixture may solidify

and must be dissolved by adding solvent and heating slightly. Mixing of the reactants at ice-bath temperature prevents discoloration. Practical grades of materials were used.

3. The use of sufficient sodium acetate is essential. If the acidic reaction products are not neutralized, the yield drops to as low as 15–20% of badly discolored product which cannot be readily purified.

4. Efficient stirring must be maintained to keep the two phases in close contact. Hydrolysis is not complete if the mixture is not heated.

5. The use of a wide-bore condenser and a simple receiver, without a stopcock, is preferable. Usually the product does not solidify at once, but occasionally it crystallizes during distillation. The use of a fraction cutter is not necessary or advisable.

3. Methods of Preparation

2-Pyrrolealdehyde has been prepared from pyrrole, chloroform, and potassium hydroxide;[2] from pyrrolemagnesium iodide and ethyl, propyl, or isoamyl formate;[3] from 2-furaldehyde dimethyl acetal and ammonium acetate;[4] and, by the method here described, from pyrrole, phosphorus oxychloride, and dimethylformamide.[5] Smith[5] has suggested a possible intermediate in this process. The method has also been applied to substituted pyrroles[6] and is similar to that described on p. 331 for the preparation of p-dimethylaminobenzaldehyde from dimethylaniline.

[1] Stanford Research Institute, Stanford, California.

[2] Bamberger and Djierdjian, *Ber.*, **33**, 536 (1900); Fischer, Beller, and Stern, *Ber.*, **61B**, 1074 (1928).

[3] Tschelinzeff and Terentjeff, *Ber.*, **47**, 2653 (1914); Putochin, *Zhur. Russ. Fiz. Khim. Obshchestva*, **59**, 809 (1927); Putochin, *Ber.*, **59B**, 1993 (1926).

[4] Elming and Clauson-Kaas, *Acta Chem. Scand.*, **6**, 867 (1952).

[5] Silverstein, Ryskiewicz, and Chaikin, *J. Am. Chem. Soc.*, **76**, 4485 (1954); Smith, *J. Chem. Soc.*, **1954**, 3842.

[6] Ryskiewicz and Silverstein, *J. Am. Chem. Soc.*, **76**, 5802 (1954); Chu and Chu, *J. Org. Chem.*, **19**, 266 (1954); Silverstein, Ryskiewicz, Willard, and Koehler, *J. Org. Chem.*, **20**, 668 (1955).

2-(1-PYRROLIDYL)PROPANOL *

(1–Pyrrolidineëthanol, β-methyl-)

$$2 \quad \boxed{} NCHCO_2C_2H_5 + LiAlH_4 \rightarrow$$
$$\quad\quad\quad \underset{CH_3}{|}$$

$$\left[\boxed{} NCHCH_2O \atop \underset{CH_3}{|} \right]_2 (C_2H_5O)_2AlLi$$

$$\left[\boxed{} NCHCH_2O \atop \underset{CH_3}{|} \right]_2 (C_2H_5O)_2AlLi + 6HCl \rightarrow$$

$$2 \quad \boxed{} NCHCH_2OH \cdot HCl + 2C_2H_5OH + AlCl_3 + LiCl$$
$$\quad\quad\quad \underset{CH_3}{|}$$

$$\boxed{} NCHCH_2OH \cdot HCl + NaOH \rightarrow$$
$$\quad \underset{CH_3}{|}$$

$$\boxed{} NCHCH_2OH + NaCl + H_2O$$
$$\quad\quad\quad \underset{CH_3}{|}$$

Submitted by ROBERT BRUCE MOFFETT.[1]
Checked by N. J. LEONARD and S. GELFAND.

1. Procedure

A 2-l. three-necked round-bottomed flask in an electric heating mantle is fitted with a mercury-sealed Hershberg stirrer, a dropping funnel, and an efficient reflux condenser topped with a tube containing soda lime and calcium chloride. In this flask are placed 21.3 g. (0.56 mole) of pulverized lithium aluminum hydride (Note 1) and 300 ml. of dry ether. The mixture is heated under reflux until most of the hydride has dissolved. A solution of 157.3 g. (0.92 mole) of ethyl α-(1-pyrrolidyl)propionate (p.466) in 200 ml. of dry ether is then added slowly with vigorous stirring at such a rate that the solvent refluxes gently. When the addition is complete and the initial reaction subsides, the mixture is stirred at the reflux temperature an additional 30 minutes. The excess lithium aluminum hydride is decomposed by adding 50 ml. of ethyl acetate slowly with stirring. This is followed by 600 ml. of 6N hydrochloric acid, added slowly with vigorous stirring.

The mixture is transferred to a separatory funnel. The water layer is separated, washed once with ether, and made strongly alkaline by the addition of 1 l. of 6N sodium hydroxide (Note 2). The mixture is returned to the original 2-l. three-necked flask. An attachment for continuous ether extraction is placed in one of the side necks,[2] and 6-in. extension columns are placed in the other two necks. The stirrer is fitted through the center column and neck. Ether is added through the column attached to the side neck until the proper ether level is attained. A stopper is then placed on this column. The stirrer is run at such a rate that gentle swirling is accomplished without hindering the separation of the ether layer, and the continuous extraction with ether is continued until test with pH paper indicates that the ether coming over contains no more basic material. The ether solution of the product is thoroughly dried over anhydrous potassium carbonate (Note 3). The drying agent is removed by filtration, and the solvent by distillation at atmospheric pressure and finally at water-pump pressure. The residue is distilled at reduced pressure through a short fractionating column; b.p. 95–96°/20 mm. (109–110°/40 mm., 80°/11 mm.); n_D^{20} 1.4780; n_D^{25} 1.4758; d_4^{25} 0.9733. The yield is 95–106 g. (80–90%).

2. Notes

1. The hydride can be pulverized rapidly and safely by breaking the large pieces with a spatula, followed by careful crushing with a mortar and pestle. Caution must be observed because the solid may inflame on prolonged grinding or abrasion. The hydride dust is caustic and irritating.

2. At this point the mixture should be a mobile milky slurry. The submitter suggests that, if the mixture is too thick to extract, more water or base may be added.

3. Sufficient drying agent should be used so that no aqueous liquid phase appears.

3. Methods of Preparation

2-(1-Pyrrolidyl)propanol has been prepared by the lithium aluminum hydride reduction of ethyl α-(1-pyrrolidyl)propionate.[3]

[1] The Upjohn Company, Kalamazoo, Michigan.
[2] *Org. Syntheses Coll. Vol.* **1**, 277 (1941).
[3] Moffett, *J. Org. Chem.*, **14**, 862 (1949).

QUINACETOPHENONE MONOMETHYL ETHER

(Acetophenone, 2'-hydroxy-5'-methoxy-)

Submitted by G. N. Vyas and N. M. Shah.[1]
Checked by William S. Johnson and R. T. Keller.

1. Procedure

In a 1-l. round-bottomed flask fitted with a reflux condenser and a calcium chloride guard tube are placed 30.0 g. (0.197 mole) of quinacetophenone [2] (Note 1) and 300 ml. of acetone (Note 2). The mixture is warmed on a steam bath to dissolve the quinacetophenone. The resulting greenish solution is cooled to room temperature under tap water, and 28 g. (0.20 mole) of anhydrous potassium carbonate is added followed by 42 g. (0.295 mole) of methyl iodide. The mixture is allowed to reflux on a water bath at 60–70° for about 6 hours (Note 3).

As much of the acetone as possible is removed by distillation on the water bath (Note 4), and the residual dark-colored liquid is cooled and acidified with 2N sulfuric acid with cooling under the water tap. The resulting mixture is steam-distilled until no oily drops are seen collecting in the condenser. The distillate, which amounts to about 2.5 l., is allowed to stand overnight at room temperature, and the greenish crystals are separated by suction filtration, washed twice with cold water, and air-dried.

The yield of quinacetophenone monomethyl ether, m.p. 48–50°, is 18–21 g. (55–64%) (Note 5).

The brown solution remaining in the distilling flask is filtered while hot; on being cooled it gives 6–7 g. of brownish needles of crude quinacetophenone.

2. Notes

1. The quinacetophenone [2] is dried in an oven at 100–110° for 2–3 hours.

2. The acetone is dried over anhydrous potassium carbonate and distilled.

3. The temperature of the water bath should not exceed 70°; otherwise serious bumping may occur.

4. The recovered acetone may be reused for another methylation.

5. The submitters state that the *dimethyl* ether of quinacetophenone is conveniently prepared by the following procedure: In a 1-l. round-bottomed flask fitted with a reflux condenser 60 g. (0.39 mole) of quinacetophenone [2] is dissolved in 300 ml. of ethanol by heating. The source of heat is then removed, and to the hot solution are alternately added in five installments with shaking a solution of sodium hydroxide (40 g. in 100 ml. of water) and dimethyl sulfate (120 g.). The heat evolved during the reaction makes the solution boil. After the addition is complete (about 20 minutes), the reaction mixture is made alkaline by the further addition of 10 g. of sodium hydroxide in 20 ml. of water and is allowed to reflux on the water bath for 3 hours. The dark mixture is distilled to remove most of the ethanol, and the residual liquid in the flask is steam-distilled. The distillate, which amounts to about 2.5 l., is cooled in an ice bath and saturated with sodium chloride, whereupon a thick oil settles to the bottom. Most of the aqueous layer is decanted, and the remaining oil is extracted with ether and dried over calcium chloride. The ether is removed by distillation, and the residue is fractionated at reduced pressure to give 50–52 g. (71–74% yield) of material boiling at 152–156°/15 mm., m.p. 20–22°.

3. Methods of Preparation

Quinacetophenone monomethyl ether has been prepared by the methylation of quinacetophenone with dimethyl sulfate and alkali,[3] by the partial demethylation of quinacetophenone dimethyl ether;[4] and by the acetylation of hydroquinone dimethyl ether with acetyl chloride in the presence of aluminum chloride.[5] It also has been obtained as a by-product in the preparation of quinacetophenone dimethyl ether.[6]

[1] M. R. Science Institute, Gujarat College, Ahmedabad, India.
[2] Org. Syntheses Coll. Vol. 3, 280 (1955).
[3] Kostanecki and Lampe, Ber., 37, 774 (1904).
[4] Baker, Brown, and Scott, J. Chem. Soc., 1939, 1926.
[5] Oliverio and Lugli, Gazz. chim. ital., 78, 16 (1948); Wiley, J. Am. Chem. Soc., 73, 4205 (1951).
[6] Kauffmann and Beisswenger, Ber., 38, 792 (1905).

SEBACIL *

(1,2-Cyclodecanedione)

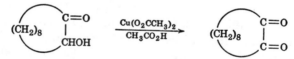

Submitted by A. T. Blomquist and Albert Goldstein.[1]
Checked by N. J. Leonard and J. C. Little.

1. Procedure

Fifty-one grams (0.30 mole) of sebacoin (Note 1), 25 ml. of methanol, 120 g. of cupric acetate monohydrate (0.60 mole), and 300 ml. of 50% aqueous acetic acid are mixed in a 1-l. flask equipped with an efficient mechanical stirrer and a reflux condenser. The mixture is heated over a free flame until refluxing occurs. The color of the mixture changes from blue to red at approximately 75°. Refluxing is continued for 1 minute. The mixture is then allowed to cool, with stirring, to about 40° (Note 2). The mixture is filtered through filter aid ("Celite") on a sintered glass funnel to remove cuprous oxide, and the filtrate is transferred to a 2-l. separatory funnel (Note 3). Saturated aqueous sodium chloride solution (310 ml.) is added to the filtrate, which is then extracted with three 150-ml. portions of ether. The combined ether extracts are washed with three 250-ml. portions of saturated salt solution, four 250-ml. portions of 5% sodium bicarbonate solution (foaming!), and once again with 250 ml. of saturated salt solution. The ether solution is then dried over 20 g. of anhydrous sodium or magnesium sulfate. The ether is removed by distillation at atmospheric pressure, and the residue is transferred to a 100-ml. flask and distilled under vacuum. The yield of sebacil is 44.4–45.1 g. (88–89%), b.p. 104–106°/10 mm. (Note 4).

2. Notes

1. The sebacoin-sebacil mixture obtained from the sebacoin preparation (p. 840) was used.

2. The sebacil may tend to crystallize during filtration if the reaction mixture is too cool.

3. The "Celite"-cuprous oxide mixture is extracted with three 50-ml. portions of ether, and the combined extracts are used as the first ether portion for the extraction of the sebacil-containing aqueous solution.

4. 1,2-Cyclononanedione has been prepared from 2-hydroxy-cyclononanone in 67–72% yield by this procedure.[2]

3. Methods of Preparation

1,2-Cyclodecanedione has also been prepared by oxidation of sebacoin with chromium trioxide in acetic acid.[3,4] Cupric acetate in acetic acid has been used for oxidation of an α-hydroxyketone by Ruggli and Zeller.[5]

[1] Cornell University, Ithaca, New York.
[2] Blomquist, Liu, and Bohrer, *J. Am. Chem. Soc.*, **74**, 3643 (1952).
[3] Blomquist, Burge, and Sucsy, *J. Am. Chem. Soc.*, **74**, 3636 (1952).
[4] Prelog, Schenker, and Günthard, *Helv. Chim. Acta*, **35**, 1610 (1952).
[5] Ruggli and Zeller, *Helv. Chim. Acta*, **28**, 741 (1945).

SEBACOIN*

(Cyclodecanone, 2-hydroxy-)

$$CH_3O_2C(CH_2)_8CO_2CH_3 \ + \ 4Na \ \longrightarrow \ \left[(CH_2)_8 \underset{C-ONa}{\overset{C-ONa}{\|}} \right] \ + \ 2NaOCH_3$$

$$\left[(CH_2)_8 \underset{C-ONa}{\overset{C-ONa}{\|}} \right] \ \xrightarrow{CH_3CO_2H} \ (CH_2)_8 \underset{CHOH}{\overset{C=O}{|}}$$

Submitted by Norman L. Allinger.[1]
Checked by N. J. Leonard, J. C. Little, and F. H. Owens.

1. Procedure

The apparatus[2] consists of a 3-l. three-necked round-bottomed creased flask, with standard ball joints and an indented cone-shaped bottom (Note 1), which is heated by means of an electric mantle and is equipped with a high-speed stirrer of stainless steel driven by a 10,000 r.p.m. motor (Note 2). One side neck is fitted with a bulb-type air-cooled condenser (Note 3), on top of which fits a 1-l. pressure-equalizing Hershberg dropping funnel (Note 4). The top of the dropping funnel is connected in turn to a U-tube containing a 1-cm. head of mercury. The entire apparatus is securely fastened to a sturdy support.

In the flask is placed 900 ml. of xylene (Note 5), and a slow stream of purified nitrogen (Note 6) is passed through the system from which the dropping funnel has been temporarily removed. The stirrer is run at slow speed, and the solvent is brought to a gentle reflux. The air stream cooling the condenser is shut off, the mercury valve is disconnected from the condenser, and a few milliliters of solvent is allowed to distil out the top of the condenser (Note 7). The dropping funnel (Note 8), containing a solution of 115 g. (0.50 mole) of dimethyl sebacate (Note 9) and 500 ml. of xylene (Note 5) is then inserted between the top of the condenser and the mercury valve. The air to the condenser is then turned on, and the electric mantle is turned off.

The solvent is allowed to cool below its boiling point, and the stirrer is gradually brought to a stop. Throughout these operations the nitrogen flow is adjusted to keep air out of the system. Through a side neck is then added 50.6 g. (2.20 g. atoms) of crust-free sodium metal cut into lumps of convenient size. The side neck is closed, and the stirrer and the heater are turned on. The sodium is dispersed by stirring at about 6000–8000 r.p.m. for 10 minutes, and, with continued heating and stirring at a rate somewhat slower (to give suitable mixing), the dropwise addition of the ester solution is begun at such a rate as to be complete in about 24 hours (Notes 10 and 11).

Heating and stirring are continued for 1 hour after the addition is completed. The stirrer is then slowed, heating is stopped, the heater is removed, and the reaction flask is allowed to cool for about 15 minutes (Note 12). The reaction flask is then cooled in a water bath, and is finally thoroughly cooled in an ice bath. A solution of 140 ml. of glacial acetic acid in an equal volume of xylene is then added dropwise during about 30 minutes with continued cooling and stirring (Note 13). After addition of 450 ml. of water, stirring is stopped, the nitrogen is turned off, and the flask is disconnected from the apparatus. The two-phase mixture is filtered through a large Büchner funnel with suction to remove a small amount of gum, and the filtrate is then poured into a 3-l. separatory funnel. The aqueous phase is drawn off, extracted with 100 ml. of xylene, and is discarded. The xylene phases are washed in series with 100 ml. of water, and are combined and dried with 10 g. of anhydrous magnesium sulfate. The solution is filtered into a 3-l. round-bottomed flask, and the bulk of the xylene is distilled with the aid of an aspirator (Notes 14 and 15). The residue is transferred to a smaller flask and is distilled through a 2-ft. Vigreux column, the fraction boiling at 134–138°/14 mm. or 124–128°/9 mm. being collected as a yellowish liquid weighing 57–63 g. (67–74%). This material solidifies on standing and is sufficiently pure for most purposes (Notes 16 and 18). For further purification it may be crystallized from 150 ml. of pentane by cooling to −10° in an ice-salt bath for several hours. The mixture is filtered, and the crystals are washed with 50 ml. of pentane which has been cooled to −80°. The pure product thus obtained is a white granular crystalline solid, m.p. 38–39°, weighing 53–56 g. (63–66%) (Notes 17 and 18).

2. Notes

1. A flask having this shape gives the most efficient mixing.[2]

2. A one-fourth-horsepower motor is adequate. A suitable motor is manufactured by Bodine Electric Company, Chicago, Illinois.

3. The use of a water-cooled glass condenser is not recommended since it might accidentally be broken and thereby cause water to flow into the flask. A metal water-cooled condenser has also been used and is satisfactory.

4. Adapted from that described in *Org. Syntheses*, Coll. Vol. **2**, 129 (1943).

5. The xylene used was purified by heating under reflux with sodium overnight and then distilling, b.p. 137–142°.

6. Linde high-purity dry nitrogen was used without further treatment.

7. The fumes may be taken off by attaching an aspirator. This procedure assures removal of all moisture from the system.

8. The funnel is dried before use with a flame and is then closed with a drying tube and allowed to cool.

9. The ester used was Eastman Kodak Company technical grade shaken with sodium carbonate, dried and distilled. The ester boils at 158–160°/11 mm.

10. The reaction time can be lengthened considerably without effect. If, however, the time is shortened appreciably, the yield may be markedly lowered.

11. Initially the reaction may take on various colors, red, purple, etc., but after a short time a dull gray-brown color appears, which is gradually replaced by a yellow-brown or olive-drab color.

12. It is important that the reaction mixture be kept out of contact with the air until it has been acidified.

13. When sufficient acetic acid has been added, the dark color of the reaction mixture is replaced by a white color, and the mixture is often quite thick. More acetic acid is not harmful.

14. Nitrogen is led through the capillary during the distillations.

15. The xylene thus recovered is purified (Note 5) and is used in the next preparation.

16. This material slowly decomposes upon standing. It may be stored for at least several months with only slight decomposition if it is kept under nitrogen in the dark and at 0°. The compound appears to be stable when pure.

17. Homologs having a ring containing 10 to 18 carbons have been prepared in an analogous manner in yields from 46 to 85%.[3]

18. This preparation has also been carried out on a 1.0 mole scale by the checkers, using a 5-l. creased flask. Comparable yields are obtainable.

3. Methods of Preparation

Sebacoin has been prepared by the cyclization of methyl or ethyl sebacate with sodium metal.[3-8]

[1] University of California, Los Angeles, California.
[2] Morton and Redman, *Ind. Eng. Chem.*, **40**, 1190 (1948).
[3] Stoll and Rouvé, *Helv. Chim. Acta*, **30**, 1822 (1947).
[4] Prelog, Frenkiel, Kobelt, and Barman, *Helv. Chim. Acta*, **30**, 1741 (1947).
[5] Stoll and Hulstkamp, *Helv. Chim. Acta*, **30**, 1815 (1947).
[6] Hansley (to E. I. du Pont de Nemours and Company), U. S. pat. 2,228,268 [*C. A.*, **35**, 2534 (1941)].
[7] Blomquist, Burge, and Sucsy, *J. Am. Chem. Soc.*, **74**, 3636 (1952).
[8] Prelog, Schenker, and Günthard, *Helv. Chim. Acta*, **35**, 1598 (1952).

SODIUM NITROMALONALDEHYDE MONOHYDRATE

(Malonaldehyde, nitro-, sodium derivative)

$$\begin{array}{c} BrC-CHO \\ \parallel \\ BrC-CO_2H \end{array} \xrightarrow[H_2O]{NaNO_2} \left[\begin{array}{c} CHO \\ | \\ C-NO_2 \\ | \\ CHO \end{array}\right]^{\ominus} Na^{\oplus} \cdot H_2O$$

Submitted by PAUL E. FANTA.[1]
Checked by CLIFF S. HAMILTON and PHILIP J. VANDERHORST.

1. Procedure

Caution! The sodium salt of nitromalonaldehyde is impact-sensitive and thermally unstable and should be handled as a potentially explosive material.

In a 2-l. three-necked round-bottomed flask, equipped with a thermometer, a dropping funnel, a mechanical stirrer and a gas vent (Note 1), are placed 258 g. (3.74 moles) of sodium nitrite and 250 ml. of water. The contents of the flask are heated and stirred to dissolve the solid. A solution of 258 g. (1 mole) of mucobromic acid (p. 688) in 250 ml. of warm 95% ethanol is placed in the dropping funnel and added dropwise with constant stirring over a period of 70–80 minutes. A mildly exothermic reaction occurs; the solution in the flask becomes deep red, and gas is evolved. During the addition, the temperature is kept at 54 ± 1° by intermittent application of an ice bath to the flask (Note 2). The mixture is stirred for an additional 10 minutes at 54 ± 1°. While being stirred continuously, it is then cooled to 0–5° by application of an ice bath. The fine, yellow precipitate is collected on a previously chilled Büchner funnel.

The slightly moist cake of crude product is transferred to a 1-l. flask and heated to boiling with a mixture of 400 ml. of 95% ethanol and 100 ml. of water. The hot solution is filtered to remove a fine yellow solid, and the clear red filtrate is cooled to 0–5°. The recrystallized product is collected on a Büchner funnel and dried in air at room temperature. The yield is 57–65 g.

(36–41%) of pink or tan needles of sodium nitromalonaldehyde monohydrate (Note 3).

2. Notes

1. The gases evolved are slightly irritating; they should be vented to a trap,[2] or the apparatus should be set up in a hood.

2. The yield is not increased and a darker, less pure product is obtained if the reaction is run at a higher temperature or for a longer time.

3. The checkers found the product to be quite pure. An almost quantitative yield of 2-amino-5-nitropyrimidine was obtained by reaction of the product with guanidine.

3. Methods of Preparation

The procedure above is a modification of the method of Hill and Torrey,[3] which was also studied by Johnson [4] and others.[5,6]

[1] Illinois Institute of Technology, Chicago, Illinois.

[2] *Org. Syntheses Coll. Vol.* **2**, 4 (1943).

[3] Hill and Torrey, *Am. Chem. J.*, **22**, 89 (1899).

[4] Johnson, P. B. Report 31092, "Preparation of Nitro Compounds" (1941) [*U. S. Bibliog. of Scientific and Industrial Reports*, **2**, 785 (1946)].

[5] Ulbricht and Price, *J. Org. Chem.*, **22**, 235 (1957).

[6] Tozaki, *Repts. Sci. Research Inst. (Japan)*, **27**, 401 (1951) [*C. A.*, **47**, 2181 (1953)].

SODIUM β-STYRENESULFONATE AND β-STYRENESULFONYL CHLORIDE

(Ethenesulfonic acid, 2-phenyl-, sodium salt)
(Ethenesulfonyl chloride, 2-phenyl-)

$$C_6H_5CH=CH_2 + dioxane \cdot SO_3 \rightarrow C_6H_5CH=CHSO_3H$$

$$C_6H_5CH=CHSO_3H + NaOH \rightarrow C_6H_5CH=CHSO_3Na + H_2O$$

$$C_6H_5CH=CHSO_3Na + PCl_5 \rightarrow$$
$$C_6H_5CH=CHSO_2Cl + POCl_3 + NaCl$$

Submitted by CHRISTIAN S. RONDESTVEDT, JR., and F. G. BORDWELL.[1]
Checked by T. L. CAIRNS and H. E. WINBERG.

1. Procedure

A. *Sodium β-styrenesulfonate*. In a 3-l. three-necked flask, equipped with a drying tube, a sealed stirrer, and a ground-glass stopper, is placed 800 ml. of ethylene chloride (Note 1). The flask with fittings is weighed and then connected through the third neck to a sulfur trioxide distillation apparatus (Note 2). Sulfur trioxide is distilled into the flask until about 300 g. has been collected. During the distillation the flask is cooled with a large pan of cold water (Note 3). The distillation is stopped, the condenser is disconnected and replaced by the glass stopper, and the flask is again weighed.

The stopper is replaced by a thermometer, and the other side neck is equipped with a two-way adapter to hold a dropping funnel and a drying tube. The sulfur trioxide solution is cooled below −5° with an ice-salt bath, then an amount of dioxane (Note 4) equivalent to the sulfur trioxide is added slowly with vigorous stirring at such a rate that the temperature does not exceed 5° (Note 5).

One molar equivalent of styrene (Note 6) dissolved in 2 volumes of ethylene chloride is added with stirring to the suspension of dioxane sulfotrioxide. The temperature is maintained below 10° during the hour required for addition, during which the suspension of colorless solid changes to a milky tan solution. The cooling bath is removed, and stirring is continued for an additional hour. After standing overnight, the nearly clear yellow solution is heated under reflux for 2 hours on the steam bath. The solution, now dark brown, is allowed to cool, is poured into 3 l. of ice water, and is stirred or shaken to ensure transfer of the sulfonic acid to the water (Note 7). The emulsion is neutralized with sodium hydroxide solution (Note 8), the layers are separated, and the aqueous phase is extracted with two 500-ml. portions of isopropyl ether (Note 9).

Several crops of crystals are now taken by evaporating the aqueous solution to the saturation point on a hot plate with stirring, then cooling. Each crop is dried for 2 hours at 100° and analyzed for unsaturation by the bromate-bromide titration method (Note 10). The first two crops usually total about 380 g. (based on 300 g. of sulfur trioxide), 49% yield, and are about 95% pure by titration. They may be further purified by one recrystallization from water, with about 80% recovery of 99+% pure salt.

The third crop generally weighs about 100 g. and is about 80% pure; the fourth crop weighs about 75 g., and is about 60% pure (Note 11). For purification, the third crop is recrystallized from the mother liquor resulting from the recrystallization of the combined first and second crops; the fourth crop is recrystallized from the third-crop mother liquor. An additional recrystallization from water is usually required to obtain material of 95+ % purity.

The total weight of 95+% pure sodium β-styrenesulfonate from 300 g. of sulfur trioxide is 450–500 g. (58–65% yield) (Notes 12 and 13).

B. *β-Styrenesulfonyl chloride.* One hundred grams of dry sodium β-styrenesulfonate (Note 14) is placed in a 500-ml. round-bottomed flask and thoroughly mixed with 120 g. of powdered phosphorus pentachloride with vigorous shaking. The flask is attached to a reflux condenser capped with a drying tube and heated on the steam bath until the reaction mixture liquefies, then for an additional 4 hours. The phosphorus oxychloride is removed by distillation at reduced pressure (water aspirator) on the steam bath. The semisolid residue is extracted three times by boiling for 15 minutes with 100-ml. portions of chloroform. The residue in the flask is dissolved in 200 ml. of ice water and extracted with 100 ml. of chloroform. The combined chloroform solutions are washed with 200 ml. of water, two 100-ml. portions of 5% sodium bicarbonate, and finally with 100 ml. of water. After drying for 2 hours over calcium chloride, the solution is treated with 3 g. of Norit, filtered, and concentrated to 150 ml., and 200 ml. of 60–68° petroleum ether is added. On cooling, 81–86 g. (83–88% yield) of β-styrenesulfonyl chloride separates as colorless needles, m.p. 88–90°. Recrystallization raises the melting point to 89–90° (Notes 15, 16, and 17).

2. Notes

1. A good grade of ethylene chloride is dried by slowly distilling a portion and rejecting the wet fore-run. Technical grade material should be carefully fractionated before use; the fraction boiling at 82.6–82.8° gives excellent results.

2. The apparatus for distillation of sulfur trioxide is constructed from a 500-ml. round-bottomed flask, connected by a ground joint to a 6-in. section of 20-mm. glass tubing. A safety tube (8-mm. tubing) reaching to the bottom of the flask is attached at the top by means of a ground joint; it is convenient to bend the top of the tube so that any acid drip can be caught in a beaker. A side arm, sealed in near the top of the wide tube, leads through a ground joint to an air condenser consisting of a 30-in. length of 12-mm. tubing, which terminates in a ground joint (with a drip tip) to fit the reaction flask. It is imperative that the apparatus be all glass, since sulfur trioxide rapidly attacks rubber or cork connections.

Sulfur trioxide is distilled by heating 60% fuming sulfuric acid contained in the flask. Spent acid may be fortified with "Sulfan B," stabilized liquid sulfur trioxide obtainable from the General Chemical Company, 40 Rector St., New York. Undiluted Sulfan B solidifies in the flask after a few heatings, and subsequent reheating may be dangerous.

3. Only slight cooling is needed at this stage. Stirring may be intermittent, if more convenient. If the sulfur trioxide solidifies above the level of the ethylene chloride from excessive cooling, it may easily be melted by warming the flask with a soft flame.

4. Dioxane is purified by refluxing for 12 hours with an excess of sodium, then distilling. The fraction boiling at 99.6–99.8° is

collected and stored over calcium chloride or, preferably, calcium hydride.

5. If the temperature is allowed to rise much above 5°, charring results and the product is discolored. A Dry Ice bath may be used to advantage. The reagent thus prepared should be used within a few hours, since dioxane is slowly attacked by sulfur trioxide.

6. Stabilized styrene of monomeric quality is used directly. The small amount of stabilizer does not interfere with the reaction.

7. In small runs, ether may be used to break the emulsion, thus permitting the separation of the layers before neutralization. In large runs, excessively large amounts of ether are required and it is preferable to neutralize first.

8. Methyl red serves as a convenient indicator. Slightly more than 1 mole of sodium hydroxide is required per mole of sulfur trioxide. Other metal salts can be prepared, using a suitable hydroxide or carbonate for the neutralization.

9. Ethyl ether may be substituted, but its greater water solubility is a disadvantage. Evaporation of the combined organic layers at reduced pressure below 40° leaves a residue of about 5% (based on sulfur trioxide) of 2,4-diphenylbutane-1,4-sultone, which may be purified (60% recovery) by recrystallization from acetone-water; m.p. 147–149°. The yield of sultone may be increased to about 25% of pure material by using 3 moles of styrene per mole of sulfur trioxide.

10. A sample (approximately 0.2 g.) is weighed accurately and dissolved in 25 ml. of water; then 25 ml. of $0.1N$ hydrochloric acid and 0.2 g. of potassium bromide are added. The solution is titrated with $0.017M$ potassium bromate until a permanent yellow color is produced. Potassium iodide (0.1 g.) is added, and the solution is backtitrated to a starch end point with $0.1N$ sodium thiosulfate. The blue color returns in about a minute since the high acidity promotes air oxidation of excess iodide. The accuracy is only slightly less if the appearance of a faint yellow bromine color is taken as the end point. One mole of potassium bromate is equivalent to 3 of sodium β-styrenesulfonate.

11. Further crops may be taken, but the recovery of pure

material from them is not worth the time required. They consist mainly of sodium 2-phenyl-2-hydroxyethane-1-sulfonate along with some sodium sulfate.

12. The procedure may be modified to prepare sodium 2-phenyl-2-hydroxyethane-1-sulfonate. The temperature during the addition of styrene is maintained below 0°. After 30 minutes of additional stirring at 0°, the mixture is hydrolyzed in ice water without heating. The crops of salts are analyzed, and those that are low in unsaturation are recrystallized from 70% aqueous ethanol. The more unsaturated crops are recrystallized from the mother liquors. The yield is 45–50% of material with less than 5% unsaturation.

13. Since the sodium sulfonates have no definite melting points, they may be converted to crystalline benzylthiuronium, p-chlorobenzylthiuronium, aniline, or p-toluidine salts for characterization.[2]

14. The salt is dried at 100° for 4 hours, then powdered.

15. An additional 6–7 g. of product, m.p. 88–89°, may be obtained by evaporation of the mother liquors and distillation of the residue as rapidly as possible at 2 mm. (bath temperature 120–140°). The total yield is thus 88–92%; in smaller runs, yields up to 96% are frequently obtained. Large batches cannot be distilled satisfactorily.

16. Carbon tetrachloride may also be used as a recrystallization solvent. The product should be stored in a tightly closed brown bottle to retard its slow decomposition.

17. The checkers used one-third quantities throughout both parts of this preparation with comparable results.

3. Methods of Preparation

Salts of β-styrenesulfonic acid have been prepared by the sulfonation of styrene with dioxane sulfotrioxide,[2] by heating styrene with ammonium sulfamate,[3] by the reaction of styrene with sodium bisulfite in the presence of oxygen,[4] and by the sulfonation of styrene with pyridine sulfotrioxide.[5]

[1] Joint contribution from Northwestern University, Evanston, Illinois, and the University of Michigan, Ann Arbor, Michigan.

[2] Bordwell, Suter, Holbert, and Rondestvedt, Jr., *J. Am. Chem. Soc.*, **68**, 139 (1946); Bordwell and Rondestvedt, Jr., *J. Am. Chem. Soc.*, **70**, 2429 (1948).
[3] Quilico and Fleischner, *Atti acad. Lincei*, **7**, 1050 (1928) [*C. A.*, **23**, 1628 (1929)].
[4] Kharasch, Schenck, and Mayo, *J. Am. Chem. Soc.*, **61**, 3092 (1939).
[5] Terent'ev and Dombrovskiĭ, *Zhur. Obshcheĭ Khim.* (*J. Gen. Chem.*), **20**, 1875 (1950) [*C. A.*, **45**, 2892 (1951)]; Terent'ev, Gracheva, and Shcherbatova, *Doklady Akad. Nauk S.S.S.R.*, **84**, 975 (1952) [*C. A.*, **47**, 3262 (1953)].

STEAROLIC ACID

$$CH_3(CH_2)_7CH{=}CH(CH_2)_7CO_2H \xrightarrow{Br_2}$$
$$CH_3(CH_2)_7CHBrCHBr(CH_2)_7CO_2H$$

$$CH_3(CH_2)_7CHBrCHBr(CH_2)_7CO_2H \xrightarrow{3NaNH_2}$$
$$CH_3(CH_2)_7C{\equiv}C(CH_2)_7CO_2Na + 2NaBr + 3NH_3$$

$$CH_3(CH_2)_7C{\equiv}C(CH_2)_7CO_2Na \xrightarrow{HCl}$$
$$CH_3(CH_2)_7C{\equiv}C(CH_2)_7CO_2H + NaCl$$

Submitted by N. A. KHAN, F. E. DEATHERAGE, and J. B. BROWN.[1]
Checked by MAX TISHLER, W. J. PALEVEDA, and E. F. SCHOENEWALDT.

1. Procedure

Bromine is added dropwise with stirring to a solution of 100 g. (0.35 mole) of oleic acid of at least 95% purity (Note 1) in 400 ml. of dry ether maintained at 0–5°, until the color of bromine persists. About 53 g. (0.33 mole) of bromine is needed; the excess is removed by addition of a few drops of oleic acid.

The sodamide required for dehydrobromination is prepared[2] in a 5-l. three-necked flask fitted with a sealed Hershberg stirrer, a gas inlet tube, and a large cold-finger condenser charged with Dry Ice-acetone. The condenser outlet is connected to a safety trap followed by a bubbler tube containing concentrated aqueous ammonia. Liquid ammonia (1.9 l., Note 2) is introduced into the flask through the inlet tube (*Hood!*); then 1.6 g. of anhydrous ferric chloride (C.P., black) is added in one portion with vigorous stirring. After 5–10 minutes, 3 g. of sodium is dropped into the brown solution to convert the iron salt into the catalytic form. After the evolution of hydrogen has ceased, the remainder of the sodium (total 43 g., 1.87 g. atoms) is added in small pieces with continued stirring. Gray, grainy crystals of sodamide settle out as the reaction proceeds.

The ethereal solution of dibromostearic acid, prepared as described above, is introduced slowly from a dropping funnel into the reaction flask. After the reaction has been allowed to proceed for 6 hours with continuous stirring, 60 g. of solid ammonium chloride (1.12 moles) is added in portions to destroy excess sodamide. The ammonia is allowed to evaporate until there remains a dry-appearing solid, and then 1 l. of water is added. The mixture is warmed to 60–70° under nitrogen (Note 3) and acidified by the addition of an excess (50 ml.) of concentrated aqueous hydrochloric acid. The aqueous layer is removed by siphoning, and the organic layer is washed with four 500-ml. portions of hot (60°) water. The aqueous layer is separated each time by siphoning. After the fourth washing, the oily product is solidified by cooling the flask in an ice bath, and the residual water is drained off. The crude product is dried on the steam bath under vacuum (Note 4).

The crude acid is dissolved in 500 ml. of petroleum ether at room temperature. The small amount of amorphous solid which may separate is removed by filtration through Supercel, and the filtrate is concentrated under reduced pressure to 300 ml. Chilling to 0–5° yields a first crop of tan crystals which is collected by suction filtration and washed with the minimum amount of ice-cold petroleum ether. Concentration of the mother liquors to 150 ml. and chilling yields a second crop of brownish crystals. The combined crops are dissolved in 300 ml. of petroleum ether, and the light-red solution is chilled to 0–5°. The almost white to light-tan crystals are collected, washed with a small amount of cold petroleum ether, and dried in a vacuum desiccator. There is obtained 51.5–61.5 g. (52–62%) of stearolic acid, m.p. 46–46.5° (Note 5).

2. Notes

1. Oleic acid of acceptable purity may be prepared from olive oil fatty acids by the method of Brown and Shinowara,[3] Wheeler and Riemenschneider,[4] Brown and Foreman,[5] or Khan, Deatherage, and Brown.[6] It may also be purchased from the Hormel Institute, Austin, Minnesota.

2. Experience has shown that about 5 g. atoms of sodium should be

used for each mole of oleic acid used as starting material. For each gram atom of sodium, 0.8–1.0 g. of ferric chloride, 1 l. of liquid ammonia, and 0.5–0.7 mole of ammonium chloride are necessary. The sodamide reaction proceeds best in comparatively dilute liquid ammonia solution.

3. The liquid stearolic acid is highly susceptible to autoxidation in the presence of air,[7] and this portion of the work-up should be conducted in a nitrogen atmosphere.

4. Alternatively, the product obtained after acidification may be taken up in ether, and the ether extract washed with water, dried over anhydrous sodium sulfate, and evaporated to dryness under reduced pressure. This procedure is reported by the submitters to be more convenient for preparations on a smaller scale, and to give a slightly improved yield.

5. This stearolic acid has been thoroughly characterized [6] by the freezing-point curve, ultraviolet and infrared spectra, ozonization, and hydrogenation. It has been shown to be free both of positional isomers and of olefinic acids such as oleic and elaidic acids. Its properties include: m.p. 46–46.5°, iodine number (Wijs titration, 30 minutes) 89.5, $n_D^{54.5}$ 1.4510, $n_D^{61.5}$ 1.4484, neutral equivalent 279.2–279.6 (theory 280.4), hydrogen uptake 95–100% of theory for a triple bond. The last trace of color is difficult to remove by recrystallization from petroleum ether. It can be removed, however, by crystallization from a 20–30% solution in acetone at −5° to −8°, or from an 8–10% solution at −20°, or by distillation (b.p. 189–190°/2mm.).

3. Methods of Preparation

Nearly all methods for preparation of stearolic acid involve dehydrohalogenation of a 9,10-dihalostearic acid, or its esters, with alcoholic potassium hydroxide; the most recent method is that of Adkins and Burks.[8] These methods employ drastic conditions, which result in poorer yields than those obtainable on dehydrohalogenation with sodamide.[6] Methyl 9,10-dibromostearate, on dehydrobromination with sodamide, yields stearolamide [6] (m.p. 82–83°) which may be hydrolyzed to stearolic acid. For preparative purposes, however, this method offers no special advantage over that described here.

[1] Ohio State University, Columbus, Ohio.
[2] Greenlee and Henne, *Inorg. Syntheses*, **2**, 128 (1946).

[3] Brown and Shinowara, *J. Am. Chem. Soc.*, **59**, 6 (1937).

[4] Wheeler and Riemenschneider, *Oil & Soap*, **16**, 207 (1939).

[5] Foreman and Brown, *Oil & Soap*, **21**, 183 (1944).

[6] Khan, Deatherage, and Brown, *J. Am. Oil Chemists' Soc.*, **28**, 27 (1951); Khan, *J. Am. Oil Chemists' Soc.*, **33**, 219 (1956).

[7] Khan, Brown, and Deatherage, *J. Am. Oil Chemists' Soc.*, **28**, 105 (1951).

[8] *Org. Syntheses Coll. Vol.* **3**, 785 (1955).

STEARONE

(18-Pentatriacontanone)

$$2C_{17}H_{35}CO_2H + MgO \rightarrow (C_{17}H_{35}CO_2)_2Mg + H_2O$$

$$(C_{17}H_{35}CO_2)_2Mg \xrightarrow{340°} C_{17}H_{35}COC_{17}H_{35} + MgO + CO_2$$

Submitted by A. G. Dobson and H. H. Hatt.[1]

Checked by R. L. Shriner and Philip R. Ruby

1. Procedure

In a 1-l. round-bottomed flask, fitted with a heated reflux condenser maintained at 100–110° (Note 1), are placed 44 g. of stearic acid (Note 2) and 20 g. (0.5 mole) of magnesium oxide (Note 3). The flask is immersed in a Wood's metal bath heated at 335–340° (Note 4). After the reaction has proceeded for 1 hour, 10-g. portions of melted stearic are added down the condenser at 15-minute intervals until an additional 240 g. (284 g., 1 mole total) has been added (Note 5). The heating is continued until the total reaction time is 10 hours.

The reaction flask is removed from the metal bath and allowed to cool to about 100°, and the liquid contents are poured with stirring into 1 l. of $4N$ sulfuric acid in a 3-l. beaker. This mixture is boiled with vigorous mechanical stirring until the frothing ceases (Note 6) and the upper layer is clear (about 2 hours). The lower aqueous layer is then siphoned off, and the upper layer is boiled for 1 hour with 1 l. of water. The water layer is separated, and the upper layer is boiled with 1 l. of a 5% sodium hydroxide solution for 1 hour with vigorous stirring. The ketone layer is then separated and boiled for 1 hour each with three successive 1 l. portions of hot water (Note 7) with good stirring.

The crude stearone is allowed to solidify and is then broken up and dried by pressing between filter paper. The crude yield is 230–240 g. (91–95%). The product melts at 84–86° and has an acid value of zero. It is purified by dissolving in 1.2 l. of a 2 : 1 mixture of benzene and absolute ethanol, filtering hot, and allowing to crystallize (Note 8). After a second crystallization from the same mixed solvents, the stearone is obtained as glistening white flakes, melting clear at 89–89.5° (shrinks at 87–88°). The yield is 204–220 g. (81–87%) (Note 9).

2. Notes

1. The condenser, attached to the flask by a ground-glass joint, is a Pyrex tube 50 cm. long and 20 mm. in internal diameter, wound with No. 26 Nichrome wire covered with asbestos paper. It is connected to a variable transformer and adjusted so as to maintain a temperature of 100–110°. A copper Liebig condenser heated by steam is also suitable provided that the inner tube has a diameter of about 20 mm.

2. The purity and melting point of the final product are dependent on the purity of the stearic acid. If Armour's Neo-Fat 1-65, m.p. 64–67° (90–95% stearic acid) is recrystallized first from 95% ethanol and then from acetone, a stearic acid melting from 67° to 68° results which yields a stearone with a melting point of 88–89° (shrinks at 86–88°).

In order to obtain stearone with the highest melting point, the checkers found it necessary to purify the above recrystallized stearic acid by converting it to the methyl ester and fractionating this ester under reduced pressure, using a 30-in. electrically heated packed fractionating column. The fraction boiling at 180–182°/4 mm., n_D^{40} 1.4364, is collected. The methyl stearate solidifies on cooling and melts at 38–39°. Saponification and acidification yielded white crystals of stearic acid melting clear at 69–69.3° (cor.).

Commercial "stearone" (Armour) is made from Neo-Fat 1-60 (which is approximately 75% stearic and 25% palmitic acid) and melts at 80–84°.

3. Merck's reagent-grade magnesium oxide is suitable. The submitters recommend 10 g. of catalyst, but this amount gave

a stearone with m.p. 87–88°. The larger amount, 20 g., gave a product melting clear at 89–89.5°.

4. The temperature of the bath should not be allowed to rise above 345°. Decomposition products are formed which lower the melting point of the final product.

5. With the rate of addition of stearic acid given, the decomposition of magnesium stearate maintains an excess of magnesium oxide in the reaction mixture. Each addition of stearic acid should take 1 to 2 minutes. Thus frothing of the reaction mixture is held under control; a brisk evolution of steam with a little entrained stearic acid follows each addition but quickly subsides and is followed by a steady effervescence due to the liberation of carbon dioxide.

6. A little unchanged magnesium stearate may cause frothing at this stage, which may be controlled by the addition of a few drops of 2-octanol and by vigorous stirring, which hastens decomposition of the soap.

7. The removal of fatty acids from neutral higher aliphatic compounds requires vigorous stirring of the hot aqueous and organic phases.

8. The solubility of stearone falls rapidly with temperature; a hot jacketed funnel is used with gentle suction.

9. This is a general method for the preparation of higher aliphatic ketones. It has been found suitable for the preparation of ketones from fatty acids containing 12 to 20 carbon atoms.

3. Methods of Preparation

Stearone has been prepared by heating the alkaline earth salts of stearic acid,[2] by reaction of liquid stearic acid with iron,[3-5] with alumina,[6] with manganous oxide and carbonate,[7] or with magnesium.[8] It has been prepared by passing stearic acid vapor over various catalysts: manganous oxide,[9] thoria aerogel,[10] or manganous oxide with chromium sesquioxide.[9] The literature contains numerous preparations from stearic acid using almost any metal or alkaline-earth oxides. It also has been obtained by the saponification of ethyl α-stearoylstearate with potassium hydroxide.[11]

Kino [12] studied the reaction of stearic acid with magnesium oxide, with alloys of magnesium, and many other metals, at high temperatures, and obtained stearone. The present method was developed from a reëxamination of this reaction.[13]

[1] Council for Scientific and Industrial Research, Melbourne, Australia.

[2] Heintz, *Jahresber. Chem.*, **8**, 515 (1855).

[3] Easterfield and Taylor, *J. Chem. Soc.*, **99**, 2298 (1911).

[4] Grün, Ulbrich, and Krczil, *Z. angew. Chem.*, **39**, 423 (1926).

[5] Piper, Chibnall, Hopkins, Pollard, Smith, and Williams, *Biochem. J.*, **25**, 2074 (1931).

[6] Kino and Kato, *Bull. Inst. Phys. Chem. Research (Tokyo)*, **19**, 179 (1940).

[7] Kino, *J. Soc. Chem. Ind. Japan*, **41**, 91 (1938); Mastagli, Lambert, and Hirigoyen, *Compt. rend.*, **248**, 1830 (1959).

[8] Kino, *J. Soc. Chem. Ind. Japan*, **40**, 194 (1937).

[9] Wortz, U. S. pat. 2,108,156 (1938) [*C. A.*, **32**, 2542 (1938)].

[10] Kistler, Swann, and Appel, *Ind. Eng. Chem.*, **26**, 388 (1934).

[11] Clement, *J. recherches centre natl. recherche sci. Lab. Bellevue (Paris)*, No. **40**, 160 (1957) [*C. A.*, **52**, 19918 (1958)]; Clement, *Compt. rend.*, **236**, 718 (1953).

[12] Kino, *J. Soc. Chem. Ind. Japan*, **40**, 194, 235, 312, 437, 464 (1937); **41**, 91, 259 (1938); **42**, 188, 362 (1939).

[13] Curtis, Dobson, and Hatt, *J. Soc. Chem. Ind.*, **66**, 402 (1947).

cis-STILBENE *

$$C_6H_5CH \quad \xrightarrow[\text{Quinoline}]{\text{Copper chromite}} \quad C_6H_5CH \\ \| \qquad\qquad\qquad\qquad\qquad \| \quad + CO_2 \\ C_6H_5CCO_2H \qquad\qquad\qquad\qquad C_6H_5CH$$

Submitted by ROBERT E. BUCKLES and NORRIS G. WHEELER.[1]
Checked by T. L. CAIRNS and J. C. LORENZ.

1. Procedure

A 500-ml. three-necked flask is fitted with a reflux condenser and a thermometer, the bulb of which reaches far enough into the flask to be covered by the liquid. A solution of 46.0 g. (0.205 mole) of α-phenylcinnamic acid (p.777) (Note 1) in 280 ml. (307 g., 2.38 moles) of quinoline (Note 2) is added to the flask along with 4.0 g. of copper chromite.[2] The reaction flask is heated by means of a mantle or an oil bath until the temperature of the reaction mixture reaches 210–220°. The mixture is kept within this temperature range for 1.25 hours. The solution is then cooled immediately and added to 960 ml. of 10% hydrochloric acid in order to dissolve the quinoline (Note 3). The product is extracted from this mixture with two 200-ml. portions of ether followed by a 100-ml. portion. The combined ether extracts are filtered to remove particles of catalyst, washed with 200 ml. of 10% sodium carbonate, and dried over anhydrous sodium sulfate. The dry solution is removed from the drying agent by filtration and heated on a steam bath to distil the ether. The residue is dissolved in a hexane fraction, b.p. 60–72° (Skellysolve B); the solution is cooled to 0° and filtered to remove *trans*-stilbene, if any. The hydrocarbon solvent is removed by distillation, and the *cis*-stilbene is distilled. The yield is 23–24 g. (62–65%), b.p. 133–136°/10 mm., 95–97°/1 mm.; n_D^{25} 1.6183–1.6193, n_D^{20} 1.6212–1.6218 (Note 4).

2. Notes

1. The isomer of α-phenylcinnamic acid of m.p. 172–173° is used (p. 777). The isomer of m.p. 137–139° yields *trans*-stilbene on decarboxylation.[3]

2. Practical grade quinoline containing about 10% of isoquinoline and quinaldine can be used. If the quinoline contains water, the desired temperature can be reached by distillation of a small amount of quinoline directly from the reaction mixture.

3. The quinoline can be recovered by neutralization of the aqueous solution, extraction of the quinoline into ether, and distillation of the dried (over barium oxide) ether extract.

4. The product obtained from this type of decarboxylation is reported to contain only about 5% of *trans*-stilbene.[4] A sample made according to the above directions can be treated with bromine in carbon tetrachloride at room temperature in the dark to give an 80–85% yield of the *dl*-dibromide which arises from *trans* addition to *cis*-stilbene. The *meso*-dibromide, which is very insoluble and easily separated, is obtained only to the extent of 10% or less. Part of the latter product may arise from the action of bromine atoms on *cis*-stilbene rather than from *trans* addition to *trans*-stilbene. The *cis*-stilbene prepared by this method is readily and completely soluble in cold absolute ethanol. It freezes solid at about −5°. Its ultraviolet absorption coefficient

(ϵ) is 1.10×10^4 at 274 mμ and 8.7×10^3 at 294 mμ, quite different from *trans*-stilbene.

3. Methods of Preparation

cis-Stilbene has been prepared by the partial hydrogenation of tolan; [5,6] by the electrolytic reduction of tolan; [7] by the reduction of tolan with a copper-zinc couple; [8] by the reduction of the low-melting isomer of α-bromostilbene with zinc dust in 90% alcohol; [9] by the illumination of *trans*-stilbene with ultraviolet light; [10] by the decarboxylation of the high-melting isomer of α-phenylcinnamic acid in the presence of barium hydroxide; [3] by heating tolan with diisobutylaluminum hydride [11] or zinc and acetic acid; [12] and by the reaction of *cis*-bromostilbene with butyllithium followed by treatment with methanol. [13] The present method is based on that of Taylor and Crawford. [14]

[1] State University of Iowa, Iowa City, Iowa.

[2] *Org. Syntheses Coll. Vol.* **2**, 142 (1943).

[3] Stoermer and Voht, *Ann.*, **409**, 36 (1915).

[4] Weygand and Rettberg, *Ber.*, **73B**, 771 (1940).

[5] Kelber and Schwarz, *Ber.*, **45**, 1946 (1912).

[6] Ott and Schröter, *Ber.*, **60**, 624 (1927).

[7] Campbell and Young, *J. Am. Chem. Soc.*, **65**, 965 (1943).

[8] Straus, *Ann.*, **342**, 238 (1905).

[9] Wislicenus and Jahrmarkt, *Chem. Zentr.*, **1901 I**, 463.

[10] Stoermer, *Ber.*, **42**, 4865 (1909).

[11] Wilke and Müller, *Chem. Ber.*, **89**, 444 (1956).

[12] Rabinovitch and Looney, *J. Am. Chem. Soc.*, **75**, 2652 (1953).

[13] Curtin and Harris, *J. Am. Chem. Soc.*, **73**, 4519 (1951).

[14] Taylor and Crawford, *J. Chem. Soc.*, **1934**, 1130.

trans-STILBENE OXIDE

(Bibenzyl, α,α'-epoxy-)

Submitted by DONALD J. REIF and HERBERT O. HOUSE.[1]
Checked by MELVIN S. NEWMAN and DONALD K. PHILLIPS.

1. Procedure

In a 1-l. three-necked flask equipped with stirrer, dropping funnel, and thermometer is placed a solution of 54 g. (0.3 mole) of *trans*-stilbene (Note 1) in 450 ml. of methylene chloride. The methylene chloride solution is cooled to 20° with an ice bath, and then the cooling bath is removed. A solution of peracetic acid (0.425 mole) in acetic acid (Note 2) containing 5 g. of sodium acetate trihydrate is added dropwise and with stirring to the reaction mixture during 15 minutes. The resulting mixture is stirred for 15 hours, during which time the temperature of the reaction mixture is not allowed to rise above 35° (Notes 3 and 4). The contents of the flask are poured into 500 ml. of water, and the organic layer is separated. The aqueous phase is extracted with two 150-ml. portions of methylene chloride, and the combined methylene chloride solutions are washed with two 100-ml. portions of 10% aqueous sodium carbonate and then with two 100-ml. portions of water. The organic layer is dried over magnesium sulfate, and the methylene chloride is distilled, the last traces being removed under reduced pressure. The residual solid is recrystallized from methanol (3 ml./g. of product) to yield 46–49 g. (78–83%) of crude *trans*-stilbene oxide, m.p. 66–69° (Note 5). An additional recrystallization from hexane (3 ml./g. of product) sharpens the melting point of the product to 68–69°. The yield is 41–44 g. (70–75%).

2. Notes

1. *trans*-Stilbene (Eastman Kodak Company) may be used

directly. A slightly higher yield is obtained if the stilbene is crystallized once from alcohol.

2. Approximately 40% peracetic acid in acetic acid is available (Becco Chemical Division, Food Machinery and Chemical Corporation, Buffalo 7, New York). Sodium acetate is added to neutralize a small amount of sulfuric acid which is present in the commercial product. The peracetic acid concentration should be determined by titration.[2] The peracetic acid solution used by the submitters contained 0.497 g. (0.00655 mole) of peracid per milliliter. Consequently 65 ml. (0.425 mole) of this solution was used in the reaction.

3. Without further cooling, the temperature of the reaction mixture usually rises to 32–35° after 1–2 hours and then gradually falls.

4. The progress of the epoxidation can be followed by measuring periodically the optical density of the reaction mixture at 295 mμ. The reaction time and temperature specified in the procedure were found to reduce the optical density of the reaction mixture at 295 mμ (and, accordingly, the *trans*-stilbene concentration) to less than 3% of its initial value. If more than this amount of unchanged *trans*-stilbene remains in the crude product, it cannot be removed by recrystallization from either methanol or hexane. Even after repeated crystallization the melting point of the product does not rise above 66–67°. Pure *trans*-stilbene oxide can be isolated from such a mixture if the mixture is treated with additional peracetic acid to convert the remaining *trans*-stilbene to *trans*-stilbene oxide.

5. This melting point and yield are obtained after the crystalline product which separates from methanol has been dried under reduced pressure for 12 hours. This drying process is unnecessary if the product is subsequently to be recrystallized from hexane.

3. Methods of Preparation

trans-Stilbene oxide has been prepared by the reaction of silver oxide with the methiodide of 1,2-diphenyl-2-dimethylaminoethanol,[3,4] by the reaction of hydrazine with hydrobenzoin,[5] and by the reaction of peracetic acid [6,7] or perbenzoic acid [8] with *trans*-

stilbene. The procedure described illustrates the use of a commercially available peracetic acid solution for the epoxidation of carbon double bonds. Since the reaction of *trans*-stilbene and other olefins conjugated with aromatic nuclei with peracids is slow, the procedure for the epoxidation of unconjugated olefins should be modified by the use of a lower reaction temperature, a shorter reaction time, and a longer period of time for the addition of the peracid.

[1] Massachusetts Institute of Technology, Cambridge, Massachusetts.

[2] Greenspan and MacKellar, *Anal. Chem.*, **20**, 1061 (1948).

[3] Read and Campbell, *J. Chem. Soc.*, **1930**, 2377.

[4] Rabe and Hallensleben, *Ber.*, **43**, 884 (1910).

[5] Müller and Kraemer-Willenberg, *Ber.*, **57B**, 575 (1924).

[6] Böeseken and Elsen, *Rec. trav. chim.*, **47**, 694 (1928).

[7] Böeseken and Schneider, *J. prakt. Chem.*, **131**, 285 (1931).

[8] Tiffeneau and Levy, *Bull. soc. chim. France*, [4]**39**, 763 (1926).

α-SULFOPALMITIC ACID

(Hexadecanoic acid, 2-sulfo-)

$$CH_3(CH_2)_{13}CH_2CO_2H + SO_3 \rightarrow CH_3(CH_2)_{13}CH(SO_3H)CO_2H$$

Submitted by J. K. Weil, R. G. Bistline, Jr., and A. J. Stirton.[1]
Checked by James Cason and Gerhard J. Fonken.

1. Procedure

Caution! Rubber gloves and a protective face shield should be worn while handling liquid sulfur trioxide, and the reaction should be carried out in a hood.

A 2-l. three-necked round-bottomed flask, with standard-taper ground-glass joints, is equipped for heating with an electric mantle or an oil bath, and fitted with a ball joint-sealed mechanical stirrer, a thermometer well, a graduated, pressure-equalizing dropping funnel, and a small vent. The dropping funnel, which should have a close-fitting stopcock well lubricated with heavy silicone grease, is placed so as to discharge sulfur trioxide well above the surface of the reaction mixture. Palmitic acid (200 g., 0.78 mole) (Note 1) and 600 ml. of carbon tetrachloride (Note 2) are added to the flask. Solution of the palmitic acid is endothermic and causes the temperature of the mixture to fall 5 to 10° below room temperature (Note 3). Stabilized liquid sulfur trioxide, 53 ml. (100 g., 1.25 moles) (Note 4), is added dropwise from the dropping funnel to the stirred mixture. Solution of palmitic acid becomes complete, the solution darkens, and the temperature rises to 45° as the sulfur trioxide is added over a 30-minute period. The reaction mixture is finally heated for 1 hour at 50–65° with continued stirring and is then chilled in an ice bath before the accessories are removed from the flask. The necks are closed with glass stoppers, and the reaction mixture is refrigerated at about −15° overnight (Note 5).

Crystallized solids are filtered by suction (Note 6), washed with cold carbon tetrachloride, and dried at room temperature in a vacuum desiccator to constant weight (1 to 2 days). The crude dark product weighs 197–223 g. (75–85%), and is usually satisfactory for the preparation of derivatives such as salts (Note 7) and esters.[2] One crystallization from acetone (7 ml./g.) at −20° yields a light gray solid, with neutralization equivalent within 1–2% of the theoretical value of 168, in a yield of 178–197 g. (68–75%). Two or three additional crystallizations (leaving a yield of only 25–30%) are required to give an almost colorless crystalline solid, m.p. 90–91° (Note 8). Once crystallized material is moderately stable to storage in a container protected from moisture, but slow darkening occurs, especially in presence of sunlight.

2. Notes

1. A purified fatty acid is recommended for the preparation of a pure α-sulfo acid. Purified palmitic acid (m.p. 60.8–61.4°, neutralization equivalent 256.2) is prepared by twice recrystallizing a good commercial grade of palmitic acid from acetone at 0°, using a solvent ratio of 10 ml. to 1 g. However, the reaction may be applied to commercial saturated higher fatty acids, if the iodine number is sufficiently low. The checkers obtained similar results with recrystallized Neo-Fat 1-56 (Armour and Company, Chicago, Ill.) or Eastman white label grade palmitic acid.

2. Other chlorinated solvents such as tetrachloroethylene or

chloroform may be used in place of carbon tetrachloride. *Caution! The reaction of sulfur trioxide with chlorinated solvents has been reported* [3] *to give phosgene and other toxic products. Adequate venting of all by-product gases is essential.*

3. The mixture which is a slurry at 15° may be used or it may be warmed to about 30° to give a clear solution. If the slurry is warmed before sulfur trioxide addition, it is usually not necessary to heat after all of the sulfur trioxide has been added.

4. Liquid sulfur trioxide may be purchased in stabilized form as Sulfan B,[4] m.p. 17°, b.p. about 45°. Caution must be exercised in handling sulfur trioxide. The liquid is highly corrosive to the skin and the vapor may cause injury if inhaled. The powerful oxidizing and dehydrating effects of sulfur trioxide should not be underestimated. The liquid may be stored in a glass-stoppered bottle if the stopper is lubricated with a heavy silicone grease. Because the liquid reacts violently with water the bottle lip should be wiped free of any moisture to avoid spattering when the liquid is poured. If solidification of sulfur trioxide should occur as the result of hydration, the vented (*Hood!*) solid may be liquefied by gentle application of heat from an infrared lamp.

5. Shorter refrigerating periods at temperatures up to +5° give slightly lower yields.

6. Since the mixture filters rather slowly, even after storage overnight, a Büchner funnel of at least 20 cm. diameter should be used, and a hard grade of filter paper (such as Whatman No. 50) is recommended. Use of a rubber dam helps to express solvent and exclude moisture from the hygroscopic product.

7. The monosodium salt $CH_3(CH_2)_{13}CH(SO_3Na)CO_2H$ may be prepared by neutralizing only the sulfonic acid group or by adding aqueous sodium sulfate to a hot aqueous solution of the crude α-sulfo acid and cooling to room temperature. The monosodium salt crystallizes in white plates, leaving most of the color in the filtrate. The disodium salt is formed by further neutralization with sodium hydroxide.

8. α-Sulfolauric, α-sulfomyristic, α-sulfostearic, and α-sulfobehenic acids have been prepared by this procedure. The table shows the melting point and neutralization equivalent after at least four recrystallizations and a 12-hour drying period at 76° under 1 mm. pressure. There is some uncertainty in the melting

points because of the hygroscopic character of the α-sulfo acids.

Neutralization Equivalent

	Found	Theoretical	M.P.
α-Sulfolauric acid	142	140.4	86.5–88°
α-Sulfomyristic acid	154	154.2	85–86.5°
α-Sulfopalmitic acid	171	168.2	90–91°
α-Sulfostearic acid	182	182.2	96–97°
α-Sulfobehenic acid	212	210.3	95–97°

3. Methods of Preparation

Other direct methods for the sulfonation of the higher fatty acids are by the use of sulfur trioxide vapor [5] or by the use of chlorosulfonic acid.[6] Indirect methods are also available for the preparation of α-sulfo fatty acids and their salts from an α-bromo fatty acid made by the Hell-Volhard-Zelinsky reaction. The bromo compound may be converted directly to the sodium salt of a sulfonic acid through the Strecker reaction [7] or may be converted to the mercaptan and oxidized to the sulfonate.[8] Sulfonation of the lower fatty acids has been studied by Backer and co-workers.[9]

α-Sulfonation with sulfur trioxide appears to be generally applicable to carboxylic acids having an α-methylene group.

[1] Eastern Regional Research Laboratory, Philadelphia 18, Pennsylvania, a laboratory of the Eastern Utilization Research Branch, Agricultural Research Service, U. S. Department of Agriculture, Philadelphia 18, Pennsylvania.

[2] Weil, Bistline, and Stirton, *J. Am. Chem. Soc.*, **75**, 4859 (1953).

[3] "Reactions of SO₃," Tech. Service Bull. SF-2, General Chemical Division, Allied Chemical and Dye Corporation, 40 Rector Street, New York 6, N. Y.

[4] "Storage, Handling and Use of Sulfan," Tech. Service Bull. SF-3, General Chemical Division, Allied Chemical and Dye Corporation.

[5] Günther (1932), PB30081, Office of Tech. Services, U. S. Department of Commerce; *Bibliography of Scientific and Industrial Reports*, **4**, 662 (1947).

[6] Günther and Hetzer (to I. G. Farbenindustrie A.-G.), U. S. pat. 1,926,442 [*C. A.*, **27**, 6001 (1933)].

[7] Mehta and Trivedi, *Melliand Textilber.*, **21**, 117, 288 (1940) [*C. A.*, **34**, 6087 (1940)].

[8] Weil, Witnauer, and Stirton, *J. Am. Chem. Soc.*, **75**, 2526 (1953).

[9] For example, de Boer, *Rec. trav. chim.*, **71**, 814 (1952).

SYRINGIC ALDEHYDE *

(Syringaldehyde)

$$CH_3O\underset{OH}{\overset{}{\bigcirc}}OCH_3 \xrightarrow[H_2SO_4]{C_6H_{12}N_4} CH_3O\underset{OH}{\overset{CHO}{\bigcirc}}OCH_3$$

Submitted by C. F. H. ALLEN and GERHARD W. LEUBNER.[1]
Checked by R. S. SCHREIBER, WM. BRADLEY REID, JR., and R. W. JACKSON.

1. Procedure

A well-stirred (Notes 1 and 2) mixture of 740 ml. of glycerol and 216 g. of boric acid, in a 2-l. three-necked round-bottomed flask fitted with a thermometer and a condenser for downward distillation, is dehydrated by heating in an oil bath to *exactly* 170°. This temperature is maintained for 30 minutes and then allowed to drop. When the temperature has fallen to 150°, a mixture of 154 g. (1 mole) of pyrogallol-1,3-dimethyl ether and 154 g. (1.1 moles) of hexamethylenetetramine (Note 3) is added as rapidly as possible through the neck holding the thermometer. The temperature drops to approximately 125°. Rapid heating is immediately started but is slowed down when the temperature begins to reach 145° and stopped at 148°. The reaction must be watched and controlled *very carefully* when this temperature is reached, since the reaction becomes exothermic at this point (Notes 4, 5, and 6). The temperature is maintained at 150–160° for approximately 6 minutes (Note 7). At the end of this reaction time the mixture is cooled to 110° as rapidly as possible (Notes 6 and 8), and a previously prepared solution of 184 ml. of concentrated sulfuric acid in 620 ml. of water is added to the reaction mixture. After being stirred for 1 hour, the mixture is cooled to 25° in an ice bath. The boric acid, which separates from the solution, is removed by filtration (Note 9) and washed free of mother liquor with 400 ml. of water. The filtrate and washings are combined and extracted with three 500-ml. portions of chloroform (Notes 10, 11, and 12).

The chloroform solution is then extracted with a filtered solution of 180 g. of sodium bisulfite in 720 ml. of water (Note 13) by stirring rapidly with a Hershberg stirrer for 1 hour. The separated bisulfite solution is washed twice with chloroform, filtered, and acidified in a hood with a solution of 55 ml. of concentrated sulfuric acid in 55 ml. of water. After careful heating on a steam bath for a short time, air is bubbled through the hot solution until all the sulfur dioxide has been expelled. The product, which separates as a mixture of crystals and oil, readily solidifies upon cooling (Note 14). The syringic aldehyde is collected by filtration, washed with cold water, and dried in an oven at 40° to give 62.5–66 g. of light-tan material, melting at 110.5–111°, which still contains a small amount of foreign material that does not melt at 300°. Recrystallization of the crude product from aqueous methanol using 30 ml. of water and 3 ml. of methanol for each 10 g. of aldehyde gives 56–59 g. (31–32%) of product melting clear at 111–112° (uncor.). A second extraction of the chloroform solution with a filtered solution of 60 g. of sodium bisulfite in 240 ml. of water gives an additional 3–4 g. of product.

2. Notes

1. The use of a Hershberg [2] stirrer is recommended.

2. It is desirable to conduct this preparation in a hood because of the large volume of ammonia liberated in the second step.

3. Eastman Kodak Company white label grade pyrogallol-1,3-dimethyl ether was used. A larger excess of hexamethylenetetramine in a small trial run did not improve the yield.

4. The reaction mixture darkens rapidly, and there is a vigorous evolution of ammonia.

5. The temperature usually rises to 160° within 5 minutes, and cooling is necessary.

6. Cooling is accomplished by playing a stream of cold water over the outside of the flask.

7. There is undoubtedly some leeway in these conditions. The same yields were obtained when the temperature was maintained at 150–160° for periods of 5 to 9 minutes. Longer reaction

times, without rapid cooling after the heating period, lowered the yield. Reaction times of 15 minutes, 30 minutes, and 1 hour gave yields of 20.8%, 10.0%, and 6.5%, respectively. Liggett and Diehl,[3] after having run a large number of other Duff reactions, have come to the conclusion that the temperature may vary between 145° and 175° without detriment to the yield.

8. About 3 to 5 minutes is required for cooling.

9. If not removed, the boric acid makes extraction of the product impossible or very difficult. Since the boric acid is finely divided, filtration is extremely slow unless large Büchner funnels, preferably with large holes, are employed. The checkers avoided the difficulty by using a filter cloth at this point.

10. The product cannot be isolated by steam distillation of the reaction mixture.

11. Syringic aldehyde is much more soluble in chloroform than in ether. Extraction is essentially complete since a fourth extraction gave only a 0.3–0.7% increase in yield.

12. If the aldehyde is isolated directly by concentration of the chloroform solution, the color is darker, and the melting point and yield are lower.

13. This represents a large excess of sodium bisulfite, but smaller amounts remove a smaller percentage of the syringic aldehyde from the chloroform solution. When larger amounts of bisulfite are employed, extraction of the product is still incomplete.

14. The product should be cooled to just 15° and filtered immediately. Longer and further cooling causes sodium sulfate to crystallize from the mixture. Very little product remains in the filtrate.

3. Methods of Preparation

This procedure is a modification of the method described by Manske and co-workers.[4] Syringic aldehyde has also been obtained by numerous other procedures from pyrogallol-1,3-dimethyl ether [5-8] from gallic acid,[9-11] and from vanillin.[12]

[1] Eastman Kodak Company, Rochester, New York.
[2] Org. Syntheses Coll. Vol. 2, 117 (1943).
[3] Liggett and Diehl, Proc. Iowa Acad. Sci., 52, 191 (1945).
[4] Manske, Ledingham, and Holmes, Can. J. Research, 23B, 100 (1945).

[5] Graebe and Martz, *Ber.*, **36**, 1031 (1903).
[6] Pauly and Strassberger, *Ber.*, **62**, 2277 (1929).
[7] Pearl, *J. Am. Chem. Soc.*, **70**, 1746 (1948).
[8] Mauthner, *Ann.*, **395**, 273 (1913).
[9] Mauthner, *J. prakt. Chem.*, **142**, 26 (1935).
[10] McCord, *J. Am. Chem. Soc.*, **53**, 4181 (1931).
[11] Sharp, *J. Chem. Soc.*, **1937**, 852.
[12] Pepper and MacDonald, *Can. J. Chem.*, **31**, 476 (1953).

TETRAACETYLETHANE

(2,5-Hexanedione, 3,4-diacetyl-)

$$CH_3COCH_2COCH_3 \xrightarrow{NaOH} [CH_3COCHCOCH_3]Na$$

$$2[CH_3COCHCOCH_3]Na + I_2 \rightarrow$$

$$(CH_3CO)_2CHCH(COCH_3)_2 + 2NaI$$

Submitted by ROBERT G. CHARLES.[1]
Checked by VIRGIL BOEKELHEIDE and HENRY FLEISCHER.

1. Procedure

A. *Sodium acetylacetonate.* A solution is prepared by dissolving 40 g. (1 mole) of sodium hydroxide in 50 ml. of water and adding to this 200 ml. of methanol. This solution is added, slowly with hand stirring, to 100 g. (1 mole) of acetylacetone (2,4-pentanedione) contained in a 500-ml. Erlenmeyer flask (Note 1). The creamy-white crystalline salt separates from solution immediately. The flask is stoppered and cooled in ice (or in a refrigerator) for 2 hours or overnight. The sodium salt is collected on a Büchner funnel and washed with two small portions of cold methanol (Note 2). After the salt is air-dried, it is dried further either by allowing it to stand in a vacuum desiccator at room temperature or by heating it in a vacuum oven at 100° for 3 hours (Note 3). The anhydrous product, which is stable and can be stored indefinitely in a stoppered jar, weighs 70–80 g. (57–66%).

B. *Tetraacetylethane.* Sodium acetylacetonate is ground to a fine powder in a mortar, and 24.4 g. (0.2 mole) of the anhydrous material or 28.9 g. of the hydrate (Note 3) is weighed into a 1-l.

Erlenmeyer flask. After 300 ml. of ether has been added, the suspension is stirred vigorously at room temperature with a magnetic stirrer. To the stirred mixture is added, dropwise from a separatory funnel, a solution of 25.4 g. (0.1 mole) of iodine dissolved in 300 ml. of ether. The rate of addition is maintained roughly constant by occasional adjustments of the stopcock, and the total addition is completed in about 2.5 hours. The reaction mixture is then poured into a large Erlenmeyer flask, and the ether is allowed to evaporate overnight at room temperature in a hood (Note 4). To the contents of the flask there is then added 500 ml. of water, and the mixture is allowed to stand for 2 hours. The remaining solid is collected on a Büchner funnel, washed several times with water, and finally dried in a vacuum desiccator. The yield (Note 5) of white solid, m.p. 185–188°, is 11–13 g. For purification, the product is taken up in 500–700 ml. of boiling methanol and the hot solution is filtered through a semi-fluted filter paper in a heated funnel. The filtrate is allowed to stand in the refrigerator for several hours. There is collected from the filtrate 8.0–11.7 g. (41–59%) of white crystals, m.p. 192–193° (cor.).

2. Notes

1. Eastman practical grade acetylacetone was found to be sufficiently pure for the preparation.

2. Washing with methanol decreases the yield somewhat but improves the purity of the product visibly. Additional, but less pure, sodium salt can be obtained if desired by combining the filtrate and washings, and evaporating.

3. The sodium salt dried in a vacuum oven is anhydrous, while that dried in a vacuum desiccator was found to contain 15.6% water.

4. Some hazard is always involved in the evaporation of ether to dryness. To minimize the hazard, peroxide-free ether should be used and the evaporation conducted behind a shield. No difficulties have been encountered in the submitter's laboratory with a number of these preparations using previously unopened cans of anhydrous ether. If desired, the ether could be recovered by distillation. The explosive hazard is probably increased, however, by such a procedure.

5. The yield is essentially the same whether anhydrous or hydrated sodium acetylacetonate is used.

3. Methods of Preparation

Tetraacetylethane has been prepared previously both by the use of sodium metal [2] and of sodium hydride [3] with acetylacetone followed by addition of iodine. Also, the compound has been prepared in low yield by the reaction of diacetyl peroxide with acetylacetone [4] and by the electrolysis of acetylacetone in an alcohol-water solution.[2] The present method, although similar to those first mentioned, is somewhat more convenient and does not require anhydrous conditions.

[1] Westinghouse Research Laboratories, Pittsburgh 35, Pennsylvania.
[2] Mulliken, *Am. Chem. J.*, **15**, 523 (1893).
[3] Mosby, *J. Chem. Soc.*, **1957**, 3997.
[4] Kharasch, McBay, and Urry, *J. Am. Chem. Soc.*, **70**, 1269 (1948).

dl-4,4′,6,6′-TETRACHLORODIPHENIC ACID

(dl-Diphenic acid, 4,4′,6,6′-tetrachloro-)

Submitted by Edward R. Atkinson, Donald M. Murphy, and James E. Lufkin.[1]

Checked by R. S. Schreiber, Wm. Bradley Reid, Jr., and R. D. Birkenmeyer.

1. Procedure

A. *3,5-Dichloro-2-aminobenzoic acid.* A solution of 45 g. (0.33 mole) of anthranilic acid, 150 ml. of concentrated hydrochloric acid, and 850 ml. of water is placed in a 2-l. three-necked flask in a hood and weighed. While the solution is stirred rapidly, chlorine is introduced until the reaction mixture gains 45 g. (0.63 mole) in weight (Note 1). The flask is surrounded by a water bath to maintain the temperature of the reaction mixture below 30° during the chlorination procedure. The reaction mixture is filtered by suction, using a large (10–12 in.) Büchner funnel; the crude product is washed with water and then dried at room temperature (Note 2). There is obtained 55–65 g. of crude product melting at about 205°. The crude product is leached with 4 ml. of boiling benzene per gram, filtered by suction, and washed on the filter with 1 ml. of cold benzene per gram. After drying at room temperature, there is obtained 46.5–53 g. (69–78%) of crude 3,5-dichloro-2-aminobenzoic acid. The melting point of this material should not be lower than 211° (Notes 3 and 4).

B. *Diazotization of 3,5-dichloro-2-aminobenzoic acid.* Fifty grams (0.24 mole) of 3,5-dichloro-2-aminobenzoic acid is dissolved in a solution of 12 g. (0.3 mole) of sodium hydroxide in 700 ml. of water. To this solution is added 20 g. (0.29 mole) of sodium nitrite, and the solution is cooled to 10° (Note 5). One hundred milliliters of concentrated hydrochloric acid (sp. gr. 1.191) and 200 ml. of water are placed in a 2-l. three-necked flask and cooled to 10°. The cold solution of sodium 3,5-dichloro-2-aminobenzoate and sodium nitrite is then added to the hydrochloric acid solution with cooling (10°) and efficient stirring at such a rate that no appreciable accumulation of undiazotized amine results (Note 6). At the conclusion of the diazotization, the resulting solution is stirred a few minutes with 2 g. of diatomaceous earth and filtered by suction (Note 7).

C. *Preparation of the reducing agent.* One hundred and twenty-five grams (0.5 mole) of cupric sulfate pentahydrate is dissolved in 500 ml. of water contained in a 3-l. three-necked flask equipped with a mechanical stirrer, and then 210 ml. of concentrated ammonium hydroxide (sp. gr. 0.90) is added with stirring. The solution is cooled to 10°. A solution of 40 g. (0.57 mole) of hydroxylamine hydrochloride in 140 ml. of water is prepared and also cooled to 10°. To the hydroxylamine hydrochloride solution there is added 95 ml. of 6N sodium hydroxide solution, and if not entirely clear, it is filtered by suction. This hydroxylamine solution is immediately added to the ammoniacal cupric sulfate solution with stirring. Reduction occurs at once with the evolution of nitrogen, and the solution becomes pale blue. If this solution is not used at once, it should be protected from the air.

D. *dl-4,4',6,6'-Tetrachlorodiphenic acid.* The reducing solution prepared above is cooled to 10° and maintained at 10–15° during the addition of the diazo solution from Part B, which is added from a dropping funnel. A feed tube having a 2-mm. opening and dipping well below the surface of the reducing solution should be attached to the stem of the dropping funnel. The feed tube should be bent upward at the end and so placed that mixing of the reducing solution occurs rapidly (Note 8). The diazo solution is added at approximately 25 ml. per minute, and excessive foaming is suppressed by the addition of small amounts of ether (Note 9). At the conclusion of the reaction (Note 10), the ammoniacal solution is transferred to two 4-l. beakers, heated to 80–90°, and rapidly acidified to litmus with concentrated hydrochloric acid with vigorous stirring (Note 11). At this point acidification is continued more carefully until the solution is acid to Congo red (Note 12). A total excess of 100 ml. of acid is then added, and the solution is allowed to stand overnight. The product is filtered by suction and washed on the filter with four 250-ml. portions of water. After drying, the yield of crude product, melting at 180–215°, is 29–38.5 g. (63–84%).

The crude product is dissolved in 3.5 ml. of concentrated sulfuric acid per gram, heated with stirring to 150° for 5 minutes, and allowed to cool overnight. The resulting product is filtered by suction through a sintered-glass funnel and washed on the filter with three 15-ml. portions of concentrated sulfuric acid at room temperature. The filter cake is removed from the funnel and boiled with 50 ml. of water to remove adherent sulfuric acid. The product is then filtered and dried. The above procedure yields 19–22 g. (41–48%) of almost colorless dl-4,4′,6,6′-tetrachlorodiphenic acid melting at 243–250° (uncor.) (Note 13). Pure acid may be obtained by a second recrystallization from concentrated sulfuric acid. Twenty grams of crude dl-4,4′,6,6′-tetrachlorodiphenic acid, m.p. 243°, recrystallized from 70 ml. of concentrated sulfuric acid yields 6.54 g. (33% recovery) of colorless product melting at 258–259°.

2. Notes

1. The rate of flow of chlorine is adjusted so that the reaction mixture is saturated with gas and some gas escapes from the surface of the solution. An indication of proper duration of

chlorination is the development of a distinct brown color in the suspension.   Further chlorination leads to a decrease in yield with the formation of polychloro products.[2]   The time required for chlorination is about 1 hour.

2. Drying at elevated temperatures gives an inferior product because of the formation of polychloro by-products at this stage.

3. Pure 3,5-dichloro-2-aminobenzoic acid [3] has a melting point of 231°.   The product described here is adequate for the subsequent step.

4. This procedure can be performed using 10 times the quantities specified.   The chlorination is carried out in a jar having a capacity of 12–14 l.   Chlorine is introduced by means of a copper tube coiled at the bottom of the jar and perforated in several places.   The time of chlorination is 2 hours.   The percentage yield is the same as that for the scale described above.

5. This solution is almost saturated with the sodium salt of 3,5-dichloro-2-aminobenzoic acid.   If crystallization occurs, additional water may be added.   Obviously, temperatures below 10° should be avoided.

6. As the salt solution enters the acid solution, there is a momentary precipitation of the amino acid, which dissolves rapidly as it is diazotized.   The checkers found that this addition took about 2 hours.

7. The diazo solution may be stored as long as 1 day at 10–15°.   The insoluble gelatinous material that forms during storage should be removed by filtration just before use.   The filter flask used at this point should be cooled in an ice bath to prevent further decomposition of diazo solution.

8. This trap arrangement prevents premature reaction of the entering diazo solution with ammonia, which otherwise would be carried up the feed tube by ascending bubbles of nitrogen.

9. The rate of addition is not a critical factor.   More rapid addition requires more vigorous stirring and may lead to troublesome foaming.

10. The solution may stand for a week before being used.

11. The checkers recommend the use of a Hershberg stirrer.

12. Basic copper salts which precipitate during the acidification redissolve before the Congo red end point is reached if the later stages of acidification are performed carefully with adequate stirring.

13. This procedure is adaptable to 10 times the quantities specified here. Diazotization is carried out in a 12-l. jar. The diazo solution is allowed to stand overnight to facilitate the separation, by decantation or siphoning, of the non-diazotizable impurities whose large-scale filtration is tedious. The main synthesis is performed in a carboy having a capacity of not less than 30 l. The metal stirrer is protected by a coat of paraffin wax. Several addition tubes are used for the diazo solution. By adding appropriate quantities of ice, the necessity of external cooling for these large vessels is avoided. The yield is 39–42% of material melting at 244–250°.

3. Methods of Preparation

3,5-Dichloro-2-aminobenzoic acid can be prepared by the chlorination of anthranilic acid in glacial acetic acid solution [3] and by the action of sulfuryl chloride on anthranilic acid.[4, 5] The procedure above is derived from a detailed study of the chlorination reaction.[2]

The method described for the preparation of dl-4,4',6,6'-tetrachlorodiphenic acid is based on the work of Atkinson and Lawler [6] but employs a more suitable reducing agent than that [7] previously used to convert diazotized anthranilic acid to diphenic acid. The product can be resolved into its optically active forms,[6] which are stable to racemization.

[1] University of New Hampshire, Durham, New Hampshire.
[2] Atkinson and Mitton, J. Am. Chem. Soc., **69**, 3142 (1947).
[3] Elion, Rec. trav. chim., **44**, 1106 (1925).
[4] Eller and Klemm, Ber., **55**, 222 (1922).
[5] Durrans, J. Chem. Soc., **123**, 1424 (1923).
[6] Atkinson and Lawler, J. Am. Chem. Soc., **62**, 1704 (1940).
[7] Org. Syntheses Coll. Vol. **1**, 222 (1941).

TETRACYANOETHYLENE*

(Ethenetetracarbonitrile)

$$4CH_2(CN)_2 + 8Br_2 + KBr \rightarrow KBr[Br_2C(CN)_2]_4 + 8HBr$$

$$KBr[Br_2C(CN)_2]_4 + 4Cu \rightarrow$$
$$2(CN)_2C{=}C(CN)_2 + 4CuBr_2 + KBr$$

Submitted by R. A. Carboni.[1]
Checked by James Cason and Edwin R. Harris.

1. Procedure

Caution! Tetracyanoethylene slowly evolves hydrogen cyanide when exposed to moist air at room temperature. This material should be handled under a hood, and contact with the skin should be avoided. The first step in the preparation should also be carried out under a hood, since bromine is used.

A. *Dibromomalononitrile-potassium bromide complex.* In a 2-l. three-necked flask equipped with an efficient stirrer, a dropping funnel, and a thermometer are placed 900 ml. of cold water, 99 g. (1.5 moles) of malononitrile (Note 1), and 75 g. (0.63 mole) of potassium bromide. The flask is then placed in an ice-water bath, the stirrer is started, and the thermometer is adjusted to extend into the liquid but not into the path of the stirrer. When the temperature of the mixture has dropped to 5–10° (much solid crystallizes), 488 g. (158 ml. at 25°, 3.05 moles) of bromine is added over a period of 2.5 hours. The stirring is continued for an additional 2 hours, while the temperature is held at 5–10°. The precipitated solid complex is collected on a Büchner funnel, washed with 150 ml. of ice-cold water and sucked as dry as possible for about 1 hour (Notes 2 and 3). The grainy product is then dried to constant weight in a vacuum desiccator over phosphorus pentoxide, at the pressure obtained with an aspirator (Notes 4 and 5). The yield of light-yellow product is 324–340 g. (85–89%) (Note 3).

B. *Tetracyanoethylene.* A mixture of 254 g. (0.25 mole) of the dibromomalononitrile-potassium bromide salt and 1 l. of dry benzene is placed in a 2-l. three-necked flask fitted with a sealed mechanical stirrer and a reflux condenser. The stirrer is started (Note 6), and 100 g. (1.57 g. atoms) of precipitated copper powder (Note 7) is added. The mixture is heated at reflux with constant stirring for 10–16 hours. The benzene layer becomes progressively deeper yellow as the reaction proceeds. At the end of the reaction period, the hot mixture is filtered by gravity, using a fluted paper. Most of the heavy solid is easily retained in the flask and is heated under reflux with 300 ml. of dry benzene, with stirring, for 30 minutes. Filtration of the hot mixture is carried out as before. Two 25-ml. portions of hot benzene are used to wash the precipitate and are decanted through the filter.

The combined filtrates are concentrated to approximately 350 ml. and cooled overnight at about 5°. The crystals are filtered by suction, washed with two 25-ml. portions of cold benzene, and dried in a vacuum desiccator (Note 8). The product weighs 35–40 g. (55–62%) and melts at 197–199° in a sealed capillary tube (Notes 9 and 10). This material is suitable for subsequent reactions if it is used within a day or two, although it gives somewhat lower yields than obtained with recrystallized material. This product may be purified, to yield material stable to storage, by recrystallization from nine times its weight of dry chlorobenzene (Note 11). There is recovered 85–90% of light beige crystals (Notes 12 and 13) melting at 199–200° (sealed capillary tube).

2. Notes

1. The malononitrile was obtained from the Winthrop-Stearns Corp., New York, N. Y., and melted at 30–31°.

2. The vapors of dibromomalononitrile-potassium bromide complex are irritating to the eyes and nose. The solid causes discoloration of the skin on contact. Manipulations should be carried out with gloves in a hood.

3. An additional few grams of product separates from the filtrate during this period, and a little more separates during 1–2 days' standing. This material amounts to 10–20 g. (2.5–5%) additional yield satisfactory for the next step.

4. The product may also be dried in a vacuum oven at 50°;

however, the deposition of some free dibromomalononitrile on the walls of the oven renders this method of drying less advantageous. If an oil pump is used to evacuate the desiccator, it should be protected by an adequate trap containing solid sodium hydroxide.

5. It is essential that the complex be thoroughly dried; otherwise the yield of tetracyanoethylene in the subsequent step is materially decreased.

6. Dryness of the complex may be assured, and checked, by attaching a distillation head to the third neck of the flask and distilling benzene until the distillate has run clear for a few milliliters. If the apparatus and reagents were properly dried, only 10–20 ml. of slightly cloudy distillate should be observed.

7. The precipitated copper powder, Grade MD 98, was obtained from Metals Disintegrating Co., Elizabeth, New Jersey.

8. The crude product retains the odor characteristic of dibromomalononitrile and will stain the skin. Pure tetracyanoethylene is practically odorless.

9. The capillary is sealed in order to prevent sublimation; it should not be evacuated unless totally immersed in the heating bath, otherwise sublimation into the cooler part of the sealed capillary will occur.

10. A small additional quantity of less pure product may be obtained by heating the mother liquor with 20 g. of fresh copper powder for 1 hour with stirring, then filtering and concentrating the filtrate to about 100 ml. An equal volume of cyclohexane is added to the hot concentrate, and the mixture is cooled at about 5° for about 30 minutes. Following this procedure in one run, the checkers obtained a yield of 2.9 g. which was recrystallized from 26 g. of chlorobenzene to give 2.1 g. of unattractive material with a poor melting point.

11. The solubility of tetracyanoethylene in chlorobenzene apparently increases sharply as the boiling point of the solvent is approached. Thus the crystals should be extracted in boiling chlorobenzene. Chlorobenzene may be conveniently dried by distilling until the distillate no longer runs cloudy (azeotrope, b.p. 90°, 28.4% water).

12. The crystals, which are yellow when wet with chlorobenzene, become light-colored as the solvent is removed during drying.

13. If an especially good product is desired, the recrystallized

material is sublimed at 130–140°/1 mm. A still better product with no trace of color may be obtained by subliming the recrystallized tetracyanoethylene through activated carbon. For example, 35 g. of tetracyanoethylene is placed in a glass thimble and covered with 20–25 g. of activated wood charcoal chips (4–8 mesh). The mouth of the thimble is covered with a coarse grade of filter paper which is held in place by wiring. The thimble is placed in a sublimer, and the sublimation is carried out at 1–2 mm. (bath temperature 175–190°). The tetracyanoethylene is recovered in 80–90% yield as a colorless, hard crystalline mass that melts at 201–202° (sealed tube).

3. Methods of Preparation

The procedure given above for the dibromomalononitrile-potassium bromide complex is essentially that of Ramberg and Wideqvist.[2] Tetracyanoethylene has also been prepared by passing malononitrile and chlorine through a hot tube at 400°.[3] The present procedure, based on that described by Cairns et al.,[3] appears to be the best preparative method. Metals, other than copper, have been used to effect the reaction.[4] Tetracyanoethylene, the first example of a percyanoölefin, has shown exceptional reactivity in a number of addition reactions. For example, it is a very active dienophile, reacting rapidly at room temperature with many 1,3-dienes to give the corresponding Diels-Alder products.[5] With aromatic hydrocarbons, it forms π-complexes of characteristic colors ranging from yellow to green,[6] and it has been used as a color-forming reagent in paper chromatography of aromatic compounds.[7]

[1] Contribution No. 480 from Central Research Department, Experimental Station, E. I. du Pont de Nemours & Co., Wilmington, Delaware.

[2] Ramberg and Wideqvist, *Arkiv Kemi, Mineral. Geol.*, **12A**, No. 22 (1937).

[3] Cairns, Carboni, Coffman, Engelhardt, Heckert, Little, McGeer, McKusick, Middleton, Scribner, Theobald, and Winberg, *J. Am. Chem. Soc.*, **80**, 2775 (1958); Heckert (to E. I. du Pont de Nemours & Co., U. S. pat. 2,794,823 [*C. A.*, **51**, 16514 (1957)].

[4] Heckert and Little (to E. I. du Pont de Nemours & Co., U. S. pat. 2,794,824 [*C. A.*, **51**, 16515 (1957)].

[5] Middleton, Heckert, Little, and Krespan, *J. Am. Chem. Soc.*, **80**, 2783 (1958).

[6] Merrifield and Phillips, *J. Am. Chem. Soc.*, **80**, 2778 (1958).

[7] Tarbell and Huang, *J. Org. Chem.*, **24**, 887 (1959).

TETRAETHYLTIN

(Tin, tetraethyl-)

$$4C_2H_5Br + 4Mg \rightarrow 4C_2H_5MgBr$$

$$SnCl_4 + 4C_2H_5MgBr \rightarrow (C_2H_5)_4Sn + 2MgCl_2 + 2MgBr_2$$

Submitted by G. J. M. Van Der Kerk and J. G. A. Luijten.[1]
Checked by M. S. Newman and L. L. Wood.

1. Procedure

A 2-l. three-necked flask is fitted with a reflux condenser (Note 1), a stirrer (Note 2), and a dropping funnel. The flask is suspended in a steam cone, which can also be used as a cooling bath. In the flask is placed 50 g. (2.05 g. atoms) of fine magnesium turnings (Note 3). In the dropping funnel is first introduced 5 ml. of a solution of 250 g. (175 ml., 2.3 moles) of ethyl bromide in 500 ml. of absolute ether. Three drops of bromine is mixed with the 5 ml., and the mixture is added to the magnesium. The Grignard reaction which starts at once (Note 4) is maintained by gradually adding the remainder of the ethyl bromide-ether solution. When the spontaneous reaction subsides, the mixture is heated gently under reflux with stirring for 30 minutes.

The flask is then cooled in ice, and in the course of about 20 minutes 83 g. of tin tetrachloride (37 ml., 0.32 mole) is added with vigorous stirring (Note 5). The mixture is heated at the reflux temperature for 1 hour, after which the condenser is set for distillation. During 1.5 hours the ether is removed by distillation while the flask is heated by an ample supply of steam (Notes 6 and 7).

The flask is again cooled in ice, the collected ether is returned to the reaction mixture, and the latter is decomposed by slowly adding first 85 ml. of ice water, then 400 ml. of ice-cold 10% hydrochloric acid. After stirring for some minutes, the contents of the flask are transferred to a separatory funnel. The layers are separated, and the ether layer is filtered through a folded filter and dried with calcium chloride (Note 8).

The ether is removed by distillation, and the crude tetra-ethyltin is distilled under water-pump vacuum, using a water bath for heating. The yield of tetraethyltin boiling at 63–65°/12 mm. is 67–72 g. (89–96%), n_D^{25} 1.4693–1.4699, d_4^{25} 1.1916 (Note 9).

2. Notes

1. A wide condenser must be employed to permit an ample reflux of ether.

2. A seal is recommended as described in *Org. Syntheses Coll. Vol.* **3**, 368 (1955), Note 1.

3. The submitters have carried out this preparation on a three-fold scale with comparable yields.

4. In general bromine starts Grignard reactions more quickly than the usually employed iodine.

5. For adding the tin tetrachloride it is advisable to use a dropping funnel which contains no ether vapor, since the latter gives troublesome formation of solid etherate.

6. The stirrer must be stopped at the beginning of the distillation or it will break, for the contents of the flask turn into a solid mass.

7. Removal of the ether is necessary to permit raising the reaction temperature. The temperature at the center of the mass reaches 60–65°. During the 1.5 hours of distillation, about 200 ml. of ether is collected, the remainder being firmly bound as etherates.

8. To obtain a product free from traces of triethyltin halide the dried ethereal solution is treated with dry ammonia, and the precipitate formed is removed by filtration.

9. The submitters report that the same yields in terms of percentages are obtained if the procedure is applied to the preparation of tetra-*n*-propyltin and tetra-*n*-butyltin.

3. Methods of Preparation

Tetraethyltin has been prepared from tin-sodium alloy and ethyl iodide;[2] from tin-sodium-zinc alloy and ethyl bromide;[3,4] from tin tetrachloride and ethylmagnesium bromide;[5-8] or triethylaluminum;[9] from tin-magnesium alloy and ethyl bromide or chloride;[10-12] and from stannous chloride and ethyllithium.[13]

The method described is essentially that of Pfeiffer and Schnurmann.[5]

[1] Organisch Chemisch Instituut T.N.O., Utrecht, The Netherlands.

[2] Löwig, *Ann.*, **84**, 317 (1852).

[3] Harada, *Sci. Papers Inst. Phys. Chem. Research Tokyo*, **35**, 290 (1939) [*Chem. Zentr.*, **1939**, II, 2912].

[4] Gilman and Arntzen, *J. Org. Chem.*, **15**, 994 (1950).

[5] Pfeiffer and Schnurmann, *Ber.*, **37**, 319 (1904).

[6] Kocheshkov, *Zhur. Obshchei Khim.*, **4**, 1359 (1934) [*Chem. Zentr.*, **1936**, II, 1707].

[7] Korsching, *Z. Naturforsch.*, **1**, 219 (1946) [*C. A.*, **41**, 1902 (1947)].

[8] Mironov, Egorov, and Petrov, *Izvest. Akad. Nauk S.S.S.R., Otdel. Khim. Nauk*, **1959**, 1400 [*C. A.*, **54**, 1266 (1960)].

[9] Zakharkin and Okhlobystin, *Izvest. Akad. Nauk S.S.S.R., Otdel. Khim. Nauk*, **1959**, 1942 [*C. A.*, **54**, 9738 (1960)].

[10] Polkinhorne and Tapley (to Albright and Wilson Ltd.), Brit. pat. 761,357 *C. A.*, **51**, 11382 (1957)].

[11] Ireland, Brit. pat. 713,727 [*C. A.*, **49**, 12530 (1955)].

[12] van der Kerk and Luijten, *J. Appl. Chem. (London)*, **4**, 307 (1954).

[13] Gilman and Rosenberg, *J. Am. Chem. Soc.*, **75**, 2507 (1953).

1,2,3,4-TETRAHYDROCARBAZOLE *

(Carbazole, 1,2,3,4-tetrahydro-)

Submitted by CROSBY U. ROGERS and B. B. CORSON.[1]
Checked by CHARLES C. PRICE, KENNETH N. CAMPBELL, and ROBERT P. KANE.

1. Procedure

A mixture of 98 g. (1 mole) (Note 1) of cyclohexanone and 360 g. (6 moles) of acetic acid contained in a 1-l. three-necked round-bottomed flask equipped with a reflux condenser, a slip-sealed stirrer, and a dropping funnel is heated under reflux and stirred while 108 g. (1 mole) of phenylhydrazine is added during 1 hour. The mixture is heated under reflux for an additional hour and poured into a 1.5-l. beaker and stirred by hand (Note 2) while it solidifies. It is then cooled to about 5° and filtered with suction, the filtrate being cooled in ice and refiltered through the filter cake. The final filtrate is discarded. The filter cake is washed with 100 ml. of water and finally with 100 ml. of 75% ethanol. Each wash is allowed to soak into the filter cake before it is sucked dry. The crude solid is air-dried overnight (Note 3) and crystallized (Note 4) from 700 ml. of methanol after treatment with de-colorizing carbon (Note 5); yield 120–135 g. of 1,2,3,4-tetrahydro-carbazole, m.p. 115–116° (Note 6). The mother liquor is con-centrated to one-fourth of its original volume and yields an ad-ditional 10 g. of product (total yield 76–85%) (Note 7).

2. Notes

1. Equivalent amounts of cyclohexanone (after suitable compensation for purity) and phenylhydrazine are used. The cyclohexanone was about 90% pure (analyzed by the procedure of Bryant and Smith[2]). Instead of analyzing the ketone, it is safe to assume 90% purity.

2. The stirring should be sufficiently vigorous to prevent the formation of lumps.

3. The crude product requires 50–70 hours of air-drying to attain constant weight (145–155 g., 85–91%). It is preferable to crystallize the partially dried product.

4. The approximate solubility of 1,2,3,4-tetrahydrocarbazole in 100 ml. of methanol at 10°, 35°, and 55° is 5, 12, and 18 g. respectively.

5. A heated funnel is desirable for filtration of the hot solution, for the product separates on slight cooling.

6. The capillary melting point of tetrahydrocarbazole ranges from 113° to 114° with slow heating and from 116° to 118° with fast heating.

7. 1,2-Benzo-3,4-dihydrocarbazole may be prepared by the same general procedure. A solution of 172 ml. (2 moles) of concentrated hydrochloric acid (sp. gr. 1.18) in 500 ml. of water is heated at the reflux temperature and stirred in a 2-l. three-necked round-bottomed flask equipped with a reflux condenser, a slip-sealed stirrer, and a dropping funnel while 108 g. (1 mole) of phenylhydrazine is added during 5 minutes. α-Tetralone (p. 898) (146 g., 1 mole or a correspondingly larger amount of material of 90% purity; see Note 1) is added in a period of 1 hour, and the mixture is stirred and heated under reflux for an additional hour. The product is cooled to room temperature with stirring, and the beadlike product is filtered, washed as above, and crystallized from 2.3 l. of methanol after treatment with decolorizing carbon. The first crop amounts to 105–110 g. and the second crop to 75–80 g., making the total yield 82–87%; m.p. 163–164°.

3. Methods of Preparation

1,2,3,4-Tetrahydrocarbazole has been prepared from cyclohexanone phenylhydrazone,[3-9] by the direct reaction of cyclohexanone with phenylhydrazine,[10] by the reaction of 2-chlorocyclohexanone with aniline,[11] by heating 2-phenylcyclohexanone oxime,[12] by the reaction of 2-hydroxycyclohexanone with aniline,[13] by treatment of cyclohexanone phenylhydrazone with a cation-exchange resin,[14] and by the hydrogenation of carbazole.[15] 1,2-Benzo-3,4-dihydrocarbazole has been prepared from the phenylhydrazone of α-tetralone [16] and by the direct reaction of α-tetralone with phenylhydrazine.[10]

[1] Mellon Institute, Pittsburgh, Pennsylvania.

[2] Bryant and Smith, *J. Am. Chem. Soc.*, **57**, 57 (1935).

[3] Drechsel, *J. prakt. Chem.*, [2] **38**, 69 (1888).

[4] Baeyer, *Ann.*, **278**, 88 (1893); Baeyer and Tutein, *Ber.*, **22**, 2178 (1889).

[5] Borsche, *Ann.*, **359**, 49 (1908).

[6] Perkin and Plant, *J. Chem. Soc.*, **119**, 1825 (1921).

[7] Hoshino and Takiura, *Bull. Chem. Soc. Japan*, **11**, 218 (1936) [*C. A.*, **30**, 5985 (1936)].

[8] Grammaticakis, *Compt. rend.*, **209**, 317 (1939).

[9] Yamada, Chibata, and Tsurui, *Pharm. Bull. (Japan)*, **1**, 14 (1953) [*C. A.*, **48**, 12078 (1954)].

[10] Rogers and Corson, *J. Am. Chem. Soc.*, **69**, 2910 (1947).

[11] Badische Anilin & Soda-Fabrik Akt.-Ges., Ger. pat. 947,068 [*C. A.*, **53**, 6250 (1959)].

[12] Löffler and Ginsburg, *Nature*, **172**, 820 (1953).

[13] Jones and Tomlinson, *J. Chem. Soc.*, **1953**, 4114.

[14] Yamada et al. (to Tanabe Drug Manufg. Co.), Jap. pat. 1284 (1954) [*C. A.*, **49**, 11720 (1955)].

[15] Adkins and Coonradt, *J. Am. Chem. Soc.*, **63**, 1563 (1941).

[16] Ghigi, *Gazz. chim. ital.*, **60**, 194 (1930).

ar-TETRAHYDRO-α-NAPHTHOL *

(1-Naphthol, 5,6,7,8-tetrahydro-)

Submitted by C. DAVID GUTSCHE and HUGO H. PETER.[1]
Checked by JOHN C. SHEEHAN, GEORGE H. BUCHI, and DWAIN M. WHITE.

1. Procedure

A 3-l. three-necked flask, equipped with a Dry Ice condenser (Note 1), a sealed Hershberg-type stirrer, and an inlet tube, is set up in a hood and charged with 108 g. (0.75 mole) of α-naphthol (Note 2). The stirrer is started, and to the rapidly stirred flask contents (Note 3) is added 1 l. of liquid ammonia as rapidly as possible (about 5 minutes). When the naphthol has gone into solution (about 10 minutes), 20.8 g. (3.0 g. atoms) of lithium metal (Note 4) is added in small pieces and at such a rate as to prevent the ammonia from refluxing too violently (Note 5). After the addition of the lithium has been completed (about 45 minutes), the solution is stirred for an additional 20 minutes and is then treated with 170 ml. (3.0 moles) of absolute ethanol which is added dropwise over a period of 30–45 minutes (Note 6). The condenser is removed, stirring is continued, and the ammonia is evaporated in a stream of air introduced through the inlet tube. The residue is dissolved in 1 l. of water, and, after the solution has been extracted with two 100-ml. portions of ether, it is carefully acidified with concentrated hydrochloric acid. The product formed is taken into ether with three 250-ml. extractions, and then the ether extract is washed with water and dried over anhydrous sodium sulfate. The ether is removed by evaporation to yield 106–108 g. (97–99%) of crude 5,8-dihydro-1-naphthol, m.p. 69–72°. This material is dissolved in 250 ml. of ethyl acetate and hydrogenated with 3.0 g. of 10% palladium on charcoal catalyst (Note 7) at 2–3. atm. pressure in a Parr apparatus until the theoretical amount of hydrogen has been absorbed (about 45 minutes). The catalyst is removed by filtration, and the solvent is removed by distillation to leave 105–107 g. of an oil which quickly solidifies, m.p. 67–69.5°. Recrystallization from 250 ml. of petroleum ether (b.p. 88–98°) gives 93–97 g. (84–88%) of almost colorless crystals, m.p. 68–68.5°.

2. Notes

1. A cold-finger type of condenser approximately 10 × 40 cm. is satisfactory.

2. Eastman's white label grade α-naphthol or equivalent is the most satisfactory starting material. Technical grade α-naphthol may be used, but it gives an inferior product that is difficult to purify.

3. Rapid stirring during the addition of the ammonia is necessary to prevent the formation of a hard cake and resultant interference with the stirrer.

4. Lithium metal strip (Metalloy Corporation, Rand Tower, Minneapolis, Minnesota) is wiped to remove the protective grease and then placed in petroleum ether (b.p. 32–37°). Pieces are cut with scissors, air-dried to remove solvent, and added to the reaction mixture.

5. During the addition of the lithium the solution turns deep blue. After this has occurred (after about one-third of the lithium has been added), the rate of addition can be increased considerably.

6. Toward the end of the addition of the alcohol, foaming may occur but may be subdued by reducing the rate of stirring.

7. See Mozingo [*Org. Syntheses Coll. Vol.* **3**, 686 (1955)] for the preparation of this catalyst.

3. Methods of Prepar .tion

ar-Tetrahydro-α-naphthol has been prepared by sodium and amyl alcohol reduction of α-naphthylamine followed by diazotization and hydrolysis;[2,3] by sodium and amyl alcohol[4–6] or lithium and ethylamine[7] reduction of α-naphthol; by sulfonation of tetralin followed by sodium hydroxide fusion;[8] and by catalytic reduction of α-naphthol.[9] *ar*-Dihydro-α-naphthol has been prepared by reduction of α-naphthol with sodium and alcohols [10] and with sodium and ammonia.[11] The use of lithium in related systems has been investigated and provides the basis for the preparation described.[12]

[1] Washington University, St. Louis, Missouri.

[2] Bamberger and Althausse, *Ber.*, **21**, 1893 (1888).

[3] Green and Rowe, *J. Chem. Soc.*, **113**, 955 (1918).

[4] Bamberger and Bordt, *Ber.*, **23**, 215 (1890).

[5] Jacobson and Turnbull, *Ber.*, **31**, 897 (1898).

[6] Bachmann and Ness, *J. Am. Chem. Soc.*, **64**, 536 (1942).

[7] Benkeser, Lambert, Ryan, and Stoffey, *J. Am. Chem. Soc.*, **80**, 6573 (1958).

[8] Schroeter, *Ann.*, **426**, 83 (1922).

[9] Musser and Adkins, *J. Am. Chem. Soc.*, **60**, 664 (1938); Gutsche and Peter, *J. Am. Chem. Soc.*, **77**, 5971 (1955).

[10] Rowe and Levin, *J. Chem. Soc.*, **119**, 2021 (1921).

[11] Birch, *J. Chem. Soc.*, **1944**, 430.

[12] Wilds and Nelson, *J. Am. Chem. Soc.*, **75**, 5360 (1953).

cis-Δ⁴-TETRAHYDROPHTHALIC ANHYDRIDE *

(4-Cyclohexene-1,2-dicarboxylic anhydride, cis-)

$$
\begin{array}{c}
\text{CH}_2 \\
\parallel \\
\text{CH} \\
\mid \\
\text{CH} \\
\diagdown \\
\text{CH}_2
\end{array}
\quad + \quad
\begin{array}{c}
\text{CH—C} \overset{O}{\underset{O}{\diagdown}} \\
\parallel \quad\quad\; \diagdown O \\
\text{CH—C} \underset{O}{\diagdown}
\end{array}
\quad \rightarrow \quad
\begin{array}{c}
\text{CH}_2 \quad \text{H} \quad\quad O \\
\text{CH} \quad\; \text{C——C} \\
\parallel \quad\quad\quad\;\; \text{H} \quad\; \diagdown O \\
\text{CH} \quad\; \text{C——C} \\
\diagdown \text{CH}_2 \quad\quad\quad\quad O
\end{array}
$$

Submitted by ARTHUR C. COPE and ELBERT C. HERRICK.[1]
Checked by CHARLES C. PRICE, KENNETH N. CAMPBELL, and ROBERT P. KANE.

1. Procedure

The apparatus shown in Fig. 16, consisting of a 2-l. three-necked round-bottomed flask fitted with an efficient stirrer (Note 1), a gas-inlet tube, a thermometer, and a reflux condenser, is assembled in a ventilated hood. Bubbler tubes containing benzene are attached to the gas inlet tube and the top of the reflux condenser, and 500 ml. of dry benzene and 196 g. (2 moles) of maleic anhydride (Note 2) are placed in the flask. Stirring is begun, the flask is heated with a pan of hot water, and butadiene is introduced (from a commercial cylinder controlled by a needle valve) at a rapid rate (0.6–0.8 l. per minute). When the temperature of the solution has reached 50° (within 3–5 minutes) the pan of water is removed. The exothermic reaction causes the temperature to reach 70–75° in 15–25 minutes. The rapid stream of butadiene is nearly completely absorbed for 30–40 minutes, after which the rate is decreased until the reaction is completed (equal rates of bubbling in the two bubbler tubes) after 2–2.5 hours. The solution is poured into a 1-l. beaker at once to avoid crystallization of the product in the reaction flask. The beaker is covered and the mixture is kept at 0–5° overnight.

The product is collected on a large Büchner funnel and washed with 250 ml. of 35–60° petroleum ether. A second crop (5–15 g.)

obtained by diluting the filtrate with an additional 250 ml. of petroleum ether is separated by filtration, combined with the first crop in a large crystallizing dish, and dried to constant weight

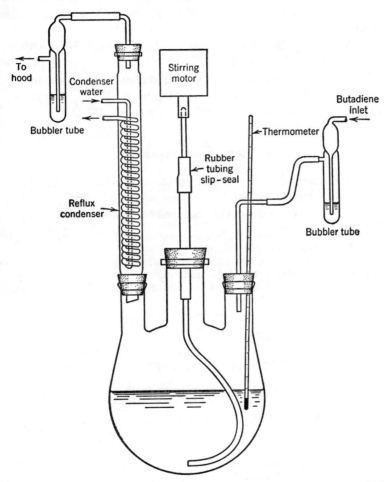

FIG. 16. Assembly of apparatus for addition of butadiene to maleic anhydride.

(1–2 hours) in an oven at 70–80°. The yield of cis-Δ⁴-tetrahydrophthalic anhydride is 281.5–294.5 g. (93–97%), m.p. 99–102° (Note 3).

2. Notes

1. Any stirrer that produces sufficiently vigorous agitation to disperse the gas through the liquid is satisfactory. It was found to be unnecessary to introduce the gas below the surface of the liquid.

2. A good grade of commercial maleic anhydride was used, m.p. 52–54°.

3. The product is analytically pure and suitable for use in preparing diethyl cis-Δ⁴-tetrahydrophthalate (p. 304). Recrystallization from ligroin [2] or ether [3] raises the m.p. to 103–104°.

3. Methods of Preparation

cis-Δ⁴-Tetrahydrophthalic anhydride has been prepared by the reaction of maleic anhydride and butadiene.[2-6] The procedure described is adapted from the one used by Kohler and Jansen.[5]

[1] Massachusetts Institute of Technology, Cambridge, Massachusetts.

[2] Diels and Alder, *Ann.*, **460**, 113 (1928).

[3] Jenkins and Costello, *J. Am. Chem. Soc.*, **68**, 2733 (1946).

[4] Farmer and Warren, *J. Chem. Soc.*, **1929**, 903.

[5] Kohler and Jansen, *J. Am. Chem. Soc.*, **60**, 2144 (1938); Cope and Herrick, *J. Am. Chem. Soc.*, **72**, 983 (1950).

[6] Fieser and Novello, *J. Am. Chem. Soc.*, **64**, 806 (1942).

TETRAHYDROTHIOPHENE *

(Thiophene, tetrahydro-)

$$ClCH_2CH_2CH_2CH_2Cl + Na_2S \rightarrow \boxed{}_S + 2NaCl$$

Submitted by J. KEITH LAWSON, WILLIAM K. EASLEY, and WILLIAM S. WAGNER.[1]
Checked by WILLIAM S. JOHNSON and W. DAVID WOOD.

1. Procedure

A 5-l. round-bottomed three-necked flask equipped with a mechanical stirrer, a reflux condenser (Note 1), and two 250-ml. dropping funnels (mounted with the aid of a Y-tube) is charged with 1.7 l. of dimethylformamide (Note 2). The flask is heated until the dimethylformamide is almost refluxing. Then, with stirring, 280 ml. (318 g., 2.5 moles) of 1,4-dichlorobutane and a solution of 359 g. (2.75 moles) of 60% sodium sulfide (Note 2) in 1 l. of hot water are added simultaneously from the dropping funnels at such a rate that the mixture refluxes without application of heat (Note 3). After the addition is complete (about 1.5 hours), the mixture is heated at reflux with stirring for an additional 2 hours. The condenser is arranged for distillation, and 600 ml. of distillate is collected (Note 4). The distillate is made alkaline by adding 20 g. of sodium hydroxide, and sodium chloride is added to the saturation point. The aqueous layer is separated and discarded, and the crude tetrahydrothiophene layer is dried over solid potassium hydroxide. Distillation through a 30-cm. Vigreux column gives, after a small fore-run (Note 5), 160–172 g. (73–78%) of colorless tetrahydrothiophene, b.p. 119–121°, n_D^{25} 1.5000–1.5014 (Notes 6 and 7).

2. Notes

1. The reaction should be carried out in a hood, or the outlet of the condenser should be connected to a fume trap.

2. Technical grade DMF is available from the Grasselli Chemicals Department of E. I. duPont de Nemours and Company. The 1,4-dichlorobutane was obtained from the Electrochemicals Department, E. I. duPont de Nemours and Company. Technical Baker and Adamson fused chip sodium sulfide assaying 60% sodium sulfide was used. The checkers obtained somewhat lower yields when the appropriate amount of reagent grade $Na_2S \cdot 9H_2O$ was employed instead of the technical material.

3. The addition of the reactants takes approximately 1.3 hours and should be carried out so that the addition of both reactants is completed at approximately the same time.

4. The dimethylformamide solution may be used repeatedly if each time the volume is reduced by the distillation of the remaining water. After two runs the accumulated sodium chloride should be removed by filtration.

5. A small amount of dimethylamine is always present in the crude product. This is removed before collection of the product

by operating the column under total reflux for 1–2 hours. The vent of the still head should be connected to a hood during this operation. If foaming does not subside after this treatment the use of a 1-l. still pot is indicated.

6. When several runs are made, an additional amount of the product can be obtained by redrying and redistilling the fore-runs. The submitters have obtained yields as high as 90% on a larger-scale operation.

7. The reaction may also be carried out in aqueous medium with an increase in the time of reaction to 4 hours. By this method the submitters obtained a 78% yield on a large-scale run. When only a single run is to be made, this procedure may be preferred as a matter of economy.

3. Methods of Preparation

Tetrahydrothiophene has been prepared by the reaction of 1,4-diiodobutane and potassium sulfide;[2] by the reaction of 1,4-dibromobutane [3,4] or 1,4-dichlorobutane [5-7] and sodium sulfide; by the reaction of tetramethylene glycol and hydrogen sulfide in the presence of alumina at high temperature;[8] by the reaction of tetrahydrofuran and hydrogen sulfide in the presence of alumina at high temperature;[9-12] by the hydrogenation of thiophene with molybdenum disulfide [13] or palladium on charcoal [14] as catalyst; and by treatment of S-(4-hydroxybutyl)iso-thiourea chloride hydrochloride with base.[15]

[1] The Chemstrand Corporation, Decatur, Alabama.

[2] von Braun and Trümpler, *Ber.*, **43**, 545 (1910).

[3] Grishkevich-Trokhimovskii, *Zhur. Russ. Fiz. Khim. Obshchestva*, **48**, 901 (1916) [*C. A.*, **11**, 785 (1917)].

[4] Tarbell and Weaver, *J. Am. Chem. Soc.*, **63**, 2940 (1941).

[5] Sumrell and Hornbaker, *J. Org. Chem.*, **23**, 1218 (1958).

[6] Reppe, *Ann.*, **596**, 80 (1955).

[7] Servigne, Szarvasi, and Neuvy, *Compt. rend.*, **238**, 2169 (1954).

[8] Yur'ev and Medovshchikov, *Zhur. Obshcheĭ Khim.*, **9**, 628 (1939) [*C. A.*, **33**, 7779 (1939)].

[9] Yur'ev and Prokina, *Zhur. Obshcheĭ Khim.*, **7**, 1868 (1937) [*C. A.*, **32**, 548 (1938)].

[10] Pipparelli, Balducci, and Floris, *Ann. chim. (Rome)*, **46**, 112 (1956) [*C. A.*, **50**, 13862 (1956)].

[11] Hirao and Hatta, *J. Pharm. Soc. Japan*, **74**, 446 (1954) [*C. A.*, **49**, 6909 (1955)].

[12] Loev and Massengale (to Pennsalt Chemicals Corp.), U. S. pat. 2,899,444 [*C. A.*, **54**, 572 (1960)].

[13] Cawley and Hall, *J. Soc. Chem. Ind.* (*London*), **62**, 116 (1943).

[14] Mozingo, Harris, Wolf, Hoffhine, Easton, and Folkers, *J. Am. Chem. Soc.*, **67**, 2092 (1945).

[15] Kienle (to American Cyanamid Co.), U. S. pat. 2,766,256 [*C. A.*, **51**, 8802 (1957)].

TETRALIN HYDROPEROXIDE

[Hydroperoxide, (1,2,3,4-tetrahydro-1-naphthyl)-]

Submitted by H. B. KNIGHT and DANIEL SWERN.[1]
Checked by JOHN C. SHEEHAN and CURT W. BECK.

1. Procedure

In a 1-l. round-bottomed three-necked flask, equipped with a thermometer, a reflux condenser, and two fritted-glass gas dispersion tubes (Note 1), is placed 600 g. (4.54 moles) of pure tetralin (Note 2). The flask is placed in a constant-temperature bath at 70°, and a finely dispersed stream of oxygen is passed through the tetralin until the peroxide content of the reaction mixture is 25–30% as shown by an active oxygen content of 2.4–2.9% (Note 3). This oxidation requires 24–48 hours. The reaction mixture is then distilled (Note 4) in an all-glass apparatus at 0.2–0.4 mm. through a 60 by 2.5 cm. Vigreux column, until a pot temperature of 70° is reached. About 370–380 g. of unoxidized tetralin, boiling at 32–45°/0.2–0.4 mm., is recovered. The pot residue is a slightly viscous amber-colored oil which weighs 225–235 g. and consists of about 80% tetralin hydroperoxide (Note 5). To obtain the pure hydroperoxide the residue is dissolved in 450 ml. of toluene and the solution is cooled to −50° with stirring (Note 6). After standing at −50° for 1 hour, the slurry is separated by suction filtration (Note 7) and the precipitate is dried at room temperature at 1–2 mm. There is thus obtained 120–125 g. of moderately pure tetralin hydroperoxide, m.p. 50.2–52.0°, active oxygen content 9.20% (Note 8). Recrystallization from 480 ml. of toluene at −30° yields 80–85 g. (44–57% yield based on the peroxide content of the oxidized tetralin) of pure tetralin hydroperoxide as a colorless solid, m.p. 54.0–54.5°, active oxygen content, 9.70% (Note 9).

2. Notes

1. Rubber connections must be avoided because rubber is rapidly attacked by tetralin. Convenient gas dispersion tubes are Pyrex No. 39533.

2. Pure tetralin was prepared from the practical grade supplied by the Eastman Kodak Company by the procedure of George and Robertson:[2] Three kilograms was fractionally distilled through a packed column, the fraction boiling at 204–207° being retained (2.7 kg.). This was gently shaken with 1 lb. of mercury, and the upper layer was carefully decanted through fluted filter paper. The crude tetralin was next shaken with saturated aqueous mercuric acetate solution, and the aqueous layer and a small amount of orange precipitate were discarded. The hydrocarbon was then shaken with 5 successive 300-ml. portions of concentrated sulfuric acid, once with 10% sodium hydroxide, and finally with water until the washings were neutral. The tetralin was dried over anhydrous calcium chloride and distilled through a packed column in a nitrogen atmosphere. In this way, 2 kg. of pure tetralin was obtained, b.p. 206–207°, n_D^{20} 1.5428.

3. Active oxygen content is determined iodometrically:[3] In an iodine flask, an accurately weighed sample (0.1–0.3 g.) is dissolved in 20 ml. of an acetic acid-chloroform solution (3:2 by volume), and 2 ml. of saturated aqueous potassium iodide solution is added. The flask is immediately flushed with nitrogen, stoppered, and allowed to stand at room temperature for 15 minutes. Fifty milliliters of water is then added with good mixing, and the liberated iodine is titrated with 0.1N sodium thiosulfate, employing starch as indicator. A blank titration, which usually does not exceed 0.2 ml., is also run. One milliliter of 0.1N sodium thiosulfate is equivalent to 0.00821 g. of tetralin hydroperoxide.

4. The distillation should be conducted behind a safety shield.

5. Peroxide loss up to this point is negligible provided an all-

glass apparatus has been used throughout and the distillation temperature has not exceeded 70°.

6. A Dry Ice-ethanol bath is convenient for cooling the solution.

7. Filtration may be carried out in a suction funnel surrounded by a Dry Ice-ethanol cooling bath, or using a cold box maintained at −50°.

8. Pure tetralin hydroperoxide has an active oxygen content of 9.74%.

9. Tetralin hydroperoxide is a convenient model compound for many studies in peroxide chemistry. It remains colorless and does not decrease in peroxide content for months if stored in the dark at or below 0°. Storage under warm summer conditions for several months results in decomposition to a dark, viscous liquid.

3. Methods of Preparation

Tetralin hydroperoxide has been prepared by the oxidation of tetralin in the presence of cobalt naphthenate,[4] manganese stearate,[5] and ceric stearate or naphthenate.[6] Tetralin hydroperoxide, produced by the oxidation of tetralin, may be separated from reaction products by formation of its sodium salt.[7] The kinetics of the liquid-phase oxidation of tetralin have been studied.[8] The present procedure is adapted from that reported by Hartmann and Seiberth [9] and Hock and Susemihl.[10]

[1] Eastern Regional Research Laboratory, U. S. Department of Agriculture, Philadelphia, Pennsylvania.

[2] George and Robertson, *Trans. Faraday Soc.*, **42**, 227 (1946).

[3] Wheeler, *J. Am. Oil Chemists' Soc.*, **9**, 89 (1932).

[4] Robertson and Waters, *J. Chem. Soc.*, **1948**, 1578.

[5] Chizhevskaya and Idel'chik, *Zhur. Obshcheĭ Khim.*, **27**, 83 (1957) [*C. A.*, **51**, 12022 (1957)].

[6] Farbenfabriken Bayer A.-G., Ger. pat. 889,443 [*C. A.*, **51**, 12974 (1957)].

[7] Johnson (to Koppers Co., Inc.), U. S. pat. 2,568,639 [*C. A.*, **46**, 5089 (1952)].

[8] Woodward and Mesrobian, *J. Am. Chem. Soc.*, **75**, 6189 (1953).

[9] Hartmann and Seiberth, *Helv. Chim. Acta*, **15**, 1390 (1932).

[10] Hock and Susemihl, *Ber.*, **66**, 61 (1933).

α-TETRALONE *

[1(2H)-Naphthalenone, 3,4-dihydro-]

I. METHOD A

$$C_6H_6 + \underset{\underset{\underset{O}{\parallel}}{\underset{C}{\diagdown}}}{\overset{CH_2-CH_2}{\underset{O}{\diagdown}\underset{CH_2}{\diagup}}} \xrightarrow{\text{AlCl}_3}$$

Submitted by CECIL E. OLSON and ALFRED R. BADER.[1]
Checked by JAMES CASON, GERHARD J. FONKEN, and WILLIAM G. DAUBEN.

1. Procedure

A 3-l. three-necked flask is fitted with a mercury-sealed stirrer, an efficient condenser capped by a drying tube filled with calcium chloride, and a wide-bore rubber tube leading to a 1-l. Erlenmeyer flask. One liter of dry, thiophene-free benzene and 104 g. (1.21 moles) of γ-butyrolactone (Note 1) are placed in the 3-l. flask. Six hundred grams (4.5 moles) of reagent grade anhydrous aluminum chloride (Note 2) is placed in the Erlenmeyer flask and is added to the stirred reaction mixture during a period of 2 hours. The mixture becomes dark brown, refluxes gently, and evolves hydrogen chloride. After addition of all the catalyst, the mixture is heated on a steam bath with continued stirring for 16 hours. It is then cooled to room temperature and poured onto 3 kg. of crushed ice drenched with 500 ml. of concentrated hydrochloric acid. The lower aqueous layer is separated and extracted with about 500 ml. of toluene. The brown, organic, upper layer and the toluene extract are combined, washed successively with water, 20% potassium hydroxide solution, and water, and distilled under reduced pressure to remove benzene, toluene, and traces of water. Distillation of the residue in a Claisen flask (Note 3) yields 160–170 g. (91–96%) of α-tetralone, b.p. 75–85°/0.3 mm., n_D^{25} 1.565–

1.568 (Notes 4 and 5). There is a residue consisting of 130–150 g. of red-purple viscous oil.

2. Notes

1. A commercial grade of butyrolactone supplied by the General Aniline and Film Corporation, 230 Park Avenue, New York 17, New York, was used by the submitters. The checkers used material from Eastern Chemical Corporation, 34 Spring Street, Newark 2, New Jersey, which was distilled before use, b.p. 208–210°.

2. The checkers found technical grade aluminum chloride, containing a little ferric chloride, equally satisfactory. After 150–200 g. of aluminum chloride is added there is usually a rather sudden and vigorous evolution of hydrogen chloride accompanied by refluxing. It is advisable to interrupt addition of aluminum chloride at this point until the mixture is refluxing smoothly.

3. The checkers distilled the product through a 0.5-meter packed column with heated jacket and partial reflux head. The yield was 148–155 g. (85–88%), b.p. 143–145°/20 mm., n_D^{25} 1.5669–1.5671. There was a fore-run of 10–15 g. and an after-run of about 8 g., b.p. 145–180°/20 mm.

4. The submitters state that varying the amounts of aluminum chloride varied the yields as shown in the table.

AlCl₃, g.	α-Tetralone, g.	Yield, %
400	119	68
500	151	86
600	165	94
700	166	95

5. Replacing the butyrolactone with 120 g. (1.20 moles) of γ-valerolactone in an otherwise identical procedure yields 150–160 g. (79–84%) of 4-methyl-1-tetralone, b.p. 108–110°/1 mm.

II. METHOD B

$$HO_2C\diagdown \text{(ring)} CH_2 CH_2 CH_2 + PCl_5 \longrightarrow ClCO\diagdown \text{(ring)} CH_2 CH_2 CH_2 + HCl + POCl_3$$

$$ClCO\diagdown \text{(ring)} CH_2 CH_2 CH_2 \xrightarrow{SnCl_4} \text{(ring)} C(=O) CH_2 CH_2 CH_2 + HCl$$

Submitted by G. Dana Johnson.[2]
Checked by James Cason, William G. Dauben, Bradford H. Walker, and Charles E. Stehr.

1. Procedure

In a dry 2-l. three-necked round-bottomed flask, fitted with a gas-tight stirrer and a reflux condenser carrying at the top a calcium chloride drying tube connected to a gas-absorption trap (a good hood is preferable), are placed 98.5 g. (0.6 mole) of γ-phenylbutyric acid [3] and 200 ml. of dry thiophene-free benzene (Note 1). After the solution has been cooled, with stirring, for a few minutes in an ice bath, 125 g. (0.6 mole) of phosphorus pentachloride is added during 5 minutes (Note 2). After the ice bath is removed the contents of the flask are heated during 20 minutes (vigorous evolution of hydrogen chloride) to the boiling point by means of a water bath, and refluxing is continued for about 5 minutes. As stirring is continued the flask is cooled in an ice-salt bath until the internal temperature (Note 3) reaches about −10°. With continued efficient cooling, there is added during 30–40 minutes a solution of 150 ml. (1.28 moles) of anhydrous stannic chloride in 150 ml. of dry thiophene-free benzene (Note 1), as the temperature is maintained below 15°. The reaction is highly exothermic, and hydrogen chloride is rapidly evolved. Stirring is continued for 1 hour at 0–10°, at the end of which time the thermometer is replaced by the condenser and the complex is decomposed by careful addition of 300 g. of ice followed by 250 ml. of concentrated hydrochloric acid. This two-phase mixture is heated to reflux, with vigorous stirring, on a water bath for about 25 minutes or until hydrogen chloride is no longer evolved (Note 4).

The cooled reaction mixture is separated in a separatory funnel, and the aqueous phase is extracted with three 50-ml. portions of benzene. These extracts are combined with each other but kept separate from the original organic phase; each wash solution is used first with the original organic phase, then the extracts. The washes are 150 ml. of water (Note 4), 100 ml. of 10% sodium carbonate solution, 100 ml. of water, and finally 50 ml. of saturated sodium chloride solution (Note 5). Solvent is distilled from the combined extracts, and the residue is distilled at reduced pressure in a Claisen flask. The yield of α-tetralone, b.p. 135–137°/15 mm., n_D^{25} 1.5671–1.5672, is 75–80 g. (85–91%).

2. Notes

1. A total of 350 ml. of dry benzene is required. It may be dried by allowing it to stand for a few days with about 1 g. of sodium wire, or by slowly distilling about 20% of a lot of benzene and cooling the residue with protection from atmospheric moisture by use of a calcium chloride tube.

2. It is convenient to weigh the phosphorus pentachloride in an Erlenmeyer flask which is then attached to a side neck of the three-necked flask by a 6-in. length of wide-bore thin-walled rubber tubing.

3. The condenser is replaced by a thermometer extending into the stirred liquid. The thermometer is inserted through a wide-bore T-tube whose side outlet is protected by a calcium chloride tube.

4. The stannic chloride complex is decomposed relatively slowly. Addition of 5–10 ml. of ether facilitates the decomposition. If the decomposition is not completed during the heating period, the wash with water gives a troublesome precipitation of stannic hydroxide. If this occurs the organic phase should either be heated for 30 minutes with, or allowed to stand overnight with, 100 ml. of 6N hydrochloric acid.

5. The wash with saturated salt solution usually gives a clean separation and removes most of the water from the organic phase. No additional drying is necessary since the remaining water is removed by azeotropic distillation with benzene.

3. Methods of Preparation

Other methods of preparation of α-tetralone have been reviewed in earlier volumes.[4, 5] Further references to the preparation of α-tetralone by the oxidation of tetralin include the use of oxygen in the presence of cobaltous chloride and acetic acid [6] and chromic anhydride.[7, 8] Beyerman [9] and Eastham [10] state that the α-tetralone obtained by the air oxidation of tetralin contains α-tetralol, which cannot be separated readily by fractional distillation. The former recommends removing the α-tetralol by treatment with chromic acid in acetic acid below 15°, while the latter investigator suggests that the reaction product be dissolved in an equal volume of 75% sulfuric acid, and the mixture extracted with hexane. The hexane extract is discarded, the acid layer is then diluted with 2 volumes of water and again extracted with hexane. The α-tetralone recovered from this hexane extract by washing, concentrating, and distilling melts at 6–7°. Eastham [10] believes that the melting point rather than the refractive index should be used as a test of purity of α-tetralone. α-Tetralone also has been obtained from tetrahydronaphthalene peroxide and sodium hydroxide or cupric chloride.[11]

4-Methyl-1-tetralone has been prepared from γ-phenylvaleric acid and sulfuric acid [12] and from γ-phenylvaleryl chloride and aluminum chloride.[13, 14]

Method I is based on the procedure of Truce and Olson.[15]

[1] Pittsburgh Plate Glass Company, Milwaukee, Wisconsin.

[2] Indiana University, Bloomington, Indiana.

[3] *Org. Syntheses Coll. Vol.* **2**, 499 (1943).

[4] *Org. Syntheses Coll. Vol.* **2**, 569 (1943).

[5] *Org. Syntheses Coll. Vol.* **3**, 798 (1955).

[6] Institut français du pétrole des carburants et lubrifiants, French pat. 1,095,348 [*C. A.*, **53**, 1289 (1959)].

[7] Staveley and Smith, *J. Inst. Petroleum*, **42**, 55 (1956).

[8] Nazarov and Burmistrova, *Zhur. Obshcheĭ Khim. (J. Gen. Chem.)*, **20**, 1304 (1950) [*C. A.*, **45**, 1562 (1951)].

[9] Beyerman, *Rec. trav. chim.*, **72**, 550 (1953).

[10] Eastham, Private communication.

[11] Johnson (to Koppers Co., Inc.), U. S. pat. 2,462,103 [*C. A.*, **43**, 3848 (1949)].

[12] Kloetzel, *J. Am. Chem. Soc.*, **62**, 1708 (1940).

[13] Mayer and Stamm, *Ber.*, **56**, 1431 (1923).

[14] v. Braun and Stuckenschmidt, *Ber.*, **56**, 1724 (1923).

[15] Truce and Olson, *J. Am. Chem. Soc.*, **74**, 4721 (1952).

β-TETRALONE *

[2(1H)-Naphthalenone, 3,4-dihydro-]

Submitted by M. D. SOFFER, M. P. BELLIS, HILDA E. GELLERSON, and ROBERTA A. STEWART.[1]

Checked by N. J. LEONARD and R. C. FOX.

1. Procedure

A. *Reduction and hydrolysis.* A solution of 129 g. (0.75 mole) of β-naphthyl ethyl ether in 1.5 l. of 95% ethanol is prepared in a 5-l. three-necked flask fitted with a mechanical stirrer, a bulb condenser topped by a Friedrichs condenser, and a Y-tube to allow for the introduction of nitrogen and sodium. The apparatus is flushed thoroughly with nitrogen, the nitrogen flow is reduced, and 225 g. (9.8 g. atoms) of sodium is added in small portions (Note 1), with efficient stirring, at a rate sufficient to maintain vigorous boiling. The hydrogen liberated passes through the condensers and a delivery tube into a hood or directly out a window. When approximately two-thirds of the sodium has been introduced, an additional 375 ml. of 95% ethanol is added to reduce the viscosity of the reaction mixture. Approximately 1.5 hours is required for the addition of the sodium. Near the end of this period, the rate of the reaction decreases and heat is supplied by means of an electric heating mantle to maintain the reflux temperature (Note 2).

After all the sodium has dissolved, the hydrogen is thoroughly swept from the system with nitrogen and the heating is discontinued. The Y-tube is replaced by a separatory funnel, and 750 ml. of water is added cautiously (Note 3), with stirring, followed by 1.5 l. of concentrated hydrochloric acid (Note 3). The acidic mixture containing precipitated sodium chloride is heated, with stirring, at the reflux temperature for 30 minutes, and then allowed to cool.

The mixture is extracted with ten 175-ml. portions of benzene or 1:1 benzene-ether mixture (Note 4). The combined extract is washed with 75-ml. portions of water until the washings are neutral to litmus. The organic solvent is removed by distillation on a steam bath. The crude oily residue is converted directly (Note 5) to the β-tetralone bisulfite addition product.

B. *Bisulfite addition product of β-tetralone.* To a solution of 325 g. (3.12 moles) of sodium bisulfite (commercial purified grade) in 565 ml. of water is added 175 ml. of 95% ethanol. The mixture is allowed to stand overnight, and the precipitated sodium bisulfite is removed by filtration. The filtrate is added to the crude β-tetralone, and the mixture is shaken vigorously. Within a few minutes the addition product separates as a voluminous precipitate. The mixture is kept cold for several hours, shaken periodically, and then filtered with the aid of suction. The precipitate is washed well, first with 125 ml. of 95% ethanol, then four times with 125-ml. portions of ether (Note 6). The colorless addition product is air-dried (Note 5) and is stored in air-tight containers. The yield is 113–131 g. (60–70% based on β-naphthyl ethyl ether) (Note 7).

C. *Regeneration of β-tetralone.* Fifty grams (0.20 mole) of β-tetralone bisulfite addition product is suspended in 250 ml. of water, and 75 g. (0.6 mole) of sodium carbonate monohydrate is added. At this point the pH of the solution is approximately 10. The mixture is extracted with five 100-ml. portions of ether (Note 8). The combined extract is washed with 100 ml. of 10% hydrochloric acid, then with 100-ml. portions of water until the washings are neutral to litmus, and is dried over anhydrous magnesium sulfate. The ether is removed by distillation, and the residue is distilled from a Claisen flask under reduced pressure, preferably in a nitrogen atmosphere. The pure β-tetralone is obtained as a colorless distillate; b.p. 70–71°/0.25 mm. (92–94°/1.8 mm., 114–116°/4.5 mm.); n_D^{20} 1.5594. The yield is 17–21 g. (40–50% based on β-naphthyl ethyl ether).

2. Notes

1. The sodium is cut into pieces about 2 in. by 0.5 in. by 0.5 in. and is kept under benzene until used.

2. The concentrated sodium ethoxide solution must be kept hot throughout the process; otherwise a voluminous precipitate separates and the mixture congeals.

3. Both the hydrolysis of the sodium ethoxide and the subsequent neutralization are very exothermic. The water and the acid may be added fast enough to maintain vigorous refluxing, but care must be exercised to keep the reaction under control. When the neutralization is complete, the refluxing subsides, and at this point the excess hydrochloric acid may be added rapidly.

4. Ether is somewhat miscible with the reaction mixture, but the benzene-ether mixture, or benzene alone, is satisfactory.

5. In this step the β-tetralone is separated from starting material and other neutral substances in the reaction mixture. β-Tetralone is sensitive to air oxidation; therefore it should not be stored in the free state. The bisulfite addition product is stable, and the dry material can be stored indefinitely without deterioration.

6. In order to obtain a product of high quality, the washing must be thorough. This is accomplished best by suspending the precipitate in the wash solvent and mixing well before filtering.

7. The yield of the bisulfite addition product is subject to variation owing to coprecipitation of sodium bisulfite.

8. The ether extract and the water layer may be tested for the presence of β-tetralone or the bisulfite addition product by the tetralone blue test.

Tetralone blue test. A few drops of the organic solvent layer or the aqueous phase is shaken in a test tube with 2 ml. of 95% ethanol, and 10 drops of 25% sodium hydroxide solution is poured down the side of the tube. In the presence of air a deep blue color appears at the interface within 1 minute.

3. Methods of Preparation

β-Tetralone has been prepared by a variety of methods, but the only practical procedures are ones involving reduction of β-naphthyl methyl ether with sodium and alcohol [2] or with sodium and liquid ammonia,[3] high-pressure catalytic hydrogenation of β-naphthol,[4] catalytic oxidation of 2-tetralol by hydrogen transfer with ethylene,[5] or the condensation of ethylene with phenylacetyl chloride in the presence of aluminum chloride.[6]

The procedure described here is an adaptation [7] of the method of Cornforth, Cornforth, and Robinson.[2]

[1] Smith College, Northampton, Massachusetts.

[2] Cornforth, Cornforth, and Robinson, *J. Chem. Soc.*, **1942**, 689.

[3] Birch, *J. Chem. Soc.*, **1944**, 430.

[4] Stork and Foreman, *J. Am. Chem. Soc.*, **68**, 2172 (1946).

[5] Adkins, Rossow, and Carnahan, *J. Am. Chem. Soc.*, **70**, 4247 (1948).

[6] Burckhalter and Campbell, *J. Org. Chem.*, **26**, 4232 (1961).

[7] Soffer, Stewart, Cavagnol, Gellerson, and Bowler, *J. Am. Chem. Soc.*, **72**, 3704 (1950).

2,2,6,6-TETRAMETHYLOLCYCLOHEXANOL

(1,1,3,3-Cyclohexanetetramethanol, 2-hydroxy-)

$$+ \ 5HCHO + H_2O \xrightarrow{\text{CaO}}$$

$$+ \ HCO_2H$$

Submitted by HAROLD WITTCOFF.[1]
Checked by R. S. SCHREIBER, WM. BRADLEY REID, JR., and JOHN L. WHITE.

1. Procedure [2]

A mixture of 196 g. (206 ml., 2 moles) of freshly distilled cyclohexanone, 332 g. (11 moles) of 99.5% paraformaldehyde (Note 1), and 1.8 l. of water is placed in a 5-l. flask equipped with a thermometer and an efficient stirrer. The mixture is cooled to 10–15° by an ice-water bath, and 70 g. (1.25 moles) of calcium oxide is added through a powder funnel over a period of 10–15 minutes. The temperature is allowed to rise slowly to 40° and is kept there by means of the cooling bath until the addition is complete. Stirring is maintained throughout. The reaction mixture is stirred for an additional 30 minutes. During this time the temperature usually falls to approximately 35°, and at this point the cooling bath is removed. The reaction mixture is then made slightly acid (pH 6–6.5) by the addition of 11–13 ml. of aqueous 87% formic acid. It is best to stir the reaction mixture for 30 minutes after neutralization in order to make sure that any suspended particles of lime are neutralized. If at the end of this time the solution is not acid, more formic acid should be added. The reaction mixture is then evaporated under reduced pressure to dryness (Note 2). The residue, which consists of a mixture of product and calcium formate, is mixed with 1 l. of absolute methanol. On warming, the organic material dissolves and the calcium formate settles to the bottom of the flask. A practically colorless solution of the product is obtained by filtration with suction (Note 3) through a heated funnel. The insoluble calcium formate is washed with about 50 ml. of methanol. Approximately one-half of the methanol is removed under reduced pressure, and the residual syrupy solution (Note 4) is allowed to crystallize in an ice chest for 24 hours. Thereafter the product is filtered and washed with 50 ml. of methanol. The mother liquor and washings are combined and set aside. With a mortar and pestle, the crystals are triturated successively with three 200–300-ml. portions of acetone, filtered, and air-dried. By continued evaporation of the mother liquor and washings, at least two successive crops of product are obtained, and these are processed as described above (Note 5). The total yield of product melting at 128–129° is 320–374 g. (73–85%). Recrystallization of 100 g. of material, m.p. 128–129°, from 175 ml. of absolute methanol yields 84 g. of pure product melting at 129–130°.

2. Notes

1. An equivalent quantity of aqueous formaldehyde free from methanol may be used.
2. The evaporation may be carried out in an ordinary distillation apparatus, but it is essential to stir the mixture to prevent bumping. It is advantageous from the point of view of speed and convenience to employ the method described below.

A 3-l. three-necked flask is fitted with an efficient rubber-sealed stirrer, an upright steam-heated condenser, the upper end of which is joined to a separatory funnel, and an outlet tube connected to a long water-cooled condenser placed for downward distillation. The end of this condenser is fitted to a 2-l. two-necked flask which is connected to a good water pump and immersed in an ice bath. The apparatus, with the exception of the separatory funnel, is placed under vacuum. The aqueous solution contained in a separatory funnel is slowly passed through the upright condenser into the three-necked flask heated on a steam bath. The syrupy residue collects in the flask, and the water vapor is removed through the condenser fitted for downward distillation. The residue is heated until no more water is removed. The total time required to strip off the water is 30 to 60 minutes.

3. The checkers found that suction filtration was very slow, owing to the very finely divided material that collected on the funnel. This was corrected by adding a filtering aid (Celite) to the crude mixture and filtering by suction, using an ordinary Büchner funnel (6-in. size). After removal of the solid by suction filtration, the methanol volume was reduced as indicated.

4. The submitter reported that it was preferable to carry out the crystallizations in metal beakers since the crystals were very hard to remove without breakage, but the checkers experienced no such difficulty.

5. Runs employing six times the quantities specified here have been carried out in 22-l. flasks with similar results.

3. Methods of Preparation

2,2,6,6-Tetramethylolcyclohexanol has been prepared by Mannich and Brose [3] by condensing cyclohexanone with formaldehyde in the presence of calcium oxide and precipitating the catalyst as calcium sulfate.

[1] General Mills, Inc., Minneapolis, Minnesota.
[2] This procedure forms a portion of the subject matter of U. S. pat. 2,462,031 [C. A., **44**, 656 (1950)]; U. S. pat. 2,493,733 [C. A., **44**, 3017 (1950)]; U. S. pat. 2,527,853 [C. A., **45**, 2262 (1951)].
[3] Mannich and Brose, Ber., **56**, 833 (1923).

TETRAPHENYLARSONIUM CHLORIDE HYDROCHLORIDE*

(Arsonium chloride, tetraphenyl-, hydrochloride)

$$3C_6H_5Cl + AsCl_3 + 6Na \rightarrow (C_6H_5)_3As + 6NaCl$$

$$(C_6H_5)_3As + H_2O_2 \rightarrow (C_6H_5)_3AsO + H_2O$$

$$(C_6H_5)_3AsO + C_6H_5MgBr \rightarrow (C_6H_5)_4AsOMgBr$$

$$(C_6H_5)_4AsOMgBr + 3HCl \rightarrow$$

$$(C_6H_5)_4AsCl \cdot HCl + MgClBr + H_2O$$

Submitted by R. L. Shriner and Calvin N. Wolf.[1]
Checked by Cliff S. Hamilton and Yao-Hua Wu.

1. Procedure

Caution! All steps in the following preparations should be performed under a hood, since volatile arsenic compounds may be liberated.

A. *Triphenylarsine.* In a 2-l. round-bottomed three-necked flask is placed 130 g. (5.65 g. atoms) of powdered sodium[2] covered with 900 ml. of benzene. The flask is fitted with an Allihn condenser, a mercury-sealed mechanical Hershberg stirrer, and a 500-ml. dropping funnel in which is placed a mixture of 170 g. (0.94 mole) of arsenic trichloride and 272 g. (2.42 moles) of chlorobenzene. About 10 ml. of the arsenic trichloride-chlorobenzene mixture is dropped into the flask, and the reaction mixture is stirred and heated on a steam bath until it darkens and boils spontaneously. The steam bath is removed, and the remainder of the arsenic trichloride-chlorobenzene mixture is added dropwise, with stirring, over a period of 1–1.5 hours at such a rate that gentle boiling is maintained (Note 1). When the addition is complete, the mixture is stirred and heated under reflux on a steam bath for 12 hours.

The reaction mixture is filtered while hot through a large Büchner funnel, and the filtrate is collected in a 3-l. filter flask. The residue (Note 2) is washed on the funnel with two 200-ml.

portions of hot benzene, pressed as dry as possible, and then transferred to a 1-l. beaker, boiled with 300 ml. of benzene, and filtered, the same funnel and flask being used. This extraction process is repeated twice.

The combined benzene filtrates are subjected to distillation from a steam bath to remove the benzene. The flask containing the residual red oil is connected to a water pump and heated under reduced pressure in an oil bath at 110–120° for 2 hours to remove unreacted starting materials. When cooled, the crude triphenylarsine solidifies to a light brown solid which melts at 57–59°. The yield is 230–240 g. (93–97%). The crude product is dissolved in 650–700 ml. of hot 95% ethanol and placed in a refrigerator overnight. The crystals are collected on a Büchner funnel and washed with 50 ml. of cold 95% ethanol. The yield is 218–225 g. (88–91%) of white crystals which melt at 61°.

B. *Triphenylarsine oxide.* In a 500-ml. round-bottomed flask equipped with a mechanical stirrer, a thermometer, and a 100-ml. dropping funnel is placed 100 g. (0.33 mole) of the recrystallized triphenylarsine (Note 3) dissolved in 200 ml. of acetone. To the solution 46 g. (0.41 mole) of 30% hydrogen peroxide is added dropwise, with stirring, over a period of 20–30 minutes. A water-ice bath is used to maintain the temperature at 25–30°. When the addition is complete, stirring is continued for 30 minutes, and the acetone is then removed by distillation.

The flask containing the residual yellow oil is fitted with a water trap and condenser, and 120 ml. of benzene is added. The water is then removed by azeotropic distillation (Note 4). When the removal of water is complete, the triphenylarsine oxide is crystallized by cooling the flask in an ice bath for 1.5–2 hours. The light-brown crystals are collected on a Büchner funnel and washed on the funnel with 25 ml. of cold benzene. The crude product weighs 97–98.5 g. (91–93%) and melts at 186–188°. The crude product is transferred to a porcelain dish and triturated with 50 ml. of benzene, collected on a Büchner funnel, pressed as dry as possible, and washed with 25 ml. of cold benzene. After drying in the air, the triphenylarsine oxide amounts to 89–92 g. (84–87%) of white crystals, melting at 189°.

C. *Tetraphenylarsonium chloride hydrochloride.* In a 2-l.

round-bottomed three-necked flask fitted with a condenser, a mercury-sealed mechanical Hershberg stirrer, and a dropping funnel is placed 40 g. (0.124 mole) of triphenylarsine oxide dissolved in 1 l. of hot benzene. To this solution there is added with vigorous stirring a solution of phenylmagnesium bromide which is prepared from 34.6 g. (0.22 mole) of bromobenzene, 6.0 g. (0.25 g. atom) of magnesium, and 200 ml. of dry ether. A brown viscous solid separates. The mixture is stirred for 15 minutes and then stirred and heated under reflux on a steam bath for 30 minutes. The solvent is removed by decantation, and the viscous solid is washed with 500 ml. of benzene. The addition product is then hydrolyzed with 100 ml. of water containing 5 ml. of concentrated hydrochloric acid.

The hydrolysis mixture is transferred to a 1-l. round-bottomed flask fitted with a reflux condenser, and 500 ml. of concentrated hydrochloric acid is added (Note 5). The mixture is heated on a steam bath for 1.5–2 hours. The flask is cooled in an ice bath; the crystals are collected on a sintered-glass funnel and washed with 200 ml. of ice-cold concentrated hydrochloric acid and then with 200 ml. of ice-cold dry ether. The crude product weighs 50–56 g. The product is dissolved in a mixture of 50 ml. of water and 150 ml. of concentrated hydrochloric acid by boiling under reflux. The tetraphenylarsonium chloride hydrochloride separates when the solution is cooled in an ice bath. The white needles are collected on a sintered-glass funnel and washed with 50 ml. of ice-cold concentrated hydrochloric acid and then with 200 ml. of ice-cold dry ether. The yield of tetraphenylarsonium chloride hydrochloride melting at 204–208° with decomposition is 42–45 g. (74–80%).

2. Notes

1. If the addition of the arsenic trichloride-chlorobenzene mixture is too rapid, the reaction becomes vigorous and must be moderated with a cooling bath.

2. Before being discarded, the residue should be treated with ethanol to destroy unreacted sodium.

3. If crude triphenylarsine is used, the final product is difficult to purify.

4. Triphenylarsine oxide is partly converted to the dihydroxide when heated with water. However, it is not hygroscopic under ordinary conditions.[3]

5. The hydrolysis product consists mainly of tetraphenylarsonium bromide, which is converted to tetraphenylarsonium chloride hydrochloride by crystallization from concentrated hydrochloric acid.

3. Methods of Preparation

Triphenylarsine has been prepared by the action of arsenic triiodide [4] or arsenic trichloride [5] on phenylmagnesium bromide and by the action of sodium and arsenic trichloride on chlorobenzene [6] or bromobenzene.[7] The method described here is essentially that of Pope and Turner.[6]

Triphenylarsine oxide has been prepared by the action of sodium hydroxide on triphenylarsine dibromide,[8] by the action of potassium permanganate on triphenylarsine,[9] and by the action of hydrogen peroxide on triphenylarsine.[10] The method described here is essentially that of Vaughan and Tarbell.[10]

Tetraphenylarsonium chloride hydrochloride has been prepared by the action of phenylmagnesium bromide on triphenylarsine oxide.[11] The method described here is essentially that of Blicke and Monroe.[11]

[1] State University of Iowa, Iowa City, Iowa.

[2] *Org. Syntheses Coll. Vol.* **1**, 252 (1941).

[3] Blicke and Cataline, *J. Am. Chem. Soc.*, **60**, 419 (1938).

[4] Burrows and Turner, *J. Chem. Soc.*, **117**, 1373 (1920).

[5] Pfeiffer and Pietsch, *Ber.*, **37**, 4621 (1904).

[6] Pope and Turner, *J. Chem. Soc.*, **117**, 1447 (1920).

[7] Michaelis, *Ann.*, **321**, 160 (1902).

[8] Philips, *Ber.*, **19**, 1031 (1886).

[9] Blicke and Safir, *J. Am. Chem. Soc.*, **63**, 575 (1941).

[10] Vaughan and Tarbell, *J. Am. Chem. Soc.*, **67**, 144 (1945).

[11] Blicke and Monroe, *J. Am. Chem. Soc.*, **57**, 720 (1935).

TETRAPHENYLETHYLENE *

(Ethylene, tetraphenyl-)

$$2(C_6H_5)_2CCl_2 + 4Cu \rightarrow (C_6H_5)_2C{=}C(C_6H_5)_2 + 2Cu_2Cl_2$$

Submitted by ROBERT E. BUCKLES and GEORGE M. MATLACK.[1]
Checked by T. L. CAIRNS and C. J. ALBISETTI.

1. Procedure

A solution of 75 g. (0.32 mole) of diphenyldichloromethane (Note 1) in 250 ml. of anhydrous benzene is placed in a 500-ml. round-bottomed flask fitted with a reflux condenser. To the solution is added 50 g. (0.78 g. atom) of powdered copper (Note 2). The mixture is boiled gently for 3 hours. The hot solution is filtered, and 250 ml. of absolute ethanol is added to the filtrate. On cooling 25–31 g. (47–60%) of light yellow crystals, m.p. 222–224°, are obtained. The mother liquor is concentrated to about 200 ml. by distillation from a 1-l. Claisen flask. Cooling the residue yields 6–12 g. of yellow product. Crystallization of this crude material from a 1:1 by volume mixture (12 ml. for each gram) of absolute ethanol and benzene gives an additional 2.5–10 g. of tetraphenylethylene, m.p. 223–224°. The total yield is 29–37 g. (55–70%).

2. Notes

1. Diphenyldichloromethane is conveniently prepared from benzophenone and phosphorus pentachloride.[2] A product of b.p. 180–181°/17 mm. is obtained in about 90% yield.

2. The checkers used bronze powder obtained from George Benda, Inc., Boonton, New Jersey. Some varieties of copper powder tended to form a dense paste which did not disperse readily and resulted in lower yields.

3. Methods of Preparation

This procedure is adapted from the method of Schlenk and

Bergmann.[3]   Tetraphenylethylene has been prepared by the reaction of diphenylmethane with diphenyldichloromethane;[4] by the reaction of diphenyldichloromethane with silver or zinc;[4] by the reaction of thiobenzophenone with copper;[5] by the reaction of diphenylmethane with sulfur;[6] by the reduction of benzophenone with amalgamated zinc in the presence of hydrochloric acid;[7] by the rearrangement of 1,2,2,2-tetraphenylethanol with acetyl chloride;[8] by the reaction of diphenylmethane with bromine, followed by treatment of the product with sodium iodide in acetone;[9] by the reaction of 1,2-difluoro-1,2-diphenylethylene with phenyllithium;[10] by the reaction of diphenylchloromethane with sodium or potassium amide in liquid ammonia;[11] and by treatment of 1,1,2,2-tetraphenylpropionitrile with potassium amide in liquid ammonia.[12]

[1] State University of Iowa, Iowa City, Iowa.

[2] *Org. Syntheses Coll. Vol.* **2**, 573 (1943).

[3] Schlenk and Bergmann, *Ann.*, **463**, 1 (1928).

[4] Norris, Thomas, and Brown, *Ber.*, **43**, 2958 (1910).

[5] Schönberg, Shütz, and Nickel, *Ber.*, **61**, 1375 (1928).

[6] Ziegler, *Ber.*, **21**, 779 (1888); Moreau, *Bull. soc. chim. France*, **1955**, 628.

[7] Steinkopf and Wolfram, *Ann.*, **430**, 113 (1923).

[8] Lévy and Lagrave, *Bull. soc. chim. France*, [4] **43**, 437 (1928).

[9] Gorvin, *J. Chem. Soc.*, **1959**, 678.

[10] Dixon, *J. Org. Chem.*, **21**, 400 (1956).

[11] Hauser, Brasen, Skell, Kantor, and Brodhag, *J. Am. Chem. Soc.*, **78**, 1653 (1956).

[12] Hauser and Brasen, *J. Am. Chem. Soc.*, **78**, 82 (1956).

2-THENALDEHYDE *

(2-Thiophenecarboxaldehyde)

$$\text{[thiophene]} + C_6H_5N(CH_3)CHO \xrightarrow{POCl_3} \text{[thiophene-CHO]} + C_6H_5NHCH_3$$

Submitted by ARTHUR W. WESTON and R. J. MICHAELS, JR.[1]
Checked by CLIFF S. HAMILTON and JOE R. WILLARD.

1. Procedure

The reaction is carried out in a 500-ml. three-necked round-bottomed flask fitted with ground-glass joints and equipped with a thermometer, a mechanical stirrer, a dropping funnel, and a calcium chloride tube. In the flask are placed 135 g. (1.0 mole) of N-methylformanilide (Note 1) and 153 g. (91 ml., 1.0 mole) of phosphorus oxychloride, and the mixture is allowed to stand for 30 minutes (Note 2). Mechanical stirring is then begun, and the flask is immersed in a cold-water bath while 92.4 g. (1.1 moles) of thiophene is added at such a rate that the temperature is maintained at 25–35° (Note 3). After the addition is complete, the reaction mixture is stirred 2 hours longer at the same temperature and is then allowed to stand at room temperature for 15 hours. The dark, viscous solution is poured into a vigorously stirred mixture of 400 g. of cracked ice and 250 ml. of water. The aqueous layer is separated and extracted with three 300-ml. portions of ether. The ether extracts are combined with the organic layer and washed twice with 200-ml. portions of dilute hydrochloric acid (Note 4) to remove all traces of N-methylaniline (Note 5). These aqueous washings are in turn extracted with 200 ml. of ether, and the ether extract is added to the ether solution of the product. The combined ether extracts are washed twice with 200-ml. portions of saturated sodium bicarbonate solution (Note 6), then with 100 ml. of water, and finally are dried over anhydrous sodium sulfate. The yellow oil obtained by concentrating the ether solution is distilled from a 100-ml. flask fitted with a satisfactory column (Note 7). The yield of 2-thiophenecarboxaldehyde boiling at 97–100°/27 mm., n_D^{23} 1.5893, is 80–83 g. (71–74%). The product darkens on standing.

2. Notes

1. Directions for preparing this intermediate have been published.[2]

2. The temperature of the mixture rises slowly to 40–45°, and a color change from yellow to red also occurs.

3. If the temperature is allowed to exceed 35°, a lower yield of aldehyde results.

4. This solution is prepared by mixing 50 g. of concentrated hydrochloric acid and 400 ml. of water. The aldehyde has appreciable solubility in strongly acidic solutions.

5. The original aqueous layer and the acidic extracts are combined, cooled, and made strongly alkaline with 450 ml. of 40% sodium hydroxide solution. The liberated N-methylaniline is extracted with three 200-ml. portions of ether. The ether extracts are combined, washed with 100 ml. of water, dried over anhydrous sodium sulfate, and concentrated. Distillation of the residue from a 200-ml. flask equipped with an 11-cm. Vigreux column gives 95.6 g. (89%) of N-methylaniline boiling at 96–100°/27 mm.; n_D^{23} 1.5717.

6. Care must be taken in adding the bicarbonate solution, as vigorous foaming occurs until neutralization is complete.

7. The submitters used an 11-cm. Vigreux column; the checkers employed a 10-in. column of the same type.

3. Methods of Preparation

In addition to the methods given earlier,[3] 2-thenaldehyde has been prepared by the decarboxylation of 2-thienylglyoxylic acid;[4] by the hydrolysis of the acetal obtained by the action of 2-thienylmagnesium iodide on ethyl orthoformate;[4] by the oxidation of 2-thenyl alcohol;[5] by treatment of N-2-thenylformaldimine with ammonium chloride and formaldehyde;[6] by the action of 2-thienylmagnesium bromide on ethyl formate;[7] by the reaction of thiophene with dimethylformamide in the presence of phosphorus oxychloride;[8] by the condensation of thiophene with 1,4-diformylpiperazine in the presence of phosphorus oxychloride;[9] and by the reaction of thiophene with dichloromethyl methyl ether in the presence of stannic chloride.[10]

The present procedure is a modification of a previously described method.[11, 12]

[1] The Abbott Laboratories, North Chicago, Illinois.

[2] *Org. Syntheses Coll. Vol.* **3**, 590 (1955).

[3] *Org. Syntheses Coll. Vol.* **3**, 811 (1955).

[4] du Vigneaud, McKennis, Simmonds, Dittmer, and Brown, *J. Biol. Chem.*, **159**, 385 (1945).

[5] Emerson and Patrick, *J. Org. Chem.*, **14**, 790 (1949); Sugasawa and Mizukami, *Pharm. Bull. (Japan)*, **3**, 393 (1955) [*C. A.*, **50**, 15534 (1956)].

[6] Hartough, Meisel, Koft, and Schick, *J. Am. Chem. Soc.*, **70**, 4013 (1948).

[7] Gattermann, *Ann.*, **393**, 215 (1912).

[8] Weston (to Abbott Laboratories), Ger. pat. 953,082 [*C. A.*, **53**, 8163 (1959)]; U. S. pat. 2,853,493 [*C. A.*, **53**, 10251 (1959)].

[9] Fujii and Yukawa, *J. Pharm. Soc. Japan*, **76**, 607 (1956) [*C. A.*, **51**, 434 (1957)].

[10] Rieche, Gross, and Höft, *Chem. Ber.*, **93**, 88 (1960).

[11] King and Nord, *J. Org. Chem.*, **13**, 635 (1948).

[12] Weston and Michaels, *J. Am. Chem. Soc.*, **72**, 1422 (1950).

3-THENALDEHYDE

(3-Thiophenecarboxaldehyde)

$$\text{Thiophene-CH}_2\text{Br} + C_6H_{12}N_4 \longrightarrow \text{Thiophene-CH}_2(C_6H_{12}N_4)\text{Br}$$

$$\text{Thiophene-CH}_2(C_6H_{12}N_4)\text{Br} \xrightarrow[\text{distil}]{\text{Steam-}} \text{Thiophene-CHO}$$

Submitted by E. CAMPAIGNE, R. C. BOURGEOIS, and W. C. MCCARTHY.[1]
Checked by CHARLES C. PRICE and E. A. DUDLEY.

1. Procedure

Hexamethylenetetramine (77 g., 0.55 mole) is dissolved in 200 ml. of chloroform, and 88 g. (0.5 mole) of 3-thenyl bromide (p. 921) is added as rapidly as possible with shaking (Note 1). A reflux condenser is attached, and the mixture is refluxed over a steam bath for 30 minutes. After being cooled, the mixture of chloroform and crystalline product (Note 2) is poured into 250 ml. of water and stirred until all the salt dissolves. The chloroform layer is separated and washed twice with 125-ml. portions of water, and the combined water extracts are steam-distilled. When the distillate comes over clear (about 1 l. of distillate is usually collected), it is acidified with a little hydrochloric acid (Note 3) and extracted with three 100-ml. portions of ether. After drying over Drierite, the ether is evaporated, and the residue is distilled. 3-Thenaldehyde is collected at 72–78°/12 mm. or 195–199°/744 mm., $n_D^{20} = 1.5860$ (Note 4). The yield is 30–40 g. (54–72%).

2. Notes

1. The reaction mixture refluxes spontaneously, and it is

necessary to be cautious in adding the reagents to prevent the chloroform from boiling over.

2. The crystalline hexamine salt may be isolated and recrystallized at this step. It softens at 120° and melts with decomposition at 150°.

3. The Sommelet procedure [2] yields a mixture of aldehyde and amines. Acidification removes the amines from the ether extract.

4. The checkers obtained the product, b.p. 80–81°/14 mm.

3. Methods of Preparation

3-Thenaldehyde has previously been prepared by Steinkopf and Schmitt [3] from 3-thienylmagnesium iodide and ethyl orthoformate in low yield. The first application of the method described here was reported by Campaigne and LeSuer.[4] 3-Thenaldehyde also has been obtained from 3-thenoic acid by the Sonn-Müller procedure [5] and from 3-bromothiophene by treatment with butyllithium and dimethylformamide.[6]

[1] Indiana University, Bloomington, Indiana.

[2] Sommelet, *Compt. rend.*, **157**, 852 (1913).

[3] Steinkopf and Schmitt, *Ann.*, **533**, 264 (1938).

[4] Campaigne and LeSuer, *J. Am. Chem. Soc.*, **70**, 1555 (1948).

[5] Nishimura, Motoyama, and Imoto, *Bull. Univ. Osaka Prefect.*, Ser. *A*, **6**, 127 (1958) [*C. A.*, **53**, 4248 (1959)].

[6] Gronowitz, *Arkiv Kemi*, 8, 441 (1955).

3-THENOIC ACID

(3-Thiophenecarboxylic acid)

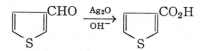

Submitted by E. CAMPAIGNE and WILLIAM M. LeSUER.[1]
Checked by CHARLES C. PRICE and E. A. DUDLEY.

1. Procedure

Silver oxide is prepared by adding a solution of 150 g. (0.88 mole) of silver nitrate in 300 ml. of water to a solution of 70 g.

(1.75 moles) of sodium hydroxide in 300 ml. of water. Continuous shaking during the addition ensures complete reaction and results in a brown semisolid mixture. To this mixture, contained in a 1-l. flask which is cooled in an ice bath, is added 47.5 g. (0.425 mole) of 3-thenaldehyde (p. 918) in small portions with stirring. The oxidation is complete in about 5 minutes after the last of the aldehyde has been added. The black silver suspension is removed by suction filtration (Note 1) and is washed with several portions of hot water. The cold combined filtrate and washings are acidified with concentrated hydrochloric acid, precipitating 49 g. of 3-thenoic acid, which melts at 136–137°. Concentration of the mother liquors gives another 3–4 g. of acid, making a total of 52–53 g. (95–97%). Recrystallization of this acid from water raises the melting point to 137–138°.

2. Note

1. This silver dissolves readily in concentrated nitric acid and may be used over and over in oxidations.

3. Methods of Preparation

3-Thenoic acid has been prepared in low yield by oxidation of 3-methylthiophene with potassium permanganate,[2–4] dilute nitric acid, chromic acid, and hydrogen peroxide,[4] and by reductive dechlorination of chloro-3-thenoic acid.[4] Starting with 3-iodothiophene, which is difficult to obtain, good yields are obtained by the Grignard procedure [5] or with cuprous cyanide and potassium cyanide in a sealed tube.[6] 3-Thenoic acid has been made also from 3-bromothiophene by reaction with cuprous cyanide in quinoline, followed by hydrolysis,[7] and by carbonation of the Grignard reagent from 3-bromothiophene which is prepared by entrainment with ethyl bromide.[8]

Oxidation of the aldehyde was not used earlier because of the difficulty of obtaining the aldehyde, which is now readily available by the Sommelet synthesis from 3-thenyl bromide. Campaigne and LeSuer [9] also used alkaline permanganate as the oxidizing agent, but the yields decreased to 40–60%. The present procedure gives an over-all yield from 3-methylthiophene of about 45%.

[1] Indiana University, Bloomington, Indiana.
[2] Muhlert, *Ber.*, **18**, 3003 (1885).
[3] Damsky, *Ber.*, **19**, 3282 (1886).
[4] Voerman, *Rec. trav. chim.*, **26**, 293 (1907).
[5] Steinkopf and Schmitt, *Ann.*, **533**, 264 (1938).
[6] Rinkes, *Rec. trav. chim.*, **55**, 991 (1936).
[7] Nishimura, Motoyama, and Imoto, *Bull. Univ. Osaka Prefect.*, Ser. A, **6**, 127 (1958) [*C. A.*, **53**, 4248 (1959)].
[8] Gronowitz, *Arkiv Kemi*, **7**, 267 (1954).
[9] Campaigne and LeSuer, *J. Am. Chem. Soc.*, **70**, 1555 (1948).

3-THENYL BROMIDE

(Thiophene, 3-bromomethyl-)

$$\text{[thiophene]}-CH_3 + C_4H_4O_2NBr \xrightarrow{\text{Peroxide}} \text{[thiophene]}-CH_2Br + C_4H_5O_2N$$

Submitted by E. Campaigne and B. F. Tullar.[1]
Checked by Charles C. Price and E. A. Dudley.

1. Procedure

Caution! This preparation should be run in a well-ventilated hood.

A 5-l. three-necked flask (Note 1) is fitted with a stirrer, an efficient reflux condenser, and a wide-mouthed funnel with a 15-cm. water-cooled stem (Note 2). A solution of 220 g. (2.24 moles) of 3-methylthiophene (Note 3) and 4 g. of benzoyl peroxide in 700 ml. of dry benzene is brought to vigorous reflux in this flask, and a mixture of 356 g. (2 moles) of N-bromosuccinimide (Note 4) and 4 g. of benzoyl peroxide is added portionwise through the wide-mouthed funnel. The dry powder is added as rapidly as the violent foaming will permit (Note 5) and is worked down through the stem of the funnel with a stirring rod. Refluxing benzene washes the lower part of the funnel continuously. The total addition requires about 20 minutes. As soon as the foaming from the last addition of N-bromosuccinimide has subsided, the flask is cooled, first with a water bath and then an ice bath. The succinimide is filtered off and washed once with dry benzene.

The filtrate is immediately transferred to a distilling flask, and the benzene is removed at reduced pressure (Note 6). The residue is distilled at 1 mm., and the fraction boiling between 75° and 78° is collected (Note 7). The major portion boils at 76° and has n_D^{25} 1.6030. This procedure yields 250–280 g. (71–79%) of a water-white product, which remains colorless for several days when stored over calcium carbonate in a refrigerator (Note 8).

2. Notes

1. The large volume of the flask is desirable to control foaming.

2. The checkers used a short, large-bore condenser.

3. Commercial 3-methylthiophene, formerly available from Socony-Vacuum Oil Company, was used. 3-Methylthiophene is now available from Winthrop-Stearns, Inc., Special Chemicals Division, New York 18, N. Y.

4. N-Bromosuccinimide can be prepared in an active state by two slight modifications of the Ziegler procedure.[2] A slight molar excess of sodium hydroxide is used, and the reaction mixture is stirred vigorously while the bromine, dissolved in an equal volume of carbon tetrachloride, is added rapidly. This produces a finely crystalline white product which is ready for use as soon as it is filtered from the water and carbon tetrachloride and thoroughly dried. The yields are generally improved by this procedure, and the N-bromosuccinimide is ready to use sooner than the product described by Ziegler et al.[2] Acetic acid is an excellent solvent for crystallization of crude N-bromosuccinimide.

5. Unless the reaction mixture is maintained at strong reflux during this addition, considerable nuclear bromination occurs with a corresponding decrease in 3-thenyl bromide yield.

6. If necessary to interrupt the procedure, the benzene solution of 3-thenyl bromide may be stored over calcium carbonate and distilled directly from the carbonate later. The addition of a little calcium carbonate before distillation avoids the formation of a difficultly removable tarry residue.

7. *Caution!* *This distillation should be carried out behind a safety shield.* Thenyl bromide is a powerful lachrymator, and some individuals may develop extensive irritation of the skin upon exposure to its vapors. The checkers observed boiling points of 78–82°/2 mm. and 80–85°/3 mm.

8. The product contains a trace of 2-bromo-3-methylthiophene, but no 2-bromo-3-thenyl bromide, and is sufficiently pure for most purposes. Samples of 3-thenyl bromide have sometimes exploded without warning, leaving deposits of black resinous material. However, storage over calcium carbonate slows this acid-catalyzed reaction. The addition of a small amount of a tertiary amine likewise increases the stability.

3. Methods of Preparation

The only method of preparative interest is that described by Campaigne and LeSuer,[3] upon which the present method is based. This method has been applied by Dittmer and others.[4]

[1] Indiana University, Bloomington, Indiana.

[2] Ziegler, Späth, Schaaf, Schumann, and Winkelmann, *Ann.*, **551**, 80 (1942).

[3] Campaigne and LeSuer, *J. Am. Chem. Soc.*, **70**, 1555 (1948); Campaigne and LeSuer (to Indiana University Foundation), U. S. pat. 2,543,544 [*C. A.*, **45**, 7601 (1951)].

[4] Dittmer, Martin, Herz, and Cristol, *J. Am. Chem. Soc.*, **71**, 1201 (1949); Shapira, Shapira, and Dittmer, *J. Am. Chem. Soc.*, **75**, 3655 (1953).

THIOBENZOIC ACID [*]

(Benzoic acid, thio)

$$KOH + H_2S \rightarrow KSH + H_2O$$

$$2KSH + C_6H_5\overset{\overset{\displaystyle O}{\|}}{C}Cl \rightarrow C_6H_5\overset{\overset{\displaystyle O}{\|}}{C}SK + KCl + H_2S$$

$$C_6H_5\overset{\overset{\displaystyle O}{\|}}{C}SK + HCl \xrightarrow{H_2O} C_6H_5\overset{\overset{\displaystyle O}{\|}}{C}SH + KCl$$

Submitted by PAUL NOBLE, JR., and D. S. TARBELL.[1]
Checked by WILLIAM S. JOHNSON and ROBERT A. KLOSS.

1. Procedure

A solution of 200 g. (3 moles) of potassium hydroxide (85%) pellets in 800 ml. of 90% ethanol (Note 1) is prepared with mechanical stirring in a 2-l. three-necked round-bottomed flask. The flask is fitted with a 500-ml. dropping funnel and a gas inlet tube extending to the bottom of the flask, and hydrogen sulfide is passed in through the inlet tube with stirring and cooling until the solution is saturated and does not give an alkaline reaction with phenolphthalein (Notes 2 and 3). The mixture is further cooled to 10–15° by means of an ice bath, and 200 g. (1.41 moles) of freshly distilled benzoyl chloride (Note 4) is introduced drop-wise over a period of about 1.5 hours with stirring while the temperature is kept below 15°. After the addition of the benzoyl chloride has been completed, the reaction mixture is stirred for an additional hour. The potassium chloride which precipitates during the addition is separated quickly by filtration through a Büchner funnel and is washed with about 200 ml. of 95% ethanol. The filtrate is placed in a 2-l. round-bottomed flask fitted with a condenser arranged for distillation and evaporated to dryness under reduced pressure on a steam bath (Note 5). The solid residue, consisting mainly of potassium thiobenzoate, is dissolved in about 700 ml. of cold water (Note 6), and the solution is filtered if considerable insoluble material is present (Note 4). The alkaline solution is extracted with 500 ml. of benzene in order to remove any neutral material. The aqueous layer is then acidified with cold 6N hydrochloric acid (Note 6) and extracted with two 500-ml. portions of peroxide-free ether (Note 7). The ether layer is washed with several portions of cold water and dried over anhydrous sodium sulfate. The ether is evaporated under reduced pressure on a steam bath (Note 5), and the residue is fractionated immediately through a short (15–20 cm.) Vigreux column at reduced pressure, dry nitrogen being admitted through the capillary (Note 8). After a very small fore-run, the yellow-orange thiobenzoic acid distils at 85–87°/10 mm. (95–97°/15 mm.); yield 120–150 g. (61–76%); n_D^{20} 1.6027. Upon refractionation, the light-yellow thiobenzoic acid, as determined by titration with standard base or alcoholic iodine, is about 99.5% pure; n_D^{20} 1.6030.

2. Notes

1. No improvement in yield was observed by substitution of absolute ethanol. The 90% is preferable to 95% ethanol, because a much smaller volume is required to dissolve the potassium hydroxide.

2. The preparation should be conducted in a well-ventilated hood or provision should be made for an exhaust tube and attachment to a gas-absorption trap.

3. The gas inlet tube should be of moderately large diameter to prevent becoming plugged with crystals during the saturation with hydrogen sulfide.

4. The use of benzoyl chloride which has not been redistilled lowers the yield by 20–30%. The use of a molar equivalent of benzoyl chloride leads to the formation of considerable benzal *bis*-thiobenzoate, $C_6H_5CH(SCOC_6H_5)_2$, which has been isolated previously as a product from the action of benzoyl chloride on potassium sulfide in ethanol.[2]

5. In order to prevent oxidation, it is inadvisable to allow the solution to stand for any appreciable time up to this point. The evaporation should be carried out without the use of a capillary. The checkers found it convenient to employ mechanical stirring (rubber-sealed stirrer) during the reduced-pressure distillation in order to prevent bumping.

6. If the temperature is allowed to rise, considerable oxidation may occur.

7. Suitable peroxide-free ether is prepared by washing ether with an equal volume of a dilute, weakly acidic solution of ferrous sulfate.

8. It is necessary to fractionate as rapidly as possible in order to prevent oxidation to the disulfide, which occurs almost completely even in the presence of nitrogen if the column is too long or if the distillation is carried out too slowly. Oil-pumped nitrogen is dried through an absorption tower containing soda lime and calcium chloride before passing to the distillation apparatus. The column should be vacuum jacketed or provided with a heated jacket.

3. Methods of Preparation

The method described is adapted from the procedures of Kym [3] and Engelhardt, Latschinoff, and Malyscheff. [4] Thiobenzoic acid has been prepared by the reaction of benzoyl chloride with potassium sulfide, [4] hydrogen sulfide in pyridine, [5,6] and magnesium bromide hydrosulfide. [7] It is formed from dibenzoyl disulfide with potassium hydrosulfide, [4] potassium hydroxide, [4,8] and ammonia. [9] It is also formed from dibenzoyl sulfide, from phenyl benzoate, and from benzoic anhydride with alcoholic potassium hydrosulfide. [4] It has been obtained from dibenzoyl sulfide and hydrogen sulfide, [10] carbon oxysulfide and phenylmagnesium bromide, [11,12] dibenzyl disulfide and sodium ethoxide, [13] benzyl chloride and sulfur in the presence of potassium hydroxide, [14] and benzylthiosulfuric acid and alkali. [15,16]

[1] University of Rochester, Rochester, New York.
[2] Bergmann, *Ber.*, **53**, 981 (1920).
[3] Kym, *Ber.*, **32**, 3533 (1899).
[4] Engelhardt, Latschinoff, and Malyscheff, *Z. Chem.*, **4**, 354 (1868).
[5] Sunner and Nilson, *Svensk Kem. Tidskr.*, **54**, 163 (1942).
[6] Lewis, *J. Chem. Soc.*, **1940**, 831.
[7] Mingoia, *Gazz. chim. ital.*, **55**, 717 (1925).
[8] Fromm and Schmoldt, *Ber.*, **40**, 2863 (1907).
[9] Busch and Stern, *Ber.*, **29**, 2150 (1896).
[10] Adkins and Thompson, *J. Am. Chem. Soc.*, **71**, 2244 (1949).
[11] Weigert, *Ber.*, **36**, 1010 (1903).
[12] Bloch, *Compt. rend.*, **204**, 1342 (1937).
[13] Fromm and Forster, *Ann.*, **394**, 338 (1912).

[14] Fromm and de Seixas Palma, *Ber.*, **39**, 3324 (1906).
[15] Price and Twiss, *J. Chem. Soc.*, **93**, 1399 (1908).
[16] Fromm and Erfurt, *Ber.*, **42**, 3818 (1909).

THIOBENZOPHENONE

(Benzophenone, thio-)

$$(C_6H_5)_2CO + H_2S \xrightarrow{HCl} (C_6H_5)_2CS + H_2O$$

Submitted by B. F. GOFTON and E. A. BRAUDE.[1]
Checked by CHARLES C. PRICE and J. FRANK GILLESPIE.

1. Procedure

Caution! This preparation should be conducted in a good hood.

A 250-ml. three-necked flask is equipped with a rubber-sealed mechanical stirrer, two gas-inlet tubes, and a mercury-sealed escape valve consisting of an outlet tube dipping into a test tube of mercury. A solution of 25 g. (0.14 mole) of benzophenone in 125 ml. of 95% ethanol is placed in the flask, which is cooled in an ice-salt freezing mixture. Hydrogen sulfide and hydrogen chloride are passed simultaneously into the stirred solution for 3 hours (Note 1). Within 1 hour the solution becomes blue. After 3 hours, the flow of hydrogen chloride is stopped and hydrogen sulfide alone is passed for a further 20 hours, with continued ice cooling. Toward the end of the reaction, the contents of the flask assume an intense violet color. The solid thiobenzophenone (23–25 g.) is filtered from the ice-cold solution in an atmosphere of carbon dioxide (Note 2), immediately dried under high vacuum (Note 3), and recrystallized twice from about 20 ml. of petroleum ether (b.p. 60–80°), giving long needles, m.p. 53–54°. The yield of purified product, is 18–21 g. (66–77%).

2. Notes

1. Hydrogen chloride was generated by dropping concentrated sulfuric acid on ammonium chloride. Hydrogen sulfide was

generated in a Kipp's apparatus from iron sulfide and hydro-chloric acid. (The checkers used tank hydrogen sulfide.)

2. A convenient filtration apparatus for this purpose may be constructed by cutting the bottom off a bottle and then inserting a Büchner funnel through a stopper in the mouth of the bottle. The space between the bottle and the funnel is then packed with Dry Ice.

3. It has been found essential to dry the crude material im-mediately after filtration; otherwise it changes into a blue oil after standing for a few hours.

3. Methods of Preparation

The method described here is adapted from the procedure of Staudinger and Freudenberger.[2] It has been found to be more convenient and to give more reproducible results than the two-stage procedure given in *Organic Syntheses*,[3] where references to other methods of preparation are also cited. Thiobenzophenone has been prepared also from benzophenone anil and thiolacetic acid.[4]

[1] Imperial College of Science and Technology, London, England.

[2] Staudinger and Freudenberger, *Ber.*, **61**, 1577 (1928).

[3] *Org. Syntheses Coll. Vol.* **2**, 573 (1943).

[4] Mihaĭlov and Savel'eva, *Izvest. Akad. Nauk S.S.S.R., Otdel. Khim. Nauk*, **1959**, 1304 [*C. A.*, **54**, 1372 (1960)].

THIOLACETIC ACID *

(Acetic acid, thiol-)

$$(CH_3CO)_2O + H_2S \rightarrow CH_3COSH + CH_3CO_2H$$

Submitted by E. K. ELLINGBOE.[1]
Checked by ARTHUR C. COPE and MALCOLM CHAMBERLAIN.

1. Procedure

Caution! All the steps of this procedure should be carried out under a hood because of the highly toxic nature of hydrogen sulfide and the probable toxicity and persistent unpleasant odor of thiolacetic acid.

A 200-ml. three-necked flask is fitted with a mercury-sealed glass stirrer, a reflux condenser, and a gas inlet tube and thermometer, both of which extend into the lower half of the flask. The top of the condenser is connected to a mercury bubbler tube, and the gas inlet tube is attached to the inlet tube of a gas-washing bottle which serves as a safety trap to prevent liquid from being drawn into the hydrogen sulfide source. The gas-washing bottle is connected through a drying tube containing anhydrous calcium sulfate (Drierite) to a T-tube. The vertical arm of the T-tube dips into mercury and forms a safety valve; the other arm is connected to a commercial cylinder (Note 1) or other source of hydrogen sulfide. To the flask are added 107 g. (100 ml., 1 mole) of 95% acetic anhydride and 1 g. (0.025 mole) of powdered sodium hydroxide. The assembly of the flask, condenser, inlet tube, and thermometer is weighed and arranged so that the amount of hydrogen sulfide introduced can be determined by subsequent weighing. The stirrer is started, and hydrogen sulfide is passed into the mixture as rapidly as possible without much loss of the gas through the bubbler connected to the top of the condenser. The temperature of the mixture rises to 55° within 30 minutes and is kept at 50–55° by intermittent cooling. The temperature begins to drop after 14–17 g. of hydrogen sulfide has been absorbed and is maintained at 50–55° by external heating. After a total reaction period of 6 hours, hydrogen sulfide ceases to be absorbed and the gain in weight amounts to about 31 g.

The reaction mixture is transferred to a 250-ml. Claisen flask (Note 2) and distilled rapidly at 200 mm. in order to separate the sodium salts (Note 3). The distillate of thiolacetic acid and acetic acid, b.p. 35–82°/200 mm., amounts to 120–124 g. It is fractionally distilled at atmospheric pressure through an efficient, variable take-off type column [2] with a 30- to 40-cm. section packed with glass helices. The fraction boiling at 86–88°, n_{D}^{25} 1.4612, is nearly pure thiolacetic acid and amounts to 55–57.5 g. (72–76%) based on the acetic anhydride (Notes 4 and 5). The residual liquid is mainly acetic acid. If distillation is continued after separation of the thiolacetic acid, the vapor temperature rises rapidly to the boiling point of acetic acid.

2. Notes

1. Cylinders of hydrogen sulfide are available from the Matheson Company, Inc., East Rutherford, New Jersey.

2. An all-glass distillation assembly equipped with ground joints is advisable because the hot vapor of thiolacetic acid rapidly softens rubber stoppers.

3. The product decomposes excessively if fractionation is attempted in the presence of the sodium salts. In larger-scale preparations the initial distillation should be conducted in small batches or, preferably, in a continuous stripping still.

4. Refractionation yields pure thiolacetic acid (with about 10% loss) with the following physical constants: b.p. 87°/760 mm., 50°/200 mm., 34°/100 mm.; n_D^{25} 1.4630; d_4^{25} 1.0634. Higher boiling points at atmospheric pressure have been reported: 88–91°,[3] 89°,[4] and 93°.[5]

5. The submitter has prepared thiolacetic acid in yields of 65–70 g. (85–92%) by a similar procedure in which the reaction mixture is placed in the glass bottle (provided with a heating jacket and thermometer or thermocouple) of a low-pressure hydrogenation apparatus.[6] The mixture is shaken and hydrogen sulfide is introduced at 25–35 p.s.i., with repressuring to 35–40 p.s.i. whenever the pressure drops to 10 p.s.i. The heat of reaction raises the temperature to 60–65° in 12–15 minutes, after which the internal temperature is maintained at 60° by heating. Hydrogen sulfide absorption becomes negligible after 4 hours, and the mixture is allowed to cool while under pressure, vented, and the product is isolated in the manner described. The reaction also was conducted in a steel autoclave or hydrogenation bomb at the full pressure of hydrogen sulfide in a commercial cylinder (about 300 p.s.i. at room temperature). Hydrogen sulfide poisons the ordinary hydrogenation catalysts, and low-pressure cylinders or hydrogenation bombs exposed to hydrogen sulfide may not be suitable for subsequent use in catalytic hydrogenations.

3. Methods of Preparation

Thiolacetic acid has been prepared from acetic acid and phosphorus pentasulfide;[5] from acetyl chloride and potassium hydrosulfide;[7] by the hydrolysis of diacetyl sulfide;[8] from acetic anhydride and hydrogen sulfide;[3,4,9,10] and from acetyl chloride and hydrogen sulfide in the presence of aluminum chloride.[11] The procedure described [9] differs from other procedures employing acetic anhydride and hydrogen sulfide [3,4] most importantly in the use of alkaline rather than acidic catalysts, which the submitter found to cause slower absorption of hydrogen sulfide under pressure and to yield considerable diacetyl sulfide in addition to thiolacetic acid. Other alkaline catalysts which have been used to effect the reaction between acetic anhydride and hydrogen sulfide are triethylamine [12] and pyridine.[13]

[1] E. I. du Pont de Nemours and Company, Inc., Wilmington, Delaware.

[2] *Org. Syntheses*, **25**, 2 (1945).

[3] Clarke and Hartman, *J. Am. Chem. Soc.*, **46**, 1731 (1924).

[4] Hands and Whitt, *J. Soc. Chem. Ind.*, **66**, 173 (1947).

[5] Kekulé, *Ann.*, **90**, 309 (1854).

[6] *Org. Syntheses Coll. Vol.* **1**, 66 (1941).

[7] Jacquemin and Vosselmann, *Compt. rend.*, **49**, 371 (1859).

[8] Davies, *Ber.*, **24**, 3551 (1891).

[9] Ellingboe, U. S. pat. 2,412,036 [*C. A.*, **41**, 2074 (1947)].

[10] McCool (to B. F. Goodrich Co.), U. S. pat. 2,568,020 [*C. A.*, **46**, 3557 (1952); U. S. pat. 2,587,580 [*C. A.*, **46**, 10192 (1952)].

[11] Arndt and Bekir, *Ber.*, **63**, 2390 (1930).

[12] Behringer and Stein, Ger. pat. 800,412 [*C. A.*, **45**, 1622 (1951)].

[13] Sjöberg, *Svensk Kem. Tidskr.*, **63**, 90 (1951) [*C. A.*, **47**, 8640 (1953)].

o-TOLUALDEHYDE

$$
\underset{\text{CH}_3}{\overset{\text{CH}_2\text{Br}}{\bigcirc}} + \left[\underset{\text{CH}_3}{\overset{\text{CH}_3}{\diagdown}} \text{C}{=}\text{NO}_2 \right]^{-} \text{Na}^+ \rightarrow
$$

$$
\underset{\text{CH}_3}{\overset{\text{CHO}}{\bigcirc}} + \underset{\text{CH}_3}{\overset{\text{CH}_3}{\diagdown}} \text{C}{=}\text{NOH} + \text{NaBr}
$$

Submitted by H. B. HASS and MYRON L. BENDER.[1]
Checked by ARTHUR C. COPE and MALCOLM CHAMBERLAIN.

1. Procedure

Eleven and one-half grams (0.5 g. atom) of sodium is dissolved in 500 ml. of absolute ethanol in a 1-l. round-bottomed flask. Forty-six grams (0.52 mole) of 2-nitropropane is added, then 92.5 g. (0.50 mole) of *o*-xylyl bromide (Note 1). The flask is attached to a reflux condenser connected to a drying tube and shaken at intervals for 4 hours. The reaction mixture, originally at room temperature, becomes warm spontaneously, and a white precipitate of sodium bromide forms (Note 2).

After a reaction period of 4 hours the sodium bromide is separated by filtration and the ethanol is removed by distillation on a steam bath. The residue of product and sodium bromide is dissolved in 100 ml. of ether and 150 ml. of water. The ether layer is washed with two 50-ml. portions of 10% sodium hydroxide solution to remove any acetoxime and excess 2-nitropropane and is then washed with 50 ml. of water. The ether layer is separated and is dried with 15 g. of anhydrous sodium sulfate, and the ether is removed by distillation on a steam bath.

The crude product is distilled from a Claisen flask under reduced pressure. The yield of *o*-tolualdehyde boiling at 68–72°/6 mm., n_D^{25} 1.5430, is 41–44 g. (68–73%) (Note 3).

2. Notes

1. o-Xylyl bromide may be obtained from the Eastman Kodak Company or may be prepared by the light-catalyzed bromination of o-xylene.[2]

2. The solution is originally supersaturated with the sodium salt of 2-nitropropane, and a precipitate of this salt may be mistaken for sodium bromide.

3. This is a general method for the preparation of substituted benzaldehydes. The following aldehydes have been prepared by the same general procedure.[3]

Aldehyde	Yield, %
p-Bromobenzaldehyde	75
Benzaldehyde	73
p-Carbomethoxybenzaldehyde	72
p-Cyanobenzaldehyde	70
p-Trifluoromethylbenzaldehyde	77

3. Methods of Preparation

A procedure for the preparation of o-tolualdehyde from o-toluanilide by the Sonn-Müller method has been published in *Organic Syntheses*.[4] In addition to the alternative methods of preparation listed there, o-tolualdehyde has been prepared from o-xylyl bromide [5] or chloride [6] and hexamethylenetetramine; by the Stephen reduction of o-tolunitrile; [7] by the reduction of the latter with lithium triethoxyaluminohydride [8] or sodium triethoxyaluminohydride; [9] by the condensation of crotonaldehyde with barium or calcium oxide at 300°; [10] by the oxidation of o-methylbenzyl alcohol with N-chlorosuccinimide; [11] by the reaction of ethyl orthoformate with o-tolylmagnesium bromide; [12] by the reaction of the o-tolylmagnesium chloride-tetrahydrofuran complex with methyl formate; [13] and by the procedure of the present preparation.[3]

[1] Purdue University, Lafayette, Indiana.

[2] Cumming, Hopper, and Wheeler, *Systematic Organic Chemistry*, p. 351, Constable and Company, London, 1937 (description of a procedure for light-catalyzed bromination which is applicable to o-xylene).

[3] Hass and Bender, *J. Am. Chem. Soc.*, **71**, 1767 (1949).

4 *Org. Syntheses Coll. Vol.* **3**, 818 (1955).

5 Sommelet, *Compt. rend.*, **157**, 852 (1913).

6 Shacklett and Smith, *J. Am. Chem. Soc.*, **75**, 2654 (1953).

7 Stephen, *J. Chem. Soc.*, **127**, 1874 (1925).

8 Brown, Shoaf, and Garg, *Tetrahedron Letters*, **3**, 9 (1959).

9 Hesse and Schrödel, *Angew. Chem.*, **68**, 438 (1956); *Ann.*, **607**, 24 (1957).

10 Wiemann, Martineau, and Tiquet, *Compt. rend.*, **241**, 807 (1955).

11 Hebbelynck and Martin, *Bull. soc. chim. Belges*, **60**, 54 (1951) [*C. A.*, **46**, 7051 (1952)].

12 Gibson, *J. Chem. Soc.*, **1956**, 808.

13 Ramsden (to Metal & Thermit Corp.), Brit. pat. 806,710 [*C. A.*, **54**, 2264 (1960)].

p-TOLUENESULFENYL CHLORIDE

$$CH_3\langle\ \rangle SH + Cl_2 \rightarrow CH_3\langle\ \rangle SCl + HCl$$

Submitted by FREDERICK KURZER and J. ROY POWELL.[1]
Checked by RICHARD S. SCHREIBER and FRED KAGAN.

1. Procedure

Caution! This preparation should be conducted in a good hood.

Chlorine (Note 1) is passed into 300 ml. of anhydrous carbon tetrachloride (Note 2) contained in a 500-ml. three-necked round-bottomed flask equipped with a mechanical stirrer, a dropping funnel protected from atmospheric moisture by a calcium chloride tube, and a gas-inlet tube equipped with a sintered-glass tip (Note 3). The amount of chlorine contained in the resulting yellow-green liquid, estimated volumetrically (Note 4), varies between 25 and 32 g. The gas-inlet tube is replaced by a calcium chloride tube, and the reaction vessel is cooled externally with ice water and protected from light by being covered with a towel.

One or two crystals of iodine are added to the chlorine solution, and a solution of *p*-toluenethiol (7.3 g. per 10 g. of dissolved chlorine, i.e., halogen in 140% excess) (Note 5) in 50 ml. of anhydrous carbon tetrachloride (Note 2) is added dropwise over a period of approximately 1 hour. The initial slight turbidity disappears gradually, a bright orange solution being formed (Note 6). After an additional 30 minutes' stirring, the solvent and excess chlorine are rapidly removed under reduced pressure at the lowest possible temperature (Note 7), the crude sulfenyl chloride being left as an orange-red mobile liquid (Note 8). For batches starting with 24.8 g. (0.2 mole) of *p*-toluenethiol, the yield varies between 27 and 30.5 g. (85–96%).

Rapid distillation under reduced pressure affords *p*-toluenesulfenyl chloride as a red mobile liquid boiling at 66–68°/0.8 mm. (74–76°/1.5 mm.; 82–84°/3.5 mm.) (Note 9). The yield of redistilled material varies between 24.5 and 28 g. (77–88%).

2. Notes

1. The checkers used tank chlorine obtained from the Ohio Chemical and Surgical Equipment Company, Detroit, Michigan. The gas was dried with concentrated sulfuric acid and used with no further purification. The rate of chlorine addition was regulated so that no chlorine escaped from the carbon tetrachloride solution. After 1.5–2.0 hours, 27–33 g. of chlorine was absorbed. The submitters generated chlorine by the action of concentrated hydrochloric acid on potassium permanganate.

2. The carbon tetrachloride is treated with a small quantity of phosphorus pentoxide, and the suspension is set aside for 1–2 hours. The clear supernatant liquid is first decanted into a dry flask and gently shaken. Any droplets of phosphoric acid are retained on the walls of this vessel; the anhydrous solvent is then easily decanted.

3. During the slow absorption of chlorine, the solution is satisfactorily protected from atmospheric moisture by means of a plug of cotton wool through which the gas-inlet tube passes. The temperature of the solution does not rise, and external cooling is not required.

4. A 2-ml. aliquot of the solution, withdrawn by means of a safety pipet, is added to 20 ml. of water containing 2 g. of potassium iodide, and the liberated iodine is titrated with good shaking with standard sodium thiosulfate solution.

5. When a smaller excess of chlorine is used poorer yields of sulfenyl chloride are obtained, a larger proportion of *p*-tolyl disulfide being formed.

6. The checkers found that the turbidity persisted throughout

the addition of the *p*-toluenethiol and that a small amount of an orange-yellow solid was present at the end of the addition. This material dissolved during the concentration under reduced pressure.

7. Since traces of sulfenyl chloride are carried over with the carbon tetrachloride, the distillate varies from light to deep yellow during the concentration. Recovery of this material by repeated distillation is not economical.

8. This material is suitable for use in further synthesis without purification by distillation, particularly if the presence of a few per cent of *p*-tolyl disulfide is of no consequence.

9. The checkers found that 76% of the product boiled at 64–66°/1.1 mm. and 24% boiled at 66–74°/1.1 mm.; however, the index of refraction of the two fractions was constant (n_D^{20} 1.6018–1.6019). As the first few drops of distillate were green, they were discarded.

3. Methods of Preparation

p-Toluenesulfenyl chloride has been prepared by the action of chlorine on a solution of *p*-toluenethiol or *p*-tolyl disulfide in anhydrous carbon tetrachloride.[2-4] Benzenesulfenyl chloride has also been obtained by the interaction of hydrogen chloride and N,N-diethylbenzenesulfenamide [5] and by reaction of benzenethiol with N-chlorosuccinimide.[6] A comprehensive review dealing with sulfenyl halides and related compounds is available.[7]

[1] University of London, London, England.

[2] Lecher, Holschneider, Koberle, Speer, and Stocklin, *Ber.*, **58**, 409 (1925).

[3] Lecher and Holschneider, Ger. pat. 423,232 (*Brit. C. A.*, **1926**, B, 386).

[4] Montanari, *Gazz. chim. ital.*, **86**, 406 (1956).

[5] Lecher and Holschneider, *Ber.*, **57**, 755 (1924).

[6] Emde, Ger. pat. 804,572 [*C. A.*, **46**, 529 (1952)].

[7] Kharasch, Potempa, and Wehrmeister, *Chem. Revs.*, **39**, 269 (1946).

p-TOLUENESULFINYL CHLORIDE

$$CH_3 \langle\!=\!\rangle SO_2Na \cdot 2H_2O + 3SOCl_2 \rightarrow$$

$$CH_3 \langle\!=\!\rangle SOCl + NaCl + 3SO_2 + 4HCl$$

Submitted by FREDERICK KURZER.[1]
Checked by RICHARD S. SCHREIBER and FRED KAGAN.

1. Procedure

Caution! This reaction should be conducted in a hood to avoid exposure to sulfur dioxide and hydrogen chloride.

To 179 g. (109 ml., 1.5 moles) of thionyl chloride (Note 1) contained in a 250-ml. round-bottomed flask, 42.8 g. (0.2 mole) of powdered sodium *p*-toluenesulfinate dihydrate [2] (Note 1) is added in portions at room temperature over a 10- to 15-minute period. A vigorous reaction occurs with the evolution of hydrogen chloride and sulfur dioxide. As the first portions of the sulfinate are added, the temperature of the reaction mixture rises, but it soon drops to approximately 0° as the addition proceeds (Note 2). The resulting reaction mixture, a clear yellow liquid containing a white opaque solid, is protected from atmospheric moisture by means of a calcium chloride drying tube and is set aside at room temperature for 1.5–2 hours. During this time slight effervescence continues (Note 3) and the white opaque suspended material gradually disintegrates to a finely divided translucent deposit.

The excess thionyl chloride is removed by distillation under reduced pressure (15–20 mm.) with the bath temperature below 50°, and the last traces are eliminated by one or two evaporations under reduced pressure after the addition of 50-ml. portions of anhydrous ether (Note 4). The residue consists of a viscous yellow oil containing a suspension of white granular inorganic solid. The crude sulfinyl chloride is readily dissolved by three successive treatments with portions of anhydrous ether (50, 30,

and 30 ml., respectively) which are decanted without difficulty
from most of the inorganic residue (Note 5). Removal of the
solvent by distillation at reduced pressure leaves the sulfinyl
chloride as a clear, pale, straw-yellow oil. The yield is 30–32 g.
(86–92%) (Note 6). Distillation at reduced pressure (Note 5)
yields p-toluenesulfinyl chloride as a deep yellow mobile oil, b.p.
113–115°/3.5 mm. or 99–102°/0.5 mm. (Note 7). The yield of
redistilled material is 23–26 g. (66–74%). A small quantity
(2–3 g.) of a dark tarry residue remains in the distilling flask.

2. Notes

1. Redistilled thionyl chloride is recommended. No difficulties
were encountered when the color of the reagent was deep yellow.
The sodium p-toluenesulfinate dihydrate should be thoroughly
air-dried to remove mechanically bound water.

2. When smaller quantities are used the thionyl chloride may
be added to the sulfinate in one portion. After an initial rise, the
temperature drops to about 0°. With larger quantities this
procedure is not practicable.

3. If the sodium p-toluenesulfinate is not finely powdered, reaction occurs more slowly with emission of slight crackling sounds.

4. The ether was dried over calcium chloride and finally over
phosphorus pentoxide. The checkers found that after the ether
evaporations the odor of thionyl chloride was still present. The
last traces of thionyl chloride were best removed at a pressure
of 1–2 mm. When free of thionyl chloride the residue no longer
has the pungent odor of either hydrogen chloride or sulfur dioxide but has a faint odor typical of sulfonyl chlorides.

5. If it is desired to distil the sulfinyl chloride subsequently,
any traces of suspended material must be removed by filtration
of the ethereal solution through a small filter, preferably into the
distilling flask. It is advantageous to concentrate the ethereal
solution to a small volume before filtering since more concentrated
solutions of sulfinyl chloride in ether have less tendency to take up
moisture from the atmosphere. The filter is rinsed with a little
anhydrous ether.

The checkers used an inverted funnel connected to a source of
dry nitrogen to provide an inert atmosphere for the filtration.

The filtration was carried out as completely as possible directly beneath the inverted funnel through which a rapid stream of nitrogen flowed.

6. The sulfinyl chloride thus obtained is satisfactory for further preparative work. The checkers found that the average yield of crude material after removal of inorganic salts by filtration was about 81%.

7. On being heated, the sulfinyl chloride undergoes transient color changes (green), and the redistilled material may be darker in color than the pale yellow residue obtained after the removal of the ether.

For distillation the checkers employed a Claisen head with a fraction cutter. Approximately 90% of the product distilled at 93–96°/1.1 mm., $n_D^{23.5}$ 1.6004, and 10% distilled at 98–100°/1.3–1.6 mm., $n_D^{23.5}$ 1.5998. The total yield of distilled material in two runs averaged 74%.

3. Methods of Preparation

Arylsulfinyl chlorides have been prepared by treating the corresponding arylsulfinic acids with an excess of thionyl chloride in the absence of solvents, either with gentle heating [3-8] or at room temperature.[9] The use of nearly equimolecular proportions of the reactants, and of ether as diluent in this reaction, has been claimed [10,11] to yield a cleaner product. *p*-Toluenesulfinyl chloride may be obtained directly from the hydrated sodium salt of *p*-toluenesulfinic acid by the action of a large excess of thionyl chloride, and the present procedure is based on this variation.[12]

[1] University of London, London, England.
[2] *Org. Syntheses Coll. Vol.* 1, 492 (1941).
[3] Hilditch and Smiles, *Ber.*, 41, 4115 (1908).
[4] von Braun and Kaiser, *Ber.*, 56, 549 (1923).
[5] Whalen and Jones, *J. Am. Chem. Soc.*, 47, 1353 (1925).
[6] Hunter and Sorenson, *J. Am. Chem. Soc.*, 54, 3368 (1932).
[7] Courtot and Frenkiel, *Compt. rend.*, 199, 557 (1934).
[8] Burton and Davy, *J. Chem. Soc.*, 1948, 528.
[9] Raiford and Hazlet, *J. Am. Chem. Soc.*, 57, 2172 (1935).
[10] Hilditch, *J. Chem. Soc.*, 97, 2585 (1910).
[11] Phillips, *J. Chem. Soc.*, 1925, 2569.
[12] Kurzer, *J. Chem. Soc.*, 1953, 549.

p-TOLUENESULFONIC ANHYDRIDE

$$2CH_3\langle\rangle SO_2OH \cdot H_2O + P_2O_5 \rightarrow$$

$$CH_3\langle\rangle SO_2OSO_2\langle\rangle CH_3 + 2H_3PO_4$$

Submitted by L. FIELD and J. W. McFARLAND.[1]
Checked by William S. JOHNSON, RAYMOND R. HINDERSINN, and A. G. FISHER.

1. Procedure

A mixture of 213 g. (1.5 moles) of phosphorus pentoxide and 14 g. of kieselguhr (Note 1) is prepared by shaking in a dry, stoppered 1-l. Erlenmeyer flask. About one-half of the mixture is then added to a mixture of 7 g. of asbestos (Note 1) and 190 g. (1.0 mole) of *p*-toluenesulfonic acid monohydrate (Note 2) in a 1-l. round-bottomed flask with one standard-taper neck protected from atmospheric moisture with a drying tube containing calcium chloride. The mixture, which soon becomes quite hot, is allowed to stand for 1 hour with occasional swirling. It is then heated in an oil bath held at about 125° for 9 hours, the remainder of the phosphorus pentoxide mixture being added in four portions during the first 3 hours. During this total 9-hour period the mass is mixed as well as possible with a metal spatula from time to time.

The drying tube is replaced by a condenser, and 400 ml. of ethylene chloride (Note 3) is added. The mixture is heated under reflux for several minutes. The flask is closed with a glass stopper and shaken well (Note 4). A thin pad of glass wool is then inserted into the lower part of the flask neck, and the extract is decanted. This extraction and decantation process is repeated with three 100-ml. portions of ethylene chloride. The extracts are combined in a 1-l. round-bottomed flask, the solvent is removed by distillation under reduced pressure, and the molten solid is swirled briefly at 3 mm. pressure until cessation of bubbling indicates complete removal of solvent. The dark oily solid, which is formed on cooling, amounts to 138–

143 g. and melts in the range 80–124° (sealed tube) (Note 5). This material is purified by dissolving it in the minimum amount (about 200 ml.) of boiling anhydrous benzene (Notes 6 and 7) and, after cooling to about 50°, adding 300 ml. of anhydrous ether (Note 7). After crystallization is well advanced, the mixture is stored at −5° overnight. The solvent is removed by forcing it under nitrogen pressure either through a 6-mm. glass tube packed with glass wool up to a constriction about 1 in. from the bottom or through a sintered-glass filter stick. Residual solvent is removed at 3-mm. pressure through a stopcock which is then closed to permit storage of the anhydride at low pressure (Note 8). The yield is 77–114 g. (47–70%) of light brown or gray prisms. The product, which softens at about 90° and melts at 110–127° (sealed tube) (Note 5), is sufficiently pure for most purposes (Note 9).

2. Notes

1. It is satisfactory to use the Super-Cel brand of kieselguhr supplied by Johns-Manville Company and asbestos of the type used in Gooch crucibles. Inclusion of these inert materials facilitates mixing and extraction of the product.

2. A good commercial grade of *p*-toluenesulfonic acid monohydrate, m.p. 104–106°, was used.

3. A good commercial grade of ethylene chloride, b.p. 83–84°, was used. It is dried by distilling a portion.

4. If shaking is too violent, the gummy residue that clings to the flask may be dislodged, and the flask may break. After extraction is complete, the phosphorus pentoxide residue can be removed from the flask by adding water and replacing this by fresh water as the mixture becomes hot.

5. Even brief exposure to moist air causes a lowering of the melting point. The melting-point tube is therefore best filled by inserting it into material ground against the flask bottom or by using a dry box; the tube is then quickly sealed. The melting point of pure *p*-toluenesulfonic anhydride is 129.5–131.5°, with softening about 120° (sealed tube).[2]

6. Commercial grades of anhydrous ether and benzene which

have been allowed to stand over sodium are satisfactory. Dry glassware should be used for transferring the solvents.

7. Usually a small amount of material remains undissolved. This material does not affect the melting point significantly but can be removed if desired by forcing the solution, kept hot by an electric heating mantle, through the filter arrangement described above into a dry flask protected from moisture by a calcium chloride tube.

8. In order to open the flask, air is admitted slowly through a calcium chloride tube. The melting point of the anhydride drops somewhat during prolonged storage with occasional opening, but the purity does not seem to be affected appreciably.[2]

9. The broad melting range apparently results from an unusual sensitivity of this property to small amounts of impurities.[2] The melting point can be raised by further recrystallizations.

3. Methods of Preparation

p-Toluenesulfonic anhydride has been prepared from the acid salt by the use of thionyl chloride [3] and phosphorus pentoxide,[2] and from the acid or sodium salt by treatment with a mixture of phosphorus pentachloride and phosphorus oxychloride.[4] It also has been obtained by heating oxime p-toluenesulfonates,[5] by reaction of p-toluenesulfonic acid with di-p-tolylcarbodiimide,[6] and by the interaction of methoxyacetylene and p-toluenesulfonic acid.[7]

[1] Vanderbilt University, Nashville 4, Tennessee.

[2] Field, *J. Am. Chem. Soc.*, **74**, 394 (1952).

[3] Meyer and Schlegl, *Monatsh.*, **34**, 573 (1913); Meyer, *Ann.*, **433**, 335 (1923).

[4] Lukashevich, *Doklady Akad. Nauk S.S.S.R.*, **114**, 1025 (1957) [*C. A.*, **52**, 3717 (1958)].

[5] Oxley and Short, *J. Chem. Soc.*, **1948**, 1524.

[6] Khorana, *Can. J. Chem.*, **31**, 585 (1953).

[7] Eglinton, Jones, Shaw, and Whiting, *J. Chem. Soc.*, **1954**, 1860.

p-TOLYLSULFONYLMETHYLNITROSAMIDE*

(*p*-Toluenesulfonamide, N-methyl-N-nitroso-)

$$CH_3\text{-}\langle \rangle\text{-}SO_2Cl + 2CH_3NH_2 \rightarrow$$

$$CH_3\text{-}\langle \rangle\text{-}SO_2\overset{H}{N}CH_3 + CH_3NH_2 \cdot HCl$$

$$(CH_3NH_2 \cdot HCl + NaOH \rightarrow CH_3NH_2 + NaCl + H_2O)$$

$$CH_3\text{-}\langle \rangle\text{-}SO_2\overset{H}{N}CH_3 + HNO_2 \rightarrow CH_3\text{-}\langle \rangle\text{-}SO_2\overset{NO}{N}CH_3 + H_2O$$

Submitted by TH. J. DE BOER and H. J. BACKER.[1]
Checked by JAMES CASON, JOHN B. ROGAN, and WM. G. DAUBEN.

1. Procedure

A total of 320 g. (1.68 moles) of *p*-toluenesulfonyl chloride (Note 1) is divided into three portions of 190, 90, and 40 g.; and a solution of alkali is prepared by dissolving 70 g. of sodium hydroxide in 70 ml. of water with cooling. The 190-g. portion of the sulfonyl chloride is added with swirling during about 5 minutes to 210 ml. (2.1 moles) of 33% aqueous methylamine (or 174 ml. of the 40% solution) contained in a 1-l. round-bottomed flask. The mixture is allowed to heat up to 80–90° in order to maintain the sulfonylmethylamide (m.p. 78°) in a molten condition (Note 2). After all this portion of the sulfonyl chloride has been added, the mixture is shaken vigorously. Boiling is prevented by mild cooling with water in order to avoid an excessive loss of methylamine.

As soon as the mixture has become acidic (Note 3), as indicated by testing a drop on litmus paper, 50 ml. of the 50% sodium hydroxide solution is added carefully with swirling. This is followed immediately by gradual addition of the 90-g. portion of the sulfonyl chloride as before. When the mixture has again become acidic (Note 3), 25 ml. of the sodium hydroxide solution is added, followed by the final 40 g. of the sulfonyl chloride. After the mixture has again become acidic, the remainder of the sodium hydroxide solution is added. The liquid phase of the final mixture should be alkaline (Note 4).

After the walls of the flask have been rinsed with a little water, the reaction is completed by heating the mixture, consisting of two layers and a precipitate of sodium chloride, on a steam bath for 15 minutes with vigorous mechanical stirring. The hot reaction mixture (Notes 2 and 5) is then poured into 1.5 l. of glacial acetic acid contained in a 5-l. round-bottomed flask, and the smaller flask is rinsed clean with 250 ml. of acetic acid (Note 6). The solution is cooled in an ice bath to about 5° (Note 7) and stirred mechanically as a solution of 124 g. (1.8 moles) of sodium nitrite in 250 ml. of water is added from a dropping funnel during about 45 minutes. The temperature of the mixture is kept below 10°, and stirring is continued for 15 minutes after addition is complete. During the reaction, the nitroso compound separates as a yellow crystalline product.

One liter of water is added to the mixture; then the precipitate is separated by suction filtration, pressed on the funnel, and washed with about 500 ml. of water. The product is transferred to a beaker, stirred well with about 500 ml. of cold water, then filtered and washed again on the funnel. This process is again repeated if necessary to remove the odor of acetic acid. After drying to constant weight in a vacuum desiccator over sulfuric acid, the product melts in the range between 55° and 60° (Note 8). The yield is 306–324 g. (85–90%) (Notes 9 and 10).

2. Notes

1. The *p*-toluenesulfonyl chloride used by the submitters was a product recrystallized from a 1:20 mixture of benzene:60–80° petroleum ether, and it melted at 67.5–69°. The checkers employed material obtained from Distillation Products Industries. Sulfonic acid may be removed from old samples of the sulfonyl chloride by thorough washing with cold water followed by immediate drying in a vacuum desiccator over sulfuric acid.

2. At lower temperatures, the sulfonylmethylamide is likely to form a hard cake. In smaller runs, e.g. 0.2 mole, the mixture heats up considerably less; therefore, the reaction mixture should

be warmed on a steam bath after the first addition of sulfonyl chloride and thereafter as necessary.

3. The reaction mixture may not become acidic after a given addition of sulfonyl chloride, especially the first one. If such is the case, no more than 5 minutes need be allowed between successive sulfonyl chloride and alkali additions. The whole procedure requires about 30 minutes.

4. If the final mixture is acidic, indicating excessive loss of methylamine, sufficient methylamine should be added to render the mixture basic. This situation is more likely to occur in smaller runs (0.1–0.2 mole).

5. The precipitate of sodium chloride need not be removed before nitrosation; it dissolves when the final product is washed with water.

6. The total amount of 1750 ml. of acetic acid is necessary to dissolve the amide completely at lower temperatures. With less acetic acid, it is difficult to secure complete nitrosation, even with very efficient stirring of the suspension.

In one experiment with 750 ml. of acetic acid, the resulting impure product was washed first with water to remove acetic acid and then with 1*N* sodium hydroxide to remove *p*-tolylsulfonylmethylamide (the nitroso compound is not attacked by cold aqueous sodium hydroxide). The yield of *p*-tolylsulfonylmethylnitrosamide was 76%, m.p. 58–60°. Acidification of the alkaline wash yielded (9% recovery) the *p*-tolylsulfonylmethylamide, m.p. 77.5–78.5°.

7. An ice-salt mixture may be used for more efficient cooling, but reaction temperatures below 0° should be avoided because the total amount of acetic acid is just sufficient to keep the sulfonylmethylamide in solution above 0°.

8. A melting point in the lower part of this range generally indicates contamination with *p*-tolylsulfonylmethylamide, which may be removed as described in Note 6; however, this purification is not necessary since the presence of this amide does not interfere in the preparation of diazomethane.

9. *Caution!* Although this material has been kept at room temperature for years without significant change, there has been reported one instance in which a sample stored for several months

detonated spontaneously. For long periods of storage, it is recommended that the material be recrystallized and placed in a dark bottle. Recrystallization is best accomplished by dissolution in boiling ether (1 ml. per g.), addition of an equal volume of low-boiling petroleum ether (technical pentane), and cooling over night in a refrigerator. Other solvents (benzene, carbon tetrachloride, chloroform) may be used for recrystallization, but the maximum temperature should not exceed 45°.

10. p-Tolylsulfonylmethylnitrosamide is a useful substance for the preparation of diazomethane (p.250). It is apparently of low toxicity. Its properties afford advantages over other nitroso compounds which have been used for preparation of diazomethane.

3. Methods of Preparation

p-Tolylsulfonylmethylnitrosamide has been prepared previously by adding sodium nitrite to an acid aqueous suspension of p-tolylsulfonylmethylamide.[2-4] The present method is more rapid and gives higher yields.[5]

[1] De Rijks-Universiteit, Groningen, the Netherlands.

[2] Friedlander, *Fortschritte der Teerfarbenfabrikation*, **10**, 1216 (1910); Bayer Co. D.R.P. 224,388 [*Chem. Zentr.*, **1910 II**, 609].

[3] Takizawa, *J. Pharm. Soc. Japan*, **70**, 490 (1950) [*C. A.*, **46**, 454 (1952)].

[4] Tomita, Kugo, and Hirai, *J. Pharm. Soc. Japan*, **73**, 1247 (1953) [*C. A.*, **48**, 13616 (1954)].

[5] de Boer and Backer, *Rec. trav. chim.*, **73**, 229 (1954).

2,4,6-TRIBROMOBENZOIC ACID

(Benzoic acid, 2,4,6-tribromo-)

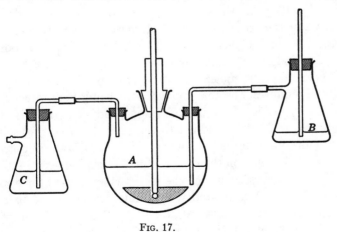

Submitted by MICHAEL M. ROBISON and BONNIE L. ROBISON.[1]
Checked by N. J. LEONARD and KENNETH CONROW.

1. Procedure

A. *3-Amino-2,4,6-tribromobenzoic acid.* The bromination apparatus consists of a 2-l. three-necked standard-taper flask *A*, equipped with a "Trubore" stirrer and attached by rubber stoppers and glass tubing to two filter flasks, *B* and *C*, as shown

FIG. 17.

in Fig. 17. The 250-ml. flask *C*, which serves as a bubbler and trap, is half-filled with water, and its side arm is connected to a water pump. The reaction flask *A* is charged with 27.4 g. (0.2 mole) of *m*-aminobenzoic acid (Note 1), 165 ml. of concentrated hydrochloric acid, and 780 ml. of water, after which it is surrounded by an ice bath, and the stirrer is started. One hundred and forty grams (45 ml., 0.88 mole) of bromine is placed in the vaporization flask *B*, which is then surrounded by a water bath heated to 40–50°. Bromine vapor is drawn through the reaction mixture at a moderate rate by applying suction at the water pump. Stirring and cooling of the mixture are continued throughout the reaction period, during which time the product precipitates (Note 2). The bromination requires about 3 hours and is complete when the slurry assumes a distinct yellow color.[2]

The cream-colored solid is separated at once by filtration and washed thoroughly with water to remove excess bromine water and acids. It is used in the next step without drying. The melting range of the crude product is usually 170–172°.

B. *2,4,6-Tribromobenzoic acid.* A 5-l. three-necked flask, equipped with a mechanical stirrer and thermometer and surrounded by an ice-salt bath, is charged with a precooled mixture of 1.02 l. of concentrated sulfuric acid and 500 ml. of water. Cooling during the reaction period is assisted by the ice-salt bath, but is effected chiefly by periodic additions of large amounts of crushed Dry Ice directly to the reaction mixture. The temperature of the stirred mixture is lowered to −5° to −10°, and 37 g. (0.54 mole) of sodium nitrite is added in small portions over a period of about 15 minutes. Cold 50% hypophosphorous acid (193 ml., 1.86 moles) (Note 3) is then added over a period of 10–15 minutes, the temperature still being maintained below −5°. A solution of the bromination product in 1.85 l. of glacial acetic acid is then added to the stirred diazotizing solution from a dropping funnel. During the 1–1.25 hours required to complete this addition the temperature is held between −10° and −15° (Note 4). Stirring of the slurry is continued for approximately 2 hours longer, and during this period the temperature is allowed to rise gradually to +5°. The loosely stoppered flask is kept in a refrigerator for 36 hours (Note 5), during which time nitrogen and some oxides of nitrogen are evolved (*Caution!*). Most of the acetic acid is then removed by volatilization in a current of steam, 3 l. of distillate being collected.

During the steam distillation the product starts to precipitate from the clear solution, and nitrogen dioxide and some hydrogen sulfide are given off. The residual liquid is cooled, and the product is collected on a sintered-glass funnel and washed with water. It is next suspended in 800 ml. of water, and approximately 9 g. of anhydrous sodium carbonate is added with shaking to bring it into solution. The yellow liquid is filtered to remove small quantities of insoluble impurities, and the acid is reprecipitated by acidification to pH 1–2 with 5% hydrochloric acid, while swirling vigorously. The cream-colored precipitate is separated by filtration, washed with water, and dried. The yield of 2,4,6-tribromobenzoic acid, sufficiently pure for preparative purposes, is 50–57.5 g. (70–80%), m.p. 192.5–194.5° (Note 6).

2. Notes

1. Matheson Company *m*-aminobenzoic acid was used without further purification.

2. Efficient stirring and cooling are necessary at this stage to obtain a final product that is not colored.

3. Baker "purified" hypophosphorous acid was used.

4. Near the end of the addition of the acetic acid solution it is usually desirable to raise the stirrer temporarily to break up the lumps which form at the surface of the solution.

5. The temperature of the refrigerator was approximately 6°.

6. The product may be recrystallized from 1:2 acetic acid-water (about 8 ml./g.), but this process effects little improvement in melting point or color, even when activated carbon is used.

3. Methods of Preparation

2,4,6-Tribromobenzoic acid has been prepared by the deamination of 2,4,6-tribromo-3-aminobenzoic acid (reagents not specified),[3] by hydrolysis of 2,4,6-tribromobenzonitrile,[4-7] and by oxidation of the tribromotoluene,[8, 9] the benzyl chloride,[10] the aldehyde,[11] and the glyoxylic acid.[12] The present method is a modification of that of Bunnett, Robison, and Pennington.[13]

---

[1] Amherst College, Amherst, Massachusetts.

[2] This is the method of Coleman and Talbot, *Org. Syntheses Coll. Vol.* **2**, 592 (1943), for 2,4,6-tribromoaniline.

[3] Hübner, *Ber.*, **10**, 1708 (1877).

[4] Sudborough, *Ber.*, **27**, 512 (1894).

[5] Wegscheider, *Monatsh.*, **18**, 218 (1897).

[6] Sudborough, Jackson, and Lloyd, *J. Chem. Soc.*, **71**, 230 (1897).

[7] Montagne, *Rec. trav. chim.*, **27**, 351 (1908).

[8] Asinger, *J. prakt. Chem.*, **142**, 291 (1935).

[9] Cohen and Dutt, *J. Chem. Soc.*, **105**, 516 (1914).

[10] Henraut, *Bull. soc. chim. Belges*, **33**, 132 (1924).

[11] Blanksma, *Chem. Weekblad*, **9**, 865 (1912) [*Chem. Zentr.*, **1912 II**, 1965].

[12] Fuchs, *Monatsh.*, **36**, 137 (1915).

[13] Bunnett, Robison, and Pennington, *J. Am. Chem. Soc.*, **72**, 2378 (1950).

TRICHLOROMETHYLPHOSPHONYL DICHLORIDE

[Phosphonic dichloride, (trichloromethyl)-]

$$CCl_4 + AlCl_3 + PCl_3 \rightarrow [Cl_3CPCl_3]^+[AlCl_4]^-$$

$$[Cl_3CPCl_3]^+[AlCl_4]^- + 7H_2O \rightarrow Cl_3C\overset{\overset{O}{\uparrow}}{\underset{\underset{Cl}{|}}{P}}Cl + AlCl_3 \cdot 6H_2O + 2HCl$$

Submitted by KENNETH C. KENNARD and CLIFF S. HAMILTON.[1]
Checked by JOHN C. SHEEHAN and J. L. YEH.

1. Procedure

Caution! This preparation should be conducted in a hood.

In a 2-l. round-bottomed three-necked flask, fitted with an efficient reflux condenser, mechanical stirrer, and dropping funnel (Note 1), are placed 133.3 g. (1 mole) of anhydrous powdered aluminum chloride, 137.4 g. (1 mole) of phosphorus trichloride, and 184.6 g. (1.2 moles) of carbon tetrachloride (Note 2). The reactants are stirred slowly until they are thoroughly mixed, and then heat is applied carefully until the reaction begins. At this point the liquid boils vigorously, and the reaction mixture becomes thicker so that faster stirring is necessary. Finally, the stirrer is stopped when the mixture becomes solid. After the reaction has cooled for 30 minutes, 1 l. of methylene chloride is run into the flask (Note 3), and the solvent is stirred vigorously until the solid is finely suspended (Note 4). The reflux condenser is replaced by a low-temperature thermometer which dips into the reaction mixture, the suspension is cooled in a Dry Ice-acetone bath, and the temperature is kept at −10 to −20° as distilled water (180 g., 10 moles) is added dropwise with vigorous stirring over a period of about 25 minutes (Note 5). After the water addition is complete, stirring is continued for 15 minutes without the cold bath.

The apparatus is dismantled, and the reaction mixture is filtered quickly by suction through a 1.5-cm. layer of filter aid on an 11-cm. Büchner funnel placed on a 2-l. filter flask. The filter cake is pressed down well and washed with three 50-ml. portions of methylene chloride. The filtrate is immediately protected from moisture by calcium chloride tubes, and the solvent is removed by distillation from a 2-l. flask. After the solution has been concentrated to about 225 ml., the hot liquid is poured into a suitable container (Note 6), and the remaining solvent is removed under reduced pressure (Note 7). The yield is 192–199 g. (81–84%) of a white, crystalline solid which melts at 155–156°.

2. Notes

1. The equipment is dried and protected from atmospheric moisture by calcium chloride tubes.

2. It is important that the reactants are pure; otherwise the complex is colored. If the containers are freshly opened, c.p. reagents are satisfactory.

3. The methylene chloride must be redistilled from aluminum chloride to prevent coloration of the product.

4. If the solid adheres to the flask, the flask is heated lightly with a free flame to loosen it.

5. Vigorous stirring is necessary to prevent caking of the aluminum chloride hydrate and to prevent localized hydrolysis. The checkers found that exactly 7 moles of water gave a somewhat improved yield, but handling was more difficult.

6. The submitters used a 500-ml. round-bottomed wide-mouthed flask.

7. The product may be warmed to aid solvent removal, but it is decomposed by prolonged heating above 60°.

3. Methods of Preparation

The aluminum chloride process [2,3] is a general method for the preparation of alkylphosphonyl dichlorides. Trichloromethylphosphonyl dichloride has been made also by the chlorination of chloromethyldichlorophosphonic dichloride in the presence of light or other catalysts [4] and by the reaction of tetrachloro(trichloromethyl)phosphorane with sulfur dioxide.[5] The procedure described here is essentially that of Kennard and Hamilton [6] and is based on the procedure of Kinnear and Perren.[2]

[1] University of Nebraska, Lincoln, Nebraska.

[2] Kinnear and Perren, *J. Chem. Soc.*, **1952**, 3437.

[3] Clay, *J. Org. Chem.*, **16**, 892 (1951).

[4] Traise and Walsh (to Victor Chemical Works), U. S. pat. 2,924,560 [*C. A.*, **54**, 11994 (1960)].

[5] Yakubovich and Ginsburg, *Zhur. Obshcheĭ Khim.*, **24**, 1465 (1954) [*C. A.*, **49**, 10834 (1955)].

[6] Kennard and Hamilton, *J. Am. Chem. Soc.*, **77**, 1156 (1955).

p-TRICYANOVINYL-N,N-DIMETHYLANILINE

(Ethenetricarbonitrile, *p*-dimethylaminophenyl-)

$$(CH_3)_2N-\langle\bigcirc\rangle + (NC)_2C{=}C(CN)_2 \rightarrow$$

$$(CH_3)_2N-\langle\bigcirc\rangle-\underset{\underset{CN}{|}}{C}{=}C(CN)_2 + HCN$$

Submitted by B. C. McKusick and L. R. Melby.[1]
Checked by James Cason and Ralph J. Fessenden.

1. Procedure

Caution! Because hydrogen cyanide is formed in this reaction, all operations up to the recrystallization of the product should be carried out in a good hood. Contact of tetracyanoethylene with the skin should be avoided.

A solution of 26.6 g. (28 ml., 0.22 mole) of N,N-dimethylaniline in 65 ml. of dimethylformamide is placed in a 250-ml. beaker clamped about 30 cm. above the base of a ring stand. The beaker is provided with a mechanical stirrer and thermometer. An iron ring is attached to the ring stand below the beaker so that the temperature of the reaction mixture can be controlled by raising or lowering an ice bath or hot water bath. Recrystallized tetra-cyanoethylene (p. 877) (25.6 g., 0.20 mole) is added in small portions over a period of about 5 minutes with good stirring. The rate of addition is such as to maintain the temperature at 45–50°, and occasional cooling with an ice bath may be necessary to keep the temperature within this range.

When all the tetracyanoethylene has been added, the reaction mixture is stirred at 45–50° for 10 minutes, and heat is supplied as needed by a water bath. *p*-Tricyanovinyl-N,N-dimethyl-aniline generally crystallizes out as a dark-blue solid during this period. At the end of the heating period, the mixture is chilled in an ice bath for 30 minutes. The tricyanovinyl compound is collected on a Büchner funnel, pressed dry with the help of a filter dam, and washed successively with 20 ml. of methanol and 40 ml. of ether. It weighs 25–30 g. after being dried in air.

The crude product is purified by recrystallization from 160–180 ml. of acetic acid. The solution (Note 1) is allowed to cool slowly to room temperature, and p-tricyanovinyl-N,N-dimethylaniline is collected on a Büchner funnel and washed successively with 20 ml. of methanol and 40 ml. of ether. The product, 23–26 g. (52–58%), is obtained as dark-blue needles, m.p. 173–175° (Note 2).

2. Notes

1. The acetic acid solution is so deep a red color that it is necessary to hold the flask over a bright light in order to determine when all the solid has dissolved. The solution will dye the skin with a fast red color.

2. Although the crystals have a very dark blue appearance, the solutions are deep red; in acetone, λ_{max}. 517 mμ (ϵ 41,500).

3. Methods of Preparation

p-Tricyanovinyl-N,N-dimethylaniline has been prepared by adding hydrogen cyanide to p-dimethylaminobenzalmalononitrile and oxidizing the adduct.[2] The present procedure, an adaptation of one that has been published,[2] is the more convenient preparative method. It can be applied to a wide variety of secondary and tertiary aromatic amines to give p-tricyanovinyl-arylamines that, like the present one, are dyes.[2] Other types of aromatic compounds also condense with tetracyanoethylene in this manner. Thus one can obtain 4-tricyanovinyl-2,6-dimethylphenol from 2,6-dimethylphenol, 2-tricyanovinylpyrrole from pyrrole, and 9-tricyanovinylphenanthrene from phenanthrene.[3]

[1] Contribution No. 484 from Central Research Department, Experimental Station, E. I. du Pont de Nemours & Co., Wilmington, Delaware.

[2] McKusick, Heckert, Cairns, Coffman, and Mower, J. Am. Chem. Soc., 80, 2806 (1958); Heckert (to E. I. du Pont de Nemours & Co.), U. S. pat. 2,889,335 [C. A., 54, 1877 (1960)].

[3] Sausen, Engelhardt, and Middleton, J. Am. Chem. Soc., 80, 2815 (1958).

TRIETHYL PHOSPHITE*

(Ethyl phosphite)

$$PCl_3 + 3C_2H_5OH + 3C_6H_5N(C_2H_5)_2 \rightarrow$$
$$(C_2H_5O)_3P + 3C_6H_5N(C_2H_5)_2 \cdot HCl$$

Submitted by A. H. Ford-Moore and B. J. Perry.[1]
Checked by William S. Johnson and James Ackerman.

1. Procedure

A solution of 138 g. (175 ml., 3 moles) of absolute ethanol (Note 1) and 447 g. (477 ml., 3 moles) of freshly distilled diethylaniline in 1 l. of dry petroleum ether (b.p. 40–60°) is placed in a 3-l. three-necked flask fitted with a sealed stirrer, an efficient reflux condenser, and a 500-ml. dropping funnel (Note 2) which is charged with a solution of 137.5 g. (87.5 ml., 1 mole) of freshly distilled phosphorus trichloride in 400 ml. of dry petroleum ether (b.p. 40–60°). The flask is cooled in a cold-water bath. With vigorous stirring (Note 3), the phosphorus trichloride solution is introduced at such a rate that the mixture boils gently towards the end of the addition. After the addition, which requires about 30 minutes, the mixture is heated under gentle reflux for about 1 hour with stirring. The suspension, containing a copious precipitate of diethylaniline hydrochloride, is then cooled and filtered with suction through a sintered-glass funnel. The cake of the amine salt is well compressed and washed with five 100-ml. portions of dry petroleum ether (b.p. 40–60°). The filtrate and washings are combined and concentrated by distillation at water-bath temperature through a 75-cm. Vigreux column. The residue is transferred to a pear-shaped flask and distilled under water-pump vacuum through a 75-cm. Vigreux column. After a small fore-run, the product is collected at 57–58°/16 mm. (51–52°/13 mm., 43–44°/10 mm.). The yield of colorless product is 138 g. (83%), n_D^{25} 1.4104–1.4106, d_4^{20} 0.963 (Notes 4 and 5).

2. Notes

1. It is important that the ethanol be thoroughly anhydrous.

The checkers employed ethanol dried over magnesium ethoxide.[2]

2. It is convenient to connect the dropping funnel to the flask by a piece of 20-mm. glass tubing about 10 cm. long which is sleeved into the neck of the flask by a section of rubber tubing. By this means, the rate of introduction of the phosphorus trichloride solution may be readily observed and clogging by the copious precipitate of diethylaniline hydrochloride is obviated.

3. If efficient mixing is not obtained, hydrogen chloride may be liberated locally and one of the ethyl groups eliminated as ethyl chloride with the resulting appearance of diethyl hydrogen phosphite in the final distillate.

4. The recovered petroleum ether and fore-run contain some of the product. By using the recovered petroleum ether in subsequent runs and adding the fore-run before the final distillation, the yield is increased to 86–90%.

5. Triisopropyl phosphite is prepared similarly, using anhydrous isopropyl alcohol in place of ethanol. It has the following properties: b.p. 43.5°/1.0 mm; n_D^{25} 1.4080; d_4^{17} 0.917.

3. Methods of Preparation

The method described here is essentially that of McCombie, Saunders, and Stacey [3] except that diethylaniline is employed in place of dimethylaniline or pyridine. Diethylaniline has the advantage that the hydrochloride formed in the reaction is very easily filtered and is non-hygroscopic. Triethyl phosphite has been prepared also from phosphorus trichloride and ethanol in the presence of other tertiary amines, such as tributyl- and triamylamine; [4] in the presence of ammonia, [5] aniline, [6] and ammonium carbamate. [7] It also has been obtained from the reaction of magnesium ethoxide with phosphorus trichloride. [8]

[1] Chemical Defence Experimental Station, Porton, Nr. Salisbury, Wilts, England.

[2] Fieser, *Experiments in Organic Chemistry*, 3rd ed., p. 286, D. C. Heath and Company, Boston, Massachusetts, 1955.

[3] McCombie, Saunders, and Stacey, *J. Chem. Soc.*, **1945**, 381.

[4] Marshall (to Monsanto Chemical Co.), U. S. pat. 2,848,474 [*C. A.*, **53**, 1144 (1959)].

[5] Boyer and Mangham (to Virginia-Carolina Chemical Corp.), U. S. pat. 2,678,940 [*C. A.*, **49**, 4704 (1955)].

[6] VEB Farbenfabrik Wolfen (by Maier-Bode and Kötz), Ger. pat. 1,028,554 [*C. A.*, **54**, 10860 (1960)].

[7] Reetz (to Monsanto Chemical Co.), U. S. pat. 2,859,238 [*C. A.*, **54**, 1299 (1960)].

[8] Mel'nikov, Mandel'baum, and Bakanova, *Zhur. Obshchei Khim.*, **28**, 2473 (1958) [*C. A.*, **53**, 3032 (1957)].

2,4,4-TRIMETHYLCYCLOPENTANONE

(Cyclopentanone, 2,4,4-trimethyl-)

Submitted by GEORGE D. RYERSON, RICHARD L. WASSON, and HERBERT O. HOUSE.[1]

Checked by JAMES CASON and RALPH J. FESSENDEN.

1. Procedure

In a 1-l. separatory funnel is placed a solution of 38.6 g. (0.25 mole) of isophorone oxide (Note 1) in 400 ml. of reagent grade benzene. To the solution is added 20 ml. (0.16 mole) of boron trifluoride etherate (Note 2). The resulting solution is mixed by swirling, allowed to stand for 30 minutes, then diluted with 100 ml. of ether and washed with 100 ml. of water (Note 3). The organic layer is shaken for 1–2 minutes with a solution of 40 g. (1.0 mole) of sodium hydroxide in 200 ml. of water (Note 4) and then washed with a second 100-ml. portion of water. The combined aqueous solutions are cooled briefly in running water and then extracted with two 50-ml. portions of ether. The ethereal extracts are added to the benzene-ether solution (Note 5), and the combined solution, after drying over anhydrous magnesium sulfate, is concentrated by distillation through a Claisen head. When the temperature of the distillate reaches about 80° (Note 6), the residual liquid is fractionally distilled (Note 7) under reduced pressure. The yield of 2,4,4-trimethylcyclopentanone, b.p.

61–62°/21 mm., n_D^{28} 1.4278–1.4288, is 17.7–19.8 g. (56–63%) (Note 8).

2. Notes

1. The preparation of isophorone oxide is described on p. 552

2. A practical grade of boron trifluoride etherate, purchased from Eastman Kodak Company, was redistilled before use. The pure etherate boils at 126°.

3. If the organic layer is dried over magnesium sulfate and fractionally distilled at this point, both 2,4,4-trimethylcyclopentanone and 2-formyl-2,4,4-trimethylcyclopentanone, b.p. 49–50° (2 mm.), n_D^{25} 1.4495, may be isolated. The pot residue from this distillation contains a small amount of the enol form of 3,5,5-trimethyl-1,2-cyclohexanedione which crystallizes from petroleum ether as white needles, m.p. 92–93°.

4. The specified period of shaking is sufficient to ensure complete deformylation of the intermediate β-ketoaldehyde. Ordinarily, no difficulty is experienced with emulsification provided that the recommended quantity (or more) of ether is added to the reaction mixture.

5. Acidification of the residual aqueous solution followed by extraction with ether permits the isolation of about 1 g. of the enol form of 3,5,5-trimethyl-1,2-cyclohexanedione (see Note 3).

6. At this point about 150–200 ml. of liquid remains in the still pot. If the distillation is continued, the peppermint-like odor of the product, 2,4,4-trimethylcyclopentanone, can be detected in the distillate.

7. The submitters used a 24-cm. jacketed Vigreux column for this distillation. The checkers used a simple type of Podbielniak column, 50 cm. in length, with heated jacket and partial reflux head; after the last of the benzene had been distilled through the column, the following fractions were received:

Wt., g.	B.P./15 mm.	n_D^{28}
0.7	35–54°	1.4288
0.7	54–55°	1.4278
7.5	55°	1.4278
9.5	55°	1.4278
4.9	Residue, m.p. 80–90°	

Thus a rather good sample of product may be obtained without use of fractionating equipment.

8. If an appreciable quantity of higher-boiling material (consisting of 2-formyl-2,4,4-trimethylcyclopentanone; see Note 4) remains after the product has been collected, the residue should be dissolved in ether and shaken with aqueous sodium hydroxide as described in the procedure. After the ethereal extract has been dried over magnesium sulfate, distillation will permit the isolation of an additional quantity of 2,4,4-trimethylcyclopentanone.

3. Methods of Preparation

2,4,4-Trimethylcyclopentanone has been prepared by the oxidation of 1-hydroxy-2,4,4-trimethylcyclopentanecarboxylic acid with lead dioxide in sulfuric acid,[2] by the hydrogenation of 2,4,4-trimethyl-2-cyclopentenone,[3,4] by the Clemmensen reduction of dimethyldihydroresorcinol,[5,6] by the distillation of powdered 2,4,4-trimethyladipic acid with sodium hydroxide,[7] and by the saponification and decarboxylation of ethyl 2-keto-1,4,4-trimethylcyclopentanecarboxylate.[8,9] The rearrangement of isophorone oxide [10] appears to represent the optimum combination of favorable yield and convenient procedure.

[1] Massachusetts Institute of Technology, Cambridge, Massachusetts.

[2] Wallach, *Ann.*, **414**, 296 (1918).

[3] Fjader, *Suomen Kemistilehti*, **5**, Suppl. 27 (1932) [*Chem. Zentr.*, **1932 I**, 3172].

[4] Nazarov and Bukhmutskaya, *Bull. acad. sci. U.R.S.S., Classe sci. chim.*, **1947**, 205 [*C. A.*, **42**, 7733 (1948)].

[5] Dey and Linstead, *J. Chem. Soc.*, **1935**, 1063.

[6] Auterinen, *Suomen Kemistilehti*, **10B**, 22 (1937) [*C. A.*, **32**, 509 (1938)].

[7] Birch and Johnson, *J. Chem. Soc.*, **1951**, 1493.

[8] Qudrati-Khuda and Mukherji, *J. Indian Chem. Soc.*, **23**, 435 (1946).

[9] Chakravarti, *J. Chem. Soc.*, **1947**, 1028.

[10] House and Wasson, *J. Am. Chem. Soc.*, **79**, 1488 (1957).

α,β,β-TRIPHENYLPROPIONIC ACID

(Propionic acid, 2,3,3-triphenyl-)

$$C_6H_5CHBrCHBrCO_2H + 2C_6H_6 \xrightarrow{AlBr_3}$$
$$(C_6H_5)_2CHCHCO_2H + 2HBr$$
$$\underset{\displaystyle C_6H_5}{|}$$

Submitted by C. P. KRIMMEL, L. E. THIELEN, E. A. BROWN, and
W. J. HEIDTKE.[1]
Checked by WILLIAM S. JOHNSON, A. L. WILDS, and J. S. JELLINEK.

1. Procedure

A 5-l. three-necked flask is fitted with a mechanical rubber-slee.′ed stirrer, a dropping funnel, and a reflux condenser capped with a calcium chloride tube leading to a gas-absorption trap.[2] The system is flame-dried, and the flask is charged with 308 g. (1 mole) of dibromohydrocinnamic acid (Note 1) and 800 ml. of dried (by distillation) thiophene-free benzene. While the dibromohydrocinnamic acid is maintained in suspension by stirring (Note 2), a freshly prepared solution of 294 g. (1.1 moles) of anhydrous aluminum bromide (Note 3) in 400 ml. of anhydrous thiophene-free benzene is added from the dropping funnel over a period of 30 minutes. The clear orange-to-red solution is then heated under reflux with stirring for 4 hours.

The mixture is cooled to room temperature and maintained there with the aid of a cooling bath while 500 ml. of concentrated hydrochloric acid is added slowly from the dropping funnel. The mixture should be stirred vigorously during this addition, which requires about 30 minutes and is accompanied by copious evolution of hydrogen bromide (Note 4) and separation of the α,β,β-triphenylpropionic acid as a thick white slurry. Stirring is continued for an additional 30 minutes, 2 l. of water is added, and the product is separated by suction filtration. The waxy filter cake is washed with two 250-ml. portions of water, then dried overnight at room temperature and finally at 75° (Note 5). The crude, almost colorless α,β,β-triphenylpropionic acid amounts to 287–302 g. (95–100% yield), m.p. 215–218°. Recrystallization by dissolution in 3 l. of isopropyl alcohol (Note 6), followed by concentration of the filtered solution to 1.5 l. before cooling, yields 200–236 g. (66–78%) of colorless needles, m.p. 220–221°.

2. Notes

1. Dibromohydrocinnamic acid is conveniently prepared by the method of Reimer.[3] The checkers employed the following procedure. To a gently boiling solution of 296 g. (2 moles) of cinnamic acid in 2 l. of c.p. carbon tetrachloride contained in a 5-l. three-necked flask fitted with a reflux condenser, a dropping funnel, and a mechanical stirrer is added slowly (1.5 hours) with stirring 320 g. (2 moles) of bromine dissolved in 200 ml. of carbon tetrachloride. After 25–50% of the bromine is added, the dibromohydrocinnamic acid begins to precipitate with evolution of heat. Stirring and heating are continued for an additional 30 minutes after all the bromine is added. The mixture is cooled to room temperature; the product is separated by suction filtration and is washed with a small amount of cold carbon tetrachloride. The air-dried product amounts to 558–580 g. (91–94% yield) of colorless crystals, m.p. 197–198° to 202–204°.

2. If a solution instead of suspension is used, troublesome gel formation may occur during the aluminum bromide addition.

3. Colorless, crystalline, anhydrous aluminum bromide supplied by the Westvaco Chemical Division, Food Machinery and Chemical Corporation, New York, New York, was used. When dissolved in dry benzene at room temperature with mechanical stirring, a perfectly clear yellow solution results, if the reagents are of high purity.

4. If stirring is not sufficiently vigorous or if the temperature is too low, the evolution of hydrogen bromide may be delayed and then may begin abruptly and be difficult to control.

5. If the wet product is introduced directly into the drying oven it may darken slightly.

6. If the solution is appreciably colored it may be treated with decolorizing carbon at this point. Toluene or dilute ethanol may also be used for the recrystallization, but these solvents are less satisfactory.

3. Methods of Preparation

α,β,β-Triphenylpropionic acid has been prepared by the alkaline hydrolysis of the addition product of phenylmagnesium bromide and methyl α-phenylcinnamate; [4] by the reaction of α-phenylcinnamic acid with benzene in the presence of aluminum chloride; [5] by the reaction of dibromohydrocinnamic acid with benzene in the presence of aluminum bromide or ferric chloride; [6] by the reaction of phenylacetic acid with benzhydryl chloride in the presence of sodamide; [7] and by the reduction of 2,3,3-triphenylacrylonitrile with benzyl alcohol and alkali, followed by hydrolysis of the crude product. [8] The procedure described here is a modification of the method of Earl and Wilson. [6]

[1] G. D. Searle and Company, Chicago, Illinois.
[2] *Org. Syntheses Coll. Vol.* **2**, 4 (1943).
[3] Reimer, *J. Am. Chem. Soc.*, **64**, 2510 (1942).
[4] Kohler and Heritage, *Am. Chem. J.*, **33**, 156 (1905).
[5] Eijkman, *Chem. Weekblad*, **5**, 655 (1908) (*Chem. Zentr.*, **1908 II**, 1100).
[6] Earl and Wilson, *J. Proc. Roy. Soc. N. S. Wales*, **65**, 178 (1932) [*C. A.*, **26**, 2976 (1932)].
[7] Hauser and Chambers, *J. Am. Chem. Soc.*, **78**, 4996 (1957).
[8] Avramoff and Sprinzak, *J. Am. Chem. Soc.*, **80**, 493 (1958).

α,α,β-TRIPHENYLPROPIONITRILE

(Propionitrile, 2,2,3-triphenyl-)

$$(C_6H_5)_2CHCN \xrightarrow[\text{Liq. NH}_3]{\text{KNH}_2} [(C_6H_5)_2CCN]^-K^+$$

$$[(C_6H_5)_2CCN]^-K^+ + C_6H_5CH_2Cl \rightarrow$$

$$(C_6H_5)_2\underset{\underset{CN}{|}}{C}{-}CH_2C_6H_5 + KCl$$

Submitted by C. R. Hauser and W. R. Dunnavant.[1]
Checked by Virgil Boekelheide and Donald R. Arnold.

1. Procedure

Caution! This preparation should be conducted in a hood to avoid exposure to ammonia.

A suspension of potassium amide (0.23 mole) in liquid ammonia is prepared in a 1-l. three-necked flask equipped with an air condenser (without drying tube), a ball-sealed mechanical stirrer, and a dropping funnel. Commercial anhydrous liquid ammonia (500 ml.) is introduced into the flask from a cylinder through an inlet tube. To the stirred ammonia is added a small piece of potassium metal. After the appearance of a blue color, a few crystals (about 0.25 g.) of ferric nitrate hydrate are added, followed by small pieces of potassium (Note 1) until 9.0 g. (0.23 g. atom) has been added. After all the potassium has been converted to the amide (Note 2), 44.6 g. (0.23 mole) of diphenylacetonitrile (Note 3) is added and the resulting greenish-brown solution is stirred for 5 minutes. To this is added, over 10 minutes, 30.5 g. (0.24 mole) of benzyl chloride (Note 4) in 100 ml. of anhydrous ether. The orange solution is stirred for 1 hour, and the ammonia is then evaporated on a steam bath as 300 ml. of anhydrous ether is being added. To the ether solution is added 300 ml. of water, whereupon the crude nitrile precipitates. The ether is then removed by distillation and the crude nitrile is collected on a Büchner funnel. The yield of crude, light-tan α,α,β-triphenylpropionitrile is 64 g. (98–99%). The nitrile is dissolved in 1.3 l. of hot ethanol, treated with Norit, and filtered. The filtrate is held at room temperature overnight, and the product collected by filtration. A second crop is obtained by concentration of the mother liquor. The total yield of α,α,β-triphenylpropionitrile, m.p. 126.5–127.5°, is 62.2–65.5 g. (95–99% yield) (Note 5).

2. Notes

1. The potassium is cut in about 0.5-g. pieces, stored under kerosene, and blotted with filter paper before addition.

2. Conversion is indicated by the discharge of the deep-blue color. This generally requires about 20 minutes.

3. Diphenylacetonitrile supplied by the Eastman Kodak Company was used without purification.

4. Eastman Kodak Company practical grade benzyl chloride was vacuum-distilled; the fraction, b.p. 63°/12 mm., was used.

5. Under comparable conditions the corresponding alkylations of diphenylacetonitrile with α-phenethyl chloride and benzhydryl

chloride have been effected to form 2,3,3-triphenylbutyronitrile and 2,2,3,3-tetraphenylpropionitrile in yields of 88 and 96% respectively.[2]

3. Methods of Preparation

The method used is that of Hauser and Brasen.[2] The benzylation of diphenylacetonitrile with benzyl chloride to form α,α,β-triphenylpropionitrile has been previously effected in 83% yield by sodium ethoxide in ethanol,[3] in 67% yield by methylmagnesium iodide in ether,[4] and in unreported yield by sodium amide in ether.[5]

[1] Duke University, Durham, North Carolina. Work supported by the Office of Ordnance Research.
[2] Hauser and Brasen, *J. Am. Chem. Soc.*, **78**, 82 (1956).
[3] Neure, *Ann.*, **250**, 140 (1889).
[4] Sisido, Nozaki, and Kurihara, *J. Am. Chem. Soc.*, **72**, 2270 (1950).
[5] Ramart, *Bull. soc. chim. France*, [4] **35**, 196 (1924).

TRIPTYCENE

(9,10-*o*-Benzenoanthracene, 9,10-dihydro-)

Submitted by GEORGE WITTIG.[1]
Checked by JOHN D. ROBERTS, M. C. CASERIO, E. S. JOHNSON, and L. SKATTEBØL.

1. Procedure

A 200-ml. three-necked flask, equipped with a ball-and-socket sealed mechanical stirrer, a pressure-compensated dropping funnel, and a reflux condenser connected to a mercury bubbler

(Note 1), is charged with 0.8 g. (0.033 g. atom) of magnesium turnings, 7.5 g. (0.042 mole) of anthracene, and 35 ml. of anhydrous tetrahydrofuran (Note 2). In the dropping funnel there is placed a solution of 5.26 g. (0.03 mole) of o-fluorobromobenzene (Note 3) in 15 ml. of tetrahydrofuran. The system is flushed with dry nitrogen for 30 minutes to remove air. The gas flow is then stopped in order to prevent extensive loss of tetrahydrofuran. The mixture is heated to and maintained at 60° (bath temperature), and one-quarter of the o-fluorobromobenzene solution is added with stirring. The appearance of a yellow color which evidences the start of reaction may not be observed immediately, and another quarter of the solution is then added dropwise over a period of about 45 minutes. When the reaction commences, the remaining solution is added dropwise over a period of 1 hour, after which the mixture is refluxed gently for 90 minutes. The almost homogeneous dark-brown mixture is poured into 100 ml. of methanol, which precipitates much of the unreacted anthracene. Without filtering, the solvents are removed under reduced pressure and the yellow residue is treated with two 50-ml. portions of 5% hydrochloric acid, filtered, and vacuum-dried. The dry, yellow residue (10 g.) is dissolved in 45 ml. of hot xylene, then 5.0 g. (0.051 mole) of maleic anhydride is added. The mixture is refluxed for 20 minutes and set aside at room temperature for 2 hours. The maleic anhydride-anthracene adduct (about 9 g.) is removed by filtration, and the brown filtrate is refluxed for 2 hours with 80 ml. of 2N sodium hydroxide solution. When cool, the organic layer is separated, washed three times with 50-ml. portions of water (Note 4), and dried over calcium chloride. The solvent is removed at reduced pressure. The brown residue is dissolved in 70 ml. of carbon tetrachloride and chromatographed on 280 g. of acid-washed alumina, using 1 l. of the same solvent to elute. After evaporation of the solvent, there remains 2.4–2.9 g. of a yellow residue which is digested with two 10-ml. portions of pentane (Note 5). The residual crude triptycene is an almost white crystalline solid of melting point 240–248°. The yield is 2.14 g. (28%) (Note 6). Recrystallization of this material from cyclohexane gives pure white crystals of melting point 255–256°.

2. Notes

1. The mercury bubbler seals the system from the air. It is connected to the top of the reflux condenser by means of 8-mm. glass tubing more than 76 cm. high. The pressure in the system may be varied by adjusting the depth of the lower end of the tube. A constant stream of dry nitrogen may be substituted for the mercury bubbler, but this inevitably results in some loss of tetrahydrofuran.

2. Tetrahydrofuran may be purified by distillation from lithium aluminum hydride.

3. o-Fluorobromobenzene as supplied by the Aldrich Chemical Company, Milwaukee, Wisconsin, may be used without further purification.

4. Small amounts of sodium chloride may be added to facilitate separation of the phases.

5. The main impurity is a yellow oil which is readily soluble in pentane.

6. The checkers report a yield of 1.35–1.68 g. (18–22%) of triptycene of melting point 245–255°.

3. Methods of Preparation

Triptycene has been prepared by Bartlett and co-workers [2] in a seven-step synthesis. It also has been obtained by the reaction of fluorobenzene with anthracene in the presence of butyllithium [3] and by the reduction of the anthracene-benzoquinone adduct with lithium aluminum hydride or sodium borohydride.[4] The present method has been published.[5]

[1] University of Tübingen, Tübingen, Germany.
[2] Bartlett, Ryan, and Cohen, J. Am. Chem. Soc., 64, 2649 (1942).
[3] Wittig and Benz, Angew. Chem., 70, 166 (1958).
[4] Craig and Wilcox, J. Org. Chem., 24, 1619 (1959).
[5] Wittig and Benz, Chem. Ber., 91, 873 (1958).

TRITHIOCARBODIGLYCOLIC ACID

(Carbonic acid, trithio-, bis[carboxymethyl] ester)

$$2KOH + H_2S \longrightarrow K_2S + 2H_2O$$

$$K_2S + CS_2 \longrightarrow (KS)_2CS$$

$$(KS)_2CS + 2ClCH_2CO_2K \longrightarrow (KO_2CCH_2S)_2CS$$

$$(KO_2CCH_2S)_2CS \xrightarrow{HCl} (HO_2CCH_2S)_2CS$$

Submitted by R. E. Strube.[1]
Checked by John D. Roberts and Stanley L. Manatt.

1. Procedure

In a 300-ml. three-necked, round-bottomed flask equipped with a magnetic stirrer and a gas-inlet tube reaching below the surface of the liquid is placed a solution of 63 g. (0.96 mole) of potassium hydroxide (Note 1) in 100 ml. of water. The solution is cooled in ice, and hydrogen sulfide is bubbled through (Note 2) with stirring until the gain of weight is 33–34 g. (Note 3). The solution is then poured into a 3-l. three-necked, round-bottomed flask provided with a stirrer, a gas-inlet tube, a reflux condenser, and a thermometer reaching into the liquid. The small flask is rinsed with 25 ml. of ice water and the rinsings added to the rest of the solution. Then 63 g. (0.96 mole) of potassium hydroxide is added and allowed to dissolve. The 3-l. flask is then well flushed with nitrogen and, at a temperature of about 30°, 76.0 g. (1.0 mole) of carbon disulfide (Note 4) is added at once. The mixture is stirred vigorously for 2 hours (Note 5) while nitrogen is passed through at a rate of about one bubble per second (Note 6) and the temperature is kept at 35–38° (Note 7). Then the gas supply is disconnected and the dark-red solution is cooled in an ice bath.

A solution of 189 g. (2.0 moles) of chloroacetic acid (Note 8) in 300 ml. of water is neutralized to litmus with a solution containing approximately 135 g. (2.1 moles) of potassium hydroxide in 300 ml. of water. The resulting potassium chloroacetate solution is placed in a dropping funnel and added to the stirred potassium thiocarbonate solution obtained above at such a rate that the temperature does not go above 40°. After the addition is complete, the stirring is continued for 1 hour at room temperature. Then 200 ml. of concentrated hydrochloric acid is added while the temperature is kept below 20° by cooling in an ice bath. Finally, the reaction mixture is stirred for 30 minutes at room temperature. The yellow precipitate is filtered and washed twice with 150-ml. portions of ice water. The crude material is dried under reduced pressure in a vacuum desiccator over calcium chloride to constant weight (about 2 days). The drying is expedited if the lumps are occasionally broken up. The yield of yellow product having m.p. 169–174° (uncor.) (Note 9) is 152–160 g. (67–71%).

2. Notes

1. Potassium hydroxide pellets, Mallinckrodt, 85% minimum KOH assay, were used.

2. A bubbler filled with mercury was placed between the gas cylinder and the gas-inlet tube. A good hood should be used throughout the procedure because hydrogen sulfide is toxic in minute concentrations.

3. Two to three hours is required to saturate the solution. The submitter used the same 3-l. flask to prepare the potassium sulfide and to carry out the subsequent reaction. The checkers found the smaller flask more convenient for following the hydrogen sulfide uptake.

4. Carbon disulfide, Mallinckrodt, analytical grade reagent, was used.

5. The carbon disulfide layer usually disappears in about 45 minutes, but longer times may be required if the stirring is not effective.

6. Contact of the reaction mixture with atmospheric oxygen is to be avoided, but the gas flow should be kept slow enough to minimize loss of carbon disulfide. A bubbler filled with water was placed between the gas cylinder and the gas-inlet tube.

7. A water bath kept at 40–43° or an electrically heated mantle may be used.

8. Chloroacetic acid, m.p. 62–64°, Eastman Kodak, was used.

9. The submitter reports yields of 160–175 g. (71–77%) of product melting at 166–172°. Recrystallization from water gives 150–165 g. (66–73%) of material melting at 174–176° (cor.).

3. Methods of Preparation

Trithiocarbodiglycolic acid can be prepared by heating an aqueous solution of the alkali salts of thiocarbonylethoxythioglycolic acid,[2,3] by heating an aqueous solution of potassium methylxanthate and sodium monochloroacetate,[3] and by heating an aqueous solution of potassium thiocarbonate and sodium monochloroacetate.[3] The compound is also formed by heating an aqueous solution of potassium ethyltrithiocarbonate and sodium monochloroacetate,[4] and by heating an aqueous solution of thiocarbonylglycolic acid-thioglycolic acid with ammonia [5] or aniline.[6] The procedure described is adapted from that of Holmberg.[7]

[1] Research Division, The Upjohn Company, Kalamazoo, Michigan.
[2] Holmberg, *J. prakt. Chem.*, **71**, 271 (1905).
[3] Biilmann, *Ann.*, **348**, 134 (1906).
[4] Holmberg, *J. prakt. Chem.*, **75**, 182 (1907).
[5] Ahlqvist, *J. prakt. Chem.*, **99**, 55 (1919).
[6] Holmberg, *J. prakt. Chem.*, **84**, 645 (1911).
[7] Holmberg, *J. prakt. Chem.*, **71**, 279 (1905).

10-UNDECYNOIC ACID

$$CH_2{=}CH(CH_2)_8CO_2H + Br_2 \rightarrow CH_2BrCHBr(CH_2)_8CO_2H$$

$$CH_2BrCHBr(CH_2)_8CO_2H + 3NaNH_2 \rightarrow$$
$$HC{\equiv}C(CH_2)_8CO_2Na + 2NaBr + 3NH_3$$

$$HC{\equiv}C(CH_2)_8CO_2Na + HCl \rightarrow HC{\equiv}C(CH_2)_8CO_2H + NaCl$$

Submitted by N. A. KHAN.[1]
Checked by T. L. CAIRNS and D. H. SMITH.

1. Procedure

Caution! All reactions involving liquid ammonia must be carried out in an effective hood.

Bromine (approximately 15 ml.) is added dropwise to 50 g. (0.271 mole) of 10-undecenoic acid (Note 1) dissolved in 210 ml. of dry ether with constant stirring while the temperature is maintained below 0° until the color of bromine persists. The excess bromine is then removed by the addition of a few drops of 10-undecenoic acid.

The sodamide for dehydrobromination of dibromohendecanoic acid is prepared according to the method of Khan et al.,[2] with some modifications (Note 2). A 3-l. three-necked flask is equipped with a stirrer and a Dry Ice-acetone cold finger reflux condenser attached through a drying tube to a trap and a bubbler filled with a saturated ammonia solution. After the condenser has been cooled with a mixture of Dry Ice and acetone, 1.5 l. of liquid ammonia is introduced into the flask through an inlet tube. Stirring is started, and 1.2–1.5 g. of ferric chloride (c.p., anhydrous, black) is added to the liquid ammonia in one portion as quickly as possible. Within 20 seconds after this addition, 6 g. of metallic sodium is dropped into the brown solution in two portions in order to convert the iron salt into catalytic iron. When the evolution of hydrogen gas ceases, the remainder of the sodium (total amount, 27.7 g., 1.2 g. atoms) is added in pieces and the solution is stirred for an additional 30 minutes (Note 3).

Finally, the solution of 10,11-dibromohendecanoic acid in absolute ether is added slowly from a separatory funnel to the reaction flask. The reaction mixture is then stirred for 6 hours. After an additional hour (Note 4) of stirring, during which time the cold-finger condenser is removed from the system, an excess of solid ammonium chloride (40 g., 0.74 mole) is introduced slowly into the reaction mixture to destroy excess sodamide. The ammonia is allowed to evaporate, 400–500 ml. of water is added, and the mixture is stirred until all the solid is broken up and dissolved (Note 5).

The water solution containing the sodium salt of the product is acidified with $6N$ hydrochloric acid and then extracted with three 200-ml. portions of ether. The ethereal extracts are combined, washed with water (until the aqueous phase shows a pH in the range 5–6), and then dried over anhydrous sodium sulfate. After removal of the solvent, the red-colored residual oil is fractionally distilled through a packed column (Note 6). The middle portion (26–28 g.) distilling at 124–130°/3 mm. is crystallized twice from petroleum ether (b.p. 30–60°) to obtain white crystals of 10-undecynoic acid. The end fractions also yield some white product after 4–5 crystallizations. A total yield of 19–24 g. (38–49%) is obtained; m.p. 42.5–43°.

2. Notes

1. 10-Undecenoic acid (undecylenic acid, commercial grade) may be obtained from the Eastman Kodak Company. The pure grade acid is also used with good results. However, for general purposes, no difficulty is encountered in preparing 10-undecynoic acid from the commercial grade acid if the final fractionation by vacuum distillation is carried out carefully.

2. The experiences with this particular preparation have led to some changes in the procedure of preparing sodamide. However, for large-scale preparation (larger than that reported here), Khan et al.[2] and Greenlee and Henne [3] have given very specific details of the methods that are to be followed. It is to be noted that the use of a Hershberg stirrer (with a rubber stopper and a powerful high-speed motor) is advantageous.

3. The complete transformation of metallic sodium into sodamide may be checked by the method of Greenlee and Henne.[3] When the evolution of hydrogen gas ceases (bubbler), it may be assumed that the above transformation is complete.

4. The checkers observed the yield to be increased somewhat by the less convenient procedure of allowing the ammonia to evaporate slowly for 12–15 hours.

5. More water may be added if needed.

6. The fractionation has also been carried out with better results through a Todd Scientific versatile column.

3. Methods of Preparation

Experiments in the submitter's laboratory have shown that the yield obtained on dehydrobromination of dibromohendecanoic acid with ethanolic potassium hydroxide [4-7] is very poor, usually lower than 30%. On the other hand, the dehydrobromination with sodamide in liquid ammonia is smooth and very satisfactory.[8] The directions employed here represent a modification of those of Lauer and Gensler.[9]

[1] Ohio State University, Columbus, Ohio.
[2] Khan, Deatherage, and Brown, *J. Am. Oil Chemists' Soc.*, **28**, 27 (1951).
[3] Greenlee and Henne, *Inorg. Syntheses*, **2**, 128 (1946).
[4] Krafft, *Ber.*, **29**, 2232 (1896).
[5] Oskerko, *Ber.*, **70**, 55 (1937).
[6] Bhattacharyya, Chakravarty, and Kumar, *Chem. & Ind.* (*London*), **1959**, 1352.
[7] Sparreboom, *Koninkl. Ned. Akad. Wetenschap., Proc., Ser. B*, **59**, 472 (1956)
[*C. A.*, **51**, 11992 (1957)].
[8] Khan, *J. Am. Oil Chemists' Soc.*, **30**, 355 (1953).
[9] Lauer and Gensler, *J. Am. Chem. Soc.*, **67**, 1171 (1945).

VANILLIC ACID*

I. SILVER OXIDE METHOD

Submitted by IRWIN A. PEARL.[1]
Checked by R. L. SHRINER and CALVIN N. WOLF.

1. Procedure

A solution of 170 g. (1.0 mole) of silver nitrate in 1 l. of water in a 2-l. beaker is treated, with stirring, with a solution of 44 g. (1.07 moles) of 97% sodium hydroxide in 400 ml. of water (Note 1). The mixture is stirred for 5 minutes, and the silver oxide is collected on an 11-cm. Büchner funnel with suction and washed free of nitrates with water (Note 2). The wet, freshly precipitated silver oxide is transferred to a 4-l. beaker (Note 1), covered with 2 l. of water, and treated with 200 g. (4.85 moles) of 97% sodium hydroxide pellets with vigorous stirring. If the temperature of the mixture at this point is below 55°, the mixture is warmed to 55–60°. With continued stirring at 55–60° (Note 3), 152 g. (1.0 mole) of vanillin (Note 4) is added; the reaction begins after a few minutes. The silver oxide is transformed to fluffy metallic silver, and considerable heat is evolved. Stirring is continued for 10 minutes, the mixture is filtered, and the precipitated silver is washed with 100 ml. of hot water. A rapid stream of sulfur dioxide gas (Note 5) is passed into the combined filtrate and washings for 2 minutes, and the resulting solution is poured into 1.1 l. of 1:1 hydrochloric acid with vigorous stirring. The resulting mixture, which should be acid to Congo red, is cooled to 15–20°. The vanillic acid is collected on a Büchner funnel, pressed to remove the mother liquor, washed with 150 ml. of ice water (Note 6), sucked as dry as possible, and air-dried. The yield is 140–160 g. (83–95%) of white needles melting at 209–210°. This product is pure enough for most purposes, but it may be purified by recrystallization from water containing a little sulfur dioxide, 1.2 l. of water being used per 100 g. of product. Pure white needles melting at 210–211° are obtained with a recovery of 90–97%.

2. Notes

1. This reaction should be performed in glass apparatus with a glass stirrer. If stainless steel is employed, the resulting vanillic acid may be dark in color.

2. The presence of nitrates in the solution will cause the formation of nitro acids when the final acidification takes place.

3. Fifty to fifty-five degrees was found to be the critical temperature for this reaction. If the reactants are mixed cold, heat must be applied to raise the temperature above 50°, at which point the reaction begins. Mixing of the reactants at temperatures much higher than 60° results in a violent reaction.

4. Commercial U.S.P. vanillin is satisfactory.

5. The treatment with sulfur dioxide prevents the product from becoming tan in color.

6. Extraction of the combined filtrate and washings with three 200-ml. portions of ether, followed by removal of the ether by distillation, yields an additional 4–20 g. of product melting at 206–208°. It may be purified by recrystallization from water containing a little sulfur dioxide to give white needles melting at 209–210°.

II. CAUSTIC FUSION METHOD

$$\text{(vanillin structure)} + 2KOH \rightarrow \text{(structure)} + H_2 + H_2O$$

$$\text{(structure)} + 2HCl \rightarrow \text{(structure)} + 2KCl$$

Submitted by Irwin A. Pearl.[1]
Checked by R. L. Shriner and Robert C. Johnson.

1. Procedure

In a stainless-steel beaker of approximately 2-l. capacity (120 mm. by 165 mm.) equipped with an efficient mechanical Nichrome or Monel stirrer (Note 1) and heated by an electric hot plate are placed 178 g. (4.3 moles) of 97% sodium hydroxide pellets, 178 g. (2.7 moles) of 85% potassium hydroxide pellets (Note 2), and 50 ml. of water. The mixture is stirred and heated to 160°, at which temperature the hot plate is turned off. Twenty-five grams of vanillin is added. The temperature drops somewhat, and after a very short time a vigorous reaction begins which raises the temperature to 180–195°. An additional 127 g. (total of 1.0 mole) of vanillin is gradually added to the reaction mixture during a period of 10–12 minutes at a rate sufficient to maintain the reaction temperature (Note 3). After all the vanillin has been added, stirring is continued for 5 minutes. The hot plate is removed, and the mixture is allowed to cool with stirring. When the mixture cools to about 150–160°, 1 l. of water is added and the mixture is stirred until all the fusion mixture is dissolved. The solution is transferred to a 4-l. beaker, 500 ml. of water being used to rinse the stirrer and metal beaker. Sulfur dioxide gas is introduced for 1 minute (Note 4), and the reaction mixture is cooled to room temperature. The mixture is acidified with about 1.2 l. of 6N hydrochloric acid using Congo red as the indicator and keeping the mixture cool during addition of the acid by stirring and cooling in an ice bath. The mixture is cooled, and the light-

tan precipitate is filtered, washed with 150 ml. of ice water, and
dried. The yield of vanillic acid melting at 206–208° is 150–160
g. (89–95%) (Note 5). The product can be recrystallized from
water, using 1.2 l. per 100 g. of vanillic acid. A 90–97% recovery
of nearly white vanillic acid melting at 209–210° is obtained.

2. Notes

1. An efficient stirrer can be made from No. 8 Nichrome or
Monel metal wire in a rectangular shape of dimensions such that
it is about two-thirds the height of the beaker and has a clearance
of 1–2 mm. with the sides of the beaker (Fig. 18). A rigid form is
maintained by spot-welding the crossed wires. The metal beaker
must be firmly clamped. It is important that there be effective
stirring throughout all the reaction mixture in order to prevent
caking, excessive local temperature rises, and foaming.

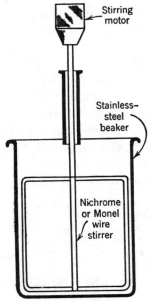

FIG. 18. Apparatus for caustic fusion of vanillin.

2. The exact proportion of sodium hydroxide to potassium hydroxide is not critical as long as the total amount of alkali is more

than 7 moles. Alkali mixtures containing more than 70% sodium hydroxide are not desirable because they are not as fluid as other mixtures.

3. The temperature of the mixture should not be allowed to rise much above 200° for any length of time because traces of protocatechuic acid will be formed and contaminate the vanillic acid.

4. The sulfur dioxide treatment prevents the product from becoming tan in color.

5. An additional few per cent may be obtained by ether extraction of the filtrate.

3. Methods of Preparation

Vanillic acid has been prepared from vanillin in small amounts by action of moist air,[2] exposure to sunlight and nitrobenzene,[3] reaction with soil bacteria,[4] ozone,[5] and by caustic fusion.[6,7] High yields of vanillic acid have been obtained from vanillin by controlled caustic fusion,[8] oxidation with silver oxide,[9,10] mercuric oxide,[11] and gold oxide,[11] and by the Cannizzaro reaction of vanillin in the presence of a silver catalyst.[12] Vanillic acid has been prepared indirectly from vanillin by the hydrolysis of acetylvanillic acid prepared by oxidation with peracetic acid of acetylvanillin,[13] by the hydrolysis of acetovanillonitrile prepared by the reaction of vanillin oxime with acetic anhydride,[14] and by the action of hydrazoic acid on vanillin, followed by hydrolysis of the intermediate 4-hydroxy-3-methoxybenzonitrile.[15] The procedures described are essentially those reported by Pearl.[8,9,16]

[1] Institute of Paper Chemistry, Appleton, Wisconsin.
[2] Tiemann, *Ber.*, 8, 1134 (1875).
[3] Ciamician and Silber, *Ber.*, 38, 3821 (1905).
[4] Robbins and Lathrop, *Soil Sci.*, 7, 475 (1919).
[5] Dorland and Hibbert, *Can. J. Research*, 18B, 33 (1940).
[6] Lock, *Ber.*, 62, 1187 (1929).
[7] Sabalitschka and Tietz, *Arch. Pharm.*, 269, 545 (1931).
[8] Pearl, *J. Am. Chem. Soc.*, 68, 2180 (1946).
[9] Pearl, *J. Am. Chem. Soc.*, 68, 429 (1946).
[10] Pearl, *J. Am. Chem. Soc.*, 68, 1100 (1946).
[11] Pearl, *J. Am. Chem. Soc.*, 67, 1628 (1945).
[12] Pearl, *J. Org. Chem.*, 12, 79 (1947).
[13] Böeseken and Greup, *Rec. trav. chim.*, 58, 528 (1939).

[14] Raiford and Potter, *J. Am. Chem. Soc.*, **55**, 1682 (1933).
[15] Schuerch, *J. Am. Chem. Soc.*, **70**, 2293 (1948).
[16] Pearl and Beyer, *Ind. Eng. Chem.*, **42**, 376 (1950); **44**, 2893 (1952).

VINYL LAURATE AND OTHER VINYL ESTERS*

(Lauric acid, vinyl ester)

$$CH_3(CH_2)_{10}CO_2H + CH_3CO_2CH{=}CH_2 \xrightarrow[\text{H}_2\text{SO}_4]{\text{(CH}_3\text{CO}_2)_2\text{Hg}}$$

$$CH_3(CH_2)_{10}CO_2CH{=}CH_2 + CH_3CO_2H$$

Submitted by DANIEL SWERN and E. F. JORDAN, JR.[1]
Checked by WILLIAM S. JOHNSON and LELAND J. CHINN.

1. Procedure

In a 500-ml. round-bottomed three-necked flask provided with a thermometer, a reflux condenser, and a gas inlet tube through which a stream of nitrogen is passed (Note 1) are placed 206 g. (2.4 moles) of freshly distilled vinyl acetate (Note 2) and 80 g. (0.4 mole) of lauric acid (Note 3). The lauric acid is dissolved by warming, and 1.6 g. of mercuric acetate is added. The mixture is shaken by hand for about 30 minutes, and 0.15 ml. of 100% sulfuric acid is added dropwise (Note 4). The solution is heated under reflux for 3 hours, then 0.83 g. of sodium acetate trihydrate is added to neutralize the sulfuric acid. The excess vinyl acetate is recovered by distillation at atmospheric pressure (vapor temperature about 70–80°) until the pot temperature reaches 125° (Note 5). The distillation is completed at 10 mm. or lower (Note 5), and, after the collection of a small quantity of low-boiling material, fairly pure vinyl laurate (Note 6) is obtained as a colorless liquid, b.p. 142–143°/10 mm. (138–139°/8 mm.; 124–126°/3 mm.). The yield is 50–57 g. (55–63%). Redistillation (Note 7) yields 48–53 g. (53–59%) of pure vinyl laurate, b.p. 142–142.5°/10 mm. (120–120.5°/2 mm.); n_D^{25} 1.4387 (Notes 8 and 9).

2. Notes

1. All operations should be conducted in a nitrogen atmosphere to minimize the formation of polymer.

2. An Eastman Kodak Company practical grade of vinyl acetate is satisfactory. It was distilled immediately before use through a 48 by ¾ in. column packed with $^3/_{32}$-in. single-turn Pyrex glass helices. The checkers employed material obtained from the Niacet Chemicals Division, Niagara Falls, New York, distilled once through a 12-in. Vigreux column, b.p. 73°/746 mm.

3. Lauric acid, m.p. 44°, was prepared from the commercial acid obtained from Armour and Company, Chicago, Illinois. The acid was recrystallized twice from acetone at −40° (10 ml. of acetone per gram of acid) and distilled under reduced pressure through a well-insulated, electrically heated 30 by 1 in. column packed with ¼-in. Berl saddles. Pure lauric acid has a boiling point of 167–168°/8 mm. and n_D^{45} 1.4316. The Eastman Kodak Company grade of lauric acid melting at 43–44° is satisfactory.

4. The 100% sulfuric acid is prepared by cautiously adding 7.3 g. of fuming sulfuric acid containing 30% sulfur trioxide to 10 g. of 95% sulfuric acid.

5. An electrically heated 18 by ½ in. Vigreux column was employed.

6. The vinyl laurate, which has an acid number of about 2, usually contains a small quantity of mercury at this stage, from which it can be separated by decantation.

7. Sufficient sodium bicarbonate is added to the pot charge to neutralize the free acid present (see Note 6).

8. Additional properties of vinyl laurate are n_D^{35} 1.4345 and d_4^{30} 0.8639. If the iodine number is determined by the Wijs method, a 200% excess of iodine chloride solution and a 1-hour reaction period should be employed in order to obtain values which are 97–99% of the theoretical value.

9. Vinyl caproate, caprylate, pelargonate, caprate, myristate, palmitate, stearate, 10-hendecenoate (undecylenate) and oleate can be prepared in a similar manner, except that in the preparation of the palmitate and stearate the fatty acids are added to a solution of mercuric acetate and sulfuric acid in vinyl acetate. Vinyl stearate is not redistilled, but the once-distilled product is recrystallized from acetone at 0° (3 ml. of acetone per gram of vinyl stearate). The amount of mercuric acetate employed was 2%, and the amount of 100% sulfuric acid was 0.3–0.4%, of the

weight of the stearic acid. Average yields and properties of these vinyl esters are given in the table.

Vinyl Ester	Yield, %	Boiling Point °C.	mm.	n_D^{30}	d_4^{30}
Caproate	40	98–99	100	1.4159	0.8837
Caprylate	55	134–135	100	1.4256	0.8719
Pelargonate	55	133–133.5	50	1.4291	0.8689
Caprate	45	148	50	1.4320	0.8670
Myristate	60	147–148	4.8	1.4407	0.8617
Palmitate (m.p. 26.7–27.1°)	35	168–169	4.5	1.4438	0.8602
Stearate (m.p. 35–36°)	30	187–188	4.3	1.4423 (at 40°)	0.8517 (at 40°)
10-Hendecenoate	70	124–124.5	10	1.4442	0.8799
Oleate	60	178	2.8	1.4533	0.8691

The acids used for preparing the vinyl esters tabulated were Eastman Kodak Company pure grade except for the following, which were obtained from the companies indicated and purified by fractionation through an efficient fractionating column: caproic acid, b.p. 96°/8 mm., Carbide and Carbon Chemicals Corporation, New York; caprylic acid, b.p. 124–125°/8 mm., capric acid, b.p. 145–146°/8 mm., Armour and Company, Chicago, Illinois; pelargonic acid, b.p. 176°/64 mm., Emery Industries, Cincinnati, Ohio; 10-hendecenoic (undecylenic) acid, b.p. 177–180°/25 mm., Baker Castor Oil Company, New York. Oleic acid was prepared from olive-oil fatty acids by low-temperature crystallization and distillation.[2]

3. Methods of Preparation

The procedure described is substantially that of Toussaint and MacDowell,[3] with minor modifications.[4] Vinyl esters of long-chain aliphatic acids have also been prepared by the reaction of acetylene with the appropriate acids,[5-8] but this reaction is not so convenient for small-scale laboratory preparations.

The interchange reaction of vinyl acetate with lauric acid in the presence of mercuric sulfate has been studied,[9] and vinyl laurate has been obtained also from lauryl chloride and acetaldehyde in the presence of pyridine.[10]

[1] Eastern Regional Research Laboratory, U. S. Department of Agriculture, Philadelphia, Pennsylvania.

[2] Brown and Shinowara, *J. Am. Chem. Soc.*, **59**, 6 (1937); Wheeler and Riemenschneider, *Oil & Soap*, **16**, 207 (1939).

[3] Toussaint and MacDowell, U. S. pat. 2,299,862 [*C. A.*, **37**, 1722 (1943)].

[4] Swern, Billen, and Knight, *J. Am. Chem. Soc.*, **69**, 2439 (1947); Swern and Jordan, *J. Am. Chem. Soc.*, **70**, 2334 (1948); Asahara and Tomita, *Yushi Kagaku Kyôkaishi (J. Oil Chemists' Soc., Japan)*, **1**, 76 (1952) [*C. A.*, **47**, 3232 (1953)].

[5] Reppe, Ger. pat. 588,352 [*C. A.*, **28**, 1357 (1934)]; U. S. pat. 2,066,075 [*C. A.*, **31**, 1037 (1937)]; Reppe et al., *Ann.*, **601**, 81 (1956).

[6] Imperial Chemical Industries, Brit. pat. 581,501 [*C. A.*, **41**, 2428 (1947)].

[7] Powers, *Ind. Eng. Chem.*, **38**, 837 (1946).

[8] Otsuka, Matsumoto, and Murahashi, *Nippon Kagaku Zasshi*, **75**, 798 (1954) [*C. A.*, **51**, 13749 (1957)].

[9] Adelman, *J. Org. Chem.*, **14**, 1057 (1949).

[10] Sladkov and Petrov, *Zhur. Obshchei Khim.*, **24**, 450 (1954) [*C. A.*, **49**, 6093 (1955)].

2-VINYLTHIOPHENE*

(Thiophene, 2-vinyl-)

$$\text{(thiophene)} + CH_3CHO + HCl \rightarrow \text{(2-(1-chloroethyl)thiophene)} + H_2O$$

$$\text{(2-(1-chloroethyl)thiophene)} + C_5H_5N \xrightarrow[\text{(2) }\Delta]{\text{(1) Quaternization}}$$

$$\text{(2-vinylthiophene)} -CH{=}CH_2 + C_5H_5N \cdot HCl$$

Submitted by W. S. EMERSON and T. M. PATRICK, JR.[1]
Checked by MAX TISHLER, P. TISHLER, and F. W. BOLLINGER.

1. Procedure

A 2-l. three-necked flask is fitted with a thermometer, a stirrer, a gas inlet tube which will reach beneath the surface of the liquid, and a vent. The flask is placed in a bath of acetone to which Dry Ice can be added. To the flask are charged 336 g. (318 ml., 4.0 moles) of thiophene (Note 1), 176 g. (177 ml., 1.33 moles) of paraldehyde, and 300 ml. of concentrated hydrochloric acid. While this mixture is stirred and maintained at 10–13° (Note 2) by means of Dry Ice, gaseous hydrogen chloride is bubbled in. At the end of 25 minutes the solution is saturated (Note 3).

The contents of the flask are poured onto 300 g. of ice, the layers are separated, and the organic portion is washed three times with 200-ml. portions of ice water (Note 4). The organic layer is added, with some cooling (Note 5), to 316 g. (322 ml., 4.0 moles) of pyridine and 2.0 g. of α-nitroso-β-naphthol in a 1-l. distilling flask. The aqueous layer is extracted with two 100-ml. portions of ether, and the combined ethereal extract is used to wash each aqueous wash in turn (Note 6). The ethereal layer is concentrated on the steam bath under a stream of nitrogen and combined with the organic mixture in the distilling flask. The mixture is allowed to stand for 1.5 hours before distillation. The distillation is performed under reduced nitrogen pressure. The distillate is collected over 1.0 g. of α-nitroso-β-naphthol in an ice-cooled receiver at successively lower pressures ending at 125°/50 mm. (Note 7). The distillate is poured onto a mixture of 400 g. of ice and 400 ml. of concentrated hydrochloric acid. The layers are separated, and the organic portion is washed successively with 100-ml. portions of 1% hydrochloric acid, water, and 2% ammonia.

The organic layer is filtered through 1 cm. of anhydrous magnesium sulfate on a sintered-glass funnel into a 500-ml. distilling flask. The aqueous layer is extracted with two 100-ml. portions of ether which are combined and used to extract the aqueous washes. The ethereal layer is washed with 50 ml. of saturated salt solution and filtered through the same funnel into a fresh suction flask. The funnel is finally washed with two 50-ml. portions of ether. The ethereal washes are concentrated on a steam bath under nitrogen (Note 8) and combined with the filtrate in the distilling flask. The filtrate is fractionally distilled under nitrogen through a 2-cm. column, 35 cm. high, packed with 6-mm. glass helices (Note 9). The column is jacketed and provided with a heater inside the jacket to minimize heat losses during distillation. The receiver is cooled in an ice bath, and the distillate is collected in three fractions, thiophene 45.6–27.9 g., b.p. 36°/150 mm.–35°/100 mm. (Note 10), intermediate 11.8–4.8 g., b.p. 35°/100 mm.–80°/98 mm., and 2-vinylthiophene (Notes 11 and 12) 191.3–224.0 g., b.p. 65–67°/50 mm., n_D^{25} 1.5701, lit.[2] b.p. 65.5–66.5°/48 mm., lit.[3] n_D^{25} 1.4698. The yield is 50–55% of the theoretical amount based on the thiophene consumed. The undistilled residue amounts to about 27 g. and the distillate in the Dry Ice-acetone trap to about 4 g.

2. Notes

1. Specific gravities at 25° were used to determine the volumes of reagents. Eastman Kodak Company thiophene, b.p. 83–85°, n_D^{25} 1.5252, and U.S.P. paraldehyde were used.

2. The submitters used an ice-salt bath, but the checkers found a Dry Ice-acetone bath more convenient and almost indispensable. Temperature control is important. Small deviations from the prescribed range result in a lower yield.

3. When the solution is saturated, copious fumes of hydrogen chloride are evolved from the vent. The reaction should not be continued beyond this point. With an ice-salt bath the time required for addition of hydrogen chloride was 35 minutes.

4. The yield will be much lower if the washing process is not carried out quickly.

5. If the mixture is too cold, the quaternization reaction will be delayed. On the other hand, if no cooling is provided, the reaction mixture may boil over spontaneously because of the exothermic nature of the reaction. Heat cracks the quaternary compound to 2-vinylthiophene which, if not removed by distillation, may undergo thermal polymerization.

6. Omission of this ethereal extraction will reduce the quoted yield by about 2 per cent.

7. At this temperature and pressure volatilization of pyridine hydrochloride occurs. Passage of vapors during the distillation through a Dry Ice-acetone trap yields 8.8 g. of thiophene contaminated with its original odoriferous impurities and 2.2 g. of an aqueous layer, both of which were discarded.

8. Omission of this ethereal extraction will reduce the quoted yield by about 6 per cent.

9. A packed column is essential to achieve the degree of fractionation required.

10. The distillation is conducted so as to keep the pot temperature below 90° until about 90% of the product has been distilled.

This is done to minimize thermal polymerization of the product. 11. If the intermediate and product fractions are not to be used immediately, α-nitroso-β-naphthol is added as a stabilizer. 12. 5-Chloro-2-vinylthiophene and 5-bromo-2-vinylthiophene have been prepared in 47% and 35% yields, respectively, by essentially the same procedure.

3. Methods of Preparation

2-Vinylthiophene has been prepared by the dehydration of α-(2-thienyl)ethanol [3-7] or β-(2-thienyl)ethanol; [8] by the condensation of vinyl chloride with 2-thienylmagnesium bromide in the presence of cobaltous chloride; [9] by the dehydrochlorination of α-(2-thienyl)ethyl chloride; [10] and by the dehydrogenation of 2-ethylthiophene. [11]

[1] Monsanto Chemical Co., Dayton 7, Ohio.

[2] Schick and Hartough, *J. Am. Chem. Soc.*, **70**, 1646 (1948).

[3] Mowry, Renoll, and Huber, *J. Am. Chem. Soc.*, **68**, 1105 (1946).

[4] Kuhn and Dann, *Ann.*, **547**, 293 (1941).

[5] Nazzaro and Bullock, *J. Am. Chem. Soc.*, **68**, 2121 (1946).

[6] Andreeva and Koton, *Zhur. Obshchei Khim.*, **27**, 997 (1957) [*C. A.*, **52**, 4598 (1958)].

[7] Van Zyl, Langenberg, Tan, and Schut, *J. Am. Chem. Soc.*, **78**, 1955 (1956).

[8] Scully and Brown, *J. Am. Chem. Soc.*, **75**, 6329 (1953).

[9] Strassburg, Gregg, and Walling, *J. Am. Chem. Soc.*, **69**, 2141 (1947).

[10] Emerson and Patrick, *J. Org. Chem.*, **13**, 729 (1948); Emerson and Patrick (to Monsanto Chemical Co.), U. S. pat. 2,547,905 [*C. A.*, **45**, 9084 (1951)].

[11] Wagner (to Phillips Petroleum Co.), U. S. pat. 2,689,855 [*C. A.*, **49**, 11720 (1955)].

o-XYLYLENE DIBROMIDE*

(o-Xylene, α,α'-dibromo-)

Submitted by EMILY F. M. STEPHENSON.[1]
Checked by WILLIAM S. JOHNSON, C. W. TAYLOR, and WILLIAM DEACETIS.

1. Procedure

Caution! o-Xylylene dibromide is a powerful and persistent lachrymator. The preparation and all subsequent handling of this substance should, therefore, be carried out in an efficient hood with adequate protection by rubber gloves. A gas mask should be at hand for emergency. All apparatus coming in contact with the dibromide should be immersed in alcoholic alkali contained in a large crock with a lid. A period of 24 hours is sufficient for decontamination. Waste substances such as filter paper and corks usually require several days of such soaking before they can be safely discarded.

A 1-l. three-necked round-bottomed flask is fitted with a rubber slip-sleeve-sealed stirrer, a dropping funnel with the tip extended to reach almost to the bottom of the flask (Note 1), and an efficient condenser leading to a gas absorption trap.[2] o-Xylene (106 g., 1 mole) (Note 2) is placed in the flask, which is heated with an oil bath and illuminated with a sun lamp (Note 3) placed 1–5 cm. from the upper portion of the flask. When the temperature of the o-xylene reaches 125°, the dropwise addition of 352 g. (2.2 moles) of bromine is commenced with stirring. The rate of addition is regulated so that all the bromine is introduced in 1.5 hours. The mixture is stirred at 125° under illumination for an additional 30 minutes. It is then allowed to cool to 60° and poured into 100 ml. of boiling 60–68° petroleum ether contained in a beaker, the transfer being assisted with small amounts of warm solvent. As the homogeneous solution cools slowly to room temperature it is stirred frequently to prevent caking of the brown crystalline product that separates. After the mixture is cool and the bulk of the dibromide has crystallized, the beaker is placed in a refrigerator for 12 hours (Note 4). The product is then separated by suction filtration, washed twice with 25-ml. portions of cold petroleum ether, and then pressed on the filter until nearly dry. Final drying is effected in a vacuum desiccator containing solid potassium hydroxide. The brown crystalline product amounts to 123–140 g. (48–53% yield), the melting point ranging between 89° and 94° (Note 5).

2. Notes

1. It is convenient to seal a short inner tube inside the stem of the dropping funnel so that the rate of addition can be observed readily. The introduction of the bromine below the surface of the *o*-xylene through an extended stem, about 4-mm. inside diameter, results in better mixing of reactants and less loss of bromine vapors.

2. The submitter used *o*-xylene obtained from Light and Company, Wraysbury, Middlesex, England. It was refluxed with sodium, then distilled from sodium, b.p. 144–144.5°, and stored over sodium.

The checkers employed the white label grade of *o*-xylene supplied by Eastman Kodak Company without further purification.

3. The submitter employed a 600-watt lamp, and the checkers used a 275-watt General Electric sun lamp.

4. Occasional stirring during the first 3–4 hours of this chilling period helps to prevent caking of the product on the side of the beaker.

5. This product is satisfactory for most preparative work. Further purification may be effected by recrystallization from 95% ethanol (3 ml./g.), to give material melting at 93–94° in 80–85% recovery. Other solvents that have been used for recrystallization are petroleum ether (British Drug House, "Analar," b.p. 60–80°) (19 ml./g.), and chloroform (1 ml./g.).

3. Methods of Preparation

o-Xylylene dibromide has been prepared from *o*-xylene by

direct bromination [3] or by treatment with N-bromosuccinimide; [4] by the direct bromination of o-xylyl bromide; [5] and by the action of concentrated hydrobromic acid on the monophenyl ether of α,α'-dihydroxy-o-xylene.[6] The present procedure is essentially that of Perkin [3] as modified by Cope and Fenton.[7]

[1] University of Melbourne, Melbourne, Australia.
[2] *Org. Syntheses Coll. Vol.* **2**, 4 (1943).
[3] Perkin, *J. Chem. Soc.*, **1888**, 5; Atkinson and Thorpe, *J. Chem. Soc.*, **1907**, 1695.
[4] Wenner, *J. Org. Chem.*, **17**, 523 (1952).
[5] von Braun and Cahn, *Ann.*, **436**, 262 (1924).
[6] von Braun and Zobel, *Ber.*, **56**, 2142 (1923).
[7] Cope and Fenton, *J. Am. Chem. Soc.*, **73**, 1668 (1951).

TYPE OF REACTION INDEX

This index lists the preparations contained in this volume in accordance with general types of reactions. Only those preparations are included which can be classified under the selected headings with some definiteness. The arrangement of types and of preparations is alphabetical.

ACETAL AND KETAL FORMATION
ACROLEIN ACETAL, 21
DICYANOKETENE ETHYLENE ACETAL, 276
ETHYL DIETHOXYACETATE, 427
β-KETOISOÖCTALDEHYDE DIMETHYL
    ACETAL, 558
MONOBENZALPENTAERYTHRITOL, 681
PHENYLPROPARGYLALDEHYDE DIETHYL-
    ACETAL, 801

ACYLATION
3-ACETAMIDO-2-BUTANONE, 5
2-AMINO-3-NITROTOLUENE, 42
BENZOYLACETANILIDE, 80
BENZOYLCHOLINE CHLORIDE, 84
β-BROMOETHYLPHTHALIMIDE, 106
p-CHLOROPHENYL SALICYLATE, 178
DIETHYL BENZOYLMALONATE, 285
N,N-DIMETHYLCYCLOHEXYLMETHYL-
    AMINE, 339
DIPHENYL SUCCINATE, 390
ETHYL AZODICARBOXYLATE, 411
ETHYL BENZOYLACETATE, 415
FLAVONE, 478
FURFURAL DIACETATE, 489
n-HEPTAMIDE, 513
HEXAMETHYLENE DIISOCYANATE, 521
6-HYDROXYNICOTINIC ACID, 532
trans-2-METHYL-2-DODECENOIC ACID, 608
2-METHYLENEDODECANOIC ACID, 616
1-METHYL-3-ETHYLOXINDOLE, 620
3-METHYLOXINDOLE, 657
o-NITROACETOPHENONE, 708
o-AND p-NITROBENZALDIACETATE, 713
PARABANIC ACID, 744
PENTAERYTHRITYL TETRABROMIDE, 753
2-PHENYLCYCLOHEPTANONE, 780
o-PHENYLENE CARBONATE, 788
PUTRESCINE DIHYDROCHLORIDE, 819
p-TOLUENESULFONYLMETHYLNITROS-
    AMIDE, 943
TRIETHYL PHOSPHITE, 955

ADDITION See also Grignard Reactions;
    Reduction
A. To C=C
    BUTYRCHLORAL, 130
    trans-2-CHLOROCYCLOPENTANOL, 157
    2-CHLORO-1,1,2-TRIFLUOROETHYL
        ETHYL ETHER, 184
    CHOLESTEROL, 195
    9,10-DIHYDROXYSTEARIC ACID, 317
    ETHYL ENANTHYLSUCCINATE, 430

ADDITION—Continued
A. To C=C—Continued
    IODOCYCLOHEXANE, 543
    STEAROLIC ACID, 851
    10-UNDECYNOIC ACID, 969
B. To C≡C
    1-ACETYLCYCLOHEXANOL, 13
    β-CHLOROVINYL ISOAMYL KETONE,
        186
C. To C=C—C=C
    3-CHLOROCYCLOPENTENE, 238
    NORBORNYLENE, 738
    cis-Δ⁴-TETRAHYDROPHTHALIC ANHY-
        DRIDE, 890
    TRIPTYCENE, 964
D. To C=O
    ATROLACTIC ACID, 58
    p-BROMOPHENYLUREA, 49
    1-CYANO-3-PHENYLUREA, 213
    1,2-CYCLOHEXANEDIONE DIOXIME,
        229
    1,2-DI-1-(1-CYANO)CYCLOHEXYL-
        HYDRAZINE, 274
    DIPHENYLACETYLENE, 377
    p-ETHOXYPHENYLUREA, 52
    1,1'-ETHYNLENE-bis-CYCLOHEXANOL,
        471
    N-METHYL-2,3-DIMETHOXYBENZYL-
        AMINE, 603
    N-METHYL-1,2-DIPHENYLETHYL-
        AMINE, 605
    1,2,3,4-TETRAHYDROCARBAZOLE, 884
    β-TETRALONE, 903
    2,2,6,6-TETRAMETHYLOLCYCLOHEX-
        ANOL, 907
    THIOBENZOPHENONE, 927
E. To C=C—C=O
    2-p-ACETYLPHENYLHYDROQUINONE,
        15
    3,4-DIHYDRO-2-METHOXY-4-METHYL-
        2H-PYRAN, 311
    β-METHYLGLUTARIC ANHYDRIDE, 630
    METHYL γ-METHYL-γ-NITROVAL-
        ERATE, 652
    β-METHYL-β-PHENYL-α,α'-DICYANO-
        GLUTARIMIDE, 662
    METHYL β-THIODIPROPIONATE, 669
    γ-PHENYLALLYLSUCCINIC ACID, 766
    PHENYLSUCCINIC ACID, 804
F. To C≡N
    N,N-DIMETHYLSELENOUREA, 359

987

ADDITION—*Continued*

F. To C≡N—*Continued*

GUANYLTHIOUREA, 502
METHYLISOUREA HYDROCHLORIDE, 645
N-PHENYLBENZAMIDINE, 769

G. *MISCELLANEOUS*

o-CHLOROPHENYLTHIOUREA, 180
2,5-DIAMINO-3,4-DICYANOTHIOPHENE, 243
2,3-DIPHENYLSUCCINONITRILE, 392

ALKYLATION The reagent employed is given in brackets.

A. *C-ALKYLATION*

[NaOC₂H₅] α-ACETYL-δ-CHLORO-γ-VALEROLACTONE, 10
p-AMINOTETRAPHENYLMETHANE, 47
[NaOC₂H₅] DL-ASPARTIC ACID, 55
[NaNH₂] n-BUTYLACETYLENE, 117
[NaOC₂H₅] DIETHYL 1,1-CYCLO-BUTANEDICARBOXYLATE, 288
[NaOC₂H₅] DIETHYL Δ²-CYCLOPEN-TENYLMALONATE, 291
[KOH] 3,4-DINITRO-3-HEXENE, 372
[Al₂O₃] HEXAMETHYLBENZENE, 520
[NaOCH₃] 9-METHYLFLUORENE, 623
[LiC₆H₅] 1-METHYLISOQUINOLINE, 641
[NaOH] TETRAACETYLETHANE, 869
p-TRICYANOVINYL-N,N-DIMETHYL-ANILINE, 953
[KNH₂] α,α,β-TRIPHENYLPROPIONI-TRILE, 962

B. *O-ALKYLATION*

ALLOXANTIN DIHYDRATE, 25
[K₂CO₃] O-METHYLCAPROLACTIM, 588
[NaOH] 3-METHYLCOUMARONE, 590
[K₂CO₃] QUINACETOPHENONE MONO-METHYL ETHER, 836
[NaOC₂H₅] o-TOLUALDEHYDE, 932

C. *N-ALKYLATION*

BENZOYLCHOLINE CHLORIDE, 84
2-BENZYLAMINOPYRIDINE, 91
BENZYLTRIMETHYLAMMONIUM ETH-OXIDE, 98
N,N'-DIETHYLBENZIDINE, 283
N,N'-DIPHENYLBENZAMIDINE, 383
N-ETHYL-p-CHLOROANILINE, 420
ETHYL N-PHENYLFORMIMIDATE, 464
ETHYL α-(1-PYRROLIDYL)PROPION-ATE, 466
o-METHYLBENZYL ALCOHOL, 582
2-METHYLBENZYLDIMETHYLAMINE, 585
1-PHENYLPIPERIDINE, 795
α-PHTHALIMIDO-o-TOLUIC ACID, 810
2-VINYLTHIOPHENE, 980

D. *S-ALKYLATION*

p-DITHIANE, 396
ETHANEDITHIOL, 401
2-FURFURYL MERCAPTAN, 491
METHYL p-TOLYL SULFONE, 674

ARBUSOV REACTION

DIISOPROPYL METHYLPHOSPHONATE, 325

ARYLATION

2-p-ACETYLPHENYLHYDROQUINONE, 15
m-NITROBIPHENYL, 718
TETRAPHENYLARSONIUM CHLORIDE HYDROCHLORIDE, 910

BLAISE REACTION

sec-BUTYL α-n-CAPROYLPROPIONATE, 120

BOORD SYNTHESIS

1,4-PENTADIENE, 746

CHLOROETHYLATION

2-VINYLTHIOPHENE, 980

CONDENSATION The term "condensation" is used here in a restricted sense and applies to those reactions in which a carbon-carbon bond is formed by the elimination of a simple molecule. Cyclization reactions are listed separately, and related reactions, such as dehydrations, may be found also under other headings. The reagent used is included in brackets.

A. *CARBONYL-ACTIVE METHYLENE CONDENSATIONS*

[β-ALANINE] 3-BENZYL-3-METHYL-PENTANOIC ACID, 93
[KOC(CH₃)₃] β-CARBETHOXY-γ,γ-DIPHENYLVINYLACETIC ACID, 132
[NaOC₂H₅] CYCLOHEPTANONE, 221
[CH₃CO₂NH₄] CYCLOHEXYLIDENE-CYANOACETIC ACID, 234
[NaOCH₃] DICYCLOPROPYL KETONE, 278
[(CH₃CO)₂O] DIETHYL ETHYL-IDENEMALONATE, 293
[C₅H₅N,C₅H₁₁N] 2,3-DIMETHOXY-CINNAMIC ACID, 327
[NaNH₂] α,β-DIPHENYLCINNAMONI-TRILE, 387
[NaCN] 2,3-DIPHENYLSUCCINONI-TRILE, 392
[C₅H₁₁N] ETHYL α-ACETYL-β-(2,3-DIMETHOXYPHENYL)PROPIONATE, 408
[CH₃CO₂NH₄] β-ETHYL-β-METHYL-GLUTARIC ACID, 441
[CH₃CO₂NH₄] ETHYL (1-PHENYL-ETHYLIDENE)CYANOACETATE, 463
[C₄H₉NH₂] o-METHOXYPHENYLACE-TONE, 573
[(C₂H₅)₃N,(CH₃CO)₂O] trans-o-NITRO-α-PHENYLCINNAMIC ACID, 730
[C₅H₅N] m-NITROSTYRENE, 731
[(C₂H₅)₃N] α-PHENYLCINNAMIC ACID, 777

B. *ESTER-ESTER CONDENSATION*

[NaOC₂H₅] CETYLMALONIC ESTER, 141

C. *ESTER-KETONE CONDENSA-TION*

[NaOCH₃] 3-CYANO-6-METHYL-2(1)-PYRIDONE, 210
[NaOC₂H₅] INDAZOLE, 536

D. *MISCELLANEOUS*

[NaOC₂H₅] α-(4-CHLOROPHENYL)-γ-PHENYLACETOACETONITRILE, 174

CONDENSATION—*Continued*
   D. MISCELLANEOUS—*Continued*
    [NaOC₂H₅] Ethyl Phenylcyano-
      acetate, 461
    3-*n*-Heptyl-5-cyanocytosine, 515
    Tetracyanoethylene, 877
    Tetraphenylethylene, 914
CYANOETHYLATION
   3-(*o*-Chloroanilino)propionitrile,
    146
   N-2-Cyanoethylaniline, 205
   α-Phenyl-α-carbethoxyglutaroni-
    trile, 776
CYCLIZATIONS These reactions include
   a variety of methods which are em-
   ployed for the formation of cyclic
   systems. The reagent used is included
   in brackets.
   N-(*p*-Acetylaminophenyl)rhodanine,
    6
   [NaOC₂H₅] α-Acetyl-δ-chloro-γ-
    valerolactone, 10
   2-Amino-4-anilino-6-(chloromethyl)-
    s-triazine, 29
   [heat] Benzofurazan Oxide, 74
   [KOH] Benzoguanamine, 78
   [H₂SO₄,SO₃] Coumalic Acid, 201
   [piperidine acetate] 3-Cyano-6-methyl-2-
    (1)-pyridone, 210
   [(CH₃CO)₂O,H₂SO₄] Diacetyl-*d*-tar-
    taric Anhydride, 242
   [C₅H₅N] 2,5-Diamino-3,4-dicyanothio-
    phene, 243
   [NaOC₂H₅] 2,4-Diamino-6-hydroxy-
    pyrimidine, 245
   [NaOC₂H₅] Diaminouracil Hydro-
    chloride, 247
   [NaOH] Dicyclopropyl Ketone, 278
   [NaOC₂H₅] Diethyl 1,1-Cyclobutane-
    dicarboxylate, 288
   [heat] 3,4-Dihydro-2-methoxy-4-
    methyl-2H-pyran, 311
   [heat] Dimethylfurazan, 342
   [H₃PO₄] 5,5-Dimethyl-2-*n*-pentyltet-
    rahydrofuran, 350
   3,5-Dimethylpyrazole, 351
   5,5-Dimethyl-2-pyrrolidone, 357
   [NaOCH₃] 1,4-Diphenyl-5-amino-1,2,3-
    triazole, 380
   p-Dithiane, 396
   [NaOH] Ethylenimine, 433
   [H₂SO₄] Ethyl Isodehydroacetate,
    549
   [CH₃CO₂NH₄] β-Ethyl-β-methylglu-
    taric Acid, 441
   [CH₃CO₂NH₄] 5-Ethyl-2-methylpyri-
    dine, 451
   [heat] Flavone, 478
   [H₂SO₄] Flavone, 479
   [AlCl₃] 9-Fluorenecarboxylic Acid,
    482
   [heat] Glutarimide, 496
   D-Gulonic-γ-lactone, 506

CYCLIZATIONS—*Continued*
   [NaOCH₃] 3-*n*-Heptyl-5-cyanocyto-
    sine, 515
   [H₂SO₄] Hexahydro-1,3,5-tripropi-
    onyl-*s*-triazine, 518
   [heat] 4-Hydroxy-1-butanesulfonic
    Acid Sultone, 529
   [*p*-C₇H₇SO₃H] 3-Hydroxytetrahydro-
    furan, 534
   Indazole, 536
   [H₂SO₄] Isodehydroacetic Acid, 549
   [NaOC₂H₅] 2-Mercapto-4-amino-5-
    carbethoxypyrimidine and 2-Mer-
    capto-4-hydroxy-5-cyanopyrimi-
    dine, 566
   Mercaptobenzimidazole, 569
   [H₂SO₄] 3-Methylcoumarone, 590
   [NaOH] Methyl Cyclopropyl Ke-
    tone, 597
   [AlCl₃] 1-Methyl-3-ethyloxindole, 620
   [(CH₂CO)₂O] β-Methylglutaric An-
    hydride, 630
   [NaOCH₃] 4-Methyl-6-hydroxypyrim-
    idine, 638
   [CaH₂] 3-Methyloxindole, 657
   [NaOC₂H₅] β-Methyl-β-phenyl-α,α'-
    dicyanoglutarimide, 662
   [H₂SO₄] 1-Methyl-3-phenylindane,
    665
   [P₄S₇] 3-Methylthiophene, 671
   [CuCr₂O₄] β-Methyl-γ-valerolac-
    tone, 677
   [NaOCH₃] Parabanic Acid, 744
   [NaOH] *o*-Phenylene Carbonate, 788
   [(CH₃CO)₂O] α-Phenylglutaric An-
    hydride, 790
   Pseudopelletierine, 816
   2,3-Pyrazinedicarboxylic Acid, 824
   [Na] Sebacoin, 840
   1,2,3,4-Tetrahydrocarbazole, 884
   Tetrahydrothiophene, 892
DARZENS REACTION
   Ethyl β,β-Pentamethyleneglyci-
    date, 459
   Methyl 3-Methyl-2-furoate, 649
DECARBONYLATION
   Cetylmalonic Ester, 141
   2,4,4-Trimethylcyclopentanone, 957
DECARBOXYLATION
   3-Acetamido-2-butanone, 5
   Coumalic Acid, 201
   1-Cyclohexenylacetonitrile, 234
   Dibromoacetonitrile, 254
   Diethyl Benzoylmalonate, 285
   4,6-Dimethylcoumalin, 337
   n-Heptamide, 513
   Laurone, 560
   3-Methylcoumarone, 590
   Methyl Cyclopropyl Ketone, 597
   3-Methylfuran, 628
   β-Methyl-β-phenylglutaric Acid,
    664
   o-Nitroacetophenone, 708
   m-Nitrostyrene, 731

DECARBOXYLATION—*Continued*
cis-STILBENE, 857
DEHALOGENATION
CHOLESTEROL, 195
1,1-DICHLORO-2,2-DIFLUOROETHYLENE, 268
DIMETHYLKETENE, 348
TRIPTYCENE, 964
DEHYDRATION The reagent used is included in brackets.
[SiO₂] AZELANITRILE, 62
[H₂SO₄] BENZHYDRYL β-CHLOROETHYL ETHER, 72
[ClSO₃H] BISCHLOROMETHYL ETHER, 101
[HCl] BUTYRCHLORAL, 130
[P₂O₅] CHLOROACETONITRILE, 144
[PCl₅] 2-CHLORONICOTINONITRILE, 166
[H₂SO₄,SO₃] COUMALIC ACID, 201
[(CH₃CO)₂O,H₂SO₄] DIACETYL-*d*-TARTARIC ANHYDRIDE, 242
[H₂SO₄] 4,4′-DICHLORODIBUTYL ETHER, 266
[MALEIC ANHYDRIDE] DIMETHYLFURAZAN, 342
[H₃PO₄] 5,5-DIMETHYL-2-*n*-PENTYLTETRAHYDROFURAN, 350
[SOCl₂] 2-ETHYLHEXANONITRILE, 436
[P₂O₅] FUMARONITRILE, 486
[heat] GLUTARIMIDE, 496
[(CH₃CO)₂O] β-METHYLGLUTARIC ANHYDRIDE, 630
[P₂O₅] NICOTINONITRILE, 706
[H₂SO₄] *trans*-1-PHENYL-1,3-BUTADIENE, 771
[(CH₃CO)₂O] α-PHENYLGLUTARIC ANHYDRIDE, 790
[P₂O₅] *p*-TOLUENESULFONIC ANHYDRIDE, 940
DEHYDROGENATION See also Oxidation. The reagent used is included in brackets.
[Pd] INDAZOLE, 536
[CuCr₂O₄] β-METHYL-δ-VALEROLACTONE, 677
DEHYDROHALOGENATION The reagent used is included in brackets.
[NaNH₂] 2-BUTYN-1-OL, 128
[KOH] 3,4-DINITRO-3-HEXENE, 372
[KOC(CH₃)₃] *trans*-2-DODECENOIC ACID, 398
[NaNH₂] ETHOXYACETYLENE, 404
[(C₂H₅)₃N] 6-KETOHENDECANEDIOIC ACID, 555
[(C₂H₅)₃N] LAURONE, 560
[collidine or LiCl in HCON(CH₃)₂] 2-METHYL-2-CYCLOHEXENONE, 162
[C₉H₇N] *trans*-2-METHYL-2-DODECENOIC ACID, 608
[NaOH] 2-METHYLENEDODECANOIC ACID, 616
[KOH] MONOVINYLACETYLENE, 683
[NaOH] 9-NITROANTHRACENE, 711
[NaNH₂] 4-PENTYN-1-OL, 755

DEHYDROHALOGENATION—*Continued*
[NaNH₂] PHENYLACETYLENE, 763
[NaNH₂] STEAROLIC ACID, 851
[NaNH₂] 10-UNDECYNOIC ACID, 969
DESULFURIZATION The reagent used is included in brackets.
[(CH₃CO)₂Pb] *o*-CHLOROPHENYLCYANAMIDE, 172
[Ni-H₂] 4-METHYL-6-HYDROXYPYRIMIDINE, 638
DIAZOTIZATION
REPLACEMENT OF DIAZONIUM GROUP
1. By Cl
1-CHLORO-2,6-DINITROBENZENE, 160
2-CHLOROPYRIMIDINE, 182
2. By H
2,4,6-TRIBROMOBENZOIC ACID, 947
3. By N₃
BENZOFURAZAN OXIDE, 74
4. By Ar
m-NITROBIPHENYL, 718
COUPLING
dl-4,4′,6,6′-TETRACHLORODIPHENIC ACID, 872
EPOXIDATION
ISOPHORONE OXIDE, 552
trans-STILBENE OXIDE, 860
ESTERIFICATION
α-CHLOROPHENYLACETIC ACID, 169
DI-*tert*-BUTYL MALONATE, 261
DIETHYL *cis*-Δ⁴-TETRAHYDROPHTHALATE, 301
DIMETHYL ACETYLENEDICARBOXYLATE, 329
ETHYL *tert*-BUTYL MALONATE, 417
6-HYDROXYNICOTINIC ACID, 532
METHYL HYDROGEN HENDECANEDIOATE, 635
EXCHANGE REACTIONS See also Acylation.
VINYL LAURATE, 977
FORMYLATION REACTIONS
p-DIMETHYLAMINOBENZALDEHYDE, 331
N-ETHYL-*p*-CHLOROANILINE, 420
INDOLE-3-ALDEHYDE, 539
2-PYRROLEALDEHYDE, 831
SYRINGIC ALDEHYDE, 866
2-THENALDEHYDE, 915
3-THENALDEHYDE, 918
FRIEDEL-CRAFTS REACTION
9-ACETYLANTHRACENE, 8
2-AMINOBENZOPHENONE, 34
3-BENZOYLPYRIDINE, 88
1-METHYL-3-ETHYLOXINDOLE, 620
NEOPHYL CHLORIDE, 702
PHENYLDICHLOROPHOSPHINE, 784
α-TETRALONE, 898
α,β,β-TRIPHENYLPROPIONIC ACID, 960
GRIGNARD REACTIONS
BENZENEBORONIC ANHYDRIDE, 68
3-BENZYL-3-METHYLPENTANOIC ACID, 93
DI-*n*-BUTYLDIVINYLTIN, 258

GRIGNARD REACTIONS—*Continued*
2-Methyl-2,5-decanediol, 601
N-Methyl-1,2-diphenylethylamine, 605
Methyl 2-Thienyl Sulfide, 667
trans-1-Phenyl-1,3-butadiene, 771
1-Phenyl-1-penten-4-yn-3-ol, 792
Tetraethyltin, 881
Tetraphenylarsonium Chloride Hydrochloride, 910
HALOGENATION
A. BROMINATION
N-Bromoacetamide, 104
4-Bromo-2-heptene, 108
p-Bromomandelic Acid, 110
2-Bromo-3-methylbenzoic Acid, 114
Dibromoacetonitrile, 254
4,4'-Dibromobiphenyl, 256
trans-2-Dodecenoic Acid, 398
trans-2-Methyl-2-dodecenoic Acid, 608
2-Methylenedodecanoic Acid, 616
Mucobromic Acid, 688
o-Phthalaldehyde, 807
3-Thenyl Bromide, 921
2,4,6-Tribromobenzoic Acid, 947
o-Xylylene Dibromide, 984
B. CHLORINATION
tert-Butyl Hypochlorite, 125
2-Chloro-2-methylcyclohexanone, 162
Diethylthiocarbamyl Chloride, 307
3-Methylcoumarone, 590
dl-4,4',6,6'-Tetrachlorodiphenic Acid, 872
p-Toluenesulfenyl Chloride, 934
C. IODINATION
Cyanogen Iodide, 207
2-Iodothiophene, 545
4-Iodoveratrole, 547
HYDROLYSIS   The subheadings indicate the types of compounds hydrolyzed.
A. AMIDE
ε-Aminocaproic Acid, 39
Atrolactic Acid, 58
N-Ethyl-*p*-chloroaniline, 420
Glutaric Acid, 496
1-Methylisoquinoline, 641
B. DIHALIDE
p-Bromomandelic Acid, 110
Ethyl Chlorofluoroacetate, 423
o-Phthalaldehyde, 807
C. ESTER
3-Benzyl-3-methylpentanoic Acid, 93
Ethyl *tert*-Butyl Malonate, 417
6-Hydroxynicotinic Acid, 532
o-Methylbenzyl Alcohol, 582
Methyl Cyclopropyl Ketone, 597
trans-2-Methyl-2-dodecenoic Acid, 608

HYDROLYSIS—*Continued*
C. ESTER—*Continued*
2-Methylenedodecanoic Acid, 616
3-Methyl-2-furoic Acid, 628
β-Methylglutaric Anhydride, 630
Methylglyoxal-ω-phenylhydrazone, 633
Methyl Hydrogen Hendecanedioate, 635
o-Nitroacetophenone, 708
D. IMIDE
dl-Aspartic Acid, 55
β-Ethyl-β-methylglutaric Acid, 441
β-Methyl-β-phenylglutaric Acid, 664
E. NITRILE
Atrolactic Acid, 58
3-Benzyl-3-methylpentanoic Acid, 93
2-Bromo-3-methylbenzoic Acid, 114
β-Ethyl-β-methylglutaric Acid, 441
Glutaric Acid, 496
D-Gulonic-γ-lactone, 506
β-Methyl-β-phenylglutaric Acid, 664
Phenylacetamide, 760
α-Phenylglutaric Anhydride, 790
Phenylsuccinic Acid, 804
F. MISCELLANEOUS
2-Aminobenzophenone, 34
2,6-Dinitroaniline, 364
Ethyl Benzoylacetate, 415
2-Furfuryl Mercaptan, 491
Glutaric Acid, 496
6-Ketohendecanedioic Acid, 555
Laurone, 560
3-Methyl-1,5-pentanediol, 660
γ-Phenylallylsuccinic Acid, 766
Pseudopelletierine, 816
β-Tetralone, 903
JAPP-KLINGEMANN REACTION
Methylglyoxal-ω-phenylhydrazone, 633
MANNICH REACTION
1-Diethylamino-3-butanone, 281
5-Methylfurfuryldimethylamine, 626
MEERWEIN REACTION
1-(*p*-Nitrophenyl)-1,3-butadiene, 727
METALATION REACTIONS
Di-*n*-butyldivinyltin, 258
Ferrocene, 473
Tetraethyltin, 881
NITRATION
2-Amino-3-nitrotoluene, 42
2,6-Dinitroaniline, 364
3-Methyl-4-nitropyridine-1-oxide, 654
9-Nitroanthracene, 711
o-Nitrocinnamaldehyde, 722
6-Nitroveratraldehyde, 735

NITROSATION
DIAMINOURACIL HYDROCHLORIDE, 247
2-PHENYLCYCLOHEPTANONE, 780
p-TOLYLSULFONYLMETHYLNITROSAMIDE, 943

OXIDATION See also Epoxidation; Dehydrogenation. The subheadings indicate the types of oxidation. The oxidizing agent employed is included in brackets.
A. CH₂ → C=O
   [CrO₃] ALLOXAN MONOHYDRATE, 23
   [Na₂Cr₂O₇] Δ⁴-CHOLESTEN-3,6-DIONE, 189
   [SeO₂] 1,2-CYCLOHEXANEDIONE DIOXIME, 229
   [Cr₂O₃,O₂] METHYL p-ACETYLBENZOATE, 579
   [CrO₃] o- AND p-NITROBENZALDIACETATE, 713
   [CrO₃] PROPIOLALDEHYDE, 813
B. CHO → CO₂H
   [Cu₂O,Ag₂O,O₂] 2-FUROIC ACID, 493
   [HNO₃] GLUTARIC ACID, 496
   [Ag₂O] 3-THENOIC ACID, 919
   [Ag₂O] VANILLIC ACID, 972
   [KOH] VANILLIC ACID, 974
C. CHOH → C=O
   [KMnO₄] ETHYL PYRUVATE, 467
   [Na₂Cr₂O₇] Δ⁴-CHOLESTEN-3,6-DIONE, 189
   [cyclohexanone] Δ⁴-CHOLESTEN-3-ONE, 192
   [Na₂Cr₂O₇] Δ⁵-CHOLESTEN-3-ONE, 195
   [Cu(O₂CCH₃)₂] SEBACIL, 838
D. R₃N → R₃N→O
   [H₂O₂] METHYLENECYCLOHEXANE, 612
   [H₂O₂] NICOTINAMIDE-1-OXIDE, 704
   [CH₃CO₃H] PYRIDINE-N-OXIDE, 828
E. MISCELLANEOUS
   [CrO₃] δ-ACETYL-n-VALERIC ACID, 19
   [Na₂S_z,NaOH] p-AMINOBENZALDEHYDE, 31
   [Br₂] 1,1'-AZO-bis-1-CYCLOHEXANENITRILE, 66
   [NaOCl] BENZOFURAZAN OXIDE, 74
   [Pb(OCOCH₃)₄] n-BUTYL GLYOXYLATE, 124
   [CH₃CO₃H] β-(o-CARBOXYPHENYL)-PROPIONIC ACID, 136
   [Na₂Cr₂O₇] CHLORO-p-BENZOQUINONE, 148
   [NaOCl] β,β-DIMETHYLGLUTARIC ACID, 345
   [O₂] p,p'-DINITROBIBENZYL, 367
   [HgO] DIPHENYLACETYLENE, 377
   [HNO₃] ETHYL AZODICARBOXYLATE, 411
   [CrO₃] 1,4-NAPHTHOQUINONE, 698
   [KMnO₄] 2,3-PYRAZINEDICARBOXYLIC ACID, 824
   [O₂] TETRALIN HYDROPEROXIDE, 895

OXIDATION—Continued
E. MISCELLANEOUS—Continued
   [H₂O₂] TETRAPHENYLARSONIUM CHLORIDE HYDROCHLORIDE, 910
OZONIZATION
   5-FORMYL-4-PHENANTHROIC ACID, 484
PRINS REACTION
   4-PHENYL-m-DIOXANE, 786
REARRANGEMENTS
   CHLORO-p-BENZOQUINONE, 148
   Δ⁴-CHOLESTEN-3-ONE, 192
   DIPHENYLACETALDEHYDE, 375
   FLAVONE, 478
   2-METHYLBENZYLDIMETHYLAMINE, 585
   METHYL CYCLOPENTANECARBOXYLATE, 594
   METHYL 3-METHYL-2-FUROATE, 649
   4-PHENYL-5-ANILINO-1,2,3-TRIAZOLE, 380
   2,4,4-TRIMETHYLCYCLOPENTANONE, 957
REDUCTION The reducing agent is given in brackets.
A. C=C → CH—CH
   [Ni–Al,NaOH] β-(o-CARBOXYPHENYL)PROPIONIC ACID, 136
   [Pt or Pd,H₂] DIETHYL cis-HEXAHYDROPHTHALATE, 304
   [Ni,H₂] DIETHYL METHYLENEMALONATE, 298
   [CuCr₂O₄,H₂] 9,10-DIHYDROPHENANTHRENE, 313
   [Pd,H₂] ETHYL α-ACETYL-β-(2,3-DIMETHOXYPHENYL)PROPIONATE, 408
   [Pd,H₂] ar-TETRAHYDRO-α-NAPHTHOL, 887
   [Na,C₂H₅OH] β-TETRALONE, 903
B. C=O → CH₂
   [H₂S] ALLOXANTIN DIHYDRATE, 25
   [Zn,HCl] CREOSOL, 203
   [H₂NNH₂,KOH] HENDECANEDIOIC ACID, 510
C. C=O → CHOH
   [CuCr₂O₄,H₂] 1,2-CYCLODECANEDIOL, 216
   [Ni,H₂] 3-METHYL-1,5-PENTANEDIOL, 660
D. CO₂R → CH₂OH
   [LiAlH₄] 2-(1-PYRROLIDYL)PROPANOL, 834
E. CONR₂ → CH₂NR₂
   [LiAlH₄] N,N-DIMETHYLCYCLOHEXYLMETHYLAMINE, 339
   [LiAlH₄] 2,2-DIMETHYLPYRROLIDINE, 354
   [LiAlH₄] LAURYLMETHYLAMINE, 564
F. NO₂ → NH₂
   [Na₂S_z,NaOH] p-AMINOBENZALDEHYDE, 31
   [Ni,H₂] CYCLOHEPTANONE, 221
   [Ni,H₂] 5,5-DIMETHYL-2-PYRROLIDONE, 357
   [Fe,HCl] o-METHOXYPHENYLACETONE, 573

REDUCTION—*Continued*
G. MISCELLANEOUS
[elect.] Chloro-*p*-benzoquinone, 148
[PCl₅] 2-Chloronicotinonitrile, 166
[Zn,HCl] Cyclodecanone, 218
[H₂S] 2,5-Diamino-3,4-dicyanothiophene, 243
[Na₂S₂O₄] Diaminouracil Hydrochloride, 247
[LiAlH₄] 2,2-Dichloroethanol, 271
[Na,NH₃] Diethyl Mercaptoacetal, 295
[Hg-Na] Hemimellitene, 508
[Ni,H₂] N-Methyl-2,3-dimethoxybenzylamine, 603
[Zn,H₂SO₄] 1,5-Naphthalenedithiol, 695
[Na,C₄H₉OH] 3-Phenyl-1-propanol, 798
[Li,C₂H₅OH] *ar*-Tetrahydro-α-naphthol, 887
[HCO₂H] 2,2,6,6-Tetramethylolcyclohexanol, 907

REFORMATSKY REACTION
4-Ethyl-2-methyl-2-octenoic Acid, 444

REISSERT REACTION
1-Methylisoquinoline, 641

REPLACEMENT REACTIONS The subheadings indicate types of replacement reactions.
A. HALOGEN BY AMINO OR SUBSTITUTED AMINO GROUPS
2-Aminobenzophenone, 34
2,6-Dinitroaniline, 364
B. HALOGEN BY CN
p-Methoxyphenylacetonitrile, 576
C. HALOGEN BY H, see Reduction
D. HALOGEN BY OH
2-Butyn-1-ol, 128
E. HYDROXYL BY HALOGEN
2-Aminobenzophenone, 34
3-Benzoylpyridine, 88
β-Bromoethylphthalimide, 106
N-Chlorobetainyl Chloride, 154
α-Chlorophenylacetic Acid, 169
Di-*tert*-butyl Malonate, 261
Dicyclopropyl Ketone, 278
1,4-Diiodobutane, 321
1,6-Diiodohexane, 323
β-Dimethylaminoethyl Chloride Hydrochloride, 333
N,N-Dimethylcyclohexylmethylamine, 339
Itaconyl Chloride, 554
Methanesulfonyl Chloride, 571
p-Methoxyphenylacetonitrile, 576
Methyl Cyclopropyl Ketone, 597
1-Methyl-3-ethyloxindole, 620
Monobromopentaerythritol, 681

REPLACEMENT REACTIONS—*Continued*
E. HYDROXYL BY HALOGEN—*Continued*
Naphthalene-1,5-disulfonyl Chloride, 693
m-Nitrobenzazide, 715
Oleoyl Chloride, 739
β-Styrenesulfonyl Chloride, 846
p-Toluenesulfinyl Chloride, 937
F. MISCELLANEOUS
3-Aminopyridine, 45
Benzoylcholine Chloride, 84
Benzyltrimethylammonium Ethoxide, 98
2-Bromo-3-methylbenzoic Acid, 114
Cyclohexene Sulfide, 232
4,4'-Dichlorodibutyl Ether, 266
1,2-Di-1-(1-cyano)cyclohexylhydrazine, 274
Diethyl Mercaptoacetal, 295
2-(Dimethylamino)pyrimidine, 336
1,4-Dinitrobutane, 368
N,N'-Diphenylbenzamidine, 383
Ethyl Isocyanide, 438
Ethyl α-Nitrobutyrate, 454
Ethyl Orthocarbonate, 457
n-Hexyl Fluoride, 525
4-Hydroxy-1-butanesulfonic Acid Sultone, 529
o-Methylbenzyl Alcohol, 582
Methyl *p*-Tolyl Sulfone, 674
m-Nitrobenzazide, 715
1-Nitroöctane, 724
Pentaerythrityl Tetrabromide, 753
Putrescine Dihydrochloride, 819
Thiobenzoic Acid, 924
Thiolacetic Acid, 928
Trichloromethylphosphonyl Dichloride, 950

RING EXPANSION REACTIONS
Cycloheptanone, 221
2-Phenylcycloheptanone, 780

SOMMELET REACTION
1-Naphthaldehyde, 690

SULFONATION
2,6-Dinitroaniline, 364
Sodium β-Styrenesulfonate, 846
α-Sulfopalmitic Acid, 862

THERMAL DECOMPOSITION REACTIONS
1,1'-Dicyano-1,1'-bicyclohexyl, 273
Diethyl Methylenemalonate, 298
Methylenecyclohexane and N,N-Dimethylhydroxylamine Hydrochloride, 612
α-Naphthyl Isothiocyanate, 700
1,4-Pentadiene, 746
Putrescine Dihydrochloride, 819
Stearone, 854
2-Vinylthiophene, 980

TYPE OF COMPOUND INDEX

Preparations are listed by functional groups or by ring systems. Phenyl, ethylenic, and acetylenic groups are not considered as substituents unless otherwise stated. Salts are included with the corresponding acids and bases.

ACETALS
ACROLEIN ACETAL, 21
CROTONALDEHYDE DIETHYL ACETAL, 22
DICYANOKETENE ETHYLENE ACETAL, 276
DIETHYL MERCAPTOACETAL, 295
ETHYL DIETHOXYACETATE, 427
β-KETOISOÖCTALDEHYDE DIMETHYL ACETAL, 558
MONOBENZALPENTAERYTHRITOL, 679
4-PHENYL-m-DIOXANE, 786
PHENYLPROPARGYLALDEHYDE DIETHYL ACETAL, 801
1,1,1′,1′-TETRAETHOXYETHYL POLYSULFIDE, 295
TIGLYLALDEHYDE DIETHYL ACETAL, 22
ACID AMIDES See Amides.

ACID ANHYDRIDES
BENZENEBORONIC ANHYDRIDE, 68
BENZOIC-CARBONIC ANHYDRIDE, 286
DIACETYL-d-TARTARIC ANHYDRIDE, 242
β-METHYLGLUTARIC ANHYDRIDE, 630
α-PHENYLGLUTARIC ANHYDRIDE, 790
cis-Δ⁴-TETRAHYDROPHTHALIC ANHYDRIDE, 890
p-TOLUENESULFONIC ANHYDRIDE, 940

ACID HALIDES
α-BROMOISOBUTYRYL BROMIDE, 348
δ-CARBOMETHOXYVALERYL CHLORIDE, 555
N-CHLOROBETAINYL CHLORIDE, 154
2,4-DI-tert-AMYLPHENOXYACETYL CHLORIDE, 742
DIETHYLTHIOCARBAMYL CHLORIDE, 307
ITACONYL CHLORIDE, 554
MALONYL DICHLORIDE, 263
METHANESULFONYL CHLORIDE, 571
NAPHTHALENE-1,5-DISULFONYL CHLORIDE, 693
m-NITROBENZOYL CHLORIDE, 715
OLEOYL CHLORIDE, 739
PALMITOYL CHLORIDE, 742
PHENYLDICHLOROPHOSPHINE, 784
RICINOLEOYL CHLORIDE, 742
β-STYRENESULFONYL CHLORIDE, 846
p-TOLUENESULFENYL CHLORIDE, 934
p-TOLUENESULFINYL CHLORIDE, 937
TRICHLOROMETHYLPHOSPHONYL DICHLORIDE, 950

ACIDS
A. UNSUBSTITUTED ACIDS
1. Monobasic
3-BENZYL-3-METHYLPENTANOIC ACID, 93
trans-2-DODECENOIC ACID, 398
3-ETHYL-3-METHYLHEXANOIC ACID, 97
4-ETHYL-2-METHYL-2-OCTENOIC ACID, 444
9-FLUORENECARBOXYLIC ACID, 482
2-METHYLDODECANOIC ACID, 618
trans-2-METHYL-2-DODECENOIC ACID, 608
2-METHYLENEDODECANOIC ACID, 616
3-METHYL-3-PHENYLPENTANOIC ACID, 97
α-PHENYLCINNAMIC ACID, 777
STEAROLIC ACID, 851
α,β,β-TRIPHENYLPROPIONIC ACID, 960
10-UNDECYNOIC ACID, 969

2. Dibasic
o-CARBOXYCINNAMIC ACID, 136
β-(o-CARBOXYPHENYL)PROPIONIC ACID, 136
β,β-DIMETHYLGLUTARIC ACID, 345
β-ETHYL-β-METHYLGLUTARIC ACID, 441
GLUTARIC ACID, 496
HENDECANEDIOIC ACID, 510
β-METHYL-β-PHENYLGLUTARIC ACID, 664
γ-PHENYLALLYLSUCCINIC ACID, 766
PHENYLSUCCINIC ACID, 804
3,4-SECO-Δ⁵-CHOLESTEN-3,4-DIOIC ACID, 191

B. SUBSTITUTED ACIDS
1. Amino Acids
ε-AMINOCAPROIC ACID, 39
3-AMINO-2,4,6-TRIBROMOBENZOIC ACID, 947
DL-ASPARTIC ACID, 55
3,5-DICHLORO-2-AMINOBENZOIC ACID, 872
6-HYDROXYNICOTINIC ACID, 532

ACIDS—*Continued*
  B. *SUBSTITUTED ACIDS*—*Continued*
    2. **Cyano Acids**
      CYCLOHEXYLIDENECYANOACETIC ACID, 234
    3. **Halogen Acids**
      3-AMINO-2,4,6-TRIBROMOBENZOIC ACID, 947
      p-BROMOMANDELIC ACID, 110
      2-BROMO-3-METHYLBENZOIC ACID, 114
      p-CHLOROMANDELIC ACID, 112
      α-CHLOROPHENYLACETIC ACID, 169
      DIBROMOHYDROCINNAMIC ACID, 961
      3,5-DICHLORO-2-AMINOBENZOIC ACID, 872
      p-IODOMANDELIC ACID, 112
      MUCOBROMIC ACID, 688
      dl-4,4',6,6'-TETRACHLORODIPHENIC ACID, 872
      2,4,6-TRIBROMOBENZOIC ACID, 947
    4. **Hydroxy and Phenol Acids**
      ATROLACTIC ACID, 58
      5,5-DIHYDROXYBARBITURIC ACID, 23
      9,10-DIHYDROXYSTEARIC ACID, 317
      6-HYDROXYNICOTINIC ACID, 532
      SODIUM 2-PHENYL-2-HYDROXY-ETHANE-1-SULFONATE, 850
      VANILLIC ACID, 972
    5. **Keto Acids**
      δ-ACETYL-n-VALERIC ACID, 19
      6-KETOHENDECANEDIOIC ACID, 555
      γ-OXOCAPRIC ACID, 432
    6. **Nitro Acids**
      m-NITROCINNAMIC ACID, 731
      trans-o-NITRO-α-PHENYLCINNAMIC ACID, 730
    7. **Miscellaneous**
      ABIETIC ACID, 1
      β-CARBETHOXY-γ,γ-DIPHENYL-VINYLACETIC ACID, 132
      COUMALIC ACID, 201
      2,3-DIMETHOXYCINNAMIC ACID, 327
      5-FORMYL-4-PHENANTHROIC ACID, 484
      2-FUROIC ACID, 493
      ISODEHYDROACETIC ACID, 549
      3-METHYLCOUMARILIC ACID, 591
      3-METHYL-2-FUROIC ACID, 628
      2-METHYL-5,6-PYRAZINEDICAR-BOXYLIC ACID, 827
      PARABANIC ACID, 744
      α-PHTHALIMIDO-o-TOLUIC ACID, 810
      2,3-PYRAZINEDICARBOXYLIC ACID, 824

ACIDS—*Continued*
  B. *SUBSTITUTED ACIDS*—*Continued*
    7. **Miscellaneous**—*Continued*
      α-SULFOPALMITIC ACID, 862
      3-THENOIC ACID, 919
      THIOBENZOIC ACID, 924
      THIOLACETIC ACID, 928
      p-TOLUENESULFONYLANTHRANILIC ACID, 34
      TRITHIOCARBODIGLYCOLIC ACID, 967
ALCOHOLS
  A. *UNSUBSTITUTED ALCOHOLS*
    1. **Primary**
      2-BUTYN-1-OL, 128
      2,3-DIMETHYLBENZYL ALCOHOL, 584
      o-METHYLBENZYL ALCOHOL, 582
      3-METHYL-1,5-PENTANEDIOL, 660
      4-PENTYN-1-OL, 755
      3-PHENYL-1-PROPANOL, 798
    2. **Secondary**
      CHOLESTEROL, 195
      1,2-CYCLODECANEDIOL, 216
      trans-METHYLSTYRYLCARBINOL, 773
      1-PHENYL-1-PENTEN-4-YN-3-OL, 792
    3. **Secondary-Tertiary**
      2-METHYL-2,5-DECANEDIOL, 601
    4. **Tertiary**
      1-ACETYLCYCLOHEXANOL, 13
      1,1'-ETHYNYLENE-bis-CYCLO-HEXANOL, 471
  B. *SUBSTITUTED ALCOHOLS* See also Acids, B,4; Esters, B,3.
      1-(AMINOMETHYL)CYCLOHEXANOL, 224
      trans-2-CHLOROCYCLOPENTANOL, 157
      CHOLESTEROL DIBROMIDE, 195
      2,2-DICHLOROETHANOL, 271
      MONOBENZALPENTAERYTHRITOL, 679
      MONOBROMOPENTAERYTHRITOL, 681
      2-(1-PYRROLIDYL)PROPANOL, 834
      SEBACOIN, 840
      2,2,6,6-TETRAMETHYLOLCYCLO-HEXANOL, 907
ALDEHYDES
  A. *ALIPHATIC ALDEHYDES*
      PROPIOLALDEHYDE, 813
  B. *ALIPHATIC, SUBSTITUTED ALDEHYDES*
      n-BUTYL GLYOXYLATE, 124
      BUTYRCHLORAL, 130
      α-CHLOROCROTONALDEHYDE, 131
  C. *AROMATIC ALDEHYDES*
      DIPHENYLACETALDEHYDE, 375
      1-NAPHTHALDEHYDE, 690
      o-TOLUALDEHYDE, 932

ALDEHYDES—*Continued*
 D. *AROMATIC, SUBSTITUTED*
   ALDEHYDES
   p-Acetamidobenzaldehyde, 32
   p-Aminobenzaldehyde, 31
   p-Dimethylaminobenzaldehyde,
    331
   5-Formyl-4-phenanthroic Acid,
    484
   o-Nitrocinnamaldehyde, 722
   6-Nitroveratraldehyde, 735
   Syringic Aldehyde, 866
 E. *DIALDEHYDES*
   β-Methylglutaraldehyde, 661
   o-Phthalaldehyde, 807
   Sodium Nitromalonaldehyde
    Monohydrate, 844
 F. *HETEROCYCLIC ALDEHYDES*
   Indole-3-aldehyde, 539
   2-Pyrrolealdehyde, 831
   2-Thenaldehyde, 915
   3-Thenaldehyde, 918
AMIDES *See also* Heterocyclic compounds (for cyclic amides).
 A. *UNSUBSTITUTED*
   2-Ethylhexanamide, 436
   Fumaramide, 486
   n-Heptamide, 513
   Phenylacetamide, 760
 B. *SUBSTITUTED*
   3-Acetamido-2-butanone, 5
   Benzoylacetanilide, 80
   2-Benzoyl-1-cyano-1-methyl-1,2-
    dihydroisoquinoline, 642
   Benzoyl-2-methoxy-4-nitroacet-
    anilide, 82
   N-Bromoacetamide, 104
   1-Cyano-2-benzoyl-1,2-dihydroi-
    soquinoline, 641
   N,N-Dimethylcyclohexanecar-
    boxamide, 339
   2,4-Dithiobiuret, 504
   N-Ethyl-*p*-chloroformanilide,
    420
   Guanylthiourea, 502
   Hexahydro-1,3,5-tripropionyl-*s*-
    triazine, 518
   Nicotinamide-1-oxide, 704
   p-Toluenesulfonylanthranilic
    Acid, 34
   p-Tolylsulfonylmethylnitros-
    amide, 943
AMIDINES
   N,N′-Diphenylbenzamidine, 383
   N-Phenylbenzamidine, 769
AMINE OXIDES
   Benzofurazan Oxide, 74
   3-Methyl-4-nitropyridine-1-oxide,
    654
   3-Methylpyridine-1-oxide, 655
   Nicotinamide-1-oxide, 704
   Pyridine-1-oxide, 828

AMINES *See also* Acids, *B*,1.
 A. *ALIPHATIC AMINES*
   1-(Aminomethyl)cyclohexanol,
    224
   1-Diethylamino-3-butanone, 281
   β-Dimethylaminoethyl Chloride
    Hydrochloride, 333
   2,3-Dimethylbenzyldimethyl-
    amine, 587
   N,N-Dimethylcyclohexylmethyl-
    amine, 339
   Laurylmethylamine, 564
   2-Methylbenzyldimethylamine,
    585
   N-Methyl-2,3-dimethoxybenzyl-
    amine, 603
   N-Methyl-1,2-diphenylethyl-
    amine, 605
   5-Methylfurfuryldimethylamine,
    626
   Pseudopellitierine, 816
 B. *AROMATIC AMINES*
   p-Aminobenzaldehyde, 31
   2-Aminobenzophenone, 34
   2-Amino-3-nitrotoluene, 42
   p-Aminotetraphenylmethane, 47
   N-Benzylaniline, 92
   3-(*o*-Chloroanilino)propionitrile,
    146
   N-2-Cyanoethylaniline, 205
   p-Dimethylaminobenzaldehyde,
    331
   2,6-Dinitroaniline, 364
   N-Ethyl-*p*-chloroaniline, 420
   4′-Methyl-2-aminobenzophenone,
    38
   p-Tricyanovinyl-N,N-dimethyl-
    aniline, 953
 C. *HETEROCYCLIC AMINES*
   2-Amino-4-anilino-6-(chloro-
    methyl)-*s*-triazine, 29
   3-Aminopyridine, 45
   Benzoguanamine, 78
   2-Benzylaminopyridine, 91
   2-Chloromethyl-4,6-diamino-*s*-
    triazine, 30
   2,5-Diamino-3,4-dicyanothiophene,
    243
   2,4-Diamino-6-hydroxypyrimidine,
    245
   2,4-Diamino-6-phenyl-*s*-triazine,
    78
   Diaminouracil Hydrochloride,
    247
   2-(Dimethylamino)pyrimidine,
    336
   1,4-Diphenyl-5-amino-1,2,3-tri-
    azole, 380
   2-Mercapto-4-amino-5-carbeth-
    oxypyrimidine, 566
   N-Methylaminopyrimidine, 336
   4-Phenyl-5-anilino-1,2,3-triazole,
    380

AMINES—*Continued*
  D. *DIAMINES*
    N,N′-DIBENZYL-*p*-PHENYLENEDI-
      AMINE, 92
    N,N′-DIETHYLBENZIDINE, 283
    PUTRESCINE DIHYDROCHLORIDE, 819
ARSENIC COMPOUNDS
  TETRAPHENYLARSONIUM CHLORIDE
    HYDROCHLORIDE, 910
  TRIPHENYLARSINE, 910
  TRIPHENYLARSINE OXIDE, 911
AZIDES
  m-NITROBENZAZIDE, 715
  o-NITROPHENYLAZIDE, 75
AZO COMPOUNDS
  1,1′-AZO-*bis*-1-CYCLOHEXANENITRILE, 66
  2,2′-AZO-*bis*-ISOBUTYRONITRILE, 67
  ETHYL AZODICARBOXYLATE, 411
  METHYL AZODICARBOXYLATE, 414
BORON COMPOUNDS
  BENZENEBORONIC ANHYDRIDE, 68
CARBOHYDRATES AND DERIVA-
TIVES
  D-GULONIC-γ-LACTONE, 506
CYANAMIDES
  o-CHLOROPHENYLCYANAMIDE, 172
  α-NAPHTHYLCYANAMIDE, 174
DIAZO COMPOUNDS
  DIAZOMETHANE, 250
  ETHYL DIAZOACETATE, 424
ESTERS *See also* Lactones.
  A. *OF UNSUBSTITUTED MONO-*
    BASIC ACIDS
    tert-BUTYL ACETATE, 263
    2,3-DIMETHYLBENZYL ACETATE, 584
    FURFURAL DIACETATE, 489
    o-METHYLBENZYL ACETATE, 582
    METHYL CYCLOPENTANECARBOX-
      YLATE, 594
    o- AND *p*-NITROBENZALDIACETATE, 713
    1,5-PENTANEDIOL DIACETATE, 748
    VINYL LAURATE, 977
  B. *OF SUBSTITUTED MONOBASIC*
    ACIDS
    1. **Halogen Esters**
      tert-BUTYL BROMOACETATE, 263
      sec-BUTYL α-BROMOPROPIONATE,
        122
      tert-BUTYL CHLOROACETATE, 263
      tert-BUTYL α-CHLOROPROPIONATE,
        263
      tert-BUTYL HYPOCHLORITE, 125
      ETHYL α-CHLOROACETOACETATE,
        592
      ETHYL CHLOROFLUOROACETATE,
        423
      ETHYL α-CHLOROPHENYLACETATE,
        169
    2. **Hydroxy Esters**
      p-CHLOROPHENYL SALICYLATE, 178
      ETHYL 4-ETHYL-2-METHYL-3-HY-
        DROXYOCTANOATE, 447
      ETHYL MANDELATE, 169

ESTERS—*Continued*
  B. *OF SUBSTITUTED MONOBASIC*
    ACIDS—*Continued*
    3. **Keto Esters**
      2-*p*-ACETYLPHENYLHYDROQUI-
        NONE DIACETATE, 16
      o-BENZOYLOXYACETOPHENONE,
        478
      tert-BUTYL *o*-BENZOYLBENZOATE,
        263
      sec-BUTYL α-*n*-CAPROYLPROPIO-
        NATE, 120
      ETHYL α-ACETYL-β-(2,3-DIMETH-
        OXYPHENYL)ACRYLATE, 408
      ETHYL α-ACETYL-β-(3,4-DIMETH-
        OXYPHENYL)ACRYLATE, 410
      ETHYL α-ACETYL-β-(2,3-DIMETH-
        OXYPHENYL)PROPIONATE, 408
      ETHYL α-ACETYL-β-(3,4-DIMETH-
        OXYPHENYL)PROPIONATE, 410
      ETHYL BENZOYLACETATE, 415
      ETHYL PYRUVATE, 467
      METHYL *p*-ACETYLBENZOATE, 579
    4. **Nitrile Esters**
      ETHYL *sec*-BUTYLIDENECYANO-
        ACETATE, 94
      ETHYL PHENYLCYANOACETATE, 461
      ETHYL β-PHENYL-β-CYANOPRO-
        PIONATE, 804
      ETHYL (1-PHENYLETHYLIDENE)-
        CYANOACETATE, 463
      α-PHENYL-α-CARBETHOXYGLU-
        TARONITRILE, 776
    5. **Miscellaneous**
      n-BUTYL GLYOXYLATE, 124
      2-CHLOROETHYL BENZOATE, 84
      N,N′-DICARBETHOXYPUTRESCINE,
        822
      ETHYL N-BENZYLCARBAMATE, 780
      ETHYL DIAZOACETATE, 424
      ETHYL DIETHOXYACETATE, 427
      ETHYL ISODEHYDROACETATE, 549
      ETHYL 3-METHYLCOUMARILATE,
        590
      ETHYL α-NITROBUTYRATE, 454
      ETHYL N-NITROSO-N-BENZYL-
        CARBAMATE, 780
      ETHYL β,β-PENTAMETHYLENE-
        GLYCIDATE, 459
      ETHYL N-PHENYLFORMIMIDATE,
        464
      ETHYL α-(1-PYRROLIDYL)PROPIO-
        NATE, 466
      2-IODOETHYL BENZOATE, 84
      2-MERCAPTO-4-amino-5-CARBETH-
        OXYPYRIMIDINE, 566
      METHYL COUMALATE, 532
      METHYL 5,5-DIMETHOXY-3-
        METHYL-2,3-EPOXYPENTANOATE,
        649
      METHYL *p*-ETHYLBENZOATE, 580
      METHYL 3-METHYL-2-FUROATE,
        649

ESTERS—*Continued*
  B. *OF SUBSTITUTED MONOBASIC*
    ACIDS—Continued
    5. **Miscellaneous**—*Continued*
      METHYL γ-METHYL-γ-NITRO-
        VALERATE, 652
      METHYL β-THIODIPROPIONATE,
        669
  C. *OF DICARBOXYLIC ACIDS*
    β-CARBETHOXY-γ,γ-DIPHENYLVINYL-
      ACETIC ACID, 132
    CETYLMALONIC ESTER, 141
    DI-*tert*-BUTYL β,β-DIMETHYLGLU-
      TARATE, 263
    DI-*tert*-BUTYL GLUTARATE, 263
    DI-*tert*-BUTYL MALONATE, 261
    DI-*tert*-BUTYL SUCCINATE, 263
    DIETHYL ACETYLENEDICARBOXYL-
      ATE, 330
    DIETHYL BENZOYLMALONATE, 285
    DIETHYL 1,1-CYCLOBUTANEDICAR-
      BOXYLATE, 288
    DIETHYL Δ²-CYCLOPENTENYLMALO-
      NATE, 291
    DIETHYL ETHYLIDENEMALONATE, 293
    DIETHYL ETHYLPHOSPHONATE, 326
    DIETHYL *cis*-HEXAHYDROPHTHALATE,
      304
    DIETHYL METHYLENEMALONATE, 298
    DIETHYL METHYLPHOSPHONATE, 326
    DIETHYL γ-OXOPIMELATE, 302
    DIETHYL *cis*-Δ⁴-TETRAHYDROPHTHAL-
      ATE, 304
    DIISOPROPYL ETHYLPHOSPHONATE,
      326
    DIISOPROPYL METHYLPHOSPHONATE,
      325
    DIMETHYL ACETYLENEDICARBOXYL-
      ATE, 329
    DIMETHYL *cis*-HEXAHYDROPHTHAL-
      ATE, 306
    DIMETHYL *cis*-Δ⁴-TETRAHYDRO-
      PHTHALATE, 306
    DIPHENYL SUCCINATE, 390
    ETHOXYMAGNESIUMMALONATE ESTER,
      285
    ETHYL AZODICARBOXYLATE, 411
    ETHYL *tert*-BUTYL MALONATE, 417
    ETHYL ENANTHYLSUCCINATE, 430
    ETHYL HYDRAZODICARBOXYLATE, 411
    METHYL AZODICARBOXYLATE, 414
    METHYL HYDRAZODICARBOXYLATE,
      413
    METHYL HYDROGEN HENDECANE-
      DIOATE, 635
    METHYL β-THIODIPROPIONATE, 669
    o-PHENYLENE CARBONATE, 788
  D. *OF TRICARBOXYLIC ACIDS*
    TRIETHYL PHOSPHITE, 955
    TRIETHYL α-PHTHALIMIDOETHANE-α,
      α,β-TRICARBOXYLATE, 55
  E. *OF TETRACARBOXYLIC ACIDS*
    ETHYL ORTHOCARBONATE, 457

ETHERS *See also* Heterocyclic compounds.
  α-ALLYL-β-BROMOETHYL ETHYL ETHER,
    750
  BENZHYDRYL β-CHLOROETHYL ETHER, 72
  BISCHLOROMETHYL ETHER, 101
  α-CHLOROETHYL ETHYL ETHER, 748
  2-CHLORO-1,1,2-TRIFLUOROETHYL ETHYL
    ETHER, 184
  CREOSOL, 203
  α,β-DIBROMOETHYL ETHYL ETHER, 749
  4,4'-DICHLORODIBUTYL ETHER, 266
  2,3-DIMETHOXYCINNAMIC ACID, 327
  ETHOXYACETYLENE, 404
  p-ETHOXYPHENYLUREA, 52
  ETHYL α-ACETYL-β-(2,3-DIMETHOXY-
    PHENYL)ACRYLATE, 408
  ETHYL α-ACETYL-β-(3,4-DIMETHOXY-
    PHENYL)ACRYLATE, 410
  ETHYL α-ACETYL-β-(2,3-DIMETHOXY-
    PHENYL)PROPIONATE, 408
  ETHYL α-ACETYL-β-(3,4-DIMETHOXY-
    PHENYL)PROPIONATE, 410
  4-IODOVERATROLE, 547
  METHOXYACETYLENE, 406
  o-METHOXYPHENYLACETONE, 573
  p-METHOXYPHENYLACETONITRILE, 576
  1-(*o*-METHOXYPHENYL)-2-NITRO-1-PRO-
    PENE, 573
  p-METHOXYPHENYLUREA, 53
  O-METHYLCAPROLACTIM, 588
  N-METHYL-2,3-DIMETHOXYBENZYL-
    AMINE, 603
  METHYL 5,5-DIMETHOXY-3-METHYL-2,3-
    EPOXYPENTANOATE, 649
  METHYLISOUREA HYDROCHLORIDE, 645
  6-NITROVERATRALDEHYDE, 735
  QUINACETOPHENONE DIMETHYL ETHER,
    837
  QUINACETOPHENONE MONOMETHYL
    ETHER, 836
  SYRINGIC ALDEHYDE, 866
  VANILLIC ACID, 972

HALOGENATED COMPOUNDS *See
  also* Acids, *B*,3; Esters, *B*,1; Ketones,
  A,2; Acid halides.
  A. *BROMO COMPOUNDS*
    α-ALLYL-β-BROMOETHYL ETHYL
      ETHER, 750
    ALLYLMAGNESIUM BROMIDE, 749
    N-BROMOACETAMIDE, 104
    β-BROMOETHYLPHTHALIMIDE, 106
    N-BROMOGLUTARIMIDE, 498
    4-BROMO-2-HEPTENE, 108
    2-BROMO-4-NITROTOLUENE, 114
    p-BROMOPHENYLUREA, 49
    CHOLESTEROL DIBROMIDE, 195
    DIBROMOACETONITRILE, 254
    4,4'-DIBROMOBIBENZYL, 257
    4,4'-DIBROMOBIPHENYL, 256
    α,β-DIBROMOETHYL ETHYL ETHER,
      749
    ETHYNYLMAGNESIUM BROMIDE, 792

HALOGENATED COMPOUNDS—
Continued
A. *BROMO COMPOUNDS—Continued*
Monobromopentaerythritol, 681
Pentaerythrityl Tetrabromide,
753
α,α,α',α'-Tetrabromo-*o*-xylene,
807
3-Thenyl Bromide, 921
o-Xylylene Dibromide, 984
B. *CHLORO COMPOUNDS*
α-Acetyl-δ-chloro-γ-valerolactone, 10
2-Amino-4-anilino-6-(chloromethyl)-s-triazine, 29
Benzhydryl β-Chloroethyl
Ether, 72
Benzoylcholine Chloride, 84
Bischloromethyl Ether, 101
Butyrchloral, 130
Chloroacetonitrile, 144
3-(*o*-Chloroanilino)propionitrile,
146
Chloro-*p*-benzoquinone, 148
N-Chlorobetainyl Chloride,
154
α-Chlorocrotonaldehyde, 130
trans-2-Chlorocyclopentanol,
157
3-Chlorocyclopentene, 239
1-Chloro-2,6-dinitrobenzene, 160
α-Chloroethyl Ethyl Ether, 748
2-Chloromethyl-4,6-diamino-s-
triazine, 30
2-Chloronicotinonitrile, 166
o-Chlorophenylcyanamide, 172
α-(4-Chlorophenyl)-γ-phenylace-
toacetonitrile, 174
o-Chlorophenylthiourea, 180
2-Chloropyrimidine, 182
3-Chlorotoluquinone, 152
2-Chloro-1,1,2-trifluoroethyl
Ethyl Ether, 184
Dichloroacetonitrile, 255
4,4'-Dichlorodibutyl Ether, 266
1,1-Dichloro-2,2-difluoroethyl-
ene, 268
2,2-Dichloroethanol, 271
2,5-Dichloroquinone, 152
β-Dimethylaminoethyl Chloride
Hydrochloride, 333
N-Ethyl-*p*-chloroaniline, 420
N-Ethyl-*p*-chloroformanilide,
420
Neophyl Chloride, 702
1-(*p*-Nitrophenyl)-4-chloro-2-
butene, 727
Phenyldichlorophosphine, 784
Tetraphenylarsonium Chloride
Hydrochloride, 910
Trichloromethylphosphonyl
Dichloride, 950

HALOGENATED COMPOUNDS—Continued
C. *FLUORO COMPOUNDS*
2-Chloro-1,1,2-trifluoroethyl
Ethyl Ether, 184
1,1-Dichloro-2,2-difluoroethyl-
ene, 268
n-Hexyl Fluoride, 525
D. *IODO COMPOUNDS*
Benzoylcholine Iodide, 84
Benzyltrimethylammonium Io-
dide, 585
n-Butyl Iodide, 322
tert-Butyl Iodide, 324
Cyanogen Iodide, 207
1,4-Diiodobutane, 321
1,6-Diiodohexane, 323
Iodocyclohexane, 324, 543
2-Iodothiophene, 545
4-Iodoveratrole, 547
Isobutyl Iodide, 324
Isopropyl Iodide, 322
n-Propyl Iodide, 324
HETEROCYCLIC COMPOUNDS
A. *THREE-MEMBERED, NITRO-
GEN*
Ethylenimine, 433
B. *THREE-MEMBERED, OXYGEN*
Ethyl β,β-Pentamethylenegly-
cidate, 459
Isophorone Oxide, 552
Methyl 5,5-Dimethoxy-3-methyl-
2,3-epoxypentanoate, 649
trans-Stilbene Oxide, 860
C. *THREE-MEMBERED, SULFUR*
Cyclohexene Sulfide, 232
D. *FIVE-MEMBERED, NITROGEN*
1,2-Benzo-3,4-dihydrocarbazole,
885
2,2-Dimethylpyrrolidine, 354
5,5-Dimethyl-2-pyrrolidone, 357
Ethyl α-(1-Pyrrolidyl)propio-
nate, 466
Indole-3-aldehyde, 539
1-Methyl-3-ethyloxindole, 620
3-Methyloxindole, 657
2-Pyrroleatdehyde, 831
2-(1-Pyrrolidyl)propanol, 834
1,2,3,4-Tetrahydrocarbazole, 884
E. *FIVE-MEMBERED, TWO
NITROGEN*
3,5-Dimethylpyrazole, 351
Indazole, 536
2-Mercaptobenzimidazole, 569
Parabanic Acid, 744
4,5,6,7-Tetrahydroindazole, 537
F. *FIVE-MEMBERED, THREE
NITROGEN*
1,4-Diphenyl-5-amino-1,2,3-tri-
azole, 380
4-Phenyl-5-anilino-1,2,3-triazole,
380

HETEROCYCLIC COMPOUNDS—*Continued*

G. FIVE-MEMBERED, ONE NITROGEN, ONE OXYGEN
2-MERCAPTOBENZOXAZOLE, 570

H. FIVE-MEMBERED, TWO NITROGEN, ONE OXYGEN
BENZOFURAZAN OXIDE, 74
DIMETHYLFURAZAN, 342

I. FIVE-MEMBERED, ONE NITROGEN, ONE SULFUR
N-(*p*-ACETYLAMINOPHENYL)RHODANINE, 6

J. FIVE-MEMBERED, OXYGEN
5,5-DIMETHYL-2-*n*-PENTYLTETRAHYDROFURAN, 350
ETHYL 3-METHYLCOUMARILATE, 590
FURFURAL DIACETATE, 489
2-FURFURYL MERCAPTAN, 491
2-FUROIC ACID, 493
3-HYDROXYTETRAHYDROFURAN, 534
3-METHYLCOUMARILIC ACID, 591
3-METHYLCOUMARONE, 590
3-METHYLFURAN, 628
5-METHYLFURFURYLDIMETHYLAMINE, 626
3-METHYL-2-FUROIC ACID, 628
METHYL 3-METHYL-2-FUROATE, 649

K. FIVE-MEMBERED, SULFUR
2,5-DIAMINO-3,4-DICYANOTHIOPHENE, 243
2-IODOTHIOPHENE, 545
METHYL 2-THIENYL SULFIDE, 667
3-METHYLTHIOPHENE, 671
TETRAHYDROTHIOPHENE, 892
2-THENALDEHYDE, 915
3-THENALDEHYDE, 918
3-THENOIC ACID, 919
3-THENYL BROMIDE, 921
2-VINYLTHIOPHENE, 980

L. SIX-MEMBERED, NITROGEN
3-AMINOPYRIDINE, 45
2-BENZOYL-1-CYANO-1-METHYL-1,2-DIHYDROISOQUINOLINE, 642
3-BENZOYLPYRIDINE, 88
4-BENZOYLPYRIDINE, 89
2-BENZYLAMINOPYRIDINE, 91
2-CHLORONICOTINONITRILE, 166
1-CYANO-2-BENZOYL-1,2-DIHYDROISOQUINOLINE, 641
3-CYANO-6-METHYL-2(1)-PYRIDONE, 210
5-ETHYL-2-METHYLPYRIDINE, 451
6-HYDROXYNICOTINIC ACID, 532
1-METHYLISOQUINOLINE, 641
3-METHYL-4-NITROPYRIDINE-1-OXIDE, 654
3-METHLYPYRIDINE-1-OXIDE, 655
NICOTINAMIDE-1-OXIDE, 704
NICOTINONITRILE, 706
1-PHENYLPIPERIDINE, 795
PYRIDINE-1-OXIDE, 828

HETEROCYCLIC COMPOUNDS—*Continued*

M. SIX-MEMBERED, TWO NITROGEN
ALLOXAN MONOHYDRATE, 23
ALLOXANTIN DIHYDRATE, 25
2-CHLOROPYRIMIDINE, 182
2,4-DIAMINO-6-HYDROPYRIMIDINE, 245
DIAMINOURACIL HYDROCHLORIDE, 247
2-(DIMETHYLAMINO)PYRIMIDINE, 336
3-*n*-HEPTYL-5-CYANOCYTOSINE, 515
2-MERCAPTO-4-AMINO-5-CARBETHOXYPYRIMIDINE, 566
2-MERCAPTO-4-HYDROXY-5-CYANOPYRIMIDINE, 566
N-METHYLAMINOPYRIMIDINE, 336
4-METHYL-6-HYDROXYPYRIMIDINE, 638
2-METHYL-5,6-PYRAZINEDICARBOXYLIC ACID, 827
2,3-PYRAZINEDICARBOXYLIC ACID, 824
QUINOXALINE, 824
2-THIO-6-METHYLURACIL, 638

N. SIX-MEMBERED, THREE NITROGEN
2-AMINO-4-ANILINO-6-(CHLOROMETHYL)-*s*-TRIAZINE, 29
BENZOGUANAMINE, 78
2-CHLOROMETHYL-4,6-DIAMINO-*s*-TRIAZINE, 30
2,4-DIAMINO-6-PHENYL-*s*-TRIAZINE, 78
HEXAHYDRO-1,3,5-TRIPROPIONYL-*s*-TRIAZINE, 518

O. SIX-MEMBERED, OXYGEN
COUMALIC ACID, 201
3,4-DIHYDRO-2-METHOXY-4-METHYL-2H-PYRAN, 311
4,6-DIMETHYLCOUMALIN, 337
ETHYL ISODEHYDROACETATE, 549
FLAVONE, 478
ISODEHYDROACETIC ACID, 549
METHYL COUMALATE, 532

P. SIX-MEMBERED, TWO SULFUR
p-DITHIANE, 396

Q. SEVEN-MEMBERED, NITROGEN
O-ETHYLCAPROLACTIM, 589
O-METHYLCAPROLACTIM, 588

HYDRAZINES
1,2-DI-1-(1-CYANO)CYCLOHEXYLHYDRAZINE, 274
1,2-DI-2-(2-CYANO)PROPYLHYDRAZINE, 275
ETHYL HYDRAZODICARBOXYLATE, 411
METHYL HYDRAZODICARBOXYLATE, 413
β-PROPIONYLPHENYLHYDRAZINE, 657

HYDROCARBONS
n-AMYLACETYLENE, 119
n-BUTYLACETYLENE, 117

HYDROCARBONS—*Continued*
CYCLOPENTADIENE, 238
9,10-DIHYDROPHENANTHRENE, 313
DIPHENYLACETYLENE, 377
HEMIMELLITENE, 508
HEXAMETHYLBENZENE, 520
n-HEXYLACETYLENE, 119
ISOAMYLACETYLENE, 119
METHYLENECYCLOHEXANE, 612
9-METHYLFLUORENE, 623
1-METHYL-3-PHENYLINDANE, 665
MONOVINYLACETYLENE, 683
NORBORNYLENE, 738
1,4-PENTADIENE, 746
PHENANTHRENE, 313
PHENYLACETYLENE, 763
trans-1-PHENYL-1,3-BUTADIENE, 771
n-PROPYLACETYLENE, 119
cis-STILBENE, 857
TETRAPHENYLETHYLENE, 914
1,1,3-TRIMETHYL-3-PHENYLINDANE, 666
TRIPTYCENE, 964
IMIDES
β-BROMOETHYLPHTHALIMIDE, 106
N-BROMOGLUTARIMIDE, 498
α,α'-DICYANO-β-ETHYL-β-METHYLGLU-
  TARIMIDE, 441
GLUTARIMIDE, 496
β-METHYL-β-PHENYL-α,α'-DICYANO-
  GLUTARIMIDE, 662
α-PHTHALIMIDO-o-TOLUIC ACID, 810
TRIETHYL α-PHTHALIMIDOETHANE-α,
  α,β-TRICARBOXYLATE, 55
ISOCYANATES
HEXAMETHYLENE DIISOCYANATE, 521
ISONITRILES
ETHYL ISOCYANIDE, 438
ISOTHIOCYANATES
α-NAPHTHYL ISOTHIOCYANATE, 700
KETENES
DIMETHYLKETENE, 348
KETONES
A. MONOKETONES
  1. Unsubstituted
    9-ACETYLANTHRACENE, 8
    Δ⁴-CHOLESTEN-3-ONE, 192
    Δ⁵-CHOLESTEN-3-ONE, 195
    CYCLODECANONE, 218
    CYCLOHEPTANONE, 221
    DICYCLOPROPYL KETONE, 278
    DI-(2-METHYLCYCLOPROPYL) KE-
      TONE, 280
    LAURONE, 560
    2-METHYL-2-CYCLOHEXENONE,
      162
    METHYL CYCLOPROPYL KETONE,
      597
    2-PHENYLCYCLOHEPTANONE, 780
    STEARONE, 854
    α-TETRALONE, 898
    β-TETRALONE, 903
    2,4,4-TRIMETHYLCYCLOPENTA-
      NONE, 957

KETONES—*Continued*
A. MONOKETONES—*Continued*
  2. Halogen Ketones
    2-CHLORO-2-METHYLCYCLOHEXA-
      NONE, 162
    5-CHLORO-2-PENTANONE, 597
    1-(*p*-CHLOROPHENYL)-3-PHENYL-
      2-PROPANONE, 176
    β-CHLOROVINYL ISOAMYL KETONE,
      186
    5α,6β-DIBROMOCHOLESTAN-3-
      ONE, 197
    1,7-DICHLORO-4-HEPTANONE, 279
    p,α,α-TRIBROMOACETOPHENONE,
      110
  3. Phenol Ketones
    2-*p*-ACETYLPHENYLHYDROQUI-
      NONE, 15
    o-HYDROXYDIBENZOYLMETHANE,
      479
    QUINACETOPHENONE MONO-
      METHYL ETHER, 836
  4. Miscellaneous
    3-ACETAMIDO-2-BUTANONE, 5
    α-ACETYL-δ-CHLORO-γ-VALERO-
      LACTONE, 10
    1-ACETYLCYCLOHEXANOL, 13
    2-*p*-ACETYLPHENYLHYDROQUI-
      NONE DIACETATE, 17
    2-*p*-ACETYLPHENYLQUINONE, 16
    2-AMINOBENZOPHENONE, 34
    BENZOYLACETANILIDE, 80
    BENZOYL-2-METHOXY-4-NITRO-
      ACETANILIDE, 82
    o-BENZOYLOXYACETOPHENONE,
      478
    3-BENZOYLPYRIDINE, 88
    4-BENZOYLPYRIDINE, 89
    1-DIETHYLAMINO-3-BUTANONE,
      281
    DIETHYL BENZOYLMALONATE, 285
    DIETHYL γ-OXOPIMELATE, 302
    ETHYL ENANTHYLSUCCINATE, 430
    FLAVONE, 478
    2-HYDROXYMETHYLENECYCLO-
      HEXANONE, 536
    ISOPHORONE OXIDE, 552
    β-KETOISOÖCTALDEHYDE DI-
      METHYL ACETAL, 558
    o-METHOXYPHENYLACETONE, 573
    METHYL *p*-ACETYLBENZOATE,
      579
    4'-METHYL-2-AMINOBENZOPHE-
      NONE, 38
    o-NITROACETOPHENONE, 708
    PHENYL 4-PYRIDYL KETONE, 89
    PSEUDOPELLETIERINE, 816
    QUINACETOPHENONE MONO-
      METHYL ETHER, 836
    SEBACOIN, 840
    THIOBENZOPHENONE, 927
B. DIKETONES
    Δ⁴-CHOLESTEN-3,6-DIONE, 189

KETONES—*Continued*
  B. DIKETONES—*Continued*
    1,2-CYCLOHEXANEDIONE, 229
    o-HYDROXYDIBENZOYLMETHANE, 479
    SEBACIL, 838
  C. TETRAKETONES
    TETRAACETYLETHANE, 869
LACTONES
  α-ACETYL-δ-CHLORO-γ-VALEROLACTONE, 10
  γ-CAPRILACTONE, 432
  COUMALIC ACID, 201
  4,6-DIMETHYLCOUMALIN, 337
  4-ETHYL-4-HYDROXY-2-METHYLOCTANOIC ACID γ-LACTONE, 447
  ETHYL ISODEHYDROACETATE, 549
  D-GULONIC-γ-LACTONE, 506
  ISODEHYDROACETIC ACID, 549
  METHYL COUMALATE, 532
  β-METHYL-δ-VALEROLACTONE, 677
NITRILES *See also* Acids, *B*,2; Esters, *B*,4.
  A. UNSUBSTITUTED
    AZELANITRILE, 62
    3-BENZYL-3-METHYLPENTANENITRILE, 95
    1-CYCLOHEXENYLACETONITRILE, 234
    1,1'-DICYANO-1,1'-BICYCLOHEXYL, 273
    2-ETHYLHEXANONITRILE, 436
    3-ETHYL-3-METHYLHEXANENITRILE, 97
    FUMARONITRILE, 486
    3-METHYL-3-PHENYLPENTANENITRILE, 97
    PALMITONITRILE, 437
    TETRACYANOETHYLENE, 877
  B. SUBSTITUTED
    1,1'-AZO-*bis*-1-CYCLOHEXANENITRILE, 66
    2,2'-AZO-*bis*-ISOBUTYRONITRILE, 67
    2-BENZOYL-1-CYANO-1-METHYL-1,2-DIHYDROISOQUINOLINE, 642
    CHLOROACETONITRILE, 144
    3-(*o*-CHLOROANILINO)PROPIONITRILE, 146
    2-CHLORONICOTINONITRILE, 166
    α-(4-CHLOROPHENYL)-γ-PHENYLACETOACETONITRILE, 174
    CYANAMIDE, 645
    1-CYANO-2-BENZOYL-1,2-DIHYDROISOQUINOLINE, 641
    N-2-CYANOETHYLANILINE, 205
    CYANOGEN IODIDE, 207
    3-CYANO-6-METHYL-2(1)-PYRIDONE, 210
    1-CYANO-3-α-NAPHTHYLUREA, 215
    1-CYANO-3-PHENYLUREA, 213
    CYCLOHEXYLIDENECYANOACETIC ACID, 234
    2,5-DIAMINO-3,4-DICYANOTHIOPHENE, 243
    DIBROMOACETONITRILE, 254
    DICHLOROACETONITRILE, 255

NITRILES—*Continued*
  B. SUBSTITUTED—*Continued*
    1,2-DI-1-(1-CYANO)CYCLOHEXYLHYDRAZINE, 274
    α,α'-DICYANO-β-ETHYL-β-METHYLGLUTARIMIDE, 441
    DICYANOKETENE ETHYLENE ACETAL, 276
    1,2-DI-2-(2-CYANO)PROPYLHYDRAZINE, 275
    α,β-DIPHENYLCINNAMONITRILE, 387
    2,3-DIPHENYLSUCCINONITRILE, 392
    3-*n*-HEPTYL-5-CYANOCYTOSINE, 515
    3-*n*-HEPTYLUREIDOMETHYLENEMALONONITRILE, 515
    2-MERCAPTO-4-HYDROXY-5-CYANOPYRIMIDINE, 566
    p-METHOXYPHENYLACETONITRILE, 576
    β-METHYL-β-PHENYL-α,α'-DICYANOGLUTARIMIDE, 662
    NICOTINONITRILE, 706
    TETRAMETHYLSUCCINONITRILE, 273
    p-TRICYANOVINYL-N,N-DIMETHYLANILINE, 953
    α,α,β-TRIPHENYLPROPIONITRILE, 962
NITRO COMPOUNDS *See also* Acids, *B*,6.
  A. UNSUBSTITUTED
    p,*p*'-DINITROBIBENZYL, 367
    1,4-DINITROBUTANE, 368
    2,3-DINITRO-2-BUTENE, 374
    1,6-DINITROHEXANE, 370
    3,4-DINITRO-3-HEXENE, 372
    1,5-DINITROPENTANE, 370
    1,3-DINITROPROPANE, 370
    9-NITROANTHRACENE, 711
    m-NITROBIPHENYL, 718
    1-NITROÖCTANE, 724
    m-NITROSTYRENE, 731
  B. SUBSTITUTED
    1. Nitro Amines
      2-AMINO-3-NITROTOLUENE, 42
      2-AMINO-5-NITROTOLUENE, 44
      2,6-DINITROANILINE, 364
    2. Miscellaneous
      BENZOYL-2-METHOXY-4-NITROACETANILIDE, 82
      2-BROMO-4-NITROTOLUENE, 114
      1-CHLORO-2,6-DINITROBENZENE, 160
      ETHYL α-NITROBUTYRATE, 454
      1-(*o*-METHOXYPHENYL)-2-NITRO-1-PROPENE, 573
      METHYL γ-METHYL-γ-NITROVALERATE, 652
      3-METHYL-4-NITROPYRIDINE-1-OXIDE, 654
      o-NITROACETOPHENONE, 708
      o- AND *p*-NITROBENZALDIACETATE, 713
      m-NITROBENZAZIDE, 715
      m-NITROBENZOYL CHLORIDE, 715

NITRO COMPOUNDS—*Continued*
B. *SUBSTITUTED—Continued*
  2. Miscellaneous—*Continued*
    o-Nitrocinnamaldehyde, 722
    o-Nitrophenylazide, 75
    1-(*p*-Nitrophenyl)-1,3-butadi-
      ene, 727
    1-(*p*-Nitrophenyl)-4-chloro-2-
      butene, 727
    1-(*m*-Nitrophenyl)-3,3-di-
      methyltriazine, 718
    6-Nitroveratraldehyde, 735
    Sodium Nitromalonaldehyde
      Monohydrate, 844
NITROSO COMPOUNDS
  Ethyl N-Nitroso-N-benzylcarba-
    mate, 780
  p-Tolylsulfonylmethylnitrosamide,
    943
ORGANO-METALLIC COMPOUNDS
  Allylmagnesium Bromide, 749
  di-*n*-Butyldivinyltin, 258
  Ethynylmagnesium Bromide, 792
  Ferrocene, 473
  Tetraethyltin, 881
OXIMES
  1,2-Cyclohexanedione Dioxime, 229
PHENOLS *See also* Acids, *B*,4; Ketones,
  A,3.
  p-Chlorophenyl Salicylate, 178
  Creosol, 203
  Syringic Aldehyde, 866
  ar-Tetrahydro-*α*-naphthol, 887
PHOSPHORUS COMPOUNDS
  Diethyl Ethylphosphonate, 326
  Diethyl Methylphosphonate, 326
  Diisopropyl Ethylphosphonate, 326
  Diisopropyl Methylphosphonate,
    325
  Phenyldichlorophosphine, 784
  Trichloromethylphosphonyl Dichlo-
    ride, 950
  Triethyl Phosphite, 955
QUATERNARY AMMONIUM COM-
  POUNDS
  Benzoylcholine Chloride, 84
  Benzoylcholine Iodide, 84
  Benzyltrimethylammonium Ethox-
    ide, 98
  Benzyltrimethylammonium Iodide,
    585
  N-Chlorobetainyl Chloride, 154
QUINONES
  2-*p*-Acetylphenylquinone, 16
  p-Benzoquinone, 152
  Chloro-*p*-benzoquinone, 148
  3-Chlorotoluquinone, 152
  2,5-Dichloroquinone, 152
  Methoxyquinone, 153
  1,4-Naphthoquinone, 698
  Phenanthrenequinone, 757
SELENIUM COMPOUNDS
  N,N-Dimethylselenourea, 359

SULFUR COMPOUNDS *See also* Hetero-
  cyclic compounds.
A. *DISULFIDES*
  Methyl 2-Thienyl Sulfide, 667
B. *MERCAPTANS*
  Diethyl Mercaptoacetal, 295
  Ethanedithiol, 401
  2-Furfuryl Mercaptan, 491
  2-Mercapto-4-amino-5-carbeth-
    oxypyrimidine, 566
  2-Mercaptobenzimidazole, 569
  2-Mercaptobenzoxazole, 570
  2-Mercapto-4-hydroxy-5-cyano-
    pyrimidine, 566
  2-Thio-6-methyluracil, 638
C. *SULFIDES*
  Methyl *β*-Thiodipropionate, 669
  1,1,1',1'-Tetraethoxyethyl Poly-
    sulfide, 295
D. *SULFONES*
  Methyl *p*-Tolyl Sulfone, 674
E. *SULFONIC ACIDS AND ESTERS*
  4-Hydroxy-1-butanesulfonic Acid
    Sultone, 529
  Sodium 2-Phenyl-2-hydroxy-
    ethane-1-sulfonate, 850
  Sodium *β*-Styrenesulfonate, 846
  α-Sulfopalmitic Acid, 862
F. *THIOPHENOLS*
  1,5-Naphthalenedithiol, 695
G. *THIOUREA DERIVATIVES*
  o-Chlorophenylthiourea, 180
  2,4-Dithiobiuret, 504
  Guanylthiourea, 502
H. *MISCELLANEOUS*
  Diethylthiocarbamyl Chloride,
    307
  Methanesulfonyl Chloride, 571
  Naphthalene-1,5-disulfonyl
    Chloride, 693
  α-Naphthyl Isothiocyanate, 700
  β-Styrenesulfonyl Chloride, 846
  Thiobenzoic Acid, 924
  Thiobenzophenone, 927
  Thiolacetic Acid, 928
  p-Toluenesulfenyl Chloride,
    934
  p-Toluenesulfinyl Chloride, 937
  p-Toluenesulfonic Anhydride,
    940
  p-Toluenesulfonylanthranilic
    Acid, 34
  p-Tolylsulfonylmethylnitros-
    amide, 943
  Trithiocarbodiglycolic Acid, 967
UREA DERIVATIVES
  p-Bromophenylurea, 49
  1-Cyano-3-*α*-naphthylurea, 215
  1-Cyano-3-phenylurea, 213
  N,N-Dimethylselenourea, 359
  asym-Dimethylurea, 361
  p-Ethoxyphenylurea, 52
  N-*n*-Heptylurea, 515

UREA DERIVATIVES—*Continued*
  3-*n*-HEPTYLUREIDOMETHYLENEMALONO-
    NITRILE, 515
  p-METHOXYPHENYLUREA, 53
  METHYLISOUREA HYDROCHLORIDE, 645
UNCLASSIFIED
  ADIPYL HYDRAZIDE, 819
  N-BENZYLIDENEMETHYLAMINE, 605

UNCLASSIFIED—*Continued*
  N,N-DIMETHYLHYDROXYLAMINE HYDRO-
    CHLORIDE, 612
  METHYLGLYOXAL-ω-PHENYLHYDRAZONE,
    633
  1-(*m*-NITROPHENYL)-3,3-DIMETHYLTRI-
    AZENE, 718
  TETRALIN HYDROPEROXIDE, 895

FORMULA INDEX

All preparations listed in the Contents are recorded in this index. The system of indexing is that used by *Chemical Abstracts*. The essential principles involved are as follows: (1) The arrangement of symbols in formulas is alphabetical except that in carbon compounds C always comes first, followed immediately by H if hydrogen is also present. (2) The arrangement of formulas is also alphabetical except that the number of atoms of any specific kind influences the order of compounds: e.g., all formulas with one carbon atom precede those with two carbon atoms, thus: CH_2I_2, CH_3NO_2, CH_5N, C_2H_2O. (3) The arrangement of entries under any heading is strictly alphabetical according to the names of the isomers. (4) Inorganic salts of organic acids and inorganic addition compounds of organic compounds are listed under the formulas of the compounds from which they are derived.

CCl_5OP TRICHLOROMETHYLPHOSPHONYL DICHLORIDE, 950

CIN CYANOGEN IODIDE, 207

CH_2N_2 DIAZOMETHANE, 250

CH_3ClO_2S METHANESULFONYL CHLORIDE, 571

$C_2Cl_2F_2$ 1,1-DICHLORO-2,2-DIFLUOROETHYLENE, 268

C_2HBr_2N DIBROMOACETONITRILE, 254

C_2H_2ClN CHLOROACETONITRILE, 144

C_2H_4BrNO N-BROMOACETAMIDE, 104

$C_2H_4Cl_2O$ BISCHLOROMETHYL ETHER, 101

$C_2H_4Cl_2O$ 2,2-DICHLOROETHANOL, 271

C_2H_4OS THIOLACETIC ACID, 928

C_2H_5N ETHYLENIMINE, 433

$C_2H_6N_2O$ METHYLISOUREA HYDROCHLORIDE, 645

$C_2H_6N_4S$ GUANYLTHIOUREA, 502

$C_2H_6S_2$ ETHANEDITHIOL, 401

C_2H_7NO N,N-DIMETHYLHYDROXYLAMINE HYDROCHLORIDE, 612

C_3H_2O PROPIOLALDEHYDE, 813

$C_3H_2N_2O_3$ PARABANIC ACID, 744

$C_3H_3NO_4$ SODIUM NITROMALONALDEHYDE MONOHYDRATE, 844

$C_3H_8N_2O$ asym-DIMETHYLUREA, 361

$C_3H_8N_2Se$ N,N-DIMETHYLSELENOUREA, 359

$C_4H_2Br_2O_3$ MUCOBROMIC ACID, 688

$C_4H_2N_2$ FUMARONITRILE, 486

$C_4H_2N_2O_4$ ALLOXAN MONOHYDRATE, 23

$C_4H_3ClN_2$ 2-CHLOROPYRIMIDINE, 182

C_4H_3IS 2-IODOTHIOPHENE, 545

C_4H_4 MONOVINYLACETYLENE, 683

$C_4H_5Cl_3O$ BUTYRCHLORAL, 130

$C_4H_6ClFO_2$ ETHYL CHLOROFLUOROACETATE, 423

$C_4H_6ClF_3O$ 2-CHLORO-1,1,2-TRIFLUOROETHYL ETHYL ETHER, 184

$C_4H_6N_2O$ DIMETHYLFURAZAN, 342

$C_4H_6N_2O_2$ ETHYL DIAZOACETATE, 424

$C_4H_6N_4O$ 2,4-DIAMINO-6-HYDROXYPYRIMIDINE, 245

$C_4H_6N_4O_2$ DIAMINOURACIL HYDROCHLORIDE, 247

C_4H_6O 2-BUTYN-1-OL, 128

DIMETHYLKETENE, 348

ETHOXYACETYLENE, 404

$C_4H_7NO_4$ dl-ASPARTIC ACID, 55

$C_4H_8I_2$ 1,4-DIIODOBUTANE, 321

$C_4H_8N_2O_4$ 1,4-DINITROBUTANE, 368

$C_4H_8O_2$ 3-HYDROXYTETRAHYDROFURAN, 534

$C_4H_8O_3S$ 4-HYDROXY-1-BUTANESULFONIC ACID SULTONE, 529

C_4H_8S TETRAHYDROTHIOPHENE, 892

$C_4H_8S_2$ p-DITHIANE, 396

C_4H_9ClO tert-BUTYL HYPOCHLORITE, 125

$C_4H_{10}ClN$ β-DIMETHYLAMINOETHYL CHLORIDE HYDROCHLORIDE, 333

$C_4H_{12}N_2$ PUTRESCINE DIHYDROCHLORIDE, 819

$C_5H_3N_3OS$ 2-MERCAPTO-4-HYDROXY-5-CYANO-PYRIMIDINE, 566

$C_5H_4Cl_2O_2$ ITACONYL CHLORIDE, 554

C_5H_4OS 2-THENALDEHYDE, 915

3-THENALDEHYDE, 918

$C_5H_4O_2S$ 3-THENOIC ACID, 919

$C_5H_4O_3$ 2-FUROIC ACID, 493

C_5H_5BrS 3-THENYL BROMIDE, 921

C_5H_5NO PYRIDINE-N-OXIDE, 828

2-PYRROLEALDEHYDE, 831

C_5H_6 CYCLOPENTADIENE, 238

$C_5H_6N_2$ 3-AMINOPYRIDINE, 45

$C_5H_6N_2O$ 4-METHYL-6-HYDROXYPYRIMIDINE, 638

C_5H_6O 3-METHYLFURAN, 628

C_5H_6OS 2-FURFURYL MERCAPTAN, 491

$C_5H_6O_4S_3$ TRITHIOCARBODIGLYCOLIC ACID, 967

C_5H_6S 3-METHYLTHIOPHENE, 671

$C_5H_6S_2$ METHYL 2-THIENYL SULFIDE, 667

C_5H_7Cl 3-CHLOROCYCLOPENTENE, 238

$C_5H_7NO_2$ GLUTARIMIDE, 496

C_5H_8 1,4-PENTADIENE, 746

$C_5H_8Br_4$ PENTAERYTHRITYL TETRABROMIDE, 753

$C_5H_8N_2$ 3,5-DIMETHYLPYRAZOLE, 351

C_5H_8O METHYL CYCLOPROPYL KETONE, 597

4-PENTYN-1-OL, 755

$C_5H_8O_3$ ETHYL PYRUVATE, 467

$C_5H_8O_4$ GLUTARIC ACID, 496

C_5H_9ClO trans-2-CHLOROCYCLOPENTANOL, 157

1007

C₅H₁₀ClNS Diethylthiocarbamyl Chloride, 307

C₅H₁₁BrO₃ Monobromopentaerythritol, 681

C₅H₁₁Cl₂NO N-Chlorobetainyl Chloride, 154

C₆H₃ClN₂ 2-Chloronicotinonitrile, 166

C₆H₃ClN₂O₄ 1-Chloro-2,6-dinitrobenzene, 160

C₆H₃ClO₂ Chloro-p-benzoquinone, 148

C₆H₄N₂ Nicotinonitrile, 706

C₆H₄N₂O₂ Benzofurazan Oxide, 74
Dicyanoketene Ethylene Acetal, 276

C₆H₄N₂O₄ 2,3-Pyrazinedicarboxylic Acid, 824

C₆H₄N₄S 2,5-Diamino-3,4-dicyanothiophene, 243

C₆H₄O₄ Coumalic Acid, 201

C₆H₅Cl₂P Phenyldichlorophosphine, 784

C₆H₅NO₃ 6-Hydroxynicotinic Acid, 532

C₆H₅N₃O₄ 2,6-Dinitroaniline, 364

C₆H₆N₂O₂ Nicotinamide-1-oxide, 704

C₆H₆N₂O₃ 3-Methyl-4-nitropyridine-1-oxide, 654

C₆H₆O₃ 3-Methyl-2-furoic Acid, 628

C₆H₆O₄ Dimethyl Acetylenedicarboxylate, 329

C₆H₆S 2-Vinylthiophene, 980

C₆H₈O₃ β-Methylglutaric Anhydride, 630

C₆H₉N₃ 2-(Dimethylamino)pyrimidine, 336

C₆H₁₀ n-Butylacetylene, 117

C₆H₁₀N₂O₂ 1,2-Cyclohexanedione Dioxime, 229

C₆H₁₀N₂O₄ 3,4-Dinitro-3-hexene, 372
Ethyl Azodicarboxylate, 411

C₆H₁₀O₂ β-Methyl-δ-valerolactone, 677

C₆H₁₀O₃ n-Butyl Glyoxylate, 124

C₆H₁₀O₆ D-Gulonic-γ-lactone, 506

C₆H₁₀S Cyclohexene Sulfide, 232

C₆H₁₁I Iodocyclohexane, 543

C₆H₁₁NO 5,5-Dimethyl-2-pyrrolidone, 357

C₆H₁₁NO₂ 3-Acetamido-2-butanone, 5

C₆H₁₁NO₄ Ethyl α-nitrobutyrate, 454

C₆H₁₂I₂ 1,6-Diiodohexane, 323

C₆H₁₃F n-Hexyl Fluoride, 525

C₆H₁₃N 2,2-Dimethylpyrrolidine, 354

C₆H₁₃NO₂ ε-Aminocaproic Acid, 39

C₆H₁₄O₂ 3-Methyl-1,5-pentanediol, 660

C₆H₁₄O₂S Diethyl Mercaptoacetal, 295

C₆H₁₅O₃P Triethyl Phosphite, 955

C₆N₄ Tetracyanoethylene, 877

C₇H₃Br₃O₂ 2,4,6-Tribromobenzoic Acid, 947

C₇H₄N₄O₃ m-Nitrobenzazide, 715

C₇H₄O₃ o-Phenylene Carbonate, 788

C₇H₅ClN₂ o-Chlorophenylcyanamide, 172

C₇H₆N₂ Indazole, 536

C₇H₆N₂O 3-Cyano-6-methyl-2(1)-pyridone, 210

C₇H₆N₂S 2-Mercaptobenzimidazole, 569

C₇H₆OS Thiobenzoic Acid, 924

C₇H₇BrN₂O p-Bromophenylurea, 49

C₇H₇ClN₂S o-Chlorophenylthiourea, 180

C₇H₇ClOS p-Toluenesulfinyl Chloride, 937

C₇H₇ClS p-Toluenesulfenyl Chloride, 934

C₇H₇NO p-Aminobenzaldehyde, 31

C₇H₈N₂O₂ 2-Amino-3-nitrotoluene, 42

C₇H₈O₃ 4,6-Dimethylcoumalin, 337

C₇H₈O₃ Methyl 3-Methyl-2-furoate, 649

C₇H₉ClO₃ α-Acetyl-δ-chloro-γ-valerolactone, 10

C₇H₉N₃O₂S 2-Mercapto-4-amino-5-carbethoxypyrimidine, 566

C₇H₁₀ Norbornylene, 738

C₇H₁₀O Dicyclopropyl Ketone, 278
2-Methyl-2-cyclohexenone, 162

C₇H₁₁ClO 2-Chloro-2-methylcyclohexanone, 162

C₇H₁₂ Methylenecyclohexane, 612

C₇H₁₂O Cycloheptanone, 221

C₇H₁₂O₂ 3,4-Dihydro-2-methoxy-4-methyl-2H-pyran, 311
Methyl Cyclopentanecarboxylate, 594

C₇H₁₂O₃ δ-Acetyl-n-valeric Acid, 19

C₇H₁₂O₄ β,β-Dimethylglutaric Acid, 345

C₇H₁₃Br 4-Bromo-2-heptene, 108

C₇H₁₃NO O-Methylcaprolactim, 588

C₇H₁₃NO₄ Methyl γ-Methyl-γ-nitrovalerate, 652

C₇H₁₄O₂ Acrolein Acetal, 21

C₇H₁₅NO n-Heptamide, 31
2-(1-Pyrrolidyl)propanol, 834

C₇H₁₇O₃P Diisopropyl Methylphosphonate, 325

C₈H₆ Phenylacetylene, 763

C₈H₆N₄O₈ Alloxantin Dihydrate, 25

C₈H₆O₂ o-Phthalaldehyde, 807

C₈H₇BrO₂ 2-Bromo-3-methylbenzoic Acid, 114

C₈H₇BrO₃ p-Bromomandelic Acid, 110

C₈H₇ClO₂ α-Chlorophenylacetic Acid, 169

C₈H₇ClO₂S β-Styrenesulfonyl Chloride, 846

C₈H₇NO₂ m-Nitrostyrene, 731

C₈H₇NO₃ o-Nitroacetophenone, 708

C₈H₇N₃O 1-Cyano-3-phenylurea, 213

C₈H₈Br₂ o-Xylylene Dibromide, 984

C₈H₈O o-Tolualdehyde, 932

C₈H₈O₃ cis-Δ⁴-Tetrahydrophthalic Anhydride, 890

C₈H₈O₃S Sodium β-Styrenesulfonate, 846

C₈H₈O₄ Isodehydroacetic Acid, 549
Vanillic Acid, 972

C₈H₈O₇ Diacetyl-d-tartaric Anhydride, 242

C₈H₉IO₂ 4-Iodoveratrole, 547

C₈H₉NO Phenylacetamide, 760

C₈H₁₀ClN N-Ethyl-p-chloroaniline, 420

C₈H₁₀N₂O₃S p-Tolylsulfonylmethylnitrosamide, 943

C₈H₁₀O o-Methylbenzyl Alcohol, 582

C₈H₁₀O₂ Creosol, 203

C₈H₁₀O₂S Methyl p-Tolyl Sulfone, 674

C₈H₁₁N 1-Cyclohexenylacetonitrile, 234
5-Ethyl-2-methylpyridine, 451

C₈H₁₂N₂O₂ Hexamethylene Diisocyanate, 521

C₈H₁₂O₄ DIETHYL METHYLENEMALONATE, 298
C₈H₁₃ClO β-CHLOROVINYL ISOAMYL KETONE, 186
C₈H₁₃NO 5-METHYLFURFURYLDIMETHYLAMINE, 626
C₈H₁₄O 2,4,4-TRIMETHYLCYCLOPENTANONE, 957
C₈H₁₄O₂ 1-ACETYLCYCLOHEXANOL, 13
C₈H₁₄O₄ β-ETHYL-β-METHYLGLUTARIC ACID, 441
C₈H₁₄O₄S METHYL β-THIODIPROPIONATE, 669
C₈H₁₅N 2-ETHYLHEXANONITRILE, 436
C₈H₁₆Cl₂O 4,4′-DICHLORODIBUTYL ETHER, 266
C₈H₁₆O₄ ETHYL DIETHOXYACETATE, 427
C₈H₁₇NO 1-DIETHYLAMINO-3-BUTANONE, 281
C₈H₁₇NO₂ 1-NITROÖCTANE, 724
C₈H₂₀Sn TETRAETHYLTIN, 881
C₉H₇NO INDOLE-3-ALDEHYDE, 539
C₉H₇NO₃ o-NITROCINNAMALDEHYDE, 722
C₉H₈O 3-METHYLCOUMARONE, 590
C₉H₈ClN₂ 3-(o-CHLOROANILINO)PROPIONITRILE, 146
C₉H₉NO p-METHOXYPHENYLACETONITRILE, 576
3-METHYLOXINDOLE, 657
C₉H₉NO₅ 6-NITROVERATRALDEHYDE, 735
C₉H₉N₅ BENZOGUANAMINE, 78
C₉H₁₀N₂ N-2-CYANOETHYLANILINE, 205
C₉H₁₀N₂O METHYLGLYOXAL-ω-PHENYLHYDRAZONE, 633
C₉H₁₀O₃ ATROLACTIC ACID, 58
QUINACETOPHENONE MONOMETHYL ETHER, 836
C₉H₁₀O₄ SYRINGIC ALDEHYDE, 866
C₉H₁₀O₅ FURFURAL DIACETATE, 489
C₉H₁₁NO p-DIMETHYLAMINOBENZALDEHYDE, 331
ETHYL N-PHENYLFORMIMIDATE, 464
C₉H₁₁NO₂ CYCLOHEXYLIDENECYANOACETIC ACID, 234
C₉H₁₂ HEMIMELLITENE, 508
C₉H₁₂N₂O₂ p-ETHOXYPHENYLUREA, 52
C₉H₁₂O 3-PHENYL-1-PROPANOL, 798
C₉H₁₄N₂ AZELANITRILE, 62
C₉H₁₄O₂ ISOPHORONE OXIDE, 552
C₉H₁₄O₄ DIETHYL ETHYLIDENEMALONATE, 293
C₉H₁₅NO PSEUDOPELLETIERINE, 816
C₉H₁₆O₄ ETHYL tert-BUTYL MALONATE, 417
C₉H₁₇NO₂ ETHYL α-(1-PYRROLIDYL)PROPIONATE, 466
C₉H₁₉N N,N-DIMETHYLCYCLOHEXYLMETHYLAMINE, 339
C₉H₂₀O₄ ETHYL ORTHOCARBONATE, 457
C₁₀H₆Cl₂O₄S₂ NAPHTHALENE-1,5-DISULFONYL CHLORIDE, 693
C₁₀H₆O₂ 1,4-NAPHTHOQUINONE, 698
C₁₀H₈BrNO₂ β-BROMOETHYLPHTHALIMIDE, 106
C₁₀H₈S₂ 1,5-NAPHTHALENEDITHIOL, 695
C₁₀H₉N 1-METHYLISOQUINOLINE, 641
C₁₀H₉NO₂ 1-(p-NITROPHENYL)-1,3-BUTADIENE, 727
C₁₀H₁₀ trans-1-PHENYL-1,3-BUTADIENE, 771
C₁₀H₁₀ClN₅ 2-AMINO-4-ANILINO-6-(CHLOROMETHYL)-s-TRIAZINE, 29
C₁₀H₁₀Fe FERROCENE, 473

C₁₀H₁₀O α-TETRALONE, 898
β-TETRALONE, 903
C₁₀H₁₀O₃ METHYL p-ACETYLBENZOATE, 579
C₁₀H₁₀O₄ β-(o-CARBOXYPHENYL)PROPIONIC ACID, 136
PHENYLSUCCINIC ACID, 804
C₁₀H₁₂O ar-TETRAHYDRO-α-NAPHTHOL, 887
C₁₀H₁₂O₂ o-METHOXYPHENYLACETONE, 573
4-PHENYL-m-DIOXANE, 786
TETRALIN HYDROPEROXIDE, 895
C₁₀H₁₂O₄ ETHYL ISODEHYDROACETATE, 549
C₁₀H₁₃Cl NEOPHYL CHLORIDE, 702
C₁₀H₁₄O₄ TETRAACETYLETHANE, 869
C₁₀H₁₅N 2-METHYLBENZYLDIMETHYLAMINE, 585
C₁₀H₁₅NO₂ N-METHYL-2,3-DIMETHOXYBENZYLAMINE, 603
C₁₀H₁₆O₂ SEBACIL, 838
C₁₀H₁₆O₃ ETHYL β,β-PENTAMETHYLENEGLYCIDATE, 459
C₁₀H₁₆O₄ DIETHYL 1,1-CYCLOBUTANEDICARBOXYLATE, 288
C₁₀H₁₈O CYCLODECANONE, 218
C₁₀H₁₈O₂ SEBACOIN, 840
C₁₀H₂₀O₂ 1,2-CYCLODECANEDIOL, 216
C₁₀H₂₀O₃ β-KETOISOÖCTALDEHYDE DIMETHYL ACETAL, 558
C₁₀H₂₀O₅ 2,2,6,6-TETRAMETHYLOLCYCLOHEXANOL, 907
C₁₁H₇NS α-NAPHTHYL ISOTHIOCYANATE, 700
C₁₁H₈O 1-NAPHTHALDEHYDE, 690
C₁₁H₁₀N₂O₂S₂ N-(p-ACETYLPHENYL)RHODANINE, 6
C₁₁H₁₀O 1-PHENYL-1-PENTEN-4-YN-3-OL, 792
C₁₁H₁₀O₃ α-PHENYLGLUTARIC ANHYDRIDE, 790
C₁₁H₁₁NO₂ ETHYL PHENYLCYANOACETATE, 461
C₁₁H₁₁NO₆ o-NITROBENZALDIACETATE, 713
p-NITROBENZALDIACETATE, 713
C₁₁H₁₂O₃ ETHYL BENZOYLACETATE, 415
C₁₁H₁₂O₄ 2,3-DIMETHOXYCINNAMIC ACID, 327
C₁₁H₁₃NO₂ 1-METHYL-3-ETHYLOXINDOLE, 620
C₁₁H₁₅N 1-PHENYLPIPERIDINE, 795
C₁₁H₁₈O₂ 10-UNDECYNOIC ACID, 969
C₁₁H₁₈O₅ DIETHYL γ-OXOPIMELATE, 302
6-KETOHENDECANEDIOIC ACID, 555
C₁₁H₂₀O₂ 4-ETHYL-2-METHYL-2-OCTENOIC ACID, 444
C₁₁H₂₀O₄ DI-tert-BUTYL MALONATE, 261
HENDECANEDIOIC ACID, 510
C₁₁H₂₂O 5,5-DIMETHYL-2-n-PENTYLTETRAHYDROFURAN, 350
C₁₁H₂₄O₂ 2-METHYL-2,5-DECANEDIOL, 601
C₁₂H₈Br₂ 4,4′-DIBROMOBIPHENYL, 256
C₁₂H₉NO 3-BENZOYLPYRIDINE, 88
C₁₂H₉NO₂ m-NITROBIPHENYL, 718
C₁₂H₁₂N₂ 2-BENZYLAMINOPYRIDINE, 91
C₁₂H₁₃N 1,2,3,4-TETRAHYDROCARBAZOLE, 884
C₁₂H₁₄O₄ β-METHYL-β-PHENYLGLUTARIC ACID, 664
C₁₂H₁₆O₄ MONOBENZALPENTAERYTHRITOL, 679
C₁₂H₁₈ HEXAMETHYLBENZENE, 520
C₁₂H₁₈ClNO₂ BENZOYLCHOLINE CHLORIDE, 84
C₁₂H₁₈INO₂ BENZOYLCHOLINE IODIDE, 84
C₁₂H₁₈N₄O 3-n-HEPTYL-5-CYANOCYTOSINE, 515

$C_{12}H_{18}O_4$ DIETHYL Δ^2-CYCLOPENTENYLMALO-
NATE, 291
DIETHYL cis-Δ^4-TETRAHYDROPHTHA-
LATE, 304
$C_{12}H_{20}O_4$ DIETHYL cis-HEXAHYDROPHTHALATE,
304
$C_{12}H_{21}NO$ BENZYLTRIMETHYLAMMONIUM ETH-
OXIDE, 98
$C_{12}H_{21}N_3O_3$ HEXAHYDRO-1,3,5-TRIPROPIONYL-s-
TRIAZINE, 518
$C_{12}H_{22}O_2$ trans-2-DODECENOIC ACID, 398
$C_{12}H_{22}O_4$ METHYL HYDROGEN HENDECANE-
DIOATE, 635
$C_{12}H_{24}Sn$ DI-n-BUTYLDIVINYLTIN, 258
$C_{13}H_9ClO_3$ p-CHLOROPHENYL SALICYLATE, 178
$C_{13}H_{10}N_4$ p-TRICYANOVINYL-N,N-DIMETHYL-
ANILINE, 953
$C_{13}H_{10}S$ THIOBENZOPHENONE, 927
$C_{13}H_{11}NO$ 2-AMINOBENZOPHENONE, 34
$C_{13}H_{12}N_2$ N-PHENYLBENZAMIDINE, 769
$C_{13}H_{13}NO_2$ ETHYL (1-PHENYLETHYLIDENE)-
CYANOACETATE, 463
$C_{13}H_{14}O_4$ γ-PHENYLALLYLSUCCINIC ACID, 766
$C_{13}H_{16}O$ 2-PHENYLCYCLOHEPTANONE, 780
$C_{13}H_{16}O_2$ PHENYLPROPARGYLALDEHYDE DI-
ETHYL ACETAL, 801
$C_{13}H_{18}O_2$ 3-BENZYL-3-METHYLPENTANOIC ACID,
93
$C_{13}H_{24}O_2$ trans-2-METHYL-2-DODECENOIC ACID,
608
2-METHYLENEDODECANOIC ACID, 616
$C_{13}H_{24}O_3$ sec-BUTYL α-n-CAPROYLPROPIONATE,
120
$C_{13}H_{29}N$ LAURYLMETHYLAMINE, 564
$C_{14}H_6Cl_4O_4$ dl-4,4',6,6'-TETRACHLORODIPHENIC
ACID, 872
$C_{14}H_8O_2$ PHENANTHRENEQUINONE, 757
$C_{14}H_9NO_2$ 9-NITROANTHRACENE, 711
$C_{14}H_{10}$ DIPHENYLACETYLENE, 377
$C_{14}H_{10}O_2$ 9-FLUORENECARBOXYLIC ACID, 482
$C_{14}H_{11}N_3O_2$ β-METHYL-β-PHENYL-α,α'-DICY-
ANOGLUTARIMIDE, 662
$C_{14}H_{12}$ 9,10-DIHYDROPHENANTHRENE, 313
9-METHYLFLUORENE, 623
cis-STILBENE, 857
$C_{14}H_{12}N_2O_4$ p,p'-DINITROBIBENZYL, 367
$C_{14}H_{12}N_4$ 1,4-DIPHENYL-5-AMINO-1,2,3-TRI-
AZOLE, 380
4-PHENYL-5-ANILINO-1,2,3-TRIAZOLE,
380
$C_{14}H_{12}O$ DIPHENYLACETALDEHYDE, 375
trans-STILBENE OXIDE, 860
$C_{14}H_{12}O_3$ 2-p-ACETYLPHENYLHYDROQUINONE,
15
$C_{14}H_{14}N_2O_2$ α-PHENYL-α-CARBETHOXYGLU-
TARONITRILE, 776
$C_{14}H_{14}O_5S_2$ p-TOLUENESULFONIC ANHYDRIDE,
940
$C_{14}H_{16}O_5$ DIETHYL BENZOYLMALONATE, 285

$C_{14}H_{20}N_2$ 1,1'-DICYANO-1,1'-BICYCLOHEXYL,
273
$C_{14}H_{20}N_4$ 1,1'-AZO-bis-1-CYCLOHEXANENITRILE,
66
$C_{14}H_{22}N_4$ 1,2-DI-1-(1-CYANO)-CYCLOHEXYL-
HYDRAZINE, 274
$C_{14}H_{22}O_2$ 1,1'-ETHYNYLENE-bis-CYCLOHEXANOL,
471
$C_{14}H_{26}O_2$ VINYL LAURATE, 977
$C_{15}H_{10}O_2$ FLAVONE, 478
$C_{15}H_{11}NO_4$ trans-o-NITRO-α-PHENYLCINNAMIC
ACID, 730
$C_{15}H_{12}O_2$ α-PHENYLCINNAMIC ACID, 777
$C_{15}H_{13}NO_2$ BENZOYLACETANILIDE, 80
$C_{15}H_{15}ClO$ BENZHYDRYL β-CHLOROETHYL
ETHER, 72
$C_{15}H_{17}N$ N-METHYL-1,2-DIPHENYLETHYLAMINE,
605
N-METHYL-1,2-DIPHENYLETHYLAMINE
HYDROCHLORIDE, 605
$C_{15}H_{20}O_5$ ETHYL α-ACETYL-β-(2,3-DIMETHOXY-
PHENYL)PROPIONATE, 408
$C_{15}H_{26}O_5$ ETHYL ENANTHYLSUCCINATE, 430
$C_{16}H_{10}O_3$ 5-FORMYL-4-PHENANTHROIC ACID,
484
$C_{16}H_{11}NO_4$ α-PHTHALIMIDO-o-TOLUIC ACID,
810
$C_{16}H_{12}ClNO$ α-(4-CHLOROPHENYL)-γ-PHENYL-
ACETOACETONITRILE, 174
$C_{16}H_{12}N_2$ 2,3-DIPHENYLSUCCINONITRILE, 392
$C_{16}H_{12}O$ 9-ACETYLANTHRACENE, 8
$C_{16}H_{14}O_4$ DIPHENYL SUCCINATE, 390
$C_{16}H_{16}$ 1-METHYL-3-PHENYLINDANE, 665
$C_{16}H_{20}N_2$ N,N'-DIETHYLBENZIDINE, 283
$C_{16}H_{32}O_5S$ α-SULFOPALMITIC ACID, 862
$C_{18}H_{15}B_3O_3$ BENZENEBORONIC ANHYDRIDE, 68
$C_{18}H_{32}O_2$ STEAROLIC ACID, 851
$C_{18}H_{33}ClO$ OLEOYL CHLORIDE, 739
$C_{18}H_{36}O_4$ 9,10-DIHYDROXYSTEARIC ACID, 317
$C_{19}H_{16}N_2$ N,N'-DIPHENYLBENZAMIDINE, 383
$C_{19}H_{18}O_4$ β-CARBETHOXY-γ,γ-DIPHENYLVINYL-
ACETIC ACID, 132
$C_{20}H_{14}$ TRIPTYCENE, 964
$C_{20}H_{30}O_2$ ABIETIC ACID, 1
$C_{21}H_{15}N$ α,β-DIPHENYLCINNAMONITRILE, 387
$C_{21}H_{17}N$ α,α,β-TRIPHENYLPROPIONITRILE, 962
$C_{21}H_{18}O_2$ α,β,β-TRIPHENYLPROPIONIC ACID,
960
$C_{23}H_{44}O_4$ CETYLMALONIC ESTER, 141
$C_{23}H_{46}O$ LAURONE, 560
$C_{24}H_{20}AsCl$ TETRAPHENYLARSONIUM CHLORIDE
HYDROCHLORIDE, 910
$C_{25}H_{21}N$ p-AMINOTETRAPHENYLMETHANE, 47
$C_{26}H_{20}$ TETRAPHENYLETHYLENE, 914
$C_{27}H_{42}O_2$ Δ^4-CHOLESTEN-3,6-DIONE, 189
$C_{27}H_{44}O$ Δ^4-CHOLESTEN-3-ONE, 192, 195
Δ^5-CHOLESTEN-3-ONE, 195
$C_{27}H_{46}O$ CHOLESTEROL, 195
$C_{35}H_{70}O$ STEARONE, 854

INDEX TO PREPARATION OR PURIFICATION OF SOLVENTS AND REAGENTS

ORGANIC SYNTHESES procedures frequently include notes describing the purification of solvents and reagents. These have been placed in a single index for convenience. The preparations of useful reagents and catalysts, as well as some techniques, determinations, and tests are included also.

Acetic acid, purification of, 712
Acetone, anhydrous, 211, 577, 837
Acetylene, purification of, 117, 187, 794
Acid number, 3
Aldehydes, stabilization against autoxidation, 447
Alizarin indicator, 416
Alumina, activation of, 796
  chromatographic, activation of, 965
  preparation for chromatography, 818
Aluminum isopropoxide, preparation of, 193
Aniline, purification of, 769
Anthracene, purification of, 9, 712
Anti-foaming agent, 138, 685, 856

Beilstein test, 199, 725
Benzaldehyde, purification of, 778
Benzene, anhydrous, 556, 592, 901
  purification of, 702
Bromine, anhydrous, 400, 619
N-Bromoacetamide, determination of purity, 105
N-Bromoglutarimide, preparation of, 498
N-Bromosuccinimide, preparation of, 922
tert-Butyl alcohol, anhydrous, 134, 460

Carbon tetrachloride, anhydrous, 935
Catalysts, β-alanine, 94
  alumina, 520, 796
  ammonium acetate, 234, 441, 451, 463
  ammonium nitrate, 21
  benzoyl peroxide, 109, 430, 921
  boron trifluoride etherate, 375, 957
  chromium oxide-calcium carbonate, 579
  copper chromite, 216, 314, 324, 337, 678, 857
  copper powder, 337, 628, 732
  cupric acetate, 146
  cuprous oxide-silver oxide, 494
  diethylamine, 205
  ferric chloride, 122, 474, 480, 573, 851, 970
  ferric nitrate, 128, 387, 404, 586, 755, 763, 963
  iron, 114
  mercuric acetate, 977
  mercuric sulfate, 13
  palladium, 538
  palladium-on-carbon, 306, 409, 537, 888
  phosphoric acid, 350

Catalysts, piperidine acetate, 210, 409
  piperidine-pyridine, 327
  platinum black, 612
  platinum oxide, 305, 306
  potassium hydroxide, 776
  pyridine, 5, 243, 381, 732
  Raney nickel, 222, 299, 314, 357, 432, 603, 639, 660, 672
  silica gel, 64
  sodium acetate, 669
  sodium ethoxide, 184
  sodium hydroxide, 929
  p-toluenesulfonic acid, 304, 534
  triethylamine, 730, 777
  Triton B, 652
  urea, 276
  zinc iodide, 801
Chlorine, anhydrous, 935
Chloroform, anhydrous, 265
Chromatographic purification, 816, 965
Chromic oxide, determination of, 20
Claisen's alkali, preparation of, 191
Cyclohexanone, anhydrous, 223

Diazomethane, preparation of, 250
Diethyl malonate, purification of, 287, 632
Dimethylamine, anhydrous, 336
Dimethylketene, determination of, 349
Dioxane, purification of, 643, 848
Distillation, azeotropic, 192, 304, 313, 416, 682, 901

Emulsion, minimizing of, 400
Ethanol, absolute, 11, 289, 428, 632, 956
  determination of water content of, 100
  paraffin-oil test for water in, 458
Ether, anhydrous, 564, 602, 938
  ethanol-free, 822
  peroxide-free, 207, 926
Ethylene glycol, anhydrous, 449
Ethylene glycol dimethyl ether, purification of, 475
Ethyl pyruvate, assay of, 469

Filtration, inverted, 561

Grignard reagent, analysis of, 750

Hexane, purification of, 610

1011

Hydrogen chloride, alcoholic, preparation
of, 428
anhydrous, 751
generation of, 171, 607, 927
Hydrogen peroxide, assay of, 319
Hydrogen selenide, generation of, 360
Hydroperoxide, quantitative estimation of,
896

Ion-exchange resin, treatment of, 39. 530
Iodine number (Wijs titration), 318, 853,
978

β-Ketoesters, test for, 122

Lithium, handling of, 888

Methylamine, generation of, 607
Methyl ethyl ketone, purification of, 86

Nitrogen, drying of, 70, 134, 926
oxygen-free, preparation of, 27
Nitromethane, anhydrous, 223

Orthophosphoric acid, preparation of, 321,
543
Oxygen content, determination of active,
896

Paraffin-oil test for water in ethanol, 458
Peroxide, test for, 830
Phenanthrene, purification of, 313
Phosgene, purification of, 524
o-Phthalaldehyde, color test for, 808
Potassium, directions for safe handling of,
134
Potassium amide, preparation of, 963
Potassium tert-butoxide, preparation of,
132, 459
Potassium fluoride, anhydrous, 526
Potassium hydroxide, methanolic, prepara-
tion, 367
Potassium hypochlorite, preparation of, 486
Potassium phthalimide, purification of, 812
Pyridine, anhydrous, 480

Quinoline, purification of, 733

Silver chloride, preparation of, 85
Silver nitrite, preparation of, 725
Silver oxide, preparation of, 548, 919, 972
Sodium amalgam, preparation of, 509
Sodium amide, preparation of, 117, 128,
296, 387, 404, 586, 755, 851, 970
Sodium ethoxide, preparation of, 11, 99,
174, 184, 221, 245, 288, 291, 396, 427,
457, 461, 536, 566, 618, 631, 662, 932
Sodium hypobromite, preparation of, 45
Sodium hypochlorite, preparation of, 74,
346
Sodium methoxide, preparation of, 29, 210,
278, 380, 382, 516, 594, 624, 638, 650,
744
Steam distillation, 45, 59, 71, 150, 193, 222,
340, 343, 355, 402, 439, 482, 492, 508,
545, 563, 573, 627, 732, 764, 778, 808,
836, 837, 918, 948
Sublimation, 696, 818, 880
Sulfur, purification of, 668
Sulfuric acid, 100%, preparation of, 978
Sulfur trioxide, distillation of, 848
liquid, 864

Tetrahydrofuran, anhydrous, 356
purification of, 259, 474, 793, 966
Tetrahydroquinone indicator, 530
Tetralin, purification of, 523, 896
Tetralone blue test, 906
Toluene, anhydrous, 194
1,2,3-Triazoles, 1,4-disubstituted-5-amino-,
titration of, 381
Triethylamine, purification of, 557, 561
Trimethylamine, generation of, 86

Unsaturation, quantitative estimation by
bromate-bromide titration method, 849

Water, deaerated, 27, 789
deoxygenated, 817
Witt-Utermann solution, 44

Xylene, purification of, 842

Zinc, amalgamated, preparation of, 204,
696

APPARATUS INDEX

A number of the procedures in Organic Syntheses describe the use of special or less common pieces of apparatus and equipment. References to many of them are recorded in this index. Illustrations are indicated by numbers in **boldface** type.

Apparatus, for acetate pyrolysis, 746
for acyloin cyclization, 840
for addition of butadiene to maleic anhydride, **891**
for bromination of *m*-aminobenzoic acid, **947**
for caustic fusion of vanillin, **975**
for constant-temperature reaction, **431**
for continuous return of heavy liquid, 580
for distillation of diethyl methylenemalonate, **300**
for evacuating and introducing nitrogen into reaction vessels, **133**, 459
for preparation of sodium amide, 117
for preparing alloxantin dihydrate, **26**
for pyrolysis of ammonium salt of azelaic acid, **63**
for pyrolytic condensation of tetrahydropyran with aniline, **795**
for reaction of ethanol with chlorotrifluoroethylene, **184**
Autoclave, high-pressure, 311, 530, 930
use of hydrogenation bomb as, 294, 452, 623, 738, 930

Bath, cooling, Dry Ice and acetone, 69, 75, 268, 747, 751
Dry Ice-ethanol, 20, 385
Dry Ice and ethylene glycol monomethyl ether, 240
Dry Ice and methylene chloride, 369
Dry Ice and trichloroethylene, 60, 387, 404
Bath, heating, salt, 498
wax, 488, 683
Wood's metal, 488, 592, 768, 854
Beaker, stainless-steel, 974
Blender, Waring-type, 111
Bottle, pressure, 261, 418

Capillary, sealed, 209, 211, 941
Capillary tube, evacuated, 199
Condenser, Dry Ice-acetone, 118, 119, 258, 269, 296, 755, 851, 970
Continuous reactor, for preparing benzoylacetanilide, **81**
for preparing oleoyl chloride, **740**

Dispersator, Duplex, 70
Distillation apparatus, for laurone, **562**
for solids, 830

Distillation apparatus, for sulfur trioxide, 848
head for, 231
steam, for solids, **397**
Distillation column, low-temperature, **686**
Distribution apparatus, cascade, 152
Dropping funnel, Hershberg, 70, 840, 964
Hershberg, modified, **747**

Electrolytic reduction, apparatus for, 149
Evaporator, rotary, 614
Extraction, countercurrent, Kies, apparatus for, 399
continuous, 46, 167, 548, 564, 645, 835
continuous, liquid-liquid, 20, 150
Wehrli, 580

Filter aid, 100, 115, 217, 283, 358, 365, 581, 660, 838, 909
Flask, addition, Erlenmeyer, 649
Claisen, modified, 109
copper, 95, 446
creased, 523, 529, 840
Dewar, 724
jacketed, **431**
Morton, 68
resin, 472
sausage, 766
stainless-steel, 95, 446, 449
Thermos, 380
Flow meter, 64, 126
Flow-rate-indicating fluid, Arochlors as, 126
Funnel, Büchner, cooling jacket for, 373, **928**
electrically heated, 811
steam-jacketed, 561, 856

Gas absorption trap, 44, 63, 157, 162, 178, 554
Gas chromatography, 614
Gas dispersion tube, 184, 523, 669, 895
Gas trap, mercury, 146, 966

Hydrogenation apparatus, 217, 222, 299, 305, 357, 409, 603, 660, 888, 930

Ion-exchange column, **40**

Kipp's apparatus, 928

Ozonizer, 485

1013

Pressure regulator, 465, 580

Rubber dam, 155, 332, 662, 864, 954

Separator, water, Barrett-type, 94
    water, Dean and Stark, 235, 573, 580, 606
    water, constant, 463
    water-benzene, 408
Soda lime, tower, 99, 387
Sodium, device for holding, 117, 763
Still, evaporative, 794
Stirrer, for caustic fusion, **975**
    centrifugal, 649
    dispersion-mill type, 70
    gas dispersion, 158
    glycerin-sealed, 291, **546**
    Hershberg, 42, 59, 69, 94, 98, 100, 124,
        163, 210, 221, 296, 314, 428, 772, 971

Stirrer, high-speed, 70, 840
    magnetic, 252, 308, 339, 340, 532, 662,
        794, 870
    Morton, 70
    Polytron, 70
    sealed, 89, 368, 862
    seal for, 650
    Stir-O-Vac, 70
    sweep-type, 573, 598
    Trubore, 89, 683, 724, 807, 814, 829, 947
    vacuum seal, 829
Sublimation apparatus, 153, 880
Swirling apparatus, 405

Tangential apparatus, **81, 740**
Tape, heating, 465, 741

Ultraviolet lamp, 807, 984

AUTHOR INDEX

Adams, A. C., 415
Ainsworth, C., 536
Albright, Charles F., 68
Alexander, Kliem, 266
Allen, C. F. H., 45, 80, 433, 739, 804, 866
Allen, C. Freeman, 398, 608, 616
Allinger, Norman L., 840
Anderson, Arthur G., Jr., 221
Angyal, S. J., 690
Archer, W. L., 331
Ashworth, P. J., 128
Atkinson, Edward R., 872

Backer, H. J., 225, 250, 943
Bader, Alfred R., 898
Bak, B., 207
Balcom, Don M., 603
Banks, Charles V., 229
Bannard, R. A. B., 393
Bantjes, A., 534
Barthel, John W., 218
Becker, Ernest I., 174, 176, 623, 657, 771
Bedell, S. F., 810
Behr, Lyell C., 342
Bell, E. W., 125
Bellis, M. P., 722, 903
Bender, Myron L., 932
Bennett, Frank, 359
Benson, Richard E., 588, 746
Berenbaum, M. B., 66, 273, 274
Berlinguet, L., 496
Bertz, R. T., 489
Bill, J. C., 807
Billig, Franklin A., 68
Bisgrove, D. E., 372
Bistline, R. G., Jr., 862
Blackwood, Robert K., 454
Blake, J., 327
Blardinelli, Albert J., 311
Blomquist, A. T., 216, 838
Boehme, Werner R., 590
Boekelheide, V., 298, 641
Bordwell, F. G., 846
Bornstein, J., 329, 810
Borum, O. H., 5
Bourgeois, R. C., 918
Bourns, A. N., 795
Boyer, J. H., 75, 532
Brasen, W. R., 508, 582, 585
Braude, E. A., 698, 927
Braun, Charles E., 8, 711
Bremer, Keith, 777
Brent, John T., 342
Brown, E. A., 960
Brown, J. B., 851

Brown, J. F., Jr., 372
Bruce, W. F., 788
Buc, Saul R., 101
Buchdahl, M. R., 106
Buchner, B., 784
Buckles, Robert E., 256, 722, 777, 857, 914
Budde, W. M., 31
Budewitz, E. P., 748
Bunnett, J. F., 114
Burckhalter, J. H., 333
Burness, D. M., 628, 649
Butler, John Mann, 486
Byers, J. R., Jr., 739

Caesar, P. D., 693, 695
Cairns, Theodore L., 588
Campaigne, E., 31, 331, 918, 919, 921
Campbell, Barbara K., 117, 763
Campbell, Kenneth N., 117, 763
Cannon, George W., 597
Carboni, R. A., 877
Carhart, Homer W., 436
Carlsmith, Allan, 828
Carmack, Marvin, 669
Cason, James, 510, 555, 630, 635
Chao, Tai Siang, 380
Chard, S. J., 520
Charles, Robert G., 869
Chinn, Leland J., 459
Chudd, C. C., 748
Ciganek, Engelbert, 339, 612
Clapp, L. B., 372
Clark, R. D., 674
Clemens, David H., 463, 662, 664
Coan, Stephen B., 174, 176
Colonge, J., 350, 601
Cook, Clinton D., 711
Cooper, F. C., 769
Cope, Arthur C., 62, 218, 234, 304, 339, 377, 612, 816, 890
Cornforth, J. W., 467
Corson, B. B., 884
Cotter, Robert J., 62, 377
Coyner, Eugene C., 727
Craig, W. E., 130
Crocker, Richard E., 278
Crovetti, Aldo J., 166, 654, 704
Croxall, W. J., 98
Cullinane, N. M., 520
Curtis, Omer E., Jr., 278
Cymerman-Craig, J., 205, 667, 700

D'Addieco, Alfred A., 234
Daub, Guido H., 390
Dauben, Hyp J., Jr., 221

1015

Davis, R. B., 392
Dawkins, C. W. C., 520
Dayan, J. E., 499
Deacon, B. D., 569
Deatherage, F. E., 851
de Boer, Th. J., 225, 250, 943
Deebel, George F., 579
Dessy, R. E., 484
DeTar, D. F., 34, 730
Dickel, Geraldine B., 317
Dickinson, C. L., 276
Diebold, James L., 254
Diehl, Harvey, 229
Dobson, A. G., 854
Donahoe, Hugh B., 157
Drummond, P. E., 810
Dryden, Hugh L., Jr., 816
Dunn, M. S., 55
Dunnavant, W. R., 962
Durham, Lois J., 510, 555, 635

Easley, William K., 892
Eastham, Jerome F., 192
Englinton, Geoffrey, 404, 755
Eliel, Ernest L., 58, 169, 626
Ellingboe, E. K., 928
Ellis, Ray C., 597
Embleton, H. W., 795
Emerson, William S., 302, 311, 579, 660, 677, 980
Endler, Abraham S., 657
English, J., Jr., 499
Englund, Bruce, 184, 423
Erickson, Floyd B., 430
Estes, Leland L., 62

Fanta, Paul E., 844
Farlow, Mark W., 521
Farmer, H. H., 441
Fawcett, J. S., 698
Feely, Wayne, 298
Fegley, Marian F., 98
Fehnel, Edward A., 669
Feldkamp, R. F., 671
Fetscher, Charles A., 735
Feuer, Henry, 368, 554
Field, L., 674
Fieser, Louis F., 189, 195
Finelli, A. F., 461, 776, 790
Fish, M. S., 327, 408
Fisk, Milton T., 169, 626
Floyd, Don E., 141
Fones, William S., 293
Fonken, Gunther S., 261
Ford-Moore, A. H., 84, 325, 955
Foster, H. M., 638
Frank, Robert L., 451
Freeman, Jeremiah P., 58
Fujiwara, Kunio, 72
Furrow, C. L., Jr., 242

Gaudry, R., 496
Gaylord, Norman G., 178

Gellerson, Hilda E., 903
Gerold, Corinne, 104
Gillis, Richard G., 396
Glickman, Samuel A., 234
Gofton, B. F., 927
Goheen, D. W., 594
Goldstein, Albert, 216, 838
Goshorn, R. H., 307
Gradsten, M. A., 518
Greenwood, F. L., 108
Grummitt, Oliver, 748, 771
Gulen, R., 679
Gunstone, F. D., 160
Guthrie, J. L., 513
Gutsche, C. David, 780, 887

Hach, Clifford C., 229
Hall, Luther A. R., 333
Hamilton, Cliff S., 950
Hamilton, Robert W., 576
Hanslick, R. S., 788
Hansuld, Mary K., 795
Harman, R. E., 148
Harris, G. C., 1
Harrisson, R. J., 493
Hart, Harold, 278
Hartman, R. J., 93
Hass, H. B., 932
Hatt, H. H., 854
Hauser, Charles R., 508, 582, 585, 708, 962
Hearst, Jerome J., 571
Heidtke, W. J., 960
Heininger, S. A., 146
Heinzelman, R. V., 573
Hennion, G. F., 683
Hering, H., 203
Herrick, Elbert C., 304, 890
Herzog, Hershel L., 753
Hexner, Peter E., 351
Hillebert, A., 207
Holmgren, A. V., 23
Hontz, Arthur C., 383
Horning, E. C., 144, 408, 461, 620, 776, 790
House, Herbert O., 367, 375, 552, 860, 957
Howard, John C., 42
Howell, Charles F., 816
Howk, B. W., 801
Huang, E. P.-Y., 93
Huang, Pao-Tung, 66, 274
Hudak, N. J., 738
Humphlett, W. J., 80, 739
Hunt, Richard H., 459
Huntress, E. H., 329

Issidorides, C. H., 679, 681

Jackson, H. L., 438
James, Philip N., 539
Janssen, Donald E., 547
Jaul, E., 307
Johnson, G. Dana, 900
Johnson, H. B., 804
Johnson, Herbert E., 780

Johnson, William S., 132, 162, 261, 390, 459
Johnston, J. Derland, 576
Jones, E. R. H., 404, 755, 792
Jones, Reuben G., 824
Jordan, E. F., Jr., 977

Kalm, Max J., 398, 608, 616
Kamath, P. M., 178
Karabinos, J. V., 506
Kaslow, C. E., 718
Kauer, J. C., 411
Kehm, Barbara B., 515
Kellert, M. D., 108
Kennard, Kenneth C., 950
Khan, N. A., 851, 969
King, Mary S., 88
Klingenberg, J. J., 110
Kluiber, Rudolph W., 261
Knight, H. B., 895
Kofod, Helmer, 491
Kogon, Irving C., 182, 336
Kohn, Earl J., 283
Koo, J., 327, 408
Kornblum, Nathan, 454, 724
Korpics, C. J., 93
Krimmel, C. P., 960
Krynitsky, John A., 436
Kurzer, Frederick, 49, 172, 180, 213, 361,
    502, 645, 934, 937

Lacey, A. B., 396
LaLonde, John, 757
Lawson, Alexander, 645
Lawson, J. Keith, 892
Leal, Joseph R., 597
Leicester, James, 525
Lesslie, T. E., 329
Leston, Gerd, 368
LeSuer, William M., 919
Leubner, Gerhard W., 866
Levens, Ernest, 68
Levis, W. W., Jr., 307
Lew, Henry Y., 545
Lieber, Eugene, 380
Lockhart, L. B., Jr., 784
Loder, J. W., 667
Longley, R. I., Jr., 302, 311, 660, 677
Lufkin, James E., 872
Luijten, J. G. A., 881

McCaleb, Kirtland E., 281
McCarthy, W. C., 918
McCloskey, Allen L., 261
McElvain, S. M., 463, 662, 664
Macey, William A. T., 525
McFarland, J. W., 940
McKeon, Thomas F., Jr., 683
McKusick, B. C., 438, 746, 953
McLaughlin, Keith C., 824
McLeod, Donald J., 510, 555, 635
McLeod, Gerald L., 345
McMahon, Robert E., 457
McRae, J. A., 393

Mallory, F. B., 74
Mansfield, G. H., 128
Marey, R., 350, 601
Mariella, Raymond P., 210, 288
Martin, D. G., 162
Mascitti, A., 788
Matar, A., 681
Matlack, George M., 914
Meinwald, J., 738
Melby, L. R., 276, 953
Merritt, Charles, Jr., 8, 711
Meyers, Cal Y., 39
Michaels, R. J., Jr., 915
Michelotti, Francis W., 29
Middleton, W. J., 243
Mikulec, Richard A., 13
Miller, Leonard E., 39
Miller, Sidney E., 141
Minin, Ronald, 182, 336
Moffett, Robert Bruce, 238, 291, 354, 357,
    427, 466, 605, 652, 834
Mosher, Harry S., 828
Mowry, David T., 486
Moyle, M., 205, 493, 700
Munch-Petersen, Jon, 715
Murphy, Donald M., 872
Murray, Joseph I., 744

Newman, M. S., 484
Nishimura, Tamio, 713
Noble, Paul, Jr., 924
Noller, C. R., 545, 571, 603
Norton, D. G., 348
Nowak, Robert M., 281

Okuda, Takuo, 566
Oliveto, Eugene P., 104
Olson, Cecil E., 898
Overberger, C. G., 29, 66, 182, 273, 274, 336

Page, G. A., 136
Pappalardo, Joseph A., 186, 558
Parham, William E., 295
Paris, G., 496
Partridge, M. W., 769
Patrick, Tracy M., Jr., 430, 980
Pearl, Irwin A., 972, 974
Pearson, David L., 221
Perkins, Edward G., 444
Perry, B. J., 325, 955
Peter, Hugo H., 887
Phillips, Donald D., 313
Pier, Stanley M., 554
Pilgrim, Frederick J., 451
Powell, J. Roy, 213, 934
Price, Charles C., 186, 558, 566, 683
Price, John A., 285
Prosser, Thomas, 169
Prout, F. S., 93

Raaen, Vernon, 130
Rabjohn, Norman, 441, 513
Raha, Chittaranjan, 263

Rao, C. N. Ramachandra, 380
Raube, Richard, 288
Rauhut, M. M., 114
Reid, E. Emmet, 47
Reif, Donald J., 375, 860
Reisner, D. B., 144
Reynolds, George A., 15, 633, 708
Rice, Rip C., 283
Richter, Henry J., 482
Riener, Edward F., 451
Rinehart, Kenneth L., Jr., 120, 444
Ringold, Howard J., 221
Ritter, E. J., 307
Roberts, John D., 457
Roberts, Royston M., 420, 464
Robison, Bonnie L., 947
Robison, Michael M., 947
Rogers, Crosby U., 884
Rondestvedt, Christian S., Jr., 766, 846
Ropp, Gus A., 130, 727
Rorig, Kurt, 576
Rosen, Milton J., 665
Rousseau, Joseph E., 711
Ruby, Philip R., 786, 798
Rutenberg, M. W., 620
Ryerson, George D., 957
Ryskiewicz, Edward E., 831

Sanderson, T. F., 1
Sandri, Joseph M., 278
Sauer, J. C., 268, 560, 801, 813
Saxton, M. R., 78
Scanlan, John T., 317
Schaefer, G. F., 31
Schaeffer, J. R., 19
Scheifele, H. J., Jr., 34
Schneider, H. J., 98
Schneider, William P., 132
Schoen, Kurt L., 623
Schoen, W., 532
Schultz, Harry P., 364
Schwartzman, Louis H., 471
Schwarz, R., 203
Searle, N. E., 424
Sedlak, J., 108
Sellas, J. T., 702
Seyferth, Dietmar, 258
Shah, N. M., 836
Shaw, B. L., 404
Shechter, Harold, 321, 323, 543
Sherman, Wm. R., 247
Short, William A., 706
Shriner, R. L., 242, 786, 798, 910
Silverstein, Robert M., 831
Simons, J. K., 78
Skattebøl, Lars, 792
Skelly, W. G., 154
Smart, B. W., 55
Smith, C. W., 348
Smith, Douglas S., 377
Smith, Newton R., 201, 337, 549, 731
Smith, P. A. S., 75, 819
Smith, Ronald Dean, 218

Smith, Walter T., Jr., 345, 702
Smolin, Edwin M., 387
Snoddy, A. O., 19, 529
Snyder, H. R., 539, 638
Soffer, M. D., 903
Soine, T. O., 106
Spangler, F. W., 433
Speziale, A. John, 401
Sprinzak, Yaïr, 91
Sroog, C. E., 271
Stacy, Gardner W., 13
Stenberg, J. F., 564
Stephens, Verlin C., 333
Stephenson, Emily F. M., 984
Stewart, Roberta A., 903
Stirton, A. J., 862
Stone, Herman, 321, 323, 543
Straley, J. M., 415
Strube, R. E., 6, 417, 967
Sugasawa, Shigehiko, 72
Summers, R. M., 718
Swern, Daniel, 317, 895, 977

Tarbell, D. S., 136, 285, 807, 924
Taylor, E. C., Jr., 166, 247, 654, 704
Taylor, G. A., 688
Teague, Peyton C., 706
Teeter, H. M., 125
Teeters, W. O., 518
Telinski, Thomas J., 576
Teranishi, Roy, 192
Tetaz, J. R., 690
Thielen, L. E., 960
Tichelaar, G. R., 93
Tipson, R. Stuart, 25
Towles, H. V., 266
Tucker, S. Horwood, 160
Tullar, B. F., 671, 921
Turner, Leslie, 828

Ulbricht, T. L. V., 566
Ungnade, H. E., 724

VanAllan, J. A., 15, 21, 245, 569, 633
Van Der Kerk, G. L. M., 881
VanderWerf, Calvin A., 157
van Tamelen, Eugene E., 10, 232
Van Zyl, G., 10
Vassel, B., 154
Vaughan, W. R., 594
Villani, Frank J., 88
Vogel, Arthur I., 525
Vogt, Paul J., 420, 464
Vyas, G. N., 836

Wade, Robert H., 221
Wagner, E. C., 383
Wagner, William S., 892
Walker, G. N., 327, 408
Ward, J. A., Jr., 392
Warnhoff, E. W., 162
Washburn, Robert M., 68
Wasson, Richard L., 552, 957

Wawzonek, Stanley, 387, 681
Webster, E. R., 433
Weijlard, John, 124
Weil, J. K., 862
Weinstock, J., 641
Wendland, Ray, 757
Wenner, Wilhelm, 23, 760
Weston, Arthur W., 915
Wheeler, Norris G., 256, 857
Wheeler, T. S., 478
White, R. A., 700
Whitehead, Calvert W., 515
Whiting, M. C., 128, 404, 755, 792
Whyte, Donald Edward, 234
Wilds, Alfred L., 281
Wiley, Richard H., 5, 201, 337, 351, 549, 731

Wilkinson, G., 473, 476
Willard, Constance, 831
Wilson, C. V., 547, 564
Wilson, J. G., 690
Wilt, James W., 254
Wittcoff, Harold, 907
Witten, Benjamin, 47
Wittig, George, 964
Wolf, Calvin N., 45, 910
Wolf, Frank J., 124
Woodburn, H. M., 271
Woods, G. Forrest, 471
Wynberg, Hans, 295, 534

Zingaro, Ralph, 359
Zuidema, G. D., 10

GENERAL INDEX

The name of a compound in SMALL CAPITAL LETTERS together with a number in boldface type indicates complete preparative directions for the substance named. The name of a compound in ordinary lightface type together with a number in boldface type indicates directions, usually adequate but not in full detail, for preparing the substance named. A name in lightface type together with a number in lightface type indicates a compound or an item mentioned in connection with a preparation.

ABIETIC ACID, **1**
 diamylamine salt of, 2
Acetaldehyde, 294
Acetamide, 104
p-Acetamidobenzaldehyde, **32**
3-ACETAMIDO-2-BUTANONE, **5**
Acetic acid, 2, 7, 11, 49, 56, 94, 149, 158, 170, 189, 197, 210, 218, 222, 318, 409, 463, 566, 570, 582, 645, 664, 698, 704, 790, 838, 922, 944
Acetic anhydride, 5, 16, 42, 242, 293, 294, 489, 631, 713, 722, 730, 748, 777, 790, 929
Acetone, 127, 210, 281, 577, 716, 727, 837
Acetonedicarboxylic acid, 816
Acetophenone, 59, 463
Acetoxime, 932
Acetylacetone, 351, 869
2-Acetylamino-3-nitrotoluene, **44**
2-Acetylamino-5-nitrotoluene, **44**
N-(*p*-ACETYLAMINOPHENYL)RHODANINE, **6**
9-ACETYLANTHRACENE, **8**
α-Acetyl-γ-butyrolactone, 597
Acetyl chloride, 8, 580
α-ACETYL-δ-CHLORO-γ-VALEROLACTONE, **10**
1-ACETYLCYCLOHEXANOL, **13**
Acetylene, 117, 186, 187, 793
2-*p*-ACETYLPHENYLHYDROQUINONE, **15**
2-*p*-ACETYLPHENYLHYDROQUINONE DI-ACETATE, **17**
2-*p*-ACETYLPHENYLQUINONE, **16**
δ-ACETYL-*n*-VALERIC ACID, **19**
Acrolein, 21, 312, 794, 817
ACROLEIN ACETAL, **21**
Acrylonitrile, 146, 147, 205, 776
Adipyl azide, 820
ADIPYL HYDRAZIDE, **819**
Alanine, 5, 94
Aldehyde-collidine, 452
ALLOXAN MONOHYDRATE, **23**, 25, 26, 28
ALLOXANTIN DIHYDRATE, **25**
Allylbenzene, 766
Allyl bromide, 749, 767
α-ALLYL-β-BROMOETHYL ETHYL ETHER, **750**
Allyl chloride, 751
Allylmagnesium bromide, **749**
Allylmagnesium chloride, **751**

Alumina, 520, 816
 activated, 244, 796, 818, 965
Aluminum bromide, 960
Aluminum chloride, 8, 89, 186, 482, 580, 621, 769, 784, 898, 950
Aluminum chloride-phosphorus oxychloride complex, 784
Aluminum isopropoxide, 193
Aluminum selenide, 360
Amberlite IR-4B resin, 39
Amberlite IR-120 resin, 530
p-Aminoacetanilide, 7
p-Aminoacetophenone, 16
2-AMINO-4-ANILINO-6-(CHLOROMETHYL)-*s*-TRIAZINE, **29**
p-AMINOBENZALDEHYDE, **31**
 azine, **33**
 oxime, **33**
 phenylhydrazone, **33**
m-Aminobenzoic acid, 948
2-AMINOBENZOPHENONE, **34**
ε-AMINOCAPROIC ACID, **39**
ε-Aminocaproic acid hydrochloride, **39**
4-Amino-3-chlorophenol, 150
N-β-Aminoethylethylenimine, 434
β-Aminoethylsulfuric acid, 433
1-(Aminomethyl)cyclohexanol, acetic acid salt, **224**
1-Amino-2-naphthol hydrochloride, 53
2-Amino-5-nitroanisole, 82
2-Amino-5-nitropyrimidine, 845
2-AMINO-3-NITROTOLUENE, **42**, 44
2-Amino-5-nitrotoluene, **44**
o-Aminophenol, 570
2-Aminopyridine, 91
3-AMINOPYRIDINE, **45**
2-Aminopyrimidine, 182
p-AMINOTETRAPHENYLMETHANE, **47**
p-Aminotetraphenylmethane hydrochloride, 47
3-AMINO-2,4,6-TRIBROMOBENZOIC ACID, **947**
6-Aminouracil, 248
Ammonia, liquid, 64, 117, 118, 128, 296, 387, 404, 441, 585, 755, 763, 851, 887, 963, 970
Ammonium acetate, 234, 394, 441, 451, 463
Ammonium chloride, 128, 259, 296, 416, 486, 506, 586, 668, 755, 773, 793, 852, 970

Ammonium hydroxide, 36, 118, 359, 365, 451, 486, 532, 764, 773, 873
Ammonium nitrate, 21
Ammonium thiocyanate, 180
n-Amylacetylene, **119**
N-Amylaniline, **284**
Amylbenzene, 522
n-Amyl fluoride, **527**
3-Amyl-4-methyl-1-phenyl-5-pyrazolone, 122
Anethole, 787
Aniline, 77, 81, 349, 383, 384, 464, 634, 764, 769, 796
Aniline benzenesulfonate, 206
Aniline hydrochloride, 47, **48**, 205
Anisaldehyde, 605
p-Anisidine hydrochloride, 53
Anisyl alcohol, **576**
Anisyl chloride, **577**
Anthracene, 8, 9, 711, 965
Anthranilic acid, 34, 872
p-toluenesulfonic acid salt, 37
Anthraquinone, 712, 758
Arsenic trichloride, 910
ARYLUREAS, **49**
Asbestos, 940
DL-ASPARTIC ACID, **55**
ATROLACTIC ACID, **58**
Azelaic acid, 64, 65
AZELANITRILE, **62**
1,1'-AZO-bis-1-CYCLOHEXANENITRILE, **66**, 273, 275
2,2'-Azo-bis-isobutyronitrile, **67**, 273, 275

Barbituric acid, 23
Barium chloride, 507, 529, 676
Barium D-gulonate, 506
Barium hydroxide, 177, 225, 506, 636
Behenic acid, 742
Benzalacetone, 312
Benzalacetophenone, 312
Benzal-bis-thiobenzoate, 925
Benzaldehyde, 392, 605, 606, 679, 777, **933**
4-Benzal-2-phenyl-5-oxazolone, 83
Benzanilide, 383, **384**
Benzene, 35, 88, 482, 556, 702, 719, 784, 898, 900, 960
BENZENEBORONIC ACID, 69, **71**
BENZENEBORONIC ANHYDRIDE, **68**
Benzenediazonium chloride, 634
Benzenesulfonyl chloride, 676, 753
Benzhydrol, 73
Benzhydryl chloride, 963
BENZHYDRYL β-CHLOROETHYL ETHER, **72**
Benzidine, 283, 284
Benzidine dihydrochloride, 284
Benzil, 377
Benzil dihydrazone, 377
Benzilic acid, 482
1,2-Benzo-3,4-dihydrocarbazole, **885**
BENZOFURAZAN OXIDE, **74**
BENZOGUANAMINE, **78**
Benzoic acid, 286

BENZOIC-CARBONIC ANHYDRIDE, **286**
Benzonitrile, 79, 123, 437, 769
Benzophenone, 133, 388, 914, 927
p-Benzoquinone, 16, **152**
BENZOYLACETANILIDE, **80**
Benzoyl chloride, 84, 384, 415, 478, 642, 924
BENZOYLCHOLINE CHLORIDE, **84**, 87
BENZOYLCHOLINE IODIDE, **84**, 85
Benzoylcholine picrate, 87
2-BENZOYL-1-CYANO-1-METHYL-1,2-DI-HYDROISOQUINOLINE, **642**
Benzoyl-2-methoxy-4-nitroacetanilide, 82
o-BENZOYLOXYACETOPHENONE, **478**, 479
Benzoyl peroxide, 109, 430, 921
3-BENZOYLPYRIDINE, **88**
4-Benzoylpyridine, **89**
Benzyl alcohol, 91
Benzylamine, 780
2-BENZYLAMINOPYRIDINE, **91**
N-Benzylaniline, **92**, 284
Benzyl chloride, 94, 99, 606, 963
Benzyl cyanide, see Phenylacetonitrile
N-BENZYLIDENEMETHYLAMINE, **605**
Benzyl isocyanide, 388
1-Benzylisoquinoline, 644
Benzylmagnesium chloride, **94**, **606**
3-BENZYL-3-METHYLPENTANENITRILE, **95**
3-BENZYL-3-METHYLPENTANOIC ACID, **93**
Benzyltrimethylammonium chloride, **99**
BENZYLTRIMETHYLAMMONIUM ETHOXIDE, **98**
Benzyltrimethylammonium hydroxide, 652
BENZYLTRIMETHYLAMMONIUM IODIDE, **585**, 586
Benzyne, 964
Betaine hydrochloride, 154
Biacetyl, 343
Bibenzyl, 96, 97, 257
Biguanide, 30
Biphenyl, 256
p-Biphenylyl isothiocyanate, 701
1-(2-Biphenylyl)urea, **51**
1-(4-Biphenylyl)urea, **51**
Bis-4-chlorobutyl ether, 529
BISCHLOROMETHYL ETHER, **101**
N,N'-Bis-2-cyanoethyl-o-phenylenedi-amine, **206**
N,N'-Bis-2-cyanoethyl-p-phenylenedi-amine, **206**
1,1-Bis(diethylaminomethyl)acetone, 281
Boric acid, 866
Boron trifluoride etherate, 375, 957
Bromine, 45, 66, 104, 110, 114, 196, 256, 257, 348, 398, 498, 526, 616, 688, 749, 807, 851, 858, 877, 881, 922, 948, 961, 969, 984
N-BROMOACETAMIDE, **104**
p-Bromoacetophenone, 110
p-Bromoaniline, 49
p-Bromobenzaldehyde, **933**
Bromobenzene, 912
α-Bromo-n-butyric acid, 621
α-Bromo-n-butyryl chloride, **621**
m-Bromocinnamic acid, **733**

2-Bromododecanoic acid, 398
β-BROMOETHYLPHTHALIMIDE, **106**
N-BROMOGLUTARIMIDE, **498**
4-BROMO-2-HEPTENE, **108**
α-BROMOISOBUTYRYL BROMIDE, **348**
p-BROMOMANDELIC ACID, **110**
2-BROMO-3-METHYLBENZOIC ACID, **114**
2-Bromo-3-methylthiophene, 923
2-BROMO-4-NITROTOLUENE, **114**
1-Bromoöctane, 724
p-Bromophenyl isothiocyanate, 701
1-(p-Bromophenyl)-3-phenyl-2-propanone, **177**
m-Bromophenylurea, **51**
o-Bromophenylurea, **51**
p-BROMOPHENYLUREA, **49**
m-Bromostyrene, **733**
N-Bromosuccinimide, 108, 254, 921
5-Bromo-2-vinylthiophene, **983**
Bronze powder, 914
Butadiene, 728, 890
Butadiene monoxide, 12
1-n-Butoxy-3-oxabicyclo(4.4.0)-3-decene, **312**
tert-Butyl acetate, **263**
n-BUTYLACETYLENE, 117
n-Butyl alcohol, 750, 798
sec-Butyl alcohol, 122
tert-Butyl alcohol, 126, 135, 776
anhydrous, 132, 264, 399, 459, 617
n-Butylamine, 284, 573
N-n-Butylaniline, 147, 284
tert-Butyl o-benzoylbenzoate, **263**
n-Butyl bromide, 118, 726
tert-Butyl bromoacetate, **263**
sec-Butyl α-bromopropionate, 121, **122**
sec-BUTYL α-n-CAPROYLPROPIONATE, **120**
n-Butyl Cellosolve, 684
tert-Butyl chloroacetate, **263**
tert-Butyl α-chloropropionate, **263**
n-Butyl cyclohexenyl ether, 312
Butyl diazoacetate, 426
tert-Butyl β,β-dimethylglutarate, **263**
9-n-Butylfluorene, **625**
tert-Butyl glutarate, **263**
n-BUTYL GLYOXYLATE, **124**
tert-BUTYL-HYPOCHLORITE, **125**
n-Butyl iodide, **322**, 726
tert-Butyl iodide, 324
1-Butylisoquinoline, 644
Butylketene dimer, 563
tert-Butyl 2-methylenedodecanoate, **617**
p-tert-Butylphenyl salicylate, **179**
n-Butylpropiolaldehyde diethyl acetal, **802**
tert-Butyl succinate, **263**
n-Butyltrivinyltin, 260
n-Butyl vinyl ether, 312
2-BUTYN-1-OL, **128**
BUTYRCHLORAL, **130**
γ-Butyrolactone, 278, 496, 898

Calcium carbide, 471
Calcium carbonate, 157, 428, 579, 922

Calcium cyanamide, 645
Calcium formate, 908
Calcium hydride, 460, 657
Calcium hypochlorite, 486
Calcium oxide, 659, 907
Calcium picrate, 87
γ-Caprilactone, **432**
ε-Caprolactam, 39, 588
n-Capronitrile, 121, 123
Caproyl chloride, 562
β-CARBETHOXY-γ,γ-DIPHENYLVINYLACETIC ACID, **132**
Carbitol, 251
p-Carbomethoxybenzaldehyde, **933**
δ-Carbomethoxyvaleryl chloride, 555, **556**
Carbon dioxide, 1, 130, 131, 296, 318, 410, 671, 927
Carbon disulfide, 25, 28, 244, 570, 967
Carbon monoxide, 142
Carbon tetrachloride, 109, 186, 285, 309, 708, 862, 922, 934, 950, 961
o-CARBOXYCINNAMIC ACID, **136**
2-Carboxy-3,5-dichlorophenyldiazonium chloride, 873
3-Carboxy-4-hydroxyquinoline, **83**
4-(o-Carboxyphenyl)-5,6-benzocoumarin, **138**
β-(o-CARBOXYPHENYL)PROPIONIC ACID, **136**
Catalysts, see "Index to Preparation or Purification of Solvents and Reagents," pp. 1011–1012
Catechol, 778, 788
Cellosolve, 115
Cerebrosterol, 200
Cetylmalonic acid, **143**
Cetylmalonic Ester, **141**
CETYLMALONIC ESTER, **141**
Chlorine, 74, 126, 130, 157, 208, 308, 346, 872, 934
Chloroacetamide, 144
Chloroacetic acid, 967
CHLOROACETONITRILE, **144**
o-Chloroacetophenone, **709**
p-Chloroacetophenone, 112
o-Chloroaniline, 146, 180
p-Chloroaniline, 420
o-Chloroaniline hydrochloride, 146
o-Chloroaniline thiocyanate, 180
3-(o-CHLOROANILINO)PROPIONITRILE, **146**
Chlorobenzene, 364, 700, 878, 910
CHLORO-p-BENZOQUINONE, **148**
N-CHLOROBETAINYL CHLORIDE, **154**
3-CHLORO-2-BUTENE-1-OL, **128**
m-Chlorocinnamic acid, **733**
o-Chlorocinnamic acid, **733**
p-Chlorocinnamic acid, **733**
α-CHLOROCROTONALDEHYDE, **131**
2-Chlorocyclohexanol, 233
2-Chlorocyclohexanone, 594
trans-2-CHLOROCYCLOPENTANOL, **157**
3-CHLOROCYCLOPENTENE, **238**, 291
1-CHLORO-2,6-DINITROBENZENE, **160**
2-CHLOROETHYL BENZOATE, **84**

α-CHLOROETHYL ETHYL ETHER, **748**
Chloroform, 265, 548, 864, 918
p-Chloromandelic acid, **112**
2-CHLORO-2-METHYLCYCLOHEXANONE, **162**
2-Chloromethyl-4,6-diamino-s-triazine, **30**
1-Chloromethylnaphthalene, 690
2-CHLORONICOTINONITRILE, **166**
m-Chloronitrobenzene, 151
o-Chloronitrobenzene, 149
9-Chloro-10-nitro-9,10-dihydroanthracene, 711
1-Chloro-1-nitroethane, 374
1-Chloro-1-nitropropane, 372
2-Chloro-2-nitropropane, 374
5-CHLORO-2-PENTANONE, **597**
p-Chlorophenol, 178
α-CHLOROPHENYLACETIC ACID, **169**
p-Chlorophenylacetonitrile, 174, 578
o-CHLOROPHENYLCYANAMIDE, **172**
o-Chlorophenylhydroxylamine, 151
o-Chlorophenyl isothiocyanate, 701
α-(4-CHLOROPHENYL)-γ-PHENYLACETO-ACETONITRILE, **174, 176**
1-(p-CHLOROPHENYL)-3-PHENYL-2-PROPANONE, **176**
o-Chlorophenyl salicylate, 179
p-CHLOROPHENYL SALICYLATE, **178**
o-CHLOROPHENYLTHIOUREA, **172, 180**
o-Chlorophenylurea, **51**
Chloropicrin, 457
2-CHLOROPYRIMIDINE, **182, 336**
m-Chlorostyrene, **733**
o-Chlorostyrene, **733**
p-Chlorostyrene, **733**
N-Chlorosuccinimide, 255
Chlorosulfonic acid, 102
m-Chlorosulfonylbenzoic acid, 697
3-Chlorotoluquinone, **152**
Chlorotrifluoroethylene, 185
2-CHLORO-1,1,2-TRIFLUOROETHYL ETHYL ETHER, **184, 423**
Chlorourea, see Monochlorourea
β-CHLOROVINYL ISOAMYL KETONE, **186, 558**
β-Chlorovinyl isobutyl ketone, **188**
β-Chlorovinyl isohexyl ketone, **188**
β-Chlorovinyl methyl ketone, **188**
5-Chloro-2-vinylthiophene, **983**
Cholestanol, 200
Δ⁴-CHOLESTEN-3,6-DIONE, **189**
Δ⁷-Cholestenol, 200
Δ⁴-CHOLESTEN-3-ONE, **192, 195, 198**
Δ⁵-CHOLESTEN-3-ONE, 195, **198**
CHOLESTEROL, 189, 192, **195**, 196
CHOLESTEROL DIBROMIDE, **195**
Cholesteryl acetate, 199
Cholesteryl chromate, 189
Choline, 148
Chromium oxide, 579
Chromium trioxide, 19, 23, 698, 713, 757, 813
Cinnamaldehyde, 312, 722, 771, 793
Cinnamic acid, 961

Cinnamic acid dibromide, 960
Claisen's alkali, 190, **191**
2,4,6-Collidine, 163
Copper, powder, 337, 628, 732, 878, 914
  sheet, 149
Copper chromite, 216, 314, 324, 337, 678, 857
Copper sulfate, 873
COUMALIC ACID, **201**, 532
CREOSOL, **203**
Crotonaldehyde, 130, 311, 312, 794
Crotonaldehyde diethyl acetal, **22**
1-(p-Cumyl)-1-propene, 787
Cupric acetate monohydrate, 146, 147, 838
Cupric bromide, 121
Cupric chloride, 728
Cuprous chloride, 160
Cuprous oxide, 493
Cyanamide, 213, **645**
Cyanoacetamide, 210, 662
Cyanoacetic acid, 234, 254
Cyanoacetylurea, 248
p-Cyanobenzaldehyde, **933**
p-Cyanobenzaldiacetate, 714
1-CYANO-2-BENZOYL-1,2-DIHYDROISOQUINOLINE, **641**
m-Cyanocinnamic acid, **733**
3-Cyano-5,6-dimethyl-2(1)-pyridone, 212
N-2-CYANOETHYLANILINE, **205**
N-2-Cyanoethyl-p-anisidine, **206**
N-2-Cyanoethyl-m-chloroaniline, **206**
2-Cyano-3-ethyl-3-methylhexanoic acid, 97
Cyanogen, 209
CYANOGEN IODIDE, **207**
N-Cyanoguanidine, 502
3-Cyano-6-isobutyl-2(1)-pyridone, 212
3-CYANO-6-METHYL-2(1)-PYRIDONE, **210**
1-Cyano-3-α-naphthylurea, 215
1-CYANO-3-PHENYLUREA, **213**
m-Cyanostyrene, **733**
Cyclobutanecarboxylic acid, 289
Cyclodecane, 219
1,2-CYCLODECANEDIOL (cis and trans), **216**
1,2-Cyclodecanedione, **838**
CYCLODECANONE, **218**, 220
Cyclodecanone semicarbazone, 220
CYCLOHEPTANONE, **221**
  bisulfite addition product, 226
Cyclohexane, 314
Cyclohexanecarboxylic acid, 339
Cyclohexanecarboxylic acid chloride, 340
1,2-CYCLOHEXANEDIONE, **229**
1,2-CYCLOHEXANEDIONE DIOXIME, **229**
Cyclohexanol, 324
Cyclohexanone, 192, 221, 225, 229, 234, 274, 459, 471, 536, 537, 781, 884, 907
Cyclohexanone 2,4-dinitrophenylhydrazone, 14
Cyclohexene, 543
Cyclohexene oxide, 232
CYCLOHEXENE SULFIDE, **232**
1-CYCLOHEXENYLACETONITRILE, **234**
CYCLOHEXYLIDENECYANOACETIC ACID, **234**

Cyclohexyl iodide, 324, **543**
Cyclohexylmethylacetylene, 802
Cyclohexylmethylpropiolaldehyde diethyl
  acetal, **802**
1,2-Cyclononanedione, **839**
Cycloöctanone, 227
CYCLOPENTADIENE, **238**, 239, 474, 476
Cyclopentene, 158

Decalin, 537
n-Decyl bromide, 618
Decyl diazoacetate, 426
n-Decyl fluoride, **527**
Decylketene dimer, 560, **561**
7-Dehydrocholesterol, 200
DIACETYL-d-TARTARIC ANHYDRIDE, **242**
DIALURIC ACID MONOHYDRATE, **28**
2,5-DIAMINO-3,4-DICYANOTHIOPHENE, **243**
2,4-DIAMINO-6-HYDROXYPYRIMIDINE, **245**
2,4-Diamino-6-phenyl-s-triazine, **78**
Diaminouracil bisulfite, 248
DIAMINOURACIL HYDROCHLORIDE, **247**
Diamylamine, 2
Di-n-amyl ketone, **562**
2,4-Di-tert-amylphenoxyacetyl chloride, **742**
DIAZOMETHANE, 225, **250**
Dibenzalpentaerythritol, 680
Dibenzeneborinic acid, 69, 70
Dibenzhydryl ether, 73
Dibenzylcadmium, 96
N,N'-Dibenzyl-p-phenylenediamine, **92**
N,N-Dibromoacetamide, 105
DIBROMOACETONITRILE, **254**
α,α'-Dibromobibenzyl, 257
4,4'-Dibromobibenzyl, **257**
4,4'-DIBROMOBIPHENYL, **256**
5α,6β-DIBROMOCHOLESTAN-3-ONE, **197**
α,β-DIBROMOETHYL ETHYL ETHER, **749**
10,11-Dibromohendecanoic acid, 970
Dibromohydrocinnamic acid, 960, **961**
DIBROMOMALONONITRILE-POTASSIUM
  BROMIDE COMPLEX, **877**
Dibromopentaerythritol, 683
Dibromostearic acid, 852
N,N'-Dibutylbenzidine, 284
Di-tert-butyl β,β-dimethylglutarate, **263**
DI-n-BUTYLDIVINYLTIN, **258**
Di-n-butyl ether, 322
Di-tert-butyl glutarate, 263
DI-tert-BUTYL MALONATE, **261**
Di-tert-butyl succinate, **263**
Di-n-butyl d-tartrate, 124
Di-n-butyltin dichloride, 258
Dibutyrolactone, 278, 279
N,N'-Dicarbethoxyputrescine, **822**
Dichloroacetic acid, 272, 427
Dichloroacetonitrile, **255**
Dichloroacetyl chloride, 271
3,5-DICHLORO-2-AMINOBENZOIC ACID, **872**
o-Dichlorobenzene, 523, 766
1,4-Dichlorobutane, **893**
1,3-Dichloro-2-butene, 128, 684
p-Di(chloro-tert-butyl)benzene, 703

2,4-Dichlorocinnamic acid, **733**
3,4-Dichlorocinnamic acid, 7?3
1,2-Dichlorocyclopentene, 159
4,4'-DICHLORODIBUTYL ETHER, **266**
1,1-DICHLORO-2,2-DIFLUOROETHYLENE, **268**
2,2-DICHLOROETHANOL, **271**
1,7-Dichloro-4-heptanone, 279
1,1-Dichloro-1-nitroethane, 374
2,5-Dichloroquinone, **152**
Dicyandiamide, 79, 502
1,1'-DICYANO-1,1'-BICYCLOHEXYL, **273**
1,2-DI-1-(1-CYANO)CYCLOHEXYLHYDRA-
  ZINE, 66, **274**, 275
α,α'-DICYANO-β-ETHYL-β-METHYLGLU-
  TARIMIDE, **441**
DICYANOKETENE ETHYLENE ACETAL, **276**
1,2-Di-2-(2-cyano)propylhydrazine, **275**
Dicyclopentadiene, 238, 475, 738
Dicyclopentadienylnickel, 477
DICYCLOPROPYL KETONE, **278**
DIELS ACID, **191**
Di-(p-ethoxyphenyl)urea, 53
Diethyl acetylenedicarboxylate, **330**
Diethyl adipate, 324, 819
Diethylamine, 147, 205, 476
Diethylamine hydrochloride, 281
1-DIETHYLAMINO-3-BUTANONE, **281**
3-Diethylaminopropionitrile, 206
N,N-Diethylaniline, 955
N,N-Diethylaniline hydrochloride, 955
DIETHYL AZODICARBOXYLATE, **411**
Diethyl benzalmalonate, 804
N,N'-DIETHYLBENZIDINE, **283**
DIETHYL BENZOYLMALONATE, **285**
Diethyl bromoacetal, 295, 297
Diethyl carbonate, 461
DIETHYL CETYLMALONATE, **141**
Diethyl chloroacetal, 404
Diethylcyanamide, 360
DIETHYL 1,1-CYCLOBUTANEDICARBOX-
  YLATE, **288**
DIETHYL Δ²-CYCLOPENTENYLMALONATE,
  291
Diethylene glycol, 313, 510, 753
Diethylene glycol monoethyl ether, 252
Diethyl ethoxymethylenemalonate, 298,
  299
DIETHYL ETHYLIDENEMALONATE, **293**
Diethyl ethylphosphonate, **326**
Diethyl fumarate, 486
DIETHYL HEPTANOYLSUCCINATE, **430**
DIETHYL cis-HEXAHYDROPHTHALATE, **304**
Diethyl hydrogen phosphite, 956
Diethyl maleate, 430
Diethyl malonate, 285, 288, 291, 293, 294,
  417, 631, 708
DIETHYL MERCAPTOACETAL, **295**
DIETHYL METHYLENEMALONATE, **298**
Diethyl methylmalonate, 300, 618
Diethyl methylphosphonate, 326
Diethyl o-nitrobenzoylmalonate, 709
Diethyl oxalate, 141, 744
DIETHYL γ-OXOPIMELATE, **302**

Diethyl phthalate, 11, 449
N,N-Diethylselenourea, 360
Diethyl sodium phthalimidomalonate, 55
Diethyl succinate, 133
DIETHYL-cis-Δ^4-TETRAHYDROPHTHALATE, 304, 305
DIETHYLTHIOCARBAMYL CHLORIDE, 307
3,4-Dihydro-2-n-butoxy-2H-pyran, 312
3,4-Dihydro-4,6-diphenyl-2-ethoxy-2H-pyran, 312
3,4-Dihydro-2-ethoxy-5-ethyl-4-n-propyl-2H-pyran, 312
3,4-Dihydro-2-ethoxy-4-furyl-2H-pyran, 312
3,4-Dihydro-2-ethoxy-6-methyl-4-phenyl-2H-pyran, 312
3,4-Dihydro-2-ethoxy-2-methyl-2H-pyran, 312
3,4-Dihydro-2-ethoxy-4-methyl-2H-pyran, 312
3,4-Dihydro-2-ethoxy-5-methyl-2H-pyran, 312
3,4-Dihydro-2-ethoxy-6-methyl-2H-pyran, 312
3,4-Dihydro-2-ethoxy-4-phenyl-2H-pyran, 312
3,4-Dihydro-2-ethoxy-2H-pyran, 312
3,4-DIHYDRO-2-METHOXY-4-METHYL-2H-PYRAN, 311, 660
3,4-Dihydro-2-methoxy-2H-pyran, 312
5,8-Dihydro-1-naphthol, 888
9,10-DIHYDROPHENANTHRENE, 313
Dihydropyran, 499
5,5-DIHYDROXYBARBITURIC ACID, 23
p-(2,5-DIHYDROXYPHENYL)ACETOPHENONE, 15
Dihydroxystearic acid, 742
9,10-DIHYDROXYSTEARIC ACID (LOW-MELTING ISOMER), 317
9,10-Dihydroxystearic acid, high-melting isomer, 320
1,4-DIIODOBUTANE, 321, 369
1,6-DIIODOHEXANE, 323, 370
1,5-Diiodopentane, 370
1,3-Diiodopropane, 370
Diisobutylthiocarbamyl chloride, 310
Diisopropyl ether, 322
Diisopropyl ethylphosphonate, 326
DIISOPROPYL METHYLPHOSPHONATE, 325
Diisopropylthiocarbamyl chloride, 310
2,3-Dimethoxybenzaldehyde, 327, 408, 603
4,4-Dimethoxy-2-butanone, 629, 649
2,3-DIMETHOXYCINNAMIC ACID, 327
3,4-Dimethoxycinnamic acid, 733
1,2-Dimethoxyethane, 475
2,3-Dimethoxyphenylacetamide, 762
3,4-Dimethoxyphenylacetamide, 762
4-(3,4-Dimethoxyphenyl)-m-dioxane, 787
1-(3′,4′-Dimethoxyphenyl)-1-propene, 787
DIMETHYL ACETYLENEDICARBOXYLATE, 329
Dimethylamine, 336, 340, 893
Dimethylamine, aqueous, 336, 361, 626, 719

p-DIMETHYLAMINOBENZALDEHYDE, 331
β-Dimethylaminoethanol, 333
β-DIMETHYLAMINOETHYL CHLORIDE HYDROCHLORIDE, 333
2-(DIMETHYLAMINO)PYRIMIDINE, 336
N,N-Dimethylaniline, 264, 331, 953
2,3-Dimethylbenzyl acetate, 584
2,3-Dimethylbenzyl alcohol, 584
N,N-Dimethylbenzylamine, 585
2,3-Dimethylbenzyldimethylamine, 508, 587
2,3-Dimethylbenzylethyldimethylammonium bromide, 583
2,3-Dimethylbenzyltrimethylammonium iodide, 508, 587
2,3-Dimethyl-2-butene, 544
Dimethyl chloroacetal, 297, 406
4,6-DIMETHYLCOUMALIN, 337
Dimethylcyanamide, 359
N,N-DIMETHYLCYCLOHEXANECARBOXAMIDE, 339, 340
N,N-DIMETHYLCYCLOHEXYLMETHYLAMINE, 339, 612
N,N-Dimethylcyclohexylmethylamine-N-oxide, 612
Di-(2-methylcyclopropyl), ketone, 280
Dimethyldivinyltin, 260
Dimethylformamide, 147, 163, 331, 382, 454, 484, 540, 811, 831, 893, 953
Dimethyl fumarate, 487
DIMETHYLFURAZAN, 342
β-β-DIMETHYLGLUTARIC ACID, 345
Dimethylglyoxime, 342
Dimethyl hendecanedioate, 635
Dimethyl cis-hexahydrophthalate, 306
N,N-DIMETHYLHYDROXYLAMINE HYDROCHLORIDE, 612
2,3-Dimethyl-2-iodobutane, 544
DIMETHYLKETENE, 348
Dimethyl mercaptoacetal, 298
2,4-Dimethyl-2-pentacosenoic acid, 611
5,5-DIMETHYL-2-n-PENTYLTETRAHYDROFURAN, 350
3,5-DIMETHYLPYRAZOLE, 351
4,6-Dimethyl-1,2-pyrone, 551
2,2-DIMETHYLPYRROLIDINE, 354
2,2-Dimethylpyrrolidine hydrochloride, 355
5,5-DIMETHYL-2-PYRROLIDONE, 355, 357
Dimethyl sebacate, 840
N,N-DIMETHYLSELENOUREA, 359
Dimethyl sulfate, 588, 674, 837
Dimethyl sulfoxide, 147, 455
Dimethyl cis-Δ^4-tetrahydrophthalate, 306
Dimethylthiocarbamyl chloride, 310
$asym$-DIMETHYLUREA, 361
Di-α-naphthylthiourea, 701
2,6-DINITROANILINE, 160, 364
p,p'-DINITROBIBENZYL, 367
1,4-DINITROBUTANE, 368
2,3-Dinitro-2-butene, 374
1,6-Dinitrohexane, 370
3,4-DINITRO-3-HEXENE, 372
1,5-Dinitropentane, 370

2,4-Dinitrophenylhydrazine, 14
1,1-Dinitropropane, 372
1,3-Dinitropropane, 370
Dioxane, 229, 578, 642, 652, 846
Dioxane sulfotrioxide, 846
4-(3,4-Dioxymethylenephenyl)-m-dioxane, 787
Dipentaerythritol, 754
Diphenic acid, 757
DIPHENYLACETALDEHYDE, 375
Diphenylacetaldehyde 2,4-dinitrophenyl-hydrazone, 376
Diphenylacetamidine, 385
Diphenylacetonitrile, 963
DIPHENYLACETYLENE, 377
1,4-DIPHENYL-5-AMINO-1,2,3-TRIAZOLE, 380
N,N'-DIPHENYLBENZAMIDINE, 383
Diphenylbenzamidine hydrochloride, 383
2,4-Diphenylbutane-1,4-sultone, 849
α,β-DIPHENYLCINNAMONITRILE, 387
Diphenyldichloromethane, 914
Diphenyldivinyltin, 260
N,N'-Diphenylformamidine, 465
1,3-Diphenyl-2-propanone, 177
DIPHENYL SUCCINATE, 390
2,3-DIPHENYLSUCCINONITRILE, 392
meso-α,α'-Diphenylsuccinonitrile, 394
Diphenyl sulfoxide, 89
Disodium monohydrogen phosphate do-decahydrate, 808, 816
p-DITHIANE, 396
2,4-DITHIOBIURET, 503, 504
Di-p-tolylacetylene, 378
Di(trimethylsilylmethyl)divinyltin, 260
Dodecanoic acid, 398
trans-2-DODECENOIC ACID, 398
3-Dodecenoic acid, 400
n-Dodecyl fluoride, 527
Dowtherm A, 672
DULCIN, 52

Elaidic acid, 320
Enanthaldehyde, 430
Epichlorohydrin, 11, 12
1,2-ETHANEDITHIOL, 396, 401
Ethanol, 225, 231, 283, 903
    anhydrous, 11, 99, 141, 142, 169, 174, 184,
        221, 245, 247, 285, 288, 291, 302, 304,
        330, 334, 399, 427, 457, 461, 536, 537,
        566, 618, 631, 662, 708, 749, 887, 932,
        955
Ethanolamine, 107
Ether, 94, 793
    anhydrous, 68, 207, 210, 272, 560, 564,
        601, 606, 667, 708, 749, 834, 881, 912,
        938
ETHOXYACETYLENE, 404
2-Ethoxy-3,4-dihydro-2H-pyran, 816
ETHOXYMAGNESIUMMALONIC ESTER, 285, 708
m-Ethoxyphenylurea, 51
o-Ethoxyphenylurea, 51
p-ETHOXYPHENYLUREA, 51, 52

Ethyl acetate, 348, 355, 409, 835, 888
Ethyl acetoacetate, 11, 408, 415, 549, 592, 634, 635, 638
p-Ethylacetophenone, 580
ETHYL α-ACETYL-β-(2,3-DIMETHOXY-PHENYL)ACRYLATE, 408
Ethyl α-acetyl-β-(3,4-dimethoxyphenyl) acrylate, 410
ETHYL α-ACETYL-β-(2,3-DIMETHOXY-PHENYL)PROPIONATE, 408
Ethyl α-acetyl-β-(3,4-dimethoxyphenyl) propionate, 410
Ethylacetylene, 118
N-Ethylaniline, 147, 284, 422
ETHYL AZODICARBOXYLATE, 411
Ethyl benzalmalonate, 83
Ethylbenzene, 580
p-Ethylbenzoic acid, 580
ETHYL BENZOLACETATE, 80, 82, 415, 635
ETHYL N-BENZYLCARBAMATE, 780
ETHYL-3-BENZYL-2-CYANO-3-METHYL-PENTANOATE, 94, 95
Ethyl bromide, 582, 792, 881
Ethyl bromoacetate, 456
Ethyl α-bromobutyrate, 454
Ethyl α-bromopropionate, 122, 445, 466
Ethyl sec-butylcyanoacetate, 96
ETHYL sec-BUTYLIDENECYANOACETATE, 94
ETHYL tert-BUTYL MALONATE, 417
O-Ethylcaprolactim, 589
Ethyl chloroacetate, 29, 30, 55, 459
Ethyl α-chloroacetoacetate, 591, 592
N-ETHYL-p-CHLOROANILINE, 420
Ethyl chlorocarbonate, see Ethyl chloro-formate
ETHYL CHLOROFLUOROACETATE, 185, 423
N-ETHYL-p-CHLOROFORMANILIDE, 420, 421
Ethyl chloroformate, 286, 411, 780
ETHYL α-CHLOROPHENYLACETATE, 169
Ethyl N-p-chlorophenylformimidate, 422
Ethyl cyanoacetate, 94, 246, 247, 441, 463, 567
Ethyl β-cyanoacrylate, 488
Ethyl 2-cyano-3-ethyl-3-methylhexanoate, 97
Ethyl 2-cyano-3-methyl-3-phenylpen-tanoate, 97
ETHYL DIAZOACETATE, 424
Ethyl dichloroacetate, 272
ETHYL DIETHOXYACETATE, 427
ETHYL ENANTHYLSUCCINATE, 430
Ethylene, 738
Ethylene chlorohydrin, 73, 84
Ethylene dibromide, 396, 402
Ethylene dichloride, 831, 846, 940
Ethylene diisothiouronium bromide, 402
Ethylene glycol, 95, 276, 449, 525, 684
Ethylene glycol dimethyl ether, 475
Ethylene glycol monomethyl ether, 240, 506
ETHYLENIMINE, 433
Ethyl α-ethoxalylstearate, sodio derivative, 142

Ethyl ethoxymethylenecyanoacetate, 566, **567**
Ethyl 4-ethyl-2-methyl-3-hydroxyocta-noate, **447**
Ethyl 4-ethyl-2-methyl-2-octenoate, 446, **448**
Ethyl 4-ethyl-2-methyl-3-octenoate, 446, **448**
9-Ethylfluorene, **625**
Ethyl formate, 210, 536, 537
Ethyl fumarate, 431
Ethyl glycidyl ether, 12
Ethyl glycinate hydrochloride, 424
Ethyl glyoxylate, 125
2-Ethylhexanal, 445
2-Ethylhexanamide, **436**
2-Ethylhexanoic acid, 436
2-ETHYLHEXANONITRILE, **436**
2-Ethylhexyl diazoacetate, 426
ETHYL HYDRAZODICARBOXYLATE, **411**, 412
Ethyl hydrogen malonate, 418
4-ETHYL-4-HYDROXY-2-METHYLOCTANOIC ACID, γ-LACTONE, **447**, 449
Ethyl iodide, 326, 438
ETHYL ISOCYANIDE, **438**
ETHYL ISODEHYDROACETATE, **549**
Ethyl isopropenyl ether, 312
Ethyl lactate, 467
Ethylmagnesium bromide, **792**
Ethyl maleate, 430
ETHYL MANDELATE, **169**
ETHYL 3-METHYLCOUMARILATE, **590**, 591
β-ETHYL-β-METHYLGLUTARIC ACID, **441**
3-Ethyl-3-methylhexanenitrile, 97
3-Ethyl-3-methylhexanoic acid, **97**
4-Ethyl-2-methyl-1,4-octanolide, **447**
4-ETHYL-2-METHYL-2-OCTENOIC ACID, **444**, 449
5-ETHYL-2-METHYLPYRIDINE, **451**
ETHYL α-NITROBUTYRATE, **454**
Ethyl α-nitrocaproate, **456**
Ethyl α-nitroisobutyrate, **456**
Ethyl α-nitroisovalerate, **456**
Ethyl α-nitrophenylacetate, 456
Ethyl α-nitropropionate, **456**
ETHYL N-NITROSO-N-BENZYLCARBAMATE, **780**
ETHYL ORTHOCARBONATE, **457**
Ethyl orthoformate, 21, 420, 465, 516, 567, 801, 802
Ethyl orthosilicate, 22
ETHYL β,β-PENTAMETHYLENEGLYCIDATE, **459**
Ethyl α-phenoxyacetoacetate, 591
Ethyl phenylacetate, 174
ETHYL PHENYLCYANOACETATE, **461**, 776
ETHYL β-PHENYL-β-CYANOPROPIONATE, **804**
ETHYL (1-PHENYLETHYLIDENE)CYANO-ACETATE, **463**, 662
ETHYL N-PHENYLFORMIMIDATE, **464**
α-Ethyl-β-n-propylacrolein, 312
ETHYL α-(1-PYRROLIDYL)PROPIONATE, **466**, 834

ETHYL PYRUVATE, **467**
sodium bisulfite addition compound, 469
Ethyl stearate, 141, **142**
Ethyl tartrate, 125
N-Ethyl-m-toluidine, **422**
Ethyl vinyl ether, 312, 817
1-Ethynylcyclohexanol, 13
1,1'-ETHYNYLENE-bis-CYCLOHEXANOL, **471**
ETHYNYLMAGNESIUM BROMIDE, **792**

Ferric alum, 648
Ferric chloride, 122, 474, 480, 573, 851, 970
Ferric nitrate, 128, 387, 404, 586, 755, 763, 963
FERROCENE, **473**
Ferrous chloride, 474, 476
FLAVONE, **478**
Fluorene, 313, 624
9-FLUORENECARBOXYLIC ACID, **482**
o-Fluorobromobenzene, 965
α-(4-Fluorophenyl)-γ-phenylacetoacetoni-trile, **175**
1-(p-Fluorophenyl)-3-phenyl-2-propanone, **177**
Formaldehyde, 469, 626, 786, 908
Formic acid, 317, 907
α-Formylethyl methyl ketone, sodium salt, 212
Formylmethyl isobutyl ketone, sodium salt, 212
5-FORMYL-4-PHENANTHROIC ACID, **484**
p-Formylstyrene, **733**
2-Formyl-2,4,4-trimethylcyclopentanone, 958
FUMARAMIDE, **486**
FUMARONITRILE, **486**
Furfural, 489, 493, 688
FURFURAL DIACETATE, **489**
Furfuryl alcohol, 491
S-2-Furfurylisothiourea, 491
2-FURFURYL MERCAPTAN, **491**
2-FUROIC ACID, **493**
β-Furylacrolein, 312
Furylacrylic acid, 302

Glutaraldehyde, 816
GLUTARIC ACID, **496**
monoamide, 497
GLUTARIMIDE, **496**, 498
Glycerol, 478, 546, 866
Glycerol-α,γ-dichlorohydrin, 12
Glycine, 5, 57
Glyoxal-sodium bisulfite, 824
Guanidine, 845
Guanidine hydrochloride, 246
Guanidine thiocyanate, 504
GUANYLTHIOUREA, **502**
D-GULONIC-γ-LACTONE, **506**

HEMIMELLITENE, **508**
HENDECANEDIOIC ACID, **510**, 636
n-HEPTAMIDE, **513**
n-Heptanoic acid, 513

2-Heptene, 108
n-Heptylamine, 515
3-n-HEPTYL-5-CYANOCYTOSINE, 515
2-Heptyl-5,5-dimethyltetrahydrofuran, 351
9-n-Heptylfluorene, 625
n-Heptyl fluoride, 527
N-n-HEPTYLUREA, 515
3-n-HEPTYLUREIDOMETHYLENEMALONONI-
TRILE, 515
n-Hexadecyl fluoride, 527
HEXAHYDRO-1,3,5-TRIPROPIONYL-8-TRI-
AZINE, 518
HEXAMETHYLBENZENE, 520
Hexamethylenediamine, 522
HEXAMETHYLENEDIAMMONIUM CHLORIDE,
522
HEXAMETHYLENE DIISOCYANATE, 521
Hexamethylenetetramine, 690, 866, 918
Hexane, 610
1,6-Hexanediol, 323
1-Hexene, 544
4-Hexen-1-yn-3-ol, 794
n-Hexylacetylene, 119
n-Hexyl alcohol, 527
N-Hexylaniline, 284
n-Hexyl bromide, 525, 726
n-Hexyl diazoacetate, 426
9-n-Hexylfluorene, 625
n-HEXYL FLUORIDE, 525
1-Hexyne, 802
Hydrazine hydrate, 377, 411, 413, 510, 537,
819
Hydrazine sulfate, 274, 351
·2,2'-Hydrazo-bis-isobutyronitrile, 275
Hydrazoic acid, 77
Hydrobromic acid, 681
Hydrochloric acid, 1, 16, 35, 39, 43, 56, 59,
75, 85, 102, 160, 170, 180, 182, 203,
213, 218, 281, 417, 421, 442, 465, 491,
497, 515, 522, 533, 573, 577, 597, 660,
679, 711, 760, 790, 805, 872, 948, 980
Hydrogen, 216, 222, 299, 305, 314, 324, 357,
409, 603, 660, 888
Hydrogen chloride, 17, 59, 169, 239, 302,
427, 580, 606, 646, 749, 828, 927, 981
Hydrogen cyanide, 274, 393
Hydrogen iodide, 324
Hydrogen peroxide, 317, 552, 612, 655, 704,
759, 911
Hydrogen selenide, 359
Hydrogen sulfide, 25, 27, 28, 244, 502, 669,
924, 927, 929, 967
Hydroquinone, 187, 311, 447, 670, 778
o-Hydroxyacetophenone, 478
4-HYDROXY-1-BUTANESULFONIC ACID
SULTONE, 529
β-Hydroxy-β-(o-carboxyphenyl)propionic
acid lactone, 138
24-Hydroxycholesterol, 200
25-Hydroxycholesterol, 200
2-Hydroxycyclodecanone, 840
2-Hydroxycyclononanone, 839
o-HYDROXYDIBENZOYLMETHANE, 479

β-Hydroxyethylphthalimide, 107
Hydroxyformoxystearic acids, 318
Hydroxylammonium chloride, 230, 873
2-HYDROXYMETHYLENECYCLOHEXANONE,
536, 537
1-(2-Hydroxy-1-naphthyl)urea, 53
6-HYDROXYNICOTINIC ACID, 532
3-HYDROXYTETRAHYDROFURAN, 534
δ-Hydroxyvaleraldehyde, 500
Hypophosphorous acid, 948

INDAZOLE, 536
Indene, 475
Indole, 540
INDOLE-3-ALDEHYDE, 539
Iodine, 207, 209, 469, 545, 548, 601, 749,
870, 934
Iodine chloride, 978
p-Iodoacetophenone, 112
IODOCYCLOHEXANE, 324, 543
2-IODOETHYL BENZOATE, 84, 85
2-Iodohexane, 544
p-Iodomandelic acid, 112
2-Iodo-5-nitrothiophene, 547
2-IODOTHIOPHENE, 545, 667
4-IODOVERATROLE, 547
Ion-exchange resins, 39, 529
Iron, filings, 225, 240
powder, 114, 474, 573
Iron selenide, 360
Iron sulfide, 928
Isoamylacetylene, 119
N-Isoamylaniline,- 284, 422
Isobutyl alcohol, 324
N-Isobutylaniline, 284
Isobutylene, 261, 418
Isobutyl iodide, 324, 726
Isobutyranilide, 349
Isobutyric acid, 348
Isocaproyl chloride, 186
ISODEHYDROACETIC ACID, 337, 549
Isonicotinic acid, 89
Isophorone, 552
ISOPHORONE OXIDE, 552, 957
Isopropyl alcohol, 956
N-Isopropylaniline, 147
Isopropyl iodide, 322, 325
p-Isopropylphenylacetamide, 762
4-(4-p-Isopropylphenyl)-5-methyl-m-
dioxane, 787
Isoquinoline, 642
Isosafrole, 787
Itaconic acid, 554, 672
ITACONYL CHLORIDE, 554

β-Ketobutraldehyde dimethyl acetal, 559,
651
γ-Ketocapric acid, 432
6-KETOHENDECANEDIOIC ACID, 510, 555
2-Ketohexamethylenimine, 39
β-KETOISOÖCTALDEHYDE DIMETHYL
ACETAL, 558
Kieselguhr, 769, 940

Lathosterol, 200
Lauric acid, 561, 977
LAURONE, 560
Lauroyl chloride, 560, **561**
LAURYLMETHYLAMINE, **564**
Lead acetate trihydrate, 173
Lead sulfide, 173
Lead tetraacetate, 124
Lithium, 887
Lithium aluminum hydride, 271, 340, 355, 474, 564, 794, 834
Lithium chloride, 163

Magnesium, 94, 258, 285, 601, 606, 667, 708, 749, 792, 881, 912, 965
Magnesium ethoxide, 11
Magnesium oxide, 419, 854
Magnesium sulfate, 467, 469
Maleic anhydride, 313, 766, 890, 965
Maleic anhydride-anthracene adduct, 965
Malic acid, 201
Malonic acid, 261, 263, 327, 732
Malonic ester, ethoxymagnesium derivative, solution of, **294**
Malononitrile, 516, 877
MALONYL DICHLORIDE, **263**, 264
Mandelic acid, 169
Manganese dioxide, 825
Melamine, 79
2-MERCAPTO-4-AMINO-5-CARBETHOXYPYRIMIDINE, **566**
2-MERCAPTOBENZIMIDAZOLE, **569**
2-Mercaptobenzoxazole, **570**
2-MERCAPTO-4-HYDROXY-5-CYANOPYRIMIDINE, **566**
Mercuric acetate, 896, 977
Mercuric chloride, 204
Mercuric oxide, 13, 378, 548
Mercury, 509, 896
Methacrolein, 312
Methallyl chloride, 702
Methanesulfonic acid, 571
METHANESULFONYL CHLORIDE, **571**
Methanol, 17, 29, 211, 268, 278, 306, 329, 367, 520, 532, 558, 580, 587, 608, 624, 636, 744
Methone, 346
Methoxyacetylene, **406**
o-Methoxybenzaldehyde, 573
4-Methoxy-3-buten-2-one, 651
p-Methoxycinnamic acid, 733
p-Methoxyphenylacetamide, **762**
o-METHOXYPHENYLACETONE, **573**
p-METHOXYPHENYLACETONITRILE, **576**
2-(m-Methoxyphenyl)cycloheptanone, **783**
2-(o-Methoxyphenyl)cycloheptanone, **783**
2-(p-Methoxyphenyl)cycloheptanone, **783**
4-(p-Methoxyphenyl)-m-dioxane, **786**
1-(o-METHOXYPHENYL)-2-NITRO-1-PROPENE, **573**
1-(p-Methoxyphenyl)-3-phenyl-2-propanone, **177**
p-Methoxyphenylurea, **51, 53**

Methoxyquinone, 153
o-Methoxystyrene, **733**
p-Methoxystyrene, **733**
METHYL p-ACETYLBENZOATE, **579**
Methyl acrylate, 669
Methylamine, 603, 606, 943
Methylamine hydrochloride, 607, 816
4'-Methyl-2-aminobenzophenone, **38**
N-Methylaminopyrimidine, **336**
N-Methylaniline, 147, **422**, 621, 916, 917
Methylaniline hydrochloride, 621
METHYL AZODICARBOXYLATE, **414**
o-METHYLBENZYL ACETATE, **582**, 583
o-METHYLBENZYL ALCOHOL, **582**
N-Methylbenzylamine, 605
2-METHYLBENZYLDIMETHYLAMINE, 582, **585**
Methyl benzyl ether, 782
2-Methylbenzylethyldimethylammonium bromide, **582**
2-Methylbenzyltrimethylammonium iodide, 587
Methyl borate, 68, 70
Methyl borate-methanol azeotrope, 70
Methyl bromide, 601
N-Methyl-α-bromo-n-butyranilide, **621**
O-METHYLCAPROLACTIM, **588**
Methyl Cellosolve, 78, 240, 506
Methyl chloride, 126, 647
Methyl chloroacetate, 649
N-Methyl-p-chloroaniline, **422**
Methyl chloroformate, 413
METHYL COUMALATE, **532**
3-METHYLCOUMARILIC ACID, **591**
3-METHYLCOUMARONE, **590**
Methyl crotonate, 631
2-Methylcyclohexanol, 19, 164
2-METHYLCYCLOHEXANONE, 19, **162**
1-Methylcyclohexene, 614
2-METHYL-2-CYCLOHEXENONE, **162**
Methylcyclopentadiene, 475
METHYL CYCLOPENTANECARBOXYLATE, **594**
METHYL CYCLOPROPYL KETONE, **597**
2-METHYL-2,5-DECANEDIOL, 350, **601**
Methyl diazoacetate, 426
N-METHYL-2,3-DIMETHOXYBENZYLAMINE, **603**
N-Methyl-3,4-dimethoxybenzylamine, 603
METHYL 5,5-DIMETHOXY-3-METHYL-2,3-EPOXYPENTANOATE, **649**
Methyl 5,5-dimethoxy-3-phenyl-2,3-epoxypentanoate, 651
N-Methyl-3,4-dioxymethylenebenzylamine, 605
N-METHYL-1,2-DIPHENYLETHYLAMINE AND HYDROCHLORIDE, **605**
2-Methyldodecanoic acid, 608, 616, **618**
2-Methyl-2-dodecenoic acid, 618
trans-2-METHYL-2-DODECENOIC ACID, **608**
2-Methyl-2-eicosenoic acid, 611
Methylene chloride, 155, 254, 425, 816, 950
METHYLENECYCLOHEXANE, **614**
2-Methylenecyclohexanone dimer, 165

2-METHYLENEDODECANOIC ACID, 616
Methylene-α-naphthylmethylamine, 691
Methyl p-ethylbenzoate, 579, 580
Methyl ethyl ketone, 84, 86, 94, 441, 794
1-METHYL-3-ETHYLOXINDOLE, 620
9-METHYLFLUORENE, 623
N-Methylformanilide, 916
2-Methylfuran, 626
3-METHYLFURAN, 628
5-METHYLFURFURYLDIMETHYLAMINE, 626
3-METHYL-2-FUROIC ACID, 628
β-Methylglutaraldehyde, 661
β-METHYLGLUTARIC ANHYDRIDE, 630
METHYLGLYOXAL-ω-PHENYLHYDRAZONE,
    633
2-Methyl-2-hexacosenoic acid, 611
METHYL HYDRAZODICARBOXYLATE, 413
Methyl hydrogen adipate, 556
METHYL HYDROGEN HENDECANEDIOATE,
    635
Methyl hydrogen β-methylglutarate, 632
4-METHYL-6-HYDROXYPYRIMIDINE, 638
Methyl iodide, 325, 326, 585, 642, 668, 836
1-METHYLISOQUINOLINE, 641
METHYLISOUREA HYDROCHLORIDE, 645
N-Methyllauramide, 564, 565
Methylmagnesium bromide, 351, 601, 771
N-Methyl-p-methoxybenzylamine, 605
Methyl cis-2-methyl-2-dodecenoate, 610
Methyl trans-2-methyl-2-dodecenoate, 610
Methyl 2-methylenedodecanoate, 610
METHYL 3-METHYL-2-FUROATE, 628, 649
METHYL γ-METHYL-γ-NITROVALERATE,
    357, 652
N-Methylmyristamide, 565
Methylmyristylamine, 565
3-Methyl-1-nitrobutane, 726
2-Methyl-1-nitropropane, 726
3-METHYL-4-NITROPYRIDINE-1-OXIDE, 654
3-Methylnonanoicnitrile, 122
Methylnonylamine, 565
Methyl orthoformate, 422
3-METHYLOXINDOLE, 657
N-Methylpelargonamide, 565
3-METHYL-1,5-PENTANEDIOL, 660, 677
3-Methyl-1-pentyn-3-ol, 794
p-Methylphenylacetamide, 762
2-(o-Methylphenyl)cycloheptanone, 783
2-(p-Methylphenyl)cycloheptanone, 783
β-METHYL-β-PHENYL-α,α'-DICYANOGLU-
    TARIMIDE, 662, 664
4-Methyl-4-phenyl-m-dioxane, 787
5-Methyl-4-phenyl-m-dioxane, 787
β-METHYL-β-PHENYLGLUTARIC ACID, 664
1-METHYL-3-PHENYLINDANE, 665
3-Methyl-3-phenylpentanenitrile, 97
3-Methyl-3-phenylpentanoic acid, 97
Methyl phenyl sulfone, 675
2-Methyl-5,6-pyrazinedicarboxylic acid,
    827
3-Methylpyridine, 655
3-Methylpyridine-1-oxide, 654, 655
2-Methylquinoxaline, 826

Methyl stearate, 855
α-Methylstyrene, 666, 787
trans-Methylstyrylcarbinol, 773
Methylsuccinic acid, disodium salt, 671
4-Methyl-1-tetralone, 899
METHYL 2-THIENYL SULFIDE, 667
METHYL β-THIODIPROPIONATE, 669
3-METHYLTHIOPHENE, 671, 921
1-(p-Methylthiophenyl)-3-phenyl-2-pro-
    panone, 177
N-Methyl-p-toluenesulfonamide, 943
N-Methyl-m-toluidine, 422
METHYL p-TOLYL SULFONE, 674
2-Methyl-2,5-undecanediol, 351
β-METHYL-δ-VALEROLACTONE, 677
Methyl vinyl ether, 311, 312
Methyl vinyl ketone, 312
Methyl violet, 382
Mineral oil, 509, 671
MONOBENZALPENTAERYTHRITOL, 679
MONOBROMOPENTAERYTHRITOL, 681
Monochlorourea, 157, 158
Monoethyl malonate, 418
MONOVINYLACETYLENE, 683
MUCOBROMIC ACID, 688, 844

Naphtha, 415
1-NAPHTHALDEHYDE, 690
Naphthalene, 691, 698
Naphthalene-1,5-disulfonic acid, disodium
    salt, 693
NAPHTHALENE-1,5-DISULFONYL CHLORIDE,
    693, 695
1,5-NAPHTHALENEDITHIOL, 695
1-Naphthol, 887
2-Naphthol, 37, 136
1,4-NAPHTHOQUINONE, 698
1-Naphthylacetamide, 762
α-Naphthylcyanamide, 174
β-Naphthyl ethyl ether, 903
α-Naphthyl isocyanate, 215, 557
α-NAPHTHYL ISOTHIOCYANATE, 700
β-Naphthyl isothiocyanate, 701
α-Naphthylphenylacetylene, 378
α-Naphthylthiourea, 700, 701
Neopentyl iodide, 726
NEOPHYL CHLORIDE, 702
Niacinamide, 705
Nicotinamide, 45, 704, 706
NICOTINAMIDE-1-OXIDE, 166, 704
Nicotinic acid, 88
NICOTINONITRILE, 706
Nicotinyl chloride, 88
Nioxime, 230
Nitric acid, 42, 412, 414, 499, 545, 711, 722,
    735, 781, 920
    fuming, 412, 414, 654
o-NITROACETOPHENONE, 708
p-Nitroacetophenone, 709
m-Nitroaniline, 718
o-Nitroaniline, 74, 75
p-Nitroaniline, 727
p-Nitroaniline hydrochloride, 727

9-NITROANTHRACENE, 711
m-Nitrobenzaldehyde, 732
o-Nitrobenzaldehyde, 730
o-NITROBENZALDIACETATE, 713
p-NITROBENZALDIACETATE, 713
m-NITROBENZAZIDE, 715
p-Nitrobenzazide, 717
Nitrobenzene, 720
m-Nitrobenzoic acid, 715
o-Nitrobenzoic acid, 709
p-Nitrobenzoic acid, 714
m-NITROBENZOYL CHLORIDE, 715, 716
o-Nitrobenzoyl chloride, 708
m-NITROBIPHENYL, 718
Nitrobutane, 726
o-NITROCINNAMALDEHYDE, 722
m-NITROCINNAMIC ACID, 731
o-Nitrocinnamic acid, 732
p-Nitrocinnamic acid, 732
Nitroethane, 573
Nitrogen, 25, 27, 64, 68, 70, 94, 98, 99, 109,
    124, 130, 132, 164, 318, 348, 374, 388,
    402, 404, 445, 457, 474, 478, 509, 642,
    649, 683, 747, 749, 771, 788, 792, 796,
    813, 816, 840, 896, 904, 925, 938, 941,
    977, 981
Nitroheptane, 726
Nitrohexane, 726
3-Nitro-4-hydroxystyrene, 733
Nitromethane, 221, 223
1-(Nitromethyl)cyclohexanol, 224
1-(Nitromethyl)cyclohexanol, sodio deriva-
    tive, 222
1-NITROÖCTANE, 724
o-NITROPHENYLAZIDE, 75, 76
1-(p-NITROPHENYL)-1,3-BUTADIENE, 727
1-(p-Nitrophenyl)-4-chloro-2-butene, 726
cis-o-Nitro-α-phenylcinnamic acid, 730
trans-o-NITRO-α-PHENYLCINNAMIC ACID,
    730
1-(m-NITROPHENYL)-3,3-DIMETHYLTRI-
    AZENE, 718
p-Nitrophenyl salicylate, 179
2-Nitropropane, 652, 932
Nitrosomethylurea, 250
α-Nitroso-β-naphthol, 981
6-Nitrosoveratric acid, 736
m-NITROSTYRENE, 731
o-Nitrostyrene, 733
p-Nitrostyrene, 733
m-Nitrotoluene, 151
o-Nitrotoluene, 151, 714
p-Nitrotoluene, 31, 114, 367, 713
Nitrourea, 361
6-NITROVERATRALDEHYDE, 735
γ-Nonanoic lactone, 601
n-Nonyl fluoride, 527
NORBORNYLENE, 738

Octadecyl alcohol, 624
9-n-Octadecylfluorene, 624
1-Octanol, 138, 724
2-Octanol, 856

n-Octyl fluoride, 527
1-Octyl nitrite, 724
Oleic acid, 317, 739, 851
Oleic acid-urea complex, 742
OLEOYL CHLORIDE, 739
Olive oil, 852
Oxalic acid, 198
γ-Oxocapric acid, 432
Oxygen, 367, 494, 579, 895
Ozone, 484

Palmitic acid, 742, 862
Palmitonitrile, 437
Palmitoyl chloride, 742
PARABANIC ACID, 744
Paraformaldehyde, 102, 281, 691, 907
Paraldehyde, 293, 451, 748, 980
1,4-PENTADIENE, 746
Pentaerythritol, 679, 681, 753
Pentaerythrityl benzenesulfonate, 753
PENTAERYTHRITYL TETRABROMIDE, 753
1,5-Pentanediol, 748
1,5-Pentanediol diacetate, 747, 748
2,4-Pentanedione, 352, 869
4-Penten-1-ol acetate, 748
4-Penten-1-yn-3-ol, 794
3-Pentyl α-bromopropionate, 123
9-n-Pentylfluorene, 625
4-PENTYN-1-OL, 755
α-naphthylurethan, 756
    silver derivative, 756
Peracetic acid, 136, 828, 860
Perchloric acid, 382
Performic acid, in situ, 317
Periodic acid, 14
Phenanthrene, 313, 314, 757
PHENANTHRENEQUINONE, 757
    sodium bisulfite adduct, 758
9-Phenanthryl isothiocyanate, 701
α-Phenethyl chloride, 963
p-Phenetidine hydrochloride, 52
Phenol, 390, 520, 592
PHENYLACETAMIDE, 760
Phenylacetic acid, 730, 760, 761, 777
N-Phenylacetimido chloride, 385
Phenylacetonitrile, 175, 380, 387, 392, 461,
    760
PHENYLACETYLENE, 763, 801, 802
γ-PHENYLALLYLSUCCINIC ACID, 766
γ-Phenylallylsuccinic anhydride, 767
4-Phenyl-5-ANILINO-1,2,3-TRIAZOLE, 380
Phenylazide, 77, 380
N-PHENYLBENZAMIDINE, 769
1-Phenylbiguanide, 29
1-Phenylbiguanide hydrochloride, 29
α-Phenyl-γ-(4-bromophenyl)acetoacetoni-
    trile, 175
trans-1-PHENYL-1,3-BUTADIENE, 771
γ-Phenylbutyric acid, 900
α-PHENYL-α-CARBETHOXYGLUTARONI-
    TRILE, 776, 790
α-PHENYLCINNAMIC ACID, 777, 857
α-Phenylcinnamonitrile, 393

2-Phenylcycloheptanone, 780
Phenyldichlorophosphine, 784
4-Phenyl-m-dioxane, 786, 798
o-Phenylene Carbonate, 788
o-Phenylenediamine, 569, 824
Phenylethynyl n-butyl dimethyl ketal, 802
Phenylethynyl methyl diethyl ketal, 802
α-Phenylglutaric acid, 790
α-Phenylglutaric Anhydride, 790
Phenyl glycidyl ether, 12
Phenylhydrazine, 657, 884, 885
2-Phenyl-2-hydroxyethane-1-sulfonic acid, sodium salt, 850
Phenyl isocyanate, 213, 561
Phenyl isothiocyanate, 701
Phenyllithium, 642
Phenylmagnesium bromide, 68, 96, 767, 912
α-Phenyl-γ-(4-methoxyphenyl)acetoacetonitrile, 175
α-Phenyl-γ-(4-methylphenyl)acetoacetonitrile, 175
α-Phenyl-γ-(4-methylthiophenyl)acetoacetonitrile, 175
Phenyl-β-naphthylamine, 772, 774
1-Phenyl-1-penten-4-yn-3-ol, 792
α-Phenyl γ-phenylacetoacetonitrile, 175
o-Phenylphenyl salicylate, 179
p-Phenylphenyl salicylate, 179
1-Phenylpiperidine, 795
3-Phenyl-1-propanol, 798
Phenylpropargyl Aldehyde Diethyl Acetal, 801
Phenyl 4-pyridyl ketone, 89
Phenylsuccinic Acid, 804
Phenylsuccinic anhydride, 806
Phenylthiourea, 181
Phenyl p-tolyl sulfone, 36, 37
Phenyltrivinyltin, 260
Phloroglucinol, 454
Phosgene, 522, 788
Phosphoric acid, 321, 323, 350, 543, 691
Phosphorus, red, 348
Phosphorus heptasulfide, 671
Phosphorus oxychloride, 166, 178, 266, 331, 384, 390, 446, 540, 784, 831, 916
Phosphorus pentachloride, 35, 166, 383, 554, 693, 847, 900, 914
Phosphorus pentoxide, 27, 70, 144, 214, 242, 321, 323, 487, 544, 706, 938, 940
Phosphorus tribromide, 107, 616
Phosphorus trichloride, 398, 784, 950, 955
o-Phthalaldehyde, 807
Phthalic acid, 56
Phthalic anhydride, 107
Phthalide, 811
Phthalideacetic acid, 138
α-Phthalimido-o-toluic Acid, 810
3-Picoline-1-oxide, 655
Piperidine, 210, 327, 409
Piperidine acetate, 210, 408
Piperonal, 605
Platinum black, in peroxide decomposition, 612

Potassium, 132, 399, 459, 617, 963
Potassium acid acetylenedicarboxylate, 329
Potassium Amide, 963
Potassium 4-amino-3,5-dinitrobenzenesulfonate, 365
Potassium bicarbonate, 412, 804
Potassium bromate, 849
Potassium bromide, 526, 877
Potassium tert-butoxide, 132, 399, 459, 617
Potassium carbonate, 211, 419, 558, 589, 781, 836
Potassium 4-chloro-3,5-dinitrobenzenesulfonate, 365
Potassium cyanate, 49
Potassium cyanide, 115, 394, 439, 496, 642, 804
Potassium ethyl malonate, 417
Potassium ethyl xanthate, 569
Potassium fluoride, 525
Potassium hydroxide, 74, 78, 91, 95, 104, 172, 226, 229, 230, 251, 253, 346, 367, 372, 402, 417, 433, 446, 471, 474, 479, 484, 498, 510, 560, 570, 591, 609, 617, 618, 627, 633, 643, 684, 728, 776, 924, 967, 974
methanolic, 367
Potassium hypochlorite, 485
Potassium iodide, 297, 319, 321, 323, 543, 896, 935
Potassium methyl sulfate, 589
Potassium nitrate, 365, 498
Potassium 1-nitropropylnitronate, 373
Potassium oxalate, 808
Potassium permanganate, 467, 825
Potassium phthalimide, 811
Potassium-sodium tartrate, 193
Potassium thiobenzoate, 924
Potassium thiocyanate, 233
Potassium trithiocarbonate, 967
Propargyl alcohol, 813
Propenylbenzene, 787
Propiolaldehyde, 813
Propionic anhydride, 657
Propionitrile, 518
β-Propionylphenylhydrazine, 657
n-Propylacetylene, 118, 119
n-Propyl alcohol, 284
N-Propylaniline, 147, 284
n-Propyl bromide, 119
N-Propylbutylamine, 284
Propylene oxide, 12
9-n-Propylfluorene, 625
n-Propyl iodide, 324
n-Propylmagnesium bromide, 96
Pseudopelletierine, 816
Pseudopelletierine hemihydrate, 817
Putrescine Dihydrochloride, 819
2,3-Pyrazinedicarboxylic Acid, 824
5-Pyrazolone of sec-butyl α-n-caproylpropionate, 122
Pyrene, 484
1-Pyrenyl isothiocyanate, 701

Pyridine, 5, 56, 198, 244, 327, 381, 383, 446, 478, 479, 732, 753, 814, 828, 981
3-Pyridinecarboxylic acid, 88
Pyridine hydrochloride, 982
PYRIDINE-N-OXIDE, **828**
Pyridine-1-oxide acetate, 829
Pyridine-1-oxide hydrochloride, **828**
Pyrogallol, 27
Pyrogallol-1,3-dimethyl ether, 866
Pyrrole, 831
2-PYRROLEALDEHYDE, **831**
Pyrrolidine, 466
2-(1-PYRROLIDYL)PROPANOL, **834**
Pyruvic acid, 468
Pyruvic aldehyde, sodium bisulfite addition product, 826

Quinacetophenone, 836, 837
Quinacetophenone dimethyl ether, **837**
QUINACETOPHENONE MONOMETHYL ETHER, **836**
Quinoline, 609, 628, 732, 857
Quinone, see p-Benzoquinone
QUINOXALINE, **824**

Raney nickel alloy, 137, 222, 283, 299, 314, 357, 432, 603, 639, 660, 672
Reissert's compound, **641**
Ricinoleoyl chloride, **742**
Rochelle salt, 194
Rosin, wood, 1

Salicylic acid, 178
Sand, 150
Sarcosine, 5
SEBACIL, 219, **838**
SEBACOIN, 216, 218, 219, 838, **840**
3,4-SECO-Δ⁵-CHOLESTEN-3,4-DIOIC ACID, **191**
Seignette salt, 194
Selenious acid, 229
Selenium, 229, 231
Selenium dioxide, 231
Silica gel, 64
Silver, 920, 973
Silver acetate, 548
Silver chloride, **85**
Silver cyanide, 438
Silver nitrate, 41, 85, 548, 614, 648, 725, 919, 972
Silver nitrite, 369, 425, **724**
Silver oxide, 493, 547, 548, 919, **972**
SILVER TRIFLUOROACETATE, **547**
Soda-lime, 27
Sodamide, see Sodium amide
Sodium, 11, 29, 99, 117, 128, 141, 174, 184, 211, 221, 245, 247, 278, 288, 291, 296, 314, 387, 396, 404, 427, 449, 457, 461, 474, 509, 536, 566, 586, 613, 618, 624, 631, 662, 744, 755, 763, 798, 841, 851, 904, 910, 932, 970
Sodium acetate, 196, 332, 489, 582, 634, 669, 728, 832, 860, 977

SODIUM ACETYLACETONATE, **869**
Sodium amalgam, **508**
Sodium amide, 117, 128, 296, 387, 404, 586, 755, 851, 970
Sodium azide, 76, 716
Sodium bicarbonate, 16, 469, 674
Sodium bisulfite, 32, 226, 469, 574, 688, 758, 826, 867
Sodium bromide, 753
Sodium carbonate, 21, 34, 128, 226, 411, 413, 719, 905
Sodium chloride, 97, 191, 302, 404, 543, 558
Sodium cyanate, 49, 515
Sodium cyanide, 59, 207, 274, 392, 506, 577
Sodium cyclopentadienide, 474
Sodium dichromate, 150, 164, 189, 197
Sodium dihydrogen phosphate, 467
Sodium diphenylketyl, 794
Sodium ethoxide, 141, 174, 184, 221, 245, 288, 291, 396, 427, 457, 461, 536, 566, 618, 631, 662, 932
Sodium formylacetone, 210
Sodium hydride, 537
Sodium hydrosulfide, 32
Sodium hydrosulfite, 16, 46, 248, 758
Sodium hydroxide, 31, 32, 37, 45, 47, 59, 74, 111, 125, 137, 182, 213, 223, 283, 345, 415, 433, 434, 491, 493, 533, 540, 548, 552, 558, 583, 598, 628, 788, 869, 957, 972, 974
Sodium hypobromite, solution of, 45
Sodium hypochlorite, solution of, 74, 346
Sodium iodide, 84, 208, 577
Sodium methoxide, 29, 210, 278, 380, 382, 516, 594, 624, 638, 650, 744
Sodium nitrate, 498
Sodium nitrite, 16, 76, 160, 170, 182, 222, 247, 454, 499, 634, 718, 725, 727, 781, 820, 844, 873, 944, 948
SODIUM NITROMALONALDEHYDE MONOHYDRATE **844**
Sodium phenoxide, 590, **592**
Sodium 2-phenyl-2-hydroxyethane-1-sulfonate, **850**
SODIUM β-STYRENESULFONATE, **846**
Sodium sulfide nonahydrate, 31, 295, 893
Sodium sulfite, 346, 529, 674
Sodium thiosulfate, 105, 158, 319, 543, 896, 935
Stannic chloride, 881, 900
Starch, 469, 896
Stearic acid, 142, 854
STEAROLIC ACID, **851**
STEARONE **854**
cis-STILBENE, **857**
trans-Stilbene, 375, 378, 858, 860
Stilbene dibromide, 858
trans-STILBENE OXIDE, 375, **860**
Styrene, 665, 786, 846
Styrene dibromide, 764, 858
Styrene oxide, 12

β-Styrenesulfonic acid; benzylthiuronium, p-chlorobenzylthiuronium, aniline, and p-toluidine salts, 850
β-Styrenesulfonyl Chloride, 846
Suberic acid, 225
Succinic acid, 390, 500
  disodium salt, 673
Succinic anhydride, 342
Succinimide, 109, 254
α-Sulfobehenic acid, 864
α-Sulfolauric acid, 864
α-Sulfomyristic acid, 864
α-Sulfopalmitic Acid, 862
  monosodium salt, 864
α-Sulfostearic acid, 864
Sulfur, 31, 244, 295, 308, 667
Sulfur dioxide, 973, 974
Sulfuric acid, 13, 19, 36, 69, 73, 149, 160, 164, 176, 201, 242, 261, 266, 318, 329, 399, 402, 418, 420, 423, 425, 432, 442, 445, 489, 518, 532, 549, 561, 591, 654, 664, 665, 695, 702, 757, 786, 813, 866, 948, 978
  fuming, 201, 364, 610, 848, 978
Sulfur trioxide, 846, 863
Sulfuryl chloride, 162, 592
Syringic Aldehyde, 866

d-Tartaric acid, 242
Tetraacetylethane, 869
α,α,α′,α′-Tetrabromo-o-xylene, 807
Tetra-n-butyltin, 882
1,1,1,2-Tetrachloro-2,2-difluoroethane, 269
dl-4,4′,6,6′-Tetrachlorodiphenic Acid, 872
Tetrachloroethylene, 863
Tetracyanoethylene, 244, 276, 877, 953
n-Tetradecyl fluoride, 527
1,1,1′,1′-Tetraethoxyethyl Polysulfide, 295, 296
Tetraethylene glycol, 614
Tetraethylthiuram disulfide, 308
Tetraethyltin, 881
1,2,3,4-Tetrahydrocarbazole, 884
Tetrahydrofuran, 258, 266, 321, 355, 474, 792, 965
Tetrahydrofurfuryl alcohol, 500, 756
Tetrahydrofurfuryl chloride, 755
cis-Δ⁴-Tetrahydrophthalic anhydride, 304, 306
4,5,6,7-Tetrahydroindazole, 537
ar-Tetrahydro-α-naphthol, 887
5,6,7,8-Tetrahydro-2-naphthylacetamide, 762
cis-Δ⁴-Tetrahydrophthalic Anhydride, 890
Tetrahydropyran, 796
Tetrahydrothiophene, 892
Tetralin, 522, 895
Tetralin Hydroperoxide, 895
α-Tetralone, 885, 898
β-Tetralone, 903
β-Tetralone, bisulfite addition product, 904

1,1,1′,1′-Tetramethoxyethyl polysulfide, 297
2,2,6,6-Tetramethylolcyclohexanol, 907
Tetramethylsuccinonitrile, 273
Tetraphenylarsonium bromide, 913
Tetraphenylarsonium Chloride Hydrochloride, 910
Tetraphenylethylene, 914
2,2,3,3,-Tetraphenylpropionitrile, 964
Tetra-n-propyltin, 882
Tetravinyltin, 260
2-Thenaldehyde, 915
3-Thenaldehyde, 918, 920
3-Thenoic Acid, 919
3-Thenyl Bromide, 918, 921
Thiobenzoic Acid, 924
Thiobenzophenone, 927
Thiolacetic Acid, 928
2-Thio-6-methyluracil, 638
Thionyl chloride, 88, 154, 169, 263, 333, 339, 436, 556, 561, 571, 621, 715, 739, 937
Thiophene, 545, 673, 916, 980
Thiourea, 401, 491, 566, 638
Tiglylaldehyde diethyl acetal, 22
o-Tolualdehyde, 932
Toluene, 38, 192, 204, 273, 461, 788
p-Toluenesulfenyl Chloride, 934
p-Toluenesulfinic acid, sodium salt, 674
  sodium salt, dihydrate, 937
p-Toluenesulfinyl Chloride, 937
p-Toluenesulfonic acid, monohydrate, 304, 306, 534, 719, 940
p-Toluenesulfonic Anhydride, 940
p-Toluenesulfonylanthranilic Acid, 34, 35
p-Toluenesulfonyl chloride, 35, 674, 943
p-Toluenethiol, 934
o-Toluidine, 42, 147
p-Tolunitrile, 714
Toluquinone, 151, 152
p-Tolyl disulfide, 936
1-(p-Tolyl)-3-phenyl-2-propanone, 177
1-m-Tolylpiperidine, 797
1-o-Tolylpiperidine, 797
1-p-Tolylpiperidine, 797
p-Tolylsulfonylmethylamide, 253
p-Tolylsulfonylmethylnitrosamide, 225, 251, 943
m-Tolylurea, 51
o-Tolylurea, 51
p-Tolylurea, 51
p,α,α-Tribromoacetophenone, 110, 111
2,4,6-Tribromobenzoic Acid, 947
Tri-n-butylvinyltin, 260
Trichloroethylene, 430
Trichloromethylphosphonyl Dichloride, 950
p-Tricyanovinyl-N,N-dimethylaniline, 953
Triethylamine, 286, 557, 560, 561, 730, 777
Triethylamine hydrochloride, 557, 560, 561

TRIETHYL 2-METHYL-1,1,3-PROPANETRI-
CARBOXYLATE, **630**
sodio derivative, 631
Triethyl orthoacetate, 802
Triethyl orthoformate, *see* Ethyl ortho-
formate
TRIETHYL PHOSPHITE, 326, **955**
TRIETHYL α-PHTHALIMIDOETHANE-α,α,-
β-TRICARBOXYLATE, **55**, 56
Triethylvinyltin, 260
Trifluoroacetic acid, 547
p-Trifluoromethylbenzaldehyde, **933**
1,2,4-Trihydroxybutane, 534
Triisopropyl phosphite, 325, 326, 956
Trimethylamine, 85, 86, 99
Trimethylamine hydrochloride, 86
Trimethylbenzene, technical, 144
3,3,5-Trimethyl-1,2-cyclohexanedione, 958
2,2,4-TRIMETHYLCYCLOPENTANONE, **957**,
958
Trimethylene chlorobromide, 288
Trimethyl n-orthovalerate, 802
1,1,3-Trimethyl-3-phenylindane, **666**
Trimethylvinyltin, 260
Trioxane, 518
TRIPHENYLARSINE, **910**
TRIPHENYLARSINE OXIDE, **911**
2,3,3-Triphenylbutyronitrile, **964**
Triphenylcarbinol, 47
α,β,β-TRIPHENYLPROPIONIC ACID, **960**
α,α,β-TRIPHENYLPROPIONITRILE, **962**
Triphenylvinyltin, 260
Tri-n-propylvinyltin, 260
TRIPTYCENE, **964**
TRITHIOCARBODIGLYCOLIC ACID, 7, **967**
Triton B, 652

γ-Undecanoic lactone, 351
10-Undecenoic acid, 969
n-Undecyl fluoride, **527**

10-UNDECYNOIC ACID, **969**
Urea, 52, 157, 247, 276, 513, 719, 744

γ-Valerolactone, 280, 899
VANILLIC ACID, **972**
Vanillin, 203, 972, 974
Veratraldehyde, 410, 605, 735
Veratrole, 548
Vinyl acetate, 977
Vinyl bromide, 258
Vinyl caprate, **978**
Vinyl caproate, **978**
Vinyl caprylate, **978**
VINYL ESTERS, **977**
Vinyl 10-hendecenoate, **978**
VINYL LAURATE, **977**
Vinylmagnesium bromide, **260**
Vinyl myristate, **978**
Vinyl oleate, **978**
Vinyl palmitate, **978**
Vinyl pelargonate, **978**
Vinyl stearate, **978**
2-VINYLTHIOPHENE, **980**
Vinyl undecylenate, **978**

Water, deaerated, 27, 789, 817
Witt-Utermann solution, 44

Xylene, 313, 474, 840
o-Xylene, 807, 933, 984
D-Xylose, 506
o-Xylyl bromide, 932
o-XYLYLENE DIBROMIDE, **984**

Zinc, amalgamated, 203, 204, 695
dust, 196, 198, 218, 268, 642, 750
foil, 121, 445
turnings, 348
Zinc chloride, 268, 750, 802
Zinc iodide, 801, 802
Zinc nitrate, 802